MODERN CONTROL ENGINEERING

D. ROY CHOUDHURY
Professor and Head
Computer Engineering Department
Delhi College of Engineering
Delhi

Prentice-Hall of India *Private Limited*
New Delhi - 110 001
2006

Rs. 350.00

MODERN CONTROL ENGINEERING
D. Roy Choudhury

© 2005 by Prentice-Hall of India Private Limited, New Delhi. All rights reserved. No part of this book may be reproduced in any form, by mimeograph or any other means, without permission in writing from the publisher.

ISBN-81-203-2196-0

The export rights of this book are vested solely with the publisher.

Third Printing ... ... **August, 2006**

Published by Asoke K. Ghosh, Prentice-Hall of India Private Limited, M-97, Connaught Circus, New Delhi-110001 and Printed by Jay Print Pack Private Limited, New Delhi-110015.

To
***The Memory of My Beloved Teacher
Professor Arun Kumar Choudhury***
(formerly of Institute of Radiophysics and Electronics,
University of Calcutta)

Contents

Preface *xiii*

1. INTRODUCTION TO CONTROL SYSTEMS 1–46

1.1 Introduction *1*
1.2 Definitions *2*
1.3 History of Development of Feedback Control Systems *4*
  1.3.1 Automatic control *4*
  1.3.2 The birth of mathematical control theory *5*
  1.3.3 The space/computer age and modern control *7*
1.4 Open-Loop Control Systems *9*
1.5 Closed-Loop (Feedback) Control Systems *10*
1.6 The Laplace Transform *14*
  1.6.1 Transfer function concept *19*
  1.6.2 Partial fraction expansion and MATLAB *21*
1.7 State Variable Formulation *26*
1.8 Properties of Transfer Function *32*
1.9 Linear Approximation of Physical Systems *34*
1.10 Case Study *37*
1.11 Unity-Feedback System *40*
1.12 Steady-State Frequency Response *42*
Summary *45*
Problems *45*

2. MATHEMATICAL MODELING OF SYSTEMS 47–124

2.1 Introduction *47*
2.2 Translational Systems *47*
2.3 Rotational Systems *48*
2.4 Electrical Analog of Mechanical Systems *49*
2.5 Mechanical Couplings *54*
2.6 Liquid-Level Systems *64*
2.7 Servomotors *66*
  2.7.1 Two-phase servomotors *67*
  2.7.2 DC motors *69*
2.8 Sensors *76*
  2.8.1 Potentiometers *76*
  2.8.2 Synchros *80*

2.8.3 Tachometers *84*
2.8.4 Linear variable differential transformer (LVDT) *86*
2.8.5 Accelerometers *88*
2.9 Magnetic Amplifier *90*
    2.9.1 Saturable core reactor *90*
2.10 Stepper Motor *94*
2.11 Block Diagram Reduction *96*
    2.11.1 Procedure for drawing the block diagram *96*
    2.11.2 Block diagram reduction techniques *97*
2.12 Signal Flow Graph *108*
2.13 Introduction to Gain Formula *109*
2.14 Mason's Gain Formula *111*
2.15 Transfer Matrix of Multivariable Systems *117*

Summary 119
Problems 119

3. CHARACTERISTICS OF FEEDBACK CONTROL SYSTEMS 125–143

3.1 Introduction *125*
3.2 Derivation of Sensitivity *126*
3.3 Sensitivity of Control Systems to Parameter Variations *127*
3.4 Effect of Disturbance Signals *137*
3.5 Alternative Design through Sensitivity Analysis *139*

Summary 141
Problems 141

4. TRANSIENT RESPONSE ANALYSIS 144–219

4.1 Introduction *144*
4.2 Typical Test Input Signals *145*
4.3 First-Order Systems *146*
    4.3.1 Unit-step response of first-order systems *146*
    4.3.2 Unit-ramp response of first-order systems *152*
    4.3.3 Unit-impulse response of first-order systems *152*
4.4 Analysis of Transient Performance from Root Pattern *153*
4.5 Second-Order Position Control System *155*
4.6 Unit-Step Response of Second-Order Systems *157*
4.7 Parameter Variation for Second-Order System *171*
4.8 Impulse Response of Second-Order System *178*
4.9 Ramp Response of Second-Order System *180*
4.10 Graphical Interpretation of Heaviside's Expansion *181*
4.11 Time-Domain Behaviour from Pole-Zero Plot *182*
4.12 Steady-State Error *188*
    4.12.1 Unit-step input *189*

Contents

    4.12.2 Unit-ramp input *189*
    4.12.3 Unit-parabolic (acceleration) input *190*
4.13 Steady-State Error in Terms of Closed-Loop Transfer Function *191*
4.14 Integral Performance Criterion *195*
4.15 Derivative Control *196*
    4.15.1 Interpretation of derivative control from root locus point of view *199*
4.16 Rate Feedback (Tachometer Feedback) Control *200*
4.17 Comparison between Derivative and Rate Feedback Control *202*
4.18 Analog Computation *210*
Summary *215*
Problems *215*

5. ROUTH STABILITY AND ROBUST CONTROL   220–249

5.1 Introduction *221*
5.2 Stability *222*
5.3 Some Results from the Theory of Equations *223*
5.4 Necessary Condition for Stability *224*
5.5 Routh Array *225*
5.6 Special Cases *227*
5.7 Application of Routh–Hurwitz Stability Criterion *232*
5.8 Stability from State-Space Representation *232*
5.9 Relative Stability *234*
5.10 Relative Stability with Minimum Damping Ratio *237*
5.11 Stability under Parameter Uncertainty: Robust Control *240*
Summary *247*
Problems *247*

6. ROOT-LOCUS TECHNIQUE   250–322

6.1 Introduction *251*
6.2 Root-Loci for Second-Order System *251*
6.3 Basic Conditions for Root Loci *254*
6.4 Rules for the Construction of Root Loci ($0 \leq K \leq \infty$) *257*
6.5 Rules for the Construction of Inverse Root Loci *298*
6.6 Effect of Adding Poles and Zeros *301*
    6.6.1 Addition of poles *301*
    6.6.2 Addition of zeros *303*
    6.6.3 Effect of varying the pole position *304*
    6.6.4 Cancellation of poles and zeros *305*
6.7 Root Contour *306*
    6.7.1 Multiple-loop system *313*
6.8 Root-locus for System with Transportation Lag *314*
6.9 Design on Root Locus *315*
Summary *318*
Problems *319*

7. PROCESS CONTROL SYSTEM   323–351

- 7.1 Introduction *323*
- 7.2 Proportional Control Action *324*
  - 7.2.1 Difficulty with proportional control *328*
- 7.3 Integral Action or Reset *328*
- 7.4 Differential Action: Derivative or Rate Control *330*
- 7.5 PID Control *331*
  - 7.5.1 Tuning rules for PID controllers *334*
- 7.6 PID Controller Design *338*
- 7.7 Ziegler–Nichols Rules for Controller Tuning *338*
  - 7.7.1 First approach *338*
  - 7.7.2 Second approach *341*
- 7.8 Adjustment According to Process Characteristics in Ziegler–Nichols Method *343*
- 7.9 Purpose of Ziegler–Nichols Tuning Method *346*
- 7.10 Designing Controller Using Root Loci *346*

Summary 350
Problems 350

8. FREQUENCY RESPONSE ANALYSIS   352–423

- 8.1 Introduction *353*
- 8.2 Bode Plot *355*
  - 8.2.1 Basic philosophy *355*
- 8.3 Response of Linear Systems *356*
- 8.4 Frequency Domain Analysis *358*
- 8.5 Construction of Bode Plot *359*
- 8.6 Log-magnitude-versus-Phase Plot *376*
  - 8.6.1 GM and PM *376*
- 8.7 Application of the Frequency Response Plot *378*
  - 8.7.1 Static position error coefficient *378*
  - 8.7.2 Static velocity error coefficient *380*
  - 8.7.3 Estimation of transfer function from Bode plot *382*
- 8.8 Minimum and Non-Minimum Phase Transfer Functions *384*
- 8.9 Stability of Control System with Time Delay *387*
- 8.10 Simulation of Time Delay *389*
- 8.11 Introduction to Polar Plot *393*
- 8.12 Construction of Polar Plot *394*
- 8.13 Relative Stability *408*
- 8.14 Phase Margin *410*
- 8.15 Gain Margin *411*
- 8.16 Conditionally Stable System *417*

Summary 420
Problems 421

9. NYQUIST STABILITY 424–484

9.1 Introduction *424*
9.2 Basic Philosophy of Stability Criterion *424*
9.3 Encircled versus Enclosed *426*
9.4 Conformal Mapping *427*
9.5 Cauchy's Theorem *430*
9.6 Relationship between GH-plane and $F(s)$-plane *433*
9.7 Nyquist Stability Criterion *433*
9.8 Conditionally Stable System *461*
9.9 Nyquist Path for Open-Loop Poles on $j\omega$-axis *466*
9.10 Closed-Loop Frequency Response *467*

    9.10.1 M-circles: constant magnitude loci *468*
    9.10.2 N-circles: constant phase-angle loci *472*
    9.10.3 Nichols chart *475*
    9.10.4 Non-unity feedback system *481*
    9.10.5 Frequency domain specifications for design *482*

Summary 483
Problems 484

10. COMPENSATION TECHNIQUES 485–521

10.1 Introduction *485*
10.2 Cascade Compensation Networks *487*
10.3 Lead Compensating Networks *488*
10.4 Characteristics of Lead Networks *490*
10.5 Lag Compensation *498*

    10.5.1 Characteristics of lag networks *499*
    10.5.2 Lag compensation techniques based on the root-locus approach *500*

10.6 Lag–lead Compensation *506*

    10.6.1 Lag–lead networks *506*
    10.6.2 Characteristics of lag–lead networks *508*

10.7 Compensation of Operational Amplifier *510*

    10.7.1 Transfer function of a practical op-amp *512*
    10.7.2 Frequency compensation *518*

Summary 521
Problems 521

11. STATE-VARIABLE FORMULATION 522–617

11.1 Introduction *522*
11.2 Concept of State *523*
11.3 Transfer Function *531*

    11.3.1 Computation of transfer function using Leverrier's algorithm *533*

11.4 State Space Representation of Multivariable Systems *543*

11.5	Set of Minimal State Variable Representation	544
11.6	State Equation from Transfer Function	546
11.7	Simulation	554
11.8	Simultaneous Equations	557
11.9	Dual Representation	559
11.10	Linear Transformation	561
11.11	Similarity Transformation	561
11.12	Invariance of Transfer Function	566
11.13	Properties of Linear Transformation	566
11.14	Diagonalization	566
11.15	Vandermonde Matrix	569
	11.15.1 Vandermonde matrix for multiple eigenvalues	573
11.16	Important Properties of Eigenvalues of a Matrix	578
11.17	Time-Domain Solution	583
11.18	State Transition Matrix	585
	11.18.1 Forced response	588
	11.18.2 Methods for evaluation of STM	590
	11.18.3 Power series method	595
	11.18.4 Cayley–Hamilton theorem	597
11.19	Power of Companion Matrix and Its Applications	604
	11.19.1 Algorithm: power of companion matrix	605
11.20	Trace of a Matrix	608
11.21	STM of Linear Time-Varying System	609
	11.21.1 Matrizant	609
	11.21.2 Kinariwala's approach	610

Summary 612
Problems 612

12. ANALYSIS AND DESIGN OF MODERN CONTROL SYSTEMS 618–676

12.1	Introduction	619
12.2	Controllability	619
12.3	Observability	623
12.4	Controllability Criterion I	625
12.5	Controllability Criterion II	626
12.6	Observability Criterion I	627
12.7	Observability Criterion II	628
12.8	Controllable and Observable Systems	629
12.9	Pole Cancellation—Stabilization	631
12.10	Concept of Transfer Function	632
12.11	Duality Property	635
12.12	Similarity Transformation Matrix from Controllability Matrix	637
12.13	Similarity Transformation Matrix from Observability Matrix	642
12.14	Pole-Placement Design through State Feedback	643

12.15 State Observer *650*

    12.15.1 Design of state observer *650*
    12.15.2 Condition for the existence of observer *651*

12.16 Eigenvalue Assignment Method for Design of G_e *652*
12.17 Transfer Function Matrix *658*

    12.17.1 Realization of transfer matrix *659*

12.18 Diagonal Form of Representation of Transfer Matrix *662*
12.19 State Space Representation in Canonical Form *667*

Summary 673
Problems 673

13. DIGITAL CONTROL SYSTEMS   677–749

13.1 Introduction *677*
13.2 Discrete-Time Systems *677*
13.3 Sampled-Data and Digital Control System *679*

    13.3.1 Digital controller vs analog controller *682*
    13.3.2 Sampling process *682*

13.4 Sample-and-Hold Device *683*
13.5 Frequency Response of Zero-Order Hold *686*
13.6 z-Transform *687*
13.7 Pulse Transfer Function of Cascaded Elements *694*
13.8 Pulse Transfer Function of Closed-Loop Systems *695*

    13.8.1 Characteristic equation *696*

13.9 Relationship between s-plane and z-plane Poles *697*
13.10 Jury's Stability Test *707*
13.11 Shifting Property of z-transform *709*
13.12 Analysis of Digital Control Systems *711*

    13.12.1 Discrete system time response *711*
    13.12.2 Time constant *713*

13.13 Steady-State Error *716*
13.14 Root Loci for Sampled-Data Control Systems *718*
13.15 Nyquist Criterion *722*
13.16 State-Variable Formulation *725*
13.17 Samuelson's Model of National Economy *729*
13.18 Solutions to State Equation by Recursive Method *730*
13.19 Direct Decomposition (Phase Variable Form) *733*
13.20 Relationship between Continuous and Discrete State Equations *735*
13.21 Cayley–Hamilton Method *738*
13.22 Discretization of Continuous Time State Equation *740*

Summary 746
Problems 746

14. NONLINEAR SYSTEMS   750–819

14.1 Introduction *750*
14.2 Common Nonlinearities *753*
14.3 Phase-Plane Analysis *755*
  14.3.1 Phase-plane method—basic concept *756*
  14.3.2 Methods for constructing trajectories *757*
14.4 Phase-Plane Analysis of Linear Control Systems *765*
14.5 Phase-Plane Analysis of Nonlinear Systems *771*
  14.5.1 Control system with nonlinear gain *772*
14.6 Describing Function Analysis *778*
  14.6.1 Derivation of describing functions *778*
14.7 Stability Analysis *794*
14.8 Liapunov Stability Analysis *798*
  14.8.1 Asymptotic stability analysis *803*

Summary 815
Problems 815

Index   *821–825*

Preface

Automatic control has become an integral part of everyday new and challenging models of dynamic systems which have to be optimized and controlled. There is thus an utmost need to understand control systems engineering on account of its multidisciplinary applications in various areas of practice (electrical, aeronautical, mechanical, chemical, industrial, and general engineering). In a nutshell, the applications of control theory are present in almost all scientific and engineering disciplines.

The main objective in writing this book has been to produce an accessible text for a two-semester course in control systems at the level of undergraduate students in any of the various branches of engineering. The book unifies the diverse methods of analysis of feedback control systems and presents the fundamentals explicitly and clearly. Highly mathematical arguments are carefully avoided in the presentation of the material. On the other hand, mathematical proofs are provided wherever they contribute to the understanding of the subject matter discussed. The theory has been supported by a variety of solved examples. Software programs have been developed in MATLAB platform for better understanding of design of control systems.

The material of this book is organized as follows. Chapters 1–10 deal with the basic material for the analysis and design of feedback control systems from the knowledge of open-loop pole-zero information using the conventional frequency domain and time-domain approaches. Chapters 11–14, provide a thorough understanding of control theory concepts such as state space, controllability, observability, and analysis and design of discrete control and nonlinear systems.

Chapter 1 serves as an introduction to control theory, narrates the history of control systems engineering, and provides a review of the mathematical background. The concepts of feedback control, the related mathematics of Laplace transform, and the linearization technique are introduced. Chapter 2 deals with developing the mathematical models of physical systems and their electrical analogue, state variable formulation concepts, the block diagram and signal flow graph representation of the system and also develops the short-cut techniques for finding the transfer function from complicated expressions. Chapter 3 explains the sensitivity to parameter variation of closed-loop control system. Chapter 4 presents the transient response analysis and steady-state error analysis. Analog computer simulation of control systems is also discussed. The role of derivative, rate feedback and PID control to improve the transient response has been thoroughly discussed. Chapter 5 deals with Routh's stability criterion. How to find stability under parameter uncertainty using Kharitonov's polynomials is also explained in this chapter. Chapter 6 presents the root-locus technique for the analysis of control systems. Chapter 7 familiarizes the students with the process control terminology and tuning action of controllers. Chapter 8 deals with the frequency domain techniques consisting of Bode plot, polar and log-magnitude-versus-frequency plot—the construction techniques and application of these plots are fully explained. The unified approach of determining the stability information of the closed-loop system from the open-loop system configuration is highlighted for frequency domain techniques by defining gain margin and phase margin. Chapter 9 deals with the investigation of stability using the Nyquist stability criterion and also deals with the closed-loop frequency response. Chapter 10 deals with the design and compensation techniques using the root locus and frequency domain response approaches.

Chapter 11 discusses the state variable formulation of the control system, be it electrical, mechanical, electro-mechanical. Minimal realization of state variable formulation using the graph theoretic approach is illustrated. Similarity transformation and its properties have been discussed. Evaluation of state transition matrix with estimation of error has been described. Solutions for linear, time-varying systems have been touched upon. Chapter 12 deals with controllability and observability of the system, design of control systems based on state space approach, and state observer design. Chapter 13 discusses the z-transformation technique, being the prerequisite for the study of discrete control systems. It analyses the discrete control systems on the same lines as discussed in the treatment of classical control theory in the earlier chapters. Chapter 14 deals with the analysis of nonlinear control systems using the phase-plane technique, describing function, and Liapunov's methods.

Applications of classical techniques based on transfer function and state variable approach for the analysis and design of control systems in MATLAB platform have been shown. User interactive programs are developed in MATLAB platform for control system designs in the relevant chapters. Answers to the drill problems have been provided. Many varied problems are included at the end of each chapter.

I deeply express my indebtedness to many of my former students, Sudeep in particular. I owe a debt of gratitude to my son Nilanjan for his critical comments and help in solving many typical problems and developing software in MATLAB platform. It is a pleasure to thank Pragati Kumar for his thoughtful comments and suggestions.

Finally, I express my appreciation to my wife Manu for her constant encouragement and support without which this project would not have been accomplished.

D. ROY CHOUDHURY

Introduction to Control Systems

1.1 Introduction

Automatic control has played a role of paramount importance in the advancement of engineering and science. Without even being familiar with the terminology of automatic control, people in the past have utilized the concept of automatic control, for instance, as in a float-control system. Besides its extreme importance in sophisticated space vehicles, missile guidance, aircraft positioning and the like, automatic control has become an integral part of our daily life starting from room temperature control, automobile steering control and extending to traffic control systems, chemical processes, economic regulation systems and so forth. Automatic control has become an integral part of modern manufacturing and industrial processes. So we ought to not only know the principles of automatic control, but also develop a good understanding of this field of control engineering.

Control engineering is not limited to any particular discipline but is equally applicable to aeronautical, chemical, mechanical, environmental, civil, electrical, electronics, instrumentation, and computer engineering, so on and so forth.

Control engineering is based on the principles of feedback theory, linear system analysis, circuit theory, and communication engineering. The basic mathematics involved comprises the Laplace transform technique, matrix algebra, and calculus.

OBJECTIVE

In this introductory chapter, we discuss all the topics which are used as tools for understanding the subsequent chapters. The study of control engineering can broadly be classified into two categories: classical control systems and modern control systems. The classical control theory, which covers a good portion of the text, requires the mathematical background in complex variables, differential equations and Laplace transforms. For modern control theory, that is, the state variable approach, we need to know matrix algebra.

The transfer function concept, physical systems, linearization of practical cases, and state variable formulation have all been touched upon. The mathematical tool Laplace trans-form is discussed with exposure to MATLAB at a preliminary level. The detailed treatment of mathematical tools has been deliberately omitted from this book as students are expected to be familiar with these tools.

CHAPTER OUTLINE

Introduction
Definitions
History of Development of Feedback Control Systems
Open and Closed-Loop Systems
Laplace Transform using MATLAB
Transfer Function Concept
State Variable Formulation
Linear Approximation
Case Study
Unity Feedback System
Steady-State Frequency Response

A control system is an interconnection of components forming a system configuration which provides the desired output response. A good understanding of the physics of different system components helps us to grasp the theory of control engineering better.

In order to understand control, analysis and design of complex systems, we must formulate the quantitative mathematical models of these systems. The systems in practice are dynamic in nature, the descriptive equations are usually differential or integro-differential. In short, the approach to dynamic system problems can be enumerated as follows:

1. Define the system and its components
2. Formulate the mathematical model
3. Write the differential equations describing the model
4. Solve the equations for the desired output variables
5. Examine the solutions and the assumptions
6. Reanalyze or proceed with the design

1.2 Definitions

Let us familiarize ourselves with the terminologies necessary to describe control systems.

Plant. A plant is a set of machine parts functioning together to perform a particular operation such as a heating furnace, a chemical reactor, and so forth.

Process. A process is one where a progressively continuing sequential operation goes on to lead towards a particular goal of an end-product or result. Examples are all kinds of systems in nature, chemical processes, economic, social, and biological processes.

System. A system is a combination of components that act together and perform a certain objective. This may imply physical, biological, economic, and the like systems.

Disturbance. A disturbance is an unwanted signal that tends to adversely affect the output of a system. The disturbance may be generated within the system or it may arise from outside the system and act as an external input.

Feedback control. The feedback control is an operation where the output is fed back to the input. The feedback output may be in-phase or out-of-phase with respect to the input and accordingly it is termed *positive* or *negative* feedback respectively. All control systems are usually negative feedback systems. The oscillators are the examples of positive feedback systems. In a negative feedback system, the difference between the reference input and the output produces an error which is reduced gradually to achieve the desired output in accordance with the reference input signal.

Servomechanism. A servomechanism is a negative feedback control system in which the output is some mechanical position and therefore a position control system is synonymous with servomechanism.

Automatic control system. An automatic control or automatic regulator system is one that maintains the actual output at the desired value in the presence of disturbances, or for a slowly varying reference input.

Process control system. An automatic control system in which the output is a variable such as temperature, pressure, flow, liquid level, or pH, is called a process control system.

Transducer. A transducer is a device which converts a signal from one form into another. Usually, the transducers employed in control systems convert a signal from any form to electrical form. This is because the electrical signal is not only easy to handle, but also easy to amplify and transmit.

Control systems may be open-loop or closed-loop. The open-loop control system is one in which the output has no effect on the control action. The output is not fed back to the input in an open-loop control system. The closed-loop control systems are feedback systems where the output is fed back to the input in out-of-phase (negative feedback) to produce an error actuating signal to drive the system in such a way so that the output becomes more closer to the desired output, that is, the reference input, at every onward instant of time (Figure 1.1). In a negative feedback control system the noise is reduced or eliminated, the gain is reduced and stability improved, and the bandwidth is increased.

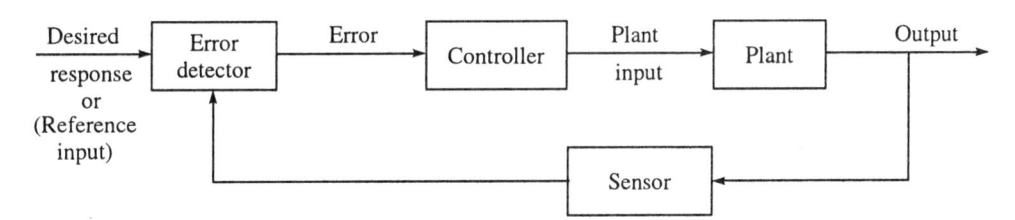

FIGURE 1.1 Closed-loop control system.

It may be mentioned here that sometimes the output quantity of interest is difficult to measure or is unmeasurable. It may then become necessary to control a secondary variable. This is termed *indirect control*. But it is always desirable to measure and control the primary variable directly, which indicates the state of the system, in order to obtain the best result. This is termed *direct control*.

Further, the dynamic characteristics of most control systems do not remain constant over time because of deterioration of components that occurs owing to aging or several other reasons. Although such small changes in the dynamic characteristics are attenuated in the feedback control system, however, if the changes in the system parameters are significant then a well-designed system must have the ability of adaptation. This kind of system is known as the *adaptive control system*.

Control systems that have the ability to learn, are called *learning control systems*. Neural networks or artificial neural networks come under this category. They are used for pattern recognition which is something different from artificial intelligence (AI). An AI system is the one that is programmed; learning systems are non-programmed. The application area in pattern recognition may be either visual recognition or speech recognition. An example of visual pattern recognition using artificial neural networks is that of fingerprint identification.

A control system may be a single-input-single-output system, or a multi-input-multi-output (multivariable) system. Further, a system may be time-varying or time invariant; linear or nonlinear.

When the parameters of a system are constant (or varying) with respect to time, it is called the time-invariant (or time varying) system. An example of a time-varying system is the booster stage of a space vehicle. As the fuel is consumed, the mass of the vehicle decreases with time. A linear system is one for which the principle of superposition holds. The system is nonlinear where the principle of superposition does not hold. All physical systems are inherently nonlinear. A nonlinear system may be linearized and proper control action may be achieved. All control system techniques developed in this book are applicable to linear time-invariant system, however, a separate chapter has been devoted to the study of nonlinear control systems.

Further, a system may be discrete-time control system or continuous-time control system. The signals in a discrete-time control system change values only at discrete instants of time, whereas signals are continuous for continuous-control systems.

1.3 History of Development of Feedback Control Systems

The objective of this book is to use modern control theory to analyse and design feedback control systems. In order to understand an area, we should examine its evolution and the reasons for its existence. Here we provide a short history of automatic control theory. Then, we give a brief discussion of the philosophies of classical and modern control theory.

1.3.1 Automatic control

There have been many developments in automatic control theory during recent years. The key developments that influenced the progress of feedback control were:

1. The Industrial Revolution in Europe whose roots can be traced back into the 1600s.
2. The development of mass communication at the beginning of the 20th century, the First World War, and the Second World War. This represented a period from about 1910 to 1945.
3. The beginning of the space/computer age in 1957.

At a point between the Industrial Revolution and the World Wars, there was an extremely important development as control theory began to acquire the language of mathematics. J.C. Maxwell provided the first rigorous mathematical analysis of a feedback control system in 1868. The period from 1868 to the early 1900s was the primitive period of automatic control. The period from 1900 until 1960 is the classical period, and the period from 1960 through present times is the modern period.

The Industrial Revolution in Europe was marked by the invention of advanced grain mills, furnaces, boilers, and the steam engine. These devices could not be adequately regulated by hand, and so arose a new requirement for automatic control systems. A variety of control devices was invented, including float regulators, temperature regulators, pressure regulators, and speed control devices. James Watt invented his steam engine in 1769, and this date marks the accepted beginning of the Industrial Revolution.

James Watt's steam engine with a rotary output motion had reached maturity by 1783, when the first one was sold. In 1788, Watt completed the design of the centrifugal flyball governor for regulating the speed of the rotary steam engine. This device employed two pivoted rotating flyballs which were flung outward by centrifugal force. As the speed of rotation increased, the flyweights swung further out and up, operating a steam flow throttling valve which slowed the engine down. Thus, a constant speed was achieved automatically.

1.3.2 The birth of mathematical control theory

The design of feedback control systems up through the Industrial Revolution was by trial-and-error together with a great deal of engineering intuition. Thus, it was more of an art than a science. In the mid 1800s, mathematics was first used to analyze the stability of feedback control systems. Since mathematics is the formal language of automatic control theory, we could call the period before this time the prehistory of control theory.

Differential equations. The British Astronomer Royal at Greenwich, G.B. Airy, was the first to discuss the instability of closed-loop systems, and the first to use differential equations in their analysis. The theory of differential equations had by then been well developed. The use of differential equations in analyzing the motion of dynamical systems was established by J.L. Lagrange (1736–1813) and W.R. Hamilton (1805–1865).

Stability theory. J.C. Maxwell analyzed the stability of Watt's flyball governor by linearizing the differential equations of motion to find the characteristic equation of the system. He studied the effect of the system parameters on stability and showed that the system is stable if the roots of the characteristic equation have negative real parts. E.J. Routh provided a numerical technique for determining when a characteristic equation had stable roots.

The Russian Vishnegradsky (1877) analyzed the stability of regulators using differential equations independently of Maxwell. Unaware of the work of Maxwell and Routh, he posed the problem of determining the stability of the characteristic equation to A. Hurwitz in 1895, who solved it independently.

The work of A.M. Liapunov was seminal in control theory. He studied the stability of nonlinear differential equations using a generalized notion of energy in 1892. Unfortunately, though his work was applied and continued in Russia, the time was not ripe in the West for his elegant theory, and it remained unknown there until approximately 1960, when its importance was finally realized.

The British engineer O. Heaviside invented operational calculus in 1892–1898. He studied the transient behaviour of systems, introducing a notion equivalent to that of the transfer function.

System theory. It is within the study of systems that feedback control theory has its place in the organization of human knowledge. Thus, the concept of a system as a dynamical entity with definite "input" and "outputs" joining it to other systems and to the environment was a key prerequisite for the further development of automatic control theory.

Mass communication system. During the development of mass communications systems at Bell Telephone Laboratories, H.S. Black demonstrated the usefulness of negative feedback in 1927 to reduce distortion in repeater amplifiers. The design problem was to introduce a phase shift at the correct frequencies in the system. Regeneration theory for the design of stable amplifiers was developed by H. Nyquist. He derived his stability criterion known as Nyquist stability criterion, based on the polar plot of a complex function. H.W. Bode in 1938 used the magnitude and phase frequency response plots of a complex function. He investigated closed-loop stability using the notions of gain margin and phase margin.

The World Wars and classical control. The development of feedback control systems became a matter of survival during the two World Wars. An important military problem during this period was the control and navigation of ships. Among the first developments was the design of sensors for the purpose of closed-loop control. In 1910, E.A. Sperry invented the gyroscope, which he used in the stabilization and steering of ships, and later in aircraft control. N. Minorsky was the first to introduce his three-term proportional-integral-derivative (PID) controller for the steering of ships. He considered nonlinear effects in the closed-loop system.

The main problem during the period of the World Wars was that of the accurate pointing of guns aboard moving ships and aircraft. With the publication of "Theory of Servomechanism" by H.L. Házen, where he coined the word servomechanisms, the use of mathematical control theory in such problems was initiated.

The Norden bombsight, developed during World War II, used synchro repeaters to relay information on aircraft altitude and velocity and wind disturbances to the bombsight, ensuring accurate delivery of weapons.

M.I.T. Radiation Laboratory. To study the control and information processing problems associated with the newly invented radar, the Radiation Laboratory was established at the M.I.T. in 1940. Much of the work in control theory during the 1940s came out of this lab. While working on an M.I.T./Sperry Corporation joint project in 1941, A.C. Hall recognized the deleterious effects of ignoring noise in control system design. He realized that the frequency-domain technology developed at Bell Labs could be employed to confront noise effects, and used this approach to design a control system for an airborne radar. Its success demonstrated the importance of frequency-domain techniques in control system design.

Using design approaches based on the transfer function, the block diagram, and frequency-domain methods, there was great success in controls design at the Radiation Lab. In 1947, N.B. Nichols developed his Nichols chart for the design of feedback systems. With the M.I.T. work, the theory of linear servomechanisms was firmly established. Working at North American Aviation, W.R. Evans presented his root-locus technique, which provided a direct way to determine the closed-loop pole locations in the s-plane. Subsequently, during the 1950s, much of controls work was focused on the s-plane, and on obtaining desirable closed-loop step-response characteristics in terms of rise time, per cent overshoot, and so on.

Stochastic analysis. At M.I.T. in 1942, N. Wiener analyzed information processing systems using models of stochastic processes. Working in the frequency domain, he developed a statistically optimal filter for stationary continuous-time signal-to-noise ratio in a communication system.

Classical control theory. By now, the automatic control theory using frequency-domain techniques had come of age, having established itself as a paradigm. On the one hand, a firm mathematical theory for servomechanisms had been established, and on the other, engineering design techniques were provided. The period after the Second World War can be called the classical period of control theory.

1.3.3 The space/computer age and modern control

With the advent of the space age, controls design in the United States turned away from the frequency-domain techniques of classical control theory to the differential equation techniques of the late 1800s, which were couched in the time domain. Frequency-domain approach of classical control theory was appropriate for linear time-invariant systems and is at its best when dealing with single-input-single-output systems, because the graphical techniques are inconvenient to apply to multivariable systems.

Classical controls design had some successes with nonlinear systems too. Using the noise-rejection properties of frequency-domain techniques, a control system can be designed that is robust to variations in the system parameters, and to measurement errors and external disturbances. Thus, classical techniques can be used on a linearized version of a nonlinear system, giving good results at an equilibrium point about which the system behaviour is approximately linear. Frequency-domain techniques can also be applied to systems with simple types of nonlinearities using the describing function approach. Unfortunately, it is not possible to design control systems for advanced nonlinear multivariable systems, such as those arising in aerospace applications, using the assumption of linearity and treating the single-input-single-output transmission pairs one at a time.

In the Soviet Union, there was a great deal of activity in nonlinear controls design. Following the lead of Liapunov, attention was focused on time-domain techniques. In 1948, Ivachenko investigated the principle of relay control, where the control signal is switched discontinuously between discrete values. Tsypkin used the phase plane for nonlinear controls design in 1955. V.M. Popov provided his circle criterion for nonlinear stability analysis.

Sputnik. Given the history of control theory in the Soviet Union, it was only natural that the first satellite, Sputnik, was launched there in 1957. The first conference of the newly formed International Federation of Automatic Control (IFAC) was fittingly held in Moscow in 1960. The launch of Sputnik engendered tremendous activity in the United States in automatic controls design. It was clear that a return was needed to the time-domain techniques of the "primitive" period of control theory, which were based on differential equations. It should be realized that the work of Lagrange and Hamilton makes it straightforward to write nonlinear equations of motion for many dynamical systems. Thus, a control theory was needed that could deal with such nonlinear differential equations. It is quite remarkable that in almost exactly 1960, major developments occurred independently on several fronts in the theory of communication and control.

Navigation. In 1960, C.S. Draper invented his inertial navigation system, which used gyroscopes to provide accurate information on the position of a body moving in space, such as a ship, aircraft, or spacecraft. Thus, the sensors appropriate for navigation and controls design were developed.

Optimal control. In the context of modern controls design, it is usual to minimize the time of transit, or a quadratic generalized energy functional or performance index, possibly with some constraints on the allowed controls. R. Bellman applied dynamic programming to the optimal control of discrete-time systems, demonstrating that the natural direction for solving optimal control problems is backwards in time. His procedure resulted in closed-loop, generally nonlinear, feedback schemes. By 1958, L.S. Pontryagin had developed his maximum principle, which solved optimal control problems relying on the calculus of variations developed by L. Euler (1707–1783). He solved the minimum-time problem, deriving an on/off relay control law as the optimal control.

Due to R. Kalman and coworkers, in 1960s a new era in control theory had begun; we call it the era of modern control. It is a time-domain approach, making it more applicable to time-varying linear systems as well as nonlinear systems. He introduced linear algebra and matrices, so that systems with multiple inputs and outputs could easily be treated. He employed the concept of the internal system state, controllability, observability and thus this approach is one that is concerned with the internal dynamics of a system, besides its input/output behaviour. The observer design is a major achievement in modern control.

In control theory, Kalman formalized the notion of optimality in control theory by minimizing a very general quadratic generalized energy function. In estimation theory, he introduced stochastic notions that applied to nonstationary time-varying systems, thus providing a recursive solution, the Kalman filter, for the least-squares approach. The Kalman filter is the natural extension of the Wiener filter to nonstationary stochastic systems. It is no accident that from this point the U.S. space programme blossomed, with a Kalman filter providing navigational data for the first lunar landing.

Nonlinear control. During 1960s, the scientists in the U.S. extended the work of Popov and Liapunov in nonlinear stability. There was an extensive application of these results in the study of nonlinear distortion in band-limited feedback loops, nonlinear process control, aircraft controls design, and eventually in robotics.

Digital control. Without computers, modern control would have had limited applications. In 1960, the second generation of computers was introduced which used solid-state technology. By 1965, Digital Equipment Corporation was building the PDP-8, and the minicomputer industry began. Finally, in 1969 W. Hoff invented the microprocessor, and a new area developed. Control systems that are implemented on digital computers must be formulated in discrete time. Therefore, the growth of digital control theory was natural at this time. The theory of sampled data systems was developed by J.R. Ragazzini, G. Franklin, and L.A. Zadeh as well as by E.I. Jury, B.C. Kuo.

With the introduction of the PC in 1983, the design of modern control systems became possible for the individual engineer. Thereafter, several software control systems design packages were developed, including PC-Matlab and others. The guaranteed performance obtained in modern control theory by solving matrix design equations means that it is often possible to design a control system that works in theory but does not give any engineering intuition about the problem. On the other hand, the frequency-domain techniques of classical control theory impart a great deal of intuition. It initiates the demand for learning both the classical control and modern control theories.

1.4 Open-Loop Control Systems

The open-loop control systems represent the simplest form of controlling devices. Their concept and functioning are illustrated here by simple examples.

The basic control problem in an open-loop system is illustrated in Figure 1.2, where the controlled variable c (i.e. output) is to be controlled in some prescribed manner by an actuating signal e through the elements of the controlled process. For instance, consider the process where the room temperature is the controlled variable c. In order to regulate c in a desired manner, an appropriate actuating signal e must be applied to the heating system.

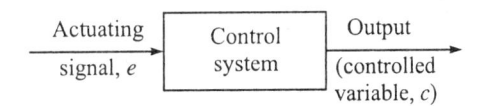

FIGURE 1.2 The basic open-loop control system.

The control adjustment of an open-loop system must depend on human intervention and estimate. In order to regulate the temperature, the human operator must estimate the amount of time of the timer for which the furnace is to be on so that the desired temperature is attained. When the preset time is up, the furnace is turned off, whether or not the room temperature has reached the desired value. The variations in environmental conditions if any, are not taken into account by this system. Therefore, an open-loop system is not truly an automatic control system.

Figure 1.3(a) illustrates a simple tank-level, open-loop control system. We wish to hold the tank level h within some reasonably acceptable limits even though the outflow through valve V_1 is varied. This can be roughly achieved by manual adjustment of the inflow rate by control valve V_2. This system does not have the capability of measuring the output flow rate through valve V_1, the input flow rate through valve V_2 or the tank level h. Figure 1.3(b) shows the block diagram relationship that exists in this system between the input (the desired tank level) and the output (the actual tank level).

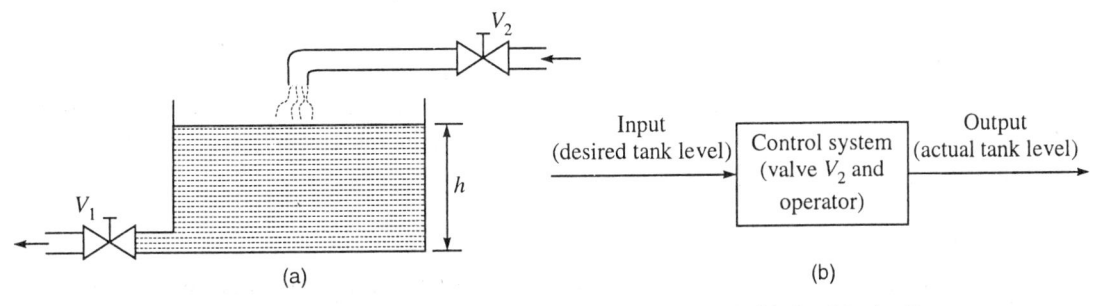

FIGURE 1.3 (a) Tank-level, open-loop control system and (b) its block diagram.

1.5 Closed-Loop (Feedback) Control Systems

What is missing in the open-loop control system for more accurate control is a link or feedback path between the output and the input of the system. A human being is probably the most complicated and sophisticated feedback control system in existence today as shown in Figure 1.4 which is self-explanatory.

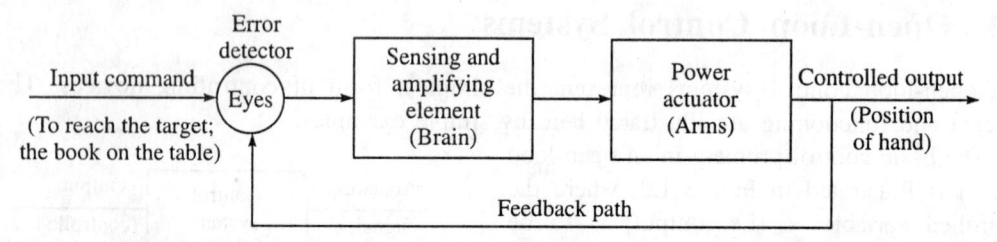

FIGURE 1.4 Human being as a closed-loop system to reach the book.

A pilot landing an aircraft is another example of a closed-loop control system. In Figure 1.5, the plant is the aircraft and the plant inputs are the pilot's manipulations of the various control surfaces and those of the throttle. The pilot with his sense of balance and motion and his visual perceptions of position, velocity, and instrument indications, acts as the sensor. The desired response is the pilot's concept of the correct flight path. The compensation or controller action is the pilot's manner of correcting the perceived errors in the flight path.

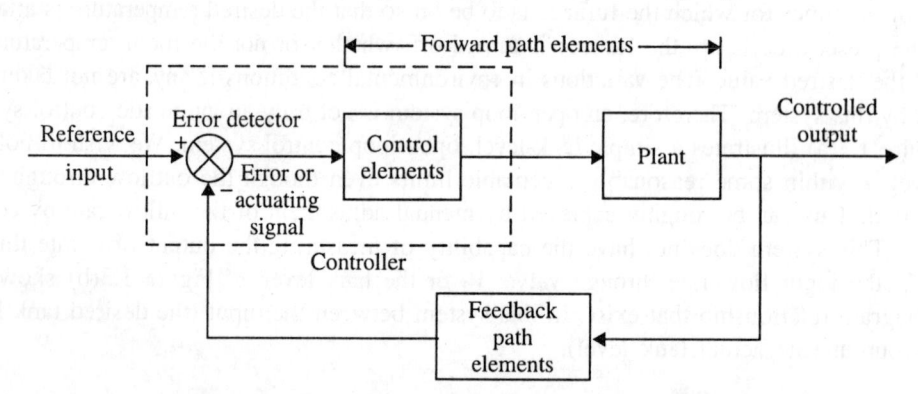

FIGURE 1.5 Block diagram of a closed-loop control system.

Closed-loop control systems derive their accurate reproduction of the reference input from feedback comparison of the output with the input. Any difference between the actual and the desired output is automatically corrected. The working principles of a few more closed-loop control systems are now discussed.

Speed control system

Historically, James Watt's flyball governor designed for controlling the speed of steam engine, was the first automatic control device used in industry. The schematic diagram of the governor

Introduction to Control Systems

is shown in Figure 1.6. Its objective is to keep the speed of the engine constant by regulating the supply of steam to it. The two flyballs on the governor rotate about a vertical axis due to the centrifugal force, at a speed proportional to the speed of the engine. The governor is directly geared to the output shaft of the engine (not shown in the figure) so that the speed of the flyballs is proportional to the output speed of the engine. The speed of the flyballs is used to control the opening of the flow control valve and thus the amount of steam entering the engine. If the speed of the engine falls below the desired value, the centrifugal force of the flyballs decreases. The lever pivoted transmits the centrifugal force from the flyballs to the top of the upper seat of the spring, causing x to decrease (move downwards) and hence by lever action the flow control valve opens more to allow more steam flow to engine, which in turn increases the speed of the engine. On the other hand, as the speed of the engine increases, the ball weights rise and move away from the shaft axis thus closing the valve more and therefore reducing the steam supply which in turn decreases the speed of the engine until equilibrium is restored.

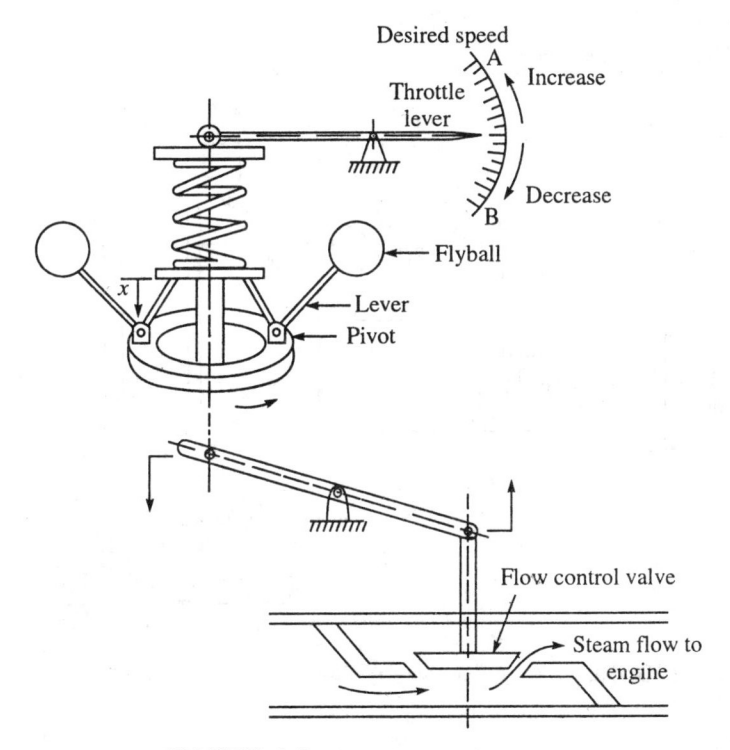

FIGURE 1.6 Speed control system.

The reference speed setting is done by the throttle lever. When the throttle lever is moved towards A, it causes x to move in the downward direction resulting in a wider opening of the flow control valve, which in turn increases the speed. The lower speed setting is achieved by moving the throttle lever in the reverse direction towards B.

Automatic tank-level control system

A tank-level control system is shown in Figure 1.7. Here, the controlled output is the liquid level h of the tank which is the plant. The valve V_1 is the output valve and V_2 is the input valve. The liquid level is sensed by a float (the sensor in the feedback path) which positions the slider arm B. Another slider arm of the potentiometric error detector is positioned according to the desired liquid level H which is the reference input. Depending upon the opening of the outlet valve V_1, the liquid level falls from the desired output level. The potentiometric error detector gives an error voltage proportional to the change in the liquid level. This error actuating signal after proper signal conditioning through the power amplifier actuates the motor which in turn regulates the valve V_2 to restore the output level of the tank to the desired liquid level at every onward instant of time until the error actuating signal is reduced to zero. Then there will be no further movement of the shaft B which implies that the output level remains at the desired level till further deviation occurs.

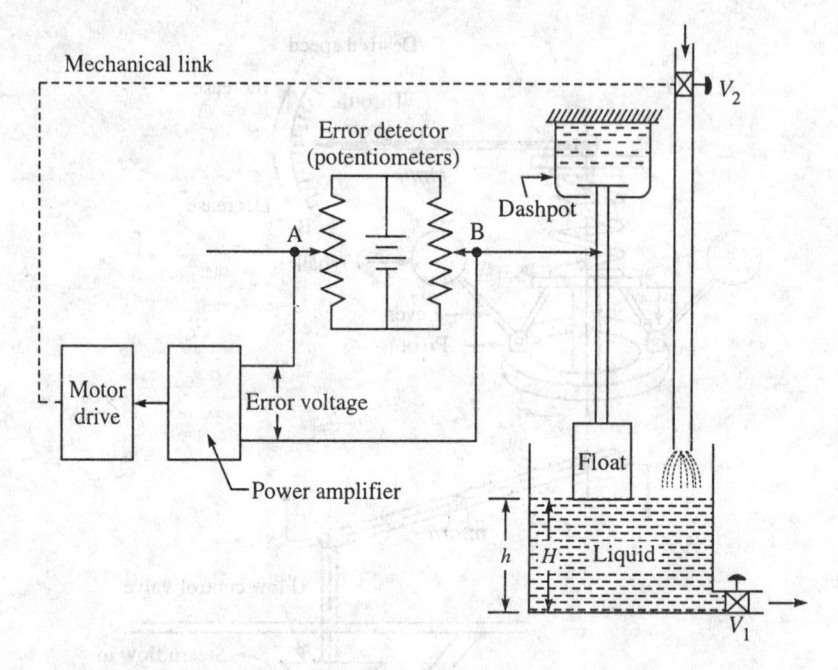

FIGURE 1.7 An automatic tank-level control system.

Servo-control system

A position-controlled servo system is shown in Figure 1.8. The servo system is used to control the position of a load shaft to which the driving motor is geared. The controlled output is denoted by θ_c and the reference desired position by θ_r. The potentiometric error detector produces an output voltage $v_e = K_p \theta_e$ where $\theta_e (= \theta_r - \theta_c)$ is the error in angular position and K_p is the proportionality constant of the error detector. The voltage v_e is amplified and used to control the field current of a dc generator which supplies the armature voltage to the driving

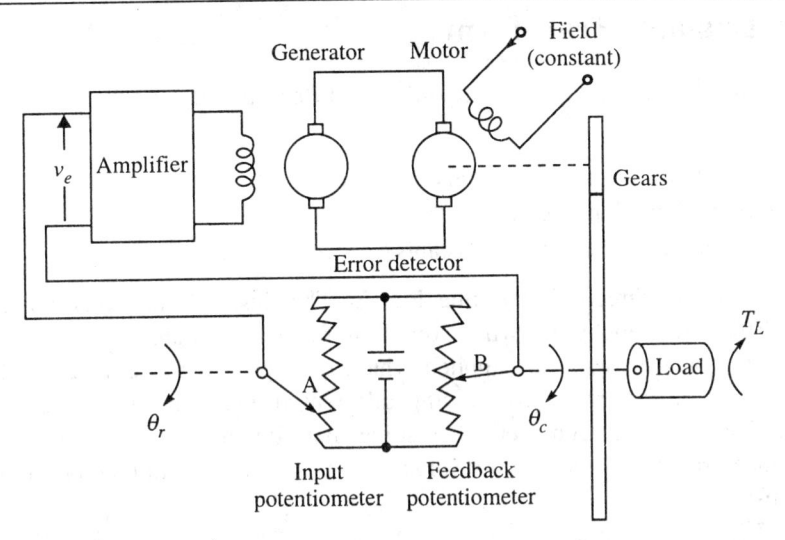

FIGURE 1.8 A position control system.

motor. The motor shaft which is coupled to the load through the gears, positions the load at the desired position θ_r so that the error is zero and then the motor shaft stops moving from the desired location.

The position control system has many applications, to mention a few, machine tool position control, missile guidance system, constant-tension control of sheet rolls in paper mills, and so forth.

Besides reducing the error between the reference input and the system output as illustrated in the above simple example, the negative feedback principle used in closed-loop control systems has the following merits and demerits:

Merits of feedback

1. It can reduce the sensitivity of the system to parameter variations.
2. It can reduce the effect of noise and disturbance on system response.
3. It produces beneficial effects on bandwidth, impedance, transient and frequency responses.

Demerits of feedback

1. It increases the number of components in the system thereby increasing complexity.
2. It affects the gain of the system.
3. It also introduces the possibility of instability in the system.

The advantages of feedback, however, far outweigh the disadvantages, as we shall see in later chapters.

The classical feedback theory, which covers a good portion of this text, requires the mathematical background in complex variables, differential equations, and Laplace transform. A brief review of Laplace transform is presented in the next section.

1.6 The Laplace Transform

Three major steps are required for solving ordinary differential equations by classical methods, namely:

(i) Determination of complementary function
(ii) Determination of particular integral
(iii) Determination of arbitrary constants

The solution is obtained directly in the time domain. The classical method is, however, difficult to apply to differential equations with excitation function which contain derivatives, and therefore the transform methods are applied which have proved to be superior. The classical method is used as the last resort for solving differential equations when the transformation method fails. The transformation method is somewhat similar to logarithmic operation.

The Laplace transform method to solve an integro-differential equation requires the following steps:

(i) Transformation of the time-domain integro-differential equation into an algebraic equation in the s (the Laplace operator) domain.
(ii) Finding the roots of the characteristic equation which is the algebraic equation in s.
(iii) Finding the inverse Laplace transform from the Laplace transformation table, to get the solution in the time domain.

The Laplace transform is one of the major mathematical tools used in the solution of ordinary integro-differential equations. For complex systems, the method of Laplace transform has a definite advantage over the classical method as explained below:

(i) The solution of linear differential equations becomes a systematic algebraic procedure by use of Laplace transforms.
(ii) The initial conditions are automatically considered in a specific transform operation.
(iii) The Laplace transform method gives the complete solution, both complementary as well as the particular solution in one operation.

The single-sided Laplace transform (commonly referred as Laplace transform) of a causal function ($f(t) = 0$ for $t < 0$) is defined as

$$F(s) = \mathscr{L}[f(t)] = \int_0^\infty f(t)e^{-st}dt$$

In the Laplace transformation we use the notation s for a complex variable, that is,

$$s = \sigma + j\omega$$

where σ is the real part and ω is the imaginary part. The properties of Laplace tranform are given in Table 1.1. The Laplace transform pairs of some commonly encountered functions are given in Table 1.2.

It may be noted that in majority of engineering applications, the time functions of interest are causal. Noncausal functions can be treated through the double-sided Laplace transform which is similar in this respect to the Fourier transform but has a stronger convergence property.

TABLE 1.1 Properties of the Laplace transform

Transform pair	$f(t) \leftrightarrow F(s)$
Linearity	$a_1 f_1(t) + a_2 f_2(t) \leftrightarrow a_1 F_1(s) + a_2 F_2(s)$
Scale change	$f\left(\dfrac{t}{a}\right) \leftrightarrow aF(as); \ a > 0$
Real translation	$f(t - t_0) \leftrightarrow e^{-st_0} F(s)$
Complex translation	$e^{-at} f(t) \leftrightarrow F(s + a)$
Real differentiation	$\dfrac{d^n}{dt^n} f(t) \leftrightarrow s^n F(s) - s^{n-1} f(0) - s^{n-2} \dfrac{d}{dt} f(0) - \cdots - \dfrac{d^{n-1}}{dt^{n-1}} f(0)$
Real integration	$\displaystyle\int_{-\infty}^{t} f(0)\, dt \leftrightarrow \dfrac{F(s)}{s} + \dfrac{\int_{-\infty}^{t} f(0)\, dt \big|_{t=0}}{s}$
Multiplication by t^n	$t^n f(t) \leftrightarrow (-1)^n \dfrac{d^n F(s)}{ds^n}$
Multiplication by $\dfrac{1}{t}$	$\dfrac{1}{t} f(t) \leftrightarrow \displaystyle\int_s^{\infty} F(s)\, ds$
Real multiplication (complex convolution)	$f_1(t) f_2(t) \leftrightarrow F_1(s) * F_2(s)$
Complex multiplication (real convolution)	$f_1(t) * f_2(t) \leftrightarrow F_1(s) F_2(s)$
Initial-value theorem	$\lim_{t \to 0} f(t) = \lim_{s \to \infty} s F(s)$
Final-value theorem	$\lim_{t \to \infty} f(t) = \lim_{s \to 0} s F(s)$

By use of the real-integration property and the final-value theorem, we obtain

$$\int_0^{\infty} f(t)\, dt = \lim_{t \to \infty} \int_0^t f(t)\, dt = \lim_{s \to 0} s \dfrac{F(s)}{s} = \lim_{s \to 0} F(s)$$

TABLE 1.2 The Laplace transform pairs

$F(s)$	$f(t); \ t \geq 0$
1	Unit impulse, $\delta(t)$
$\dfrac{1}{s}$	Unit step, $u(t)$
$\dfrac{n!}{s^{n+1}}$	$t^n \quad (n = \text{integer})$
$\dfrac{1}{s+a}$	e^{-at}
$\dfrac{1}{(s+a)^n}$	$\dfrac{1}{(n-1)!} t^{n-1} e^{-at} \quad (n = \text{integer})$

(Contd.)

TABLE 1.2 The Laplace transform pairs (*Contd.*)

$F(s)$	$f(t); t \geq 0$
$\dfrac{a}{s(s+a)}$	$1 - e^{-at}$
$\dfrac{1}{(s+a)(s+b)}$	$\dfrac{1}{b-a}(e^{-at} - e^{-bt})$
$\dfrac{s}{(s+a)(s+b)}$	$\dfrac{1}{b-a}(be^{-bt} - ae^{-at})$
$\dfrac{s}{s(s+a)(s+b)}$	$\dfrac{1}{ab}\left[1 + \dfrac{1}{a-b}(be^{-at} - ae^{-bt})\right]$
$\dfrac{\omega}{s^2 + \omega^2}$	$\sin \omega t$
$\dfrac{s}{s^2 + \omega^2}$	$\cos \omega t$
$\dfrac{\omega}{(s+a)^2 + \omega^2}$	$e^{-at} \sin \omega t$
$\dfrac{s+a}{(s+a)^2 + \omega^2}$	$e^{-at} \cos \omega t$
$\dfrac{1}{s^2(s+a)}$	$\dfrac{1}{a^2}(at - 1 + e^{-at})$
$\dfrac{\omega_n^2}{s^2 + 2\zeta\omega_n s + \omega_n^2}$	$\dfrac{\omega_n}{\sqrt{1-\zeta^2}} e^{-\zeta\omega_n t} \sin(\omega_n \sqrt{1-\zeta^2}\, t); \quad \zeta < 1$
$\dfrac{s}{s^2 + 2\zeta\omega_n s + \omega_n^2}$	$\dfrac{-1}{\sqrt{1-\zeta^2}} e^{-\zeta\omega_n t} \sin(\omega_n \sqrt{1-\zeta^2}\, t - \phi)$
	$\phi = \tan^{-1} \dfrac{\sqrt{1-\zeta^2}}{\zeta}; \quad \zeta < 1$
$\dfrac{\omega_n^2}{s(s^2 + 2\zeta\omega_n s + \omega_n^2)}$	$1 - \dfrac{1}{\sqrt{1-\zeta^2}} e^{-\zeta\omega_n t} \sin(\omega_n \sqrt{1-\zeta^2}\, t + \phi)$
	$\phi = \tan^{-1} \dfrac{\sqrt{1-\zeta^2}}{\zeta}; \quad \zeta < 1$

EXAMPLE 1.1 Draw the waveforms of time functions (a) and (b) and find their Laplace transforms.

(a) $f(t) = 0$, for $t < 0$
$= e^{-0.4t} \cos 2t$, for $t \geq 0$

(b) $f(t) = 0$, for $t < 0$
$= \sin\left(4t + \dfrac{\pi}{3}\right)$, for $t \geq 0$

Solution: (a) $F(s) = \dfrac{s + 0.4}{(s + 0.4)^2 + (2)^2} = \dfrac{s + 0.4}{s^2 + 0.8s + 4.16}$

Multiplication of waveforms of $e^{-0.4t}$ as in (i) and $\cos 2t$ as in (ii) below, graphically gives the curve of $e^{-0.4t} \cos 2t$ for $t \geq 0$, which is shown in (iii).

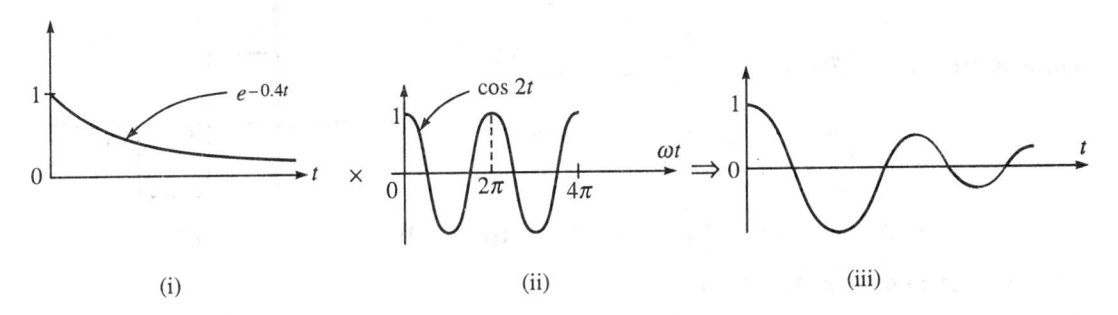

(i)   (ii)   (iii)

(b) We have drawn the curve (iv) with $\omega = 4 \Rightarrow 2\pi f = 2\pi/T = 4 \Rightarrow T = \pi/2$ and $\theta = \pi/3 = 60°$. Hence, $f(t) = 0.5 \sin 4t + 0.866 \cos 4t$. Then,

$$F(s) = \dfrac{0.5 \times 4}{s^2 + 4^2} + \dfrac{0.866s}{s^2 + 4^2} = \dfrac{2 + 0.866s}{s^2 + 16}$$

EXAMPLE 1.2 Obtain the inverse Laplace transform of $F(s)$ given by

$$F(s) = \dfrac{2 + 0.866s}{s^2 + 16}$$

(iv)

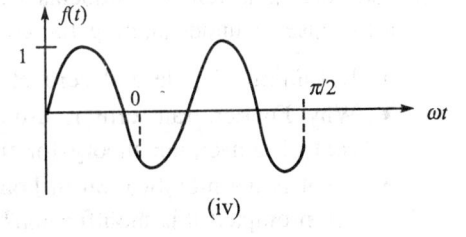

Solution: $F(s) = \dfrac{2 + 0.866s}{s^2 + 16} = \dfrac{(1/2)4}{s^2 + 4^2} + \dfrac{0.866s}{s^2 + 4^2}$

Therefore, $f(t) = \mathscr{L}^{-1}F(s) = 0.5 \sin 4t + 0.866 \cos 4t = \sin\left(4t + \dfrac{\pi}{3}\right)$

EXAMPLE 1.3 Given, $F(s) = \dfrac{\omega \cos \theta + s \sin \theta}{s^2 + \omega^2}$, derive the time function $f(t)$ and draw its waveform.

Solution: $f(t) = \mathscr{L}^{-1}F(s)$

$$= \cos\theta \, \mathscr{L}^{-1}\frac{\omega}{s^2+\omega^2} + \sin\theta \, \mathscr{L}^{-1}\frac{s}{s^2+\omega^2}$$

$$= \cos\theta \sin\omega t + \sin\theta \cos\omega t$$

$$= \sin(\omega t + \theta)$$

The time function is drawn in (v).

EXAMPLE 1.4 Given,

$$F(s) = \frac{2}{s}(1 - 2e^{-as} + e^{-2as}),$$

derive the time function $f(t)$ and draw its waveform.

Solution: $f(t) = \mathscr{L}^{-1}F(s) = \mathscr{L}^{-1}\frac{2}{s}(1 - 2e^{-as} + e^{-2as})$

$$= \mathscr{L}^{-1}\frac{2}{s} - \mathscr{L}^{-1}\frac{4e^{-as}}{s} + \mathscr{L}^{-1}\frac{2e^{-2as}}{s}$$

$$= (2 \times 1) - [4 \times 1(t-a)] + [2 \times 1(t-2a)]$$

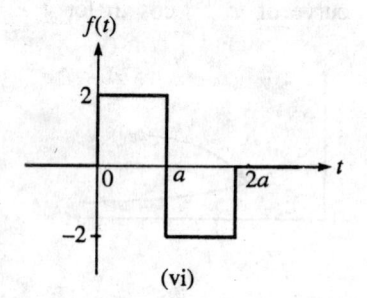

The time function is drawn in (vi).

Students are expected to be familiar with Laplace transform as this topic is covered in courses in mathematics and circuit theory. The Laplace transform theory is therefore not repeated here. It is felt that students can brush up this topic with reference to the following issues for ease in understanding the control theory presented in this book.

- Definition of pole and zero of rational functions.
- Why Fourier transform is used in communication theory and Laplace transform in control and circuit theory for the analysis of continuous linear systems.
- What is the modification in Fourier transform for Laplace transform to emerge. What has prompted this modification? What is the advantage involved in this modification?
- What is the advantage of the Laplace transform method in solving ordinary linear time-invariant differential equations over the classical method of solving differential equations.
- Definition of convolution integral.
- What is the advantage of the final-value theorem in Laplace transform.
- If $\mathscr{L}[f_1(t)] = F_1(s)$ and $\mathscr{L}[f_2(t)] = F_2(s)$, what is $\mathscr{L}[f_1(t)f_2(t)]$ in terms of $F_1(s)$ and $F_2(s)$.
- Given $f_1(t) = e^{-t}$, $t > 0$ and $f_2(t) = 1$, $t > 0$ and zero otherwise, evaluate $\mathscr{L}[f_1(t)f_2(t)]$ by the Laplace transform method. Compare the result obtained with that obtained by convolution integral.

- When the Laplace transform method is available, where is the need for convolution integral? Discuss.
- Definitions of rational and irrational functions.
- For inverse Laplace transform, the knowledge of roots of the characteristic equation is essential. Discuss.
- What is Dirichlet condition?
- What is the condition under which the final-value theorem is valid?

1.6.1 Transfer function concept

The transfer function is defined as the ratio of the Laplace transform of output to the Laplace transform of input with all initial conditions as zero. The concept of transfer function is applicable to single-input-single-output, linear time-invariant systems. The dynamics of a linear time-invariant system (whose coefficients are constant) are represented by a linear differential equation such as:

$$a_n \frac{d^n y(t)}{dt^n} + a_{n-1} \frac{d^{n-1} y(t)}{dt^{n-1}} + \cdots + a_1 \frac{dy(t)}{dt} + a_0 y(t) = b_m \frac{d^m u(t)}{dt^m} + \cdots + b_1 \frac{du(t)}{dt} + b_0 u(t) \quad (1.1)$$

The Laplace transform of the above equation with zero initial conditions leads to

$$(a_n s^n + a_{n-1} s^{n-1} + \cdots + a_1 s + a_0) Y(s) = (b_m s^m + b_{m-1} s^{m-1} + \cdots + b_1 s + b_0) U(s) \quad (1.2)$$

Designating the transfer function of the system as $G(s)$, we get

$$\text{Transfer function, } G(s) = \left. \frac{\text{Laplace transform of output}}{\text{Laplace transform of input}} \right|_{\text{with zero initial conditions}}$$

$$= \frac{Y(s)}{U(s)} = \frac{b_m s^m + b_{m-1} s^{m-1} + \cdots + b_1 s + b_0}{a_n s^n + a_{n-1} s^{n-1} + \cdots + a_1 s + a_0} \quad (1.3)$$

In general, the transfer function $G(s)$ is the ratio of two polynomials in s such as

$$G(s) = \frac{P(s)}{Q(s)} \quad (1.4)$$

The points at which the numerator polynomial $P(s)$ equals zero are called the zeros of the system $G(s)$. A zero is indicated by a small circle 'o'. The points at which the denominator polynomial $Q(s)$ equals zero are called the poles of the system $G(s)$. A pole is indicated by a small cross '×'. At zeros the transfer function $G(s)$ becomes zero and at poles it becomes infinite. The denominator polynomial for the transfer function is called the characteristic polynomial. The roots of the characteristic equation (characteristic polynomial equated to zero) are the poles and characterize the system behaviour.

The transfer function is a rational function. It means that its number of zeros and poles are equal taking poles and zeros at infinity into consideration. For example, if for a transfer function, the numerator polynominal is of degree 2 and the denominator polynomial is of degree 5, then there are 5 finite poles, 2 finite zeros and 3 zeros at infinity.

With zero initial conditions, the Laplace transform of the output of the system shown in Figure 1.9 is

$$C(s) = G(s)R(s) \qquad (1.5)$$

For an impulse input, i.e. $R(s) = 1$, the impulse response is the transfer function in time domain, i.e.

$$c(t) = \mathscr{L}^{-1}C(s) = \mathscr{L}^{-1}G(s) = g(t)$$

FIGURE 1.9

EXAMPLE 1.5 Figure 1.10 represents an electrical system which is used for integration or to provide phase lag. Derive the transfer function of the system.

Solution: The application of KVL to the circuit of Figure 1.10 gives the loop equation as

$$Ri + \frac{1}{C}\int i\,dt = e_i(t)$$

and the output

$$e_o(t) = \frac{1}{C}\int i\,dt$$

$$\therefore \qquad G(s) = \frac{E_o(s)}{E_i(s)} = \frac{1}{RCs + 1}$$

(with zero initial conditions)

FIGURE 1.10 Example 1.5: Integrating circuit.

EXAMPLE 1.6 Figure 1.11(a) shows a spring-mass-damper system. Derive the transfer function of the system.

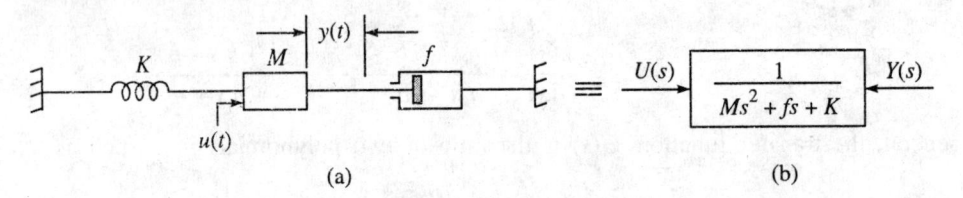

FIGURE 1.11 Example 1.6: spring-mass-damper system.

Solution: The force due to spring is

$$F_S = -Ky$$

where K is the spring constant. The negative sign is due to recoiling of the spring in the direction opposite to that of applied force $u(t)$.

The force due to viscous damping is

$$F_f = -f\frac{dy}{dt}$$

where f is the viscous damping coefficient. The negative sign is due to the direction of viscous damping force being opposite to that of $u(t)$.

Introduction to Control Systems

The force due to inertial mass M is

$$F_M = M\frac{d^2y}{dt^2}$$

The force balance equation gives

$$M\left(\frac{d^2y}{dt^2}\right) = u(t) - Ky - f\left(\frac{dy}{dt}\right)$$

or

$$M\left(\frac{d^2y}{dt^2}\right) + f\left(\frac{dy}{dt}\right) + Ky = u(t) \tag{1.6}$$

or

$$G(s) = \frac{Y(s)}{U(s)} = \frac{1}{Ms^2 + fs + K} \tag{1.7}$$

as shown in Figure 1.11(b). The transfer function is thus a complex quantity having both magnitude and phase (see Example 1.9).

1.6.2 Partial fraction expansion and MATLAB

Computation (MATLAB) approach in control system analysis is widely accepted and its understanding in this book is presented with the help of examples as we proceed onward in different chapters. Now we explain the partial fraction expansion of $P(s)/Q(s)$.

Consider the rational function $F(s)$ as the ratio of two polynomials of s as $P(s)/Q(s)$, which can be after partial fraction expansion obtained as

$$F(s) = \frac{P(s)}{Q(s)} = \frac{b_m s^m + b_{m-1} s^{m-1} + \cdots + b_1 s + b_0}{s^n + a_{n-1} s^{n-1} + \cdots + a_1 s + a_0}$$

$$= k(s) + \frac{r_1}{s - p_1} + \frac{r_2}{s - p_2} + \cdots + \frac{r_n}{s - p_n} \tag{1.8}$$

where p_i's are the poles and r_i's are the corresponding residues and $k(s)$ is the direct term.

As an example in the line of Eq. (1.8), we consider $F(s)$ as

$$F(s) = \frac{s^5 + 8s^4 + 23s^3 + 35s^2 + 28s + 3}{s^3 + 6s^2 + 8s} = (s^2 + 2s + 3) + \frac{0.375}{s+4} + \frac{0.25}{s+2} + \frac{0.375}{s}$$

For this function $F(s)$,

```
num = [1  8  23  35  28  3];
den = [0  0  1  6  8  0];
```

the command

$$[r, p, k] = \text{residue (num, den)}$$

gives the following result:

```
num=[1 8 23 35 28 3];
den=[0 0 1 6 8 0];
[r,p,k]=residue(num,den)

r =
    0.3750
    0.2500
    0.3750

p =
   -4
   -2
    0

k =
    1   2   3
```

If the degree of the numerator polynomial is less than that of the denominator polynomial, then $F(s)$ is strictly proper, and $k(s) = 0$. The order of the denominator polynomial is the number of poles. The poles may be distinct or repetitive. If the degree of the numerator polynomial is equal to or greater than that of the denominator polynomial, then $F(s)$ is improper.

For a strictly proper function $F(s)$, let

$$F(s) = \frac{K(s+z_1)(s+z_2)\cdots(s+z_m)}{(s+p_1)(s+p_2)\cdots(s+p_n)}; \quad n > m \tag{1.9}$$

$$= \frac{A_1}{s+p_1} + \frac{A_2}{s+p_2} + \cdots + \frac{A_j}{s+p_j} + \cdots + \frac{A_n}{s+p_n}$$

where the residue, A_j ($j = 1, 2, \ldots, n$) can be evaluated as

$$A_j = [(s+p_j)F(s)]_{x=-p_j} \tag{1.10}$$

For distinct poles, consider

$$F(s) = \frac{6}{s^3 + 6s^2 + 11s + 6}$$

For this function $F(s)$,

```
num = [0 0 0 6];
den = [1 6 11 6];
```

the command

$$[r, p, k] = \text{residue(num, den)}$$

gives the following result:

$$F(s) = \frac{3}{s+3} + \frac{-6}{s+2} + \frac{3}{s+1}$$

```
num=[0 0 0 6];
den=[1 6 11 6];
[r,p,k]=residue(num,den)
r =
    3.0000
   -6.0000
    3.0000
p =
   -3.0000
   -2.0000
   -1.0000
k =
   [ ]
```

It may be noted that the direct term k is zero and is indicated in MATLAB output as $k = [\ ]$. The original function $F(s)$ is obtained from r,p,k with the following program:

```
           [num,den]=residue(r,p,k);
           printsys(num,den,'s')
num/den =
   -3.9968e-015s^2 - 1.4211e-014 s + 6
   ─────────────────────────────────────
          s^3 + 6s^2 + 11s + 6
```

It effectively becomes

```
num/den =
         6
   ──────────────────
   s^3 + 6s^2 + 11s + 6
```

If a pole p_j has multiplicity $r > 1$, the partial fraction contains additional terms of the form

$$F(s) = \frac{A_{j,1}}{s+p_j} + \frac{A_{j,2}}{(s+p_j)^2} + \cdots + \frac{A_{j,r}}{(s+p_j)^r} \qquad (1.11)$$

where

$$A_{j,r} = \left[(s+p_j)^r F(s)\right]$$

$$A_{j,r-1} = \frac{d}{ds}\left[(s+p_j)^r F(s)\right]_{s=-p_j}$$

$$\vdots \qquad \vdots$$

$$A_{j,r-i} = \frac{1}{(r-i)!}\frac{d^{r-i}}{ds^{r-i}}\left[(s+p_j)^r F(s)\right]_{s=-p_j}$$

for $i = 2$ to $r - 1$

EXAMPLE 1.7 Consider:

$$F(s) = \frac{10(s+2)}{(s+1)^2(s+3)} = \frac{A_{1,1}}{s+1} + \frac{A_{1,2}}{(s+1)^2} + \frac{A_2}{s+3} = \frac{2.5}{s+1} + \frac{5}{(s+1)^2} + \frac{-2.5}{s+3}$$

where $A_{1,2} = \left[(s+1)^2 F(s)\right]_{s=-1} = 5$; $A_{1,1} = \frac{d}{ds}\left[(s+1)^2 F(s)\right]_{s=-1} = 2.5$; $A_2 = \left[(s+3)F(s)\right]_{s=-3} = -2.5$

```
den1=conv([1 1],[1 1])
den=conv([den1],[1 3])
den1 =
        1   2   1
den =
        1   5   7   3
```

```
num=[0 0 20 40];
den1=conv([den1],[1 1]);
den=con([den1],[1 3]);
printsys(num,den,'s')
[r, p, k]=residue(num,den)
den =
        1   2   1
num/den =
          20 s + 40
     ─────────────────────
     s^3 + 5s^2 + 7s + 3
r =
       -5.0000
        5.0000
       10.0000
p =
       -3.0000
       -1.0000
       -1.0000
k =
       [ ]
```

Finding zeros and poles of transfer function

MATLAB has a command

$$[z,p,K] = tf2zp(num,den)$$

to obtain the zeros, poles and gain K of the transfer function.

Consider the function defined by

$$F(s) = \frac{20s + 20}{s^3 + 5s^2 + 7s + 3}$$

then the MATLAB program

```
num=[0 0 20 20];
den=[1 5 7 3];
[z,p,K]=tf2zp(num,den)
```

will produce the output as

```
z =
    -1
p =
    -3.0000
    -1.0000 + 0.0000i
    -1.0000 - 0.0000i
K =
    20
```

A zero is at $s = -1$. The poles are at $s = -3, -1$ and -1. The gain K is 20.

Further, if the zeros, poles and gain K are given, then the following MATLAB program will give the numerator and denominator polynomials of the transfer function as

```
z = [-1];
p=[-3;-1;-1];
K=20;
[num,den]=zp2tf(z,p,K);
printsys(num,den, 's')
num/den =
            20s + 20
        ---------------------
        s^3 + 5s^2 + 7s + 3
```

1.7 State Variable Formulation

Let us define state, state variable, state vector and state space.

State. The state of a dynamic system is the minimal amount of information required, together with the initial condition at time $t = t_0$ and input excitation, to completely specify the future behaviour of the system for any time $t > t_0$.

State variables. These are the smallest set of variables which determine the state of the dynamic system. If at least n variables $x_1(t)$, $x_2(t)$, ..., $x_n(t)$ are needed to completely describe the future behaviour of the system, together with the initial state and input excitation, then these n variables $[x_1(t), x_2(t), ..., x_n(t)]$ are a set of state variables. Note that the state variables need not be physically measurable or observable quantities.

State vector. The n state variables can be considered the n components of the state vector $X(t)$ described in n-dimensional vector-space called the state space. For example, the state vector

$$X(t) = [x_1(t) \quad x_2(t) \quad \cdots \quad x_n(t)]^T$$

The differential equation (1.6) represents the dynamics of the spring-mass-damper system of Figure 1.11. Let us choose the output as one of the state variables, i.e.

$$y = x_1$$

and
$$\dot{x}_1 = x_2 = \dot{y}$$

In order to represent the system dynamics in terms of state variables, substituting the above state variables in Eq. (1.6), we obtain

$$M\frac{dx_2}{dt} + fx_2 + Kx_1 = u(t) \tag{1.12}$$

Rewriting the above two state variable equations by a set of two first-order differential equations as

$$\dot{x}_1 = x_2 = 0.x_1 + 1.x_2 + 0.u(t) \tag{1.13}$$

and from Eq. (1.12), we get

$$\dot{x}_2 = -\frac{K}{M}x_1 - \frac{f}{M}x_2 - \frac{1}{M}u(t) \tag{1.14}$$

Equations (1.13) and (1.14) can be written in matrix form as

$$\dot{X} = AX + Bu \tag{AB}$$

and output
$$y = CX + Du \tag{CD}$$

where
$$A = \begin{bmatrix} 0 & 1 \\ -\dfrac{K}{M} & -\dfrac{f}{M} \end{bmatrix}, \quad B = \begin{bmatrix} 0 \\ \dfrac{1}{M} \end{bmatrix}, \quad C = [1 \quad 0], \quad D = [0]$$

Introduction to Control Systems

The block diagram representation of Eqs. (AB) and (CD) is shown in Figure 1.12(a).

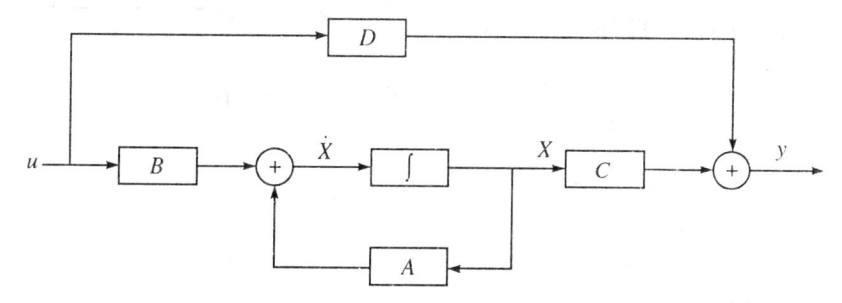

FIGURE 1.12(a) Block diagram representation of Eqs. (AB) and (CD).

The simulation of the spring-mass-damper system is shown in Figure 1.12(b).

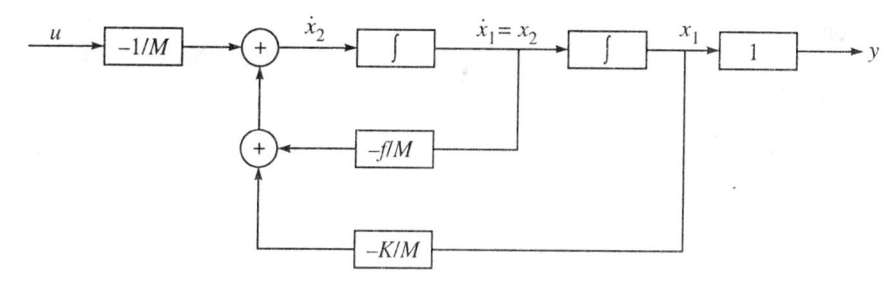

FIGURE 1.12(b) Block diagram of the spring-mass-damper system shown in Figure 1.11.

Derivation of transfer function

Let us first consider the problem of determining the transfer function of a system having state variable representation as in Eqs. (AB) and (CD). Taking the Laplace transforms, we get

$$sX(s) - X(0) = AX(s) + BU(s) \tag{1.15}$$

and
$$Y(s) = CX(s) + DU(s) \tag{1.16}$$

As by definition,

$$\text{Transfer function} = \left.\frac{\text{Laplace transform of output}}{\text{Laplace transform of input}}\right|_{\text{Initial conditions}=0}$$

we put $X(0) = 0$, and grouping the two $X(s)$ terms, we get

$$[sI - A]X(s) = BU(s)$$

where the identity matrix I has been introduced to allow the indicated factoring as the Laplace operator s is a scalar. If both sides of this equation are now premultiplied by $[sI - A]^{-1}$, we obtain

$$X(s) = (sI - A)^{-1}BU(s) \tag{1.17}$$

By substituting Eq. (1.17) into Eq. (1.16), we obtain

$$Y(s) = [C[sI - A]^{-1}B + D]U(s)$$

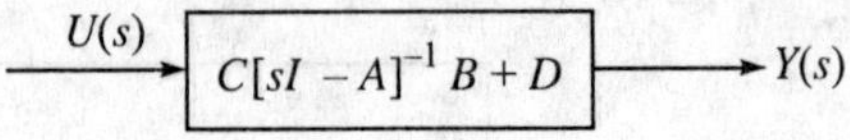

The transfer matrix $G(s)$ [see Figure 1.12(c)] for a multivariable system is therefore given by

FIGURE 1.12(c) Transfer matrix.

$$G(s) = \frac{Y(s)}{U(s)} = C[sI - A]^{-1}B + D = C\frac{\text{Adjoint }[sI - A]}{|sI - A|}B + D$$

If there is no direct coupling between input and output (which is the usual case), i.e. $D = 0$, in that case, we have

$$G(s) = C[sI - A]^{-1}B = C\frac{\text{Adjoint}[sI - A]}{|sI - A|}B$$

In the case of the single-input-single-output (SISO) system, $G(s)$ is a scalar and is termed the transfer function.

For illustration, let us take the dynamics of the spring-mass-damper system using the differential equation

$$\ddot{y} + 3\dot{y} + 2y = u \qquad (1.18)$$

Now, let the choice of the states be

$$x_1 = y$$

and

$$x_2 = \dot{x}_1 = \dot{y}$$

Then the differential equation (1.18) becomes

$$\ddot{x}_1 + 3\dot{x}_1 + 2x_1 = u$$

or

$$\frac{d}{dt}(\dot{x}_1) + 3\dot{x}_1 + 2x_1 = u$$

or

$$\dot{x}_2 + 3x_2 + 2x_1 = u$$

or

$$\dot{x}_2 = -2x_1 - 3x_2 + u$$

Now, we can rewrite these derivatives of the states as

$$\dot{x}_1 = x_2 = 0.x_1 + 1.x_2 + 0.u$$

and

$$\dot{x}_2 = -2.x_1 - 3.x_2 + 1.u$$

From the preceding two equations we can write the state variable formulation in matrix form as

$$\begin{bmatrix} \dot{x}_1 \\ \dot{x}_2 \end{bmatrix} = \begin{bmatrix} 0 & 1 \\ -2 & -3 \end{bmatrix} \begin{bmatrix} x_1 \\ x_2 \end{bmatrix} + \begin{bmatrix} 0 \\ 1 \end{bmatrix} u$$

that is,

$$\dot{X} = AX + Bu \qquad \text{(AB)}$$

where
$$A = \begin{bmatrix} 0 & 1 \\ -2 & -3 \end{bmatrix}, \quad B = \begin{bmatrix} 0 \\ 1 \end{bmatrix}, \quad X = \begin{bmatrix} x_1 \\ x_2 \end{bmatrix}$$

and as we have chosen output $y = x_1$, then the output in matrix form can be written as

$$y = \begin{bmatrix} 1 & 0 \end{bmatrix} \begin{bmatrix} x_1 \\ x_2 \end{bmatrix} + [0]u$$

that is,
$$y = CX + Du \quad \text{(CD)}$$
where
$$C = [1 \ 0], \quad D = [0]$$

Hence the dynamics of the spring-mass-damper system can be represented by the vector-matrix differential equations (AB) and (CD) where X is the state vector of order (2×1); u is the input of order (1×1), a scalar; y is the output of order (1×1), a scalar; A is the system matrix of order (2×2); B is the input-coupling matrix of order (2×1); C is the output-coupling matrix of order (1×2) and D is the input-output coupling matrix of order (1×2).

Substituting the values of the matrices A, B, C, D in the transfer function expression derived from state variable formulation, we get the transfer function of the system as

$$G(s) = C[sI - A]^{-1}B + D$$

$$= [1 \ 0] \left\{ s \begin{bmatrix} 1 & 0 \\ 0 & 1 \end{bmatrix} - \begin{bmatrix} 0 & 1 \\ -2 & -3 \end{bmatrix} \right\}^{-1} \begin{bmatrix} 0 \\ 1 \end{bmatrix} + [0]$$

$$= [1 \ 0] \left\{ \begin{bmatrix} s & 0 \\ 0 & s \end{bmatrix} - \begin{bmatrix} 0 & 1 \\ -2 & -3 \end{bmatrix} \right\}^{-1} \begin{bmatrix} 0 \\ 1 \end{bmatrix}$$

$$= [1 \ 0] \frac{\begin{bmatrix} s+3 & 1 \\ -2 & s \end{bmatrix}}{s^2 + 3s + 2} \begin{bmatrix} 0 \\ 1 \end{bmatrix} = \frac{1}{s^2 + 3s + 2}$$

The transfer function of the spring-mass-damper system obtained from state variable formulation is the same as that obtained from the basic definition of transfer function by the usual Laplace transform method. The simulation of the system is shown in Figure 1.12(d).

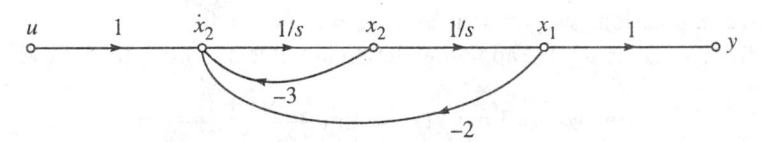

FIGURE 1.12(d) System simulation.

A *transfer function may only be defined for a linear, time-invariant (constant parameter) single-input-single-output system*. Transfer function is an input-output description of the behaviour of a system where the information about the initial conditions is lost. Thus, the transfer function description does not include any information concerning the internal structure of the system and its behaviour. On the other hand, the state variable approach can handle multivariable systems and take into account of the initial conditions as well.

In order to illustrate the usefulness of the Laplace transformation technique and the steps involved in determining the solution of a linear differential equation representing the dynamics of the system, reconsider the spring-mass-damper system which is rewritten as

$$M\left(\frac{d^2y}{dt^2}\right) + f\left(\frac{dy}{dt}\right) + Ky = u(t)$$

with the initial conditions

$$y(0) = y(t)|_{t=0} = y_0 \quad \text{and} \quad \dot{y}(0) = \frac{dy}{dt}\bigg|_{t=0} = 0$$

Taking Laplace transform, we get

$$M[s^2Y(s) - sy(0) - \dot{y}(0)] + f[sY(s) - y(0)] + KY(s) = U(s)$$

or

$$Y(s) = \frac{(Ms+f)y_0}{Ms^2 + fs + K} = \frac{P(s)}{Q(s)}$$

The denominator polynomial $Q(s) = 0$ is the characteristic equation, since the roots of this equation determine the time response and hence the characteristics of the system. The roots of the characteristic equation $Q(s) = 0$ are called the poles or the points of singularities of the system. The numerator polynomial $P(s) = 0$ gives the zeros of the system. At poles the function $Y(s)$ becomes infinity.

For the homogeneous case, $u(t) = 0$, then considering $K/M = 2$, $f/M = 3$ and $y(0) = y_0 = 1$, $\dot{y}(0) = 0$, we get

$$Y(s) = \frac{(s+3)y_0}{s^2 + 3s + 2} = \frac{s+3}{s^2 + 3s + 2} = \frac{s+3}{(s-s_1)(s-s_2)} = \frac{s+3}{(s+1)(s+2)}$$

$$= \frac{2}{s+1} - \frac{1}{s+2} \quad \text{(by partial fraction expansion)}$$

Therefore, $\quad y(t) = \mathscr{L}^{-1}Y(s) = \mathscr{L}^{-1}\frac{2}{s+1} - \mathscr{L}^{-1}\frac{1}{s+2} = 2e^{-t} - e^{-2t}$

From the above expression, we get that as $t \to \infty$, $y(\infty) = 0$.

Looking from another angle, applying the final-value theorem to $Y(s)$, we get

$$\lim_{t \to \infty} y(t) = \lim_{s \to 0} sY(s) = \lim_{s \to 0} \frac{s(s+3)}{(s+1)(s+2)} = 0$$

that is, the final position for the mass is the normal equilibrium position $y = 0$.

EXAMPLE 1.8 Consider the linear differential equation

$$\frac{d^2y}{dt^2} + \frac{dy}{dt} = e^{4t}$$

with the initial conditions

$$y(0) = 2 \quad \text{and} \quad \dot{y}(0) = 0$$

Taking Laplace transform of both sides, we get

$$[s^2Y(s) - sy(0) - \dot{y}(0)] + [sY(s) - y(0)] = \frac{1}{s-4}$$

Substituting the initial conditions,

$$Y(s) = \frac{2s^2 - 6s - 7}{s(s+1)(s-4)} \qquad (1.19)$$

The partial fraction expansion gives

$$Y(s) = \frac{7/4}{s} + \frac{1/5}{s+1} + \frac{1/20}{s-4}$$

The inverse Laplace transform gives

$$y(t) = \frac{7}{4} + \frac{1}{5}e^{-t} + \frac{1}{20}e^{4t}$$

It is obvious that the final value of $y(t)$ is infinite. However, if one were to apply the final value theorem to Eq. (1.19), the incorrect final value of (7/4) would be obtained. This example, therefore, clearly illustrates *that the final-value theorem cannot be applied when the function* is not *analytic in the right-half of s-plane.*

For a multivariable system having m-inputs and n-outputs, the input-output relationship is shown in the matrix form as

$$\begin{bmatrix} Y_1(s) \\ \vdots \\ Y_n(s) \end{bmatrix} = \begin{bmatrix} G_{11}(s) & \cdots & G_{1m}(s) \\ \vdots & & \vdots \\ G_{n1}(s) & \cdots & G_{nm}(s) \end{bmatrix} \begin{bmatrix} U_1(s) \\ \vdots \\ U_m(s) \end{bmatrix} \qquad (1.20)$$

where

$y_i (i = 1, 2, \ldots, n)$ are the n-outputs
$u_j (j = 1, 2, \ldots, m)$ are the m-inputs

$G_{ij}(s)$ is the transfer function (scalar) relating the ith output variable to the jth input variable. In matrix form, we can write

$$Y(s) = G(s)U(s)$$

where $G(s)$ is the transfer matrix and $Y(s)$ is the Laplace transform of output vector and $U(s)$ is the Laplace transform of input vector.

For example, the multivariable system shown in Figure 1.13(a) has two inputs and two outputs. Using the transfer matrix relation, we can write the simultaneous equations for the output variables as

FIGURE 1.13(a) System transfer matrix.

$$Y_1(s) = G_{11}(s)U_1(s) + G_{12}(s)U_2(s)$$
$$Y_2(s) = G_{21}(s)U_1(s) + G_{22}(s)U_2(s)$$
(1.21)

Further, using the superposition principle, we can get the transfer function as

$$G_{ij}(s) = \left. \frac{Y_i(s)}{U_j(s)} \right|_{U_k(s)=0,\, k \neq j}$$

The block diagram representation of the set of equations (1.21) is shown in Figure 1.13(b).

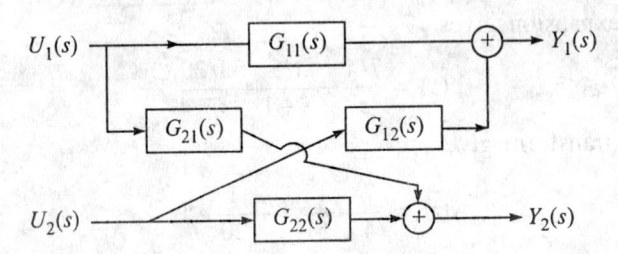

FIGURE 1.13(b) Multivariable system transfer function connectivity.

1.8 Properties of Transfer Function

We introduced the concept of transfer function in control theory in Section 1.6.1. We stated that the concept of transfer function is limited to linear, time-invariant, differential equation systems. Below, we summarize some important properties of the transfer function.

(i) The transfer function of a system is the Laplace transform of its impulse response.
(ii) The transfer function concept is applicable only to linear, time-invariant systems.
(iii) The transfer function does not take care of the initial conditions of the system. This is the greatest disadvantage. This is clear from the definition of the transfer function, stated in Section 1.6.1.
(iv) The transfer function is applicable to single-input-single-output systems, though for multivariable systems, the transfer matrix can be obtained by using the principle of superposition.
(v) The degree of the denominator polynomial of the transfer function is the order of the system. The denominator polynomial gives the poles and the numerator polynomial gives the zeros.

Before we proceed to the next property, let us see what a pole is and what a zero is.

What is a pole?

A pole is the most common type of singularity* and plays a very important role in the study of classical control theory.

A pole can be stated as: If a function $G(s)$ is analytic and single-valued in the neighbourhood of s_i, it is said to have a pole of order q at $s = s_i$ if the limit

$$\lim_{s \to s_i} [(s - s_i)^q G(s)] \tag{1.22}$$

has a finite, nonzero value.

What is a zero?

A zero is stated as: If a function $G(s)$ is analytic at $s = s_i$, it is said to have a zero of order p at $s = s_i$ if the limit

$$\lim_{s \to s_i} [(s - s_i)^p G(s)] \tag{1.23}$$

is finite or nonzero.

(vi) The transfer function can be written in the following forms:

(a) *The ratio of two polynomials form as:*

$$G(s) = \frac{Y(s)}{U(s)} = \frac{s^m + b_{m-1} s^{m-1} + \cdots + b_1 s + b_0}{s^n + a_{n-1} s^{n-1} + \cdots + a_1 s + a_0} ; \quad m \leq n \tag{1.24}$$

(b) *In time constant form as:*

$$\frac{K(sT_a + 1) \cdots (sT_m + 1)}{s^N (sT_1 + 1)(sT_2 + 1) \cdots (sT_l + 1)} ; \quad N + l \geq m \tag{1.25}$$

(c) *In pole-zero form as:*

$$G(s) = \frac{Y(s)}{U(s)} = \frac{K(s + z_1)(s + z_2) \cdots (s + z_m)}{s^N (s + p_1)(s + p_2) \cdots (s + p_l)} ; \quad N + l \geq m \tag{1.26}$$

where K is the gain factor. The poles and zeros may be complex conjugate or real. At poles the function becomes infinite and at zeros the function becomes zero. This transfer function is of type-N system and order of the system is $(N + l)$. Obviously $m \leq (N + l)$. Forms (b) and (c) are almost alike.

(vii) It is emphasized that not all transfer functions are rational algebraic expressions. The transfer function of a system including time delays contains terms of the form e^{-Ts} where T is the time delay in units of time. See Eq. (1.28).

(viii) The transfer function may be classified as (a) *strictly proper* when the degree of the denominator polynomial n is greater than that of the numerator polynomial m; i.e. $n > m$, (b) *proper* when $n \leq m$. We will confine our discussion to systems having $n \geq m$ from the physically realizable point of view.

---

*Singularities of a function are the points at which the function or its derivatives do not exist.

EXAMPLE 1.9 The system dynamics containing a time delay is represented as

$$\frac{dy(t)}{dt} + y(t) = u(t-T) \tag{1.27}$$

The transfer function of the system can be obtained by taking the Laplace transform of both sides with all initial conditions zero as

$$sY(s) + Y(s) = e^{-sT}U(s)$$

Then the transfer function becomes

$$G(s) = \frac{Y(s)}{U(s)} = \frac{e^{-sT}}{s+1} \tag{1.28}$$

EXAMPLE 1.10 The unit-step response of a given system is obtained as

$$y(t) = 1 - (7/3)e^{-t} + (3/2)e^{-2t} - (1/6)e^{-4t}$$

Determine the transfer function.

Solution: Since the time derivative of step is an impulse, the impulse response of the system is

$$g(t) = \frac{dy}{dt} = (7/3)e^{-t} - 3e^{-2t} + (2/3)e^{-4t}$$

The Laplace transform of $g(t)$ gives the desired transfer function as

$$G(s) = \frac{7/3}{s+1} + \frac{-3}{s+2} + \frac{2/3}{s+4} = \frac{s+8}{(s+1)(s+2)(s+4)}$$

1.9 Linear Approximation of Physical Systems

A system is linear if and only if the properties of superposition and homogeneity are satisfied. When a system at *rest* is subjected to an excitation $x_1(t)$, it provides a response $y_1(t)$. Furthermore, when the system is subjected to an excitation $x_2(t)$, it provides a corresponding response $y_2(t)$. For a linear system it is *necessary* that the excitation $[x_1(t) + x_2(t)]$ results in a response $[y_1(t) + y_2(t)]$. This is called the *principle of superposition.*

It is necessary that the magnitude scale factor is preserved in a linear system. Consider a system with an input x which results in an output y. It is necessary that the response of a linear system to a constant multiple α of an input x is equal to the response to the input multiplied by the same constant α so that the output is equal to αy. This is called the property of *homogeneity.*

A system characterized by the relation $y = x^2$ is not linear since the principle of superposition is not valid. A system represented by the relation $y = mx + c$ ($c \neq 0$) is not linear since it does not satisfy the homogeneity property, whereas for the system represented by the relation $y = mx$ the homogeneity property is satisfied. In fact, the system represented by the relation $y = mx$ is linear as both the properties, i.e. the principle of superposition and the homogeneity property are satisfied. However the device represented by the relation $y = mx + c$ may be considered linear about an operating point (x_0, y_0) for small changes Δx and Δy. When $x = x_0 + \Delta x$ and $y = y_0 + \Delta y$, we have

$$y_0 + \Delta y = mx_0 + m\Delta x + c \qquad (1.29)$$

and therefore $\Delta y = m\Delta x$, i.e. $m = \Delta y/\Delta x$, which satisfies the necessary conditions required for a linear system. In this way, we can linearize the nonlinear elements assuming small-signal conditions as done in electronic circuits.

A majority of physical systems are linear within some range of variables. For example, the spring-mass-damper system as described by Eq. (1.6), as long as subjected to small deflections $y(t)$, is linear. However, if the deflection continues to increase beyond limits, the spring would be overextended and break. Therefore, the question of linearity within the range of applicability only must be considered for each system.

Let us consider the input-output relationship of the system depicted in Figure 1.14 and represented by the relationship

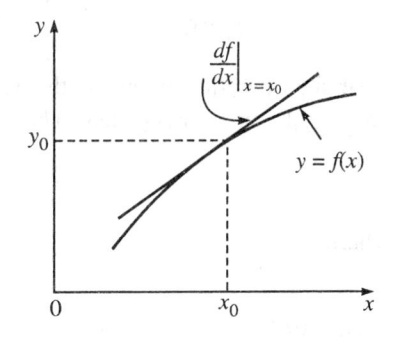

$$y(t) = f(x(t)) \qquad (1.30)$$

FIGURE 1.14 Linearization.

If the normal operating point is designated by x_0 and the corresponding output point is y_0, then Eq. (1.30) may be expanded into a Taylor's series about this operating point (x_0, y_0) as follows:

$$y = f(x) = f(x_0) + \left.\frac{df}{dx}\right|_{x=x_0} \frac{x - x_0}{1!} + \left.\frac{d^2 f}{dx^2}\right|_{x=x_0} \frac{(x - x_0)^2}{2!} \qquad (1.31)$$

where the derivatives df/dx, d^2f/dx^2, ... are evaluated at $x = x_0$. If the variation $x = x - x_0$ is small, we may neglect the higher order terms in $(x - x_0)$. Then Eq. (1.31) may be written as

$$y = y_0 + m(x - x_0) \qquad (1.32)$$

where $\qquad y_0 = f(x_0) \qquad$ and $\qquad m = \left.\dfrac{df}{dx}\right|_{x=x_0}$

Equation (1.32) can be written as

$$y - y_0 = m(x - x_0)$$

or
$$\Delta y = m\Delta x \qquad (1.33)$$

which indicates that Δy is proportional to Δx. Equation (1.33) gives a linear mathematical model for the nonlinear system of Eq. (1.30).

Now, if the dependent variable y depends upon several excitation variables, such as x_1 and x_2, then the functional relationship is written as

$$y = f(x_1, x_2) \qquad (1.34)$$

In order to obtain a linear approximation to this nonlinear system, we may expand Eq. (1.34) into a Taylor series about the normal operating point (x_{10}, x_{20}). Then Eq. (1.34) becomes

$$y = f(x_{10}, x_{20}) + \left[\frac{\partial f}{\partial x_1}(x_1 - x_{10}) + \frac{\partial f}{\partial x_2}(x_2 - x_{20})\right]$$

$$+ \frac{1}{2!}\left[\frac{\partial^2 f}{\partial x_1^2}(x_1 - x_{10})^2 + 2\frac{\partial^2 f}{\partial x_1 \partial x_2}(x_1 - x_{10})(x_2 - x_{20}) + \frac{\partial^2 f}{\partial x_2^2}(x_2 - x_{20})^2\right] + \cdots \qquad (1.35)$$

Near the operating point, as the variations are small, the higher order terms in $(x_1 - x_{10})$ and $(x_2 - x_{20})$ can be neglected. Then the linearized equation becomes

$$y - y_0 = m_1(x_1 - x_{10}) + m_2(x_2 - x_{20}) \qquad (1.36)$$

where

$$y_0 = f(x_{10}, x_{20}), \quad m_1 = \left.\frac{\partial f}{\partial x}\right|_{x_1 = x_{10}}, \quad m_2 = \left.\frac{\partial f}{\partial x}\right|_{x_2 = x_{20}}$$

EXAMPLE 1.11 Consider a pendulum oscillator shown in Figure 1.15(a). The torque on the mass is

$$T = mgl \sin \theta \qquad (1.37)$$

where g is the gravitational constant, m is the mass, l is the length, and θ is the angular displacement.

The equilibrium condition for the mass is $\theta_0 = 0°$. The nonlinear graphical relationship between torque T and angle θ is shown in Figure 1.15(b). The first derivative evaluated at the operating (equilibrium) point leads to the linear approximation as

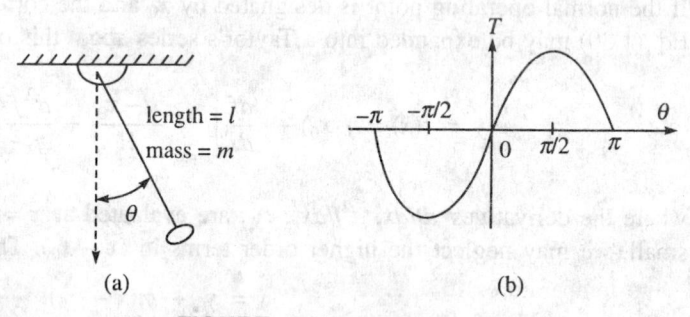

FIGURE 1.15 Example 1.11.

$$T = mgl \left.\frac{d \sin \theta}{d\theta}\right|_{\theta = \theta_0} (\theta - \theta_0)$$

$$= mgl (\cos \theta_0)(\theta - \theta_0) = mgl\theta \qquad (1.38)$$

This approximation is reasonably accurate for $-\pi/4 < \theta < \pi/4$.

1.10 Case Study

Let us consider for proper understanding a precision temperature control system as shown in Figure 1.16 and analyze the dynamics of this system.

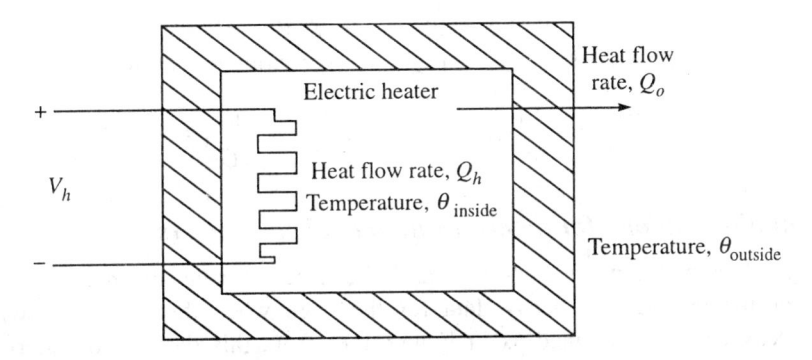

FIGURE 1.16 Precision temperature control system.

The rate of change of difference in temperature between inside and outside the oven is related to heat flow Q_h supplied by the heater and the flow rate through the oven insulation Q_o as

$$C\frac{d}{dt}(\theta_{inside} - \theta_{outside}) = Q_h - Q_o \qquad (1.39)$$

For constant (or very slowly varying) outside temperature, this relation becomes

$$C\frac{d}{dt}(\theta_{inside}) = Q_h - Q_o \qquad (1.40)$$

The constant of proportionality C is the thermal capacity of the oven in J/K (joules per degree kelvin). Q_h and Q_o are the heat flow rates in joules/sec (J/s), that is, watt (W). θ_{inside} and $\theta_{outside}$ are the temperatures in degree kelvin.

Further, the heat flow rate supplied by the heater is

$$Q_h = \frac{V_h^2}{R} \text{ J/s (or W)} \qquad (1.41)$$

where
  V_h is the heater voltage in volts
  R is the electrical resistance of the heater in ohms.

Again, the rate of heat loss through the oven insulation is proportional to the temperature difference across the insulation. That is,

$$Q_o = \frac{\theta_{inside} - \theta_{outside}}{r} \tag{1.42}$$

where r is the thermal resistance of the insulation in K-s/J, or K/W.

Hence substituting the values of Q_h and Q_o, we get from Eq. (1.40) as

$$\frac{d\theta_{inside}}{dt} + \frac{1}{rC}\theta_{inside} = \frac{1}{rC}\theta_{outside} + \frac{1}{RC}V_h^2 \tag{1.43}$$

Experimental observations for determining model parameters

The numerical values of the parameters rC and RC can be estimated from the measurements. Table 1.3 gives the temperature decay data for the oven when the heater is switched off. Exponential decay curve is obtained as in Figure 1.17 from this data. In order to determine the time constant, we know

$$120e^{-t/rC} = 120e^{-T/rC} = 120e^{-1} \approx 44 \quad \text{(as } e = 2.71\text{)}$$

TABLE 1.3 Oven temperature decay data

Time (s)	Temperature (°C)
0000	120
0470	108
0980	92
1525	80.5
2520	63
4260	40.5
⋮	⋮
5000	30
6000	30
⋮	⋮

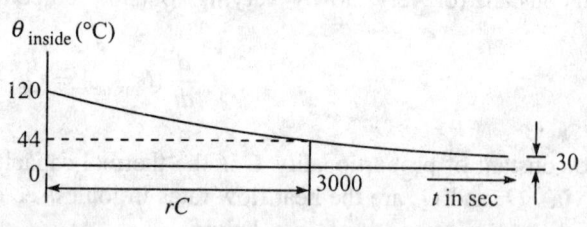

FIGURE 1.17 Plot of oven temperature decay obtained from Table 1.3.

From the curve of Figure 1.17, we get the time to reach temperature 44°C as 3000 seconds. Hence the exponential curve has a time constant obtained from the curve as

$$rC = 3000$$

The outside temperature $\theta_{outside} = 30°C$ can be inferred from the curve.

Table 1.4 gives the steady-state oven temperature (°C) versus the heater voltage (V) and the corresponding graph is drawn in Figure 1.18. The outside temperature $Q_{outside} = 30°C$ is again obtained from the graph.

Introduction to Control Systems

TABLE 1.4 Steady-state oven temperature data

Heater voltage, V_h (V)	Steady-state oven temperature, θ_{inside} (°C)
9.8	34.1
20.0	50.5
29.5	74.0
40.0	110

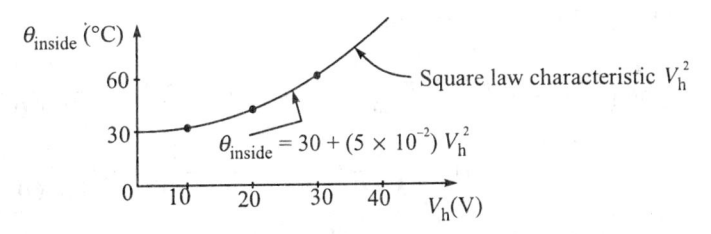

FIGURE 1.18 Plot of steady-state oven temperature obtained from Table 1.4.

The equation

$$\theta_{inside} = \theta_{outside} + (r/R)V_h^2 \tag{1.44}$$

has been fitted with the curve obtained from the data of Table 1.4 and the equation of the curve becomes

$$\theta_{inside} = 30 + (5 \times 10^{-2})V_h^2 \tag{1.45}$$

That is, $r/R = 5 \times 10^{-2}$. This gives

$$RC = (rC)(R/r) = \frac{3000}{5 \times 10^{-2}} = 60{,}000 \text{ s}$$

By substituting the values in Eq. (1.43), the oven system is therefore approximately modelled as

$$\frac{d\theta_{inside}}{dt} + \frac{1}{3000}\theta_{inside} = \frac{1}{3000}\theta_{outside} + \frac{1}{60000}V_h^2 \tag{1.46}$$

Linearization about the operating point

For sufficiently small changes about an operating point, the nonlinear equation representing the dynamics of the system can be linearized by expressing each of the signals involved as the sum of a constant nominal value plus a deviation from the nominal value as

$$V_h = \overline{V}_h + v_h(t)$$
$$\theta_{inside} = \overline{\theta}_{inside} + \theta_{inside}(t)$$
$$\theta_{outside} = \overline{\theta}_{outside} + \theta_{outside}(t)$$

where the symbols with overbars are the constant nominal values and the lowercase symbols with time represent the deviations from the nominal values.

Suppose the nominal values are chosen as:

$$\bar{V}_h = 20 \text{ V}, \quad \bar{\theta}_{inside} = 50°C, \text{ and } \bar{\theta}_{outside} = 30°C$$

Then
$$V_h = 20 + v_h(t)$$
$$\theta_{inside} = 50 + \theta_{inside}(t)$$
$$\theta_{outside} = 30 + \theta_{outside}(t)$$

Now substituting in Eq. (1.44) the oven model becomes as

$$\frac{d\theta_{inside}(t)}{dt} + \frac{1}{3000}[\theta_{inside}(t) + 50] = \frac{1}{3000}[\theta_{outside}(t) + 30] + \frac{1}{60,000}[20 + v_h(t)]^2$$

or
$$\frac{d\theta_{inside}(t)}{dt} + \frac{1}{3000}\theta_{inside}(t) = \frac{1}{3000}\theta_{outside}(t) + \frac{1}{1500}v_h(t) + \frac{1}{60,000}v_h^2(t) \quad (1.47)$$

For small perturbations, the term $v_h^2(t)$ is negligible compared to $v_h(t)$ and therefore we obtain the linearized first order differential equation of the oven as

$$\dot{\theta}_{inside}(t) = -\frac{1}{3000}\theta_{inside}(t) + \frac{1}{3000}\theta_{outside}(t) + \frac{1}{1500}v_h(t) \quad (1.48)$$

1.11 Unity-Feedback System

Figure 1.19 shows the closed-loop motor speed control system where a voltage corresponding to a measure of the actual speed is compared with a reference voltage, r. The difference between these two voltages is then amplified and applied as actuating signal to the motor to maintain the speed as per the reference input.

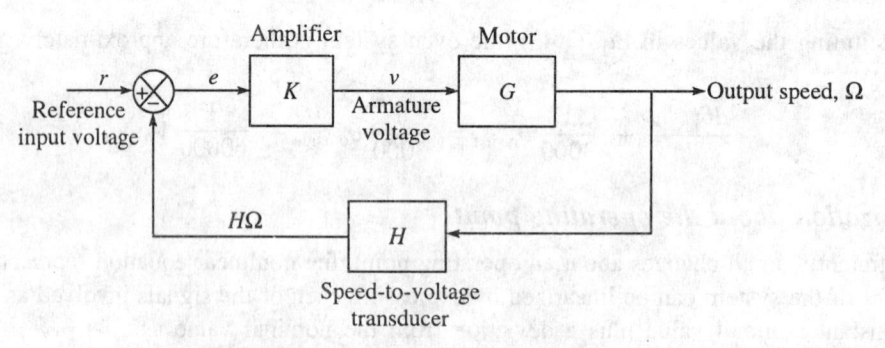

FIGURE 1.19 A generalized control loop.

From Figure 1.19, we get the relationships as

$$\Omega = G \times v \quad (1.49)$$

$$e = r - H\Omega \tag{1.50}$$

$$\Omega = KGe = KG(r - H\Omega) = KGr - KGH\Omega \tag{1.51}$$

Rearranging, we get

$$\Omega = \frac{KG}{1+KGH}r \tag{1.52}$$

It is often more convenient to work with a modified version of Figure 1.19 as shown in Figure 1.20 in order to obtain an expression directly relating the actual speed to the desired speed rather than to the reference input voltage. In this case we can imagine an input, 'desired speed', variable which is then multiplied by a gain exactly equivalent to that of the transducer, in order to give an appropriate reference input voltage as in Figure 1.20. Figure 1.21 is equivalent to Figure 1.20. It makes no difference whether the gain H is applied before or after the comparator, so long as it is applied to both the signals being compared. Figure 1.21 is the unity-feedback system form of the closed-loop system. Further, gain K can be modified accordingly by taking $H = 1$.

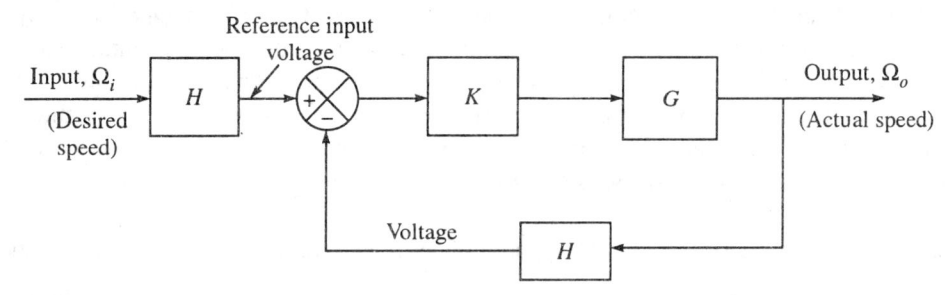

FIGURE 1.20 Modified version of Figure 1.19.

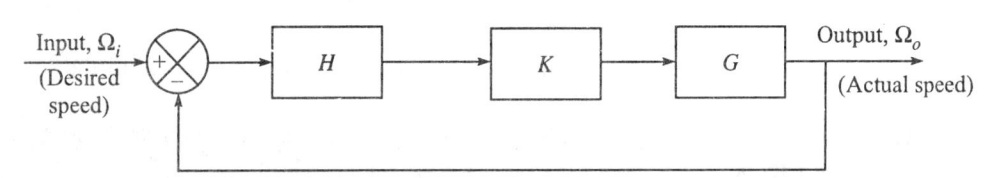

FIGURE 1.21 Unity-feedback form of closed-loop system of Figure 1.19.

Hence conceptually for all discussion, to consider the unity-feedback system as in Figure 1.22 is no loss of generality, though in this book we have taken the open-loop system as $G(s)H(s)$ instead of $G(s)$ for the unity-feedback system.

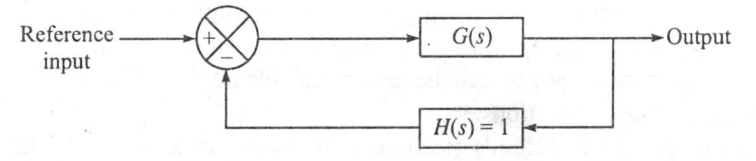

FIGURE 1.22 Unity-feedback system.

1.12 Steady-State Frequency Response

Let us consider the system shown in Figure 1.23 whose dynamics is represented by the differential equation

$$\frac{d^2y}{dt^2} + 3\frac{dy}{dt} + 2y(t) = 2u(t)$$

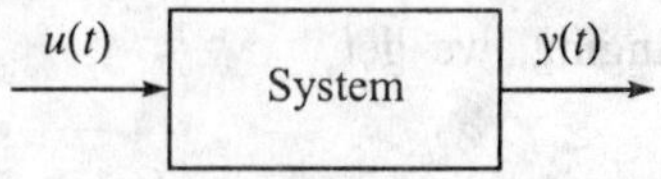

FIGURE 1.23 Forced open-loop system.

Taking the Laplace transform and rearranging, we get

$$Y(s) = \frac{2}{s^2 + 3s + 2} U(s) + \frac{(s+3)\, y(0) + \dot{y}(0)}{s^2 + 3s + 2}$$

In general, we can write

$$Y(s) = \frac{P(s)}{Q(s)} U(s) + \frac{I(s)}{Q(s)} \qquad (1.53)$$

where $I(s)$ is a function of the initial conditions on $y(t)$ and $G(s) = P(s)/Q(s)$ is the system's open-loop transfer function. The output solution $y(t)$ of an ordinary differential equation is given by the complementary function plus the particular integral. The complementary function involves $I(s)/Q(s)$ and upon performing partial fraction expansion and taking the inverse Laplace transform gives the transient response which dies down for a stable system. The particular integral function involves $[P(s)/Q(s)]U(s)$ and upon performing convolution integral gives the steady-state solution.

Let us consider the system's step response. The input forcing function producing the step response is given by

$$u(t) = 0 \text{ for } t < 0$$
$$= K \text{ for } t \geq 0$$

If the system is asymptotically stable, all its poles and zeros are contained within the left-half s-plane.

The steady-state response may be obtained from the residue associated with the pole at $s = 0$, and is found to be

$$y(\infty) = \lim_{s \to 0} sY(s) = K \left.\frac{P(s)}{Q(s)}\right|_{s=0} = \left.KG(s)\right|_{s=0}$$

It may be noted that all the poles in the left-half s-plane produce the decaying exponential terms in the time domain, and that their contribution to $y(t)$ therefore diminishes with increasing time. Hence the steady-state response to a forcing step input is simply the systems's gain, $\left.G(s)\right|_{s=0}$ times the amplitude K of the step. Since the residue associated with the pole at the origin is required, the steady-state response can be easily calculated. Also, note that this residue is dependent on the pole-zero locations.

A detailed analysis of a system's steady-state response to a sinusoidal harmonic forcing input is now developed. In fact, this will give you the answer as to why we put $s = j\omega$ for

Introduction to Control Systems

steady-state in the frequency response analysis as we will later see while studying Bode plot, polar plot, etc. in Chapter 9.

Let the input function be of the form

$$u(t) = K \sin \omega t$$

which, on Laplace transformation, becomes

$$U(s) = \frac{\omega K}{s^2 + \omega^2}$$

Substituting for $U(s)$ in Eq. (1.53) gives

$$Y(s) = \frac{P(s)}{Q(s)} \frac{\omega K}{s^2 + \omega^2} + \frac{I(s)}{Q(s)}$$

Since the system is assumed to be stable, the effect of the initial conditions diminishes with time and

$$\mathcal{L}^{-1}\left[\frac{I(s)}{Q(s)}\right] \to 0 \quad \text{as} \quad t \to \infty$$

Therefore,

$$y(t) = \mathcal{L}^{-1}\left[\frac{P(s)}{Q(s)} \frac{\omega K}{s^2 + \omega^2}\right] \quad \text{as} \quad t \to \infty$$

Partial fraction expansion of the term within the brackets leads to

$$\frac{P(s)}{Q(s)} \frac{\omega K}{s^2 + \omega^2} = \frac{A}{s - j\omega} + \frac{B}{s + j\omega} + \text{(all terms arising from } Q(s)\text{)}$$

As the system is stable, all the terms arising from the system's characteristic polynomial $Q(s)$ must be functions which disappear with time as t tends to infinity Hence, the steady-state response may be found by solving for A and B. Using Heaviside formula,

$$A = \left(\frac{P(s)}{Q(s)} \frac{\omega K (s - j\omega)}{(s - j\omega)(s + j\omega)}\right)\Bigg|_{s = j\omega} = \frac{K}{2j} \frac{P(j\omega)}{Q(j\omega)} = \frac{K}{2j} G(j\omega)$$

Similarly,

$$B = -\frac{K}{2j} G(-j\omega)$$

Further, as $G(j\omega)$ and $G(-j\omega)$ are complex, these may be written in polar form as

$$G(j\omega) = M(\omega) e^{j\phi(\omega)} \quad \text{or} \quad G(j\omega) = M e^{j\phi}$$

and

$$G(-j\omega) = M(\omega) e^{-j\phi(\omega)} \quad \text{or} \quad G(j\omega) = M e^{-j\phi}$$

Substituting for $G(j\omega)$ and $G(-j\omega)$, we get A and B as

$$A = \frac{K}{2j} M e^{j\phi} \quad \text{and} \quad B = -\frac{K}{2j} M e^{-j\phi}$$

Now we can write the steady-state response as

$$y(t)_{ss} = \mathcal{L}^{-1}\left[\frac{A}{s-j\omega} + \frac{B}{s+j\omega}\right]$$

or
$$y(t)_{ss} = Ae^{j\omega t} + Be^{-j\omega t}$$

or
$$y(t)_{ss} = \frac{K}{2j}Me^{j\phi}e^{j\omega t} - \frac{K}{2j}Me^{-j\phi}e^{-j\omega t}$$

$$= \frac{K}{2j}M(e^{j(\phi+\omega t)} - e^{-j(\phi+\omega t)})$$

$$= KM \sin(\omega t + \phi)$$

Therefore, for sinusoidal input, the steady-state output is also sinusoidal, with a magnification factor M, but shifted by an amount ϕ as shown in Figure 1.24. In general, both the magnitude M and phase ϕ are functions of frequency and can be represented by $M(\omega)$ and $\phi(\omega)$, respectively. For any given frequency, both magnitude $M(\omega)$ and phase $\phi(\omega)$ may be determined from the transfer function $G(s)$.

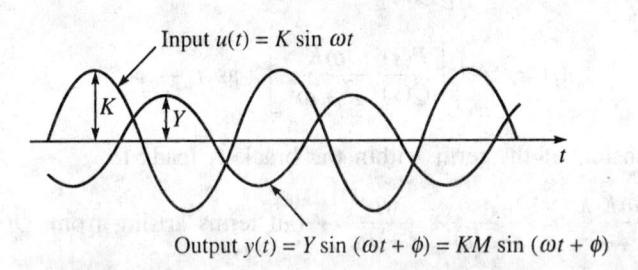

FIGURE 1.24 Input and output sinusoidal signals.

The reason for putting $s = j\omega$ for steady-state in frequency response analysis is thus justified, though s is a complex quantity, i.e. $s = \sigma + j\omega$

By using the graphical technique we can find the magnification factor $M(\omega)$ and phase shift $\phi(\omega)$. This is illustrated below with an example.

For the transfer function,

$$G(s) = \frac{s+4}{s^2+5s+6}$$

the pole-zero map of the system is shown in Figure 1.25.

The function $G(j\omega_1)$ at point $j\omega_1$ is evaluated as

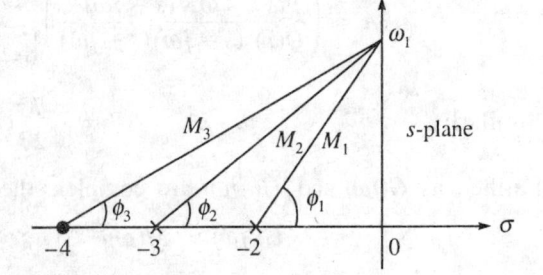

FIGURE 1.25 Pole-zero map for the system $G(s) = (s+4)/(s^2 + 5s + 6)$.

$$G(j\omega_1) = M(\omega_1)e^{j\phi(\omega_1)}$$

where
$$M(\omega_1) = \frac{M_3}{M_1 M_2} = \frac{\sqrt{(\omega_1^2 + 4^2)}}{\sqrt{(\omega_1^2 + 2^2)}\sqrt{(\omega_1^2 + 3^2)}}$$

and
$$\phi(\omega_1) = \phi_3 - (\phi_1 + \phi_2)$$

Note that when ω_1 is set equal to zero, $M(0)$ gives the system's gain called the dc gain and $\phi(0)$ is zero.

Summary

Feedback control is a fundamental fact of modern industry and society. Feedback control systems are extensively used in industrial applications. To obtain quantitative mathematical models useful for engineering analysis and design, we need to understand the feedback control mechanism which we have explained in this chapter. Further, in this chapter we explained concepts such as state variable formulation, transfer function, pole-zero map, and linearization to facilitate their detailed discussion in the subsequent chapters. The mathematical tool used for the analysis of control systems, namely the Laplace transform, was reviewed. We also presented the MATLAB approach to obtain the partial-fraction expansion of $P(s)/Q(s)$ and also to obtain the zeros and poles of $P(s)/Q(s)$.

Problems

1.1 Draw the block diagram of a thermostatically controlled home-heating system indicating the function of each element.

1.2 Draw the block diagram of a traffic light control system indicating the function of each element and assuming that the light duration depends on the density of traffic.

1.3 Draw the block diagram of a student-teacher learning process which is inherently a feedback control system and indicate the function of each block.

1.4 The toilet cistern is basically a feedback control system. Draw the block diagram of the cistern indicating the function of each element.

1.5 The government has introduced free economy assuming the basic law of demand and supply. Analyze all the attributes of the market and develop the block diagram of the demand and supply system.

1.6 Find the Laplace transform of the following functions.
  (a) $\theta(t) = 4 \sin(2 - 4t)$
  (b) $\theta(t) = 7e^{-0.2t} \sin 10.5t$
  (c) $\theta(t) = 8 \cos\left(6t - \frac{\pi}{4}\right)$
  (d) $f(t) = 0$ for $t < 0$
      $= te^{-t} \sin 5t$ for $t > 0$

1.7 Find the inverse Laplace transforms of the following.

(a) $F(s) = \dfrac{1}{s(s^2 + \omega^2)}$

(b) $F(s) = \dfrac{5e^{-s}}{s+1}$

(c) $F(s) = \dfrac{1}{s^2(s^2 + \omega^2)}$

(d) $F(s) = \dfrac{\omega_n^2}{s(s^2 + 2\zeta\omega_n s + \omega_n^2)}$

1.8 Obtain the partial-fraction expansion of the following functions with MATLAB.

(a) $F(s) = \dfrac{2}{(s+1)(s+3)^2}$

(b) $F(s) = \dfrac{2s+10}{s^2+10s+24}$

(c) $F(s) = \dfrac{s+2}{s(s+1)(s^2+9)}$

1.9 A function $P(s)/Q(s)$ consists of the following zeros, poles, and gain K:

zeros at $s = -1$, $s = -2$
poles at $s = 0$, $s = -4$, $s = -6$
gain $K = 5$

Obtain the expression $P(s)/Q(s)$ with MATLAB.

Mathematical Modeling of Systems

OBJECTIVE

This chapter provides the student with the basic tools and experience to create models of physical systems found in electrical, mechanical, electromechanical, and fluid applications. All sections are intended to be self-contained and are also supplemented with numerous examples. The chapter begins with an introduction to simulation and modeling techniques used for physical systems. Block diagrams are used for modeling, and simulation environments are restricted to those that are visual or block diagram-based, as opposed to text based. Following a brief description of block diagrams and the simulation process, modeling derivation using a modified analogy approach, extended to produce a block diagram instead of an impedance diagram, is discussed. The remainder of the chapter uses the method to derive subsystem models commonly encountered in electrical, mechanical translation, mechanical rotation, electromechanical and fluid applications. Component modeling, which is the derivation of mathematical equations suitable for computer simulation, plays a critical role during the design stage of control systems.

CHAPTER OUTLINE

Introduction
Translational and Rotational Systems
Analogous Systems
Liquid-Level Systems
Servomotors
Sensors
Magnetic Amplifier
Stepper Motor
Block Diagram Reduction
Signal Flow Graph
Multi-Variable Systems

2.1 Introduction

The first step in the analysis of a dynamic system is to derive its mathematical model. The derivation of a mathematical model with reasonable accuracy is of paramount importance. The mathematical model of a physical system, for instance, electrical, mechanical, electromechanical, acoustic, and so on can be obtained provided the dynamics of the system under investigation is known. We are already familiar with the mathematical model of an electrical system. The concept of an analogous system is useful in practice since one type of system may be easier to handle experimentally than another. If the response of one physical system to a given excitation is determined, the response of all other systems which can be described by the same set of equations can then be derived for the same excitation function. Systems remain analogous as long as their differential equations or transfer functions are of identical form.

To find the electrical analogy of a mechanical system, the parameters for (i) translational and (ii) rotational systems have to be first defined.

2.2 Translational Systems

A translational system has three types of forces due to passive elements.

1. The *inertial force* f_M due to inertial mass M, as shown in Figure 2.1(a), is by Newton's second law given by

$$f_M = Ma = M\frac{du}{dt} = M\frac{d^2x}{dt^2}$$

where x is the displacement, u the velocity, and a the acceleration.

2. The *viscous damping force* f_D, as shown in Figure 2.1(b), due to viscous damping coefficient D is a retarding force proportional to velocity u, i.e.

$$f_D = Du = D\frac{dx}{dt}$$

3. The *spring force* f_K is proportional to the displacement (deformation) x. Let K as shown in Figure 2.1(c) be the compliance of the spring which is the reciprocal of its stiffness or the spring constant. Then

$$f_K = \frac{1}{K}x = \frac{1}{K}\left[\int_0^t u\,dt + x(0)\right]$$

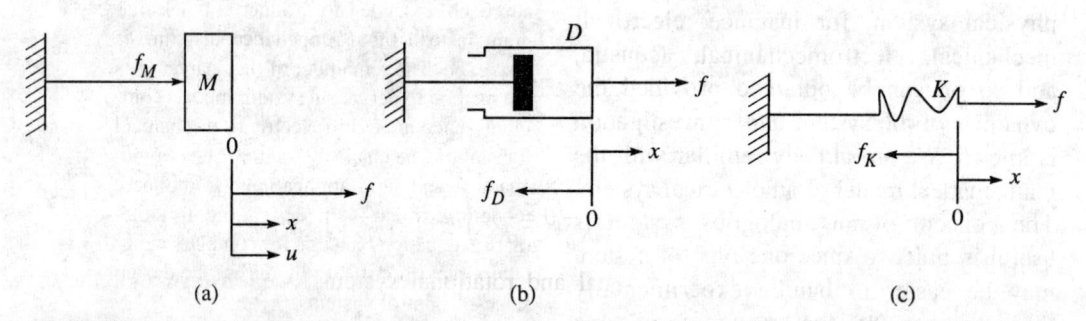

FIGURE 2.1 Mechanical passive elements: (a) mass element, (b) damper element, and (c) compliance of the spring.

2.3 Rotational Systems

Let us now consider the components of the rotational mechanical system shown in Figure 2.2(a). This system consists of a rotatable disc of moment of inertia I_θ and a shaft of stiffness $1/K_\theta$, that is, shaft compliance K_θ. The disc rotates in a viscous medium with viscous friction coefficient D_θ.

Let T be the applied torque which tends to rotate the disc. The free-body diagram is shown in Figure 2.2(b). A more realistic rotational mechanical system is shown in Figure 2.2(c) where compliance of $2K_\theta$ from each side of the disc is shown. Effectively, the system of Figure 2.2(c) boils down to that shown in Figure 2.2(a).

The following are the three types of torques due to rotational elements that resist rotational motion.

1. The *intertial torque* T_I is equal to the moment of inertia I_θ times the angular acceleration α, i.e.

Mathematical Modeling of Systems

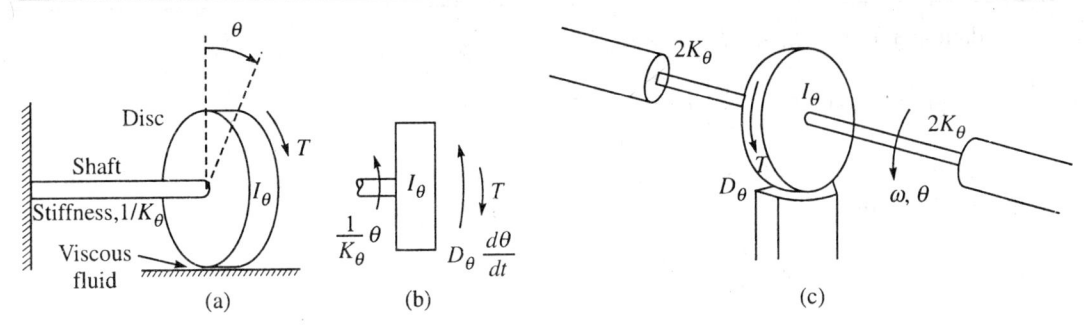

FIGURE 2.2 Rotational mechanical system.

$$T_I = I_\theta \alpha = I_\theta \frac{d\omega}{dt} = I_\theta \frac{d^2\theta}{dt^2}$$

where ω is the angular velocity and θ the angular displacement.

2. The *damping torque* T_D is equal to the rotational damping coefficient D_θ times the angular velocity ω in a linear system, i.e.

$$T_D = D_\theta \omega = D_\theta \frac{d\theta}{dt}$$

3. The *spring torque* T_K is equal to θ times the torsional stiffness of the spring, which is the reciprocal of torsional compliance K_θ, i.e.

$$T_K = \left(\frac{1}{2K_\theta} + \frac{1}{2K_\theta}\right)\theta = \frac{1}{K_\theta}\theta = \frac{1}{K_\theta}\int \omega dt = \frac{1}{K_\theta}\left[\int_0^t \omega dt + \theta(0)\right]$$

Comparing the equations of translational and rotational systems, we observe that these two systems are analogous.

2.4 Electrical Analog of Mechanical Systems

In translational mechanical systems, the D'Alembert's principle states that:

For any body, the algebraic sum of the externally applied forces and the forces resisting the motion in any given direction is zero.

The equilibrium equation of a translational mechanical (spring-mass-damper) system subjected to an external force f as shown in Figure 2.3(a), by D'Alembert's principle is

$$f + f_M + f_D + f_K = 0 \qquad (2.1)$$

where

$$\text{inertial force, } f_M = -M\frac{du}{dt} = -M\ddot{x}$$

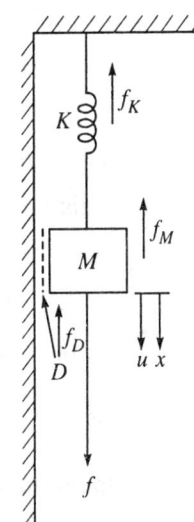

FIGURE 2.3(a) Spring-mass-damper system.

damping force, $f_D = -Du = -M\dot{x}$

spring force, $f_K = \dfrac{1}{K}\left[\displaystyle\int_0^t u\,dt + x(0)\right]$

Note that the directions of forces due to inertia, damping and spring are all opposite to that of the applied external force f. Then, the equilibrium equation becomes

$$M\frac{du}{dt} + Du + \frac{1}{K}\int u\,dt = f$$

or
$$M\frac{du}{dt} + Du + \frac{1}{K}\left[\int_0^t u\,dt + x(0)\right] = f \tag{2.2}$$

or
$$M\ddot{x} + D\dot{x} + \frac{x}{K} = f$$

A systematic way of analyzing is to draw the free-body diagram as shown in Figure 2.3(b) assuming that the gravitational effect is negligible.

Again by Newton's law,

$$\sum \text{forces} = Ma$$

or
$$f - D\frac{dx}{dt} - \frac{x}{K} = M\frac{d^2x}{dt^2}$$

or
$$M\ddot{x} + D\dot{x} + \frac{x}{K} = f$$

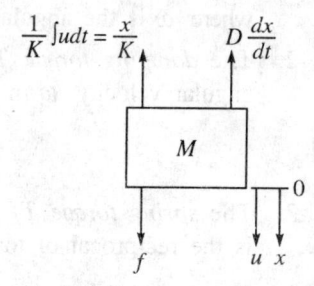

FIGURE 2.3(b) Free-body diagram of spring-mass-damper system.

which is the same equation as we obtained from the mathematical model of the spring-mass-damper system by applying the D'Alembert's principle.

Similarly, for the rotational mechanical system the D'Alembert's principle can be stated as:

For any body, the algebraic sum of the externally applied torques and the torque resisting the rotation about any axis is zero.

The equilibrium equation for Figure 2.2(a) by D'Alembert's principle is

$$T + T_I + T_D + T_K = 0 \tag{2.3}$$

where

inertial torque, $T_I = -I_\theta \dfrac{d\omega}{dt}$ (2.4)

damping torque, $T_D = -D_\theta \omega$ (2.5)

spring torque, $T_K = -\dfrac{1}{K_\theta}\left[\displaystyle\int_0^t \omega\,dt + \theta(0)\right]$ (2.6)

The following rule for drawing the force–current (force–voltage) analogous electrical circuits from mechanical systems will prove useful:

Each junction in the mechanical system corresponds to a node or junction (closed loop) which joins (consists of) electrical excitation sources and passive elements, analogous to the mechanical driving sources and passive elements, connected to the junction. All points on a rigid mass are considered the same junction.

The electrical analogous circuit of the mechanical (translational as well as rotational) system with force–current analogy is shown in Figure 2.4(a) and that with force–voltage analogy is shown in Figure 2.4(b).

The equations describing these two circuits are:

For Figure 2.4(a):

$$C\frac{dv}{dt} + Gv + \frac{1}{L}\left[\int_0^t v\,dt + v(0)\right] = i$$

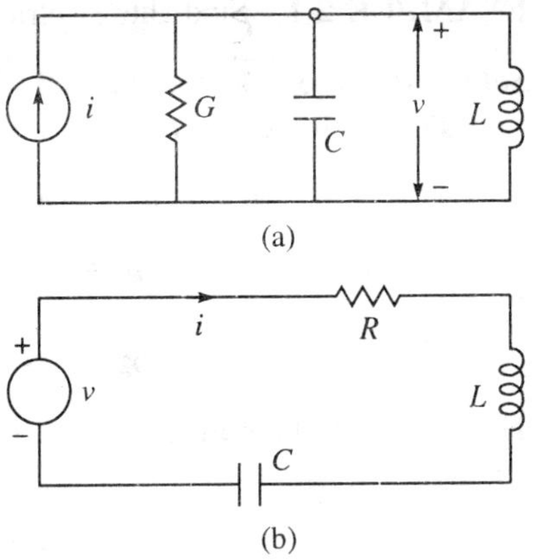

For Figure 2.4(b): $L\frac{di}{dt} + Ri + \frac{1}{C}\left[\int_0^t i\,dt + q(0)\right] = v$

FIGURE 2.4 (a) Force–current analogy and (b) force–voltage analogy.

Table 2.1 gives the list of analogous quantities. The relationship between the analogies force–current and force–voltage and between the mechanical and electrical system components is given in Table 2.2. The units of translational and rotational motion are also given in Table 2.1.

TABLE 2.1 Analogous quantities in electrical and mechanical systems

Mechanical system		Electrical system	
Translational (unit)	Rotational (unit)	Force–current	Force–voltage
Force, f (N)	Torque, T (N-m)	Current, i	Voltage, v
Velocity, u (m/s)	Angular velocity, ω (rad/s)	Voltage, v	Current, i
Displacement, x (m)	Angular displacement, θ (rad)	Flux linkage, ϕ	Charge, q
Mass, M (kg)	Moment of inertia, I_θ (kg-m^2)	Capacitance, C	Inductance, L
Viscous damping coefficient, D (N-m/m/s)	Rotational damping coefficient, D_θ (N-m/rad/s)	Conductance, G	Resistance, R
Compliance, K	Torsional compliance, K_θ	Inductance, L	Capacitance, C

TABLE 2.2 Relationship between, through and across variables of analogous system components

Electrical system		Mechanical system	
f–i analogy	f–v analogy	Translational	Rotational
$i = C(dv/dt)$	$v = L(di/dt)$	$f = M(du/dt)$	$T = I_\theta(d\omega/dt)$
$i = Gv$	$v = Ri$	$f = Du$	$T = D_\theta \omega$
$i = \frac{1}{L}\int v\,dt$	$v = \frac{1}{C}\int i\,dt$	$f = \frac{1}{K}\int u\,dt$	$T = \frac{1}{K}\int \omega\,dt$

EXAMPLE 2.1 Find the electric analog of the mechanical system shown in Figure 2.5(a).

Solution: The reference directions are as shown by the arrows in the given Figure 2.5(a). Consider the forces on mass M_1. The external force on M_1 is f. The resisting forces in M_1 are:

(i) Inertial force, $f_{M1} = -M_1 \ddot{x}_1 = -M_1 \dot{u}_1$ (2.7)

(ii) Damping force, $f_{D1} = -D_1(\dot{x}_1 - \dot{x}_2) = -D_1(u_1 - u_2)$

and
$$f_{D2} = -D_2 \dot{x}_1 = -D_2 u_1 \qquad (2.8)$$

The free-body diagram is shown in Figure 2.5(b). By D'Alembert's rule, the force equilibrium for mass M_1 is

$$M_1 \frac{du_1}{dt} + (D_1 + D_2)u_1 - D_1 u_2 = f \qquad (2.9)$$

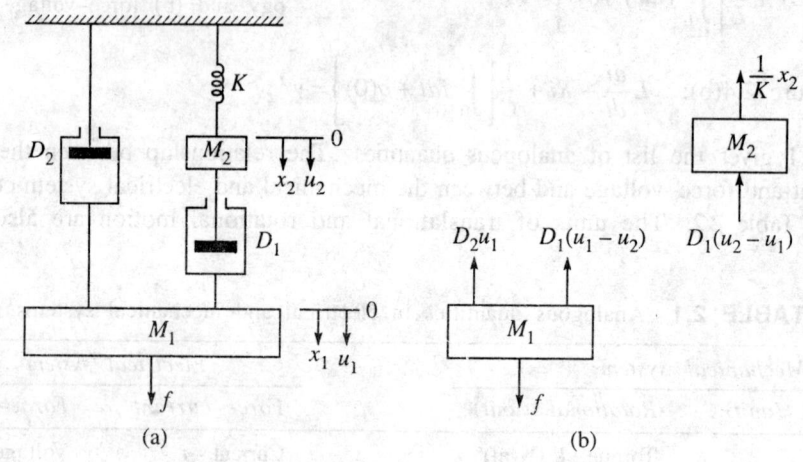

FIGURE 2.5 Example 2.1: (a) mechanical system of two degrees of freedom and (b) free-body diagram.

Similarly by D'Alembert's rule, the force balance equation for mass M_2 is

$$-D_1 u_1 + M_2 \frac{du_2}{dt} + D_1 u_2 + \frac{1}{K}\left[\int_0^t u_2\, dt + x_2(0)\right] = 0 \qquad (2.10)$$

By the f–v analogy, the electrical analog circuit of Eqs. (2.9) and (2.10) is as shown in Figure 2.6 with parameter equivalences mentioned therein. The loop equations are:

$$L_1\left(\frac{di_1}{dt}\right) + (R_1 + R_2)i_1 - R_1 i_2 = v \qquad (2.11)$$

and
$$-R_1 i_1 + L_2\left(\frac{di_2}{dt}\right) + R_1 i_2 + \left(\frac{1}{C}\right)\left[\int_0^t i_2\, dt + q_2(0)\right] = 0 \qquad (2.12)$$

Mathematical Modeling of Systems

The conversion into electrical parameters is in accordance with Table 2.1. Equations (2.11) and (2.12) are identical to Eqs. (2.9) and (2.10) respectively. Since corresponding to the two coordinates x_1 and x_2, the mechanical system has two junctions, similarly, we have two loops in the f–v analogous electrical circuit.

Further, in order to draw the electrical equivalent of the mechanical system by the f–v analogy just by looking at the mechanical system, the rule for drawing the f–v analogous electric circuit from the mechanical system can be restated as follows:

> *Each junction in the mechanical system corresponds to a closed-loop which consists of electrical excitation sources and passive elements analogous to the mechanical driving sources and passive elements connected to the junction. All points on a rigid mass are considered the same junction.*

Elaborating this, we can write that each junction on M_1 and M_2 corresponds to two closed loops. All points on rigid mass M_1 are considered the same junction and all points on the rigid mass M_2 are also considered the same junction. There will be two loops. The first loop consists of electrical voltage source v, inductance L_1 analogous to M_1, resistance R_2 analogous to D_2, and R_1 analogous to D_1. The second loop consists of inductance L_2 analogous to M_2, capacitance C analogous to compliance K and resistance R_1 analogous to D_1. Further, R_1 is common to both the loops so that current $(i_1 - i_2)$ flows through R_1 just as the force on D_1 depends on $(\dot{x}_1 - \dot{x}_2)$. This is how we have drawn the f–v analogous electric circuit as shown in Figure 2.6.

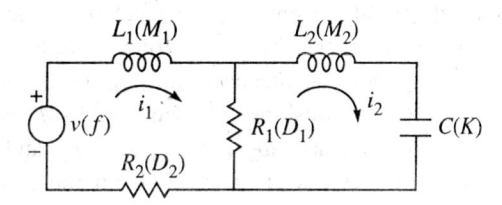

FIGURE 2.6 Example 2.1: electric analog circuit by f–v analogy.

Similarly, by the f–i analogy using Table 2.1, the electric analog circuit is shown in Figure 2.7 with parameter equivalences mentioned therein. The node equations using KCL are

$$C_1\left(\frac{dv_1}{dt}\right) + (G_1 + G_2)v_1 - G_1 v_2 = i \qquad (2.13)$$

and

$$-G_1 v_1 + C_2\left(\frac{dv_2}{dt}\right) + G_1 v_2 + \left(\frac{1}{L}\right)\left[\int_0^t v_2 \, dt + \phi(0)\right] = 0 \qquad (2.14)$$

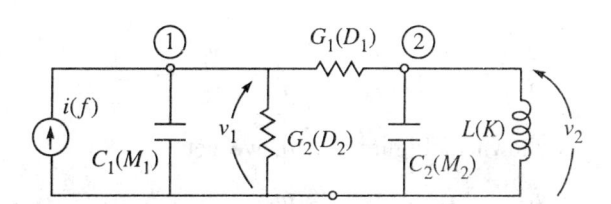

FIGURE 2.7 Example 2.1: electric analog circuit by f–i analogy.

Equations (2.13) and (2.14) are analogous with Eqs. (2.9) and (2.10) and have one-to-one correspondence with f–i analogy from Table 2.1.

Corresponding to two coordinates x_1 and x_2, there are two independent nodes in the electrical analog circuit of Figure 2.7. The first node joins the current source $i(f)$, a capacitance $C_1(M_1)$ and two conductances $G_1(D_1)$ and $G_2(D_2)$. The second node joins $C_2(M_2)$, $L(K)$ and $G_1(D_1)$. Note that $G_1(D_1)$ is common to both nodes. The electrical analogous circuits of Figures 2.6 and 2.7 are dual to each other.

The f–i analogous electric circuit can also be drawn at a glance following the rule stated earlier for drawing the f–i analogous electric circuit together with making use of Table 2.2 for conversion of parameters.

2.5 Mechanical Couplings

The common mechanical coupling devices, i.e. friction wheels, gear trains, levers, etc. also have electrical analogs. These mechanical coupling devices act as matching devices like transformers in electrical systems. These mechanical devices transmit energy from one part of a system to another in such a way that force, torque, speed and displacement are altered. The inertia and friction of these mechanical coupling devices are neglected in the ideal case considered.

The relationships between torques T_1 and T_2, angular displacements θ_1 and θ_2, the angular velocities ω_1 and ω_2, radii r_1 and r_2, and the number of teeth n_1 and n_2 of the mechanical coupling devices are derived from the following facts.

Friction wheels

In nonslipping friction wheels of Figure 2.8, the points of contact P_1 on wheel 1 and P_2 on wheel 2 must have the same linear velocity because they move together and experience equal and opposite forces. Being a rotational system, it is convenient to use angular velocity and torques. The following relations hold good

$$\frac{T_1}{T_2} = \frac{r_1}{r_2} \quad \text{and} \quad \frac{\omega_1}{\omega_2} = \frac{r_2}{r_1}$$

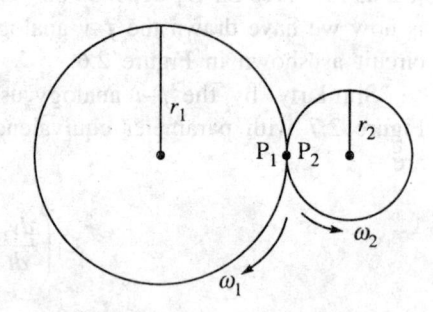

FIGURE 2.8 Friction wheels in pair.

The electrical analog of the friction wheel-pair is an ideal transformer having the turns ratio $n_1 : n_2$. The f–v analogy (see Figure 2.9(a)) can be written as

$$\frac{\omega_1}{\omega_2} \Rightarrow \frac{i_1}{i_2}; \quad \frac{r_1}{r_2} \Rightarrow \frac{n_1}{n_2}; \quad \frac{v_1}{v_2} \Rightarrow \frac{T_1}{T_2}$$

From the f–i analogy, as shown in Figure 2.9(b), we get

$$\frac{i_1}{i_2} \Rightarrow \frac{T_1}{T_2}; \quad \frac{v_1}{v_2} \Rightarrow \frac{\omega_1}{\omega_2}; \quad \frac{r_2}{r_1} \Rightarrow \frac{n_1}{n_2}$$

The reversal of current directions and voltage polarities in the secondaries of Figure 2.9 corresponds to reversal of directions of both torque and angular velocity due to coupling. This is equivalent to putting dots on the opposite ends of primary and secondary windings of the transformer.

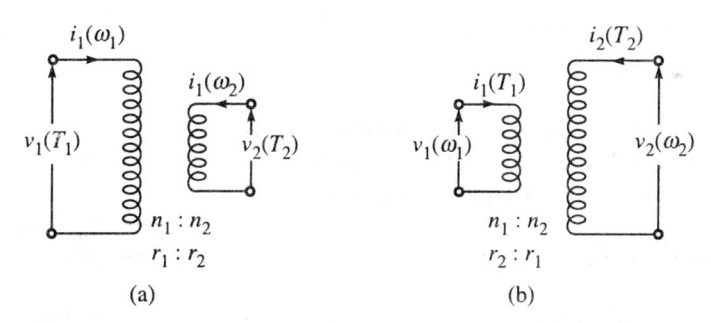

FIGURE 2.9 (a) f–v analogy and (b) f–i analogy of a friction-wheel pair.

Gear train

Gear trains are used in control systems to attain the mechanical matching of motor to load. Usually, the servomotor operates at a high speed but has low torque. To drive a load with high torque and low speed by such a motor, speed reduction and torque magnification are achieved by gear trains. Two gears are shown coupled together in Figure 2.10.

1. The number of teeth on the surface of the gears is proportional to radii r_1 and r_2 of the gears, that is,
$$r_1 n_1 = r_2 n_2 \tag{2.15}$$

2. The distance travelled along the surface of each gear is the same. Therefore,
$$\theta_1 r_1 = \theta_2 r_2 \tag{2.16}$$

3. The work done by one gear is equal to that done by the other since there is assumed to be no loss. Thus,
$$T_1 \theta_1 = T_2 \theta_2 \tag{2.17}$$

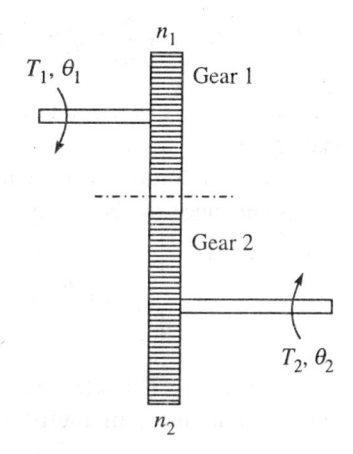

FIGURE 2.10 Gear train.

If the angular velocities ω_1 and ω_2 of the gears are brought into picture, Eqs. (2.15) to (2.17) lead to

$$\frac{T_1}{T_2} = \frac{\theta_2}{\theta_1} = \frac{n_1}{n_2} = \frac{\dot{\theta}_2}{\dot{\theta}_1} = \frac{\ddot{\theta}_2}{\ddot{\theta}_1} \tag{2.18}$$

Figure 2.11 shows a motor driving a load through a gear train which consists of two gears coupled together. The moment of inertia and viscous friction of motor and gear 1 are denoted by $I_{\theta1}$ and $D_{\theta1}$ and those of gear 2 and load are denoted by $I_{\theta2}$ and $D_{\theta2}$, respectively.

For the first shaft, the differential equation is

$$I_{\theta1}\ddot{\theta}_1 + D_{\theta1}\dot{\theta}_1 + T_1 = T_M \quad (2.19)$$

where T_M is the torque developed by the motor and T_1 is the load torque on gear 1 due to the rest of the gear train.

For the second shaft, the differential equation is

$$I_{\theta2}\ddot{\theta}_2 + D_{\theta2}\dot{\theta}_2 + T_L = T_2 \quad (2.20)$$

where T_2 is the torque transmitted to gear 2 and T_L is the load torque. Here the stiffness of the shafts of the gear train is assumed to be infinite.

Eliminating T_1 and T_2 from Eqs. (2.19) and (2.20) with the help of Eq. (2.18), we get

$$I_{\theta1}\ddot{\theta}_1 + D_{\theta1}\dot{\theta}_1 + \frac{n_1}{n_2}(I_{\theta2}\ddot{\theta}_2 + D_{\theta2}\dot{\theta}_2 + T_L) = T_M \quad (2.21)$$

The equivalent electric analog on $T-v$ (torque–voltage) analogy from Eq. (2.21) is shown in Figure 2.12(a) with n as the turns ratio (n_1/n_2) which is less than unity.

Eliminating θ_2 from Eq. (2.21) with the help of Eq. (2.18) yields

$$\left[I_{\theta1} + \left(\frac{n_1}{n_2}\right)^2 I_{\theta2}\right]\ddot{\theta}_1 + \left[D_{\theta1} + \left(\frac{n_1}{n_2}\right)^2 D_{\theta2}\right]\dot{\theta}_1 + \left(\frac{n_1}{n_2}\right)T_L = T_M \quad (2.22)$$

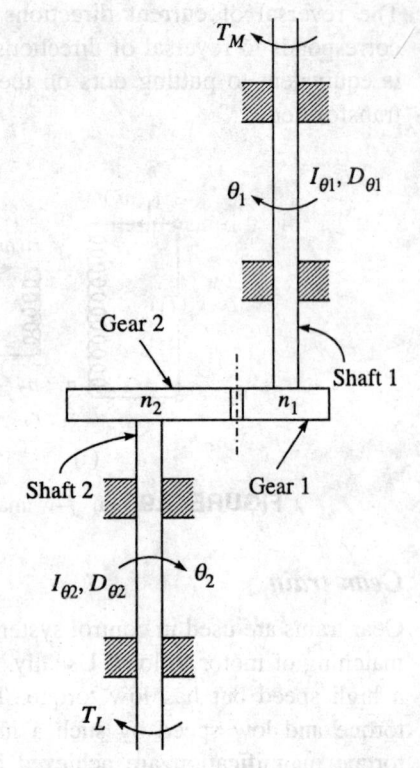

FIGURE 2.11 Gear train between motor and load.

The equivalent $T-v$ electric analog of Eq. (2.22) is shown in Figure 2.12(b). Thus the equivalent moment of inertia and viscous friction of the gear train referred to shaft 1 are

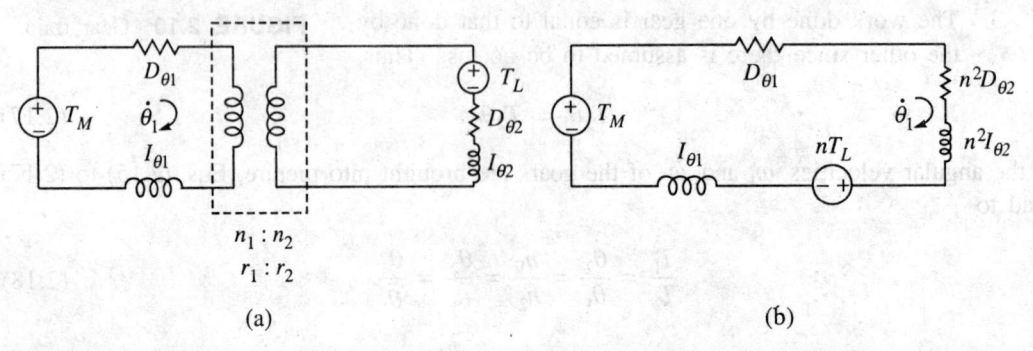

FIGURE 2.12 (a) $T-v$ analogy of Eq. (2.21) and (b) $T-v$ analogy of Eq. (2.22) of gear train.

$$I_{\theta 1}(\text{eq}) = I_{\theta 1} + \left(\frac{n_1}{n_2}\right)^2 I_{\theta 2} \qquad (2.23)$$

$$D_{\theta 1}(\text{eq}) = D_{\theta 1} + \left(\frac{n_1}{n_2}\right)^2 D_{\theta 2} \qquad (2.24)$$

Hence the torque equation referred to shaft 1 is

$$I_{\theta 1}(\text{eq})\ddot{\theta}_1 + D_{\theta 1}(\text{eq})\dot{\theta}_1 + \left(\frac{n_1}{n_2}\right)T_L = T_M \qquad (2.25)$$

where $(n_1/n_2)T_L$ is the load torque referred to shaft 1.

Similarly, the torque equation referred to load shaft may be expressed as

$$I_{\theta 2}(\text{eq})\ddot{\theta}_2 + D_{\theta 2}(\text{eq})\dot{\theta}_2 + T_L = \left(\frac{n_2}{n_1}\right)T_M \qquad (2.26)$$

where $I_{\theta 2}(\text{eq}) = I_{\theta 2} + \left(\frac{n_2}{n_1}\right)^2 I_{\theta 1}$ and $D_{\theta 2}(\text{eq}) = D_{\theta 2} + \left(\frac{n_2}{n_1}\right)^2 D_{\theta 1}$

The electric analog circuit based on T–i (torque–current) analogy is shown in Figure 2.13. Students are advised to derive this circuit on the same lines as the T–v analogous circuit.

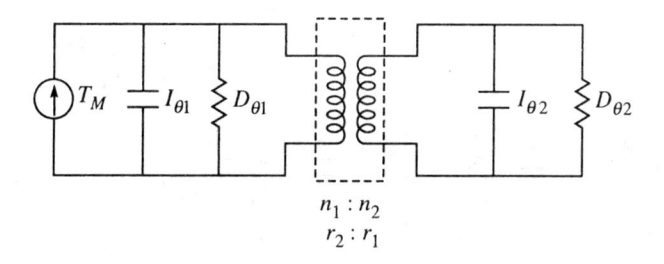

FIGURE 2.13 T–i analogy of gear train.

EXAMPLE 2.2 Draw the electric analog of the mechanical system shown in Figure 2.14(a) by both the f–v analogy and f–i analogy. Write the equilibrium equation of the mechanical system.

Solution: By D'Alembert's principle, the equilibrium equation for mass M_1 is

$$M_1\left(\frac{du_1}{dt}\right) + M_2\left(\frac{du_1}{dt}\right) + D_1(u_1 - u_2) + \frac{1}{K_1}\left[\int_0^t (u_1 - u_2)dt + x_1(0) - x_2(0)\right] = f \qquad (2.27)$$

Similarly for mass M_3,

$$M_3\left(\frac{du_2}{dt}\right) + \frac{1}{K_0}\left[\int_0^t u_2 dt + x_2(0)\right] + \frac{1}{K_1}\left[\int_0^t (u_2 - u_1)dt + x_2(0) - x_1(0)\right] + D_1(u_2 - u_1) = 0 \qquad (2.28)$$

The electric circuit by f–v analogy is shown in Figure 2.14(b) and the loop equations are

$$(L_1 + L_2)\frac{di_1}{dt} + R_1(i_1 - i_2) + \frac{1}{C_1}\left[\int_0^t (i_1 - i_2)dt + q_1(0) - q_2(0)\right] = v \qquad (2.29)$$

and $$L_3\frac{di_2}{dt} + \frac{1}{C_0}\left[\int_0^t i_2 dt - q_2(0)\right] + R_1(i_2 - i_1) + \frac{1}{C_1}\left[\int_0^t (i_2 - i_1) + q_2(0) - q_1(0)\right] = 0 \qquad (2.30)$$

The f–i analogous electric circuit is shown in Figure 2.14(c) and the node equations are

$$(C_1 + C_2)\left(\frac{dv_1}{dt}\right) + G_1(v_1 - v_2) + \frac{1}{L_1}\left[\int_0^t (v_1 - v_2)dt + \phi_1(0) - \phi_2(0)\right] = i \qquad (2.31)$$

and

$$C_3\left(\frac{dv_2}{dt}\right) + \frac{1}{L_0}\left[\int_0^t v_2 dt - \phi_2(0)\right] + G_1(v_2 - v_1) + \frac{1}{L_1}\left[\int_0^t (v_2 - v_1)dt + \phi_2(0) - \phi_1(0)\right] = 0 \qquad (2.32)$$

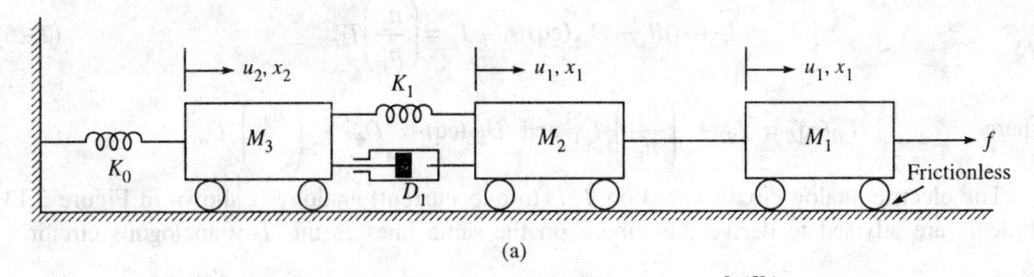

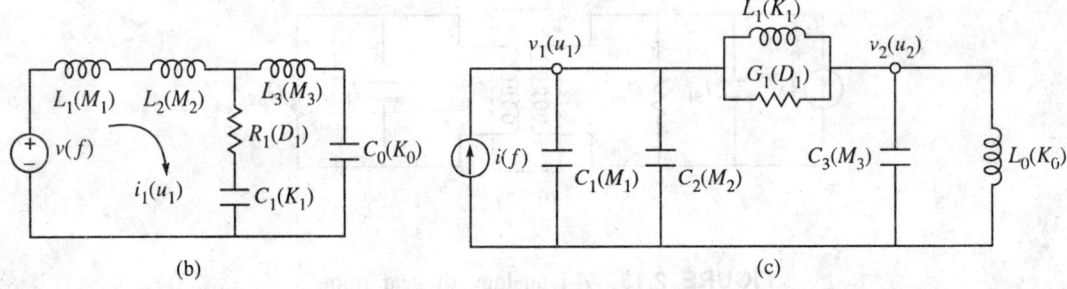

FIGURE 2.14 Example 2.2: (a) mechanical system, (b) f–v analogy, and (c) f–i analogy.

EXAMPLE 2.3 Draw the electric analog by both the f–v analogy and f–i analogy of the mechanical system shown in Figure 2.15(a). Write the equilibrium equations of the mechanical system.

Solution: The free-body diagram is shown in Figure 2.15(b). Using the D'Alembert's principle, the equilibrium equation for mass M_3 is

$$M_3\frac{du_3}{dt} + \frac{1}{K_3}\left[\int_0^t (u_3 - u_2)dt + x_3(0) - x_2(0)\right] = f = F\sin\omega t \qquad (2.33)$$

and that for M_2 is

$$M_2\frac{du_2}{dt} + \frac{2}{K_2}\left[\int_0^t (u_2 - u_1)dt + x_2(0) - x_1(0)\right] + \frac{1}{K_3}\left[\int_0^t (u_2 - u_1)dt + x_2(0) - x_1(0)\right] = 0 \qquad (2.34)$$

Mathematical Modeling of Systems

and that for mass M_1 is

$$M_1 \frac{du_1}{dt} + \frac{2}{K_2}\left[\int_0^t (u_1 - u_2)dt + x_1(0) - x_2(0)\right] + \frac{1}{K_1}\left[\int_0^t u_1 dt + x_1(0)\right] + D_1 u_1 = 0 \qquad (2.35)$$

The electric analog circuit by f–v analogy is shown in Figure 2.15(c) and the loop equations are

$$L_3 \frac{di_3}{dt} + \frac{1}{C_3}\left[\int_0^t (i_3 - i_2)dt + q_3(0) - q_2(0)\right] = v \qquad (2.36)$$

$$L_2 \frac{di_2}{dt} + \frac{2}{C_2}\left[\int_0^t (i_2 - i_1)dt + q_2(0) - q_1(0)\right] + \frac{1}{C_3}\left[\int_0^t (i_2 - i_3)dt + q_2(0) - q_3(0)\right] = 0 \qquad (2.37)$$

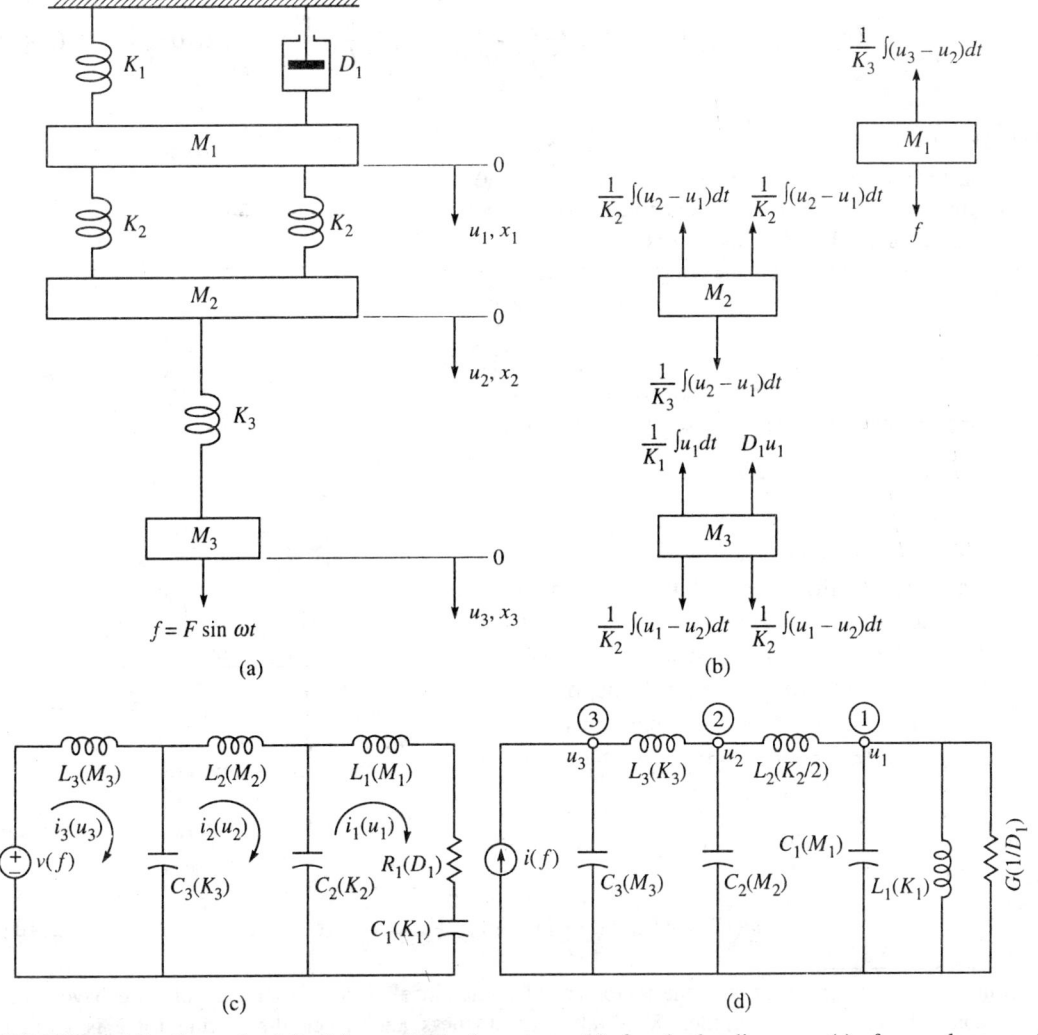

FIGURE 2.15 Example 2.3: (a) mechanical system, (b) free-body diagram, (c) f–v analogy, and (d) by f–i analogy.

and $\quad L_2 \dfrac{di_2}{dt} + \dfrac{2}{C_2}\left[\int_0^t (i_2 - i_1)dt + q_2(0) - q_1(0)\right] + \dfrac{1}{C_3}\left[\int_0^t (i_2 - i_3)dt + q_2(0) - q_3(0)\right] = 0 \quad (2.38)$

The electric analog circuit by f–i analogy is shown in Figure 2.15(d) and the node equations are

$$C_3 \dfrac{dv_3}{dt} + \dfrac{1}{L_3}\left[\int_0^t (v_3 - v_2)dt + \phi_3(0) - \phi_2(0)\right] = i \quad (2.39)$$

$$C_2 \dfrac{dv_2}{dt} + \dfrac{1}{L_3}\left[\int_0^t (v_2 - v_3)dt + \phi_2(0) - \phi_3(0)\right] + \dfrac{2}{L_2}\left[\int_0^t (v_2 - v_1)dt + \phi_2(0) - \phi_1(0)\right] = 0 \quad (2.40)$$

and $\quad \dfrac{2}{L_2}\left[\int_0^t (v_2 - v_1)dt + \phi_1(0) - \phi_2(0)\right] + C_1 \dfrac{dv_1}{dt} + G_1 v_1 + \dfrac{1}{L_1}\left[\int_0^t v_1 dt + \phi_1(0)\right] = 0 \quad (2.41)$

Drill Problem 2.1

The dynamic vibration absorber shown in Figure 2.15(a) is simplified where $K_2 = M_2 = 0$ and force f is applied at M_1 instead of at M_3. This system is a representative of many situations involving the vibration of machines containing unbalanced components. The parameters M_3 and K_3 may be chosen so that the main mass M_1 does not vibrate when $f = F \sin \omega t$. (a) Obtain the differential equations describing the system. (b) Draw the analogous electrical circuit based on the force–current analogy.

EXAMPLE 2.4 An important consideration in the design of the suspension system for an automobile is to increase the comfort level of the passengers by absorbing the vibrations caused by the terrain of the road. A model for the vertical suspension is shown in Figure 2.16. Obtain an equivalent electrical network, then write a set of state equations for the system. Determine the transfer function matrix using x_1 and x_2 to be the inputs u_1 and u_2.

Solution: The force–balance equation* for M_2 in Figure 2.17(a) is

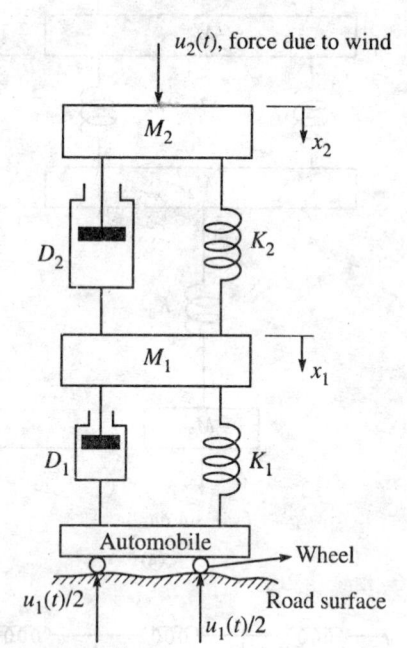

FIGURE 2.16 Example 2.4: suspension system of an automobile.

$$u_2(t) = M_2 \ddot{x}_2 + D_2(\dot{x}_2 - \dot{x}_1) + K_2(x_2 - x_1) \quad (2.42)$$

---

* Spring constant K as stiffness is the reciprocal of compliance. In an earlier example, we have used K for compliance. In this example, K is taken as stiffness and hence the spring force is written differently. This is to get familiarity with both the terminologies of spring component.

Mathematical Modeling of Systems

The force–balance equation for M_1 in Figure 2.17(b) is

$$M_1\ddot{x}_1 + D_1\dot{x}_1 + K_1x_1 = D_2(\dot{x}_2 - \dot{x}_1) + K_2(x_2 - x_1) \qquad (2.43)$$

The force–balance equation in Figure 2.17(c) is

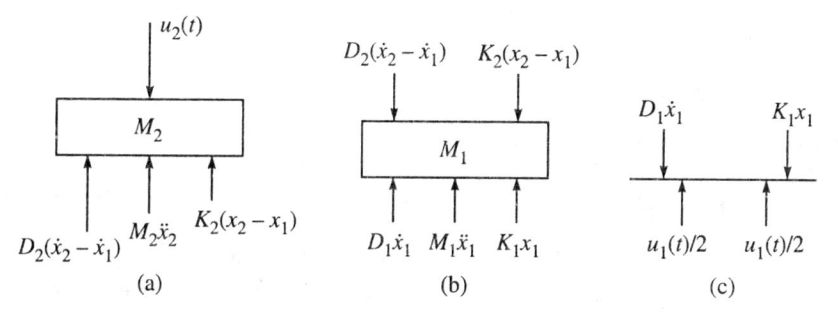

$$u_1(t) = D_1\dot{x}_1 + K_1x_1 \qquad (2.44)$$

FIGURE 2.17 Example 2.4: force–balance equations.

EXAMPLE 2.5 Draw the analogous electric circuit for the mechanical system shown in Figure 2.18(a) where x_i is the input displacement, x_o is the output displacement, D is the viscous damping coefficient, and K is the compliance of the spring.

Solution: By D'Alembert's principle, we can write the following equation

$$D\dot{x}_o = \frac{1}{K}(x_i - x_o)$$

Taking the Laplace transform and assuming all initial conditions to be zero (for derivation of transfer function), we get

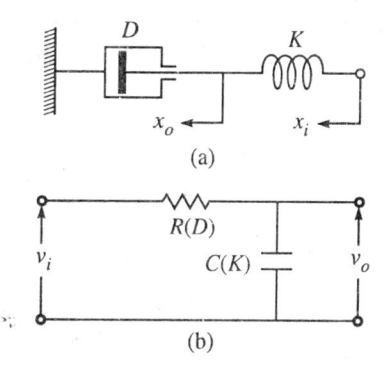

FIGURE 2.18 Example 2.5: (a) mechanical system and (b) analogous electric circuit.

$$\frac{X_o(s)}{X_i(s)} = \frac{1}{DKs + 1}$$

Obviously the analogous electric equivalent circuit is a low-pass RC circuit as shown in Figure 2.18(b), whose transfer function becomes

$$\frac{V_o(s)}{V_i(s)} = \frac{1}{RCs + 1}$$

EXAMPLE 2.6 Draw the analogous electric circuit for the mechanical system shown in Figure 2.19(a), where x_i is the input displacement, x_o is the output displacement, D is the viscous damping coefficient, and K is the compliance of the spring.

Solution: By D'Alembert's principle, we can write the following equation

$$D(\dot{x}_i - \dot{x}_o) = \frac{1}{K}x_o$$

Taking the Laplace transform and assuming all initial conditions to be zero (for derivation of transfer function), we get the transfer function as

$$\frac{X_o(s)}{X_i(s)} = \frac{DKs}{DKs+1}$$

The analogous electric equivalent circuit is a high-pass RC circuit as shown in Figure 2.19(b), whose transfer function becomes

$$\frac{V_o(s)}{V_i(s)} = \frac{RCs}{RCs+1}$$

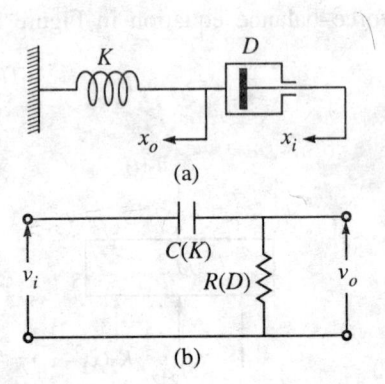

FIGURE 2.19 Example 2.6: (a) mechanical system (b) analogous electric circuit.

EXAMPLE 2.7 Draw the electrical equivalent circuit of the mechanical system shown in Figure 2.20(a), where x_i is the input displacement, x_o is the output displacement, y is the displacement of the spring, D_1, D_2 are the viscous damping coefficients, and K_1, K_2 are the compliances of the springs.

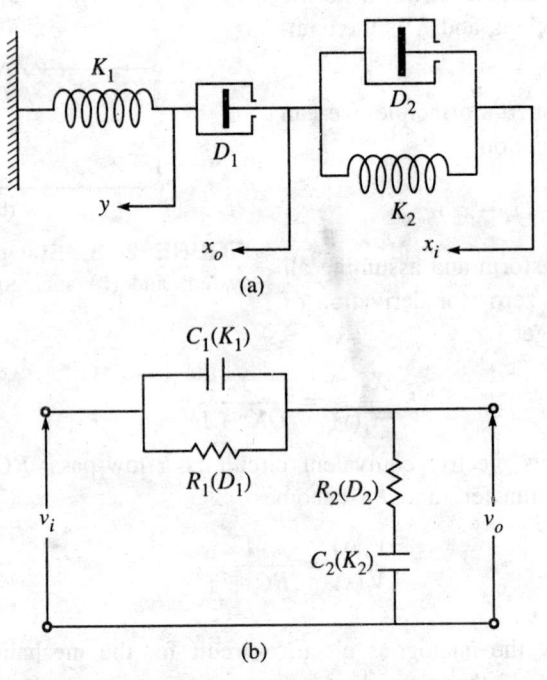

FIGURE 2.20 Example 2.7: (a) mechanical system and (b) analogous electric circuit.

Solution: By D'Alembert's principle, we can obtain the following equations:

$$D_2(\dot{x}_i - \dot{x}_o) + \frac{1}{K_2}(x_i - x_o) = D_1(\dot{x}_o - \dot{y}) \tag{2.45}$$

and

$$D_1(\dot{x}_o - \dot{y}) = \frac{1}{K_1} y \tag{2.46}$$

where $D_2(\dot{x}_i - \dot{x}_o)$ and $D_1(\dot{x}_o - \dot{y})$ are the damping forces; $(x_i - x_o)/K_2$ and y/K_1 are the spring forces.

Taking the Laplace transform and assuming zero initial conditions (for derivation of transfer function), we obtain from Eq. (2.46)

$$Y(s) = \frac{K_1 D_1 s}{K_1 D_1 s + 1} X_o(s) \tag{2.47}$$

Putting Eq. (2.47) in Eq. (2.45), we get

$$X_o(s) \left[\frac{D_1 s}{K_1 D_1 s + 1} + D_2 s + \frac{1}{K_2} \right] = X_i(s) \left[D_2 s + \frac{1}{K_2} \right]$$

From which the transfer function is obtained as

$$\frac{X_o(s)}{X_i(s)} = \frac{D_2 + \dfrac{1}{K_2 s}}{D_2 + \dfrac{1}{K_2 s} + \dfrac{D_1}{1 + K_1 D_1 s}}$$

The analogous electric circuit is as shown in Figure 2.20(b), whose transfer function is

$$\frac{V_o(s)}{V_i(s)} = \frac{R_2 + \dfrac{1}{C_2 s}}{R_2 + \dfrac{1}{C_2 s} + \dfrac{R_1}{1 + R_1 C_1 s}}$$

Note that K is the compliance of the spring which is the reciprocal of its stiffness.

EXAMPLE 2.8 Draw the electric analog and derive the transfer function of a mechanical lead network as shown in Figure 2.21(a), where x_i, x_o, y, D_1, D_2, and K have their usual significance.

Solution: By D'Alembert's rule, we can obtain the following equations:

$$D_2(\dot{x}_i - \dot{x}_o) = D_1(\dot{x}_o - \dot{y})$$

and

$$D_1(\dot{x}_o - \dot{y}) = \frac{1}{K} y$$

where $D_2(\dot{x}_i - \dot{x}_o)$ and $D_1(\dot{x}_o - \dot{y})$ are the damping forces and y/K is the spring force.

Taking Laplace transform and assuming all initial conditions to be zero (for derivation of transfer function), we obtain

$$\frac{X_o(s)}{X_i(s)} = \frac{D_1 Ks + 1}{\left(\dfrac{D_2}{D_1 + D_2}\right) D_1 Ks + 1} \times \frac{D_2}{D_1 + D_2} = \frac{s + 1/T}{s + 1/\alpha T}$$

where
$$T = D_1 K$$

and
$$\alpha = \frac{D_2}{D_1 + D_2}$$

The analogous electric circuit by f–v analogy is shown in Figure 2.21(b). The transfer function is given by

$$\frac{V_o(s)}{V_i(s)} = \frac{Z_2}{Z_1 + Z_2} = \frac{s + 1/T}{s + 1/\alpha T}$$

where $Z_1 = \dfrac{R_1}{R_1 Cs + 1}$, $Z_2 = R_2$, $T = R_1 C$, and $\alpha = \dfrac{R_2}{R_1 + R_2}$

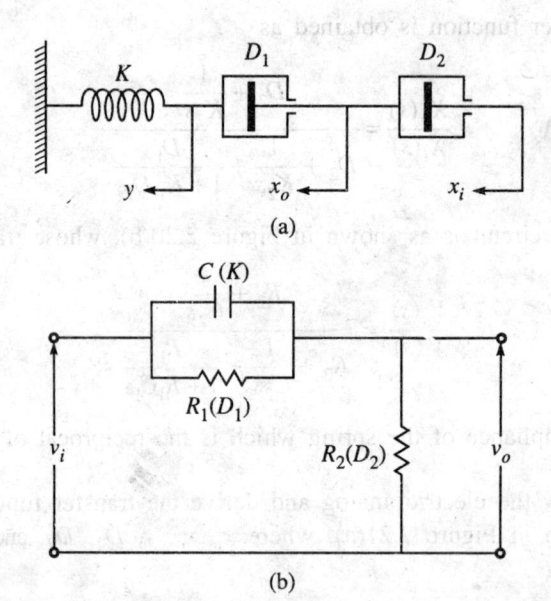

FIGURE 2.21 Example 2.8: (a) mechanical system and (b) f–v analogy.

2.6 Liquid-Level Systems

The concept of resistance and capacitance to describe the dynamics of a liquid-level system is introduced here. Consider the flow of a liquid through a short pipe into a tank. The resistance R to liquid flowing into the tank can be defined as

$$R = \frac{\text{change in liquid level}}{\text{change in flow rate}}$$

For laminar flow, the resistance

$$R = \frac{dH}{dQ} = \frac{H}{Q}$$

is constant and analogous to electrical resistance, where

H is the steady-state head (height), in m

Q is the steady-state volumetric flow rate, in m^3/s

R is the valve resistance, in S/m^3.

Consider the system shown in Figure 2.22(a), where q_i, q_o are the small deviations in the respective inflow and outflow rates from the steady-state flow rate Q, and h is the small deviation of the head from the steady-state head H.

Assume laminar flow to get a linear equation. The rate of change of fluid volume in the tank is equal to the flow-in less flow-out, i.e.

$$\frac{dV}{dt} = q_i - q_o$$

Further, since $dV = Cdh$, where C is the fluid capacitance of the tank defined as the ratio of change in volume stored to the change in head, we get

$$Cdh = (q_i - q_o)dt$$

Again,

$$q_o = \frac{h}{R}$$

The differential equation for a constant value of R becomes

$$RC\frac{dh}{dt} + h = Rq_i$$

With q_i as the input and h as the output head, the transfer function becomes

$$\frac{H(s)}{Q_i(s)} = \frac{R}{RCs+1} \qquad (2.48)$$

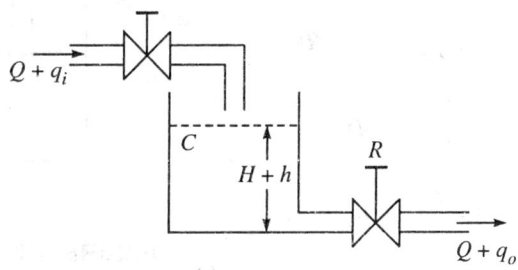

FIGURE 2.22(a) Liquid-level system.

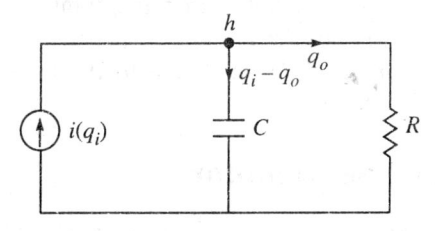

FIGURE 2.22(b) Electric circuit analog.

The electric circuit analog is shown in Figure 2.22(b). In this analysis, the resistance R includes the resistance due to exit and entrance of the tank. The effects of compliance and inertia have been neglected.

However, if q_o is taken as the output with the input q_i being the same and where the relation used is $q_o = \dfrac{h}{R}$ or $Q_o(s) = \dfrac{1}{R} H(s)$, then the transfer function becomes

$$\frac{Q_o(s)}{Q_i(s)} = \frac{1}{RCs + 1} \tag{2.49}$$

Liquid-level system with interaction

Consider the liquid-level system where two tanks interact as shown in Figure 2.23(a), for which we can obtain the following equations (assuming small variations of the variables from the steady-state values) as

$$\frac{h_1 - h_2}{R_1} = q_1$$

$$C_1 \frac{dh_1}{dt} = q - q_1$$

$$\frac{h_2}{R_2} = q_2$$

$$C_2 \frac{dh_2}{dt} = q_1 - q_2$$

FIGURE 2.23(a) Liquid-level system with interactions.

Considering q and q_2 as the respective input and output, we get the transfer function as

$$\frac{Q_2(s)}{Q(s)} = \frac{1}{R_1 R_2 C_1 C_2 s^2 + (R_1 C_1 + R_2 C_2 + R_2 C_1)s + 1} \tag{2.50}$$

The electric analog of the liquid-level system of Figure 2.23(a) is shown in Figure 2.23(b), where R_1 includes the resistance of the exit from tank 1 and entrance to tank 2 and R_2 is the resistance of exit from tank 2. The compliance and inertia effects have been neglected.

It may be noted that the transfer function of the system is not the product of the two isolated first-order transfer functions. It is due to the loading effect.

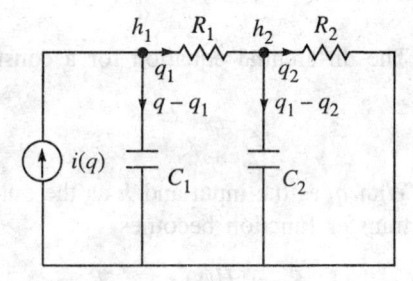

FIGURE 2.23(b) Electric circuit analog.

2.7 Servomotors

The electromechanical systems that we will discuss here are servomotors. The servomotors are of tow types: two-phase servomotors and dc servomotors. Further, the dc servomotors are either of armature-controlled type or of field-controlled type. The most important characteristic of the servomotor is the maximum acceleration. For a given available torque, the moment of inertia of the rotor must be minimum.

Let J_m and f_m be, respectively, the moment of inertia and friction of the motor, J_L and f_L be, respectively, the moment of inertia and friction of the load on the output shaft. Assume that the moment of inertia and friction of the gear train are either negligible or included in J_L and f_L, respectively. Then, the equivalent moment of inertia $J = J_{eq}$ and the equivalent friction $f = f_{eq}$ referred to the motor shaft can be rewritten from Eqs. (2.23) and (2.24) as

$$\left. \begin{array}{l} J = J_{eq} = J_m + n^2 J_L \quad (n<1) \\ f = f_{eq} = f_m + n^2 J_L \quad (n<1) \end{array} \right\} \quad (2.51)$$

where n is the gear ratio between the motor and the load. If neither J_m nor $n^2 J_L$ is negligibly small compared with the other, then the equivalent moment of inertia $J = J_{eq}$ and the equivalent friction $f = f_{eq}$ must be used for evaluating the transfer function of the motor-load combination. *It may be noted that the more commonly used symbols are used in different sections, for example, the symbols f and D are the same and used for either viscous friction or damping coefficient. Similarly for moment of inertia, I_θ and J are used interchangeably.*

2.7.1 Two-phase servomotors

A two-phase servomotor, commonly used for instrument servomechanisms, is similar to a conventional two-phase induction motor except for its special design considerations. It uses a squirrel-cage rotor. This rotor has a small diameter-to-length ratio to minimize the moment of inertia and to obtain a good accelerating characteristic. The two-phase servomotor is very rugged and reliable. Its power range is between a fraction of a watt and only a few hundred watts.

The schematic diagram of a two-phase servomotor is shown in Figure 2.24(a). Here one phase (fixed field) of the motor is continuously excited from the reference voltage, the frequency of which is usually 50 or 400 Hz (normally 50 Hz is for general purpose and 400 Hz is for military use); and the other phase (control field) is driven with the control voltage—a suppressed carrier signal which is 90° phase-shifted in time with respect to the reference voltage.

Since the control phase voltage is made 90° out of phase with respect to the voltage of the fixed phase, the stator windings for the fixed and control phases are placed 90° apart in space. These considerations are based on the fact that torque is produced most efficiently on the shaft when the phase-winding axes are in space quadrature and voltages in the two phases are in time quadrature.

The two stator windings are normally excited by a two-phase power supply. If a two-phase power supply is not available, however, then the fixed-phase winding may be connected to a single-phase power supply through a capacitor, which will provide the 90° phase shift. The control-phase winding is connected from the same single-phase power supply.

In the two-phase servomotor, the polarity of the control voltage determines the direction of rotation. Then the instantaneous control voltage e_c is of the form

$$\begin{aligned} e_c(t) &= E_c(t) \sin \omega t & \text{for } E_c(t) > 0 \\ &= |E_c(t)| \sin (\omega t + \pi) & \text{for } E_c(t) < 0 \end{aligned}$$

Since the reference voltage is constant, the torque T and the angular speed $\dot{\theta}$ are also functions of the control voltage $E_c(t)$. If variations in $E_c(t)$ are also compared with the ac supply frequency, the torque developed by the motor is proportional to $E_c(t)$. Figure 2.24(b) shows the curves $\theta_c(t)$ versus t, $E_c(t)$ versus t, and torque $T(t)$ versus t. The angular speed at steady state is proportional to the control voltage $E_c(t)$.

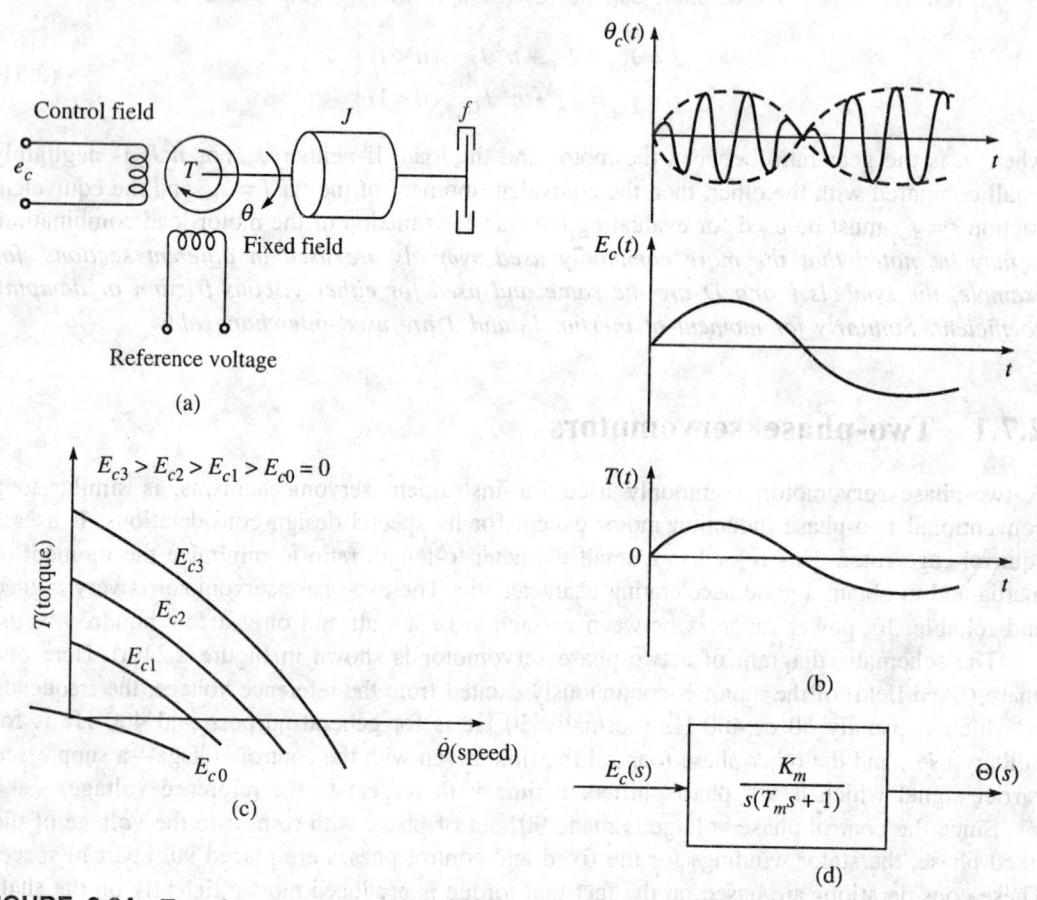

FIGURE 2.24 Two-phase servomotor: (a) schematic diagram, (b) curves showing $\theta_c(t)$ versus t, $E_c(t)$ versus t, and $T(t)$ versus t, (c) torque-speed curves, and (d) block diagram.

A family of torque-speed curves, when the rated voltage is applied to the fixed-phase winding and various voltages are applied to the control-phase winding, gives the steady-state characteristics of the two-phase servomotor. The transfer function of a two-phase ac servomotor may be obtained from such torque-speed curves if they are parallel and equidistant straight lines. Generally, the torque-speed curves are parallel for a relatively wide speed range but may not be equidistant, i.e. for a given speed, the torque may not vary linearly with respect to the control voltage. In a low-speed region, however, the torque-speed curves are usually straight lines and equidistant in a region of low control voltages. Since the servomotor seldom operates at high speeds, the linear portions of the torque-speed curves may be extended to the high-speed region. If the assumption is made that they are equidistant for all control voltages, then the servomotor may be considered linear.

Figure 2.24(c) shows a set of torque-speed curves for various values of control voltages. The torque-speed curve corresponding to zero control voltage passes through the origin. Since the slope of this curve is normally negative, if the control-phase voltage becomes equal to zero, the motor develops that torque which is necessary to stop the rotation.

The servomotor provides a large torque at zero speed. This torque is necessary for rapid acceleration. From Figure 2.24(c), we see that the torque T generated is a function of the motor-shaft angular speed $\dot{\theta}$ and the control voltage E_c. The equation for any torque-speed line is

$$T = -K_n \dot{\theta} + K_c E_c \qquad (2.52)$$

where K_n and K_c are positive constants.

The torque-balance equation for the two-phase servomotor is

$$T = J\ddot{\theta} + f\dot{\theta} \qquad (2.53)$$

where J is the moment of inertia of the motor and that of the load referred to the motor shaft and f is the viscous-friction coefficient of the motor and that of the load referred to the motor shaft as mentioned in Eq. (2.51). From Eqs. (2.52) and (2.53), we obtain the following equation:

$$J\ddot{\theta} + (f + K_n)\dot{\theta} = K_c E_c \qquad (2.54)$$

Noting that the control voltage E_c is the input and the angular displacement θ of the motor shaft is the output, we see that the transfer function of the system is given by

$$\frac{\Theta(s)}{E_c(s)} = \frac{K_c}{Js^2 + (f + K_n)s} = \frac{K_m}{s(T_m s + 1)} \qquad (2.55)$$

where, $K_m = \dfrac{K_c}{f + K_n}$ = motor gain constant and $T_m = \dfrac{J}{f + K_n}$ = motor time constant.

Figure 2.24(d) shows the block diagram for this system. From the dynamics in Eq. (2.55), we can see that $(f + K_n)$ is a viscous-friction term produced together by both the motor and the load. Here, K_n is the negative of the slope of the torque-speed curves. Hence for steeper torque-speed characteristics, the damping of the motor is higher. If the motor inertia is sufficiently low, then for most of the frequency range we have $|T_m s| \ll 1$ and the servomotor acts as an integrator.

The transfer function given by Eq. (2.55) is based on the assumption that the servomotor is linear. In practice, however, it is not quite so. For torque-speed curves not quite parallel and equidistant, the value of K_n is not constant and, therefore, the values and K_m and T_m are also not constant—they vary with the control voltage.

2.7.2 DC motors

A dc motor is often employed in a control system where an appreciable amount of shaft power is required. DC motors that are used in control systems are called dc servomotors. The dc motors are much more efficient than two-phase ac servomotors. The dc motors have separately excited fields. They are either armature-controlled with fixed field (i.e. the magnetic field is produced by a permanent magnet) or field-controlled with fixed armature current.

Armature-controlled dc servomotor

Armature-controlled d.c. servomotors used in instruments employ a fixed permanent-magnet field, and the control signal is applied to the armature terminals.

Consider the armature-controlled dc servomotor shown in Figure 2.25(a). In this system:

- R_a is the armature resistance ($\approx 0.2\ \Omega$)
- L_a is the armature inductance, in henries (negligibly small)
- i_a is the armature current, in amperes
- i_f is the field current, in amperes
- e_a is the applied armature voltage, in volts
- e_b is the back emf, in volts
- θ is the angular displacement of the motor shaft, in radians
- T is the torque delivered by the motor, in newton-metre (N-m)
- J is equivalent moment of inertia of the motor plus that of the load referred to the motor shaft, in kg-m^2
- f is the equivalent viscous-friction coefficient of the motor plus that of the load referred to the motor shaft, in N-m/rad/s.

The torque T developed by the motor is proportional to the product of the armature current i_a and the air-gap flux ψ, which in turn is proportional to the field current i_f, or

$$\psi = K_f i_f$$

where K_f is a constant. The torque T can, therefore, be written as

$$T = K_f i_f K_1 i_a$$

where K_1 is a constant.

In the armature-controlled dc motor, the field current is held constant. For a constant field current, the flux becomes constant, and the torque becomes directly proportional to the armature current so that

$$T = K i_a$$

where K is the motor-torque constant.

When the armature is rotating, a voltage proportional to the product of the flux and angular velocity is induced in the armature. For a constant flux, the induced voltage is directly proportional to the angular velocity $d\theta/dt$. Thus,

$$e_b = K_b \frac{d\theta}{dt} \tag{2.56}$$

where K_b is a back emf constant.

The speed of an armature-controlled dc motor is controlled by the armature voltage e_a supplied by an amplifier. The differential equation following KVL for the armature circuit is

Mathematical Modeling of Systems

$$L_a \frac{di_a}{dt} + R_a i_a + e_b = e_a \tag{2.57}$$

The armature current produces the torque, which is applied to the inertial mass and friction; hence the force balance equation is

$$J \frac{d^2\theta}{dt^2} + f \frac{d\theta}{dt} = T = K i_a \tag{2.58}$$

For the definition of the transfer function, assume that all initial conditions are zero, and taking the Laplace transforms of Eqs. (2.56), (2.57), and (2.58), we obtain

$$K_b s \Theta(s) = E_b(s) \tag{2.59}$$

$$(L_a s + R_a) I_a(s) + E_b(s) = E_a(s) \tag{2.60}$$

$$(Js^2 + fs)\Theta(s) = T(s) = K I_a(s) \tag{2.61}$$

Considering $E_a(s)$ as the input and $\Theta(s)$ as the output, we can construct the block diagram, as shown in Figure 2.25(b), from Eqs. (2.59) to (2.61). The effect of the back emf is seen to be the feedback signal proportional to the speed of the motor. This back emf thus increases the effective damping of the system. The transfer function of this system is obtained as

$$\frac{\Theta(s)}{E_a(s)} = \frac{K}{s[L_a J s^2 + (L_a f + R_a J)s + R_a f + K K_b]} \tag{2.62}$$

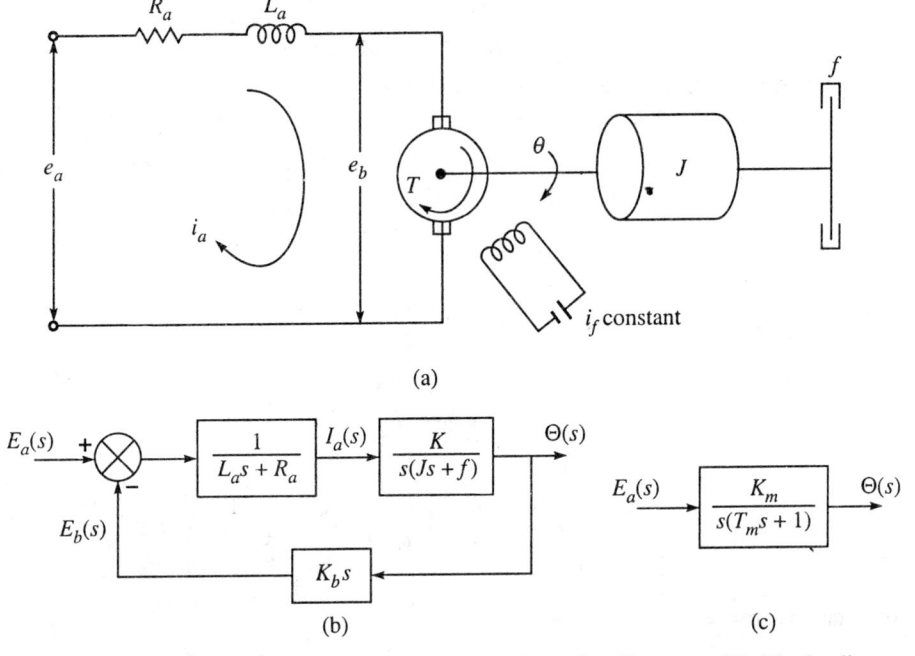

FIGURE 2.25 Armature-controlled dc motor: (a) schematic diagram, (b) block diagram, and (c) simplified block diagram.

The inductance L_a in the armature circuit is usually small and may be neglected. If L_a is neglected, then the transfer function given by Eq. (2.62) reduces to

$$\frac{\Theta(s)}{E_a(s)} = \frac{K_m}{s(T_m s + 1)} \qquad (2.63)$$

where

$$K_m = K/(R_a f + KK_b) = \text{motor gain constant}$$
$$T_m = R_a J/(R_a f + KK_b) = \text{motor time constant}$$

State-space formulation

From Eq. (2.63) the differential equation for the system is

$$\ddot{\theta} + \frac{1}{T_m}\dot{\theta} = \frac{K_m}{T_m} e_a \qquad (2.64)$$

Let us define the state variables as

$$x_1 = \theta \quad \text{and} \quad x_2 = \dot{\theta} = \dot{x}_1$$

Substituting these in Eq. (2.64), we get a new set of equations as

$$\dot{x}_1 = 0.x_1 + 1.x_2 + 0.u$$

$$\dot{x}_2 = 0.x_1 - \frac{1}{T_m}.x_2 + \frac{K_m}{T_m}.u$$

where the input variable $u = e_a$ and the output variable

$$y = \theta = x_1 = 1.x_1 + 0.x_2 + 0.u$$

Then the state variable formulation of the dynamics of the dc servomotor represented by the differential equation (2.64) can be written in vector-matrix differential equation form as

$$\begin{bmatrix} \dot{x}_1 \\ \dot{x}_2 \end{bmatrix} = \begin{bmatrix} 0 & 1 \\ 0 & -\frac{1}{T_m} \end{bmatrix} \begin{bmatrix} x_1 \\ x_2 \end{bmatrix} + \begin{bmatrix} 0 \\ \frac{K_m}{T_m} \end{bmatrix} u$$

and output

$$y = \begin{bmatrix} 1 & 0 \end{bmatrix} \begin{bmatrix} x_1 \\ x_2 \end{bmatrix} + 0.u$$

or, in a general way

$$\dot{X} = AX + Bu$$

and output

$$y = CX + Du$$

where

$$A = \begin{bmatrix} 0 & 1 \\ 0 & -\frac{1}{T_m} \end{bmatrix}, \quad B = \begin{bmatrix} 0 \\ \frac{K_m}{T_m} \end{bmatrix}; \quad C = \begin{bmatrix} 1 \\ 0 \end{bmatrix}^T$$

and X, the state vector $= \begin{bmatrix} x_1 \\ x_2 \end{bmatrix}$.

The scalar output $y = \theta$ and the scalar input $u = e_a$.

Mathematical Modeling of Systems

The state flow graph (SFG) of Eq. (2.63) is shown in Figure 2.26.

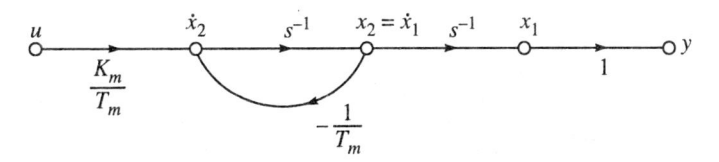

FIGURE 2.26 State flow graph.

Drill Problem 2.2

The practical setup of a position-control servomechanism is shown in Figure 2.27. Assume that the input and output of the system are the input shaft position and the output shaft position respectively. Obtain the closed-loop transfer function after going through the detailed individual block diagram of the system with the following numerical values for the system constants.

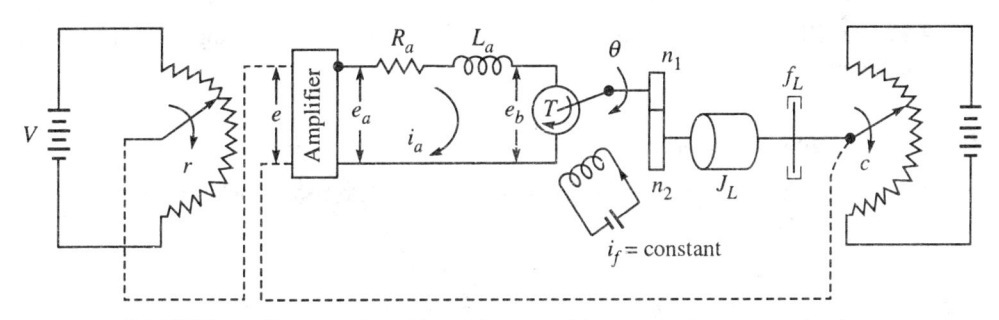

FIGURE 2.27 Drill Problem 2.2: a position-control servomechanism.

r, c, θ are the angular displacement of input, output and motor shaft, in rad

K_1 is the gain of the potentiometric error detector (= 7.64 V/rad)

K_p is the amplifier gain (= 10 V/V)

e_a is the applied armature voltage, in volts

e_b is the back emf in volts with back emf constant, K_b = 0.055 V-s/rad

R_a, L_a and i_a are respectively the resistance (= 0.2 Ω), inductance (negligible) and current of armature winding

K is the motor torque constant (= 6×10^{-5} N-m/V)

J_m is the inertia of the motor (= 1×10^{-5} kg-m^2) and f_m (negligible) is the viscous friction of the motor

J_L is the inertia of load (= 4.4×10^{-3} kg-m^2) and f_L is viscous friction of the load (= 4×10^{-2} N-m/rad/s).

n is the gear ratio, with $n_1/n_2 = 0.1$

Answer:
Figures 2.28(a) and (b) depict the block diagram of the system shown in Figure 2.27. Finally, we obtain in Figure 2.28(c) the closed-loop transfer function.

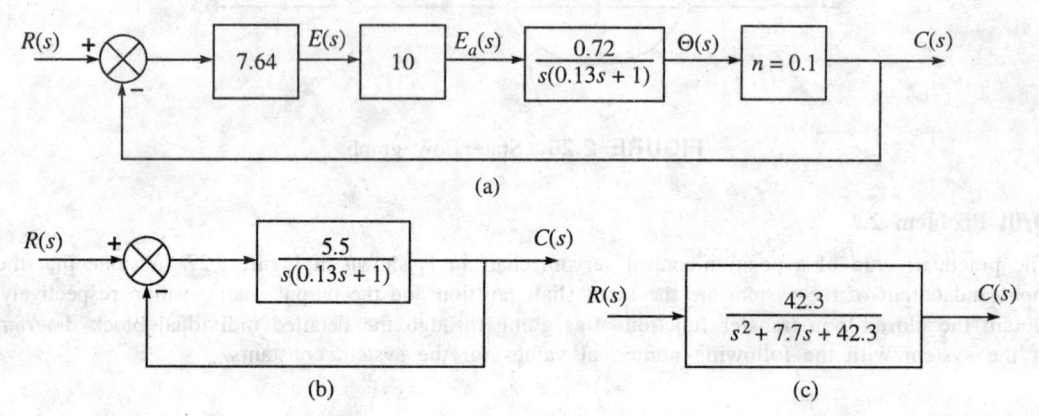

FIGURE 2.28 (a) Block diagram of the system shown in Figure 2.27, (b) simplified block diagram, and (c) closed-loop transfer function.

Field-controlled dc servomotor

Figure 2.29(a) is a schematic diagram of a field-controlled dc motor, where the various parameters are defined as follows:

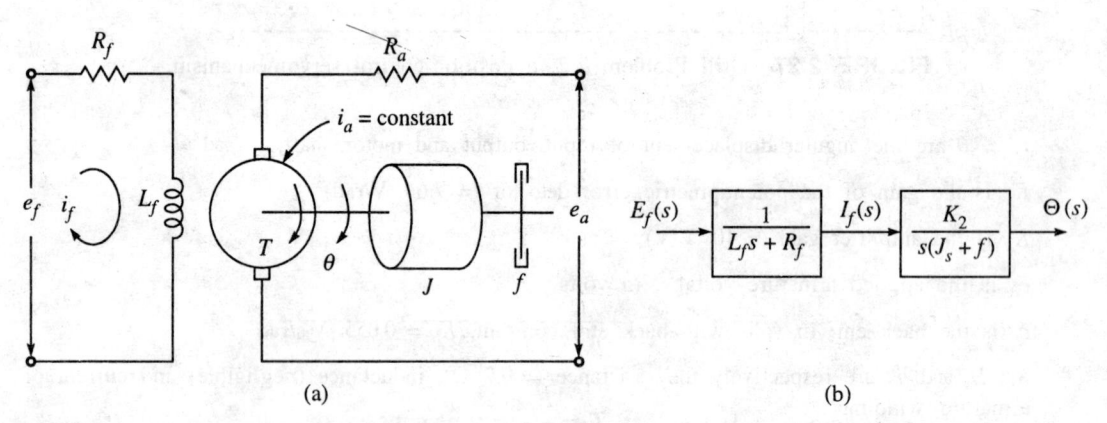

FIGURE 2.29 (a) Schematic diagram of a field-controlled dc motor and its (b) block diagram.

Field winding: R_f is the resistance in ohms, L_f the inductance in henries, i_f the current in amperes, and e_f is the applied field voltage in volts.

Armature winding: R_a is the resistance in ohms, i_a the current in amperes, and θ is the angular displacement of the motor shaft in radians.

Motor: T is the torque developed in N-m, J is the equivalent moment of inertia of the motor and that of the load referred to the motor shaft in kg-m^2, f is the equivalent viscous-friction coefficient of the motor and that of the load referred to the motor shaft in N-m/rad/s.

In this system, the field voltage e_f is the control input supplied by an amplifier. The armature current is maintained constant; this may be accomplished by applying a constant voltage source to the armature and inserting a very large resistance in series with the armature. The efficiency is necessarily low, but such a field-controlled dc motor may be used for a speed-control system. Note that maintaining a constant armature current i_a is more difficult than maintaining a constant field current i_f because of the back emf in the armature circuit.

The torque T developed by the motor is proportional to the product of the air-gap flux ψ and armature current i_a, so we get

$$T = K_1 \psi i_a$$

where K_1 is a constant. Since the air-gap flux ψ and the field current i_f are proportional for the usual operating range of the motor and i_a is assumed to be constant, we can rewrite the above equation as

$$T = K_2 i_f$$

where K_2 is a constant. The equations for this system are

$$L_f \frac{di_f}{dt} + R_f i_f = e_f \quad (2.65)$$

and

$$J \frac{d^2\theta}{dt^2} + f \frac{d\theta}{dt} = T = K_2 i_f \quad (2.66)$$

Taking the Laplace transform of Eqs. (2.65) and (2.66), assuming zero initial conditions for evaluating the transfer function, we obtain

$$(L_f s + R_f) I_f(s) = E_f(s) \quad (2.67)$$

$$(J s^2 + f s) \Theta(s) = K_2 I_f(s) \quad (2.68)$$

Considering $E_f(s)$ as the input and $\Theta(s)$ as the output, the block diagram of Eqs. (2.67) and (2.68) is shown in Figure 2.29(b). From this block diagram, the transfer function of this system is obtained as

$$\frac{\Theta(s)}{E_f(s)} = \frac{K_2}{s(L_f s + R_f)(J s + f)} = \frac{K_m}{s(T_f s + 1)(T_m s + 1)} \quad (2.69)$$

where, $K_m = \dfrac{K_2}{R_f f}$ = motor gain constant; $T_f = \dfrac{L_f}{R_f}$ = time constant of field circuit; and $T_m = \dfrac{J}{f}$
= time constant of inertia-friction element.

Since the field inductance L_f is not negligible, the transfer function of a field-controlled dc motor is of the third order and hence stability is poorer compared to that of an armature-controlled dc motor.

2.8 Sensors

Sensors can be divided into two basic classes: (i) internal state sensors and (ii) external state sensors. To discuss all of them is beyond the scope of this book, but some of them are discussed from application point of view. Under the first category exists:

- Potentiometers
- Synchros
- Tachometers
- Differential transformers (i.e. LVDTs and RVDTs)
- Accelerometers
- Optical interrupters
- Resolvers
- Linear inductive scales
- Optical encoders (absolute and incremental)

And the second category includes:

- Strain gauges
- Pressure transducers
- Proximity devices
- Ultrasonic sensors
- Electromagnetic sensors
- Elastometric sensors

2.8.1 Potentiometers

A potentiometer may be either of linear or rotational type, as shown in Figures 2.30(a) and (b) respectively, commonly made with wire-wound or conductive plastic resistance material. Rotary potentiometers are either in single revolution or in multi revolution form, and with limited or unlimited rotational motion. Refer to an N-turn rotary potentiometer having a fixed reference voltage E as shown in Figure 2.30(b); the output voltage $v_o(t)$ being proportional to the shaft position $\theta_c(t)$ can be written as

$$v_o(t) = K_S \theta_c(t) \quad (2.70)$$

where K_S is the proportionality constant and given by

$$K_S = \frac{E}{2\pi N} \text{ V/rad} \quad (2.71)$$

Two potentiometers connected in parallel as shown in Figure 2.31 allow the comparison of positions of two remotely located shafts. The output voltage is

$$e(t) = K_S[\theta_r(t) - \theta_c(t)] \quad (2.72)$$

and the block diagram representation is shown in Figure 2.32. This arrangement of Figure 2.31 may be recalled as potentiometric error detector and is often used in the dc motor control

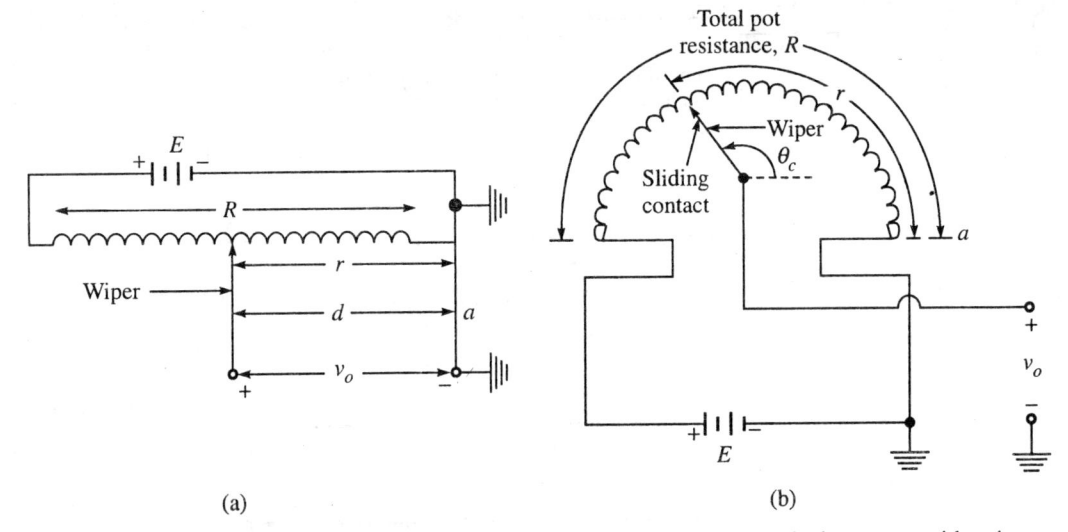

FIGURE 2.30 Wire-wound potentiometer (pot). The wiper makes physical contact with wires on the resistive coil. Point *a* corresponds to zero output (i.e. zero resistance): (a) linear output proportional to *d* and (b) rotary output proportional to θ_c.

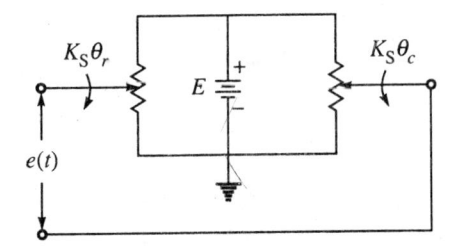

FIGURE 2.31 Potentiometer as error detector.

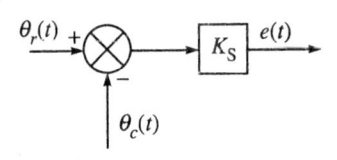

FIGURE 2.32 Block diagram representation of Figure 2.31.

system (that is, slowly varying system) such as that in Figure 2.33. In case of the dc servomotor used in a dc position control servo system, the motor has to have a small diameter-to-length ratio to minimize the moment of inertia in order to get good accelerating torque.

The schematic diagram of an ac control system having a two-phase servomotor is shown in Figure 2.34(a). This is similar to the dc control system of Figure 2.33(a) except that the signal voltage applied to the potentiometric error detector is sinusoidal. Here one phase (fixed field) of the motor is continuously excited from the reference voltage $v(t)$ whose frequency is, usually 50 or 400 Hz, much higher than that of the signal that is being transmitted through the system. The other phase (control field) is driven with the control voltage (suppressed carrier signal).

In a two-phase servomotor, the rotor has a small diameter-to-length ratio in order to minimize the moment of inertia and to obtain a good accelerating torque characteristic, otherwise it is similar to the conventional two-phase induction motor.

The phase windings are in space quadrature and the voltages in the two phases are in time quadrature. This makes the torque produced as maximum. The signal $v(t)$ is referred to as the

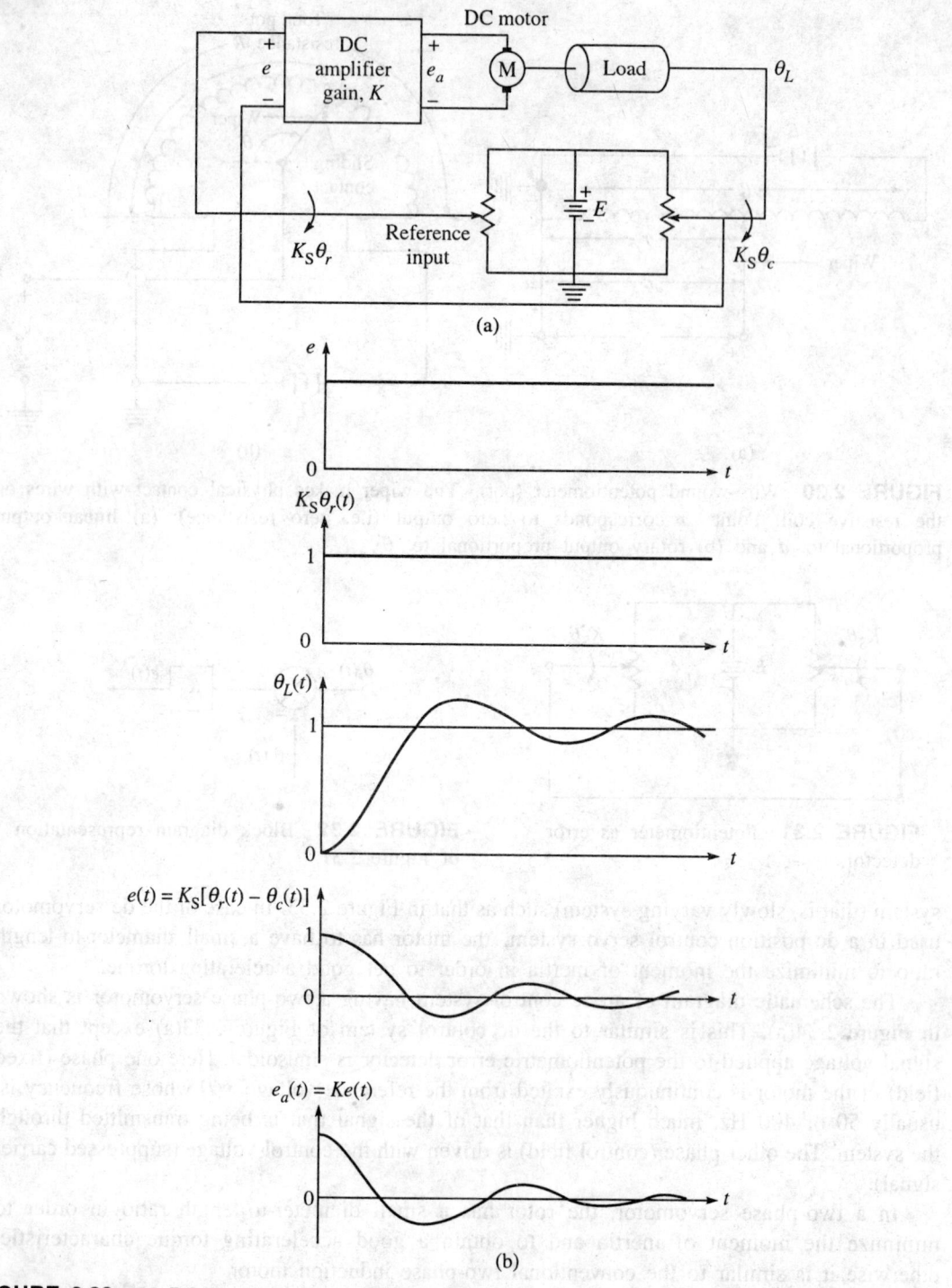

FIGURE 2.33 (a) DC motor position-control system with potentiometers as error sensors (detector) and (b) typical waveforms of signals in the control system of figure (a).

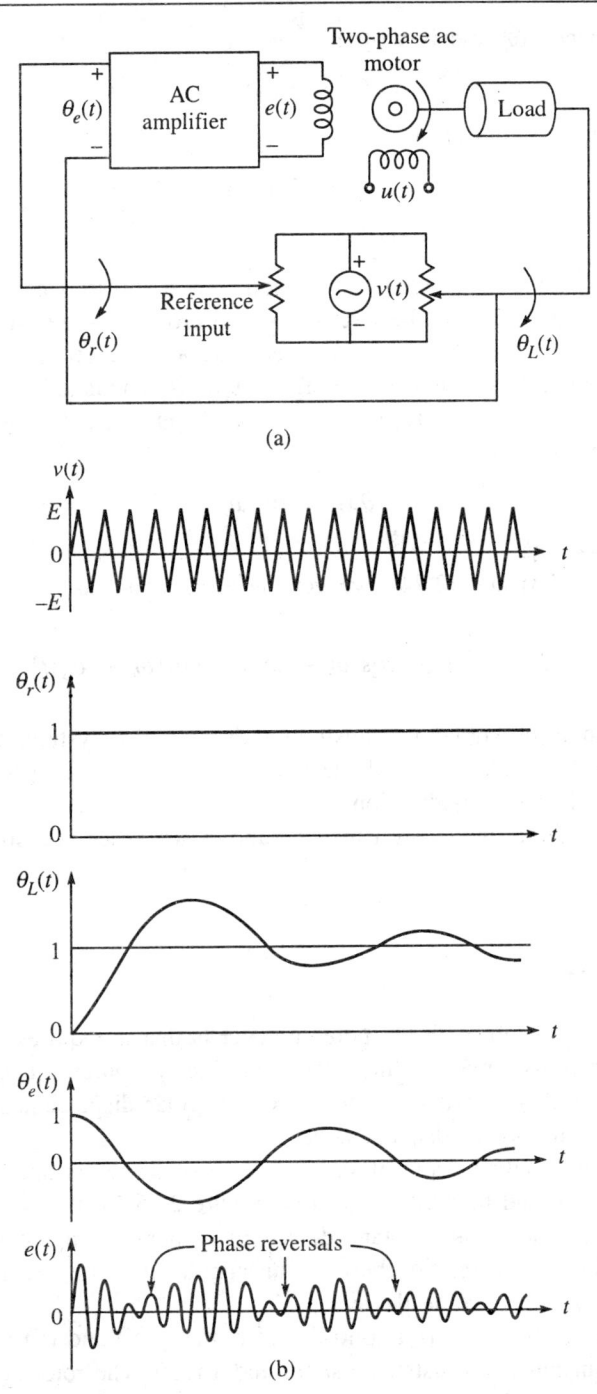

FIGURE 2.34 (a) AC control system with potentiometers as error detectors and (b) typical waveforms of signals in the control system of figure (a).

carrier, having frequency ω_c and is written as

$$v(t) = E \sin \omega_c t \qquad (2.73)$$

Analytically, the output of the error detector is given by

$$e(t) = K_S \theta_e(t) v(t) \qquad (2.74)$$

where
$$\theta_e(t) = \theta_r(t) - \theta_L(t)$$

For $\theta_e(t)$ shown in Figure 2.34(b), $e(t)$ becomes a suppressed carrier modulated signal. A reversal in phase of $e(t)$ occurs whenever the signal crosses the zero-axis and causes the ac motor to reverse direction according to the desired sense of correction of the error signal $\theta_e(t)$.

Refer to the ac control system where a signal $\theta_e(t)$ is modulated by a carrier signal $v(t)$, the resultant signal $e(t)$ contains only the two sidebands $\omega_c \pm \omega_s$. Let us assume that $\theta_e(t)$ is also a sinusoid given by

$$\theta_e(t) = \sin \omega_s t$$

where normally $\omega_s \ll \omega_c$.

Then from Eqs. (2.73) and (2.74), we get the error signal as

$$e(t) = \frac{1}{2} K_S E [\cos(\omega_c - \omega_s)t - \cos(\omega_c + \omega_s)t]$$

When the modulated error signal is transmitted through the system, the motor acts as a demodulator, so that the displacement of the load will be of the same form as that of the dc (low frequency) signal before modulation.

Control systems with ac signals are usually found in aerospace systems that are susceptible to noise.

2.8.2 Synchros

A significant practical problem with the potentiometer is that is requires a physical contact of the wiper with the resistance material in order to produce an output. The synchro overcomes that difficulty. It is a rotary transducer that converts angular displacement into an ac voltage or an ac voltage into an angular displacement.

Synchros are used widely in control systems as error detectors and encoders because of their rugged construction and high reliability (see Figure 2.35 for the schematic diagram of a synchro). Basically, a synchro is a rotary device that operates on the same principle as a transformer and produces a correlation between an angular position and a voltage or a set of voltages. Therefore, synchros are ac devices.

A synchro system consists of (i) a control transmitter (CX) and (ii) a control transformer (CT). The control transmitter consists of a stator and a rotor. The rotor is a dumb-bell-shaped magnetic structure. The supply is given to the rotor by means of slip rings, which are actually mounted on the stator housing. The secondaries are in the skewed slots all along the periphery of the stator and are 120° apart because of their mechanical displacement.

Mathematical Modeling of Systems

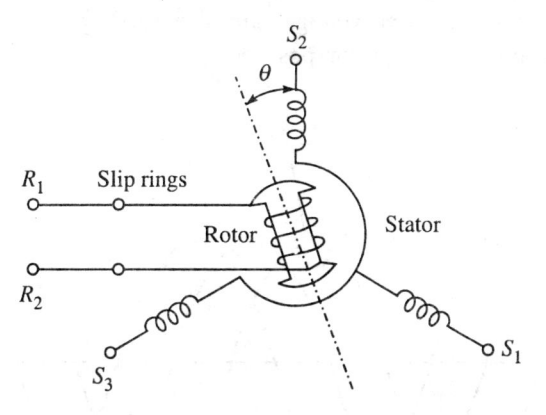

FIGURE 2.35 Schematic diagram of a synchro transmitter.

The induced secondary voltage will depend upon the angle of the rotor shaft. For reference, the zero degree position of the shaft is defined when the rotor is in alignment with the coil S_2. In this position, the voltage in coil S_2 is maximum, and similaly the maximum voltage in coils S_1 and S_3 will result at 120° and 240° positions respectively. The voltage in S_2 is a function of θ and so is the voltage in S_1 and S_3. Thus,

$$E_{0s2} = A \cos \theta$$

$$E_{0s1} = A \cos (\theta - 120°)$$

$$E_{0s3} = A \cos (\theta - 240°)$$

The connections of the synchro are made between the terminals and hence

$$E_{s1s2} = E_{0s1} - E_{0s2}$$

$$= A \cos (\theta - 120°) - A \cos \theta$$

$$= A \left[-\frac{1}{2} \cos \theta + \frac{\sqrt{3}}{2} \sin \theta - \cos \theta \right]$$

$$= \sqrt{3} A \left[-\frac{\sqrt{3}}{2} \cos \theta + \frac{1}{2} \sin \theta \right]$$

Therefore,
$$E_{s2s1} = \sqrt{3} A \left[\frac{\sqrt{3}}{2} \cos \theta - \frac{1}{2} \sin \theta \right]$$

$$= \sqrt{3} A \cos (\theta + 30°) \quad (2.75)$$

Similarly,
$$E_{s3s2} = \sqrt{3} A \cos (\theta + 150°) \quad (2.76)$$

and
$$E_{s1s3} = \sqrt{3} A \cos (\theta + 270°) \quad (2.77)$$

It is thus observed that the three line voltages are also 120° apart. These voltages are plotted in Figure 2.36 as a function of the rotor position θ.

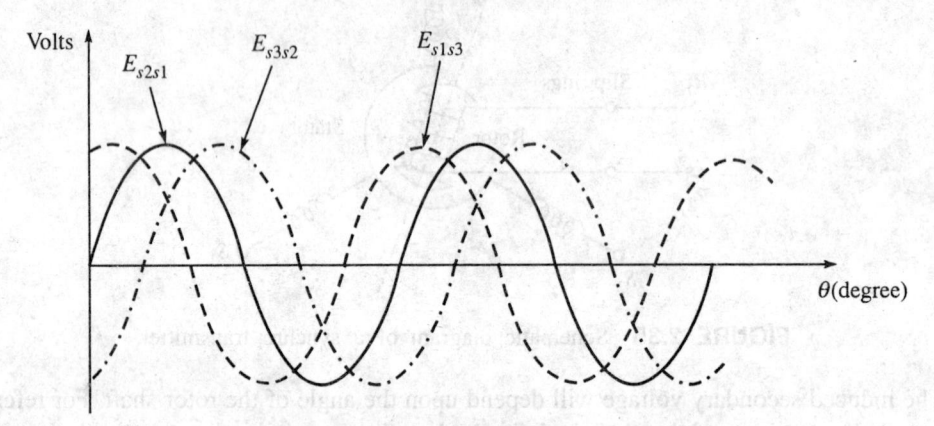

FIGURE 2.36 Line voltages in a synchro.

The control transformer is similar in appearance to that of a control transmitter except that its rotor is round and has distributed rotor windings. The rotor is made round so that it may have a high input impedance and hence not load the supply voltage.

Synchro-pair error detector

This is the most commonly used error detector in ac systems and is known by several names such as Selsyn, Telesyn Circutrol, Dichloyn, Teletorque and Autosyn, etc. The rotation here is not limited to 360° and it has high sensitivity and infinite resolution.

The schematic diagram of a synchro-pair error detector is shown in Figure 2.37. The stator windings of the control transformer are connected electrically to the control transmitter and hence the magnetic field established in the control transformer depends upon the terminal voltages of the control transmitter, which are functions of the angular position of the transmitter rotor. Since the voltage induced in the rotor of the control transformer depends upon the angle at which its turns are cut by the magnetic field of the stator, this induced voltage is determined both by the angular position of the transformer rotor and the angular direction of the stator magnetic field (and hence depends on the angular position of the transmitter rotor). Numerically, the voltage induced in the transformer rotor is equal to the sine of the difference angle between θ_r and θ_L multiplied by the maximum voltage induced. The transformer generally has higher impedance windings than those of the transmitter for loading reasons.

When the control transformer is connected as a component of the servo system, the correct zero position must be selected. This is done as follows:

(i) Set the control transformer shaft to one of the zero output positions.
(ii) Rotate the shaft slightly in the counterclockwise direction.
(iii) If the voltage now induced in the transformer rotor is in phase with the excitation applied to the transmitter rotor the position selected in step (i) is correct, otherwise it is at 180° phase shift position.

Referring to the arrangement shown in Figure 2.37, the voltages given by Eqs. (2.75), (2.76) and (2.77) are now impressed across the corresponding stator terminals of the control transformer. When the rotor positions of the two synchros are in perfect alignment, the voltage generated across the terminals of the CT rotor windings is zero. When the two rotor shafts are not in alignment, the rotor voltage of the CT is approximately a sine function of the difference between the two shaft angles, as shown in Figure 2.38.

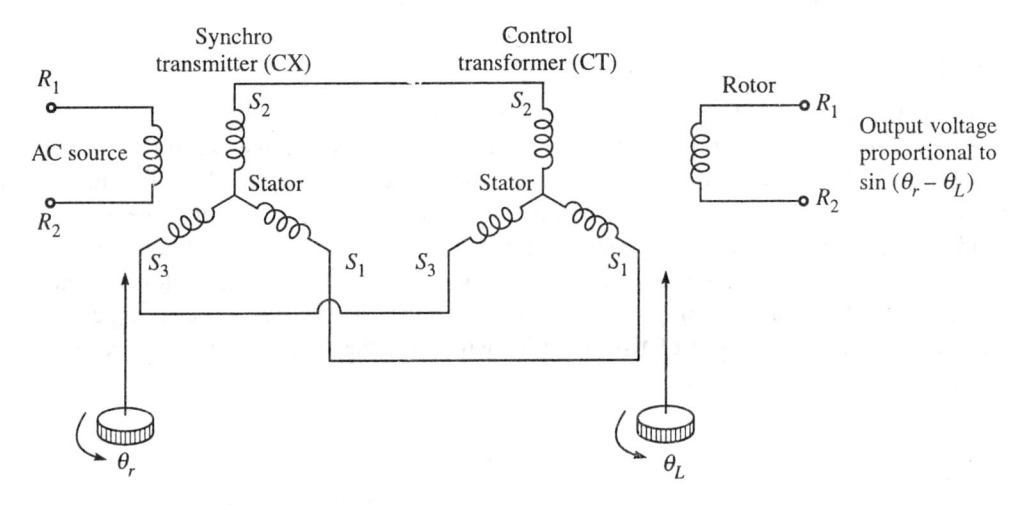

FIGURE 2.37 Synchro error detector.

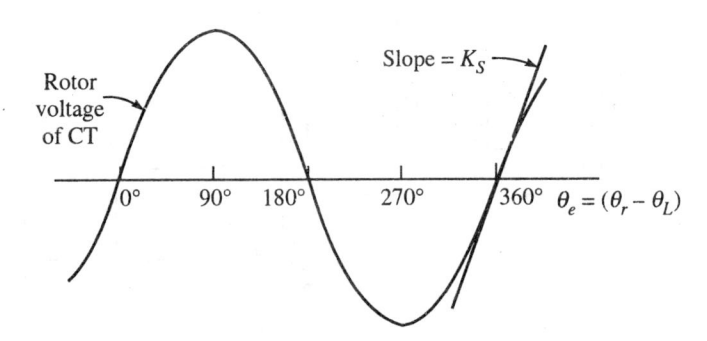

FIGURE 2.38 Rotor voltage of control transformer (CT) as a function of the difference of the rotor positions of the two synchros.

From Figure 2.38 it is apparent that the synchro error detector is a nonlinear device. However, for small angular deviations of up to 15 degrees in the vicinity of the two null positions, the rotor voltage of the control transformer is approximately proportional to the difference between the positions of the rotors of the transmitter and the control transformer. Therefore, for small deviations, the transfer function of the synchro error detector can be approximated by a constant K_S, i.e.

$$K_S = \frac{E}{\theta_r - \theta_L} = \frac{E}{\theta_e} \qquad (2.78)$$

where

  E is the error voltage
  θ_r is the shaft position of the synchro transmitter (CX), in degrees
  θ_L is the shaft position of the synchro control transformer (CT), in degrees
  θ_e is the error between the two shaft positions, in degrees
  K_S is the sensitivity of the error detector, in volts per degree.

Applications

The schematic diagram of a position-control system employing a synchro error detector is shown in Figure 2.39. The purpose of the control system is to make the controlled shaft follow the angular displacement of the reference-input shaft as closely as possible. The rotor of the control transformer is mechanically connected to the controlled shaft, and the rotor of the synchro transmitter is mechanically connected to the reference-input shaft. When the controlled shaft is aligned with the reference shaft, the error voltage is zero and the motor does not turn. When an angular misalignment exists, an error voltage of relative polarity appears at the amplifier input, and the output of the amplifier will drive the motor in such a direction so as to reduce the error.

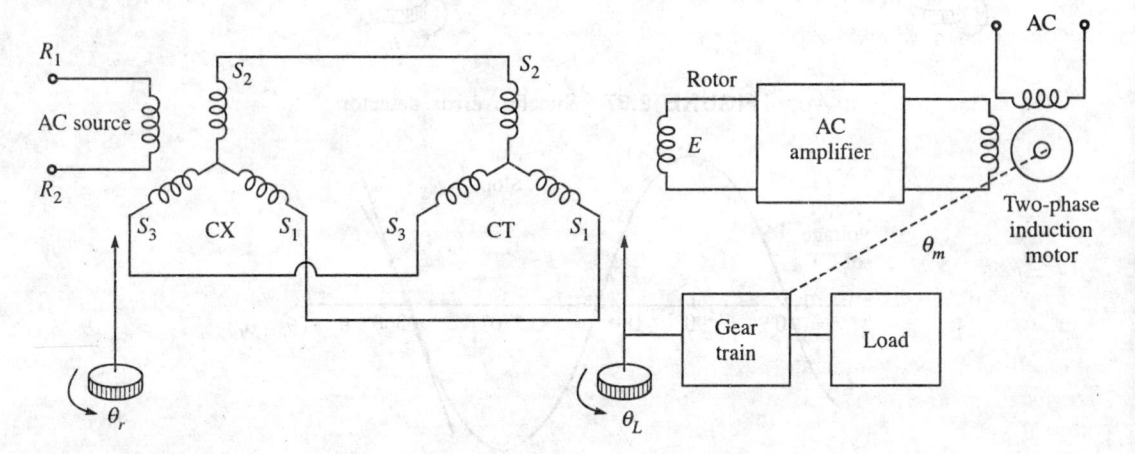

FIGURE 2.39 Position-control system.

2.8.3 Tachometers

Tachometers are electromechanical devices that convert mechanical energy into electrical energy. The device works essentially as a generator with the output voltage proportional to the magnitude of the angular velocity.

The most common type of dc tachometer contains an iron-core rotor. The magnetic field is provided by a permanent magnet, and no external supply voltage is necessary. The windings on the rotor (armature) are connected to the commutator segments and the output voltage is taken across a pair of brushes that ride on the commutator segments.

Mathematical Modeling of Systems

To reduce the inertia of the rotor assembly, the moving-coil tachometers contain a rotor that is iron-less. In this case the rotor with the armature windings is shaped in the form of a cup and is cantilevered between the permanent magnet poles and the inner iron structure; the later is needed for the completion of the flux paths.

For an ac tachometer, a sinusoidal voltage of rated value is applied to the primary winding. A secondary winding is placed at a 90-degree angle mechanically with respect to the primary winding. When the rotor shaft is rotated, the magnitude of the sinusoidal output voltage is proportional to the rotor speed. The phase of the voltage is dependent on the direction of rotation. An ac tachometer can be used in a dc control system by using a phase-sensitive demodulator to convert the ac output to dc. On the other hand, the output of a dc tachometer can be modulated for the control of ac system components, if necessary.

Mathematical modeling of tachometers

Regardless of the type of a tachometer, its basic characteristic is that the output voltage is proportional to the rotor speed. Thus, the dynamics of the tachometer can be represented by the equation

$$e_t(t) = K_t \frac{d\theta(t)}{dt} = K_t \omega(t) \tag{2.79}$$

where $e_t(t)$ is the output voltage, $\theta(t)$ the rotor displacement in radians, $\omega(t)$ the rotor velocity in rad/s and K_t the tachometer constant in V/rad/s. The value of K_t is usually given as a catalogue parameter in volts per 1000 rpm (V/krpm). A typical value for K_t is 6 V/krpm.

The transfer function of the tachometer is obtained by taking the Laplace transform of Eq. (2.79),

$$\frac{E_t(s)}{\Theta(s)} = K_t s$$

In control systems most of the tachometers used are of the dc variety (i.e. the output voltage is a dc signal). The dc tachometers are used in control systems in many ways; they can be used as velocity indicators to provide shaft-speed readout or to provide velocity feedback for speed control or stabilization. Figure 2.40 shows the block diagram of a typical velocity-control system.

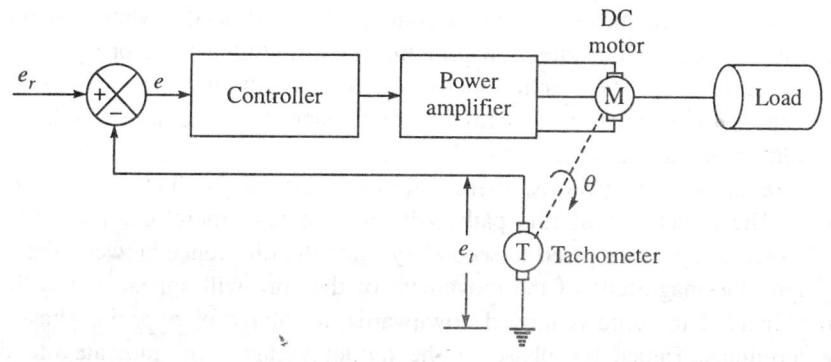

FIGURE 2.40 Velocity-control system with tachometer feedback.

In a position-control system, velocity feedback is often used to improve the stability or the damping of the overall system. Figure 2.41 shows the block diagram of such an application. In this case the tachometer feedback forms an inner loop to improve the damping characteristics of the system and the accuracy of the tachometer is not so critical for this type of application.

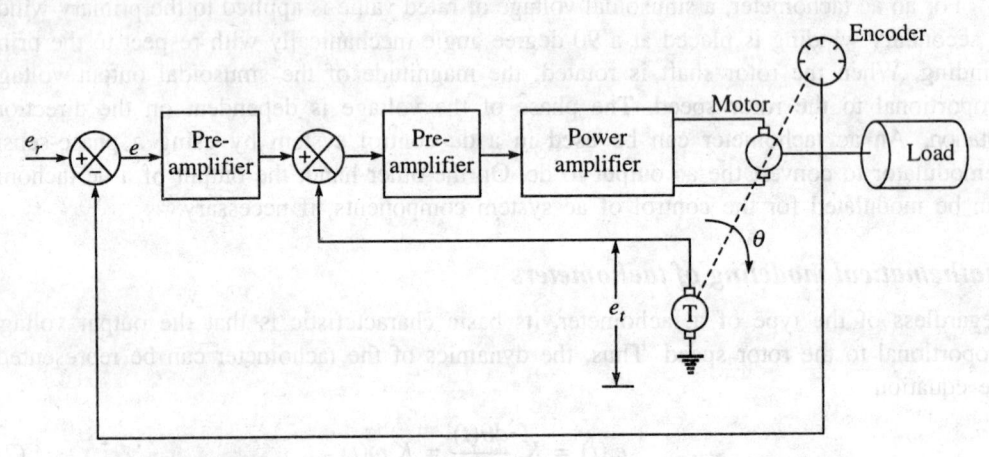

FIGURE 2.41 Position-control system with tachometer feedback.

The third and most traditional use of dc tachometers is in providing visual speed readout of a rotating shaft. Tachometers used in this capacity are generally connected directly to a voltmeter calibrated in rpm.

2.8.4 Linear variable differential transformer (LVDT)

The differential transformer employs the principle of electromagnetic induction and hence is usable only for alternating signals. Such a transformer however has a primary winding, two secondary windings I and II, and a movable core. The secondary windings are identical in respect of their number of turns as well as in respect of their placement on both sides of the primary winding as shown in Figure 2.42(a).

The secondary windings are connected in series opposition, so that the voltages in the two secondaries subtract. The movable core is connected to the shaft, whose position is to be controlled. Figure 2.42(a) illustrates the principle of differential transformer. If the movable core is in the centre or middle position, equal voltages will be induced in both the secondary windings because of the symmetry. Because of series opposition, the net secondary voltage will be zero as illustrated in Figure 2.42(b).

If the core is moved upwards, there will be more air-gap between the primary and secondary II. The reluctance of this path will increase and, therefore, less voltage will be developed in secondary II compared to secondary I and the difference between the two voltages depending upon the magnitude of the movement of the core will appear across the terminals. On the other hand if the core is moved downwards, a voltage of opposite phase will appear across the terminals. Hence the phase of the output voltage will indicate the direction of movement of the core while the magnitude of output voltage will be proportional to the displacement of the core from the centre position. This is illustrated in Figure 2.42(c).

This is the most popular magnetic type of error detector. It can be used as 'mechanical displacement'-to-'electrical voltage' type transducer. The characteristic is shown in Figure 2.42(d). When the core is exactly at the central position, the voltage is not zero because of residual magnetism. This is linear characteristic, symmetrical about the vertical axis. The output loses its linear relationship with displacement beyond some limits and this property

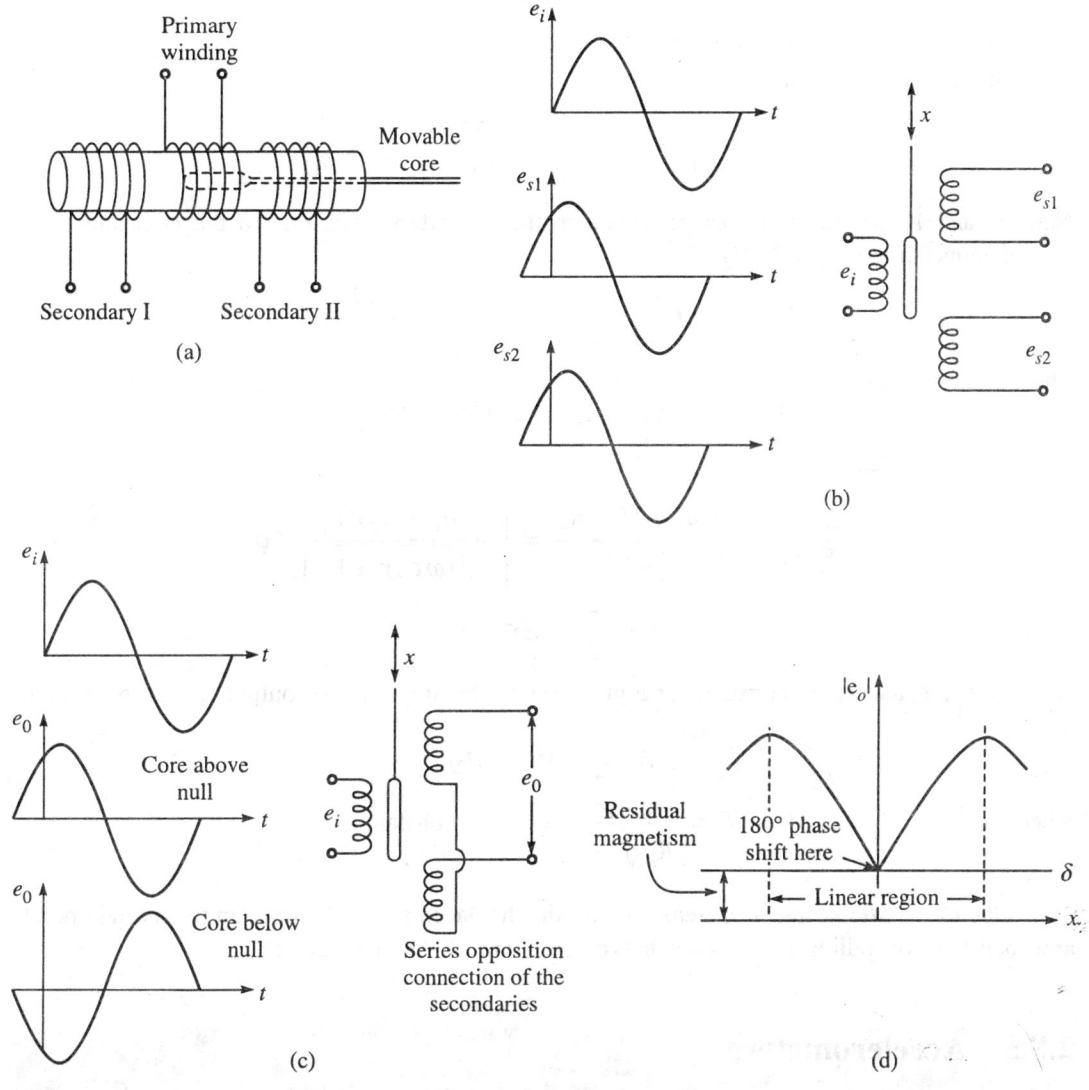

FIGURE 2.42 (a) Construction of LVDT, (b) LVDT core in null position and the corresponding voltages, (c) series opposition connection of secondaries and voltages for different core positions, and (d) magnitude of output voltage for core displacement of an LVDT.

restricts the range of the LVDT. The drooping occurs because of the core going out of bounds. This transducer can be used for measuring pressure indirectly. Weighing machines, load cells can use this type of transducer.

Circuit analysis

When the secondaries of LVDT are open-circuited, the equation for the primary becomes

$$i_p R_p + L_p \frac{di_p}{dt} = e_i$$

where symbols have their usual meaning.

Taking Laplace transform,

$$I_p(s) = \frac{E_i(s)}{sL_p + R_p} = \frac{E_i(s)/R_p}{\tau_p s + 1}; \quad i_p = \frac{L_p}{R_p}$$

Now e_{s_1} and e_{s_2} are the voltages generated in the secondary coils due to the coefficients of mutual inductances M_1 and M_2. Thus,

$$e_{s_1} = M_1 \frac{di_p}{dt} \quad \text{and} \quad e_{s_2} = M_2 \frac{di_p}{dt}$$

The output voltage,

$$E_o(s) = E_{s_1}(s) - E_{s_2}(s) = (M_1 - M_2) s I_p(s)$$

Substituting for $I_p(s)$, we get

$$\frac{E_o(s)}{E_i(s)} = \frac{s(M_1 - M_2)/R_p}{s\tau_p + 1} = \left| \frac{\omega(M_1 - M_2)/R_p}{\sqrt{(\omega\tau_p)^2 + 1}} \right| \angle \phi$$

where

$$\phi = \frac{\pi}{2} - \tan^{-1} \omega\tau_p$$

since ω, R_p, τ_p and e_i are constant for a given setup, the amplitude of output A_o can be written as

$$A_o = K(M_1 - M_2)$$

where

$$K = \frac{\omega e_i}{R_p \sqrt{(\omega\tau_p)^2 + 1}} = \text{constant}$$

The value $(M_1 - M_2)$ keeps on increasing with the displacement of the core up to a certain point and then it starts falling as the core moves past one of the secondaries.

2.8.5 Accelerometers

Let us briefly consider how a device that can be used to obtain linear acceleration and referred to as accelerometer operates. From Figure 2.43 it can be seen that an accelerometer consists

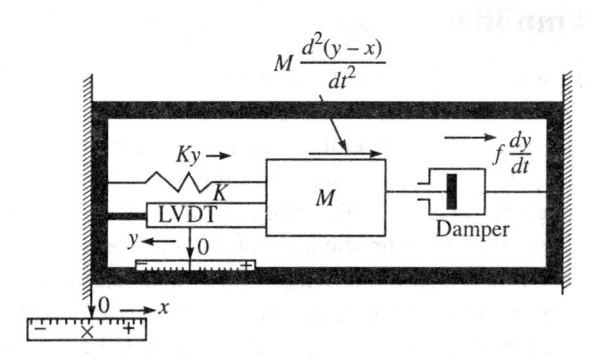

FIGURE 2.43 Basic diagram of a linear accelerometer.

of three basic elements:

- A mass M
- Some type of linear displacement sensor (e.g. an LVDT)
- A set of spring, damper having spring constant (reciprocal of compliance) K and viscous damping coefficient f

The accelerometer is used in seismic measurements. Because of the seismic jerk (which can be simulated in the laboratory experiment as if putting the frame on a moving vehicle), the frame of the accelerometer is accelerated, the spring deflects until it produces enough force to accelerate the mass at the same rate as the frame. The deflection of the spring measured by a linear displacement sensor like LVDT is a direct measure of acceleration. Let

x be the displacement of the frame with respect to a fixed reference frame
y be the displacement of the frame M with respect to a fixed reference frame

The force balance equation becomes

$$M \frac{d^2(y-x)}{dt^2} + f \frac{dy}{dt} + Ky = 0$$

or

$$M \frac{d^2 y}{dt^2} + f \frac{dy}{dt} + Ky = M \frac{d^2 x}{dt^2} = Ma$$

where a is the input acceleration.

If a constant acceleration is applied to the accelerometer, the output displacement y becomes constant under steady-state. Then $\frac{dy}{dt} = \frac{d^2 y}{dt^2} = 0$, that is,

$$Ma = Ky \quad \text{or} \quad a = \left(\frac{K}{M}\right) y$$

From this expression, it is apparent that the acceleration of the mass is proportional to the distance, that is, the output of the sensor LVDT will be proportional to the actual acceleration. The accelerometer must include enough damping so that the spring-mass combination does not 'ring' significantly (i.e. does not produce damped sinusoidal oscillations).

2.9 Magnetic Amplifier

Magnetic amplifier is a misnomer. There is no amplification as such in a magnetic amplifier. The gate winding and the load are connected in series. The output signal is distributed between the gate winding and the load. When operating near the saturation region of the nonlinear saturation characteristics of the magnetic core, the output signal across the gate winding is negligibly small and hence significantly high across the load. And in that sense, it acts as an amplifier though as such there is no amplification in the true sense. Such a magnetic device is called the *saturable core reactor*. The efficiency of this magnetic device is greatly enhanced by the addition of a silicon rectifier in the output circuit. This combination of a saturable core reactor with rectifiers is called magnetic amplifier or self-saturated amplifier. The beauty of the magnetic amplifier is its robustness and control of a large output quantity (i.e. voltage across load) by the variation of a small input quantity (dc bias in the control winding) as is illustrated by the saturable core reactor principle.

Linear stabilized power supplies are being built with magnetic amplifiers because of their robustness. Before we proceed further, the working principle of the saturable core reactor is explained.

2.9.1 Saturable core reactor

A saturable core reactor acts as a variable inductance connected in series with a load across an input ac power supply. It looks like a transformer having two windings wound on a core of steel as shown in Figure 2.44. One of these windings named control winding (CW) receives a small dc current controlled by a potentiometer and a fixed dc voltage source. This acts as an input bias signal that controls the amount of alternating current that can flow through the other winding, namely the gate winding (GW) and the load in series. The alternating current in the gate winding produces an objectionable ac voltage in the control winding that may disturb the fixed dc bias in it. Therefore, most saturable reactors include two identical cores *A* and *B* as shown in Figure 2.45; each steel core has its own ac winding named GW1 and GW2, while the dc coil surrounds one leg of each core. Here the two ac windings GW1 and GW2 are connected in parallel and in phase opposition to each other. Thus within the dc coil, that is, CW, two flux movements are in opposite directions at the same instant of time; therefore, no ac voltage

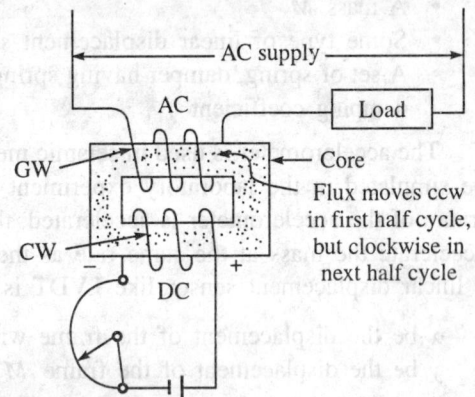

FIGURE 2.44 Saturable reactor with one gate winding.

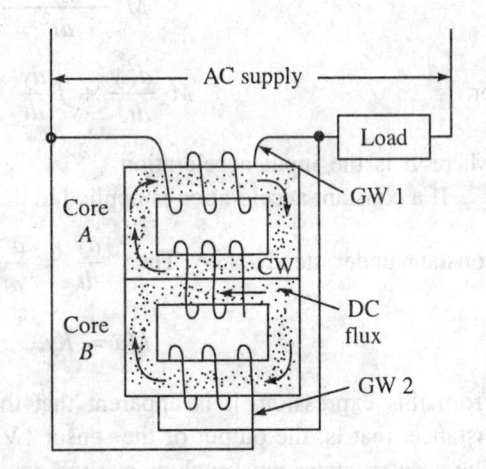

FIGURE 2.45 Saturable reactor with two gate windings.

is superimposed on the dc coil. There is no offset, the circuit is balanced, so that no objectionable ac voltage is induced in the dc circuit. Suppose, the dc current I in CW having N number of turns is high enough to put the dc bias near the knee region of the flux ϕ vs NI magnetization curve or B-H curve.

As $L = d\phi/dI$, the inductance L will be very less in the saturation region. The reactance ωL of the output ac winding (in this case GW) is significant less for saturation region operation. The output is absorbed in GW and load. The output current and load voltage are increased as the dc input bias current in CW is raised. The inductance of ac winding is proportional to the incremental permeability, which in turn, is proportional to the slope of the magnetization curve. As the dc current in CW increases, the inductance of the ac winding (GW) decreases and it becomes almost zero at the saturation point B or E as shown in Figure 2.46(a), which represents the B-H or magnetization (ϕ-NI) curve of ordinary magnetic core. Silicon-iron and nickel-iron alloy cores have a very small hysteresis width and a sharp knee break as shown in Figure 2.46(b). Note that the reactance of GW will be very less in the saturation region. Application of ac input voltage limited within YZ leads the current, that is, inductance L of GW is quite high and hence reactance ωL of GW is high, the output current is low and hence the output voltage across load is low, whereas the voltage across GW is relatively high. Hence application of an ac input in the linear range of operation of the magnetization curve is not able to transfer much of input share to output load. The output ac voltage is distributed as the voltage across GW and the load. Observations with fixed value of input ac supply voltage are as follows:

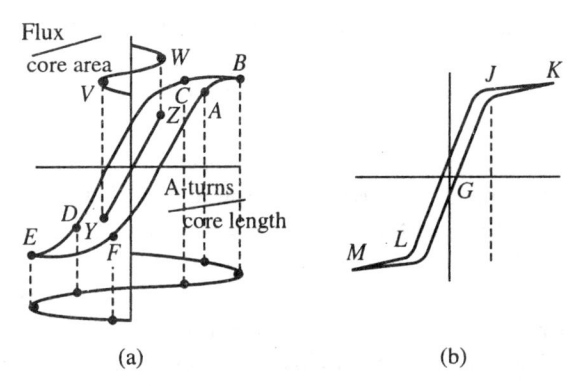

FIGURE 2.46 (a) B-H magnetization curve of ordinary steel and (b) B-H magnetization curve of alloy steel.

With N as the number of turns of CW, when dc current I in CW gives NI at point B, on flux vs ampere-turns curve, that is, (ϕ-NI) curve in Figure 2.47(a), for this portion of ac input operating in linear range, L is high, ωL is high, output voltage across GW is high and voltage across load is less.

In the same line of thought as the dc current in CW is increased, the dc bias also shifts towards right at point C, D and E of Figure 2.47(a) and accordingly the ac output voltage across GW decreases as ωL becomes less and less, whereas the ac voltage across load increases as shown in Figures 2.47(b) to (d). A typical variation of ac load output versus dc current in CW is shown in Figure 2.48 from which one can conclude that by providing a small dc current in CW, a large amount of ac input supply voltage can be transferred to the load as output ac

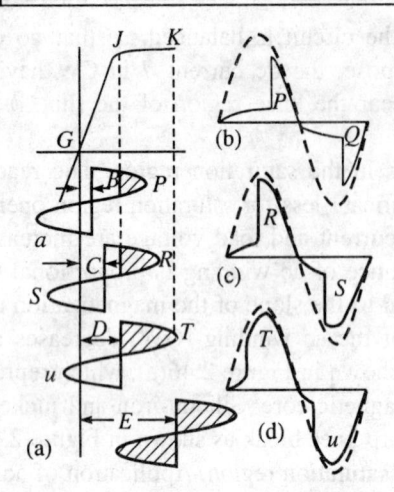

FIGURE 2.47 Large value of direct current increases saturation and ac output.

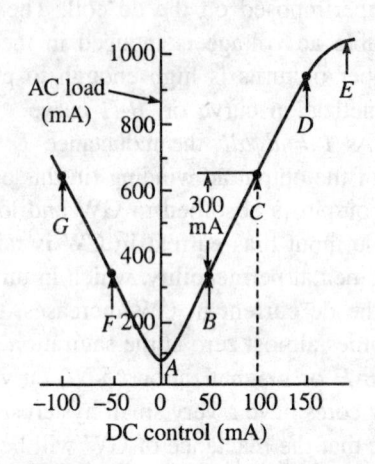

FIGURE 2.48 AC output of saturable reactor versus dc control input.

voltage, which obviously will be less than the ac input voltage. Figure 2.48 shows the typical current amplification characteristic, obviously the output current will be higher, the reactance ωL of GW will be less with higher dc input in CW because L decreases with increased bias NI.

A small amount of dc voltage with a high number of turns of CW can produce the necessary amount of dc bias to produce a large ac output across the load. It may be noted that a dc bias current of 5 mA in CW having a 400-turn coil produces the same amount of flux as is caused by 2 A flowing in a 1000-turn coil at the same location; the mmf in both cases is said to be 2000 ampere-turns.

It may be noted that most saturable reactors are made of special steel (a spiral winding of grain-oriented silicon-iron or nickel-iron or nickel-iron alloys) that has a narrow magnetization curve as shown in Figure 2.49(a) and a sharp knee at J and L. Near the saturation region the inductance reduces drastically which in turn reduces the output voltage in GW significantly and thus the voltage across the load increases significantly as shown in Figure 2.49(b).

Further note that, part of the dc voltage in CW can be generated from the input ac supply by providing a silicon (high current tolerability) rectifier as shown in Figure 2.50.

Hence, part of dc bias to provide saturation is obtained by providing a rectifier and therefore a magnetic device with a rectifier attachment can be called the *self-saturable reactor*. Such an arrangement is called the magnetic amplifier.

For ac load, the magnetic amplifier having rectifier along with the saturable core reactor is as shown in Figure 2.50.

For dc load, the magnetic amplifier with a full-wave bridge rectifier arrangement is as shown in Figure 2.51. Further, an alternative arrangement, a centre-tapped transformer for dc load is as shown in Figure 2.52 where only two diodes are required.

The use of magnetic amplifier in the feedback control mode for designing stabilized power supply is in vogue for its robustness. One should recall the working principle of linear regulated power supply where the nonlinear characteristic of the series active power transistor (active device) is replaced by the nonlinear saturation characteristic of the magnetic amplifier. The

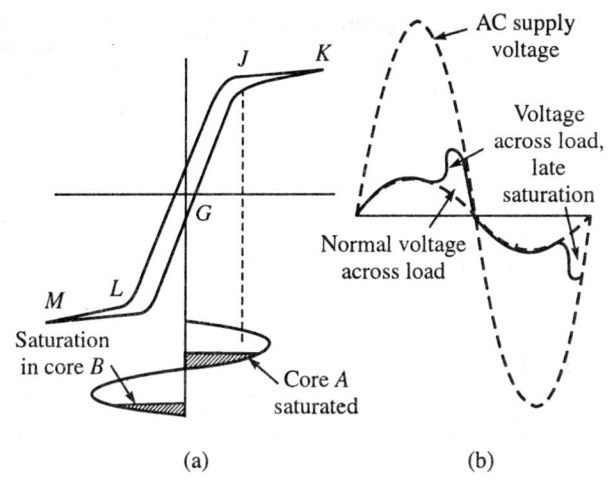

FIGURE 2.49 (a) Magnetization curve of alloy steel and (b) ac voltage waves showing saturation.

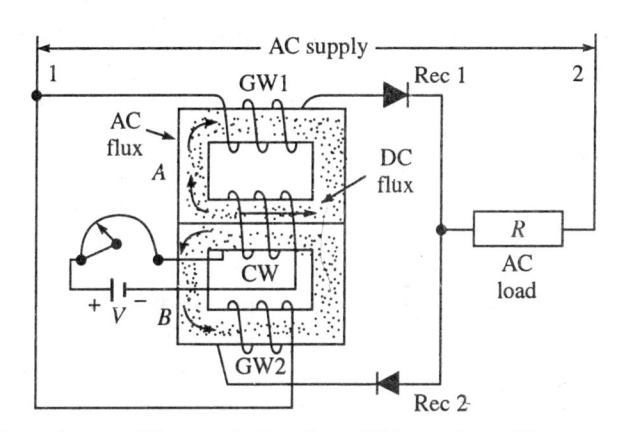

FIGURE 2.50 Magnetic amplifier made by the addition of rectifiers causing self-saturation.

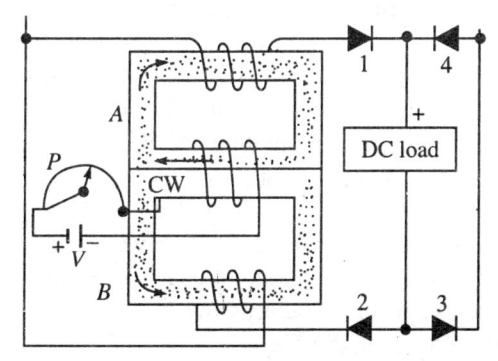

FIGURE 2.51 Magnetic amplifier with rectifier bridge permits control of dc load.

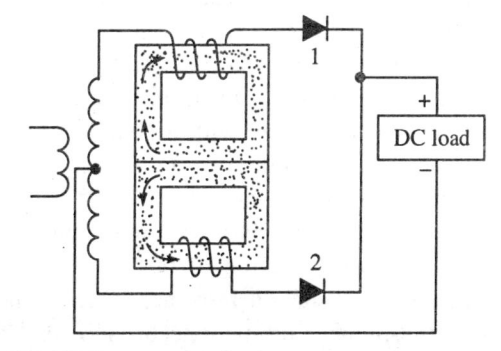

FIGURE 2.52 Centre-tap transformer with magnetic amplifier to control dc load.

magnetic amplifier having bias and feedback windings for feedback control application is shown in Figure 2.53. The bias winding is for fine control and the feedback winding is for course control of dc bias in control winding so as to permit variation of the point of operation on the magnetization curve in order to have better control of output across the load.

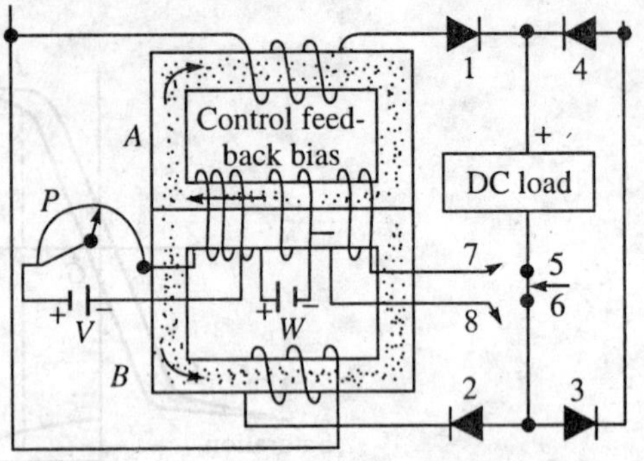

FIGURE 2.53 Magnetic amplifier with bias and feedback windings. The rectifier bridge permits control of dc load.

2.10 Stepper Motor

The two most widely used types of stepper motors are:

(i) Variable reluctance (VR) type
(ii) Permanent magnet (PM) type

The basic operational features of the variable reluctance motor are only being described here, being easy to understand.

The structure of a typical VR stepper motor is shown in Figure 2.54. It may be noted that unlike servomotor, both the stator and rotor of stepper are toothed structures and they do not have the same number of teeth. For example, the stator has eight teeth (located every 45°) and the rotor has six (located every 60°). In addition, each stator tooth has a coil wound on it, and the oppositely placed coils (e.g. A and A') are grouped together and referred to as a phase. In this case, there are four phases (labeled A, B, C, and D).

The operation of the stepper motor is based on the principle of *minimum reluctance* whereby a magnetic structure always attempts to reorient itself so as to minimize the length of any air-gap in the magnetic path. When phase A is energized, rotor teeth 1 and 4 (R4, 1) will align with stator teeth 1 and 5 (S5, 1) as shown in Figure 2.54(a) and will remain in this position as long as the coils in the same phase are energized. As long as the excitation remains on coils A-A' there is a *holding torque,* so that if any applied external torque is less than this value, no motion will occur. Suppose that the excitation is removed from phase A and placed on phase B, then R6,3 will align with S8,4 as shown in Figure 2.54(b). It can be seen that the rotor has moved clockwise through an angle of 15°(60° − 45° = 15°). This process can be repeated for phases C, D and then back to A as shown in Figures 2.54(c), (d), and (e). In each case, a 15° step occurs with a complete sequence of phase excitations (e.g. A, B, C, D, and A), producing a rotation of 60°. Consequently in this case, it requires six such cycles to cause one complete rotor revolution of 360° and therefore this case can be referred as a "24-step/revolution" motor. Further, making the phase excitation sequence as A, D, C, B, A will produce counterclockwise rotation of 60°. Translator design plays an important role in exercising a proper phase excitation sequence which is not in the scope of our discussion.

If the number of stator teeth is N_s and that of rotor teeth N_r, then

$$\text{step angle} = 360° \frac{|N_s - N_r|}{N_s N_r}$$

Mathematical Modeling of Systems 95

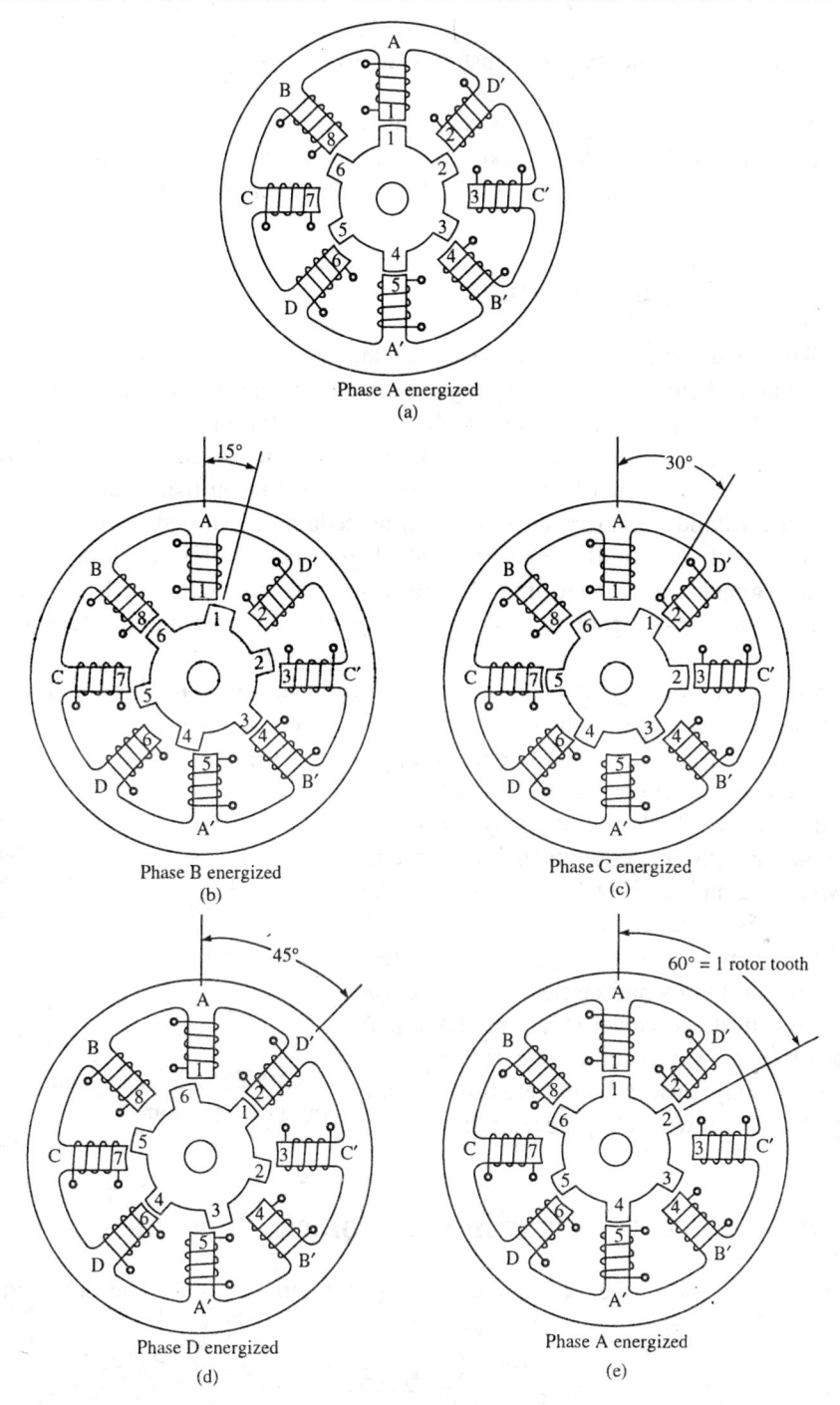

FIGURE 2.54 Basic structure of a VR-type stepper motor excited in the sequence ABCD: (a) Phase A is energized, (b) Phase B is energized, (c) Phase C is energized, (d) Phase D is energized, and (e) Phase A is energized.

and
$$\text{number of steps/revolution} = \frac{N_s N_r}{|N_s - N_r|}$$

The stepper motor is used as the digital actuator element in incremental motion control systems such as computer peripherals like printers, tape drives, etc. CNC machines, machine tools and process control systems also use stepper motors for this function.

2.11 Block Diagram Reduction

The block diagram representation of a system is adequate to shown the interrelationships of controlled output and input variable. Though the block diagram reduction technique can be suitably utilized for simplifying a complex block diagram, the procedure is cumbersome. An alternative method for representing the relationship between system variables is that of by signal flow graph. Mason's gain formula developed by S.J. Mason makes the transfer function calculation easier without requiring any reduction procedure or manipulation of the signal flow graph. This is a straightforward method for calculating the transfer function.

Signal flow graph is a directed line diagram consisting of nodes which are connected by several directed branches. Signal flow graph is a graphical representation of a set of linear relations.

The block diagram of a system consists of unidirectional functional blocks. They show the direction of flow and the operations on the system variables, in such a way that a relationship is established between the input and output as a path is traced through the diagram. Each functional block has entry and exit terminals indicated by arrows.

Before going into reduction techniques, let us describe what a summing point and a pick-off point or a branch point is. A summing point is one at which several system variables are added or subtracted, usually represented by a small circle with a cross (or without a cross) in it, as shown in Figure 2.55(a). A pick-off point or branch point is one at which the input variables may proceed unaltered to several different paths, as shown in Figure 2.55(b).

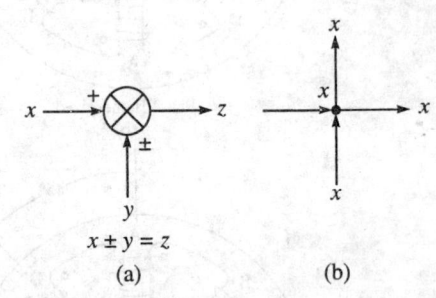

FIGURE 2.55 (a) Summing point and (b) pick-off point.

2.11.1 Procedure for drawing the block diagram

As an illustration, consider the RC circuit shown in Figure 2.56(a). The equations for this circuit are

$$i = \frac{e_i - e_o}{R} \tag{2.80}$$

and

$$e_o = \frac{1}{C}\int i\,dt \tag{2.81}$$

Mathematical Modeling of Systems

Taking the Laplace transform with zero initial conditions, we get

$$I(s) = \frac{E_i(s) - E_o(s)}{R} \tag{2.82}$$

$$E_o(s) = \frac{I(s)}{sC} \tag{2.83}$$

Equation (2.82) represents a summing operation and the corresponding diagram is shown in Figure 2.56(b). Equation (2.83) is represented by the block diagram in Figure 2.56(c). Assembling these two elements, we obtain the overall block diagram for the system as shown in Figures 2.56(d) and 2.56(e).

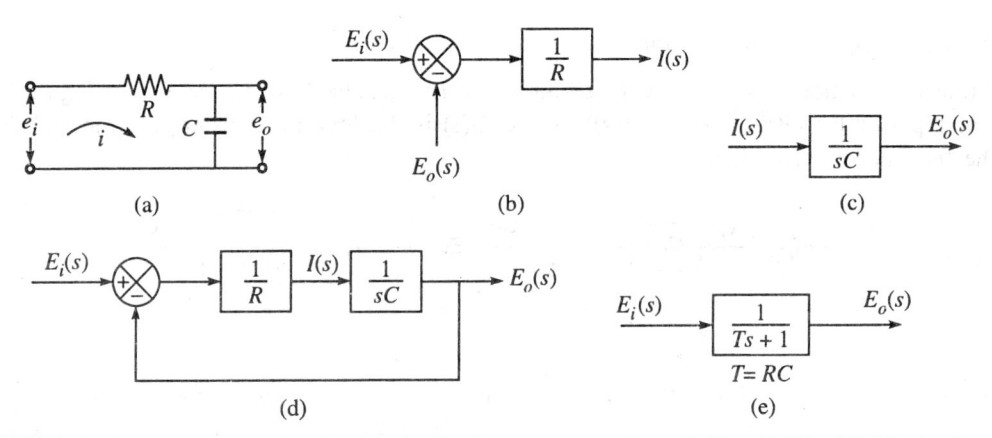

FIGURE 2.56 (a) RC circuit, (b) block diagram representation of Eq. (2.82), (c) block diagram representation of Eq. (2.83), and (d) and (e) overall block diagram representations of the system in (a).

2.11.2 Block diagram reduction techniques

In block diagram reduction techniques, a total of seven rules need to be considered. The seven rules are as follows:

1. Combining blocks in cascade

Blocks connected in cascade can be replaced by a single block with transfer function equal to the product of the individual transfer functions provided there is *no loading*. This is illustrated in Figure 2.57.

FIGURE 2.57 Combining blocks in cascade.

2. Parallel blocks

The transfer function of a parallel combination is simply the sum of the individual transfer functions. This is illustrated in Figure 2.58.

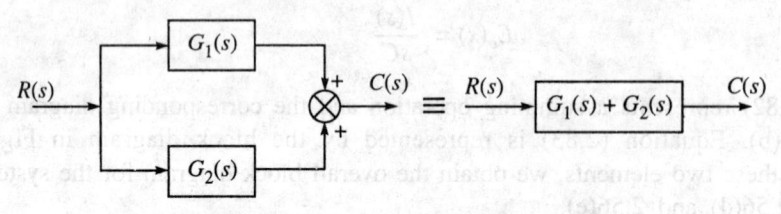

FIGURE 2.58 Combining parallel blocks.

3. Eliminating a feedback loop

The transfer function of a system with a simple negative feedback loop as shown in Figure 2.59 can be replaced by $G(s)/(1 + G(s)H(s))$, where $G(s)$ is the forward transfer function and $H(s)$ is the feedback transfer function.

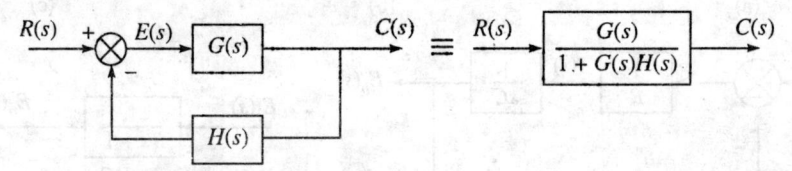

FIGURE 2.59 Eliminating a feedback loop.

Here,
$$C(s) = G(s)E(s)$$
$$E(s) = R(s) - H(s)C(s)$$

or
$$C(s) = G(s)[R(s) - H(s)C(s)]$$

Therefore,
$$\frac{C(s)}{R(s)} = \frac{G(s)}{1 + G(s)H(s)}$$

4. Moving a summing point ahead of a block

When a summing point is moved ahead of block G, the output is unaltered (see Figure 2.60).

$$Z = GX + Y$$

$$Z = G\left(X + Y\frac{1}{G}\right) = GX + Y$$

FIGURE 2.60 Moving a summing point ahead of a block.

Mathematical Modeling of Systems

This rule is illustrated with the following example in Figure 2.61(a) when the summing point is shifted ahead of block G_1 and the resulting simplification leads to block diagram as shown in Figure 2.61(b). Then the transfer function from Figure 2.61(b) becomes

$$G(s) = \left(1 + \frac{G_2}{G_1}\right)\left(\frac{G_1}{1+G_1}\right) = \frac{G_1 + G_2}{1 + G_1}$$

which is equal to the transfer function of Figure 2.61(a).

5. Moving a summing point behind a block

When summing point is moved behind the block G as shown Figure 2.62, the output is unlatered.

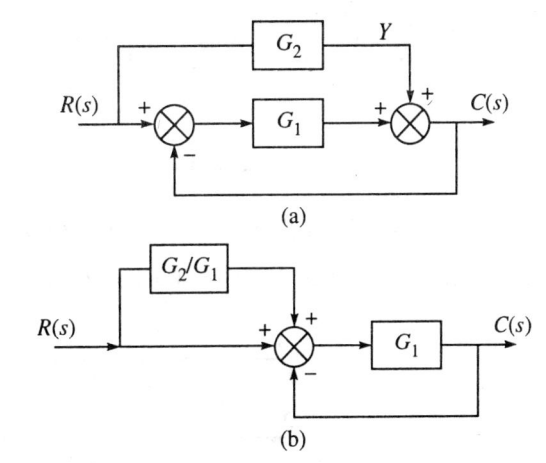

FIGURE 2.61 Example of moving a summing point ahead of a block.

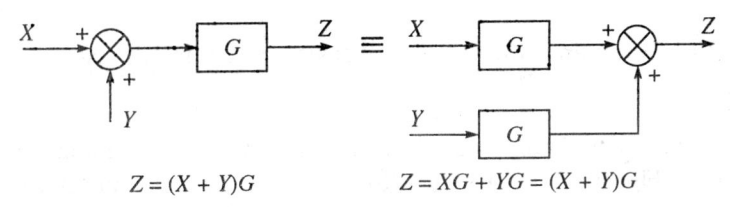

$$Z = (X + Y)G \quad\quad Z = XG + YG = (X + Y)G$$

FIGURE 2.62 Moving a summing point behind a block.

This rule is illustrated with an example in Figure 2.63. The transfer function of Figure 2.63(b) is

$$C(s) = \left(\frac{G_1}{1+G_1} + \frac{G_2}{1+G_1}\right) R(s)$$

$$= \left(\frac{G_1 + G_2}{1+G_1}\right) R(s)$$

which is equal to the transfer function of Figure 2.63(a).

6. Moving a pick-off point ahead of a block

The pick-off point in Figure 2.64 has been moved ahead of block G.

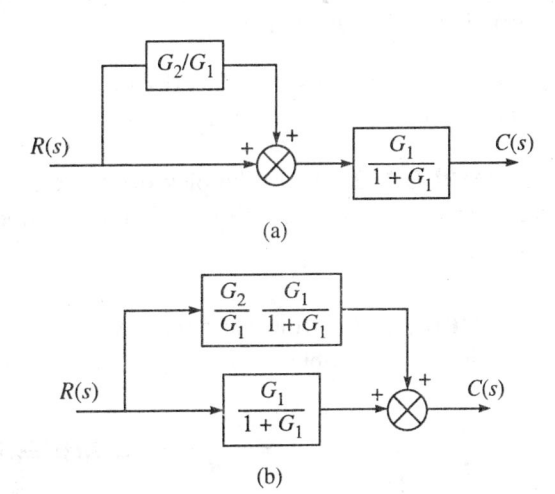

FIGURE 2.63 Example of moving a summing point behind a block.

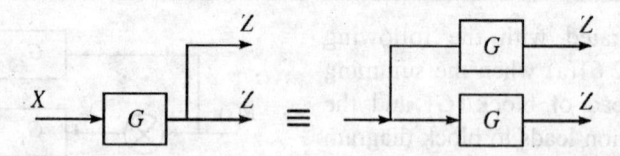

FIGURE 2.64 Moving a pick-off point ahead of a block.

7. *Moving a pick-off point behind a block*

The pick-off point in Figure 2.65 has been moved behind the block G.

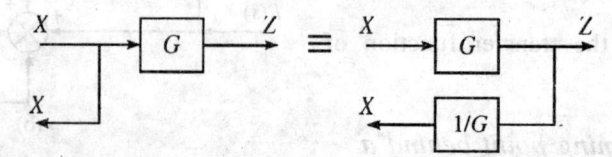

FIGURE 2.65 Moving a pick-off point behind a block.

EXAMPLE 2.9 Follow the simplification of the block diagram through Figures 2.66(a) to (e) and determine the overall transfer function.

EXAMPLE 2.10 Determine the overall transfer function of the block diagram in Figure 2.67(a).

Solution: Move the second summing junction in Figure 2.67(a) behind the block G_2, and follow the resulting Figures 2.67(b) to (d), to obtain the overall transfer function.

EXAMPLE 2.11 Move the pick-off point X in Figure 2.68(a) ahead of G_3 and then find the overall transfer function.

Solution: Follow Figures 2.68(b) through (d) for the determination of the overall transfer function.

EXAMPLE 2.12 Move the pick-off point Y in Figure 2.68(a) behind the block G_3 as shown in Figure 2.69(a). Then through simplification, obtain the overall transfer function as in Figure 2.69(b).

EXAMPLE 2.13 From the circuit diagram of Figure 2.70(a), we can write the equations in time domain and s-plane as

$$\frac{v_i - v_1}{R_1} = i_1 \; ; \; \frac{V_i(s) - V_1(s)}{R_1} = I_1(s) \Rightarrow \text{Block diagram in Figure 2.70(b)}$$

$$v_1 = \frac{1}{C_1}\int (i_1 - i_2)\,dt \; ; \; V_1(s) = \frac{1}{C_1 s}[I_1(s) - I_2(s)] \Rightarrow \text{Block diagram in Figure 2.70(c)}$$

Mathematical Modeling of Systems

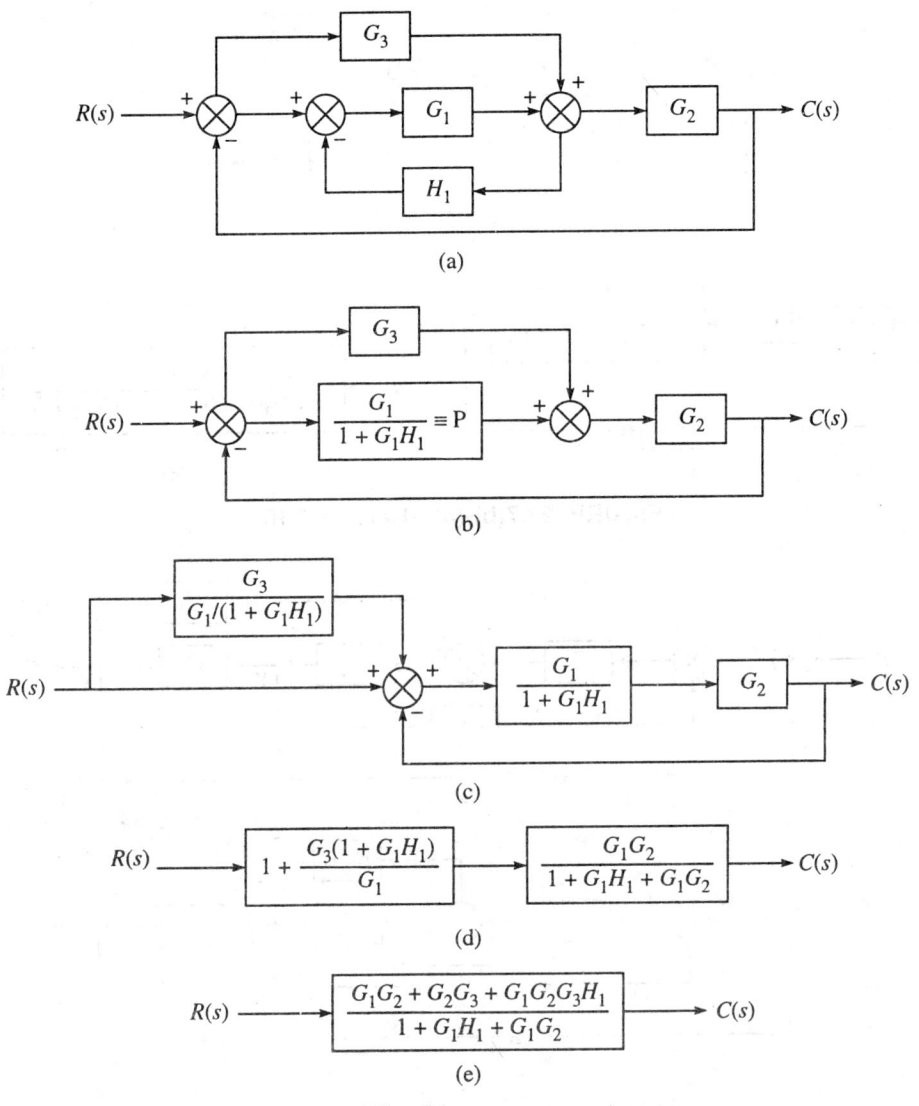

FIGURE 2.66 Example 2.9.

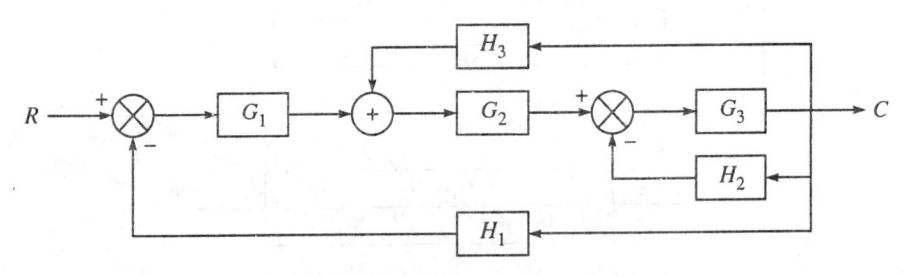

FIGURE 2.67(a) Example 2.10.

FIGURE 2.67(b)-(d) Example 2.10.

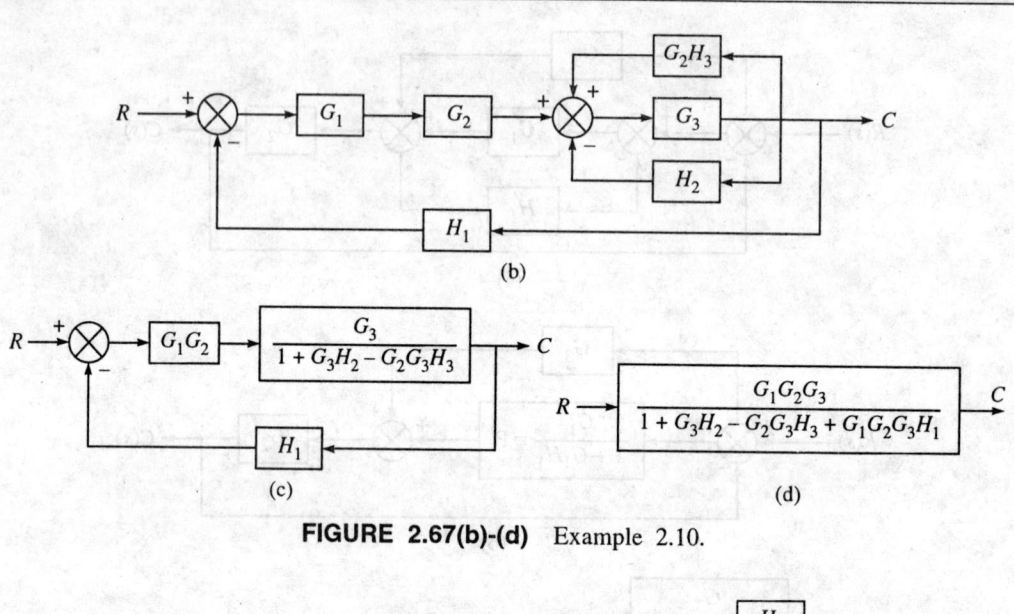

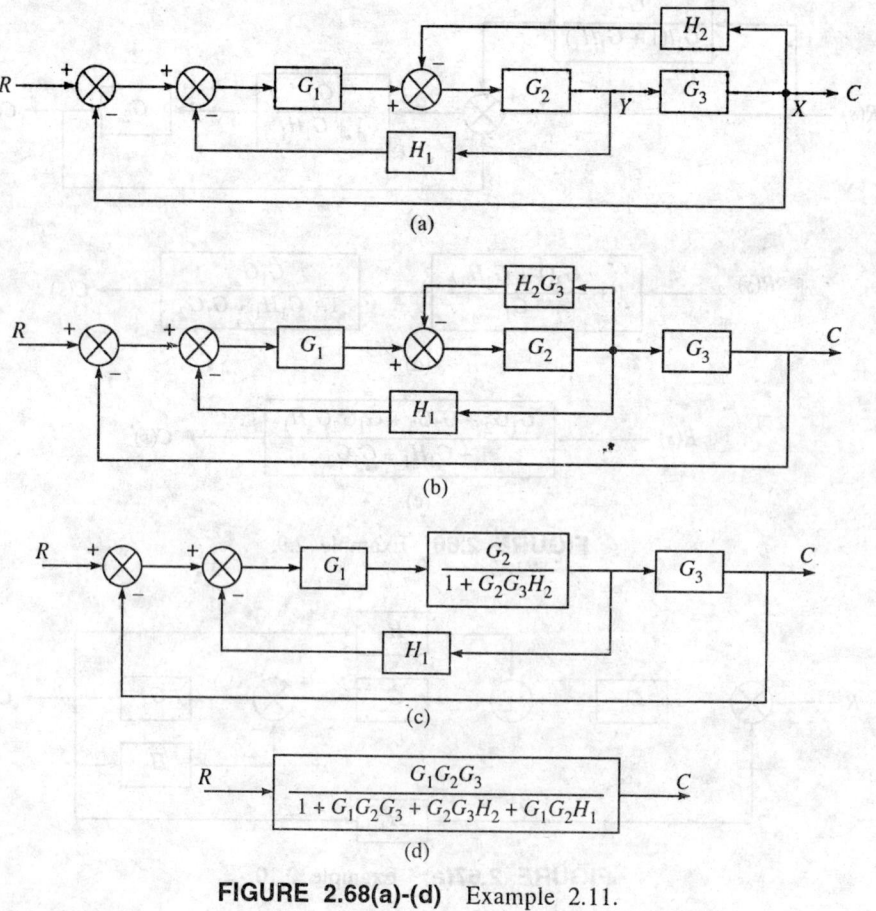

FIGURE 2.68(a)-(d) Example 2.11.

Mathematical Modeling of Systems

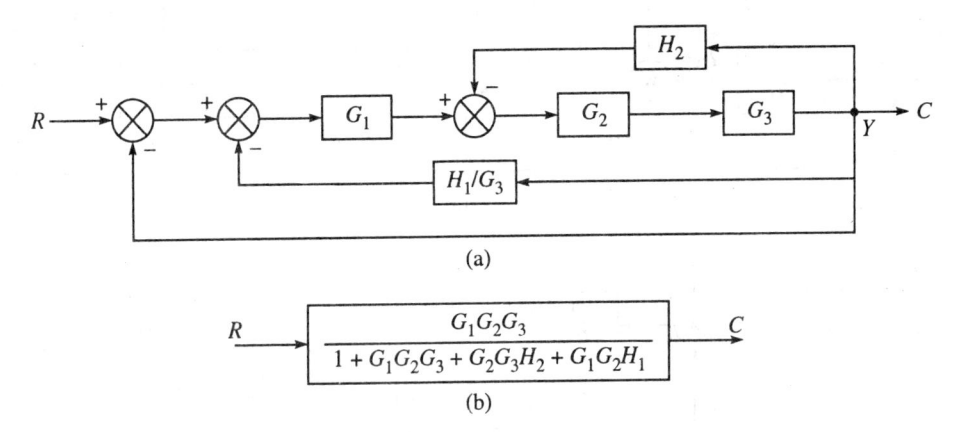

FIGURE 2.69 Example 2.12.

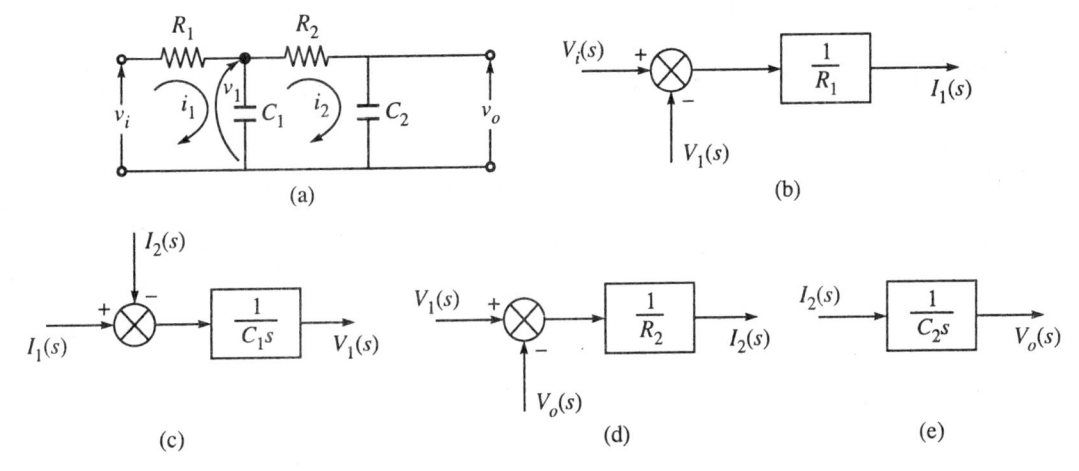

FIGURE 2.70 Example 2.13.

$$\frac{v_1 - v_o}{R_2} = i_2 \ ; \ \frac{V_1(s) - V_o(s)}{R_2} = I_2(s) \Rightarrow \text{Block diagram in Figure 2.70(d)}$$

$$v_o = \frac{1}{C_2}\int i_2 dt \ ; \ V_o(s) = \frac{I_2(s)}{C_2 s} \Rightarrow \text{Block diagram in Figure 2.70(e)}$$

The complete system block diagram is shown in Figure 2.71(a). Move the summing point X behind the block $1/R_1$ and then move the pick-off point Y ahead of the block $1/C_2 s$. Then obtain Figure 2.71(b). Then by simple manipulation, you will ultimately get the transfer function as shown in Figure 2.71(e) through steps (c) and (d).

(a)

(b)

(c)

(d)

$$V_i(s) \longrightarrow \boxed{\dfrac{1}{R_1R_2C_1C_2s^2 + (R_1C_1 + R_2C_2 + R_1C_2)s + 1}} \longrightarrow V_o(s)$$

(e)

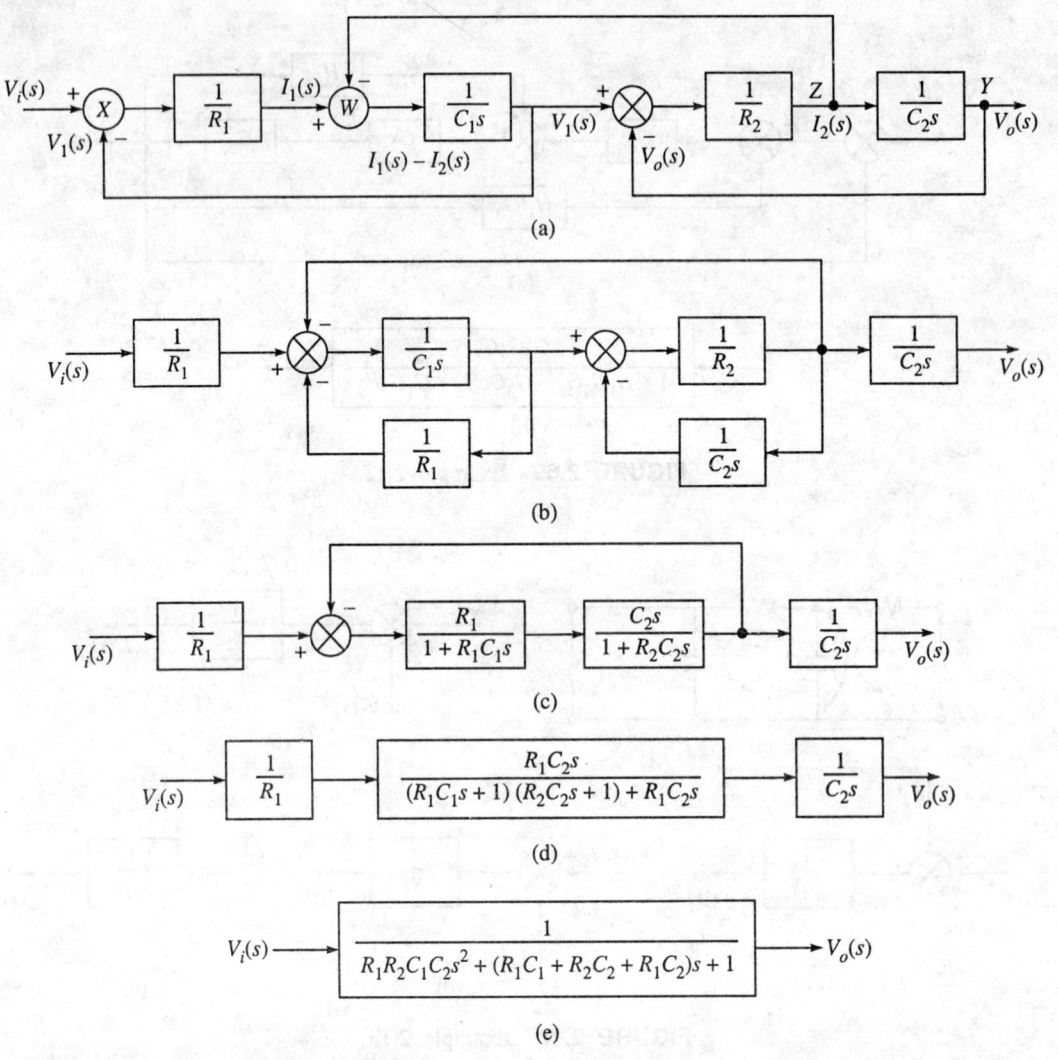

FIGURE 2.71(a)-(e) Example 2.13: system block diagram and overall transfer function.

EXAMPLE 2.14 In order to futher illustrate the shifting technique, the block diagram in Figure 2.72(a) may be considered.

(a)

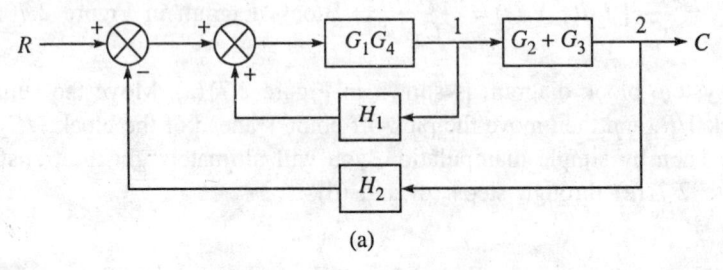

FIGURE 2.72(a) Example 2.14.

Moving the pick-off point '1' to behind the block ($G_2 + G_3$), we get Figure 2.72(b), and then through sequence of Figures 2.72(c) through (e), we finally obtain the transfer function as shown in Figure 2.72(e).

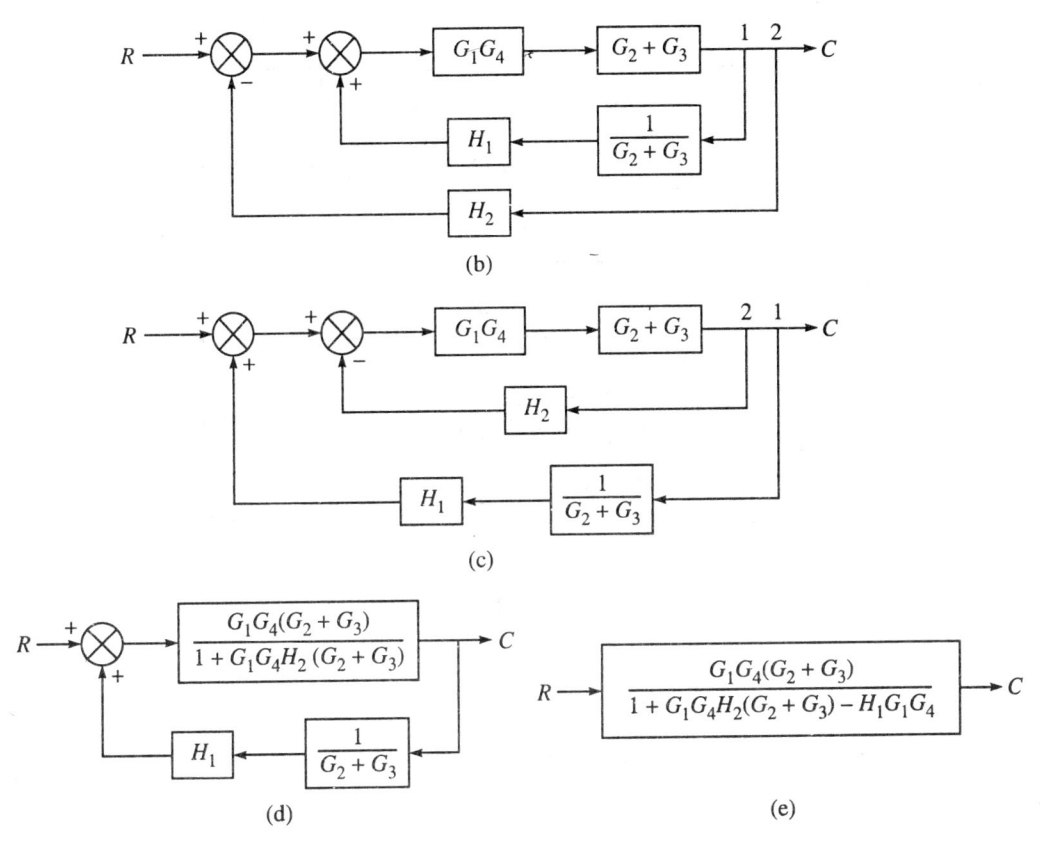

FIGURE 2.72(b)-(e) Example 2.14.

EXAMPLE 2.15 Simplify the block diagram shown in Figure 2.73(a) to open-loop form.

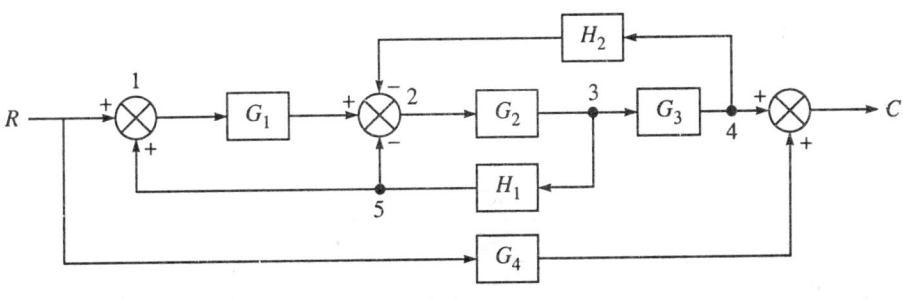

FIGURE 2.73(a) Example 2.15.

Solution: Moving the summing point behind (to the right of) block G_1 of Figure 2.73(a), we get Figure 2.73(b). Then through the sequence of Figures 2.73(c), (d), and (e), we finally reach Figure 2.73(f) which gives the transfer function of Figure 2.73(a).

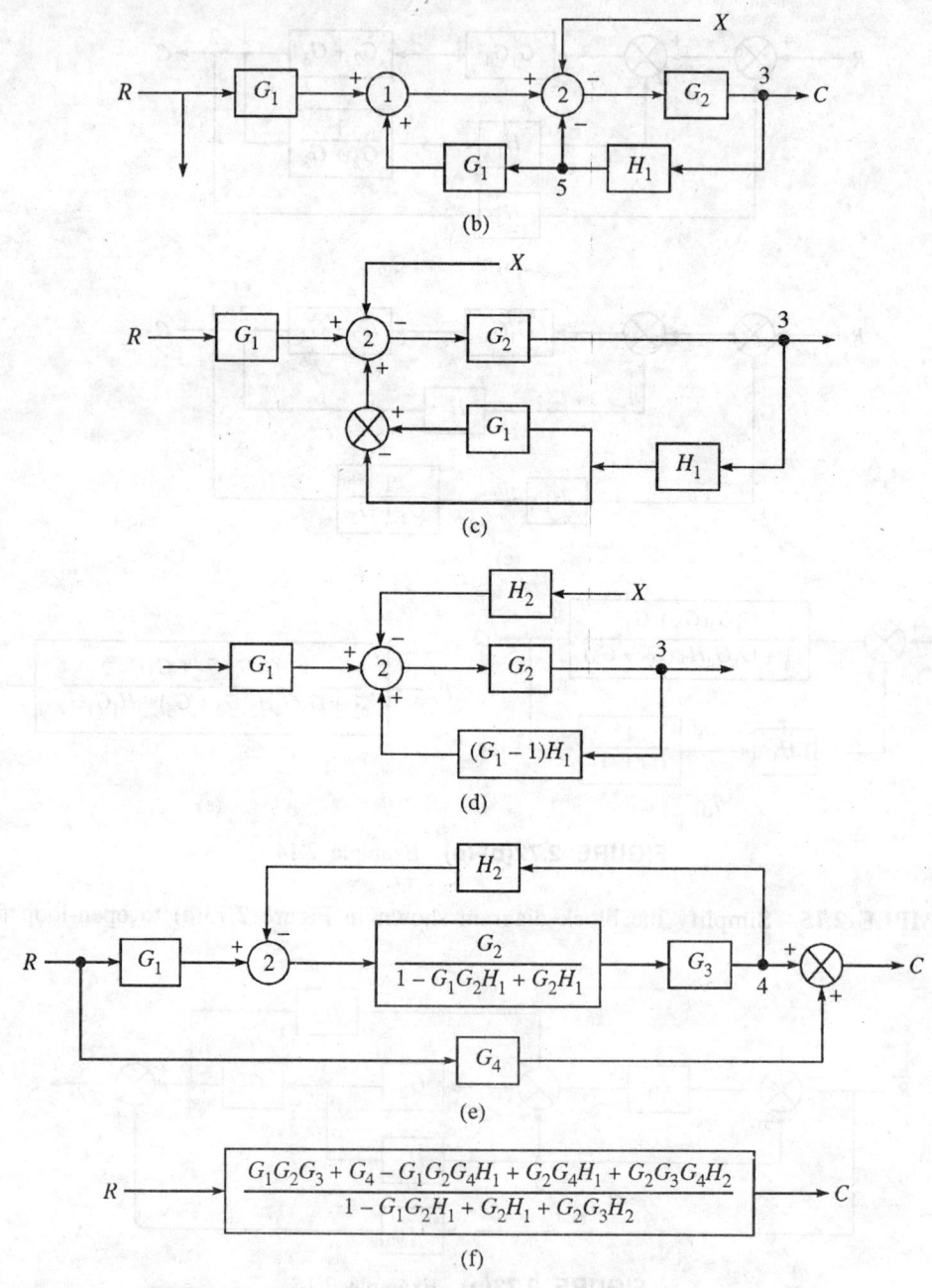

FIGURE 2.73(b)-(f) Example 2.15: overall transfer function.

Mathematical Modeling of Systems

EXAMPLE 2.16 Determine the transfer function matrix of Figure 2.74 using the block diagram reduction technique.

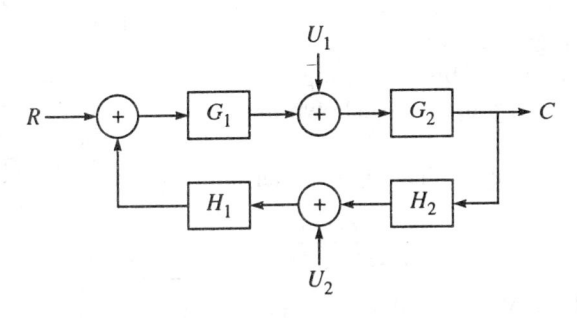

FIGURE 2.74 Example 2.16.

Solution: We need to get the expression,

$$C(s) = [G_{CR}(s) \quad G_{CU1}(s) \quad G_{CU2}(s)] \begin{bmatrix} R(s) \\ U_1(s) \\ U_2(s) \end{bmatrix}$$

where

$$G_{CR}(s) = \frac{C(s)}{R(s)}\bigg|_{U_1=U_2=0} = \frac{G_1 G_2}{1 - G_1 G_2 H_1 H_2}$$

$$G_{CU1}(s) = \frac{C(s)}{U_1(s)}\bigg|_{R=U_2=0} = \frac{G_2}{1 - G_1 G_2 H_1 H_2}$$

$$G_{CU2}(s) = \frac{C(s)}{U_2(s)}\bigg|_{R=U_1=0} = \frac{G_1 G_2 H_1}{1 - G_1 G_2 H_1 H_2}$$

EXAMPLE 2.17 For the multivariable system shown in Figure 2.75(a), determine G_{ij} where the input-output relationship is described as $C_i = G_{ij} R_j$

that is,

$$\begin{bmatrix} C_1 \\ C_2 \end{bmatrix} = \begin{bmatrix} G_{11} & G_{12} \\ G_{21} & G_{22} \end{bmatrix} \begin{bmatrix} R_1 \\ R_2 \end{bmatrix}$$

By definition,

$$G_{11} = \left(\frac{C_1}{R_1}\right)\bigg|_{R_2=0}, \quad G_{12} = \left(\frac{C_1}{R_2}\right)\bigg|_{R_1=0}$$

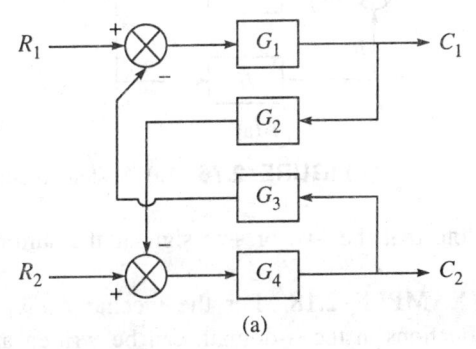

FIGURE 2.75(a) Example 2.17.

$$G_{21} = \left(\frac{C_2}{R_1}\right)\bigg|_{R_2=0}, \quad G_{22} = \left(\frac{C_2}{R_2}\right)\bigg|_{R_1=0}$$

Make $R_2 = 0$ to get G_{11} and G_{21}. The resultant figure is as shown in Figure 2.75(b).

Now G_{11} can be obtained by neglecting C_2 and the redrawn block diagram is as shown in Figure 2.75(c).

Again, $G_{21} = C_2/R_1$ can be obtained in a similar way by neglecting C_1 to get $G_{21} = \dfrac{-G_1G_2G_4}{1-G_1G_2G_3G_4}$

Now from the definition of G_{12} and G_{22}, make $R_1 = 0$ and proceed in a similar way to get G_{12}, by neglecting C_1, and G_{22} by neglecting C_2.

Hence the multivariable system can be written as

$$\begin{bmatrix} C_1 \\ C_2 \end{bmatrix} = \frac{1}{1-G_1G_2G_3G_4} \begin{bmatrix} G_1 & -G_1G_3G_4 \\ -G_1G_2G_4 & G_4 \end{bmatrix} \begin{bmatrix} R_1 \\ R_2 \end{bmatrix}$$

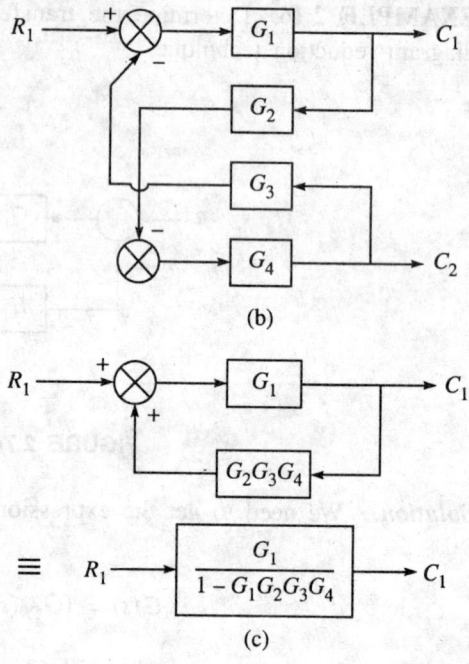

FIGURE 2.75(b)-(c) Example 2.17.

2.12 Signal Flow Graph

From the block diagram of the system, we can draw the signal flow graph (SFG) easily where each variable of the block diagram becomes a node and each block becomes a branch having the same transmittance as the gain of the concerned block. The construction of SFG is illustrated in Figure 2.76(b) using the block diagram of the feedback control system shown in Figure 2.76(a).

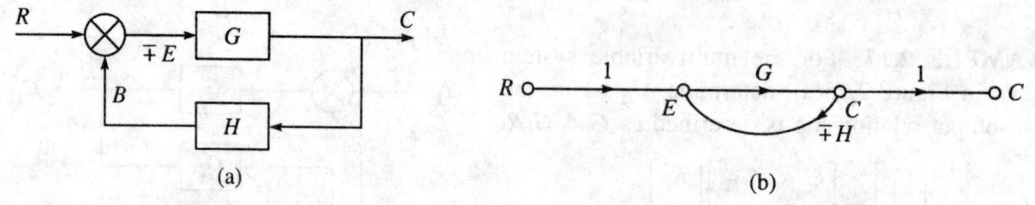

FIGURE 2.76 (a) System block diagram and (b) its signal flow graph.

Note that the −ve or +ve sign of the summing point is associated with H in SFG.

EXAMPLE 2.18 For the mechanical system shown in Figure 2.77(a), the force-balance equations in the s-domain can be written as

$$F(s) + K_1X_2(s) = (M_1s^2 + f_1s + K_1)X_1(s)$$

and
$$K_1 X_1(s) = (M_2 s^2 + f_2 s + K_1 + K_2) X_2(s)$$

where $F(s)$ is the Laplace transform of the force applied, K_1 and K_2 are the spring constants, f_1 and f_2 are the viscous friction (damping) coefficients and $X_1(s)$ and $X_2(s)$ are the displacements. The objective of this example is to show how one approaches the Mason's gain formula for obtaining the transfer function. Let us put

$$A = M_1 s^2 + f_1 s + K_1$$
and
$$B = M_2 s^2 + f_2 s + K_1 + K_2$$

Then, we can write

$$\left(\frac{1}{A}\right) F(s) + \left(\frac{K_1}{A}\right) X_2(s) = X_1(s)$$

and
$$\left(\frac{K_1}{B}\right) X_1(s) = X_2(s)$$

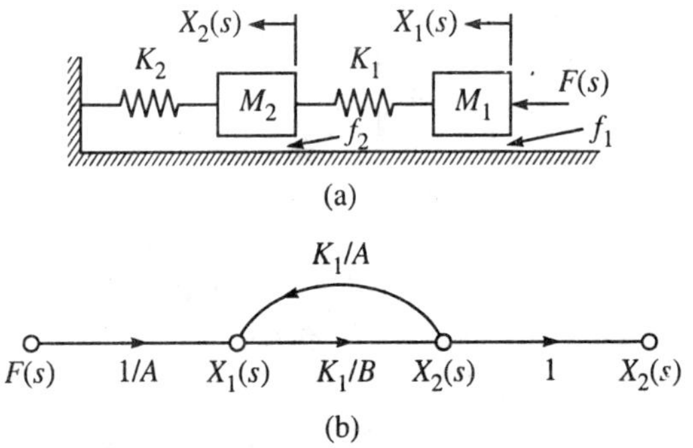

The signal flow graph can be drawn as in Figure 2.77(b).

Here the forward path gain $P_1 = K_1/AB$ and the loop gain $P_{11} = K_1^2/AB$.

$$\Delta = 1 - \frac{K_1^2}{AB}; \quad \Delta_1 = 1$$

FIGURE 2.77 (a) Mechanical system and (b) its signal flow graph.

Therefore, the transfer function

$$\frac{X_2(s)}{F(s)} = \frac{K_1}{(M_1 s^2 + F_1 s + K_1)(M_2 s^2 + f_2 s + K_1 + K_2) - K_1^2}$$

$$= \frac{K_1}{AB - K_1^2} = \frac{(K_1/AB) \cdot 1}{1 - \frac{K_1^2}{AB}} = \frac{P_1 \Delta_1}{\Delta} \qquad (2.84)$$

2.13 Introduction to Gain Formula

Consider the following set of algebraic equations

$$a_{11} x_1 + a_{12} x_2 + r_1 = x_1 \qquad (2.85)$$
$$a_{21} x_1 + a_{22} x_2 + r_2 = x_2 \qquad (2.86)$$

where r and x are the input and output respectively. Rewriting the above equations in matrix form, we get

$$\begin{bmatrix} 1 - a_{11} & -a_{12} \\ -a_{21} & 1 - a_{22} \end{bmatrix} \begin{bmatrix} x_1 \\ x_2 \end{bmatrix} = \begin{bmatrix} r_1 \\ r_2 \end{bmatrix} \qquad (2.87)$$

Using Cramer's rule, we get

$$x_1 = \frac{(1-a_{22})r_1 + a_{12}r_2}{\Delta} = \frac{(1-a_{22})}{\Delta}r_1 + \frac{a_{12}}{\Delta}r_2 = g_{11}r_1 + g_{12}r_2$$

$$x_2 = \frac{a_{21}r_1 + (1-a_{11})r_2}{\Delta} = \frac{a_{21}}{\Delta}r_1 + \frac{(1-a_{11})}{\Delta}r_2 = g_{21}r_1 + g_{22}r_2$$

where

$$\Delta = \begin{vmatrix} 1-a_{11} & -a_{12} \\ -a_{21} & 1-a_{22} \end{vmatrix} = 1 - (a_{11} + a_{22} + a_{12}a_{21} + a_{11}a_{22})$$

and

$$\left.\frac{x_1}{r_1}\right|_{r_2=0} = g_{11} = \frac{1-a_{22}}{\Delta}; \quad \left.\frac{x_1}{r_2}\right|_{r_1=0} = \frac{a_{12}}{\Delta} = g_{12}$$

$$\left.\frac{x_2}{r_1}\right|_{r_2=0} = g_{21} = \frac{a_{21}}{\Delta}; \quad \left.\frac{x_2}{r_2}\right|_{r_1=0} = g_{22} = \frac{1-a_{11}}{\Delta}$$

Now let us draw the signal flow graph from the given simultaneous equations as in Figure 2.78.

In this signal flow graph, r_1 and r_2 are the two inputs; x_1 and x_2 are the two outputs. The individual loop gains are a_{11}, a_{22}, and $a_{12}a_{21}$. The gain product of two nontouching loops is only $a_{11}a_{22}$.

From the solution, the denominator is equal to 1 minus sum of each of the self-loops a_{11}, a_{22}, and $a_{12}a_{21}$ plus the product of two nontouching loops a_{11} and a_{22}.

Further, g_{11}, that is, the ratio of x_1 with input r_1 only is $(1 - a_{22})$, which is the value of Δ not touching the path 1 from r_1 and x_1. Again for g_{12}, the ratio of x_1 with input r_2 only is a_{12}. The numerator for x_2 also will follow the same line.

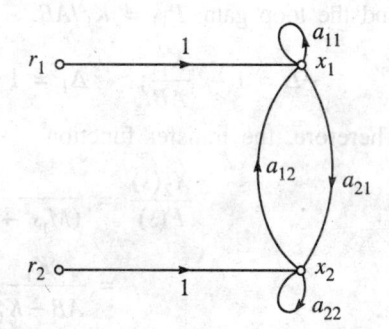

FIGURE 2.78 Signal flow graph of simultaneous Eqs. (2.85) and (2.86).

With this background, the Mason's gain formula has been developed to determine the linear dependance T_{ij} between the independent variable (often called the input variable) x_i and a dependent variable x_j. Before we proceed further, let us define the following terms with reference to Figure 2.78.

Node. A node is a point representing a variable or signal.

Branch. A branch is a directed line segment joining two nodes.

Transmittance. The transmittance is the gain between two nodes. The gain of a branch is therefore the transmittance.

Input node or source. An input node or source is a node which has only the outgoing branches. This corresponds to an independent variable.

Output node or sink. An output node or sink is a node which has only the incoming branches. This corresponds to a dependent variable.

Mixed node. A mixed node is a node which has both incoming and outgoing branches. See Figure 2.79.

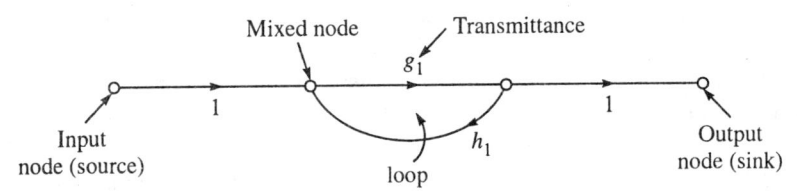

FIGURE 2.79 Mixed node.

Path. A path is a traversal of connected branches in the direction of branch arrows. The path may be open if no node is crossed more than once. If the path ends at the same node from which it began and does not cross any other node more than once, it is closed. If a path crosses some node more than once but ends at a different node from which it began, it is neither open nor closed.

Loop. A loop is a closed path.

Loop gain. The product of all the branch transmittances of a loop is the loop gain.

Nontouching loops. Loops are nontouching if they do not possess any common node.

Forward path. A forward path is a path from an input node (source) to an output node (sink) which does not cross any nodes more than once.

Forward path gain. A forward path gain is the product of all the branch transmittances of a forward path.

2.14 Mason's Gain Formula

The gain formula is often used to relate the output variable $C(s)$ to the input variable $R(s)$ and is given as

$$T(s) = \frac{C(s)}{R(s)} = \frac{\sum_k P_k \Delta_k}{\Delta} \qquad (2.88)$$

where

P_k = path gain or transmittance of the kth forward path
Δ = determinant of the graph
 = 1 − (sum of all the different individual loop gains)
  + (sum of gain products of all the possible combinations of two nontouching loops)
  − (sum of gain products of all the possible combinations of three nontouching loops)
  + ...
Δ_k = the cofactor of the path P_k
 = the determinant with loops touching the kth forward path

Modern Control Engineering

The summation is taken over all the possible k-paths from $R(s)$ to $C(s)$. The determinant

$$\Delta = 1 - \sum_m P_{m1} + \sum_m P_{m2} - \sum_m P_{m3} + \cdots$$

where P_{mr} = gain product of the mth possible combinations or r nontouching loops.

For example, P_{12} signifies as follows: The subscript 2 denotes two nontouching loops and the subscript 1 denotes as the first numbered element in that category (of '2'). It may be noted that the loops are nontouching if they do not have any common nodes.

EXAMPLE 2.19 Consider the SFG of Figure 2.80. Determine the transfer function $C(s)/R(s)$ using the Mason's gain formula.

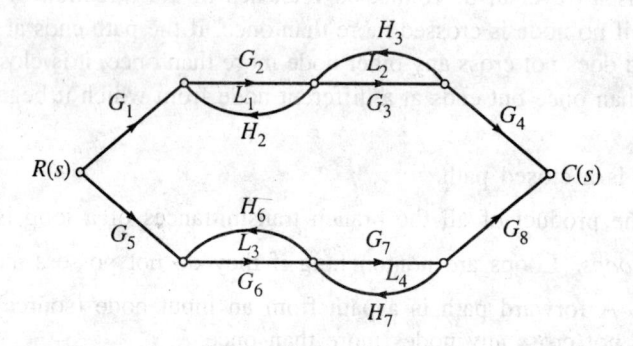

FIGURE 2.80 Example 2.19: signal flow graph.

Solution: 1. The forward paths connecting the input $R(s)$ and output $C(s)$ are

$$P_1 = G_1G_2G_3G_4 \quad \text{and} \quad P_2 = G_5G_6G_7G_8$$

2. There are four individual self-loops such as

$$P_{11} = G_2H_2; \quad P_{21} = H_3G_3; \quad P_{31} = G_6H_6; \quad P_{41} = G_7H_7$$

3. The loops P_{11} and P_{21} do not touch P_{31} and P_{41}. Hence there are four possible combinations of two nontouching loops with loop gain products as

$$P_{12} = P_{11}P_{31} = G_2H_2H_6G_6; \quad P_{22} = P_{11}P_{41} = G_2H_2H_7G_7$$
$$P_{32} = P_{21}P_{31} = G_3H_3H_6G_6; \quad P_{42} = P_{21}P_{41} = G_3H_3H_7G_7$$

4. There are **no** combinations of three nontouching loops, etc. Therefore,

$$P_{m3} = P_{m4} = \ldots = 0$$

Now, $\quad \Delta = 1 - (P_{11} + P_{21} + P_{31} + P_{41}) + (P_{21} + P_{22} + P_{32} + P_{42})$

5. Now, the cofactor of Δ along path P_1 is evaluated by removing the loops that touch path

P_1 from Δ. Therefore,
$$\Delta_1 = 1 - (P_{31} + P_{41})$$
Similarly,
$$\Delta_2 = 1 - (P_{11} + P_{21})$$
Therefore, the transfer function from Eq. (2.88) becomes

$$T(s) = \frac{C(s)}{R(s)} = \frac{P_1\Delta_1 + P_2\Delta_2}{\Delta}$$

$$= \frac{G_1G_2G_3G_4(1-P_{31}-P_{41}) + G_5G_6G_7G_8(1-P_{11}-P_{21})}{1-P_{11}-P_{21}-P_{31}-P_{41}+P_{12}+P_{22}+P_{32}+P_{42}}$$

EXAMPLE 2.20 Draw the SFG of the circuit shown in Figure 2.81(a) and find the transfer function by the Mason's gain formula.

Solution: The circuit equations can be written as

$$i_1 = \frac{v_1 - v_2}{R_1} = \frac{v_1}{R_1} - \frac{v_2}{R_1}$$

$$i_2 = \frac{v_2 - v_3}{R_2} = \frac{v_2}{R_2} - \frac{v_3}{R_2}$$

$$v_2 = i_1 R_3 - i_2 R_3$$

$$v_3 = i_2 R_4$$

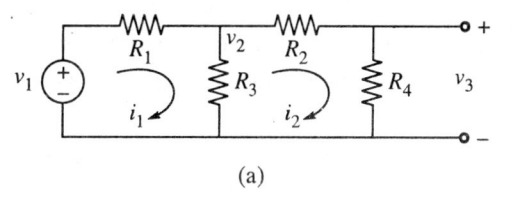

FIGURE 2.81(a) Example 2.20.

The SFG can be drawn as in Figure 2.81(b).

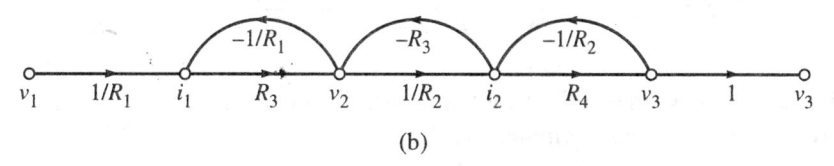

FIGURE 2.81(b) Example 2.20: signal flow graph.

In SFG, the number of forward paths is 1, and the forward path gain as in Figure 2.81(c) can be written as $P_1 = R_3R_4/R_1R_2$.

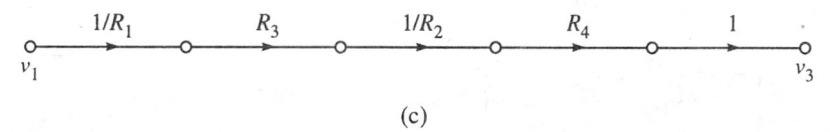

FIGURE 2.81(c) Example 2.20: forward path gain.

The individual loops as in Figure 2.81(d) are:

$$P_{11} = -\frac{R_3}{R_1}$$

$$P_{21} = -\frac{R_3}{R_2}$$

$$P_{31} = -\frac{R_4}{R_2}$$

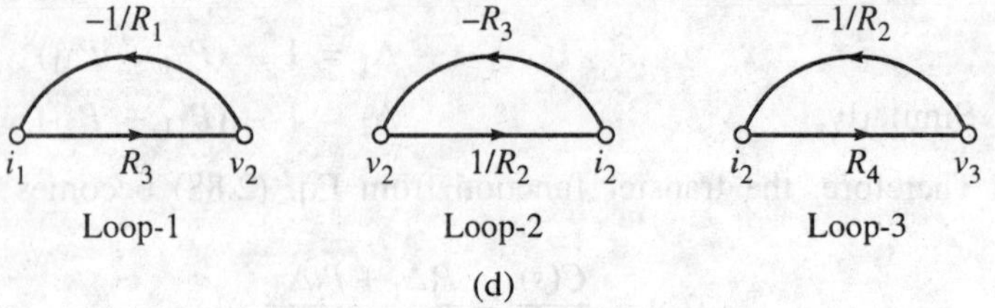

FIGURE 2.81(d) Example 2.20: individual loops.

There is only one possibility for the combination of two nontouching loops, i.e. loops 1 and 3. Hence

$$P_{12} = \text{gain product of the two nontouching loops}$$

$$= P_{11}P_{31} = \frac{R_3 R_4}{R_1 R_2}$$

There are no three loops that do not touch, i.e. there is no possible combination of three nontouching loops; hence $P_{m3} = 0$. Therefore,

$$\Delta = 1 - (P_{11} + P_{21} + P_{31}) + P_{12} = 1 + \left(\frac{R_3}{R_4} + \frac{R_3}{R_2} + \frac{R_4}{R_2}\right) + \left(\frac{R_3 R_4}{R_1 R_2}\right)$$

$$= \frac{R_1 R_2 + R_1 R_3 + R_1 R_4 + R_2 R_3 + R_3 R_4}{R_1 R_2}$$

Since all loops touch the forward path, $\Delta_1 = 1$. Finally, the transfer function

$$T(s) = \frac{V_3(s)}{V_1(s)} = \frac{P_1 \Delta_1}{\Delta} = \frac{R_3 R_4}{R_1 R_2 + R_1 R_3 + R_1 R_4 + R_2 R_3 + R_3 R_4}$$

EXAMPLE 2.21 Draw the SFG of the circuit shown in Figure 2.82(a) and find the transfer function by the Mason's gain formula.

Solution: From KCL and KVL, the network equations in the s-domain can be written as

$$I_1 = \frac{V_1 - V_2}{R_1} = \frac{V_1}{R_1} - \frac{V_2}{R_1}$$

$$I_2 = \frac{V_2 - V_3}{R_2} = \frac{V_2}{R_2} - \frac{V_3}{R_2}$$

$$I_3 = \frac{I_1 - I_2}{sC_1} = \frac{I_1}{sC_1} - \frac{I_2}{sC_1}$$

$$I_4 = \frac{I_2}{sC_2}$$

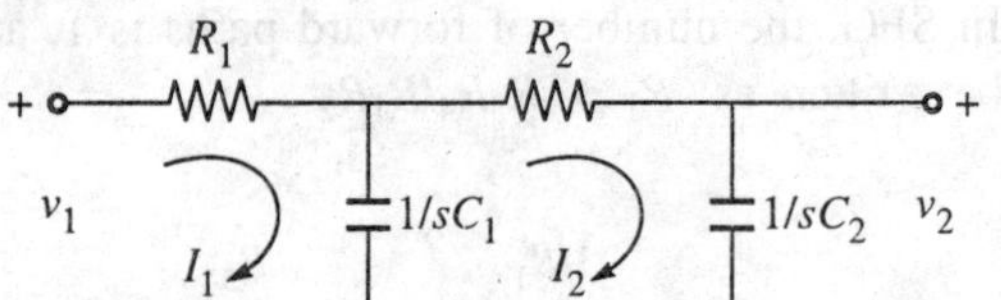

FIGURE 2.82(a) Example 2.21.

Mathematical Modeling of Systems

The SFG can be drawn as in Figure 2.82(b).

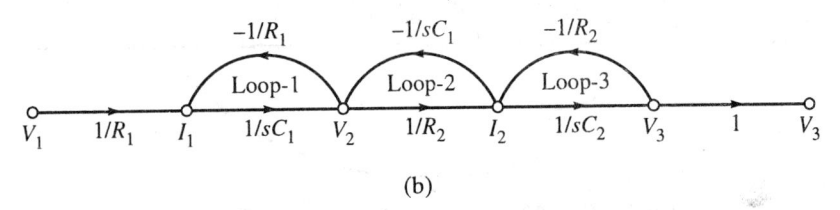

FIGURE 2.82(b) Example 2.21: signal flow graph.

The only forward path gain is

$$P_1 = \frac{1}{s^2 R_1 R_1 C_1 C_2}$$

The individual loop gains are

$$P_{11} = \frac{-1}{sC_1 R_1}, \quad P_{21} = \frac{-1}{sC_1 R_2} \quad \text{and} \quad P_{31} = \frac{-1}{sC_2 R_2}$$

There is only one possible combination of two nontouching loops with loops 1 and 3. Hence

$$P_{12} = \text{gain product of the two nontouching loops} = P_{11}P_{31} = \frac{1}{s^2 R_1 R_2 C_1 C_2}$$

There are no possible combinations of three nontouching loops, i.e. $P_{m3} = 0$. Therefore,

$$\Delta = 1 - (P_{11} + P_{21} + P_{31}) + P_{12} = 1 + \frac{1}{s}\left(\frac{1}{C_1 R_1} + \frac{1}{C_1 R_2} + \frac{1}{C_2 R_2}\right) + \frac{1}{s^2 R_1 R_2 C_1 C_2}$$

Since all loops touch the forward path, $\Delta_1 = 1$.
Finally, the overall transfer function is

$$T(s) = \frac{V_3(s)}{V_1(s)} = \frac{P_1 \Delta_1}{\Delta} = \frac{1}{s^2 R_1 R_2 C_1 C_2 + s(C_1 R_1 + C_2 R_2 + C_1 R_2) + 1}$$

EXAMPLE 2.22 Determine the transfer function by the Mason's gain formula of the SFG shown in Figure 2.83.

Solution: There are two forward paths: $P_1 = G_1 G_2 G_4$ and $P_2 = G_1 G_3 G_4$.
There are three individual loops. The loop gains are:

$$P_{11} = G_1 G_4 H_1; \quad P_{21} = -G_1 G_2 G_4 H_2; \quad P_{31} = -G_1 G_3 G_4 H_2$$

There is no possible combination of two nontouching loops, i.e. $P_{m2} = 0$.
All loops touch both the forward paths; hence

$$\Delta_1 = 1; \quad \Delta_2 = 1$$

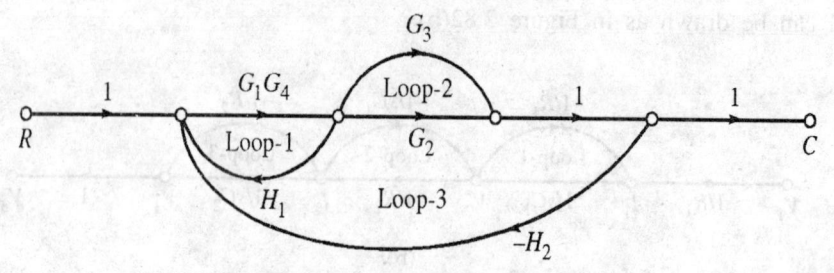

FIGURE 2.83 Example 2.22: signal flow graph.

Therefore, the transfer function by the Mason's gain formula becomes

$$T(s) = \frac{C(s)}{R(s)} = \frac{P_1 \Delta_1 + P_2 \Delta_2}{\Delta}$$

$$= \frac{G_1 G_2 G_4 + G_1 G_3 G_4}{1 - G_1 G_4 H_1 + G_1 G_2 G_4 H_2 + G_1 G_3 G_4 H_2} = \frac{G_1 G_4 (G_2 + G_3)}{1 - G_1 G_4 H_1 + G_1 G_4 H_2 (G_2 + G_3)}$$

EXAMPLE 2.23 Determine the transfer function $C(s)/R(s)$ of Figure 2.84 using the Mason's gain formula.

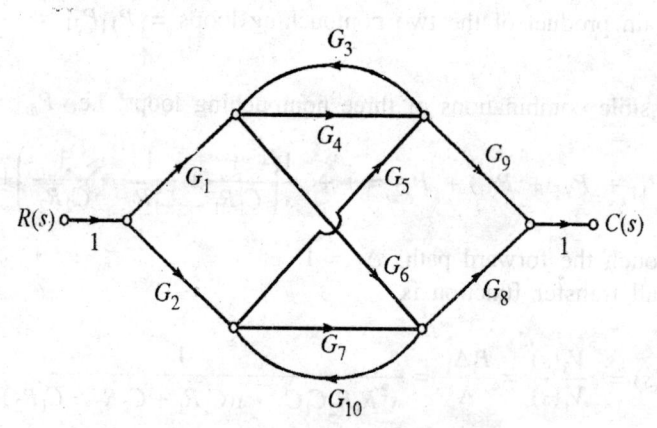

FIGURE 2.84 Example 2.24: signal flow graph.

Solution: In the given SFG, there are six forward paths and their forward path gains are:
$P_1 = G_1 G_4 G_9$; $P_2 = G_2 G_7 G_8$; $P_3 = G_2 G_5 G_9$; $P_4 = G_1 G_6 G_8$; $P_5 = G_2 G_5 G_3 G_6 G_8$ and $P_6 = G_1 G_6 G_{10} G_5 G_9$

The individual loops are

$$P_{11} = G_3 G_4; \qquad P_{21} = G_7 G_{10}; \qquad P_{31} = G_5 G_3 G_6 G_{10}$$

There is one possible combination of two nontouching loops with loop gains

$$P_{12} = G_1 G_3 G_7 G_{10}$$

Mathematical Modeling of Systems

There is no possible combination of three nontouching loops, i.e. $P_{m3} = 0$
The determinant of the graph is

$$\Delta = 1 - G_4G_3 - G_7G_{10} - G_5G_3G_6G_{10} + G_4G_3G_7G_{10}$$

and $\quad \Delta_1 = 1 - G_7G_{10}; \quad \Delta_2 = 1 - G_3G_4; \quad \Delta_3 = 1; \quad \Delta_4 = 1; \quad \Delta_5 = 1; \quad \Delta_6 = 1$

Hence the transfer function from the Mason's gain formula is

$$T(s) = \frac{C(s)}{R(s)} = \sum_{i=1}^{6} \frac{P_i \Delta_i}{\Delta} = \frac{1}{1 - G_3G_4 - G_7G_{10} - G_5G_3G_6G_{10} + G_3G_4G_7G_{10}} [G_1G_4G_9(1 - G_7G_{10})]$$

$$+ G_2G_7G_8(1 - G_3G_4) + G_2G_5G_9 + G_1G_6G_8 + G_2G_3G_5G_6G_8 + G_1G_5G_6G_9G_{10}]$$

2.15 Transfer Matrix of Multivariable Systems

EXAMPLE 2.24 Consider the multiple-input-multiple-output feedback system shown in Figure 2.85, where G_1, G_2, H_1 and H_2 are the transfer function matrices and R, E_1, V, W, C, D, F_1 and F_2 are the column matrices where the order of vectors R is $r \times 1$, V is $m \times 1$, C is $n \times 1$ and that of vector D is $p \times 1$. Determine the dimensions of all other matrices in the diagram.

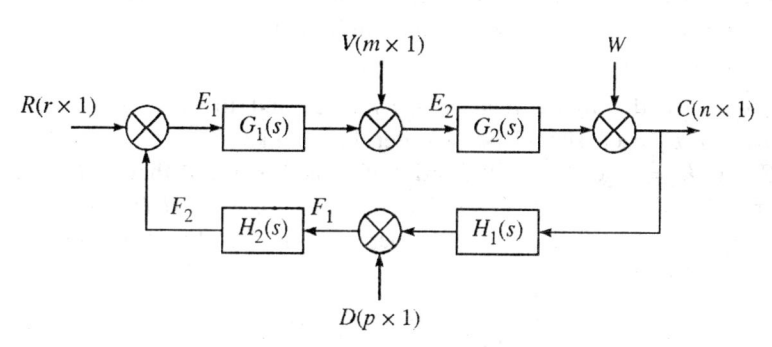

FIGURE 2.85 Example 2.24: multivariable system.

Solution: We observe from the given figure, $R - F_2 = E_1$ and to make sense, as R is of order $r \times 1$ then F_2 and E_1 each should be of order $r \times 1$ like R. If G_1E_1 is to be conformable, then G_1 should be of r columns. Further, V is of order $m \times 1$ and from the relation $G_1E_1 + V = E_2$, we need to get the order of G_1 as $m \times r$. Obviously, the order of E_2 must be $m \times 1$. In a similar way, $G_2E + W = C$ and C is of order $n \times 1$ (given), then obviously W should be of order $n \times 1$, G_2 must be of order $n \times m$. Similar reasoning for $H_1C + D = F_1$ leads to conclude that for the given order of matrices D of $p \times 1$, C of $n \times 1$, F_1 has to be a column matrix of order $p \times 1$; then H_1 has to be of order $p \times n$. With similar reasoning we can conclude that H_2 has to be of order $r \times p$ and then F_2 would be of order $r \times 1$. This serves as a check as well.

EXAMPLE 2.25 In continuation of Example 2.24, let us determine the transfer matrix for the multivariable system of Figure 2.85 with the following modifications: all inputs R, V, W and D are scalar; the output C is also scalar and the transfer functions G_1, G_2, H_1, and H_2 are also scalar.

Solution: Here we have 4 inputs and 1 output. The transfer matrix should be of order 1×4. The input-output relationship can be written as

$$[C] = [G_{11} \quad G_{12} \quad G_{13} \quad G_{14}] \begin{bmatrix} R \\ V \\ W \\ D \end{bmatrix}$$

where the scalars are given by

$$G_{11} = \left.\frac{C}{R}\right|_{V,W,D=0} = \frac{G_1 G_2}{1+G_1 G_2 H_1 H_2}; \quad G_{12} = \left.\frac{C}{V}\right|_{R,W,D=0} = \frac{G_2}{1+G_1 G_2 H_1 H_2}$$

$$G_{13} = \left.\frac{C}{W}\right|_{V,R,D=0} = \frac{G_2}{1+G_1 G_2 H_1 H_2}; \quad G_{14} = \left.\frac{C}{D}\right|_{V,R,R=0} = \frac{H_2 G_1 G_2}{1+G_1 G_2 H_1 H_2}$$

EXAMPLE 2.26 Consider Figure 2.85 with the following modifications: inputs V, W, and D are absent. The feedback system is with only single scalar input R (i.e. order 1×1) and two outputs C (i.e. order 2×1).

We observe from the figure that, $R - F_2 = E_1$ and to make sense, as R is of order 1×1 then F_2 and E_1 each should be of order 1×1 like R.

Again $C = G_2 E_2 = G_2 G_1 E_1$. As the order of C is 2×1 and that of E_1 is 1×1, then we can get the order of $[G_2 G_1]$ as 2×1. Similar reasoning for $F_2 = H_2 H_1 C$ leads to conclude that the order of $[H_2 H_1]$ is 1×2 because the order of C is 2×1 and F_2 is 1×1.

The given data is as

$$G_2 G_1 = \begin{bmatrix} \dfrac{1}{s+1} & \dfrac{1}{s+2} \end{bmatrix}^T$$

We now find the characteristic equation and the transfer function matrix relating R and C.

Using the relation

$$E_1 = R - F_2 = R - H_2 H_1 C = R - H_2 H_1 G_2 G_1 E_1$$

leads to $[I + H_2 H_1 G_2 G_1]E_1 = R$, that is, $E_1 = [I + H_2 H_1 G_2 G_1]^{-1} R$

As $C = G_2 G_1 E_1$, we get $C = G_2 G_1 [I + H_2 H_1 G_2 G_1]^{-1} R$

Hence the transfer function matrix is

$$G_2 G_1 [I + H_2 H_1 G_2 G_1]^{-1} \tag{2.89}$$

and the characteristic equation is

$$\left|I + H_2 H_1 G_2 G_1\right| = 1 + \frac{s}{s+1} + \frac{1}{s+2} = 0 \qquad (2.90)$$

Using Eq. (2.89), the characteristic equation is, $\left|I + G_2 G_1 H_2 H_1\right| = 0$; identity matrix I is of order 2×2. Then, we get

$$\left|\begin{bmatrix} 1 & 0 \\ 0 & 1 \end{bmatrix} + \begin{bmatrix} \dfrac{s}{s+1} & \dfrac{1}{s+1} \\ \dfrac{s}{s+2} & \dfrac{1}{s+2} \end{bmatrix}\right| = \left(1 + \frac{s}{s+1}\right)\left(1 + \frac{1}{s+2}\right) - \frac{s}{(s+1)(s+2)} = 1 + \frac{s}{s+1} + \frac{1}{s+2} = 0$$

(2.91)

which is identical to Eq. (2.90), but Eq. (2.89) is without matrix inversion.

Summary

In this chapter we discussed quantitative mathematical models of control components and systems. From the knowledge of the physics of the system the mathematical model is obtained. The systems in practice are nonlinear in nature. One has to linearize the system using the Taylor series approximation as illustrated in Chapter 1. For linear systems, one may utilize the Laplace transformation and its related input-output relationship, that is, the transfer function. In fact, we have very limited mathematical tools for system analysis and for this reason, the transfer function has got wide acceptability for linear systems.

Further, one is more familiar with one's own discipline for system analysis. The system may be electrical, mechanical, pneumatic, so on and so forth. Hence we talked about the analogous system. The transfer function concept goes with single-input-single-output (SISO) linear systems. For multi-input-multi-output (MIMO) systems, that is, multivariable systems, the input-output relationship is not the scalar transfer function, but it is the transfer matrix.

Using the transfer function notations, the block diagram models of systems of interconnected components were developed. Transfer functions of complicated systems were simplified using the block diagram reduction techniques. An alternative use of transfer function models in signal flow graph form was investigated where using the Mason's gain formula, the transfer function can be easily obtained. We discussed some of the sensors used in control systems. Some of the widely-used control components are servomotors (both ac and dc), stepper motors, and magnetic amplifiers. We also discussed the industrial applications of these components.

Problems

2.1 Make an experimental setup to study the LVDT. Indicate the industrial applications of LVDT.

2.2 Make an experimental setup to study the magnetic amplifier. Indicate its industrial application for stabilized power supply.

2.3 Make an experimental setup to study the synchro pair. Indicate its application in remote control.

2.4 Obtain a set of simultaneous integro-differential equations representing the network shown in Figure P.2.4.

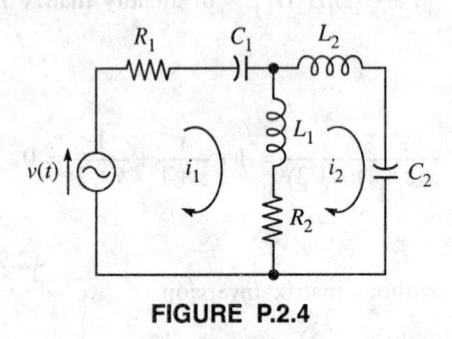

FIGURE P.2.4

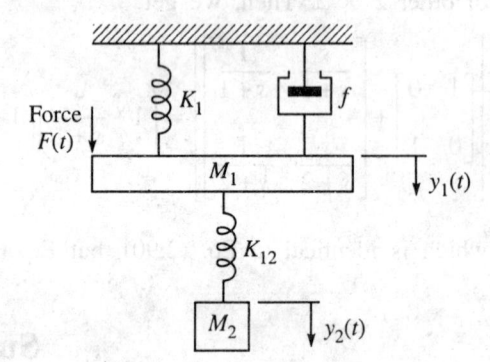

FIGURE P.2.5

2.5 The dynamic vibration absorber system of Figure P.2.5 is representative of many situations involving the vibration of machines containing unbalanced components. The parameters M_2 and K_{12} may be chosen so that the main mass M_1 does not vibrate when the force applied is $F(t) = a \sin \omega_0 t$. (a) Obtain the dynamics of the system in terms of differential equations. (b) Draw the analogous electrical circuit based on the force–current analogy.

2.6 A coupled spring-mass system is shown in Figure P.2.6. The masses and springs are assumed to be equal. (a) Obtain the dynamics of the system. (b) Draw an analogous electric circuit based on the force–current analogy.

2.7 A multivariable feedback control system is shown in Figure P.2.7 with transfer matrices of the system given as

$$G(s) = \begin{bmatrix} \dfrac{1}{s} & \dfrac{1}{s+2} \\ 5 & \dfrac{1}{s+1} \end{bmatrix}; \quad H(s) = \begin{bmatrix} 1 & 0 \\ 0 & 1 \end{bmatrix}$$

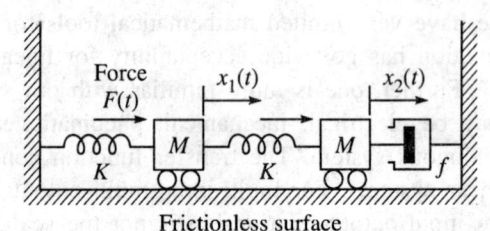

FIGURE P.2.6

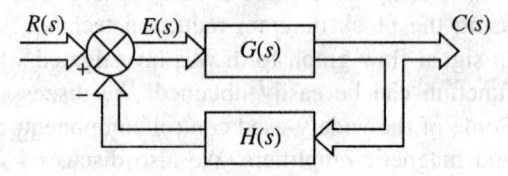

FIGURE P.2.7

Determine the closed-loop transfer function matrix for the system.

2.8 Determine the following transfer relationships from the block diagram shown in Figure P.2.8.

$$\left.\dfrac{C_2(s)}{R_1(s)}\right|_{R_2=0}, \quad \left.\dfrac{C_2(s)}{R_2(s)}\right|_{R_1=0}, \quad \left.\dfrac{C_1(s)}{R_1(s)}\right|_{R_2=0}, \quad \left.\dfrac{C_1(s)}{R_2(s)}\right|_{R_1=0}$$

Write the transfer function matrix for the system showing the connectivity in the form of

$$C(s) = G(s)R(s)$$

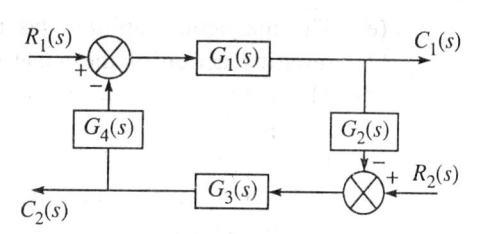

2.9 Obtain the transfer function of the circuit shown in Figure P.2.9.

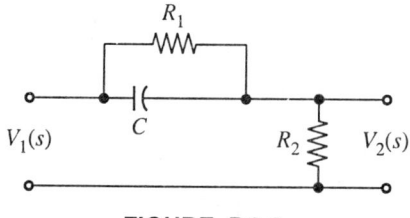

FIGURE P.2.8

FIGURE P.2.9

2.10 In a dc servomotor used in the field-control mode, the armature current is maintained constant and the input voltage v is applied to the field winding, for which the schematic diagram is shown in Figure P.2.10. Determine the transfer function $\Theta(s)/V(s)$.

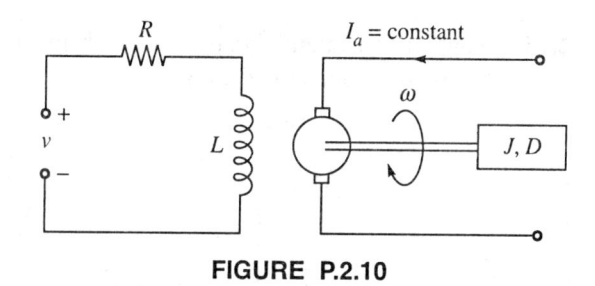

FIGURE P.2.10

2.11 An algebraic set of equations may be rewritten in matrix form as

$$\begin{bmatrix} x_1 \\ x_2 \\ x_3 \end{bmatrix} = \begin{bmatrix} 3 & -1 & 0 \\ 0 & 1 & 4 \\ 2 & 0 & 1 \end{bmatrix} \begin{bmatrix} x_1 \\ x_2 \\ x_3 \end{bmatrix}$$

Draw the signal flow graph of the matrix equation.

2.12 For the circuit of Figure P.2.12, write the dynamics of the system and find the transfer function. Draw the SFG and verify the transfer function obtained by using the Mason's gain formula.

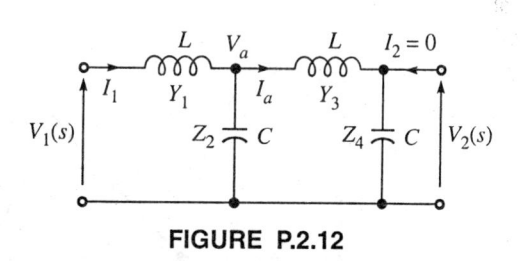

FIGURE P.2.12

2.13 State true or false. Explain why?

(a) A magnetic amplifier can use a magnetic circuit that has a core consisting entirely of air or wood.

(b) A steeper magnetization of *B-H* curve indicates greater inductance.

(c) A saturable-core reactor (without rectifiers) permits the same flow of ac through it if the dc is reversed in its control winding.

(d) A coil wound around a brass core retains the same inductance at all values of current in the coil.

(e) The magnetic amplifier has greater gain than the saturable reactor because the resetting action of the negative half wave is blocked.
(f) The addition of a bias winding changes the available gain of a magnetic amplifier.
(g) A feedback magnetic amplifier uses no fewer than four windings in its reactor.
(h) If 1 mA flows in the dc control winding of a magnetic amplifier while its ac output is 1 A, this amplifier has a gain of 1000.
(i) A separate dc supply is always needed for the control of a magnetic amplifier.
(j) Positive feedback can cause faster response in a magnetic amplifier.
(k) Magnetic amplifiers can be used for providing stabilized power supply.

2.14 Draw a signal flow graph for the following set of linear equations:

$$3y_1 + y_2 + 5y_3 = 0$$
$$y_1 + 2y_2 - 4y_3 = 2$$
$$-y_2 - y_3 = 0$$

2.15 Draw an equivalent signal flow graph for the block diagram in Figure P.2.15 and find the transfer function $C(s)/R(s)$.

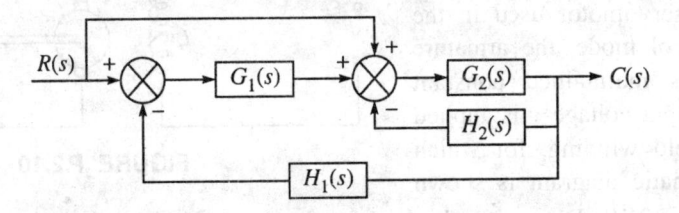

FIGURE P.2.15

2.16 Determine the transfer function of Figure P.2.16 using the Mason's gain formula.

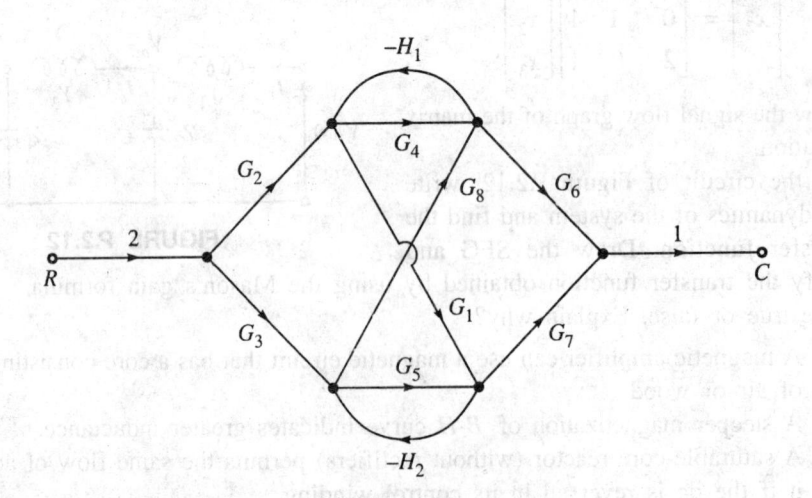

FIGURE P.2.16

2.17 For the block diagrams shown in Figures P.2.17(a) and P.2.17(b), determine the overall transfer function $C(s)/R(s)$ by block diagram simplification and verify by using the Mason's rule.

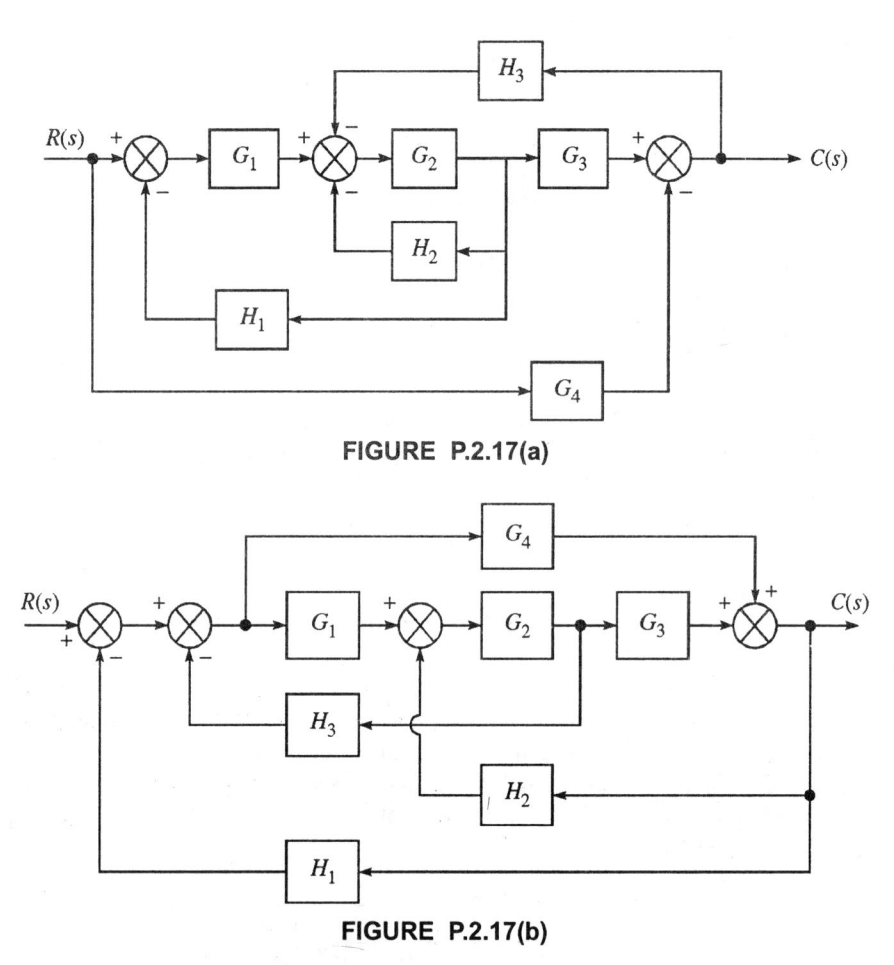

FIGURE P.2.17(a)

FIGURE P.2.17(b)

2.18 Obtain a signal-flow graph to represent the following set of algebraic equations where x_1 and x_2 are to be considered the dependent variables and 8 and 13 are the inputs:

$$x_1 + 2x_2 = 8$$
$$2x_1 + 3x_2 = 13$$

Determine the value of each dependent variable by using the gain formula. After solving for x_1 by Mason's formula, verify the solution by using Cramer's rule.

2.19 Find the gains, y_6/y_1, y_3/y_1, and y_5/y_2 for the signal flow graph shown in Figure P.2.19.

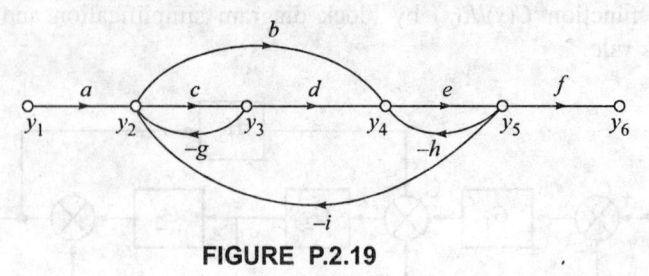

FIGURE P.2.19

2.20 Find the gain y_5/y_1 for the signal flow graph shown in Figure P.2.20.

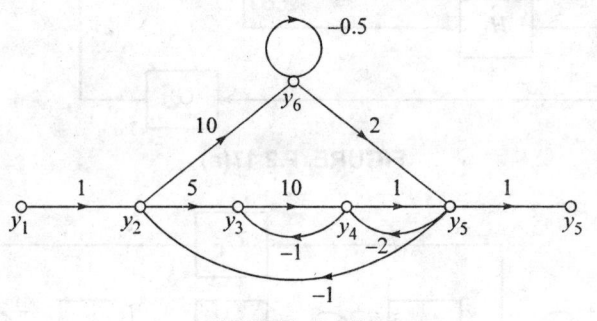

FIGURE P.2.20

2.21 A mechanical system is shown in Figure P.2.21, which is subjected to a known displacement $x_3(t)$ with respect to the reference. (a) Determine the two independent equations of motion. (b) Obtain the equations of motion in terms of the Laplace transform assuming that the initial conditions are zero. (c) Draw a signal-flow graph representing the system of equations. (d) Obtain the relationship between $X_1(s)$ and $X_3(s)$, $T_{13}(s)$, by using the Mason's gain formula.

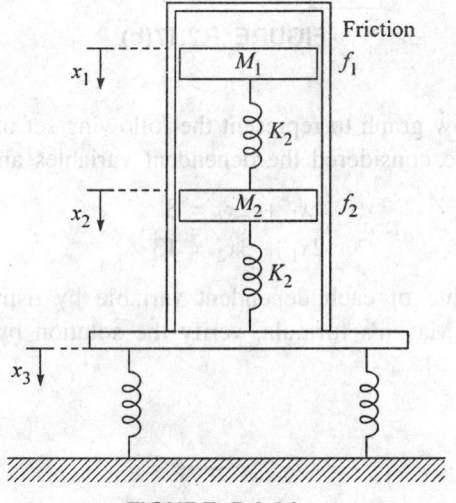

FIGURE P.2.21

Characteristics of Feedback Control Systems

OBJECTIVE

In the open-loop control system, there is no provision for supervision of the output and no mechanism to correct the system behaviour for any lack of proper performance of system components. The concept of feedback revolutionized the design methodology of control systems as it provides a self-correcting mechanism. One of the primary purposes of using feedback in control systems is to reduce the sensitivity of the system to parameter variations. This has made the industrial production much more cost effective.

CHAPTER OUTLINE

Introduction
Derivation of Sensitivity
Sensitivity of Control Systems to Parameter Variation
Effect of Disturbance Signals
Alternative Design through Sensitivity Analysis

3.1 Introduction

In this chapter we will study some of the important characteristics of the feedback control system. Now that we are able to derive the mathematical models of the components of control systems, a control system may be defined as an interconnection of control components forming a system configuration which will provide a desired system response.

All through this text, we will be concerned with only negative feedback systems (without mentioning it explicitly), until stated otherwise. In a negative feedback control system, an actuating signal proportional to the error between the desired and actual response is generated to drive the process or plant, so that at every onward instant of time the actual response approaches towards the desired response. In fact, we can say that a negative feedback system is a proportional control system. The feedback signal gives the system the capability to act as a self-correcting mechanism.

To illustrate the characteristics and advantages of negative feedback, let us consider a simple, single-loop feedback control system. It is true that many practical control systems are not single-loop in character, but a single-loop system is being used here only as an illustration. A thorough understanding of the benefits of feedback is obtained from a study of the single-loop system, which study can then be extended to multi-loop systems.

3.2 Derivation of Sensitivity

In a negative feedback closed-loop control system as shown in Figure 3.1, the error signal $E(s)$ can be written as

$$E(s) = R(s) - C(s)H(s) \tag{3.1}$$

and the output is written as

$$C(s) = G(s)E(s) = G(s)[R(s) - H(s)C(s)]$$

and therefore,

$$C(s) = \frac{G(s)}{1 + G(s)H(s)} R(s) \tag{3.2}$$

The error-actuating signal from Eq. (3.1) is

$$E(s) = \frac{1}{1 + G(s)H(s)} R(s) \tag{3.3}$$

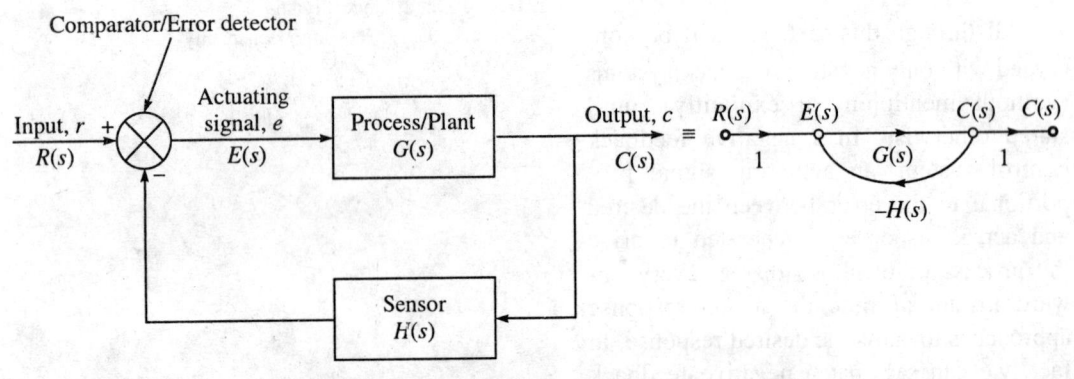

FIGURE 3.1 A closed-loop system and its SFG.

It is thus evident that in order to reduce the error, the magnitude of $[1 + G(s)H(s)]$ should be greater than one, that means, $|G(s)H(s)| \gg 1$ holds over the range of frequencies under consideration.

It may be noted that due to the low-pass nature of control systems, this inequality $|G(s)H(s)| \gg 1$ holds, but will not be satisfied at high frequencies. Also, if $G(s)H(s)$ turns out to be a negative number close to unity at some value of the frequency, then we may have a very large error.

EXAMPLE 3.1 A unity-feedback system is shown in Figure 3.2. Express the steady-state error as a percentage of the desired speed if the controller gain K is (i) 1, (ii) 5, and (iii) 10.

Characteristics of Feedback Control Systems

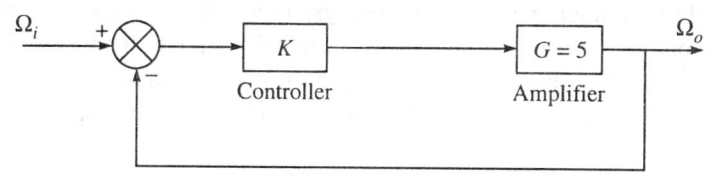

FIGURE 3.2 Example 3.1: unity-feedback system.

Solution: The closed-loop gain is $\dfrac{KG}{1+KG}$, where $G = 5$.

For $K = 1, 5$ and 10, the closed-loop system has gain of $5/6 = 0.83$, $25/26 = 0.96$, and $50/51 = 0.98$ respectively.

$$\text{The \% error for } K = 1 \text{ is } \dfrac{1-0.83}{1} \times 100\% = 17\%$$

$$\text{The \% error for } K = 5 \text{ is } \dfrac{1-0.96}{1} \times 100\% = 4\%$$

$$\text{The \% error for } K = 10 \text{ is } \dfrac{1-0.98}{1} \times 100\% = 2\%$$

Thus the steady-state error is reduced as the controller gain (or the forward path gain in general) is increased.

EXAMPLE 3.2 Suppose in Example 3.1, the gain of the amplifier increases by 10 per cent as a result of parameter changes. What is the new closed-loop gain for $K = 1, 5,$ and 10?

Solution: A 10 per cent increase in the gain of the amplifier gives the new process (amplifier) gain of 5.5.
The closed-loop gain becomes

$$\text{for } K = 1; \quad \text{as} \quad \dfrac{1(5.5)}{1+1(5.5)} = \dfrac{5.5}{6.5} = 0.85$$

$$\text{for } K = 5; \quad \text{as} \quad \dfrac{5(5.5)}{1+5(5.5)} = \dfrac{27.5}{28.5} = 0.96$$

$$\text{for } K = 10; \quad \text{as} \quad \dfrac{10(5.5)}{1+10(5.5)} = \dfrac{55}{56} = 0.98$$

This shows the relative insensitivity of the closed-loop gain to changes in the forward path gain and also shows that the closed-loop system becomes progressively less sensitive to such forward path parameter variations as the controller gain K is continued to be increased.

3.3 Sensitivity of Control Systems to Parameter Variations

The process or plant represented by the transfer function $G(s)$ is subject to change under changing environment, aging, etc. The change in output due to change in a parameter of the process $G(s)$ is obvious in an open-loop control system whereas in a closed-loop control system there is an attempt to correct the output because of negative feedback. The reduction in sensitivity of the control system to parameter variations is a primary advantage of the feedback in a control system. Rewriting the output expression for a closed-loop control system,

$$C(s) = \frac{G(s)}{1 + G(s)H(s)} R(s)$$

For $G(s)H(s) \gg 1$,

$$C(s) \cong \frac{1}{H(s)} R(s) \tag{3.4}$$

and for a unity-feedback system, $H(s) = 1$, then

$$C(s) = R(s)$$

that is, the output follows the input. This is the ideal desired condition. But the practical difficulty is that the condition $G(s)H(s) \gg 1$ may cause the system response highly oscillatory and even unstable. Also the fact remains that as we approach the condition $G(s)H(s) \gg 1$, we reduce the effect of $G(s)$ on output. This is important, and hence the primary advantage of the feedback control system over the open-loop control system lies in the reduced effect of variations in parameters of the process $G(s)$.

The system sensitivity is defined as the ratio of the percentage change in the system transfer function to the percentage change in the plant transfer function.

The system transfer function $T(s)$ is defined as

$$T(s) = \frac{C(s)}{R(s)}$$

Our objective is to derive the sensitivity in the case of open-loop and closed-loop systems under parameter variations of the process $G(s)$.

$$\text{For the open-loop case, } T(s) = G(s) \tag{3.5}$$

$$\text{For the closed-loop case, } T(s) = \frac{G(s)}{1 + G(s)H(s)} \tag{3.6}$$

The sensitivity of T with respect to G is defined as

$$S_G^T = \frac{\frac{\Delta T(s)}{T(s)}}{\frac{\Delta G(s)}{G(s)}} = \frac{\text{percentage change in } T(s)}{\text{percentage change in } G(s)} \tag{3.7}$$

Hence the sensitivity for the open-loop system becomes

$$S_G^G = \frac{\frac{\Delta G(s)}{G(s)}}{\frac{\Delta G(s)}{G(s)}} = 1 \quad [\because T(s) = G(s)]$$

That is, the sensitivity of the open-loop system is unity.

The sensitivity expression for the closed-loop system can be rewritten, in the limit, for small incremental changes, as

$$S_G^T = \frac{\frac{\partial T}{T}}{\frac{\partial G}{G}} = \frac{\partial \ln T}{\partial \ln G}$$

Now substituting $T(s) = \frac{G}{1+GH}$, the sensitivity of the closed-loop system given by is

$$S_G^T = \frac{\partial T}{\partial G} \frac{G}{T} = \frac{1}{(1+GH)^2} \frac{G}{\frac{G}{(1+GH)}}$$

or $$S_G^T = \frac{1}{1+GH} \qquad (3.8)$$

Hence sensitivity for the closed-loop system has improved by a factor $\frac{1}{1+GH}$. Equation (3.8) states that if $GH \gg 1$, a 10 per cent change in the forward transfer function $G(s)$ due to transistor replacement, temperature changes, etc. will appear as a change in the overall closed-loop transfer function $T(s)$ of approximately $\left[\frac{10}{G(s)H(s)}\right]$ per cent.

As an illustration, consider a 10 per cent increase in $G(s)$, i.e. let us assume $G'(s) = 1.1G(s)$ and the corresponding closed-loop transfer function is

$$T'(s) = \frac{G'}{1+G'H} = \frac{1.1G}{1+(1.1G)H}$$

whereas the closed-loop transfer function $T(s)$ corresponding to the forward transfer function $G(s)$ is

$$T(s) = \frac{G}{1+GH}$$

Therefore, $\Delta T(s) = T'(s) - T(s)$

$$= \frac{1.1G}{1+1.1GH} - \frac{G}{1+GH} = \frac{0.1G}{(1+1.1GH)(1+GH)}$$

The percentage change in $T(s) = \frac{\Delta T(s)}{T(s)} \times 100 = \frac{10}{1+1.1GH} \approx \frac{10}{GH}$ per cent (as $GH \gg 1$).

Similarly, for a 10 per cent decrease in $G(s)$, we get

$$T'(s) = \frac{0.9G}{1+0.9GH}$$

Therefore, $$\Delta T(s) = T(s) - T'(s) = \frac{0.1G}{(1+GH)(1+0.9GH)}$$

The percentage change in $T(s) = \dfrac{\Delta T(s)}{T(s)} \times 100 = \dfrac{10}{0.9GH} \approx \dfrac{10}{GH}$ per cent (as $GH \gg 1$).

EXAMPLE 3.3 A feedback amplifier is to be designed to have an overall gain of a feedback system as 40 dB and a sensitivity of 5 per cent to internal variations in the gain of the amplifier. Find the required gain of the open-loop transfer function $G(s)H(s)$ and also the forward transfer function $G(s)$.

Solution: From the definition of dB as $20 \log_{10}\left(\dfrac{V_2}{V_1}\right)$, we get

$$T = \frac{G}{1+GH} \approx \frac{G}{GH} = 100 = 40 \text{ dB, assuming } GH \gg 1.$$

The sensitivity of the overall gain to internal variations in the gain of the amplifier given as 5 per cent (= 0.05) means that

$$S_G^T = \frac{1}{1+GH} \approx \frac{1}{GH} = 0.05$$

so that $GH = 20$ and $G = \dfrac{20}{H} = (20)(100) = 2000 = 66$ dB

Thus, in order to achieve an overall gain of 100 (i.e. 40 dB), we must start with an amplifier having a gain of 2000 (i.e. 66 dB). The sacrifice of gain has resulted in considerable degree of stability.

The sensitivity of the closed-loop system may be reduced below that of the open-loop system by increasing $G(s)H(s)$ over the frequency range of interest. Now we will see the effect on sensitivity of the variation in parameter of the feedback element $H(s)$. Let $H(s) = \beta$. Then the sensitivity,

$$S_\beta^T = \frac{\dfrac{\partial T}{T}}{\dfrac{\partial \beta}{\beta}} = \frac{\beta}{T} \cdot \frac{\partial T}{\partial \beta} = \left(\frac{\beta}{H} \cdot \frac{dH}{d\beta}\right)\left(\frac{H}{T} \cdot \frac{dT}{dH}\right)$$

$$= S_\beta^H \left(\frac{H}{T} \cdot \frac{(-G)G}{(1+GH)^2}\right) = S_\beta^H \frac{H}{\dfrac{G}{1+GH}} \cdot \frac{-G^2}{(1+GH)^2}$$

$$= -S_\beta^H \frac{GH}{1+GH} = -\frac{GH}{1+GH} \quad (\because S_\beta^H = 1) \qquad (3.9)$$

When $GH \gg 1$, the magnitudes of S_β^T and S_β^H are nearly equal, both are unity in this case. The changes in $H(s)$ directly affect the output response. *Therefore, it is important to ensure that the feedback components do not vary with environmental changes, or in other words should be maintained constant.*

Hence we can conclude that though we can use components of higher tolerance (say 10 per cent tolerance resistors) in the forward path, we should use only high precision components in the feedback path (say 1 per cent tolerance resistors).

It may thus be observed that the advantage of negative feedback is the reduction in the sensitivity to variations in the parameters of the forward path.

Historically, the development of feedback amplifiers was done by the engineers at Bell Telephone Laboratories. In the old days, for a long distance telephone system, repeaters made of vacuum tubes were used. Because of mass-scale production of repeaters, the gain varied considerably between different repeaters, primarily due to the variation in amplification factors of vacuum tubes. For cost-effective products, less precision components were used to keep the cost of production low. The primary objective of introducing negative feedback was to make the gain of amplifiers in repeaters relatively insensitive to variations in the parameters of the vacuum tubes. This led to the design of feedback control systems where the transfer function of the closed-loop system was made relatively insensitive to small changes in the values of the parameters of the components in the forward path of the system.

In order to have the mathematical support, let us consider a change in the process so that the new process is $G(s) + \Delta G(s)$ due to ageing effect. As the output $C(s) = G(s)R(s)$, then, the change in output in the open-loop case is calculated as

$$C(s) + \Delta C(s) = [G(s) + \Delta G(s)]R(s)$$

or

$$\Delta C(s) = \Delta G(s)R(s)$$

whereas in closed-loop system, as we have

$$C(s) = \frac{G(s)}{1+G(s)H(s)}R(s)$$

Therefore,

$$C(s) + \Delta C(s) = \frac{G(s)+\Delta G(s)}{1+[G(s)+\Delta G(s)]H(s)}R(s)$$

Substituting the value of $C(s)$, we get

$$\frac{G(s)R(s)}{1+G(s)H(s)} + \Delta C(s) = \frac{G(s)+\Delta G(s)}{1+[G(s)+\Delta G(s)]H(s)}R(s)$$

or

$$\Delta C(s) = \left[\frac{G(s)+\Delta G(s)}{1+[G(s)+\Delta G(s)]H(s)} - \frac{G(s)}{1+G(s)H(s)}\right]R(s)$$

$$= \frac{\Delta G(s)}{[1+G(s)H(s)+\Delta G(s)H(s)][1+G(s)H(s)]}R(s) \quad \text{(after simplification)}$$

When $G(s)H(s) \gg \Delta G(s)H(s)$, as is often the case, we get

$$\Delta C(s) = \frac{\Delta G(s)}{[1+G(s)H(s)]^2} R(s) \qquad (3.10)$$

It is evident from the expression (3.10) that the change in output under parameter variation of the process due to ageing, in a closed-loop system is reduced by a factor $[1 + G(s)H(s)]$ [see Eq. (3.2) for comparison], which is much greater than unity over the range of frequencies of interest. This emphasizes the advantage of negative feedback in the reduction of sensitivity to variations in parameters of the forward path.

If α be a parameter of $G(s)$, then the sensitivity of the open-loop system having transfer function $G(s)$ with respect to the parameter α is defined as

$$S_\alpha^G = \frac{\dfrac{dG}{G}}{\dfrac{d\alpha}{\alpha}} = \frac{d\ln G}{d\ln \alpha} = \frac{\alpha}{G}\frac{dG}{d\alpha} \qquad (3.11)$$

Rewriting the above equation as

$$S_\alpha^G = \lim_{\Delta\alpha \to 0} \frac{\dfrac{\Delta G}{G}}{\dfrac{\Delta\alpha}{\alpha}}$$

which can be considered as the fractional change in G due to a very small fractional change in α. This is the sensitivity of the transfer function of the open-loop system to variations in α.

Now we will find the sensitivity of the closed-loop system $T(s)$ to variations in α, which can be written as

$$S_\alpha^T = \frac{\alpha}{T}\frac{dT}{d\alpha} = \frac{\alpha}{\dfrac{G}{1+GH}}\frac{dT}{dG}\frac{dG}{d\alpha}$$

$$= \frac{\alpha(1+GH)}{G} \frac{1}{(1+GH)^2} \frac{dG}{d\alpha}$$

$$= \frac{\alpha}{G}\frac{dG}{d\alpha}\frac{1}{1+GH} \qquad (3.12)$$

Substituting Eq. (3.11) into Eq. (3.12), we get

$$S_\alpha^T = \frac{S_\alpha^G}{1+GH} \qquad (3.13)$$

Expression (3.8) is in tune with the expression (3.13).

Characteristics of Feedback Control Systems

Now we will see the effect on sensitivity caused by the variation in the feedback element $H(s)$. The sensitivity is

$$S_H^T = \frac{dT}{dH}\frac{H}{T} = \left(\frac{G}{1+GH}\right)^2 \left(\frac{-H}{\frac{G}{1+GH}}\right)$$

$$= \frac{-GH}{1+GH} \approx 1 \quad \text{as } GH \gg 1 \tag{3.14}$$

It is evident from the expression (3.14) that the sensitivity changes directly with $H(s)$ and the output is affected. Hence it is necessary to ensure that the feedback components do not vary under environmental changes. Needless to say that when GH is large, the sensitivity approaches unity.

It is interesting to know that the feedback does not reduce the sensitivity to variations in parameters in the feedback path. Let β be a parameter of $H(s)$. Then,

$$S_\beta^T = \frac{\beta}{T}\frac{dT}{d\beta} = \left(\frac{\beta}{H}\frac{dH}{d\beta}\right)\left(\frac{H}{T}\frac{dT}{dH}\right)$$

$$= S_\beta^H \frac{H}{T}\frac{(-G)G}{(1+GH)^2} = -S_\beta^H \frac{H(1+GH)}{G}\frac{G^2}{(1+GH)^2}$$

$$= -S_\beta^H \frac{GH}{1+GH} \approx \left|S_\beta^H\right| \quad \text{as } GH \gg 1 \tag{3.15}$$

It therefore follows that the magnitudes of S_β^T and S_β^H are nearly equal if the loop gain $G(s)H(s) \gg 1$.

Usually the transfer function can be written as the ratio of the numerator polynomial to the denominator polynomial. It can be written as follows when α is a parameter and subject to variation.

$$T(s, \alpha) = \frac{N(s, \alpha)}{D(s, \alpha)}$$

Then the sensitivity to α can be written as

$$S_\alpha^T = \frac{\partial \ln T}{\partial \ln \alpha} = \frac{\partial \ln N}{\partial \ln \alpha}\bigg|_{\alpha_0} - \frac{\partial \ln D}{\partial \ln \alpha}\bigg|_{\alpha_0}$$

$$= S_\alpha^N - S_\alpha^D \tag{3.16}$$

where α_0 is the normal value of the parameter.

A simple example will illustrate the value of feedback in reducing the sensitivity.

EXAMPLE 3.4 Consider a closed-loop position control system with armature-controlled dc servomotor in the forward path having transfer function

$$G(s) = \frac{K}{s(s+\alpha)}$$

and the feedback transfer function

$$H(s) = \beta$$

Solution: Suppose the normal value of $K = 20$, $\alpha = 4$ and $\beta = 1$.
Hence the given transfer functions become

$$G(s) = \frac{K}{s(s+\alpha)} = \frac{20}{s(s+4)} \quad \text{and} \quad H(s) = \beta = 1$$

Now the sensitivities of the open-loop transfer function to parameter variations in K and α are as follows:

$$S_K^G = \frac{K}{G}\frac{dG}{dK} = s(s+\alpha)\frac{1}{s(s+\alpha)} = 1$$

and

$$S_\alpha^G = \frac{\alpha}{G}\frac{dG}{dK} = \frac{\alpha}{\frac{K}{s(s+\alpha)}} \frac{-K}{s(s+\alpha)^2} = \frac{-\alpha}{s+\alpha} = -\frac{4}{s+4}$$

The sensitivity of the feedback transfer function $H(s)$ to parameter variation in β is given as

$$S_\beta^H = \frac{\beta}{H}\frac{dH}{d\beta} = 1$$

The sensitivities of the closed-loop system transfer function can now be evaluated as

$$S_K^T = \frac{S_K^G}{1+G(s)H(s)} = \frac{1}{1+G(s)H(s)} = \frac{s(s+\alpha)}{s(s+\alpha)+K} = \frac{s^2+4s}{s^2+4s+20}$$

$$S_\alpha^T = \frac{S_\alpha^G}{1+G(s)H(s)} = \frac{-\alpha}{s+\alpha}\frac{s(s+\alpha)}{s(s+\alpha)+K} = \frac{-4s}{s^2+4s+20}$$

$$S_\beta^T = -S_\beta^G \frac{G(s)H(s)}{1+G(s)H(s)} = \frac{-K}{s^2+\alpha s+K} = \frac{-20}{s^2+4s+20}$$

The sensitivities of the closed-loop transfer function $T(s)$, that is, S_K^T and S_α^T with respect to the variation in parameters in the forward transfer function, are seen to be small at low frequencies and very small at dc(zero frequency). On the other hand, the sensitivity of $T(s)$, that is, S_β^T, with respect to variation in parameter β in the feedback transfer function $H(s)$ has the magnitude 1 at zero frequency. *Thus, the feedback has not reduced the sensitivity to variation of a parameter in the feedback path.*

Let us consider the effect of a 10 per cent change in the parameter K. Then the resulting change in the closed-loop transfer function can be obtained from

$$\frac{\Delta T(s)}{T(s)} = S_K^T \cdot \frac{\Delta K}{K} = \left(\frac{s^2 + 4s}{s^2 + 4s + 20}\right)(0.1)$$

or

$$\Delta T(s) = \left(\frac{s^2 + 4s}{s^2 + 4s + 20}\right)(0.1)\left(\frac{20}{s^2 + 4s + 20}\right) = \frac{2s(s+4)}{(s^2 + 4s + 20)^2}$$

The advantage of a feedback system is that the effect of the variations in the parameters of the process in the forward transfer function $G(s)$ is reduced. In the case of an open-loop system, the components of the open-loop transfer function must be very carefully selected to meet the exact specifications. Slight changes in the components will affect the output drastically for an open-loop system, whereas in the case of a feedback system the sensitivity to the parameter variation in the forward path will be reduced by a factor of $(1 + GH)$. This benefit of the feedback system is a profound advantage which is gainfully utilized in the design of the electronic amplifier in communication industry. Historically, the engineers at the Bell Telephone Laboratories developed the feedback amplifier, and in the design utilized the idea of feedback suitably, as mentioned earlier as well.

EXAMPLE 3.5 The op-amp inverting amplifier is shown in Figure 3.3. Assume open-loop gain $A = 10^4$, $R_f = 10$ kΩ and $R_i = 1$ kΩ; $\beta^\dagger = R_i/R_f = 0.1$. Calculate the sensitivity S_A^T and S_β^T.

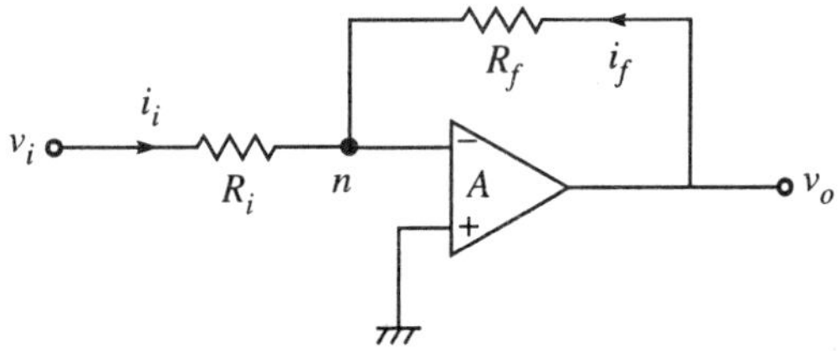

FIGURE 3.3 Example 3.5: inverting amplifier.

Solution: Applying KCL at virtual ground node n of operational amplifier having high input impedance so that the current through the operational amplifier is zero, i.e.

$$i_i + i_f = 0$$

or

$$\frac{v_i - v_n}{R_i} + \frac{v_o - v_n}{R_f} = 0 \qquad (3.17)$$

After simplification and rearranging, we get

$$\frac{v_o}{v_i} = -\frac{A}{1 + (1+A)\left(\frac{R_i}{R_f}\right)} = -\frac{A}{1 + A\beta} \quad \left(\because \frac{v_o}{v_n} = -A, \ \beta = \frac{R_i}{R_f} \text{ and } 1 \ll A\right) \qquad (3.18)$$

For $A \gg 1$, $\dfrac{v_o}{v_i} = -\dfrac{1}{\dfrac{1}{A} + \beta} = -\dfrac{1}{\beta} = -\dfrac{R_f}{R_i}$. This is usually the case.

---

†For the circuit of Figure 3.3, the feedback current $i_f = v_o/R_f$. Again KCL at node n leads to $i_f = -i_i$ which implies $v_o/R_f = -v_i/R_i$ and leads to $v_i = (R_i/R_f)v_o = -\beta v_o$, that is, the portion of the output feedback is $\beta = |R_i/R_f|$.

From Eq. (3.18), we get
$$v_o + A\beta v_o = -Av_i$$
or
$$v_o = (v_i + \beta v_o)(-A) \tag{3.19}$$

Following Eq. (3.19), the signal flow graph is drawn in Figure 3.4. The closed-loop transfer function now becomes

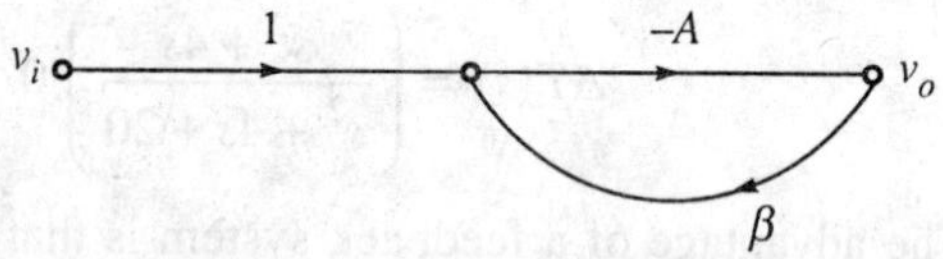

$$T(s) = \frac{G(s)}{1 - G(s)H(s)} = \frac{-A}{1 + A\beta} = \frac{-A}{1 - (-A)(\beta)}$$

FIGURE 3.4 Example 3.5: signal flow graph of circuit of Figure 3.3.

The feedback factor in the diagram is $H(s) = \beta$; the forward transfer function $G(s) = -A$. The op-amp is subject to variations in amplification, A. The sensitivity of the open-loop system is

$$S_A^G = \frac{A}{G}\frac{\partial G}{\partial A} \approx 1$$

The sensitivity of the closed-loop system is

$$S_A^T = \frac{A}{T}\frac{\partial T}{\partial A} = \frac{A}{\frac{A}{1-A\beta}}\frac{1}{(1-A\beta)^2} = \frac{1}{1-A\beta}$$

For $A = 10^4$, $\beta = 0.1$, we get

$$S_A^T = \frac{1}{1 - 10^3} = 0.001$$

The change in gain in the forward transfer function affects the output performance of the closed-loop system insignificantly by a factor of 0.001; the greater the value of the open-loop gain A, the less significant is the effect of its change on the closed-loop system.

The sensitivity due to variations in the feedback parameter β (say, changes in R_f due to ageing) is

$$S_\beta^T = \frac{\beta}{T}\frac{\partial T}{\partial \beta} = \frac{\beta}{\frac{A}{1-A\beta}}\frac{A^2}{(1-A\beta)^2} = \frac{A\beta}{1-A\beta}$$

Substituting the values of $A = 10^4$, $\beta = 0.1$, we get

$$S_\beta^T = 1$$

i.e. the sensitivity to β is approximately equal to unity.

Drill Problem 3.1

For a unity-feedback system having open-loop transfer function $K/(1 + 4s)(1 + s)$, with nominal value of $K = 10$ and a tolerance of 5 per cent, find the output percentage tolerance for a step input.

Ans. ± 0.413

3.4 Effect of Disturbance Signals

Most control systems are subjected to unwanted disturbance (noise) signals. For example, noise may be generated in electronic amplifiers, or due to load fluctuations, or due to gusts of winds affecting radar antennas, and so forth. The block diagram of such a system subjected to noise is shown in Figure 3.5(a) where $N(s)$ is the disturbance or noise signal.

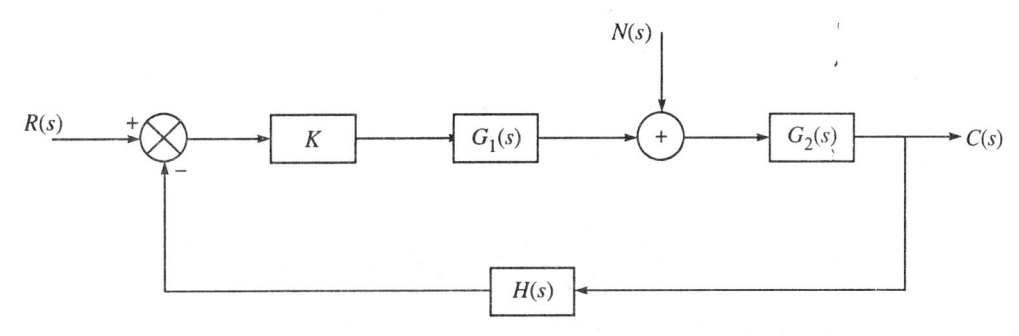

FIGURE 3.5(a) Control system subjected to noise.

The output may be written as

$$C(s) = \frac{KG_1(s)G_2(s)}{1+KG_1(s)G_2(s)H(s)} R(s) + \frac{G_2(s)}{1+KG_1(s)G_2(s)H(s)} N(s)$$

Hence, due to feedback, the effect of noise is reduced by $\dfrac{1}{KG_1(s)H(s)}$ over the frequency range of interest. Thus by increasing the loop gain, we can reduce the effect of noise occurring at the output level. Load disturbance is an example of this kind. Now let us talk about a practical example of load disturbance.

The important effect in a feedback control system is the partial limitation of the effect of disturbance signals. Feedback systems have the beneficial aspect that the effect of distortion, noise and unwanted disturbances can be effectively reduced. In the steel rolling mill of Figure 3.5(b), rolls passing steel bars through them are subjected to large load changes or disturbances. As the steel bar approaches the rolls, the rolls are turning unloaded. However, when the steel bar engages the rolls, the load on the rolls increases immediately to a large value. This loading effect can be approximated by a step change in disturbance D, then noise can be written as $N(s) = D/s$ in Figure 3.5(a).

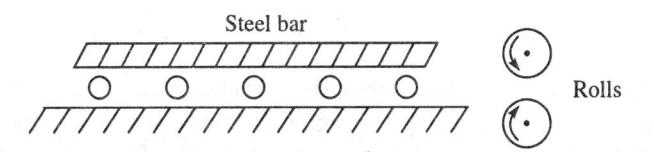

FIGURE 3.5(b) Steel rolling mill operation.

EXAMPLE 3.6 The output of the speed control system shown in Figure 3.6 is given by

$$C(s) = \frac{2}{s^2 + 1.2s + 2.2} R(s) + \frac{0.2(s+1)}{s^2 + 1.2s + 2.2} N(s)$$

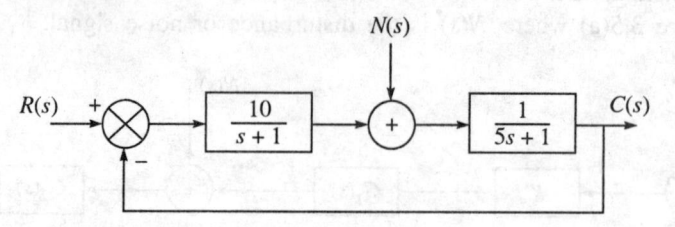

FIGURE 3.6 Example 3.6: speed control system.

The reduction in the disturbance is evident by noting that $C(s)/R(s)$ is 10 times more than $C(s)/N(s)$ which is precisely the value of $KG(s)$ at $s = 0$. The reduction in disturbance is possible at the output level. If the disturbance occurs at the input level, the effect of input and noise would be the same at the output.

In general, the primary reason for introducing the feedback is the ability to reduce the effect of disturbances and noise signals generated within the feedback loop. For the noise or disturbance $N(s)$ generated by the measurement sensor as shown in Figure 3.7, the output $C(s)$ is

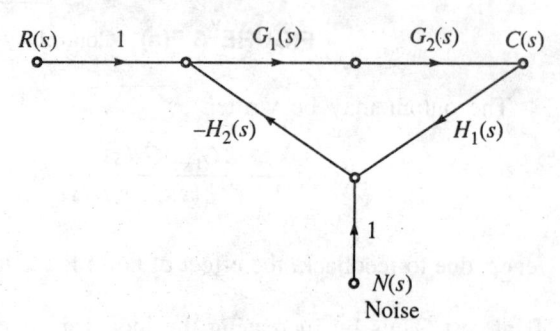

FIGURE 3.7 Example 3.6.

$$C(s) = \frac{G_1(s)G_2(s)}{1 + G_1G_2H_1H_2(s)} R(s) + \frac{-G_1G_2H_2(s)}{1 + G_1G_2H_1H_2(s)} N(s) \quad (3.20)$$

The effect of noise on the output is

$$C(s) = \frac{-G_1G_2H_2(s)}{1 + G_1G_2H_1H_2(s)} N(s) \quad (3.21)$$

which is approximately

$$C(s) = -\frac{1}{H_1(s)} N(s) \quad (3.22)$$

Thus, for optimum performance of the system, the designer must obtain a maximum value of $H_1(s)$, which is equivalent to maximizing the signal-to-noise ratio of the measurement sensor. This necessity is equivalent to requiring that the quality of the feedback sensor be as good as $S_H^T = 1$.

EXAMPLE 3.7 Obtain the expression for the output temperature θ_o in terms of the demanded temperature θ_i, the disturbances d_1 and d_2 and the gain K and G being as shown in Figure 3.8.

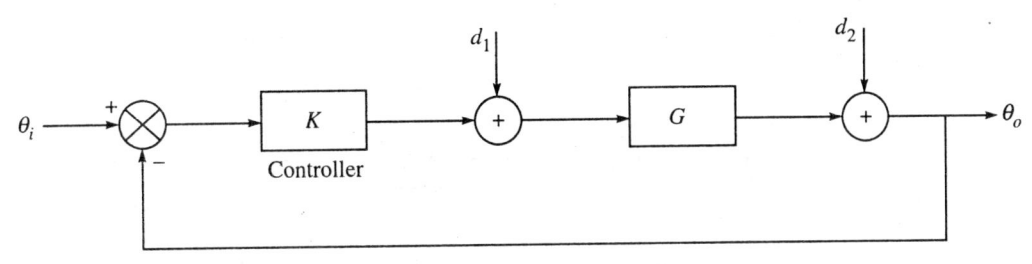

FIGURE 3.8 Example 3.7.

Solution: The output temperature θ_o can be written as

$$\theta_o = \frac{KG}{1+KG}\theta_i + \frac{G}{1+KG}d_1 + \frac{1}{1+KG}d_2$$

In the absence of the feedback loop the disturbances d_1 and d_2 would give rise to errors of Gd_1 and d_2, respectively, in the output. The factor $(1 + KG)$ is therefore a measure of how much the effects of disturbance are modified by the feedback action. This property is often known as *disturbance rejection*.

3.5 Alternative Design through Sensitivity Analysis

Suppose the plant transfer function is given. We select the closed-loop transfer function $T(s)$ such that the given design specifications are met by a suitable controller design. The closed-loop transfer function $T(s)$ may be realized by different feedback structures. Now the question is which feedback structure is better? Sensitivity analysis allows a quantitative comparison of these feedback structures to decide in favour of the best. This is illustrated with the following example.

Let the given plant transfer function be

$$G(s) = \frac{K}{s(s+1)} \quad \text{with nominal value of } K = 1$$

Let the desired closed-loop transfer function to satisfy the design specification be

$$T(s) = \frac{25}{s^2 + 5s + 25}, \quad \text{for the frequency of interest } \omega = 5 \text{ rad/s.}$$

Now the transfer function $T(s)$ may be realized by a single-loop configuration as shown in Figure 3.9, with nominal value of $K = 1$.

Alternatively, the same $T(s)$ can be realized by a two-loop structure of Figure 3.10 (nominal value of $K = 1$).

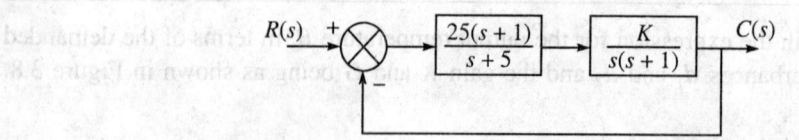

FIGURE 3.9 Closed-loop system with single-loop configuration.

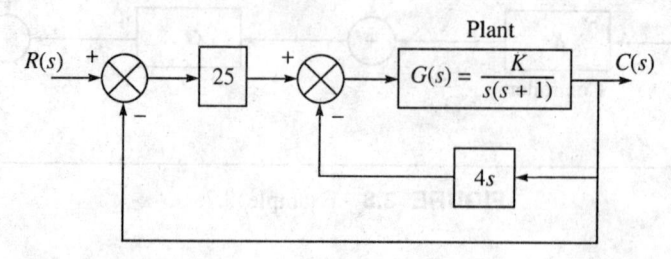

FIGURE 3.10 Closed-loop system with two-loop configuration.

Now we will perform the sensitivity analysis for the systems in both of these figures. Referring to Figure 3.9, the closed-loop transfer function

$$T(s) = \frac{C(s)}{R(s)} = \frac{25K}{s^2 + 5s + 25K}$$

The sensitivity of the closed-loop system to variation in parameter K having nominal value $K = 1$ is found as

$$S_K^T = \frac{\partial T}{\partial K}\frac{K}{T} = \frac{s(s+5)}{s^2 + 5s + 25K} = \frac{s(s+5)}{s^2 + 5s + 25}$$

The magnitude of the sensitivity at the given frequency $\omega = 5$ is

$$\left|S_K^T\right| = 1.41$$

Now referring to Figure 3.10, the closed-loop transfer function is

$$T(s) = \frac{25K}{s^2 + (1 + 4K)s + 25K}$$

and the sensitivity

$$S_K^T = \frac{s(s+1)}{s(s+1) + K(4s+25)} = \frac{s(s+1)}{s^2 + 5s + 25}; \text{ for } K = 1$$

which at frequency $\omega = 5$ becomes

$$\left|S_K^T\right| = \sqrt{\frac{26}{25}} \approx 1$$

The sensitivity analysis done thus far shows the superiority of the two-loop system of Figure 3.10 over the single-loop system of Figure 3.9. Thus, the sensitivity analysis gives a firm basis for comparison of alternative designs to satisfy the design specifications.

Summary

In this chapter we studied the various aspects of feedback control systems.
The main advantages of feedback are:

1. Reduction in sensitivity to variations in parameters in the forward path.
2. Control over both the transient as well as the steady-state response and hence the accuracy obtainable by adjusting the loop gain.
3. Reduction in the effect of noise or disturbance at the output level, leading to improved stability.

The disadvantages of feedback are:

1. Feedback decreases the overall gain; hence an increase in the loop gain of the system to offset the reduction in gain due to feedback requires additional hardware that causes increased complexity.
2. Feedback does not reduce the sensitivity to variations in components in the feedback path and hence the components in the feedback path must be made more precise. This results in an increase in the overall cost of the system.
3. Sensors required for the feedback path may introduce noise in the system and thereby reduce the overall accuracy.
4. The introduction of feedback may lead to the possibility of instability of the closed-loop system due to phase lag in the feedback loop, even though the open-loop system may be stable. The result is that, what was intended to be negative feedback may turn out to be positive feedback at some higher frequencies*.

In concluding remarks, it is fair to say that the advantages of feedback are much more significant than the disadvantages.

Problems

3.1 For the system shown in Figure P.3.1, determine the sensitivity of the transfer function of the closed-loop system to changes in parameters K, p and α if the normal values of K, p and α are 10, 2, and 0.14 respectively.

Ans. $S_K^T = \dfrac{s(s+2)}{s^2 + 3.4s + 10}$; $S_p^T = \dfrac{-2s}{s^2 + 3.4s + 10}$; $S_\alpha^T = \dfrac{-1.4s}{s^2 + 3.4s + 10}$

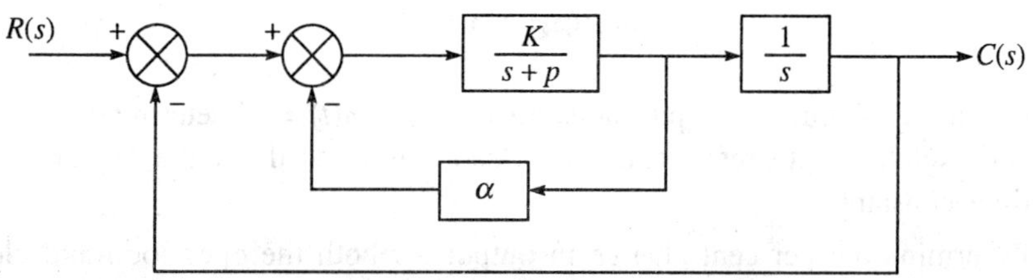

FIGURE P.3.1

---

*D. Roy Choudhury and Shail B. Jain, *Linear Integrated Circuits*, Second Edition, New Age International, New Delhi, 2003.

3.2 Consider Figure P.3.1, assume the input $r(t) = t$, $t > 0$; otherwise zero. Determine the percentage change in e_{ss} for a 5 per cent change in the value of (i) K, (ii) p, and (iii) α.

Ans. (i) -0.01; (ii) 0.01; (iii) 0.007

3.3 The transfer function of a unity-feedback system is

$$\frac{K(s+2)}{s(s+1)(1+sT)}$$

Determine the relative change in $\Delta C/C$ output for (a) the relative change in time constant $\Delta T/T$ and (b) the relative change in the forward path gain $\Delta K/K$ of the system

Ans. (a) $\dfrac{Ts^2(s+1)}{s(s+1)(1+sT)+K(s+2)} \dfrac{\Delta T}{T}$; (b) $\dfrac{s(s+1)(1+sT)}{s(s+1)(1+sT)+K(s+2)} \dfrac{\Delta K}{K}$

3.4 For the closed-loop control system shown in Figure P.3.4: (i) determine the sensitivity of the closed-loop system transfer function to a small change in the value of K where the normal value of K is assumed to be 400. Assume $H(s) = 1$. (ii) Assume $H(s) = \dfrac{1}{(1+0.1s)}$ and repeat (i).

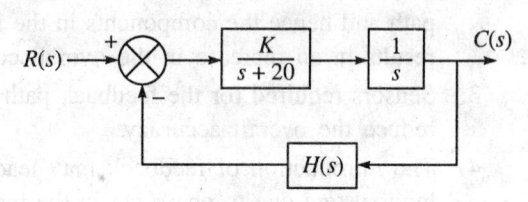

FIGURE P.3.4

3.5 The block diagrams of open- and closed-loop systems shown in Figures P.3.5(a) and (b) have nominal value of $K = 1$. Do the followings:

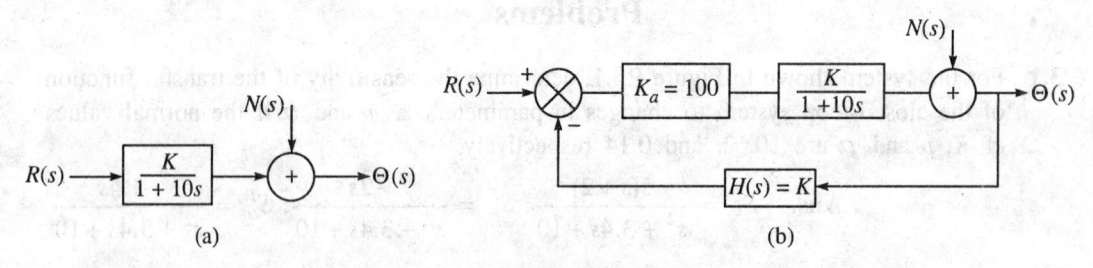

FIGURE P.3.5

(i) Given $r(t) = $ unit step input, noise disturbance $n(t) = 0$, determine the output $\theta(t)$ and sketch it with respect to time. Also determine the value of output after one time constant.

(ii) Determine the per cent change in output for both the open-loop and closed-loop cases for a 5 per cent change in K where the nominal value of $K = 1$.

(iii) Determine and sketch the output $\theta(t)$ subject to a disturbance $n(t) = $ step input of size 0.1 after the output $\theta(t)$ has reached the steady-state value.

3.6 The transfer function of the forward path of a unity-feedback system is

$$G(s) = \frac{K(s+2)}{s^2(s+4)}$$

(a) For $K = 8$, determine the steady-state error when the input to the system is

$$r(t) = 4 - t - 2t^2$$

(b) What will be the change in the steady-state error if there is a 5 per cent change in the value of K?

(c) What will be the change in the steady-state error if there is a 5 per cent change in the location of the zero at $s = -2$?

(d) What will be the change in the steady-state error if there is a 5 per cent change in the location of the pole at $s = -4$?

3.7 The forward path transfer function of a position control system with velocity feedback is given by

$$G(s) = \frac{K}{s(s+p)}$$

The transfer function of the feedback path is

$$H(s) = 1 + \alpha s$$

Determine the sensitivity of the transfer function of the closed-loop system to changes in K, p, and α if the nominal values are $K = 10$, $p = 2$, and $\alpha = 0.14$.

3.8 What will be the percentage change in the steady-state error of the system in Problem 3.7 for a 5 per cent change in the value of (a) K, (b) p, and (c) α, if the input is a unit-ramp function?

Transient Response Analysis

4.1 Introduction

The time response of a control system is usually divided into two parts: the transient response and the steady-state response. If $c(t)$ is a time response, then, in general,

$$c(t) = c_t(t) + c_{ss}(t) \qquad (4.1)$$

where $c_t(t)$ denotes the transient portion and $c_{ss}(t)$ the steady-state portion of the response.

In control systems the steady-state response is simply the response when time reaches infinity. Therefore, a sine wave is considered a steady-state because its behaviour does not alter at $t = \infty$.

Transient response is defined as that part of the response which goes to zero as time becomes large. Therefore, $c_t(t)$ has the property

$$\lim_{t \to \infty} c_t(t) = 0 \qquad (4.2)$$

In feedback control systems, since inertia and friction are unavoidable, the output response cannot instantaneously follow a sudden change in the input and therefore a transient is usually observed. However, it may be noted that a control system is basically a low-pass system. The steady-state in a feedback control system, when compared with input, gives an indication of the accuracy of system. If the steady-state output does not exactly agree with the input, the system is said to have a steady-state error.

OBJECTIVE

The basic objective of this chapter is to determine the time response of a control system in transient phase. Even if a system is stable in steady-state, it may have an intolerable overshoot in the transient state, causing the system to break down. It is therefore necessary to learn about the transient response in time domain so that design specification in time domain can be understood. The relationship between the time-domain and frequency-domain specification has been developed to better appreciate the design specifications. Steady-state error has been discussed. Suitable controller design through proper understanding of a Proportional + Integral + Derivative (PID) controller has been touched upon. In the design of the closed-loop control system, the role of root-locus technique has been highlighted. MATLAB approach to the solution of transient response problems has been touched upon. After studying this chapter, you will be proficient in the following topics.

CHAPTER OUTLINE

Introduction
First-Order Systems
Second-Order Systems
Step Response
Parameter Variation
Impulse Response
Ramp Response
Graphical Interpretation
Analysis from Pole-Zero Plot
Steady-State Error
Integral Performance Criterion
Derivative and Rate Feedback Control
Analog Computation

4.2 Typical Test Input Signals

In the time-domain analysis, the following test signals are often used:

(i) *Step displacement input (step function).* This is the instantaneous change in the reference input variable, for example, a sudden rotation of a shaft. The mathematical representation of a step function is

$$r(t) = R \quad \text{for} \quad t > 0, \ R \text{ is a constant}$$
$$= 0 \quad \text{for} \quad t < 0$$

or $\quad r(t) = Ru(t) \quad$ (4.3)

where $u(t)$ is the unit-step function. The function $r(t)$ is not defined at $t = 0$. The step function $r(t)$ is shown in Figure 4.1(a). The unit-step function or $u(t)$ has the property

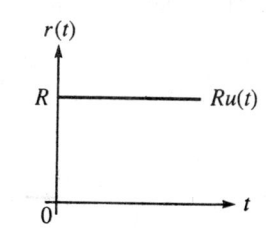

$$u(t) = 1 \quad \text{for} \quad t > 0$$
$$= 0 \quad \text{for} \quad t < 0$$

FIGURE 4.1(a) Step function.

(ii) *Velocity input (ramp function).* In this case, the reference-input variable has a constant change in position with respect to time. Mathematically, a ramp function is represented by

$$r(t) = Rt \quad \text{for} \quad t > 0$$
$$= 0 \quad \text{for} \quad t < 0 \quad (4.4)$$

or $\quad r(t) = Rtu(t) \quad$ (4.5)

The ramp function is shown in Figure 4.1(b).

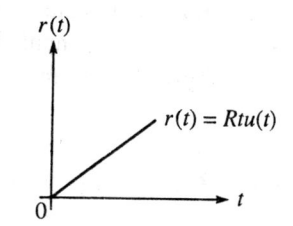

FIGURE 4.1(b) Ramp function.

(iii) *Acceleration input (parabolic function).* The mathematical representation of an acceleration input is

$$r(t) = Rt^2 \quad \text{for} \quad t > 0$$
$$= 0 \quad \text{for} \quad t < 0 \quad (4.6)$$

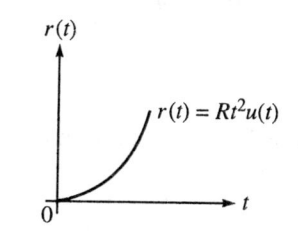

or $\quad r(t) = Rt^2 u(t) \quad$ (4.7)

FIGURE 4.1(c) Acceleration (parabolic) function.

The acceleration function is shown in Figure 4.1(c).

Under the above category of 'typical test input signals' we have not considered the sinusoidal input though it is very much a standard input signal. The sinusoidal input is discussed in the chapter on frequency response analysis. The obvious reason is that, for the sinusoidal input the output response of a linear time-invariant system remains sinusoidal but with magnitude scaled and a shift in phase. Hence for the sinusoidal input, we consider only the steady-state analysis where we substitute $s = j\omega$ in frequency response analysis.

4.3 First-Order Systems

Consider the first-order system shown in Figure 4.2. Physically, this system may represent an RC circuit, or a thermal system, and so forth. The input-output relationship is given by

$$\frac{C(s)}{R(s)} = \frac{1}{Ts+1} \qquad (4.8)$$

In the following, we shall analyze the system response to inputs such as unit-step, unit-ramp, and unit-impulse functions. The initial conditions will be assumed zero.

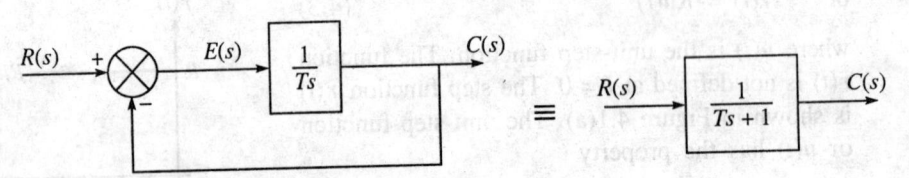

FIGURE 4.2 First-order system.

Note that all systems having the same transfer function will exhibit the same output in response to the same input. For any given physical system, the mathematical response can be given a physical interpretation.

4.3.1 Unit-step response of first-order systems

Since the Laplace transform of the unit-step function is $1/s$, substituting $R(s) = 1/s$ into Eq. (4.8), we get

$$C(s) = \frac{1}{s(Ts+1)} \qquad (4.9)$$

Expanding $C(s)$ into partial fractions,

$$C(s) = \frac{1}{s} - \frac{1}{Ts+1} = \frac{1}{s} - \frac{1}{s+(1/T)} \qquad (4.10)$$

Taking the inverse Laplace transform of Eq. (4.10),

$$c(t) = 1 - e^{-t/T}; \quad t \geq 0 \qquad (4.11)$$

The error,

$$e(t) = r(t) - c(t) = e^{-t/T} \qquad (4.12)$$

The steady-state error is

$$e_{ss} = \lim_{t \to \infty} e(t) = 0$$

Thus the system tracks the step input with zero steady-state error.

Equation (4.11) states that initially the output $c(t)$ is zero and finally it becomes unity. One of the important characteristics of such an exponential response curve $c(t)$ is that at $t = T$, the value of $c(t)$ is 0.632 or the response $c(t)$ has reached 63.2% of its total change. This may be easily seen by substituting $t = T$ in $c(t)$, namely

$$c(t) = 1 - e^{-T/T} = 1 - e^{-1} = 0.632$$

It is well known that T is the time constant of the system. The smaller the time constant, the faster is the system response.

Another important characteristic of the exponential response is that the slope or the tangent line at $t = 0$ is $1/T$ since

$$\left.\frac{dc}{dt}\right|_{t=0} = \left.\frac{1}{T}e^{-t/T}\right|_{t=0} = \frac{1}{T} \qquad (4.13)$$

The output would reach the final value at $t = T$ if it maintained its initial speed of response. From Eq. (4.11), we see that the slope of the response curve $c(t)$ decreases monotonically from $1/T$ at $t = 0$ to zero at $t = \infty$.

The exponential response curves $c(t)$ and $e(t)$ of Eqs. (4.11) and (4.12) are shown in Figure 4.3. In one time constant, the exponential curve has gone through changes from 0 to 63.2% of the final value. In two time constants, the response reaches 86.5% of the final value. At $t = 3T$, $4T$ and $5T$ the response reaches 95%, 98.2% and 99.3% respectively of the final value. Thus, for $t \geq 4T$, the response settles within 2% of the final value. As seen from Eq. (4.11), the steady-state is reached mathematically only after an infinite time. In practice, however, a reasonable estimate of the *settling time* is the length of time the response curve needs to reach the 2% line of the final value, i.e. four time constants. Similarly, another reasonable estimate of the settling time in vogue is the length of time the response curve $c(t)$ needs to reach the 5% line of the final value, i.e. three time constants.

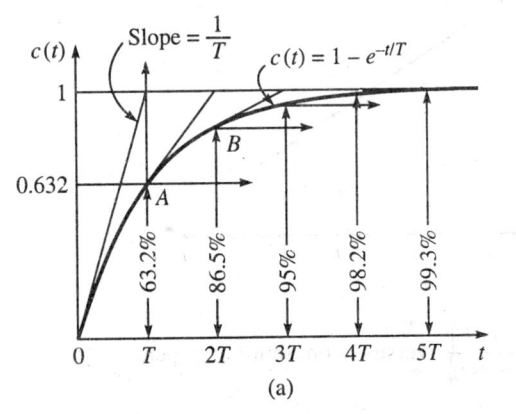

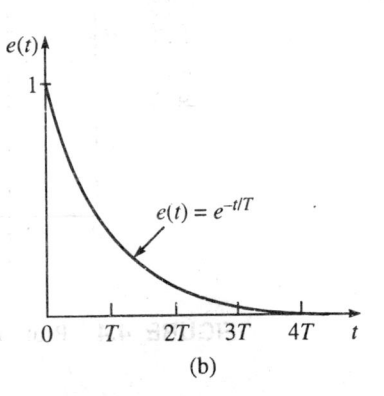

FIGURE 4.3 First-order system: (a) output curve and (b) error curve.

Further, another important parameter of system response is *delay time* which is a measure of the response time to reach 50% of the steady-state or final value of the response.

In a similar way, another important estimate of system response is the *rise time* which is defined as the time duration of the response to reach from 10% to 90% of the steady-state value. Usually, this duration of time is used to define rise time in case of monotonic response (quite obvious in case of first-order system or for the system having only real poles) whereas for the underdamped second-order system it is the time to reach from 0% to 100% of the steady-state response.

Consider the system whose order has to be determined experimentally. In order to determine experimentally whether or not the system is of first-order, plot the curve log $[c(t) - c(\infty)]$ as function of t. If the curve turns out to be a straight line, the system is of the first order. The time constant T can be read from the graph as the time T which satisfies the following equation

$$c(T) - c(\infty) = 0.368[c(0) - c(\infty)]$$

But this method is cumbersome. Note that instead of plotting log $\left[\dfrac{|c(t) - c(\infty)|}{|c(0) - c(\infty)|}\right]$ versus t on ordinary graph paper, it is much more convenient to plot $\dfrac{|c(t) - c(\infty)|}{|c(0) - c(\infty)|}$ versus t, on semilog paper. Refer to Figure 4.4.

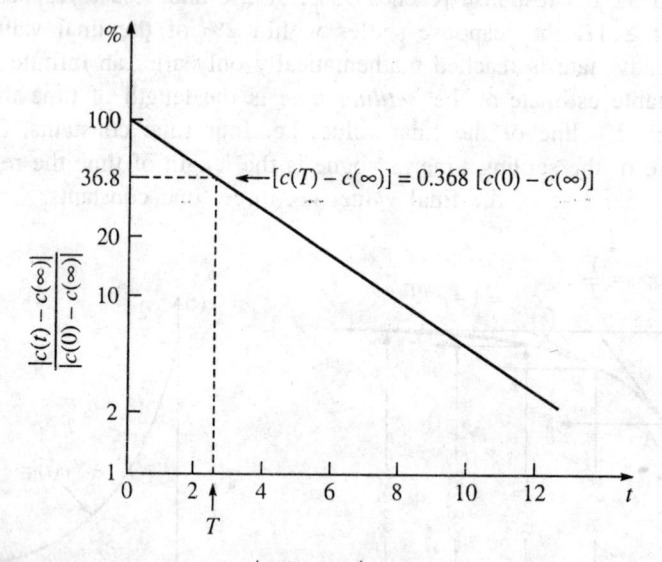

FIGURE 4.4 Plot of $\dfrac{|c(t) - c(\infty)|}{|c(0) - c(\infty)|}$ versus t on semilog paper.

Liquid-level control system

The closed-loop control system of a first-order liquid-level control is shown in Figure 4.5(a). It is a proportional controller just like the cistern in the toilet of our homes. The sensor is the float. Assume that:

$\overline{H}$ is the steady-state head (height), in m

$\overline{Q}$ is the steady-state volumetric flow rate, in m^3/s; Q_i is referred to incoming and Q_o to outgoing flow rates

$\overline{R}$ is the steady-state value of valve resistance, in S/m^2.

The resistance to liquid flow or the restriction is defined as the change in the level difference necessary to cause a unit change in flow rate, that is,

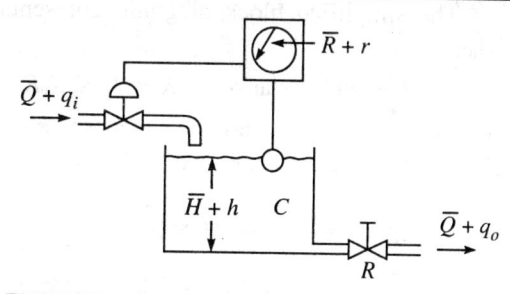

FIGURE 4.5(a) Liquid-level control system.

$$R = \frac{\text{change in level difference, in m}}{\text{change in flow rate, in m}^3/\text{s}}$$

The small deviations r, q_i, h and q_o are (assumed to be sufficiently small so that the system can be approximated by a linear mathematical model) from their respective steady-state values $\overline{R}$, $\overline{Q}_i$, $\overline{H}$, and $\overline{Q}_o$. Assuming the controller to be a proportional one, inflow q_i is proportional to the actuating error signal e, that is,

$$q_i = K_p K_v e$$

where

K_p is the gain of the controller
K_v is the gain of the control valve
K_b is the gain of the float in the feedback path.

The block diagram representation of the system is shown in Figure 4.5(b).

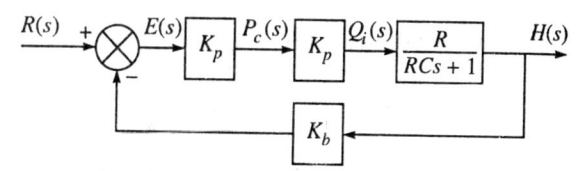

FIGURE 4.5(b) Block diagram of liquid level control system.

The closed-loop transfer function can be written as

$$\frac{H(s)}{R(s)} = \frac{\dfrac{K_p K_v R}{RCs+1}}{1+\dfrac{K_b K_p K_v R}{RCs+1}} = \frac{K_p K_v R}{RCs+1+K_b K_p K_v R}$$

In order to realize the closed-loop system of Figure 4.5(b) as unity-feedback system, we rewrite the above expression using the block diagram algebra as

$$\frac{X(s)}{R(s)} \times \frac{H(s)}{X(s)} = \frac{1}{K_b} \times \frac{K_b K_v K_b R}{RCs+1+K_b K_p K_v R}$$

or

$$\frac{X(s)}{R(s)} \times \frac{H(s)}{X(s)} = \frac{1}{K_b} \times \frac{K}{Ts+1+K}$$

where $X(s)$ is a dummy variable.

The simplified block diagram representation of Figure 4.5(b) is shown in Figure 4.5(c), where

$$T = RC \quad \text{and} \quad K = K_b K_p K_v R$$

Now our task is to find the response $h(t)$ to a change in reference input. We shall assume a unit-step change in scaled down input $x(t) = \dfrac{1}{K_b} r(t)$.

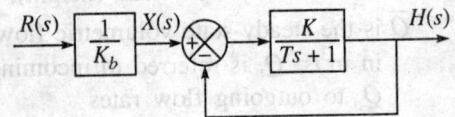

FIGURE 4.5(c) Simplified block diagram.

Unit-step response of the closed-loop transfer function between $H(s)$ and $X(s)$ is

$$h(t) = \mathscr{L}^{-1}[H(s)] = \mathscr{L}^{-1}\left[\frac{K}{Ts+1+K} \times \frac{1}{s}\right] = \frac{K}{1+K}(1-e^{-t/T_1}) \quad \text{for } t \geq 0$$

where
$$T_1 = \frac{T}{1+K}$$

The response curve $h(t)$ is shown in Figure 4.5(d). The steady-state value of $h(t)$ is

$$h(\infty) = \frac{K}{1+K}$$

The steady-state error is

$$e_{ss} = 1 - \frac{K}{1+K} = \frac{1}{1+K}$$

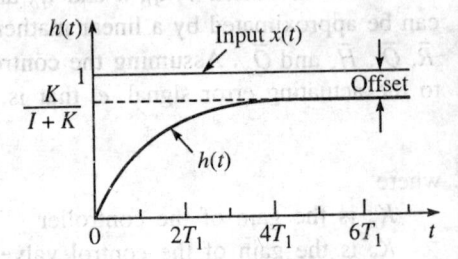

FIGURE 4.5(d) Curve of $h(t)$ versus t.

This error is called *offset*. The value of the offset is reduced as the gain K becomes larger.

Parameter variation for first-order system

The response of the first-order transfer function $1/(1 + Ts)$ subjected to a unit-step input for various values of time constant T is computed by MATLAB program. It is obvious from the step response curve of the MATLAB Program 4.1 that as the value of T increases, the delay time, the rise time and the settling time of the step response all increase.

MATLAB Program 4.1

```
% Calculation of Td, Tr, Ts
% for varying time constant

%program1.m
clear;
pack;
clc;
j=1;
while(j<6)
```

Transient Response Analysis

```
    T(j)=input('Enter value of "T": ');
       num=[1];
       den=[T(j) 1];
t=0:.0001:14;
[y1,x1,T1]=step(num,den,t);
i=1;
% calculating delay time,Td
while y1(i)<.5,
  i=i+1;
end
Td(j)=T1(i);
i=1;
while y1(i)<.1,
  i=i+1;
end
ten=T1(i);
% calculating rise time,Tr
  while y1(i)<.90,
  i=i+1;
end
ninety=T1(i);
Tr(j)=ninety-ten;
i=1;
% calculating settling time,T
  while y1(i)<.95,
  i=i+1;
end
Ts(j)=T1(i);
step(num,den,t);
axis([0 14 0 1.2]);
title('Unit step response of a first-order system for varying time constant');
pop1=sprintf('%g',T(j));
text(T1(10000),y1(10000),pop1);
j=j+1;
hold on;
end
j=1;
while(j<6)
  pop=sprintf('%6g %6g %6g %6g',T(j),Td(j),Tr(j),Ts(j));
text(6,.6,' T    td    tr    ts');
text(6,.6–(.1*j),pop);
j=j+1;
end

Enter Value of "T": 3
Enter Value of "T": 2
Enter Value of "T": 1
Enter Value of "T": .5
Enter Value of "T": .25
```

T	td	tr	ts
3	2.0795	6.5917	8.9872
2	1.3863	4.3944	5.9915
1	0.6932	2.1972	2.9958
0.5	0.3466	1.0986	1.4979
0.25	0.1733	0.5493	0.749

4.3.2 Unit-ramp response of first-order systems

Since the Laplace transform of the unit-ramp function is $1/s^2$, we obtain the output of the system of Figure 4.2 as

$$C(s) = \frac{1}{Ts+1}\frac{1}{s^2} \qquad (4.14)$$

Expanding $C(s)$ into partial fractions,

$$C(s) = \frac{1}{s^2} - \frac{T}{s} + \frac{T^2}{Ts+1} \qquad (4.15)$$

Taking the inverse Laplace transform of Eq. (4.15),

$$c(t) = t - T + Te^{-t/T} \; ; \quad t \geq 0 \qquad (4.16)$$

The error,

$$e(t) = r(t) - c(t) = T(1 - e^{-t/T}) \qquad (4.17)$$

As t approaches infinity, $e^{-t/T}$ approaches zero, and thus the error signal $e(t)$ approaches T, that is, $e(\infty) = T$. The unit-ramp input and system output are shown in Figure 4.6. The error, in following the unit-ramp input is equal to T for sufficiently large time t. The smaller the time constant T, the smaller is the steady-state error in following the ramp input.

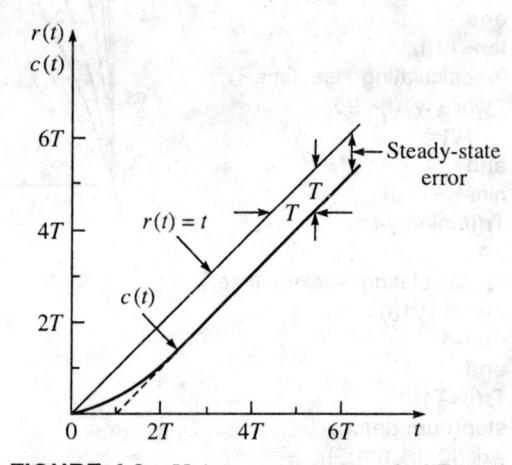

FIGURE 4.6 Unit-ramp response of $1/(Ts + 1)$.

4.3.3 Unit-impulse response of first-order systems

For unit impulse, that is, $r(t) = \delta(t)$, the Laplace transform is $R(s) = \mathscr{L}[\delta(t)] = 1$, then we can write, as usual

$$\frac{C(s)}{R(s)} = \frac{1}{Ts+1}$$

For unit-impulse input, the output response

$$c(t) = \mathscr{L}^{-1}[C(s)] = \mathscr{L}^{-1}\left[\frac{R(s)}{Ts+1}\right] = \mathscr{L}^{-1}\left[\frac{1}{Ts+1}\right] = \frac{1}{T}e^{-t/T} \; ; \quad t \geq 0 \qquad (4.18)$$

is shown in Figure 4.7.

We can conclude from Eqs (4.11), (4.16) and (4.18) and that the unit-impulse response is the derivative of the unit-step response and this can be seen from the unit-impulse input which is the derivative of the unit-step input. Comparison of the system response to these three inputs

Transient Response Analysis

clearly indicates that the response to the derivative of an input signal can be obtained by differentiating the response of the system to the original signal. It can also be seen that the response to the integral of the original signal can be obtained by integrating the response of the system to the original signal and by determining the integration constants from the zero output initial condition.

Drill Problem 4.1

Determine the time constant of the closed-loop system of Figure 4.8 for two cases with (a) $K_1 = 5$, $K_2 = 2$ and (b) $K_1 = 2$, $K_2 = 5$.

Ans. 1.6

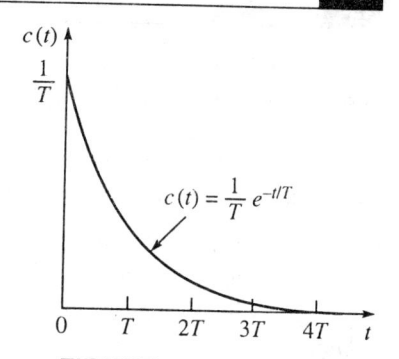

FIGURE 4.7 Unit-impulse response of $1/(Ts + 1)$.

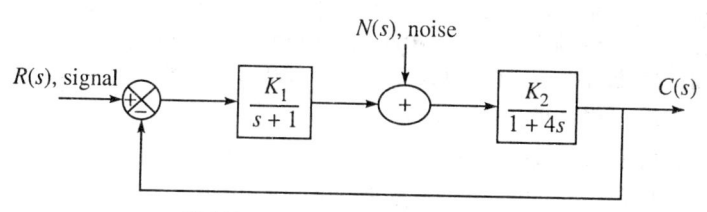

FIGURE 4.8 Drill Problem 4.5

Drill Problem 4.2

Refer to Figure 4.8. Determine the reduction in noise for both cases (a) and (b) at dc.

Ans. (a) 5 (b) 2

Drill Problem 4.3

Refer to Figure 4.8. Determine the frequency response of the output with respect to the input signal and that of the output with respect to the disturbance, both at $\omega = 0.5$ rad/s.

Ans. (a) $\frac{C}{R}(j0.5) = 0.97 \exp(-j0.244)$; $\frac{C}{N}(j0.5) = 0.217 \exp(j0.219)$.

(b) $\frac{C}{R}(j0.5) = 0.97 \exp(-j0.244)$; $\frac{C}{N}(j0.5) = 0.542 \exp(j0.219)$.

4.4 Analysis of Transient Performance from Root Pattern

The output response of a system, represented by a linear differential equation, subjected to a step input can be written as

$$c(t) = A_0 + A_1 e^{r_1 t} + A_2 e^{r_2 t} + A_3 e^{r_3 t} + \cdots$$

For a stable system the values of all r_is are negative real or complex conjugate with negative real part. Thus as t increases, all of the exponential terms decrease.

Case I. For all real roots and if $|r_1| < |R_2| < |r_3| \ldots$ and $A_1 > A_2 > A_3 > \ldots$, which is usually true, then as time t increases the higher numbered terms approach zero and their contribution to the response $c(t)$ becomes insignificant; essentially all of the response is due to one or two (r_1, r_2) of the roots and these roots are said to be dominant. That is, the real parts of the dominant roots are very small and they are close to origin so far as real roots are concerned.

Case II. For a pair of complex roots having negative real parts close to origin, it is almost always true that the contribution from all other roots having negative real parts away from the origin will disappear quickly. Then the complex pole pair near the origin is said to be *dominant*.

With the concept of dominant poles in higher-order models, it is reasonably clear, just from these considerations, that the transient performance of most systems depends largely on the dominant complex conjugate poles. Here we will choose an armature-controlled dc servomotor as an example of a second-order system for analyzing its transient response.

Dominant-pole analysis requires that the transient contribution from the non-dominant system poles be small. This implies:

(i) The non-dominant poles are well to the left of the dominant pole(s), so that the corresponding transients die away rapidly.

(ii) Any pole near the dominant pole(s) is close to a zero, so that the magnitude of its transient response will be very small.

The visual inspection of the pole-zero maps of Figure 4.9 shows two typical dominant-pole systems—one having a dominant single pole and the other having a dominant pair of complex

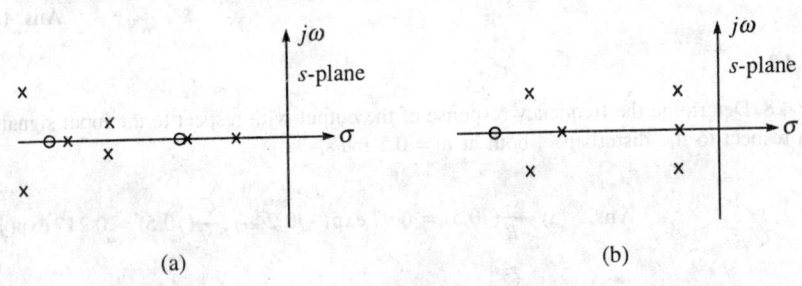

FIGURE 4.9 Typical dominant-pole systems: (a) a single dominant pole and (b) a dominant pair of complex conjugate poles.

conjugate poles, and correspondingly the time-domain responses of systems having similar pole-zero maps appear predominantly those of first- or second-order systems. This is illustrated with the following example.

EXAMPLE 4.1 The following transfer function is subjected to unit-step input:

$$\frac{C(s)}{U(s)} = \frac{2810(s+4)}{(s+3.8)(s+6)(s^2+2s+17)(s^2+10s+29)}$$

The pole-zero map is shown in Figure 4.10(a).

Taking inverse Laplace transform, the time domain expression is

$$c(t) = 1 + e^{-t}(0.622 \cos 4t - 0.571 \sin 4t) - 0.518 e^{-3.8t}$$
$$+ e^{-5t}(0.973 \cos 2t - 3.33 \sin 2t) - 2.08 e^{-6t}$$

The unit-step response for the full system is shown in Figure 4.10(b). The dominant poles are $-1 \pm j4$ and the response of the second-order dominant underdamped system becomes

$$c(t) = 1 + e^{-t}(0.622 \cos 4t - 0.571 \sin 4t)$$

The dominant system response is also shown in Figure 4.10(b). Note that for small values of t the two responses are distinguishable, but as time increases the two responses become indistinguishable.

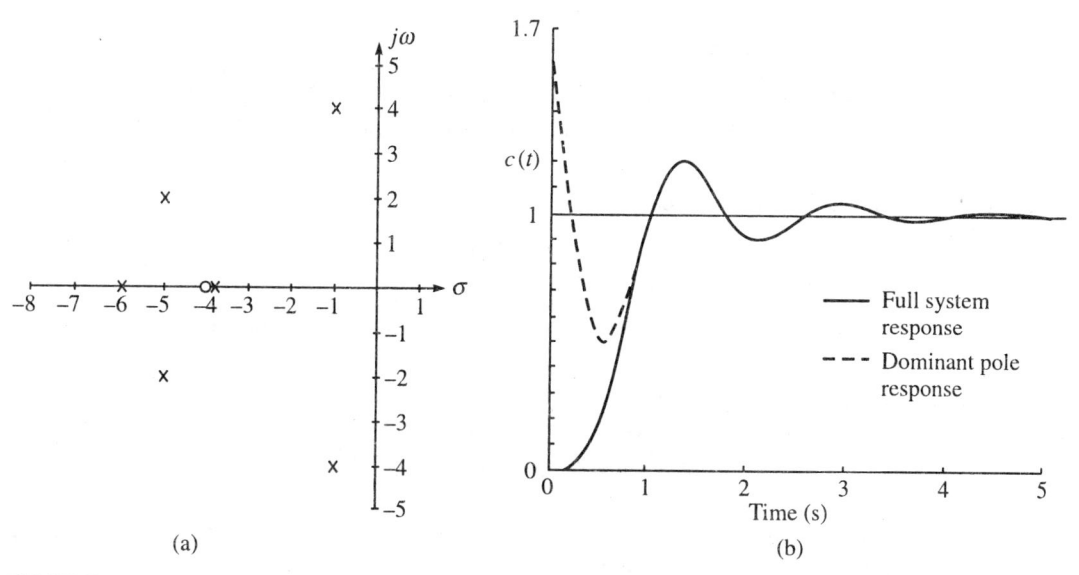

FIGURE 4.10 Example 4.1: (a) pole-zero map and (b) step response for the full system and the dominant system.

4.5 Second-Order Position Control System

Now we will find the response of a second-order control system subjected to a step input. Here we take a practical position control system comprising mechanical loading with gear ratio, the potentiometric error detector, an amplifier and an armature-controlled dc servomotor. This system is illustrated in Figure 4.11(a).

For constant field current, the torque developed by the motor is

$$T = K_2 i_a$$

where K_2 is the motor torque constant and i_a the armature current.

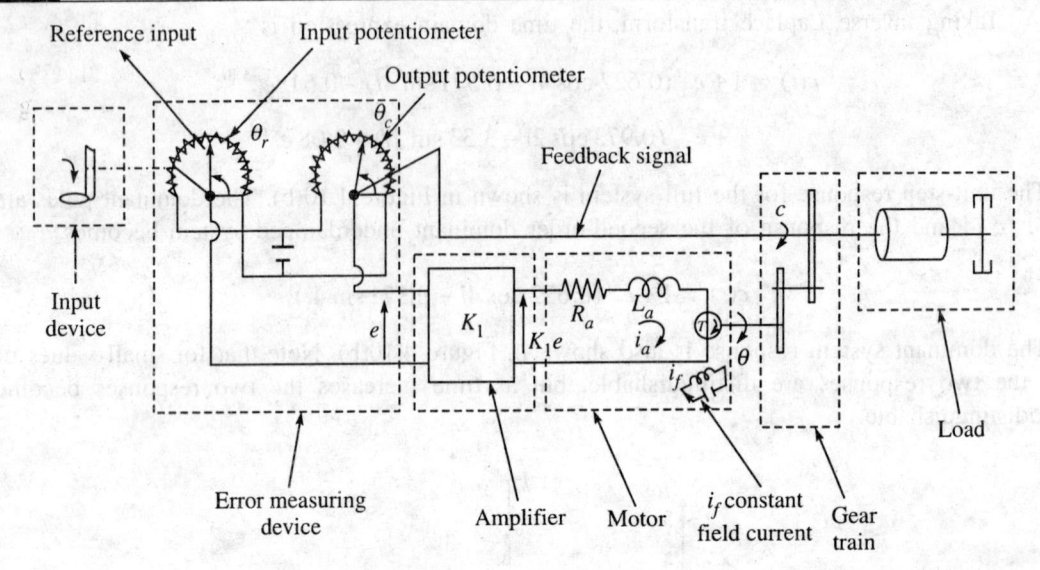

FIGURE 4.11(a) Second-order position control system.

For the armature circuit, we get

$$L_a \frac{di_a}{dt} + R_a i_a + K_3 \frac{d\theta}{dt} = K_1 e$$

where K_3 is the back emf constant of the motor, and θ the angular displacement of the motor shaft.

The torque-equilibrium equation is

$$J_\theta \frac{d^2\theta}{dt^2} + f_\theta \frac{d\theta}{dt} = T = K_2 i_a$$

where J_θ and f_θ are the inertia and viscous friction respectively of the combination of motor, load and gear ratio referred to the motor shaft.

Taking the Laplace transform of the ratio of motor shaft displacement $\Theta(s)$ to the error signal $E(s)$, the forward transfer function is

$$G(s) = \frac{\Theta(s)}{E(s)} = \frac{K_1 K_2}{s(L_a s + R_a)(J_\theta s + f_\theta) + K_2 K_3}$$

Again for the gear ratio n, we get

$$C(s) = n\Theta(s)$$

Further,

$$E(s) = K_0[R(s) - C(s)]$$

Therefore,

$$G(s) = \frac{K_0 K_1 K_2 n}{s(L_a s + R_a)(J_\theta s + f_\theta) + K_2 K_3 s} \qquad (4.19)$$

Since L_a is usually small and can be neglected, then the forward transfer function $G(s)$ becomes

Transient Response Analysis

$$G(s) = \frac{K_0 K_1 K_2 n}{s[R_a(J_\theta s + f_\theta) + K_2 K_3]} = \frac{K_0 K_1 K_2 n/R_a}{J_\theta s^2 + (f_\theta + K_2 K_3/R_a)s} \quad (4.20)$$

Let $J = J_\theta/n^2$ = moment of inertia referred to the output shaft

$$f = \frac{f_\theta + (K_2 K_3/R_a)}{n^2} = \text{viscous friction coefficient referred to the output shaft}$$

and $K = K_0 K_1 K_2 / n R_a$

Then the forward transfer function

$$G(s) = \frac{K}{Js^2 + fs} = \frac{K}{s(Js + f)} \quad (4.21)$$

The block diagram of the system is shown in Figure 4.11(b). This is a type-1 system.

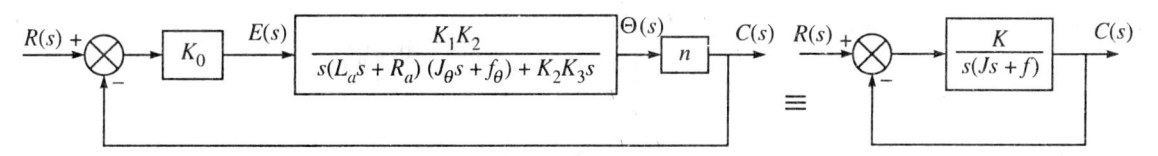

FIGURE 4.11(b) Block diagram of the system in Figure 4.11(a).

4.6 Unit-Step Response of Second-Order Systems

Consider the position-controlled servomechanism shown in Figure 4.11(b) where $G(s) = \frac{K}{s(Js + f)}$ is the forward transfer function of an armature-controlled dc servomotor. The closed-loop transfer function is

$$\frac{C(s)}{R(s)} = \frac{K}{Js^2 + fs + K} = \frac{K/J}{s^2 + (f/J)s + (K/J)} \quad (4.22)$$

In transient response analysis, it is convenient to write

$$\frac{K}{J} = \omega_n^2, \quad \frac{f}{J} = 2\zeta\omega_n$$

where ω_n is the undamped natural frequency and ζ the damping ratio of the system. With this notation, Figure 4.11(b) can be modified to that of Figure 4.12. Then the closed-loop transfer function is

$$\frac{C(s)}{R(s)} = \frac{\omega_n^2}{s^2 + 2\zeta\omega_n s + \omega_n^2}$$

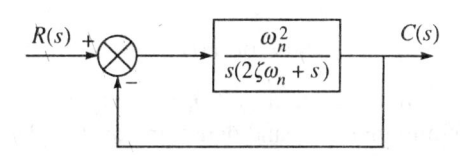

Figure 4.12 Generalized block diagram of the system in Figure 4.11(b).

$$= \frac{\omega_n^2}{(s+\zeta\omega_n+j\omega_d)(s+\zeta\omega_n-j\omega_d)}$$

where $\quad \omega_d = \omega_n\sqrt{1-\zeta^2} \quad (0 < \zeta < 1)$ (4.23)

is called the *damped natural frequency*.

For unit-step input, $R(s) = 1/s$, therefore

$$C(s) = \frac{\omega_n^2}{s(s^2+2\zeta\omega_n s+\omega_n^2)} = \frac{1}{s} - \frac{s+2\zeta\omega_n}{(s+\zeta\omega_n)^2+\omega_d^2} - \frac{\zeta\omega_n}{(s+\zeta\omega_n)^2+\omega_d^2} \quad (4.24)$$

Taking the inverse Laplace transform, we get the time-domain solution as

$$c(t) = 1 - e^{-\zeta\omega_n t}\cos\omega_d t - \frac{\zeta}{\sqrt{1-\zeta^2}}e^{-\zeta\omega_n t}\sin\omega_d t$$

$$= 1 - e^{-\zeta\omega_n t}\left[\cos\omega_d t + \frac{\zeta}{\sqrt{1-\zeta^2}}\sin\omega_d t\right] \quad (4.25)$$

$$= 1 - \frac{e^{-\zeta\omega_n t}}{\sqrt{1-\zeta^2}}\left[\frac{\omega_n\sqrt{1-\zeta^2}}{\omega_n}\cos\omega_d t + \frac{\zeta\omega_n}{\omega_n}\sin\omega_d t\right]$$

$$= 1 - \frac{e^{-\zeta\omega_n t}}{\sqrt{1-\zeta^2}}(\sin\theta\cos\omega_d t + \cos\theta\sin\omega_d t)$$

$$= 1 - \frac{e^{-\zeta\omega_n t}}{\sqrt{1-\zeta^2}}\sin(\omega_d t + \theta) \quad (4.26)$$

where $\quad \theta = \tan^{-1}\frac{\sqrt{1-\zeta^2}}{\zeta}$

The output expression of Eq. (4.26) is depicted in Figure 4.13(a) and the phase shift θ is illustrated in Figure 4.13(b).

It can be seen that the frequency of transient oscillation is the damped frequency ω_d as in Eq. (4.23) and thus varies with the damping ratio ζ. In the range $0 < \zeta < 1$, with decreasing damping ratio ζ, the frequency of oscillation increases as is evident from Figure 4.14. The steady-state value of $c(t)$ is given by $c_{ss}(t) = \lim_{t\to\infty} c(t) = 1$.

In expression (4.26), the magnitudes of sinusoidal terms have the damping ratio term ζ in both numerator and denominator. For decreasing value of ζ in the range $0 < \zeta < 1$, the negative exponential term in the numerator is more predominant than the denominator term $\sqrt{1-\zeta^2}$. Hence with decreasing value of ζ in the range $0 < \zeta < 1$, the swings of the output expression

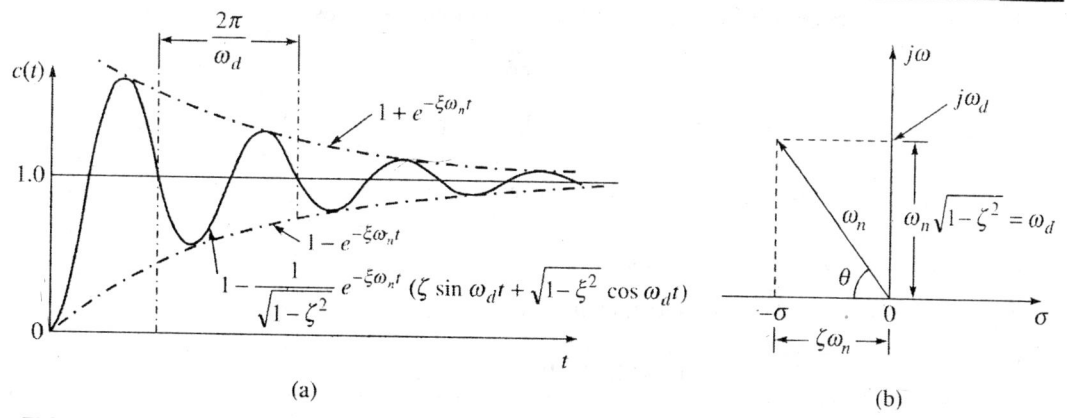

FIGURE 4.13 (a) Response of second-order system to unit-step input and (b) phase shift θ.

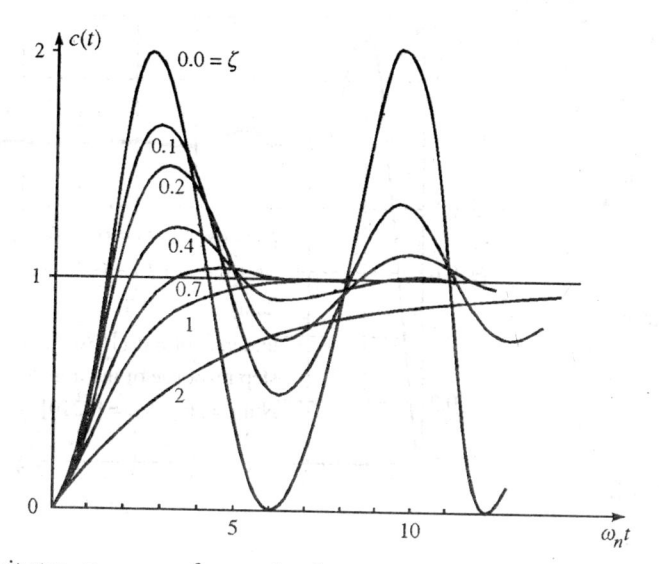

FIGURE 4.14 Unit-step response of second-order system with varying values of damping ratio ζ.

$c(t)$ in Eq. (4.26) are increasingly larger as shown in Figure 4.14.

The error signal is

$$e(t) = r(t) - c(t) = \frac{e^{-\zeta\omega_n t}}{\sqrt{1-\zeta^2}} \sin(\omega_d t + \theta) \qquad (4.27)$$

The error signal exhibits a damped sinusoidal oscillation as shown in Figure 4.15.

At steady-state, i.e. at $t = \infty$, the error becomes zero. Thus far we have discussed the underdamped case where the damping ratio is in the range $0 < \zeta < 1$.

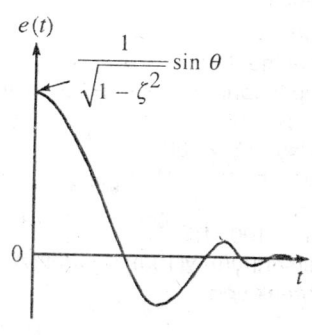

FIGURE 4.15 Error signal.

For $\zeta = 0$, the response is undamped and the output becomes

$$c(t) = 1 - \cos \omega_n t \qquad (4.28)$$

The output gives sustained oscillations, as shown in Figure 4.14, of undamped natural frequency ω_n. It may be observed that with damping factor ζ increasing in the range $0 < \zeta < 1$, the damped frequency of oscillation ω_d decreases, and obviously $\omega_n > \omega_d$. Note that for $0 < \zeta < 1$, the roots become complex conjugate and the response gives oscillations. For $\zeta > 1$, the oscillations do not exist, the response becomes overdamped and the roots become real.

MATLAB program **TransientResponse.m** has been made as follows:

MATLAB Program 4.2

```
% TransientResponse.m
clear;
pack;
clc;
num=[10];
den=[1 2 10];
t=.1:0.1:10;
[y,x,t]=step(num,den,t);
for i=2:1:100
if y(i)<y(i-1)
break,end
end
overshoot=y(i)-y(100);
peaky=y(i);
peaktime=t(i);
while y(i)<y(i-1)
  i=i+1;
end
undershoot=y(100)-y(i);
undery=y(i);
undertime=t(i);
pop4=sprintf('undershoot=%g
at time %g seconds',undershoot,undertime);
i=1;
while y(i)<y(100)
  i=i+1;
end
risetime=t(i);
pop2=sprintf('risetime=%g',risetime);
slope=100;
delay=0.5*y(100);
epsilon=.000001;
i=2;
for i=100:-1:2
  if(sqrt((y(i)-y(100))^2)>delay)
  break,end
```

Step Response plot annotations:
- peak overshoot = 0.346199 at time 1.1
- risetime = 0.7
- delay time = 0.3
- undershoot = 0.116803 at time 2.2 seconds
- step response of the transfer function with Num = [10] den = [1 2 10]

```
  slope=(y(i)-y(i-1))/(t(i)-t(i-1));
end
delaytime=t(i);
pop3=sprintf('delay time=%g',delaytime);
pop1=sprintf('peak overshoot=%g
at time %g',overshoot,peaktime);
step(num,den,t);
gtext(pop1);
gtext(pop2);
gtext(pop3);
gtext(pop4);
gtext('step response of the
transfer function with');
gtext('Num=[10] den=[1 2 10]')
grid on;
```

Critically-damped case ($\zeta = 1$)

In this case, repetitive roots occur. For unit-step input, $R(s) = 1/s$, then $c(t)$ can be written as

$$c(t) = \mathcal{L}^{-1} C(s) = \mathcal{L}^{-1} \frac{\omega_n^2}{s(s+\omega_n)^2} \qquad (4.29)$$

$$= 1 - e^{-\omega_n t}(1 + \omega_n t)$$

Refer Figure 4.13(a).

Overdamped case ($\zeta > 1$)

In this case two poles of the closed-loop transfer function $C(s)/R(s)$ are distinct real negative. For unit-step input, $R(s) = 1/s$, $C(s)$ becomes

$$C(s) = \frac{\omega_n^2}{s(s+\zeta\omega_n + \omega_n\sqrt{\zeta^2-1})(s+\zeta\omega_n - \omega_n\sqrt{\zeta^2-1})}$$

or

$$c(t) = 1 + \frac{\omega_n}{2\sqrt{\zeta^2-1}}\left(\frac{e^{-s_1 t}}{s_1} - \frac{e^{-s_2 t}}{s_2}\right) \qquad (4.30)$$

where roots $s_1, s_2 = \omega_n(\zeta \pm \sqrt{\zeta^2-1})$.

The response $c(t)$ becomes non-oscillatory and includes two exponentially decaying terms as shown in Figure 4.13(a).

We have thus seen that the dynamic behaviour of a second-order system can be described in terms of two parameters, namely, ζ and ω_n. If $0 < \zeta < 1$, the closed-loop poles are complex conjugate and lie in the left-half of the s-plane. The system is underdamped and the response is oscillatory. If $\zeta = 1$, the system is critically damped. An overdamped system corresponds to $\zeta > 1$. The transient responses of critically damped and overdamped systems do not oscillate. If $\zeta = 0$, the transient response does not die out and gives sustained oscillations.

The roots of the characteristic equation vary widely for different values of the damping ratio ζ.

For $0 < \zeta < 1$: $s_1, s_2 = -\zeta\omega_n \pm j\omega_n\sqrt{1-\zeta^2}$ (Re $s_1, s_2 < 0$)
(complex conjugate roots in underdamped case)

For $\zeta = 1$: $s_1, s_2 = -\zeta\omega_n$
(negative real repetitive roots in critically-damped case)

For $\zeta > 1$: $s_1, s_2 = -\zeta\omega_n \pm \omega_n\sqrt{1-\zeta^2}$ (Re $s_1, s_2 < 0$)
(real negative roots in overdamped case)

For $\zeta = 0$: $s_1, s_2 = \pm j\omega_n$
(sustained oscillations)

For $\zeta < 0$: $s_1, s_2 = -\zeta\omega_n \pm j\omega_n\sqrt{1-\zeta^2}$ (Re $s_1, s_2 > 0$)
(positive roots with negative damping ratio, and the response diverges out)

Referring to Figure 4.11(b), we can say that if the value of motor gain constant K is increased, it amounts to increase in ω_n; this in turn reduces the damping factor ζ; the response will be faster as ω_d increases. The magnitude of response is more, and the stability reduces. Instead, if the viscous friction coefficient f is reduced, effectively ζ is reduced and the response becomes faster. If f is increased, ζ is increased and the response becomes sluggish.

The variation in design parameters has the effect on performance indices. Quantitatively, the performance indices are related to:

(i) How fast the system moves to follow the input (rise time)?

(ii) How long does it take to reach the final value (settling time)?

(iii) How oscillatory the system is (indicative of the damping factor)?

Performance indices

The performance indices for specifying the transient response of an underdamped system to a unit-step input are the following with reference to Figure 4.16:

(a) *Delay time, t_d:* It is the time required for the response to reach 50% of the final value the very first time.

(b) *Rise time, t_r:* It is the time required for the response to rise from 10% to 90% of the final value in cases of overdamped or critically-damped second-order systems, or 0 to 100% of the final value in case of the underdamped second-order system.

(c) *Peak time, t_p:* It is the time required for the response to reach the first peak of the overshoot.

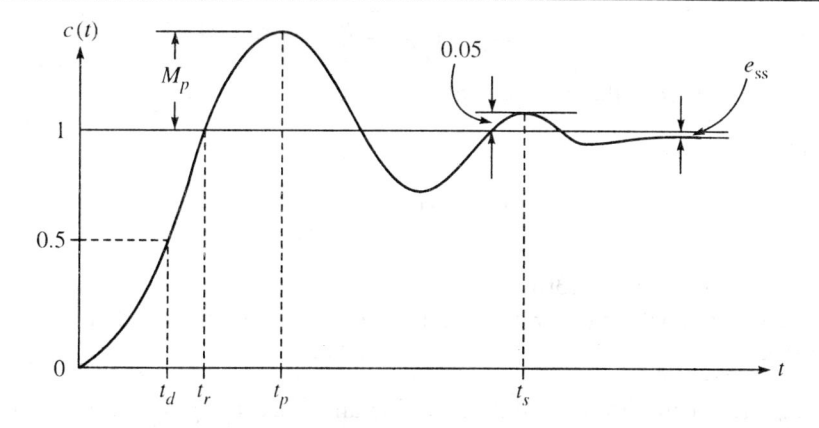

FIGURE 4.16 Unit-step response curve showing performance indices.

(d) *Maximum overshoot, M_p:* It is the normalized difference between the maximum peak value and the final steady-state output and is expressed in per cent value. It is defined as

$$M_p, \text{ maximum per cent overshoot} = \frac{c(t_p) - c(\infty)}{c(\infty)} \times 100\% \tag{4.31}$$

(e) *Settling time, t_s:* The time required for the response to reach and stay within a specified tolerance band (usually 2% to 5%) of its final value.

Derivation of performance indices

Rise time, t_r. Referring to Eq. (4.25), substituting $t = t_r$, we obtain

$$c(t_r) = 1 = 1 - e^{-\zeta\omega_n t_r}\left[\cos\omega_d t_r + \frac{\zeta}{\sqrt{1-\zeta^2}}\sin\omega_d t_r\right]$$

or

$$e^{-\zeta\omega_n t_r}\left[\cos\omega_d t_r + \frac{\zeta}{\sqrt{1-\zeta^2}}\sin\omega_d t_r\right] = 0$$

Since $e^{-\zeta\omega_n t_r} \neq 0$ for any infinite time, we obtain

$$\left[\cos\omega_d t_r + \frac{\zeta}{\sqrt{1-\zeta^2}}\sin\omega_d t_r\right] = 0$$

or

$$\tan\omega_d t_r = -\frac{\sqrt{1-\zeta^2}}{\zeta} = -\frac{\omega_d}{\sigma}$$

where
$$\omega_d = \omega_n\sqrt{1-\zeta^2}$$

and $\sigma = \zeta\omega_n$, the real part of complex root, i.e. attenuation

Hence, rise time

$$t_r = \frac{1}{\omega_d}\tan^{-1}\left(\frac{\omega_d}{-\sigma}\right) = \frac{\pi - \theta}{\omega_d} \qquad (4.32)$$

where θ is defined in Figure 4.13(b).

Clearly for fast response, i.e. for t_r to be less, ω_n should be high and/or ζ should be less in the range $0 < \zeta < 1$.

Peak time, t_p. Referring to Eq. (4.26), we obtain peak time by solving the equation $dc/dt = 0$; i.e.

$$\left.\frac{dc}{dt}\right|_{t=t_p} = \sin\omega_d t_p \frac{\omega_n}{\sqrt{1-\zeta^2}} e^{-\zeta\omega_n t_p} = 0$$

or $\sin\omega_d t_p = 0$

or $\omega_d t_p = 0,\ \pi,\ 2\pi,\ \ldots$

Since peak time corresponds to the first peak overshoot, i.e. $\omega_d t_p = \pi$

Hence
$$t_p = \frac{\pi}{\omega_d} \qquad (4.33)$$

The peak time corresponds to one-half cycle of the frequency of the damped oscillation. The second peak, i.e. the first undershoot occurs at $t = 2\pi/\omega_d$. The third peak, i.e. the second overshoot occurs at $t = 3\pi/\omega_d$, and so on.

Maximum overshoot, M_p. The maximum overshoot occurs at the peak time or at $t = t_p = \pi/\omega_d$. Thus, from Eq. (4.25), M_p is obtained as

$$M_p = c(t_p) - 1$$

$$= -e^{-\zeta\omega_n t_p}\left(\cos\pi + \frac{\zeta}{\sqrt{1-\zeta^2}}\sin\pi\right) \qquad (\because \omega_d t_p = \pi)$$

$$= -e^{-\zeta\omega_n\pi/\omega_d} = e^{-(\zeta/\sqrt{1-\zeta^2})\pi} \qquad (4.34)$$

The maximum per cent overshoot is $e^{-(\zeta/\sqrt{1-\zeta^2})\pi} \times 100\%$.

It is important to note that for the underdamped second-order system, the maximum overshoot for a step input depends only on the value of the damping coefficient ζ and is independent of the value of ω_n. Since from the definition of ζ [from Figure 4.13(b)]

$$\zeta = \cos\theta \qquad (4.35)$$

It is clear that ζ defines a radial line on the s-plane and therefore complex roots anywhere on this line produce the same percentage overshoot regardless of the value of ω_n (distance from origin).

Subsidence ratio. In a decaying oscillation this is the ratio of the amplitudes of successive cycles. A subsidence ratio and the peak overshoot would provide a practical optimal response information for many process control systems.

A subsidence ratio is usually expressed as $R:1$, which signifies that the first peak at time t_{p1}, i.e. the first maximum peak overshoot $M(t_{p1}) = [c(t_{p1}) - c_{ss}]$ is R times greater than the second peak at time t_{p2}, i.e. the second peak overshoot $M(t_{p2}) = [c(t_{p2}) - c_{ss}]$. It may be shown that for any two adjacent peaks, the subsidence ratio of a second-order underdamped system will be the same. The first peak occurs at $t_{p1} = \pi/\omega_d$ and the second peak (for overshoot) at $t_{p2} = 3\pi/\omega_d$. Hence

$$M(t_{p1}) = RM(t_{p2})$$

Substituting for $M(t_{p1})$ and taking natural logarithm yields

$$\frac{\zeta}{\sqrt{1-\zeta^2}} = \frac{\ln(R)}{2\pi}$$

which indicates that the subsidence ratio is a function of damping ratio ζ alone. The logarithm of subsidence ratio is called the logarithmic decrement as illustrated below.

The unity-feedback system of Figure 4.12 is subjected to a unit-step input, the first maximum overshoot at first peak time $t = t_p$ is M_p and let us denote it by M_{p1}. Then,

$$M_{p1} = \exp\left[\left(-\frac{\zeta}{\sqrt{1-\zeta^2}}\right)\pi\right]$$

and the second maximum overshoot M_{p2} at $t = t_p + \frac{2\pi}{\omega_d} = \frac{3\pi}{\omega_d}$ is

$$M_{p2} = \exp\left[\left(-\frac{\zeta}{\sqrt{1-\zeta^2}}\right)3\pi\right]$$

Thus the logarithmic decrement is defined as

$$\ln\left[\frac{M_{p1}}{M_{p2}}\right] = \frac{2\pi\zeta}{\sqrt{1-\zeta^2}}$$

Similarly, the nth maximum overshoot M_{pn} at $t = t_p + (n-1)T$ where $T = 2\pi/\omega_d$, the period of oscillation, is

$$M_{pn} = \exp\left[-\frac{\zeta}{\sqrt{1-\zeta^2}}(2n-1)\pi\right]$$

Then,
$$\frac{M_{p1}}{M_{pn}} = \exp\left[\frac{\zeta}{\sqrt{1-\zeta^2}}(n-1)2\pi\right]$$

or
$$\ln\left[\frac{M_{p1}}{M_{pn}}\right] = (n-1)\frac{2\pi\zeta}{\sqrt{1-\zeta^2}}$$

In this process, we can determine the damping ratio ζ experimentally as

$$\zeta = \frac{\frac{1}{n-1}\ln\left[\frac{M_{p1}}{M_{p2}}\right]}{\sqrt{\left[4\pi^2 + \left(\frac{1}{n-1}\ln\left[\frac{M_{p1}}{M_{p2}}\right]\right)^2\right]}} \qquad (4.36)$$

Settling time, t_s. The transient response of Eq. (4.26) is rewritten as

$$c(t) = 1 - \frac{e^{-\zeta\omega_n t}}{\sqrt{1-\zeta^2}}\sin(\omega_d t + \theta)$$

The curves of $1 \pm \dfrac{e^{-\zeta\omega_n t}}{\sqrt{1-\zeta^2}}$ are the envelope curves of the transient response to a unit-step input.

The response is within the envelope curves as shown in Figure 4.17. The time constant of these envelope curves is T. The rate of decay depends on the value of the time constant which is $1/\zeta\omega_n$.

The unit-step response of the first-order low-pass RC circuit reaches 98% of the final value after four time constants, i.e. $4T$, where $T = RC$ is the time constant.

Hence for a second-order underdamped system the settling time (with 2% criterion) is defined as the four times the time constant, i.e. $4T$ or $4/\zeta\omega_n$.

The settling time (with 5% criterion) is defined as the three times the time constant, i.e. $3T$ or $3/\zeta\omega_n$. It is clear that for smaller values of damping coefficient ζ, the rise time t_r reduces but the settling time t_s increases.

For 2% settling time,
$$0.02 = \exp(-\zeta\omega_n t_s)$$

Taking natural logarithm, we get
$$t_s = \frac{4}{\zeta\omega_n}$$

For 5% settling time,
$$0.05 = \exp(-\zeta\omega_n t_s)$$

Transient Response Analysis

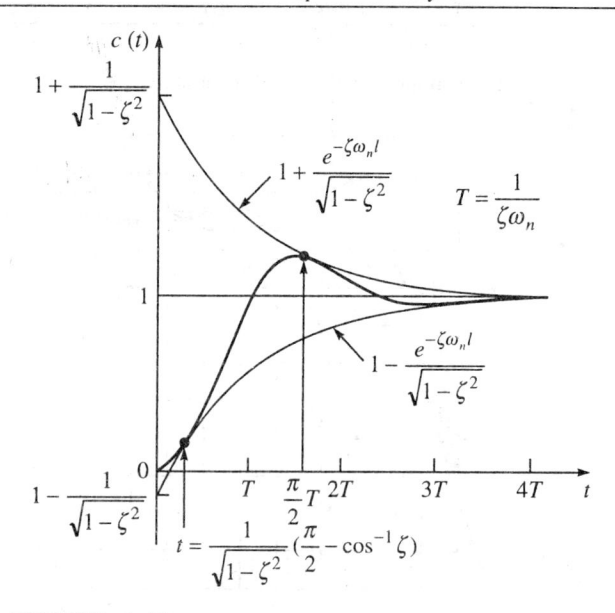

FIGURE 4.17 Envelope curves of transient response.

Taking natural logarithm, we get

$$t_s = \frac{3}{\zeta \omega_n}$$

Settling time is a function of the real part of the dominant poles. The reciprocal of $\zeta \omega_n$ has units of time and is referred as time constant of the equivalent second-order system.

Now, if we want to see the variation of response with the variation of damping ratio, let us draw the response $c(t)$ versus normalized time $\omega_n t$ for various values of ζ as well as the tolerance band, which may be with 2% criterion where $4T = 4/\zeta \omega_n$ or with 5% criterion where $3T = 3/\zeta \omega_n$.

Let us consider the tolerance band of 5% criterion. Figure 4.18 shows the unit-step response versus normalized time $\omega_n t$ for various values of damping ratio ζ as well as the tolerance band.

We can observe from the plot of Figure 4.18 the following:

The output response curve for $\zeta = 1$ touches the 5% band, that is, $3T$. Then as we decrease the damping ratio from 1 till 0.68 (approximately), the normalized time $\omega_n t$ decreases monotonically. It is observed that as the response $c(t)$ for $\zeta = 0.68$ just enters the upper limit of 5% band, there is a jump in the value of $\omega_n t$, i.e. a discontinuity occurs as shown in Figure 4.18. Then again, as we decrease the damping ratio ζ from 0.68 till 0.43 where the response is oscillatory because of lower damping ratio, the normalized time $\omega_n t$ increases monotonically; then as the response for $\zeta = 0.43$ just enters the 5% band, there is a jump in the value of $\omega_n t$, i.e. another discontinuity occurs at $\omega_n t = 0.68$. Then again as we decrease the damping ratio ζ from 0.43 till 0.28, the normalized time $\omega_n t$ increases monotonically till another jump occurs (not shown in the response plot).

Hence our observation in a nutshell is that as the damping is gradually reduced from unity (critically damped) in the range $1 > \zeta \geq 0.68$, the normalized settling time $\omega_n t_s$ decreases monotonically.

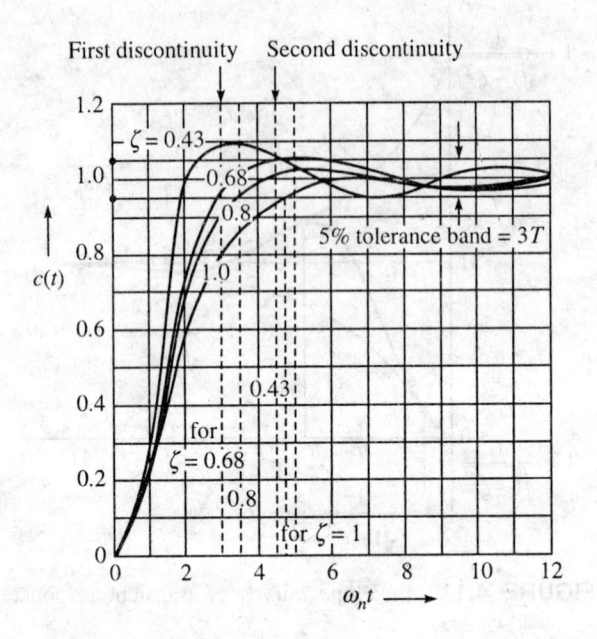

FIGURE 4.18 Settling time for various values of ζ.

As the damping ratio is reduced further, the normalized settling time $\omega_n t_s$ suddenly jumps up, i.e. settling time has a discontinuity which occurred for damping ratio of 0.68 (shown in the plot) and then again increases slowly with the decrease in ζ which is in the usual norm in the corresponding band.

The response plot has another discontinuity corresponding to the first undershoot for damping ratio of 0.43 touching the lower limit of 5% tolerance band and the normalized settling time $\omega_n t_s$ suddenly jumps up. Similar discontinuities occur with further decrease in damping ratio.

Hence from the response $c(t)$ versus the normalized settling time $\omega_n t_s$, we get the different discontinuity bands with damping ratio ζ as the parameter. A rough estimate has been made in Table 4.1.

In Figure 4.19, a plot of normalized settling time $\omega_n t_s$ versus damping ratio from the data obtained in Figure 4.18 has been made. Table 4.1 is the rough estimate for formulating the plot of normalized settling time vs. damping ratio from Figure 4.18.

Following the same line of treatment, one can use 2% band, that is, $4T$, and draw the plot of normalized settling time versus damping ratio.

MATLAB program **SettleTimeVaryingZeeta.m** with its graphical output is given in the following:

MATLAB Program 4.3

```
%SettleTimeVaryingZeeta.m
% Plot of Normalized settling time (with 5% criterion) vs Damping coefficient
zeta=0;
```

Transient Response Analysis

```
while zeta<1
num=[1];
den=[1 2*zeta 1];
t=0:.1:14;
[y1,x1,T1]=step(num,den,t);
i=2;
settle=.05;
for i=140:-1:2
  if(abs(y1(i)-y1(140))>=settle)
  break,end
end
Ts=T1(i);
text(zeta,Ts,'-')
ylabel('Normalized settling time')
xlabel('Damping coefficient')
axis([0 1 0 14])
zeta=zeta+.01;
zeta
Ts
end
title('Normalized settling time vs
damping coefficient');
grid on;
```

Normalized settling time vs damping coefficient

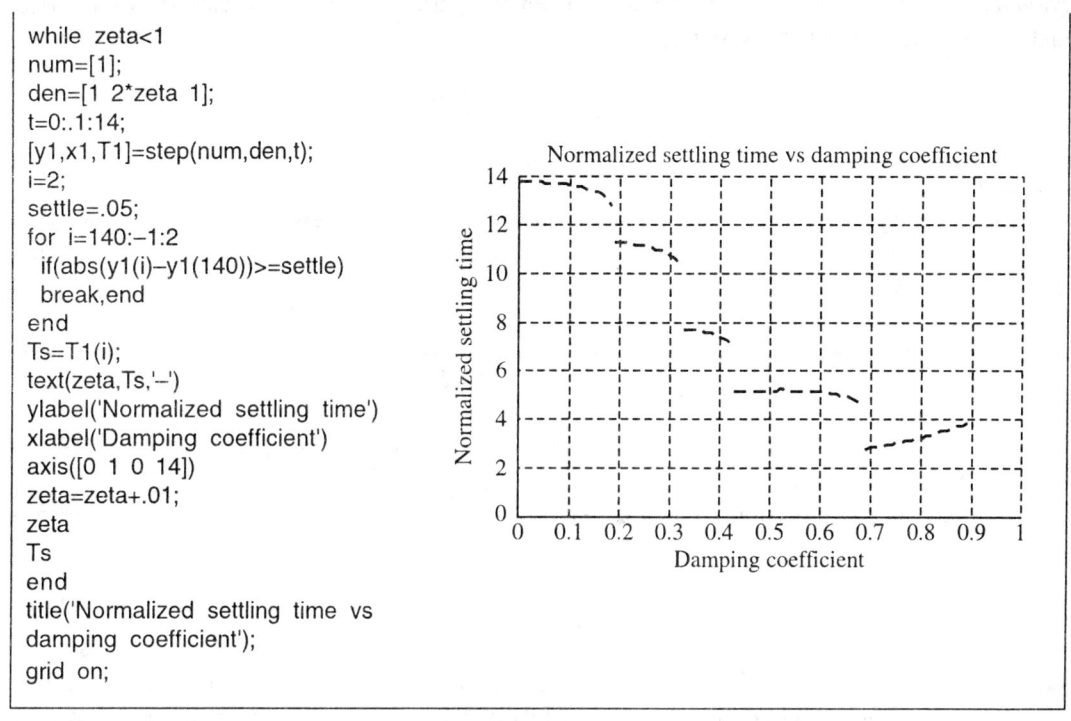

TABLE 4.1 Comparison of normalized settling time of second-order system with 5% tolerance band

ζ	Settling time, $\omega_n t_s$	Different discontinuity band
0.24	11	Another band
0.27	10	large oscillation exists
0.28	7.9	
0.3	7.8	Another band
0.4	7.0	oscillation exists
0.43	6.5	
0.44	4.5	
0.5	4.5	Another band
0.6	4.2	overshoot exists
0.64	4.0	
0.68	3.8	
0.7	3.2	
0.8	3.33	One band for no oscillation
0.9	4.0	
1.0	4.73	

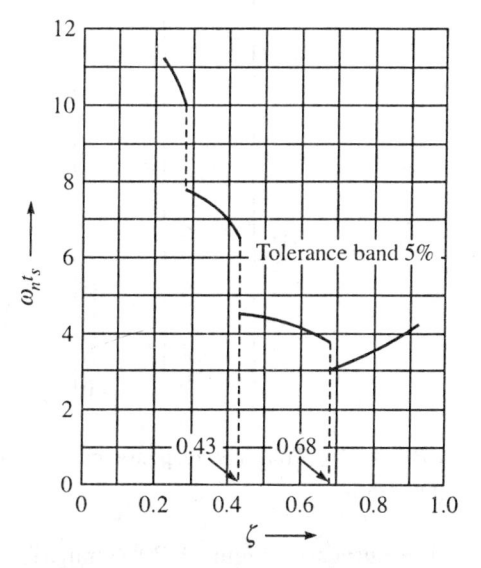

FIGURE 4.19 Normalized settling time versus damping ratio for 5% tolerance band.

Number of oscillations to settling time. Given the periodic time of the damped oscillations and the settling time of the system, then

$$\text{number of oscillations} = \frac{\text{settling time}}{\text{periodic time}}$$

and is a function of the damping ratio.

Summarizing the results of the performance indices, we can rewrite these expressions, in terms of the damping ratio of the second-order underdamped system subjected to unit-step input, as follows:

Equation (4.33) is rewritten as $\qquad \omega_n t_p = \dfrac{\pi}{\sqrt{1-\zeta^2}}$ (4.37)

Equation (4.34) is rewritten as $\qquad M_p(\%) = \exp(-\pi\zeta/\sqrt{1-\zeta^2}) \times 100$ (4.38)

Settling time, t_s (5% criterion) is written as $\omega_n t_s = \dfrac{4}{\zeta}$ (4.39)

Damped frequency as $\qquad \dfrac{\omega_d}{\omega_n} = \sqrt{1-\zeta^2}$ (4.40)

Equations (4.37) to (4.40) define important characteristics of the transient response in terms of the root location as defined by ζ and ω_n. Plots of the transient characteristics are shown in Figure 4.20 with ζ as the variable parameter.

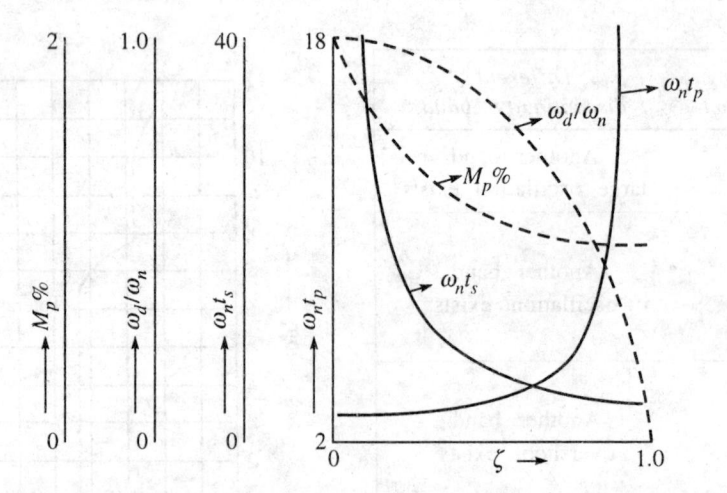

FIGURE 4.20 Transient response characteristics of the second-order system as a function of the damping ratio.

The curves of Figure 4.20 summarize the dynamic step response characteristics for the second-order underdamped system. These are of paramount importance in the analysis and design. Usually the design specifications put bounds on performance indices, more specifically, if the bound is on maximum overshoot $M_p\%$ and settling time t_s, then one can enter the curves of Figure 4.20 and for the specified bound value of $M_p\%$, evaluate the value of the damping coefficient range. Then after knowing ζ, the curve yields a value for $t_s = 4/\zeta\omega_n$ (for 2% criterion) and thus the required value of ω_n can be obtained and again the peak time t_p can be evaluated. The curves of Figure 4.20 will all the more be appreciated when used in conjunction with the frequency response curves.

EXAMPLE 4.2 The response of a system is obtained as

$$c(t) = 1 + 0.2 \exp(-60t) - 1.2 \exp(-10t)$$

when subjected to a unit-step input. Obtain the closed-loop transfer function. Determine ω_n and ζ.

Solution: Laplace transform of the unit-impulse response is the transfer function of the system. Hence, the impulse response is

$$g(t) = \frac{dc(t)}{dt} = 12 \exp(-10t) - 12 \exp(-60t)$$

The closed-loop transfer function becomes

$$G(s) = \frac{12}{s+10} - \frac{12}{s+60} = \frac{600}{s^2 + 70s + 600}$$

Comparing the characteristic polynomial $s^2 + 70s + 600$ with the standard equation

$$s^2 + 2\zeta\omega_n s + \omega_n^2$$

we get $\quad \omega_n = 10\sqrt{6} \quad$ and $\quad \zeta = \dfrac{3.5}{\sqrt{6}}$

4.7 Parameter Variation for Second-Order System

For the second-order stable system having the closed-loop transfer function

$$\frac{\omega_n^2}{s^2 + 2\zeta\omega_n s + \omega_n^2}$$

the unit-step response is computed using a MATLAB program with undamped natural frequency $\omega_n = 1$ rad/s and several values of damping coefficient ζ, i.e. 0.2, 0.5, 0.707, 1, and 2. The corresponding unit-step response obtained through MATLAB Program 4.4 illustrates the significance of the damping coefficient ζ. The inference is that when ζ increases, delay time and rise time all increase. The settling time is large for small and large values of ζ. The maximum overshoot decreases as ζ increases in the range $0 < \zeta < 1$ at the expense of delay time, rise time and settling time. For those reasons, in classical design, the optimum value for ζ is 0.707 because it gives maximum overshoot as 4.3% and yet the delay time, the rise time, and the settling time are all reasonable.

MATLAB Program 4.4

```
% Calculation of Tr,Td,Tp,Mp,Ts with
% varying zeta keeping Wn constant
clear;
pack;
clc;

j=1;
wn=1;

while(j<6)
  zeta(j)=input('Enter value of damping ratio: ');
    num=[wn*wn];
    den=[1 2*zeta(j)*wn wn*wn];
t=0:.1:26;
[y1,x1,T]=step(num,den,t);
i=1;
while y1(i)<.1,
  i=i+1;
end
ten=T(i);
% calculating rise time,Tr:10% to 90% criteria
  while y1(i)<.90,
   i=i+1;
end
ninety=T(i);
value=ninety-ten;
digits(6);
Tr(j)=value;
i=1;
% calculating delay time,Td
while y1(i)<.5,
  i=i+1;
end
Td(j)=T(i);

for i=2:1:260
if y1(i)<=y1(i-1)
break,end
end
over=y1(i)-y1(260);
% calculating peak time,Tp and overshoot,Mp
Mp(j)=over;
Tp(j)=T(i);
% calculating settling time,Ts using 5% criteria
settle=0.05;
i=2;
for i=260:-1:2
 if(abs(y1(i)-y1(260))>=settle)
  break,end
```

Step response

Zeta	Td	Tr	Ts	Mp	Tp
2	2.9	8.2	11.3	0	25.9
1	1.7	3.3	4.7	0	25.9
0.707	1.5	2.1	2.9	0.0432	4.5
0.5	1.3	1.7	5.2	0.163	3.7
0.2	1.2	1.2	13.6	0.53	3.3

```
end
Ts(j)=T(i);
step(num,den,t);
axis([0 30 0 1.6]);
title('Unit Step Response of 2nd Order
System for varying zeta with Wn constant');
pop1=sprintf('%6.3g',zeta(j));
text(T(40),y1(40),pop1);

hold on;
j=j+1;
end

j=1;
while(j<6)
pop=sprintf('%6.3g %6.3g %6.3g %6.3g %6.3g
%6.3g',zeta(j),Td(j),Tr(j),Ts(j),Mp(j),Tp(j));
text(6,.6, 'Zeta Td Tr Ts Mp Tp');
text(6,.6-(.1*j),pop);
j=j+1;
end

Enter value of damping ratio: 2
Enter value of damping ratio: 1
Enter value of damping ratio: .707
Enter value of damping ratio: .5
Enter value of damping ratio: 0.2
```

In a similar way while keeping the damping ratio constant at $\zeta = 0.5$ and varying the undamped natural frequency ω_n, i.e. 5, 10 and 20, the unit-step response of the second-order system of Figure 4.12 is computed using MATLAB Program 4.5 and the plot obtained can be used to investigate the effects of varying ω_n. From these it may be concluded that the maximum overshoot does not vary with ω_n because of the constant value of the damping coefficient ζ in the underdamped range. As the value of ω_n increases, the value of delay time, rise time, settling time and t_p all decrease and the maximum overshoot remains constant. Thus classical design calls for a high value of undamped natural frequency ω_n in order to produce a response that rises and settles fast without penalizing the overshoot. However, from the practical point of view, high ω_n requires high loop gain which is a costly proposition and may also cause instability.

MATLAB Program 4.5

```
% Calculation of Td,Tr,Ts,Mp,Tp for
% varying Wn keeping zeta constant

clc;
clear;
pack;
```

```
clg;
j=1;
zeta=0.5;
while(j<4)
  wn(j)=input('Enter value of natural freq: ');
       num=[wn(j)*wn(j)];
       den=[1 2*zeta*wn(j) wn(j)*wn(j)];
  t=0:.01:2;
[y1,x1,T]=step(num,den,t);
i=1;
while y1(i)<.1,
  i=i+1;
end
ten=T(i);
  while y1(i)<.90,  % calculating rise time,Tr
  i=i+1;
end
ninety=T(i);
value=ninety-ten;
Tr(j)=value;
i=1;
while y1(i)<.5,  % calculating delay time,Td
  i=i+1;
end
Td(j)=T(i);
for i=2:1:200
if y1(i)<=y1(i-1)
break,end
end
over=y1(i)-y1(200);
Mp(j)=over;  % calculating peak time,Tp
overshoot,Mp
Tp(j)=T(i);
settle=0.05;  % using 5% criteria
i=2;
for i=200:-1:2
  if(abs(y1(i)-y1(200))>=settle)
  break,end
end
Ts(j)=T(i);
step(num,den,t);
axis([0 2 0 1.2]);
title('Unit Step Response of 2nd
Order System
for varying Wn keeping zeta
constant');
pop1=sprintf('%g',wn(j));
text(T(40),y1(40),pop1);
hold on;
j=j+1;
end
```

Step response

Wn	Td	Tr	Ts	Mp	Tp
20	0.07	0.08	0.26	0.16075	0.19
10	0.13	0.17	0.52	0.162857	0.37
5	0.26	0.33	1.05	0.160725	0.74

```
j=1;
while(j<4)
pop=sprintf('%6g %6g %6g %6g %6g
%6g',wn(j),Td(j),Tr(j),Ts(j),Mp(j),Tp(j));
text(.45,.5, 'Wn Td Tr Ts Mp Tp');
text(.4,.5-(.1*j),pop);
j=j+1;
end

Enter value of natural freq: 20
Enter value of natural freq: 10
Enter value of natural freq: 5
```

Simulation of an electrical system (having series *RLC* circuit driven by a voltage source and output taken across the capacitor) done in SPICE leads to the same result as above. The transfer function becomes

$$G(s) = \frac{1/LC}{s^2 + (R/L)s + 1/LC} = \frac{\omega_n^2}{s^2 + 2\zeta\omega_n s + \omega_n^2}$$

For variation of ζ keeping ω_n constant, putting (i) $L = 10$ μH, $C = 1$ nF, $R = 180$ Ω leads to $\zeta = 0.9$, (ii) $L = 10$ μH, $C = 1$ nF, $R = 100$ Ω leads to $\zeta = 0.5$, (iii) $L = 10$ μH, $C = 1$ nF, $R = 50$ Ω leads to $\zeta = 0.25$, and (iv) $L = 10$ μH, $C = 1$ nF, $R = 15$ Ω leads to $\zeta = 0.075$. Keeping ω_n constant in all cases gives the step-response curve having the same undamped natural frequency but different overshoots because of change of damping coefficient. In the same line, ω_n can be varied keeping ζ constant by changing the parameters in a suitable fashion to show the laboratory demonstration of electrical systems.

EXAMPLE 4.3 The output response of a spring-mass-damper system is shown in Figure 4.21. From the output response, calculate the per cent overshoot that occurs at 2 seconds after the step force of 2 newtons is applied and the mass M shows a final displacement of 0.1 metre. Obtain the value of mass M, viscous friction coefficient f and the spring constant K.

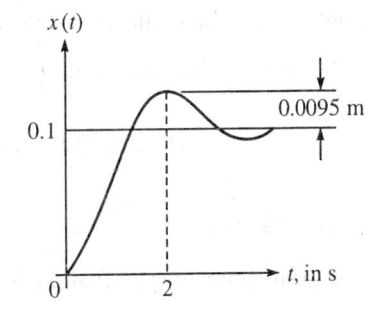

FIGURE 4.21 Example 4.3: output response.

Solution: Laplace transform of the system dynamics is

$$(Ms^2 + fs + K)X(s) = U(s) = \frac{2}{s}$$

Hence, $$X(s) = \frac{2}{s(Ms^2 + fs + K)}$$

The characteristic equation is, $s^2 + \left(\dfrac{f}{M}\right)s + \left(\dfrac{K}{M}\right) = 0$

Further, $\lim_{t \to \infty} x(t) = \lim_{s \to 0} sX(s) = \dfrac{2}{K} = 0.1$ or $K = \dfrac{2}{0.1} = 20$ N/m

Using the relation,

$$M_p = 0.0095 = \exp\left(-\dfrac{\zeta \pi}{\sqrt{1-\zeta^2}}\right)$$

gives $\zeta = 0.6$.

Further, $t_p = 2 = \dfrac{\pi}{\omega_d} = \dfrac{\pi}{\omega_n\sqrt{1-0.36}} = \dfrac{\pi}{0.8\omega_n}$ gives $\omega_n = 1.96$ rad/s

Also, $\omega_n^2 = \dfrac{K}{M} = \dfrac{20}{M} = (1.96)^2$ gives $M = 5.2$ kg

From, $2\zeta\omega_n = \dfrac{f}{M}$, we get $f = 2 \times 0.6 \times 1.96 \times 5.2 = 12.2 \dfrac{\text{N-m}}{\text{rad/s}}$

EXAMPLE 4.4 A unity-feedback system having forward transfer function

$$G(s) = \dfrac{K}{s(Ts+1)}$$

is subjected to a unit-step input. Determine the values of K and T from the output response curve shown in Figure 4.22. Find the rise time and settling time.

Solution: Using the relation,

$$M_p = 0.254 = \exp\left(\dfrac{-\zeta\pi}{\sqrt{1-\zeta^2}}\right)$$

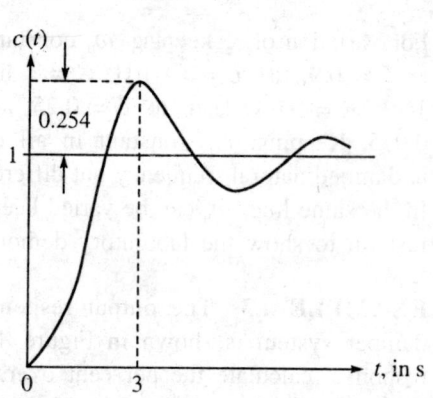

FIGURE 4.22 Example 4.4: output response.

yields $\zeta = 0.4$.

The peak time is specified as 3 seconds, therefore,

$$t_p = 3 = \dfrac{\pi}{\omega_n\sqrt{1-\zeta^2}} = \dfrac{3.14}{\omega_n\sqrt{(1-0.4^2)}} = \dfrac{3.427}{\omega_n} \quad \text{or} \quad \omega_n = \dfrac{3.427}{3} = 1.142 \text{ rad/s}$$

The characteristic equation is $1 + GH = 0$; that is, $Ts^2 + s + K = 0$.

Comparing with the standard form of equation, $s^2 + 2\zeta\omega_n s + \omega_n^2 = 0$; we get $\omega_n = \sqrt{\dfrac{K}{T}}$
$= 1.142$ rad/s

Again, $2\zeta\omega_n = \dfrac{1}{T}$, therefore, $T = 1.09$

As $\omega_n = \sqrt{\dfrac{K}{T}}$; hence $K = \omega_n^2 T = (1.142)^2(1.09) = 1.42$

Substituting $\omega_d = \omega_n\sqrt{1-\zeta^2} = 1.016$ and $\theta = \tan^{-1}\dfrac{\sqrt{1-\zeta^2}}{\zeta} = 66.4°$, rise time, $t_r = \dfrac{\pi - \theta}{\omega_d}$

≈ 2 s. Settling time, t_s (for 2% criterion) is, $t_s = \dfrac{4}{\zeta\omega_n} = 6.97$ s

EXAMPLE 4.5 Consider the unity-feedback control system having forward transfer function

$$G(s) = \dfrac{K}{s(Ts+1)}$$

with K and T as usual being the motor gain constant and motor time constant. It is desired to select K and T so that the time-domain specifications will be satisfied. The transient response to a unit-step input should be as fast as possible and with an overshoot less than 5%. Further, the settling time should be less than 4 seconds for 2% criterion. The minimum damping ratio for an overshoot of 4.3% is 0.707.

Solution: For 2% criterion, the settling time is $t_s = \dfrac{4}{\zeta\omega_n} \leq 4$ s; which gives $\zeta\omega_n \geq 1$.

Further, $\zeta\omega_n = 1$ and $\zeta = 0.707$ satisfy our overshoot requirement.
The closed-loop transfer function becomes

$$\dfrac{C(s)}{R(s)} = \dfrac{K}{Ts^2 + s + K} = \dfrac{K/T}{s^2 + s/T + K/T}$$

Comparing with the standard form, we get

$$\omega_n^2 = \dfrac{K}{T} \quad \text{and} \quad 2\zeta\omega_n = \dfrac{1}{T}$$

From $\zeta = 0.707$; $\zeta\omega_n = 1$, we get $\omega_n = \sqrt{2}$; Also, we get $T = 0.5$ and $K = 1$.

EXAMPLE 4.6 For a unity-feedback system having open-loop transfer function $G(s)$ as

$$\dfrac{25}{s(s+5)}$$

determine the rise time t_r, settling time t_s, first peak time t_{p1} and maximum per cent overshoot when the closed-loop system is subjected to unit-step input.

Solution: The characteristic equation $s^2 + 5s + 25 = 0$ leads to $\omega_n = 5$ and $\zeta = 0.5$.

Now, $\omega_d = \omega_n\sqrt{1-\zeta^2} = 4.33$ and $\sigma = \zeta\omega_n = 1$

Phase angle, $\beta = \tan^{-1}\dfrac{\omega_d}{\sigma} = 1.344$ rad and rise time, $t_r = \dfrac{\pi - \beta}{\omega_d} = 0.416$ s

Settling time, t_s:

With 2% criterion, $t_s = \dfrac{4}{\zeta\omega_n} = 4$ s; with 5% criterion, $t_s = \dfrac{3}{\zeta\omega_n} = 3$ s

The first peak time, $t_{p1} = \dfrac{\pi}{\omega_d} = 0.726$ s, and maximum % overshoot,

$$M_p = \exp\left(-\dfrac{\zeta\pi}{\sqrt{1-\zeta^2}}\right) \times 100 = 16.3\%$$

Drill Problem 4.4

A unity-feedback system having closed-loop transfer function $\omega_n^2/(s^2 + 2\zeta\omega_n s + \omega_n^2)$ should have a maximum overshoot as 20% and settling time ≤ 2 s. Determine the required values of the damping ratio ζ and ω_n.

Ans. $\zeta = 0.4559$ and $\omega_n = 4.3864$

Drill Problem 4.5

For a unity-feedback system having forward transfer function $25/s(s+10\zeta)$, determine the response of the closed-loop system subjected to a unit-step input having damping coefficient ζ as (i) 1 and (ii) 1.5.

Ans. (i) $c(t) = 1 - e^{-5t} - 5te^{-5t}$; (ii) $c(t) = 1 - 1.1708e^{-13.09t} + 0.1708e^{-1.9t}$

Drill Problem 4.6

For the Drill Problem 4.5, determine the value of the damping ratio in order to get 10% maximum overshoot.

Ans. 0.591

4.8 Impulse Response of Second-Order System

For unit-impulse input $r(t)$, the corresponding $R(s) = 1$. The unit impulse response of the system in Figure 4.12 can be written as

$$c(t) = \mathcal{L}^{-1}C(s) = \mathcal{L}^{-1}\dfrac{\omega_n^2}{s^2 + 2\zeta\omega_n s + \omega_n^2} \tag{4.41}$$

(i) For underdamped case, i.e. $0 < \zeta < 1$ the time response is

$$c(t) = \dfrac{\omega_n}{\sqrt{1-\zeta^2}}e^{-\zeta\omega_n t}\sin\omega_d t \tag{4.42}$$

The response is shown in Figure 4.23 for different values of ζ.

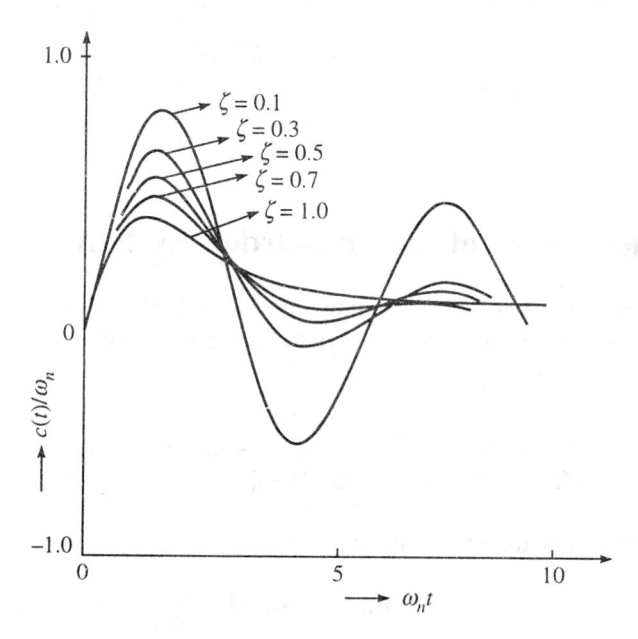

FIGURE 4.23 Unit impulse response of second-order system for different values of damping ratio.

(ii) For critically-damped case, i.e. $\zeta = 1$, the response is

$$c(t) = \omega_n^2 \, t \, e^{-\omega_n t} \qquad (4.43)$$

(iii) For overdamped case, i.e. $\zeta > 1$, the response is

$$c(t) = \frac{\omega_n}{2\sqrt{\zeta^2 - 1}} \left[e^{-(\zeta - \sqrt{\zeta^2 - 1})\omega_n t} - e^{-(\zeta + \sqrt{\zeta^2 - 1})\omega_n t} \right] \qquad (4.44)$$

It may be noted that as unit-impulse function is the time derivative of unit-step function, the unit-step response can be obtained by differentiating the unit-ramp response of the system.

It is apparent from the analysis that for an overdamped or critically-damped system where the roots are real, the corresponding step response does not overshoot but increases or decreases monotonically and approaches a constant value.

In order to derive the maximum overshoot, refer to Eq. (4.42) and find $dc/dt = 0$ (for underdamped case) to get the time at which the peak occurs, i.e.

$$t = \frac{\tan^{-1}\dfrac{\sqrt{1-\zeta^2}}{\zeta}}{\omega_n\sqrt{1-\zeta^2}}; \qquad 0 < \zeta < 1 \qquad (4.45)$$

and the maximum overshoot is

$$c(t)|_{max} = \omega_n \exp\left[-\frac{\zeta}{\sqrt{1-\zeta^2}} \tan^{-1}\frac{\sqrt{1-\zeta^2}}{\zeta}\right]; \quad \text{for } 0 < \zeta < 1 \qquad (4.46)$$

4.9 Ramp Response of Second-Order System

For a unit-ramp input $r(t)$, the corresponding Laplace transform is $R(s) = 1/s^2$.

The inverse Laplace transform of $C(s)$ yields the time domain solution for the response $c(t)$ as

$$c(t) = t - \frac{2\zeta}{\omega_n} + \frac{e^{-\zeta\omega_n t}}{\omega_n\sqrt{1-\zeta^2}} \sin(\omega_d t + \theta) \qquad (4.47)$$

The steady-state error for unit-ramp input is

$$e_{ss} = \lim_{t \to \infty}[r(t) - c(t)] = \frac{2\zeta}{\omega_n} \qquad (4.48)$$

Referring to Figure 4.11(b), the steady-state error is

$$e_{ss} = \lim_{s \to \infty} sE(s)$$

or
$$e_{ss} = \lim_{s \to 0} s \cdot \left(\frac{Js^2 + fs}{Js^2 + fs + K}\right) \cdot \frac{1}{s^2} = \frac{f}{K} \qquad (4.49)$$

$$= \frac{2\zeta}{\omega_n}$$

where $\quad \zeta = \dfrac{F}{2\sqrt{KJ}} \quad$ and $\quad \omega_n = \sqrt{\dfrac{K}{J}}$

It is apparent from the above expression that in order to have an acceptable steady-state error in following a ramp input, the damping coefficient ζ must not be too small and ω_n must be sufficiently large. Further, referring to Figure 4.11(b) we can conclude that to make e_{ss} small the gain K has to be large. A large value of K would, however, make the value of damping coefficient small and increase the maximum overshoot which is undesirable. To overcome this difficulty, two schemes to improve the behaviour have been proposed: (i) proportional-plus-derivative controller and (ii) tachometer feedback control, both of which have been discussed in a later section.

The response $c(t)$ of the second-order system (underdamped) subjected to unit-ramp input for various values of damping coefficient ζ is plotted in Figure 4.24. For a lower value of ζ in the underdamped case, the response is more oscillatory as in the case of unit-step input.

Transient Response Analysis

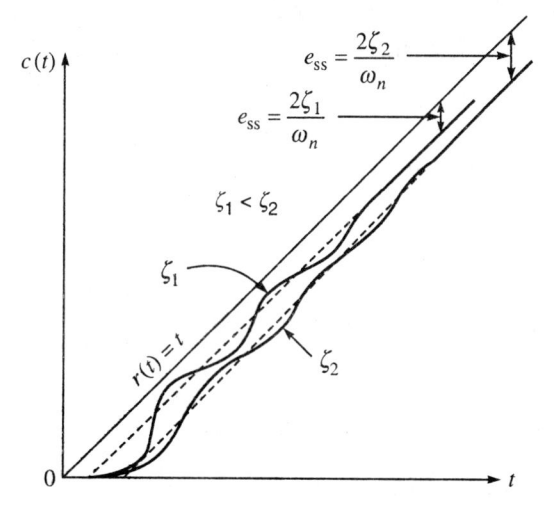

FIGURE 4.24 Unit-ramp response of underdamped second-order system.

4.10 Graphical Interpretation of Heaviside's Expansion

There is an alternative way to obtain the coefficients of partial fraction expansion graphically from the pole-zero map in s-plane as illustrated by the following example.

EXAMPLE 4.7 Determine the inverse Laplace transform by the graphical technique using the pole-zero map in s-plane of the transfer function

$$G(s) = \frac{4(s+1)}{(s+2)(s^2+2s+2)} = \frac{4(s+1)}{(s+2)(s+1-j)(s+1+j)}$$

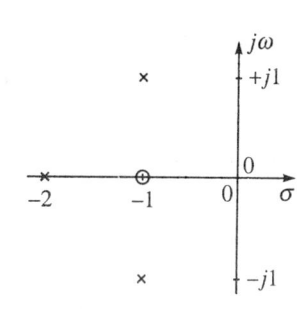

FIGURE 4.25(a) Pole-zero map.

The pole-zero map is shown in Figure 4.25(a).
The partial fraction expansion gives

$$G(s) = \frac{C_1}{s+2} + \frac{C_2}{s+1-j} + \frac{C_3}{s+1+j}$$

From Figure 4.25(b), the coefficient of the pole at -2 is

$$C_1 = \frac{4(-2-(-1))}{\sqrt{2}\sqrt{2}} = -2$$

Since an odd number of real poles and zeros exist to the right of this pole, its sign is negative. Similarly, from Figure 4.25(c), the coefficient of the pole at $-1+j1$ is

$$C_2 = \frac{4\{(-1+j1)-(-1)\}}{\{(-1+j1)-(-2)\}\{(-1+j1)-(-1-j1)\}}$$

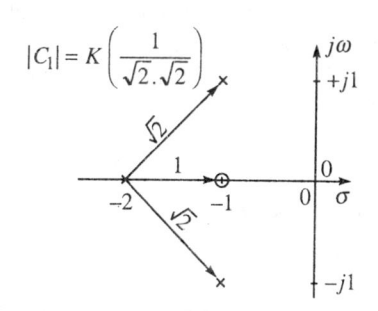

FIGURE 4.25(b) Residue at pole $s = -2$.

$$= \frac{4j}{(1+j1)(2j)} = 1 - j1$$

$$C_2 = K\left(\frac{1}{2\sqrt{2}}\right)e^{-j45°}$$

Obviously, $\quad C_3 = 1 + j1$

Taking the Laplace inverse transform, we get

$$g(t) = -2e^{-2t} + 2e^{-t}(\cos t + \sin t)$$

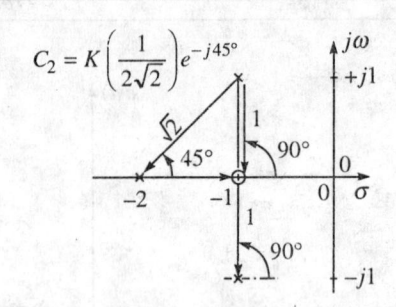

FIGURE 4.25(c) Residue at complex pole $s = -1 + j$.

4.11 Time-Domain Behaviour from Pole-Zero Plot

In general, for an array of poles in the s-plane, as shown in Figure 4.26, the poles s_b and s_d are quite different from the conjugate pole pairs. Further, s_b and s_d are the real poles and the response due to the poles s_b and s_d converges monotonically. The poles s_b and s_d may correspond to the quadratic function,

$$s^2 + 2\zeta\omega_n s + \omega_n^2 = 0; \quad \zeta > 1 \qquad (4.50)$$

and the roots

$$s_b, s_d = -\zeta\omega_n \pm \omega_n\sqrt{\zeta^2 - 1}; \quad \zeta > 1 \qquad (4.51)$$

The contribution to the total response, due to poles s_b and s_d is

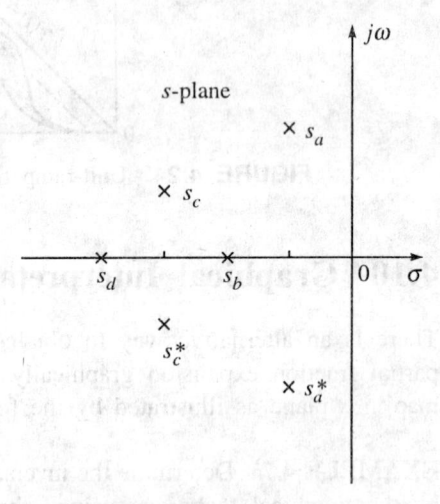

$$K_b e^{s_b t} + K_d e^{s_d t} \qquad (4.52)$$

FIGURE 4.26 Array of poles.

The contribution to response due to the pole s_b is predominant compared to that of s_d as $|s_d| \gg |s_b|$. In this case, s_b is dominant compared to s_d. The pole s_b is the dominant pole amongst these two real poles. The response due to pole at s_d dies down faster compared to that due to pole at s_b.

Now, let us discuss the complex conjugate poles s_a and s_a^* which belong to the quadratic factor

$$s^2 + 2\zeta\omega_n s + \omega_n^2; \quad \zeta > 1 \qquad (4.53)$$

The roots are

$$s_a, s_a^* = -\zeta\omega_n \pm j\omega_n\sqrt{1 - \zeta^2} \qquad (4.54)$$

Contribution to the total response by the complex conjugate poles s_a and s_a^* is

$$K_a e^{-\zeta\omega_n t} e^{j\omega_n\sqrt{1-\zeta^2}\, t} + K_a^* e^{-\zeta\omega_n t} e^{-j\omega_n\sqrt{1-\zeta^2}\, t} \qquad (4.55)$$

Transient Response Analysis

The factor $\exp(-\zeta\omega_n t)$ gives the monotonically decreasing function, whereas the factor $\exp(j\omega_n\sqrt{1-\zeta^2}\,t)$ gives sustained oscillation. The resultant will give the damped sinusoid waveform, as shown in Figure 4.27. The rate of decay of the sinusoidal waveform depends on the real part $\zeta\omega_n$ of the complex poles.

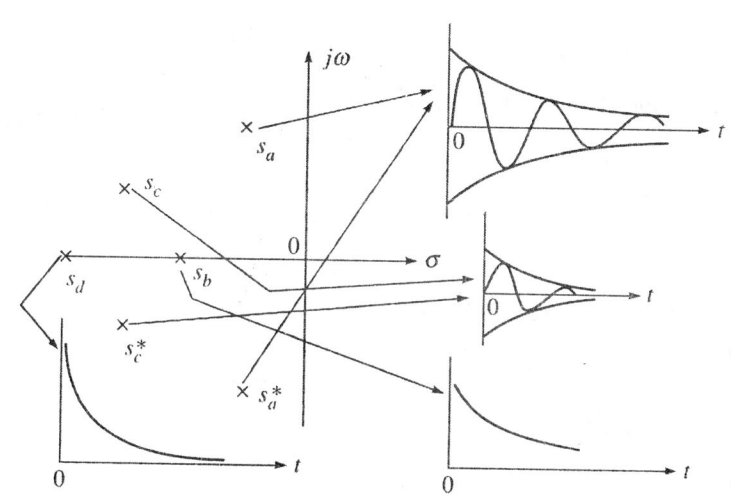

FIGURE 4.27 Nature of response with arbitrary magnitude corresponding to all poles.

Similarly, s_c and s_c^* are also the complex conjugate pole pair. The response due to s_c and s_c^* will die down faster than that due to s_a and s_a^*. Hence s_a and s_a^* are the dominant complex conjugate pole pair compared to the complex pole pair s_c and s_c^*.

The time-domain response can be obtained by taking the inverse Laplace transform after the partial fraction expansion, i.e.

$$\mathscr{L}^{-1}\left[\frac{K_a}{s-s_a} + \frac{K_a^*}{s-s_a^*} + \frac{K_b}{s-s_b} + \frac{K_c}{s-s_c} + \frac{K_c^*}{s-s_c^*} + \frac{K_d}{s-s_d}\right] \quad (4.56)$$

where the residues K_a and K_a^* are complex conjugate, as also the residues K_c and K_c^*.
Any residue K_r can be obtained as

$$K_r = H \times \frac{(s-s_1)\cdots(s-s_n)}{(s-s_a)\cdots(s-s_r)\cdots(s-s_m)}(s-s_r)\Bigg|_{s=s_r} \quad (4.57)$$

The above equation is composed of factors of general form $(s_r - s_n)$, where both s_n and s_r are known complex numbers. The term $(s_r - s_n)$ is also a complex number expressed in polar coordinates as

$$s_r - s_n = M_{rn}\exp(j\phi_{rn}) \quad (4.58)$$

Hence, $\quad K_r = H \times \dfrac{M_{r1}M_{r2}\cdots M_{rn}}{M_{ra}M_{rb}\cdots M_{rm}}\exp j(\phi_{r1}+\phi_{r2}+\cdots+\phi_{rn}-\phi_{ra}-\phi_{rb}-\cdots-\phi_{rm}) \quad (4.59)$

For the output to converge, all poles of the system must lie in the left-half of the s-plane and can never occur in the right-half of the s-plane. Poles can be on the boundary (the imaginary axis) subject to the limitation that such poles are simple. Consider the transform pair with pole-zero locations as in Figure 4.28 and we get

$$\frac{s}{(s^2 + \omega_0^2)^2} = \frac{t}{2\omega_0} \sin \omega_0 t \quad (4.60)$$

The response is unbounded as shown in Figure 4.29. Response for multiple poles at other places is bounded since such terms give rise to terms of the form $t^m \exp(-at)$, and in the limit

$$\lim_{t \to \infty} t^m e^{-at} = 0 \quad (4.61)$$

Hence the response is bounded.

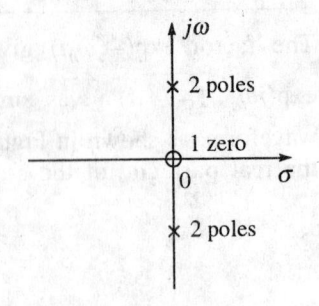

FIGURE 4.28 Pole pattern

Figure 4.30 illustrates the relationship between the roots of the characteristic equation and α, ζ, ω_n, and ω_d.

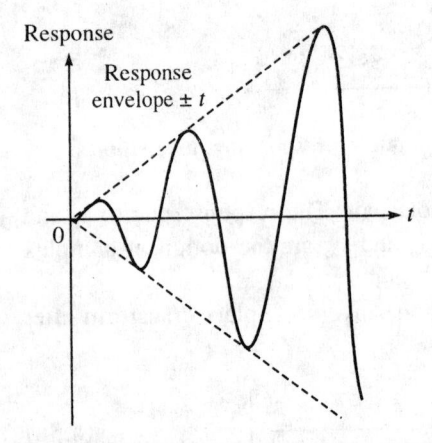

FIGURE 4.29 The inverse Laplace transform of the function having poles and zeros of Figure 4.28.

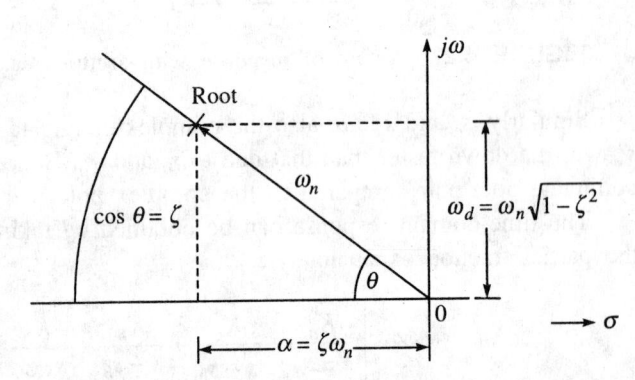

FIGURE 4.30 Relationship between the roots of the characteristic equation and α, ζ, ω_n, and ω_d.

For complex conjugate roots, ω_n is the radial distance from the root to the origin of the s-plane, α is the real part of the complex roots and is the inverse of the time constant. The damping ratio ζ is equal to the cosine of the angle between the radial line to the root and the negative real axis, i.e. $\zeta = \cos \theta$.

This signifies that if roots are changed due to the variation of ω_n, keeping ζ = constant (< 1), then the roots will lie on the line $\zeta = \cos \theta$.

Figure 4.31 shows the constant ω_n-loci, the constant ζ-loci, the constant $\alpha(=\zeta\omega_n)$-loci and the constant ω_d-loci. The effect of roots on damping ratio is shown in Figure 4.32 which shows that $0 < \zeta < \infty$ corresponds to negative-half of the s-plane and $-\infty < \zeta < 0$ corresponds to positive-half of the s-plane, the step response will settle to its constant steady-state value of the

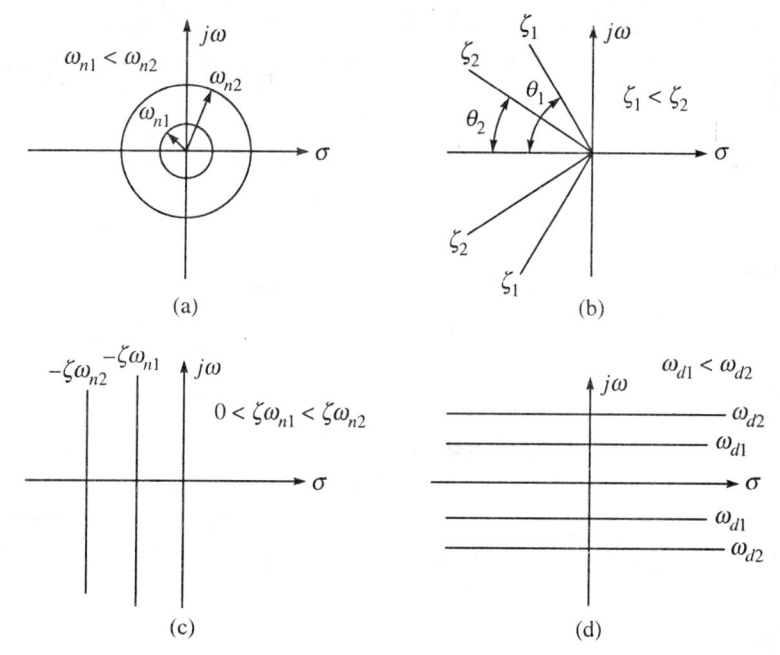

FIGURE 4.31 Performance lines in the s-plane: (a) constant ω_n, (b) constant ζ, (c) constant $\zeta\omega_n$, and (d) constant ω_d.

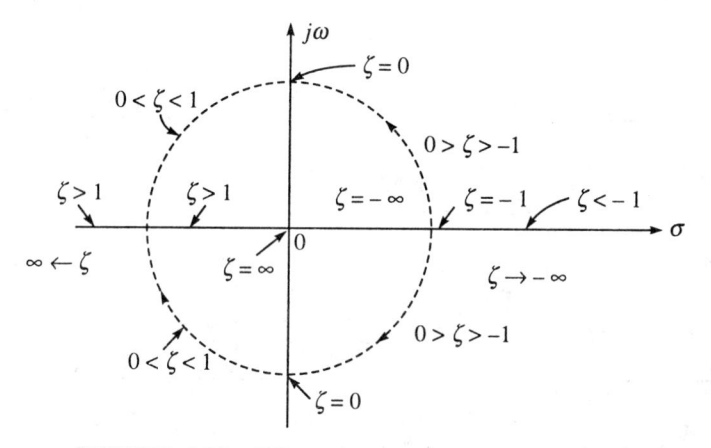

FIGURE 4.32 Effect of roots on damping ratio ζ.

negative exponent, i.e. $\exp(-\zeta\omega_n t)$. Negative damping corresponds to a response that grows without bound. The zero value of damping corresponds to sustained oscillation. The effect of roots position on system response is shown in Figure 4.33.

It is likely that for optimal step response, one has to impose the limits on the value of ω_d; this defines bands on the rise time, the peak time and the damped frequency ω_d. A minimum and maximum time constant defines the peak overshoot and probably more important the settling time.

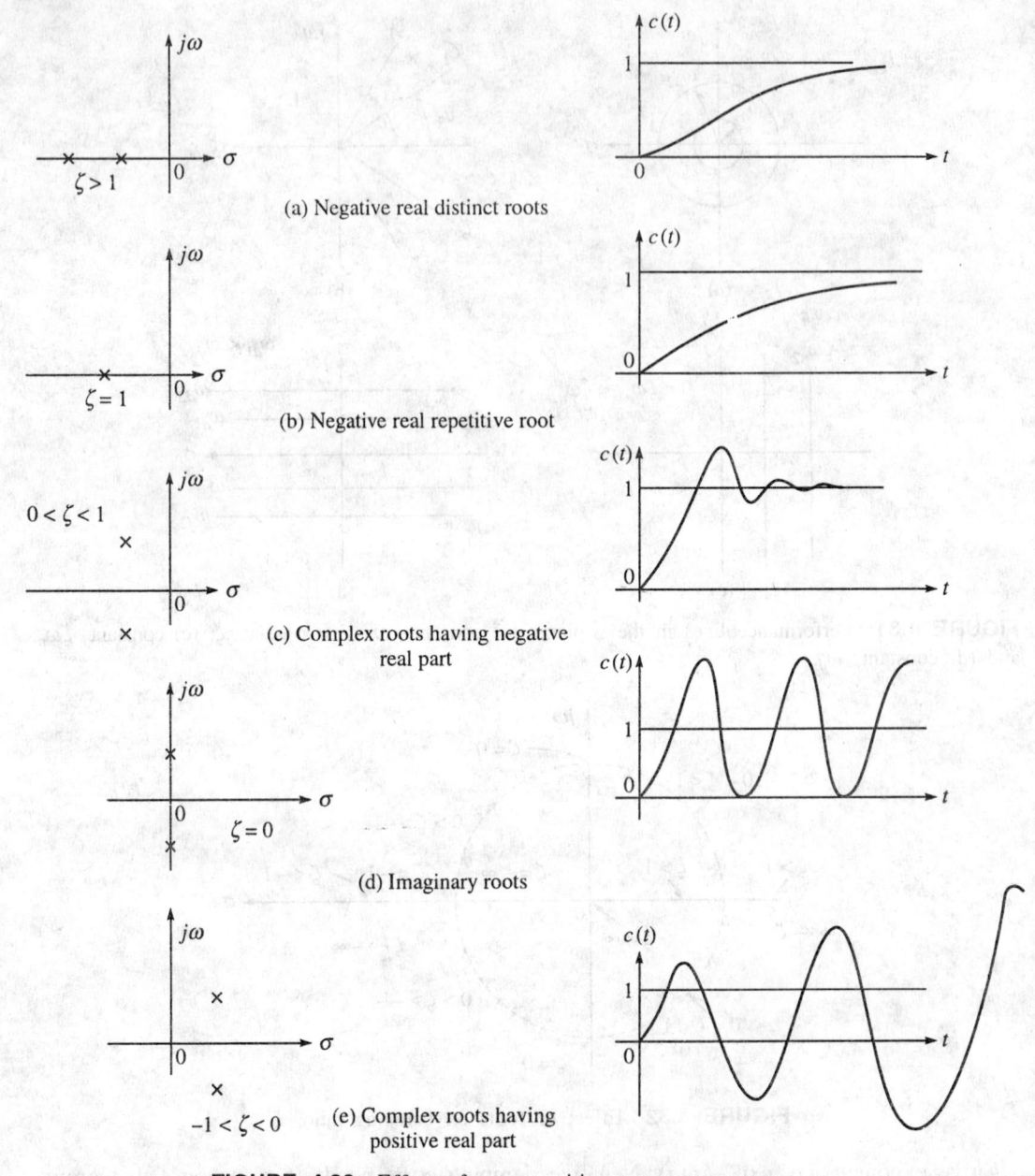

FIGURE 4.33 Effect of roots-position on system response.

(a) Negative real distinct roots, $\zeta > 1$
(b) Negative real repetitive root, $\zeta = 1$
(c) Complex roots having negative real part, $0 < \zeta < 1$
(d) Imaginary roots, $\zeta = 0$
(e) Complex roots having positive real part, $-1 < \zeta < 0$

Using the s-plane performance criterion, it can be ensured that the dominant poles lie within the shaded area (see Figure 4.34). Again from the design point of view, it is advised to keep the specification to the minimum. Over-specification may create unnecessary difficulties. Removal of the constraints on the maximal value of ζ and/or on the minimum value of ω_d would ease the design problem considerably.

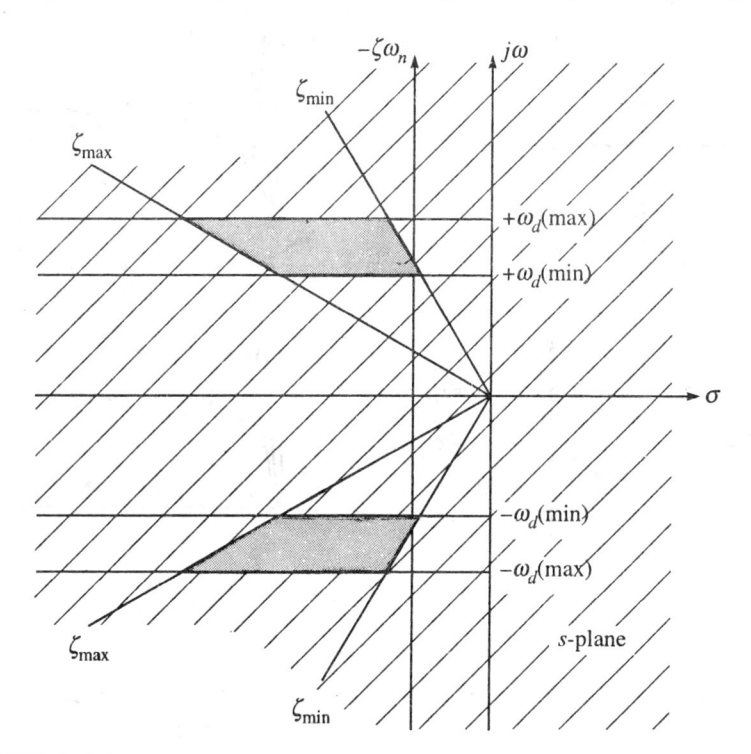

FIGURE 4.34 Dominant pole positions and s-plane performance criteria.

It will be useful to consider the effect on the time response of shifting the pole-position along the s-plane performance lines. Figure 4.35 illustrates this shift for two second-order underdamped systems.

When the systems have the same $\zeta\omega_n$ value, the responses are contained within the same exponential envelopes, but the frequency of oscillation is greater for the system having the larger ω_d value as shown in Figure 4.35(a). Keeping the $\zeta\omega_n$ value constant (that is, time constant) for both the systems, the effect on time response with the vertical movement on the constant $\zeta\omega_n$-axis is shown which indicates that the rate of decay remains the same though the damped frequency changes but effectively the settling time of response remains unaffected despite variation of poles on the constant $\zeta\omega_n$-line.

The same value of damping ratio ζ as in Figure 4.35(b) shows the same peak overshoot but different damping frequencies and settling times. The settling time is much faster for the system whose poles have the more negative real part. Movement of poles on the constant ζ-line keeps the overshoot unaffected though the settling time and damped frequency change.

Figure 4.35(c) shows two systems with the same value of ω_d; the more negative poles indicate higher value of damping ratio ζ and obviously more heavily damped response. The peak times for both responses are the same as ω_d is unaltered for both pairs of poles. The more negative poles produce the faster and more heavily damped response.

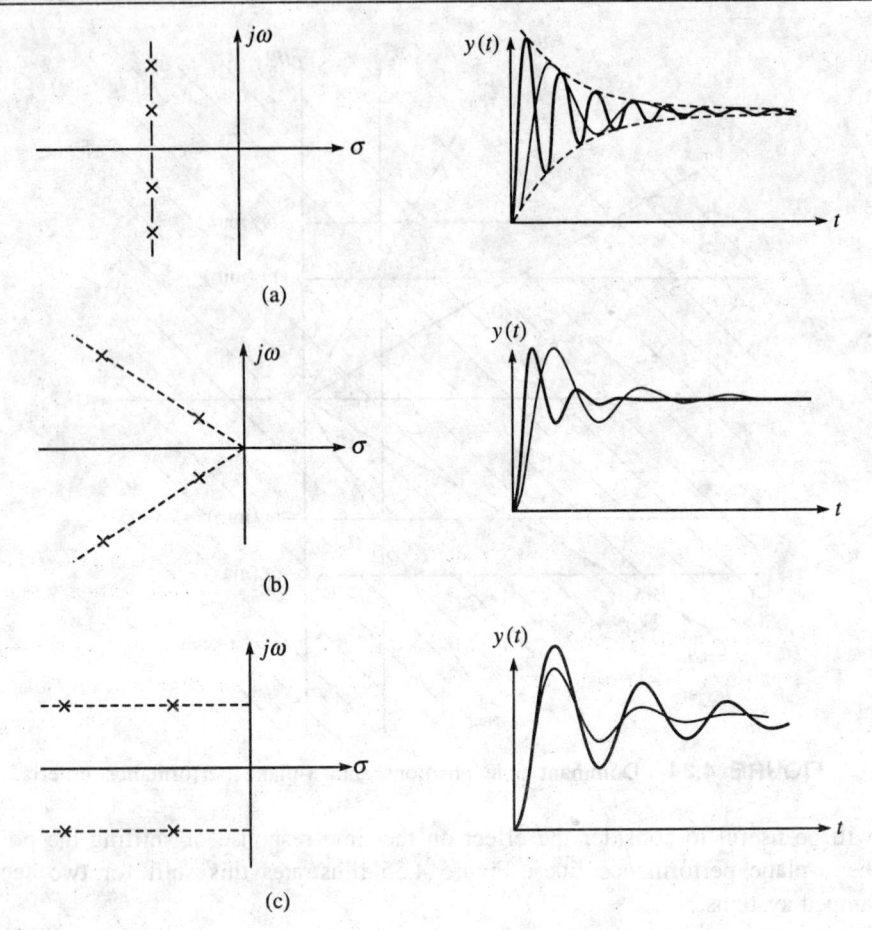

FIGURE 4.35 A system's pole positions and its step response for (a) the same $\zeta\omega_n$ value, (b) the same ζ value, and (c) the same ω_d value.

4.12 Steady-State Error

The use of feedback control system is resorted to for accuracy, usually in the steady-state; occasionally during dynamic operation, we have to have an estimate about the accuracy of the system over long periods of operation, that is, the steady-state error which may be impractical for a human operator to evaluate because of fatigue. We can get the error expression in the frequency domain of the closed-loop system having forward transfer function $G(s)$ and feedback transfer function $H(s)$, as shown in Figure 4.36, i.e.

$$E(s) = \frac{R(s)}{1+G(s)H(s)}$$

which implies the time variation of the error.

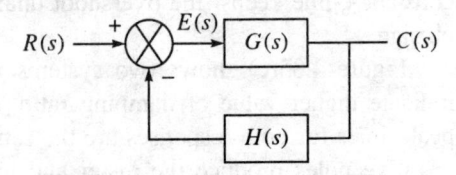

FIGURE 4.36 Closed-loop system.

Transient Response Analysis

By taking the inverse Laplace transform, the transient variation in e(t) can be obtained. But as we are interested in the steady-state error, we will not take the inverse Laplace transform but find the steady-state error by using the final-value theorem of the Laplace transformation for which we need a more explicit expression of the open-loop transfer function as the ratio of two polynomials, and to be physically realizable this expression must have more poles than zeros.

Consider the following open-loop transfer function

$$G(s)H(s) = \frac{K(T_a s + 1) \cdots (T_m s + 1)}{s^N (T_1 s + 1) \cdots (T_p s + 1)} \quad (4.62)$$

The terms s^N in the denominator indicates the pole at origin of multiplicity N.

A system is called Type 0 if $N = 0$, i.e. no pole at the origin. A system is called Type 1 if $N = 1$, i.e. one pole at the origin, Type 2 if $N = 2$, and so on.

By final-value theorem the steady-state error e_{ss} is obtained as

$$e_{ss} = \lim_{t \to \infty} e(t) = \lim_{s \to 0} sE(s) = \lim_{s \to 0} \frac{sR(s)}{1 + G(s)H(s)} \quad (4.63)$$

4.12.1 Unit-step input

Input, $\quad r(t) = u(t); \quad R(s) = \dfrac{1}{s}$

$$e_{ss} = \lim_{s \to 0} \frac{s}{1 + G(s)H(s)} \frac{1}{s} = \frac{1}{1 + \lim_{s \to 0} G(s)H(s)} = \frac{1}{1 + K_p} \quad (4.64)$$

where $\quad K_p = \lim_{s \to 0} G(s)H(s) = G(0)H(0) = G(0) \quad (4.65)$

is defined as the static position error coefficient for unity-feedback systems.

4.12.2 Unit-ramp input

Input, $r(t) = t; \quad R(s) = \dfrac{1}{s}$

$$e_{ss} = \lim_{s \to 0} \frac{s}{1 + G(s)H(s)} \frac{1}{s^2} = \frac{1}{\lim_{s \to 0} sG(s)H(s)} = \frac{1}{1 + K_v} \quad (4.66)$$

where the static velocity error coefficient, $K_v = \lim_{s \to 0} sG(s)H(s)$.

Now, $\quad e_{ss} = \dfrac{1}{K_v} = \infty \quad$ (for Type 0 system); $\quad e_{ss} = \dfrac{1}{K_v} = \dfrac{1}{K} \quad$ (for Type 1 system) and $e_{ss} = \dfrac{1}{K_v} = 0$ (for Type 2 or a higher-order system)

Type 0 system with unity feedback is incapable of following a ramp input in the steady state. Type 1 system with unity feedback can follow a ramp input with a constant finite error, i.e. at steady state, the output velocity will be equal to the input velocity but there would exist a constant finite positional error. A finite velocity error implies that after transients have died out, the input and output move at the same velocity but have a finite position difference.

4.12.3 Unit-parabolic (acceleration) input

Input, $\quad r(t) = \dfrac{t^2}{2}; \quad R(s) = \dfrac{1}{s^3}$

$$e_{ss} = \lim_{s \to 0} \dfrac{1}{1+G(s)H(s)} \dfrac{1}{s^3} = \lim_{s \to 0} \dfrac{1}{s^2 G(s)H(s)} = \dfrac{1}{K_a} \qquad (4.67)$$

where $\quad K_a = \lim_{s \to 0} s^2 G(s)H(s) \qquad (4.68)$

is defined as the static acceleration error coefficient.

The steady-state error for unit parabolic input is

$e_{ss} = \dfrac{1}{K_a} = \infty$ (for Type 0 and Type 1 systems); $\quad e_{ss} = \dfrac{1}{K_a} = \dfrac{1}{K}$ (for Type 2 system)

and $e_{ss} = 0$ (for Type 3 or a higher-order system)

Type 0 and Type 1 systems are incapable of following the parabolic input, Type 2 with unity feedback can follow with a finite error and Type 3 or a higher-order system with unity feedback, can follow a parabolic input without any error.

The error coefficients K_p, K_v and K_a are thus indicative of the steady-state performance. Table 4.2 summarizes the steady-state error for Type 0, Type 1 and Type 2 systems when they are subjected to various inputs.

TABLE 4.2 Steady-state error in terms of gain K

System type	Step input $r(t) = 1$	Ramp input $r(t) = t$	Acceleration input $r(t) = t^2/2$
Type 0	$\dfrac{1}{1+K}$	∞	∞
Type 1	0	$\dfrac{1}{K}$	∞
Type 2	0	0	$\dfrac{1}{K}$

It may thus be observed that from the knowledge of e_{ss} for a given input, we can find the error coefficient which, in turn, tells us about the type of the system. For example, for a velocity (ramp) input if e_{ss} is finite, it gives an estimate for velocity error coefficient $K_v = \lim_{s \to 0} sG(s)H(s)$ to be finite which is only possible for a Type 1 system. For acceleration (parabolic) input, if the steady-state error e_{ss} is infinite, what conclusion can we draw about the open-loop transfer function $G(s)H(s)$? Obviously, it could be either a Type 1 or a Type 0 system.

4.13 Steady-State Error in Terms of Closed-Loop Transfer Function

The steady-state error e_{ss} is often more conveniently obtained from the expression of the transfer function of the closed-loop system. Let us define the transfer function of the closed-loop system as

$$T(s) = \frac{C(s)}{R(s)} \tag{4.69}$$

The Laplace transform of the error $e(t)$ for unity-feedback system is

$$E(s) = R(s) - C(s) = R(s) - T(s)R(s) = R(s)[1 - T(s)] \tag{4.70}$$

The steady-state error e_{ss} is

$$e_{ss} = \lim_{s \to 0}[sE(s)] = \lim_{s \to 0}\left[sR(s)\{1 - T(s)\}\right] \tag{4.71}$$

Using the expression in Eq. (4.71), we will derive the steady-state error e_{ss} for different standard inputs.

Case I: For unit-step input, $R(s) = 1/s$; then the steady-state error is

$$e_{ss} = \lim_{s \to 0}[sE(s)] = \lim_{s \to 0}\left[sR(s)\{1-T(s)\}\right] = \lim_{s \to 0}\left[1-T(s)\right] = 1 - T(0) \tag{4.72}$$

Thus the steady-state error for unit-step input is zero if $T(0) = 1$, that is, if the dc gain of the closed-loop transfer function is unity.

As an illustration, consider the closed-loop transfer function

$$T(s) = \frac{10s + 50}{s^3 + 7s^2 + 35s + 50}$$

From Eq. (4.72), we get the steady-state error $e_{ss} = 0$ to a unit-step input as $T(0) = 1$.

Case II: For unit-ramp input, $R(s) = 1/s^2$; then the steady-state error

$$e_{ss} = \lim_{s \to 0}[sE(s)] = \lim_{s \to 0}[sR(s)\{1-T(s)\}] = \lim_{s \to 0}\left[\frac{1-T(s)}{s}\right] \tag{4.73}$$

It is evident from Eq. (4.73) that $e_{ss} = \infty$ unless $T(0) = 1$ for which we have to use L'Hospital's rule to obtain

$$e_{ss} = \lim_{s \to 0} \left[\frac{-dT(s)}{ds} \right] \qquad (4.74)$$

Let the closed-loop transfer function be

$$T(s) = \frac{b_m s^m + b_{m-1} s^{m-1} + \cdots + b_1 s + b_0}{a_n s^n + a_{n-1} s^{n-1} + \cdots + a_1 s + a_0} = K \cdot \frac{(s + z_1)(s + z_2) \cdots (s + z_m)}{(s + p_1)(s + p_2) \cdots (s + p_n)} \qquad (4.75)$$

where $m \leq n$.

Since differentiation of $T(s)$ here is a difficult task, we shall use the natural logarithm of $T(s)$ for finding e_{ss} using Eq. (4.74).

From Eq. (4.75), after taking natural logarithm, we get

$$\ln[T(s)] = \ln K + \sum_{i=1}^{m} \ln(s + z_i) - \sum_{j=1}^{n} \ln(s + p_j) \qquad (4.76)$$

Taking the derivative of Eq. (4.76), we get

$$\frac{d}{ds}[\ln T(s)] = \frac{1}{T(s)} \frac{dT(s)}{ds} = \sum_{i=1}^{m} \frac{1}{s + z_i} - \sum_{j=1}^{n} \frac{1}{s + p_j}$$

or

$$\frac{dT(s)}{ds} = T(s) \left[\sum_{i=1}^{m} \frac{1}{s + z_i} - \sum_{j=1}^{n} \frac{1}{s + p_j} \right] \qquad (4.77)$$

Coming back to Eq. (4.74), since $T(0) = 1$, the steady-state error for unit-ramp input using Eq. (4.77) becomes

$$e_{ss} = \lim_{s \to 0} \left[-\frac{dT(s)}{ds} \right] = \lim_{s \to 0} \left[-T(s) \left\{ \sum_{i=1}^{m} \frac{1}{s + z_i} - \sum_{j=1}^{n} \frac{1}{s + p_j} \right\} \right]$$

$$= -T(0) \left[\sum_{i=1}^{m} \left(\frac{1}{z_i} \right) - \sum_{j=i}^{n} \left(\frac{1}{p_j} \right) \right]$$

$$= -\sum_{i=1}^{m} \left(\frac{1}{z_i} \right) + \sum_{j=1}^{n} \left(\frac{1}{p_j} \right); \quad \text{(as } T(0) = 1\text{)} \qquad (4.78)$$

Thus we have got a very simple expression (4.78) for the steady-state error for unit-ramp input which is valid only if $T(0) = 1$.

It may further be noted that the knowledge of locations of poles and zeros may not be available; in that case when the closed-loop transfer function $T(s)$ is expressed as a ratio of polynomials then by using the properties of polynomials, we get

$$-\sum_{i=1}^{m}\left(\frac{1}{z_i}\right) = -\left(\frac{b_1}{b_m}\right) \quad \text{and} \quad -\sum_{j=1}^{n}\left(\frac{1}{p_j}\right) = -\left(\frac{a_1}{a_n}\right)$$

Then the expression for e_{ss} becomes

$$e_{ss} = -\left(\frac{b_1}{b_m}\right) + \left(\frac{a_1}{a_n}\right) \qquad (4.79)$$

As an illustration, consider the closed-loop transfer function

$$T(s) = \frac{10s + 50}{s^3 + 7s^2 + 35s + 50}$$

Using Eq. (4.79), we get the steady-state error for a unit-ramp input, $e_{ss} = 0.5$ as $T(0) = 1$. Further, if $T(s)$ is given in the factored form as

$$T(s) = \frac{10(s+5)}{(s+2)(s+2.5+j4.33)(s+2.5-j4.33)}$$

then using Eq. (4.79), we get steady-state error, $e_{ss} = 0.5$ to a unit-ramp input.

Case III: For unit-parabolic input, $R(s) = 1/s^3$ and the steady-state error e_{ss} can be written as

$$e_{ss} = \lim_{s \to 0}[sE(s)] = \lim_{s \to 0}[sR(s)\{1 - T(s)\}] = \lim_{s \to 0}\left[\frac{1 - T(s)}{s^2}\right] \qquad (4.80)$$

Hence the steady-state error will be infinite unless

$$T(0) = 1 \qquad (4.81)$$

and

$$\lim_{s \to 0} \frac{d}{ds}[\ln T(s)] = 0 \qquad (4.82)$$

Equation (4.82) implies

$$\sum_{i=1}^{m} \frac{1}{z_i} = \sum_{j=1}^{n} \frac{1}{p_j} \qquad (4.83)$$

When both the conditions of Eqs. (4.81) and (4.82) are satisfied, then applying the L'Hospital's rule twice, we get

$$e_{ss} = \lim_{s \to 0}\left[-\left(\frac{1}{2}\right)\frac{d^2}{ds^2}\{T(s)\}\right] \qquad (4.84)$$

In order to evaluate Eq. (4.84), we differentiate Eq. (4.76) twice to get

$$\frac{d^2}{ds^2}[\ln T(s)] = -\sum_{i=1}^{m}\frac{1}{(s+z_i)^2} + \sum_{j=1}^{n}\frac{1}{(s+p_j)^2} \quad (4.85)$$

Then the steady-state error e_{ss} to a unit-parabolic input is therefore given by

$$e_{ss} = \frac{1}{K_a} = \left(\frac{1}{2}\right)\left[\sum_{i=1}^{m}\left(\frac{1}{z_i^2}\right) - \sum_{j=1}^{n}\left(\frac{1}{p_j^2}\right)\right] \quad (4.86)$$

Hence to find e_{ss} it is not required to know the locations of poles and zeros of $T(s)$.

EXAMPLE 4.8 An automobile speed control system is shown in Figure 4.37.

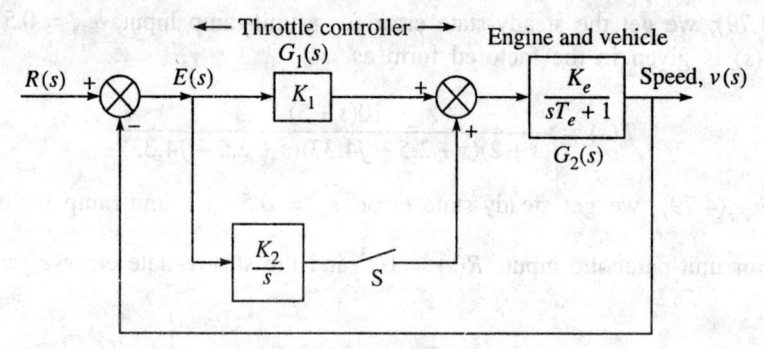

FIGURE 4.37 Example 4.8: an automobile speed control system.

The throttle controller, $G_1(s)$, is given as

$$G_1(s) = K_1 + \frac{K_2}{s} = \frac{K_1 s + K_2}{s}$$

when the switch S is closed, otherwise $G_1(s) = K_1$, that is, when the switch S is open. The given plant, that is, the engine and the vehicle,

$$G_2(s) = \frac{K_e}{sT_e + 1}$$

When the switch S is open, then the positional error coefficient K_p (for a step input) is

$$K_p = \lim_{s \to 0} G_1(s)G_2(s) = \lim_{s \to 0} \frac{K_1 K_e}{sT_e + 1} = K_1 K_e$$

The steady-state error e_{ss} becomes

$$e_{ss} = \frac{1}{1 + K_p} = \frac{1}{1 + K_1 K_e} = \text{finite value}$$

When the switch S is closed, then the positional error coefficient K_p (for a step input) is

$$K_p = \lim_{s \to 0} G_1(s)G_2(s) = \lim_{s \to 0} \frac{K_1 s + K_2}{s} \frac{K_e}{sT_e + 1} = \infty$$

and the steady-state error e_{ss} becomes zero for a step input.
Similarly, the velocity error coefficient K_v is

$$K_v = \lim_{s \to 0} sG_1(s)G_2(s) = \lim_{s \to 0} s\frac{K_1 s + K_2}{s} \frac{K_e}{sT_e + 1} = K_2 K_e$$

The steady-state error e_{ss} becomes

$$e_{ss} = \frac{1}{K_v} = \text{finite quantity (a constant)}$$

The error coefficients K_p, K_v and K_a of a control system describe the ability of the system to reduce or to eliminate the steady-state error. These are indicators for numerical measures of the steady-state performance. The designer determines the error constants for a given system and attempts to determine the methods of increasing the error constants while maintaining an acceptable transient response. For example, for the speed control system, to reduce e_{ss}, we have to increase K_v, that means increasing the gain factor $K_e K_2$. Any increase in $K_e K_2$ decreases the damping ratio ζ and the system response to step input will be more oscillatory with a high overshoot. Hence a compromise would have to be determined which would provide the largest K_v based on the smallest ζ allowable.

Drill Problem 4.7
Verify the steady-state errors given in Table 4.2.

4.14 Integral Performance Criterion

A performance index is a qualitative measure of the performance of a system and is chosen so that the emphasis is on important system specifications. A qualitative measure of the performance of a system is necessary where one can adjust certain parameters for the design of an optimum system. Here the system parameters are adjusted so that the index reaches an extremum value, commonly a minimum value. For this purpose, a performance index must be chosen and measured. This performance index is also called the *cost function*.

A suitable cost function is the integral of the square of the error (ISE) defined as

$$J_1 = \int_0^T e^2(t)dt \qquad (4.87)$$

where $e(t) = r(t) - c(t)$ is the error between the desired output (i.e. the reference input) and the actual output for a unity-feedback system. The upper limit T is a finite time chosen somewhat arbitrarily so that the integral approaches a steady-state value. It is usually convenient to choose

T as the settling time, T_s. The step response to a unity-feedback underdamped system is shown in Figure 4.38(a); the error in Figure 4.38(b); the square of error in Figure 4.38(c) and the integral of the square of error in Figure 4.38(d). As the squaring circuit is readily available, the integral of the square of error is a widely accepted performance index. Furthermore, the squared error is mathematically convenient for analytical and computational purposes.

Another popular criterion is the integral of the absolute value of the error (IAE) defined as

$$J_2 = \int_0^T |e(t)| dt \qquad (4.88)$$

It can be measured easily using an analog computer and that is the main advantage.

Another criterion is the integral of the time multiplied by the square of the error (ITSE), defined as

$$J_3 = \int_0^T t e^2(t) dt \qquad (4.89)$$

Still another criterion is the time multiplied by the absolute value of the error (ITAE), defined as

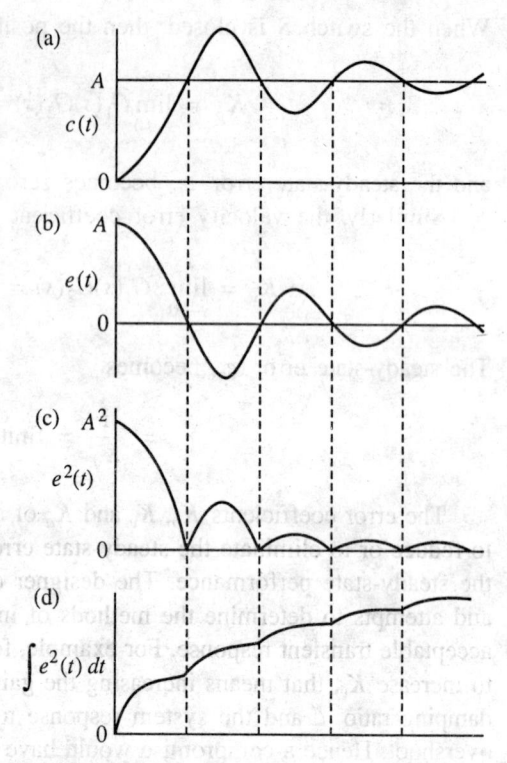

FIGURE 4.38 The calculation of the integral squared error.

$$J_4 = \int_0^T t |e(t)| dt \qquad (4.90)$$

4.15 Derivative Control

A negative feedback control system of Figure 4.39 is one which develops a correcting effort proportional to the magnitude of the actuating signal. Thus the negative feedback system is a proportional type of feedback system. The limitation of the proportional control system is that in the underdamped case if the forward loop gain is increased, the damping factor ζ is reduced further in the range $0 < \zeta < 1$ and, in turn, the overshoot increases and the rise time decreases. There should be a compromise between the rise time and overshoot with the proper choice of gain to attain the proper value of damping factor ζ. Sometimes a compromise cannot be reached to have the overshoot within an acceptable tolerance or the system may even be unstable. Further, there may be a steady-state error as well.

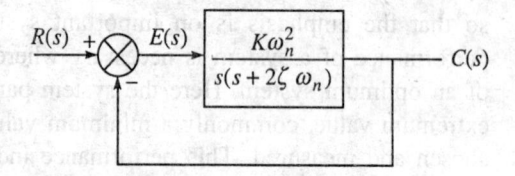

FIGURE 4.39 Feedback control system.

Transient Response Analysis

The situation demands that we consider some other type of control, in addition to the proportional type of control, which may improve the transient and steady-state behaviour of an ordinary negative feedback control system. The following three basic control schemes are adapted to modify the performance of the underdamped negative feedback control systems. (i) derivative control, (ii) integral control, and (iii) rate feedback control.

Let us describe the basic principle of D-control. The description can be made in two parts: philosophy of derivative control and analytical explanation.

The typical unit-step response $c(t)$ of an underdamped (proportional type servomechanism) second-order negative feedback system is shown in Figure 4.40(a). The corresponding error curve $e(t) = 1 - c(t)$ and the time rate of change of error, i.e. de/dt are shown in Figures 4.40(b) and (c), respectively. Assume that the system has only proportional type of control. In fact, all negative feedback control systems are basically proportional control systems. In the negative feedback control system when error $e(t) = 0$, the ouput $c(t)$ is equal to the reference input. As $e(t) = 0$, the actuating signal to drive the motor will be zero and the motor shaft will remain stationary. Therefore, the curve is swinging from positive to negative and then again to positive at the instants $t = t_1, t_3, t_5, \ldots$ passing through zero. The large overshoot is entirely due to excessive amount of positive torque developed by the motor in the time interval $0 < t < t_1$ during which the error signal $\int_0^{t_1} e\,dt$ is positive. For the time interval $t_1 < t < t_3$, the error signal $\int_{t_1}^{t_3} e\,dt$ is negative and the motor torque is in the reverse direction, i.e. the motor develops a retarding torque bringing the overshooting output back. The retarding torque was in excess because of the excess amount of negative error signal in the interval $t_1 < t < t_3$ and therefore the undershoot occurs. Again during the interval $t_3 < t < t_5$, the error signal $\int_{t_3}^{t_5} e\,dt$ is positive. It results in a positive torque which tends to reduce the undershoot in the response caused by the negative torque in the interval $t_2 < t < t_3$. The process is repeated and for a stable system, a steady-state is reached.

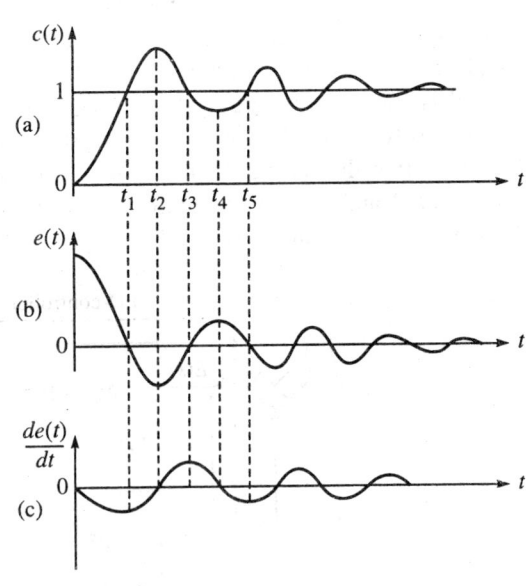

FIGURE 4.40 Waveforms of $c(t)$, $e(t)$ and $de(t)/dt$ showing the effect of derivative control.

The cause for the overshoot and undershoot and its remedy can be summarized as follows:

(i) The positive correcting torque in the interval $0 < t < t_1$ is too large. This should be reduced.

(ii) The retarding torque in the interval $t_1 < t < t_2$ is inadequate. This should be increased.

(iii) The negative correcting torque in the interval $t_2 < t < t_3$ is in excess. This should be reduced.

(iv) The positive correcting torque in the interval $t_3 < t < t_4$ is not sufficient. This should be increased.

Now let us incorporate derivative control in addition to the existing proportional control. This results in an actuating signal for the motor. See the de/dt curve in Figure 4.40(c). The actuating signal for the motor is now $[e(t) + T_d(de/dt)]$ which improves the resulting motor torque as prescribed in the remedy. The block diagram of the proportional plus derivative control is shown in Figure 4.41. The signal supplied to the motor consists of two components: one proportional to the instantaneous error, the other the time derivative of the error signal. It is seen that during the interval $0 < t < t_1$, the de/dt is negative. Hence it is seen that with proportional and derivative control, the response of the closed-loop control system improves. This has been explained in Table 4.3.

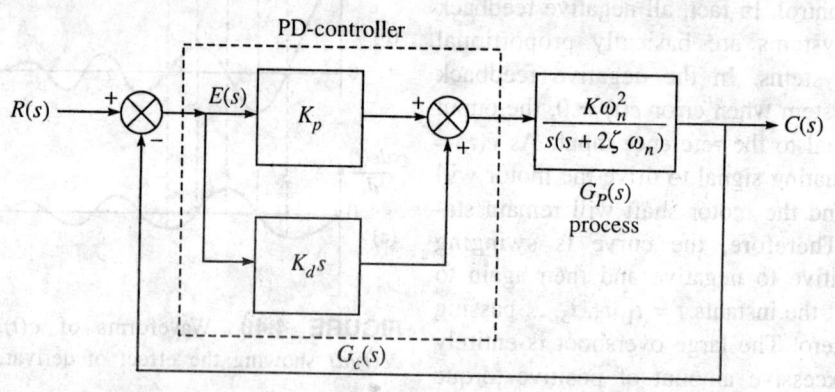

FIGURE 4.41 Feedback control system with proportional and derivative control.

TABLE 4.3 Effect of proportional and derivative control

Time interval	Error, $e(t)$	de/dt	Net $(e + de/dt)$	Correcting torque developed	Effect
$0 < t < t_1$	+ve	−ve	Reduced to less +ve	Less +ve	Overshoot is less
$t_1 < t < t_2$	−ve	−ve	Increased to more −ve	Retarding torque	Overshoot is less
$t_2 < t < t_3$	−ve	+ve	Reduced to less −ve	Negative correcting torque reduced	Undershoot reduced
$t_3 < t < t_4$	+ve	+ve	Increased to more +ve	Positive retarding torque increases	Undershoot reduced
$t_4 < t < t_5$	+ve	−ve	Less +ve	Retarding torque reduced	Less overshoot

This is to note that derivative control is effective in the transient part of the response as error is varying, whereas in the steady-state, usually if any error is there, it is constant and does not vary with time. In this aspect, derivative control is not effective in the steady-state part of the response.

In the steady-state part, if any error is there, integral control will be effective to give proper correcting torque to minimize the steady-state error. An integral controller is basically a low-pass circuit and hence will not be effective in transient part of the response where error is fast changing. Hence for the whole range of time response both derivative and integral control actions should be provided in addition to the in-built proportional control action for negative feedback control systems and altogether named as PID (proportional-integral-derivative) control.

4.15.1 Interpretation of derivative control from root locus point of view

We have seen that with proportional and derivative control, the transient response of the closed-loop control system improves. It may be noted that for constant steady-state error, derivative control is not effective as the time derivative of a constant is zero. However, derivative control affects the steady-state error of the system only if the steady-state error varies with time. The effect of D-control on time response of a feedback control system can be explained by analytical means as follows.

The derivative control introduces a zero and the order of the system remains the same. From the root locus point of view we can show that due to derivative control, the stability has improved. The open-loop transfer function without derivative control is

$$\frac{K\omega_n^2}{s(s+2\zeta\omega_n)} \qquad (4.91)$$

and the corresponding root loci when the open-loop gain K is varying from zero to infinity is shown in Figure 4.42(a).

The open-loop transfer function with derivative feedback is

$$\frac{K\omega_n^2(1+sT_d)}{s(s+2\zeta\omega_n)} \qquad (4.92)$$

and the corresponding root loci is shown in Figures 4.42(b) and (c) for two different values of T_d. From these root-locus diagrams, the effect of derivative control on damping factor of the transient response is apparent. The effect of adding a zero at $-1/T_d$ is to cause the original root loci to bend towards the left. In Figure 4.42(b) for gain $0 < K < K_1$ the system is overdamped and as the gain K is increased further, the system becomes underdamped in the range $K_1 < K < K_2$ and the closed-loop poles become complex conjugate. For gain increased beyond K_2 the system becomes overdamped again as shown in Figure 4.42(b). Further, if T_d is such that $-2\zeta\omega_n < -1/T_d < 0$, then the system is overdamped for all values of K in the range $0 < K < \infty$ and overshoot never occurs. as depicted in Figure 4.42(c). It may be concluded from

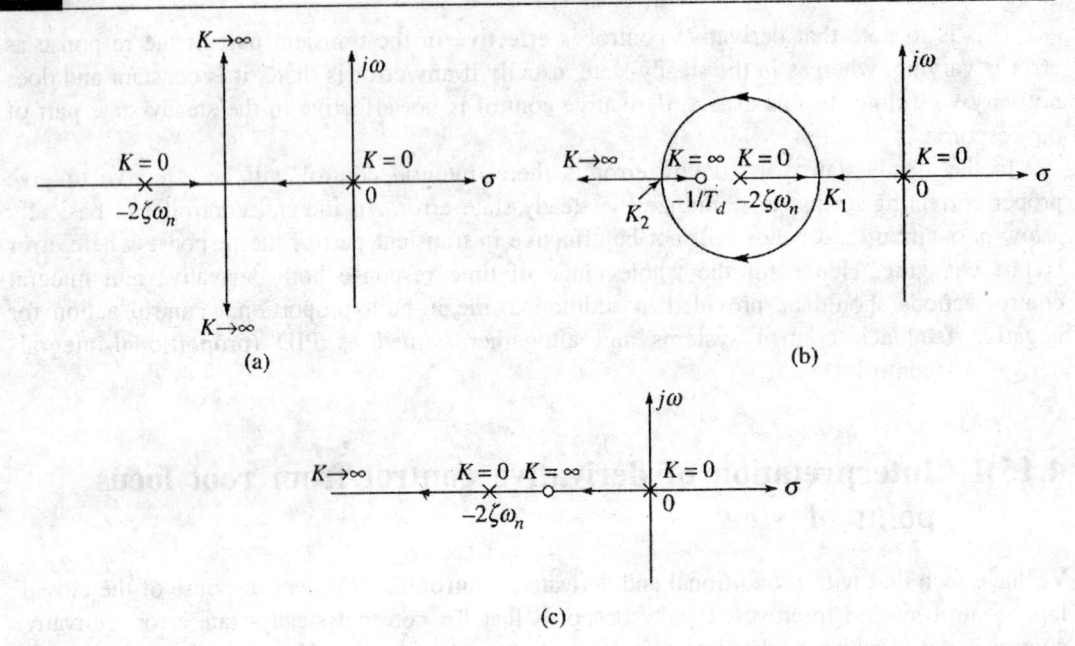

FIGURE 4.42 Effect of derivative control from root locus point of view: (a) negative feedback system with proportional control only, (b) derivative control with $|1/T_d| > |2\zeta\omega_n|$, and (c) derivative control with $|1/T_d| < |2\zeta\omega_n|$.

the root-loci diagram that the system with derivative control is more stable compared to that without derivative control. However in this particular example both the systems (with and without derivative control) are stable. It may be noted that a system with derivative control in the forward path the error signal is of low voltage level and that may not be able to drive the motor, hence a power amplifier is required.

4.16 Rate Feedback (Tachometer Feedback) Control

The philosophy of derivative control can be extended by feeding back the derivative of the output signal through tachogenerator and comparing it with the reference input as shown in Figure 4.43. This figure shows the block diagram of a simple second-order system with a secondary path feeding back the output velocity of the system. The closed-loop transfer

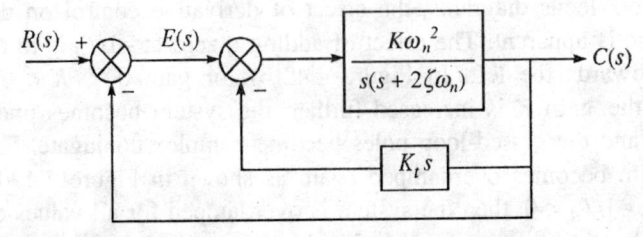

FIGURE 4.43 Rate feedback (tachogenerator control).

function of the system is given by

$$\frac{C(s)}{R(s)} = \frac{K\omega_n^2}{s^2 + (2\zeta\omega_n + KK_t\omega_n^2)s + K\omega_n^2} \tag{4.93}$$

and the characteristic equation is

$$s^2 + (2\zeta\omega_n + KK_t\omega_n^2)s + K\omega_n^2 = 0 \tag{4.94}$$

Comparing this equation with the characteristic equation derived from Eq. (4.92), we see that they are of the same form. In fact, if T_d in Eq. (4.92) were replaced by K_t, both the characteristic equations would be identical. Therefore, we can conclude that the rate feedback (velocity feedback or tachometer feedback) also improves (changes) the damping coefficient of the time response of a control system. This is to mention here that, for this same damping ratio, the amount of overshoot for a second-order system will be the same for derivative control in the forward path and rate feedback, but the two responses are not the same. In general, because of the added zero in the open-loop transfer function, the rise time of the derivative control system in the forward path will be faster.

The derivative control has its own effect on the transient response because in this part of the response, the error is varying. The derivative control has no effect on the steady-state response because in the steady-state the error is usually constant and not varying.

However, if the steady-state error increases with time, a torque is again developed proportional to *de/dt*, which reduces the magnitude of the error.

The effect of derivative control on time response of a feedback control system can also be explained by analytical means as in the following. The open-loop transfer function of the system with proportional plus derivative control shown in Figure 4.43 is

$$G(s) = \frac{C(s)}{E(s)} = \frac{K(1+T_d s)\omega_n^2}{s(s+2\zeta\omega_n)} \tag{4.95}$$

Hence the closed-loop transfer function is

$$\frac{C(s)}{R(s)} = \frac{K(1+T_d s)\omega_n^2}{s^2 + (2\zeta\omega_n + KT_d\omega_n^2)s + K\omega_n^2} \tag{4.96}$$

The characteristic equation of the closed-loop system with PD-control is

$$s^2 + (2\zeta w_n + KT_d\omega_n^2)s + K\omega_n^2 = 0 \tag{4.97}$$

The form of the last equation is similar to that of the characteristic equation of the proportional type system, except that the coefficient of s is increased by the quantity $KT_d\omega_n^2$. This actually means that the damping coefficient of the system is increased. This implies that the rise time has increased, that is, the system will have a longer delay. The system becomes more stable. The proportional plus derivative control introduces a zero in numerator of the transfer function. Again, with the introduction of zero by PD-control, the system becomes more stable from the root locus point of view.

4.17 Comparison between Derivative and Rate Feedback Control

The rate feedback is provided by the tachogenerator. The output level is high and that is why a tachogenerator can be applied in the feedback path to have derivative control action. On the other hand, as the error level is low, we have to provide a high-pass *RC* circuit in the forward path to perform the derivative action. (Note that differentiation amplifies the noise effects.) In fact, if discontinuous noises are present, differentiation amplifies the discontinuous noises more than the useful signal. For example, the output of a potentiometer is a discontinuous voltage signal due to potentiometer brush moving on the winding.)

We have to have the power amplifier for derivative control in the forward path, as the error signal level is low and is not sufficient enough to drive the plant. When we use the tachogenerator feedback we do not need the power amplifier because the output signal level is quite high. Hence cost-wise both the control schemes are of the same order; the one used only depends on the availability of the components.

EXAMPLE 4.9 The given feedback system is as shown in Figure 4.44. Determine the value of gain K and that of velocity feedback constant K_1 so that the poles of the closed-loop system lie on a $\pm 60°$ line from the negative-real axis in the s-plane and the output reaches within $\pm 2\%$ of the steady-state value not before 2 s. Also determine the percentage overshoot and occurrence of peak time when the system is subjected to a unit-step input.

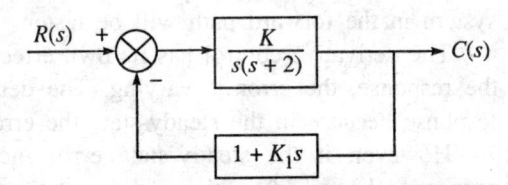

FIGURE 4.44 Example 4.9.

Solution: Given $\zeta = \cos 60° = 0.5$. Again with 2% criterion, $t_s = 2$ s $= 4/\zeta\omega_n$. Hence, $\omega_n = 4/2(0.5) = 4$ rad/s.

The closed-loop transfer function,

$$\frac{C(s)}{R(s)} = \frac{K}{s^2 + (KK_1 + 2)s + K}$$

Comparing with the standard form,

$$\frac{\omega_n^2}{s^2 + 2\zeta\omega_n s + \omega_n^2} = \frac{16}{s^2 + 4s + 16}$$

We get the value of $K = 16$, $K_1 = 0.125$. From the formula, we calculate

$$\% \text{ overshoot} = 100 \exp\left[-\frac{\zeta}{\sqrt{1-\zeta^2}}\pi\right] = 16.3\%$$

and

$$t_p = \frac{\pi}{\omega_n\sqrt{1-\zeta^2}} = 0.906 \text{ s}$$

Drill Problem 4.8

Refer to Figure 4.44, determine the values of K and K_1 so that for the closed-loop system subjected to unit-step input the % overshoot is 20% and the peak time is 1 s. Determine the values of t_s, t_r with 5% criterion.

Ans. $K = 12.5$, $K_1 = 0.097$, $t_r = 0.65$ s, and $t_s = 2.48$ s

EXAMPLE 4.10 When the system shown in Figure 4.45(a) is subject to a unit-step input, the system output response is shown in Figure 4.45(b). Determine the values of K and k from the output response curve. Obtain the rise time and settling time.

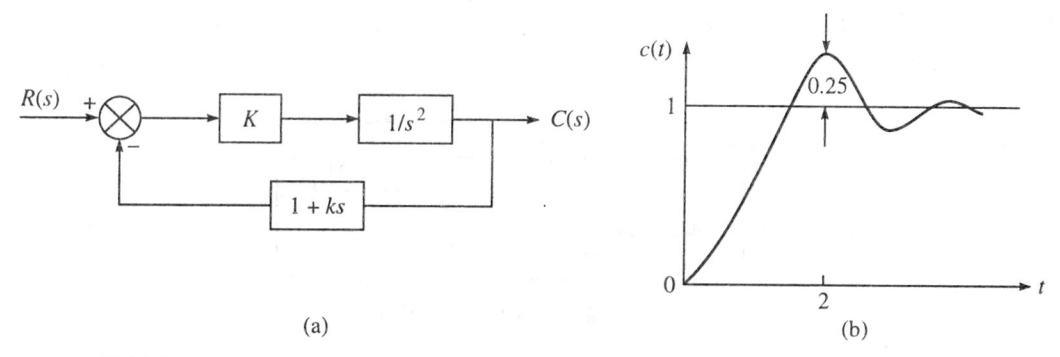

FIGURE 4.45 Example 4.10: (a) closed-loop system and (b) output response.

Solution: Given $G(s) = K/s^2$, $H(s) = 1 + ks$. The closed-loop transfer function is

$$\frac{C(s)}{R(s)} = \frac{G}{1+GH} = \frac{K}{s^2 + Kks + K},$$

where the characteristic equation is $s^2 + Kks + K = 0$

Comparing with the standard form of equation, $s^2 + 2\zeta\omega_n s + \omega_n^2 = 0$, we obtain

$$\omega_n = \sqrt{K} \quad \text{and} \quad 2\zeta\omega_n = Kk$$

Maximum % overshoot $M_p = 25\%$ (as specified in Figure 4.45(b)). Therefore,

$$\exp\left(\frac{-\pi\zeta}{\sqrt{1-\zeta^2}}\right) = M_p = 0.25, \text{ which yields } \zeta = 0.404.$$

The given peak time (as shown in Figure 4.45(b)), $t_p = 2$ s

Therefore, $t_p = 2 = \dfrac{\pi}{\omega_d} = \dfrac{\pi}{\omega_n\sqrt{1-\zeta^2}}$, which yeilds $\omega_n = 1.72$.

From comparison with the characteristic equation, we get

$$K = \omega_n^2 = (1.72)^2 = 2.95 \text{ N-m}, \quad \text{and} \quad k = \frac{2\zeta\omega_n}{K} = \frac{2(0.404)(1.72)}{2.95} = 0.471$$

The rise time t_r can be obtained as

$$t_r = \frac{\pi - \theta}{\omega_d} = 1.26 \text{ s, where } \theta = \tan^{-1}\left(\frac{\omega_n\sqrt{1-\zeta^2}}{\zeta\omega_n}\right) = 66.17° = 1.154 \text{ rad}$$

Settling time t_s can be obtained as

$$t_s(\text{for 2\% criterion}) = \frac{4}{\zeta\omega_n} = 5.76 \text{ s}, \quad \text{and} \quad t_s(\text{for 5\% criterion}) = \frac{3}{\zeta\omega_n} = 4.32 \text{ s}$$

EXAMPLE 4.11 For the system shown in Figure 4.46, analyze the system from all aspects with two varying parameters K_p and K_d.

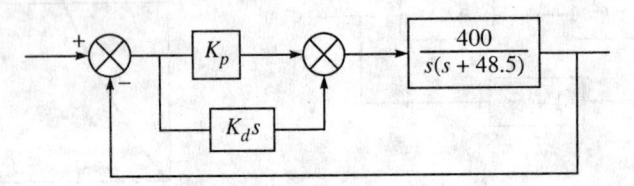

FIGURE 4.46 Example 4.11.

Solution: Case I: $K_p = 2.94$, $K_d = 0$
The characteristic equation is, $1 + G(s)H(s) = 0$

or $\quad s^2 + 48.5s + (400 \times 2.94) = s^2 + 48.5s + 1176 = 0$

Comparing with the standard equation, $s^2 + 2\zeta\omega_n s + \omega_n^2 = 0$, we get

$$\omega_n = \sqrt{1176} \text{ rad/s} \quad \text{and} \quad \zeta = 0.707$$

The unit-step response for Case I is the upper curve shown in Figure 4.47(a).

Case II: $K_p = 2.94$, $K_d = 0.05$. The open-loop transfer function is

$$G(s)H(s) = \frac{400(K_p + K_d s)}{s(s + 48.5)}$$

The characteristic equation is

$$s^2 + (48.5 + 400K_d)s + 400K_p = 0$$

or $\quad s^2 + (48.5 + 400 \times 0.05)s + 400 \times 2.94 = 0$

Comparing with standard equation $s^2 + 2\zeta\omega_n s + \omega_n^2 = 0$, we get $\omega_n = \sqrt{1176}$ rad/s and $\zeta = 1$.

The system becomes critically damped and sluggish. The unit-step response for Case II is the lower curve shown in Figure 4.47(a).

Case III: $K_p = 100$, $K_d = 0$. Then, $\zeta = 0.12125$ which gives, % overshoot = 68%
In order to reduce overshoot, increase K_d to 0.05 with same $K_p = 100$ which gives $\zeta = 0.1715$ and the overshoot reduces as is evident from Figure 4.47(b). Again for $K_p = 100$ and $K_d = 0.8788$ gives $\zeta = 1$ and the response is shown in Figure 4.47(b).

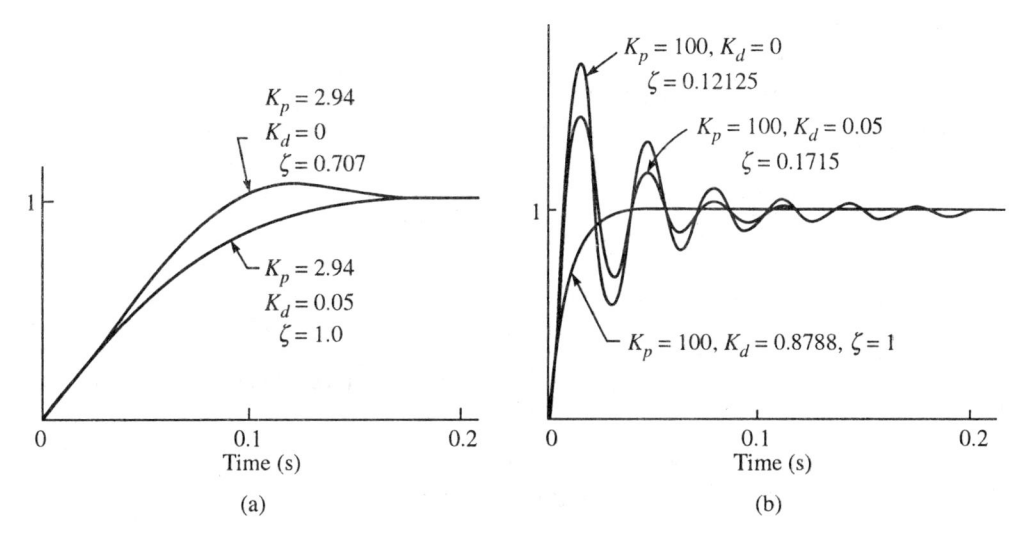

FIGURE 4.47 Example 4.11: (a) unit-step response with PD control and with $K_p = 2.94$ and (b) unit-step response with PD control and with $K_p = 100$.

Further, the problem may be looked from a different angle. Suppose $K_d \neq 0$; with $K_p = 2.94$, adjust K_d in such a way that $\zeta = 1$. Then proceed as follows.

The open-loop transfer function

$$G(s)H(s) = \frac{400(K_p + K_d s)}{s(s + 48.5)}$$

The characteristic equation, $s^2 + (48.5 + 400K_d)s + 400K_p = 0$
Putting $K_p = 2.94$, we get

$$s^2 + (48.5 + 400K_d)s + 1176 = 0$$

Comparing with the standard equation, $s^2 + 2\zeta\omega_n s + \omega_n^2 = 0$, we get

$$\omega_n = \sqrt{1176} \text{ rad/s} \quad \text{and} \quad 2\zeta\omega_n = 48.5 + 400K_d$$

For ζ to be 1 and $\omega_n = \sqrt{1176}$, K_d becomes 0.05.

The effect of introducing D-control is evident from Figures 4.47(a) and (b).

Analytically, the design of a PD controller can be carried out by the root-contour method since the characteristic equation involves two parameters K_d and K_p. Let us first put $K_d = 0$, then the characteristic equation becomes

$$s^2 + 48.5s + 400K_p = 0$$

which can be rewritten as

$$1 + \frac{400K_p}{2(s+48.5)} \Rightarrow 1 + GH = 0$$

with the variation of K_p in the range $0 < K_p < \infty$; the root loci is shown in Figure 4.48.

Now with $K_d \neq 0$, the characteristic equation becomes

$$1 + G(s)H(s) = 1 + \frac{400(K_p + K_d s)}{s(s+48.5)} = 0$$

or $s^2 + (48.5 + 400K_d)s + 400K_p = 0$

or $s^2 + 48.5s + 400K_p + 400K_d s = 0$

We rewrite this equation as

$$1 + \frac{400K_d s}{s^2 + 48.5s + 400K_p} = 0$$

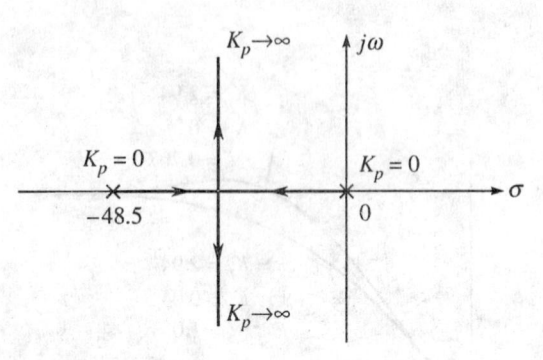

FIGURE 4.48 Example 4.11: root loci.

The root contours with K_p = constant, as K_d varies in the range $0 < K_d < \infty$, are constructed as shown in Figure 4.49. For K_p = 2.94 and K_d = 0, the roots of the characteristic equation are at $-24.25 \pm j24.25$. When K_d increases, the two roots move towards the real axis along a circular arc. When K_d = 0.05, the two roots are real and repetitive and equal to -34.29 which indicates critical damping. Hence for $0 < K_d < 0.05$ and K_p = constant = 2.94, the two roots are complex conjugate as is evident from the root-contour diagram.

Similarly, when K_p = 100 and K_d = 0, the roots are $-24.25 \pm j198.5$. As K_d varies in the range $0 < K_d < \infty$, keeping a constant value of K_p = 100, the root contour is as in Figure 4.49. For $0 < K_d < 0.8788$ the roots are complex conjugate. For K_d = 0.8788, the system is critically damped. For $0.8788 < K_d < \infty$, the roots are real.

Integral control

We want to start our discussion by explaining as to why do we need an integral controller together with a proportional one. Or, in other words, what makes us think about the necessity of an integral controller. In this regard we would like to say that the proportional control of a plant that does not possess an integrating property (which means that the plant transfer function does not include the factor $1/s$) suffers offset in response to step inputs. This is evident from Figure 4.50. For the case when $N = 0$, the positional error coefficient $K_p = \lim_{s \to 0} G(s)H(s) = K$ and the steady-state error $e_{ss} = \frac{1}{1+K} = a$ finite quantity. For the case when $N \neq 0$, $K_p = \infty$ and hence $e_{ss} = 0$.

Figure 4.51 shows the block diagram of a feedback control system that has a second-order process with transfer function $G_p(s)$ and a controller with proportional and integral control

Transient Response Analysis

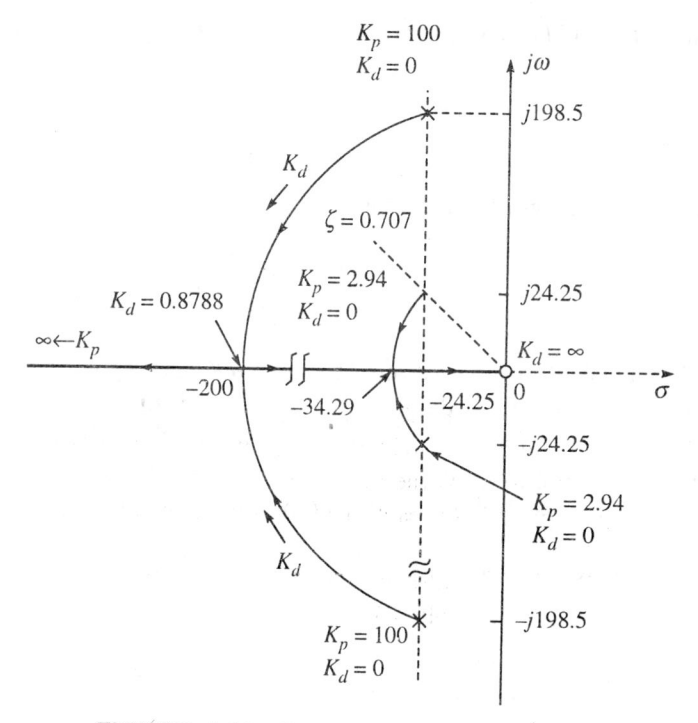

FIGURE 4.49 Example 4.11: root contours.

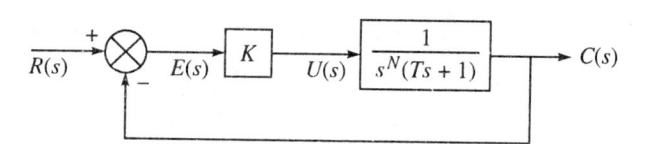

FIGURE 4.50 Integral control.

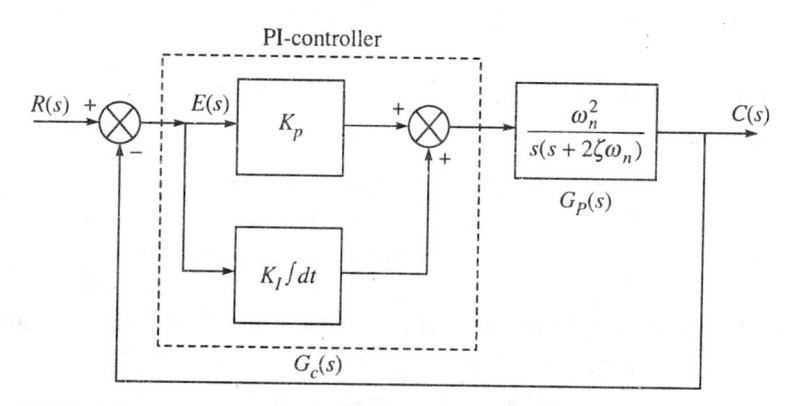

FIGURE 4.51 Feedback control system with proportional-integral control.

whose transfer function is $G_c(s)$ and which can be written as

$$G_P(s) = \frac{\omega_n^2}{s(s+2\zeta\omega_n)} \quad \text{and} \quad G_c(s) = K_p + \frac{K_I}{s} = \frac{K_p s + K_I}{s} = \frac{K_p\left(s + \frac{K_I}{K_P}\right)}{s}$$

The open-loop transfer function of the overall system is

$$G_c(s)G_P(s) = \frac{\omega_n^2(K_p s + K_I)}{s^2(s+2\zeta\omega_n)}$$

The PI-controller is equivalent to adding a zero at $s = -K_I/K_p$ and a pole at $s = 0$ to the transfer function $G_P(s)$. One obvious effect is that it increases the order of the original system without integral control by one. Therefore, the steady-state error of the original system (which is without integral control) is improved by one order; that is, if the steady-state error to a given input is constant, the integral control reduces it to zero (provided that the final system is stable, of course).

As an illustration, consider Figure 4.51 having the same plant transfer function as used in the case of derivative control of Example 4.11 rewritten as

$$G_P(s) = \frac{400}{s(s+48.5)} \quad \text{and} \quad K_p = 100, \quad K_I = 10$$

The closed-loop transfer function of the system becomes

$$\frac{40000(s+0.1)}{s^3 + 48.5s^2 + 40000s + 4000} = \frac{40000(s+0.1)}{(s+0.10001)(s^2 + 48.5s + 40000)}$$

The poles are at -0.10001, $-24.2 \pm j198.5$ and a zero at -0.1. The zero is very close to the real pole of the closed-loop transfer function and thus for all practical purposes, the closed-loop transfer function can be approximated as

$$\frac{40000}{s^2 + 48.5s + 40000}$$

The unit-step response is shown in Figure 4.52 for

(i) $K_p = 100$, $K_I = 10$; (ii) $K_p = 10$, $K_I = 1$; (iii) $K_p = 2$, $K_I = 0.2$

It may be observed that the controller configurations highlighted, show somewhat that K_p is proportional to K_I. This is not absolutely necessary, but was done so that the real pole will be close to its zero, then the transient response can be predicted by two complex poles only.

With PI-controller, the basic characteristic equation of the closed-loop system becomes

$$1 + \frac{400(K_p s + K_I)}{s^2(s+48.5)} = 0$$

Transient Response Analysis

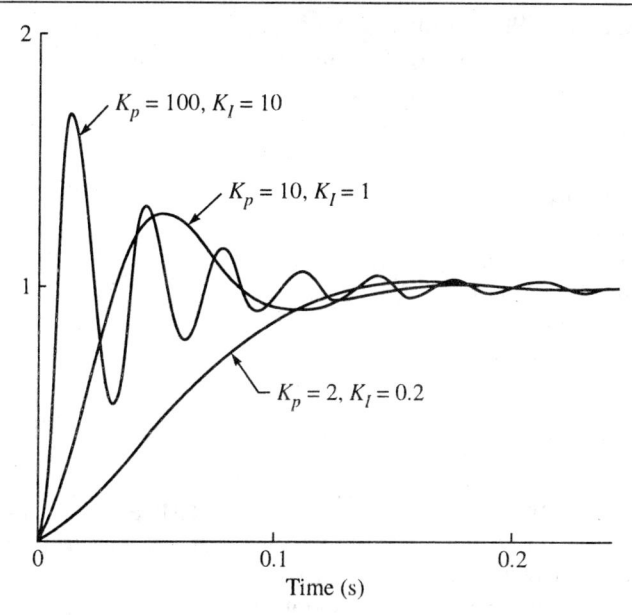

FIGURE 4.52 Unit-step response of the control system in Figure 4.51 with PI-controller.

or
$$s^3 + 48.5s^2 + 400K_p s + 400K_I = 0$$

Applying Routh's test, the stability range is obtained as $0 \geq K_I \geq 48.5K_p$ to show the effect of K_I on stability.

With $K_p = 100$, the root-loci plots for $0 < K_I < \infty$ are constructed from pole-zero configuration of

$$\frac{400K_I}{s(s^2 + 48.5s + 40000)}$$

and are shown in Figure 4.53. This indicates the adverse effect of integral control as the closed-loop system becomes unstable for $K_I > 4850$ ($= 48.5K_p$).

For design purposes, it is better to highlight the open-loop transfer function rewritten as

$$\frac{400K_p \left(s + \dfrac{K_I}{K_p} \right)}{s^2(s + 48.5)}$$

whose root-loci plot is shown in Figure 4.54 for $0 < K_I < \infty$, but with $K_I/K_p \ll 48.5$. The key to

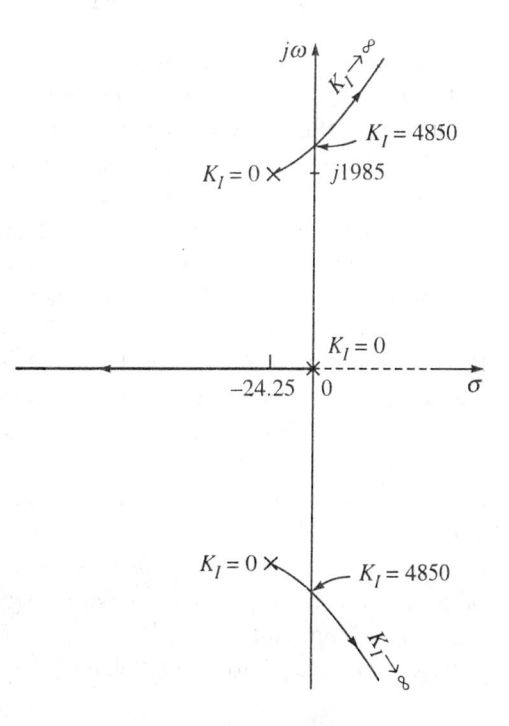

FIGURE 4.53 Root contours.

the successful design of a PI-controller lies in the satisfaction of inequality condition $0 < K_I < 48.5 K_p$.

Thus far we have learned about PD and PI controllers. The PD controller is effective in the transient part and ineffective in the steady-state part of the response while the PI controller is effective in the steady-state part and ineffective in the transient part of the response. Hence for the total response, both PI and PD control actions have to be incorporated in the PID controller.

The suggested PID-controller is

$$G_c(s) = K_p + K_d s + \frac{K_I}{s}$$

with $K_p = 100$, $K_d = 0.8788$ and $K_I = 10$. The unit-step response for the same plant is shown in Figure 4.55. The transient response is quite satisfactory with fast rise time, and the peak overshoot is only just 7%.

FIGURE 4.54 Root loci with K_I/K_p equal to a small number.

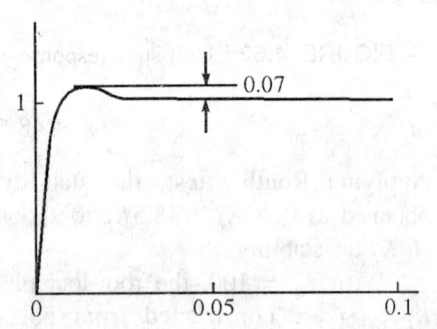

FIGURE 4.55 Unit-step response of the control system with PID controller.

4.18 Analog Computation

Analog computer was the most useful engineering tool available for the analysis and design of both linear and nonlinear systems till about 1970s. With the advent of digital computers and particularly with the easy accessibility of PCs, people now prefer digital computer simulation rather than analog simulation. We will discuss analog computation here only from the academic point of view. Further, operational amplifier is the heart of an analog computer. An operational amplifier can perform various mathematical operations, such as sign inversion, summation and integration. Differentiators are not being used for analog computation because of noise. A step of noise after the differentiator becomes an impulse which may saturate the op-amp, making the system go into saturation. This makes the closed-loop system open-loop. Operational amplifier is a differential amplifier of high gain of order $10^6 \sim 10^8$ and very high input impedance. Hence the current drawn at the input of an operational amplifier is negligibly small. The output voltage is usually limited to ±10 volts. The schematic diagram of an operational amplifier is shown in Figure 4.56(a) with open-loop gain $K = 10^6 \sim 10^8$. The output voltage e_o is related to input voltage e_g as

FIGURE 4.56(a) Schematic diagram of op-amp.

$$e_o = -K e_g$$

A feedback amplifier is shown in Figure 4.56(b) where Z_f is the feedback impedance inserted in the feedback path and Z_i is the input impedance in series with the op-amp. The KCL at the summing junction SJ gives

$$i_1 = i_2 + i_g$$

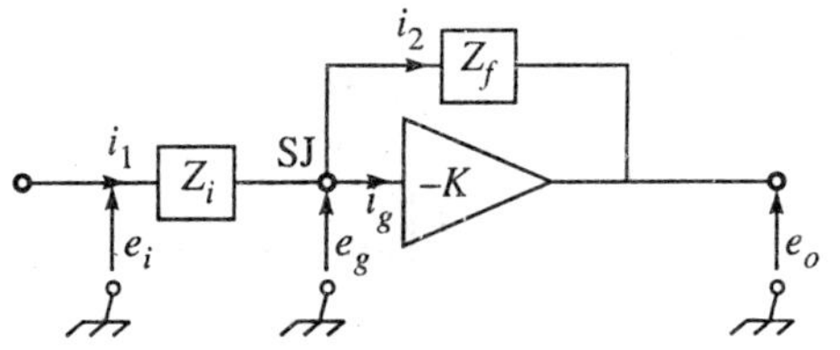

FIGURE 4.56(b) Feedback amplifier.

Again, $i_g \approx 0$, as the internal impedance of the amplifier is very high and it takes practically negligible input current. Hence the summing junction SJ is at virtual ground. Therefore,

$$i_1 = i_2$$

or

$$\frac{e_i - e_g}{Z_i} = \frac{e_g - e_o}{Z_f}$$

As K, the open-loop gain is very high and for finite output $e_o = -Ke_g$, practically $e_g \approx 0$, then, we get

$$\frac{e_o}{e_i} = -\frac{Z_f}{Z_i}$$

or in Laplace domain

$$E_o(s) = -\frac{Z_f}{Z_i} E_i(s)$$

Sign inversion: Putting $Z_i = R_i$, $Z_f = R_f$, we get $e_o = -(R_f/R_i)e_i$

The output voltage is equal to the (R_f/R_i) times the input voltage. Figure 4.57 shows a circuit giving the sign inversion of gain unity with $R_i = R_f = 1$ kΩ. It may be noted that though (R_f/R_i) is the ratio, but a low value of the resistance of the order of 1 Ω is not permitted because of loading effect.

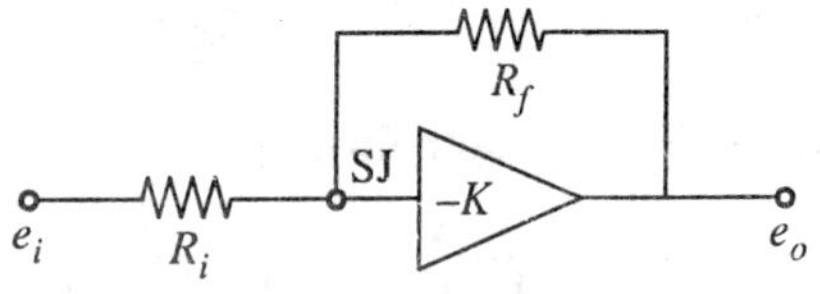

FIGURE 4.57 Sign inversion with op-amp.

What is loading effect? In Figure 4.57, the input impedance of the circuit is R_i as the terminal SJ is virtual ground. Hence the function generator providing the input voltage e_i will be loaded heavily if we put R_i of very low value. Current drawn from the input function generator will be very high and heat generation will be so high that the function generator will get burnt.

Summation amplifier: Figure 4.58(a) shows a schematic diagram of a summation amplifier which adds n-inputs with different gains using the principle of superposition as

$$e_o = -\left(\frac{R_f}{R_1}e_1 + \frac{R_f}{R_2}e_2 + \cdots + \frac{R_f}{R_n}e_n\right)$$

As an illustration, in Figure 4.56(b), $R_f = 1$ MΩ, $R_1 = 0.25$ MΩ, $R_2 = 1$ MΩ and $R_3 = 0.1$ MΩ. Then, $e_o = -(4e_1 + e_2 + 10e_3)$, whose symbolic representation is shown in Figure 4.58(c).

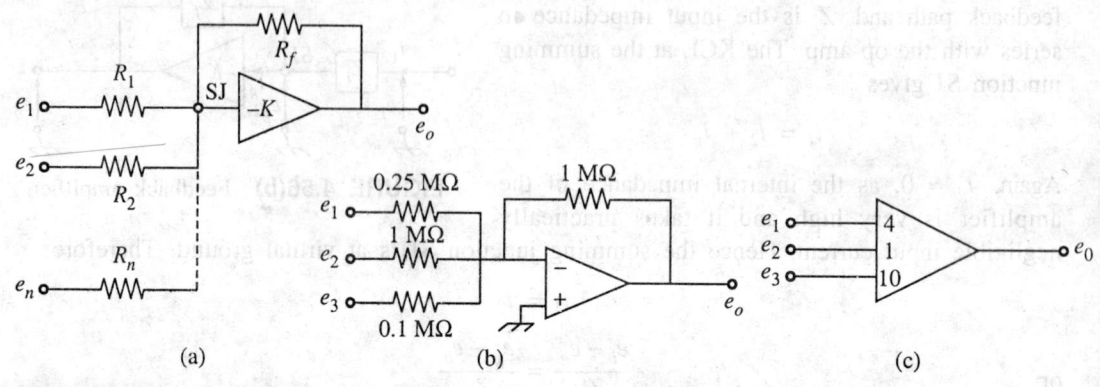

FIGURE 4.58 Summation amplifier: (a) schematic representation, (b) an illustration, and (c) symbolic representation.

Integrator: If a capacitor C is used in the feedback path and a resistor R as input impedance as shown in Figure 4.59, the op-amp becomes an integrator. The integrator symbol is shown in Figure 4.59(b). Here,

$$Z_f = \frac{1}{sC}; \quad Z_i = R$$

then, $\quad E_o(s) = -\frac{1}{s}\frac{1}{RC}E_i(s)$

which in time domain becomes

$$\int_0^t de_o = -\frac{1}{RC}\int_0^t e_i\, dt$$

or $\quad e_o(t) - e_o(0) = -\frac{1}{RC}\int_0^t e_i\, dt$

or $\quad e_o(t) = -\frac{1}{RC}\int_0^t e_i\, dt + e_o(0)$

where $e_o(0)$ is the initial condition of the integrator given by some dc voltage, i.e. the capacitor C is initially charged to $e_o(0)$ volts. The circuit arrangement to provide initial condition is as shown in Figure 4.60(a). The symbolic representation of an integrator with initial condition is shown in Figure 4.60(b).

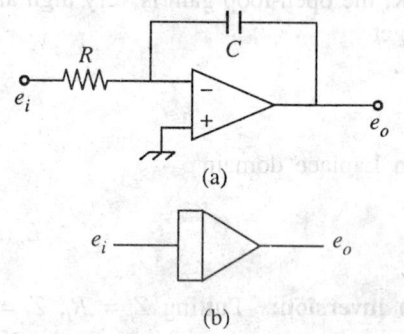

FIGURE 4.59 Integrator: (a) circuit diagram and (b) symbolic representation.

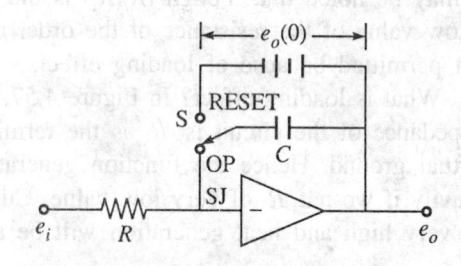

FIGURE 4.60(a) Setting up of initial control condition.

Transient Response Analysis

As an illustration, for $R = 1\text{ M}\Omega$, $C = 1\text{ }\mu\text{F}$

$$e_o(t) = -\int_0^t e_i\, dt + e_o(0)$$

The initial condition $e_o(0)$ is indicated by the side of the circle in Figure 4.60(b). Dimensionally, RC is in seconds.

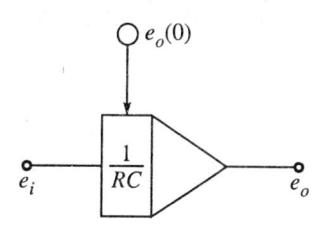

FIGURE 4.60(b) Symbolic representation of integrator with initial condition.

Setting up of initial condition: Refer to Figure 4.60(a). The dc voltage source has its inherent internal impedance r (not shown in the figure) with is very small. The capacitor will charge to 98% of $e_o(0)$ volts with a time constant rC in 4 time constants with the output terminal as positive polarity, when the switch S is in position RESET. When the switch S is in the OP (operation) position, normal integration takes place as $-\dfrac{1}{RC}\int e_i\, dt$ starting with the initial condition $e_o(0)$. Figure 4.60(b) is the commonly used symbol for this case.

As an illustration, the op-amp set-up for

$$e_o(t) = -\int_0^t 5\, dt - 4 = -5t - 4$$

is shown in Figure 4.61 indicating the input step of 5 volt and the integrated output ramp starting with the initial condition of -4 volts.

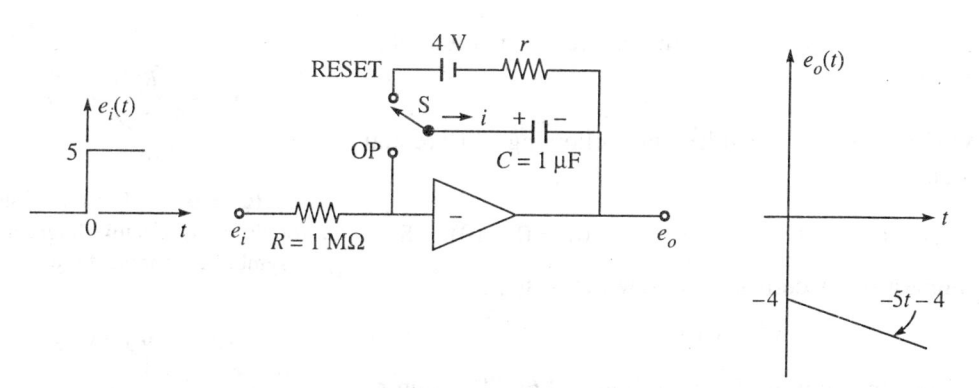

FIGURE 4.61 An example of setting up of initial condition.

Summing integrator: Figure 4.62(a) shows a schematic diagram of a summing integrator which integrates n inputs. Using superposition theorem, the output $e_o(t)$ is

$$e_o(t) = -\left(\dfrac{1}{R_1 C}\int e_1\, dt + \cdots + \dfrac{1}{R_n C}\int e_n\, dt\right) + e_o(0)$$

For example, $n = 3$; $C = 1\text{ }\mu\text{F}$, $R_1 = 1\text{ M}\Omega$, $R_2 = 0.25\text{ M}\Omega$ and $R_3 = 0.1\text{ M}\Omega$

Then,

$$e_o(t) = -\left(\int e_1\, dt + 4\int e_2\, dt + \cdots + 10\int e_3\, dt\right) + e_o(0)$$

Figure 4.62(b) is the commonly used symbol for such a case.

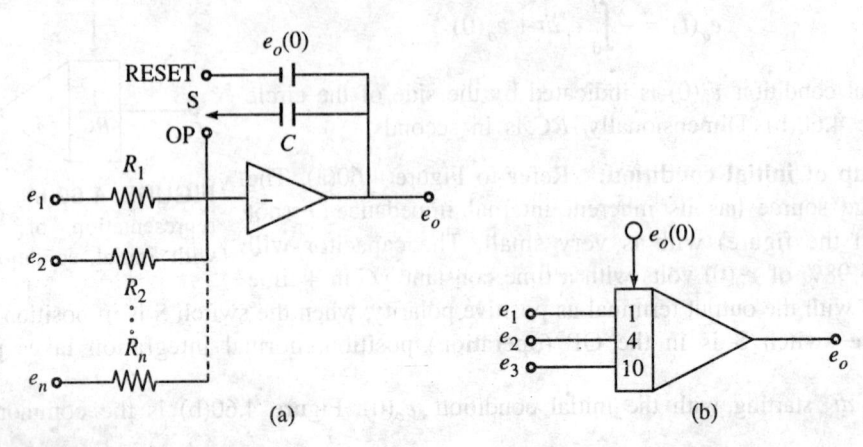

FIGURE 4.62 Summar integrator: (a) schematic diagram and (b) symbolic representation.

Multiplication by a fraction: The multiplication of e_i by a constant α where $0 < \alpha < 1$ can be accomplished by the potentiometer shown in Figure 4.63(a). The output

$$e_o = \frac{R_o}{R_i} e_i$$

Figure 4.63(b) shows the symbolic representation of a potentiometer.

EXAMPLE 4.12 Consider the following differential equation

$$\ddot{x} + 10\dot{x} + 16x = 0 \quad \text{with} \quad x(0) = 0, \dot{x}(0) = 8$$

The highest-order derivative is rewritten as

$$\ddot{x} = -10\dot{x} - 16x$$

The SFG simulation is shown in Figure 4.64. The sum of the inputs to the first integrator is the highest derivative term that originally was assumed to be available. A sign change is associated with each operational amplifier. The number of operational amplifiers in any loop must be odd, otherwise the op-amp output will saturate and lead to positive feedback.

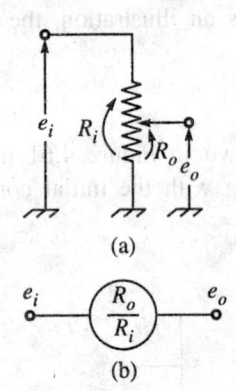

FIGURE 4.63 Multiplication by a fraction: (a) circuit diagram and (b) symbolic representation.

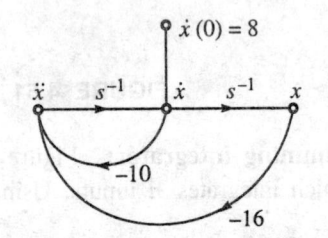

FIGURE 4.64 Example 4.12: SFG simulation.

EXAMPLE 4.13 Set up the analog computer simulation for generating a sinusoidal function $x(t) = 2 \sin 3t$.

Solution: Let $x(t) = 2 \sin 3t$; the initial condition becomes $x(0) = 0$. Then the derivative becomes

$$\dot{x}(t) = 6 \cos 3t; \text{ the initial condition becomes } \dot{x}(0) = 6$$

and

$$\ddot{x}(t) = -18 \sin 3t = -9 \times 2 \sin 3t = -9x(t)$$

Hence the dynamics of the system becomes

$$\ddot{x} + 9x = 0 \text{ with } x(0) = 0 \text{ and } \dot{x}(0) = 6$$

whose output becomes the desired sinusoidal signal $x(t) = 2 \sin 3t$. The SFG simulation is shown in Figure 4.65.

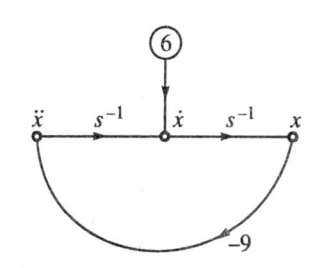

FIGURE 4.65 Example 4.13: SFG simulation.

Summary

In this chapter we talked about the performance index and the usefulness of standard test signals. Then several performance measures for a standard step input test signal were delineated. For example, the overshoot, peak time, and settling time of the response of the system under test for a step input signal were considered. One has to be very thorough about these performance measures for the design of control of a closed-loop control system. The relationship between the location of the s-plane roots of the system transfer function and the system response was discussed. The most important measure of the system performance is the steady-state error for specific test input signals. In this regard, the final-value theorem plays an important role. The relationship of the steady-state error of a system in terms of the system parameters was developed by utilizing the final-value theorem. The discussion of proportional, derivative and integral controllers was presented. MATLAB programs were developed for performance measures.

Problems

4.1 The block diagram of an open-loop temperature-control system is shown in Figure P.4.1(a) and that of a closed-loop system in Figure P.4.1(b) with nominal value of $K = 1$.

(a) If $r(t)$ is a unit step and the disturbance $d(t)$ is zero, then determine the temperature $e(t)$ and plot it. Determine the time for the temperature to reach 63% of its final vlaue in each case.

(b) Determine the % change in $\theta(t)$ for 5% change in K.

(c) After the steady-state of $\theta(t)$ has reached, a disturbance $d(t)$ of step 0.1 size is observed due to a sudden increase in the volume of the fluid in the tank. Determine the effect of this disturbance on the temperature for both the open- and closed-loop cases and plot them.

Heating system

Reference temperature $r(t)$ → $\dfrac{K}{1+10s}$ → (+ with $d(t)$) → Normalized temperature $\theta(t)$

FIGURE P.4.1(a)

Reference temperature $r(t)$ → (+/−) → Amplifier $K_a = 100$ → Heating system $\dfrac{K}{1+10s}$ → (+ with $d(t)$) → Normalized temperature $\theta(t)$; feedback $K_t = 1$

FIGURE P.4.1(b)

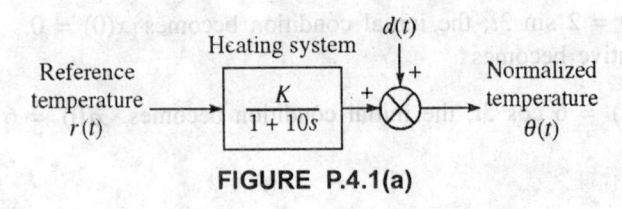

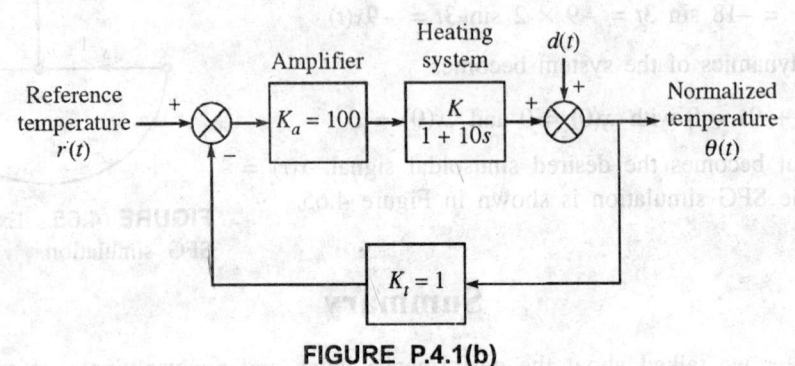

Ans. *For open-loop system* 　　　　*For closed-loop system*
  (a) $\theta(t) = 1 - \exp(-0.1t)$ 　　(a) $0.99(1 - e^{-10.1t})$
  　Time = 10 s 　　　　　　　　　Time = 0.99 s
  (b) 5% 　　　　　　　　　　　　(b) 0.0495%
  (c) Output $\theta(t)$ changes by 0.1 (c) Output $\theta(t)$ changes by 0.001 in 4 s

4.2 For an automobile control system shown in Figure P.4.2:

(a) Determine the steady-state error when the system is subjected to a unit-step input and also express it in %.

(b) After attaining a steady speed along a level road, suddenly the gradient changes. This corresponds to a load torque disturbance of step 0.1. Determine the new steady-state error.

Ans. (a) $e_{ss} = 0.001996$ or 0.1996%; (b) $e_{ss} = 0.005988$

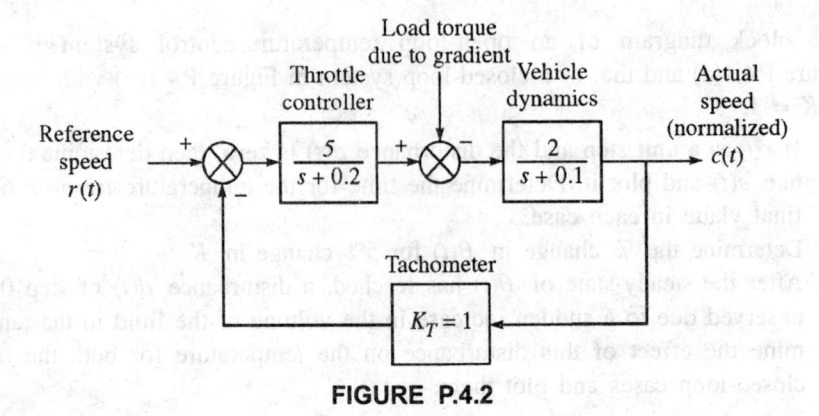

Reference speed $r(t)$ → (+/−) → Throttle controller $\dfrac{5}{s+0.2}$ → (+ with Load torque due to gradient, −) → Vehicle dynamics $\dfrac{2}{s+0.1}$ → Actual speed (normalized) $c(t)$; Tachometer $K_T = 1$ feedback

FIGURE P.4.2

Transient Response Analysis

4.3 A position-control system with velociy feedback damping is shown in Figure P.4.3.

(a) Determine the settling time and the maximum overshoot in the respone to a unit-step input as well as the steady-state error to a unit-ramp input in the absence of velocity feedback, i.e. for $\alpha = 0$.

(b) Repeat if $\alpha = 2$.

(c) Determine the value of α so that the damping ratio of the closed-loop system transfer function may be increased to 0.6. What are the values of the settling time, maximum overshoot, and the steady-state error to a unit-ramp input in this case?

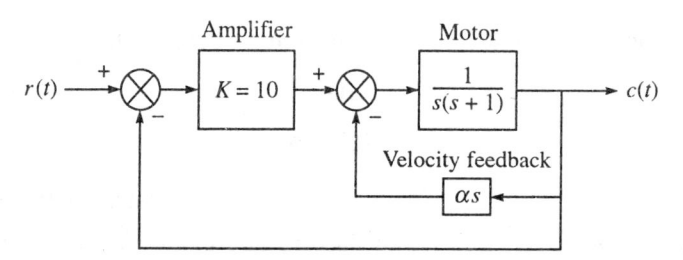

FIGURE P.4.3

4.4 The block diagram of the altitude control system for a missile is shown in Figure P.4.4. Determine the values of K_1 and K_2 in order that the closed-loop system may have undamped natural frequency $\omega_n = 10$ rad/s and damping ratio $\zeta = 0.5$. What will be the settling time and the maximum overshoot for a unit-step input?

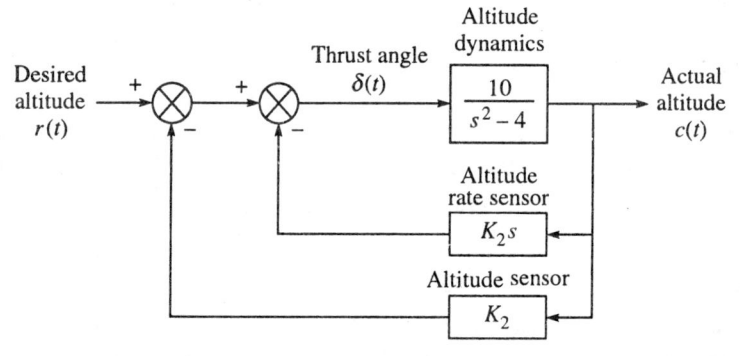

FIGURE P.4.4

4.5 The block diagram of a position-control system with velocity feedback is shown in Figure P.4.5. Determine the value of α so that the step response has a maximum overshoot of 10%. What is the steady-state error to a unit-ramp input for this value of α?

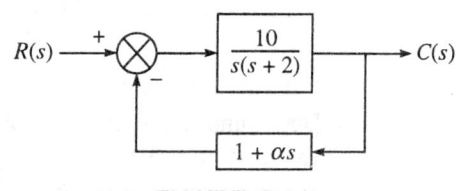

FIGURE P.4.5

4.6 Find the steady-state error for a unity-feedback system having transfer function

$$\frac{20}{s(s+2)(s^2+2s+2)}$$

subjected to (a) unit step, (b) unit ramp, and (c) parabolic input $t^2/2$.

Ans. (a) 0; (b) 0.2; and (c) ∞

4.7 Determine the steady-state error of the transfer function of Problem 4.6 for the following inputs: (a) $(8 + 3t)$ and (b) $(2 + 5t + 2t^2)$.

Ans. (a) 0.8 and (b) ∞

4.8 Repeat Problem 4.6 for the transfer function of the system

$$\frac{108}{s^2(s+4)(s^2+3s+12)}$$

Ans. (a) 0; (b) 0; and (c) 4/9

4.9 For a unity-feedback control system having the forward transfer function

$$G(s) = \frac{K}{s(s+a)}$$

the overshoot of the closed-loop system to a unit-step input is less than 30%. Draw the closed-loop frequency response and then determine the resonant frequency ω_r. Determine the bandwidth of the closed-loop system.

4.10 The open-loop transfer function of a unity feedback system is

$$\frac{K}{s(s+2)}$$

It is specified that the response of the system to a unit-step input should have a maximum over shoot of 10%, and the settling time should be less than one second.

(a) Is it possible to satisfy both the specifications simultaneouly by adjusting K?
(b) If not, determine the value of K that will satisfy the first specification. What will be the settling time and the time to reach the first peak for this case?

4.11 The measurement of heart rate is proposed as in Figure P.4.11. An electronic pacemaker regulates the speed of the heart pump. It is desired to design the amplifier gain K to yield a tightly controlled system with a settling time to a step disturbance of less than one second. The overshoot to a step in desired heart rate should be less than 15%

(a) Determine the suitable range of gain K.
(b) Determine the sensitivity of the system to small changes in K for the nominal value of $K = 10$.
(c) Determine the sensitivity at dc.
(d) Determine the magnitude of the sensitivity at normal heart rate of 72 **beats/s**.

Transient Response Analysis

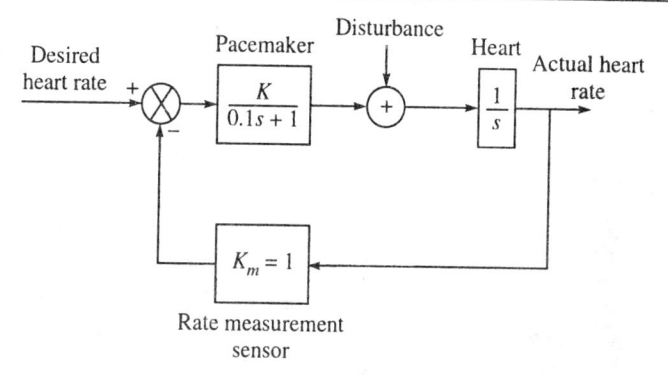

FIGURE P.4.11

(Block diagram: Desired heart rate → summing junction → Pacemaker $\frac{K}{0.1s+1}$ → summing junction with Disturbance → Heart $\frac{1}{s}$ → Actual heart rate; feedback through Rate measurement sensor $K_m = 1$)

4.12 Given the function subjected to unit-step input,

$$\frac{C(s)}{U(s)} = \frac{2810(s+4)}{(s+3.8)(s+6)(s^2+2s+17)(s^2+10s+29)}$$

find the time domain response through MATLAB program. Also, draw the response considering the dominant pole and compare the two results.

4.13 An analog computer is often used for simulating dynamic systems and studying the effect of adjusting the values of certain parameters. Its basic components are operational amplifiers, which can be used as adders, integrators, inverters, and potentiometers. The simulation diagram for a particular system is shown in Figure P.4.13. Determine the transfer function.

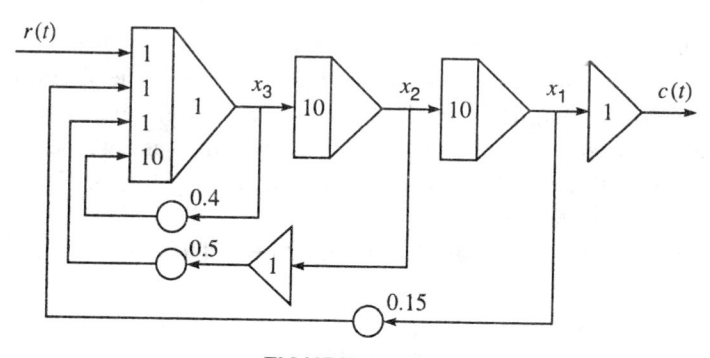

FIGURE P.4.13

Edward John Routh was born on 20 January, 1831 in Quebec, Canada.

Edward entered University College, London, in 1847 having won a scholarship. There he studied under De Morgan whose influence led to him deciding on a career in mathematics. Routh obtained his M.A. from London in 1853 and was awarded the Gold medals for Mathematics and Natural Philosophy. In January 1854 Routh graduated with a B.A. from Cambridge. He was Senior Wrangler in the Mathematical Tripos examinations and was ranked first. In 1855, Routh was elected a fellow of Peterhouse and was appointed as a College lecturer in Mathematics. In the following year, he was appointed as assistant tutor at Peterhouse.

1831–1907

In 1864, Routh married Hilda Airy, the Astronomer Royal. They had five sons and one daughter. One of his sons, Rupert John Routh, served in the Indian Civil Service.

Routh became the most famous of the Cambridge coaches for the Mathematical Tripos. During his career he coached about 700 pupils, of whom about 480 were Wranglers out of around 900 Wranglers over these 30 years.

He contributed to mathematics with some excellent papers and some outstanding texts. His research interests areas were geometry, dynamics, astronomy, waves vibrations and harmonic analysis.

E.J. Routh provided a numerical techniques for determining when a characteristic equation has stable roots.

He published famous advance treatises which became standard applied mathematics texts. He was elected a fellow of the Cambridge Philosophical Society. He became a founder member of the London Mathematical Society; he was elected a fellow of the Royal Astronomical Society and of the Royal Society. He was awarded honorary degrees from a number of universities including Glasgow and Dublin. He was made an honorary fellow of Peterhouse.

On June the 7th 1907 in Cambridge, Cambridgeshire, Edward Routh breathed his last.

Routh Stability and Robust Control

5.1 Introduction

A linear system will be defined stable if and only if all the poles of the system transfer function have negative real parts. This definition is stronger, in as much as it does not allow simple poles on the imaginary axis. All the components of the natural response will then decrease with time. This requires the evaluation of the roots of the characteristic equation. Further, there is no analytical method for finding the roots of polynomials of order greater than five. However, it is not necessary to determine the actual location of all the poles of the transfer function. In fact, for determining the absolute stability we only need to know the number of right-hand poles, if any. Routh–Hurwitz criterion was developed independently by Hurwitz (1895) in Germany and E.J. Routh (1892) in the United Kingdom. This method is to find the number of roots having positive real parts without actually solving for them.

Let the characteristic polynomial (i.e. the denominator of the transfer function after the pole-zero cancellation of factors common with the numerator) be given as

$$F(s) = a_n s^n + a_{n-1} s^{n-1} + \cdots + a_1 s + a_0$$

For stability, all the roots must be in the negative-half of the s-plane. The following observations can be made:

OBJECTIVE

Information about the stability of a control system is of paramount importance for any design problem. The beauty of the Routh–Hurwitz algorithm lies in finding the absolute stability of the system without determining the roots of the system. Then comes the problem of stability of a control system under parameter variations. Kharitonov's polynomial comes to rescue here. One has to generated only four Kharitonov polynomials for Routh–Hurwitz test for stability irrespective of the order of the system. Only from four polynomials generated from maximum and minimum variations of parameters, Routh–Hurwitz test can be performed to ensure stability under parameter variations.

CHAPTER OUTLINE

Introduction
Stability
Some Results from the Theory of Equations
Routh Array
Stability from State-Space Representation
Relative Stability
Parameter Uncertainty: Robust Control

(i) If all the roots have negative real parts, then it is necessary for all the coefficients of the polynomial $F(s)$ to have the same sign.

(ii) No coefficients of the polynomial $F(s)$ can have zero value. The existence of one or more roots at the origin is obvious by the absence of the term a_0 or both the terms a_0 and a_1, and so forth.

For example, the equation $s^2 + 1 = 0$, has two imaginary roots and represents a neutrally stable system. The equation $s^2 - 1$, represents an unstable system. Similarly, $s^2 - 3s + 1 = 0$ is also unstable. On the other hand as in the equation

$$s^4 + 2s^3 + 3s^2 + 4s + 5 = 0$$

all the coefficients are present and positive but the system is unstable as will be seen in Example 5.2. Observations (i) and (ii) are necessary but not sufficient conditions for stability. The Routh–Hurwitz criterion is the sufficient condition for stability which we are going to discuss here.

5.2 Stability

The stability of a control system can be defined as follows:

Definition 1. A system is stable if its impulse response (i.e. the transfer function) approaches zero as time approaches infinity.

Definition 2. A system is stable if every bounded input produces a bounded output.

Further, if the system has distinct roots on the imaginary axis, the system is said to be marginally stable; this system is not unstable as there is no root in the right-half of s-plane. But in this instance, the impulse response does not decay to zero although it is bounded. In addition, certain bounded inputs (such as $\sin t$) produce unbounded output. Therefore, marginally stable systems are unstable from Routh's stability criterion point. We can strictly define stability only for that system which has roots in the negative-half of the s-plane.

The system described by the Laplace transformed differential equation as

$$(s^2 + 1)Y(s) = U(s)$$

has the characteristic equation

$$s^2 + 1 = 0$$

The two roots are $\pm j$. This system is marginally stable and has no roots having positive real parts. The output response for sinusoidal input $u(t) = \sin t$ will be

$$y(t) = t \sin t$$

which is unbounded.

Complex conjugate poles of the form $\dfrac{c}{s^2 + \omega_0^2}$ lead to the conjugate poles on the $j\omega$-axis.

The corresponding time function is $\left(\dfrac{c}{\omega_0}\right)\sin \omega_0 t$. The response does not die out as time t increases because there is no exponential damping. Now if the system is subjected to a sinusoid of the same frequency ω_0, then a double conjugate pair of poles results and the output in the s-domain becomes of the form $\dfrac{1}{(s^2 + \omega_0^2)^2}$ which in time domain becomes $\left(\dfrac{1}{2\omega_0^3}\right)(\sin \omega_0 t - \omega_0 t \cos \omega_0 t)$ and increases without bound as time t increases. Physically, we are exciting a natural resonance of the system with an input at precisely the resonant frequency. Because there is no loss ($\zeta = 0$, no damping) associated with this mode of the system, the output grows without bound.

For the system with a simple pole at the origin and driven by a step, the output results into a ramp, and becomes unbounded. Repeated poles on the $j\omega$-axis lead to the terms $\dfrac{1}{(s^2 + \omega^2)^2}$ and the time function becomes unbounded.

To summarize, a *causal*, time-invariant system having transfer function $G(s)$ is stable if:

1. All the poles of $G(s)$ are in the left-half of s-plane.
2. The degree of denominator polynomial of $G(s)$ is greater or equal to that of numerator polynomial, i.e. the transfer function has to be rational.

The last condition is required in order to eliminate terms like s^{m-n}, a differentiator of degree $(m - n)$. If a system is excited by an input like $\sin \omega t$, then the output of the system would include the term like $\omega^{m-n} \sin \omega t$. This term could be made as large as desired by increasing the input frequency ω.

5.3 Some Results from the Theory of Equations

The roots of the polynomial are the functions of its coefficients and vice versa. Consider a monic[1] polynomial having roots r_1, r_2, r_3, and r_4 as

$$s^4 + As^3 + Bs^2 + Cs + D$$
$$= (s + r_1)(s + r_2)(s + r_3)(s + r_4)$$
$$= s^4 + (r_1 + r_2 + r_3 + r_4)s^3 + (r_1r_2 + r_1r_3 + r_1r_4 + r_2r_3 + r_2r_4 + r_3r_4)s^2$$
$$+ (r_1r_2r_3 + r_1r_2r_4 + r_2r_3r_4 + r_1r_3r_4)s + r_1r_2r_3r_4$$
$$= s^4 + \sum_{i=1}^{4} r_i s^{4-1} + \sum_{\substack{i=1, j=1 \\ i \neq j}}^{4} r_i r_j s^{4-2} + \sum_{\substack{i=1, j=1, k=1 \\ i \neq j \neq k}}^{4} r_i r_j r_k s^{4-3} + \prod_{i=1}^{4} r_i s^0$$

For the nth order polynomial, we can generalize the above expression as

---

[1] A monic polynomial is one for which the coefficient of the highest power term is unity.

$$s^n + As^{n-1} + Bs^{n-2} + \cdots + Cs + Ds^0 = s^n + \sum_{i=1}^{n} r_i s^{n-1} + \sum_{\substack{i=1, j=1 \\ i \neq j}}^{n} r_i r_j s^{n-2}$$

$$+ \sum_{\substack{i=1, j=1, k=1 \\ i \neq j \neq k}}^{n} r_i r_j r_k s^{n-3} + \cdots + \prod_{i=1}^{n} r_i s^0$$

(5.1)

It is clear that no coefficient can be zero or negative unless there is at least one root in the right-half of s-plane. We inspect the coefficient terms of the polynomial and if any term is missing, this means that the coefficient is zero and there is at least one root lying at the origin or in the right-half of s-plane, and the system is said to be unstable. In a like manner, if any coefficient is negative, i.e. if there is a change of sign, the system is unstable. Unfortunately, the converse is not true; when all terms are present and all signs are positive, there may still be roots in the right-half of s-plane; so a stability test is needed. So for the system to be stable, the necessary condition is:

There should not be any missing term of the coefficients of the characteristic equation and there should not be any change of sign of the coefficients.

5.4 Necessary Condition for Stability

In general, a characteristic equation may be written in the form

$$F(s) = \sum_{i=0}^{n} a_i s^i = 0$$

The output response of a system represented by a linear differential equation subjected to a step input can be written as

$$c(t) = A_0 + A_1 e^{r_1 t} + A_2 e^{r_2 t} + A_3 e^{r_3 t} + \cdots$$

For a stable system the values of all r's are negative real or complex with negative real parts. Thus as t increases, all of the exponential terms decrease.

The value of the coefficient of the s^0 term is the product of all the roots. For example, when we increase the forward gain in a feedback control system, the value of the coefficient of the s^0 term increases. Clearly, if the product of the roots is unchanged, then for most systems, other roots must get smaller. So we can say that as we increase the gain we can expect some roots to move to left, others to right. If the gain increases too high, some of the roots moving to the right will move into the right-half of s-plane and the system will become unstable. This method of analysis is also applicable when any parameter other than the gain parameter is varied.

If $F(s)$ has real coefficients, then its complex roots will occur in conjugate pairs. Let $F(s)$ have p-real roots $(-\alpha_i, i = 1, 2, \cdots, p)$ and q-pair of complex roots $(-\beta_l + j\phi_l, l = 1, 2, \cdots, q)$, so that $n = p + 2q$. Then in factored form, $F(s)$ can be written as

$$F(s) = a_0 \prod_{i=1}^{p}(s + \alpha_i) \prod_{l=1}^{q}(s^2 + 2\beta_l s + \beta_l^2 + \phi_l^2)$$

For a system to be stable, all the α's and β's must be positive. By expansion, it is found that all the α's must exist and must be of the same sign. This is a necessary condition but not sufficient. For sufficient condition, we will talk about Routh's stability.

5.5 Routh Array

Arrange the coefficients of the characteristic polynomial in descending order, in rows and columns, as in Eq. (5.2) below:

$$\begin{array}{c|ccccc}
s^n & a_n & a_{n-2} & a_{n-4} & a_{n-6} & \cdots \\
s^{n-1} & a_{n-1} & a_{n-3} & a_{n-5} & a_{n-7} & \cdots \\
s^{n-2} & c_1 & c_2 & c_3 & c_4 & \cdots \\
s^{n-3} & d_1 & d_2 & d_3 & d_4 & \cdots \\
\vdots & \vdots & \vdots & \vdots & & \cdots \\
s^3 & e_1 & e_2 & e_3 & & \\
s^2 & f_1 & f_2 & & & \\
s^1 & g_1 & & & & \\
s^0 & h_1 & & & &
\end{array} \qquad (5.2)$$

The powers of s (i.e. $s^n, s^{n-1}, \cdots, s_1, s^0$) are used for indexing purpose. The entries in s^n and s^{n-1} rows are filled up from the coefficients of the characteristic polynomial of Eq. (5.1). The entries in the successive rows are obtained using the following procedure:

$$c_1 = \frac{-\begin{vmatrix} a_n & a_{n-2} \\ a_{n-1} & a_{n-3} \end{vmatrix}}{a_{n-1}} = \frac{a_{n-1}a_{n-2} - a_n a_{n-3}}{a_{n-1}}$$

$$c_2 = \frac{-\begin{vmatrix} a_n & a_{n-4} \\ a_{n-1} & a_{n-5} \end{vmatrix}}{a_{n-1}} = \frac{a_{n-1}a_{n-4} - a_n a_{n-5}}{a_{n-1}}$$

$$c_3 = \frac{-\begin{vmatrix} a_n & a_{n-6} \\ a_{n-1} & a_{n-7} \end{vmatrix}}{a_{n-1}} = \frac{a_{n-1}a_{n-6} - a_n a_{n-7}}{a_{n-1}}$$

and so on.

Moving down a row, the d_i entries are obtained as

$$d_1 = \frac{-\begin{vmatrix} a_{n-1} & a_{n-3} \\ c_1 & c_2 \end{vmatrix}}{c_1} = \frac{c_1 a_{n-3} - c_2 a_{n-1}}{c_1}$$

$$d_2 = \frac{-\begin{vmatrix} a_{n-1} & a_{n-5} \\ c_1 & c_3 \end{vmatrix}}{c_1} = \frac{c_1 a_{n-5} - c_3 a_{n-1}}{c_1}$$

and so forth until the $(n + 1)$th row is obtained.

After forming the Routh array, apply the following rules for the fulfillment of the Routh–Hurwitz criterion.

Rule 1. The number of roots of $F(s)$ with positive real parts is equal to the number of sign changes in the entries in the first column of the Routh array (5.2).

Rule 2. The sufficient condition for all roots of $F(s)$ to lie in the left-half of s-plane (a stable system) is that the first column of the array contains no sign changes.

EXAMPLE 5.1 Consider

$$F(s) = a_3 s^3 + a_2 s^2 + a_1 s + a_0 \tag{5.3}$$

The Routh array is:

$$\begin{array}{c|cc} s^3 & a_3 & a_1 \\ s^2 & a_2 & a_0 \\ s^1 & \dfrac{a_1 a_2 - a_3 a_0}{a_2} & \\ s^0 & a_0 & \end{array}$$

The condition for all roots to have negative real parts is

$$(a_1 a_2 - a_0 a_3) > 0$$

and

$$a_i > 0, \ i = 0, 1, 2, 3$$

EXAMPLE 5.2 Consider the system whose characteristic equation is

$$s^4 + 2s^3 + 3s^2 + 4s + 5 = 0 \tag{5.4}$$

The Routh array is:

$$\begin{array}{c|ccc} s^4 & 1 & 3 & 5 \\ s^3 & \cancel{2} & \cancel{4} & \\ s^3 & 1 & 2 & \rightarrow \text{The second row is divided by 2} \\ s^2 & 1 & 5 & \\ s^1 & -3 & & \\ s^0 & 5 & & \end{array}$$

As the number of changes of sign in the first column of the array is 2, so two roots of the given characteristic equation must have positive real parts. Hence the system is unstable. It is to be noted that the s^3-row is divided by 2, thus considerably simplifying the subsequent algebra.

EXAMPLE 5.3 Consider the system whose characteristic equation is

$$s^4 + (2 \times 10^3)s^3 + (3 \times 10^6)s^2 + (4 \times 10^9)s + (5 \times 10^{12}) = 0 \tag{5.5}$$

Is the system stable?

Before setting up the Routh array, normalize the polynomial coefficients as nearly as possible. Setting $s = 10^3 p$ in Eq. (5.5) leads to

$$p^4 + 2p^3 + 3p^2 + 4p + 5 = 0 \qquad (5.6)$$

The Routh array is:

$$\begin{array}{c|ccc}
p^4 & 1 & 3 & 5 \\
p^3 & \cancel{2} & \cancel{4} & \\
p^3 & 1 & 2 & \\
p^2 & 1 & 5 & \\
p^1 & -3 & & \\
p^0 & 5 & &
\end{array}$$

$\rightarrow$ The second row is divided by 2

There are two changes of sign in the first column of the array, so there are two roots in the right-hand side of s-plane and hence the system is unstable.

Drill Problem 5.1

Determine the number of roots of each of the following polynomials in the right-half of s-plane using Routh's array:

(a) $s^4 + 5s^3 + 20s^2 + 40s + 50$
(b) $s^3 + s^2 + 2s + 24$
(c) $s^7 + s^6 + s^5 + 2s^4 + 3s^3 + 4s^2 + 5s + 6$

Ans. (a) none; (b) two; (c) four

5.6 Special Cases

Rule 3. If a first column term of any row is zero, but the remaining terms are not zero or there is no remaining term, then the zero term is replaced by a very small positive number ε and the rest of the array is evaluated as ε approaches zero.

EXAMPLE 5.4 Consider the characteristic equation

$$s^3 + 2s^2 + s + 2 = 0 \qquad (5.7)$$

The array of coefficients is:

$$\begin{array}{c|cc}
s^3 & 1 & 1 \\
s^2 & 2 & 2 \\
s^1 & 0 \approx \varepsilon & \\
s^0 & 2 &
\end{array}$$

$\rightarrow$ The zero in the third row is replaced by ε

If the sign of the coefficient above the zero ($\approx \varepsilon$) is the same as that of below it, then it indicates that there is a pair of imaginary roots. Actually, Eq. (5.7) has two roots at $s = \pm j$. As there is no change of sign in the first column of the array, no root lies in the right-half of the s-plane.

We can use this information to determine all the three roots easily. Let the imaginary root factors be written as $(s^2 + a)$, where $a > 0$. Let the real root factor be $(s + b)$, where $b > 0$. Then

$$(s + b)(s^2 + a) = s^3 + bs^2 + as + ab = s^3 + 2s^2 + s + 2$$

Comparing the two forms shows that
$$b = 2, \ a = 1$$

The roots are, $s = -2, \ \pm j1$

EXAMPLE 5.5 Consider the characteristic equation

$$s^4 + s^3 + 2s^2 + 2s + 3 = 0 \tag{5.8}$$

The Routh array is:

s^4	1	2	3
s^3	1	2	
s^2	$0 \ (\approx \varepsilon)$	3	
s^1	$2 - \dfrac{3}{\varepsilon}$		
s^0	3		

For small $\varepsilon > 0$, the fourth element of the first column is negative, indicating two changes of sign. Thus, there are two roots with positive real parts.

EXAMPLE 5.6 Consider the polynomial

$$F(s) = s^5 + 2s^4 + 3s^3 + 6s^2 + 10s + 15 \tag{5.9}$$

The Routh array is:

s^5	1	3	10
s^4	2	6	15
s^3	$0(\approx \varepsilon)$	$\dfrac{5}{2}$	
s^2	$6 - \dfrac{5}{\varepsilon}$	15	
s^1	$\dfrac{30\varepsilon - 25 - 30\varepsilon^2}{12\varepsilon - 10}$		
s^0	15		

There are two sign changes in the first column of the array as ε approaches zero. This indicates two roots in the right-half of s-plane.

The other alternative to Rule 3 is to put $s = 1/p$ in the original characteristic equation and then apply the Routh's test on the modified equation in terms of p. The number of positive roots in the p-plane is equal to that in the s-plane. This is simple but may not be always possible as is evident from the following example.

Routh Stability and Robust Control

EXAMPLE 5.7 Consider the characteristic polynomial

$$s^3 + 2s^2 + s + 2 \tag{5.10}$$

Substitute $s = 1/p$, then the modified polynomial in p becomes

$$2p^3 + p^2 + 2p + 1$$

The Routh array formulation will not be complete till we put $0 \approx \varepsilon$ and proceed. However, we take a new characteristic polynomial:

$$s^5 + 2s^4 + 3s^3 + 6s^2 + 2s + 1$$

Formulation of the Routh array will be held up in the s^3-row. We then put a small positive quantity ε (≈ 0) and proceed with the array and conclude with two right-half plane roots. The system is therefore unstable. The other alternative approach is to put $s = 1/p$ in the polynomial and then apply the Routh test to arrive at the same result.

The third alternative for this type of difficulty is to multiply the characteristic equation by $(s + 1)$, say, and proceed with the Routh's test for the modified polynomial. We will then arrive at the same conclusion as obtained previously. The polynomial for the Routh's test becomes

$$(s^5 + 2s^4 + 3s^3 + 6s^2 + 2s + 1)(s + 1) = s^6 + 3s^5 + 5s^4 + 9s^3 + 8s^2 + 3s + 1$$

From the first column of the Routh's array the same conclusion as earlier about the number of positive roots is obtained, i.e. two roots have positive real parts.

Drill Problem 5.2

Determine the number of roots of the following polynomials by Routh's array:

(a) $s^5 + 2s^4 + 3s^3 + 6s^2 + 10s + 15$

(b) $s^3 + Ks + 5K;\quad K > 0$

Ans. (a) two; (b) two

Rule 4. When all the elements in any one row of the Routh's array are zero, the condition indicates that there are roots of equal magnitude lying radially opposite in the s-plane (i.e. pair of real roots with opposite signs and/or pair of conjugate roots on the imaginary axis and/or complex conjugate roots forming quadrates in the s-plane). Such roots can be found by solving the auxiliary equation $A(s)$ which is formed with coefficients of the row, just above the row of zeros in the Routh's array. This polynomial $A(s)$ gives the number and location of root pairs of the characteristic equation, which are symmetrically located in the s-plane. Because of a zero-row in the array, the Routh's test breaks down. In such a case, the evaluation of the rest of the array can be continued by using the coefficients of the derivative of the auxiliary polynomial $A(s)$ in the next row. This has been illustrated below by an example.

EXAMPLE 5.8 Consider the polynomial

$$F(s) = s^6 + 5s^5 + 11s^4 + 25s^3 + 36s^2 + 30s + 36 \tag{5.11}$$

The array of coefficients is:

s^6	1	11	36	36
s^5	5	25	30	
s^4	6	30	36	← Auxiliary polynomial $A(s)$
s^3	0	0		

The terms in the s^3-row are all zero. The auxiliary polynomial is then formed from the coefficients of the s^4-row, as

$$A(s) = 6s^4 + 30s^2 + 36 \qquad (5.12)$$

The roots of the auxiliary equation are $\pm j\sqrt{2}$ and $\pm j\sqrt{3}$.

Now the derivative of the auxiliary equation with respect to s is

$$\frac{dA(s)}{ds} = 24s^3 + 60s \qquad (5.13)$$

The terms in the s^3-row of the Routh array are replaced by the coefficients of $dA(s)/ds$, namely 24 and 60. The array of the coefficients then becomes:

s^6	1	11	36	36
s^5	5	25	30	
s^4	6	30	36	
s^3	24	60		← Coefficients of $dA(s)/ds$
s^2	15	36		
s^1	2.4			
s^0	36			

As there is no change of sign in the first column of the new array, the original polynomial of Eq. (5.11) has no roots lying in the right-half of the s-plane. In fact, we get

$$\frac{F(s)}{A(s)} = (s+2)(s+3) \qquad (5.14)$$

Clearly, the original polynomial has no roots lying in the positive-half of s-plane and hence the system having the characteristic polynomial as in Eq. (5.11) is stable. This provides a way for finding the roots of the sixth order polynomial without recourse to numerical methods.

The auxiliary polynomial $A(s)$ obtained is of even order, but it may be noted that the auxiliary polynomial is not always of even order. For example, the characteristics equation

$$s(s^2 - 1)(s^2 + 2s + 4) = 0 \qquad (5.15)$$

produces the auxiliary polynomial

$$A(s) = 4s^3 - 4s \qquad (5.16)$$

EXAMPLE 5.9 Consider

$$F(s) = s^6 + 4s^5 + 12s^4 + 16s^3 + 41s^2 + 36s + 72 \qquad (5.17)$$

The Routh array is:

s^6	1	12	41	72
s^5	4	16	36	
s^4	8	32	72	
s^3	0	0		

The auxiliary polynomial

$$A(s) = 8s^4 + 32s^2 + 72 \tag{5.18}$$

and

$$\frac{dA(s)}{ds} = 32s^3 + 64s \tag{5.19}$$

The terms in the s^3-row are replaced by the coefficients of $dA(s)/ds$, namely 32 and 64. The array of coefficients then becomes:

s^6	1	12	41	72
s^5	4	16	36	
s^4	8	32	72	
s^3	32	64		
s^2	16	72		
s^1	−80			
s^0	72			

As there are two changes of sign in the first column of the Routh array, two roots lie in the positive-half of s-plane.

In fact, the auxiliary equation

$$A(s) = 8s^4 + 32s^2 + 72 = 8(s^2 + \sqrt{2}s + 3)(s^2 - \sqrt{2}s + 3)$$

has the roots in the right-hand side of s-plane.

Now,

$$q(s) = \frac{F(s)}{A(s)} = s^2 + 4s + 8$$

The Routh table of $q(s)$, shown below, indicates that it has no root in the right-half of s-plane.

s^2	1	8
s^1	4	
s^0	8	

Hence $F(s)$ has two roots in the right-half of s-plane and it is the characteristic polynomial of an unstable system.

Drill Problem 5.3

How many roots of each of the following polynomials are in the right-half of s-plane? Comment on stability as per Routh's definition.

(a) $s^5 + 6s^4 + 15s^3 + 30s^2 + 44s + 24$
(b) $s^6 + 4s^5 + 12s^4 + 16s^3 + 41s^2 + 36s + 72$
(c) $s^3 + 10s^2 + 50s + 500$

Ans. (a) none (two on the $j\omega$-axis); (b) two, unstable; (c) none, stable

5.7 Application of Routh–Hurwitz Stability Criterion

By the application of the Routh algorithm, we can find the stability range of a parameter value. Consider the system shown in Figure 5.1.

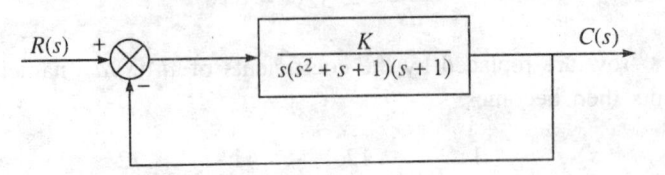

FIGURE 5.1 Closed-loop system.

Let us determine the range of the parameter K for stability. The closed-loop system transfer function is

$$\frac{C(s)}{R(s)} = \frac{K}{s(s^2 + s + 1)(s + 2) + K} \qquad (5.20)$$

The characteristic polynomial is

$$F(s) = s(s^2 + s + 1)(s + 2) + K$$
$$= s^4 + 3s^3 + 3s^2 + 2s + K$$

The array of coefficients becomes:

s^4	1	3	K
s^3	3	2	
s^2	$\dfrac{7}{3}$	K	
s^1	$2 - \dfrac{9K}{7}$		
s^0	K		

For stability, $K > 0$ and all the elements in the first column of the array must be positive. Therefore, the range of stability of the parameter K is $0 < K < 14/9$

5.8 Stability from State-Space Representation

The dynamics of the system can be represented by the vector-matrix differential equation

$$\dot{X} = AX + Bu \qquad \text{(AB)}$$

and output

$$y = CX + Du$$

The characteristic equation is given as

$$\Delta(s) = |sI - A| = 0 \qquad (5.21)$$

Routh Stability and Robust Control

The system described by Eq. (AB) will be stable if all the roots of the characteristic Eq. (5.21), that is the eigenvalues of the system matrix A, have negative real parts.

Instead of evaluating the roots of the characteristic equation, we may use Routh–Hurwitz criterion to determine the number of roots lying in the right-half of s-plane. This is illustrated below with an example.

EXAMPLE 5.10 The state variable representation of a third-order system is given below. Use Routh's criterion to determine the stability of the system whose state-space representation is given as

$$\frac{d}{dt}\begin{bmatrix} x_1 \\ x_2 \\ x_3 \end{bmatrix} = \begin{bmatrix} 0 & 1 & 0 \\ 0 & 0 & 1 \\ -6 & -11 & -6 \end{bmatrix}\begin{bmatrix} x_1 \\ x_2 \\ x_3 \end{bmatrix} + \begin{bmatrix} 0 \\ 0 \\ 1 \end{bmatrix} u$$

and output

$$y = \begin{bmatrix} 1 & 0 & 0 \end{bmatrix}\begin{bmatrix} x_1 \\ x_2 \\ x_3 \end{bmatrix}$$

Here the characteristic equation

$$\Delta(s) = |sI - A| = \begin{vmatrix} s & -1 & 0 \\ 0 & s & -1 \\ 6 & 11 & s+6 \end{vmatrix} = s^3 + 6s^2 + 11s + 6 = 0$$

The Routh array is as follows:

s^3	1	11
s^2	6	6
s^1	10	
s^0	6	

Since there is no change of sign in the first column of the Routh's array, the system is stable.

Drill Problem 5.4

Use Routh's criterion to determine the stability of the system represented by the vector-matrix differential equation

$$\frac{d}{dt}\begin{bmatrix} x_1 \\ x_2 \\ x_3 \end{bmatrix} = \begin{bmatrix} 0 & 1 & -8 \\ 1 & 0 & -5 \\ 0 & 2 & -6 \end{bmatrix}\begin{bmatrix} x_1 \\ x_2 \\ x_3 \end{bmatrix} + \begin{bmatrix} 2 \\ 1 \\ 0 \end{bmatrix} u$$

and output
$$y = \begin{bmatrix} 1 & 0 & 0 \end{bmatrix} \begin{bmatrix} x_1 \\ x_2 \\ x_3 \end{bmatrix}$$

Ans. The system is stable as there is no change of sign in the first column of the Routh's array of the characteristic equation.

5.9 Relative Stability

Application of Routh's criterion, as discussed above, only tells us whether a linear system is stable or not? It gives the absolute stability information. We often like to know the degree of instability, i.e. how far from the $j\omega$-axis is the real part of the dominant pole. Routh's criterion can be utilized for obtaining this relative stability information by shifting the vertical axis in the s-plane to obtain p-plane as shown in Figure 5.2. Hence in the polynomial $F(s)$, if we replace s by $(p - a)$ we get a new polynomial $F(p)$. Routh's test on the polynomial $F(p)$ gives the indication of the roots of $F(s)$ to the right of the line $s = -a$ in the s-plane.

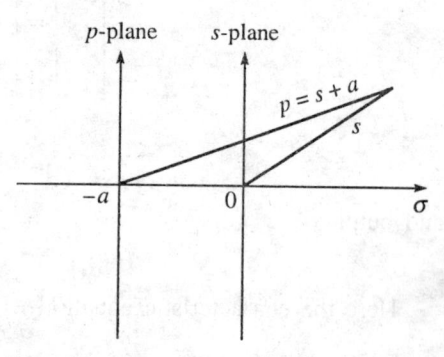

FIGURE 5.2 Shift of the vertical axis to the left by a.

Further, the relative stability depends on the settling time of a system. The settling time is inversely proportional to the real part of the dominant roots. Then the relative stability can be specified by requiring that all roots must be to the left of the line $s = -a$.

EXAMPLE 5.11 Consider the open-loop transfer function of a unity-feedback system

$$G(s) = \frac{K}{s(1 + sT)} \tag{5.22}$$

It is desired that all roots of the closed-loop system lie in a region towards the left of the line $s = -a$ so that the damping coefficient has a minimum value. Determine the values of K and T.

Solution: The characteristic equation is

$$F(s) = 1 + G(s)H(s) = 1 + \frac{K}{s(1 + sT)} = 0$$

or

$$F(s) = Ts^2 + s + K = 0 \tag{5.23}$$

The settling time being inversely proportional to the real part of the dominant roots, the relative stability can be specified by requiring that all roots of the characteristic equation must lie to the left of the line $s = -a$. To ensure that all roots lie to the left of the line $s = -a$, let us put $s = p - a$ in $F(s)$ to get

$$T(p - a)^2 + (p - a) + K = 0$$

or
$$Tp^2 + (1 - 2T)p + (a^2T - a + K) = 0 \qquad (5.24)$$

The Routh array is:

p^2	T	$(a^2T - a + K)$
p^1	$(1 - 2T)$	
p^0	$(a^2T - a + K)$	

To ensure that there is no change of sign in the first column of the Routh's array, we need

$$T > 0, \quad (1 - 2T) > 0, \quad \text{and} \quad (a^2T - a + K) > 0$$

The first two inequalities give $0 < T < 0.5$. The third inequality gives $K > a(1 - aT)$.

To ensure that all roots lie to the left of the line $s = -a$, the desired value of K can be obtained for any given value of a and the value of time constant T can be chosen keeping in view the already obtained bounds on T.

EXAMPLE 5.12 Consider

$$F(s) = s^3 + 9s^2 + 26s + K$$

Find the value of K required so that the dominant time constant is less than or equal to 0.5. The system should be designed to be slightly underdamped.

Solution: The largest time constant = 0.5. This means that no root can lie to the right of the line $s = (-2)$. Putting $s = p - 2$ gives:

$$F(p) = (p - 2)^3 + 9(p - 2)^2 + 26(p - 2) + K$$
$$= p^3 + 3p^2 + 2p + (K - 24) \qquad (5.25)$$

The Routh array is:

p^3	1	2
p^2	3	$K - 24$
p^1	$10 - \dfrac{K}{3}$	
p^0	$K - 24$	

No change of sign occurs for $24 \leq K \leq 30$. This guarantees that the dominant time constant is less than or equal to 0.5.

Now for $K = 24$, there exists a single root at origin in the p-plane (i.e. at $p = 0$) as can be seen from $F(p)$, or equivalently a root at $s = -2$. If $K = 30$, the third row of the array is zero, with positive entries above and below. This gives a pair of imaginary roots in p-plane, that is, for $K = 30$, the two roots are $s = -2 \pm jb$, where b is unknown at this point. Putting $K = 30$ in $F(p)$, we get

$$p^3 + 3p^2 + 2p + 6 = (p + a)(p^2 + b^2) = p^3 + ap^2 + b^2 p + ab^2 \qquad (5.26)$$

Comparing the two forms, we get

$$a = 3, \quad b^2 = 2, \quad ab^2 = 6 \quad \text{or} \quad a = 3, \quad b = \sqrt{2}$$

From Eq. (5.26), $p = -a$ and $p = \pm jb$ will satisfy $F(p)$ for $K = 30$.

In terms of s, this gives $s = p - 2 = -a - 2 = -5$ and $s = p - 2 = \pm j\sqrt{2} - 2 = -2 \pm j\sqrt{2}$

The damping factor ζ can be obtained from Figure 5.3 as

$$\zeta = \cos \theta = 0.816$$

EXAMPLE 5.13 For the system shown in Figure 5.4 where K is the gain of the power amplifier and B is the gain of the velocity feedback, determine the stability range in the B-K plane by Routh's stability criterion.

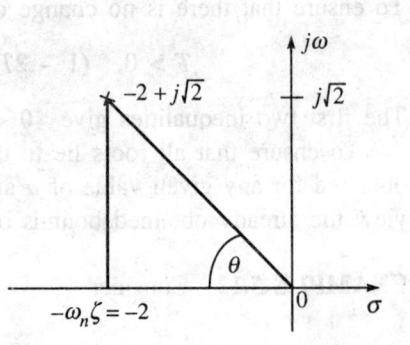

FIGURE 5.3 Example 5.12.

Solution: The closed-loop transfer function is

$$\frac{C(s)}{R(s)} = \frac{K}{s(s + 5)(s + 10) + Bs + K}$$

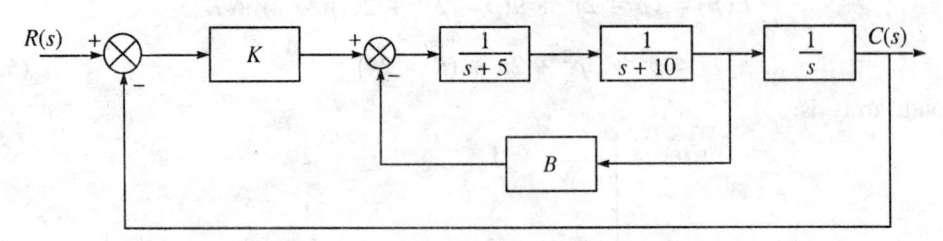

FIGURE 5.4 Example 5.13: block diagram of a positioning system.

The characteristic equation is

$$s(s + 5)(s + 10) + Bs + K = s^3 + 15s^2 + (50 + B)s + K = 0 \qquad (5.27)$$

This third-order polynomial contains two adjustable variables.

The Routh array is:

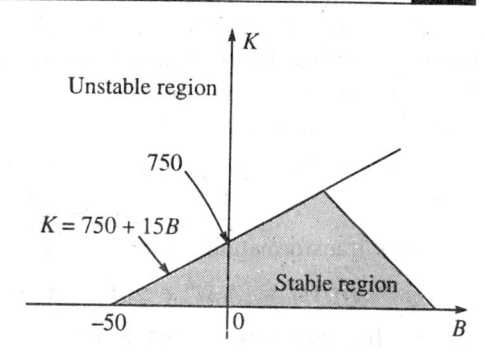

s^3	1	$50 + B$
s^2	15	K
s^1	$\dfrac{750 + 15B - K}{15}$	
s^0	K	

(5.28)

FIGURE 5.5 Example 5.13: stability boundary.

Note that the stability limit is now a function of both K and B. Clearly, K must be positive and it is necessary that $K \leq (750 + 15B)$. This equation defines a straight line on B-K plane as shown in Figure 5.5. It is clear that the stability range would be $0 < K < 750$ if B were zero (i.e. no velocity feedback); but with velocity feedback, the stability range can be above 750 if one adjusts B to preserve stability. The stability range is marked as the shaded region.

5.10 Relative Stability with Minimum Damping Ratio

Suppose we want to know about the stability of the system in addition to knowing whether all the roots lie within the radiating lines indicated by the minimum damping ratio, i.e. $\zeta_{min} = \cos \theta$ as shown shaded in Figure 5.6. Is it possible? We are going to discuss this. The condition for a stable system is that the effective damping ratio must be greater than some minimum value, where the minimum damping ratio is specified by the cosine of the angle between the negative axis and either of the radiating lines.

Rotate the s-plane axes by an angle $\phi = (\pi/2) - \theta$ counterclockwise as shown in Figure 5.7. With this rotation, a system having all its poles lying to the left of $\bar{s}$-plane, where $\bar{s} = \Sigma + j\Omega$, will satisfy a minimum damping ratio criterion $\zeta = \cos \theta = \sin \phi$; this is simply the coordinate transformation.

Putting $s = e^{j\phi}\bar{s}$, where $\bar{s} = \Sigma + j\Omega$, in $F(s)$, then

$$F(s) = F(e^{j\phi}\bar{s}) = H(\bar{s}) \quad (5.29)$$

Again,

$$s^n = e^{jn\phi}(\bar{s})^n = [\cos n\phi + j \sin n\phi](\bar{s})^n \quad (5.30)$$

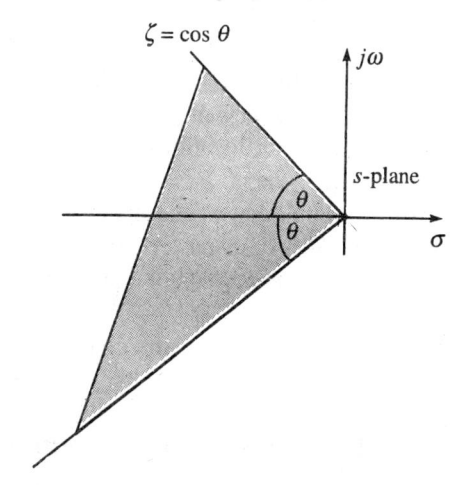

FIGURE 5.6 Stability region in shaded area.

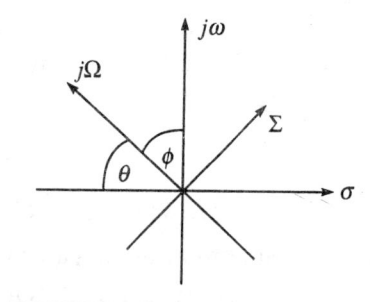

FIGURE 5.7 Rotation of s-plane.

The new characteristic polynomial (transformed), $H(\bar{s})$, has complex coefficients and hence the normal Routh criterion technique cannot be applied. Here is the necessity to adapt Routh algorithm for polynomials having complex coefficients. The results are given below without proof and derivation.

Consider the transformation of a complex characteristic polynomial $H(\bar{s})$ such that

$$H(\bar{s}) = H(j\Omega) \tag{5.31}$$

With this transformation the characteristic polynomial becomes

$$H(j\Omega) = (A_0\Omega^n + A_1\Omega^{n-1} + \cdots + A_n) + j(B_0\Omega^n + B_1\Omega^{n-1} + \cdots + B_n) \tag{5.32}$$

and the first two rows of the Routh array may be written as

$$\begin{array}{c} \text{Row} \\ \downarrow \\ 0 \quad A_0 \quad A_1 \quad \cdots \quad A_n \\ 1 \quad B_0 \quad B_1 \quad \cdots \quad B_n \end{array} \tag{5.33}$$

provided $B_0 \neq 0$ and the array may be completed in the normal way. For, $B_0 = 0$ is the special case to be treated as

$$G(\bar{s}) = jH(\bar{s}) \tag{5.34}$$

This new transformation swaps the first two rows of the Routh array and multiplies the new 0th row by −1. This method removes the first column zero from row 1. As $H(\bar{s})$ and $G(\bar{s})$ have the same root locations, without loss of generality, we can use this new transform when in $H(j\Omega)$ the element $B_0 = 0$ and then proceed in a normal way to compute the Routh array.

After completion of the array, we take the product of rows 1 and 2, of rows 3 and 4, of rows 5 and 6, and so on. All these products will be nonzero and of the same sign for roots to lie in the negative-half of the s-plane. This is illustrated below with an example.

EXAMPLE 5.14 Given the characteristic equation as

$$F(s) = s^3 + 4s^2 + 6s + 4 = 0$$

test whether the system has an effective minimum damping ratio of 0.5?

Solution: The system's absolute stability has been checked by going through the normal Routh's test and hence we will now check the relative stability.

An effective damping ratio of 0.5, that is (cos $\theta = 0.5 \to \theta = 60° \to \phi = 90° - \theta = 30°$ and $\zeta = \cos\theta = \sin\phi = 0.5$) means that the s-plane requires rotation by $\phi = \sin^{-1}(0.5) = 30°$. Hence $s = e^{j30°}\bar{s}$, and therefore $H(\bar{s})$ becomes

$$H(\bar{s}) = F(e^{j30°}\bar{s}) = (\cos 90° + j\sin 90°)(\bar{s})^3 + 4(\cos 60° + j\sin 60°)(\bar{s})^2 + 6(\cos 30° + j\sin 30°)\bar{s} + 4 = 0$$

or
$$H(\bar{s}) = j\bar{s}^3 + (2 + j3.46)\bar{s}^2 + (5.2 + j3)\bar{s} + 4 = 0$$

Consider the transformation of a complex characteristic polynomial such that

$$H(j\Omega) = (\Omega^3 - 2\Omega^2 - 3\Omega + 4) + j(-3.46\Omega^2 + 5.2\Omega) = 0$$

Now we have to arrange the Routh's array. By inspection it is clear that in row 1, we have $B_0 = 0$, so we cannot proceed further. Routh array for this system is obtained by considering

$$G(\bar{s}) = jH(\bar{s})$$

that is, $G(j\Omega) = (3.46\Omega^2 - 5.2\Omega) + j(\Omega^3 - 2\Omega^2 - 3\Omega + 4) = 0$

which gives the Routh array as

Row ↓

Row					
0		0	3.46	−5.2	0
1	+ve	1	−2	−3	4
2		3.46	−5.2	0	
3	+ve	−0.49	−3	4	
4		−26.38	28.24		
5	+ve	−3.52	4		
6		−1.74			

(5.35)

The stability check is carried out by taking the product of rows 1 and 2, of rows 3 and 4, and of rows 5 and 6 and these first column products are all found to be positive. Hence the system has an effective minimum damping ratio of at least 0.5 as all these products are nonzero and of the same sign.

EXAMPLE 5.15 Take the same characteristic equation as that of the previous example, i.e.

$$F(s) = s^3 + 4s^2 + 6s + 4 = 0$$

and test whether the system has an effective minimum damping ratio of 0.866.

Solution: For minimum damping ratio of 0.866, the s-plane has to be rotated by an angle of 60°, and then

$$H(j\Omega) = (2\Omega^2 - 5.2\Omega + 4) + j(\Omega^3 - 3.46\Omega^2 + 3\Omega) = 0$$

As $B_0 \neq 0$, we can formulate the Routh's array from $H(j\Omega)$ as

Row ↓

Row					
0		0	2	−5.2	4
1	+ve	1	−3.46	3	0
2		2	−5.2	4	
3	+ve	−0.86	1		
4		−2.87	4		
5	−ve	−0.20			
6		4			

(5.36)

The stability check may be carried out by taking the product of rows 1 and 2, of rows 3 and 4, and of rows 5 and 6. As the first column products contain a sign change, this system does not have the required effective minimum damping ratio of 0.866. It is obvious that $F(s)$ has the roots -2 and $-1 \pm j1$; from this root information the damping ratio comes out to be 0.707. Hence instead of rotating the s-plane by 60° if we rotate it by 45°, then the relative stability could have been ensured.

Do not get confused. Our aim is to know the relative stability without solving for the roots, otherwise it would have been a trivial problem.

Drill Problem 5.5

A unity-feedback system has the forward transfer function as

$$G(s) = \frac{K}{s(s+3)(s+10)}$$

(a) Determine the range of K for the system to be stable.
(b) What should be the upper limit of K if all poles of the closed-loop system are to lie to the left-half of the line $\sigma = -1$?

Ans. (a) $0 < K < 390$; (b) 13.86

5.11 Stability under Parameter Uncertainty: Robust Control

The robust control problem, that is, the problem of designing accurate control systems in the presence of significant plant uncertainties, is classical. However, over the last few years, significant new theory on this problem has been developed.

Routh's stability is an absolute stability criterion. Now the parameter of the characteristic equation is subject to variation within its minimum and maximum levels. In that case, it is practically impossible to perform Routh's stability test for innumerable possibilities of parameter combinations in the characteristic equation. Fortunately, the Russian mathematician, Kharitonov, came to rescue in the late 1970s. He published a paper on purely mathematical point of view in some mathematical journal in 1978 without knowing the demanding problem of robust control. Using the Kharitonov's method, only four polynomials have to be generated from the parameter variations within the maximum and minimum levels. The Routh's stability test is performed on only these four polynomials to declare the system stability under the parameter variations. Here, we will discuss the Kharitonov's method.[1,2,3]

Let us assume the polynomial $p(s)$ as

$$p(s) = p_0 + p_1 s + p_2 s^2 + \cdots + p_{n-1} s^{n-1} + p_n s^n = p_e(s) + p_o(s) \qquad (5.37)$$

---

[1] Kharitonov, V.L. "Asymptotic stability of an equilibrium position of a family of systems of linear differential equations", *Differential'nya Uraveniya*, Vol. **14**, pp. 2086–88, 1978.
[2] Minnichelli, R.J., J.J. Anagnost, and C.A. Desoer, "An elementary proof of Kharitonov's stability theorem with extensions", *IEEE Trans. Auto. Contr.*, Vol. **AC-34**, 1989.
[3] Chapellat, H. and S.P. Bhattacharyya, "Generalization of Kharitonov's Theorem: Robust Stability of Interval Plants", *IEEE Trans. Auto. Contr.*, Vol. **AC-34**, No. 3, March 1989, pp. 306–11.

Routh Stability and Robust Control

where the even part $p_e(s)$ is

$$p_e(s) = p_0 + p_2 s^2 + p_4 s^4 + \cdots \quad (5.38)$$

and the odd part $p_o(s)$ is

$$p_o(s) = p_1 s + p_3 s^3 + p_5 s^5 + \cdots \quad (5.39)$$

Then find,

$$p_e(\omega) = p_e(s)\big|_{s=j\omega} = p_0 - p_2 \omega^2 + p_4 \omega^4 - p_6 \omega^6 + \cdots \quad (5.40)$$

and

$$p_o(\omega) = \frac{p_o(s)}{s}\bigg|_{s=j\omega} = p_1 - p_3 \omega^2 + p_5 \omega^4 - p_7 \omega^6 + \cdots \quad (5.41)$$

Now $p(s)$ is Hurwitz if and only if:
  (i) $p_e(\omega)$ and $p_o(\omega)$ have the highest coefficient terms of the same sign.
  (ii) the roots of $p_e(\omega)$ and $p_o(\omega)$ are all real and interlacing.

Suppose the roots of $p_e(\omega)$ are $\omega_{e1}, \omega_{e2}, \ldots, \omega_{em}$ and those of $p_o(\omega)$ are $\omega_{o1}, \omega_{o2}, \ldots, \omega_{o(m-1)}$. Roots of $p_e(\omega)$ and $p_o(\omega)$ are interlacing means that (see Figure 5.8)

$$\omega_{e1} < \omega_{o1} < \omega_{e2} < \omega_{o2} < \omega_{e3} < \omega_{o3} < \omega_{e4} < \cdots \quad (5.42)$$

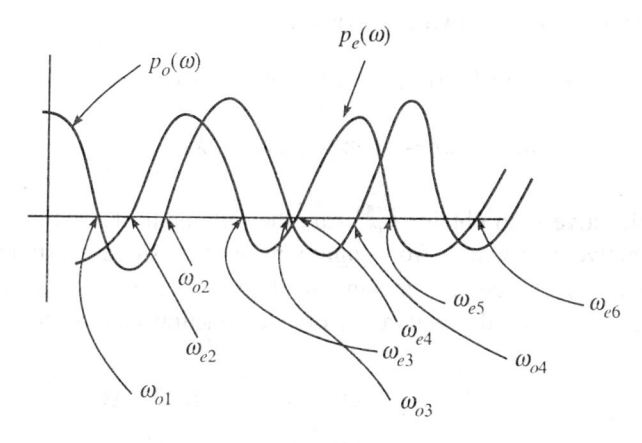

FIGURE 5.8 Roots of $p_e(\omega)$ and $p_o(\omega)$.

If $p(s)$ is Hurwitz, then:
  (i) All coefficients are of the same sign.
  (ii) The phase of $p(s)|_{s=j\omega}$ decreases monotonically from $-n\pi/2$ at $\omega = -\infty$ to $n\pi/2$ at $\omega = \infty$ (see Figure 5.9).

Now,

$$p(j\omega) = p_e(\omega) + j p_o(\omega) \quad (5.43)$$

Let us define the even function $p_e(\omega)$ as K_e and the odd function $p_o(\omega)$ as K_o. Further, suppose that the parameters of the polynomial $p(s)$ vary within limits of minimum and maximum as

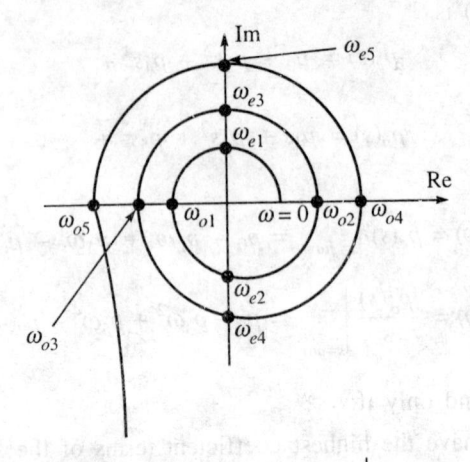

FIGURE 5.9 Phase of $p(s)|_{s=j\omega}$.

X and Y respectively. That is, the parameter p_0 varies between its minimum value X_0 and its maximum value Y_0. Similarly, the parameter p_1 varies between X_1 and Y_1. The parameter p_2 varies between its minimum X_2 and maximum Y_2, and so on. In general, the parameter p_i varies between its minimum X_i and maximum Y_i for $i = 0, 1, 2, \cdots$.

Now let us formulate the following polynomials

$$K_e^{\min}(\omega), \quad K_e^{\max}(\omega), \quad K_o^{\min}(\omega) \quad \text{and} \quad K_o^{\max}(\omega)$$

as

$$K_e^{\min}(\omega) = X_0 - Y_2\omega^2 + X_4\omega^4 - Y_6\omega^6 + X_8\omega^8 - Y_{10}\omega^{10} + \cdots \quad (5.44)$$

In Eq. (5.44), the objective is to obtain $K_e^{\min}(\omega)$, the minimum of even function; i.e. subtract maximum of all negative coefficients from minimum of all positive coefficients of Eq. (5.40). In a similar argument, $K_e^{\max}(\omega)$, the maximum of even function, $K_o^{\min}(\omega)$, the minimum of odd function and $K_o^{\max}(\omega)$, the maximum of odd function can be generated as (see also Figure 5.10)

$$K_e^{\max}(\omega) = Y_0 - X_2\omega^2 + Y_4\omega^4 - X_6\omega^6 + Y_8\omega^8 - X_{10}\omega^{10} + \cdots \quad (5.45)$$

$$K_o^{\min}(\omega) = X_1 - Y_3\omega^2 + X_5\omega^4 - Y_7\omega^6 + X_9\omega^8 - Y_{11}\omega^{10} + \cdots \quad (5.46)$$

$$K_o^{\max}(\omega) = Y_1 - X_3\omega^2 + Y_5\omega^4 - X_7\omega^6 + Y_9\omega^8 - X_{11}\omega^{10} + \cdots \quad (5.47)$$

Following Eq. (5.40), we can get

$$K_e^{\min}(s) \quad \text{from} \quad K_e^{\min}(\omega) \quad \text{and} \quad K_e^{\max}(s) \quad \text{from} \quad K_e^{\max}(\omega)$$

as

$$K_e^{\min}(s) = X_0 + Y_2 s^2 + X_4 s^4 + Y_6 s^6 + X_8 s^8 + Y_{10} s^{10} + \cdots \quad (5.48)$$

and

$$K_e^{\max}(s) = Y_0 + X_2 s^2 + Y_4 s^4 + X_6 s^6 + Y_8 s^8 + X_{10} s^{10} + \cdots \quad (5.49)$$

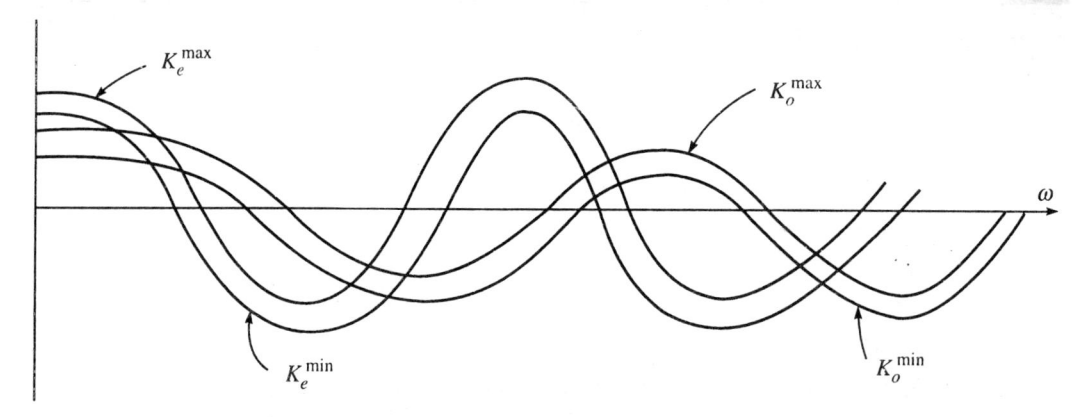

FIGURE 5.10 Interlacing of even and odd polynomials.

Following Eq. (5.41), we can get

$K_o^{min}(s)$ from $K_o^{min}(\omega)$ and $K_e^{max}(s)$ from $K_o^{max}(\omega)$

as
$$K_o^{min}(s) = X_1 s + Y_3 s^3 + X_5 s^5 + Y_7 s^7 + X_9 s^9 + Y_{11} s^{11} + \cdots \qquad (5.50)$$

and
$$K_e^{max}(s) = Y_1 s + X_3 s^3 + Y_5 s^5 + X_7 s^7 + Y_9 s^9 + X_{11} s^{11} + \cdots \qquad (5.51)$$

Four possible polynomials are to be ultimately formulated out of these four polynomials given in Eqs. (5.48) through (5.51) and named as $K^1(s)$, $K^2(s)$, $K^3(s)$, and $K^4(s)$ as follows:

$$K^1(s) = K_e^{min}(s) + K_o^{min}(s)$$
$$= X_0 + X_1 s + Y_2 s^2 + Y_3 s^3 + X_4 s^4 + X_5 s^5 + Y_6 s^6 + \cdots \qquad (5.52)$$

$$K^2(s) = K_e^{min}(s) + K_o^{max}(s)$$
$$= X_0 + Y_1 s + Y_2 s^2 + X_3 s^3 + X_4 s^4 + Y_5 s^5 + \cdots \qquad (5.53)$$

$$K^3(s) = K_e^{max}(s) + K_o^{min}(s)$$
$$= Y_0 + X_1 s + X_2 s^2 + Y_3 s^3 + Y_4 s^4 + X_5 s^5 + \cdots \qquad (5.54)$$

$$K^4(s) = K_e^{max}(s) + K_o^{max}(s)$$
$$= Y_0 + Y_1 s + X_2 s^2 + X_3 s^3 + Y_4 s^4 + Y_5 s^5 + \cdots \qquad (5.55)$$

Now these four polynomial Eqs. (5.52) through (5.55) obtained from the characteristic polynomial $p(s)$ with the variation of parameter p_i within limits X_i and Y_i for $i = 0, 1, 2, \ldots$ need to pass the Routh's stability test individually, for the system to be stable under perturbation or uncertainty conditions.

In order to understand the theory behind the Kharitonov's method, the following examples are given here.

The concept of uncertainty is briefly explained by the following four rules.

$$X = [a, b] \text{ means } \forall\, x \in R: a \leq x \leq b$$

Assume two intervals of confidence in R:

$$A = [a_1, \; a_2]$$
$$B = [b_1, \; b_2]$$

Then, Addition : $A + B = [a_1 + b_1, \; a_2 + b_2]$
Subtraction : $A - B = [a_1 - b_2, \; a_2 - b_1]$
Multiplication : $A \cdot B = [a_1 \cdot b_1, \; a_2 \cdot b_2]$
Division : $A/B = [a_1/b_2, \; a_2/b_1]$

EXAMPLE 5.16 Suppose the parameter variations are as follows:

$a = 4 \pm 1$ implies $3 \leq a \leq 5$ implies $a = [3, 5]$
$G = 0.15 \pm 0.1$ implies $0.05 \leq G \leq 0.25$ implies $G = [0.05, 0.25]$
$K = 2.5 \pm 0.5$ implies $2 \leq K \leq 3$ implies $K = [2, 3]$

Now we have to calculate $(1 + aG)/G$. The question is, should we calculate $(1 + aG)/G$ directly? No, then linearity will not be maintained because of addition of constant 1.

So, $\dfrac{1 + aG}{G} = \left(a + \dfrac{1}{G}\right)$ should be considered.

Now, $\left(a + \dfrac{1}{G}\right) = [3, 5] + \dfrac{1}{[0.05, 0.25]} = [3,5] + [4,20] = [7,25]$

EXAMPLE 5.17 In a position control system of a damped rotating gun turret of Figure 5.11, the nominal system parameters and their uncertainty ranges are

$$a = 4 \pm 1, \quad T = 0.15 \pm 0.1 \quad \text{and} \quad K = 2.5 \pm 0.5$$

Determine the stability of the closed-loop system.

Solution: From Figure 5.11, the closed-loop transfer function becomes

$$\frac{\Theta(s)}{R(s)} = \frac{K}{s(s+a)(1+Ts) + K}$$

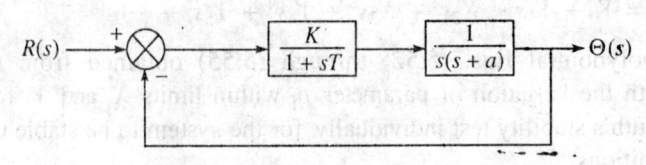

FIGURE 5.11 Example 5.17.

The characteristic polynomial becomes

$$p(s) = s(s+a)(1+Ts) + K = s^3 + \frac{1+aT}{T}s^2 + \frac{a}{T}s + \frac{K}{T}$$

Routh Stability and Robust Control

$$= c_3 s^3 + c_2 s^2 + c_1 s + c_0$$

where $\quad c_3 = 1, \quad c_2 = \dfrac{1+aT}{T} = a + \dfrac{1}{T}, \quad c_1 = \dfrac{a}{T}, \quad c_0 = \dfrac{K}{T}$

The parameter variations are as follows:

$a = 4 \pm 1$	implies $3 \leq a \leq 5$	implies $a = [3, 5]$
$T = 0.15 \pm 0.1$	implies $0.05 \leq T \leq 0.25$	implies $T = [0.05, 0.25]$
$K = 2.5 \pm 0.5$	implies $2 \leq K \leq 3$	implies $K = [2, 3]$

Hence, $\quad c_0 = \dfrac{K}{T} = \dfrac{[2,3]}{[0.05, 0.25]} \in [8, 60]$

$c_1 = \dfrac{a}{T} = \dfrac{[3,5]}{[0.05, 0.25]} \in [12, 100]$

$c_2 = a + \dfrac{1}{T} = [3, 5] + \dfrac{1}{[0.05, 0.25]} = [3, 5] + [4, 20] \in [7, 25]$

and $\quad c_3 = 1$

The four Kharitonov polynomials as per Eqs. (5.52) to (5.55) are

$$K^1(s) = 8 + 12s + 25s^2 + s^3$$
$$K^2(s) = 8 + 100s + 25s^2 + s^3$$
$$K^3(s) = 60 + 12s + 7s^2 + s^3$$
$$K^4(s) = 60 + 100s + 7s^2 + s^3$$

For each $K^i(s)$, reading alternate coefficients from right to left, we determine the following Routh arrays:

\multicolumn{8}{c}{Four Kharitonov polynomials}
\multicolumn{2}{c}{$K^1(s)$}	\multicolumn{2}{c}{$K^2(s)$}	\multicolumn{2}{c}{$K^3(s)$}	\multicolumn{2}{c}{$K^4(s)$}
1	12	1	100	1	12	1	100
25	8	25	8	7	60	7	60
$\dfrac{292}{25}$	0	$\dfrac{2492}{25}$	0	$\dfrac{24}{7}$	0	$\dfrac{640}{7}$	0
8		8		60		60	

Since column 1 for each Kharitonov polynomial contains no change of sign, we can conclude that all of the roots for each $K^i(s)$ ($i = 1, 2, 3, 4$) have negative real parts. Thus, the control system is stable for all parameter values in the specified ranges. That is, the feedback control system is guaranteed asymptotically stable.

EXAMPLE 5.18 Consider a unity-feedback system having forward transfer function

$$G(s) = \frac{K}{s(s+a)(s+b)}$$

with uncertainty ranges of the parameters K, a, b as given below:

$$K = 200 \pm 20; \qquad a = 4 \pm 0.5; \qquad b = 6 \pm 1$$

We have to determine the closed-loop system stability under the given parameter variation using the Kharitonov's four polynomials.

Solution: The characteristic polynomial

$$P(s) = K + abs + (a+b)s^2 + s^3$$
$$= P_0 + P_1 s + P_2 s^2 + P_3 s^3$$

From the given ranges on the parameters, the coefficients of the characteristic polynomial are

$$P_3 = 1$$
$$8.5 \leq P_2 = (a+b) \leq 11.5$$
$$17.5 \leq P_1 = ab \leq 31.5$$
$$180 \leq P_0 = K \leq 220$$

We now obtain the four Kharitonov polynomials as

$$K^1(s) = 180 + 17.5s + 11.5s^2 + s^3$$
$$K^2(s) = 180 + 31.5s + 11.5s^2 + s^3$$
$$K^3(s) = 220 + 17.5s + 8.5s^2 + s^3$$
$$K^4(s) = 220 + 31.5s + 8.5s^2 + s^3$$

Apply the Routh–Hurwitz criterion to all these four polynomials $K^i(s)$ ($i = 1, 2, 3, 4$). In this particular case, the first column of the Routh array for the three polynomials $K^1(s)$, $K^2(s)$ and $K^4(s)$ are all positive (that is, no change of sign in the first column). But the polynomial $K^3(s)$ has two changes of sign in the first column of the Routh array as shown in the following table.

	Four Kharitonov polynomials
	$K^1(s)$		$K^2(s)$		$K^3(s)$		$K^4(s)$
1	17.5	1	31.5	1	17.5	1	31.5
11.5	180	11.5	180	8.5	220	8.5	220
1.84		15.84		−8.38		5.6	
180		180		220		220	

This shows that the closed-loop system will be unstable for the given set of parameter variations.

Drill Problem 5.6

For a unity-feedback system having open-loop transfer function

$$\frac{K}{s(s+a)(s+b)}$$

the uncertainty ranges of the parameters are as follows:

$8 \leq K \leq 12$; $2.5 \leq a \leq 3.5$ and $3.8 \leq b \leq 4.2$

Determine the stability of the closed-loop system using Kharitonov's method.

Ans. The closed-loop is stable for all parameter variations in the specified ranges.

Summary

The absolute stability of a linear system can be determined by Routh–Hurwitz criterion without solving for the roots of the system. The criterion may be suitably utilized for determining the range of open-loop gain for the closed-loop system to be stable. For the state variable formulation of the system, the criterion can be utilized for finding the number of eigenvalues of the system matrix lying in the right-half of the s-plane. Further, in case of uncertainties of the parameter variation, if their ranges of variations are known, we can investigate the robust parametric stability by using Routh–Hurwitz criterion on the four polynomials formed using Kharitonov's theorem.

Problems

5.1 Using Routh–Hurwitz criterion, determine the number of roots lying in the right-half s-plane of the following characteristic equations and comment on the stability of each system.

(a) $s^3 + 20s^2 + 9s + 100 = 0$
(b) $s^3 + 20s^2 + 9s + 200 = 0$
(c) $3s^4 + 10s^3 + 5s^2 + s + 2 = 0$
(d) $s^4 + 2s^3 + 6s^2 + 8s + 8 = 0$
(e) $s^6 + 2s^5 + 8s^4 + 12s^3 + 20s^2 + 16s + 16 = 0$

5.2 From the following characteristic equations, determine the value of K for the system to be stable.

(a) $s^4 + 22s^3 + 10s^2 + 2s + K = 0$
(b) $s^4 + 20Ks^3 + 5s^2 + (10 + K)s + 15 = 0$
(c) $s^3 + (K + 0.5)s^2 + 4Ks + 50 = 0$

5.3 Using Routh–Hurwitz stability criterion, determine the stability of the following characteristic polynomials:

(a) $s^2 + 3s + 1$

(b) $s^3 + 4s^2 + 5s + 6$
(c) $s^3 + 3s^2 - 6s + 10$
(d) $s^4 + s^3 + 2s^2 + 6s + 8$
(e) $s^4 + s^3 + 2s^2 + 2s + K$
(f) $s^5 + s^4 + 2s^3 + s + 3$
(g) $s^5 + s^4 + 2s^3 + s^2 + s + K$

Determine the range of K for the system to be stable. Further, determine the number of roots, if any, in the right-hand plane.

5.4 A feedback control system has the forward transfer function $k(s + 40)/s(s + 10)$ and the feedback transfer function $1/(s + 20)$.

(a) Determine the range of gain K for the system to be stable.
(b) Determine the number of imaginary roots.
(c) Reduce the gain K to one-half the magnitude of that of the borderline value and determine the relative stability using Routh–Hurwitz criterion.

5.5 Determine the range of stability of the systems with the following characteristic equations:

(a) $s^3 + s^2 + 2s + 0.5 = 0$
(b) $s^4 + 9s^3 + 30s^2 + 42s + 20 = 0$
(c) $s^3 + 19s^2 + 110s + 200 = 0$

5.6 Determine the relative stability of the following unity-feedback control systems (a) by determining the location of the roots in the s-plane, and (b) by using Routh–Hurwitz criterion. The open-loop transfer functions are given as:

(i) $G(s) = \dfrac{65 + 33s}{s^2(s + 20)}$; (ii) $G(s) = \dfrac{24}{s(s^3 + 10s + 35s + 50)}$;

(ii) $G(s) = \dfrac{2(s+4)(s+8)}{s(s+5)^2}$

5.7 For the given open-loop transfer function

$$G(s)H(s) = \dfrac{K(s+1)}{s(1+Ts)(1+2s)}$$

determine the region in the K-T plane in which the closed-loop system is stable.

5.8 The open-loop transfer function of a unity-feedback control system is

$$G(s) = \dfrac{K(s+5)(s+40)}{s^3(200+s)(1000+s)}$$

Discuss the stability of the closed-loop system as a function of K. Determine (a) the values of K for which the closed-loop system gives sustained oscillations and (b) the frequency of oscillation.

5.9 The open-loop transfer function of a unity-feedback system is

$$G(s) = \frac{K}{s(1+Ts)}$$

It is desired that all the roots of the system's characteristic equation lie in the region to the left of the line $s = -a$. This will assure that not only is a stable system obtained, but also that the system has a minimum amount of damping. Extend the Routh–Hurwitz criterion to this case, and determine the values of K and T required so that there are no roots to the right of the line $s = -a$.

5.10 Medical science reveals that humans use visual signals, otoliths, and signals from their semicircular canals for feedback to control their posture and retain their balance. Posture control of a unity-feedback system has the open-loop transfer function as

$$G(s)H(s) = \frac{K(s+2)}{s(s+5)(s+2s+5)}$$

Determine the value of gain K when a person's balance is lost on account of the system just starting to oscillate, and also calculate the roots when the system oscillates.

5.11 A feedback control system has the following characteristic equation

$$s^3 + (4+K)s^2 + 6s + 8(K+2) = 0$$

What is the maximum value of K before the system starts to oscillate? Determine the frequency of oscillation.

5.12 The characteristic equation of a feedback control system is

$$s^6 + 2s^5 + 5s^4 + 8s^3 + 8s^2 + 8s + K = 0$$

What is the maximum value of K before the system starts to oscillate? Determine (a) the frequency of oscillation and (b) the value of the roots when the system oscillates.

5.13 The dynamics of a motorcycle and an average rider can be represented by an open-loop transfer function

$$G(s)H(s) = \frac{K(s+30s+1125)}{(s+20)(s+10s+25)(s^2+60s+3400)}$$

(a) Calculate the range of K for the system to be stable.
(b) Calculate the range of K for the system to be stable when the numerator polynomial and the denominator polynomial $(s^2 + 60s + 3400)$ are neglected.

5.14 A system has the characteristic equation

$$1 + \frac{K(s^2+4s+a)}{s(s+T)(s^2+2s+2)} = 0$$

Use Kharitonov's technique to determine the range(s) of parameter values for K for guaranteed asymptotic stability with $10 \le a \le 20$ and $5 \le T \le 8$.

Walter R. Evans was born in 1920. He was the recipient of the 1987 American Society of Mechanical Engineers Rufus Oldenburger Medal and the 1988 AACC Richard E. Bellman Control Heritage Award. He passed away at the age of 79 on July 10, 1999 in Whittier, CA.

Walter Evans' principal contribution to the field of automatic control was his invention of the Evans Root Locus Method in 1948 and his subsequent invention of the Spirule, a tool used in conjunction with the root-locus method. Because it codifies very useful frequency information about a feedback system in such an intuitive and appealing graphical form, Evan's root-locus method has enjoyed widespread use in the design of control systems and is now a standard chapter in texts on feedback control systems.

1920–1999

Evans received his B.S. in Electrical Engineering from Washington University in St. Louis, Missouri in 1941 and the M.S. degree in Electrical Engineering from the University of California at Los Angeles in 1951. During his lifetime, he worked as an engineer at several companies, including General Electric, Autonetics, and Ford Aeronautic Company. He also served as an instructor at Washington University for few years.

Root-Locus Technique

6.1 Introduction

The roots of the characteristic equation of the closed-loop system determine the dynamic behaviour of the overall system. Thus, for the analysis of problems of control engineering, it is important to study the location of poles and zeros of the closed-loop system in the s-plane, with the variation of a system parameter.

For a higher-order polynomial it is a tedious job to find the roots. The classical technique for factoring the polynomial is not convenient because the computations must be repeated as a single parameter of the open-loop transfer function is varied. Using the root-locus technique, the roots of the closed-loop system (corresponding to a particular value of the variable parameter) can be located on the open-loop pole-zero configuration. It may be noted that the variable parameter is usually the gain K in the present text unless otherwise stated, but any other variable of the open-loop transfer function may also be used.

OBJECTIVE

The root-locus technique is a graphical method for drawing the locus of roots in the s-plane as a system parameter is varied and hence it is of paramount importance in the design of control systems. Since the root-locus method provides graphical information, an approximate sketch can be used to obtain qualitative information about the stability and performance of the system. The beauty of the root-locus technique is that it gives knowledge about the roots of the closed-loop system from the open-loop pole-zero configuration when one system parameter is varied in the positive range, and all this without solving the roots of the characteristic polynomial of the closed-loop system.

CHAPTER OUTLINE

Introduction
Basic Conditions for Root Loci
Construction Rules for Root Loci
Construction Rules for Inverse Root Loci
Effect of Adding Poles and Zeros
Root Contour
Root Locus with Delay
Design on Root Locus

6.2 Root-Loci for Second-Order System

Consider the second-order system shown in Figure 6.1 which represents a typical position control system. For the negative feedback control system having forward transfer function

$$G(s) = \frac{K}{s(s+4)} \qquad (6.1)$$

where K is a gain parameter and feedback transfer function $H(s) = 1$ for a unity-feedback system, the closed-loop transfer function is

$$\frac{C(s)}{R(s)} = \frac{G(s)}{1 + G(s)H(s)} = \frac{K}{s^2 + 4s + K} \tag{6.2}$$

Since the dynamic behaviour of a system is controlled by the roots of the characteristic equation, it is important to investigate the variation of the roots of the closed-loop system with the variation of the parameter K. The characteristic equation of the system is

$$s^2 + 4s + K = 0 \tag{6.3}$$

which is analogous to

$$s^2 + 2\zeta\omega_n s + \omega_n^2 = 0$$

where ζ is the damping coefficient and ω_n is the undamped natural frequency of the closed-loop system. The roots of the closed-loop system are

$$s_1 = -2 + \sqrt{4 - K} \quad \text{and} \quad s_2 = -2 - \sqrt{4 - K} \tag{6.4}$$

The movement of the roots s_1, s_2 when K is varied in the range $0 \leq K < +\infty$ is shown in Figure 6.2.

(i) For $K = 0$: The two roots are at $s_1 = 0$ and $s_2 = -4$. The two roots are the same as the poles of the open-loop transfer function $G(s)H(s) = K/s(s + 4)$ at points A and B of Figure 6.2.

(ii) For $0 < K < 4$: Both the roots s_1 and s_2 are negative real.

(iii) For $K = 4$: Both the roots are repetitive and $s_1 = s_2 = -2$ as at point C.

(iv) For $4 < K < +\infty$: The two roots are complex conjugate pairs with the negative real part equal to -2. $s_1 = -2 + j\sqrt{K - 4}$ and $s_2 = -2 - j\sqrt{K - 4}$. The two roots move towards points D and E respectively as gain K increases from 4.

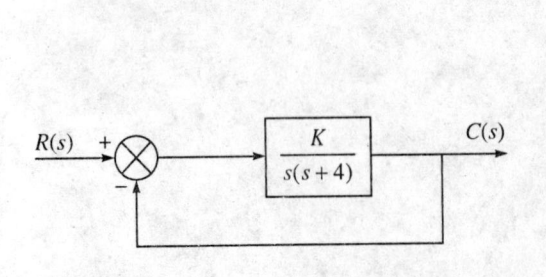

FIGURE 6.1 Typical position control system.

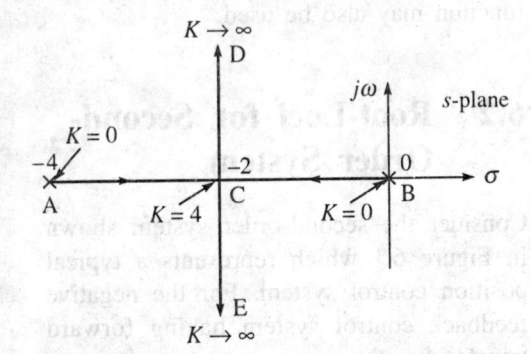

FIGURE 6.2 Root loci for the system shown in Figure 6.1.

The root loci therefore consist of straight lines AB and DE intersecting at C. The loci start from two open-loop poles at $s = 0$ (i.e. at B where $K = 0$) and at $s = -4$ (i.e. at A where $K = 0$). The root-locus starting from A moves towards the right and that starting from B moves towards the left for $0 < K < 4$ till they intersect at C for $K = 4$. For values of $K > 4$, the loci move along the lines CD and CE as shown in Figure 6.2.

From the root loci, the following information about the system behaviour may be inferred.

(i) *Stability:* The closed-loop system is stable for all positive values of gain $0 \leq K < +\infty$.

(ii) *Transient response:* For all values of gain K between 0 and 4, the system is overdamped (the damping coefficient $\zeta > 1$); for $4 < K < +\infty$, the system is underdamped ($\zeta < 1$); for $K = 4$, the system is critically damped ($\zeta = 1$) having repetitive roots.

For $\zeta < 1$, the damping coefficient is given by

$$\zeta = 2/\sqrt{K} \tag{6.5}$$

The undamped natural frequency ω_n increases with the increase in K. For all values of $K > 4$, the settling time of the step response is constant since the real parts of the two roots are fixed and are equal to -2.

We have considered an example of a second-order system where the root-locus can be obtained from analytical considerations. Every time with the new parameter value K, we have to determine the roots of the closed-loop system analytically which is a formidable task especially for higher-order systems. The root-locus technique is a systematic graphical method by which we can draw the root-locus diagram of the closed-loop system from open-loop pole-zero information when one variable parameter (say, gain K) is varied in the range $0 \leq K < +\infty$.

The root-locus technique was introduced by W.R. Evans in 1948 and the art has since been developed and extensively applied to the analysis and design of control systems. The beauty of the technique is that it gives the complete information of the closed-loop poles from the open-loop pole-zero information without solving for the roots of the closed-loop system when one of the parameters (say, gain K) of the open-loop transfer function is varied.

In control system problems, the complete root-locus diagram is a plot of the loci of the poles of the closed-loop transfer function (or roots of the characteristic equation) when one parameter of the open-loop transfer function is varied from $-\infty$ to $+\infty$. We have in our definition, root loci, inverse root loci and complete root loci plots.

(i) Usually when only one parameter, say gain K, is varied between 0 to $+\infty$ the plot is called the root-locus diagram.

(ii) When K is varied between 0 to $-\infty$, the plot is called the inverse root-locus diagram.

(iii) The plot is called the complete root loci when K is varied in the range $-\infty < K < \infty$, that is, the complete root loci is made up of root loci and inverse root loci plots.

Further, when more than one parameter is considered to be variable, the plot is referred to as a root-contour diagram. Our point of interest is the root loci for the variable parameter K in the range $0 \leq K < \infty$.

Drill Problem 6.1

Determine the poles and zeros including those at infinity (if any) of each of the following rational transfer functions $G(s)$. Mark each finite pole by a small cross (×) and each finite zero by a small circle (o) in the s-plane. Graphically, measure the magnitude and phase at $\omega = 2$ rad/s for each of the transfer functions.

(i) $\dfrac{10(s+2)}{s^2(s+1)(s+10)}$ (ii) $\dfrac{10s(s+1)}{(s+2)(s^2+3s+2)}$ (iii) $\dfrac{10(s+2)}{s(s^2+2s+2)}$

Drill Problem 6.2

A unity-feedback system has forward transfer function, $G(s) = K/s(s+2)$. Determine the value of K for the closed-loop poles having damping ratio of 0.5.

Ans. 4

6.3 Basic Conditions for Root Loci

Let us consider the negative feedback system of Figure 6.3 where $G(s)$ is the forward transfer function and $H(s)$ is the feedback ratio. The closed-loop transfer function can be written as

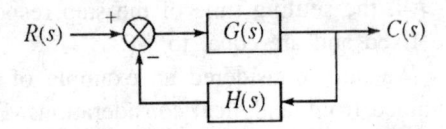

$$\frac{C(s)}{R(s)} = \frac{G(s)}{1+G(s)H(s)} \qquad (6.6)$$

FIGURE 6.3 Negative feedback system.

In a feedback control system without transportation lag, the open-loop transfer function $G(s)H(s)$ is a rational algebraic function and can be written as

$$G(s)H(s) = \frac{K(s^m + b_{m-1}s^{m-1} + \cdots + b_1 s + b_0)}{s^n + a_{n-1}s^{n-1} + \cdots + a_1 s + a_0} = K\frac{P(s)}{Q(s)} = \frac{K(s+z_1)(s+z_2)\cdots(s+z_m)}{(s+p_1)(s+p_2)\cdots(s+p_n)}$$

$$= \frac{K\prod_{i=1}^{m}(s+z_i)}{\prod_{j=1}^{n}(s+p_j)} = KG_1(s)H_1(s); \quad m \le n \qquad (6.7)$$

where K is the variable parameter, say, gain, a real quantity varying in the range $0 \le K < +\infty$, and $G_1(s)H_1(s)$ is a complex quantity. $Q(s)$ is an nth order polynomial of s as

$$Q(s) = s^n + a_{n-1}s^{n-1} + \cdots + a_1 s + a_0$$

and $P(s)$ is an mth order polynomial of s as

$$P(s) = s^m + b_{m-1}s^{m-1} + \cdots + b_1 s + b_0$$

and n and m are positive integers and $m \le n$.

Root-Locus Technique

The characteristic equation is

$$1 + G(s)H(s) = 0 \quad \text{or} \quad G(s)H(s) = -1 \tag{6.8}$$

i.e.
$$KG_1(s)H_1(s) = -1 \quad \text{or} \quad \frac{K\prod_{i=1}^{m}(s+z_i)}{\prod_{j=1}^{n}(s+p_j)} = -1 \tag{6.9}$$

This leads to the following two conditions:

(i) The magnitude condition for root loci becomes

$$|G(s)H(s)| = 1 \tag{6.10}$$

i.e.
$$|G_1(s)H_1(s)| = \frac{\prod_{i=1}^{m}(s+z_i)}{\prod_{j=1}^{n}(s+p_j)} = \frac{1}{|K|}; \quad 0 \le K < \infty \tag{6.11}$$

(ii) The phase condition or angle condition for root loci becomes

$$\angle G(s)H(s) = \angle K + \sum_{i=1}^{m}\angle(s+z_i) - \sum_{j=1}^{n}\angle(s+p_j)$$

$$= (2k+1)\pi; \quad k = 0, 1, 2, \cdots \tag{6.12}$$

or
$$\angle G_1(s)H_1(s) = \sum_{i=1}^{m}\angle(s+z_i) - \sum_{j=1}^{n}\angle(s+p_j)$$

$$= (2k+1)\pi; \quad k = 0, 1, 2, \cdots \tag{6.13}$$

as K is a real quantity, $\angle K = 0$.

It may be noted that for any given value of K, say K_1, between 0 and $+\infty$, any point P (i.e. $s = s_1$) in the s-plane which satisfies the magnitude and phase conditions of Eqs. (6.11) and (6.13), respectively, is a point on the root loci. Alternatively, the loci of the roots corresponding to all values of K in the range plotted in Figure 6.2 must satisfy both the magnitude and phase conditions, i.e. for any point on the root loci, one can immediately determine the value of the variable parameter K from the magnitude condition that will yield a closed-loop pole at a desired point on the root loci. Similarly, it can be shown that any desired point on the root loci must satisfy the angle condition given by Eq. (6.13).

As an illustration, let

$$G(s)H(s) = \frac{K(s+z_1)}{s(s+p_2)(s+p_3)} \tag{6.14}$$

The poles and zeros are located in the s-plane as shown in Figure 6.4. Let $s = s_1$ be any desired point P on the root loci, then the magnitude condition

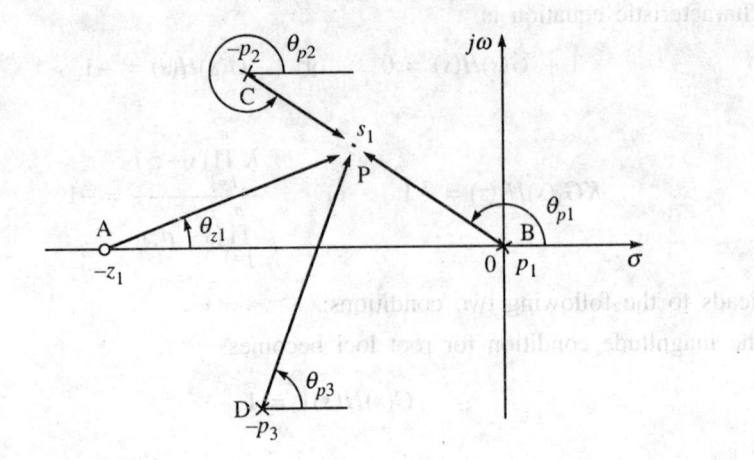

FIGURE 6.4 Pole-zero configuration of $G(s)H(s) = \dfrac{K(s+z_1)}{s(s+p_2)(s+p_3)} \equiv \dfrac{K(s+z_1)}{(s-0)(s+p_2)(s+p_3)}$.

$$\left| G(s)H(s) \right|_{s=s_1} = \left| KG_1(s)H_1(s) \right|_{s=s_1} = 1$$

or
$$\left| G_1(s_1)H_1(s_1) \right| = \frac{1}{|K|} \qquad (6.15)$$

i.e.
$$\frac{|(s_1+z_1)|}{|(s_1-0)||(s_1+p_2)||(s_1+p_3)|} = \frac{AP}{BP \cdot CP \cdot DP} = \frac{1}{|K|} \qquad (6.16)$$

The factor $|(s_1 + z_1)|$ is recognized as the length of the vector drawn from the zero at $-z_1$ to the point s_1 and the vector length is represented by AP. The factor $|(s_1 + p_2)|$ is the length of the vector drawn from the pole at $-p_2$ to s_1 and the vector length is represented by CP. Similarly, the vector length for $|(s_1 + p_3)|$ is represented by DP and the vector length for the pole at the origin, i.e. $|s_1 - 0|$ is represented by the vector length BP. Similarly, any point P at $s = s_1$ on the root loci must satisfy the phase condition, i.e.

$$\angle G_1(s_1)H(s_1) = (2k+1)\pi; \; k = 0, 1, 2, \cdots \qquad (6.17)$$

or
$$\angle \frac{s_1 + z_1}{s_1(s_1 + p_2)(s_1 + p_3)} = (2k+1)\pi \qquad (6.18)$$

or
$$\angle(s_1 + z_1) - (\angle(s_1 - 0) + \angle(s_1 + p_2) + \angle(s_1 + p_3)) = (2k+1)\pi \qquad (6.19)$$

or
$$\theta_{z_1} - (\theta_{p_1} + \theta_{p_2} + \theta_{p_3}) = (2k+1)\pi \qquad (6.20)$$

where the angles $\theta_{z_1}, \theta_{p_1}, \theta_{p_2},$ and θ_{p_3} are the arguments of the vectors measured with the positive real axis as the zero reference. The angle in the clockwise direction is negative and the angle in the counterclockwise direction is positive with respect to the positive real axis taken as the reference. It is to be noted that K is a real quantity and its angle contribution is zero.

Root-Locus Technique

Consequently, given the pole-zero configuration of an open-loop transfer function, the construction of the root-locus diagram of the closed-loop system involves a search for the s_1 point which will satisfy both the magnitude and angle conditions specified by Eqs. (6.15) and (6.17).

Although searching for all the points s_1 is apparently an almost impossible task, the actual procedure is not so complex. Normally the root loci can be sketched following certain rules of construction as described in the following.

6.4 Rules for the Construction of Root Loci ($0 \leq K \leq \infty$)

The rules for the construction of root loci are enumerated below:

Rule 1. *The $K = 0$ or the starting points on the root loci are at the poles of $G(s)H(s)$.*

Proof: Rewrite Eq. (6.11) as

$$|G_1(s)H_1(s)| = \frac{\prod_{i=1}^{m}|(s+z_i)|}{\prod_{j=1}^{n}|(s+p_j)|} = \frac{1}{|K|} \tag{6.21}$$

As K approaches zero, the value of $G_1(s)H_1(s)$ in Eq. (6.21) approaches infinity. Again the value of $G_1(s)H_1(s)$ become infinity at poles $s = -p_j$. It means that the starting point (i.e. $K = 0$) of each root loci is a pole (i.e. $s = -p_j$) of $G(s)H(s)$, the open-loop transfer function.

Rule 2. *The $K = +\infty$ or the terminating points on the root loci are the zeros of $G(s)H(s)$.*

Proof: With reference to Eq. (6.21) as K approaches $+\infty$, the value of Eq. (6.21) approaches zero. Again the value of Eq. (6.21) becomes zero at zeros of $G_1(s)H_1(s)$. It means that the terminating point of each root loci (i.e. $K = \infty$) is a zero of $G_1(s)H_1(s)$, which is the same as a zero of the open-loop transfer function $G(s)H(s)$. It may be noted that for any given $G(s)H(s)$, if the number of finite zeros Z is less than the number of finite poles P, then $(P - Z)$ zeros lie at infinity. This is because, for a rational function the total number of poles and zeros must be equal if the poles and zeros at infinity are counted as well.

To illustrate the preceding rules for the construction of root loci, let us consider the open-loop transfer function of the system as

$$G(s)H(s) = \frac{K(s+3)}{s(s+5)(s+6)(s^2+2s+2)} \tag{6.22}$$

where the gain parameter K varies in the range $0 < K < \infty$. Now let

$$G_1(s)H_1(s) = \frac{s+3}{s(s+5)(s+6)(s^2+2s+2)} = \frac{1}{K}$$

It may be mentioned that the poles and zeros of both $G_1(s)H_1(s)$ and $G(s)H(s)$ are the same. The pole-zero configuration is shown in Figure 6.5. The finite poles are located at $s = 0$,

$s = -5$, $s = -6$ and $s = -1 \pm j1$. The finite zero is located at $s = -3$. Therefore, these five poles are the starting points ($K = 0$ points) of the five root loci. As $G(s)H(s)$ is a rational function, the total number of poles must be equal to the total number of zeros, if the poles and zeros at infinity are counted. Hence, obviously four zeros are located at infinity. One of five root loci terminates at the finite zero of $G(s)H(s)$ at $s = -3$ and the other four root loci terminate at zeros located at infinity.

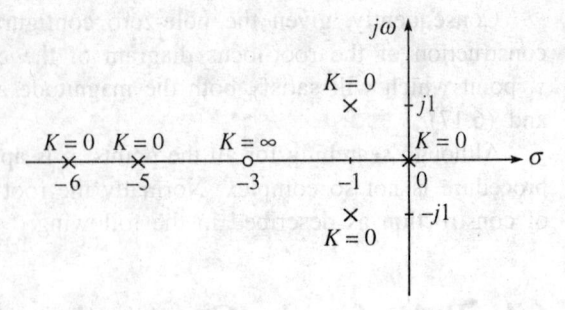

FIGURE 6.5 Pole-zero configuration.

Rule 3. *The number of separate root loci N is given by*

$$N = P \text{ if } P > Z \tag{6.23}$$

$$N = Z \text{ if } Z > P \tag{6.24}$$

Since a complete root loci is formed between a pair of a pole and a zero of $G(s)H(s)$, the total number of root loci of a given system must be equal to P or Z, whichever is greater. In fact, $N = P$ because for a physically realizable system $P \geq Z$. For example, in the open-loop transfer function of Eq. (6.22), the finite number of poles, $P = 5$ and the finite number of zeros, $Z = 1$. Hence the number of separate root loci $N = P = 5$ as $P > Z$.

Rule 4. *Symmetry of the root loci:* The root loci are symmetrical about the real axis in the s-plane. The proof is self evident as the complex roots occur in conjugate pairs and hence root loci are symmetrical about the real axis in s-plane.

Rule 5. *Angles of asymptotes and the intersection of asymptotes on the real axis:* If the number of finite zeros Z is less than the number of finite poles P, then $(P - Z)$ is the number of root loci that must terminate at zeros at infinity. The asymptotes to these $(P - Z)$ number of root loci intersect at a common point (called centroid) σ_A on the real axis, given by

$$\sigma_A = \frac{\text{sum of poles} - \text{sum of zeros}}{P - Z} \tag{6.25}$$

and are inclined to the real axis at angles ϕ_k given by

$$\phi_k = \frac{(2k+1)\pi}{P - Z}; \quad k = 0, 1, 2, \ldots, (P - Z - 1) \tag{6.26}$$

Equations (6.25) and (6.26) can be proved by rewriting Eq. (6.9) as

$$-K = \frac{\prod_{j=1}^{n}(s + p_j)}{\prod_{i=1}^{m}(s + z_i)} = \frac{s^n + a_{n-1}s^{n-1} + \ldots}{s^m + b_{m-1}s^{m-1} + \ldots} = \frac{s^n - \left(\sum_{j=1}^{n} p_j\right)s^{n-1}}{s^m - \left(\sum_{i=1}^{m} z_i\right)s^{m-1}} \tag{6.27}$$

As lim $s \to \infty$ is the condition of asymptote for large s, we can neglect the lower-order terms of s and Eq. (6.27) is approximated as

$$-K = s^{n-m} - \left(\sum_{j=1}^{n} p_j - \sum_{i=1}^{m} z_i\right) s^{n-m-1} + \ldots$$

by long division and can be approximated to

$$-K \approx s^{n-m}\left(1 - \frac{\sum_{j=1}^{n} p_j - \sum_{i=1}^{m} z_i}{s}\right)$$

or

$$s\left(1 - \frac{\sum_{j=1}^{n} p_j - \sum_{i=1}^{m} z_i}{s}\right)^{\frac{1}{n-m}} = (-K)^{\frac{1}{n-m}} \tag{6.28}$$

Binomial expansion leads to

$$s\left(1 - \frac{\sum_{j=1}^{n} p_j - \sum_{i=1}^{m} z_i}{(n-m)s} + \ldots\right) = (-K)^{\frac{1}{n-m}} \tag{6.29}$$

Again, if the terms higher than the second are neglected for large value of s, we get

$$s - \frac{\sum_{j=1}^{n} p_j - \sum_{i=1}^{m} z_i}{n-m} = (-K)^{\frac{1}{n-m}} \tag{6.30}$$

Substituting $s = \sigma + j\omega$ into Eq. (6.30) and using Demoivre's algebraic theorem, yields

$$(\sigma + j\omega) - \frac{\sum_{j=1}^{n} p_j - \sum_{i=1}^{m} z_i}{n-m} = \left|K^{\frac{1}{n-m}}\right|\left(\cos\frac{(2k+1)\pi}{n-m} + j\sin\frac{(2k+1)\pi}{n-m}\right) \tag{6.31}$$

for $\quad k = 0, 1, 2, \ldots, n - m - 1$

Equating real and imaginary parts, we get

$$\sigma - \frac{\sum_{j=1}^{n} p_j - \sum_{i=1}^{m} z_i}{n-m} = \left|K^{\frac{1}{n-m}}\right|\cos\frac{(2k+1)\pi}{n-m} \tag{6.32}$$

and

$$\omega = \left|K^{\frac{1}{n-m}}\right|\sin\frac{(2k+1)\pi}{n-m} \tag{6.33}$$

Solving for $K^{\frac{1}{n-m}}$ from Eqs. (6.32) and (6.33), we get

$$\left|K^{\frac{1}{n-m}}\right| = \frac{\omega}{\sin\frac{(2k+1)\pi}{n-m}} = \frac{\sigma - \frac{\sum_{j=1}^{n}p_j - \sum_{i=1}^{m}z_i}{n-m}}{\cos\frac{(2k-1)\pi}{n-m}}$$

and solving for ω, we get

$$\omega = \tan\frac{(2k+1)\pi}{n-m}\left[\sigma - \frac{\sum_{j=1}^{n}p_j - \sum_{i=1}^{m}z_i}{n-m}\right] \qquad (6.34)$$

Equation (6.34) represents an equation of straight line in the s-plane which is of the form

$$\omega \approx m(\sigma - \sigma_A) \qquad (6.35)$$

where m is the slope and σ_A is the intersection on the σ-axis. Thus,

$$m = \tan\frac{(2k+1)\pi}{n-m} = \frac{(2k+1)\pi}{P-Z}; \qquad k = 0, 1, \ldots, P-Z-1$$

and

$$\sigma_A = \frac{\sum_{j=1}^{n}p_j - \sum_{i=1}^{n}z_i}{n-m} \qquad (6.36)$$

The intersection of the asymptotes on the real axis is therefore given by

$$\sigma_A = \frac{\sum\text{finite poles of }G(s)H(s) - \sum\text{finite zeros of }G(s)H(s)}{P-Z} \qquad (6.37)$$

$$= \frac{\sum\text{real part of finite poles of }G(s)H(s) - \sum\text{real part of finite zeros of }G(s)H(s)}{P-Z}$$

The angles of asymptotes are given by

$$\phi_k = \tan^{-1}m = \frac{(2k+1)\pi}{P-Z} \qquad \text{for} \qquad k = 0, 1, 2, \ldots, (P-Z-1) \qquad (6.38)$$

Considering again the $G(s)H(s)$ of Eq. (6.22), the intersection of asymptotes on the real axis as shown in Figure 6.6 is given by

$$\sigma_A = \frac{\sum\text{real part of the finite poles of }G(s)H(s) - \sum\text{real part of the finite zeros of }G(s)H(s)}{P-Z}$$

$$= \frac{(0-5-6-1-1)-(-3)}{5-1} = -2.5$$

The angles of asymptotes shown in Figure 6.6 are given by

$$\phi_k = \frac{(2k+1)\pi}{P-Z}; \quad k = 0, 1, 2, \ldots, P-Z-1$$

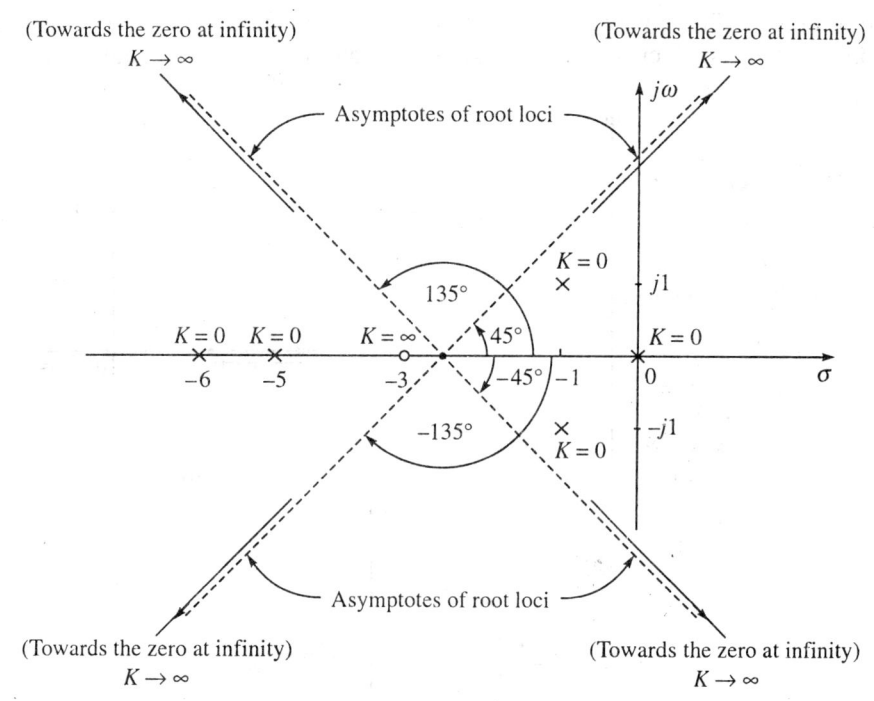

FIGURE 6.6 Angles of asymptotes and intersection of asymptotes on the real axis for
$$GH(s) = \frac{K(s+3)}{s(s+5)(s+6)(s^2+2s+2)}.$$

i.e. for $\quad k = 0, \ \phi_0 = \pi/4 = 45°; \quad\quad k = 1; \ \phi_1 = 3\pi/4 = 135°;$
$k = 2, \ \phi_2 = 5\pi/4 = -3\pi/4 = -135°; \quad k = 3; \ \phi_3 = 7\pi/4 = -\pi/4 = -45°.$

Drill Problem 6.3

Given the open-loop transfer function as $K/(s^2 + 4s + 5)s$, the root loci has one break-in point for $K = 1.852$, justify. Is there any break-away point? Determine the angles of asymptotes and the point of intersection of asymptotes on the real axis. Determine also the angle of departure from the open-loop complex pole in the upper-half of the s-plane.

Ans. 60°, 180°, −60°; −1.33°; −63.43°

Rule 6. *Existence of root loci on the real axis:* Any point on the real axis is a part of the root locus if and only if the number of finite poles and zeros of $G(s)H(s)$ on the real axis to the right of the point is odd.

Proof: This property follows by applying the angle condition at any point on the real axis of the s-plane. At any point s_1 on the real axis, the angles of the vectors from each pair of complex poles or zeros of $G(s)H(s)$ occuring in conjugate pairs will add up to zero (see Figure 6.7(a)). We need consider only the poles and zeros on the real axis. Further, the angle subtended at any point s_1 on the real axis by a real pole or zero of $G(s)H(s)$ to the left of the point is always zero (see Figure 6.7(b)). Hence we need to consider the poles and zeros of $G(s)H(s)$ on the real axis that are to the right of that point. Since each of these subtends an angle π (see Figure 6.7(c)), we must have an odd number of finite poles and zeros to the right of the point on the root loci on the real axis of s-plane to satisfy the angle condition.

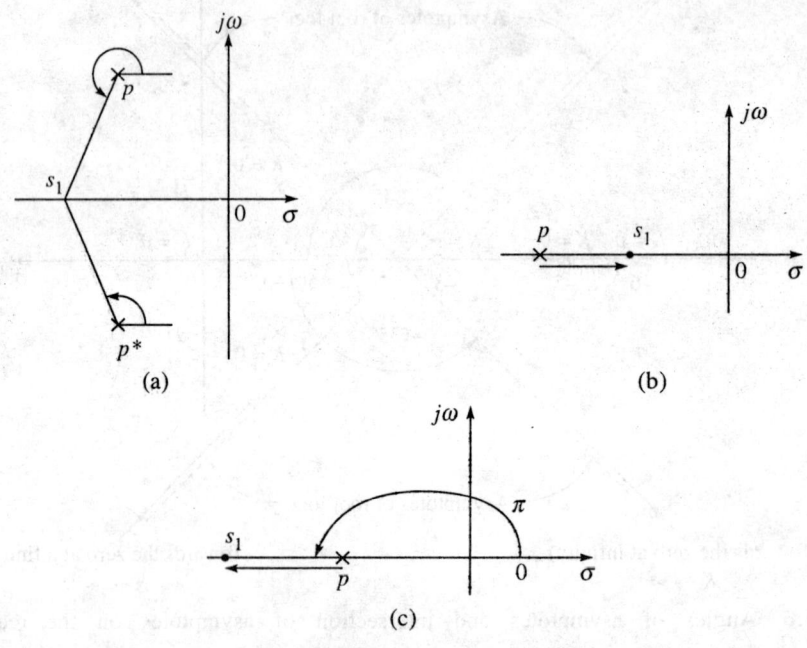

FIGURE 6.7 (a) Sum of the angles subtended at s_1 by complex conjugate poles/zeros is zero, (b) the angle subtended at s_1 by a real pole/zero to the left of the point is 0°, and (c) the angle subtended at s_1 by a real pole/zero to the right of the point is π.

Considering again the open-loop transfer function of Eq. (6.22), the existence of the root loci on the real axis is shown in Figure 6.8. It is apparent that the occurrence of the loci on the real axis is not affected by the complex poles and zeros of $G(s)H(s)$. There are root loci, i.e. for $0 < K < \infty$ on the real axis between (a) $s = 0$ and $s = -3$ and between (b) $s = -5$ and $s = -6$.

FIGURE 6.8 Existence of root loci on the real axis.

Rule 7. *Intersection of the root loci on the imaginary axis:* If the root-locus crosses the $j\omega$-axis for some values of K, this can be readily determined through the Routh–Hurwitz criterion. From the Routh table we can determine the value of K that makes the system just unstable, as well as the locations of the resulting roots on the $j\omega$-axis. This is illustrated again with the $G(s)H(s)$ of Eq. (6.22), whose characteristic equation is

$$s^5 + 13s^4 + 54s^3 + 82s^2 + (60 + K)s + 3K = 0 \tag{6.39}$$

The Routh tabulation is

s^5	1	54	$60 + K$
s^4	13	82	$3K$
s^3	47.7	$60 + 0.769K$	
s^2	$65.6 - 0.212K$	$3K$	
s^1	$\dfrac{3940 - 105K - 0.163K^2}{65.6 - 0.212K}$		
s^0	$3K$		

In order to have the closed-loop poles in the left-half of the s-plane, i.e. for the closed-loop system to be stable, the quantities in the first column of the Routh tabulation array should be of the same sign, i.e. positive. Therefore, the following inequalities must be satisfied:

(i) $\qquad (65.6 - 0.212K) > 0 \qquad$ or $\qquad K < 309$

(ii) $\qquad (3940 - 105K - 0.163K^2) > 0 \qquad$ or $\qquad K < 35 \tag{6.40}$

(iii) $\qquad K \geq 0$

Hence all the roots of the characteristic Eq. (6.39) will stay in the left-half of the s-plane if K lies in the range $0 < K < 35$. This means that the root loci just crosses the imaginary axis when $K \geq 35$. The coordinate at the crossover point on the imaginary axis that corresponds to $K = 35$ can be determined from the auxiliary equation to be formed from Routh's array as

$$A(s) = (65.6 - 0.212K)s^2 + 3K = 0$$

Substituting $K = 35$, we get

$$58.2s^2 + 105 = 0$$

which yields $\qquad s = \pm j1.34, \qquad$ i.e. $\qquad \omega = \pm 1.34$

Hence the intersection of the root loci on the imaginary axis will take place at points $\omega = 1.34$ rad/s and $\omega = -1.34$ rad/s for gain $K = 35$.

Drill Problem 6.4

Determine the point of intersection of the root loci with the imaginary axis for a unity-feedback system with forward transfer function $G(s) = K/s(s + 4)(s + 16)$. Find the value of K at this crossover frequency.

Ans. $s = +j8;\ K = 1280$

Rule 8. *Angles of departure (from poles) and angles of arrival (at zeros) of the root loci:* The angle of departure (arrival) of root loci at a pole (zero) of $G(s)H(s)$ denotes its behaviour near the pole (zero). For the root loci ($0 \leq K \leq \infty$) these angles can be determined by use of Eq. (6.13). This is illustrated with the open-loop transfer function of Eq. (6.22). Suppose it is desired to determine the angle at which the root locus leaves the pole at $s = -1 + j1$ (see Figure 6.9). Note that the unknown angle θ is measured with respect to the real axis with anticlockwise movement as positive and clockwise movement as negative angle. Let us assume that s_1 is a point on the root locus leaving the pole at $s = -1 + j1$ and s_1 is very close to the pole at $s = -1 + j1$. Then the point s_1 must satisfy Eq. (6.13). Thus,

$$\angle G_1(s_1)H_1(s_1) = \angle(s_1 + 3) - [\angle s_1 + \angle(s_1 + 1 + j1) + \angle(s_1 + 5) + \angle(s_1 + 6) + \angle(s_1 + 1 - j1)]$$
$$= (2k + 1)\pi \quad \text{for } k = 0, 1, 2 \tag{6.41}$$

Since s_1 is very close to the pole at $-1 + j1$, the angles of the vectors are drawn from the finite poles and zero to the point s_1 in Figure 6.9 and then Eq. (6.41) becomes

$$26.6° - (135° + 90° + 14° + 11.4° + \theta) = (2k + 1)180°; \quad k = 0, 1, 2$$

Therefore,
$$\theta = -43.8° = 316.2°$$

Hence the angle of departure of the root locus from the pole at $(-1 + j1)$ is $316.2°$. It may be noted that the angle of departure of the root locus from the complex conjugate pole at $(-1 - j1)$ is also $-316.2°$ as the root loci about the real axis is symmetric.

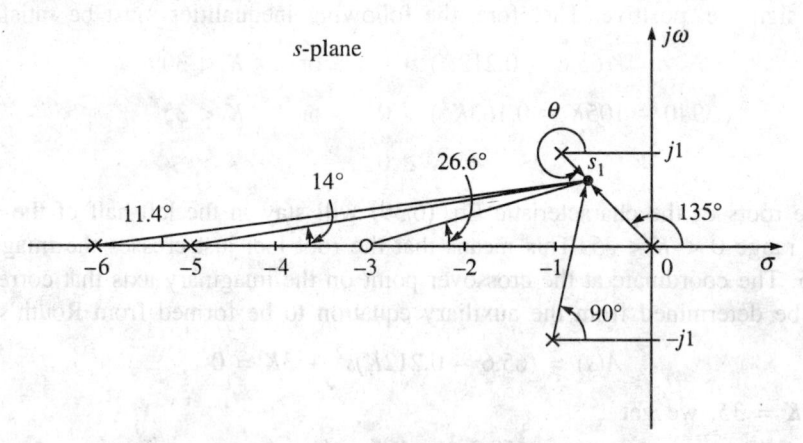

FIGURE 6.9 Determination of angle at which the root locus leaves the pole at $s = -1 + j1$.

Rule 9. *Break-away or break-in points:* Break-away points on the root loci of an equation correspond to multiple-order roots of the equation. A root-locus diagram can, of course, have more than one break-away point. Moreover, the break-away points need not always be on the real axis. However, because of the conjugate symmetry of the root loci, the break-away points must either be real or in complex conjugate pairs. If r number of root loci branches meet at a point, they break away at an angle of $\pm 180°/r$.

Two branches of the root loci on the real axis meet at a point and depart at right angles to it as depicted in Figure 6.10(a). This is called the *break-away* point and is characterized by the fact that the characteristic polynomial has multiple roots at the break-away point. The dual situation occurs when the branches of the root loci enter the real axis. This may be called the *break-in* or *re-entry* point as depicted in Figure 6.10(b). Normally, a break-away point occurs on a segment of the root locus on the real axis commencing from two poles as in Figure 6.10(a), and a re-entry point occurs between two zeros as in Figure 6.10(b). Further, there may be points on the real axis as in Figure 6.10(c) where more than two branches of the root loci meet. The angles at which the branches will leave the real axis are given by $\pm 180°/r$, where r is the number of root loci branches approaching the break-away point. Again, the root-locus may have more than one break-away point as in Figure 6.13. Moreover all the break-away points need not be always on the real axis (see Example 6.1, Figure 6.13). Obviously, if the break-away points are in the complex plane, they must be in complex conjugate pairs because of the property of symmetry of root loci about the real axis.

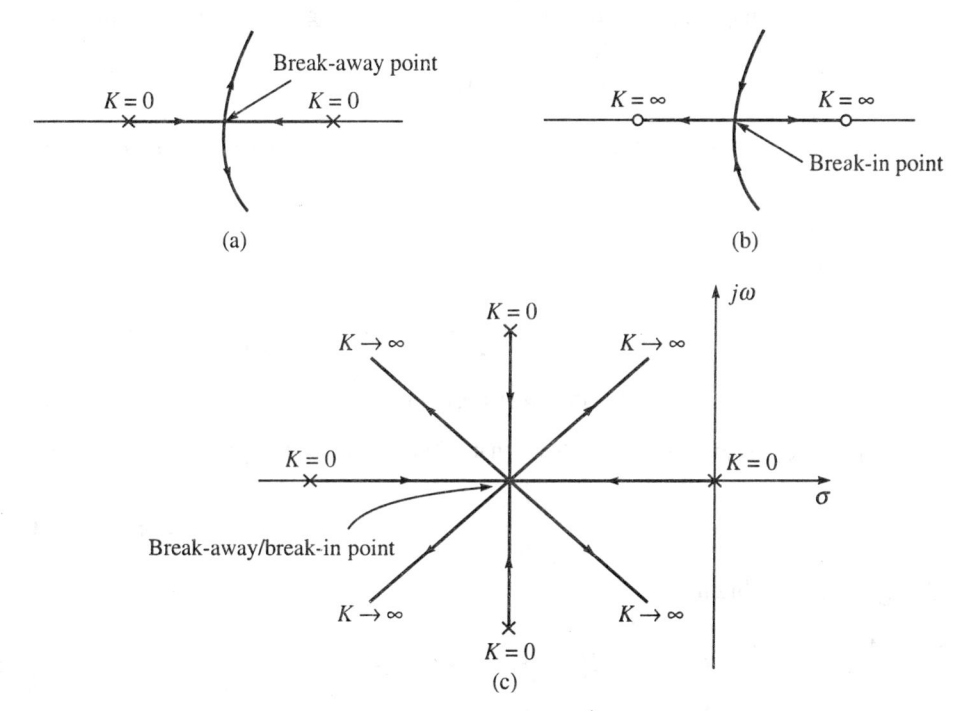

FIGURE 6.10 (a) Break-away point, (b) break-in point, and (c) break-away/break-in point.

Method I

Note that $F(s) = 1 + G(s)H(s) = 0$ has multiple roots at points where

$$\frac{d}{ds} F(s) = 0$$

The break-away or break-in points represent occurrence of multiple roots. Then the break-away/break-in points on the root loci of the characteristic equation

$$F(s) = 1 + G(s)H(s) = 1 + KG_1(s)H_1(s) = 0$$

must satisfy

$$\frac{d}{ds}[G(s)H(s)] = \frac{d}{ds}[G_1(s)H_1(s)] = 0 \tag{6.42}$$

It is important to point out that the condition for the break-away/break-in point given by the above equation is a necessary one but not sufficient. In other words, all break-away/break-in points must satisfy Eq. (6.42), but all solutions of this equation are not necessarily the break-away/break-in points. To be a break-away/break-in point, the solution of Eq. (6.42) must also satisfy the characteristic equation or must be a factor of the characteristic equation for some real value of the parameter K in the range $0 \le K < \infty$.

Break-away/break-in points are thus determined by solving the following equation:

$$\frac{d}{ds}[G(s)H(s)] = 0$$

For higher-order systems the method is laborious, sometimes a formidable task.

Method II

Let us take the characteristic equation as

$$F(s) = 1 + G(s)H(s) = 1 + KG_1(s)H_1(s) = 1 + K\frac{P(s)}{Q(s)} = 0$$

or

$$F(s) = Q(s) + KP(s) = 0 \tag{6.43}$$

where $Q(s)$ and $P(s)$ do not contain K. Note that $F(s) = 0$ has multiple roots at points where

$$\frac{dF(s)}{ds} = 0 \tag{6.44}$$

From Eq. (6.43), we obtain

$$\frac{dF(s)}{ds} = Q'(s) + KP'(s) = 0 \tag{6.45}$$

where

$$Q'(s) = \frac{d}{ds}Q(s) \quad \text{and} \quad P'(s) = \frac{d}{ds}P(s)$$

The particular value of K which will yield multiple roots of the characteristic equation as obtained from Eq. (6.45) is

$$K = -\frac{Q'(s)}{P'(s)} \tag{6.46}$$

Substituting the value of K in Eq. (6.43), we get

$$F(s) = Q(s) - \frac{Q'(s)}{P'(s)} \times P(s) = 0$$

or
$$Q(s)P'(s) - Q'(s)P(s) = 0 \qquad (6.47)$$

Again from Eq. (6.43),

$$K = -\frac{Q(s)}{P(s)}$$

Then
$$\frac{dK}{ds} = -\frac{Q'(s)P(s) - Q(s)P'(s)}{P^2(s)} \qquad (6.48)$$

If dK/ds is set equal to zero, we get the equation which is same as Eq. (6.47), which is the Eq. (6.45) itself and is the condition for occurrence of multiple roots of the characteristic equation $F(s) = 0$ of the closed-loop system. Therefore, the break-away/break-in points can be simply determined from the roots of

$$\frac{dK}{ds} = 0 \qquad (6.49)$$

Further, it may be noted that to be the break-away/break-in points, the solution of Eq. (6.49) should satisfy the characteristic equation. In fact, for a break-away point

$$\frac{d^2K}{ds^2} < 0 \qquad (6.50)$$

and for a break-in point

$$\frac{d^2K}{ds^2} > 0 \qquad (6.51)$$

The value of K at break-away points

As we have the value of s at the break-away point $s = s_1$ from either of the methods I and II, there is a necessity for knowing the value of gain K at the break-away point as

$$K = \left.\frac{1}{G_1(s)H_1(s)}\right|_{s=s_1} \qquad (6.52)$$

This will give us important information concerning the design of the system.

As an example, consider $G(s)H(s)$ of Eq. (6.22), having the characteristic equation as

$$\frac{K(s+3)}{s(s+5)(s+6)(s^2+2s+2)} = -1$$

The break-away points can be obtained by solving $dK/ds = 0$, i.e.

$$s^5 + 13.5s^4 + 66s^3 + 142s^2 + 123s + 45 = 0 \tag{6.53}$$

To determine the break-away points, we have to determine the roots of Eq. (6.53). From common sense and from the structure of root loci, a break-away point should lie between $-6 < s < -5$.

After a few trial-and-error calculations using the bisection theorem, the root of Eq. (6.53) that corresponds to the break-away point is found to be $s = -5.53$. As a check, the point $s = -5.53$ lies on the admissible range on the real axis where root loci can exist (Rule 6). Hence the break-away point is $s = -5.53$ as shown in Figure 6.8. The other four roots of Eq. (6.53) are $s = -0.656 \pm j0.468$ and $s = -3.33 \pm j1.204$. These four points are not the break-away points as they do not satisfy the characteristic equation. After going through all the rules for the construction of root loci, the root loci for the feedback system having the open-loop transfer function as

$$G(s)H(s) = \frac{K(s+3)}{s(s+5)(s+6)(s^2+2s+2)}$$

can be drawn as shown in Figure 6.11.

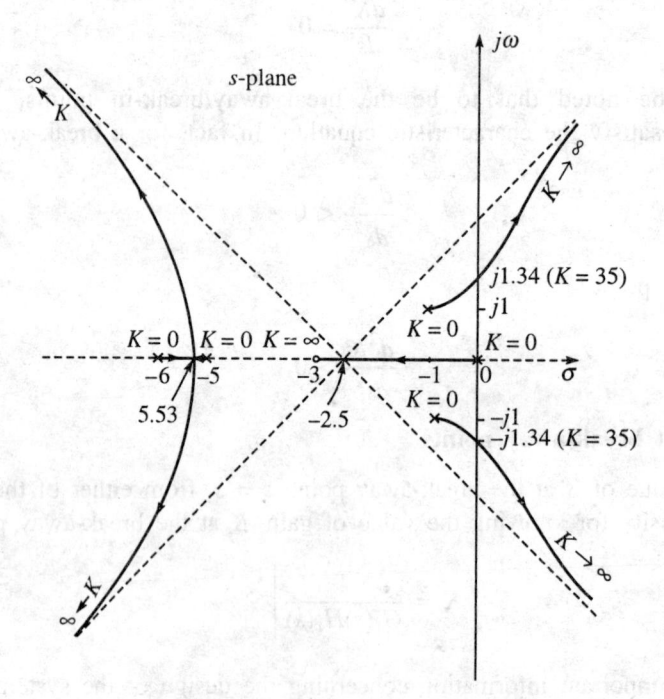

FIGURE 6.11 Root loci of $\dfrac{K(s+3)}{s(s+5)(s+6)(s^2+2s+3)}$.

Drill Problem 6.5

Determine the break-away and break-in points of the root loci plot of a unity-feedback system having the open-loop transfer function as

$$G(s)H(s) = \frac{K(s+3)}{s(s+2)}$$

Ans. $-1.268;\ -4.732$

Let us now summarize the steps utilized in evaluating the loci of roots of characteristic equation:

1. Write the characteristic equation in the pole-zero form so that the parameter of interest K appears as $1 + KG_1(s)H_1(s) = 0$.
2. Locate the open-loop poles and zeros of $G_1(s)H_1(s) = 0$.
3. Locate the segments of the real axis where root loci exist.
4. Determine the number of separate root loci.
5. Locate the angles of the asymptotes and the intersection of the asymptotes.
6. Determine the break-away point on the real axis (if any).
7. By utilizing the Routh–Hurwitz criterion, determine the point at which the locus crosses the imaginary axis (if it does so).
8. Estimate the angle of departure of root loci from complex poles and the angle of arrival of root loci at complex zeros.

Note that for solving any particular problem of root loci, we do not need all the rules of construction, but we have to be familiar with all the rules in order to construct the root loci of all systems.

We have studied in detail the construction procedures of root loci. Let us now test our knowledge. Draw the root-locus plot for the transfer function $G(s)H(s) = K/s$. Here the number of root locus is 1. The angle of asymptote is π. The existence of the root locus on the real axis is along the negative real axis where the angle condition is satisfied. See Figure 6.12(a).

Similarly, the root loci for transfer function $G(s)H(s) = K/s^2$, is as shown in Figure 6.12(b). Here the number of root loci is two. The angles of asymptotes are $\pi/2$ and $-\pi/2$. The existence of root loci on the real axis is nil, i.e. no root loci exists on the real axis because of the concerned property, that no root loci exists on the real axis if the sum of finite poles and zeros to the right of the point in s-plane is even.

The angle of departure condition is satisfied only on the imaginary axis.

Now we should be able to draw the rough sketch of root loci at a glance for the following open-loop transfer functions for $0 < K < \infty$.

$GH(s) = K/s$ as in Figure 6.12(a)

$GH(s) = K/s^2$ as in Figure 6.12(b)

$GH(s) = K/(s + 1)$ as in Figure 6.12(c)

$GH(s) = K/(s - 1)$ as in Figure 6.12(d)

$$GH(s) = K/(s+2)^2 \text{ as in Figure 6.12(e)}$$

$$GH(s) = K(s+1)/(s+2)^2 \text{ as in Figure 6.12(f)}$$

$$GH(s) = \frac{K(s+2)}{(s+1)^2} \text{ as in Figure 6.12(g)}$$

$$GH(s) = \frac{K}{s^2 + 2s + 2} \text{ as in Figure 6.12(h)}$$

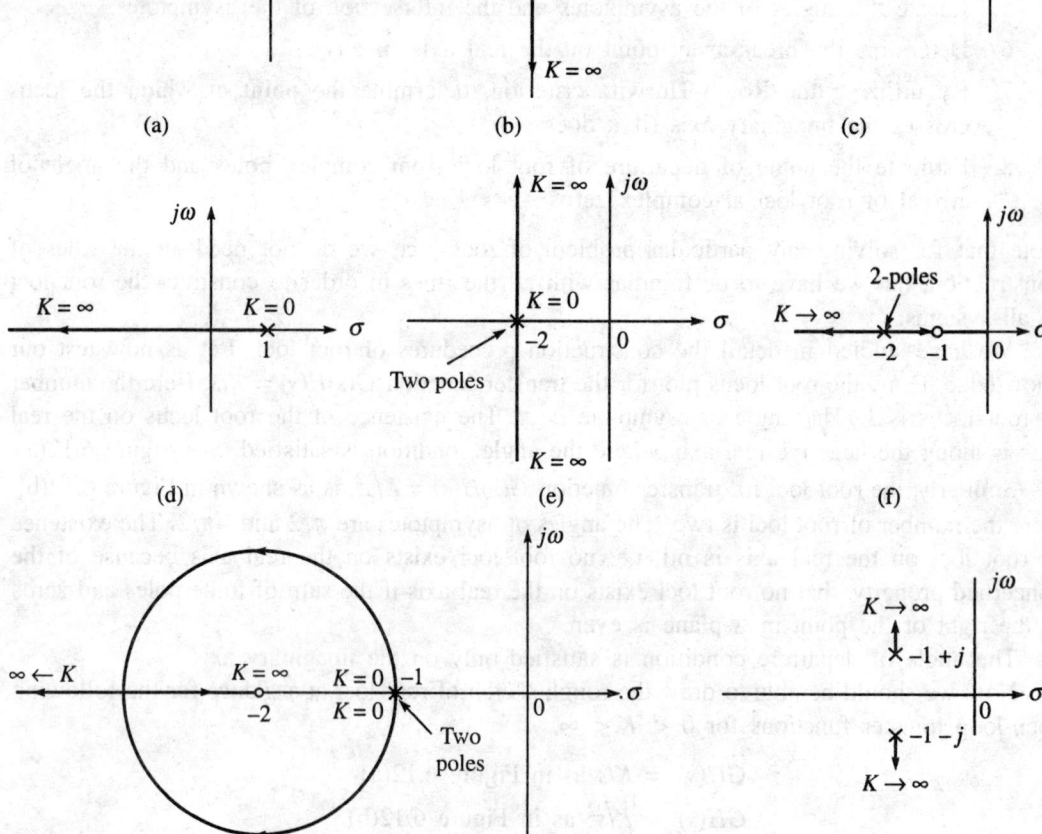

FIGURE 6.12 Root loci: (a) K/s, (b) K/s^2, (c) $K/(s+1)$, (d) $K/(s-1)$, (e) $K/(s+2)^2$, (f) $K(s+1)/(s+2)^2$, (g) $K(s+2)/(s+1)^2$, and (h) $K/(s^2+2s+2)$.

Root-Locus Technique

Drill Problem 6.6

Several pole-zero locations of open-loop transfer functions are shown in Figure 6.13. Draw the rough sketch of root loci plot.

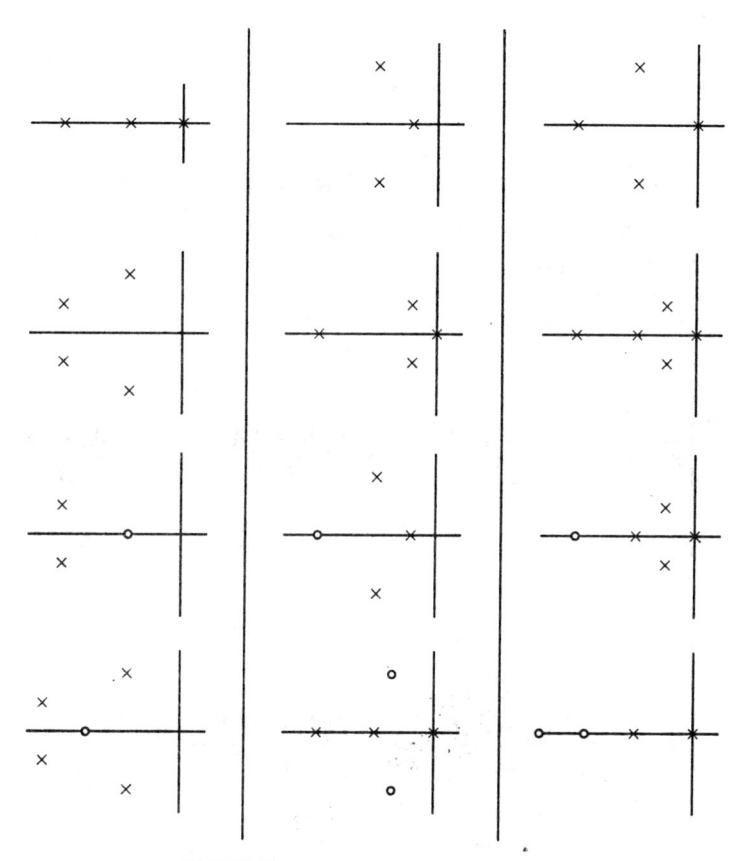

FIGURE 6.13 Drill Problem 6.6.

Rule 10. *Alternative method for obtaining break-away points:* For very high-order systems, solving $dK/ds = 0$ for obtaining the break-away points may be a tedious job, sometimes impossible. An algorithm for finding the break-away points on the root loci ($0 < K < \infty$) was introduced by Remec which is explained as follows:

Let the characteristic equation be

$$F(s) = A_n s^n + A_{n-1} s^{n-1} + \cdots + A_1 s + A_0 = 0 \tag{6.54}$$

Differentiating Eq. (6.54) and equating the result to zero,

$$F'(s) = B_n s^{n-1} + B_{n-1} s^{n-2} + \cdots + B_2 s + B_1 = 0 \tag{6.55}$$

Arranging the coefficients of $F(s)$ and $F'(s)$ in two rows,

$$\begin{matrix} A_n & A_{n-1} & A_{n-2} & \cdots & A_1 & A_0 \\ B_n & B_{n-1} & B_{n-2} & \cdots & B_1 & 0 \end{matrix}$$ (6.56)

As an example, consider a third-order characteristic equation,
$$F(s) = A_3 s^3 + A_2 s^2 + A_1 s + A_0$$
Then
$$F'(s) = B_3 s_2 + B_2 s + B_1$$

The tabulation of array is (note that it involves cross-multiplication as in Routh tabulation but not exactly as in Routh's array)

$$\begin{array}{c|cccc} s^3 & A_3 & A_2 & A_1 & A_0 \\ s^2 & B_3 & B_2 & B_1 & 0 \\ s^2 & C_3 = \dfrac{B_3 A_2 - B_2 A_3}{B_3} & C_2 = \dfrac{B_3 A_1 - B_1 A_3}{B_3} & C_1 = \dfrac{B_3 A_0 - 0 \times A_3}{B_3} & \\ s^2 & B_3 & B_2 & B_1 & \\ s^1 & D_3 = \dfrac{B_3 C_2 - B_2 C_3}{B_3} & D_2 = \dfrac{B_3 C_1 - B_1 C_3}{B_3} & & \\ s^1 & E_3 = \dfrac{D_3 B_2 - D_2 B_3}{D_3} & E_2 = \dfrac{D_3 B_1 - 0 \times B_3}{D_3} & & \\ s^1 & D_3 & D_2 & & \end{array}$$

$$s^0 \quad F_3 = \frac{D_3 E_2 - D_2 E_3}{D_3}$$

If $F(s)$ has multiple-order roots, which means that the root loci will have break-away points, a row of the tabulation shown above will contain all zero elements, that is, premature termination will occur. The multiple-order roots, which are the break-away points are obtained by solving the equation formed by using the row of coefficients just preceding the row of zeros.

The above alternative method for obtaining the break-away point is illustrated here by taking the open-loop transfer function of the feedback control system as

$$G(s)H(s) = \frac{K}{s(s+4)(s^2+4s+20)}$$

The characteristic equation is

$$F(s) = 1 + G(s)H(s) = 0$$

or

$$F(s) = s^4 + 8s^3 + 36s^2 + 80s + K = 0$$

Then

$$F'(s) = s^3 + 6s^2 + 18s + 20 = 0 \tag{6.57}$$

The following tabulation is made:

$$\begin{array}{c} s^4 \\ \begin{bmatrix} s^3 \\ s^3 \\ s^3 \end{bmatrix} \end{array} \quad \begin{array}{cccc} 1 & 8 & 36 & 80 & K \\ 1 & 6 & 18 & 20 \\ 2 & 18 & 60 & K \\ 1 & 6 & 18 & 20 \end{array}$$

$$\begin{bmatrix} s^2 & 6 & 24 & K-40 \\ s^2 & 2 & \left(18 - \dfrac{K-40}{6}\right) & 20 \\ s^2 & 6 & 24 & K-40 \end{bmatrix} \tag{6.58}$$

$$\begin{bmatrix} s^1 & \left(10 - \dfrac{K-40}{6}\right) & \left(20 - \dfrac{K-40}{3}\right) \\ s^1 & 12 & K-40 \\ s^1 & \left(10 - \dfrac{K-40}{6}\right) & \left(20 - \dfrac{K-40}{3}\right) \end{bmatrix}$$

$$s^0 \quad K - 64$$

All the elements of the first two of s^1-group would be zero for $K = 100$. Putting this value of $K = 100$ in the equation of the preceding row, i.e.

$$6s^2 + 24s + (K - 40) = 0 \tag{6.59}$$

or

$$s^2 + 4s + 10 = 0 \tag{6.60}$$

Therefore, the break-away points occur at

$$s = -2 \pm j2.45 \tag{6.61}$$

which are the solutions of Eq. (6.60).

For the other multiple roots (i.e. for the break-away points) any other value of $K \neq 100$, i.e. for $K = 64$ all the elements of the s^0-row will be zero. Similarly, putting the value $K = 64$ in the equation formed from the preceding row, we get

$$\left(10 - \frac{K-40}{6}\right)s + \left(20 - \frac{K-40}{3}\right) = 0 \quad \text{which gives} \quad s = -2$$

Hence $s = -2$ is the other break-away point which, in fact, lies on the admissible range of existence of root loci on the real axis.

EXAMPLE 6.1 Draw the root loci of the characteristic equation of the closed-loop system given as

$$s(s + 4)(s^2 + 4s + 20) + K = 0$$

Solution: Dividing both sides by the terms that do not contain K, we get

$$1 + \frac{K}{s(s+4)(s^2+4s+20)} = 1 + G(s)H(s) = 0$$

The poles of $G(s)H(s)$ are at $s = 0, -4$, and $-2 \pm j4$ which are the starting points, i.e. $K = 0$.

The number of finite poles, $P = 4$ and finite zeros $Z = 0$; so the number of root loci, $N = 4$. The root loci are symmetrical about the real axis.

The angles of asymptotes are $45°, 135°, 225°$, and $315°$. The point of intersection of the asymptotes on the real axis is at $s = -2$.

The existence of asymptotes is on the real axis between $s = 0$ to $s = -4$.

The limiting value for the gain parameter K for stability is obtained as $K = 260$ by Routh–Hurwitz criterion from the characteristic equation $s^4 + 8s^3 + 36s^2 + 80s + K = 0$.

The points of intersection of root loci on the imaginary axis are obtained as $\omega = \pm 3.16$ by formulating the auxiliary equation and putting the limiting value of K in it.

The angle of departure of the root locus from the complex pole $-2 + j4$ is $-90°$.

For the break-away point, with

$$K = -s(s + 4)(s^2 + 4s + 20)$$

$$\frac{dK}{ds} = \frac{d}{ds}(s^4 + 8s^3 + 36s^2 + 80s) = 0$$

or $\quad s^3 + 6s^2 + 18s + 20 = (s + 2)(s + 2 + j2.45)(s + 2 - j2.45) = 0$

The break-away points are at $s = -2$ and at $s = -2 \pm j2.45$. The root loci is shown in Figure 6.14.

The break-away points obtained here by solving $dK/ds = 0$, with $H(s) = 1$ and $K = -s(s + 4)(s^2 + 4s + 20)$ of the said open-loop transfer function are the same as those obtained by the tabular method.

Root-Locus Technique

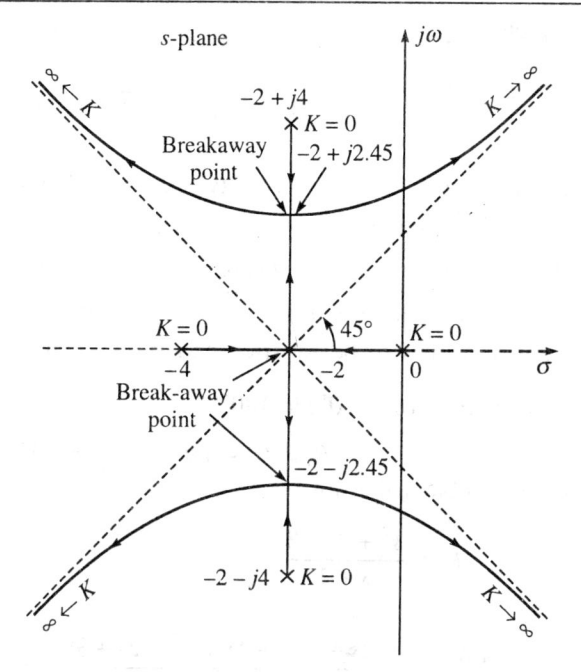

FIGURE 6.14 Example 6.1: root loci.

EXAMPLE 6.2 Draw the root loci of a unity-feedback system having open-loop transfer function $G(s)H(s) = \dfrac{K(s+2)}{s^2+2s+3}$. Determine:

(a) The value of K for which repetitive roots occur.
(b) The range of K for which the closed-loop system becomes underdamped.
(c) The value of K for which the system will have damping ratio of 0.7.

Solution: $G(s)H(s) = \dfrac{K(s+2)}{s^2+2s+3} = \dfrac{K(s+2)}{(s+1-j\sqrt{2})(s+1+j\sqrt{2})}$

In order to draw the root loci, let us follow the following rules:

1. $K = 0$ are the starting points, i.e. the open-loop poles are at, $s = -1 \pm j\sqrt{2}$
2. $K = \infty$, are the terminating points, i.e. one finite zero is at $s = -2$ and the other zero is at $s = \infty$.
3. The number of finite poles, $P = 2$ and the number of finite zeros, $Z = 1$. Hence the number of root loci, $N = 2$. One root locus will terminate at $s = \infty$. Hence only one angle of asymptote is required.
4. The root loci will be symmetric about the real axis.

5. Angles of asymptote are

$$\theta_k = \frac{(2k+1)\pi}{P-Z} \; ; \quad k = 0, 1, \cdots, \overline{P-Z-1} \Rightarrow \theta_0 = \pi$$

6. Existence of root loci on the real axis is between $s = -2$ and $s = -\infty$.
7. Intersection of root loci with the imaginary axis does not occur.
8. Angle of departure: See Figure 6.15(a) where we choose a test point s_1 in the vicinity of the complex open-loop pole at $(-1 + j\sqrt{2})$. Then if the test point lies on the root loci, the angle condition has to be satisfied and from Figure 6.15(a), we get

$$\theta_z - (\theta_p + \theta_p^*) = 180°$$

or
$$\theta_p = \theta_z - 180° - \theta_p^* = 55° - 180° - 90° = 145°$$

9. The break-in point is obtained from

$$K = \frac{s^2 + 2s + 3}{s + 2}$$

or
$$\frac{dK}{ds} = -\frac{(2s+2)(s+2) - (s^2 + 2s + 3)}{(s+2)^2} = 0$$

which gives $s^2 + 4s + 1 = 0$ or $s = -3.732$ and $s = -0.268$

The point $s = -3.732$ is the break-in point and the other point at $s = -0.268$ is not because it is not within the permissible range of root loci on the real axis. The root loci is drawn in Figure 6.15(b).

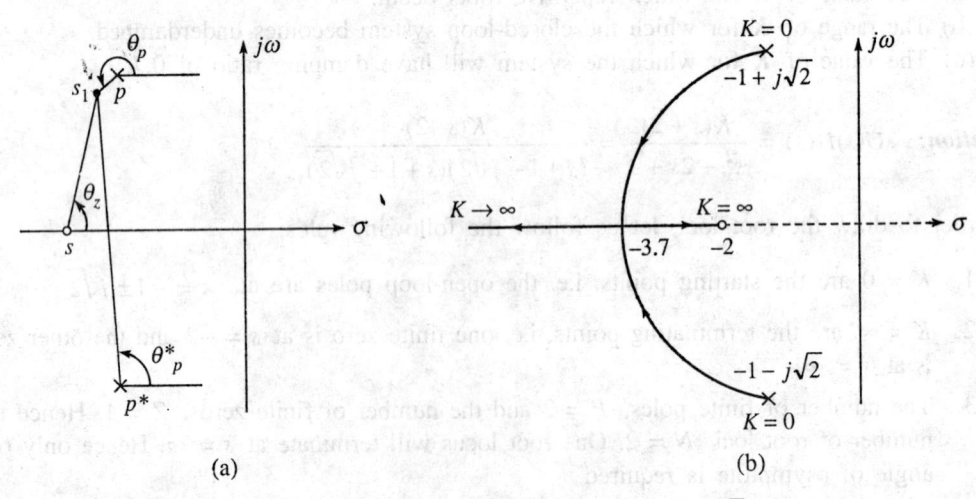

FIGURE 6.15 Example 6.2: (a) angle of departure from $p = -1 + j\sqrt{2}$ to s_1 on the root loci of $G(s)H(s) = K(s + 2)/(s^2 + 2s + 3)$ and (b) root loci of $G(s)H(s) = K(s + 2)/(s^2 + 2s + 3)$.

Root-Locus Technique

To show the occurrence of a circular root loci in the present system, we need to derive the equation for the root locus. From the angle condition,

$$(\angle s + 2) - (\angle s + 1 - j\sqrt{2}) - (\angle s + 1 + j\sqrt{2}) = \pm(2k+1)\pi$$

Putting $s = \sigma + j\omega$, we get

$$\tan^{-1}\left(\frac{\omega - \sqrt{2}}{\sigma + 1}\right) + \tan^{-1}\left(\frac{\omega + \sqrt{2}}{\sigma + 1}\right) = \tan^{-1}\left(\frac{\omega}{\sigma + 2}\right) \pm (2k+1)\pi$$

Using the relation, $\tan(x \pm y) = \dfrac{\tan x \pm \tan y}{1 \mp \tan x \tan y}$, we get

$$\tan\left[\tan^{-1}\left(\frac{\omega - \sqrt{2}}{\sigma + 1}\right) + \tan^{-1}\left(\frac{\omega + \sqrt{2}}{\sigma + 1}\right)\right] = \tan\left[\tan^{-1}\left(\frac{\omega}{\sigma + 2}\right) \pm (2k+1)\pi\right]$$

or

$$\frac{\dfrac{\omega - \sqrt{2}}{\sigma + 1} + \dfrac{\omega + \sqrt{2}}{\sigma + 1}}{1 - \dfrac{(\omega - \sqrt{2})(\omega + \sqrt{2})}{(\sigma + 1)^2}} = \frac{\dfrac{\omega}{\sigma + 2} \pm 0}{1 \mp \left(\dfrac{\omega}{\sigma + 2}\right) \times 0}$$

or $\quad \omega[(\sigma + 2)^2 + \omega^2 - 3] = 0$

or $\quad \omega = 0, \quad (\sigma + 2)^2 + \omega^2 = (\sqrt{3})^2$

The first equation $\omega = 0$ is the real axis equation from $s = -2$ to $s = -\infty$ and the second one is the equation of a circle with centre at $(\sigma = -2, \omega = 0)$ and radius equal to $\sqrt{3}$.

It is to be noted that for complicated systems having many poles and zeros, any attempt to derive the equation for root loci is discouraged. Let us now find the answers to the given problem.

(a) The value of gain K at the break-in point (where the repetitive root occurs, that is, the damping factor is just unity) is obtained as

$$\left|\frac{K(s+2)}{s^2 + 2s + 3}\right|_{s = -3.732} = 1 \quad \text{or} \quad K = 5.4641$$

(b) The range of K for which the system becomes underdamped (i.e. complex conjugate poles) is

$$0 < K < 5.4641$$

(c) Here the damping ratio $= 0.7$, i.e. $\cos\theta = 0.7$, $\theta = 45.57°$

$\therefore \qquad \tan\theta = 1.0202 \quad \text{or} \quad \omega = 1.0202\sigma$

Putting this value of ω in the equation of the circle,

$$(\sigma + 2)^2 + \omega^2 = (\sqrt{3})^2, \text{ we get } \sigma = -1.6659$$

Again, $\sigma = \zeta\omega_n = 0.7\omega_n$ and $\omega = \omega_n\sqrt{1-\zeta^2} = 1.6995$

Hence the value of K at $s = -1.6659 + j1.6995$ can be obtained as

$$K = \left|\frac{(s+1-j\sqrt{2})(s+1+j\sqrt{2})}{s+2}\right|_{s=-1.6659+j1.6995} = 1.3318$$

Therefore, $K = 1.3318$ is the value of gain where the complex conjugate closed-loop poles will have the damping ratio $\zeta = 0.7$.

EXAMPLE 6.3 Draw the root loci of the closed-loop system for the given open-loop transfer function

$$G(s)H(s) = \frac{K}{s(s+4)(s+4+j4)(s+4-j4)}$$

as K varies from zero to infinity.

Solution:

1. The poles (i.e. $K = 0$ are the starting points of root loci) are located at $s = 0, -4, -4 + j4, -4 - j4$.
2. There is no finite zero.
3. The number of finite poles, $P = 4$ and the number of finite zeros, $Z = 0$, so the number of separate root loci, $N = 4$.
4. The root loci are symmetric about the real axis.
5. Angles of asymptotes are:

$$\theta_k = \frac{(2k+1)\pi}{P-Z}; \quad k = 0, 1, \ldots, P-Z-1$$

$$= \frac{(2k+1)\pi}{4}, \quad k = 0, 1, \ldots, 3$$

$$= 45°, 135°, 225°, 315°$$

The intersection of asymptotes is

$$\sigma_A = \frac{-4-4-4}{4} = -3$$

The asymptotes are drawn as shown in Figure 6.16(a).

6. Existence of root loci on the real axis is between $s = 0$ and $s = -4$.

Root-Locus Technique

7. The characteristic equation is rewritten as

$$s(s+4)(s^2+8s+32) + K = s^4 + 12s^3 + 64s^2 + 128s + K = 0$$

Therefore, the Routh array is:

s^4	1	64	K
s^3	12	128	
s^2	b_1	K	
s^1	c_1		
s^0	K		

where $b_1 = \dfrac{12 \times 64 - 128}{12} = 53.33$ and $c_1 = \dfrac{53.33 \times 128 - 12K}{53.33}$

Hence the limiting value of gain for stability is $K = 568.85$ and the roots of the auxiliary equation are

$$53.33s^2 + 570 = 53.33(s + j3.25)(s - j3.25)$$

The points where the root loci cross the imaginary axis are obtained from

$$53.33s^2 + K = 0 \quad \text{which leads to} \quad \omega = \pm 3.266$$

The limiting value $K = 568.85$ signifies that the closed-loop system is stable for $0 < K < 568.85$. If the gain K is increased beyond $K = 568.85$, the closed-loop system will become unstable. Now if you are asked, what would be the gain margin for the value of gain $K_1 = 100$. The question means that the margin of gain at $K_1 = 100$ would be $20\log_{10}(K/K_1)$ in dB before the system goes to instability.

An alternative approach is to let $s = j\omega$ in the characteristic equation, equate both the real part and the imaginary part to zero, and then solve for ω and K. For the present system, the characteristic equation with $s = j\omega$, is

$$(j\omega)^4 + 12(j\omega)^3 + 64(j\omega)^2 + 128(j\omega) + K = 0$$

or

$$(\omega^4 - 64\omega^2 + K) + j4\omega(32 - 3\omega^2) = 0$$

Equating both the real and imaginary parts of this last equation to zero, we obtain

$$32 - 3\omega^2 = 0 \quad \text{and} \quad \omega^4 - 64\omega^2 + K = 0$$

from which, we obtain $\omega = 3.2666$ and $K = 568.85$.

8. The angle of departure at the complex pole $p_1 = -4 + j4$ can be estimated by utilizing the angle criterion as

$$\theta_{p_1} = -135° = +225°$$

9. The break-away point is obtained from $dK/ds = 0$ which gives

$$s^3 + 9s^2 + 32s + 32 = 0$$

To solve this equation, i.e. to get the roots is a tedious job. The break-away point is therefore estimated by evaluating

$$K = P(s) = -s(s + 4)(s + 4 + j4)(s + 4 - j4)$$

between $s = -4$ and $s = 0$. We expect the break-away point to lie between $s = -3$ and $s = -1$ and therefore we search for a maximum value of $P(s)$ in that region. The maximum of $P(s)$ is found to lie at approximately $s = -1.5$. A more accurate estimate of the break-away point is normally not necessary or worthwhile. A closer estimate is in fact -1.57 as indicated in Figure 6.16(a). The root loci are shown in Figure 6.16(b).

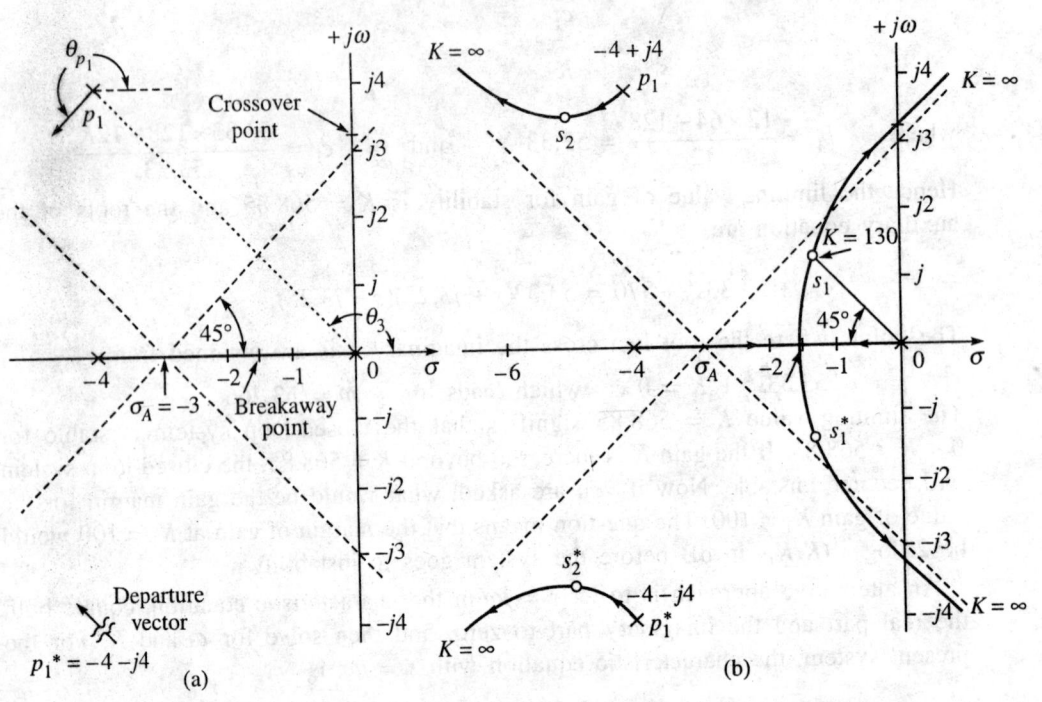

FIGURE 6.16 Example 6.3: (a) asymptotes and (b) root loci.

Suppose you are now asked to find the value of K at $\zeta = \sqrt{2}$. The gain K can be determined graphically as shown in Figure 6.16(b). Draw a line $\theta = 45°$ (as $\zeta = \sqrt{2} = \cos \theta$ gives $\theta = 45°$). The intersection of root loci and $\theta = 45°$ line gives s_1 and s_1^* points graphically. The vector lengths to the root location s_1 from the open-loop poles are evaluated and result in a gain at s_1 of

$$K = |s_1||s_1 + 4||s_1 - p_1||s_1 - p_1^*| = (1.9)(3)(3.8)(6) \approx 130$$

The remaining pair of complex conjugate roots of closed-loop system occurs at s_2 and s_2^* when $K = 130$. It is obvious that at $K = 130$, s_1, s_1^* are the dominant pole pair compared to s_2 and s_2^* pole pair.

EXAMPLE 6.4 Draw the root loci of the open-loop transfer function of the feedback control system

$$G(s)H(s) = \frac{K}{s(s+4)(s+5)}$$

Solution: We have finite open-loop poles at $s = 0, -4, -5$ and no finite zero. The number of root loci is $N = 3$, as $P = 3$, $Z = 0$. The root loci on the real axis will exist between 0 and -4, and between -5 and ∞. All the three root loci will terminate at zeros at ∞. The angles of asymptotes are: $\theta_0 = \pi/3$, $\theta_1 = \pi$ and $\theta_2 = 5\pi/4 = -\pi/3$.

The intersection of the asymptotes on the real axis is: $\sigma_A = -3$

Now, $\qquad K = -s(s+4)(s+5) = -s^3 - 9s^2 - 20s$

The break-away point is obtained from $dK/ds = 0$ which gives

$$-3s^2 - 18s - 20 = 0$$

that is, $s = -1.4725$ and -4.5275 of which the only one at $s = -1.4725$ is the admissible break-away point as it lies on the segment between $s = 0$ and -4. The point -4.5275 does not exist on the admissible segment of the root loci on the real axis and hence need not be considered.

The value of K at break-away point $s = -1.4725$ is obtained as

$$K = -s(s+4)(s+5)\big|_{s=-1.4725} = 13.128$$

The intersection of root loci with the imaginary axis is determined from Routh's table, with the characteristic equation $s^3 + 9s^2 + 20s + K = 0$. The Routh's table is

s^3	1	20
s^2	9	K
s^1	$\frac{180-K}{9}$	0
s^0	K	

Hence the critical gain before the closed-loop system goes to instability is $K_c = 180$, and the auxiliary equation is $9s^2 + K = 0$. Putting $K = 180$, we get

$$s = \pm j2\sqrt{5} \quad \text{i.e.} \quad \omega = \pm 2\sqrt{5}$$

A sketch of the complete root loci is shown in Figure 6.17.

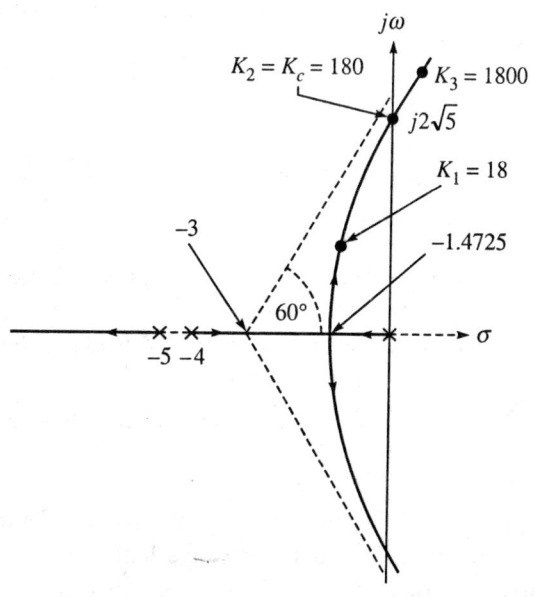

FIGURE 6.17 Example 6.4: root loci.

Now suppose you are asked to find the gain margin for the following gains:

(i) For $K_1 = 18$, the margin of gain is $20 \log(K_c/K_1) = 20$ dB, the closed-loop system is stable and margin of gain is 20 dB before the system goes to instability.

(ii) For $K_2 = 180$, the margin of gain is $20 \log(K_c/K_2) = 0$ dB; the closed-loop system is criticality (just) stable, no margin of gain is left to play with before the system goes to instability.

(iii) For $K_3 = 1800$, the gain margin is -20 dB, the closed-loop is unstable.

These observations are evident from Figure 6.17. This concept is useful for understanding the design parameters.

EXAMPLE 6.5 Sketch the root loci of the open-loop transfer function of the feedback control system

$$G(s)H(s) = \frac{K}{s(s^2 + 6s + 25)}; \quad 0 < K < \infty$$

Determine the gain margin at $K = K_1 = 15$ and at $K = K_2 = 1500$.

Solution: The poles are at $s = 0$ and $s = -3 \pm j4$. Here $P = 3$, $Z = 0$. The number of root loci, $N = 3$, each terminating on a zero at infinity. The root loci on the real axis exist on the entire negative real axis. The angles of asymptotes are

$$\theta_0 = \pi/3, \quad \theta_1 = \pi \quad \text{and} \quad \theta_2 = 5\pi/3 = -\pi/3$$

The intersection of asymptotes on the real axis is

$$\sigma_A = \frac{0 - 3 + j4 - 3 - j4}{3} = -2$$

The characteristic polynomial is

$$s^3 + 6s^2 + 25s + K$$

Routh's array becomes:

s^3	1	25
s^2	6	K
s^1	$\frac{150-K}{6}$	0
s^0	K	

Hence $K = 150$ when root loci just cross the imaginary axis before going to the positive-half s-plane. The auxiliary equation is $6s^2 + K = 0$.

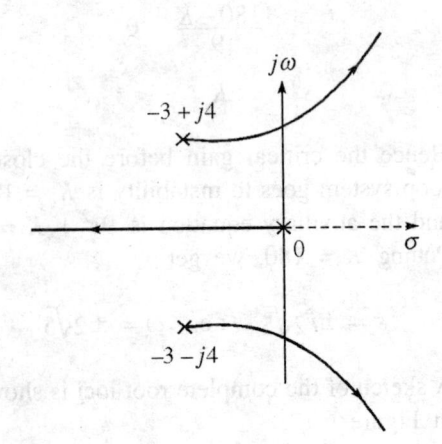

FIGURE 6.18 Example 6.5: root loci.

Putting $K = 150$ in the auxiliary equation, we get $s = \pm j5$, i.e. $\omega = \pm 5$ as the cross-over point on the imaginary axis.

The angle of departure from the upper complex pole $(-3 + j4)$ is calculated as

$$\theta = 180° - (90° + 126.87°) = -36.87°$$

The sketch of the root loci is shown in Figure 6.18. Note that the determination of break-away point is not necessary here, though it may give the value of $s = 2 \pm j2.0817$ from $dK/ds = 0$, which does not satisfy the characteristic equation. We do not have any break-away point on real axis.

Now the gain margin at $K_1 = 15$, is +20 dB, the closed-loop system is therefore stable. For $K_2 = 1500$ the gain margin is -20 dB; the closed-loop system is therefore unstable.

It may be noted that all the rules of construction may not be necessary for drawing the root-locus plot, as can be in this case where the break-away point is not necessary.

EXAMPLE 6.6 Draw the root loci of the open-loop transfer function of the feedback control system

$$G(s)H(s) = \frac{K}{s(s+2)(s^2+6s+25)}$$

Solution: Here $P = 4$, $Z = 0$, $N = 4$. Poles are at $s = 0, -2$ and $-3 \pm j4$. The root loci on the real axis exist only between $s = 0$ and -2. All the four root loci will terminate on zeros at infinity. The intersection of asymptotes on the real axis is $\sigma_A = -2$.

The angles of asymptotes are

$$\theta_0 = \pi/4, \ \theta_1 = 3\pi/4, \ \theta_2 = 5\pi/4 = -3\pi/4, \text{ and } \theta_3 = 7\pi/4 = -\pi/4$$

Here,
$$K = -s(s+2)(s^2+6s+25)$$

Therefore,
$$\frac{dK}{ds} = -(4s^3 + 24s^2 + 74s + 50) = 0$$

The break-away point is searched between $s = 0$ and $s = -2$ and is found to be at $s = -0.8981$ and the other two are in the non-admissible range on the real axis of root loci and hence ignored. The value of K at the break-away point $s = -0.8981$ is obtained as

$$K = -s(s+2)(s^2+6s+25)\big|_{s=-0.8981} = 20.206$$

The characteristic polynomial is

$$s^4 + 8s^3 + 37s^2 + 50s + K$$

Routh's array is:

s^4	1	37	K
s^3	8	50	
s^2	30.75	K	
s^1	$\dfrac{1537.5 - 8K}{30.75}$		
s^0	K		

Hence $K = 1537.5/8 = 192.1875$.

The auxiliary equation is

$$30.75 s^2 + K = 0$$

Putting $K = 192.1875$ in the auxiliary equation, we get $s = \pm j2.5$, i.e. $\omega = \pm 2.5$. The angle of departure from the upper complex pole $(-3 + j4)$ is given by

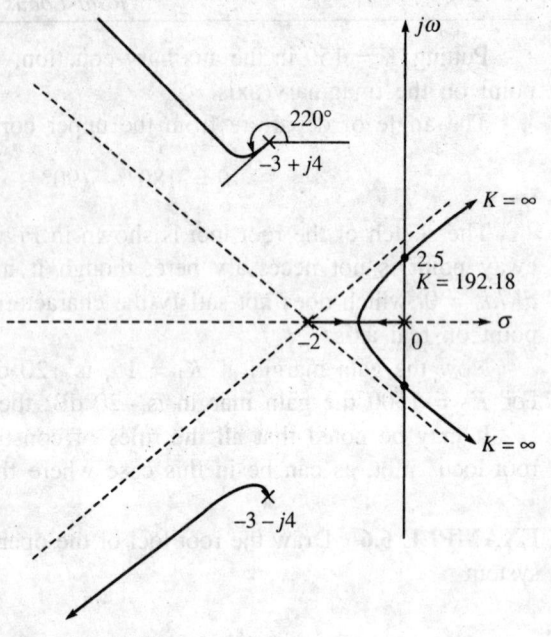

FIGURE 6.19 Example 6.6: root loci.

$$\theta = 180° - (90° + 126.87° + 104.04°) = -140.91°$$

The root loci are shown in Figure 6.19.

EXAMPLE 6.7 Draw the root loci of open-loop transfer function of the feedback control system given as

$$G(s)H(s) = \frac{K(s+3)}{s(s+2)}; \quad 0 < K < \infty$$

Solution: The poles are at $s = 0$ and -2. The zero is at $s = -3$. The number of poles $P = 2$, and the number of finite zeros $Z = 1$. The number of root loci, $N = 2$. The root loci on the real axis lie between $s = 0$ and -2, and between $s = -3$ to infinity. The angles of asymptotes are: $\theta_0 = \pi$, $\theta_1 = 2\pi = 0°$.

From the characteristic equation the magnitude condition $|GH| = 1/K$ leads to

$$K = -\frac{s(s+2)}{s+3}$$

Then,

$$\frac{dK}{ds} = -(s^2 + 6s + 6) = 0$$

gives $s_1 = -4.1732$ and $s_2 = -1.268$ as the break-away/break-in points. In fact, $s_1 = 1.268$ is the break-away point as $\left.\dfrac{d^2K}{ds^2}\right|_{s=s_1} > 0$, and $s_2 = -4.1732$ is the break-in point as $\left.\dfrac{d^2K}{ds^2}\right|_{s=s_2} < 0$.

Clearly, for finding the break-away/break-in point, dK/ds is the preferred option than dG/ds because for all physically realizable systems the degree of denominator polynomial of open-loop transfer function is higher or equal to that of the numerator polynomial.

The root loci will be circular in the complex s-plane. This may be proved as follows: On the root loci, the angle condition becomes

$$\angle G(s)H(s) = \angle s+3 - \angle s - \angle s+2 = 180° \qquad \text{(i)}$$

Putting $s = \sigma + j\omega$ in Eq. (i), we have

$$\angle \sigma + j\omega + 3 - \angle \sigma + j\omega - \angle \sigma + j\omega + 2 = 180° \text{ which can be rewritten as}$$

$$\tan^{-1}\left(\frac{\omega}{\sigma+3}\right) - \tan^{-1}\left(\frac{\omega}{\sigma}\right) = 180° + \tan^{-1}\left(\frac{\omega}{\sigma+2}\right)$$

or

$$\tan\left[\tan^{-1}\frac{\omega}{\sigma+3} - \tan^{-1}\frac{\omega}{\sigma}\right] = \tan\left[180° + \tan^{-1}\frac{\omega}{\sigma+2}\right]$$

or

$$\frac{-3\omega}{\sigma(\sigma+3)+\omega^2} = \frac{\omega}{\sigma+2} \quad \left(\text{Using } \tan(x \pm y) = \frac{\tan x \pm \tan y}{1 \mp \tan x \tan y} \text{ and simplifying}\right)$$

or

$$(\sigma+3)^2 + \omega^2 = (\sqrt{3})^2$$

The above equation represents a circle with centre at ($\sigma = -3$, $\omega = 0$) and radius $\sqrt{3}$. These can be rechecked from the break-away and break-in points. The root loci are shown in Figure 6.20. The gain K at the break-away point -1.268 is obtained as $K_1 = 0.5359$ and at the break-in point -4.1732 the gain is obtained as $K_2 = 7.73$.

Hence the closed-loop system is overdamped for gain K in the ranges, $0 < K < 0.5359$ and $7.73 < K < \infty$ and the closed-loop system is underdamped in the range $0.05359 < K < 7.73$.

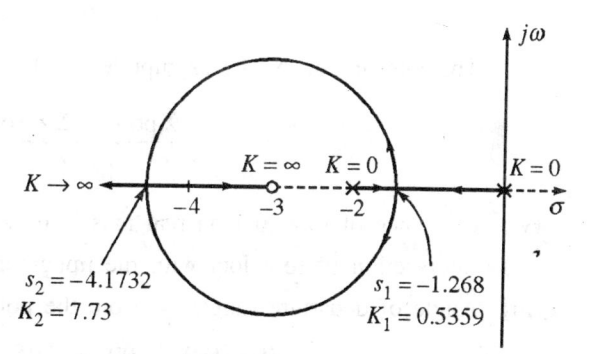

$s_2 = -4.1732$
$K_2 = 7.73$

$s_1 = -1.268$
$K_1 = 0.5359$

FIGURE 6.20 Example 6.7: root loci.

EXAMPLE 6.8 For the open-loop transfer function $20(1 + ks)/s(s + 1)(s + 4)$, (a) draw the root loci and (b) determine the value of k such that the damping ratio of one of the closed-loop poles is 0.4.

Solution: (a) $G(s)H(s) = \dfrac{20(1+ks)}{s(s+1)(s+4)}$

The characteristic equation is

$$1 + G(s)H(s) = 0$$

or

$$s^3 + 5s^2 + 4s + 20 + 20ks = 0$$

Assuming $20k = K$, we can rewrite

$$1 + \dfrac{Ks}{s^3 + 5s^2 + 4s + 20} = 0$$

or

$$1 + \dfrac{Ks}{(s+j2)(s-j2)(s+5)} = 0$$

where $G(s)H(s) = \dfrac{Ks}{(s+5)(s+j2)(s-j2)}$; for $0 < (K = 20k) < \infty$

For the construction of the root loci, we follow the following procedure:

 (i) Starting points ($K = 0$) are at finite poles, $s = -5$, $s = -j2$, $s = +j2$.
 (ii) Terminating points ($K = \infty$) are at $s = 0$, $s = \infty$, $s = \infty$.
 (iii) Number of root loci, $N = 3$, the number of finite poles, $P = 3$ and that of finite zeros, $Z = 1$.
 (iv) Symmetry of root loci about the real axis holds.
 (v) Angles of asymptotes are

$$\theta_0 = \pi/2, \ \theta_1 = -\pi/2$$

The intersection of the asymptotes on the real axis is

$$\dfrac{\Sigma \text{poles} - \Sigma \text{zeros}}{P - Z} = -\dfrac{5}{2}$$

 (vi) Existence of root loci on real axis is between $s = 0$ to $s = -5$.
 (vii) Intersection of root loci with the imaginary axis does not occur in this example.
 (viii) Angle of departure (angle θ) from the pole at $s = j2$ is obtained as

$$\theta = 180° - 90° - 21.8° + 90° = 158.2°$$

 (ix) Break-away/break-in point does not occur in this example.

The root loci diagram for the system is shown in Figure 6.21.

 (b) The damping ratio $\zeta = 0.4$ means that $\zeta = \cos\theta = 0.4$, i.e. $\theta = \pm 66.42°$ with the negative-real axis. There are two intersections of the root loci branch in the upper-half of

s-plane with the straight line of angle 66.42°. Thus, these two values of K will give the damping ratio of the closed-loop poles equal to 0.4. At point P the value of K is

$$K = \frac{(s+j2)(s-j2)(s+5)}{s}\bigg|_{s=-1.05+j2.4} = 8.98$$

or
$$k = K/20 = 0.449 \text{ at point P.}$$

At point Q, the value of K is

$$K = \frac{(s+j2)(s-j2)(s+5)}{s}\bigg|_{s=-2.15+j4.95} = 28.26$$

or
$$k = K/20 = 1.413 \text{ at point Q.}$$

Hence for $k = 0.449$, the three closed-loop poles are

$$s = -1.05 \pm j2.4, \quad s = -2.902$$

The closed-loop transfer function

$$\frac{C(s)}{R(s)} = \frac{20}{s^3 + 5s^2 + 12.98s + 20}$$

The unit-step response becomes

$$c(t) = 1 - 0.747e^{-2.902t} - 0.253e^{-1.05t}\cos(2.4t) - 1.0113e^{-1.05t}\sin(2.4t)$$

For $k = 1.413$, the three closed-loop poles are at $s = -2.15 \pm j4.95$ and $s = -0.6823$.
The closed-loop transfer function

$$\frac{C(s)}{R(s)} = \frac{20}{s^3 + 5s^2 + 32.26s + 20}$$

The unit-step response is

$$c(t) = 1 - 1.0924e^{-0.6823t} + 0.0924e^{-2.15t}\cos(4.95t) - 0.1102e^{-2.15t}\sin(4.95t)$$

The response is shown in Figure 6.22. Clearly, the oscillatory terms damp out much faster than the purely exponential term. The system with $k = 0.449$ which exhibits faster response with small overshoot has much better characteristic than the system with $k = 1.413$ which exhibits a slow overdamped response. Therefore, we should choose $k = 0.4490$ for the present system.

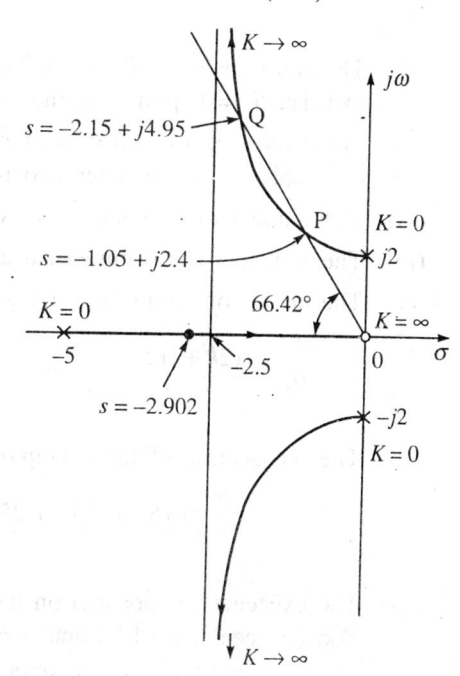

FIGURE 6.21 Example 6.8: root loci.

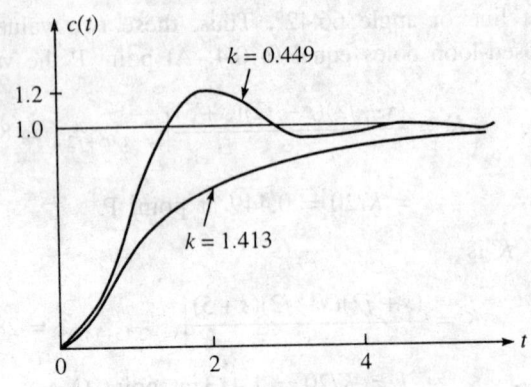

FIGURE 6.22 Example 6.8: system response.

EXAMPLE 6.9 Consider the open-loop system transfer function

$$G(s)H(s) = \frac{K(s^2 + 1.5s + 1.5625)}{(s - 0.75)(s + 0.25)(s + 1.25)(s + 2.25)}; \quad K > 0$$

Draw the root loci diagram. Determine the values of gain for a stable closed-loop system.

Solution: By inspection, the denominator and the numerator polynomials are of proper form. The coefficient of s^2 in the numerator polynomial and that of s^4 in the denominator polynomial is unity.

(i) The poles are at $s = 0.75, -0.25, -1.25$ and -2.25 which are the starting points of root loci, i.e. $K = 0$ points of the root loci.

(ii) Finite zeros at $s = -0.75 \pm j1$ are the terminating points of the two root loci, i.e. $K = \infty$ points and the other two root loci terminate at infinity.

(iii) The number of root loci is 4 as $P = 4$ and $Z = 2$.

(iv) The root loci are symmetrical about the real axis.

(v) The angles of asymptotes are given by

$$\theta_k = \frac{(2k+1)\pi}{P - Z}; \quad k = 0, 1 \quad \text{i.e.} \quad \theta_0 = \pi/2; \quad \theta_1 = 3\pi/2$$

The intersection of the asymptotes with the real axis is

$$= \frac{(0.75 - 0.25 - 1.25 - 2.25) - (-0.75 - 0.75)}{4 - 2} = -0.75$$

(vi) The existence of root loci on the real axis is in between $s = 0.75$ and $s = -0.25$ and also between $s = -1.25$ and $s = -2.25$.

(vii) The point of intersection of the root loci with the imaginary axis is obtained from Routh–Hurwitz criterion. The characteristic equation of the closed-loop system is

Root-Locus Technique

obtained from

$$1 + G(s)H(s) = 0$$

i.e. $s^4 + 3s^3 + (0.875 + K)s^2 + (1.5K - 2.063)s + (1.563K - 0.527) = 0$

The Routh array for this equation is formulated as

s^4	1	$0.875 + K$	$1.563K - 0.527$
s^3	3	$1.5K - 2.063$	
s^2	$\dfrac{3(0.875 + K) - (1.5K - 2.063)}{3}$	$1.563K - 0.527$	
s^1	A		
s^0	$1.563K - 0.527$		

To be on the imaginary axis, the s^1 row must be an all-zero row, i.e.

$$A = \frac{3(0.875 + K) - (1.5K - 2.063)}{3} \times (1.5K - 2.063) - 3(1.563K - 0.527) = 0$$

which may be rearranged to give

$$0.75K^2 - 3.377K - 1.643 = 0 \quad \text{or} \quad K = 2.251 \pm 2.694$$

Since only positive values of K are required, then $K = 4.945$.
Note that the stable range of the value of K is $\infty > K > 4.945$ for the closed-loop system to be stable.

Putting this value of $K = 4.945$ in the equation obtained from the second row of Routh array, i.e. $3s^2 + (1.5K - 2.063) = 0$, we get

$$\omega = \sqrt{\left(\frac{1.5 \times 4.945 - 2.063}{3}\right)} = \pm 1.34$$

(viii) The angle of arrival at the complex zero $s = -0.75 \pm j1$ is obtained as in Figure 6.23. From the angle criterion, we get

$$(90° + \phi) - [(180° - 34°) + 34° + (180° - 63°) + 69°] = (2k + 1)\pi; \text{ where } k \text{ is an integer}$$

i.e. $\phi = 90°$

(ix) The break-away/break-in points are obtained from $dK/ds = 0$, as

$$s = 0.26, -1.76 \quad \text{and} \quad -0.75 \pm j1.74$$

From the above rules, the root loci may now be drawn as shown in Figure 6.24. It may be noted that it would be a good exercise for the reader to find the value of gain K at the break-away and break-in points. The root loci obtained by MATLAB is also shown (see MATLAB Program 6.1).

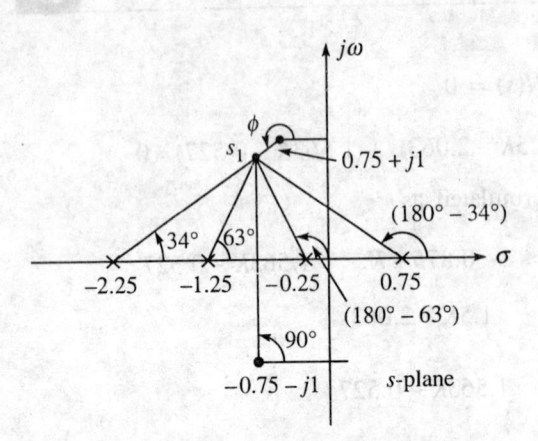

FIGURE 6.23 Example 6.9: application of angle criterion.

FIGURE 6.24 Example 6.9: root loci.

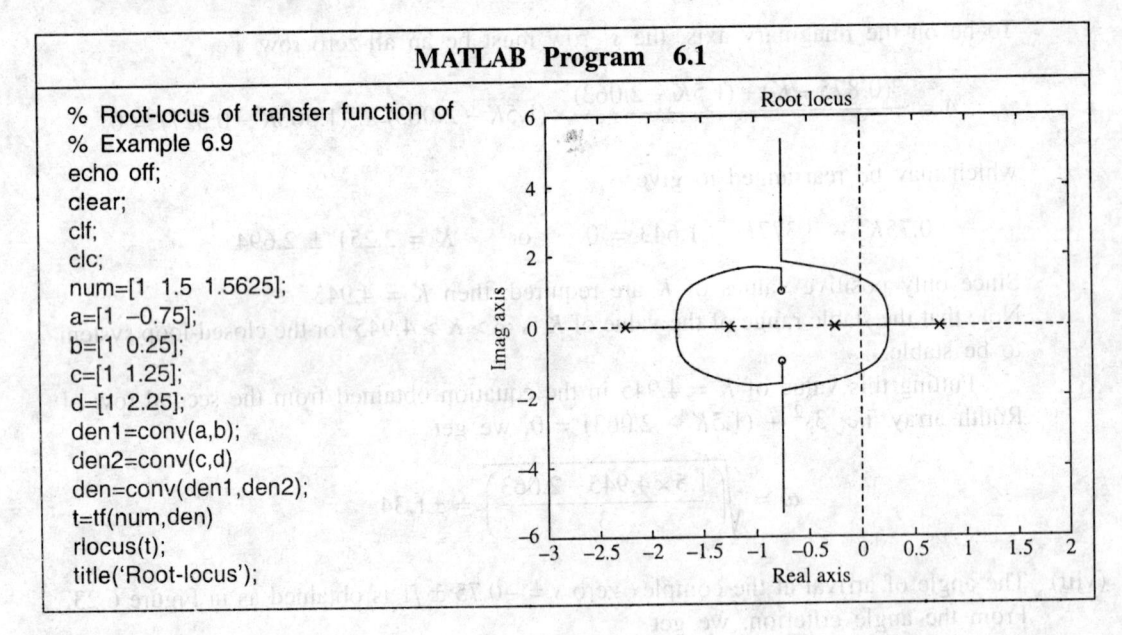

```
                    MATLAB Program   6.1
% Root-locus of transfer function of
% Example 6.9
echo off;
clear;
clf;
clc;
num=[1 1.5 1.5625];
a=[1 −0.75];
b=[1 0.25];
c=[1 1.25];
d=[1 2.25];
den1=conv(a,b);
den2=conv(c,d);
den=conv(den1,den2);
t=tf(num,den)
rlocus(t);
title('Root-locus');
```

EXAMPLE 6.10 Draw the root loci of the closed-loop system having open-loop transfer function $G(s)H(s)$ as

$$G(s)H(s) = \frac{Ks}{(s^2+4)(s^2+16)}; \quad 0 < K < \infty$$

Solution: The gain $K = 0$ points are at $s = j2$, $s = -j2$, $s = j4$, and $s = -j4$. The gain $K = \infty$ points are at $s = 0$ and three other at $s = \infty$. The number of root loci are 4, as $P = 4$ and $Z = 1$. Symmetry of root loci exists about the real axis. Root loci exist between $s = 0$ to $-\infty$ on the negative real axis.

The angles of asymptotes are

$$\theta_k = \frac{(2k+1)\pi}{P-Z}; \quad k = 0, 1, 2, \text{ i.e. } 60°, 180° \text{ and } -60°$$

The intersection of asymptotes is at $s = 0$. The break-in point occurs at $s = -\infty$ on the negative real axis.

The angle of departure from pole at $s = j4$ is calculated as $0°$ and that from pole at $s = j2$ is calculated as $180°$.

The root loci is drawn in Figure 6.25 for $0 < K < \infty$.

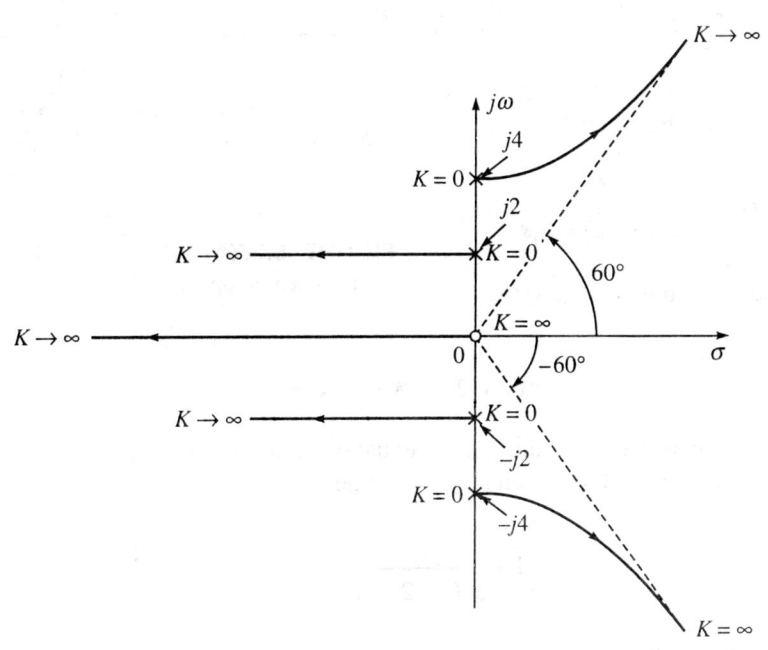

FIGURE 6.25 Example 6.10: root loci.

EXAMPLE 6.11 The block diagram of a feedback control system is shown in Figure 6.26(a). Sketch the root loci for $K \geq 0$ when the switch S is open. Determine the stability of the system as a function of K. Close the switch S so that the minor feedback loop is in effect. Set $K = 1$ and show by the root-locus plot how the system is stabilized when K_t varies. What is

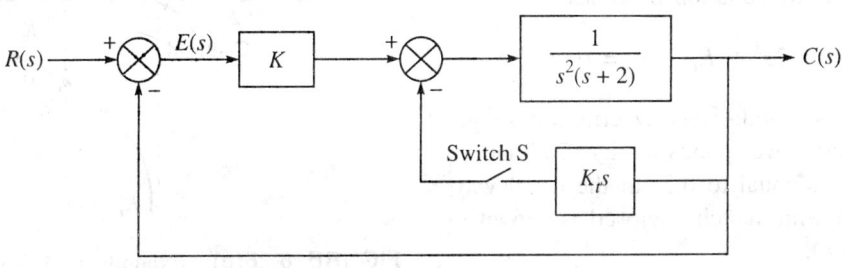

FIGURE 6.26(a) Example 6.11: control system.

the minimum value of K_t which stabilizes the system? What is the system's damped natural frequency corresponding to this value of K_t?

Solution: When switch S is open, the open-loop transfer function becomes

$$GH(s) = \frac{K}{s^2(s+2)}$$

The root-loci plot for $0 < K < \infty$ is shown in Figure 6.26(b).

When the switch S is closed, the open-loop transfer function becomes:

$$G(s)H(s) = \frac{K}{s^2(s+2) + K_t s}$$

The characteristic equation $1 + G(s)H(s) = 0$ becomes

$$s^2(s+2) + K_t s + K = 0$$

Now for $K = 1$, we can rearrange the system equation in terms of K_t as variable in the range $0 < K_t < \infty$, then the characteristic equation becomes

$$1 + \frac{K_t s}{s^2(s+2) + 1}$$

Then the pole-zero of the open-loop transfer function $\frac{K_t s}{s^2(s+2)+1}$ becomes $s_1 = -2.2$ and $s_{2,3} = 0.1 \pm j0.656$. Now we proceed to draw the root-loci plot for $0 < K_t < \infty$ while $K = 1$. The characteristic equation becomes

$$s^3 + 2s^2 + K_t s + 1 = 0$$

from which by Routh–Hurwitz criterion we get the $j\omega$-axis crossover points as $\pm j0.707$ and K_t at that point is equal to 0.5. In the usual way, the root loci with switch S closed is shown in Figure 6.26(c).

FIGURE 6.26(b) Example 6.11: root-loci plot with switch S open.

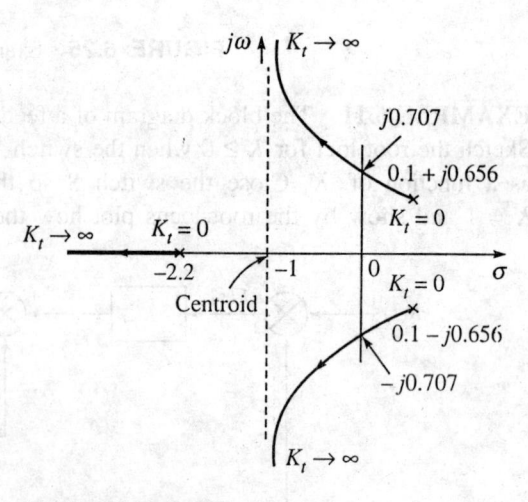

FIGURE 6.26(c) Example 6.11: root-loci plot with switch S closed.

EXAMPLE 6.12 The open-loop transfer function of the given system is

$$G(s)H(s) = \frac{100K(s+5)(s+40)}{s^3(s+100)(s+200)}$$

Draw the root loci for K in the range $0 < K < \infty$.

The same question can be written in a different form as: for the given characteristic equation

$$s^3(s+100)(s+200) + 100K(s+5)(s+40)$$

draw the root loci for $0 < K < \infty$.

Solution: The starting points of root loci are at 0, 0, 0, –100, and –200. The terminating point of root loci are at –5, –40, ∞, ∞, and ∞. The number of root loci is 5. The root loci on real axis exist from origin to –5, and from –100 to –40, and from –200 to infinity. The root loci are symmetrical about the real axis. The angles of asymptotes are $\pi/3$, π, and $-\pi/3$. The intersection of asymptotes on real axis is at –85. The intersection of root loci on $j\omega$-axis is obtained from Routh-tabulation of the characteristic equation as

$$\omega = \pm 25.8 \text{ for } K = 2818 \text{ and } \omega = \pm 77.7 \text{ for } K = 18837$$

The root loci is drawn in Figure 6.27.

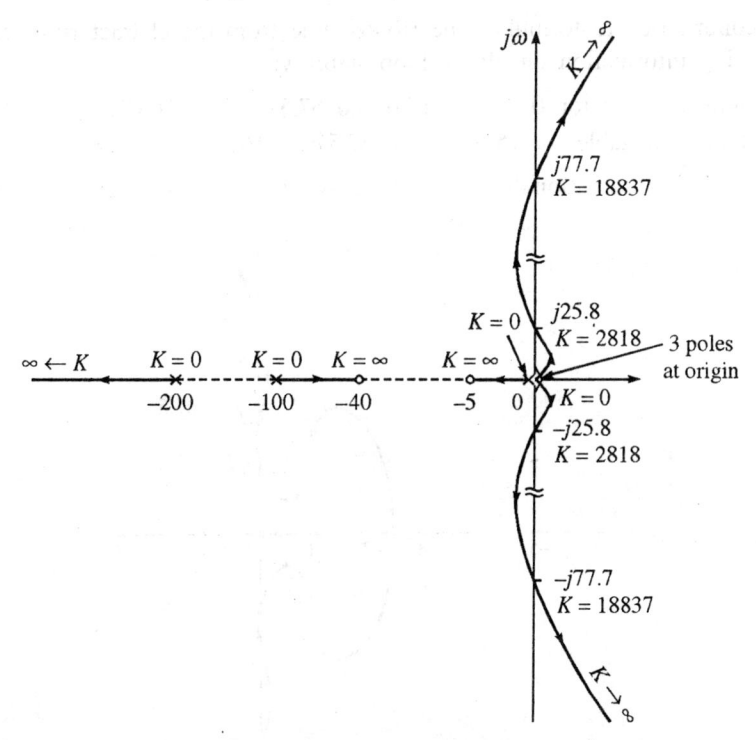

FIGURE 6.27 Example 6.12: root loci.

The concluding comment is that the closed-loop system is conditionally stable for gain K in the positive range. That is, for the range $0 < K < 2818$, the closed-loop system is unstable. Again for the range of gain $18837 < K < \infty$, the closed-loop system is unstable. For the range of gain $2818 < K < 18837$, the closed-loop system is stable. Hence the closed-loop system is conditionally stable for variable gain K in the range $0 < K < \infty$.

EXAMPLE 6.13 For the system shown in Figure 6.28, draw the root loci and then determine its stability.

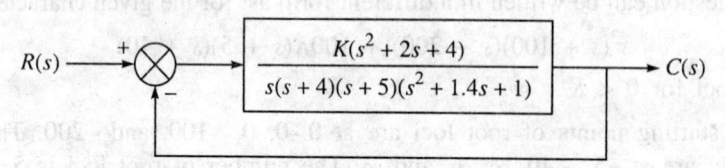

FIGURE 6.28 Example 6.13: control system.

Solution: The characteristic equation of the closed-loop system is

$$1 + G(s)H(s) = 1 + \frac{K(s^2 + 2s + 4)}{s(s+4)(s+6)(s^2 + 1.4s + 1)} = 0$$

or

$$s(s+4)(s+6)(s^2 + 1.4s + 1) + K(s^2 + 2s + 4) = 0$$

Students are recommended to formulate the Routh table from the characteristic equation and verify the following information on closed-loop stability:

(i) The system is stable for $0 < K < 15.6$ and $67.5 < K < 163.6$.
(ii) The system is unstable for $15.6 < K < 67.5$ and $163.6 < K < \infty$.

The system is hence conditionally stable. The root loci is shown in Figure 6.29.

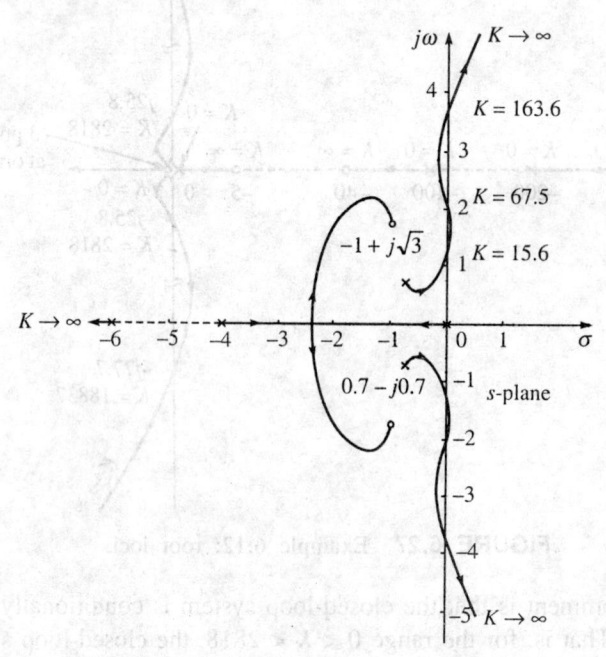

FIGURE 6.29 Example 6.13: root loci.

Root-Locus Technique

The same stability information for the conditionally stable system can be obtained from the root-loci plot as that obtained from Routh–Hurwitz criterion. Problems could arise if the designer stipulated a gain between $67.5 < K < 163.6$ as this gain would dampen the transient response but in this case reduction of gain would make the closed-loop system unstable.

The problem can be overcome by using a suitable compensating network $G_c(s)$ in the forward path in cascade as shown in Figure 6.30. With $G_c(s) = (s + 3)/(s + 5)$, the root-loci plot of the compensated system is shown in Figure 6.31. The closed-loop system is now stable for all values of gain K, i.e. $0 < K < \infty$.

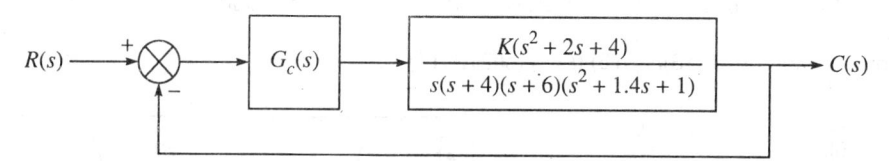

FIGURE 6.30 Example 6.13: addition of compensating network.

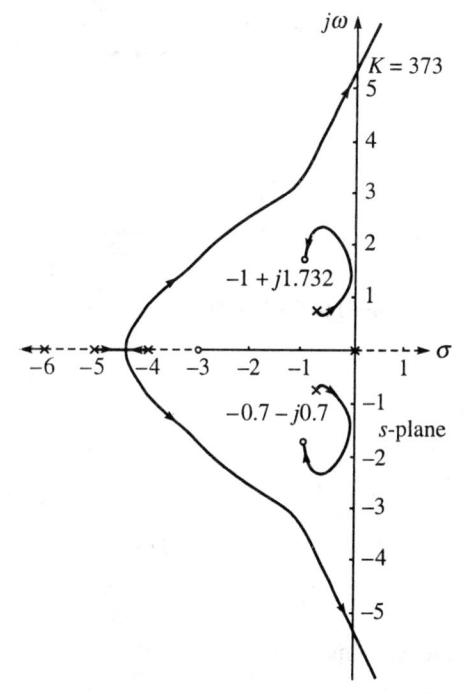

FIGURE 6.31 Example 6.13: root loci with compensator.

We have seen that root loci may be constructed in MATLAB platform. But we are yet to know how to evaluate the salient features (such as break-away/break-in points, intersection point with imaginary axis, upper limit of gain before the system goes to instability, etc.) of root-loci plot which are required for design of control systems. Obviously, the professional version of MATLAB is equipped with all the salient features of root-loci plot. The program for determining all the salient features of root loci required for design purposes is incorporated as an executable file with root-loci plot in MATLAB platform. For a better perspective and perception a few relevant programs in C language in MATLAB platform have been given (see MATLAB Program 6.2).

MATLAB Program 6.2

```
% Root loci.m
clear;
pack;
clc;
ch=menu('Root-locus','Demo1','Demo2','Demo3','Enter Tr. Fn.','Quit');
if ch==1,
% Root-loci plot of k(s+3)/s(s+5) (s+6) (s^2+2s+1)
    num=[1 3];
    a = [1 0];
    b = [1 5];
    c = [1 6];
    d = [1 2 2];
    den 1 = conv (a, b);
    den 2 = conv (c, d);
    den = conv (den 1, den 2);
    stp=.8;
end
if ch==2,
    num=[1 1];
    den=[1 3 12 -16 0];
    stp=.8;
end
if ch==3,
% Root-loci plot of G(s)H(s)=k/s(s+4) (s^2+4s+20)
    num=[1];
    a=[1 0];
    b=[1 4];
    den 1=conv [a, b];
    den 2=conv [1 4 20];
    den=conv [den1 den2];
    den=[1 8 36 80 0];
    stp=0.9;
end
if ch==4,
    stp=.8;
    coe=input('input the degree of numerator');
c=1;
coe1=coe+1;
while c<=coe1,
    num(c)=input('enter the value');
    c=c+1;
end
    doe=input('input the degree of numerator');
c=1;
doe1=doe+1;
```

Root locus:

Cross pt.-0.035785-1.3091ik=32.1

brk pt = -5.7953 + 0i

Cross pt.-0.035785-1.3091ik=32.1

Root-Locus Technique

```
while c<=doe1,
   den(c)=input('enter the value');
   c=c+1;
end
 end
 if ch==5,
   quit;
end
[r,k]=rlocus(num,den);
r
k
rlocus(num,den);
title('Root-locus');
len=length(den)-1;
%No of Cols in r
len1=len-1;
l=1;
while l<=len1,
   j=l+1;
   while j<=len,
       z=1;
           while z<=length(r),
%Difference between points
       dif=r(z,l)-r(z,j)
       x=real(dif);
       y=imag(dif);

% Testing brk. Points.
if abs(x)<=stp & abs(y)<=stp
       x1=real(r(z,l));
       y1=imag(r(z,l));
       br1=num2str(x1);
       br2=num2str(y1);
       br=['brk pt= ',br1];
       bq=[' + ',br2,'i'];
       bf=[br,bq];
       text(x1,y1,bf);
         while abs(x)<=stp & abs(y)<=stp,
           dif=r(z,l)-r(z,j);
           x=real(dif);
           y=imag(dif);
           z=z+1;
         end
end
       z=z+1;
       end
                   j=j+1;
           end
               l=l+1;
            end
               a=1;
```

Root locus:
crosspt.-0.014149+2.586ik=36.1
crosspt.-0.022372+1.6125ik=24.1
brk pt = -2.2257 + 0.37614i   brk pt = 0.18058 + 0i
crosspt.-0.022372+1.6125ik=24.1
crosspt.-0.014149+2.586ik=36.1

Root locus:
cross pt. –0.017575 + 3.1512ik =256
brk pt = -2 + 2.0063i
brk pt = -1.6047 + 0i
brk pt = -2 + -2.824i
cross pt. –0.017575 – 3.1512ik =256

```
while a<=len,
   b=1;
      while b<=length(r),
         m=r(b,a);
 % Testing the crossover points.
 if abs(real(r(b,a)))<0.7 & imag(r(b,a))~=0,
 m11=num2str(r(b,a));
            m33=num2str(k(b));
            m44=['k=',m33];
            s=['cross pt.',m11,m44];
            text(real(r(b,a)),imag(r(b,a)),s);
         while abs(real(r(b,a)))<0.7 & imag(r(b,a))~=0,
            b=b+1;
         end
            end
      b=b+1;
   end
   a=a+1;
end

%The brk points of the algorithm are calculated using
%the logic that at brk points diff of the roots is
%nearly zero(Threshold value is taken as 0.8 or 0.9).
%So the value calculated manually and using MATLAB
%may not come out to be exactly equal.
```

6.5 Rules for the Construction of Inverse Root Loci

The rules for constructing inverse root loci (IRL) for one parameter, say, gain K varying in the range $-\infty < K \leq 0$ are listed in the following:

(i) *$K = 0$ points:* The $K = 0$ points on the IRL are the poles of the open-loop transfer function $G(s)H(s)$. These are the terminating points of IRL.

(ii) *$K = -\infty$ points:* The $K = -\infty$ points on the IRL are the zeros of $G(s)H(s)$ (including zeros at infinity). These are the starting points of IRL.

(iii) *Number of separate IRL:* The total number of IRL is equal to the number of finite poles P or the number of finite zeros Z, whichever is greater.

(iv) *Symmetry of IRL:* IRL are symmetrical about the real axis.

(v) *Asymptotes of IRL:* For large values of s, the angles of asymptotes are

$$\theta_k = \frac{2k\pi}{P-Z}; \quad k = 0, 1, 2, \ldots, P-Z-1 \qquad (6.62)$$

The point of intersection of the asymptotes on the real axis is given by

$$\sigma_1 = \frac{\sum \text{real parts of poles of } G(s)H(s) - \sum \text{real parts of zeros of } G(s)H(s)}{P-Z}$$

Root-Locus Technique

(vi) *IRL on real axis:* On a given section on the real axis in the s-plane, the existence of IRL is found for $-\infty < K \leq 0$ in the section only if the total number of finite poles and zeros to the right of the section is even.

(vii) *Intersection of IRL on the imaginary axis:* The values of ω and K at the crossing points of the IRL on the imaginary axis of the s-plane may be obtained by Routh–Hurwitz criterion, similar to the root loci case.

(viii) *Angle of departure and arrival:* The angle of departure of inverse root locus for $-\infty < K \leq 0$ from a zero or the angle of arrival at a pole of $G(s)H(s)$ can be determined by assuming a point s_1 on the inverse root locus that is associated with the zero or pole and which is very close to the zero or pole. The angle of departure or arrival of an inverse root locus is determined from

$$G(s_1)H(s_1) = \sum_{i=1}^{m}(s_1 - z_i) - \sum_{j=1}^{n}(s_1 - p_j) = 2k\pi, \text{ where } k = 0, \pm 1, \pm 2, \ldots \quad (6.63)$$

(ix) *Break-away and break-in points:* The break-away or break-in points on the IRL are determined by finding the roots of

$$\frac{dK}{ds} = 0 \quad \text{or} \quad \frac{d}{ds}[G(s)H(s)] = 0 \quad (6.64)$$

Otherwise, we can use the alternative method as discussed in the context of the construction of root loci.

(x) *Calculation of values of K on the IRL:* The absolute value of K at any point s_1 on the inverse root loci can be determined from the equation

$$|K| = \frac{1}{G(s_1)H(s_1)}$$

$$= \frac{\text{products of lengths of vectors drawn from the poles of } G(s)H(s) \text{ to } s_1}{\text{products of lengths of vectors drawn from zeros of } G(s)H(s) \text{ to } s_1} \quad (6.65)$$

EXAMPLE 6.14 Consider the open-loop transfer function

$$G(s)H(s) = \frac{K(s+3)}{s(s+5)(s+6)(s^2+2s+2)}$$

Draw the inverse root loci for $-\infty < K < 0$.

Solution: The pole-zero configuration of the open-loop transfer function $G(s)H(s)$ is shown in Figure 6.32. The construction rules for inverse root loci are:

 (i) The starting points ($K = -\infty$) are at $s = -3$, $s = -\infty$, $s = -\infty$, $s = -\infty$ and $s = -\infty$
 (ii) The terminating points ($K = 0$) are at $s = 0$, $s = -1 \pm j1$, $s = -5$ and $s = -6$.
 (iii) The number of separate IRL = 5 as $P = 5$, $Z = 1$ and $P > Z$.
 (iv) There is symmetry of IRL about the real axis.

(v) The angles of asymptotes of IRL for large values of s are given by

$$\theta_k = \frac{2k\pi}{P-Z}; \quad k = 0, 1, \ldots, P - Z - 1$$

i.e. $\theta_0 = 0°, \; \theta_1 = \pi/2, \; \theta_2 = \pi, \; \theta_3 = 3\pi/2$

The asymptotes intersect the real axis at -2.5.

(vi) IRL will exist on the real axis if the number of poles and zeros to the right is even. Hence IRL exists between $+\infty$ to 0 and between -3 to -5 and between -6 to $-\infty$.

(vii) There is no intersection of IRL with the imaginary axis for $-\infty < K < 0$ as found out by Routh–Hurwitz criterion.

(viii) The angle of departure or arrival as determined from Eq. (6.65) is $\theta = 180° - 43.8° = 136.2°$.

(ix) The break-away points are obtained by solving $dK/ds = 0$, which gives

$$s^5 + 13.5s^4 + 66s^3 + 142s^2 + 123s + 45 = 0$$

i.e. $s = -5.53; \; -0.656 \pm j0.468; \; -3.33 \pm j1.204$

None of these points are on IRL and hence IRL has no break-away points.

Now the inverse root loci is drawn as shown in Figure 6.32 with the dotted lines for the variable parameter K in the range $-\infty \leq K \leq 0$. Further, for variable parameter K in the range $0 \leq K < \infty$, the root loci is already drawn in Figure 6.11. Now for complete root loci for

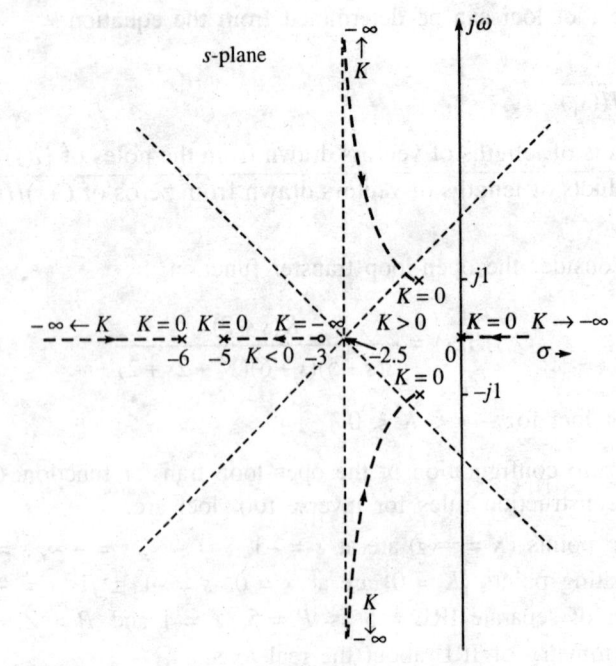

FIGURE 6.32 Example 6.14: inverse root loci for $-\infty \leq K \leq 0$.

$-\infty < K \le \infty$, the inverse root loci and root loci have to be added together and are shown in Figure 6.33.

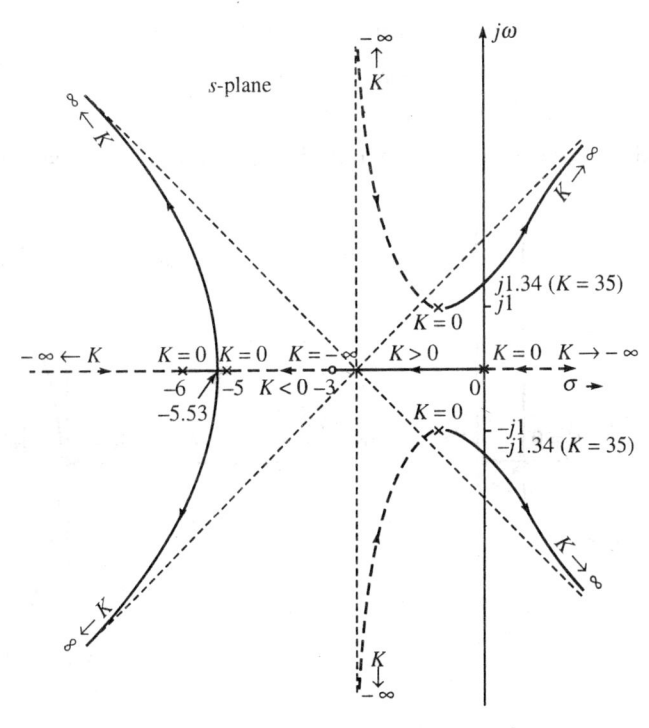

FIGURE 6.33 Example 6.14: complete root loci.

6.6 Effect of Adding Poles and Zeros

The root-locus technique is used to design a control system. In this connection, some of the salient features of the root-locus technique are now highlighted.

6.6.1 Addition of poles

Consider
$$G(s)H(s) = \frac{K}{s(s+a)}; \quad a > 0 \qquad (6.66)$$

The root-locus diagram is shown in Figure 6.34(a).
Now a pole at $-b$ is added so that the new transfer function becomes

$$G(s)H(s) = \frac{K}{s(s+a)(s+b)}; \quad |b| > |a| \qquad (6.67)$$

and the corresponding root loci is shown in Figure 6.34(b). The complex closed-loop poles are going towards the unstable (right-half of s-plane) region for $K > K_1$. The stable range is $0 < K < K_1$.

Next another pole at $-c$ is added so that the new transfer function becomes

$$G(s)H(s) = \frac{K}{s(s+a)(s+b)(s+c)}; \quad |c| > |b| > |a| \tag{6.68}$$

and the corresponding root loci is shown in Figure 6.34(c). The stable range becomes

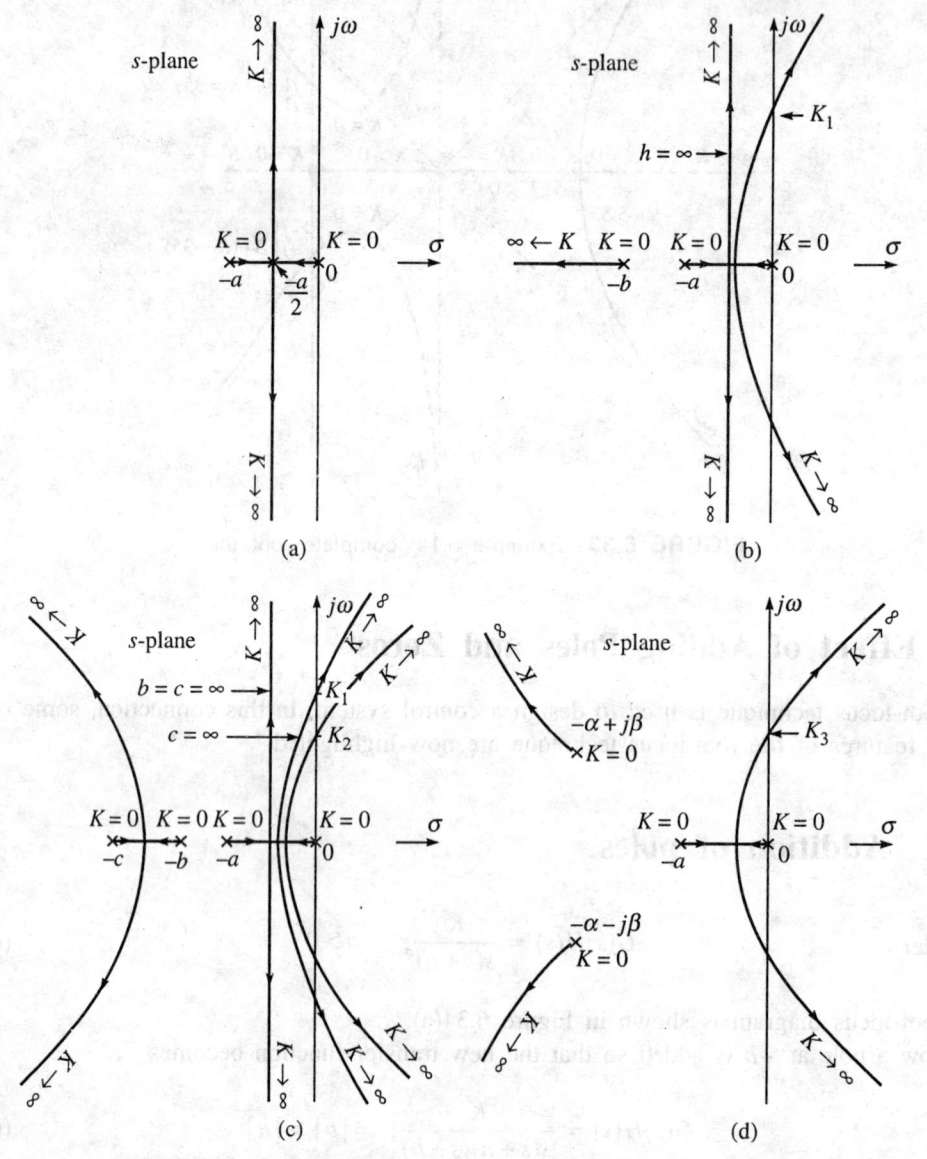

FIGURE 6.34 Effect of addition of poles on system stability.

Root-Locus Technique

$0 < K < K_2 < K_1$, i.e. less than the previous case. The closed-loop system is stable up to gain K_2 which is obviously less than K_1.

Next we add two complex conjugate poles to the original system of Figure 6.34(a) so that the new open-loop transfer function of the system becomes

$$G(s)H(s) = \frac{K}{s(s+a)(s+\alpha+j\beta)(s+\alpha-j\beta)}; \quad |\alpha| > |a| \qquad (6.69)$$

The pole-zero configuration is shown in Figure 6.34(d). The root-locus plot is drawn. The system is now stable up to the value of gain $K_3 < K_2$ and hence less stable compared to the original system. If we move the complex poles further, the system becomes even more restricted. Hence we can conclude from the root-locus technique that the system becomes less stable as we add the poles. Addition of poles to $G(s)H(s)$ is the same as the increase in the order of the system. Addition of poles to the function $G(s)H(s)$ has the effect of moving the root loci towards the right-half of s-plane.

6.6.2 Addition of zeros

In a similar fashion, adding zeros to $G(s)H(s) = K/s(s + 2)$ [see Figure 6.35(a)] has the effect of moving the root loci towards the left making the system more stable.

Adding a zero at $-b$ where $|b| > |a|$, the root-locus plot is shown in Figure 6.35(b). The root-locus plot has moved towards the left. The closed-loop system becomes more stable compared to the previous case. Move zero at $-b$ towards right so that $|b| < |a|$ as in Figure 6.34(b). The closed-loop poles become real, the system becomes overdamped. Stability improves further. Add two complex zeros at $-b \pm j\beta$. The root-locus plot is shown in Figure 6.35(c). The system stability improves compared to that of Figure 6.35(b).

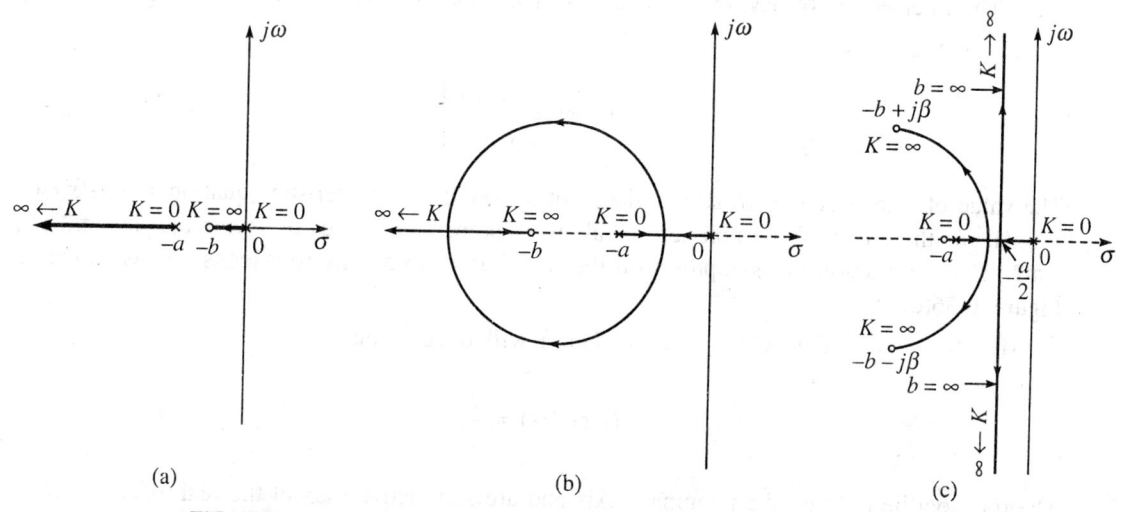

FIGURE 6.35 Effect of addition of zeros on system stability.

6.6.3 Effect of varying the pole position

Consider
$$G(s)H(s) = \frac{K(s+1)}{s^2(s+a)} \qquad (6.70)$$

Therefore,
$$K = -\frac{s^2(s+a)}{(s+1)} \qquad (6.71)$$

Now $dK/ds = 0$ gives the break-away points as

$$\frac{-(a+3)}{4} \pm \frac{1}{4}\sqrt{a^2 - 10a + 9} \qquad (6.72)$$

(i) Let $a = 10$. Then,
$$G(s)H(s) = \frac{K(s+1)}{s^2(s+10)} \qquad (6.73)$$

Now $dK/ds = 0$ gives the break-in point at $s = -2.5$ and the break-away point at -4.0.

The root loci on the real axis exist between -1 and ∞ on the negative real axis. The angles of asymptotes are $\pi/2, -\pi/2$. The intersection of asymptotes is at -4.5. The root-loci plot is shown in Figure 6.36(a).

(ii) Let the pole at $-a$ be shifted from -10 to -9. Then

$$G(s)H(s) = \frac{K(s+1)}{s^2(s+9)} \qquad (6.74)$$

Now $dK/ds = 0$ gives the breakaway point at -3. The angles of asymptotes are $\pi/2$ and $-\pi/2$. The intersection of asymptotes is at -4. The root-locus plot is shown in Figure 6.36(b).

(iii) Putting $a = 8$,

$$G(s)H(s) = \frac{K(s+1)}{s^2(s+8)} \qquad (6.75)$$

The value of s in equation $dK/ds = 0$ does not satisfy the characteristic equation $1 + G(s)H(s) = 0$. Hence there is no finite non-zero break-away point. The angles of asymptotes are $\pi/2$ and $-\pi/2$. The intersection of asymptotes on the real axis is -3.5. The root-locus plot is shown in Figure 6.36(c).

(iv) For $a = 1$, the pole-zero cancellation will occur. Then

$$G(s)H(s) = \frac{K}{s^2}$$

The root loci lie only on the imaginary axis and are symmetrical about the real axis. The root-locus plot is shown in Figure 6.36(d).

(a) $a = 10$

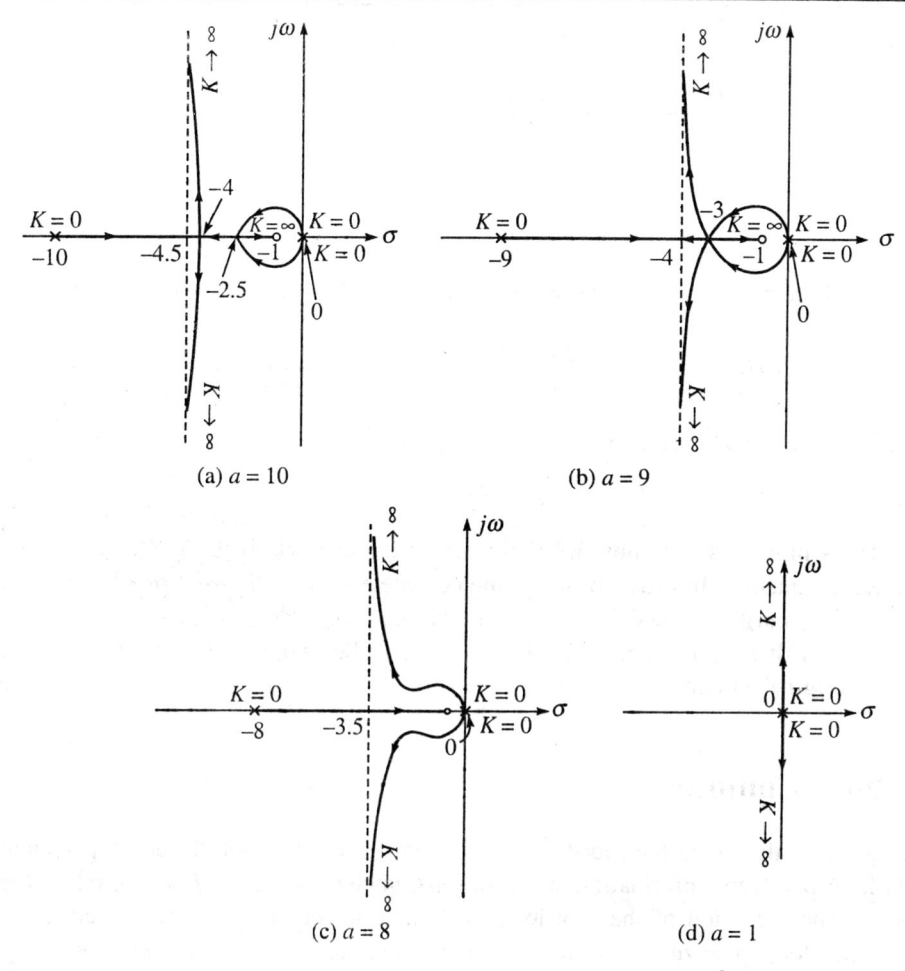

(b) $a = 9$

(c) $a = 8$

(d) $a = 1$

FIGURE 6.36 Root loci for the variable pole of $G(s)H(s) = K(s + 1)/s^2(s + a)$; $1 \leq a \leq 10$, $0 < K < \infty$.

6.6.4 Cancellation of poles and zeros

It is important to note that if the denominator and numerator of $G(s)H(s)$ involve common factors, then the open-loop poles and zeros will cancel each other and the degree of the characteristic equation is reduced by one or more. To illustrate this, consider a negative feedback system having forward transfer function

$$G(s) = \frac{K}{s(s+2)(s+3)}$$

and feedback ratio $H(s) = (s + 2)$. Then the characteristic equation from the closed-loop transfer function

$$\frac{C(s)}{R(s)} = \frac{G(s)}{1+G(s)H(s)} = \frac{K}{s(s+2)(s+3)+K(s+2)}$$

can be written as

$$s(s+2)(s+3) + K(s+2) = 0$$

But because of the cancellation of the common term $(s + 2)$, we have the characteristic equation

$$1+G(s)H(s) = 1 + \frac{K}{s(s+2)(s+3)} \times (s+2) = 1 + \frac{K}{s(s+3)} = \frac{s(s+3)+K}{s(s+3)}$$

The reduced characteristic equation is

$$s(s + 3) + K = 0$$

The root-locus plot does not show all the roots of the characteristic equation, only the roots of the reduced equation. In order to obtain the complete set of closed-loop poles, the cancelled pole of $G(s)H(s)$ must be added to those of the closed-loop poles obtained from the root-locus plot of $G(s)H(s)$. It is to be remembered that the cancelled pole of $G(s)H(s)$ is a closed-loop pole of the original system.

6.7 Root Contour

Thus far we have discussed the root-locus technique for the study of closed-loop systems from the open-loop pole-zero information when one parameter, say, gain K is varied in the range $0 < K < \infty$. The extension of the root-locus technique to the study of the closed-loop system from the open-loop pole-zero information when more than one parameter is varying in the range from 0 to ∞, is known as the root contour. The root contour can be constructed following the same rules that we earlier explained in the context of construction of root loci. This is being illustrated with examples for better understanding.

Consider the system shown in Figure 6.37 where K and α are varying in the range 0 to infinity. The characteristic equation is

$$s^2 + \alpha s + K = 0$$

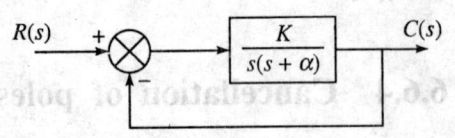

FIGURE 6.37

The root-contours are to be plotted by allowing both α and K to vary in the range 0 to ∞. Rewriting the characteristic equation as

$$1 + \alpha\left(\frac{s}{s^2 + K}\right) = 0 \qquad (6.76)$$

Root-Locus Technique

such that α appears as a variable (say, gain) parameter and is varying in the range 0 to ∞ for different, particular values of K, i.e. as if the open-loop transfer function is

$$\frac{\alpha s}{s^2 + K} \tag{6.77}$$

For various values of K by allowing α to vary in the range 0 to ∞, the root contour can be constructed. The root contour of Eq. (6.77) originates from poles at $s = \pm j\sqrt{K}$ and terminates on zeros at $s = 0$ and at $s = -\infty$.

Now,
$$\frac{d\alpha}{ds} = -\frac{-(s^2 - K)}{s^2} = 0 \tag{6.78}$$

gives the break-away point at $s = \pm\sqrt{K}$.

It is evident from Figure 6.37 that the break-away point has to lie on the real axis between $s = 0$ to $s = -\infty$. Therefore, only $s = -\sqrt{K}$ corresponds to the break-away point. The root contours for various values of K with α varying from 0 to ∞ are shown in Figure 6.38.

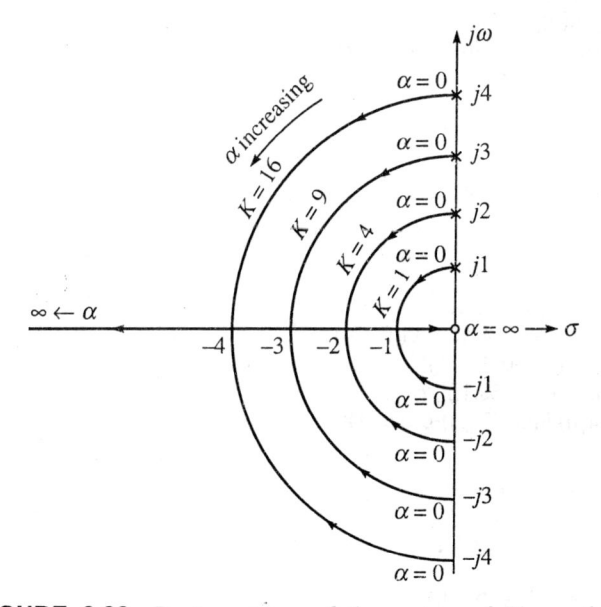

FIGURE 6.38 Root contours of the system of Figure 6.37.

It can be shown that the complex root branches are circular as we proved earlier in Example 6.2. MATLAB program to plot the root contours of a characteristic equation is given next.

MATLAB Program 6.3

```
%   rootcontour.m
%   To plot the root-contour of the characteristic equation
%           s^2 + αs + K = 0
clear;
pack;
clc;

axis([-5,2,-5,5]);
hold on;
for K = 1 : 4,
   num = [1 0];
   den = [1 0 K^2];
   rlocus(num,den);
   l = length(den)-1;
   for α = 0:0.1:10,
      [R]= rlocus(num,den,α);
%   To find the intersection with the imaginary axis.
      for n = 1:l,
         if abs(real(R(n)))<=0.001 & imag(R(n))~=0,
            str1 = sprintf(' %.1fj',imag(R(n))) ;
         text(0,imag(R(n)),str1);
         str1 = sprintf('         (α = %.1f)',α) ;
         text(0,imag(R(n)),str1);
            end;
         end;

%   To find the Breakaway or Breakin Point.

      for p = 1:l-1,
      for q = p+1:l,
         if abs(R(p)-R(q))<= 0.1,
            str1 = sprintf(' %.1f',real(R(p))) ;
text(real(R(p)),imag(R(p))-0.4,str1);
         str1 = sprintf('(α = %.1f)',α) ;
         text(real(R(p)),imag(R(p))-0.8,str1);
            end;
            end;
         end;
      end;

   pop1 = sprintf(' K = %g',K^2);
   text(-K,0.5,pop1);
   hold on;

end;
title('Root-contour for the characteristic eqn. s^2 + αs + K = 0');
hold off;
```

EXAMPLE 6.15 Consider the feedback control system with open-loop transfer function

$$G(s)H(s) = \frac{K}{s(s+1)(s+\alpha)}$$

in which both K and α are variable in the range 0 to ∞.

Manipulating the characteristic equation

$$1 + G(s)H(s) = 1 + \frac{K}{s(s+1)(s+\alpha)} = 0 \quad \text{or} \quad s^2(s+1) + \alpha(s+1)s + K = 0$$

in the form where α appears as a variable (gain) parameter of root-loci plot, we can write

$$1 + \frac{\alpha s(s+1)}{s^2(s+1) + K} = 0 \tag{i}$$

This is of the form where the root locus with respect to the parameter α can be drawn for an open-loop transfer function

$$\frac{\alpha s(s+1)}{s^2(s+1) + K}$$

The root-contours of Eq. (i) originate (i.e. for $\alpha = 0$) at open-loop poles of the reduced characteristic equation

$$s^2(s+1) + K = 0$$

and terminate (i.e. for $\alpha = \infty$) at 0, -2 and $-\infty$. The reduced characteristic equation is rewritten as

$$1 + \frac{K}{s^2(s+1)} = 0$$

The root locus of the reduced characteristic equation with K as a variable parameter is plotted in Figure 6.39 with open-loop poles 0, 0 and -1.

The root contours for various values of K with varying α in the range $0 < K < \infty$ are drawn in Figure 6.40 following the root-locus technique. The value of α at which the root contours will cross the $j\omega$-axis are obtained by Routh criterion from the characteristic equation

$$s^3 + (\alpha + 1)s^2 + \alpha s + K = 0$$

The Routh array is:

s^3	1	α
s^2	$\alpha + 1$	K
s^1	$\dfrac{\alpha(\alpha+1) - K}{\alpha + 1}$	
s^0	K	

310 Modern Control Engineering

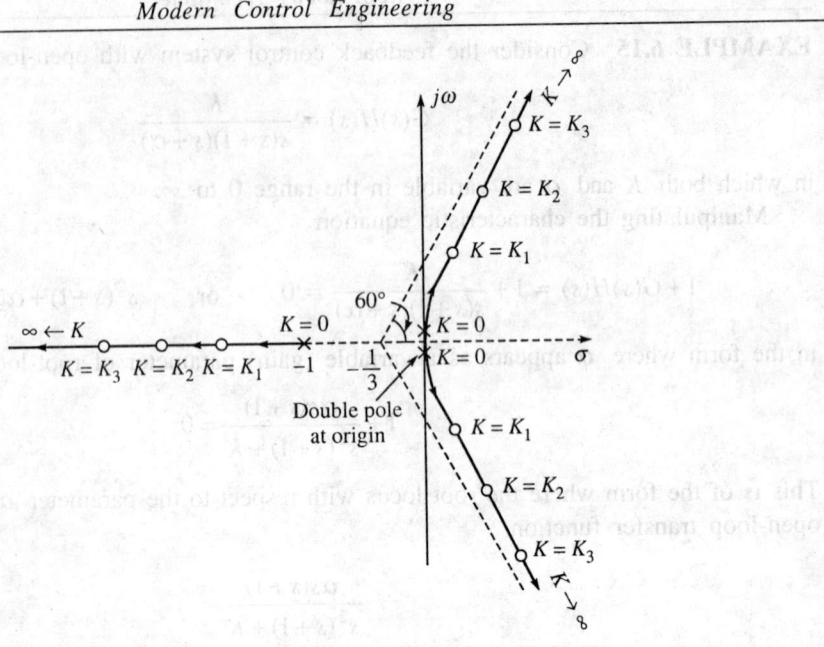

FIGURE 6.39 Example 6.15: root loci of $G(s)H(s) = K/s^2(s+1)$.

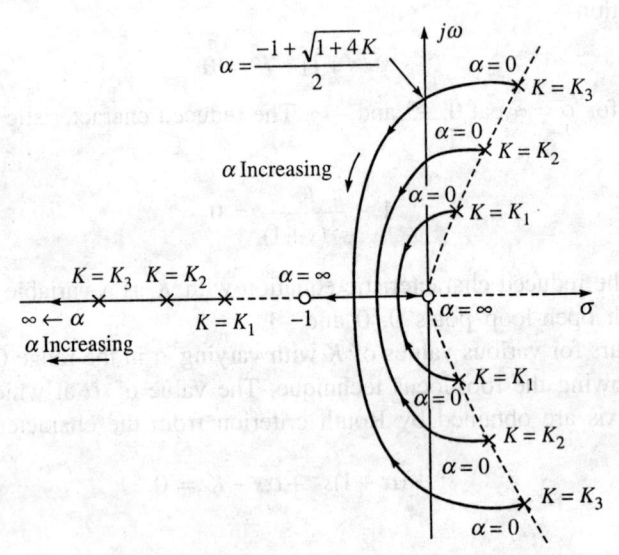

FIGURE 6.40 Example 6.15: root loci of $G(s)H(s) = \alpha s(s+1)/[s^2(s+1) + K]$.

The root contours cross the $j\omega$-axis for

$$\alpha(\alpha + 1) - K = 0 \quad \text{or} \quad \alpha = \frac{-1 + \sqrt{1 + 4K}}{2}$$

since only the positive value is permitted.

Root-Locus Technique

EXAMPLE 6.16 (i) For the system given in Figure 6.41, draw the root-loci first with $K_h = 0$ and K as variable. (ii) Obtain the value of K so that the system damping ratio is 0.158. (iii) For this value of the system gain, draw the root loci with K_h as variable. (iv) Find the value of K_h that improves the system damping ratio to 0.5, obtained using the root-loci plot. Grapical approximations are acceptable. Make qualitative plots of unit-step and unit-ramp responses of the above system with and without K_h feedback being operative.

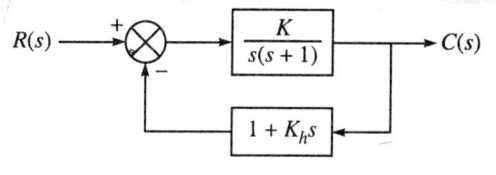

FIGURE 6.41 Example 6.16.

Solution: (i) $G(s)H(s) = \dfrac{K(1 + K_h s)}{s(s + 1)}$

The characteristic equation is

$$s(s + 1) + K(1 + K_h s) = 0$$

or

$$s^2 + s + KK_h s + K = 0$$

For part (i), $K_h = 0$. The reduced characteristic equation is therefore

$$s^2 + s + K = 0$$

and the equivalent characteristic equation $1 + G_1(s) = 0$

yields

$$G_1(s) = \dfrac{K}{s(s+1)}$$

The root-loci plot is shown in Figure 6.42.

(ii) For $\zeta = 0.158 = \cos\theta$ yields $\theta = 80.9°$. Further,

$$s^2 + s + K = 0 \text{ yeilds } \omega_n^2 = K$$

This is equivalent to the standard form of equation with $\zeta = 0.158$, yielding

$$s^2 + 2(0.158)\sqrt{K}\, s + K = 0$$

Comparing, we get

$$2(0.158)\sqrt{K} = 1 \quad \text{or} \quad K = 10$$

Then $s^2 + s + 10 = 0$ yields $s_1 = -0.5 \pm j3.122$. The point s_1, in order to lie on root-loci needs the magnitude condition to be satisfied for a check.

(iii) Root-loci for $K = 10$, the characteristic equation for variable K_h is

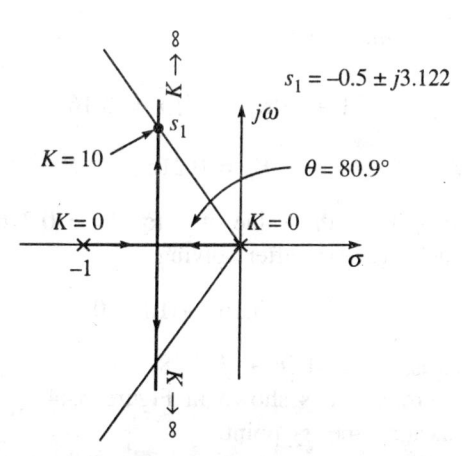

FIGURE 6.42 Example 6.16: root-loci with $K_h = 0$.

$$s^2 + (1 + 10K_h)s + 10 = 0$$

or
$$s^2 + s + 10 + 10K_h s = 0$$

or
$$1 + \frac{10K_h s}{s^2 + s + 10} = 0$$

Hence we have to find the root-loci for the variable K_h in $0 < K_h < \infty$ where the open-loop transfer function is

$$\frac{sK_h}{s^2 + s + 10}$$

The poles are at $-0.5 \pm j3.122$ and zero at origin. The number of root-loci = 2.
The break-in point $dK_h/ds = 0$ yeilds $s = -\sqrt{10} = -3.16$. The root-loci, for $K = 10$ are shown in Figure 6.43.

(iv) $\zeta = 0.5 = \cos\theta$ or $\theta = 60°$
The characteristic equation for the variable K_h with $K = 10$ is

$$s^2 + (1 + 10K_h)s + 10 = 0$$

Hence $\omega_n = \sqrt{10}$

Now with $\zeta = 0.5$ and $\omega_n = \sqrt{10}$, the characteristic equation becomes

$$s^2 + 2(0.5)\sqrt{10}\, s + 10 = 0$$

Comparing, we get

$$1 + 10K_h = \sqrt{10} = 3.16$$

or $K_h = 0.216$

Let us have the value of s for $K_h = 0.216$ which becomes after solving

$$s^2 + 3.16s + 10 = 0$$

that is, $s_2 = -1.56 + j2.738$.
The root-loci is shown in Figure 6.44 indicating the s_2 point.

(v) Qualitative sketches for unit step and ramp responses are shown in Figures 6.45(a) and (b) respectively.

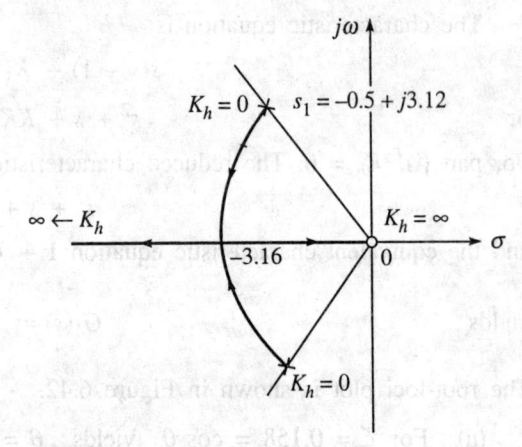

FIGURE 6.43 Example 6.16: root-loci with K_h as variable.

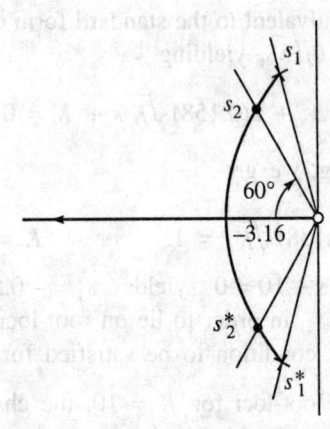

FIGURE 6.44 Example 6.16: root-loci with $\zeta = 0.5$.

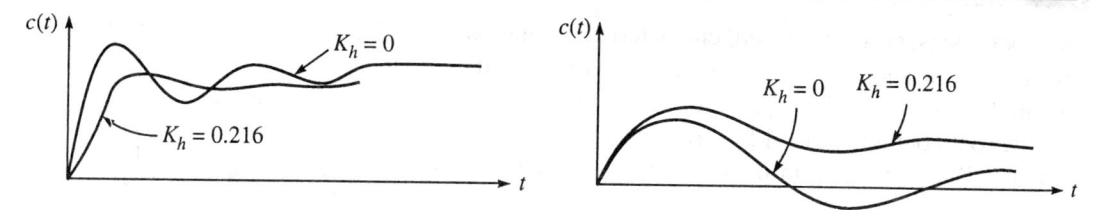

FIGURE 6.45 Example 6.16: (a) unit-step response and (b) unit-ramp response with and without K_h.

6.7.1 Multiple-loop system

Consider the system shown in Figure 6.46(a) where two feedback loops exist. The equivalent system is shown in Figure 6.46(b), where the open-loop transfer function

$$G(s) = \frac{C(s)}{E(s)} = \frac{\dfrac{K}{s(s+2)}}{1 + \dfrac{sKK_t}{s(s+2)}}$$

$$= \frac{K}{s(s+2) + sKK_t}$$

The characteristic equation of the system is $1 + GH = 0$, i.e.

$$s(s+2) + sKK_t + K = 0$$

FIGURE 6.46 (a) System with two feedback loops and (b) its equivalent system.

Rewriting this as

$$1 + \frac{sKK_t}{s(s+2) + K} = 1 + \frac{\alpha s}{s(s+2) + K} = 0 \qquad (6.79)$$

where $\alpha = KK_t$.

The root contours can be plotted for various values of K with α varying from 0 to ∞. The root contours of Eq. (6.79) originate (i.e. for $\alpha = 0$) at open-loop poles of the reduced characteristic equation

$$s(s+2) + K = 0$$

The reduced characteristic equation is rewritten as

$$1 + \frac{K}{s(s+2)} = 0$$

The root locus of the reduced characteristic equation with K as a variable parameter can be plotted as dotted line in Figure 6.47 with open-loop poles at 0 and -2. The root contours plotted for various values of K with $\alpha = KK_t$ varying from 0 to infinity are shown in Figure 6.47.

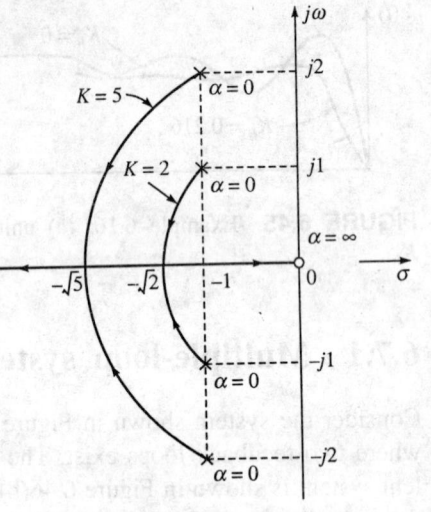

6.8 Root-locus for System with Transportation Lag

Consider a system having open-loop transfer function

$$G(s)H(s) = \frac{Ke^{-sT}}{s(s+2)}$$

FIGURE 6.47

where T is the transportation delay in seconds and is given as 1 s. Let us draw the root loci for K varying in the range $0 < K < \infty$.

If the transportation delay is small, then we can assume

$$e^{-sT} = 1 - sT$$

Rewriting $G(s)H(s)$ as

$$G(s)H(s) = \frac{K(1-s)}{s(s+2)}; \quad T = 1$$

i.e.

$$G(s)H(s) = \frac{-K(s-1)}{s(s+2)}$$

The characteristic equation becomes

$$1 - G(s)H(s) = 0$$

Here the following two rules have to be modified:

(i) The angle condition instead of Eq. (6.13) becomes: $\angle \dfrac{K(s-1)}{s(s+2)} = 2k(\pi);\ k = 0, 1, 2, \ldots$

(ii) The existence of root loci for even (instead of odd as in Rule 6) number of poles and zeros to the right-hand side on the real axis, for the open-loop transfer function

$$\frac{K(s-1)}{s(s+2)}$$

The other rules are not to be modified.
The root loci is drawn as shown in Figure 6.48.

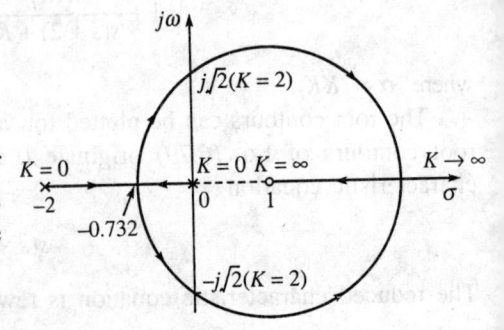

FIGURE 6.48 Root-loci plot of $1 + [Ke^{-s}/s(s+2)] = 0$ with e^{-s} is approximated as $(1 - s)$.

6.9 Design on Root Locus

To achieve a desired speed of response, the time constant of the system would have to be set below some specified value. Or, to obtain a desired oscillatory response, the damping characteristics of the system would have to be specified. The speed of response of the system is determined by the largest time constant appearing in the characteristic equation. On the root-locus plot, lines parallel to the imaginary axis represent lines that have constant $1/\tau$ value, where τ is the time constant. Hence, these can be considered as lines of fixed time constant value. Further, the undamped natural frequency ω_n and the damped natural frequency ω_d relate to the oscillatory behaviour of the system response. Lines that are parallel to the real axis, as in Figure 6.49, are lines of constant ω_d.

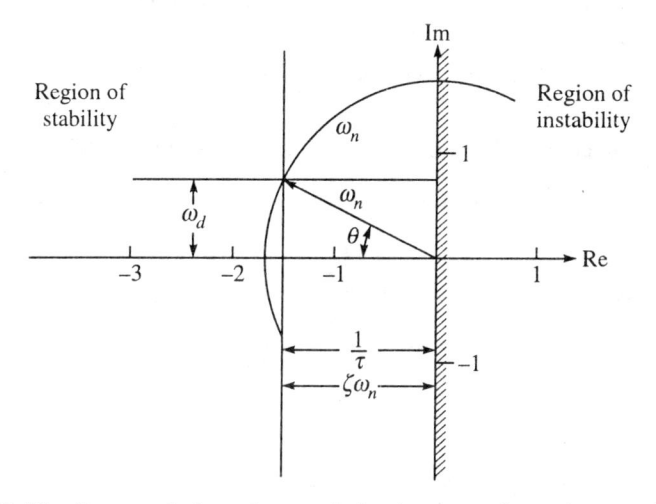

FIGURE 6.49 System design characteristics interpreted on the root-locus plot.

Constant ω_n values are indicated on the root locus plot by circles with the origin as the centre. We have already derived some important relations which are repeated here for ready reference such as the relation between ω_d and ω_n as

$$\omega_d = \omega_n \sqrt{1-\zeta^2}$$

Natural frequency can be related to the time constant by means of the damping ratio as

$$\frac{1}{\tau} = \zeta \omega_n$$

That means the vertical lines in complex plane represent constant $1/\tau$ or constant $\zeta \omega_n$ values. Further, the angle that relates the inclination of the damping ratio is given by

$$\cos \theta = \zeta$$

It is obviously clear that all the above three equations are applicable to underdamped systems, i.e. to systems with values of ζ less than unity.

EXAMPLE 6.17 For a feedback system of Figure 6.50:

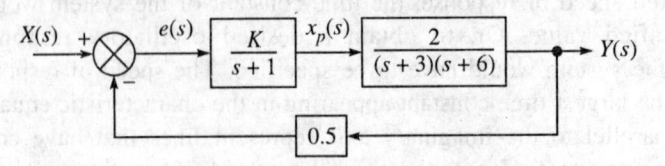

FIGURE 6.50 Example 6.17: control system.

(i) It is desired to have the maximum value of the time constant $\tau = 0.8$ s (for relatively fast response) without any oscillation of the closed-loop system response. Determine the value of gain K.

(ii) Further, check the response of the closed-loop system with the modified gain K for a step input $x(t) = 5$.

(iii) For $K = 10$, what will be the effect on system performance?

Solution: (i) The closed-loop transfer function

$$\frac{Y(s)}{X(s)} = \frac{G(s)}{1 + G(s)H(s)} = \frac{\frac{2K}{(s+1)(s+3)(s+6)}}{1 + \frac{K}{(s+1)(s+3)(s+6)}} = \frac{2K}{s^3 + 10s^2 + 27s + 18 + K}$$

The open-loop transfer function is

$$G(s)H(s) = \frac{K}{(s+1)(s+3)(s+6)}$$

The time constant $\tau = 0.8$ s means that $1/\tau = 1.25$. To satisfy the design specification, the roots of the characteristic equation must lie on the root-locus plot to the left of $\sigma = -1.25$ line. Since the non-oscillatory response is required, the roots must lie on the real axis only. From the root-locus plot drawn as shown in Figure 6.51, the break-away point lies at -1.88, it means that a double pole occurs at -1.88 on the real axis for the particular value of gain K(say K_1). Further, it is quite obvious that the third pole of gain K_1 will be on the real axis away from -6. Now our task is to find the time constant τ corresponding to $\sigma = -1.88$ which comes out to be $1/1.88 = 0.53$ s which is within the design constraint of $\tau \leq 0.8$ s and hence acceptable.

In order to determine the value of $K(= K_1)$ that corresponds to $s_1 = -1.88$, we make use of the magnitude condition as

$$G(s)H(s)\Big|_{s=s_1} = \frac{K}{(s+1)(s+3)(s+6)} = -1$$

which leads to

$$K = K_1 = -4.06$$

Root-Locus Technique

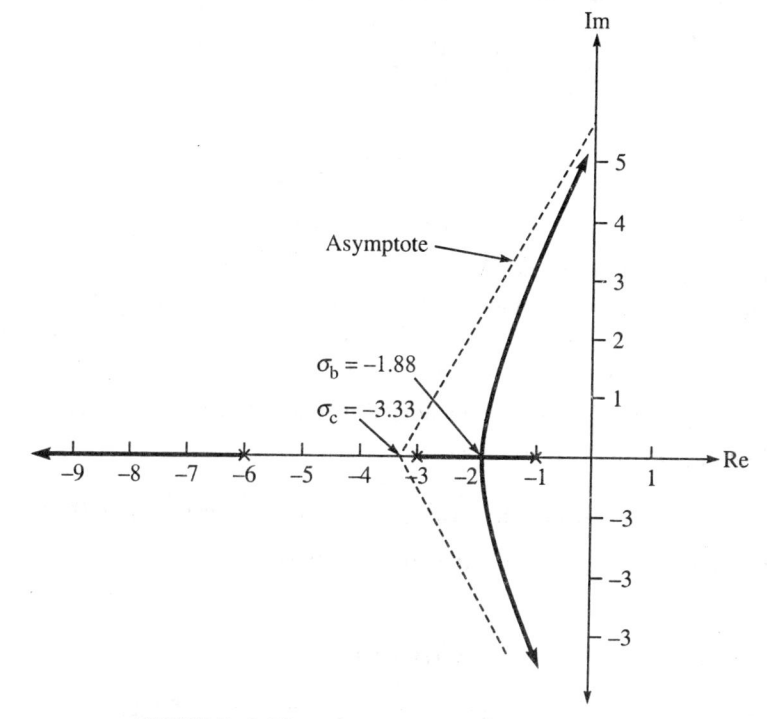

FIGURE 6.51 Example 6.17: root-loci plot.

The placement of the third root will be s_3 for $K = 4.06$ (obviously, s_3 will be to the left of the line $\sigma = -6$) and satisfy the magnitude condition and can be found out as

$$(s_3 + 1)(s_3 + 3)(s_3 + 6) = -1$$

which leads to $s_3 = -6.25$.

Now with $K = 4.06$, the double poles are at -1.88 ($t = 0.53$ s) and the third one at -6.24 ($t = 0.16$ s) which is < 0.8 s. Hence the speed of response is satisfactory as per the design constraint for $K = 4.06$.

(ii) Now the second part of the problem is to determine the closed-loop system response for a step of magnitude $x(t) = 5$.

$$\frac{Y(s)}{X(s)} = \frac{2K}{s^3 + 10s^2 + 27s + 18 + K} = \frac{2(4.06)}{s^3 + 10s^2 + 27s + 18 + 4.06} = \frac{8.12}{s^3 + 10s^2 + 27s + 22.06}$$

Hence the output response $y(t)$ becomes

$$y(t) = \mathscr{L}^{-1}\left[\frac{5(8.12)}{s(s^3 + 10s^2 + 27s + 22.06)}\right] = 1.84 - 4.95te^{-1.88t} - 1.5e^{-1.88t} - 0.34e^{6.24t}$$

(iii) Now with $K = 10$, the gain condition to be satisfied, the characteristic equation is obtained as

$$\frac{10}{(s+1)(s+3)(s+6)} = -1$$

or
$$1 + G(s)H(s) = 1 + \frac{10}{(s+1)(s+3)(s+6)} = 0$$

leads to the roots as $-1.72 \pm j1.123$ and -6.515.

The closed-loop response is oscillatory and slower than case (i). Then nature of this response is determined by the complex roots whose imaginary part $\pm j1.123$ tells that the response is oscillatory, while the real part -1.732 determines the speed of response. The time constant associated with these roots is $\tau = 1/1.72 = 0.57$s and the time constant associated with the third root is $\tau = 1/6.515 = 0.15$ s. The undamped natural frequency $\omega_n = \sqrt{(1.742)^2 + (1.123)^2}$ = 2.07 rad/s. Again, ω_d can be determined as

$\omega_d = \omega_n\sqrt{1-\zeta^2}$ which leads to, $1.123 = 2.07\sqrt{1-\zeta^2}$, hence $\zeta = 0.84$

The closed-loop system is stable as all the roots have negative real part.

Summary

Root loci are the curves on the *s*-plane along which the roots of the characteristic equation move as a parameter (such as gain) is varied in the range from 0 to ∞. There is one locus for each root, that is, the root locus is unique. The beauty of the root-locus technique is that we can get the information about the roots of the characteristic equation of the closed-loop system from the open-loop pole-zero configuration when one parameter is varied in the positive range from 0 to ∞, without solving the characteristic equation of the closed-loop system.

The rules for constructing the root loci have been derived and their use has also been demonstrated, when one parameter (say gain) is varied in the positive range. Families of loci for more than one parameter varying can also be constructed as discussed in the construction of root contours. From the academic point of view, the construction of the inverse range from 0 to −∞ has also been discussed.

The root loci may be used for analyzing stability and transient performances. The effects of parameter adjustment can be nicely visualized and hence the root-locus technique is of paramount importance in the design of control systems. It provides the designer with considerable insight into the system's stability, performance, and response characteristics. From the general shape of a root-loci diagram, one can get the idea of the type of controller or compensator needed to meet the particular design criterion.

The limitations of the root-locus technique lie in its inability to deal with more than one variable at a time, and the difficulty in dealing with time delays. Time delay or transportation delay cannot be handled conveniently using the standard Evan's rules of construction of root loci.

The root loci may be constructed by computer programming in the MATLAB platform. In order to evaluate the salient features of root-locus plots, such as break-away/break-in points, the intersection point with the imaginary axis, etc. one can develop the relevant program in C language in MATLAB platform as has been done here for better perception. Otherwise, using the Control Tool Box, these features can be evaluated.

Root-Locus Technique

The analytical expression for the break-away point cannot be obtained in MATLAB platform. One has to develop the relevant algorithm to sort out the problem. Each locus starts on a pole and terminates on a zero either at a finite point or at infinity. At overlapping points, multiplicity of root occurs and at that point loci depart. The root loci clashing at a point give the break-away or break-in point. This concept is useful in determining the break-away or break-in point in MATLAB platform for drawing the root loci.

Problems

6.1 Draw the rough sketch of the root loci for a unity-feedback system for each of the following forward transfer functions. Label all the critical points.

(a) $\dfrac{K}{s}$ (b) $\dfrac{K}{s^2}$

(c) $\dfrac{K}{s^3}$ (d) $\dfrac{K}{s(s+2)}$

(e) $\dfrac{K}{s(s+5)(s+9)}$ (f) $\dfrac{K}{s(s+4)(s^2+2s+2)}$

(g) $\dfrac{K}{s(s^2+2s+2)}$

6.2 The position control system with velocity feedback is shown in Figure P.6.2. Draw the root-loci for α varying in the range 0 to ∞. Then determine the value of α so that the damping coefficient of the closed-loop system may be 0.5.

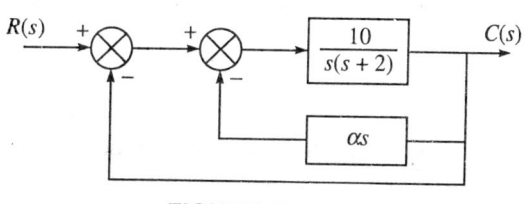

FIGURE P.6.2

6.3 For each of the unity-feedback systems having the transfer function $G(s)$ given below, where the variable parameter K is varying in the positive range from 0 to infinity, draw the root loci. Calculate the values of centroid of the asymptotes, the number of asymptotes and the values of ω at which root loci cross the imaginary axis.

(a) $\dfrac{K}{s(s+5)^2}$ (b) $\dfrac{K}{s(s+1)(s+2)}$

(c) $\dfrac{K(s+1)^2}{s^3}$ (d) $\dfrac{K(s+5)}{s^2(s+3)}$

(e) $\dfrac{K(s+7)}{s(s+1)(s+2)}$ (f) $\dfrac{K(s+2)}{s(s+1)(s+3+j10)(s+3-j10)}$

(g) $\dfrac{K(s^2+2s+25)}{s(s+1)(s^2+4s+25)}$ (h) $\dfrac{K(s+15)8}{(s+5)(s+2+j6)(s+2-j6)}$

6.4 For the open-loop transfer function

$$G(s)H(s) = \frac{K(s+10)}{s(s+5)(s+25)(s+50)}$$

draw the root loci. Determine the value of K for the system having damping coefficient as 0.707, and also find the value of the complex conjugate roots.

6.5 Using the root-locus technique, show how to estimate the roots of the characteristic polynomial

$$s^4 + 100s^2 + 10{,}000 = 0$$

6.6 Draw the the root loci for the system shown in Figure P.6.6.

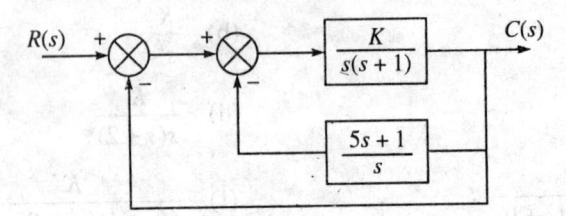

FIGURE P.6.6

6.7 For the open-loop transfer function

$$G(s)H(s) = \frac{K(s+1)(s+5)}{s^2(s+2)}$$

draw the root loci. Determine the values of s at which root loci leave the real axis and re-enter for the system to be underdamped.

6.8 For the open-loop transfer function

$$G(s)H(s) = \frac{K(s+2)}{s(s+1)(s+5)(s+8)}$$

draw the root loci. At some point the imaginary coordinate is $\omega = 1.5$, what is then the real value of s.

6.9 For the open-loop transfer function

$$G(s)H(s) = \frac{K}{s(s+1)(s+2)}$$

draw the root loci. Determine the value of K for the closed-loop pole to be at $s = -5$. Determine the other two complex conjugate poles at this value of K.

6.10 For the open-loop transfer function

$$G(s)H(s) = \frac{K(s+1)}{s(s+2)(s+3)(s+4)}$$

draw the root loci. Determine the roots from the root loci for the value of $K = 10$.

6.11 For the open-loop transfer function

$$G(s)H(s) = \frac{K(s+5)}{s(s+1)}$$

draw the root loci. Then determine the location of roots for $K = 9$. Determine the range of K for which the closed-loop system is overdamped. Repeat the same for the system to be underdamped.

6.12 For a velocity feedback system shown in Figure P.6.12, draw the root loci for velocity constant K varying in the positive range from 0 to ∞. Determine the value of K for the closed-loop system to have damping coefficient as 0.707. Determine the roots at this value of K.

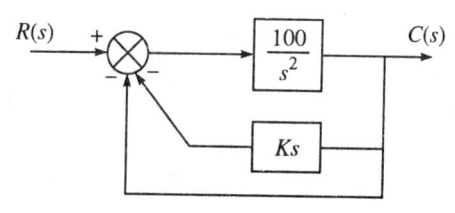

FIGURE P.6.12

6.13 For the system shown in Figure P.6.13, draw the root loci considering z as the variable parameter.

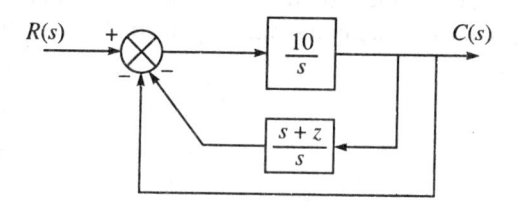

FIGURE P.6.13

6.14 For the system shown in Figure P.6.14, draw the root loci considering K as the variable parameter.

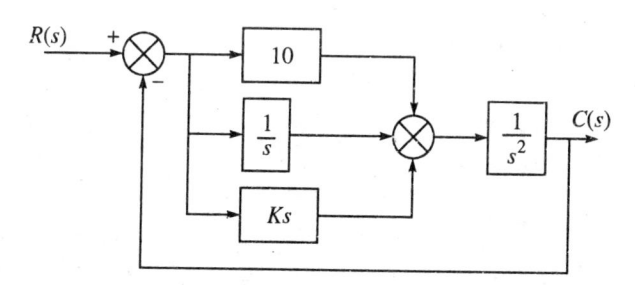

FIGURE P.6.14

6.15 For a unity-feedback system having the open-loop transfer function

$$\frac{20(s^2 + as + 10)}{s^2(s+1)}$$

draw the root loci considering the variable parameter as a.

6.16 From the root loci point of view, show the effect of variable parameter K on the transient response of the system shown in Figure P.6.16.

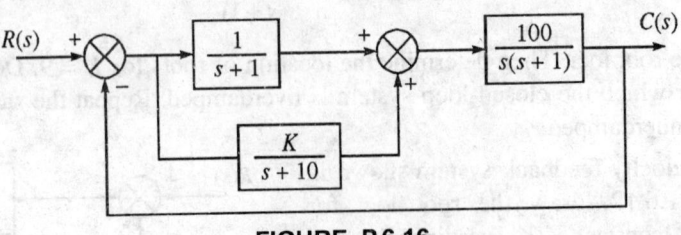

FIGURE P.6.16

6.17 For a unity-feedback system having the open-loop transfer function $G(s)$ as

$$G(s) = \frac{K(s+9)}{s(s^2 + 4s + 11)}$$

plot the root loci. Locate the closed-loop poles on the root loci such that the dominant closed-loop poles have a damping ratio equal to 0.5. Determine the range of gain K for stability.

Process Control System

7.1 Introduction

One of the oldest and best known examples of a typical continuous controller is the original steam engine speed governor used by James Watt as depicted in Figure 7.1.

The engine drives the governor mechanism via a pulley. As the engine speed increases the centrifugal force tends to move the flyweights in an outward direction, thus actuating a lever in such a manner that the steam valve begins to close as the engine speed begins to increase.

As the steam valve can take up any position between fully open and fully closed extremities, such adjustment is termed *continuous control*.

Further, amongst the disciplines of continuous control there are two schools of thought. One school advocates pneumatic control and the other electronic control. However, depending upon the control application, one has to adopt either on-off control or continuous control. In the case of a process with large inertia, the continuous controller cannot react instantaneously to a fast changing input, whereas an on-off controller reacts immediately and brings the control output near to the set value.

Here we will discuss the philosophy of feedback control in process control systems. Any negative feedback control system is obviously a proportional control system. Our objective is to explain the continuous controller such as proportional (P),

OBJECTIVE

By now, we have acquired some amount of knowledge about the classical control systems. The process control terminology though not introduced so far is mainly structured in the light of classical control theory. This chapter provides a glimpse of the design methodology of process control systems so that students get the confidence in handling the same.

CHAPTER OUTLINE

Introduction
Proportional Control Action
Integral Action or Reset
Differential Action
PID Control
Ziegler–Nicholer Rules
Process Characteristics
Ziegler–Nichols Tuning
Controller Design

Proportional and Integral (PI), Proportional, Integral and Derivative (PID) for controlling a process control system. To begin with, we will consider the steam engine speed governor indeed as the process.

7.2 Proportional Control Action

The fly-ball governor was used for controlling the speed of James Watt's steam engine in 1770. After understanding its working, it should be an easy jump to the electronic equivalent, also shown in Figure 7.1.

In Figure 7.2 (the simplified block diagram version of Figure 7.1), the speed of the Watt's steam engine is detected by a tachogenerator which supplies voltage proportional to the speed. This input signal to the controller is compared with the set-value, which can be adjusted

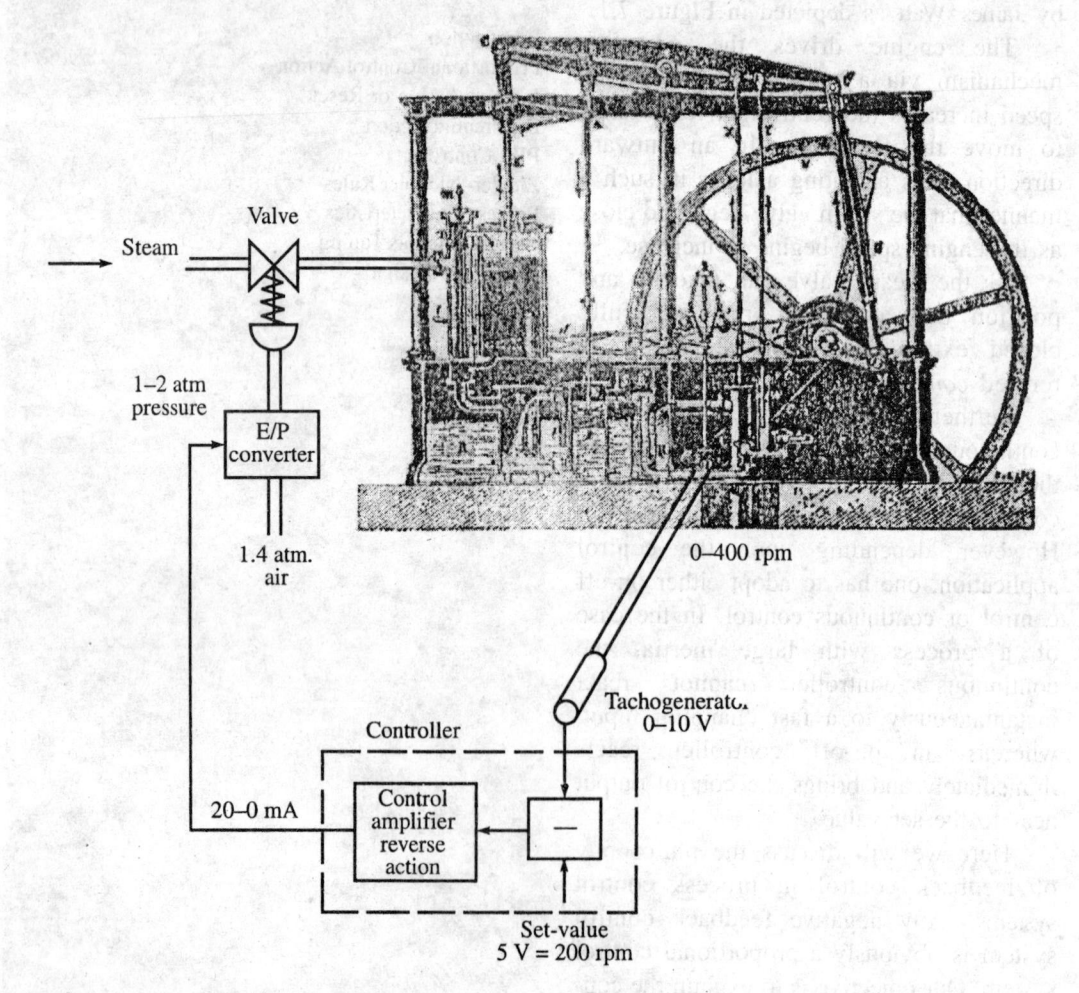

FIGURE 7.1 James Watt's steam engine with electronic speed controller.

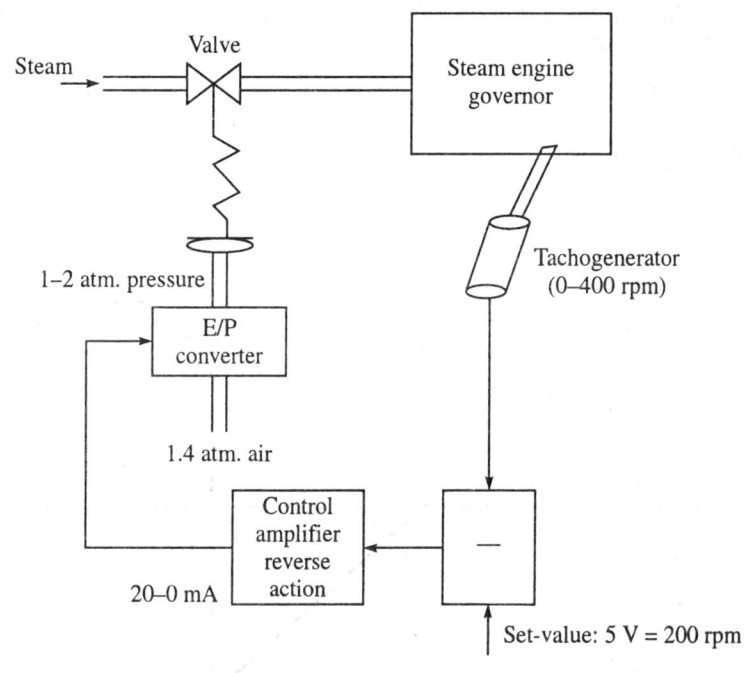

FIGURE 7.2 Simplified version of Figure 7.1.

manually to any required value. If the speed of the engine (the output of the tachogenerator) does not match the required set-value (the voltage represented by the set-point selected), the controller which supplies a correcting signal, moves the pneumatic valve in the steam line to another position.

Let us assume that the controller can supply as output a signal between 0 and 20 mA, the actual value depends on the difference existing between the set-value and the actual speed of the engine. Let us assume that the set-point speed of the engine is 200 rpm, and that the maximum permissible speed is 400 rpm. We adjust the controller in such a way that if the actual speed of the engine (i.e. the input to the controller) and the set-point coincide (i.e. no deviation), then the output signal of the controller will be 10 mA. This value of 10 mA is chosen in order to have sufficient correcting control action in both directions. This 10 mA is converted into a 0.6 kg/cm^2 gauge pressure signal by an electro-pneumatic (E/P) converter and then used to position a pneumatic valve to 50% opening. If, due to increased load, the speed of the engine decreases, the tachogenerator will produce less voltage and the controller will detect too low a signal at its input (negative deviation). The result will be that the controller will supply an increased output signal to the E/P converter. This increased signal will result in a higher pressure on the valve and this will increase its opening, allowing more steam to flow to the engine. The increased steam supply will again increase the speed of the engine. In principle, this is continuous proportional control and forms the basis of all feedback control schemes.

The control action may be either *reverse* or *direct*. When the controller's decreased input results in increased output, this is termed *reverse action*. However, there exist many applications which require that the increased controller's input should result in increased controller output. This is termed *direct action*.

Referring to Figure 7.3, we have the control valve with reverse control action where we assumed that at maximum speed (400 rpm) the controller will tend to close the steam valve completely and that at the nominal speed of 200 rpm (= controller set-point) the steam valve will be 50% open and that at 0 rpm the controller will tend to completely open the steam valve.

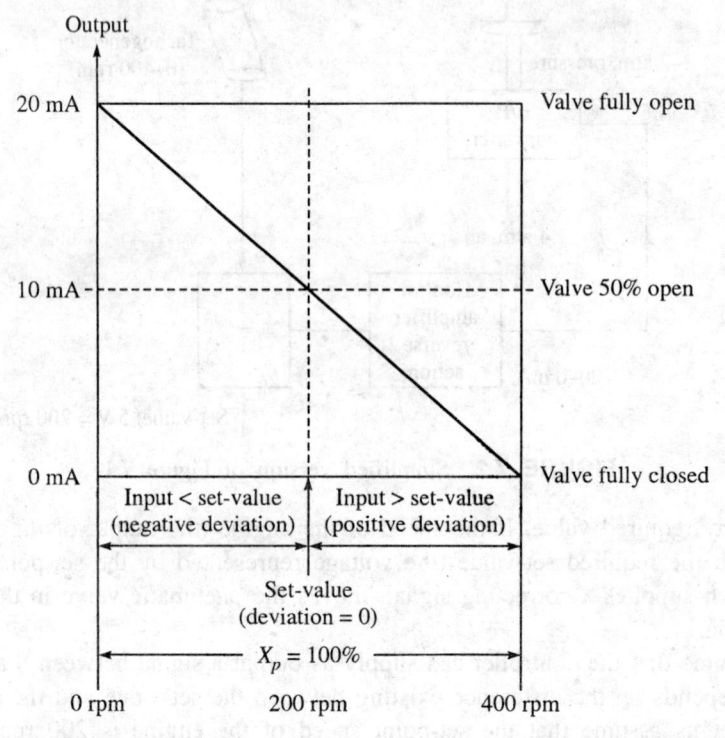

FIGURE 7.3 Control valve with reverse control action.

However, for different operating requirements, it might be necessary to have the valve fully closed at 300 rpm and to have it fully opened at 100 rpm. In this case, the output of the controller should be 0 mA at 300 rpm and 20 mA at 100 rpm. This is illustrated in Figure 7.4. The proportional action line of the controller is now steeper. In other words, the proportional band has become narrower (X_p small).

For some applications it might be necessary that the valve be never completely closed or never completely opened. In this case, for example, the output should be only 15 mA at minimum speed and 5 mA at maximum speed. In this case the proportional band becomes wider (X_p large) as illustrated in Figure 7.4.

In Figure 7.3, the proportional band is 100%, whereas in Figure 7.4, the proportional band is 50% as indicated by the continuous line and 200% as indicated by the dotted line.

It is also possible to make the proportional band 0% as shown in Figure 7.5(a) when the smallest deviation from the set-point would result in sudden complete opening or complete closing of the valve. This becomes the ON-OFF control. At the other extreme, the proportional

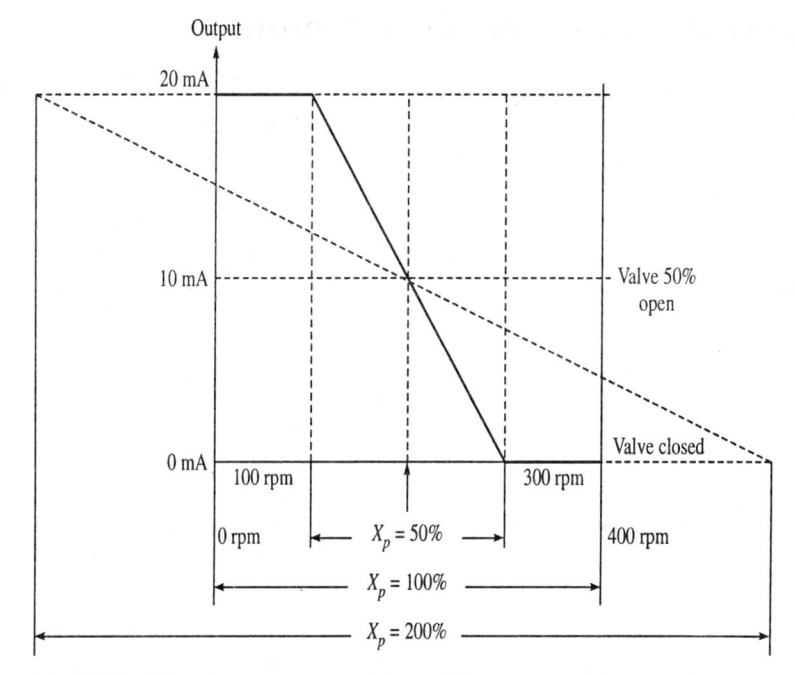

FIGURE 7.4 Control valve with a different operating requirement.

band would be infinite as shown in Figure 7.5(b), where, whatever the deviation may be from the set-point, the output of the controller would remain 10 mA. Thus there would be no control action at all.

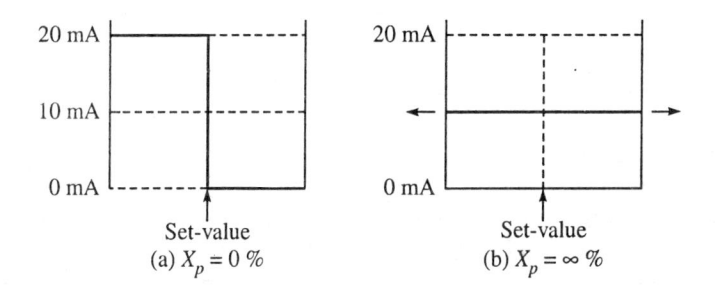

FIGURE 7.5 Proportional band: (a) ON-OFF control and (b) Nil control.

The proportional band is commonly used to refer to the sensitivity or gain of the proportional controller. Proportional band is the % of change in the input to the controller which is the error signal, required to cause 100% change in the output of the actuator. Thus a small (high) proportional band means high (small) gain of the controller or the controller with high (low) sensitivity. Thus a gain of 4%/1% means that there is a change of 4% in the output if the change in the input is 1%. Then

$$\text{Proportional band} = \frac{100\%}{\text{gain in percentage}/\%} = \frac{100\%}{4\%/1\%} = 25\%$$

7.2.1 Difficulty with proportional control

Proportional control seems to be most ideal, however, there is one major disadvantage invariably coupled with proportional control, i.e. offset or proportional offset. Under nominal load conditions, steam engine will run at 200 rpm when the steam valve is 50% opened (controller output is 10 mA). If the load on the engine is increased, the speed will drop. This results in negative control deviation which, in turn, will force the controller to increase its output signal because of the reverse action in the controller. Consequently, the steam valve will be opened wider and the engine will accelerate again. This, however, has the effect that the speed comes closer to the set-value and the control deviation decreases. If the proportional controller could manage to restore the process to the set-value, this would mean that the deviation would become zero again, which would however by definition coincide with an output signal of 10 mA. The increase in speed due to the larger output signal of the controller reduces the control deviation. The controller sensing this, tends now to close the valve again.

In fact, a proportional-controller can only supply sufficient steam to the engine for as long a duration as the load remain nominal. Any other load condition will force the controller to seek a new operating point along its proportional action line, automatically resulting in a deviation from the required set-point. This deviation is termed *proportional offset* or just *offset*. This offset increases in magnitude as the load conditions of the process differ more and more from their nominal value. The offset also increases as the width of the proportional band of the controller is made greater.

7.3 Integral Action or Reset

Offset is inherent in a proportional controller, which will either give too much or too little output as soon as the load conditions depart from nominal.

Therefore the P-controller must be equipped with additional means in order to get rid of this undesirable effect. We have to consider the inclusion of the unit which supplies sufficient additional output current to the control valve so that the offset is eliminated and the control deviation becomes zero. Let us try to describe this in detail with the help of Figure 7.6.

Due to an increased load on the engine, the P-controller tries to supply sufficient output current so that the set-value speed could be reached again. Let us assume that this would require 19 mA output. Due to its characteristics the P-controller cannot succeed in reaching this and will stabilize the process at 15 mA [Figure 7.6(b), resulting in a proportional offset of 4 mA (which is equivalent to loss of some engine speed).

The additional unit (the integral or I-unit) will also detect the control deviation and will start to supply an ever increasing current to the control valve for as long as a deviation exists. The larger the deviation, the faster the increase in output current will be [Figure 7.6(d)].

This additional current will give the valve additional opening, resulting in an increased engine speed and a decreased deviation. This, in turn, will cause the output current of the P-controller to become smaller [Figure 7.6(c)]. After a certain period, due to the steadily increasing current from the I-unit, the offset will get nearer to zero and the P-controller will again reach its nominal 10 mA output current. In the meantime, the I-unit has constantly increased its output current and already produces some 9 mA. As soon as the set-point speed of the engine is reached, the I-unit will stop increasing the output current. Then, the sum of the currents from P and I-units would be exactly 19 mA, which is the current required to reach the set-point speed under the new load conditions [Figure 7.6(e)]

(a) — Process deviation / Proportional offset

(b) Output of controller 15 mA / 10 mA — Controller P-action (No I-action)

(c) 10 mA / 19 mA — Controller P-action (I-action available)

(d) 0 mA — Controller I-action

(e) 19 mA / 10 mA — Controller P + I action

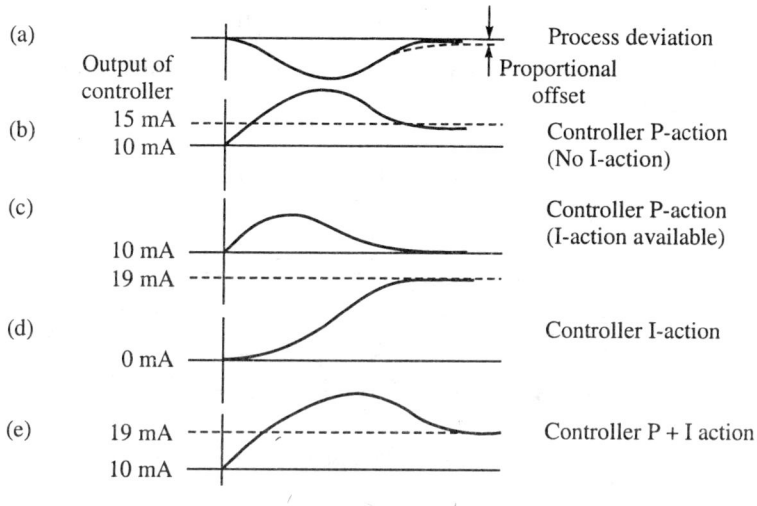

FIGURE 7.6 Response of P-I controller.

The response fully depends on the speed of the increase in current from the I-unit, i.e. whether it really stops at zero offset or whether the current continues to increase for a while. In the latter case we will be confronted with an overshoot and the same phenomenon takes place all over again in the other direction. It is important to understand that the combined action of P and I-unit, if adjusted to optimal values, can result in a smooth control action without overshoot and without proportional offset.

Briefly recapitulating, we see that:

- The P-unit tries to correct the position of the valve rapidly and proportionally to the control deviation but does not succeed completely.
- The I-unit starts to give a steadily increasing current as long as a control deviation continues to exist. The larger the deviation, the faster the increase or decrease in extra current. In the new balanced situation, the I-unit together with the 10 mA from the P-unit will supply all the current (positive or negative) necessary to form the required total output current to eliminate the control deviation.
- The more I-action (the faster the increase in current per time unit for a given deviation value), the less stable the control loop. At one extreme the current increases almost instantaneously to its maximum regardless of the magnitude of the deviation; on the other extreme the I-unit will not produce any current regardless of the size of deviation. Somewhere in between will be the optimum setting. For practical reasons, the value for the "I-action" is usually denoted by its reciprocal value $1/I$—action which is called T_i. Thus T_i represents the time interval necessary for integral action to cause

the same amount of correcting movement as that produced by proportional action due to a steady deviation from the set-value. This time interval T_i is usually expressed in minutes. Thus,

$$T_i = \infty \text{ means no I-action}$$

$$T_i = 0 \text{ means infinite I-action}$$

For steady process control, a continuous controller should include both types of control actions, i.e. P and I as shown in Figure 7.7.

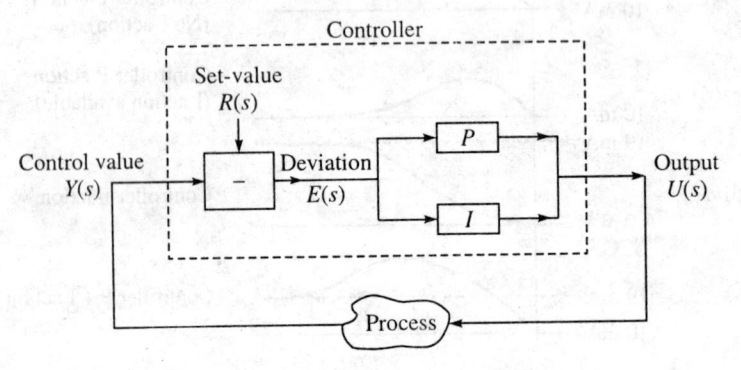

FIGURE 7.7 Continuous controller.

7.4 Differential Action: Derivative or Rate Control

As long as the process is running under steady-state conditions, the PI-controller will do fine. Even if the process disturbances occur, the PI-controller will handle these very well. However, the wider the proportional band (X_p) and the smaller the integral action (T_i), the less responsive the control system will become. Depending on the characteristics of the process, for instance, when running under steady-state conditions with small disturbances, the optimal PI setting would give optimum results, but as soon as too large a disturbance occurs, then the PI setting would be too slow. This especially can occur under starting-up conditions.

What in fact is required, is the extra amount of push in the beginning. As long as the speed of the steam engine is far below the set-value, it does no harm to give the control valve an additional lift; much more than is actually required for the set-value speed. This would result in the engine obtaining an excess of steam, which would accelerate the engine rapidly, enabling it to reach the set-value sooner. This situation must not, however, last longer than just a few moments, otherwise the speed of the engine might become too high and we would be confronted with an overshoot again. This extra push in the beginning is given by the derivative unit. It detects at its input the change in increase or decrease of the deviation. Thus it is not the magnitude of the deviation as such (as was the case with the P and the I control mode), but the speed with which the deviation value increases or decreases is what counts. The higher this speed (the quicker the control value runs away from the set-value) the higher the extra push signal or the D-action at the output as shown in Figure 7.8. This implies, on the other hand,

Process Control System

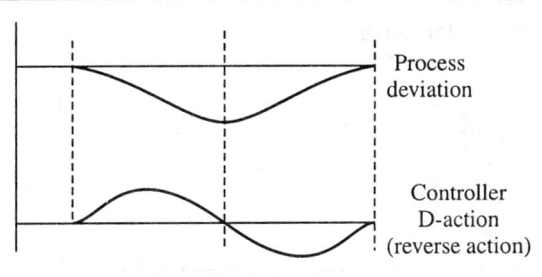

FIGURE 7.8 D-action.

that an existing deviation which remains constant will not cause any D-action. D-action is normally denoted as T_d in time-units and thus $T_d = 0$ implies no D-action and $T_d = \infty$ implies infinite D-action.

7.5 PID Control

In order to obtain optimal control for different processes, a continuous controller should include three different control modes: proportional action P, integral action I, and derivative action D as shown in Figure 7.9. The controller is of PID type. In Figure 7.10, it is shown how the process will react when:

(a) The controller acts as P-controller ($T_i = \infty$, $T_d = 0$)
(b) The controller acts as PI-controller ($T_d = 0$ and optimal setting for X_p and T_i)
(c) The controller acts as PID-controller (optimal settings for X_p, T_i and T_d)

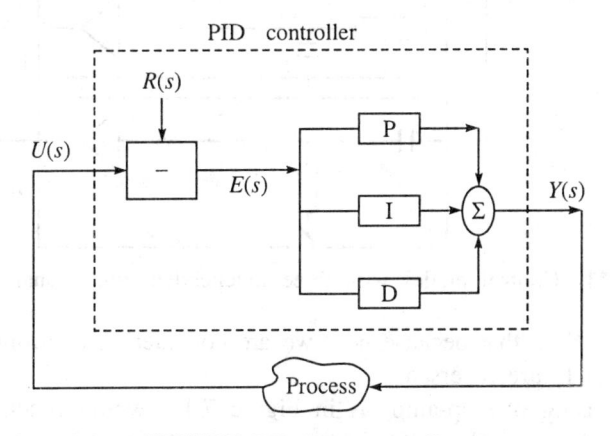

FIGURE 7.9 PID controller.

All three modes should be separately adjustable. It would require three independent operational amplifiers with control modes as shown in Figure 7.11. This scheme is costly and requires complex electronic circuitry. However, it is possible to achieve the same result cost-effectively by using one operational amplifier with the control modes as feedback circuit, as

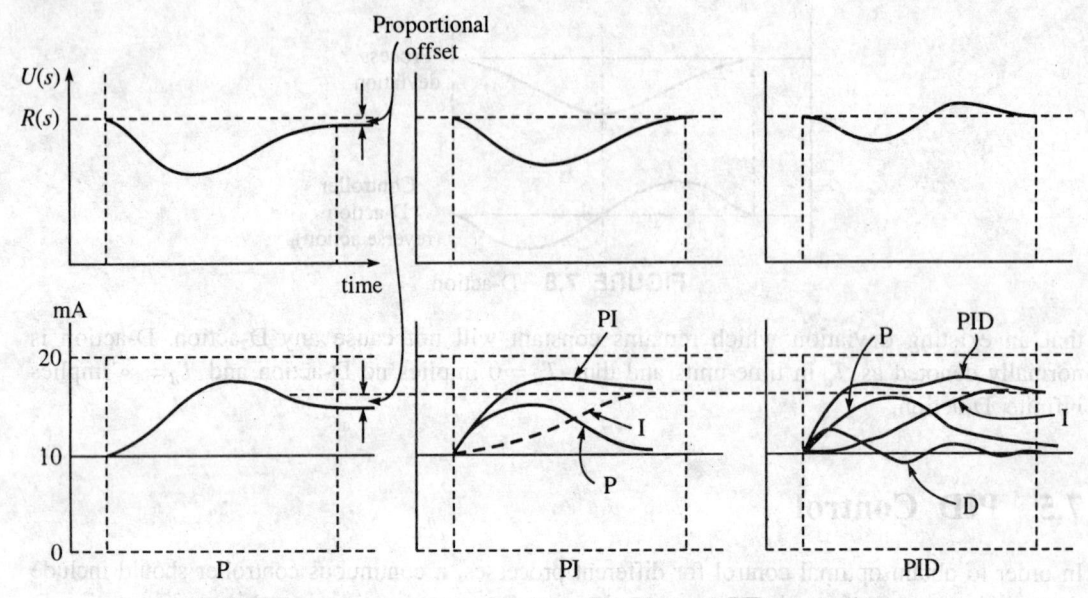

FIGURE 7.10 Process response to P-, PI-, and PID controllers.

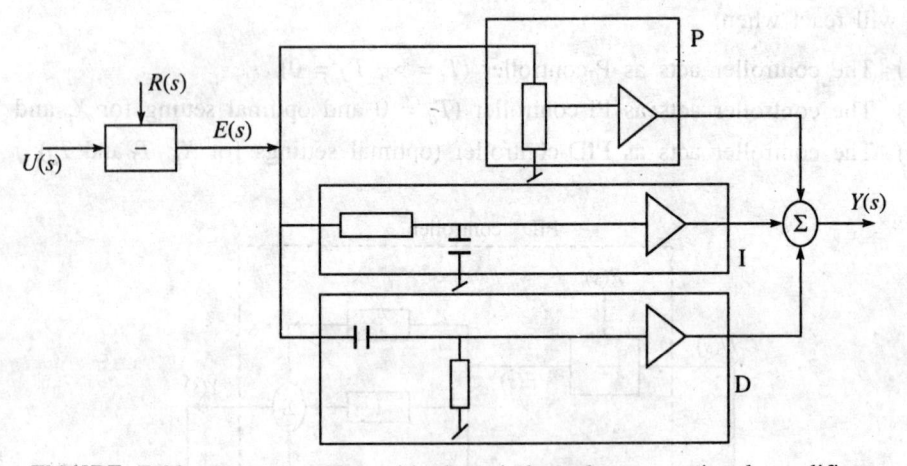

FIGURE 7.11 Control modes with three independent operational amplifiers.

shown in Figure 7.12. Note that because now we are considering controller in feedback, the *RC* networks for I and D are reversed.

The PID-control using one op-amp, as in Figure 7.12, would function well in normal operating conditions. However, the difficulty arises in the following cases:

(1) At start-up, when it takes a long time before the control valve reaches the set-value.

(2) When there is a sudden increase or decrease in the set-value, which is equivalent to case (1).

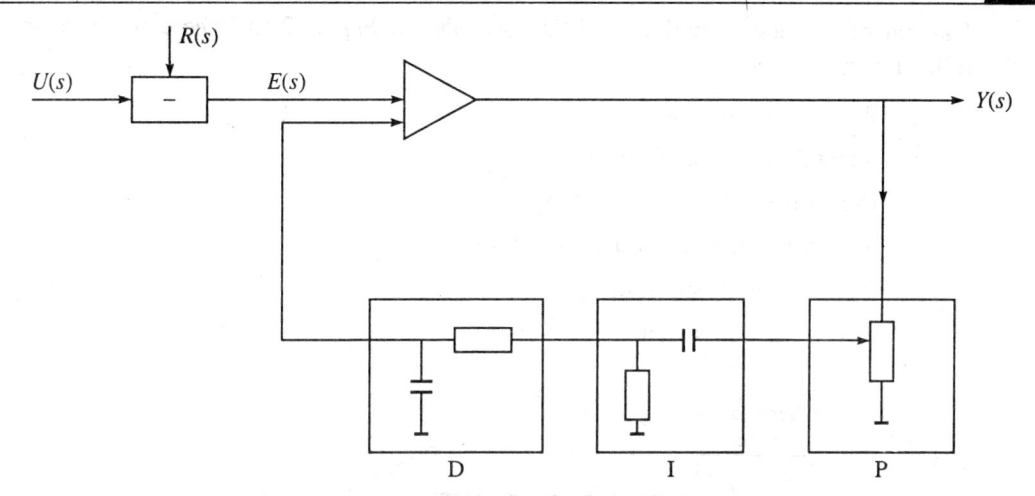

FIGURE 7.12 Control modes as feedback circuit.

Bearing in mind that we are now operating with a feedback arrangement, the P, I and D unit can no longer react to the control deviation as appearing at the input of the controller but can only be influenced by the output signal of the controller. A process normally requires a long period of time from start-up to reach the set-value. Due to the persistence of the control deviation, the controller will, after a while, produce maximum output (20 mA). This is due to the fact that the control deviation is larger than the proportional band and due to the action of the I-unit which has been given plenty of time to reach 20 mA. This saturated condition will continue until the process actually has reached the set-value. Only then will the controller start to reduce its output signal (much too late) and then only will the P, I and D unit be able to detect that some thing has happened at the input of the controller. The result will be that a serious overshoot would occur. In other words, if the deviation conditions last too long, the P, I and D unit will be unable to initiate corrective action.

There are two ways to overcome this drawback:

1. During start-up the I-action is switched off manually or automatically. By doing so, the I-unit cannot drive the controller output to its saturated condition, and thus the proportional action remains active.

2. The D-unit is not included in the feedback circuit but operates through a separate amplifier directly on the deviation signal at the input circuit of the controller. Note that the input signal to the controller keeps changing as long as the control valve does not reach the set-point. We agreed that the output signal to the controller during this period cannot change because it produced full scale signal (20 mA) Thus, with the D-unit acting in the input section, the PI section of the controller receives the sum of deviation signal at its input. Under normal running conditions, this works as decried before but at start-up conditions it offers a real advantage. We know that as long as the control valve changes, the D-unit will produce a signal. This signal will result in a breaking action, an early reduction in controller output signal, so that the control valve can reach the set-value smoothly without undue overshoot. We finally come to the definite block

diagram of a modern continuous PID controller in Figure 7.13, where we have the following components:

- Proportional unit with X_p-setting
- Integral unit with T_i-setting
- Derivative unit with T_d-setting
- Switch to eliminate I-action (start-up)
- Rate before Reset (start-up)
- Reverse action switch

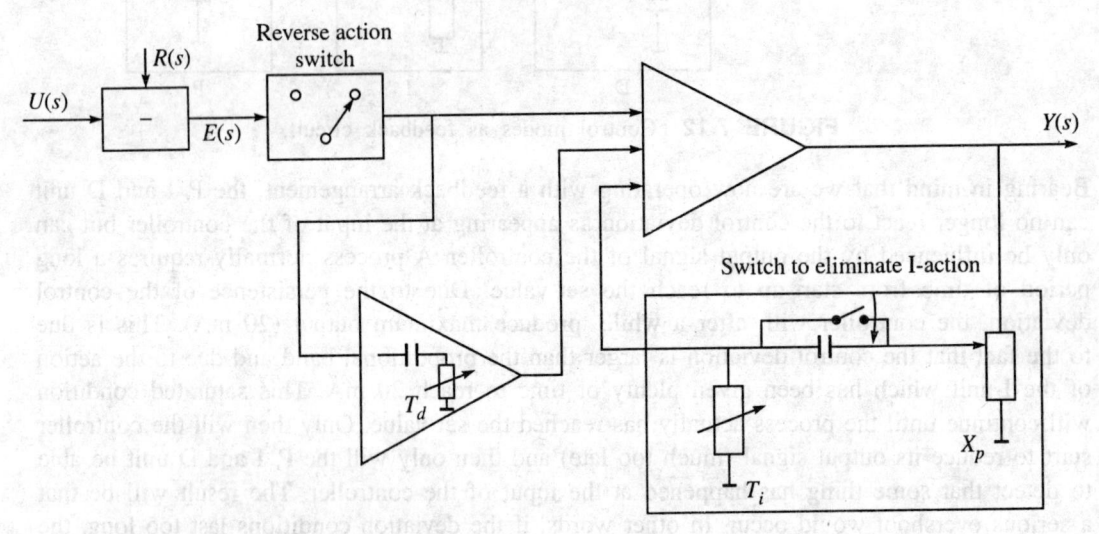

FIGURE 7.13 Block diagram of a modern PID controller.

The response of ideal controllers to standard inputs is given in Table 7.1.

7.5.1 Tuning rules for PID controllers

The PID controllers are very frequently used in industrial control systems:

Proportional (P) action   : $u(t) = K_p e$ \hfill (7.1)

Integral (I) action   : $u(t) = K_i \int e \, dt$ \hfill (7.2)

Derivative (D) action   : $u(t) = K_d \dfrac{de}{dt}$ \hfill (7.3)

(P + I + D) action   : $u(t) = K_p e + K_i \int e \, dt + K_d \dfrac{de}{dt}$ \hfill (7.4)

where, K_p is the proportional gain, K_i is the integral gain, and K_d is the derivative gain.

TABLE 7.1 Response of ideal controllers to standard inputs

Standard input	Impulse	Step	Ramp	Sinusoidal
Proportional $\tilde{m} = K_c \tilde{e}$	$\frac{1}{K_c}$, ∞	$\frac{1}{K_c}$	ramp	$\tilde{e}$, $K_c\tilde{e}$
Integral or reset $\tilde{m} = \frac{1}{\tau_i}\int_\theta^t \tilde{e}\,dt$	$\frac{1}{\tau_i}$	ramp	parabolic	sinusoid
Derivative or rate $\tilde{m} = \tau_d \frac{d\tilde{e}}{dt}$	τ_d, ∞, $-\infty$	∞	τ_d	sinusoid
Proportional + Reset $\tilde{m} = K_c(\tilde{e} + \frac{1}{\tau_i}\int_\theta^t \tilde{e}\,dt)$	K_c, $\frac{1}{\tau_i}$	K_c	ramp	sinusoid

In this case K_p, K_i and K_d become the controller parameters, that is, they are constants (adjustable on controller).

Let X_p be the proportional band where

$$K_p = \frac{1}{X_p} \tag{7.5}$$

The transfer function $G_c(s)$ of the PID controller is

$$G_c(s) = \frac{Y(s)}{E(s)} = K_p\left(1 + \frac{1}{T_i s} + T_d s\right) \tag{7.6}$$

And the output of the PID controller can be written as

$$y = \frac{1}{X_p}e + \frac{1}{X_p T_i}\int e\,dt + \frac{T_d}{X_p}\frac{de}{dt} \tag{7.7}$$

where
  y is the output signal from the PID controller
  e is the control deviation
  X_p is the proportional band (= $1/K_p$) (7.8)
  T_i is the integral action time (= K_i/X_p) (7.9)
  T_d is the derivative action time (= $K_d X_p$) (7.10)

It is noted that in actual PID controllers, instead of adjusting the proportional gain K_p, we adjust the proportional band X_p. The proportional band is $1/K_p$ and is expressed in per cent, for example, 25% band corresponds to $K_p = 4$. The control action of individual controllers is given in Table 7.2.

TABLE 7.2 P-, I-, and D-control actions

	Control action	Main purpose		Stability	Reaction time control loop
X_p (%) P	A control deviation will produce controller output proportional to the magnitude of the deviation	Produces bulk of corrective effort	$X_p >$ (proportional band larger)	better	slower
			$X_p <$ (proportional band smaller)	worse	quicker
T_i (time) I	The moment a control deviation exists, the controller starts to produce an ever-increasing output current. The larger the deviation, the faster this increase or decrease in output current	Elimination of proportional offset	$T_i >$ (I-action smaller)	better	slower
			$T_i <$ (I-action larger)	worse	quicker
T_d (time) D	Any *change* in control deviation will *result* in an output, proportional to the speed at which the deviation is changing	Extra push for quick return to set-value	$T_d >$ (D-action larger)	worse	quicker
			$T_d =$ optimal	better	quicker
			$T_d <$ (D-action smaller)	worse	slower

Figure 7.14 shows an electronic PID controller using operational amplifiers. The transfer function $E(s)/E_i(s)$ is given by

$$\frac{E(s)}{E_i(s)} = -\frac{Z_2}{Z_1} = -\left(\frac{sR_2C_2+1}{sC_2}\right)\left(\frac{sR_1C_1+1}{R_1}\right) \quad (7.11)$$

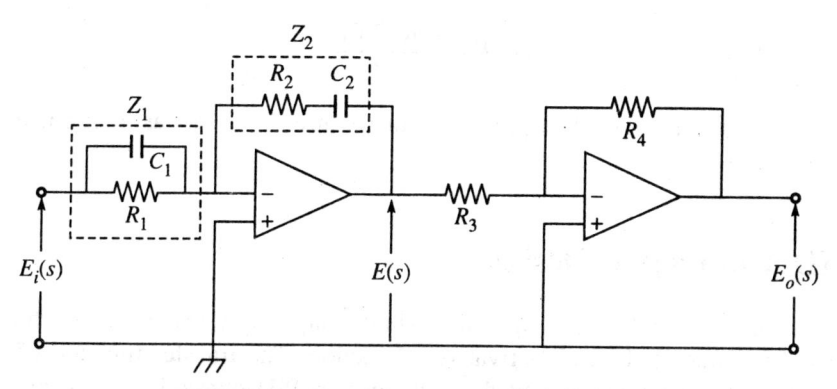

FIGURE 7.14 Electronic PID controller.

where
$$Z_1 = \frac{R_1}{sR_1C_1 + 1} \quad \text{and} \quad Z_2 = \frac{sR_2C_2 + 1}{sC_2}$$

Further noting that

$$\frac{E_o(s)}{E(s)} = -\frac{R_4}{R_3} \tag{7.12}$$

we have

$$\frac{E_o(s)}{E_i(s)} = \frac{E_o(s)}{E(s)} \frac{E(s)}{E_i(s)}$$

$$= \frac{R_4(R_1C_1 + R_2C_2)}{R_1R_3C_2}\left[1 + \frac{1}{(R_1C_1 + R_2C_2)s} + \frac{R_1C_1R_2C_2}{R_1C_1 + R_2C_2}s\right] \tag{7.13}$$

Thus comparing with Eq. (7.6), we get

$$K_p = \frac{R_4(R_1C_1 + R_2C_2)}{R_1R_3C_2} \tag{7.14}$$

$$T_i = R_1C_1 + R_2C_2 \tag{7.15}$$

$$T_d = \frac{R_1C_1R_2C_2}{R_1C_1 + R_2C_2} \tag{7.16}$$

Using Eq. (7.7), we can rewrite in terms of the proportional gain, integral gain, and derivative gain as

$$K_p = \frac{R_4(R_1C_1 + R_2C_2)}{R_1R_3C_2} \tag{7.17}$$

$$K_i = \frac{R_4}{R_1R_3C_2} \tag{7.18}$$

$$K_d = \frac{R_2 R_4 C_1}{R_3} \qquad (7.19)$$

Note that the second operational amplifier circuit (Figure 7.14) acts as a sign inverter as well as a gain adjuster.

7.6 PID Controller Design

We have seen that controllers for single-input-single-output systems consist of three elements: proportional (P), Integral (I), and Derivative (D) action. The transfer function of a controller which includes all three terms is called the three-term PID controller, given by

$$G_c = K_p \left(1 + T_d s + \frac{1}{T_i s} \right)$$

where, K_p, T_d, and T_i, have usual meaning as already explained.

Based on the three-term PID controller, there may be derived a number of others controllers. The majority of the industrial control elements are of the P or PI-type. These controllers are derived from the three-term PID controller $G_c(s)$ by making adjustments to T_d and T_i as

$T_d = 0$ and $T_i = \infty$ gives a P-controller

$T_d = 0$ and $T_i =$ finite gives a PI-controller

Commercially available pneumatic or electronic controllers may be of non-interacting or interacting type depending on the principles of action. Only the derivative action is never implemented in practice because of noise problem. It may be noted that interacting controller means that an adjustment of any one parameter affects the other parameters, whereas non-interacting means the other way round.

7.7 Ziegler–Nichols Rules for Controller Tuning

When the process transfer function model is available, the following results can be used. They are named after J.G. Ziegler and Nathaniel Burgess Nichols, who developed them in the 1940s.

7.7.1 First approach

The process to be controlled is as shown in Figure 7.15. Under pure proportional control, the system is asymptotically stable in the range $0 \leq K_p < K_c$, and goes unstable in an oscillatory manner when $K_p > K_c$. Now observe the following:

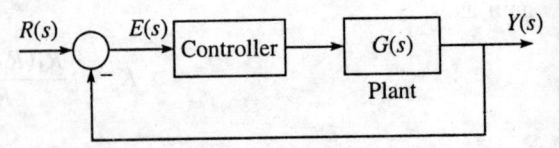

FIGURE 7.15 Controller tuning with provisions for P, PI, and PID controller settings.

1. Increase the gain K_p from 0 to K_c (decrease the proportional band X_p until the process starts to oscillate). At this critical gain K_c the closed-loop system is marginally stable so any gain adjustments must be carried out with extreme care. (If the output does not exhibit sustained oscillations for whatever value K_p may take, then this method does not apply).
2. Note the value K_c and the period of oscillation T.
3. The recommended settings of K_p, T_i, and T_d are given in Table 7.3 for different types of controller design.

TABLE 7.3 Ziegler–Nichols tuning rule based on critical gain K_c and critical period T

Type of controller	K_p	T_i	T_d
P	$0.5K_c$	∞	0
PI	$0.45K_c$	$0.83T$	0
PID	$0.6K_c$	$0.5T$	$0.125T$

Note that the PID controller as per Eq. (7.6) gives

$$G_c(s) = K_p\left(1 + \frac{1}{T_i s} + T_d s\right) = 0.6K_c\left(1 + \frac{1}{0.5Ts} + 0.125Ts\right)$$

$$= 0.075K_c T \frac{(s + 4/T)^2}{s} \qquad (7.20)$$

Thus, the PID controller has a pole at the origin and double zeros at $s = -4/T$.

As the transfer function model of the plant is available, Routh's array may be used to establish the critical gain K_c and the corresponding period of oscillation T. The procedure is:

1. Find the system's closed-loop characteristic equation under pure proportional control.
2. Form Routh's array and establish the critical gain K_c that produces an all-zero row. If the system goes unstable in an oscillatory manner, the all-zero row will be the row associated with s^1, the auxiliary polynomial will be of second order and there will be no roots of the remainder polynomial with positive real parts. Note that the system should remain stable for all positive values of K_p below the critical value.
3. Use the auxiliary polynomial to find the period of oscillation T, and apply the recommended settings given above.

EXAMPLE 7.1 As an illustration, the use of Ziegler–Nichols rules for finding the P, PI and PID controller settings for a plant whose transfer function model is available, is best understood by the following example. The unity-feedback system of Figure 7.15 having the plant transfer function

$$G(s) = \frac{6}{(s+1)(s+2)(s+3)}$$

and the controller in the forward path has provisions for P, PI and PID controller settings. Perform the controller settings and find the corresponding closed-loop transfer function and the response of the closed-loop system subjected to a unit-step input.

Solution: The characteristic equation of the closed-loop system under pure proportional control having proportional controller gain K_p as per the procedure discussed earlier, is

$$1 + K_p G(s) = 0$$

$$F(s) = 1 + G(s)H(s) = 1 + \frac{6 \times K_p}{(s+1)(s+2)(s+3)} = 0$$

or

$$s^3 + 6s^2 + 11s + 6(1 + K_p) = 0$$

The Routh array is:

s^3	1	11
s^2	6	$6(K_p + 1)$
s^1	$11 - (K_p + 1)$	
s^0	$6(K_p + 1)$	

The range of stability is $0 < K_p \leq 10$. The critical gain $K_c = 10$. The auxiliary polynomial is

$$6s^2 + 6(10 + 1) = s^2 + 11 = 0$$

Then the frequency of oscillation ω is obtained as

$$\omega = \sqrt{11}$$

The critical period, $T = 2\pi/\omega = 1.895$.

Therefore, the recommended settings as per Table 7.3 are as follows:

For P-control, $K_p = 0.5 K_c = 0.5(10) = 5$, which gives the closed-loop transfer function as

$$\frac{Y(s)}{R(s)} = \frac{G'}{1 + G'} = \frac{30}{s^3 + 6s^2 + 11s + 36}$$

where

$$G' = K_p G(s) = \frac{30}{(s+1)(s+2)(s+3)}$$

For PI-control, as per Table 7.3, $K_p = 0.45 K_c = 4.5$ and $T_i = 0.83T = 0.83(1.895) = 1.572$, which gives the closed-loop transfer function as

$$\frac{Y(s)}{R(s)} = \frac{G'}{1 + G'} = \frac{42.5s + 27}{1.572 s^4 + 9.434 s^3 + 17.3 s^2 + 51.89 s + 27}$$

where

$$G' = K_p \left(1 + \frac{1}{T_i s}\right) G(s) = \frac{4.5\left(1 + \dfrac{1}{1.572 s}\right) \times 6}{(s+1)(s+2)(s+3)}$$

For PID-control, $K_p = 0.6K_c = 0.6(10) = 6$, $T_i = 0.5T = 0.5(1.895) = 0.947$ and
$$T_d = 0.125T = 0.125(1.895) = 0.237$$
which gives the closed-loop transfer function as
$$\frac{Y(s)}{R(s)} = \frac{G'}{1+G'} = \frac{8.525s^2 + 34.1s + 36}{0.947s^4 + 5.683s^3 + 18.5s^2 + 39.78s + 36}$$
where
$$G' = K_p\left(1 + \frac{1}{T_i s} + T_d s\right)G(s) = \frac{6\left(1 + \frac{1}{0.947s} + 0.237s\right) \times 6}{(s+1)(s+2)(s+3)}$$

Figure 7.16 shows the unit-step response under P, PI and PID control. Under proportional control, the response is relatively fast, and for reducing the steady-state error, a PI controller may be used. The introduction of integral action reduces the stability of the system. The introduction of derivative action has a stabilizing effect on the plant, but derivative action cannot be used if the controller is being fed with measurement noise (spurious signals).

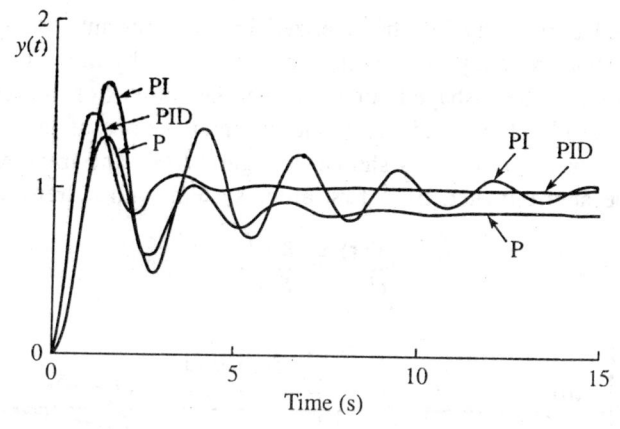

FIGURE 7.16 Time response under P PI, and PID control.

However, if the plant is complicated such that its mathematical model cannot be easily obtained, then analytical approach to the design of the PID controller is not possible. Then we must resort to experimental approaches to the design of PID controllers.

7.7.2 Second approach

Ziegler–Nichols tuning rules have been widely used to tune PID controllers in process control systems where the plant dynamics are not precisely known. Over many years, such tuning rules proved to be very useful. Ziegler–Nichols tuning rules can, of course, be applied to plants whose dynamics are known. Further, it may be noted that if plant dynamics are known, many analytical and graphical approaches to the design of PID controllers are available, in addition to Ziegler–Nichols tuning rules.

The test procedure is carried out as follows:

1. Obtain the unit-step transient response for the open-loop plant shown in Figure 7.17. If the plant involves neither the integrator nor the dominant complex conjugate poles, then a unit-step response curve may look like an S-shaped curve as shown in Figure 7.18. It may be noted that if the response does not exhibit an S-shaped curve, this method does not apply.

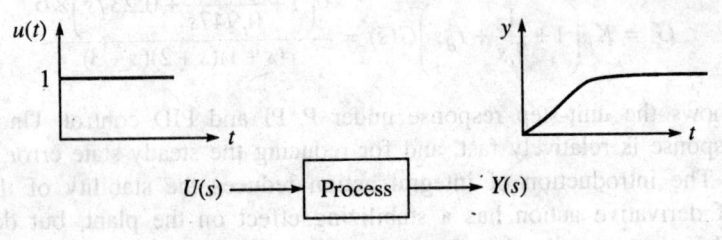

FIGURE 7.17 Open-loop plant.

2. The S-shaped curve may be characterized by the constants, delay time L and time constant T. Measure delay time L and time constant T by drawing a tangent line at the inflection point of the S-shaped curve (so that the slope R of the response becomes the maximum possible slope) and determine the intersections of the tangent line with the time axis and line $y(t) = K$, as shown in Figure 7.18. The transfer function $Y(s)/U(s)$ may then be approximated by a first-order system with a transportation lag. Then

$$\frac{Y(s)}{U(s)} = \frac{Ke^{-Ls}}{Ts+1} \tag{7.21}$$

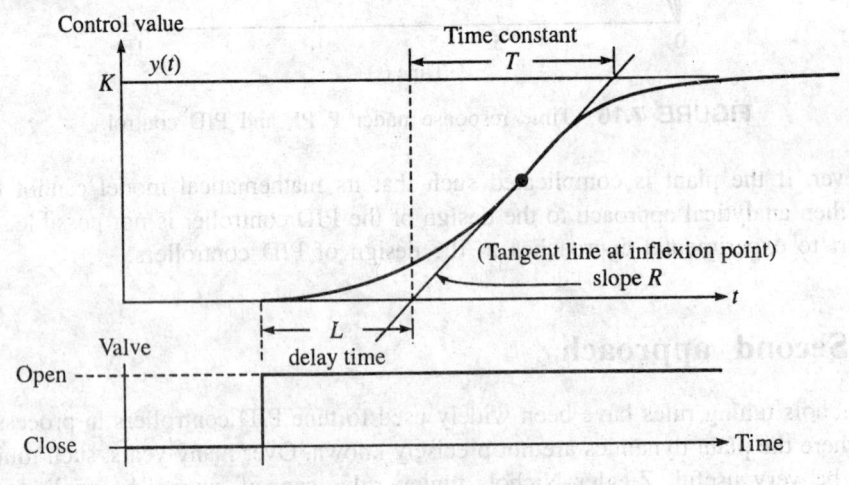

FIGURE 7.18 Unit-step response curve.

The transient response of a process to a step change of input is defined by two characteristics—delay time L and time constant T. Both characteristics together, more or less, define whether the controller can keep the systems within limits. As long as time constant T is relatively large compared with delay time L, the process is easily controllable with a simple controller. Difficulty arises with processes where the delay time L is large compared with time constant T.

3. The recommended Ziegler and Nichols settings to set the values of K_p, T_i and T_d are according to the formulae shown in Table 7.4.

TABLE 7.4 Ziegler–Nichols tuning rule based on step response of plant

Type of controller	$K_p = 1/X_p$	T_i	T_d
P	T/L	∞	0
PI	$0.9(T/L)$	$L/0.3$	0
PID	$1.2(T/L)$	$2L$	$0.5L$

Note that the PID controller as per Eq. (7.6) gives

$$G_c(s) = K_p \left(1 + \frac{1}{T_i s} + T_d s\right)$$

$$= \frac{1.2T}{L}\left(1 + \frac{1}{2Ls} + \frac{Ls}{2}\right)$$

$$= 0.6T \frac{\left(s + \frac{1}{L}\right)^2}{s} \quad (7.22)$$

Thus the PID controller has a pole at the origin and double zeros at $s = -1/L$.

7.8 Adjustment According to Process Characteristics in Ziegler–Nichols Method

Before being able to use any formula we have to measure the reaction of the complete loop to a disturbance. In Figure 7.19 such a control loop is shown. The output current u of the controller operates a valve via an electro-pneumatic (E/P) converter. The steam heats the process, the temperature of which is detected by a thermocouple. The mV signal from the thermocouple is converted by a transmitter to a mA signal which then serves as input signal to the controller.

Now we have to determine the combined response of all the elements which constitute the loop, i.e. the E/P converter, the valve, the process, the sensing element (transducer, the thermocouple) and the transmitter all in cascade as shown in Figure 7.19. This response is defined

[Figure 7.19: Process loop diagram showing a control loop with Controller (E, 0–20 mA input; U, 0–20 mA output), E/P converter receiving Air, Valve with Steam, Process with thermocouple sensing element, Transmitter (100–800°C) producing Y(s), and reference R(s).]

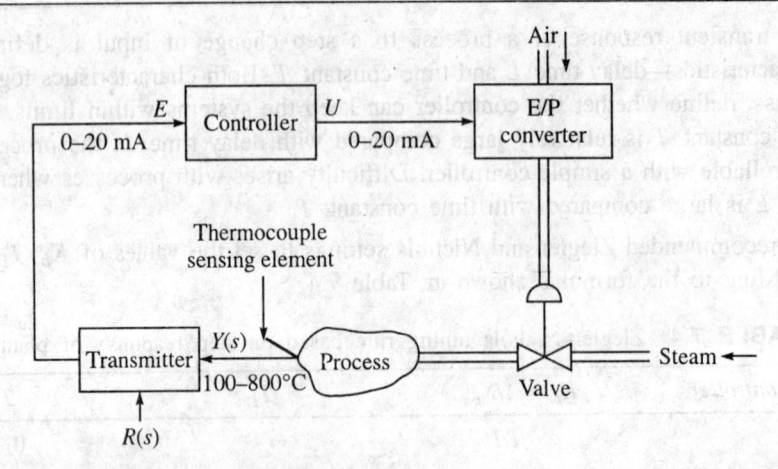

FIGURE 7.19 Process loop.

as the amplification factor of the control loop, $A = \Delta E/\Delta U$. If the time constant T, delay-time or dead-time L and amplification factor A of the loop are known, it is possible to calculate the approximate settings for P-control (i.e. K_p or X_p), I-control (T_i') and D-control (T_d) according to Table 7.4 as:

$$K_p = 1.2\left(\frac{1}{A}\right)\left(\frac{T}{L}\right)\% \qquad (7.23)$$

or

$$X_p = 83(A)\left(\frac{L}{T}\right)\% \qquad (7.24)$$

For unity amplification factor, K_p becomes $1.2(T/L)$ as given in Table 7.4.

$$K_p = \left(\frac{1.2}{LR}\right)\%; \; R \text{ is the slope of the S-shaped response curve.}$$

$$T_i = 2L \text{ (min)} \qquad (7.25)$$

$$T_d = 0.5L \text{ (min)} \qquad (7.26)$$

The practical method for determining T, L, and A is given below:

For this test we will operate the controller in the manual mode setting, in order to be able to artificially create a step change in process input and then observe the reaction of the output of the process (i.e. input to the controller), by means of a recorder. Before doing so, however, we have to make sure that the process remains within the safe limits, that is, to say that we have to choose an operating level (controller output current I_n = nominal valve position) which still permits sufficient further output current change in either direction ($\pm I_d$) allowing the process to still remain within the safe operating limits but large enough to produce a noticeable effect on the recorder.

With this in mind, we continue as follows:

1. Record the output signal of the process (= output of the transmitter).

2. Select a nominal controller output current I_n.
3. Manually reduce the controller output current to a value $I_n - I_d$ and let the process recover.
4. Manually increase the controller output to the value $XI_n + I_d$ and let the process recover again.
5. Calculate $\Delta U = (I_n + I_d) - (I_n - I_d) = 2I_d$.
6. Determine from the recorder the jump in the process output current (= transmitter output current) = ΔE.
7. Calculate $A = \Delta E / \Delta U$.
8. Determine from the recording, T and L.
9. Calculate $X_p (= 1/K_p)$, T_i, T_d.

EXAMPLE 7.2 In order to illustrate the practical settings and calculations of X_p, T_i and T_d, let us note the data from the experimental observation, as shown in Figure 7.20.

- Output signal for nominal conditions = 12.5 mA
- A variation of ±2.5 mA is optimal for the process and the result can be clearly followed on the recorder.

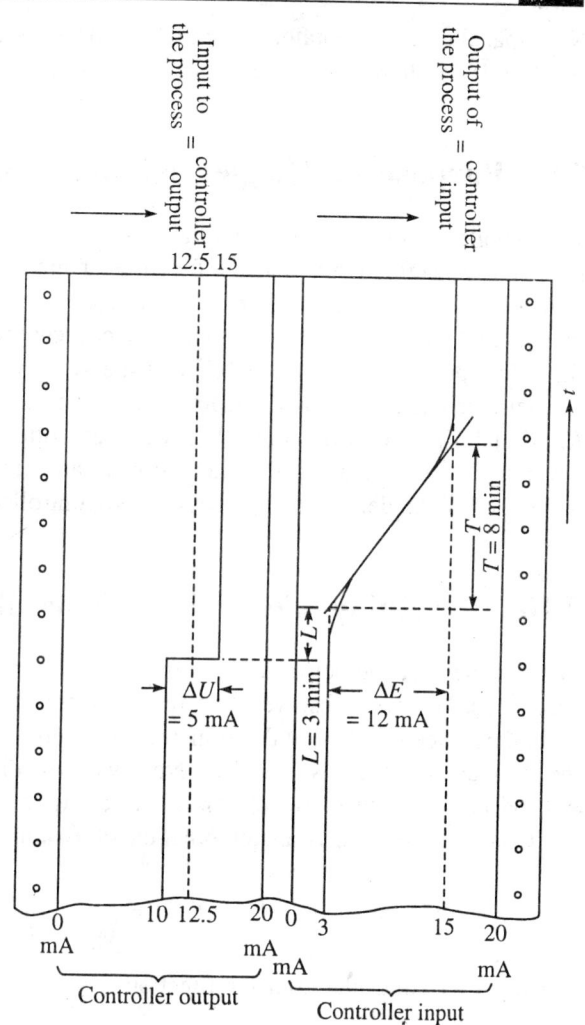

FIGURE 7.20 Input-output response.

- From the recording curve it is found that:
$$L = 3 \text{ min}, \quad T = 8 \text{ min}$$

Thus

$$\Delta E = 15 - 3 = 12 \text{ mA}; \quad \Delta U = 2 \times 2.5 = 5 \text{ mA}, \quad \text{and} \quad A = \frac{12}{5} = 2.4$$

From Eq. (7.24), $X_p = 83 \times 2.4 \times (3/8) = 74.7\% \approx 75\%$ or $K_p = 1.33$

From Eq. (7.26), $T_d = 0.5 \times 3 = 1.5$ min

From Eq. (7.25), $T_i = 2 \times 3 = 6$ min

Note that the values obtained in this way are only approximate and can therefore only serve as a guide to shorten the trial and error procedure.

7.9 Purpose of Ziegler–Nichols Tuning Method

The Ziegler–Nichols tuning method for PID controller for continuous-time systems is particularly useful if the transfer function of the system is not known. It was noted that the proportional part of the compensatory network improves the sensitivity to parameter variations, the integral part improves the steady-state accuracy if any steady-state error exists and the derivative part improves the stability of the system by increasing the damping. In practice, the Ziegler–Nichols tuning method provides a good set of initial values of the various constants K_p, T_i and T_d of the PID controller, which can often be improved by on-line tuning. The main advantage is that by using this method one can completely bypass the need to obtain an exact mathematical model for the process to be controlled.

7.10 Designing Controller Using Root Loci

In Example 7.2, the settings recommended by Ziegler and Nichols for PID controllers were used. It is, however, observed that for some applications, the resulting time response tends to be underdamped and no indication for the improvement of the performance is given through the Ziegler and Nichols methods. Now we will discuss through an example, how root loci method can be used to select controller settings.

Consider the plant transfer function of Example 7.1

$$G(s) = \frac{6}{(s+1)(s+2)(s+3)}$$

and the PID controller transfer function

$$G_c(s) = 1 + T_d s + \frac{1}{T_i s}$$

The scheme is shown in Figure 7.21.

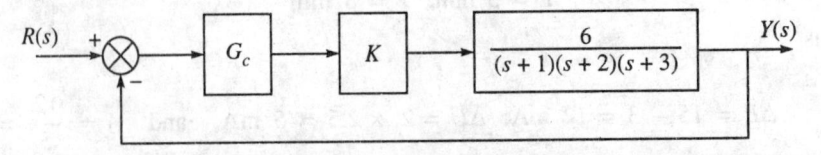

FIGURE 7.21 Plant and the PID controller.

Case A (Root loci for P control)

For proportional control, $T_d = 0$, $T_i = \infty$. The closed-loop characteristic equation with pure proportional controller is

$$1 + \frac{6K}{(s+1)(s+2)(s+3)} = 0$$

The root loci is drawn in Figure 7.22(a).

In the root loci, the closed-loop complex pole positions are identical for critical gain $K_c = 10$, Ziegler and Nichols gain (from Table 7.3) as $K_{ZN} = 0.5K_c = 5$. Clearly the response dynamics would be improved by reducing the controller gain K to 0.6. However, this increases the offset as the position error coefficient (as per the symbol used in earlier chapters)

$$K_p = \lim_{s \to 0} G(s)H(s) = 0.6$$

and the steady-state error, $e_{ss} = \dfrac{1}{1+K_p} = 0.625$. See Figure 7.22(b).

Case B (Root loci for PI control)

Under PI control, the closed-loop characteristic equation becomes

$$1 + \frac{6K\left(1 + \dfrac{1}{T_i s}\right)}{(s+1)(s+2)(s+3)} = 1 + \frac{6K\left(s + \dfrac{1}{T_i}\right)}{s(s+1)(s+2)(s+3)} = 0$$

In the root-loci method either K or T_i can be varied with the other fixed. If T_i is fixed at the recommended Ziegler–Nichols value $T_i = 0.83T$ (from Table 7.2) where, from Routh–Hurwitz criterion we get $T_i = 1.57$, then the characteristic equation becomes

$$1 + \frac{6K(s + 0.637)}{s(s+1)(s+2)(s+3)} = 0$$

The root loci is shown in Figure 7.22(c) for the variable parameter gain K in the range $0 < K < \infty$. Again the position error coefficient K_p with the recommended Ziegler–Nichols setting for $T_i = 1.57$ becomes

$$K_p = \lim_{s \to 0} \frac{6K(s + 0.637)}{s(s+1)(s+2)(s+3)} = \infty$$

Hence the steady-state error for a unit-step input is

$$e_{ss} = \frac{1}{1+K_p} = 0$$

It may be noted that the PI controller introduces a pole at the origin and a zero at $s = -0.637$. The PI-controller's pole at the origin reduces the error offset for unit step input to zero.

With only P-control the critical gain $K_c = 10$. For PI-control the Ziegler–Nichols gain $K_{ZN} = 0.45K_c = 4.5$ and at $K_{ZN} = 4.5$, the pair of dominant complex conjugate roots are also shown.

The response for PI-controlled system is shown in Figure 7.22(d) for $K_{ZN} = 4.5$. From this plot it appears that the response could be improved by reducing the gain K to 2 as shown. Further reduction in gain K will reduce the effective dominance of the complex poles and in turn, damping coefficients of complex poles will increase as is shown in Figure 7.22(d). For two values of $K = K_{ZN} = 4.5$ and $K = 2$, the step responses are shown.

Now we will draw the root loci keeping K fixed at $K = 0.45$ and varying T_i in the range $0 < T_i < \infty$. To see the effect of adjustments to T_i, proceed as follows: Set the controller gain fixed at the recommended value $K = 0.45K_c = 0.45(10) = 4.5$ as specified in Ziegler–Nichols settings for PI control. Modify the characteristic equation as

$$1 + G(s)H(s) = 0$$

or

$$1 + \frac{(4.5)(6)(s + 1/T_i)}{s(s+1)(s+2)(s+3)} = 1 + \frac{27(s + 1/T_i)}{s^4 + 6s^3 + 11s^2 + 6s} = 0$$

Thus, the characteristic equation becomes

$$s^4 + 6s^3 + 11s^2 + 33s + \frac{27}{T_i} = 0$$

or

$$1 + \frac{27/T_i}{s^4 + 6s^3 + 11s^2 + 33s} = 1 + \frac{27/T_i}{s(s+5.11)(s+0.4666 \pm j5.109)} = 0$$

If $K_1 = 27/T_i$, then draw the root loci for the open-loop transfer function

$$\frac{K_1}{s(s+5.11)(s+0.4666 + j5.109)(s+0.4666 - j5.109)}$$

for $0 < K_1 < \infty$ and the root loci is shown in Figure 7.22(e). On the root loci it is indicated with arrows the increasing values of K_1, that is, decreasing values of T_i. The critical value of T_i is $(108/121)$. Further, $T_i < (108/121)$ and with $K = 4.45$, the system becomes unstable. The root loci of Figure 7.22(e) for $T_i = 1.57$ (as stipulated by Ziegler–Nichols PI controller settings) should coincide with the root loci of Figure 7.22(c) for $K = 4.45 = K_{ZN}$.

Case C (Root loci for PID control)

The analysis of the PID controller using the root-locus approach where three parameters are to be adjusted, demands that two variables be fixed (T_d and T_i) and the other parameter K may be variable. Set $T_d = 0.125T$ and $T_i = 0.5T$ as recommended in Ziegler–Nichols Table 7.3 and for this particular example $T = 1.895$. Then $T_d = 0.237$ and $T_i = 0.947$.

The characteristic equation is

$$1 + \frac{6K(1 + 0.237s + 1/0.947s)}{(s+1)(s+2)(s+3)} = 1 + \frac{6K(s + 2.11 \pm j0.07)}{s(s+1)(s+2)(s+3)} = 0$$

The root-loci plot for variations of K in the range $0 < K < \infty$ is drawn in Figure 7.22(f), from which the closed-loop system is seen to be stable for all values of gain K. For the proposed

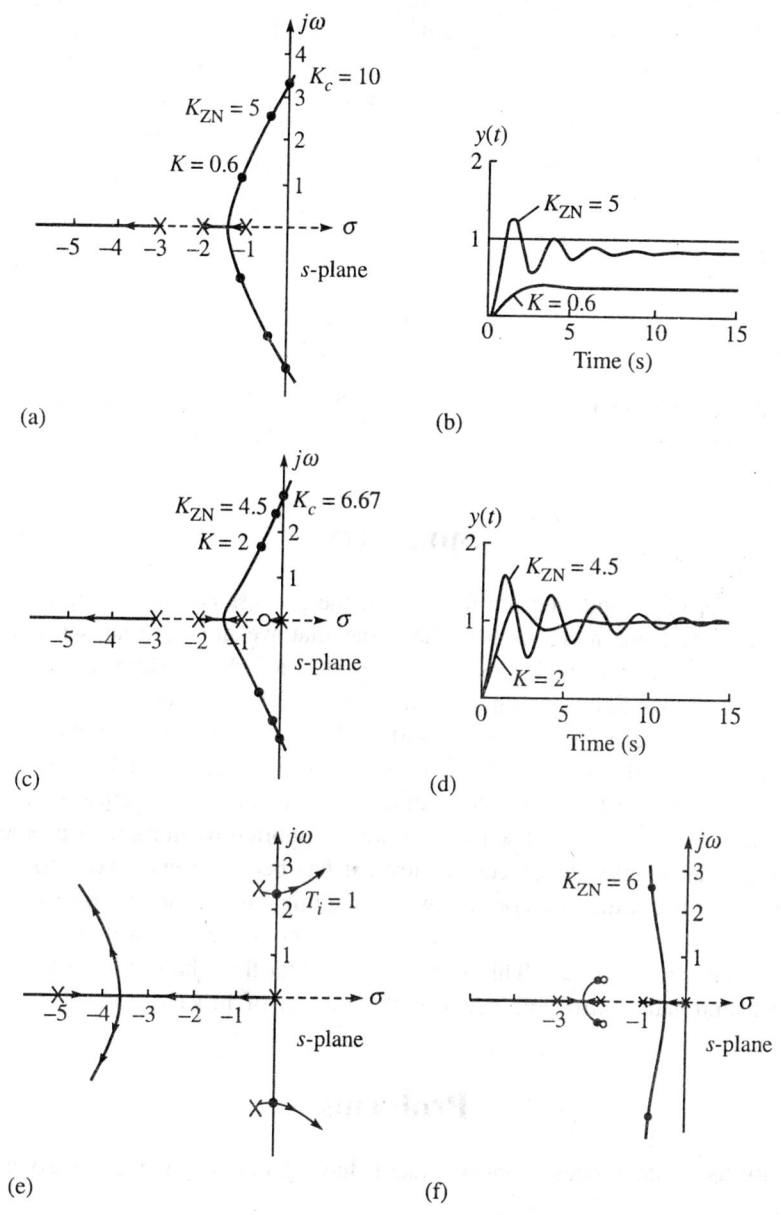

FIGURE 7.22 Designing controller using root loci.

Ziegler–Nichols gain $K = K_{ZN} = 0.6K_c = 0.6(10) = 6$ and the already proposed values set of $T_i = 0.5T$ and $T_d = 0.125T$, the unit-step response is shown in Figure 7.23. For comparison, see the response to a step input for P, PI and PID control as shown in Figure 7.16 (repeated below as Figure 7.24).

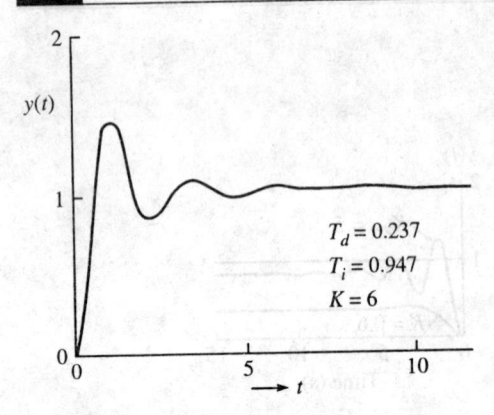

FIGURE 7.23 Time response under PID control.

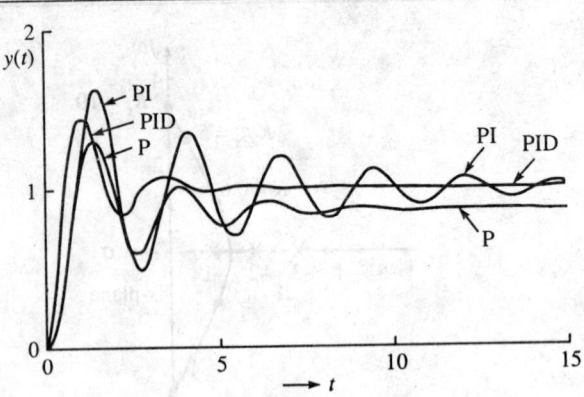

FIGURE 7.24 Time response under P, PI and PID control.

Summary

In this chapter we briefly explained the features of the process control system. The techniques used in process control are more or less the same that we are familiar with from classical control systems, only the terminologies are different somewhere. Here an attempt has been made to unify these two lines of control. The concept of feedback, the action of feedback principle in process control systems, the proportionality band and its relationship with dc gain, necessity of PI control and finally the PID control, and the circuitry have been discussed. The tuning procedure of PID-controller by the Ziegler–Nichols method, together with its analytical approach, has been discussed. The discussion is well supported by numerical problems. Though this chapter is an introduction to process control, it has been the endeavour to cover most of the basic principles of continuous control, without going into complex control loops such as multi-element control, ratio control, cascade control, computer control, fractioning column control, reactor control, and so on. This is only to say that this chapter is an humble approach to motivate conventional control engineers to the subject of process control systems.

Problems

7.1 A unity-feedback process control system having plant dynamics is given by

$$G_P(s) = \frac{4}{(s^2 + 8s + 80)(s + 1)}$$

and a PID (three-term) controller with transfer function is given by

$$G_c(s) = \frac{20(T_i T_d s^2 + T_i s + 1)}{T_i s}$$

Establish the values of T_i and T_d for the closed-loop system to be stable.

Ans. $T_i > 0$ and $T_d > -79/90$

7.2 Use the Ziegler–Nichol rules to design a three-term controller for a plant model having open-loop transfer function

$$G(s) = \frac{1}{s(s+1)(s+2)(s+3)}$$

Show that the resulting closed-loop system is stable.

Ans. $K_c = 10$, $T = 2\pi$
PID settings are: $K = 6$, $T_i = \pi$, and $T_d = \pi/4$
The closed-loop characteristic equation of the model and PID controller is
$$s^5 + 6s^4 + 11s^3 + (6 + 1.5\pi)s^2 + 6s + 6/\pi = 0$$

and by Routh's array it is stable.

7.3 Consider the electronic PID controller of Figure 7.14. Determine the values of R_1, R_2, and R_4 for the given values of $C_1 = C_2 = 10$ μF and $R_3 = 10$ kΩ.

Ans. $R_1 = R_2 = 153.85$ kΩ and $R_4 = 197.1$ kΩ

1905–1982

Hendrik Wade Bode was born on 24 December 1905, in Madison, Wisconsin. He began his career at Bell Telephone Laboratories in 1926, where he specialized in research pertaining to electrical network theory and to its application to long distance communication facilities. While employed at Bell Laboratories, he attended Columbia University Graduate School, and received the Ph.D. degree in 1935.

With the outbreak of World War II, Bode turned to the development of electronic fire control devices, and in recognition of his contributions in this field, he was awarded the Presidential Certificate of Merit (in 1948) for his work in the field of fire control under the NDRC of the OSRD.

Bode's work in electric filters and equalizers led to broader aspects of communication transmission, resulting in the publication in 1945 of his book *Network Analysis and Feedback Amplifier Design*, which is considered a classic in its field.

In 1944, Bode was placed in charge of the Mathematics Research Group at Bell Laboratories, and in 1952 became Director of Mathematical Research. In 1955 he was appointed Director of Research in the Physical Sciences, and in 1958 assumed the responsibilities of Vice-President in charge of one of two vice-presidential areas devoted to military development.

Bode retired from Bell Telephone Laboratories in October 1967, at the age of 61, after 41 years service in a distinguished career. He was immediately elected Gordon McKay Professor of Systems Engineering at Harvard University, whose he particularly worked on military projects in areas where there were difficulties concerned with decision making.

While Bode's chief interests have been science and engineering, he found time to pursue two major hobbies. One of these was reading, in which he was known as an avid reader on a wide variety of subjects. His other interest was sail-boating on Long Island Sound during his early career in the New York area, and after World War II, he acquired a converted LCT, which he operated on the upper reaches of the Chesapeake Bay adjacent to the Eastern Shore of Maryland. He was also a "do-it-your-selfer," interested particularly in gardening around his home. Bode's wife is the former Barbara Poore. They have two children, Katherine Anne Bode and Beatrice Anne Hathaway Bode.

Frequency Response Analysis

8.1 Introduction

Earlier we learned about a variety of input test signals such as impulse, step, ramp in Chapter 4 on transient response analysis. Here we discuss the frequency response analysis with the sinusoidal input signal. When a linear time-invariant system is subjected to a pure sinusoidal input, for example, $A \sin \omega t$, the output response contains the transient part which dies down (because of the roots with negative real parts for a stable system) in steady state, i.e. in the limit $t \to \infty$, only the steady-state part of the response exists. Hence there is no need to put $s = \sigma + j\omega$, instead we substitute $s = j\omega$ for steady-state analysis. In a nutshell, the frequency response of a system is defined as the steady-state response of the system to a sinusoidal input signal. The sinusoid is a unique input signal, and the resulting output signal for a linear system as well as the signals throughout the system, are sinusoidal in the steady-state; the output waveform differs from the input waveform only in amplitude and phase angle. Look back at Figure 1.24.

In Chapter 5 we discussed stability of closed-loop systems using the Routh–Hurwitz criterion which does not require the knowledge of the roots of the characteristic polynomial. On the other hand, the root-locus method shows how the roots of the characteristic polynomial move in the s-plane when one parameter is varied from zero to infinity. The basic requirement for these analyses is that the transfer function

OBJECTIVE

The knowledge of transfer function is a prerequisite for stability analysis, however, the transfer function may not be available in all practical situations. Hence the frequency response analysis is the only way out. The obvious question may be why should one know all the techniques for stability analysis? This is because, as a design engineer, one has to have all shots in arm to understand the design specifications. Techniques to draw the frequency response of open-loop systems should be learnted first, followed by a study of stability of the closed-loop systems. Then and then only one can know how to design the control system as per design specifications. The genesis of frequency response analysis is to determine the stability of a closed-loop system from the frequency response of the open-loop system, be it Bode plot, polar plot, or log-magnitude-versus-phase plot.

CHAPTER OUTLINE

Introduction
Bode Plot
Non-minimum Phase Function
System with Time Delay
Polar Plot
Relative Stability
Conditionally Stable System

of the open-loop system has to be known precisely, which may not be the case in practical situations. The remedy lies in obtaining the frequency response of a stable open-loop system experimentally, and then it is not necessary to know the transfer function.

Frequency response analysis may be of open-loop frequency response or closed-loop frequency response. The design engineer has to analyse both the open-loop frequency response and the closed-loop frequency response, though, sometimes one kind of analysis can be derived from the other kind. Further, frequency response analysis can be applied to systems that do not have rational transfer function such as a system with transportation lag.

Here in this chapter we want to understand the open-loop frequency response. We will deal with the closed-loop frequency response in the next chapter. The open-loop frequency response analysis covers the techniques for the construction and interpretation of the Bode plot, the polar plot, and the log-magnitude versus phase plot. The basic philosophy or beauty of the open-loop frequency response analysis is that we get the information about the relative stability of the closed-loop system from any of these open-loop frequency response plots. What do we mean by this? The closed-loop characteristic equation is $1 + G(s)H(s) = 0$, i.e. the open-loop transfer function, $G(s)H(s) = -1$. Obviously, $\angle G(s)H(s)$ is a complex quantity, hence the magnitude $|G(s)H(s)| = 1$ and phase $\angle G(s)H(s) = -180°$ for a just stable system.

In Bode plot, the frequency response plot consists of two curves: magnitude in dB versus frequency and phase versus frequency, where frequency is plotted in logarithmic scale. After drawing the frequency response plot of the open-loop transfer function $G(s)H(s)$ in case of Bode plot, then comparing with the right-hand side of the basic equation $G(s)H(s) = -1$, we find that the magnitude curve of 1 is the 0-dB line and the phase curve is the $-180°$ line. Comparing the magnitude curve of Bode plot of open-loop transfer function $G(s)H(s)$ with that of 0-dB line and comparing the phase curve of Bode plot of open-loop transfer function $G(s)H(s)$ with that of $-180°$ line, will give information about the relative stability of the closed-loop control system. Relative stability information is specified in terms of gain margin and phase margin which we will soon discuss in detail.

In the same way, comparing the polar plot of the open-loop transfer function with the $(-1, j0)$ point, gives the information about the relative stability of the closed-loop system, again in terms of the gain margin and phase margin.

It may be mentioned that the log-magnitude-versus-phase plot is an extension of the Bode plot and again in the same way, the stability of the closed-loop system can be obtained from the open-loop plot of the log-magnitude-versus-phase angle.

Thus the objective of frequency domain analysis is to find the stability of the closed-loop system from the open-loop transfer function in the steady state (putting $s = j\omega$) by assigning

$$G(j\omega)H(j\omega) = -1$$

i.e. $\quad |G(j\omega)H(j\omega)| = 1 \quad$ and $\quad \angle G(j\omega)H(j\omega) = -180°$

which is the condition for just stability of the closed-loop system. The same philosophy is applied in root-locus approach, though the root-locus plot is a time domain approach as it ultimately talks about the pole positions of the closed-loop system, again from the open-loop pole-zero information with one parameter varying. The frequency response analysis presents a qualitative picture of the transient response, the correlation between the frequency and transient responses are indirect except for the case of the second-order system.

8.2 Bode Plot

One of the main attractions of the Bode diagram (named after Hendrik W. Bode born in 1905) is the ease with which it can be produced and modified by using pencil-and-paper methods. The open-loop transfer function represented in factored form is approximated by a number of asymptotic straight line functions (one for each factor) which may be added graphically to produce plots accurate enough for most design studies. The Bode plot is used to determine the gain margin and phase margin for analysing the stability of the closed-loop system. Nowadays, simple computer programs are also available to produce accurate Bode plots of open-loop transfer functions. Thus, the design problems have become easier today.

This section considers the traditional method of producing Bode plots based on asymptotic (straight line) approximations to the various elements within the loop, then cascading them all for the overall open-loop transfer function.

8.2.1 Basic philosophy

For a feedback control system, the open-loop transfer function in the frequency domain (putting $s = j\omega$, for steady state) is

$$GH(j\omega) = |GH(\omega)| e^{j\phi(\omega)} \tag{8.1}$$

Taking the natural logarithm, we get

$$\ln GH(j\omega) = \ln GH(\omega) + j\phi(\omega) \tag{8.2}$$

where $\ln (GH)$ is the magnitude in nepers. The logarithm of the magnitude is normally expressed in terms of the logarithm to the base 10, so that we use

$$\text{logarithmic gain} = 20 \log_{10} GH(\omega) \tag{8.3}$$

where the units are in decibels (dB).

Similarly, the phase angle ϕ is in degrees.

One may ask that since the Laplace operator s is a complex quantity, i.e. $s = \sigma + j\omega$, then why should we put $s = j\omega$ only?

A linear time-invariant system having transfer function $GH(s)$ when subjected to a sinusoidal input $\sin \omega t$, has the steady-state output $|GH(j\omega)| \sin[\omega t + \theta(\omega)]$ which is also sinusoidal having a phase difference $\theta(\omega)$ with respect to input, where $|GH(j\omega)|$ and $\theta(\omega)$ are the amplitude and phase angle of $GH(s)$ obtained by putting $s = j\omega$ in the steady state. Hence in order to get the frequency response of the system having transfer function $GH(s)$, we put $s = j\omega$ in the steady state.

The basic advantage of logarithmic representation is that multiplication and division for the pole-zero representation of transfer function are replaced by addition and subtraction. The logarithmic scale is nonlinear, as a result, it enables us to cover a greater range of frequencies. Semilog paper comes in one, two, three or four cycles, indicating the range of coverage in decades. It may be noted that we cannot locate $\omega = 0$ on log scale since $\log 0 = -\infty$.

Let us now introduce a logarithmic frequency variable,

$$u = \log_{10} \omega \quad \text{or} \quad \omega = 10^u$$

The two frequency intervals can be written as

$$u_2 - u_1 = \log_{10} \omega_2 - \log_{10} \omega_1 = \log_{10}\left(\frac{\omega_2}{\omega_1}\right) \tag{8.4}$$

These frequency intervals are either the octave for which $\omega_2 = 2\omega_1$ or the decade for which $\omega_2 = 10\omega_1$. The slopes of the straight lines in the Bode plot will be expressed in terms of these two frequency intervals, i.e. in decade or octave. The number of decades between any two frequencies ω_1 and ω_2 is given by

$$\text{Number of decades} = \frac{\log_{10}\left(\frac{\omega_2}{\omega_1}\right)}{\log_{10} 10} = \log_{10}\left(\frac{\omega_2}{\omega_1}\right) \tag{8.5}$$

$$20 \text{ dB/decade} = 20 \log_{10}\left(\frac{\omega_2}{\omega_1}\right) = (20 \times 0.301) \log_2\left(\frac{\omega_2}{\omega_1}\right)$$

$$= 6 \log_2\left(\frac{\omega_2}{\omega_1}\right) = 6 \text{ dB/octave} \tag{8.6}$$

8.3 Response of Linear Systems

Let us write the output $y(t)$ in the complex domain s, when a system with transfer function $g(t)$ is subjected to input $u(t)$, i.e.

$$Y(s) = G(s) U(s)$$

Let the transfer function of the system be

$$G(s) = \frac{N_1(s)}{D_1(s)} = \frac{N_1(s)}{(s - p_1)(s - p_2)\ldots(s - p_n)}$$

and the transformed input be

$$U(s) = \frac{N_2(s)}{D_2(s)} = \frac{N_2(s)}{(s - q_1)(s - q_2)\ldots(s - q_m)}$$

then $Y(s)$ can be written as

$$Y(s) = \sum_{i=1}^{n} \frac{c_i}{s - p_i} + \sum_{i=1}^{m} \frac{k_i}{s - q_i}$$

Frequency Response Analysis

The inverse Laplace transform is

$$y(t) = \underbrace{\sum_{i=1}^{n} c_i e^{p_i t}}_{\text{transient part}} + \underbrace{\sum_{i=1}^{m} k_i e^{q_i t}}_{\text{steady-state part}}$$

The complete response of a system to any forcing function is thus composed of a transient portion that results from natural modes of the system, and a steady-state portion that depends on both the system and the forcing function. In a stable system, the terms in the first sum die away with increasing time t, while the steady-state portion may contain terms that are nonzero indefinitely.

EXAMPLE 8.1 A low-pass RC network with $R = 2\ \Omega$ and $C = 0.5$ F is driven by the input voltage 10 cos $4t$. Determine the output voltage v_o across C.

Solution: The transfer function $G(s)$ is

$$G(s) = \frac{1}{s+1}$$

The Laplace transform of the input waveform is

$$U(s) = \mathscr{L}(10 \cos 4t) = \frac{10s}{s^2 + 16}$$

The transformed output is

$$V_o(s) = G(s)U(s) = \frac{10s}{(s^2 + 16)(s+1)}$$

$$= \frac{-10/17}{s+1} + \frac{(10/17)s}{s^2 + 16} + \frac{(160/17)}{s^2 + 16}$$

Therefore,

$$v_o(t) = \underbrace{-\frac{10}{17} e^{-t}}_{\text{transient part}} + \underbrace{\left(\frac{10}{17} \cos 4t + \frac{40}{17} \sin 4t\right)}_{\text{steady-state part}}$$

We can obtain the steady-state component directly via the Fourier transform. Putting $s = j\omega$ for steady state, we get

$$G(j\omega) = \frac{1}{1 + j\omega}$$

For angular frequency of $\omega = 4$,

$$G(j\omega) = \frac{1}{1 + j4} = \frac{1 - j4}{17}$$

$$= \frac{1}{\sqrt{17}} \angle -76°$$

Hence, the steady-state response to an input of $10 \cos 4t$ is

$$v_o(t) = \frac{10}{\sqrt{17}} \cos(4t - 76°) = \frac{10}{17} \cos 4t + \frac{40}{17} \sin 4t$$

which is the same result as obtained by the Laplace transform method.

The obvious conclusion is that the frequency response of a system with transfer function $G(s)$ is simply $G(s)$ evaluated along the $j\omega$-axis. If the input $u(t)$ is $\sin \omega t$, then the steady-state output $y(t)$ is

$$y(t) = |G(j\omega)| \sin[\omega t - \theta(\omega)]$$

where the functions $|G(j\omega)|$ and $\theta(\omega)$ are the amplitude and angle functions obtained from the transfer function $G(s)$ by substituting $j\omega$ for s. This is an alternative way of establishing that the frequency response of a linear time-invariant system is the magnitude and phase of $G(j\omega)$. This relationship between the frequency response of a system and the transfer function of a system makes it easy to obtain the steady-state response for sinusoidal input.

8.4 Frequency Domain Analysis

Before we explain how a Bode plot is drawn, let us talk about why the Bode plot needs to be drawn, i.e. what the objective of the Bode plot is. For a single-loop control system as shown in Figure 8.1, the characteristic equation is

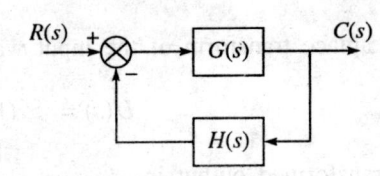

FIGURE 8.1 Single-loop control system.

$$1 + G(s)H(s) = 0 \quad \text{or} \quad G(s)H(s) = -1 \quad (8.7)$$

which can be written as:

(i) The magnitude condition

$$\text{dB} = 20 \log |G(s)H(s)| = 20 \log |1| = 0 \text{ dB} \quad (8.8)$$

(ii) The phase condition

$$\angle G(s)H(s) = (2k + 1)\pi; \quad k = 0, 1, 2,\ldots \quad (8.9)$$

where $G(s)H(s)$ is the open-loop transfer function.

The intersection of the magnitude curve of the open-loop transfer function $G(s)H(s)$ in dB with the 0-dB line [as is apparent from Eq. (8.8)] gives the *gain crossover frequency*. This is the critical frequency, that is, the critical stable point so far as only the gain of the closed-loop system is concerned. Similarly, the intersection of the phase curve of the open-loop transfer function $G(s)H(s)$ with the $-180°$ line [as is apparent from Eq. (8.9)] gives the *phase crossover frequency*. A measure of the magnitude at the phase crossover frequency to the magnitude curve from the 0-dB line gives the *gain margin* of the closed-loop system. Similarly, a measure of the phase at the gain cross-over frequency to the phase curve from the $-180°$ line gives the *phase margin* of the closed-loop system.

Now the objective of the Bode plot seems to be clear, i.e. the application of Bode plot is to compare the magnitude curve of the open-loop transfer function $G(s)H(s)$ in dB with the 0-dB line at the phase crossover frequency in order to let the designer know how much stable the closed-loop system is, gain-wise. Similarly, the comparison of the phase curve of the open-loop transfer function $G(s)H(s)$ with the $-180°$ line at the gain crossover frequency tells the designer as to how stable the closed-loop system is, phase-wise. Thus the gain and phase margins constitute the relative stability information of a closed-loop system obtained from the open-loop transfer function.

8.5 Construction of Bode Plot

To illustrate the construction of Bode plot, let us write the generalized expression for the open-loop transfer function $G(s)H(s)$ as

$$G(s)H(s) = \frac{K_s \prod_{k=1}^{Q} (s^2 + 2\zeta_k \omega_{nk} s + \omega_{nk}^2) \prod_{i=1}^{M} (sT_i + 1)}{s^N \prod_{m=1}^{v} (s^2 + 2\zeta_m \omega_{nm} s + \omega_{nm}^2) \prod_{l=1}^{u} (sT_l + 1)} \quad (8.10)$$

where K_s, T_i, T_l, ζ_k, ζ_m, ω_{nk}, and ω_{nm} are real constants.

Putting $s = j\omega$, in the steady state, the open-loop transfer function can be written as

$$G(j\omega)H(j\omega) = \frac{K \prod_{i=1}^{M} (j\omega T_i + 1) \prod_{k=1}^{Q} (1 + j2\zeta_k u_{nk} - u_{nk}^2)}{(j\omega)^N \prod_{l=1}^{u} (j\omega T_l + 1) \prod_{m=1}^{v} (1 + j2\zeta_m u_{nm} - u_{nm}^2)} \quad (8.11)$$

where

$$K = \frac{K_s \prod_{k=1}^{Q} \omega_{nk}^2}{\prod_{m=1}^{v} \omega_{nm}^2} \quad u_{nk} = \frac{\omega}{\omega_{nk}}; \quad u_{nm} = \frac{\omega}{\omega_{nm}}$$

and ζ is the damping coefficient, and ω_n is the undamped natural frequency.

Interpreting Eq. (8.11) and expressing the magnitude in dB gives

$$|G(j\omega)H(j\omega)|$$
$$= 20 \log |K| + \sum_{k=1}^{Q} 20 \log |(1 + j2\zeta_k u_{nk} - u_{nk}^2)| + \sum_{i=1}^{M} 20 \log |(1 + j\omega T_i)|$$
$$- \begin{cases} 20 \log (j\omega)^N \\ \text{or} \\ 20 N \log |j\omega| \end{cases} - \sum_{l=1}^{u} 20 \log |(j\omega T_l + 1)| - \sum_{m=1}^{v} 20 \log |(1 + j2\zeta_m u_{nm} - u_{nm}^2)| \quad (8.12)$$

and the angle equation is

$$\angle[G(j\omega)H(j\omega)] = \sum_{i=1}^{M} \tan^{-1}(\omega T_i) + \sum_{k=1}^{Q} \tan^{-1}\left(\frac{2\zeta_k u_{nk}}{1-u_{nk}^2}\right)$$

$$- N\left(\frac{\pi}{2}\right) - \sum_{l=1}^{u} \tan^{-1}(\omega T_l) - \sum_{m=1}^{v} \tan^{-1}\left(\frac{2\zeta_m u_{nm}}{1-u_{nm}^2}\right) \quad (8.13)$$

In general, the open-loop transfer function $G(j\omega)H(j\omega)$ may contain four simple types of factors:

(i) Constant K
(ii) Zeros or poles at the origin $(j\omega)^{\pm N}$
(iii) Simple zero or pole $(1 + j\omega T)^{\pm 1}$
(iv) Complex zeros or poles $(1 + j2\zeta u - u^2)^{\pm 1}$

These four different kinds of factors will now be investigated separately. All logarithms are to the base 10.

(i) **The constant term K**

$$K(\text{dB}) = 20 \log|K| = \text{constant} \quad (8.14)$$
$$\text{Arg } K = 0° \quad (8.15)$$

The plot of the constant term K is shown in Figure 8.2.

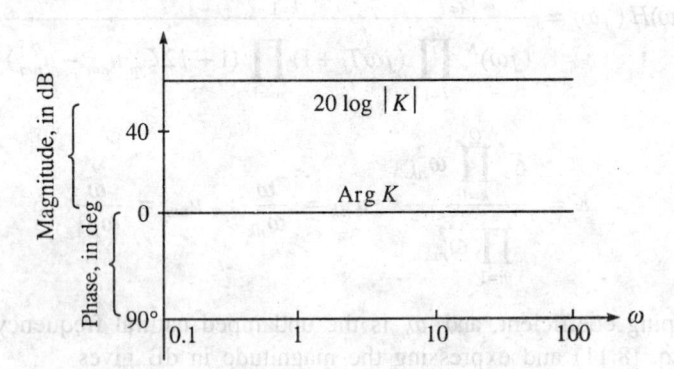

FIGURE 8.2 Magnitude and phase curve of the constant term K.

(ii) **Zeros or poles at origin**

For the factor $(j\omega)^{\pm N}$, the log magnitude is given by

$$20 \log|(j\omega)^{\pm N}| = \pm 20N \log \omega \quad \text{dB} \quad (8.16)$$

which is the equation of a straight line with slope $\pm 20N$ dB/decade or $\pm 6N$ dB/octave, passing through the 0-dB line at point $\omega = 1$. The term $(j\omega)^{+N}$ signifies that there are N zeros at origin, and the term $(j\omega)^{-N}$ signifies that there are N poles at origin. To draw this straight line, we need one reference point on it, since the slope is known. Let it be $\omega = 1$ on the 0-dB line, through which we draw a straight lines of $\pm 20N$ dB/decade.

The phase angle of $(j\omega)^{\pm N}$ is

$$\text{Arg } (j\omega)^{\pm N} = (\pm N) \times 90° \tag{8.17}$$

which is constant over the entire frequency range.

The magnitude and phase curves of the terms $(j\omega)^{\pm N}$ are shown in Figures 8.3(a) and 8.3(b), respectively.

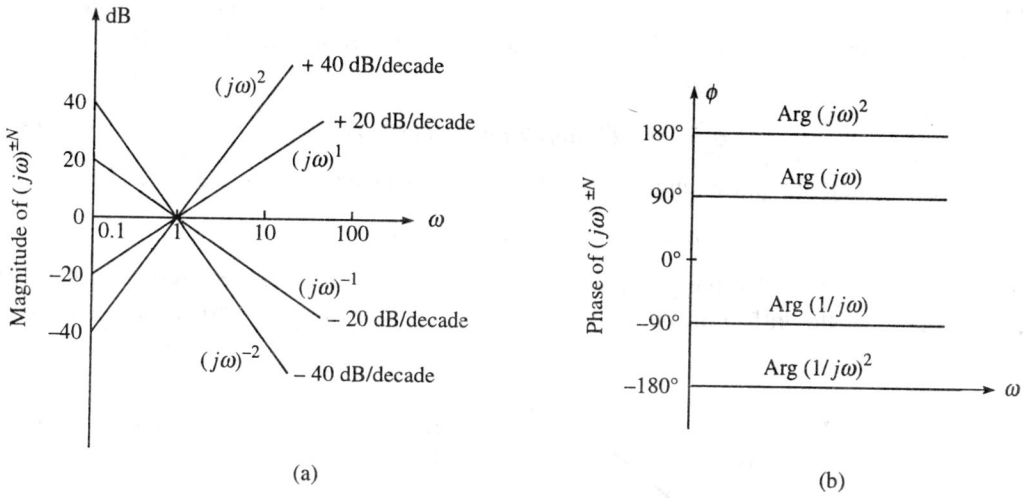

FIGURE 8.3 Magnitude and phase curves of the term $(j\omega)^{\pm N}$.

(iii) **Simple zero or pole: $(1 + j\omega T)^{\pm 1}$**

(a) *Simple zero:* Let $\quad G(j\omega)H(j\omega) = 1 + j\omega T$

The magnitude in dB is

$$20 \,|\log G(j\omega)H(j\omega)| = 20 \log \sqrt{1^2 + \omega^2 T^2} \tag{8.18}$$

At very low frequencies, $\omega \ll 1/T$, the log magnitude is approximated as

$$20 \log |G(j\omega)H(j\omega)| = 20 \log 1 = 0 \text{ dB} \tag{8.19}$$

The logarithmic plot of $20 \log |G(j\omega)H(j\omega)|$ versus $\log \omega$ is, therefore, a 0-dB straight line coincident with the horizontal axis at $\omega \ll 1/T$.

At very high frequencies, $\omega \gg 1/T$, the log magnitude is approximated as

$$20 \log |G(j\omega)H(j\omega)| = 20 \log \sqrt{\omega^2 T^2} = 20 \log \omega T = 20 \log \omega + 20 \log T \tag{8.20}$$

which is the equation of a straight line with slope +20 dB/decade or +6 dB/octave, passing through the 0-dB line at $\omega = 1/T$. To draw this straight line, we need one reference point on it, say $\omega = 1/T$, since the slope is known.

Therefore the log-magnitude versus log-frequency curve of simple zero $(1 + j\omega T)^{+1}$ can be approximated by two straight line asymptotes; a straight line at 0-dB for the frequency range $0 < \omega < 1/T$, and the other, a straight line with a slope +20 dB/decade or +6 dB/octave for the frequency range $1/T < \omega < \infty$ as shown in Figure 8.4(a). The frequency $\omega = \omega_c = 1/T$ at which the intersection of the low-frequency and the high-frequency asymptotes occurs is called the *corner* frequency or *break* frequency.

At corner frequency $\omega = 1/T$, the log magnitude equals $20 \log \omega T = 20 \log 1 = 0$ dB; at $\omega = 10/T = 10\omega_c$, the log magnitude is $20 \log \omega T = 20 \log 10 = 20$ dB. Thus the value of $20 \log \omega t$ dB increases by 20 dB for every decade of ω. For $\omega >> 1/T$, the log-magnitude curve is thus asymptotic to a straight line with a slope of +20 dB/decade (or +6 dB/octave).

The phase angle ϕ of simple zero is

$$\phi = \text{Arg } G(j\omega)H(j\omega) = \tan^{-1} \omega T \qquad (8.21)$$

At zero frequency, the phase angle is 0°. At corner frequency, the phase angle is

$$\phi = \tan^{-1} \omega T = \tan^{-1} 1 = 45°$$

At infinity, $\phi = 90°$. Since the phase angle is given by an inverse tangent function, it is skew symmetric about the inflection point at $\phi = 45°$. The phase curve is shown in Figure 8.4(b).

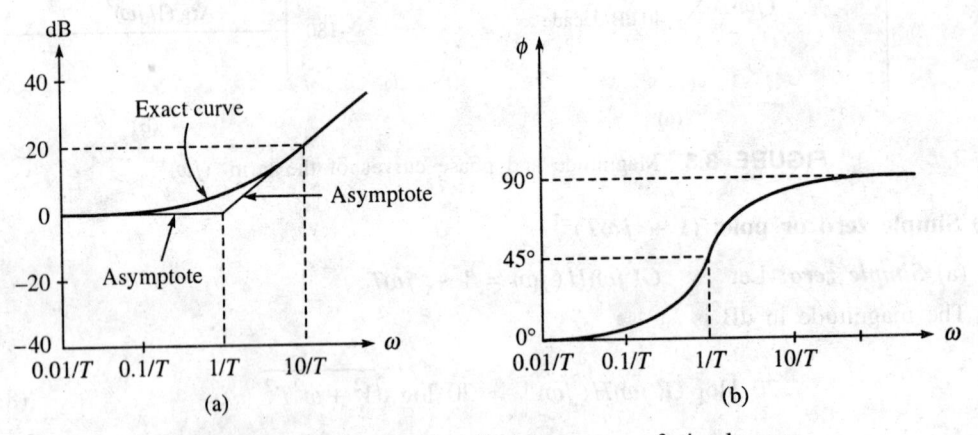

FIGURE 8.4 Magnitude and phase curves of simple zero.

The actual curves deviate slightly from straight line asymptotes and the errors in magnitude and phase are given in Table 8.1(a).

(b) **Simple pole:** Let $\quad G(j\omega)H(j\omega) = (1 + j\omega T)^{-1}$

Now the magnitude in dB is given by

$$20 \log |G(j\omega)H(j\omega)| = 20 \log|(1 + j\omega T)^{-1}| = -20 \log \sqrt{(1^2 + \omega^2 T^2)} \qquad (8.22)$$

Frequency Response Analysis

At very low frequencies, i.e. $\omega T \ll 1$, from Eq. (8.22) we get

$$20 \log |G(j\omega)H(j\omega)| = 0 \text{ dB} \tag{8.23}$$

And at very high frequencies, i.e. $\omega T \gg 1$, we get

$$20 \log |G(j\omega)H(j\omega)| = -20 \log \omega T = -20 \log \omega - 20 \log T \tag{8.24}$$

Equation (8.24) represents a straight line with a slope of –20 dB/decade (or –6 dB/octave) and Eq. (8.23) reprints a 0-dB straight line coincident with the horizontal axis. The intersection of the low frequency and the high frequency asymptotes gives the corner frequency $\omega = \omega_c = 1/T$.

The phase angle of a simple pole is

$$\text{Arg } G(j\omega)H(j\omega) = -\tan^{-1} \omega T \tag{8.25}$$

The linear asymptotic approximate curve and the phase curve are shown in Figure 8.5.

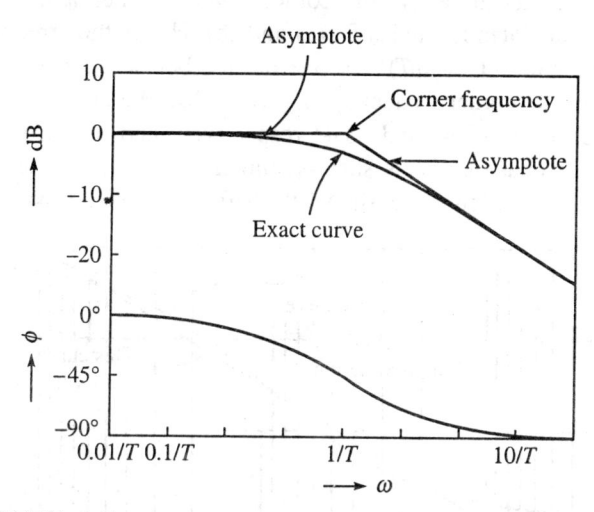

FIGURE 8.5 Magnitude and phase curves of simple pole.

Hence for transfer function $(1 + j\omega T)^{\pm 1}$, at frequency one octave below the corner frequency, i.e. at $\omega = 0.5\omega_c$, the error in magnitude is

$$\pm 20 \log \sqrt{1^2 + \left(0.5 \frac{\omega_c}{\omega_c}\right)^2} - (\pm 20 \log 1) = \pm 0.97 \text{ dB} \approx \pm 1 \text{ dB} \tag{8.26}$$

Similarly at one octave above the corner frequency, i.e. $\omega = 2\omega_c$, the error in magnitude is

$$\pm 20 \log \sqrt{(1^2 + 2^2)} - (\pm 20 \log 2) = \pm 0.97 \text{ dB} \approx \pm 1 \text{ dB} \tag{8.27}$$

For simple zero or pole, i.e. $(1 + j\omega T)^{\pm 1}$, at corner frequency, $\omega = \omega_c$, the error in magnitude is

$$\pm 20 \log \sqrt{(1^2 + 1^2)} - (\pm 20 \log 1) = \pm 3.03 \text{ dB} \approx \pm 3 \text{ dB} \tag{8.28}$$

Further, for simple zero or pole, the error at one decade below the corner frequency, i.e. at $\omega = 0.1\omega_c$ is

$$\pm 20 \log \sqrt{1^2 + \left(0.1\frac{\omega_c}{\omega_c}\right)^2} - (\pm 20 \log 1) \approx \pm 0.04 \text{ dB} \qquad (8.29)$$

Similarly, at one decade above the corner frequency i.e. at $\omega = 10\omega_c$, the error is

$$\pm 20 \log \sqrt{(1^2 + 10^2)} - (\pm 20 \log 10) = \pm 0.04 \text{ dB} \qquad (8.30)$$

The actual phase angle of $(1 + j\omega T)^{\pm 1}$ at $\omega = \omega_c$ is $\pm 45°$ and at one octave below the corner frequency, i.e. at $\omega = 0.5\omega_c$, it is $\pm 26.6°$; and at frequency one octave above the corner frequency, i.e. at $\omega = 2\omega_c$, it is $\pm 63.4°$. Hence, the error in phase is the difference between the actual and the approximate values.

Similarly, the actual phase angle at one decade below the corner frequency, i.e. at $\omega = 0.1\omega_c$ is $\pm 5.7°$ and at one decade above the corner frequency, i.e. at $\omega = 10\omega_c$, it is $\pm 84.3°$. The errors in phase are tabulated in Tables 8.1(a) and (b). Hence the error in phase of a simple zero, i.e. for transfer function $(1 + j\omega T)^{+1}$ at one decade below the corner frequency is $+5.7°$ and at one decade above the corner frequency is $-5.7°$. Also, the error at one octave above and below the corner frequency is $+3.4°$ and $-3.4°$ respectively and the error at corner frequency is $0°$. Hence the error in phase curve is skew-symmetric about the inflection point at corner frequency for a simple pole. Figure 8.6 shows the straight line approximations and the true

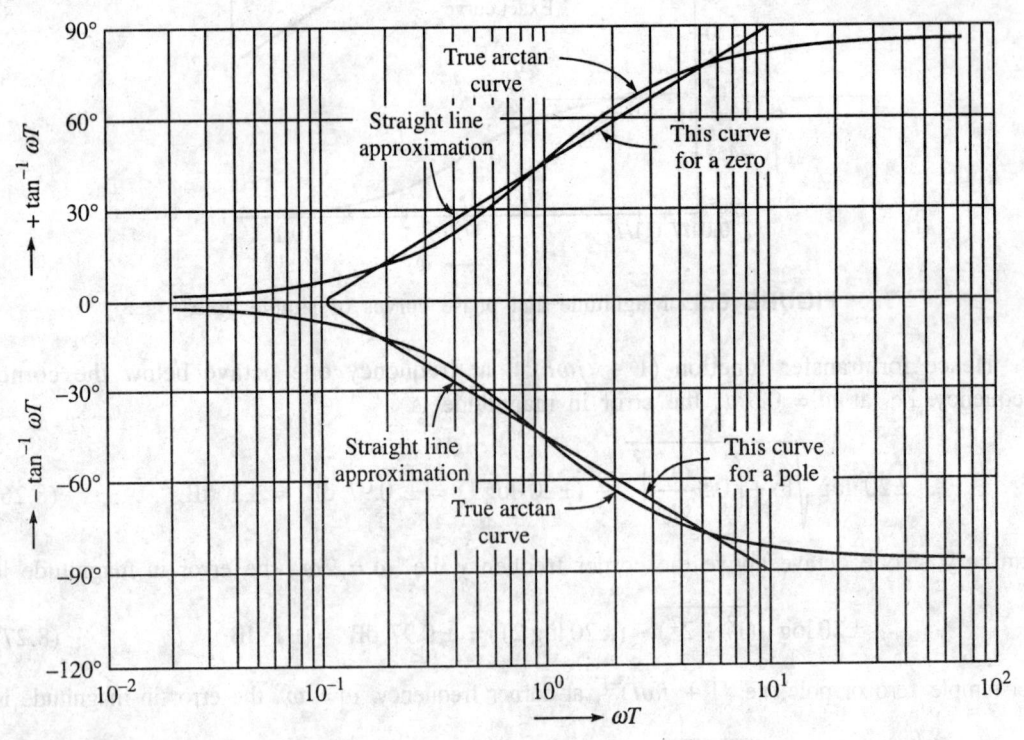

FIGURE 8.6 Straight line approximations and the true arctan curves for simple pole and zero.

Frequency Response Analysis

TABLE 8.1(a) Errors in magnitude and phase of simple zero $(1 + j\omega T)^{+1}$

Corner frequency $\omega_c = 1/T$		Actual magnitude (dB)	Asymptotic value of magnitude (dB)	Error in magnitude (dB)	Actual phase	Asymptotic value of phase	Error in phase
$\omega = 0.1/T$	One decade below ω_c	0.303	0	+0.04	5.7°	0°	+5.7°
$\omega = 0.5/T$	One octave below ω_c	0.97≈1	0	+1	26.6°	30°	−3.4°
$\omega = 1/T$	At corner frequency ω_c	3	0	+3	45°	45°	0°
$\omega = 2/T$	One octave above ω_c	7	6	+1	63.4°	60°	+3.4°
$\omega = 10/T$	One decade above ω_c	20.303	20	+0.04	84.3°	90°	−5.7°

TABLE 8.1(b) Errors in magnitude and phase of simple pole $(1 + j\omega T)^{-1}$

Corner frequency $\omega_c = 1/T$		Actual magnitude (dB)	Asymptotic value of magnitude (dB)	Error in magnitude (dB)	Actual phase	Asymptotic value of phase	Error in phase
$\omega = 0.1\omega_c$	One decade below ω_c	−0.303	0	−0.04	−5.7°	0°	−5.7°
$\omega = 0.5\omega_c$	One octave below ω_c	−0.97≈1	0	−1	−26.6°	−30°	+3.4°
$\omega = \omega_c$	At corner frequency ω_c	−3	0	−3	−45°	−45°	0°
$\omega = 2\omega_c$	One octave above ω_c	−7	−6	−1	−63.4°	−60°	−3.4°
$\omega = 10\omega_c$	One decade above ω_c	−20.303	20	−0.04	−84.3°	−90°	+5.7°

arctan curves with errors taken into account for simple pole and zero with normalized frequency axis ω/ω_c which is ωT.

For simple pole and zero, the actual phase curve is skew symmetric about the straight line approximation as shown in Figure 8.6. The upper one is for simple zero and the lower one is for simple pole.

(iv) **Complex zeros or poles**

(a) *Quadratic poles:* Now consider the second-order transfer function having complex conjugate poles with damping factor $\zeta < 1$, i.e.

$$G(s)H(s) = \frac{\omega_n^2}{s^2 + 2\zeta\omega_n s + \omega_n^2}; \quad \zeta < 1$$

Putting $s = j\omega$ and $\mu = \dfrac{\omega}{\omega_n}$, we get

$$G(j\omega)H(j\omega) = \frac{1}{1 + j2\zeta\mu - \mu^2} \tag{8.31}$$

If $\zeta \geq 1$, it results in two real poles that can be treated like the case of real poles. Rewriting Eq. (8.31) as

$$G(j\omega)H(j\omega) = \frac{1}{\left[1 - \left(\dfrac{\omega}{\omega_n}\right)^2\right] + j2\zeta\left(\dfrac{\omega}{\omega_n}\right)} \tag{8.32}$$

the magnitude in dB is

$$20 \log|G(j\omega)H(j\omega)| = -20 \log\sqrt{[1 - (\omega/\omega_n)^2]^2 + 4\zeta^2(\omega/\omega_n)^2} \tag{8.33}$$

At low frequency, i.e. at $\omega/\omega_n \ll 1$, we get

$$20 \log|G(j\omega)H(j\omega)| = -20 \log 1 = 0 \text{ dB}$$

Thus the low frequency asymptote of the magnitude plot is a straight line, i.e. line on the 0-dB axis of the Bode plot.

At very high frequencies, i.e. at $\omega/\omega_n \gg 1$, we get

$$20 \log|G(j\omega)H(j\omega)| = -20 \log\sqrt{(\omega/\omega_n)^4} = -40 \log(\omega/\omega_n) \tag{8.34}$$

This represents an equation of a straight line with a slope -40 dB/decade (i.e. -12 dB/octave).

The intersection of the two asymptotes at $\omega = \omega_n = \omega_c$ gives the corner frequency since at this frequency

$$-40 \log(\omega/\omega_n) = -40 \log 1 = 0 \text{ dB} \tag{8.35}$$

The magnitude in dB versus frequency in logarithmic scale is shown in Figure 8.7. The actual magnitude curve may differ strikingly from asymptotes near the corner frequency ω_n and the deviation will depend upon the value of damping ratio ζ. The actual deviation is quite large for small values of ζ.

From Eq. (8.32), the magnitude of $G(j\omega)H(j\omega)$ can be written as

$$|G(j\omega)H(j\omega)| = \frac{1}{\sqrt{[1 - (\omega/\omega_n)^2]^2 + [2\zeta(\omega/\omega_n)]^2}} \tag{8.36}$$

Frequency Response Analysis

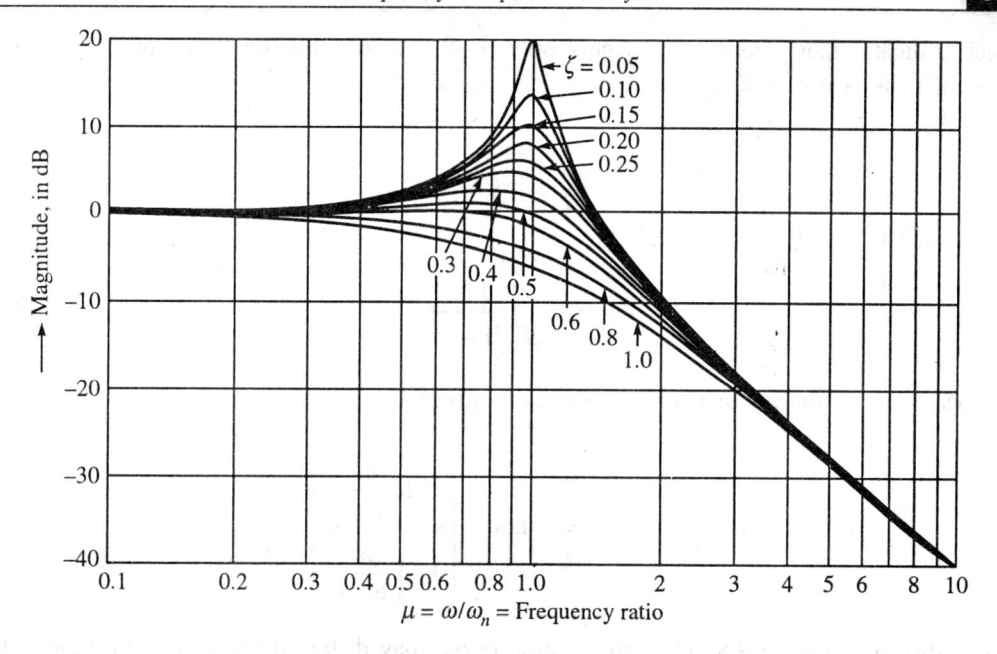

FIGURE 8.7 Magnitude curves of quadratic poles.

If $|G(j\omega)H(j\omega)|$ has a peak value at some frequency, this frequency is called the *resonant frequency*. Since the numerator of $|G(j\omega)H(j\omega)|$ is constant, the peak value of $|G(j\omega)H(j\omega)|$ will occur when

$$\left(1 - \frac{\omega^2}{\omega_n^2}\right)^2 + \left(2\zeta\frac{\omega}{\omega_n}\right)^2 = g(\omega) \quad \text{(say)}$$

is a minimum which can be rewritten as

$$g(\omega) = 1 + \left(\frac{\omega^2}{\omega_n^2}\right)^2 - 2\frac{\omega^2}{\omega_n^2} + 4\frac{\omega^2}{\omega_n^2}\zeta^2$$

$$= \left[\left(\frac{\omega^2}{\omega_n^2}\right) - (1 - 2\zeta^2)\right]^2 + 4\zeta^2(1 - \zeta^2) = \left[\frac{\omega^2 - \omega_n^2(1 - 2\zeta^2)}{\omega_n^2}\right]^2 + 4\zeta^2(1 - \zeta^2)$$

Now, for $g(\omega)$ to be minimum, $\frac{d}{d\omega}[g(\omega)] = 0$ gives

$$\left[\frac{\omega^2 - \omega_n^2(1 - 2\zeta^2)}{\omega_n^2}\right] 4\omega = 0 \quad \text{or} \quad \omega = \omega_n\sqrt{1 - 2\zeta^2}$$

Thus the resonant frequency ω_r is

$$\omega_r = \omega_n\sqrt{1 - 2\zeta^2}\,; \quad 0 \leq \zeta \leq 0.707$$

which indicates that a maximum occurs only if $\zeta < 0.707$ and the maximum value of the resonant peak is obtained as

$$\frac{1}{2\zeta\sqrt{1-\zeta^2}}$$

and obviously the peak gain in dB is

$$20 \log \frac{1}{2\zeta\sqrt{1-\zeta^2}} \tag{8.37}$$

The phase of $G(j\omega)H(j\omega)$ from Eq. (8.32) is given by

$$\angle G(j\omega)H(j\omega) = -\tan^{-1}\left(\frac{2\zeta\frac{\omega}{\omega_n}}{1-\left(\frac{\omega}{\omega_n}\right)^2}\right) \tag{8.38}$$

and is shown in Figure 8.8. The actual phase curve may differ strikingly for different values of damping coefficient ζ.

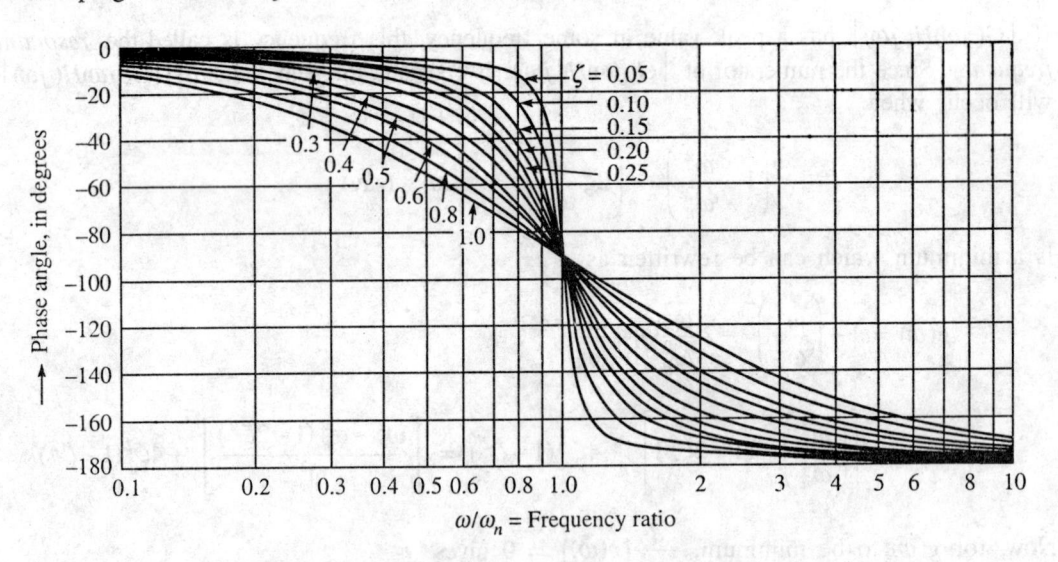

FIGURE 8.8 Phase angle curves of quadratic poles.

(b) *Quadratic zeros:* The analysis of the Bode plot of the second-order transfer function with complex conjugate zeros can be made in the same way but by inverting the curves of Figures 8.7 and 8.8. Here the low frequency asymptote of the curve of magnitude in dB vs normalized frequency ω/ω_n on the logarithmic scale is a 0-dB line. The high frequency asymptote is a straight line of +40 dB/decade (or +12 dB/octave) slope beginning at the corner frequency $\omega_c = \omega_n$. The phase is

$$+\tan^{-1}\left[\frac{2\zeta(\omega/\omega_n)}{1-(\omega/\omega_n)^2}\right] \qquad (8.39)$$

Obviously, the phase and magnitude differ widely at the corner frequency from the true values. Further, for quadratic poles and zeros, the magnitude curve is with errors incorporated for different values of ζ, that is, it is the true magnitude curve. A family of master curves such as those shown in Figure 8.7 can be used to plot the magnitude of the quadratic poles. For complex poles, the slope is −40 dB/decade whereas for that of complex zeros, the slope is +40 dB/decade.

When complex poles or zeros occur in the transfer function, the specific arctan curve depends on the value of ζ. A family of master curves such as those shown in Figure 8.8 can be used to plot the angle of the quadratic zeros. For quadratic poles the phase is from 0° to −180° as shown in Figure 8.8 whereas for quadratic zeros the phase is from 0 to +180°.

The overall gain of Eq. (8.11) is now obtained as the algebraic sum of the gains for the different factors of Eq. (8.12). For the asymptotic plot, the summation can be carried out very easily if we realize that any contribution to the gain is made only for the frequencies above the corner frequency for that factor. Hence, if we start with the plot for ω less than the lowest corner frequency, then all we have to do is to account for the change in slope at the various corner frequencies.

EXAMPLE 8.2 Draw the logarithmic magnitude plot of the transfer function

$$G(s)H(s) = \frac{K_1(s+a)}{s(s+b)(s^2+2\zeta\omega_n s+\omega_n^2)}; \qquad a = 1/T_1 \text{ and } b = 1/T_2 \qquad (8.40)$$

Solution: Rewriting Eq. (8.40) by putting $s = j\omega$ (in the steady state), we obtain

$$GH(j\omega) = \frac{K\left(1+\dfrac{j\omega}{a}\right)}{j\omega\left(1+\dfrac{j\omega}{b}\right)\left[1-\left(\dfrac{\omega}{\omega_n}\right)^2+j2\zeta\dfrac{\omega}{\omega_n}\right]}; \qquad K = \frac{K_1 a}{b\omega_n^2} \qquad (8.41)$$

For the given open-loop transfer function $GH(j\omega)$, the terms for the logarithmic magnitude plot are as follows:

1. Constant, $K \Rightarrow 20 \log K$
2. Pole at origin $\Rightarrow \dfrac{1}{j\omega}$

3. (a) Simple zero $\Rightarrow 1 + \dfrac{j\omega}{a} \Rightarrow 1 + j\omega T_1$

   (b) Simple pole $\Rightarrow \dfrac{1}{1 + \dfrac{j\omega}{b}} \Rightarrow \dfrac{1}{1 + j\omega T_2}$

5. Quadratic poles

$$\dfrac{1}{1 - \left(\dfrac{\omega}{\omega_n}\right)^2 + j2\zeta\left(\dfrac{\omega}{\omega_n}\right)}$$

For the transfer function in Eq. (8.40), the overall gain is shown in Figure 8.9, where it is assumed that

$$a < b < \omega_n, \text{ i.e. } \dfrac{1}{T_1} < \dfrac{1}{T_2} < \omega_n$$

For $\omega < 1/T_1$, only the first and second factors are significant, and they give us a straight line of slope -20 dB/decade corresponding to $(20 \log K - 20 \log \omega)$.

As the frequency is increased, we first get the effect of the simple zero at $s = -a$, i.e. the term $20 \log (1 + j\omega T_1)$ when ω reaches the value a. The linear factor causes the slope to be increased by $+20$ dB/decade, resulting in a straight line of zero slope. As the frequency is further increased, we get the effect of simple pole at $s = -b$, i.e. the term $-20 \log (1 + j\omega T_2)$, when ω reaches the value b. At the corner frequency $\omega = b$, i.e. at $\omega = 1/T_2$, the effect of the simple pole is to reduce the slope by -20 dB/decade. Finally, at the corner frequency ω_n, the effect of the complex poles is to further reduce the slope by -40 dB/decade. See Figure 8.9.

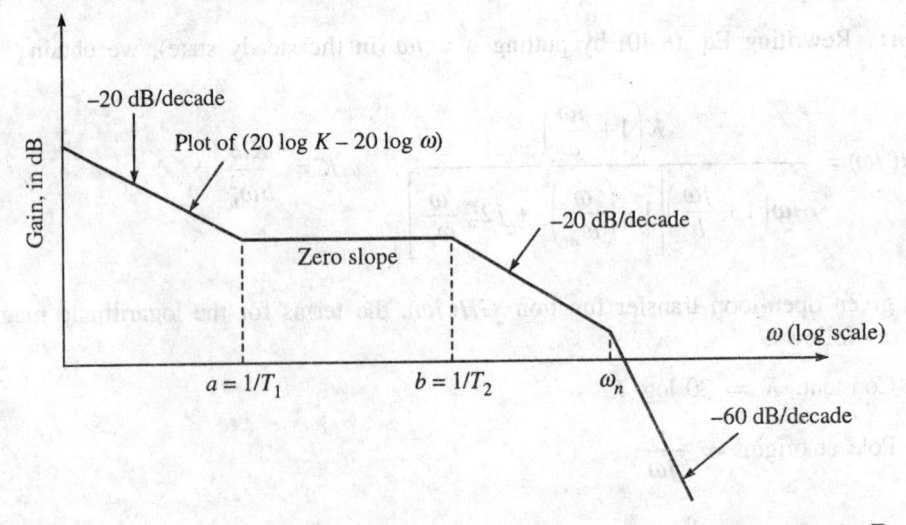

FIGURE 8.9 Example 8.2: asymptotic plot of gain for the transfer function given in Eq. (8.41).

Frequency Response Analysis

EXAMPLE 8.3 Given the open-loop transfer function

$$G(s)H(s) = \frac{13653.3 \times 10^8 (s+150)}{s(s+40)(s+800)(s^2+2000s+4\times 10^6)} \quad (8.42)$$

draw the asymptotic magnitude curve of Bode plot on semilog paper.

Solution: In the steady state, putting $s = j\omega$ and rewriting in proper form, we get

$$G(j\omega)H(j\omega) = \frac{1600\left(1+\dfrac{j\omega}{150}\right)}{j\omega\left(1+\dfrac{j\omega}{40}\right)\left(1+\dfrac{j\omega}{800}\right)\left[1-\left(\dfrac{\omega}{2000}\right)^2+\dfrac{j\omega}{2000}\right]}$$

The open-loop transfer function $GH(j\omega)$ contains four simple types of factors as follows:

(i) Constant $K = 1600$; dB $= 20 \log 1600 = 64$ dB
(ii) Pole at the origin $(j\omega)^{-1}$; slope -20 dB/decade passing through the 0-dB line at $\omega = 1$
(iii) Simple poles and zeros:

  (a) Factor $\left(1+\dfrac{j\omega}{40}\right)^{-1}$; real pole having corner frequency $\omega_c = 40$; asymptotic lines are:

  0-dB line for $\omega < \omega_c = 40$ and line with slope -20 dB/decade for $\omega > \omega_c = 40$. Both the asymptotes meet at $\omega_c = 40$.

  (b) Factor $\left(1+\dfrac{j\omega}{150}\right)$; real zero having corner frequency $\omega_c = 150$; asymptotic lines are:

  0-dB line for $\omega < \omega_c = 150$ and line with slope $+20$ dB/decade for $\omega > \omega_c = 150$. Both the asymptotes meet at $\omega_c = 150$.

  (c) Factor $\left(1+\dfrac{j\omega}{800}\right)^{-1}$; real pole having corner frequency $\omega_c = 800$; asymptotic lines, are: 0-dB line for $\omega < \omega_c = 800$ and line with slope -20 dB/decade for $\omega > \omega_c = 800$. Both the asymptotes meet at $\omega_c = 800$.

(iv) Complex poles: factor

$$\frac{1}{\left[1-\left(\dfrac{\omega}{2000}\right)^2+j\dfrac{\omega}{2000}\right]^{-1}};$$

or, corner frequency $\omega_c = 2000$; asymptotic lines are: 0-dB line for $\omega \ll \omega_c = 2000$ and line with slope -40 dB/decade for $\omega \gg \omega_c = 2000$. Both the asymptotes meet at $\omega_c = 2000$.

The procedure to draw the magnitude (or gain) plot of $G(j\omega)H(j\omega)$ is summarized as follows:

(i) Prepare the semilog graph sheet, choosing scales and marking all corner frequencies.

(ii) Construct the lowest frequency asymptote to start the plot.
(iii) Complete the asymptotic plot by extending the asymptote to higher frequencies while changing the slope at each corner frequency. This is demonstrated in Figure 8.10.
(iv) If necessary (not done in this problem), obtain the actual curve by incorporating error corrections as per Table 8.1 at appropriate corner frequencies.

The given transfer function is of type 1, so the lowest frequency asymptote has a slope of -20 dB/decade. The velocity error coefficient, $K_v = 1600$. Locate $\omega = K_v = 1600$ on the 0-dB axis, draw a straight line with slope of -20 dB/decade through this point and extend this line to low frequencies. The complete diagram is now constructed. See Figure 8.10. Starting with this low frequency line as $|G(j\omega)H(j\omega)|$ (in dB), proceed along it to higher frequencies, encountering the lowest corner frequency at $\omega = p = 40$ where the slope is made more negative by -20 dB/decade, that is, the slope is -40 dB/decade till $\omega = 150$. At $\omega = 150$, the slope changes by $+20$ dB/decade and hence for $150 < \omega < 800$, the slope is -20 dB/decade. For $800 < \omega < 2000$, the slope is -40 dB/decade.

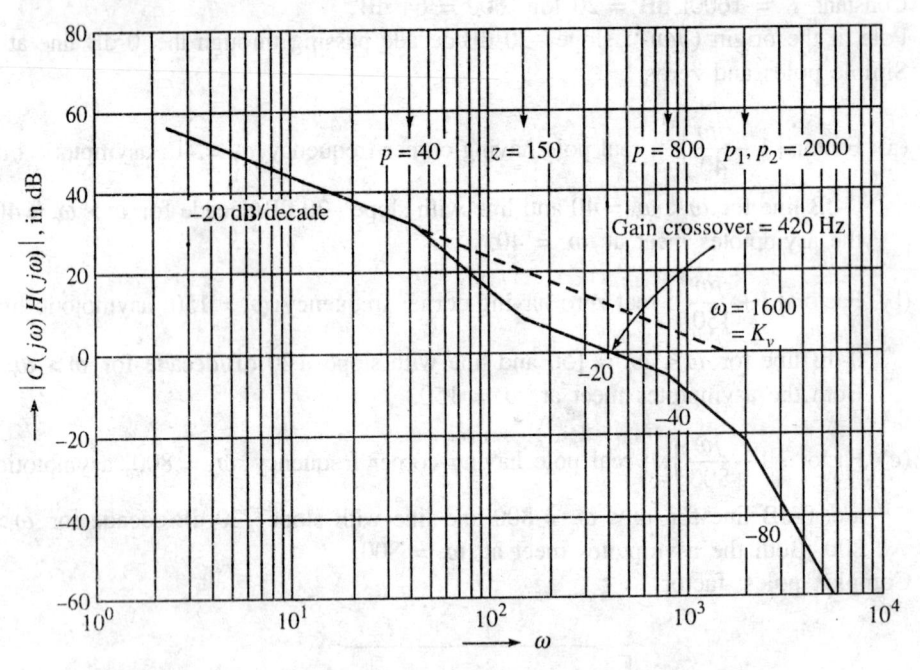

FIGURE 8.10 Example 8.3: construction of magnitude asymptotes.

At $\omega = 2000$, the corner frequency of the quadratic poles, the line slope changes by another -40 dB/decade and hence for $2000 < \omega < \infty$, the slope is -80 dB/decade. This competes the straight line (or asymptotic) representation of the Bode magnitude curve. Note that the value of $\zeta = 0.5$ as defined for the complex poles. It is needed only for the true curve, not for the asymptotic curve.

Here an indication of the gain-crossover frequency is important for determining the phase margin for stability analysis of the closed-loop system from the phase plot of the open-loop

Frequency Response Analysis

transfer function. The gain-crossover frequency is the frequency at which the gain curve crosses the 0-dB line; and from the magnitude curve it is found to be 420 Hz.

EXAMPLE 8.4 Draw the phase curve of the Bode magnitude plot for the transfer function of Example 8.3.

Solution: For convenience, let us rewrite the transfer function as

$$G(H)(j\omega) = \frac{1600\left(1 + \dfrac{j\omega}{150}\right)}{j\omega\left(1 + \dfrac{j\omega}{40}\right)\left(1 + \dfrac{j\omega}{800}\right)\left[1 - \left(\dfrac{\omega}{2000}\right)^2 + \dfrac{j\omega}{2000}\right]}$$

In general, the phase curve consists of four simple types of factors:

(i) Constant, $K = 1600$; Arg $K = 0°$
(ii) Pole at the origin, $(j\omega)^{-1}$; the phase shift of $(j\omega)^{-1}$ is Arg $(j\omega)^{-1} = -1 \times 90 = -90°$
(iii) Simple poles and zeros:

   (a) The phase angle asymptote of simple pole factor $\left(1 + \dfrac{j\omega}{40}\right)^{-1}$ is, Arg $\left(1 + \dfrac{j\omega}{40}\right)^{-1} =$ $-\tan^{-1}\dfrac{\omega}{40}$, i.e. at corner frequency, $\omega = \omega_c = 40$, it is $-45°$ and one decade below ω_c, i.e. at $\omega = 0.1\omega_c = 4$, it is $0°$ and one decade above ω_c, that is, at $\omega = 10\omega_c = 400$, it is $-90°$. For the true phase curve, we incorporate the errors as in Table 8.1.

   (b) The phase shift asymptote for simple pole factor $\left(1 + \dfrac{j\omega}{800}\right)^{-1}$ is $-\tan^{-1}\left(\dfrac{\omega}{800}\right)$, i.e. $-45°$ at corner frequency $\omega = \omega_c = 800$ and $0.1°$ at $\omega = 0.1\omega_c = 80$ and $-90°$ at $\omega = 10\omega_c = 8000$. For the true phase curve, we incorporate errors as in Table 8.1.

   (c) The phase shift asymptote for simple zero factor $\left(1 + \dfrac{j\omega}{150}\right)$ is $\tan^{-1}\left(\dfrac{\omega}{150}\right)$, i.e. at corner frequency $\omega = \omega_c = 150$ it is $45°$ and at $\omega = 0.1\omega_c = 15$, it is $0°$ and at $\omega = 10\omega_c = 1500$, it is $90°$. For the true phase curve, we incorporate errors from Table 8.1(a).

(iv) Quadratic poles: The phase shift for quadratic pole factor

$$\frac{1}{1 - \left(\dfrac{\omega}{2000}\right)^2 + \dfrac{j\omega}{2000}} \quad \text{is} \quad -\tan^{-1}\left[\frac{\dfrac{\omega}{2000}}{1 - \left(\dfrac{\omega}{2000}\right)^2}\right]$$

i.e. at corner frequency $\omega = \omega_c = 2000$, it is $-90°$ and for $\omega \ll \omega_c$, i.e. in practical sense $\omega = 0.1\omega_c = 200$, it is $0°$ and for $\omega \gg \omega_c$, i.e. in practical sense $\omega = 10\omega_c = 20000$, it is $-180°$.

The complete phase-angle curve can be obtained by adding the ordinates of the phase-angle curves for all factors at selected values of ω; plotting the points obtained, and drawing a smooth curve through the calculated points. See Figure 8.11.

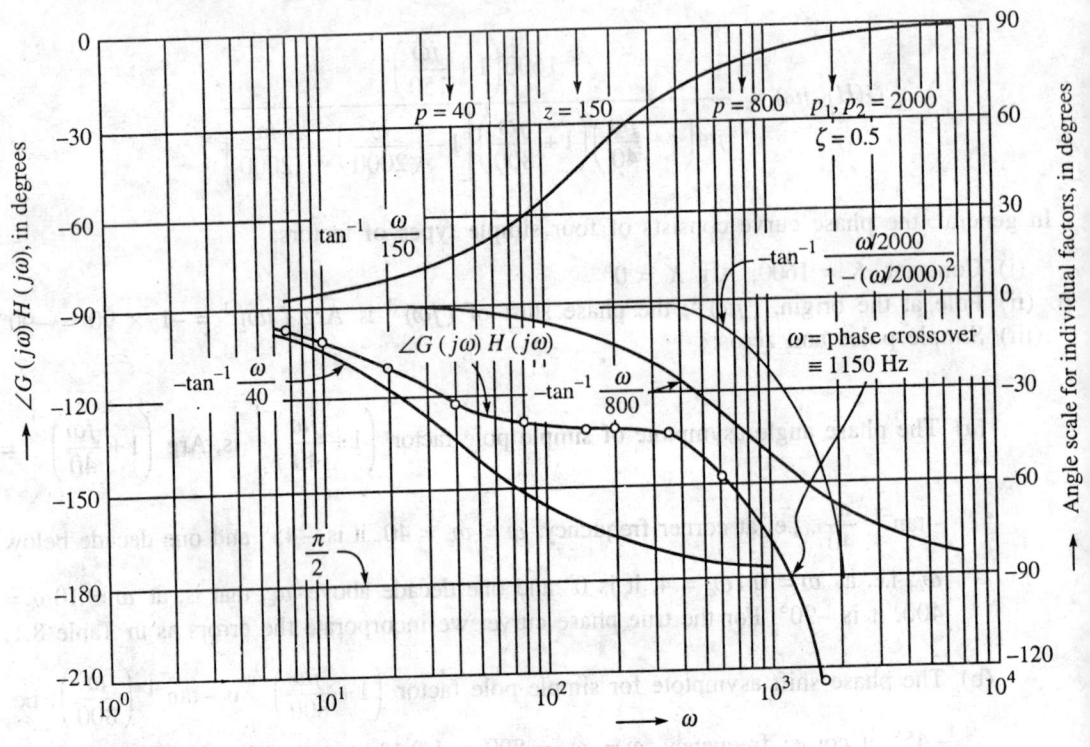

FIGURE 8.11 Example 8.4: phase angle curve.

Here an indication of the phase-crossover frequency is important for determining the gain margin for stability analysis of the closed-loop system from the gain magnitude curve of the open-loop system transfer function. The phase-crossover frequency is the frequency at which the phase curve crosses the $-180°$ line; and from the phase curve it is found to be 1150 Hz.

EXAMPLE 8.5 For the open-loop transfer function of Example 8.3 which for convenience is rewritten as

$$G(j\omega)H(j\omega) = \frac{1600\left(1 + \dfrac{j\omega}{150}\right)}{j\omega\left(1 + \dfrac{j\omega}{40}\right)\left(1 + \dfrac{j\omega}{800}\right) \times \left[\left(1 - \dfrac{\omega}{2000}\right)^2 + \dfrac{j\omega}{2000}\right]}$$

Frequency Response Analysis

draw both the magnitude and phase curves on the same graph paper to find the following:
  (i) Gain-crossover frequency, ω_g
 (ii) Phase margin, PM
(iii) Phase-crossover frequency, ω_p
 (iv) Gain margin, GM

Draw the conclusion about the stability of the closed-loop system.

Solution: It is convenient to plot both curves—Bode magnitude curve and Bode phase curve on the same sheet with the same scale for determining the stability of the closed-loop system. Note that both are completely independent, so the ordinate scales are chosen as convenient. It is recommended, however, that the horizontal axis for both the Bode magnitude curve and Bode phase curve be made the same line. This simplifies the interpretation of phase margin and gain margin as shown in Figure 8.12. On this composite diagram:

1. Locate the gain-crossover point and from that point draw a perpendicular line to touch the phase curve and measure the phase margin.
2. Locate the phase-crossover point and from that point draw a perpendicular to touch the magnitude curve and measure the gain margin.

The gain-crossover frequency is found to be 420 Hz. In Figure 8.12 the vertical line drawn from the magnitude plot at $\omega = 420$ Hz to the phase curve gives the phase margin which happens to be $+44°$.

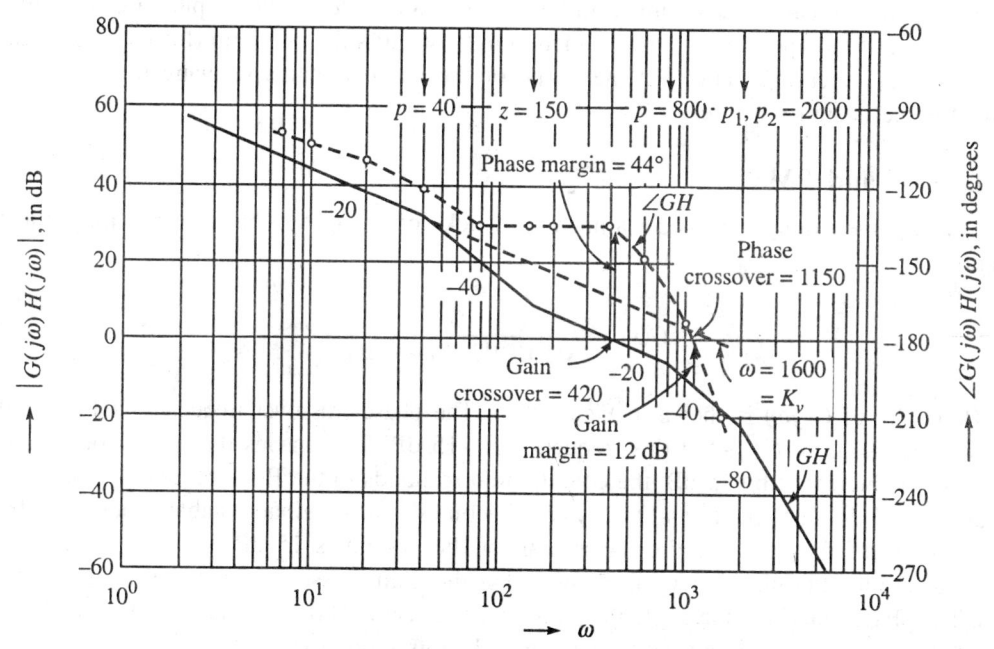

FIGURE 8.12 Example 8.5: magnitude and phase curves.

Phase-crossover frequency is found to be 1150 Hz from Figure 8.12. The vertical line drawn from the phase plot at $\omega = 1150$ Hz to the magnitude plot is the measure of the magnitude in dB at the intersection point of the magnitude curve with the 0-dB line which gives the gain margin as 12 dB.

The closed-loop system is stable as GM > 0 and PM > 0. For a stable system, both GM and PM should be positive.

If the phase margin is positive, the system is stable. In a like manner, when the gain curve is below the phase crossover, the gain margin is positive, which also indicates a stable system.

8.6 Log-magnitude-versus-Phase Plot

The log-magnitude-versus-phase plot is a plot of the logarithmic magnitude in dB versus the phase angles for a frequency range of interest. Such plots are commonly called the Nichols plots. The two curves in the Bode diagram are combined into one by eliminating the common frequency terms to construct the log-magnitude-versus-phase plot. In fact, such a plot can easily be constructed by reading the values of the log-magnitude (dB) and phase angle from the Bode diagram at every frequency point of the frequency range of interest. This is illustrated with an example given below.

The advantages of the log-magnitude-versus-phase plot are that the relative stability of the closed-loop system can be determined. In the log-magnitude-versus-phase plot, a change in the gain constant of the open-loop transfer function merely shifts the curve up (for increasing gain) or down (for decreasing gain), but the shape of the curve remains unchanged. This is very useful for the design purpose.

8.6.1 GM and PM

As an illustrative example, consider a unity-feedback control system having open-loop transfer function as

$$G(s) = \frac{10}{s(1+0.02s)(1+0.2s)}$$

The Bode plot of $G(j\omega)$ is shown in Figure 8.13. The phase-crossover frequency is 15.88 rad/s, and the magnitude of $G(j\omega)$ at this frequency is -15 dB. This means that if the loop gain of the system is increased by 15 dB, the condition corresponds to the Bode plot of $G(j\omega)$ passing through the (0-dB line) point, and the system becomes just marginally stable. Therefore, from the definition of the gain margin, the margin of the system is 15 dB.

To determine the phase margin, we note that the gain-crossover frequency is at $\omega = 6.22$ rad/s. The phase margin is the angle the phase curve must be shifted so that it will pass through the $-180°$ axis at the gain-crossover frequency. In this case,

$$\text{Phase Margin} = 180° - 150° = 30°$$

Frequency Response Analysis

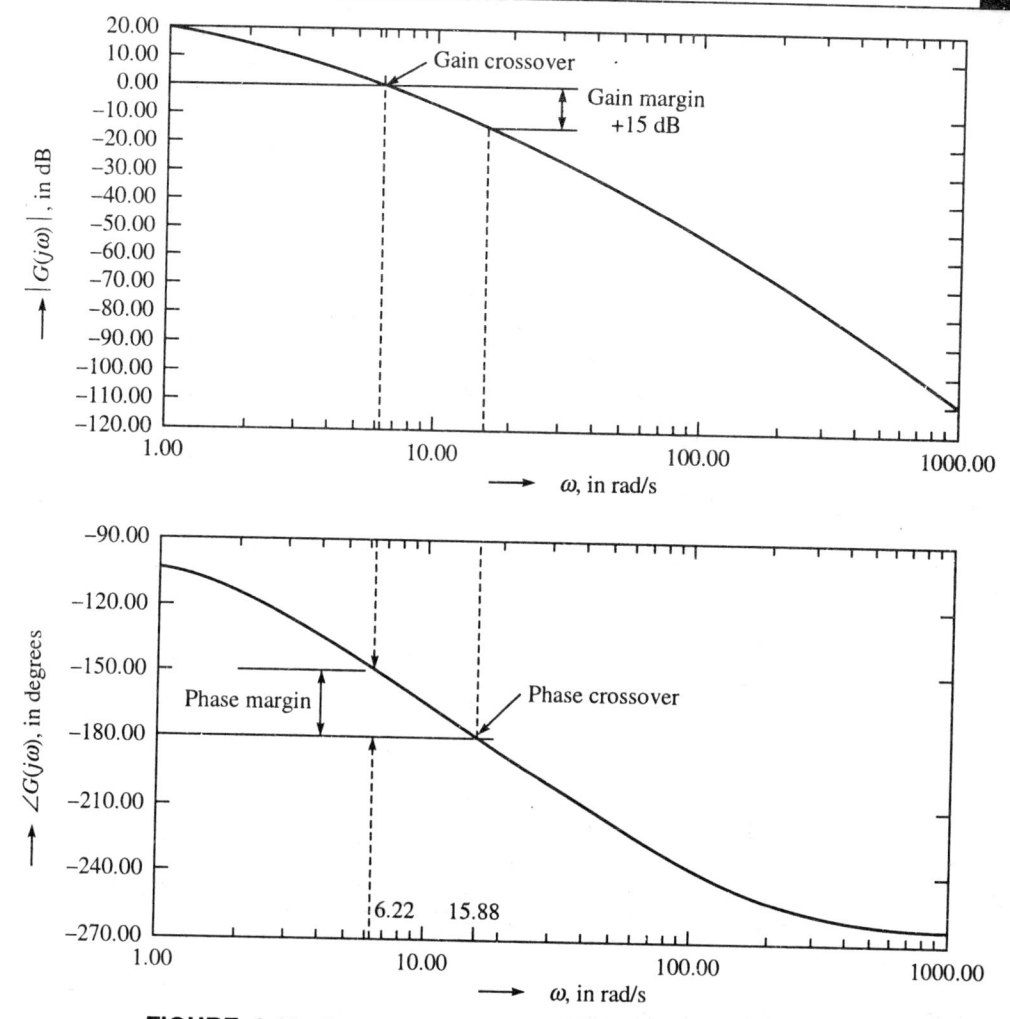

FIGURE 8.13 Bode plot of $G(s) = 10/[s(1 + 0.2s)(1 + 0.02s)]$.

In general, the procedure of determining the gain margin and the phase margin from the Bode plot may be outlined as follows:

1. The gain margin is measured at the phase-crossover frequency ω_p, i.e.

$$\text{Gain Margin} = -|G(j\omega_p)H(j\omega_p)| \text{ dB}$$

2. The phase margin is measured at the gain-crossover frequency ω_g, i.e.

$$\text{Phase Margin} = 180° + GH(j\omega_g)$$

The gain and phase margins are illustrated on the log-magnitude-versus-phase plot as shown in Figure 8.14. On the magnitude-versus-phase plot of the unity-feedback system having open-loop transfer function $G(j\omega)$, the phase-crossover point is, where the locus intersects the $-180°$ axis, and the gain-crossover point is where the locus intersects the 0-dB axis.

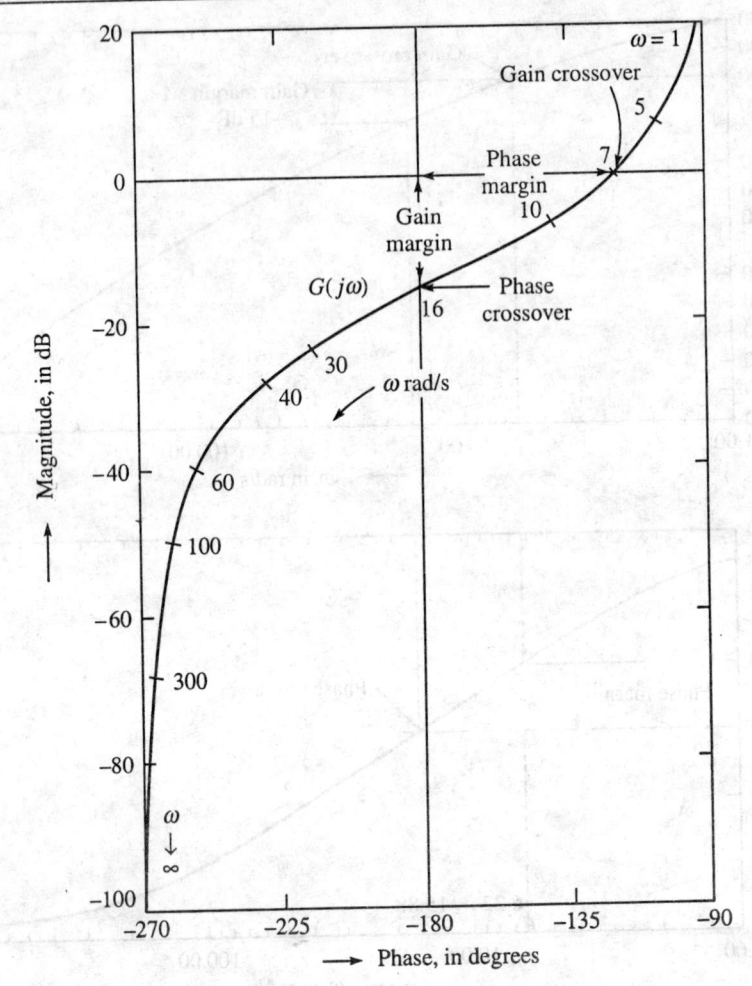

FIGURE 8.14 Magnitude-versus-phase plot of $G(s) = 10/[s(1 + 0.2s)(1 + 0.02s)]$.

Table 8.2 gives some more examples of log-magnitude-versus-phase plots.

8.7 Application of the Frequency Response Plot

8.7.1 Static position error coefficient

For a type-0 system, the open-loop transfer function is

$$G(s)H(s) = \frac{K}{(s+1)(s+10)} \qquad (8.43)$$

TABLE 8.2 Log-magnitude-versus-phase plots of unity-feedback systems

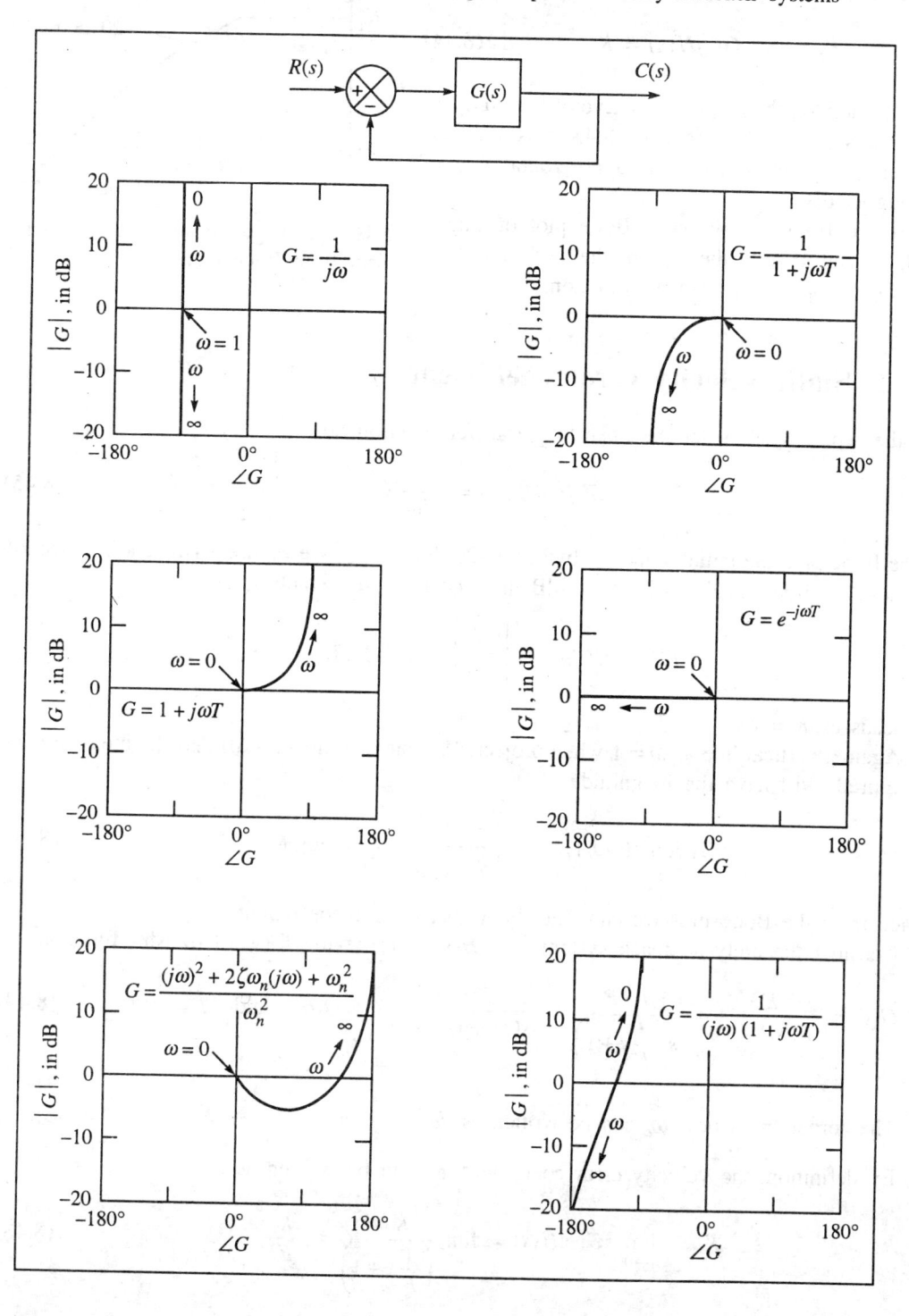

$G = \dfrac{1}{j\omega}$

$G = \dfrac{1}{1 + j\omega T}$

$G = 1 + j\omega T$

$G = e^{-j\omega T}$

$G = \dfrac{(j\omega)^2 + 2\zeta\omega_n(j\omega) + \omega_n^2}{\omega_n^2}$

$G = \dfrac{1}{(j\omega)(1 + j\omega T)}$

By definition, the positional error coefficient K_p is

$$K_p = \lim_{s \to 0} G(s)H(s) = K \qquad (8.44)$$

From Bode plot, the magnitude curve of type-0 open-loop transfer function of Eq. (8.43) has the low frequency asymptote which is a horizontal line of $(20 \log K)$ dB.

Hence from the observed Bode plot of Figure 8.15, we can identify the type of the system and the gain of the open-loop transfer function.

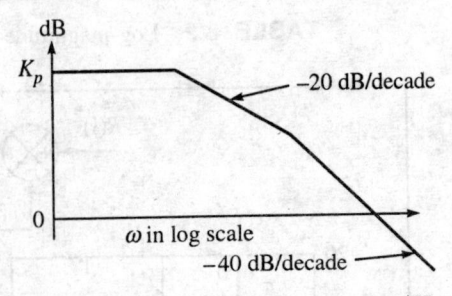

FIGURE 8.15 Position error coefficient from Bode plot.

8.7.2 Static velocity error coefficient

For the type-1 system, let the open-loop transfer function be

$$G(j\omega)H(j\omega) = \frac{K}{j\omega} \qquad (8.45)$$

In the Bode plot, the initial slope is obviously -20 dB/decade (see Figure 8.16). The intersection of the -20 dB/decade line with the 0-dB line occurs at ω_1 which gives

$$G(j\omega)H(j\omega) = \left|\frac{K}{j\omega}\right|_{\omega=\omega_1} = 0 \text{ dB}$$

and leads to $K = \omega_1$.

Again a vertical line at $\omega = 1$ when projected to intersect the -20 dB/decade line (extended if required) will have the magnitude

$$|G(j\omega)H(j\omega)|_{\omega=1} = \left|\frac{K_v}{j\omega}\right|_{\omega=\omega_1} = 20 \log K_v \qquad (8.46)$$

Hence from the Bode plot we can find the velocity error coefficient.

Consider the unity-feedback system (i.e. $H(s) = 1$) having forward transfer function,

$$G(s) = \frac{K}{s(Js+F)} = \frac{K/F}{s\left(\frac{J}{F}s+1\right)} = \frac{K'}{s(Ts+1)}; \quad K' = K/F \text{ and } T = J/F. \qquad (8.47)$$

The corner frequency ω_c can be written as $\omega_c = \dfrac{1}{T} = \dfrac{F}{J}$.

By definition, the velocity error coefficient K_v can be written as

$$K_v = \lim_{s \to 0} sG(s)H(s) = \lim_{s \to 0} \frac{K/F}{\left(\frac{J}{F}s+1\right)} = \frac{K}{F} \qquad (8.48)$$

Frequency Response Analysis

Hence the velocity error coefficient K_v is K/F. The -20 dB/decade line (extended if required) intersects the 0-dB line at $\omega = \omega_1$ will be

$$20 \log \left|\frac{K/F}{s}\right|_{s=j\omega} = 20 \log \left|\frac{K/F}{s}\right|_{\omega=\omega_1} = 0 \text{ dB} = 20 \log 1 \qquad (8.49)$$

and gives $\omega_1 = \dfrac{K}{F}$

Therefore, $\omega_1 = K_v = \dfrac{K}{F}$.

Hence the extension of the -20 dB/decade line to intersect the 0-dB line at frequency ω_1 (say) gives $\omega_1 = K/F$. See Figure 8.16. Further, the intersection of the -40 dB/decade line with the 0-dB line gives ω_3. By rewriting the open-loop transfer function $G(s)H(s)$ as

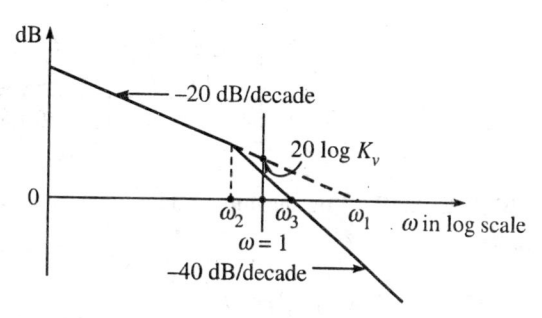

FIGURE 8.16 Velocity error coefficient from Bode plot.

$$20 \log \left|\frac{K/J}{s\left(s+\frac{F}{J}\right)}\right|_{s=j\omega} = 20 \log \left|\frac{K/J}{j\omega\left(j\omega+\frac{F}{J}\right)}\right|_{\omega=\omega_3} = 0 \text{ dB}$$

gives
$$\omega_3^2 = K/J \quad \text{or} \quad \omega_3 = \sqrt{K/J} \qquad (8.50)$$

For the first-order part, the corner frequency

$$\omega_2 = \frac{F}{J} \qquad (8.51)$$

Therefore, $\omega_1 \omega_2 = \omega_3^2$ or $\dfrac{\omega_1}{\omega_3} = \dfrac{\omega_3}{\omega_2}$

On logarithmic plot, $\log \omega_1 - \log \omega_3 = \log \omega_3 - \log \omega_2$, i.e. ω_3 is just the midway point between ω_1 and ω_2.

Again comparing the open-loop transfer function $\dfrac{K/J}{s(s+F/J)}$ with the standard form $\dfrac{\omega_n^2}{s(s+2\zeta\omega_n)}$, we get the damping coefficient as

$$\zeta = \frac{F}{2\sqrt{KJ}} = \frac{\omega_2}{2\omega_3}$$

The intersection of the initial -20 dB/decade segment (or its extension) with $\omega = 1$ line has the magnitude of $20 \log K_v$ as stated earlier.

The log magnitude curve of type-2 system is shown in Figure 8.17. The intersection of the initial −40 dB/decade segment (or its extension) with $\omega = 1$ line has the magnitude $20 \log K_a$.

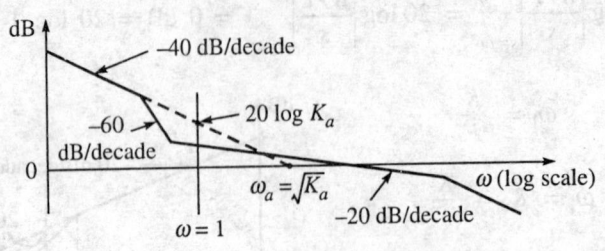

FIGURE 8.17 Log-magnitude plot of type-2 system (acceleration error coefficient from Bode plot).

Since at low frequency

$$G(j\omega)H(j\omega) = \frac{K_a}{(j\omega)^2} \tag{8.52}$$

it follows that

$$20 \log \left| \frac{K_a}{(j\omega)^2} \right| = 20 \log K_a \tag{8.53}$$

The frequency ω_a at the intersection of the initial −40 dB/decade segment (or its extension) with the 0-dB line gives the square root of K_a numerically, as can be seen from the following:

$$20 \log \left| \frac{K_a}{(j\omega)^2} \right| = 20 \log 1 = 0$$

which yields

$$\omega_a = \sqrt{K_a} \tag{8.54}$$

8.7.3 Estimation of transfer function from Bode plot

Approximately, the magnitude curve (i.e. gain in dB versus frequency in logarithmic scale) is given by straight line segments of slope $20n$-dB/decade, where n is an integer. Note the corner frequencies at the poles and zeros. The negative slope is for poles and the positive slope for zeros. For change of slope by 40 dB/decade, i.e. 2×20 dB/decade leads to the conclusion that the break frequency may be due to

(i) a double real pole or zero
(ii) a pair of complex poles or zeros with the undamped natural frequency given by the corner or break frequency.

Frequency Response Analysis

For case (ii) the damping coefficient may be estimated by examining the frequency response near the break frequency. Further, only the magnitude curve is not sufficient to determine the exact transfer function. The magnitude curves of $K/(s+1)$ and $K/(s-1)$ are the same [see Figure 8.18(a)]. But with the given phase curve in Figure 8.18(b) (lagging in nature), we can ascertain the exact transfer function as $K/(1+s/a)$. So in general, the magnitude and phase curves together can determine the exact transfer function of the system.

As another example, the straight line approximation of the magnitude plot in Figure 8.19(a) gives the estimate of nature of the transfer function of the form

$$\frac{K}{s^2 + 2\zeta\omega_n s + \omega_n^2}$$

The value of K is determined from low frequency gain and ω_n is determined from break frequency in the magnitude

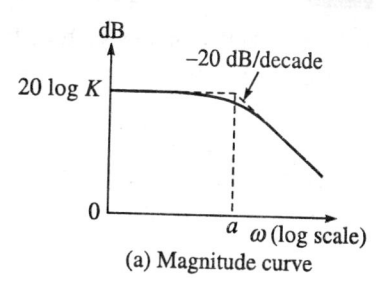

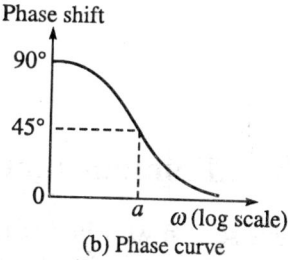

FIGURE 8.18 Bode plot.

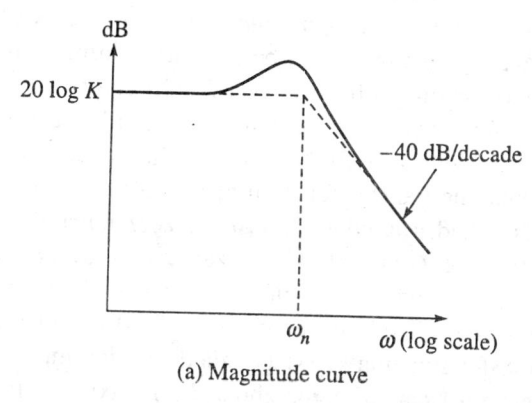

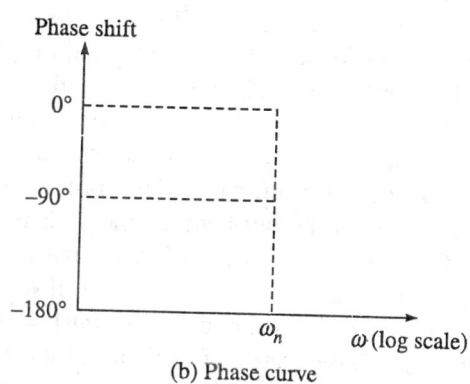

FIGURE 8.19 Bode plot.

plot, and ω_n is also the frequency at which the phase shift is 90° as obtained from the phase curve in Figure 8.19(b). Finally, the damping coefficient ζ is obtained from the following:

$$G(j\omega)\Big|_{\omega=\omega_n} = \frac{K}{2\zeta\omega_n^2}$$

Looking at another example, from the magnitude plot of Figure 8.20, the corner frequencies are ascertained as 2 and 10 from the straight line approximations. The corner frequency 2 gives the estimate for zero and 10 gives the estimate for pole. From the initial slope of −20 dB/decade, it

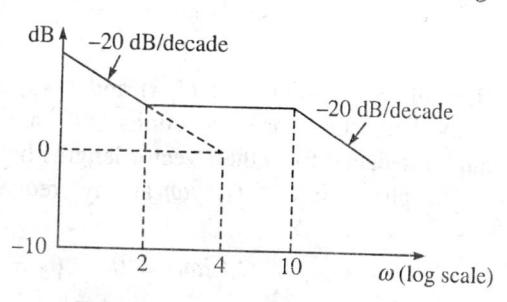

FIGURE 8.20 Magnitude plot.

is ascertained that the system transfer function is of type-1. Further, the intersection of the extended projection (dotted) of initial -20 dB/decade line with the 0-dB line gives

$$20 \log 4 = 20 \log K_v = 20 \log K \quad \text{or} \quad K = 4$$

Hence the transfer function is

$$\frac{4\left(1 + \frac{s}{2}\right)}{s\left(1 + \frac{s}{10}\right)}$$

8.8 Minimum and Non-Minimum Phase Transfer Functions

In this section we will discuss the minimum and non-minimum phase transfer functions. From the name it is clear that it is of comparative nature. We are talking about the stable system only, i.e. when all the poles lie in the left-half of the s-plane. A system may have zeros located in the right-hand s-plane and may still be stable. The stable transfer functions with zeros in the right-hand s-plane are classified as *non-minimum phase transfer functions*. If the zeros of a transfer function are all reflected about the $j\omega$-axis, then there will be no change in magnitude of the transfer function, and the only difference arising will be in phase shift characteristics. If the phase characteristics of two system transfer functions are compared, it can be seen that the net phase shift over the frequency range from zero to infinity is less in the system having all its zeros in the left-hand of the s-plane. Thus, the transfer function of a stable $G_1(s)$ having all its zeros in the left-hand of the s-plane is called a *minimum phase transfer function*.

Consider another system having transfer function $G_2(s)$ with all the zeros of $G_1(s)$ reflected about the $j\omega$-axis into the right-hand of the s-plane and all its poles being the same as those of $G_1(s)$, then the magnitude $|G_2(j\omega)| = |G_1(j\omega)|$. The transfer function $G_1(s)$ is called the minimum phase transfer function and the transfer function $G_2(s)$ is called the non-minimum phase transfer function. Reflection of any zero or a pair of zeros about the $j\omega$ axis results in a non-minimum phase transfer function.

Consider the transfer functions $G_1(s)$ and $G_2(s)$ as

$$G_1(s) = \frac{s+z}{s+p} \quad \text{and} \quad G_2(s) = \frac{s-z}{s+p}$$

The pole-zero patterns of $G_1(s)$ and $G_2(s)$ are shown in Figures 8.21(a) and 8.21(b) respectively. Both the transfer functions $G_1(s)$ and $G_2(s)$ have the same magnitude characteristics as can be deduced from their vector lengths but their phase characteristics are obviously different.

The phase shift of $G_1(j\omega)$ at any frequency $\omega = \omega_1$ can be written as

$$\angle G_1(j\omega_1) = \theta_1 - \theta_2 = \angle \tan^{-1}\frac{\omega_1}{z} - \angle \tan^{-1}\frac{\omega_1}{p}$$

Frequency Response Analysis

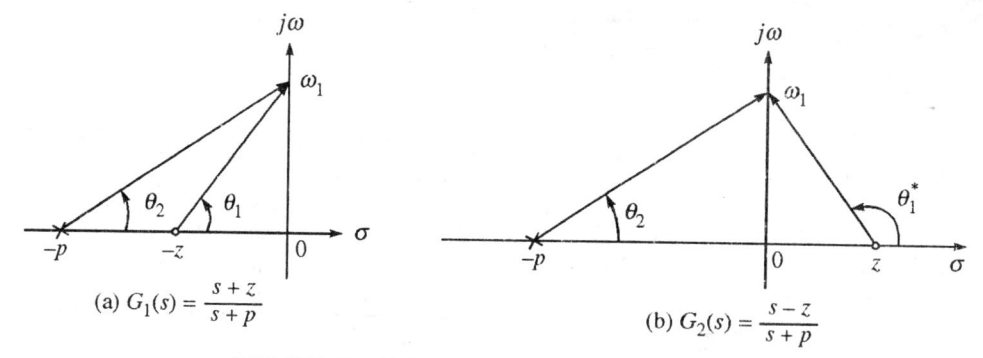

(a) $G_1(s) = \dfrac{s+z}{s+p}$

(b) $G_2(s) = \dfrac{s-z}{s+p}$

FIGURE 8.21(a) and (b) Pole-zero patterns.

which is an acute angle for $\omega > 0$ and clearly ranges over less than 90° (as is apparent from the figure).

The phase shift of $G_2(j\omega)$ at $\omega = \omega_1$ can be written as

$$\angle G_2(j\omega_1) = \theta_1^* - \theta_2 = \left(\pi - \angle\tan^{-1}\dfrac{\omega_1}{z}\right) - \angle\tan^{-1}\dfrac{\omega_1}{p}$$

which is an obtuse angle for $\omega > 0$ and ranges from 0° to 180° for $\omega = \infty$ to $\omega = 0$ respectively in the frequency range $0 < \omega < \infty$. Strictly speaking, the phase shift of $G_1(j\omega)$ is acute and that of $G_2(j\omega)$ obtuse.

It means $G_1(s)$ has minimum phase characteristic while $G_2(s)$ has non-minimum phase characteristic. The range of phase shift of a minimum phase transfer function is the least possible or minimum corresponding to a given amplitude curve, while the range of a non-minimum phase curve is greater than the minimum possible for the given amplitude curve. See Figure 8.21(c).

Further, it may be mentioned that we have got acute angle for minimum phase function and obtuse angle for non-minimum phase function. The angles are +ve as shown in Figure 8.21(c) for $|z| < |p|$, as for $G_1(s) = \dfrac{s+1}{s+2}$ and $G_2(s) = \dfrac{s-1}{s+2}$. If $|z| > |p|$, as for $G_1(s) = \dfrac{s+2}{s+1}$ and $G_2(s) = \dfrac{s-2}{s+1}$, then the angles are in the negative range as shown in Figure 8.21(d).

An all-pass network is realized with symmetrical lattice network as in Figure 8.22(a). A symmetrical pattern of poles and zeros is obtained for this network as shown in Figure 8.22(b). Again the magnitude remains constant with frequency (in this case becomes unity) as in Figure 8.22(c), whereas the phase characteristic varies from 0° to −360°. It is evident from Figure 8.22(b) that $\theta_2 = 180° - \theta_1$ and $\theta_2^* = 180° - \theta_1^*$. The phase shift is $\phi(\omega) = -2(\theta_1 + \theta_1^*)$. Obviously, this is a non-minimum phase symmetrical lattice network and has wide applications as repeater in telephone communication networks.

Earlier we have used the casual definition of minimum and non-minimum phase transfer functions by considering a stable system, i.e. having poles in the left-half of the s-plane. Strictly speaking, from academic point, transfer functions having no poles or zeros in the right-

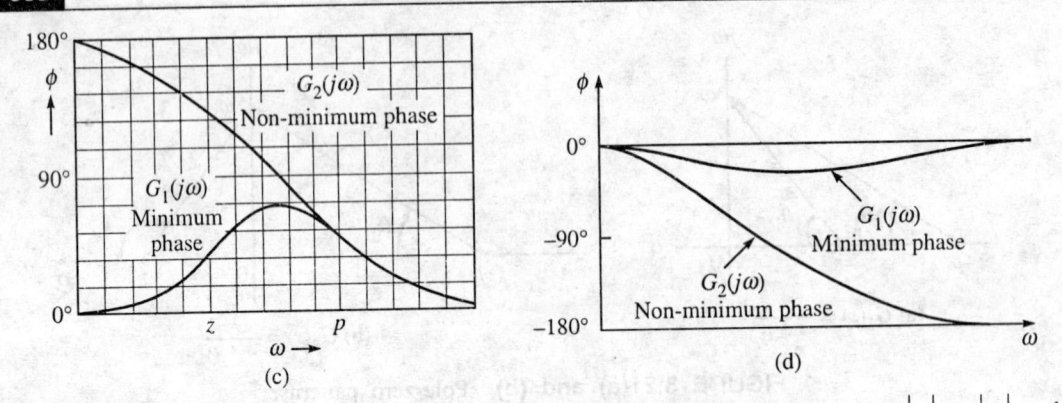

FIGURE 8.21(c) and (d) (c) Phase angle characteristics of $G_1(j\omega)$ and $G_2(j\omega)$ for $|z| < |p|$ and (d) phase angle charateristics of $G_1(j\omega)$ and $G_2(j\omega)$ for $|z| > |p|$.

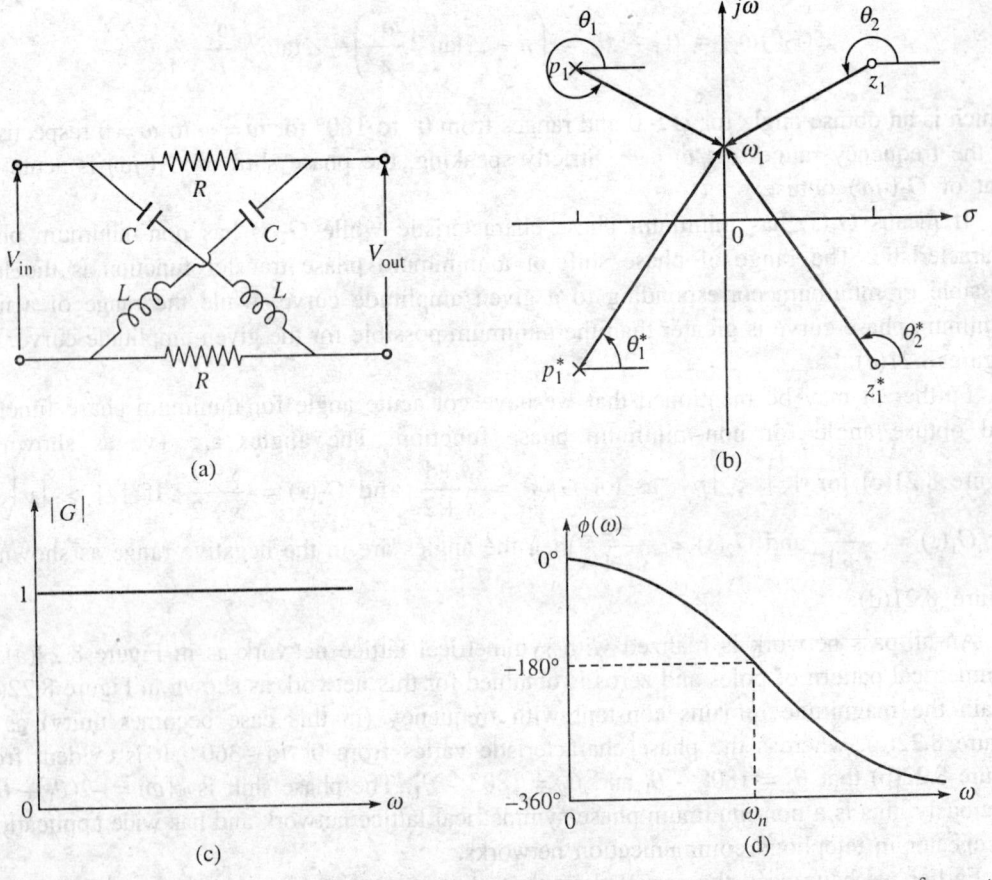

FIGURE 8.22 (a) All-pass network, (b) pole-zero pattern (c) frequency response of constant magnitude curve, and (d) frequency response of phase curve.

half of the s-plane are minimum-phase transfer functions; whereas those having poles and/or zeros in the right-half of the s-plane are non-minimum phase transfer functions. We have, for all practical purpose, considered poles only on the left-hand side because of stable system. Systems with minimum phase transfer functions are called minimum phase systems, whereas those with non-minimum phase transfer functions are called non-minimum phase systems. Systems having the same magnitude characteristic have the range in phase angle as minimum for minimum-phase transfer functions, while the range in phase angle for the non-minimum phase transfer functions is greater than that for the minimum-phase transfer functions.

Non-minimum-phase systems are slow in response because of their faulty behaviour at the start of response (see the MATLAB exercise in this chapter). A common example of a non-minimum phase system is the presence of transportation lag in control systems. In control systems, the presence of transportation lag produces excessive phase lag which is instrumental for the system to become unstable, and thus in practical control systems the excessive phase lag should be avoided.

For a minimum-phase system with n-poles and m-zeros, excluding poles or zeros at s = origin, the net phase angle change as ω decreases from ω = infinity to ω = 0, is given by $(n - m)\pi/2$.

In Bode plot, we require both magnitude and phase diagrams in order to determine the transfer function of the system correctly. The transfer function can be determined either from the magnitude plot or from the phase plot alone if the desired system is a minimum phase system. In order to determine the transfer function of a non-minimum phase system, both magnitude and phase plots are required.

Consider now the magnitude plot as shown in Figure 8.23. The gain $K = 1$ and the corner frequency $\omega_c = 1$. The transfer function may be either $1/(s + 1)$ or $1/(s - 1)$, because the magnitude for both the cases is $1/\sqrt{(1+\omega^2)}$. Hence from the given magnitude curve, the transfer function should be written as $1/(s \pm 1)$ unless it is stated that the system is a minimum phase system for which case the transfer function will be only $1/(s + 1)$.

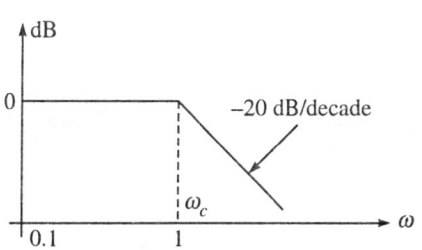

FIGURE 8.23 Magnitude plot.

Hence for a minimum phase system, the Bode magnitude plot alone completely and unambiguously determines the transfer function. The same is also true for the phase plot alone for a minimum phase system. But the magnitude plot has the advantage as it can be easily constructed using the approximate straight-line asymptotes.

8.9 Stability of Control System with Time Delay

There are many control systems which have a time delay within the closed loop of the system and which affects the stability of the system. Fortunately, the Nyquist criterion can be utilized to determine the effect of the time delay on the relative stability of the feedback system. A pure time delay, without attenuation, is represented by the transfer function, $G_d(s) = e^{-sT}$, where T is the delay time. The Nyquist criterion remains valid for a system with a time delay, since the factor e^{-sT} does not introduce any additional poles or zeros within the contour. This factor, however, adds a phase shift to the frequency response without altering the magnitude curve.

We can plot the Bode diagram including the delay factor, and investigate the stability relative to the 0-dB, −180° point. The delay factor $e^{-j\omega T}$ results in a phase shift, $\phi(\omega) = -\omega T$, and is readily added to the phase shift resulting from $G_a G(j\omega)$; see Figure 8.24 of Example 8.6. Example 8.6 also shows the simplicity of this approach for constructing the Bode diagram.

EXAMPLE 8.6 A level control system is shown in Figure 8.24 and its block diagram in Figure 8.25(a) with typical transfer function of each of the components of the level control system such as for

(i) Hydraulic actuator, transfer function, $G_a(s) = \dfrac{10}{s+1}$

(ii) Water tank, transfer function, $G(s) = \dfrac{3.15}{30s+1}$

(iii) Float transfer function, $G_f(s) = \dfrac{s^2}{9} + \dfrac{s}{3} + 1$

(iv) Transfer function of delay element, $G_d(s) = e^{-sT}$, with $T = 1$ s

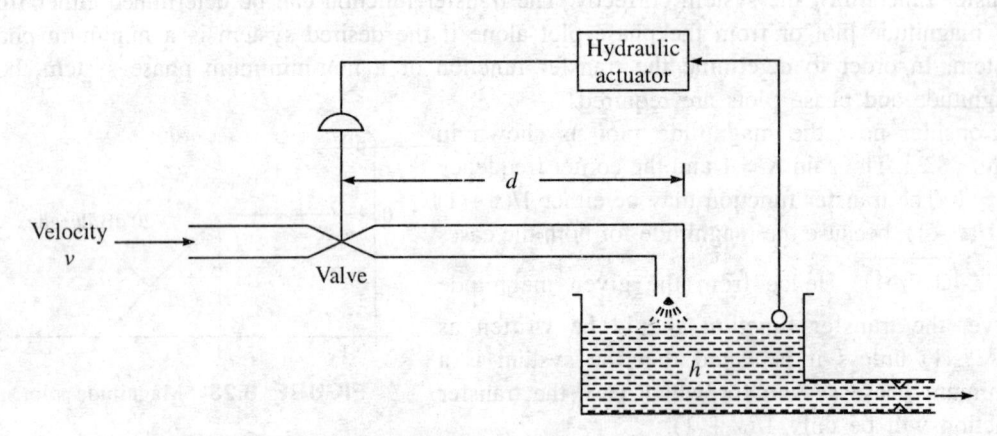

FIGURE 8.24 Example 8.6: a liquid level control system.

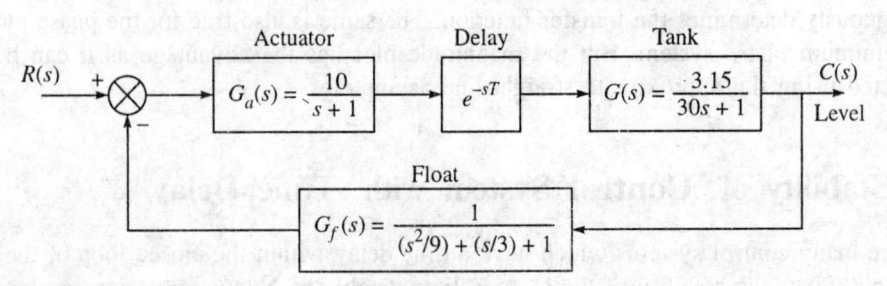

FIGURE 8.25(a) Example 8.6: block diagram of the liquid level control system.

The time delay between the valve adjustment and the fluid output is, $T = d/v$.

Frequency Response Analysis

Now, if the flow rate is 5 m³/s, the cross-sectional area of the pipe is 1 m², and the distance d is equal to 5 m, we will have the time delay $T = 1$ s. The open-loop transfer function is then

$$GH(s) = G_a(s)G(s)G_f(s)e^{-sT} = \frac{31.5\, e^{-s}}{(s+1)(30s+1)\left(\frac{s^2}{9}+\frac{s}{3}+1\right)}$$

Figure 8.25(b) shows the Bode diagram for the level control system.

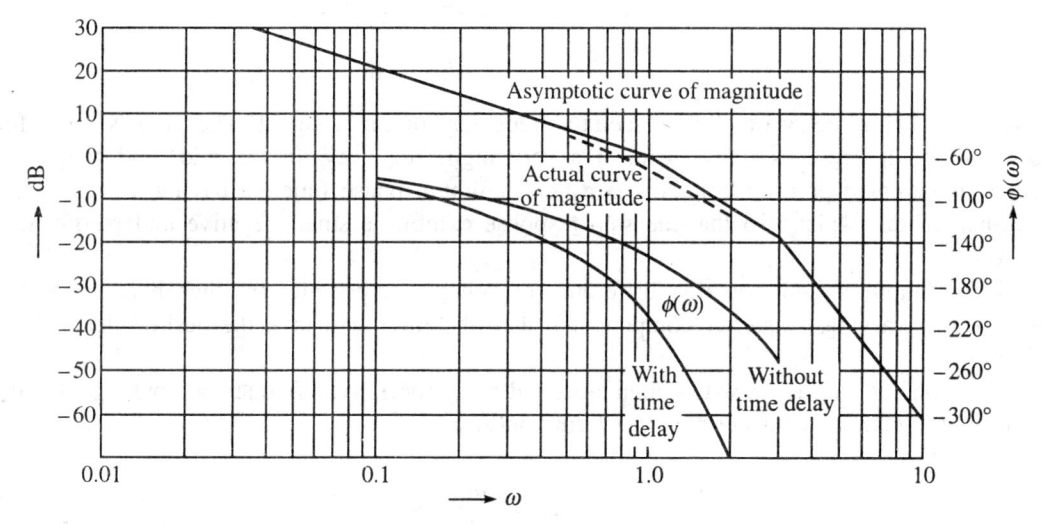

FIGURE 8.25(b) Example 8.6: Bode diagram for the level control system.

The logarithmic gain curve crosses the 0-dB line at $\omega = 0.8$. Therefore the phase margin of the system without time delay would be 40°. However, with the time delay added, we find that the phase margin is equal to −3°, and the system is unstable.

Therefore the system gain must be reduced in order to provide a reasonable phase margin of 30°. The gain would have to be decreased by a factor of 5 dB to $K = 31.5/1.78 = 17.7$.

A time delay, e^{-sT}, in a feedback system introduces an additional phase lag and results in a less stable system. Therefore as pure time delays are unavoidable in many systems, it is often necessary to reduce the loop gain in order to obtain a stable response. However, the penalty for stability is the resulting increase in the steady-state error of the system as the loop gain has to be reduced.

8.10 Simulation of Time Delay

Systems described inherently by transcendental transfer functions are difficult to handle. There are many ways of approximating $\exp(-sT)$ by a rational function. It is known that:

$$e^{-sT} = 1 - sT + \frac{(sT)^2}{2!} - \frac{(sT)^3}{3!} + \dots$$

and
$$e^{-sT} = \frac{1}{e^{sT}} = \frac{1}{1 + sT + \frac{(sT)^2}{2!} + \frac{(sT)^3}{3!} + \cdots}$$

From Pade approximation, we get

$$e^{-sT} = \frac{e^{\frac{-sT}{2}}}{e^{\frac{sT}{2}}} = \frac{1 - \frac{sT}{2} + \frac{s^2T^2}{8} - \cdots}{1 + \frac{sT}{2} + \frac{s^2T^2}{8} + \cdots}$$

This gives a time delay of T for a certain frequency of input signal. The gain is unity for $\omega \ll 1$ and it becomes unity again at a very high frequency. In the mid-band frequency range, variation of gain occurs. This is a non-minimum phase function having a zero in the right-half of the s-plane so that the step response exhibits a small negative undershoot near $t = 0$.

If the input signal contains different frequency components (neither harmonics nor very widely varying frequency components), then the filter will give the desired delay of T seconds.

Now we will discuss how to obtain a desired delay for a given frequency input signal, with Pade approximation (considering three terms only) as

$$e^{-sT} = \frac{\left(1 - T^2 \frac{\omega^2}{8}\right) - j\frac{\omega}{2}T}{\left(1 - T^2 \frac{\omega^2}{8}\right) + j\frac{\omega}{2}T}$$

Since we know the signal frequency ω, i.e. how much time does it take for 1 cycle, the phase shift $\phi = -\tan^{-1}\omega T$ is calculated from the desired delay and the input signal frequency. So we can calculate what should be the value of T, say, for $\omega = 1$. Thus,

$$e^{-Ts} = \frac{\mathscr{L} x(t-T)}{\mathscr{L} x(t)} = \frac{\frac{T^2}{8}s^2 - \frac{T}{2}s + 1}{\frac{T^2}{8}s^2 + \frac{T}{2}s + 1} = \frac{\left(1 - \frac{T^2}{8}\right) - j\frac{T}{2}}{\left(1 - \frac{T^2}{8}\right) + j\frac{T}{2}}$$

See MATLAB simulation where the negative undershoot is seen at the initial stage. Further, for $T = 20$ s, with the 1st order Pade approximation, MATLAB simulation Delay.m has been made which gives the value of delay time as 15.7 s. Further with the 2nd order Pade approximation the delay time has improved to 16.6 s in MATLAB simulation. It is expected that the time delay T would approach 20 s with a higher order Pade approximation.

MATLAB Program 8.1

```
% Delay.m
clear;
pack;
clc;
% Original 1st order System
num=[10];
den=[210 1];
% Original System cascaded
% with 1st order Pade's approximation.
num_1=[-100 10];
den_1=[2100 220 1];
% Original System cascaded
% with 2nd order Pade's approximation,
num_2=[500 -100 10];
den_2=[10500 2150 220 1];
% otherwise use Matlab Pade function

t=0:10:1490;

[Y1,X1,T1]=step(num_1,den_1,t);
i=1;
while Y1(i)<9.95 % with 5% criterion.
 i=i+1;
end
settletime=T1(i);
pop2=sprintf('Settle Time = %g',settletime);
step(num_1, den_1,t);
hold on;
step(num, den,t);
text(1100,4,pop2);
pop3=sprintf('--->Function with time delay');
text(T1(30),Y1(30),pop3);
grid on;
hold off;
pause;
clf;
t=0:.1:149;
[Y1,X1,T1]=step(num_1,den_1,t);
i=2;
while Y1(i)>0
 i=i+1;
end
while Y1(i)<0
 i=i+1;
end
delaytime=T1(i);
pop1=sprintf('Delay Time =%g \n with 1st order \n Pade approximation',delaytime);
step(num_1,den_1,t);
```

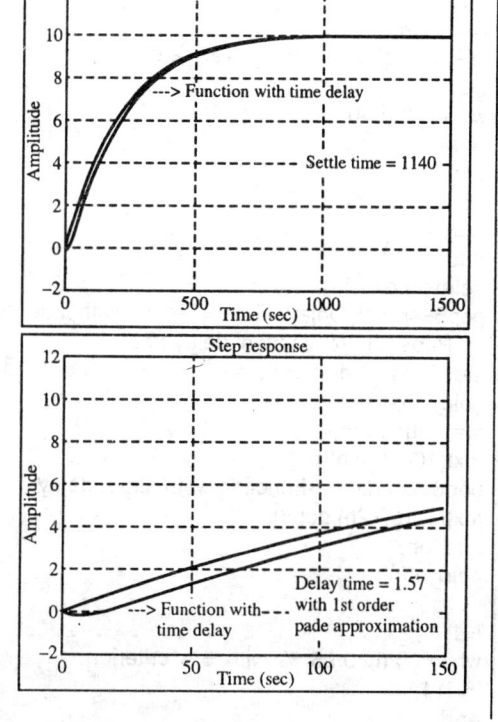

```
hold on;
step(num,den,t);
text(100,1,pop1);
pop4=sprintf('--->Function with time delay');
text(T1(i),Y1(i),pop4);
grid on;

pause;
clf;
t=0:10:1490;
[Y2,X2,T2]=step(num_2,den_2,t);
i=1;
while Y2(i)<9.95 % with 5% criterion.
 i=i+1;
end
settletime=T2(i);
pop2=sprintf('Settle Time = %g',settletime);
step(num_2, den_2,t);
hold on;
step(num, den,t);
text(1100,4,pop2);
pop3=sprintf('--->Function with time delay');
text(T2(30),Y2(30),pop3);
grid on;
hold off;
pause;
clf;
t=0:.1:149;
[Y2,X2,T2]=step(num_2,den_2,t);
i=2;
while Y2(i)>0
 i=i+1;
end
while Y2(i)<0
 i=i+1;
end
delaytime=T2(i);
pop5=sprintf('Delay Time =%g \n with 2nd order
\n Pade approximation',delaytime);
step(num_2,den_2,t);
hold on;
step(num,den,t);
text(100,1,pop5);
pop6=sprintf('--->Function with time delay');
text(T2(i),Y2(i),pop6);
grid on;
hold off;

i=1;
while Y2(i)<9.95 % with 5% criterion.
 i=i+1;
end
```

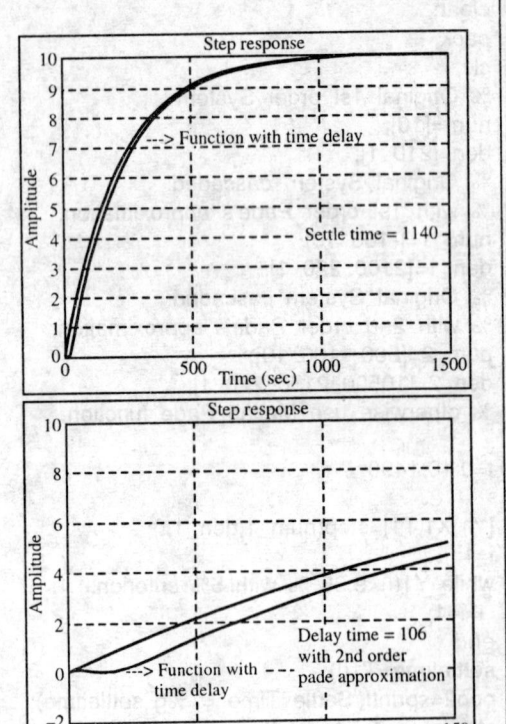

```
settletime=T2(i);
pop2=sprintf('Settle Time = %g',settletime);
step(num_2, den_2,t);
hold on;
step(num, den,t);
text(1100,4,pop2);
pop3=sprintf('--->Function with time delay');
text(T2(30),Y2(30),pop3);
grid on;
hold off;
pause;
clf;
t=0:.1:149;
[Y2,X2,T2]=step(num_2,den_2,t);
i=2;
while Y2(i)>0
 i=i+1;
end
while Y2(i)<0
 i=i+1;
end
delaytime=T2(i);
pop5=sprintf('Delay Time =%g \n with 2nd order
\n Pade approximation',delaytime);
step(num_2,den_2,t);
hold on;
step(num,den,t);
text(100,1,pop5);
pop6=sprintf('--->Function with time delay');
text(T2(i),Y2(i),pop6);
grid on;
hold off;
```

8.11 Introduction to Polar Plot

The disadvantage of the Bode plot is that we have two separate curves, namely the magnitude and phase plots, showing the variation of gain and phase shift with frequency. The information contained in these two plots can be combined into one curve called the *polar plot*. The polar plot of a sinusoidal transfer function $G(j\omega)$ is a plot of the magnitude of $G(j\omega)$ versus the phase angle of $G(j\omega)$, on the polar coordinates, as ω is varied from zero to infinity. Thus the polar plot is the locus of vectors $|G(j\omega)| \angle G(j\omega)$ as ω is varied from zero to infinity. An example of such a plot is shown in Figure 8.26. It is usual to use the frequency range from 0 to ∞ rather than $-\infty$ to ∞.

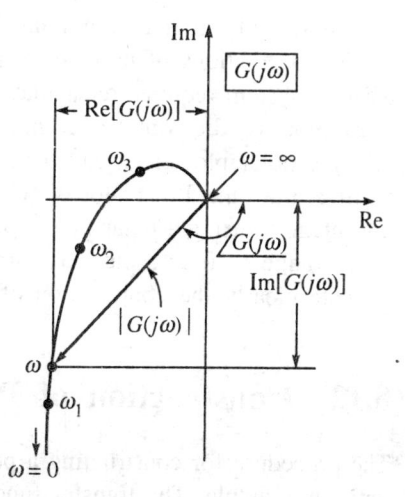

FIGURE 8.26 Polar plot.

In a casual way, the polar plot is often called the Nyquist plot. The polar plot is strictly for the frequency range $0 < \omega < \infty$ while the Nyquist plot is in the frequency range $-\infty < \omega < \infty$. The information on negative frequency is redundant because the magnitude and the real part of $G(j\omega)$ are even functions, while the phase and imaginary part of $G(j\omega)$ are odd functions of ω, as illustrated in Figure 8.27. Note that, for real systems

$$G(-j\omega) = \overline{G(j\omega)} = -G(j\omega) \qquad (8.55)$$

(a) Magnitude of $G(j\omega)$

(b) Phase of $G(j\omega)$

(c) Real part of $G(j\omega)$

(d) Imaginary part of $G(j\omega)$

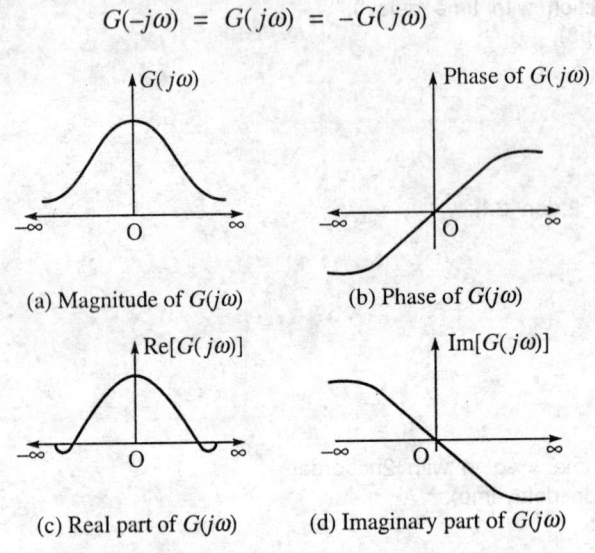

FIGURE 8.27 Four parts of a network function for $s = j\omega$.

and so we consider the positive range of frequencies only.

Both the magnitude $|G(j\omega)|$ and phase angle $\angle G(j\omega)$ must be calculated directly for each value of frequency ω in order to construct the polar plot. In polar plots, a positive (negative) phase angle is measured counterclockwise (clockwise) from the positive real axis. Further, it may be mentioned that the polar plot can be utilized to determine the closed-loop system stability after determining the gain and phase margins.

The limitations of polar plots are readily apparent. The addition of poles or zeros to an existing system requires recalculations. Furthermore, the calculation of frequency response by polar plot is tedious and does not indicate the effect of the individual poles or zeros.

If a polar plot of cascaded transfer functions $G_1(j\omega) G_2(j\omega)$ is desired, it is convenient to first draw the Bode plot of $G_1(j\omega) G_2(j\omega)$ and then convert it into the polar plot. The advantage of using a polar plot is that it depicts the frequency response characteristics over the entire range of frequencies in a single plot. The disadvantage is that the polar plot does not indicate clearly the contribution of each of the individual factors of the transfer function.

8.12 Construction of Polar Plot

The procedure for constructing a polar plot for frequency range $0 < \omega < \infty$ is illustrated here with an example. The transfer function $G(j\omega)$ is a complex quantity having magnitude and phase as a function of frequency. The plot in polar coordinates is replaced by the real and imaginary parts of the transfer function in Cartesian coordinates.

EXAMPLE 8.7 For a low-pass *RC* circuit (lag network) of Figure 8.28(a), the transfer function which has a negative real pole becomes

$$G(s) = \frac{V_2(s)}{V_1(s)} = \frac{1}{1+sT}, \quad \text{where } T = RC \tag{8.56}$$

For sinusoidal steady state, putting $s = j\omega$,

$$G(j\omega) = \frac{1}{1+j\omega T} = \frac{1}{\sqrt{1+\omega^2 T^2}} \angle -\tan^{-1} \omega T = |G(j\omega)| \angle G(j\omega) \tag{8.57}$$

The following terminal points on the locus are determined by varying ω from zero to infinity.

ω	$|G(j\omega)|$	$\angle G(j\omega)$
0	1	0
1/T	0.707	$-45°$
∞	0	$-90°$

The polar plot of this transfer function as shown in Figure 8.28(b) is a semicircle as the frequency ω is varied from zero to infinity. The centre is located at 0.5 on the real axis, and radius is equal to 0.5.

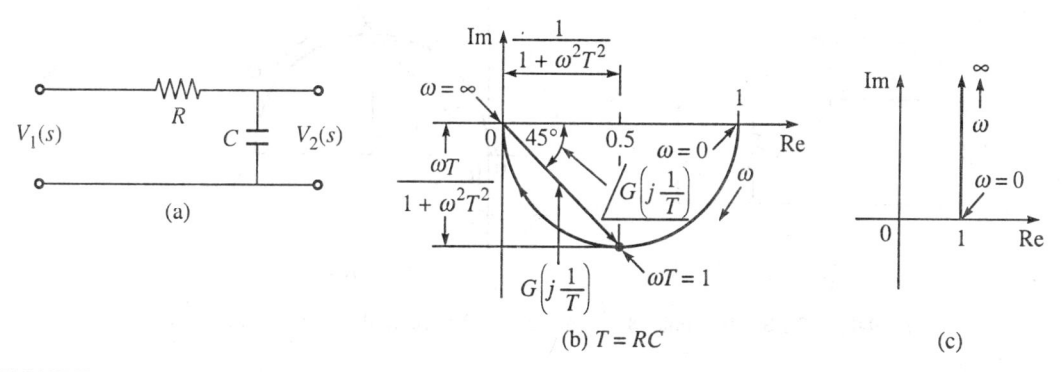

FIGURE 8.28 Example 8.7: (a) lag network, (b) its polar plot, and (c) the polar plot of $(1 + j\omega T)$.

To prove that the polar plot is a semicircle, we define $G(j\omega) = X + jY$

where X = real part of $G(j\omega) = \dfrac{1}{1+\omega^2 T^2}$; Y = imaginary part of $G(j\omega) = -\dfrac{\omega T}{1+\omega^2 T^2}$

Then, $(X - 1/2)^2 + Y^2 = \dfrac{1}{4}\left(\dfrac{1-\omega^2 T^2}{1+\omega^2 T^2}\right)^2 + \left(\dfrac{-\omega T}{1+\omega^2 T^2}\right)^2 = \left(\dfrac{1}{2}\right)^2$

Thus, in the X-Y plane, $G(j\omega)$ is a semicircle with centre at $X = 1/2$, $Y = 0$ and with radius = 1/2 as shown in Figure 8.28(b) for the frequency range of variation $0 < \omega < \infty$. The output lags behind the input for all ω in the range $0 < \omega < \infty$.

The polar plot of any function $G(j\omega) = (1 + j\omega T)$ is simply the upper half of the straight line passing through the point $(1, j0)$ in the complex $G(j\omega)$ plane and parallel to the imaginary axis as shown in Figure 8.28(c). The polar plot of $(1 + j\omega T)$ will be a straight line in the first quadrant of $G(j\omega)$ plane having value 1 at $\omega = 0$ and moving on the straight line as $\omega \to \infty$. The polar plot of $(1 + j\omega T)$ has an appearance completely different from that of $1/(1 + j\omega T)$, that is, $(1 + j\omega T)^{-1}$. The impedance function of a series RL circuit is $(R + j\omega L)$ and it has a similar expression to that of $(1 + j\omega T)$ expression.

EXAMPLE 8.8 Given the transfer function of a high-pass RC circuit (lead network) as in Figure 8.29(a), we have

$$G(s) = \frac{V_2(s)}{V_1(s)} = \frac{s}{s + 1/T}, \quad \text{where } T = RC \qquad (8.58)$$

For sinusoidal steady state, putting $s = j\omega$,

$$G(j\omega) = \frac{j\omega}{j\omega + 1/T} = \left| \frac{\omega T}{\sqrt{1 + \omega^2 T^2}} \right| \angle(90° - \angle\tan^{-1} \omega T) = |G(j\omega)| \angle G(j\omega) \qquad (8.59)$$

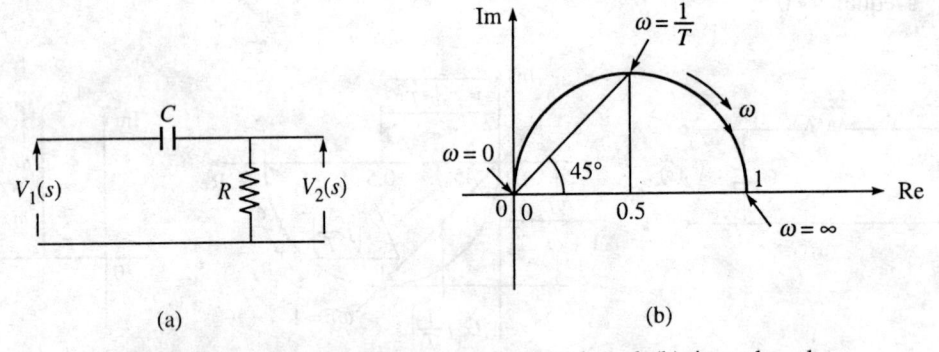

FIGURE 8.29 Example 8.8: (a) lead network and (b) its polar plot.

Following the same procedure as in Example 8.7, we get the points on the locus, determined by varying ω from zero to infinity.

The polar plot of this transfer function is a semicircle as the frequency ω is varied from zero to infinity as shown in Figure 8.29(b). The centre is located at 0.5 on the real axis, and the radius is equal to 0.5. The phase of the output voltage leads the input for all ω. This network is called a lead network and is used for compensation in the design of systems.

ω	$|G(j\omega)|$	$\angle G(j\omega)$
0	0	90°
1/T	0.707	45°
∞	1	0°

Frequency Response Analysis

Drill Problem 8.1

Prove that the polar plot of high-pass *RC* circuit is a semicircle.

Integrator

The polar plot of the transfer function having a pole at the origin, i.e. $G(s) = 1/s$ is shown in Figure 8.30. At steady state,

$$G(j\omega) = \frac{1}{j\omega} = -j\frac{1}{\omega} = \frac{1}{\omega}\angle -90° = |G(j\omega)|\angle G(j\omega)$$

The polar plot is obviously the negative imaginary axis. The points on the polar plot are also shown below.

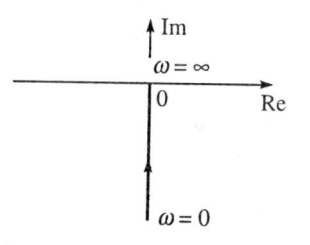

ω	$|G(j\omega)|$	$\angle G(j\omega)$
0	∞	0
Any finite +ve frequency	Finite +ve value	$-90°$
∞	0	$-90°$

FIGURE 8.30 Polar plot of transfer function $G(j\omega) = 1/j\omega$ (integrator).

Derivative factor

The polar plot of a zero at the origin, $G(j\omega) = j\omega$, is on the positive imaginary axis as the magnitude $|G(j\omega)| = \omega$ and phase $\angle G(j\omega) = +90°$. The polar plot is shown in Figure 8.31.

EXAMPLE 8.9 For an *RLC* series network, the output is taken across the capacitor. The transfer function is obtained as

$$G(s) = \frac{\omega_n^2}{s^2 + 2\zeta\omega_n s + \omega_n^2} \qquad (8.60)$$

FIGURE 8.31 Polar plot of transfer function $G(j\omega) = j\omega$ (derivative factor).

where the undamped natural frequency, $\omega_n = 1/\sqrt{LC}$ and the damping coefficient, $\zeta = \frac{R}{2}\sqrt{\frac{C}{L}}$.

For sinusoidal steady state,

$$G(j\omega) = \frac{1}{1 + j2\zeta(\omega/\omega_n) - (\omega/\omega_n)^2} \qquad (8.61)$$

The low-frequency and high-frequency portions of the polar plot of the sinusoidal transfer function $G(j\omega)$ are given by

$$\lim_{\omega \to 0} G(j\omega) = 1\angle 0° \quad \text{and} \quad \lim_{\omega \to \infty} G(j\omega) = 0\angle 180°$$

respectively. The polar plot of this sinusoidal transfer function starts at $1\angle 0°$ and terminates at $0\angle 180°$ as ω increases from zero to infinity. Thus the high frequency portion of $G(j\omega)$ is tangential to the negative real axis. The values of $G(j\omega)$ at any intermediate frequency range of interest can be calculated directly or by the use of a logarithmic plot. The polar plot of the given transfer function is shown in Figure 8.32(a). The exact shape of a polar plot depends on the value of the damping coefficient but the general shape of the plot is the same for both the underdamped case ($0 < \zeta < 1$) and the overdamped case ($\zeta > 1$).

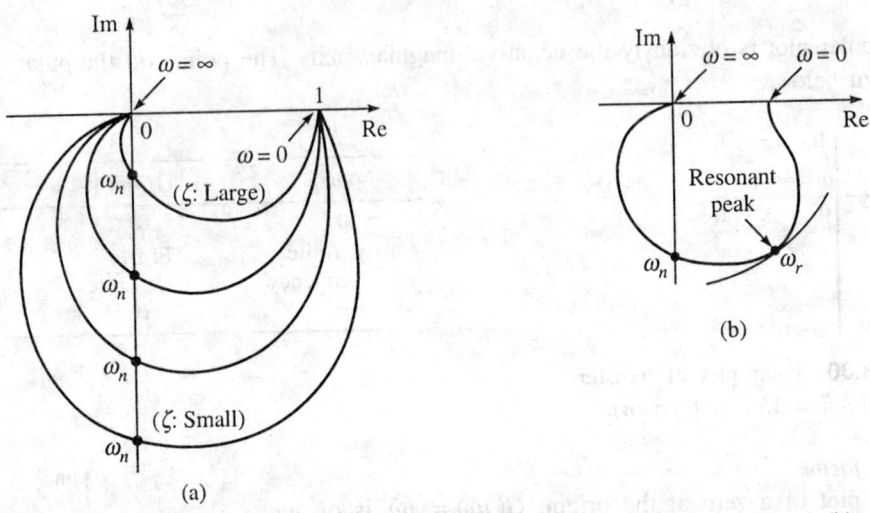

FIGURE 8.32 Example 8.9: (a) polar plot of transfer function of Eq. (8.61) and (b) polar plot showing the resonant peak and resonant frequency ω_r.

For the underdamped case, at $\omega = \omega_n$, we get $G(j\omega_n) = 1/j2\zeta$. The phase angle at $\omega = \omega_n$ is $-90°$. Therefore, it can be seen that the frequency at which the $G(j\omega)$ locus intersects the imaginary axis is the undamped natural frequency ω_n. In the polar plot, the frequency point whose distance from the origin is maximum corresponds to the resonant frequency ω_r. The peak value of $G(j\omega)$ is obtained as the ratio of the magnitude of the vector at resonant frequency ω_r to the magnitude of the vector at $\omega = 0$. The resonant frequency ω_r is indicated in the polar plot shown in Figure 8.32(b).

For the overdamped case, as the damping coefficient ζ increases beyond unity, the $G(j\omega)$ locus approaches a semicircle. This may be observed from the fact that for a heavily damped system, the characteristic roots are real and one is much smaller than the other. Since for a sufficiently large value of ζ, the effect of a larger root on the response becomes very small, the system behaves like a first-order transfer function such as $1/(1 + j\omega T)$.

For a series RLC network, the impedance function becomes

$$Z(j\omega) = 1 + j2\zeta(\omega/\omega_n) - (\omega/\omega_n)^2 \tag{8.62}$$

with
$$\omega_n = 1/\sqrt{LC} \quad \text{and} \quad \zeta = \frac{R}{2}\sqrt{\frac{C}{L}}$$

The polar plot is shown in Figure 8.33.

EXAMPLE 8.10 Sketch the polar plot of the transfer function

$$G(s) = \frac{1}{s(sT+1)}$$

Solution: The sinusoidal transfer function is obtained by putting $s = j\omega$ as

$$G(j\omega) = \frac{1}{j\omega(1+j\omega T)} = -\frac{T}{1+\omega^2 T^2} - j\frac{1}{\omega(1+\omega^2 T^2)}$$

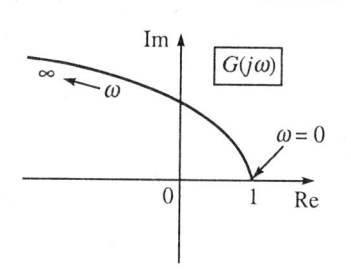

FIGURE 8.33 Polar plot of transfer function of Eq. (8.61).

The low-frequency portion of the polar plot becomes

$$\lim_{\omega \to 0} G(j\omega) = -T - j\infty = \infty \angle -90°$$

and the high-frequency portion becomes

$$\lim_{\omega \to \infty} G(j\omega) = 0 - j0 = 0 \angle -180°$$

At any intermediate frequency in the range $0 < \omega < \infty$, $|G(j\omega)|$ is of finite value and phase is in the range $-90° < \angle G(j\omega) < 0°$.

The general shape of the polar plot of $G(j\omega)$ is shown in Figure 8.34(a). The $G(j\omega)$ plot is asymptotic to the vertical line passing through the point $(-T, 0)$.

The rough sketch of a polar plot can be constructed from the magnitude and phase information at the terminal points and at any other intermediate point in the range $0 < \omega < \infty$. However, a more detailed polar plot can be drawn as shown in Figure 8.34(b) by calculating the magnitude M and phase θ of the given transfer function at different frequencies in the range $0 < \omega < \infty$ as shown in Table 8.3. Substituting $T = 1$, in the given transfer function, we get

$$G(s) = \frac{1}{s(s+1)} \tag{8.63}$$

$$M = |G(j\omega)| = \frac{1}{\omega\sqrt{1+\omega^2}} \quad \text{and} \quad \theta = \angle G(j\omega) = -90° - \tan^{-1}\omega$$

It may be mentioned that for the given transfer function $G_1(j\omega)$, as of Eq. (8.63), which is a type-1 system, the polar plot is shown in Figure 8.35(a). However if we rotate the type-1 transfer function by $-90°$, then the polar plot would be for

$$G_2(j\omega) = \frac{1}{(j\omega)^2(j\omega+1)}$$

which is also shown in Figure 8.35(a), where

$$M = \frac{1}{\omega^2\sqrt{\omega^2+1^2}} \quad \text{and} \quad \theta = -180° - \tan^{-1}\omega$$

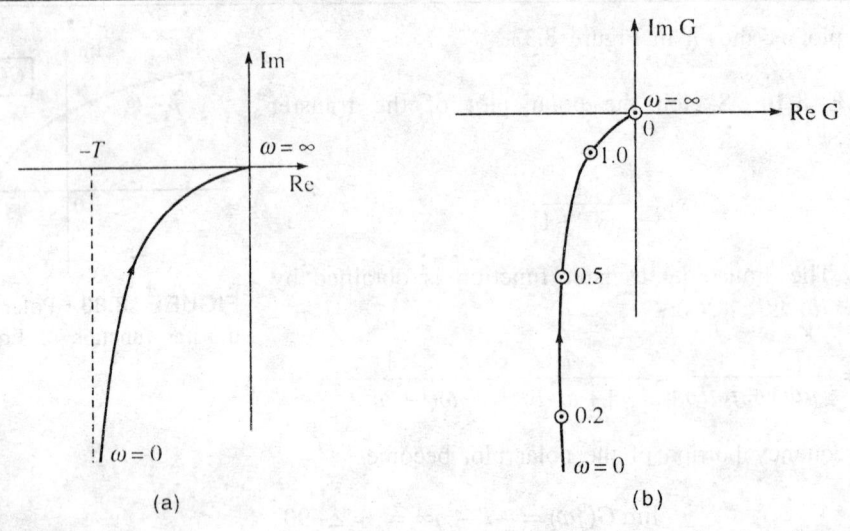

FIGURE 8.34 Example 8.10: (a) polar plot of $1/j\omega(1 + j\omega T)$ and (b) polar plot of transfer function $1/j\omega(1 + j\omega)$.

TABLE 8.3 Frequency response of the transfer function $G(s) = 1/s(s + 1)$

ω	M	θ, in degrees
0	∞	−90
0.1	9.95	−95.71
0.2	4.903	−101.31
0.5	1.789	−116.56
1.0	0.707	−135
2.0	0.204	−153.43
5.0	0.039	−168.7
7.0	0.0202	−171.87
10	0.01	−174.89
20	0.0025	−177.14
50	0.0004	−178.85
70	0.0002	−179.18

This means that the addition of a pole at the origin (i.e. integrator) rotates the polar plot by −90°. Hence the general shape of the low-frequency portion of the polar plot of type-1 will be asymptotic to the −ve imaginary axis (−90°) and that for the type-2 system will be asymptotic to −ve real axis (i.e. −180°), and so on. See Figure 8.35(a).

If the degree of the denominator polynomial is greater than that of the numerator, then the polar plot converges to the origin clockwise and at $\omega = \infty$ the plot is tangent to one of the axes at origin.

When both numerator and denominator are of the same degree, the polar plot starts at a finite distance on the real axis and ends at a finite point on the real axis.

The general shapes of polar plots of type-0, 1, 2 systems are shown in Figure 8.35(b).

Frequency Response Analysis

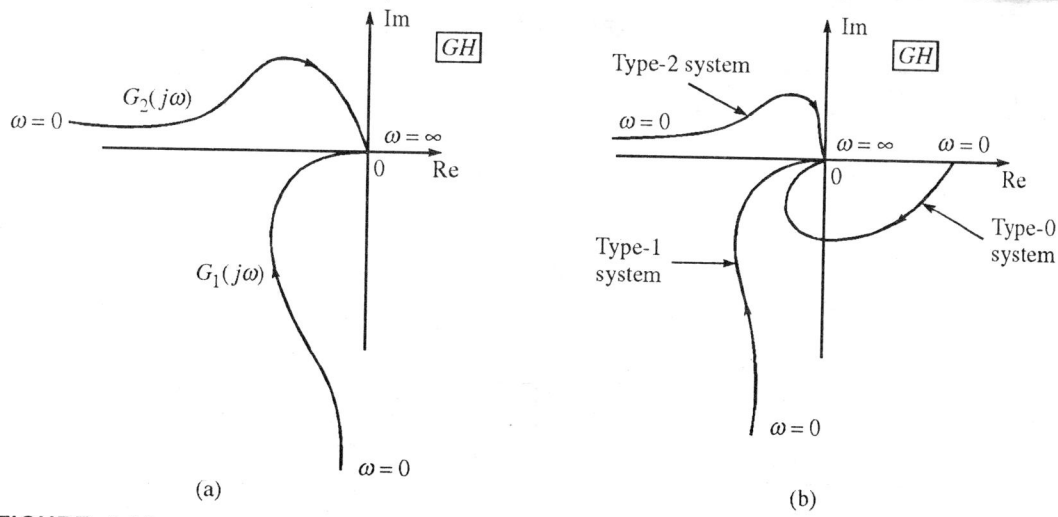

FIGURE 8.35 (a) Polar plots of $G_1(j\omega)$ and $G_2(j\omega)$ and (b) polar plots of type-0, type-1, and type-2 systems.

The polar plots of some simple transfer functions with their pole-zero patterns are given in Table 8.4.

TABLE 8.4 Polar plots of simple transfer functions

Transfer function $G(j\omega)$	Pole-zero pattern	Polar plot	Remarks
$\dfrac{1}{j\omega}$			
$\dfrac{1}{(j\omega)^2}$	2 poles		By rotating the polar plot of $1/j\omega$ by $-90°$
$1 + j\omega T$			
$\dfrac{1}{(1+j\omega T_1)(1+j\omega T_2)(1+j\omega T_3)}$			

(Contd.)

TABLE 8.4 (Contd.)

Transfer function G(jw)	Pole-zero pattern	Polar plot	Remarks
$\dfrac{1}{(j\omega)^2(1+j\omega T_1)(1+j\omega T_2)(1+j\omega T_3)}$			By rotating the polar plot of serial 4 by $-180°$
$\dfrac{1}{1+j\omega T}$	$-1/T$		
$\dfrac{1}{j\omega(1+j\omega T)}$	$-1/T$		By rotating the polar plot of serial 6 by $-90°$

EXAMPLE 8.11 The transportation lag

$$G(j\omega) = \exp(-j\omega T)$$

can be written as

$$G(j\omega) = 1\angle(\cos\omega T - j\sin\omega T) = 1\angle-\omega T$$

since the magnitude of $G(j\omega)$ is always unity and the lagging phase angle varies linearly with ω, the polar plot shown in Figure 8.36(a) is a unit circle. The corresponding Bode plot is shown in Figure 8.36(b).

EXAMPLE 8.12 Consider the transfer function

$$G(s) = \dfrac{\exp(-sL)}{1+sT}$$

The sinusoidal transfer function $G(j\omega)$ can be written as

$$G(j\omega) = [\exp(-j\omega L)]\left(\dfrac{1}{1+j\omega T}\right)$$

The magnitude and phase angle are, respectively

$$|G(j\omega)| = \left|\exp(-j\omega L)\cdot\dfrac{1}{1+j\omega T}\right| = \dfrac{1}{\sqrt{1+\omega^2 T^2}}$$

Frequency Response Analysis

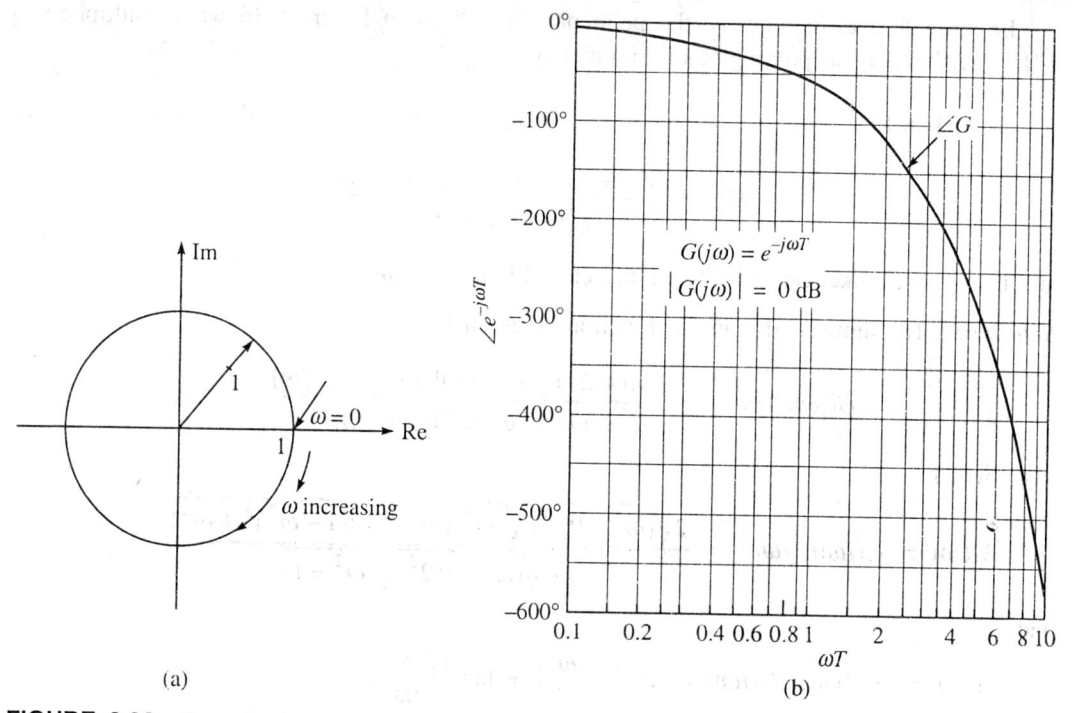

FIGURE 8.36 Example 8.11: (a) polar plot of transportation lag and (b) Bode plot of transportation lag.

and
$$\angle G(j\omega) = \angle \exp(-j\omega L) + \angle \frac{1}{1+j\omega T}$$
$$= -\omega L - \tan^{-1} \omega T$$

Since the magnitude decreases monotonically from unity and the phase angle also decreases monotonically with increasing ω, the polar plot is a spiral as shown in Figure 8.37.

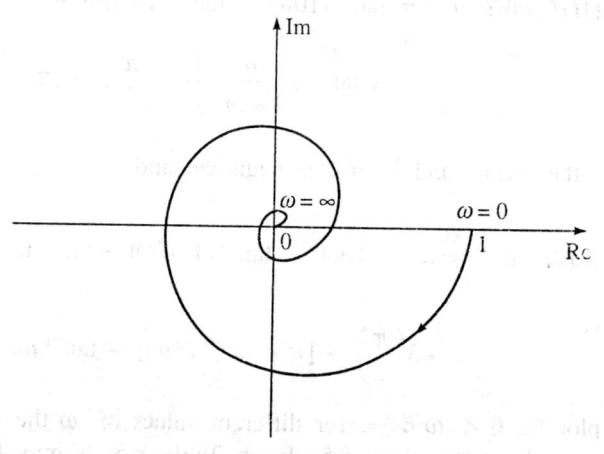

FIGURE 8.37 Example 8.12: polar plot of $\exp(-j\omega L)/(1+j\omega T)$.

Looking in a different way, the polar plot of e^{-sL} as in Figure 8.36 when multiplied by $1/(1 + sT)$, gives the polar plot of Figure 8.37.

EXAMPLE 8.13 Draw the polar plot for the system having the open-loop transfer function

$$G(s)H(s) = \frac{2(s+0.1)(s+0.6)(s^2+s+1)}{s^3(s-0.2)(s+1)}$$

and hence determine the stability of the closed-loop system.

Solution: The sinusoidal transfer function is given by

$$G(j\omega)H(j\omega) = \frac{2(j\omega+0.1)(j\omega+0.6)[(1-\omega^2)+j\omega]}{(j\omega)^3(j\omega-0.2)(j\omega+1)}$$

from which

$$M(\omega) = |G(j\omega)H(j\omega)| = \frac{2\sqrt{(\omega^2+0.1^2)}\sqrt{(\omega^2+0.6^2)}\sqrt{[(1-\omega^2)^2+\omega^2]}}{\omega^3\sqrt{(\omega^2+0.2^2)}\sqrt{(\omega^2+1)}}$$

and

$$\theta(\omega) = \arg[G(j\omega)H(j\omega)] = \tan^{-1}\left(\frac{\omega}{0.1}\right) + \tan^{-1}\left(\frac{\omega}{0.6}\right)$$

$$+ \tan^{-1}\left(\frac{\omega}{1-\omega^2}\right) - 3\tan^{-1}\left(\frac{\omega}{0}\right) - \tan^{-1}\left(\frac{\omega}{-0.2}\right) - \tan^{-1}\omega$$

Remember that the term $(1/s^3)$ gives an argument contribution of $3\tan^{-1}(\omega/0)$ or $-3(90°) = -270°$.

Note that for $0 < \omega < 1$, the term $\tan^{-1}\left(\frac{\omega}{1-\omega^2}\right)$ is positive, so

$$\theta(\omega) = \arg[G(j\omega)H(j\omega)] = \tan^{-1}(10\omega) + \tan^{-1}(1.67\omega) +$$

$$+ \tan^{-1}\left(\frac{\omega}{1-\omega^2}\right) - 3\left(\frac{\pi}{2}\right) - [\pi - \tan^{-1}(5\omega)] - \tan^{-1}\omega$$

and for $1 < \omega < \infty$, the term $[\omega/(1-\omega^2)]$ is negative, and

$$\phi(\omega) = \arg[G(j\omega)H(j\omega)] = \tan^{-1}(10\omega) + \tan^{-1}(1.67\omega) + \left[\pi - \tan^{-1}\left(\frac{\omega}{\omega^2-1}\right)\right]$$

$$-3\left(\frac{\pi}{2}\right) - [\pi - \tan^{-1}(5\omega)] - \tan^{-1}\omega]$$

To draw the polar plot for $0 < \omega < \infty$, for different values of ω the corresponding values of $M(\omega)$ and $\theta(\omega)$ are shown in Table 8.5. From Table 8.5, it may be observed that for

TABLE 8.5 Example 8.13: points for $GH(j\omega)$ locus for $0 < \omega < \infty$

ω	$M(\omega)$	$\theta(\omega)$
0	∞	$-\pi/2$
0.1	761.8	$-8.915°$
0.2	120.2	$+37.33°$
0.5	9.54	$+105.6°$
1.2	2.2	$+199.2°$
$\vdots$	$\vdots$	$\vdots$
∞	0	$-\pi/2$

$0.5 < \omega < 1.2$, the polar plot just crosses the 180° line and we get the value of ω as ω_p at the crossing-over point (otherwise, equating the phase to +180° line we get the frequency ω_p) and then putting the value of ω_p into the magnitude equation, we get $M(\omega_p) > 1$. The polar plot is shown in Figure 8.38.

FIGURE 8.38 Example 8.13: polar plot.

EXAMPLE 8.14 Draw the polar plot of

$$G(s) = \frac{10}{s(s+1)(s+2)}$$

Solution: Substituting $s = j\omega$ for steady state, we have

$$G(j\omega) = \frac{10}{j\omega(j\omega+1)(j\omega+2)}$$

Then,

$$\lim_{\omega \to 0} \angle G(j\omega) = \lim_{\omega \to 0} \angle \frac{10}{j\omega} = -90° \text{ and } \lim_{\omega \to 0} |G(j\omega)| = \infty$$

and

$$\lim_{\omega \to \infty} \angle G(j\omega) = \lim_{\omega \to \infty} \angle \frac{10}{(j\omega)^3} = -270° \text{ and } \lim_{\omega \to \infty} |G(j\omega)| = 0$$

Rationalizing $G(j\omega)$ gives

$$G(j\omega) = \frac{10[-3\omega^2 - j\omega(2-\omega^2)]}{[-3\omega^2 - j\omega(2-\omega^2)][-3\omega^2 + j\omega(2-\omega^2)]} \quad (8.64)$$

$$= \frac{-30\omega^2}{9\omega^4 + \omega^2(2-\omega^2)} - \frac{j10\omega(2-\omega^2)}{9\omega^4 + \omega^2(2-\omega^2)}$$

Thus,

$$\text{Re}[G(j\omega)] = \frac{-30\omega^2}{9\omega^4 + \omega^2(2-\omega^2)} = 0 \quad \text{gives} \quad \omega = \infty \quad (8.65)$$

which means that the $G(j\omega)$ plot intersects the imaginary axis only at the origin. Also,

$$\text{Im}[G(j\omega)] = \frac{-10(2-\omega^2)}{9\omega^3 + \omega(2-\omega^2)} = 0 \quad \text{gives} \quad \omega^2 = 2 \quad (8.66)$$

which gives the intersection on the real axis of $G(j\omega)$ at $\omega = \pm\sqrt{2}$ rad/s. Substituting $\omega = \sqrt{2}$ into Eq. (8.65) gives the intersect at

$$G(j\omega) = \frac{-30 \times 2}{9 \times 4} = -\frac{5}{3}$$

Hence $|G(j\sqrt{2})| = -\frac{5}{3}$ and $\angle G(j\sqrt{2}) = -180°$.

The polar plot is drawn as shown in Figure 8.39.

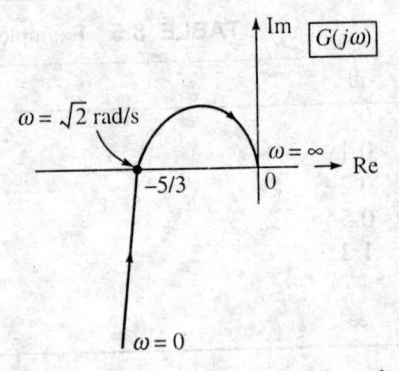

FIGURE 8.39 Example 8.14: polar plot of $G(j\omega) = 10/j\omega(j\omega + 1)(j\omega + 2)$.

EXAMPLE 8.15 Draw the polar plot of the given open-loop transfer function

$$GH(s) = \frac{100K(s+5)(s+40)}{s^3(s+100)(s+200)}$$

and discuss the stability of the closed-loop system.

Solution: Substituting $s = j\omega$, we have

$$GH(j\omega) = \frac{100K(j\omega+5)(j\omega+40)}{(j\omega)^3(j\omega+100)(j\omega+200)}$$

$$M(\omega) = |GH(j\omega)| = \left| \frac{100K\sqrt{\omega^2+5^2}\sqrt{\omega^2+40^2}}{\omega^3\sqrt{\omega^2+100^2}\sqrt{\omega^2+200^2}} \right|$$

$\theta(\omega) = \angle GH(j\omega) = \angle[\tan^{-1}(\omega/5) + \tan^{-1}(\omega/40) - \tan^{-1}(\omega/100) - \tan^{-1}(\omega/200) - 270°]$

$GH(j\omega) = \text{Re}(\omega) + j\text{Im}(\omega)$ (through rationalization)

Now $\text{Im}(\omega) = 0$ occurs at $\omega = 25.8$ rad/s, 77.7 rad/s and at ∞. Then we get $\text{Re}(\omega)$ at these frequencies as

$\text{Re}(25.8) = (-3.497 \times 10^{-4})K = M(25.8)$, magnitude of $GH(j\omega)$ at $\omega = 25.8$

$\text{Re}(77.7) = (-5.3 \times 10^{-5})K = M(77.7)$, magnitude of $GH(j\omega)$ at $\omega = 77.7$

$\text{Re}(\infty) = 0 = M(\infty) = $ magnitude of $GH(j\omega)$ at $\omega = \infty$

ω	$M(\omega)$	$\theta(\omega)$
0	∞	$-270°$
$0 < \omega < 25.8$	finite	$-180° < \theta < -270°$
25.8	$(-3.497 \times 10^{-4})K$	$-180°$
$25.8 < \omega < 77.7$	finite	$-180° < \theta < -90°$
77.7	$(-5.3 \times 10^{-5})K$	$-180°$
$77.7 < \omega < \infty$	finite	$-180° < \theta < -270°$
∞	0	$-270°$

Frequency Response Analysis

The polar plot is drawn as shown in Figure 8.40. The following observations are made.

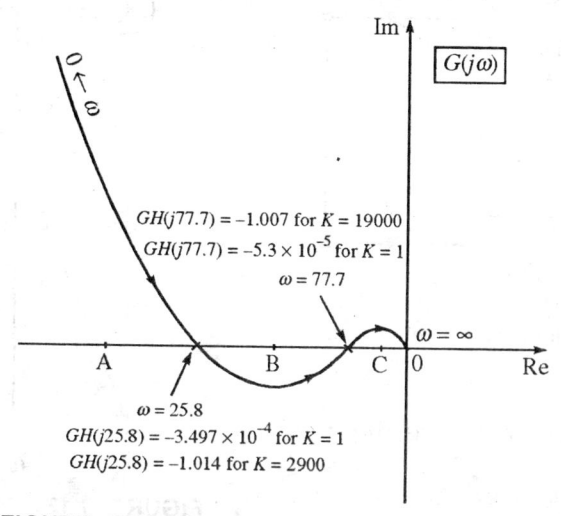

FIGURE 8.40 Example 8.15: polar plot of $GH(s)$.

Key values on the plot:
- $GH(j77.7) = -1.007$ for $K = 19000$
- $GH(j77.7) = -5.3 \times 10^{-5}$ for $K = 1$
- $\omega = 77.7$
- $\omega = 25.8$
- $GH(j25.8) = -3.497 \times 10^{-4}$ for $K = 1$
- $GH(j25.8) = -1.014$ for $K = 2900$

For gain K in the range $0 < K < 2818$, the $(-1, j0)$ point is somewhere in the region 'A' which is encircled (see the definition of encirclement) by the polar plot of the open-loop transfer function $GH(j\omega)$. Hence the closed-loop transfer function is unstable for gain in the range $0 < K < 2818$.

Further, for gain K in the range $2818 < K < 18837$, the $(-1, j0)$ point is somewhere in the region 'B' which is not encircled by the polar plot of $GH(j\omega)$. Hence the closed-loop system is stable in the range of gain $2818 < K < 18837$.

Again for gain K in the range $18837 < K < \infty$, the $(-1, j0)$ point is somewhere in the region 'C' which is again encircled by the polar plot of the open-loop transfer function $GH(j\omega)$ and hence the closed-loop system is unstable for the range of gain $18837 < K < \infty$.

This shows that the given closed-loop system is conditionally stable. It may be mentioned here that the same problem may be solved through different approaches and we would arrive at the same result.

EXAMPLE 8.16 Consider the symmetrical lattice network shown in Figure 8.41.

Here, $$G(s) = \frac{V_2(s)}{V_1(s)} = \frac{V_P(s) - V_Q(s)}{V_1(s)} = \frac{s^2 - s + 1}{s^2 + s + 1} = \frac{(s + z_1)(s + z_2)}{(s + p_1)(s + p_2)}$$

where the complex conjugate poles and zeros are

$$z_{1,2} = \frac{1}{2} \pm j\frac{\sqrt{3}}{2}$$

$$p_{1,2} = -\frac{1}{2} \pm j\frac{\sqrt{3}}{2}$$

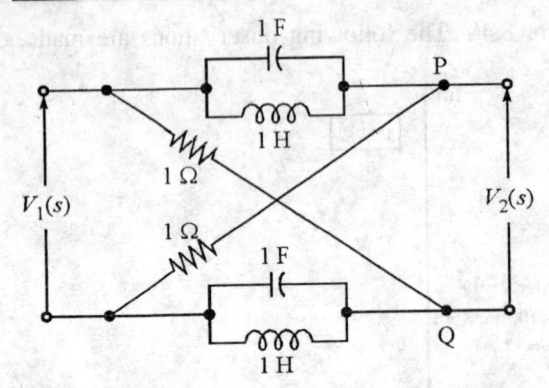

FIGURE 8.41 Example 8.16: symmetrical lattice network.

The poles and zeros form a quad as shown in Figures 8.42.

Now, the magnitude of $G(j\omega)$ is

$$|G(j\omega)| = \left|\frac{V_2(j\omega)}{V_1(j\omega)}\right| = 1$$

The phase of $G(j\omega)$

$$\angle G(j\omega) = \phi_1 + \phi_2 - \theta_1 - \theta_2$$

i.e. the output magnitude always remains equal to the input magnitude for all frequencies, and the phase varies. The phase curve is shown in the Bode plot of Figure 8.43(a). The polar plot is shown in Figure 8.43(b), where the phase varies between 0° and 360° as ω varies between zero and infinity whereas the magnitude remains constant.

Such a network has wide applications in the compensation of telephone lines. The network with sinusoidal input has the property that the magnitude of the output remains the same as that of the input, but there is distortion in phase as ω varies.

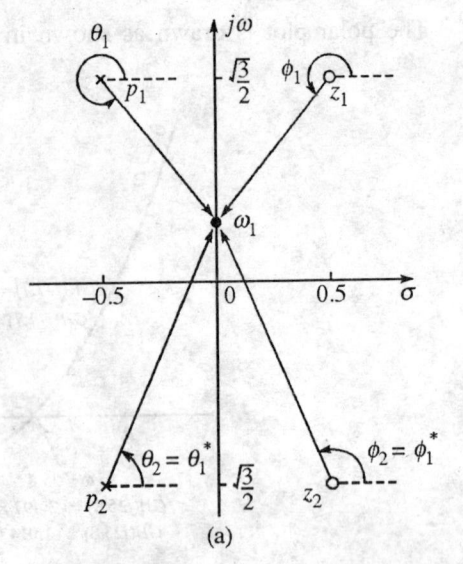

FIGURE 8.42 Example 8.16: pole-zero positions.

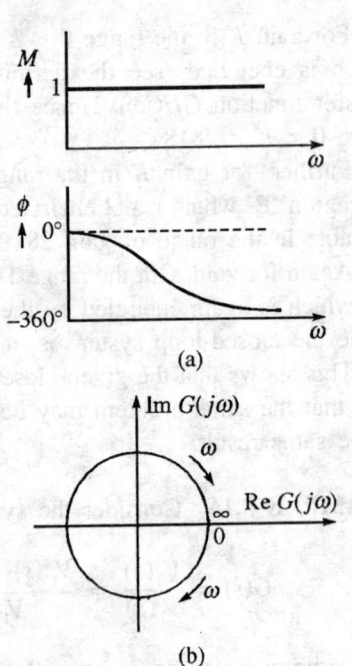

FIGURE 8.43 Example 8.16: (a) Bode plot and (b) polar plot.

8.13 Relative Stability

We have already seen in root-locus technique that as the gain K is increased the closed-loop system has a tendency to become relatively less stable. In transient response for the underdamped system, the overshoot increases with the increase in gain K, i.e. the stability deteriorates. We made the same observations from the Routh's stability criterion, i.e. after a certain value of gain K, the system goes to instability. In the polar plot shown in Figure 8.44 for the open-loop transfer function

$$G(s)H(s) = \frac{K}{s(sT_1 + 1)(sT_2 + 1)} \quad (8.67)$$

there is a tendency for the polar plot to encircle the critical point $-1 + j0$ as the gain K is increased. The closeness of the open-loop transfer function $G(j\omega)H(j\omega)$ to the $-1 + j0$ point can be used as a measure of the margin of stability. This does not apply to a conditionally stable system. In the polar plot of Figure. 8.44 we know that for gain $K = K_1$, the intersection of $G(j\omega)H(j\omega)$ curve with the negative real axis occurs closer to the origin than that of the polar plot of $G(j\omega)H(j\omega)$ having gain K_2, where $K_2 > K_1$. The polar plot for the open-loop transfer function GH with gain K_2 just passes through the $-1 + j0$ point. So the closed-loop system at this stage is just at the verge of instability, i.e. critically stable. The system with gain K_1 is obviously stable as seen from the corresponding polar plot, that means, for proper design we can increase the gain beyond K_1 before the closed-loop system goes to instability. In other words, we have the margin of gain to play with before the system goes to instability. In this case, the gain margin is positive. If the gain is still increased from K_2 to K_3, then the corresponding polar plot encircles the $-1 + j0$ point and the gain margin becomes negative, or in other words, the closed-loop system becomes unstable.

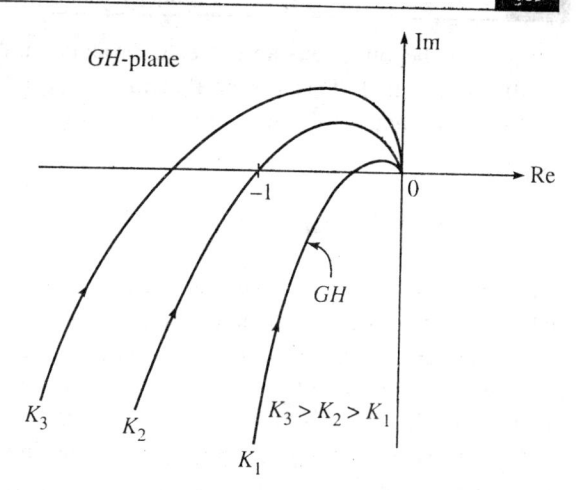

FIGURE 8.44 Polar plot of $G(j\omega)H(j\omega)$.

The open-loop transfer function $G(s)H(s)$ is a complex quantity and it has both magnitude (gain) and phase. The closed-loop system stability depends on both gain and phase. The relative stability of the system under consideration has to be discussed in terms of the gain and phase, i.e. gain margin and phase margin.

Some of the important queries as follows may arise while discussing the stability of feedback control systems.

(i) If the system is stable, then how much stable it is, i.e. what the degree of stability is? This means how much gain can we increase before the system goes to instability.
(ii) If the system is unstable then by what margin of gain the system is unstable, i.e. to make the system just stable, by what amount should the gain be reduced?

We know that for a negative feedback system if the output becomes unbounded, i.e. the closed-loop system is unstable, then from the characteristic equation, we get

$$G(s)H(s) = -1 \quad (8.68)$$

This gives the magnitude condition as

$$|G(s)H(s)| = 1 \quad (8.69)$$

and the phase condition as

$$\angle G(s)H(s) = -180° \quad (8.70)$$

The system becomes unstable, i.e. it produces unbounded output if the two Eqs. (8.69) and (8.70) are satisfied. This means that in the closed-loop system stability, the point $-1 + j0$ or the 0-dB line and $-180°$ line are the important factors to be considered.

8.14 Phase Margin

If the magnitude of GH is just unity, i.e. 0-dB, then the closed loop system is just critically stable. This means that 0-dB is the critical point for determining the system stability (gain-wise). Now when Eq. (8.69) is satisfied, then to make the closed-loop system just critically stable phase-wise, a measure of phase angle of GH from $-180°$ is the phase margin, that is, the margin of phase we can play with (phase-wise) to make the closed-loop system just critically stable. If $\angle GH$ is less than $-180°$, then the phase margin is negative, otherwise positive. For a positive phase margin the closed-loop system is stable, and for a negative phase margin the closed-loop system is unstable keeping gain at the critical point, i.e. $|GH| = 1$ or 0-dB. The beauty of the frequency domain technique is, that from the open-loop phase characteristic, i.e. GH-plot, we can measure the relative stability (phase-wise) of the closed-loop system by finding the margin of phase that the system posseses before going to the critically stable point.

The frequency at which $|GH| = 1$ is called the gain crossover frequency, ω_g. The measure of phase of open-loop transfer function at the gain crossover frequency ω_g from the $-180°$ line is known as the phase margin (PM). Then

$$\text{Phase margin (PM)} = \angle G(j\omega_g) H(j\omega_g) - 180° \tag{8.71}$$

The analytical procedure for computing the PM involves finding the phase of $G(j\omega)H(j\omega)$ at the gain crossover frequency ω_g and then subtracting $180°$ from this phase. The analytical method involves tedious computations, so a graphical method is preferred.

The phase margin is a measure of the angle between the gain crossover point to the critical point $(-1, j0)$. Phase margin is therefore defined as the angle through which the polar plot of $G(s)H(s)$ must be rotated in order that the gain crossover point on the polar plot passes through the critical point $(-1, j0)$. Figure 8.45 shows the phase margin which is positive, i.e. phase-wise the closed-loop system is stable. Thus, we note that the relative stability (phase-wise) information can be obtained from the polar plot of the open-loop transfer function GH.

In a polar plot, a line may be drawn from origin to the point at which the unit circle crosses the $G(s)H(s)$ locus. The angle γ from the negative real axis to this line is the phase margin. The phase margin is positive for $\gamma > 0$ and negative for $\gamma < 0$ with the existing sign convention of angles.

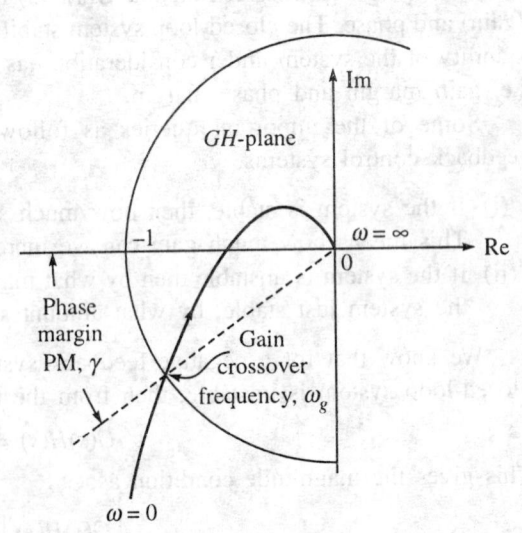

FIGURE 8.45 Definition of phase margin from polar plot.

8.15 Gain Margin

Again from the relation $\angle GH = -180°$ when the open-loop transfer function GH crosses the negative real axis, i.e. the 180° line, then the closed-loop system is phase-wise critically stable. The frequency at which the open-loop transfer function GH crosses the 180° line is called the phase crossover frequency ω_p. Now a measure of magnitude of $|GH|$ at phase crossover frequency from the critical point $(-1, j0)$ gives the margin of gain that we can play with for design purposes before the system becomes just unstable or critically stable (gain-wise). The gain margin is a measure of closeness of the phase crossover frequency point to the critical point $-1 + j0$. The physical significance of gain margin is the amount of gain in dB that can be allowed to increase in the loop before the system reaches instability. Refer to Figure 8.46 where the phase crossover frequency is ω_p and the magnitude of GH at ω_p is $|G(j\omega_p)H(j\omega_p)|$. Then the gain margin (GM) in dB is defined as

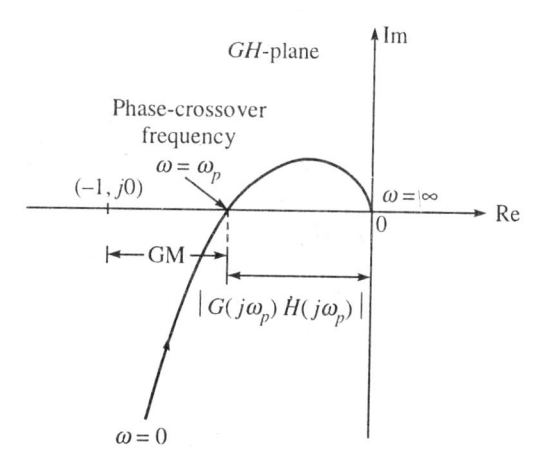

FIGURE 8.46 Definition of gain margin from polar plot.

$$\text{Gain Margin (GM)} = 20 \log_{10}\left[-\{G(j\omega)H(j\omega)|_{\omega=\omega_p}\} - (-1)\right]$$

$$= 20 \log_{10}(1 - |G(j\omega_p) H(j\omega_p)|) \text{ dB}$$

$$= 20 \log_{10}\left(\frac{1}{|G(j\omega_p) H(j\omega_p)|}\right) \text{ dB} \qquad (8.72)$$

Obviously if the GM is positive then the closed-loop system is stable. For a first-order or a second-order system, the GM is always infinite and hence it is stable for all gain since the polar plots for such systems do not cross the negative real axis. When the critical point $(-1, j0)$ is enclosed by the GH-plot, the magnitude of $G(j\omega_p)H(j\omega_p)$ is greater than unity and the gain margin (GM) in dB becomes negative. It should be noted that negative GM in dB does not always correspond to an unstable system. The stability condition can be ascertained from the Nyquist plot that encircles the $(-1, j0)$ point as will be discussed in the next chapter.

In polar plots a measure of gain magnitude of the open-loop transfer function at phase crossover frequency, i.e. $G(j\omega)H(j\omega)|_{\omega=\omega_p}$ from the $-1 + j0$ point as shown in Figure 8.46 is the gain margin as given by Eq. (8.71). Here GM is positive.

In Bode plots, a vertical line is drawn from the phase crossover frequency point of the phase curve to the magnitude curve of $G(j\omega)H(j\omega)$. The gain margin in dB is obtained by subtracting $G(j\omega)H(j\omega)|_{\omega=\omega_p}$ in dB from 0-dB.

For minimum-phase systems, if both the gain margin and phase margin are positive, then the closed-loop system will be stable. The closed-loop system may be unstable if either of the phase margin or gain margin is negative.

For satisfactory performance of minimum phase systems, the gain margin should be greater than 6-dB and phase margin should be within 30° to 60°.

Illustrations in Figure 8.47 show positive and negative phase and gain margins through polar plots in Figure 8.47(a) and through Bode plots in Figure 8.47(b) and through magnitude-phase plots in *GH*-plane as in Figure 8.47(c).

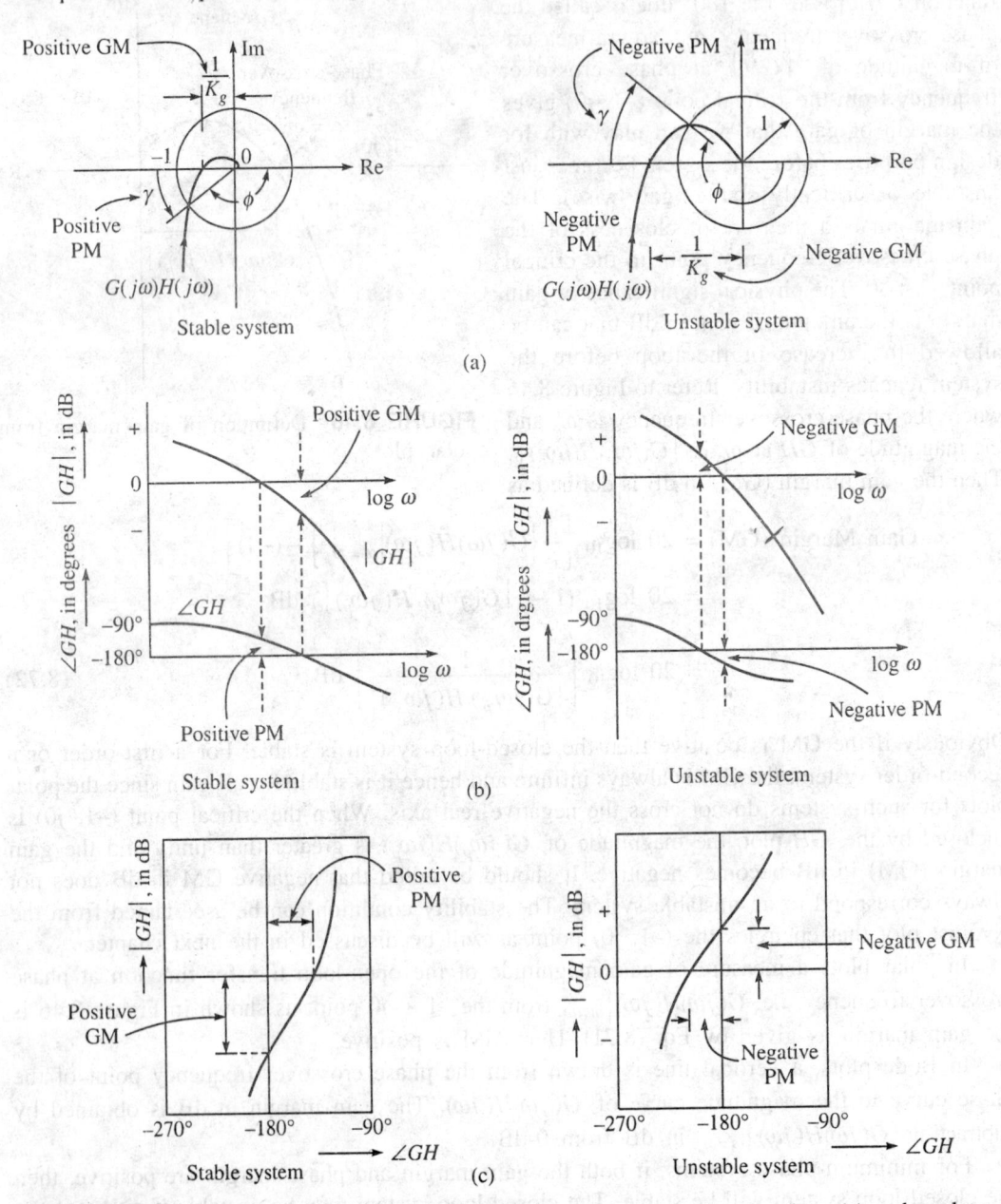

FIGURE 8.47 Illustrations of positive and negative phase and gain margins: (a) polar plots, (b) Bode plots, and (c) magnitude-phase plots.

Gain margin and phase margin are the two common design criteria related to the open-loop frequency response. We are explaining here from different angles to arrive at the same definition. The gain margin being the amount of gain by which the gain of a stable system must be increased for the polar plot to pass through the $(-1, j0)$ point, it may be defined as

$$GM = \frac{K_c}{K} = \frac{K_c|GH(j\omega_p)|}{K|GH(j\omega_p)|} = \frac{1}{K|GH(j\omega_p)|} = \frac{1}{a}$$

where K_c is the critical loop gain for marginal stability and K is the actual loop gain (see Figure 8.48).

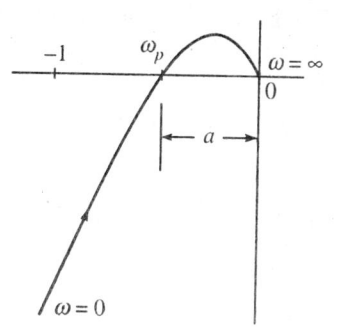

FIGURE 8.48 Polar plot.

In terms of dB, the gain margin is

$$\text{GM in dB} = 20 \log_{10}(\text{GM}) = -20 \log_{10} GH(j\omega_p) = -20 \log_{10} a$$

EXAMPLE 8.17 Obtain the polar plot for the unity feedback control system having open-loop transfer function as

$$GH(s) = \frac{K}{s(s+2)(s+50)}; \quad K = 1300$$

Solution: The polar plot is shown in Figure 8.49(a) and from the plot we get, $a = 0.25$. With $K = 1300$, the GM $= 1/a = 1/0.25 = 4$, or GM in dB $= 20 \log 4 = 12.04$ dB. Thus the critical loop gain is, $K_c = (\text{GM})K = 4(1300) = 5200$. See Figure 8.49(b).

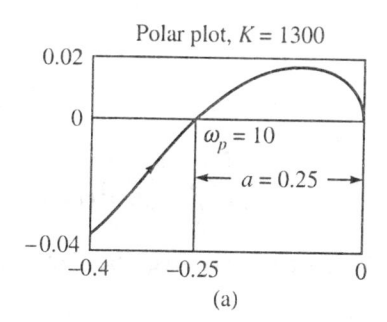

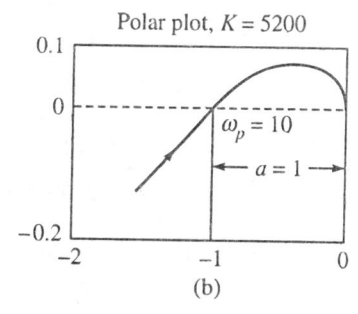

FIGURE 8.49 Example 8.17: (a) polar plot with $K = 1300$ and (b) polar plot with $K = 5200$.

The gain margin alone is inadequate to indicate relative stability when the system parameters affecting the phase of $GH(j\omega)$ are subject to variation. The phase margin is required to indicate the degree of stability.

$$\text{PM} = \angle GH(j\omega_g) - (-180°)$$

In this example, for

$$K = 1300, \quad \omega_g = 4.89$$

and

$$\text{PM} = -163.36° - (-180°) = 16.64°$$

The gain and phase margins along with ω_p and ω_g are obtained more easily from the Bode plot. The phase margin may be read directly from the Bode plot at the frequency ($\omega_g = 4.89$) at which the amplitude curve crosses the 0-dB line. The gain margin may be read (in dB) at the frequency ($\omega_p = 10$) at which the phase angle curve crosses the $-180°$ line. From the Bode plot of Figure 8.50, these margins are: GM = 12.04 dB and PM = 16.64°.

It may be noted that the polar plot, that is, the Nyquist plot and the Bode plot can be obtained by using MATLAB. The MATLAB function [re, Im] = nyquist (num, den, ω) returns the real and imaginary parts of a transfer function for the specified range of frequencies.

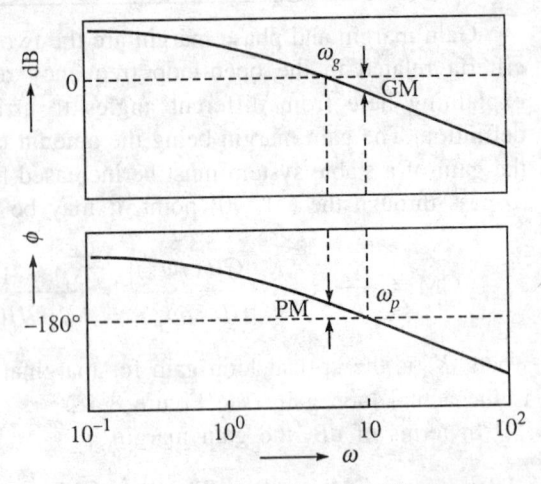

FIGURE 8.50 Example 8.17: Bode plots.

The program has been developed in MATLAB platform to find ω_p, GM, ω_g, and PM. This has been treated at the end of the chapter as an MATLAB application.

EXAMPLE 8.18 Given the open-loop transfer function

$$G(s)H(s) = \frac{K}{(s+1)(2s+1)(3s+1)}$$

and substituting $K = 5$, and $s = j\omega$, we have

$$G(j\omega)H(j\omega) = \frac{5}{(1-11\omega^2) + j6\omega(1-\omega^2)}$$

$$= \frac{5(1-11\omega^2)}{(1-11\omega^2)^2 + 36\omega^2(1-\omega^2)^2} - j\frac{5 \times 6\omega(1-\omega^2)}{(1-11\omega^2)^2 + 36\omega^2(1-\omega^2)^2}$$

Equating the imaginary part equal to zero, we get the gain crossover frequency $\omega_p = \pm 1$ rad/s. Substituting the value of ω_p, we get

$$|G(j\omega_p)H(j\omega_p)| = 0.5$$

Then, the gain margin (GM) in dB = 20 log 1 − 20 log 0.5 = 20 log (1/0.5) = 20 log 2 = 6 dB. It signifies that we can increase the gain K by 6 dB or by a factor of 2 before the closed-loop system goes to instability. That means the limiting factor of increasing the gain K is (2 × 5) or 10. Further by Routh–Hurwitz criterion from the characteristic equation, $1 + GH = 0$, we can find the stability range as $0 < K \leq 10$, which tallies with the result obtained from the gain margin information.

Frequency Response Analysis

EXAMPLE 8.19 Given the open-loop transfer function

$$G(s)H(s) = \frac{20}{s(s+2)(s+10)}$$

draw the Bode plot. Then determine the gain margin and phase margin. Determine the value of gain to be increased such that the system may have (a) a gain margin of 6 dB and (b) a phase margin of 45°.

Solution: The Bode plot is drawn in Figure 8.51 following Table 8.6.

TABLE 8.6 Example 8.19: frequency response data

ω	Gain (dB)	Phase shift (degree)
0.1	20	−93.4
0.5	5.75	−106.9
1.0	−1.01	−122.3
2.0	−9.2	−146.3
3.0	−15.0	−163
5.0	−23.7	−184.8
10.0	−37.2	−213.7

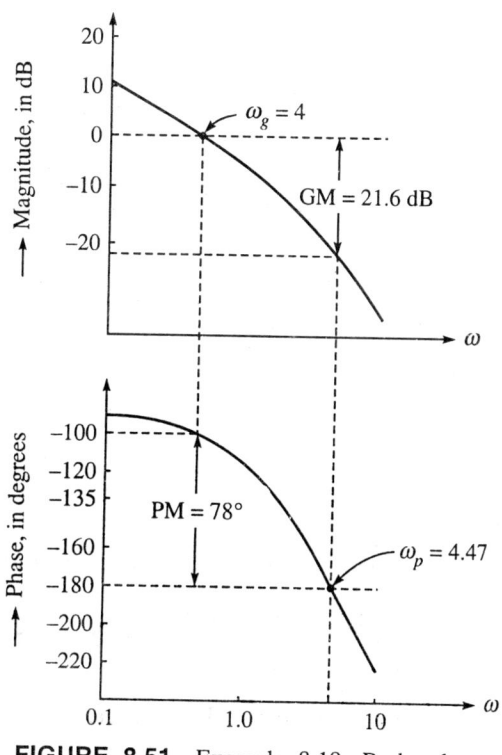

FIGURE 8.51 Example 8.19: Bode plots.

From the Bode plot, we get the phase crossover frequency $\omega_p = 4.47$ and the gain margin as 21.6 dB. The gain crossover frequency $\omega_g = 0.4$ and the phase margin is 78°.

(a) To obtain gain margin as 6 dB, increase the gain in dB by 21.6 − 6 = 15.6 dB. Then the gain is to be increased by a factor 6.03 as $20 \log_{10}$ (gain factor) = 15.6 and we know that if $\log_b a = c$, then $a = b^c$.

(b) To obtain a phase margin of 45°, first find the frequency at which the phase shift is 45° = (180° + γ) − 180°; or $\gamma = 45°$ or −135° and from the plot, we get it as 1.5 rad/s where the gain is seen to be −5.5 dB. Thus, the gain has to be increased by a factor $10^{5.5/20} = 1.88$. These values can be checked from theoretical calculations as well.

EXAMPLE 8.20 Given the open-loop transfer function as

$$GH(s) = \frac{K}{s(T_1 s + 1)(T_2 s + 1)}; \quad K > 0$$

Determine the condition of stability from the polar plot. Verify this condition using Routh–Hurwitz criterion.

Solution: In the steady-state, we put $s = j\omega$ to obtain

$$GH(j\omega) = \frac{K}{j\omega(j\omega T_1 + 1)(j\omega T_2 + 1)}$$

$$= X + jY = \frac{-K(T_1 + T_2)}{1 + \omega^2(T_1^2 + T_2^2) + \omega^4 T_1^2 T_2^2} + j\frac{-K(1 - \omega^2 T_1 T_2)/\omega}{1 + \omega^2(T_1^2 + T_2^2) + \omega^4 T_1^2 T_2^2}$$

$$= \frac{K}{\sqrt{\left[\omega^4(T_1 + T_2)^2 + \omega^2(1 - \omega^2 T_1 T_2)^2\right]}} \angle(-\tan^{-1}\omega T_1 - \tan^{-1}\omega T_2 - \pi/2)$$

The polar plot can be drawn with the following data. The nature of the plot will be as shown in Figure 8.48.

ω	M	ϕ
0	∞	$-90°$
$0 < \omega < \infty$	Finite value	$-90° > \phi > -270°$
∞	0	$-270°$

For negative real axis crossing, i.e. $Y = 0$ leads to

$$\frac{-K(1 - \omega^2 T_1 T_2)/\omega}{1 + \omega^2(T_1^2 + T_2^2) + \omega^4 T_1^2 T_2^2} = 0 \quad \text{or} \quad 1 - \omega^2 T_1 T_2 = 0 \quad \text{or} \quad \omega = 1/\sqrt{T_1 T_2}$$

The magnitude of the real part at $\omega = 1/\sqrt{T_1 T_2}$ becomes

$$\left.\frac{-K(T_1 + T_2)}{1 + \omega^2(T_1^2 + T_2^2) + \omega^4 T_1^2 T_2^2}\right|_{\omega = 1/\sqrt{T_1 T_2}} = \frac{-K T_1 T_2}{T_1 + T_2}$$

Therefore, for the closed-loop system to be stable

$$\frac{-K T_1 T_2}{T_1 + T_2} > -1 \quad \text{or} \quad K < \frac{T_1 + T_2}{T_1 T_2}$$

The characteristic equation is $1 + GH(s) = 0$. That is,

$$T_1 T_2 s^3 + (T_1 + T_2)s^2 + s + K = 0$$

The Routh array is

s^3	$T_1 T_2$	1
s^2	$T_1 + T_2$	K
s	$\dfrac{T_1 + T_2 - K T_1 T_2}{T_1 + T_2}$	
s^0	K	

For the closed-loop system to be stable, the first column of the Routh array should not have any change of sign for which the condition is $0 < K < \dfrac{T_1 + T_2}{T_1 T_2}$. This is in accordance with the polar plot condition derived above.

8.16 Conditionally Stable System

The left-hand rule states that if the $(-1 + j0)$ point lies to the left of the locus, then the closed-loop system is stable. Consider an open-loop transfer function of a unity-feedback system as

$$GH(s) = \frac{K(s^2 + 2s + 4)}{s(s+4)(s+6)(s^2 + 1.4s + 1)}$$

The polar plot is drawn for $K = 1$ and is shown in Figure 8.52. The $(-1 + j0)$ point is to the left of $GH(j\omega)$ locus, i.e. point C, and therefore from the left-hand rule the closed-loop system is stable. The GM = 2.6.

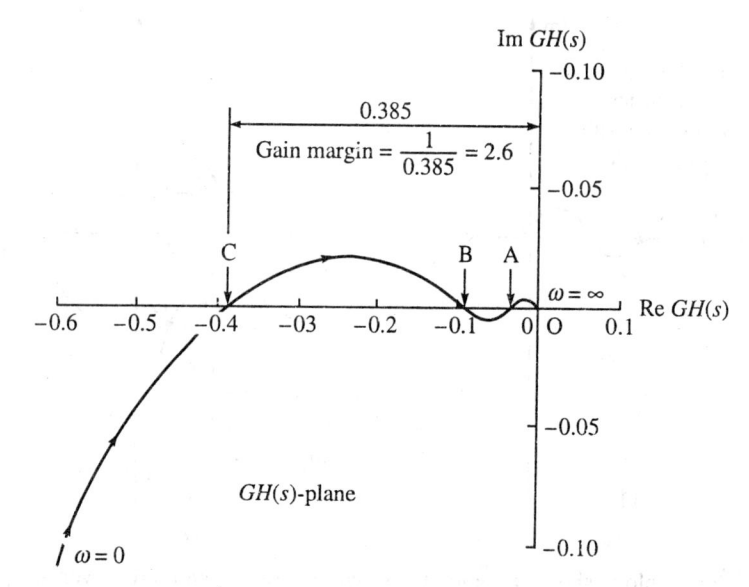

FIGURE 8.52 Polar plot of $\dfrac{K(s^2 + 2s + 4)}{s(s+4)(s+6)(s^2 + 1.4s + 1)}$ with $K = 1$.

Any increase in K affects only the magnitude ratio and not the phase angle. If the magnitude ratio is increased by a factor 6, the $GH(j\omega)$ locus passes through the $(-1 + j0)$ point, that is, the closed-loop system is stable for all values of gain less than $2.6K$ or $2.6(6) = 15.6$. The unstable range of K is $15.6 < K < 67.5$, when the $(-1 + j0)$ point will lie in between the C and B points which indicates from the left-hand rule the unstable range of gain of the system.

For a further increase in the value of K from 67.5 in the range $67.5 < K < 163.6$, the $(-1 + j0)$ point lies in between the points B and A, which from the left-hand rule indicates that the closed-loop system is stable for this range of K.

For a still further increase in the value of K from 163.6 in the range $163.6 < K < \infty$, the $(-1 + j0)$ point lies in between the point A and the origin O, which by the left-hand rule indicates that the closed-loop system is unstable for this range of K.

Hence the system is conditionally stable. In the range of gain K in $0 < K < 15.6$ and $67.5 < K < 163.6$, the closed-loop system is stable, but unstable in the range $15.6 < K < 67.5$ and $163.6 < K < \infty$.

It may be mentioned here that these results are in total agreement with those obtained in root loci plot of Example 6.13.

It is important to note that the conditionally stable system will have two or more phase crossover frequencies as in Figure 8.53(a). Further, some higher order systems with complicated numerator dynamics may have two or more gain-crossover frequencies as in Figure 8.53(b) with $\omega_1 > \omega_2 > \omega_3$. The question may arise that for which gain-crossover frequency the phase margin has to be measured. It is obvious that safe margin to operate is that phase margin which is measured at the highest gain-crossover frequency and with reference to Figure 8.53(b), it is ω_1.

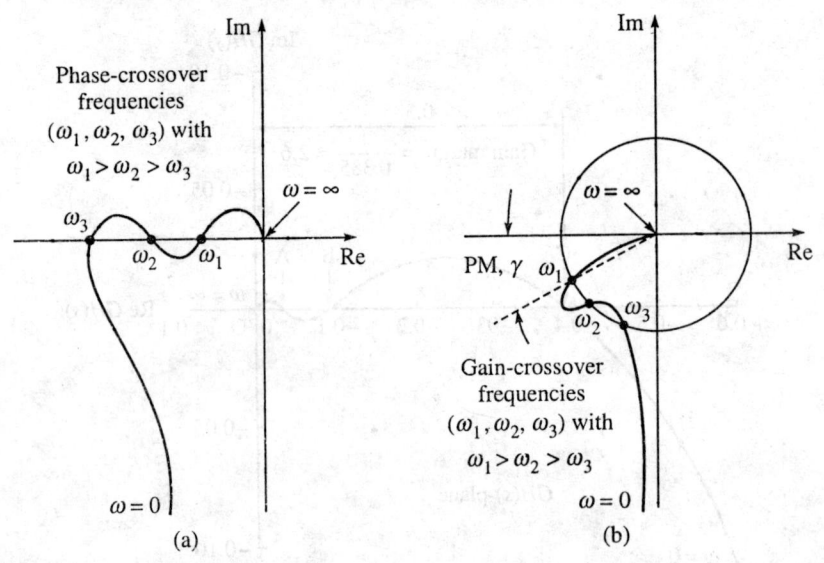

FIGURE 8.53 Polar plots showing more than two phase or gain-crossover frequencies where $\omega_1 > \omega_2 > \omega_3$.

Drill Problem 8.2

The open-loop transfer function is given by

$$\frac{K}{s(s+1)(s+2)}$$

for (i) $K = 1$ and (ii) $K = 10$.

(a) Obtain the Bode plot. Then determine the gain margin (GM) and the phase margin (PM). Draw your conclusions about the stability of the closed-loop system.

(b) Draw the polar plot and comment about the stability of the closed-loop system.

(c) Obtain the dB-magnitude-versus-phase plot. Find GM and PM. Comment about the stability of the closed-loop system.

(d) Verify the result in MATLAB.

Ans.

(a) See Figure 8.54. For $K = 1$: GM = 15.6 dB, PM = 53.4°. The closed-loop system is stable. For $K = 10$: GM = −4.4 dB, PM = −13°. The closed-loop system is unstable.

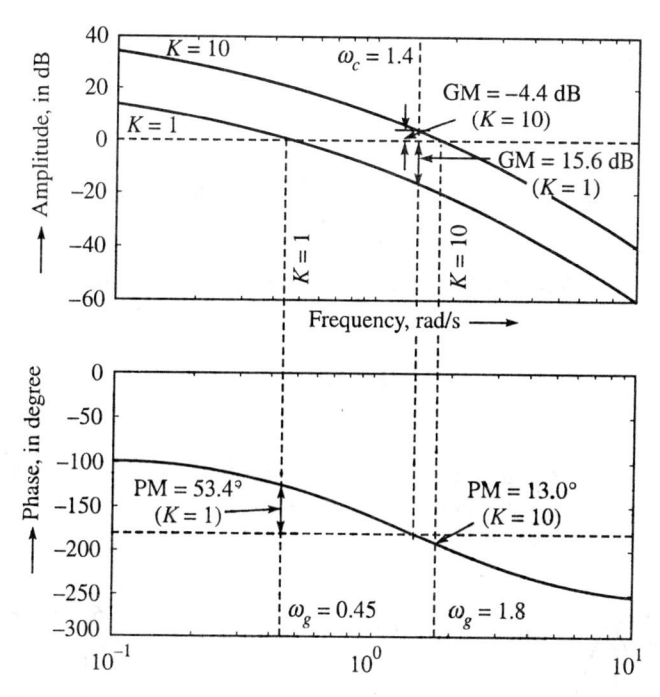

FIGURE 8.54 Drill Problem 8.2: Bode diagrams for $K = 1$ and $K = 10$.

(b) See Figure 8.55. For $K = 1$, the closed-loop system is stable. For $K = 10$, the closed-loop system is unstable.

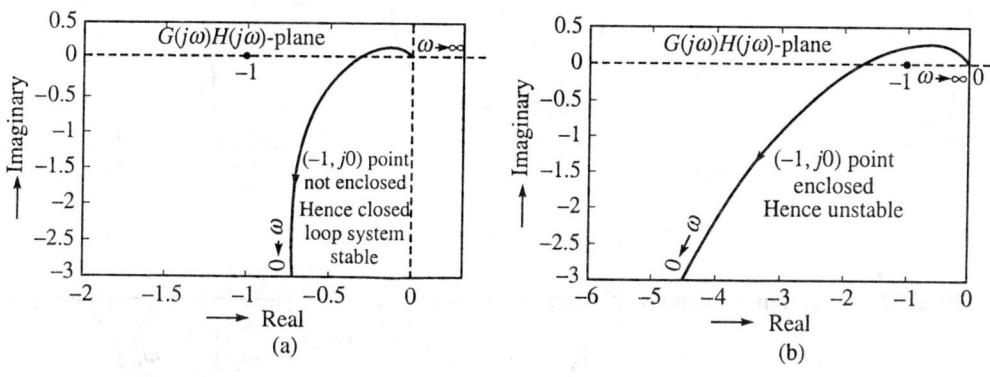

FIGURE 8.55 Drill Problem 8.2: polar plots for $K = 1$ and $K = 20$.

(c) See Figure 8.56. For (i) $K = 1$: GM = 15.6 dB and PM = 53.4°. The closed-loop system is stable. For (ii) $K = 10$: GM = −4.4 dB, PM = −13°. The closed-loop system is unstable.

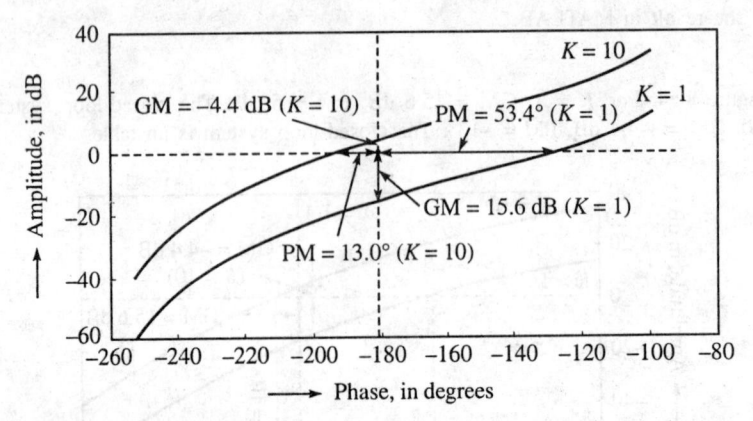

FIGURE 8.56 Drill Problem 8.2: magnitude-versus-phase plots for $K = 1$ and $K = 10$.

Summary

We studied in detail the construction of all the frequency response plots, namely the Bode plot, the log-magnitude-versus-phase plot and polar plots, including their applications. Three representations of underdamped second-order all pole systems are shown in Figure 8.57 for better clarity and comparison.

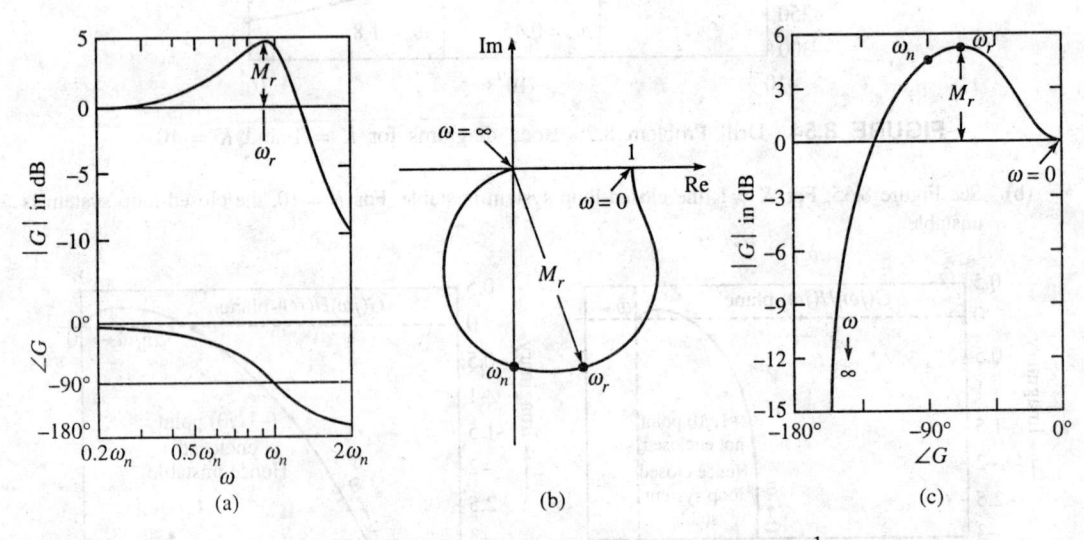

FIGURE 8.57 Three representations of frequency response of $\dfrac{1}{1 + 2\zeta\left(j\dfrac{\omega}{\omega_n}\right) + \left(j\dfrac{\omega}{\omega_n}\right)^2}$, for $\zeta > 0$:

(a) Bode diagram, (b) polar plot, and (c) log-magnitude-versus-phase plot.

Bode plot can be obtained without much computation, and computer programs for Bode plots can easily be made. The polar plot requires comparatively larger computations. Identification of an unknown system, i.e. an estimation of the transfer function can be made from the Bode plot.

However, it may be noted that except for the minimum-phase transfer function (i.e. for the stable transfer function having all zeros in the left-half of the *s*-plane) the transfer function cannot be determined from the magnitude curve alone.

Problems

8.1 Draw the asymptotic magnitude and phase plots of the following open-loop transfer functions:

(a) $\dfrac{100\left(\dfrac{s}{7}+1\right)}{(s+1)^2\left(\dfrac{s}{70}+1\right)\left(\dfrac{s}{200}+1\right)}$

(b) $\dfrac{(300)^2(0.1s+1)(0.0025s+1)}{s^2(0.01s+1)(0.00025s+1)(0.0001s+1)}$

(c) $\dfrac{2\times 10^4(s+10)}{s(s+5)(s+200)}$

(d) $\dfrac{4}{(s+1)^3}$

8.2 For the following open-loop transfer functions

(a) $\dfrac{225(0.1j\omega+1)}{(j\omega)^2(0.01j\omega+1)}$

(d) $\dfrac{1000}{(j\omega+1)(0.01j\omega+1)(0.0005j\omega+1)}$

draw the Bode plot, determine the gain-crossover frequency, phase-crossover frequency, gain margin, and phase margin and then comment upon the stability of each system.

8.3. (a) Draw the Bode plot for the following open-loop transfer function:

$$\dfrac{100(0.02s+1)}{(s+1)(0.1s+1)(0.01s+1)^2}$$

(b) Mark the following on the Bode diagram: (i) gain-crossover frequency, (ii) phase margin, (iii) phase-crossover frequency, and (iv) gain margin.

8.4. Draw the Bode diagram for the following open-loop transfer function:

$$\dfrac{200(0.1s+1)}{s(0.2s+1)(0.05s+1)}$$

Is the system stable?

8.5 Draw the Bode plot of the system shown in Figure P.8.5. Find the gain and phase margins. Comment upon the stability of the system.

8.6 The two magnitude diagrams of Bode plot of the open-loop transfer function

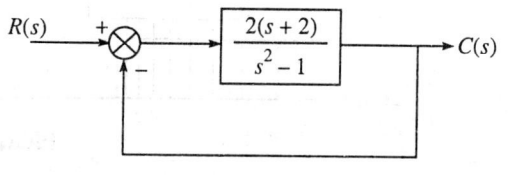

FIGURE P.8.5

of minimum-phase function are shown in Figures P.8.6(a) and (b). Determine the transfer function for each system. Sketch the curves for each system.

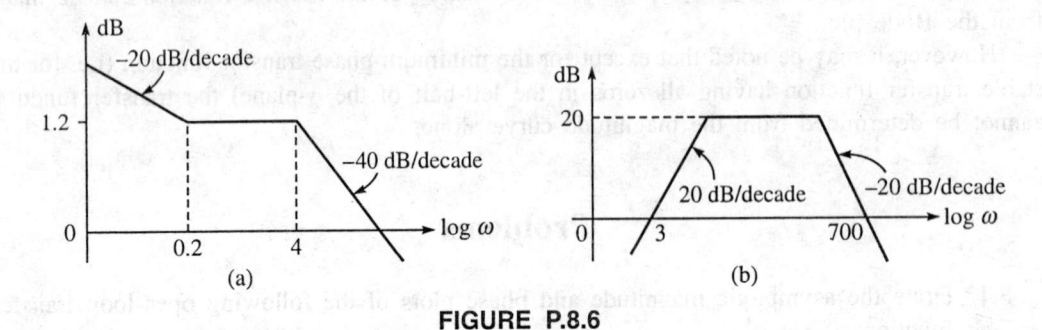

FIGURE P.8.6

8.7 The Bode plot shown in Figure P.8.7 is for a unity-feedback system, with a minimum-phase transfer function $G(s)$. Determine the gain margin GM (in dB) and phase margin PM (in degrees). Is the closed-loop system stable? Determine $G(s)$.

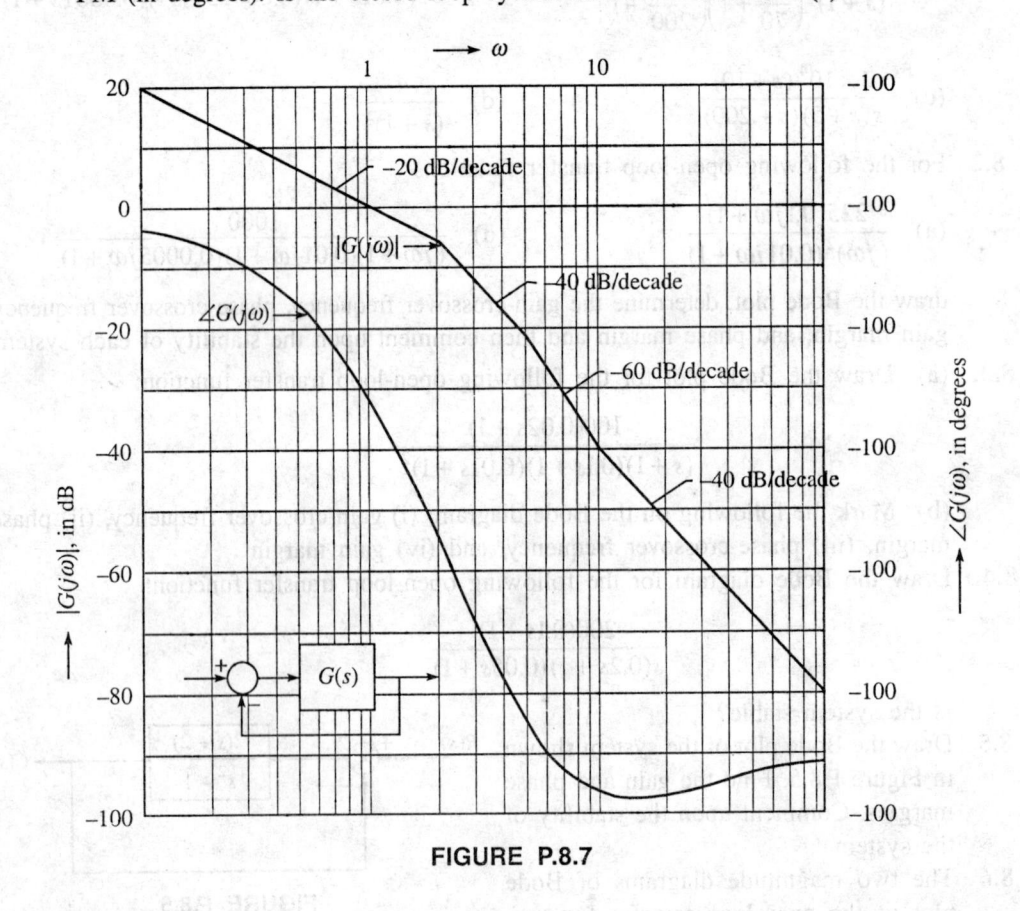

FIGURE P.8.7

8.8 Consider a unity-feedback system with the open-loop transfer function

$$\frac{K}{s(s+1)(s+2)}, \quad K = 2$$

Draw the polar plot. Determine GM and PM and comment upon the closed-loop system stability.

8.9 The polar plot shown in Figure P.8.9 is of a minimum-phase transfer function of a unity-feedback system.

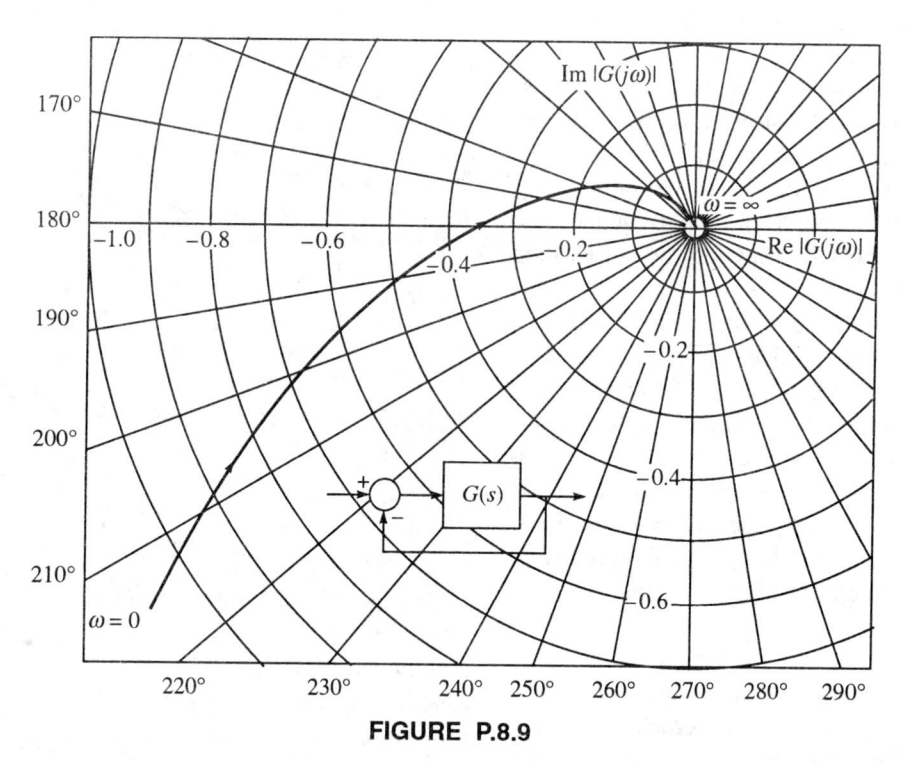

FIGURE P.8.9

(a) Determine the gain margin GM (in dB) and the phase margin PM (in degrees). Is the closed-loop system stable?
(b) Determine the value of K. Evaluate the open-loop transfer function.
(c) Determine the value of K so that the closed-loop system has phase margin, PM = 20°. Determine the corresponding gain margin GM (dB).

Nyquist Stability

9.1 Introduction

A frequency domain stability criterion was developed by H. Nyquist in 1932 and remains a fundamental approach to the investigation of stability of linear control systems. The Nyquist stability criterion is useful for determining the stability of the closed-loop system from the open-loop frequency response without determining the closed-loop poles. Further, the Nyquist stability criterion is very convenient because many a times the mathematical model of the physical system to be designed is not available, only the frequency response is obtainable. It may be noted that the absolute stability of the system can be obtained from its characteristic polynomial by using the Routh–Hurwitz criterion. Nyquist stability criterion is useful for the design of control systems.

The Nyquist stability criterion is based upon a theorem in the theory of the functions of complex variables due to Cauchy. Cauchy's theorem is concerned with mappings of contours in the complex s-plane, which uses complex variable theory.

9.2 Basic Philosophy of Stability Criterion

In this chapter, we will discuss the Nyquist stability criterion and the related mathematics. The overall transfer function of the closed-loop system shown in Figure 9.1 is given by

$$\frac{C(s)}{R(s)} = \frac{G(s)}{1+G(s)H(s)} \quad (9.1)$$

OBJECTIVE

Strictly speaking, the Nyquist plot, a graphical technique, is for the frequency range $-\infty < \omega < \infty$ and is different from the polar plot which is for the frequency range $0 < \omega < \infty$. Using Cauchy's theorem of complex variables, we can determine the stability of closed-loop systems from the knowledge of, the number of encirclements of the $(-1, j0)$ point by the Nyquist contour of the open-loop frequency response. In this regard, the technique for drawing the Nyquist contour of the given open-loop transfer function for the whole frequency range $-\infty < \omega < \infty$, in parts, has to be learnt. Then the stability of the closed-loop system can be determined using Cauchy's theorem.

The closed-loop stability from Nichols' chart for the design aspect of control system is touched upon, for which the prerequisite fundamentals concerning M and N circles have also been highlighted.

CHAPTER OUTLINE

Introduction
Basic Philosophy of Stability Criterion
Conformal Mapping
Nyquist Stability Criterion
Conditionally Stable System
Closed-Loop Frequency Response
M and N-Circles
Nichols' Chart
Frequency Domain Specification for Design

Nyquist Stability

The characteristic equation is

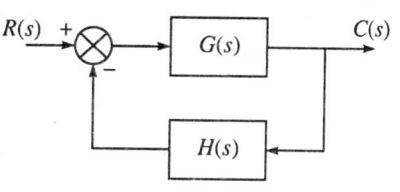

$$F(s) = 1 + G(s)H(s) = 0 \qquad (9.2)$$

Further, $\quad G(s)H(s) = \dfrac{N(s)}{D(s)} \qquad (9.3)$

FIGURE 9.1 Closed-loop system.

Then, $\quad F(s) = 1 + G(s)H(s) = 1 + \dfrac{N(s)}{D(s)} = \dfrac{N'(s)}{D(s)} \qquad (9.4)$

Now we can conclude:

1. Closed-loop poles = zeros of $F(s)$ = zeros of $[1+ G(s)H(s)]$
   = roots of the characteristic equation
2. Poles of $F(s)$ = poles of the open-loop transfer function $G(s)H(s)$
3. For the closed-loop system to be asymptotically stable, there is no restriction on the poles and zeros of the open-loop transfer function, but the poles of the closed-loop system or the roots of the characteristic equation must all be located in the left-half of the s-plane.

In order to determine whether the closed-loop system is stable, we must find out whether

$$F(s) = 1 + G(s)H(s) = 0$$

has any roots in the right-half of the s-plane (including the $j\omega$-axis). Let

$$F(s) = 1 + G(s)H(s) = 1 + \dfrac{K \prod_{i=1}^{n} (s+s_i)}{\prod_{k=1}^{M} (s+s_k)} \qquad (9.5)$$

Finding the zeros of $F(s)$, i.e. the poles of the closed-loop system, is a difficult task but using the Nyquist graphical technique we can find the stability of the closed-loop system from the open-loop pole information and principle of argument. As $s = \sigma + j\omega$, a complex quantity, the function $F(s)$ is also a complex quantity.

Let $F(s)$ be defined as

$$F(s) = u + jv \qquad (9.6)$$

and represented on the complex $F(s)$-plane with coordinates u and v. Equation (9.5) indicates that for every point s in the s-plane at which $F(s)$ is analytic, we can find a corresponding $F(s)$ in the $F(s)$-plane.

A function $F(s)$ is analytic in the s-plane provided the function and all its derivatives exist. The points in the s-plane where the function (or its derivatives) does not exist, are called singular points. The poles of a function are singular points.

Since any number of points of analyticity in the s-plane can be mapped on to the $F(s)$-plane, it follows that for a contour in the s-plane which does not go through any singular point, there corresponds a contour in the $F(s)$-plane.

We choose a contour T_s in the s-plane. The contour T_s in s-plane moving in a prescribed direction (in this text, it is taken as clockwise) is mapped on the $F(s)$-plane and is defined as T_F in the $F(s)$-plane. Then we determine the number of encirclements of the origin of $F(s)$ by T_F in the same direction as T_s and suppose it is N. Then from the principle of argument, we get

$$N = Z - P \qquad (9.7)$$

where P is the number of poles of $F(s)$ within T_s-contour, i.e. the poles of the open-loop transfer function $G(s)H(s)$ lying in the right-hand side of s-plane, if any, and Z is the number of zeros of $F(s)$ within T_s-contour, i.e. the unstable zeros of $F(s)$ or the poles of the closed-loop system lying in the right-hand side of the s-plane. For stability information, we are concerned with the right-hand side closed-loop poles only. The number of encirclements of the origin of the $F(s)$-plane, i.e. of the $(-1 + j0)$ point of the GH-plane gives the information about the stability of the closed-loop system as $Z = N + P$, i.e. the value of Z gives the number of poles of the closed-loop system lying in the right-hand side of the s-plane.

9.3 Encircled versus Enclosed

The term *encircled* and *enclosed* are frequently used in the interpretation of the Nyquist criterion and need clarification.

Encircled. A point is said to be encircled by a closed path if it is found inside the path. In Figure 9.2, point A is said to be encircled by the closed path T in the prescribed clockwise direction since point A is inside the closed path whereas the point B is not encircled by the closed path T since it is outside the path.

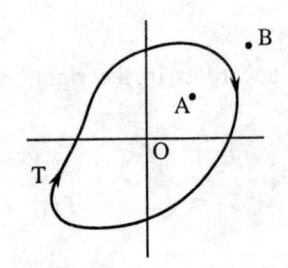

FIGURE 9.2 Encirclement of point A by the closed path T.

Enclosed. A point or region is said to be enclosed by a closed path if it is found to be to the right of the path when the path is traversed in the prescribed clockwise direction. The point A is enclosed by T in Figure 9.3(a) but not in Figure 9.3(b). The point B or the region outside T in Figure 9.3(b) is considered to be enclosed.

Number of encirclements and enclosures. When the point A is encircled or enclosed by a closed path, a number N may be assigned to the number of encirclements or enclosures (in the prescribed direction), as the case may be. For clockwise direction, it is $+N$ and for counter-clockwise direction it is $-N$. In Figure 9.4(a), the encirclement is clockwise, as point A is encircled once, hence $N = +1$ and point B is encircled twice, hence $N = +2$. In Figure 9.4(b), the encirclement is counterclockwise as point A is encircled once, hence $N = -1$ and point B is encircled twice, hence $N = -2$.

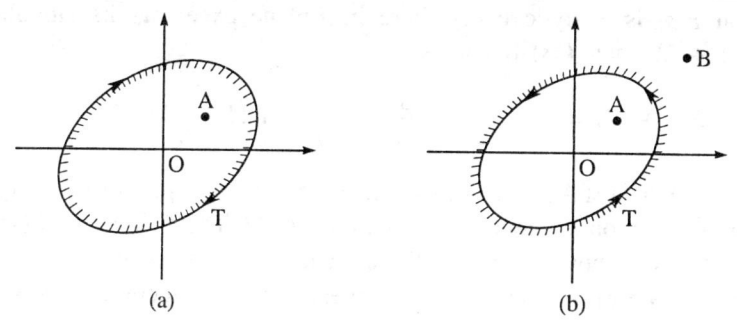

FIGURE 9.3 (a) Point A is considered enclosed by the closed path T and (b) point B and the region outside T are considered enclosed by T (Point A here is not considered enclosed by T).

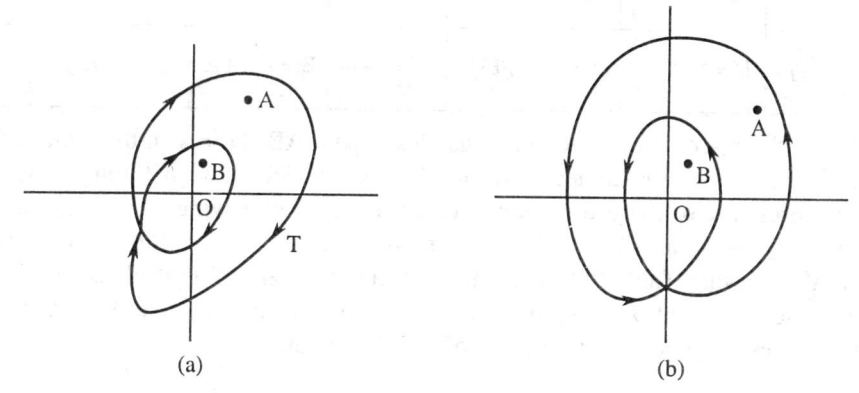

FIGURE 9.4 (a) Clockwise encirclement and (b) counterclockwise encirclement.

9.4 Conformal Mapping

Here we will explain conformal mapping due to Cauchy's theorem considering the entire s-plane, while Nyquist applied this technique to stability analysis where only the right-hand s-plane is the point of interest and thus is considered accordingly. The characteristic equation of the system is

$$F(s) = 1 + G(s)H(s) = 0$$

Consider, for example, the open-loop transfer function

$$G(s)H(s) = \frac{6}{(s+1)(s+2)}$$

Then

$$F(s) = 1 + \frac{6}{(s+1)(s+2)} = 0 \tag{9.8}$$

or

$$F(s) = \frac{(s+1.5+j2.4)(s+1.5-j2.4)}{(s+1)(s+2)} = 0 \tag{9.9}$$

The zeros of $F(s)$ are $-1.5 \pm j2.4$ and the poles of $F(s)$ are -1 and -2.

The function $F(s)$ is analytic everywhere in s-plane except at its singular points. For example, if $s = 1 + j2$, then $F(s)$ becomes

$$F(1 + j2) = 1 + \frac{6}{(2 + j2)(3 + j2)} = 1.11 - j0.577$$

The T_s-contour ABCDEFA in the s-plane enclosing the two poles of $F(s)$ lying at -1 and -2 in the clockwise direction is mapped as T_F-contour A'B'C'D'E'F'A' in the $F(s)$-plane following Table 9.1, and this is shown in Figure 9.5(a). The encirclement of origin of $F(s)$-plane by T_F-contour in Figure 9.5(a) is twice in the counterclockwise direction, that is, $N = -2$.

TABLE 9.1

s-plane $\sigma+j\omega$	$A = -3+j0$	$B = -3+j2$	$C = 1+j2$	$D = 1+j0$	$E = 1-j2$	$F = -3-j2$	$A = -3+j0$
$F(s)$-plane $u+jv$	$A' = 4+j0$	$B' = 0.7+j0.9$	$C' = 1.11-j0.577$	$D' = 2+j0$	$E' = 1.11+j0.57$	$F' = 0.7-j0.9$	$A' = 4+j0$

In Figure 9.5(b), the T_s-contour is a continuous closed path ABCDEFA in the s-plane (marked by the boundary points as mentioned in Table 9.2) which does not go through any singular points and encloses in clockwise direction two poles (P) of $F(s)$ lying at -1 and -2 and two zeros (Z) of $F(s)$ lying at $-1.5 \pm j2.4$. It is mapped as T_F-contour in the $F(s)$-plane A'B'C'D'E'F'A' following Table 9.2. The encirclement of origin of $F(s)$-plane by T_F-contour in the same sense as that of T_s-contour is $(Z - P)$ times, i.e. in this case $N = 0$, that means the T_F-contour does not encircle the origin of the $F(s)$-plane.

TABLE 9.2

s-plane $\sigma+j\omega$	$A=-3+j0$	$B=-3+j3$	$C=1+j3$	$D=1+j0$	$E=1-j3$	$F=-3-j3$	$A=-3+j0$
$F(s)$-plane $u+jv$	$A'=4+j0$	$B'=0.677+j0.45$	$C'=0.923-j0.385$	$D'=2+j0$	$E'=0.923+j0.385$	$F'=0.677-j0.415$	$A'=4+j0$

In Figure 9.5(c), the T_s-contour EFGHE in the clockwise direction encloses neither any pole nor any zero of $F(s)$ and is mapped as T_F-contour E'F'G'H'E' in $F(s)$-plane following Table 9.3. There is no encirclement of origin in the $F(s)$-plane, that is, $N = 0$.

TABLE 9.3

s-plane $\sigma + j\omega$	$E = -4 - j1$	$F = -4 - j0$	$G = -3 + j0$	$H = -3 - j1$	$E = -4 - j1$
$F(s)$-plane $u + jv$	$E' = 1.05 - j1.35$	$F' = 2 + j0$	$G' = 4 + j0$	$H' = 1.6 - j1.8$	$E' = 1.05 - j1.35$

In Figure 9.5(d), the T_s-contour ABCDA enclosing one zero at $(-1.5 + j2.4)$ of $F(s)$ in the clockwise direction is mapped as T_F-contour A'B'C'D'A' in the $F(s)$-plane, following

Nyquist Stability

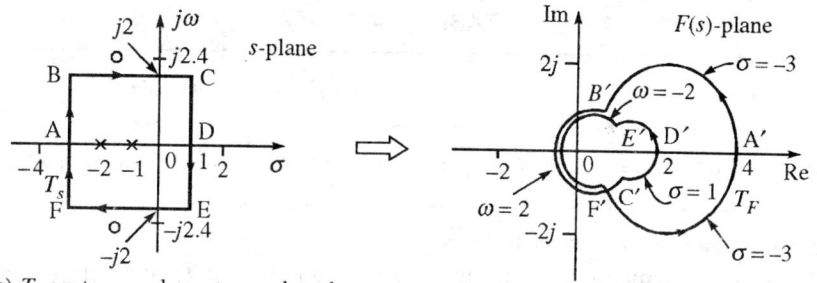

(a) T_s-contour encloses two poles, the corresponding T_F-contour encircles the orgin twice.

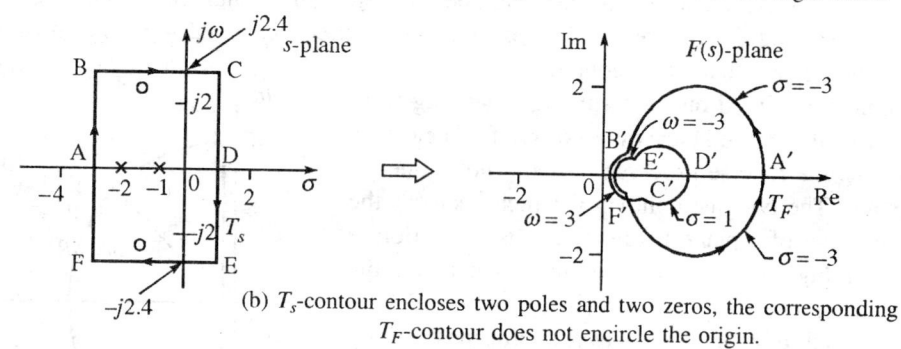

(b) T_s-contour encloses two poles and two zeros, the corresponding T_F-contour does not encircle the origin.

(c) T_s-contour does not enclose any pole or zero in s-plane, the corresponding T_F-contour does not encircle the origin.

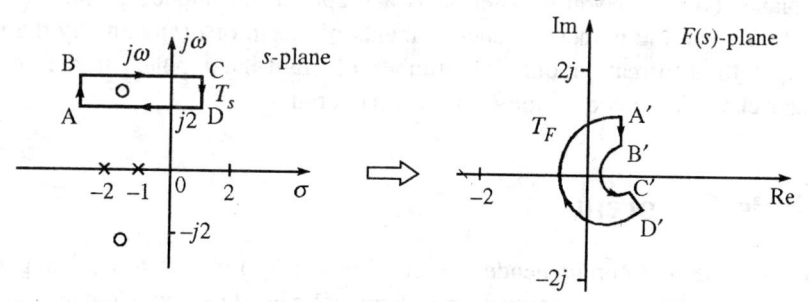

(d) T_s-contour encloses one zero, the corresponding T_F-contour encircles the origin once.

FIGURE 9.5 Conformal mapping for the function $F(s)$ of Eq. (9.9).

Table 9.4. The encirclement of origin of the $F(s)$-plane is one in the clockwise direction, that is, $N = +1$.

TABLE 9.4

s-plane $\sigma + j\omega$	$A = -3 + j2$	$B = -3 + j3$	$C = 1 + j3$	$D = 1 + j2$	$A = -3 + j2$
$F(s)$-plane $u + jv$	$A' = 0.7 + j0.9$	$B' = 0.677 + j0.45$	$C' = 0.923 - j0.385$	$D' = 1.1 - j0.577$	$A' = 0.7 + j0.9$

The conclusions can be drawn as follows:

The number of encirclements of origin in the $F(s)$-plane by the T_F-contour, corresponding to T_s-contour in the s-plane enclosing Z number of zeros and P number of poles of $F(s)$ in arbitrarily prescribed (in this text, taken as clockwise) direction, is $(Z - P)$, i.e. in the same sense of direction as that of T_s-contour.

It should be pointed out that although mapping from the s-plane to the $F(s)$-plane is one-to-one for a rational function $F(s)$, the reverse process is usually not a one-to-one mapping. The Nyquist contour T_s which encloses the entire right-hand of s-plane in the clockwise direction is as shown in Figure 9.6. The contour passes along the $j\omega$-axis from $-j\infty$ to $+j\infty$. The contour is completed by a semicircular path of radius R where R approaches infinity thus covering the entire right-hand side of the s-plane. However, it may be noted that if $F(s)$ has a pole at the origin of the s-plane or at some points on the $j\omega$-axis, we must exclude the point of singularity of $F(s)$ in the T_s-contour by making a detour along an infinitesimal semicircle. See Figure 9.12.

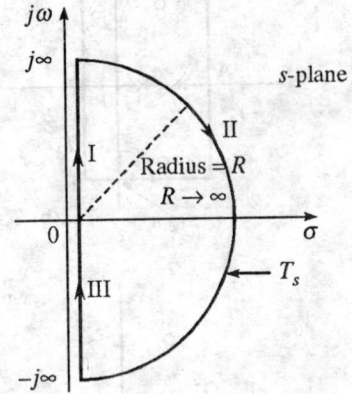

FIGURE 9.6 Nyquist contour T_s.

The obvious question may arise that why the T_s-contour covering only the right-hand side of s-plane? This is because our intention is to find the stability of the closed-loop system, which in other words, is the information about the number of poles of the closed-loop system in the right-hand of s-plane. Further, Nyquist criterion is a graphical technique to find the stability from the information about the number of encirclements of origin of $F(s)$-plane by the mapped T_F-contour along with information only of number of right-hand poles of the open-loop system, which aspect is elaborated in the Cauchy's Theorem.

9.5 Cauchy's Theorem

Cauchy's theorem can be best comprehended by considering $F(s)$ in terms of the angle due to each pole and zero as the contour T_s is traversed in the clockwise direction. Now let us consider the function

$$F(s) = \frac{(s + z_1)(s + z_2)}{(s + p_1)(s + p_2)} \tag{9.10}$$

Nyquist Stability

where z_i is a zero of $F(s)$ and p_k is a pole of $F(s)$. Then,

$$F(s) = |F(s)| \angle F(s) = \frac{|s+z_1||s+z_2|}{|s+p_1||s+p_2|}(\angle s+z_1 + \angle s+z_2 - \angle s+p_1 - \angle s+p_2)$$

$$= |F(s)|(\phi_{z1} + \phi_{z2} - \phi_{p1} - \phi_{p2}) \tag{9.11}$$

Figure 9.7(a) shows an arbitrarily chosen trajectory T_s in the s-plane with an arbitrary point s_1 on its path. The function $F(s)$ evaluated at $s = s_1$ is given by

$$F(s_1) = \frac{(s_1+z_1)(s_1+z_2)}{(s_1+p_1)(s_1+p_2)}$$

The factor $(s_1 + z_1)$ can be represented graphically by the vector drawn from $-z_1$ to s_1. Similar vectors can be defined for $(s_1 + z_2)$, $(s_1 + p_1)$ and $(s_1 + p_2)$.

Thus $F(s_1)$ is represented by the vectors drawn from the given poles and zeros to the point s_1 as shown in Figure 9.7(a). Now, considering the vectors as shown for a specific contour T_s, we can determine the angles as s_1 traverses the contour in the prescribed clockwise direction until it returns to the starting point. Clearly the angles generated by the vectors drawn from the poles and zeros that are not encircled by T_s when s_1 completes one round trip are zero, whereas the vector $(s_1 + z_1)$ drawn from the zero at $-z_1$, which is encircled by T_s, generates an angle (clockwise sense) of 2π radians. If Z number of zeros were enclosed within T_s, then the angle would be equal to $2\pi(Z)$ radians. Following this reasoning, if Z zeros and P poles are encircled as contour T_s is traversed in the clockwise direction, then

$$2\pi(Z) - 2\pi(P) \tag{9.12}$$

is the net resultant angle of $F(s)$. Thus the net angle of T_F of the contour in the $F(s)$-plane [Figure 9.7(b)], is simply, $\phi_F = \phi_Z - \phi_P$

or
$$2\pi N = 2\pi Z - 2\pi P \tag{9.13}$$

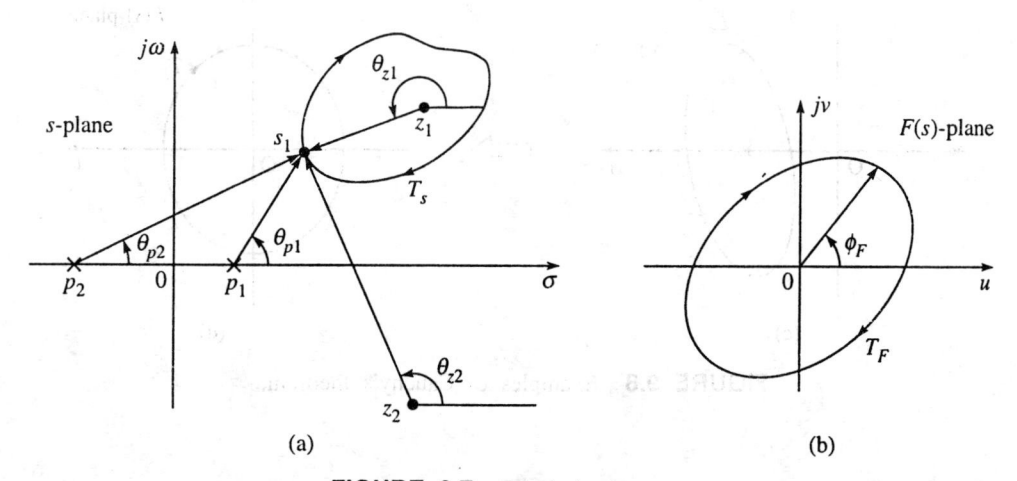

FIGURE 9.7 Cauchy's theorem.

and the net number of encirclements of the origin of the $F(s)$-plane in the clockwise direction is

$$N = Z - P \tag{9.14}$$

In the pole-zero pattern of $F(s)$ shown in Figure 9.8(a), the contour T_s encloses three zeros and one pole. Therefore, $N = 3 - 1 = 2$, and T_F completes two clockwise encirclements of the origin in the $F(s)$-plane as in Figure 9.8(b).

For the pole and zero pattern and the contour T_s as shown in Figure 9.8(c), one pole is encircled and no zeros are encircled. Therefore we have

$$N = Z - P = -1$$

and we expect one encirclement of the origin by the contour T_F in the $F(s)$-plane. However, since the sign of N is negative, we find that the encirclement moves in the counterclockwise direction as shown in Figure 9.8(d).

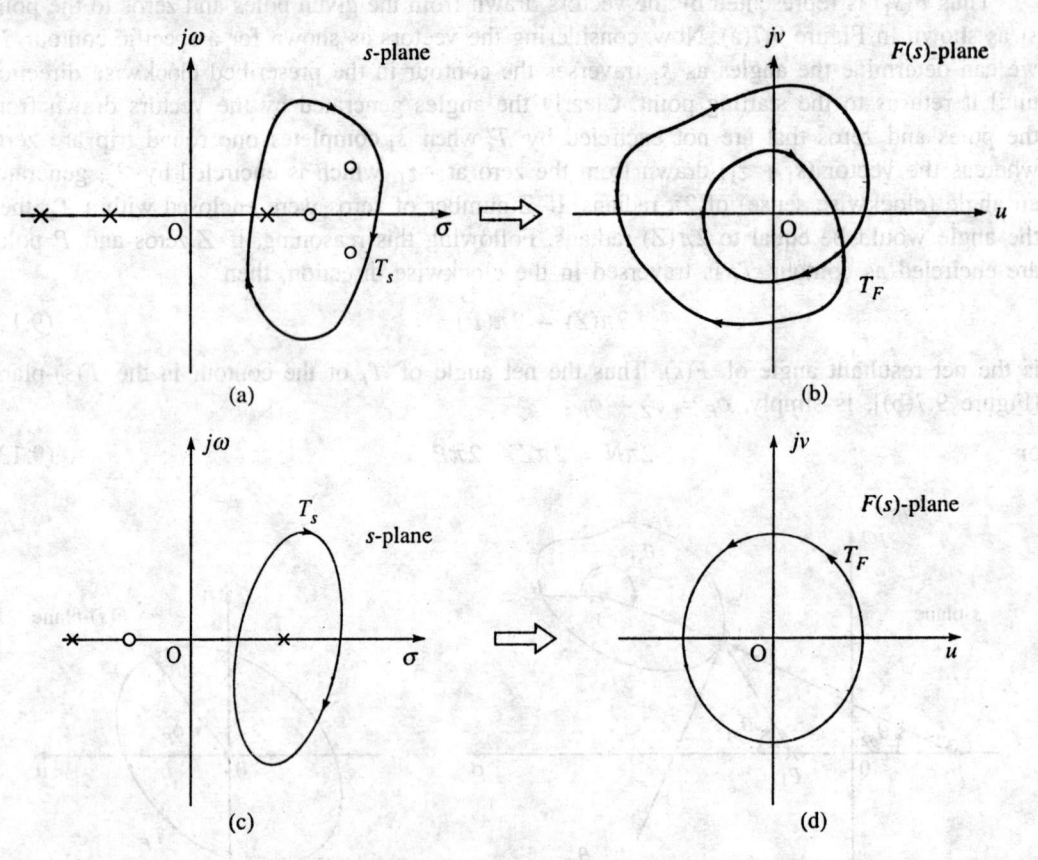

FIGURE 9.8 Examples of Cauchy's theorem.

9.6 Relationship between *GH*-plane and *F(s)*-plane

It may be noted that

$$G(s)H(s) = [1 + G(s)H(s)] - 1 = F(s) - 1 \qquad (9.15)$$

That is, the contour T_F of $F(s)$-plane is obtained from the contour T_{GH} of *GH*-plane through coordinate shifting by $(-1 + j0)$. The origin of $F(s)$-plane is $(-1 + j0)$ of *GH*-plane. Thus the encirclement of the origin by the contour T_F is equivalent to the encirclement of the point $(-1 + j0)$ by the contour T_{GH} as shown in Figure 9.9(a). The relationship between the *GH*-plane and the $F(s)$-plane is shown in Figure 9.9(b).

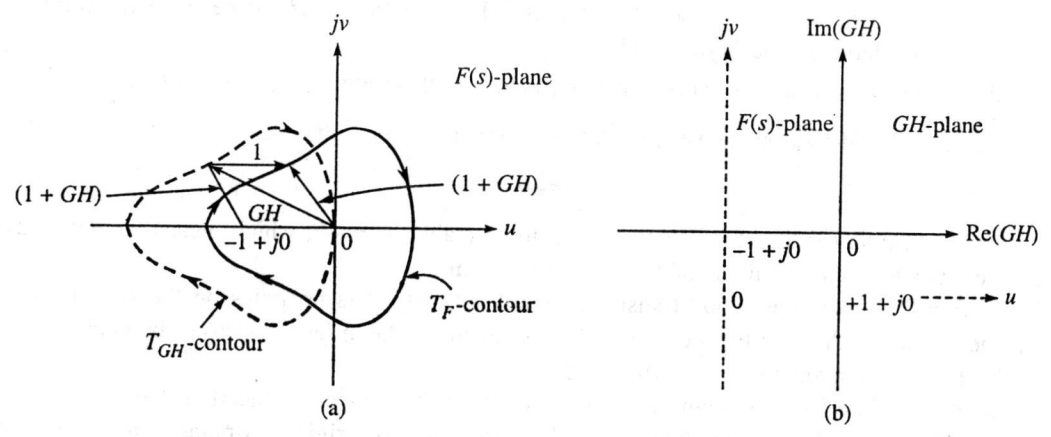

FIGURE 9.9 Relationship between *GH*-plane and $F(s)$-plane.

The number of roots (or zeros) of $[1 + G(s)H(s)]$ in the right-hand *s*-plane is represented by the expression

$$Z = N + P \qquad (9.16)$$

Clearly, if the number of poles of $G(s)H(s)$ in the right hand *s*-plane is zero ($P = 0$), we require for a stable system that be $N = 0$ and the contour T_{GH} must not encircle the $(-1 + j0)$ point. Also, if $P \neq 0$ and we require for a stable system that $Z = 0$, then we must have $N = -P$, i.e. have P number of counterclockwise encirclements. More about this in the next section.

9.7 Nyquist Stability Criterion

If the open-loop transfer function $G(s)H(s)$ has k poles in the right-half of the *s*-plane, then for closed-loop stability the $G(s)H(s)$ locus, as a representative point traces out the Nyquist contour in the prescribed (clockwise) direction, must encircle the $(-1 + j0)$ point k times in the reverse of the prescribed (that is, counterclockwise) direction. It may be noted that when the open-loop transfer function $G(s)H(s)$ has poles and/or zeros on the $j\omega$-axis, the Nyquist contour has to be modified as per the restriction of conformal mapping in the theory of complex variables.

The Nyquist stability criterion provides a convenient method for finding the number of zeros of $F(s) = 1 + G(s)H(s)$ in the right-half of the s-plane directly from the Nyquist plot of open-loop transfer function $G(s)H(s)$. The Nyquist stability criterion is defined in terms of the $(-1, j0)$ point on the Nyquist plot. The Nyquist plot is obtained by mapping the function $G(s)H(s)$. The Nyquist contour is chosen so that it encircles the entire right-half of s-plane. When the s-plane locus T_s is the Nyquist contour, the Nyquist stability criterion is given by

$$Z = N + P$$

where

P = number of poles of $G(s)H(s)$ or $F(s)$ in the right-half of the s-plane.

N = number of clockwise encirclements of $(-1, j0)$ point in GH-plane or of the origin in $F(s)$-plane by the Nyquist plot.

Z = number of zeros of $F(s) = 1 + G(s)H(s)$ in the right-half of the s-plane.

For the closed-loop system to be stable, Z must be zero, that is,

$$N = -P$$

If $P \neq 0$, then for the closed-loop system to be stable, $N = -P$, which means we must have counterclockwise encirclement of the $(-1 + j0)$ point.

If $P = 0$, i.e. the open-loop transfer function $G(s)H(s)$ has no poles on the right-hand of s-plane, then, $Z = N$, thus for a closed-loop system to be stable there must not be encirclement of the $(-1 + j0)$ point by the $G(s)H(s)$ locus.

Since multiple loops systems may or may not include poles in the right-half of s-plane, we should be careful. A simple inspection of encirclement of origin of s-plane is not sufficient, Routh's stability test has then to be performed.

EXAMPLE 9.1 Given the open-loop transfer function

$$G(s)H(s) = \frac{K(s+2)}{(s+1)(s-1)} \qquad (9.17)$$

where $K = 1$. Draw the complete Nyquist plot and determine the stability of the closed-loop system.

Solution: From the open-loop transfer function, it is observed that there is one open-loop pole in the right-half of s-plane, therefore, $P = 1$. The T_s-contour in the s-plane is shown in Figure 9.6 and traversing in the prescribed (clockwise) direction covering the entire right-hand side of s-plane gives the following three sections:

(i) **Section I:** This section, from 0 to $+\infty$ along the $j\omega$-axis of the T_s-contour, when mapped in the GH-plane, becomes the polar plot as depicted in Figure 9.10(a). The polar plot of

$$G(s)H(s) = \frac{s+2}{(s+1)(s-1)}$$

Nyquist Stability

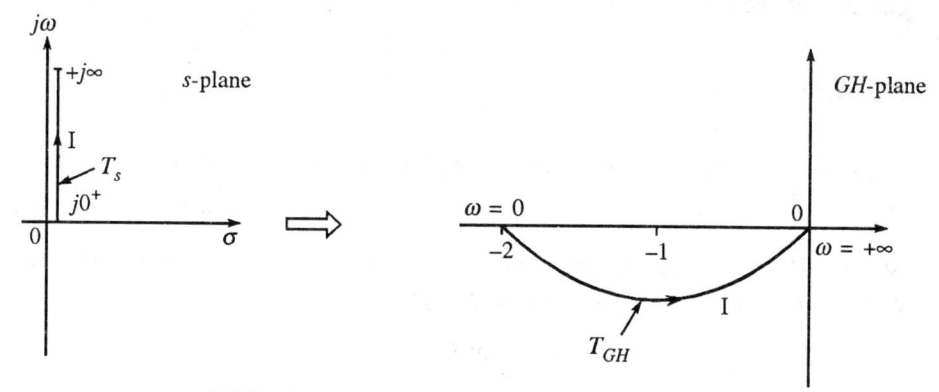

(a) Mapping of section I from T_s-contour to T_{GH}-contour

(b) Mapping of section II from T_s-contour to T_{GH}-contour

(c) Mapping of section III from T_s-contour to T_{GH}-contour

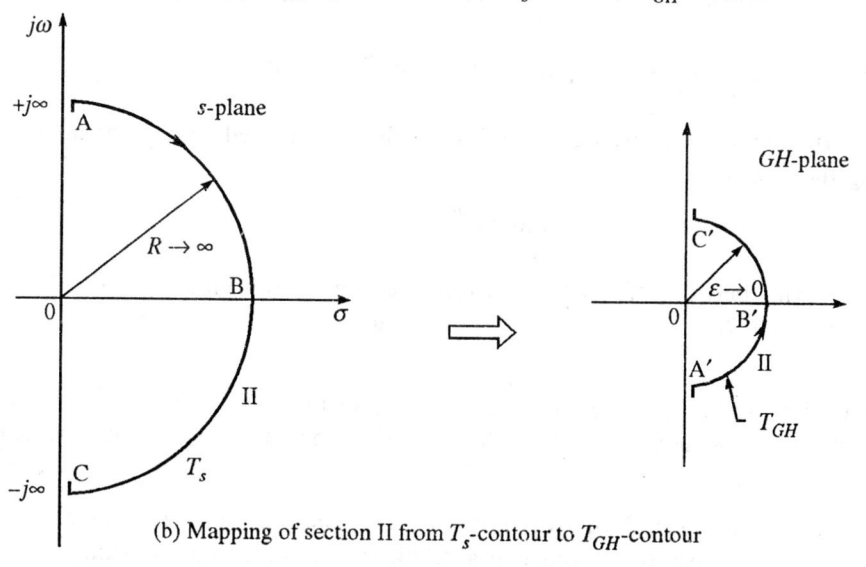

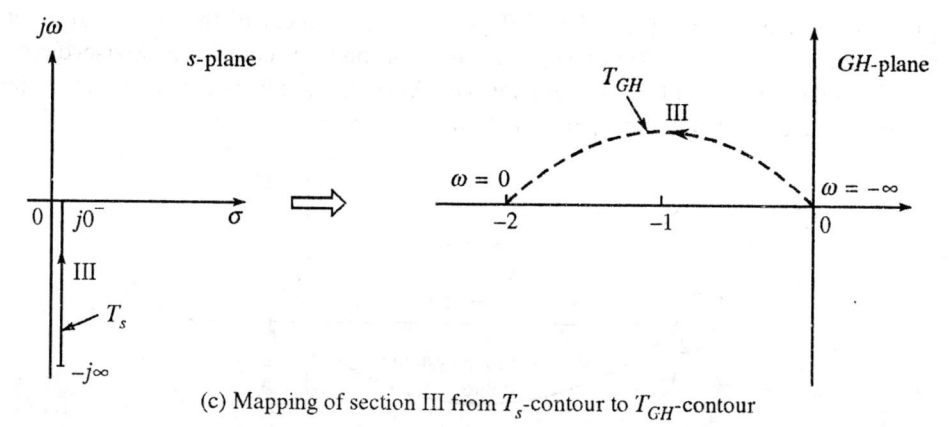

FIGURE 9.10 Example 9.1: Nyquist plots.

can be obtained as follows:

Magnitude is: $\left|\dfrac{j\omega + 2}{(j\omega + 1)(j\omega - 1)}\right|$

When θ varies from $+\pi/2$ to $-\pi/2$ through $0°$ in the T_s-plane, we have

$\angle\tan^{-1}(\omega/2) - \angle\tan^{-1}\omega - \angle(\pi - \tan^{-1}\omega) = -\pi + \tan^{-1}(\omega/2)$; for $\omega = 0$ to $+\infty$

(ii) **Section II:** From $+j\infty$ through $j0$ to $-j\infty$ with infinite radius ($R \to \infty$) covering the entire right-hand side of the s-plane, we put

$$s = \lim_{R\to\infty} Re^{j\phi} \qquad (9.18)$$

The entire right-hand side of the s-plane of T_s is mapped in T_{GH}-contour as

$$\lim_{R\to\infty} G(s)H(s)\Big|_{s=Re^{j\phi}} = \lim_{R\to\infty}\left(\frac{2}{R}e^{-j\phi}\right) = 0 \cdot e^{-j\phi}$$

i.e. for ϕ varying between $\pi/2$ to $-\pi/2$ through 0 is mapped to a point at the origin having the radius as

$$\lim_{\varepsilon\to 0} \varepsilon\, e^{j\theta} \qquad (9.19)$$

with θ moving between $\pi/2$ to $-\pi/2$ through 0 as depicted in Figure 9.10(b). The points A, B, and C of T_s-contour correspond to the mapped T_{GH} plot at points A′, B′ and C′ respectively.

(iii) **Section III:** From $-j\infty$ to $j0$ along the $j\omega$-axis of the T_s-contour mapped in s-plane becomes the inverse polar plot when mapped as T_{GH}-contour in the GH-plane as depicted in Figure 9.10(c).

In order to get the complete Nyquist plot T_{GH} in GH-plane corresponding to the T_s-contour in s-plane, we add part-by-part of T_{GH}-contour as in Figure 9.11 of the corresponding T_s-contour in s-plane. From the Nyquist plot (Figure 9.11), the encirclement of the $(-1 + j0)$ point in the GH-plane (that is, origin of $F(s)$-plane) is seen to be once in counterclockwise direction, so $N = -1$. From the open-loop transfer function $G(s)H(s)$, the number of poles lying in the right-half of the s-plane is $P = 1$. So $Z = P + N = 1 - 1 = 0$.

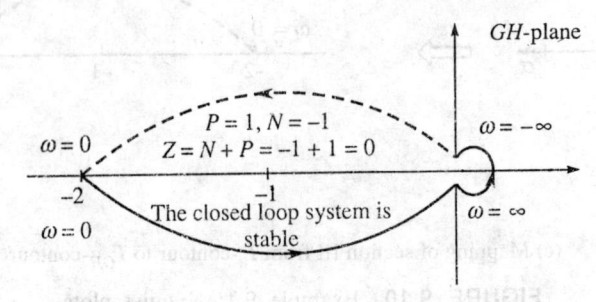

FIGURE 9.11 Example 9.1: Nyquist plot.

Thus, there is no zero of $F(s)$, that is, no pole of closed-loop system that lies in the right-hand side of the s-plane. Hence the closed-loop system is stable.

EXAMPLE 9.2 A single-loop feedback control system has the open-loop transfer function

$$G(s)H(s) = \frac{K}{s(s+a)} \tag{9.20}$$

where K and a are positive constants. Draw the Nyquist plot and comment on the stability of the closed-loop system.

Solution: $G(s)H(s)$ has a pole at the origin. The Nyquist contour T_s in the s-plane is shown in Figure 9.12 where an infinitesimal small detour around the pole at the origin is effected by a small semicircle of radius ε where $\varepsilon \to 0$. This detour is a consequence of the condition of Cauchy's theorem which requires that the contour T_s cannot pass through the point of singularity of $F(s)$, i.e. the pole at the origin.

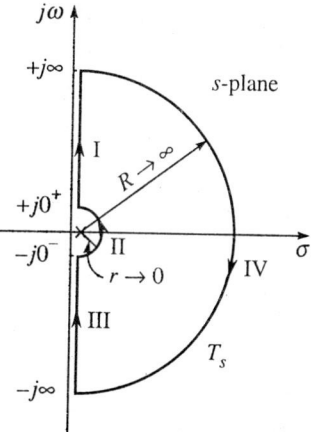

FIGURE 9.12 Example 9.2: Nyquist contour.

The entire Nyquist contour is divided into four sections as shown in Figure 9.12, such as: $-j\infty$ to $-j0^-$ as section III, $-j0^-$ to $+j0^+$ as section II, $+j0^+$ to $+j\infty$ as section I and $+j\infty$ to $-j\infty$ in clockwise direction as section IV.

(i) **Section I:** The Nyquist plot for the section $+j0^+$ to $+j\infty$ is (section I) basically the polar plot for real frequency and the section $-j\infty$ to $-j0^-$ (section III) is the inverse polar plot because the real part of $G(j\omega)H(j\omega)$ is even and the imaginary part is odd; it therefore follows that

$$\text{Im } G(-j\omega)H(-j\omega) = -\text{Im } G(+j\omega)H(+j\omega) \tag{9.21}$$

and

$$\text{Re } G(-j\omega)H(-j\omega) = +\text{Re } G(+j\omega)H(+j\omega) \tag{9.22}$$

The polar plot [Figure 9.13(b)] for negative values of ω can be made by reflecting the polar plot [Figure 9.13(a)] for positive frequency, i.e. polar plot upon the real axis of the GH-plane.

Section I of the Nyquist contour, i.e. the portion from $\omega = 0_+$ to $\omega = +\infty$, mapping of the positive imaginary axis is obtained by calculating $K/j\omega(j\omega + a)$ at various values of ω and plotting them in the $G(s)H(s)$-plane. In other words, section I of T_s-contour is mapped by the function $G(s)H(s)$ as the polar plot since $s = j\omega$ and Eq. (9.20) becomes after rationalization

$$G(j\omega)H(j\omega) = \frac{K(-\omega^2 - ja\omega)}{\omega^4 + a^2\omega^2} \tag{9.23}$$

The intersect of $G(j\omega)H(j\omega)$ on the real axis is determined by equating the imaginary part of $G(j\omega)H(j\omega)$ to zero. Thus the frequency at which $G(j\omega)H(j\omega)$ intersects the real axis is found from

$$\text{Im } G(j\omega)H(j\omega) = \frac{Ka\omega}{\omega^4 + a^2\omega^2} = -\frac{Ka}{\omega(\omega^2 + a^2)} = 0 \tag{9.24}$$

which gives $\omega = \infty$. This means that the only intersect on the real axis in the GH-plane is at

438   Modern Control Engineering

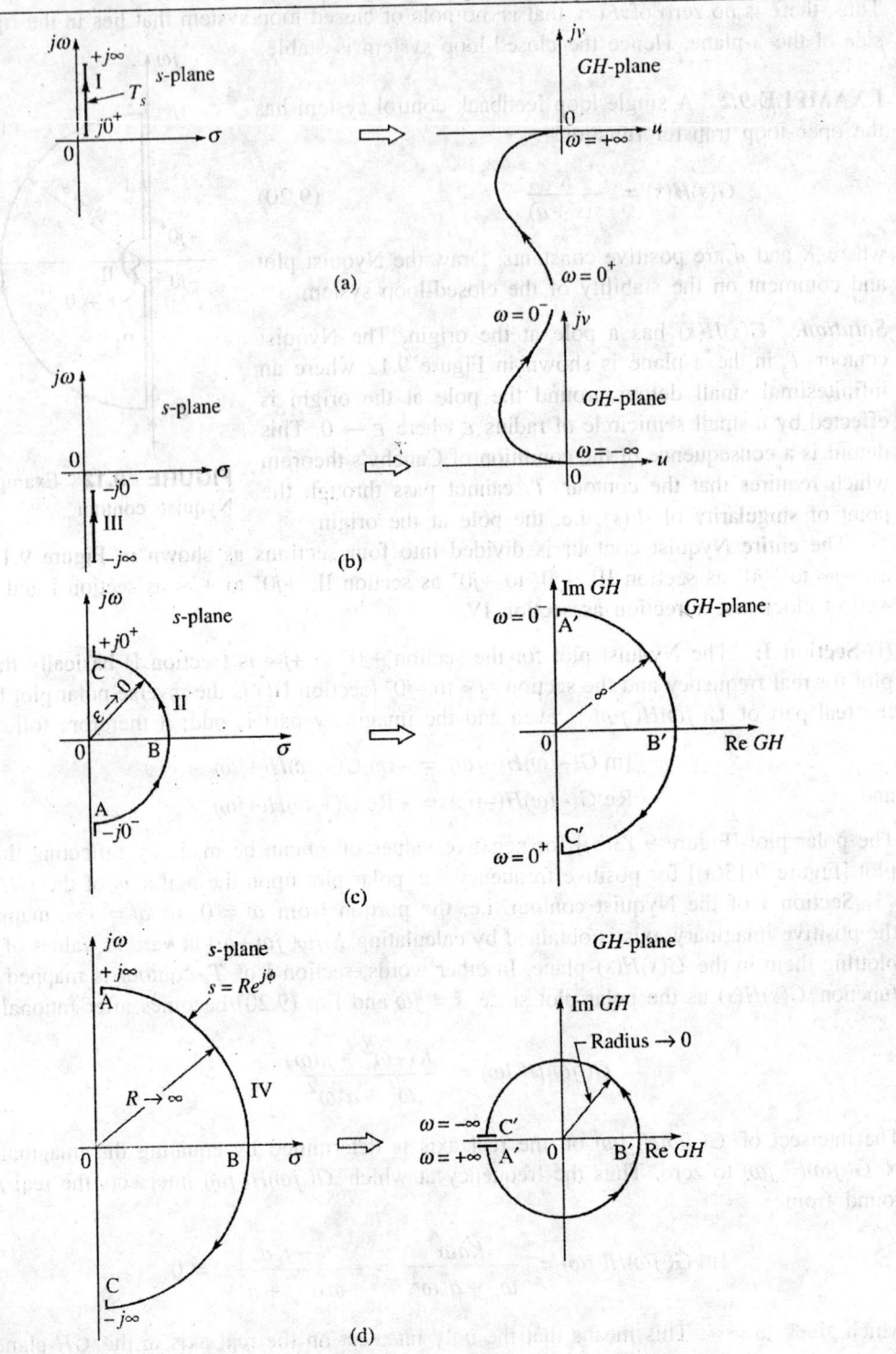

FIGURE 9.13(a)–(d)   Example 9.2: Nyquist plots.

Nyquist Stability

the origin with $\omega = \infty$. The polar plot in the GH-plane is drawn in Figure 9.13(a) following Table 9.5 given below:

TABLE 9.5 Example 9.2

ω	0	$0 < \omega < \infty$	∞
$G(j\omega)H(j\omega)$	∞	finite	$0°$
$\angle G(j\omega)H(j\omega)$	0	$-180° > \phi > -90°$	$-180°$

(ii) Section II: The portion from $-j0^-$ to $+j0^+$ of the Nyquist contour is magnified as shown in Figure 9.13(c). The points on the section are represented by

$$s = \lim_{\varepsilon \to 0} \varepsilon e^{j\theta} \tag{9.25}$$

where ε ($\varepsilon \to 0$) and θ denote the magnitude and phase of the phasor respectively. As the Nyquist contour is traversed from $-j0^-$ to $+j0^+$ along section II, the phasor of Eq. (9.20) rotates in the counterclockwise direction through 180°. In going from $-j0^-$ to $+j0^+$, the phase angle θ varies from $-90°$ through $0°$ to $+90°$. The corresponding Nyquist plot of $G(s)H(s)$ can be determined as

$$G(s)H(s)\Big|_{s=\lim_{\varepsilon \to 0}\varepsilon e^{j\theta}} = \lim_{\varepsilon \to 0} \frac{K}{\varepsilon e^{j\theta}(\varepsilon e^{j\theta} + a)} \approx \lim_{\varepsilon \to 0} \frac{K}{(a\varepsilon)e^{j\theta}} = \infty e^{-j\theta} \tag{9.26}$$

which indicates that all points on the Nyquist plot T_{GH} of $G(s)H(s)$ that correspond to section II of the Nyquist contour T_s have an infinite magnitude and the corresponding phase is opposite to that of the s-plane locus. Since the Nyquist contour in s-plane varies from $-90°$ through $0°$ to $+90°$ in counterclockwise direction, the minus sign in the phase relation of Eq. (9.26) indicates that the corresponding $G(s)H(s)$ plot should have a phase that varies from $+90°$ through $0°$ to $-90°$ in the clockwise direction as shown in Figure 9.13(c). Points A, B, C on the s-plane contour T_s of infinitesimally small radius are mapped into the respective points A', B', and C' on the $G(s)H(s)$ locus T_{GH} of infinite radius.

(iii) Section IV: The same technique may also be used to determine the behaviour of $G(s)H(s)$ plot T_{GH} in GH-plane corresponding to section IV of Nyquist contour (from $+j\infty$ to $-j\infty$) T_s in s-plane in Figure 9.12. The T_{GH} plot is as shown in Figure 9.13(d). The points on the semicircle may be represented by the phasor

$$s = \lim_{R \to \infty} R e^{j\phi} \tag{9.27}$$

Substituting Eq. (9.27) in Eq. (9.20) yields

$$G(s)H(s)\Big|_{s=\lim_{R \to \infty} R e^{j\phi}} = \lim_{R \to \infty} \frac{K}{R^2 e^{j2\phi}} = 0 \times e^{-j2\phi} \tag{9.28}$$

It is evident from Eq. (9.28) that the infinite semicircular arc of the Nyquist contour T_s in the s-plane represented by $s = \lim_{R \to \infty} R e^{j\phi}$ (ϕ varies from $+90°$ through $0°$ to $-90°$ in clockwise direction) maps into a point at the origin of the GH-plane and is described by a phasor with infinitesimally small magnitude which rotates around the origin by $2 \times 180° = 360°$ in counterclockwise direction (i.e. from $-180°$ through $0°$ to $180°$ in counterclockwise direction).

Thus the $G(s)H(s)$-plot T_{GH} that corresponds to section IV of the Nyquist path is sketched as shown in Figure 9.13(d). The points A, B, C in T_s-contour are mapped to the corresponding points A', B' and C' in T_{GH}-contour as shown in Figure 9.13(d).

The complete Nyquist plot T_{GH} is plotted in the GH-plane corresponding to the s-plane T_s-contour by integrating all the parts of Figures 9.13(a) to (d) in GH-plane and is shown in Figure 9.13(e). We know $P = 0$, from the Nyquist plot T_{GH} in GH-plane, the number of clockwise encirclement of the $-1 + j0$ point is zero; $N = 0$. Hence $Z = N + P = 0$, i.e. there is no pole of the closed-loop system having positive real part. The closed-loop system is therefore stable.

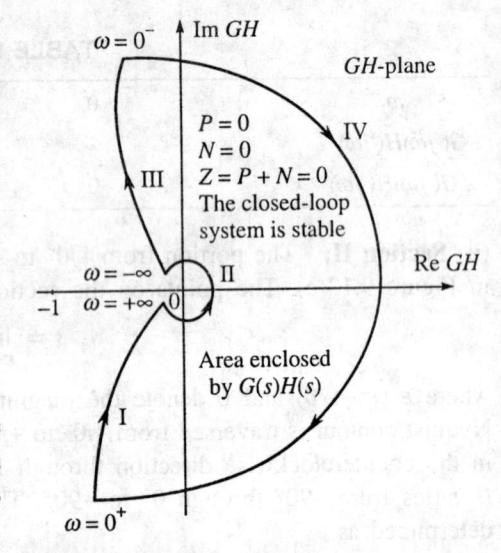

FIGURE 9.13(e) Example 9.2: complete Nyquist plot of $G(s)H(s)$.

EXAMPLE 9.3 Consider the open-loop transfer function

$$G(s)H(s) = \frac{K(T_2 s + 1)}{s^2(T_1 s + 1)}; \quad T_1 > T_2$$

Draw the Nyquist plot and comment on the stability of the system.

Solution: We have a double pole at the origin. The Nyquist contour T_s must make a small detour around the poles while still attempting to enclose the entire right-half of the s-plane including the $j\omega$-axis. The mapping of the Nyquist contour T_s (Figure 9.14) is obtained as follows:

1. Semicircular indent, that is, the portion from $-j0^-$ to $+j0^+$ as in Figure 9.15(a) represented by

$$s = \lim_{\varepsilon \to 0} \varepsilon\, e^{j\theta} \quad (9.29)$$

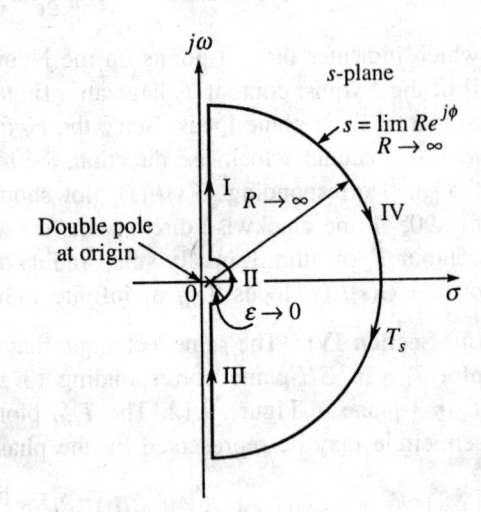

FIGURE 9.14 Example 9.3: Nyquist contour.

where θ varying from $-90°$ through $0°$ to $+90°$ is mapped into

$$\lim_{\varepsilon \to 0} \left[\frac{K(T_2\, \varepsilon\, e^{j\theta} + 1)}{\varepsilon^2\, e^{j2\theta}(T_1\, \varepsilon\, e^{j\theta} + 1)} \right] = \lim_{\varepsilon \to 0} \left[\frac{K}{\varepsilon^2\, e^{j2\theta}} \right] = \infty \times e^{-j2\theta}$$

$$= \infty \times [\angle 180° \to \angle 0° \to \angle -180°] \quad (9.30)$$

This part of the map is an infinite circle shown in Figure 9.15(b).

Nyquist Stability

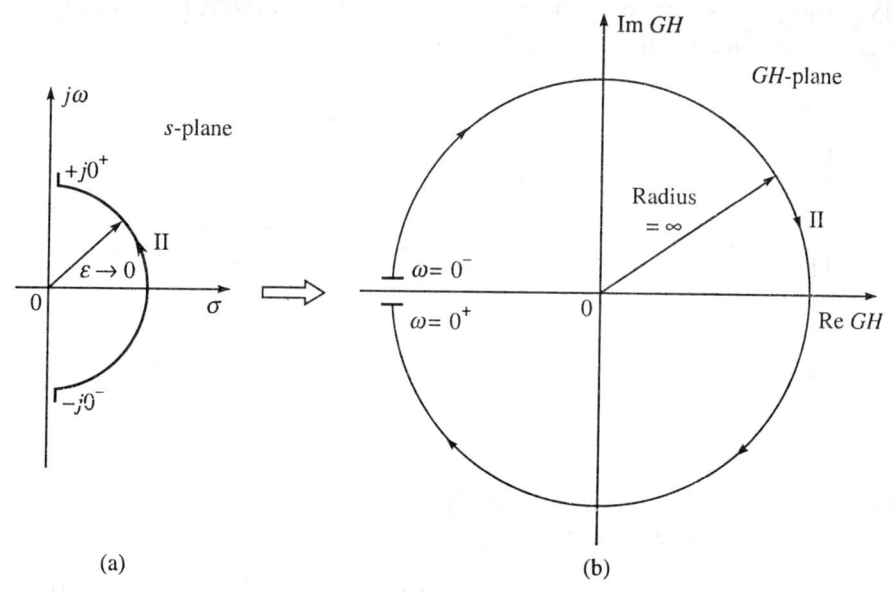

FIGURE 9.15 Example 9.3: (a) section II of Nyquist contour of Figure 9.14 and (b) Nyquist plot of $G(s)H(s)$ that corresponds to section II.

2. Along the $j\omega$-axis, that is, the portion from $j0^+$ to $+j\infty$

$$G(j\omega)H(j\omega) = \frac{K(jT_1\omega + 1)}{(j\omega)^2 (jT_2\omega + 1)} \qquad (9.31)$$

For various values of ω, the magnitude and phase of $G(j\omega)H(j\omega)$ are calculated and plotted as shown in Figure 9.16, which is nothing but the polar plot. Further, it may be noted that the $G(j\omega)H(j\omega)$ locus does not intersect the real axis as

$$-270° < \angle G(j\omega)H(j\omega) < -180° \qquad \text{for } 0^+ < \omega < +\infty \qquad (9.32)$$

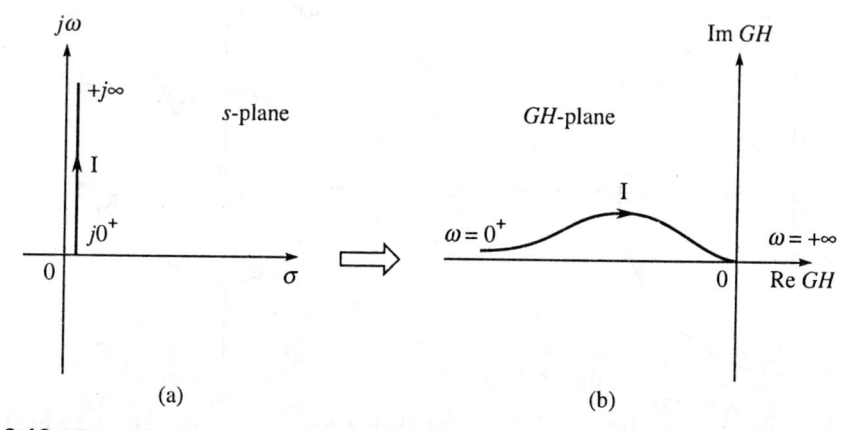

FIGURE 9.16 Example 9.3: (a) section I of Nyquist contour of Figure 9.14 and (b) Nyquist plot of $G(s)H(s)$ that corresponds to section I.

3. Along the $j\omega$-axis, for the portion $-j\infty$ to $-j0^-$, the $G(j\omega)H(j\omega)$ locus becomes the inverse polar plot as shown in Figure 9.17(b).

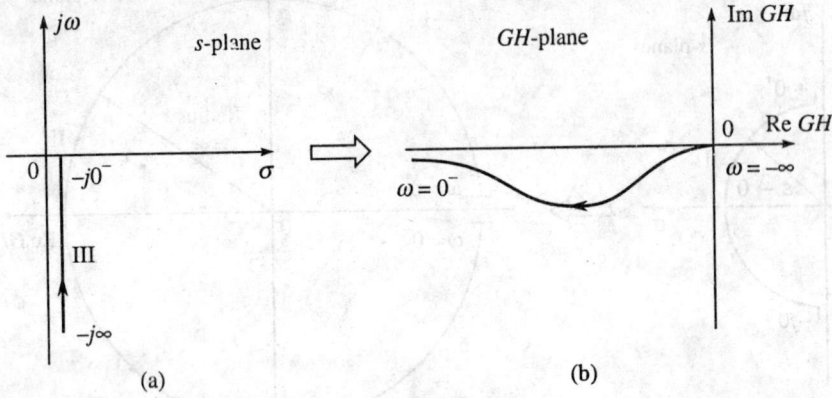

FIGURE 9.17 Example 9.3: (a) section III of Nyquist contour of Figure 9.14 and (b) Nyquist plot of $G(s)H(s)$ that corresponds to section III.

4. The portion $+j\infty$ to $-j\infty$, that is, the infinite semicircle of the Nyquist contour (Figure 9.14) where s can be represented by $s = \lim_{R\to\infty} R e^{j\phi}$ (ϕ varies from $+90°$ through $0°$ to $-90°$) is mapped into

$$\lim_{R\to\infty}\left[\frac{K(T_2 R e^{j\phi} + 1)}{R^2 e^{j2\phi}(T_1 R e^{j\phi} + 1)}\right] = 0 \cdot e^{-j2\phi} = 0 \; (\angle -180° \text{ through } \angle 0° \text{ to } \angle +180°) \quad (9.33)$$

This part of the map is an infinitesimally small circle (as a point) shown in Figure 9.18(b). The points A, B, C of Figure 9.18(a) correspond to the points A′, B′ and C′ of Figure 9.18(b).

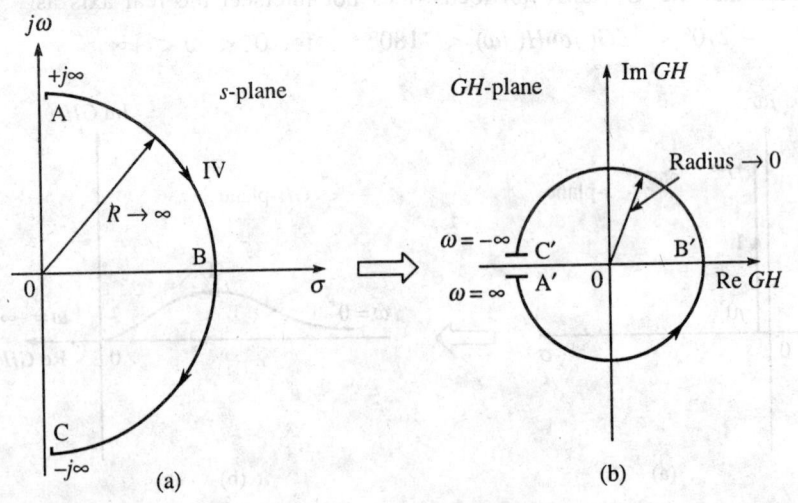

FIGURE 9.18 Example 9.3: (a) section IV of Nyquist contour of Figure 9.14 and (b) Nyquist plot of $G(s)H(s)$ that corresponds to section IV.

Nyquist Stability

The complete mapped plot corresponding to the Nyquist contour of Figure 9.14 is obtained by joining the curves (b) of Figures 9.15 through 9.18 and is shown in Figure 9.19. From this plot, it is observed that the $(-1 + j0)$ point is encircled twice in the clockwise direction, therefore, $N = +2$. From the given transfer function it is seen that no pole of $G(s)H(s)$ lies in the right-half s-plane, i.e. $P = 0$. Hence $Z = P + N$, that is, two zeros of $F(s)$ lie in the right-half s-plane from which we conclude that the system is unstable.

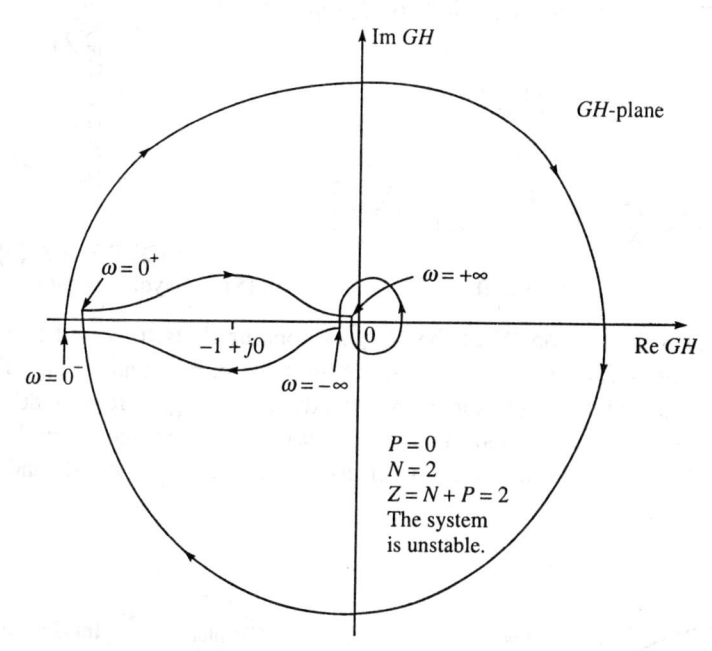

FIGURE 9.19 Example 9.3: complete Nyquist plot.

EXAMPLE 9.4: Consider the control system shown in Figure 9.20(a). Draw the complete Nyquist plot and determine the closed-loop stability.

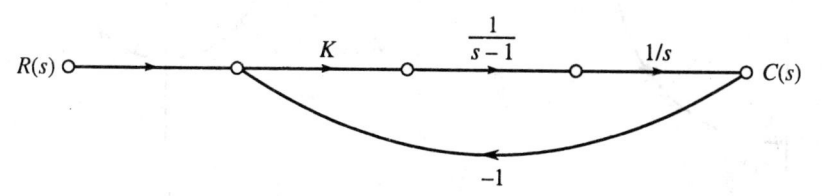

FIGURE 9.20(a) Example 9.4: second-order feedback control system.

Solution: The open-loop transfer function becomes

$$G(s)H(s) = \frac{K}{s(s-1)} \qquad (9.34)$$

The Nyquist contour T_s in s-plane is shown in Figure 9.20(b) and has the detour at the origin of infinitesimal radius as $G(s)H(s)$ has a pole at origin. The prescribed direction of traversing in T_s is clockwise. The construction of Nyquist plot of $G(s)H(s)$ is outlined as follows:

(i) **Section IV:** The infinite semicircle of T_s represented by

$$s = \lim_{R \to \infty} Re^{j\phi}$$

is mapped as

$$\lim_{R \to \infty} \left[\frac{K}{Re^{j\phi}(Re^{j\phi}-1)} \right] = \lim_{R \to \infty} \left[\frac{K}{R^2 e^{2j\phi}} \right]$$

$$= 0 \cdot e^{-j2\phi} \quad (9.35)$$

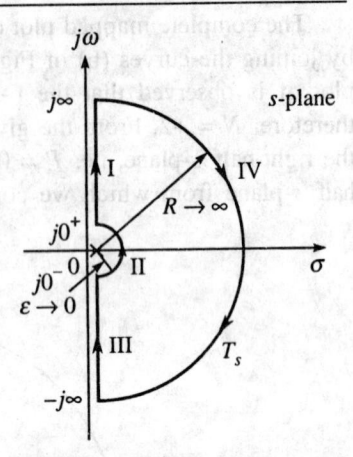

FIGURE 9.20(b) Example 9.4: Nyquist contour in s-plane.

As the phasor for section IV of the Nyquist contour T_s is traversed from $+90°$ through $0°$ to $-90°$ clockwise, Eq. (9.35) indicates that the corresponding Nyquist plot T_{GH} of $G(s)H(s)$ is traced by a phasor of practically zero length from $-180°$ through $0°$ to $+180°$ through a total of $360°$ in the clockwise sense as shown in Figure 9.21. The points A, B and C in Figure 9.21(a) correspond to the points A', B' and C' respectively in Figure 9.21(b).

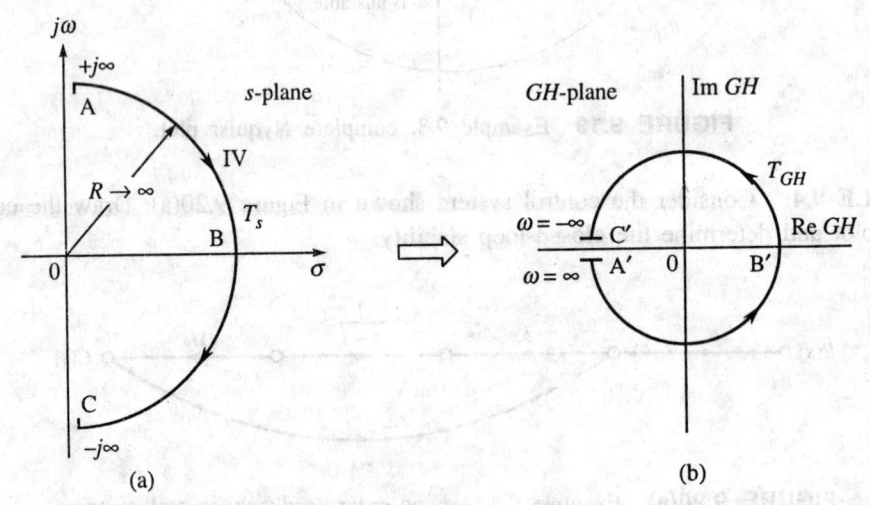

FIGURE 9.21 Example 9.4: (a) section IV of Nyquist contour of Figure 9.20(b) and (b) Nyquist plot of $G(s)H(s)$ that corresponds to section IV.

(ii) **Section I:** Considering section I of Nyquist contour T_s [Figure 9.20(b)], along the $j\omega$-axis for the portion $j0^+$ to $+j\infty$, the polar plot is given by

$$G(j\omega)H(j\omega) = \frac{K}{j\omega(j\omega - 1)}$$

Nyquist Stability

$$= \frac{K}{\sqrt{\omega^2 + \omega^4}} \angle (-\pi/2) - \tan^{-1}(-\omega)$$

$$= \frac{K}{\sqrt{(\omega^2 + \omega^4)}} \angle -(\pi/2) - (\pi - \tan^{-1}\omega)$$

$$= \frac{K}{\sqrt{\omega^2 + \omega^4}} \angle (\pi/2) + \tan^{-1}\omega = M\angle\phi$$

where $\quad M = \dfrac{K}{\omega\sqrt{(1+\omega^2)}} \quad$ and $\quad \phi = \dfrac{\pi}{2} + \tan^{-1}\omega \quad\quad (9.36)$

Plotting the values for $0^+ < \omega < +\infty$ as per Table 9.6, we get the polar plot T_{GH} as shown in Figure 9.22 in the *GH*-plane. This is the Nyquist plot of *GH* corresponding to section I of Nyquist contour T_s.

TABLE 9.6 Example 9.4: section I

ω	0	finite +ve value	$+\infty$
M	∞	finite +ve value	0
ϕ	90°	0° < ϕ < 180°	+180°

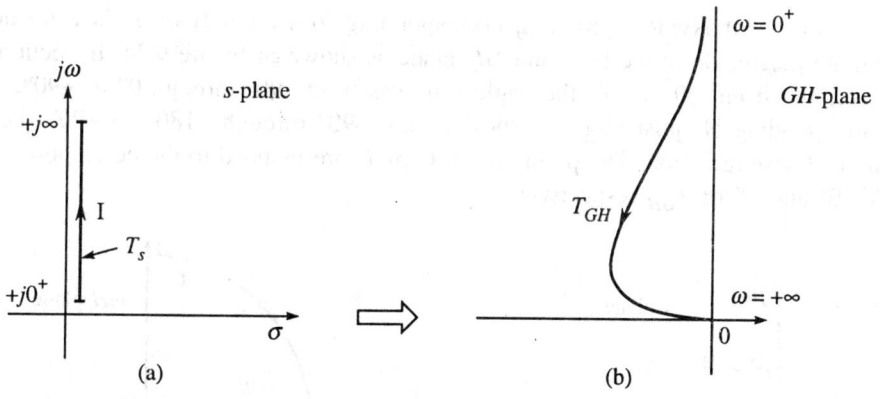

FIGURE 9.22 Example 9.4: (a) section I of Nyquist contour of Figure 9.20(b) and (b) Nyquist plot of $G(s)H(s)$ that corresponds to section I.

(iii) **Section III:** For the portion from $-j\infty$ to $-j0^-$, the Nyquist plot T_{GH} for section III of T_s of Figure 9.20(b) will be the inverse polar plot and is shown in Figure 9.23.

(iv) **Section II:** In the semicircular indent (section II of T_s), that is, portion $-j0^-$ to $+j0^+$, the *s* is represented by $s = \varepsilon e^{j\theta}$, with $\varepsilon \to 0$ and θ varying from $-90°$ through $0°$ to $+90°$. Now replace, $s = \lim_{\varepsilon \to 0} \varepsilon e^{j\theta}$ in $G(s)H(s)$ as

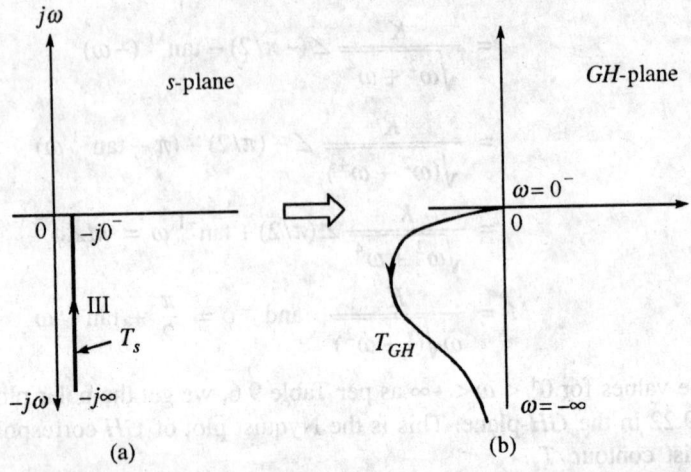

FIGURE 9.23 Example 9.4: (a) section III of Nyquist contour of Figure 9.20(b) and (b) Nyquist plot of $G(s)H(s)$ that corresponds to section III.

$$\lim_{\varepsilon \to 0}\left[\frac{K}{\varepsilon e^{j\theta}(\varepsilon e^{j\theta}-1)}\right] = \lim_{\varepsilon \to 0}\left[\frac{-K}{\varepsilon e^{j\theta}}\right] = \lim_{\varepsilon \to 0}\left[\frac{-K}{\varepsilon}e^{-j\theta}\right]$$

$$= \lim_{\varepsilon \to 0}\left[\frac{K}{\varepsilon}\right]e^{-j(\pi+\theta)} = \infty \angle -(\pi+\theta) \qquad (9.37)$$

The portion of Nyquist plot T_{GH} corresponding to section II of T_s is a semicircle of infinite magnitude in the left-hand GH-plane as shown in Figure 9.24. In section II of T_s traversing from $-j0^-$ to $j0^+$, the angle θ moves from $-90°$ through $0°$ to $+90°$. Then the corresponding Nyquist plot T_{GH} moves from $-90°$ through $-180°$ to $+90°$ (i.e. $-270°$) in clockwise direction. The points A, B, C of T_s are mapped to the corresponding points A′, B′ and C′ of T_{GH} respectively.

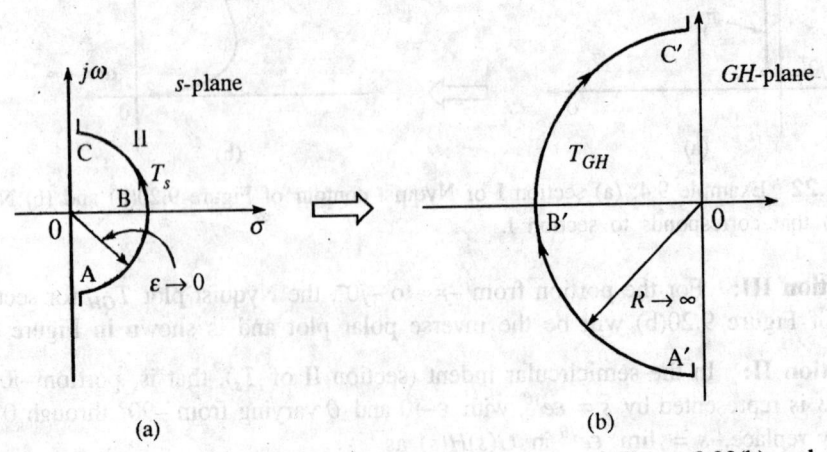

FIGURE 9.24 Example 9.4: (a) section II of Nyquist contour of Figure 9.20(b) and (b) Nyquist plot of $G(s)H(s)$ that corresponds to section II.

Now the complete Nyquist plot, obtained by simply adding the different sections of the Nyquist plot T_{GH} as per the sequence of corresponding T_s contours, is shown in Figure 9.25. It is seen that the number of clockwise rotation about the $(-1 + j0)$ point is one, i.e. $N = 1$. As $P = 1$, then $Z = N + P = 1 + 1 = 2$, hence the closed-loop system is unstable.

EXAMPLE 9.5 Plot the Nyquist diagram for the system having the open-loop transfer function $G(s)H(s)$ as

$$\frac{2(s+0.1)(s+0.6)(s^2+s+1)}{s^3(s-0.2)(s+1)}$$

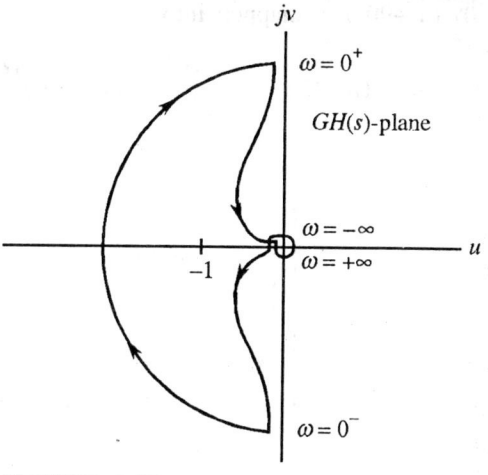

FIGURE 9.25 Example 9.4: Nyquist plot.

and hence, determine the stability of the closed-loop system.

Solution: For the given open-loop transfer function, there is one open-loop pole at $s = 0.2$ within the Nyquist contour T_s covering the entire right-hand side of the s-plane, and therefore $P = 1$.

$G(s)H(s)$ has three poles at origin. The Nyquist contour T_s in the s-plane is shown in Figure 9.26.

The Nyquist plot T_{GH} for section I of T_s is the polar plot as shown in Figure 9.27. The Nyquist plot T_{GH} for section III of T_s is the inverse polar plot and is as shown in Figure 9.28.

Section II of the Nyquist contour is magnified as shown in Figure 9.29. The points in this section are represented by

$$s = \lim_{\varepsilon \to 0} \varepsilon e^{j\theta} \quad (9.38)$$

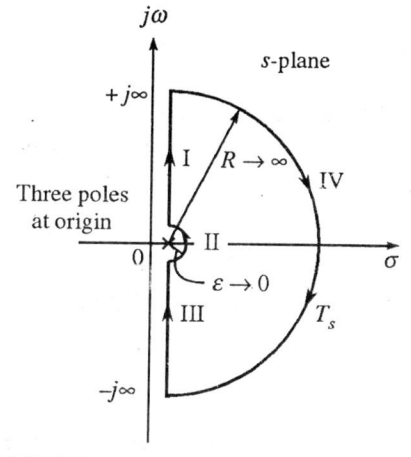

FIGURE 9.26 Example 9.5: Nyquist contour in s-plane.

where ε and θ denote the magnitude and phase of the phasor respectively. As the Nyquist path is traversed from $-j0^-$ to $+j0^+$ along section II, the phasor of Eq. (9.38) rotates in the counterclockwise direction through $180°$. While going from $-j0^-$ to $+j0^+$, θ varies from $-90°$ through $0°$ to $+90°$. For small values of s, the open-loop transfer function $G(s)H(s)$ may be approximated as

$$G(s)H(s) \approx \frac{0.6}{-s^3}$$

The semicircular indent represented by $s = \lim_{\varepsilon \to 0} \varepsilon e^{j\theta}$ (and with θ varying from $-90°$ through

0° to +90°) is mapped into

$$\lim_{s \to 0} G(s)H(s) = \frac{-0.6}{s^3} = \lim_{\varepsilon \to 0}\left(-\frac{0.6}{\varepsilon^3 e^{j3\theta}}\right) = \left(\lim_{\varepsilon \to 0} \frac{0.6}{\varepsilon^3}\right) e^{-j(\pi+3\theta)} = \infty \angle -\pi - 3\theta$$

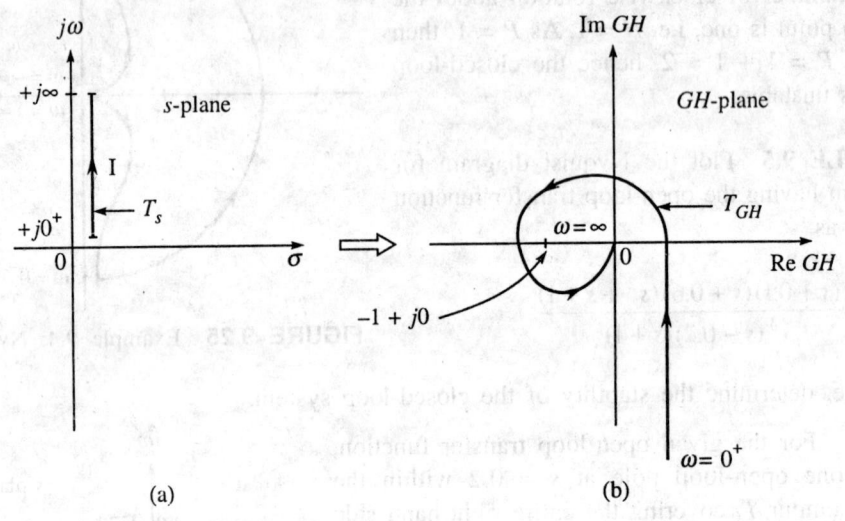

FIGURE 9.27 Example 9.5: (a) section I of Nyquist contour of Figure 9.26 and (b) Nyquist plot of $G(s)H(s)$ that corresponds to section I.

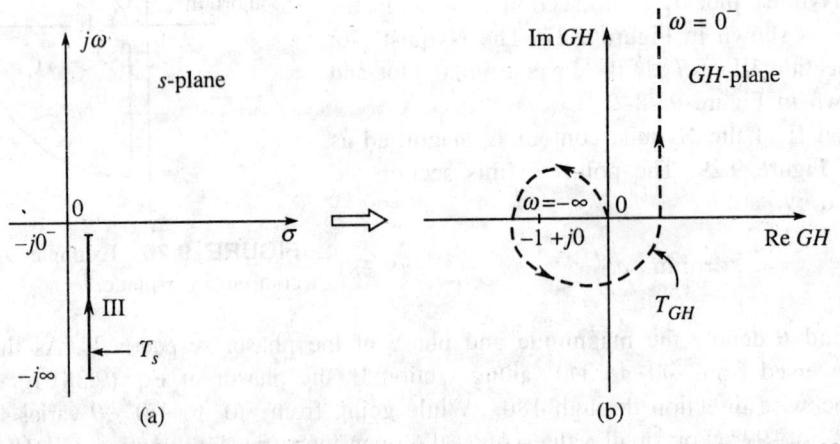

FIGURE 9.28 Example 9.5: (a) section III of Nyquist contour of Figure 9.26 and (b) Nyquist plot of $G(s)H(s)$ that corresponds to section III.

The portion of Nyquist plot T_{GH} corresponding to section II of T_s is a one-half circle of infinite magnitude covering the entire GH-plane starting from $+(\pi/2)$ through $-\pi$ to $-(\pi/2)$ covering $6(\pi/2)$ in clockwise direction and is shown in Figure 9.29. Further, it may be noted that the points A, B, and C of T_s-contour are mapped as points A′, B′ and C′ of T_{GH}-plot.

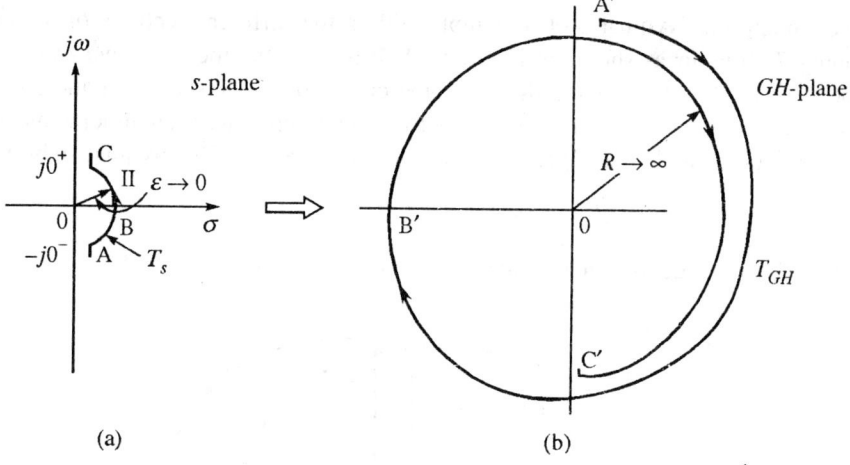

FIGURE 9.29 Example 9.5: (a) section II of Nyquist contour of Figure 9.26 and (b) Nyquist plot of $G(s)H(s)$ that corresponds to section II.

The infinite semicircular arc (section IV) of the Nyquist contour T_s as represented by

$$s = \lim_{R \to \infty} Re^{j\phi}$$

where ϕ varies from $+90°$ through $0°$ to $-90°$, is mapped as T_{GH} in GH-plane as

$$\lim_{R \to \infty} \frac{2}{Re^{j\phi}} = \lim_{R \to \infty} \left(\frac{2}{R}\right) e^{-j\phi} = 0 \ [-90° \text{ through } 0° \text{ to } +90°]$$

i.e. mapped at the origin of GH-plane. The Nyquist plot T_{GH} in GH-plane corresponding to section IV of Nyquist contour T_s of s-plane is shown in Figure 9.30. The points A, B and C of T_s are mapped to the corresponding points A′, B′ and C′ of T_{GH}.

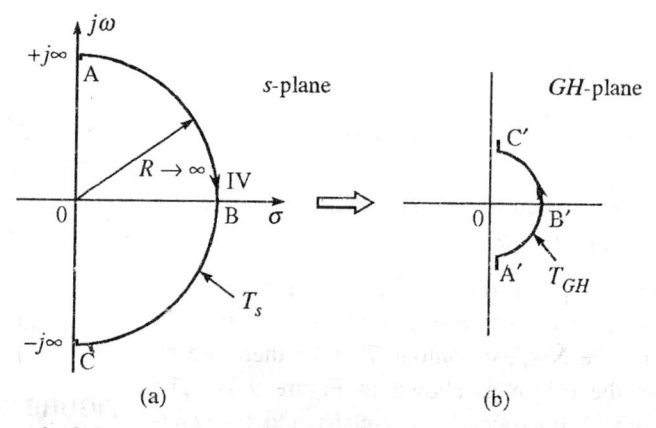

FIGURE 9.30 Example 9.5: (a) section IV of Nyquist contour of Figure 9.26 and (b) Nyquist plot of $G(s)H(s)$ that corresponds to section IV.

Now complete the Nyquist plot by simply adding the different sections of T_{GH} as per the corresponding T_s-contour as shown in Figure 9.31. It is seen that there are two counterclockwise encirclements of the $(-1 + j0)$ point, the two inner circles of the Nyquist plot and one clockwise encirclement of infinite radius. The net number of counterclockwise encirclements is therefore plus one (two counterclockwise minus one clockwise), so $N = 1$. The Nyquist stability formula,

$$Z = P + N = 1 + (-1) = 0$$

Hence it may be concluded that the closed-loop system is stable.

GH-plane

$-1 + j0$

$P = 1$
$N = +1 - 2 = -1$
$Z = N + P = -1 + 1 = 0$
The closed-loop system is stable

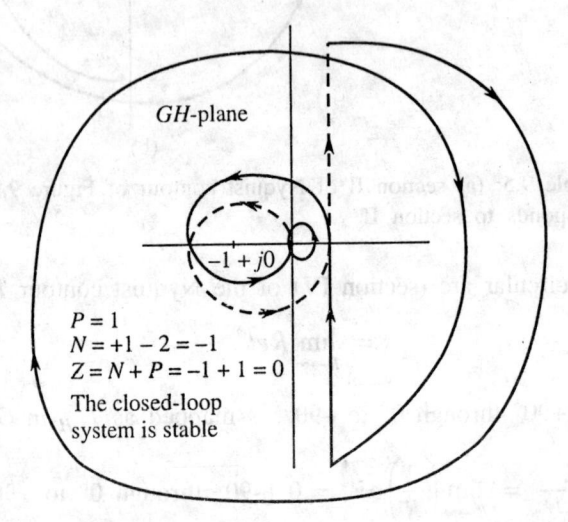

FIGURE 9.31 Example 9.5: complete Nyquist plot.

It would be a useful exercise for the readers to check the closed-loop stability of the various systems in Examples 9.1 to 9.5 using the Routh criterion.

EXAMPLE 9.6 Consider the open-loop transfer function

$$G(s)H(s) = \frac{K(s+3)}{s(s-1)} \quad \text{with } K = 10$$

and draw the complete Nyquist plot and then determine the stability of the closed-loop system.

Solution: The open-loop transfer function has a pole in the positive-half of s-plane, therefore $P = 1$. Further, $G(s)H(s)$ has a pole at the origin. The Nyquist contour T_s must therefore be indented to bypass the origin as shown in Figure 9.32. The mapping of the Nyquist contour T_s in s-plane into the GH-plane may be carried out as follows:

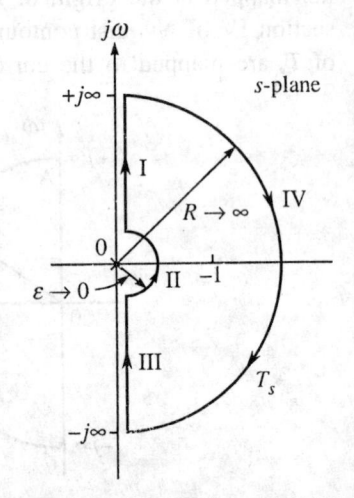

FIGURE 9.32 Example 9.6: Nyquist contour in s-plane.

1. The semicircular indent (section II) around the pole at the origin represented by

$s = \lim_{\varepsilon \to 0} \varepsilon e^{j\theta}$ (θ varying from $-90°$ through $0°$ to $+90°$ as s moves along the T_s-contour) maps into

$$\lim_{s \to 0} G(s)H(s) = \lim_{\varepsilon \to 0} -\frac{3 \times 10}{\varepsilon e^{j\theta}} = \infty \cdot e^{-j(\pi+\theta)} \qquad (9.39)$$

The value of $30/\varepsilon$ approaches infinity as ε approaches zero and in GH-plane the Nyquist plot for the corresponding section II varies from $-90°$ through $180°$ to $-270°$ ($+90°$) as s moves along the semicircle. Thus an infinitesimal semicircular indent around the origin of s-plane maps into a semicircular arc of infinite radius in $G(s)H(s)$-plane as shown in Figure 9.33.

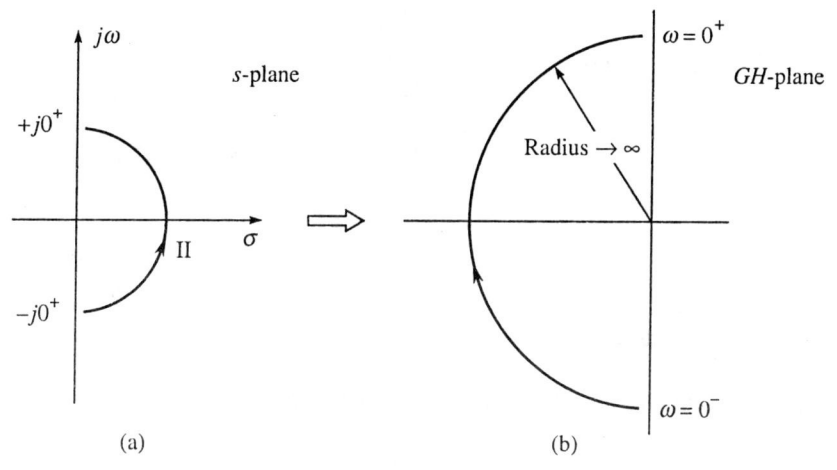

FIGURE 9.33 Example 9.6: (a) section II of Nyquist contour of Figure 9.32 and (b) Nyquist plot of $G(s)H(s)$ that corresponds to section II.

2. For section I ($+j0^+$ to $+j\infty$), the corresponding Nyquist plot in the GH-plane is actually the polar plot for positive real frequencies. Thus:

$$G(j\omega)H(j\omega) = \frac{10(j\omega+3)}{j\omega(j\omega-1)} = Me^{j\phi} \qquad (9.40)$$

where $\quad M = |G(j\omega)H(j\omega)|$

and $\quad \phi = \angle G(j\omega)H(j\omega) = -3(\pi/2) + \tan^{-1}(\omega/3) + \tan^{-1}\omega \qquad (9.41)$

For the imaginary part to become zero, we need to have $\phi = \pi$. Using the formula

$$\tan^{-1} A + \tan^{-1} B = \tan^{-1}\left(\frac{A+B}{1-AB}\right)$$

we get
$$\tan^{-1}\left(\frac{\frac{\omega}{3}+\omega}{1-\frac{\omega^2}{3}}\right) = \frac{\pi}{2}$$

which leads to $\omega = \sqrt{3}$ and then $M = |GH(j\omega)|_{\omega=\sqrt{3}} = |GH(j\sqrt{3})| = -10$. The polar plot is drawn in Figure 9.34 from the following table using Eq. (9.41) as:

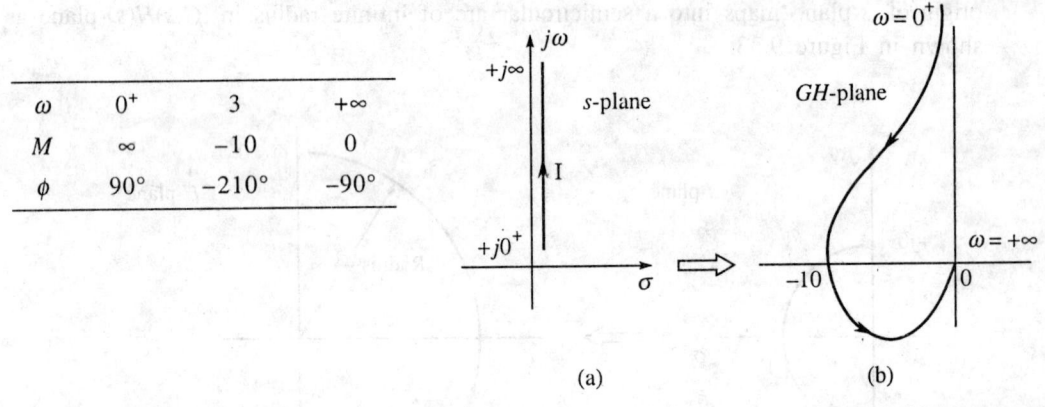

ω	0^+	3	$+\infty$
M	∞	-10	0
ϕ	$90°$	$-210°$	$-90°$

(a) (b)

FIGURE 9.34 Example 9.6: (a) section I of Nyquist contour of Figure 9.32 and (b) Nyquist plot of $G(s)H(s)$ that corresponds to section I.

3. The infinite semicircular arc (section IV) of the Nyquist contour T_s represented by

$$s = \lim_{R \to \infty} R e^{j\phi}$$

is mapped into the GH-plane with ϕ varying from $+90°$ through $0°$ to $-90°$ traversing in clockwise direction as:

$$\lim_{R \to \infty} \frac{10}{R e^{j\phi}} = \lim_{R \to \infty} \left(\frac{10}{R}\right) e^{-j\phi} = 0 \cdot e^{-j\phi}$$

i.e. the origin of the GH-plane. The $G(s)H(s)$ locus thus turns at the origin with zero radius from $-90°$ through $0°$ to $+90°$. The Nyquist plot corresponding to section IV of Nyquist contour is shown in Figure 9.35.

4. The mapping of the negative imaginary axis from $-j\infty$ to $-j0^-$ is the mirror image of that for the positive imaginary axis and is shown as the inverse polar plot in Figure 9.36.

The complete Nyquist plot for $G(s)H(s)$ is shown in Figure 9.37. The Nyquist plot indicates that the $(-1 + j0)$ point is encircled by the $G(s)H(s)$ locus once in the counter-

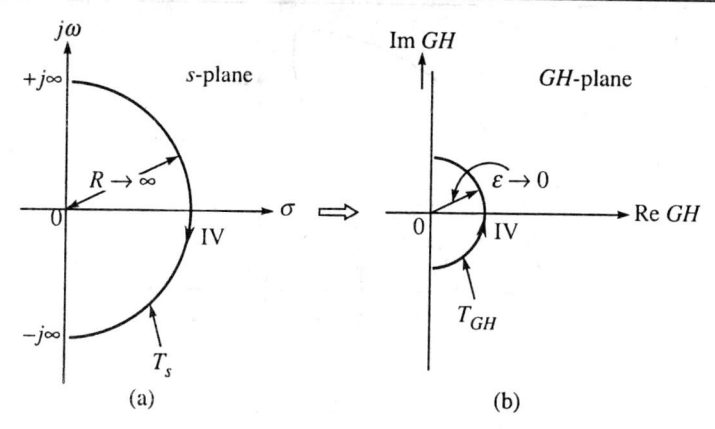

FIGURE 9.35 Example 9.6: (a) section IV of Nyquist contour of Figure 9.32 and (b) Nyquist plot of $G(s)H(s)$ that corresponds to section IV.

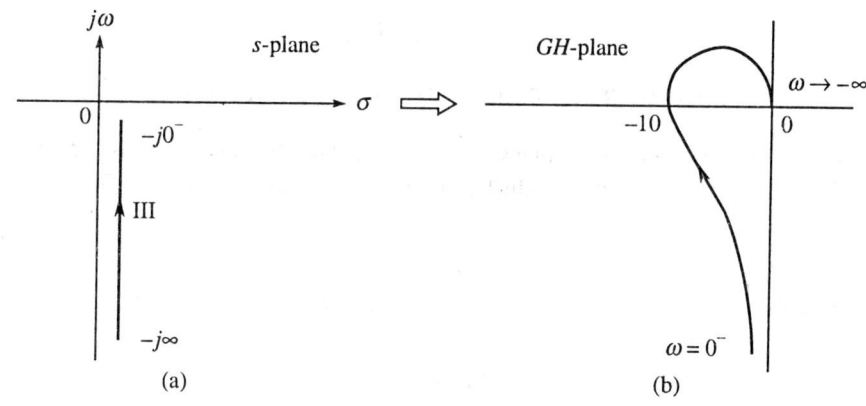

FIGURE 9.36 Example 9.6: (a) section III of Nyquist contour of Figure 9.32 and (b) Nyquist plot of $G(s)H(s)$ that corresponds to section III.

clockwise direction. Therefore $N = -1$. We are already having $P = 1$. Thus Z is found from $Z = N + P = 0$ which indicates that there is no zero of $1 + G(s)H(s)$ in the right half s-plane and the closed-loop system is stable. This is one of the examples where an unstable open-loop system becomes stable when the loop is closed. Further, it may be noted that by varying gain K the unstable open-loop system may be stable and unstable as well. The same conclusion can be drawn from Routh's algorithm.

EXAMPLE 9.7 Consider the open-loop transfer function

$$G(s)H(s) = \frac{K(s-1)}{s(s+1)}; \quad K > 0 \tag{9.42}$$

Here the characteristic equation

$$F(s) = 1 + G(s)H(s) = s^2 + (1+K)s - K = 0 \tag{9.43}$$

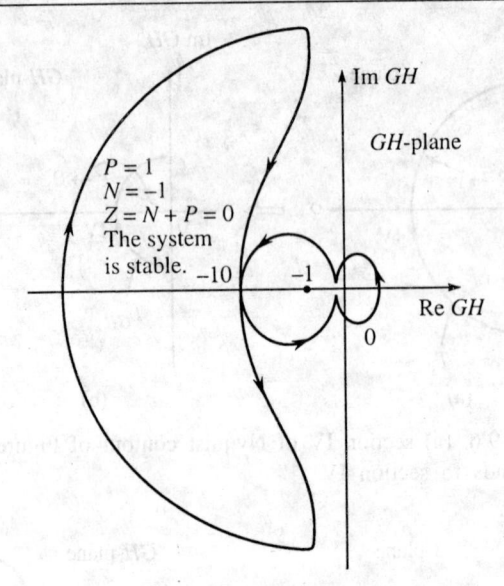

FIGURE 9.37 Example 9.6: complete Nyquist plot.

has one root in positive-half of s-plane, so $P = 1$. The Nyquist contour T_s is shown in Figure 9.38(a). The corresponding Nyquist plot T_{GH} as shown in Figure 9.38(b) is drawn as follows.

1. The small semicircular detour (section II) around the pole at origin $s = 0$ is to exclude the singular point pole at the origin of s-plane. The origin of the s-plane represented by setting

$$s = \lim_{\varepsilon \to 0} \varepsilon \, e^{j\theta} \qquad (9.44)$$

where θ is varying from $-90°$ through $0°$ to $+90°$, is mapped in $G(s)H(s)$ and it becomes

$$\lim_{s \to 0} G(s)H(s) = \lim_{s \to 0} \frac{-K}{s} = \lim_{\varepsilon \to 0} \frac{-K}{\varepsilon \, e^{j\theta}}$$

$$= \lim_{\varepsilon \to 0} \frac{K}{\varepsilon} \cdot e^{j(\pi+\theta)}$$

$$= \infty \cdot e^{-j(\pi-\theta)} \qquad (9.45)$$

which indicates that all points on Nyquist plot of $G(s)H(s)$ that correspond to section II of the Nyquist contour T_s have an infinite magnitude and the corresponding phase is opposite to that of the s-plane locus. Since the Nyquist contour in s-plane varying from $-90°$ through $0°$ to $+90°$ for section II is in counterclockwise direction, the phase relation of Eq. (9.45) indicates that the corresponding $G(s)H(s)$ plot should have a phase that varies from $-90°$ through $-180°$ to $-270°$ or $+90°$ in the clockwise direction [Figure 9.38(b)].

2. The portion from $+j\infty$ through 0 to $-j\infty$ (section IV) covers the entire right-hand side by a semicircle of infinite radius.

Nyquist Stability

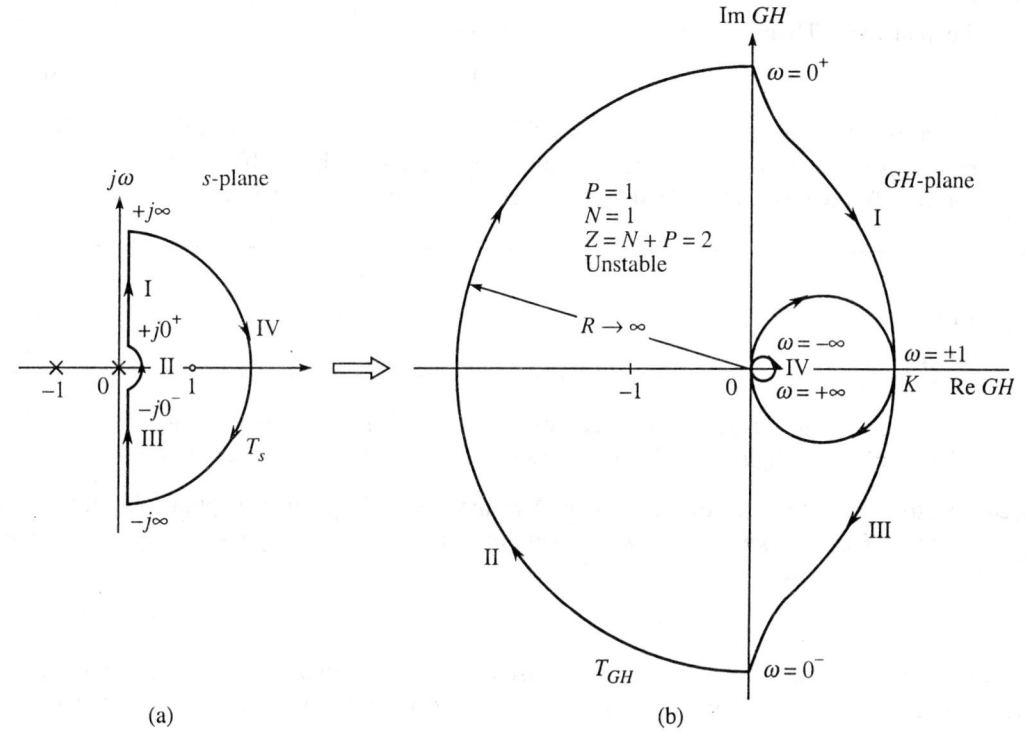

FIGURE 9.38 Example 9.7: (a) Nyquist contour T_s in s-plane and (b) Nyquist plot T_{GH} in GH-plane.

Putting
$$s = \lim_{R \to \infty} Re^{j\phi} \qquad (9.46)$$

in $G(s)H(s)$, we have

$$\lim_{s \to \infty} \frac{K}{s} = \lim_{R \to \infty} \frac{K}{Re^{j\phi}} = 0 \cdot e^{-j\phi} \qquad (9.47)$$

where ϕ varies from $+90°$ through $0°$ to $-90°$ in clockwise direction in s-plane. With Eq. (9.47), the infinite semicircular arc of Nyquist contour T_s maps into the origin of GH-plane and is described by a phasor with infinitesimally small magnitude which moves around the origin correspondingly from $-90°$ through $0°$ to $+90°$ in counterclockwise direction. Thus the Nyquist plot of $G(s)H(s)$ corresponding to section IV goes around the origin $180°$ in counterclockwise direction with zero magnitude.

3. Putting $s = j\omega$, rationalizing and equating the imaginary part of $G(j\omega)H(j\omega)$ to zero, we get the frequency at which $\text{Im}(GH)$ is zero and the value of the intercept on the real axis in GH-plane as

$$\text{Im}[G(j\omega)H(j\omega)] = K \frac{2\omega + j(1+\omega^2)}{\omega(\omega^2+1)} = 0 \qquad (9.48)$$

and which gives $\omega = \pm 1$ rad/s as the frequencies at which the $G(j\omega)H(j\omega)$ locus crosses the real axis. Then

$$G(j1)H(j1) = K \qquad (9.49)$$

Based on this information at $\omega = \pm 1$ and from the table below, we can draw the Nyquist plot T_{GH} in the GH-plane corresponding to section III. This is basically the polar plot for real frequencies. This is drawn as section III in Figure 9.38(b).

ω	0	any +ve value	∞
$G(j\omega)H(j\omega)$	∞	finite	0
$\angle G(j\omega)H(j\omega)$	$+90°$	$90° > \phi < -90°$	$-90°$

4. Putting $s = -j\omega$, we get the inverse polar plot which is the mirror image of section III of T_{GH} plot in Figure 9.38(b) and drawn as section I in the same figure.

Based on this information, the complete Nyquist plot T_{GH} in GH-plane is drawn in Figure 9.38(b). The number of clockwise encirclements of $(-1 + j0)$ point by T_{GH}-contour in the GH-plane is 1. Also,

$$Z = N + P = 1 + 1 = 2$$

This means that the closed-loop system is unstable as it has two poles on the right-half of s-plane. The Nyquist plot further shows that the closed-loop system becomes unstable for all $K > 0$.

EXAMPLE 9.8 For the open-loop system

$$G(s)H(s) = \frac{K}{s(T_1s + 1)(T_2s + 1)}$$

Determine the closed-loop system stability through the Nyquist plot.

Solution: In this example, the open-loop system has a pole at the origin. The Nyquist contour T_s is therefore indented to bypass the origin as shown in Figure 9.39.

The mapping of the Nyquist contour is obtained as follows:

1. Semicircular indent (section II) is represented by

$$s = \lim_{\varepsilon \to 0} \varepsilon e^{j\theta}$$

where θ varying from $-90°$ through $0°$ to $+90°$ is mapped into

$$\lim_{\varepsilon \to 0}\left[\frac{K}{\varepsilon e^{j\theta}(T_1\varepsilon e^{j\theta} + 1)(T_2\varepsilon e^{j\theta} + 1)}\right]$$

$$= \lim_{\varepsilon \to 0}\left[\frac{K}{\varepsilon e^{j\theta}}\right] = \infty \cdot e^{-j\theta}$$

$$= \infty \angle +90° \text{ through } \angle 0° \text{ to } \angle -90°$$

Nyquist Stability

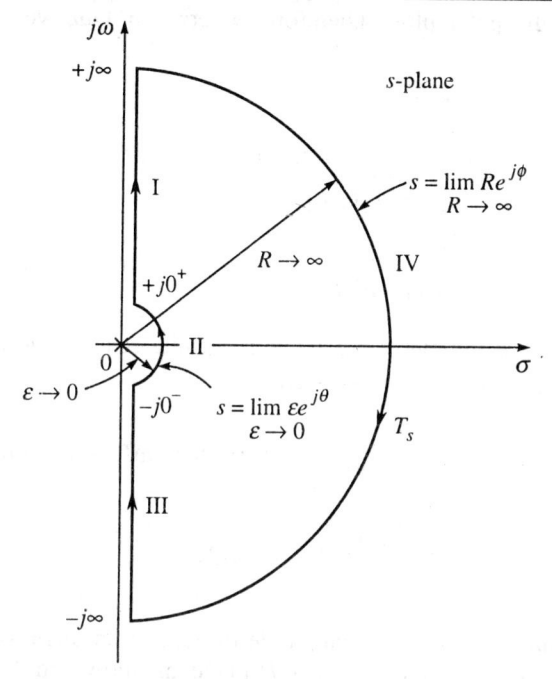

FIGURE 9.39 Example 9.8: Nyquist contour.

This part of the Nyquist plot is an infinite semicircle shown in Figure 9.40(a). The points A, B, C of section II of T_s-contour are mapped as points A', B', and C' respectively in the GH-plane as T_{GH}.

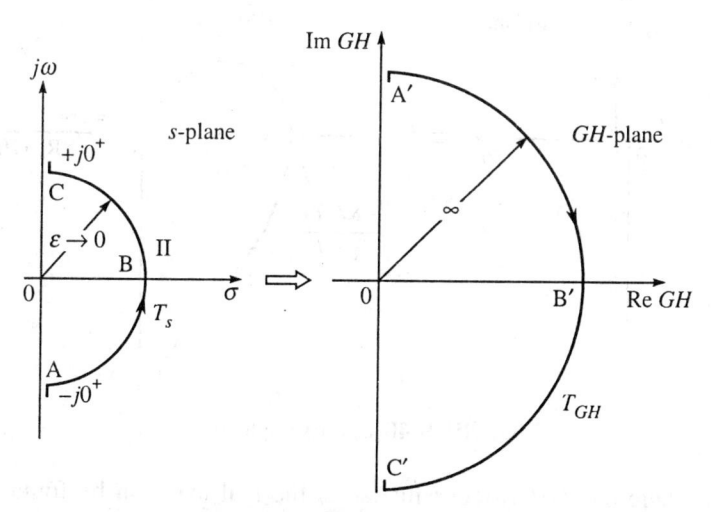

FIGURE 9.40(a) Example 9.8.

2. Along the $j\omega$-axis (section I), for various positive real values of ω, the $G(j\omega)H(j\omega)$ is calculated and plotted as shown in Figure 9.40(b). The $G(j\omega)H(j\omega)$ locus along the

$j\omega$-axis becomes the polar plot. Therefore, where $s = +j\omega$, we have

$$GH(j\omega) = \frac{K}{j\omega(j\omega T_1 + 1)(j\omega T_2 + 1)}$$

$$= \frac{-K(T_1 + T_2) - jK(1/\omega)(1 - \omega^2 T_1 T_2)}{1 + \omega^2(T_1^2 + T_2^2) + \omega^4 T_1^2 T_2^2}$$

$$= \frac{K}{[\omega^4(T_1^2 + T_2^2)^2 + \omega^2(1 - \omega^2 T_1 T_2)^2]^{1/2}} \angle[-\tan^{-1}\omega T_1 - \tan^{-1}\omega T_2 - (\pi/2)] \quad (9.50)$$

When $\omega = 0^+$, the magnitude of the locus is infinite at an angle of $-90°$ in the $GH(s)$-plane. When ω approaches $+\infty$, we have

$$\lim_{\omega \to \infty} GH(j\omega) = \lim_{\omega \to \infty} \left| \frac{1}{\omega^3} \right| \angle[-(\pi/2) - \tan^{-1}\omega T_1 - \tan^{-1}\omega T_2]$$

$$= \left(\lim_{\omega \to \infty} \left| \frac{1}{\omega^3} \right| \right) \angle -(3\pi/2) \quad (9.51)$$

Therefore $GH(j\omega)$ approaches a magnitude of zero at an angle of $-270°$ and the locus T_{GH} must cross the real GH-axis in GH-plane as shown in Figure 9.40(b) at point $\left(\frac{-K T_1 T_2}{T_1 + T_2} \right)$.

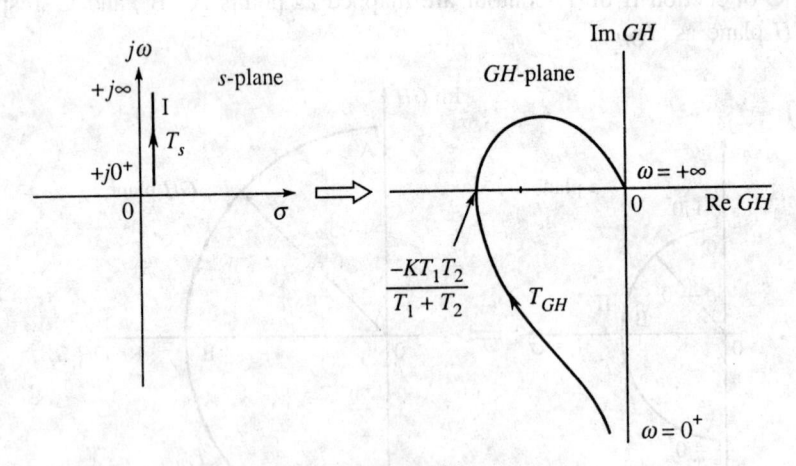

FIGURE 9.40(b) Example 9.8.

The point where the $GH(s)$-locus intersects the real axis can be found by setting the imaginary part of $GH(j\omega)$ equal to zero. Then, we have from Eq. (9.50)

$$\text{Im}[GH(j\omega)] = \frac{-K(1/\omega)(1 - \omega^2 T_1 T_2)}{1 + \omega^2(T_1^2 + T_2^2) + \omega^4 T_1^2 T_2^2} = 0$$

which leads to

$$1 - \omega^2 T_1 T_2 = 0 \quad \text{or} \quad \omega = \frac{1}{\sqrt{T_1 T_2}} \tag{9.52}$$

The magnitude of the real part of $GH(j\omega)$ at this frequency is

$$\text{Re}[GH(j\omega)] = \frac{-K(T_1 + T_2)}{1 + \omega^2(T_1^2 + T_2^2) + \omega^4 T_1^2 T_2^2}\bigg|_{\omega^2 = 1/T_1 T_2}$$

$$= \frac{-K T_1 T_2}{T_1 + T_2} \tag{9.53}$$

3. Along the $j\omega$-axis (section III), the $G(j\omega)H(j\omega)$ becomes the inverse polar plot as shown in Figure 9.40(c). The mapping is symmetrical for $GH(j\omega)$ and is the inverse polar plot.

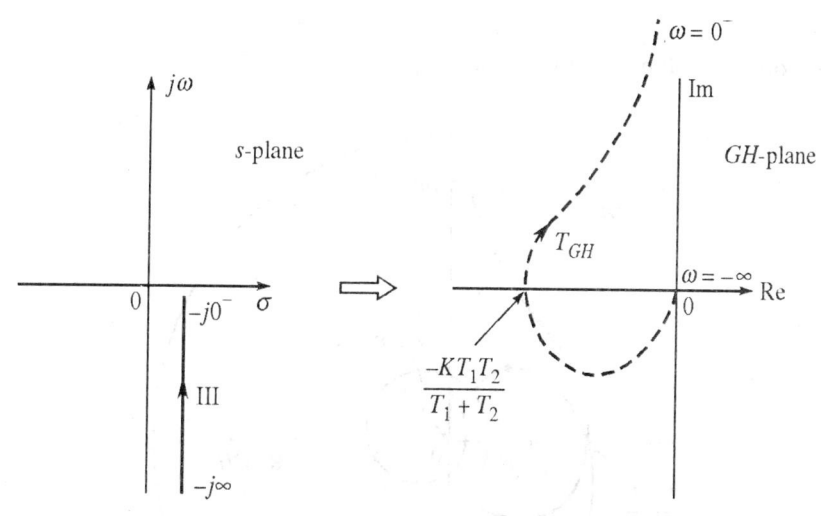

FIGURE 9.40(c) Example 9.8.

4. The infinite semicircle (section IV) of the Nyquist contour represented by $s = \lim_{R \to \infty} Re^{j\phi}$ (ϕ varying from $+90°$ through $0°$ to $-90°$) is mapped into

$$\lim_{R \to \infty}\left[\frac{K}{T_1 T_2 R^3} \frac{1}{e^{j3\phi}}\right] = \frac{1}{\infty}\left(e^{-j3\phi}\right)$$

$$= 0 \ [\angle -270° \text{ through } \angle 0° \text{ to } \angle +270°]$$

This part of the map as shown in Figure 9.40(d) is a point of infinitesimally small radius $\varepsilon \to 0$ of $540°$ revolution counterclockwise. The points A, B, C of section IV of T_s contour are mapped as points A′, B′, and C′ respectively in GH-plane as T_{GH}.

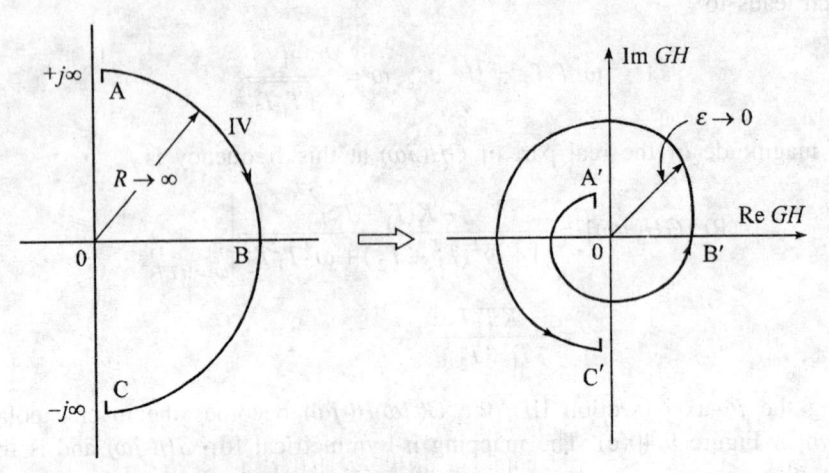

FIGURE 9.40(d) Example 9.8

The complete Nyquist plot is shown in Figure 9.41.

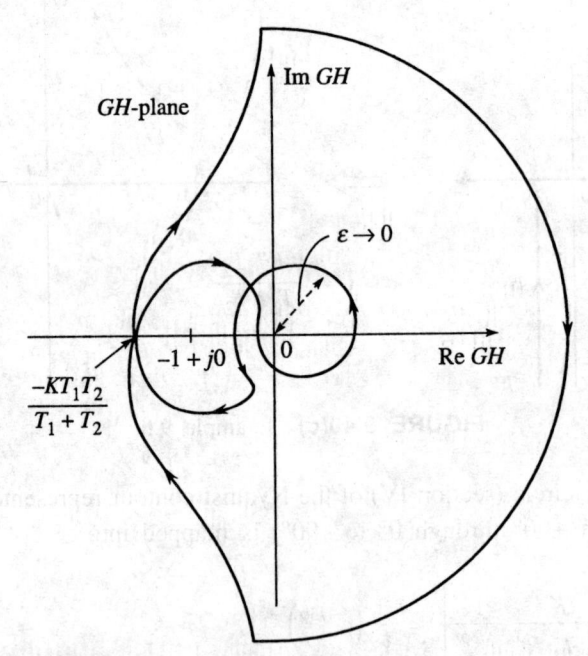

FIGURE 9.41 Example 9.8: complete Nyquist plot.

If $\dfrac{K T_1 T_2}{T_1 + T_2} > 1$, then $N = 2$, $P = 0$, hence $Z = P + N = 2$ and the system is unstable with two roots of the closed-loop system being in right-half s-plane. If $\dfrac{K T_1 T_2}{T_1 + T_2} < 1$, then $N = 0$ and $Z = P + N = 0$ the closed-loop system becomes stable.

9.8 Conditionally Stable System

Figure 9.42 shows an example of minimum phase function $GH(j\omega)$ locus, for which the closed-loop system can be made unstable by varying the open-loop gain. In the figure, if the open-loop gain is chosen such that the critical point $(-1 + j0)$ lies in between A and B, then $N = 0$ and $P = 0$; hence $Z = 0$ which means that the closed-loop system is stable.

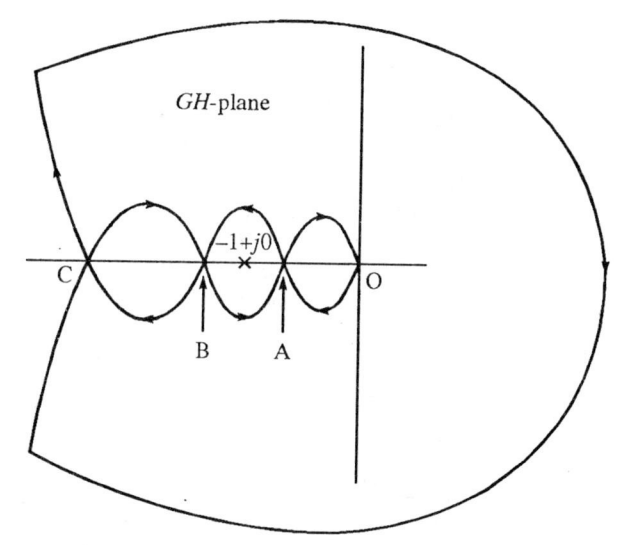

FIGURE 9.42 Conditionally stable system.

Now the open-loop gain is modified such that the critical point $(-1 + j0)$ lies in between O and A or between B to C, then in both the cases $N = 2$ and $P = 0$, so $Z = 2$ which means that the closed-loop system is unstable. Hence the system becomes conditionally stable.

In fact, as an illustration, consider the open-loop system

$$G(s)H(s) = \dfrac{K(s^2 + 2s + 4)}{s(s + 4)(s + 6)(s^2 + 1.4s + 1)}$$

The open-loop gain K is chosen to vary in the range $67.5 < K < 16.36$, the critical point $(-1 + j0)$ will be in between A and B for which, by the Nyquist criterion, the closed-loop system becomes stable. Further K varying in the range $16.36 < K < \infty$ or $0 < K < 67.5$, the critical point $(-1 + j0)$ will be between O and A and between B and C, for which, in both the cases the closed-loop system becomes unstable by Nyquist criterion.

Further, it may be noted that the same result can be derived from the polar plot as well as from the root-locus plot. We can get the same information by Routh's criterion, but for relative stability information for the design of a control system, Nyquist criterion is a useful graphical technique.

EXAMPLE 9.9 Draw the Nyquist plot for the open-loop transfer function

$$GH(s) = \frac{100K(s+5)(s+40)}{s^3(s+100)(s+200)}$$

Solution: In this example, the open-loop system has three poles at the origin. The Nyquist contour is therefore indented to bypass the origin as shown in Figure 9.43.

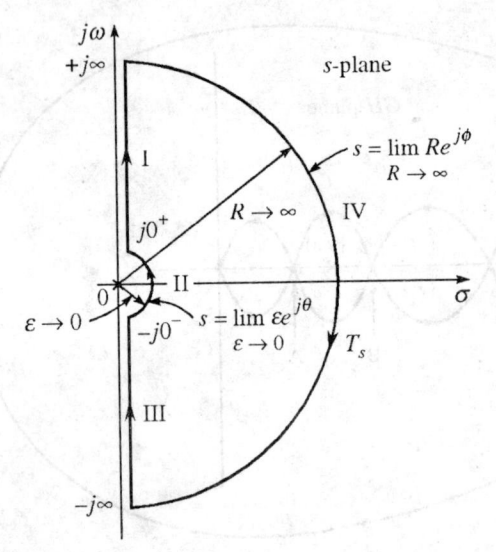

FIGURE 9.43 Example 9.9: Nyquist contour.

The mapping of the Nyquist plot corresponding to sections of Nyquist contour of Figure 9.43 is obtained as follows.

1. Semicircular indent (section II) represented by

$$s = \lim_{\varepsilon \to 0} \varepsilon e^{j\theta}$$

where θ varies from $-90°$ through $0°$ to $90°$, is mapped into

$$\lim_{\varepsilon \to 0} \frac{100K(\varepsilon e^{j\theta}+5)(\varepsilon e^{j\theta}+40)}{(\varepsilon e^{j\theta})^3(\varepsilon e^{j\theta}+100)(\varepsilon e^{j\theta}+200)}$$

$$= \lim_{\varepsilon \to 0} \frac{K}{\varepsilon^3 e^{j3\theta}} = \infty \cdot e^{-j3\theta} = \infty \; \angle+270° \to \angle 0° \to \angle -270°$$

This part of the map in the GH-plane is an infinite semicircle shown in Figure 9.44(a).

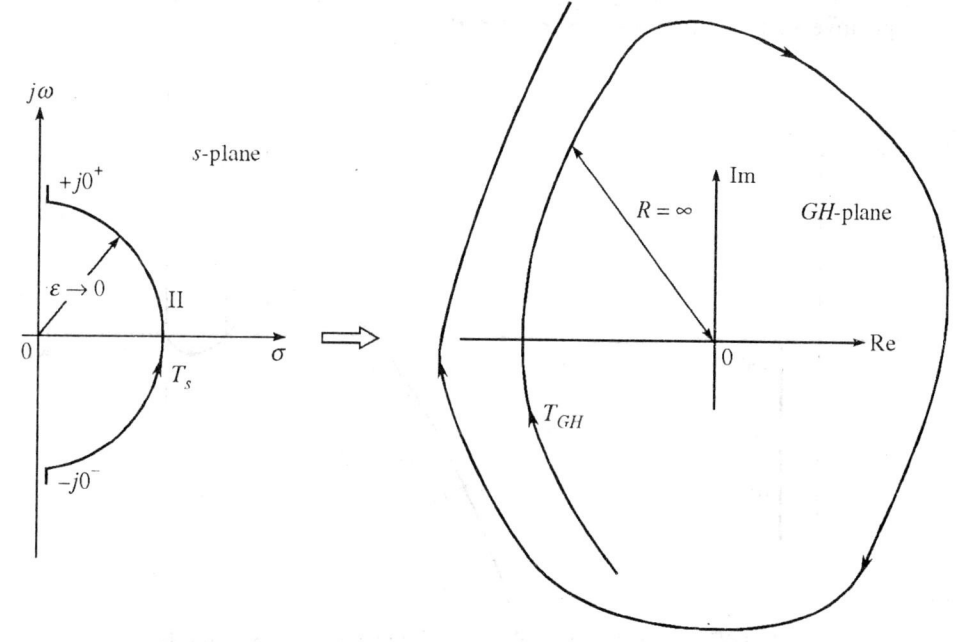

FIGURE 9.44(a) Example 9.9: Nyquist plot of section II.

2. Along the $j\omega$-axis (section I), for various positive real values of ω, the $GH(j\omega)$ is calculated and plotted as shown in Figure 9.44(b). The $GH(j\omega)$ locus along the $j\omega$-axis becomes the polar plot.

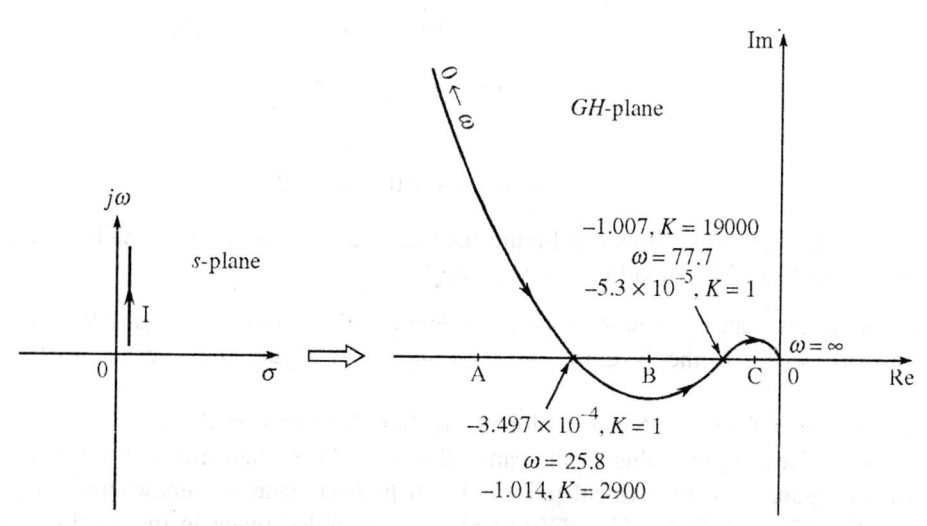

FIGURE 9.44(b) Example 9.9: Nyquist plot of section I.

3. Along the $j\omega$-axis (section III), the $GH(j\omega)$ locus T_{GH} corresponding to T_s-contour along the $j\omega$-axis from $-j\infty$ to $-j0^-$ becomes the inverse polar plot as shown in

Figure 9.44(c). The mapping is symmetrical for $GH(j\omega)$ and $GH(-j\omega)$, so that $GH(j\omega)$ is the inverse polar plot for $-\infty < \omega < 0^-$.

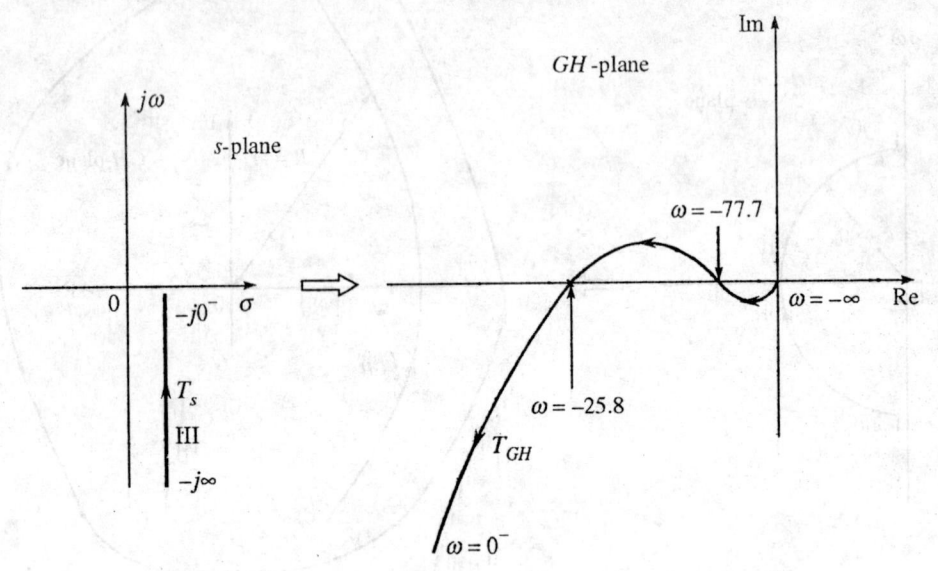

FIGURE 9.44(c) Example 9.9: Nyquist plot of section III.

4. The infinite semicircle (section IV) of the Nyquist contour represented by $s = \lim_{R\to\infty} Re^{j\phi}$ (ϕ varies from $+90°$ through $0°$ to $-90°$) is mapped into

$$\lim_{R\to\infty} \frac{100K(Re^{j\phi}+5)(Re^{j\phi}+40)}{(Re^{j\phi})^3(Re^{j\phi}+100)(Re^{j\phi}+200)}$$

$$= \lim_{R\to\infty} \frac{100K}{R^3 e^{j3\phi}} = \lim_{R\to\infty} \frac{100K}{R^3} e^{-j3\phi}$$

$$= 0 \angle{-270°} \to \angle{0°} \to \angle{+270°}$$

This part of the map shown in Figure 9.44(d) is a point of infinitesimally small radius, $\lim \varepsilon \to 0$ of 540° revolution clockwise.

The complete Nyquist plot is obtained by adding the T_{GH} contours in Figure 9.44(a) through Figure 9.44(d) which is the Nyquist plot T_{GH} shown in Figure 9.45 corresponding to the T_s-contour.

The conclusion from the Nyquist plot in the GH-plane can be drawn as follows.

For gain K having any value in the range $0 < K < 2818$, then drawing the Nyquist plot as shown in Figure 9.45, the critical point $(-1, j0)$ in $F(s)$-plane is somewhere in the region A, that is, the origin in $F(s) = [1 + GH(s)]$ plane is encircled twice in the clockwise (which is the chosen reference direction of T_s contour) direction and hence $N = 2$; the number of poles of $GH(s)$ lying in positive-half of s-plane, that is, $P = 0$. In fact, these ranges of k may be obtained from root loci plot as in Figure 6.27.

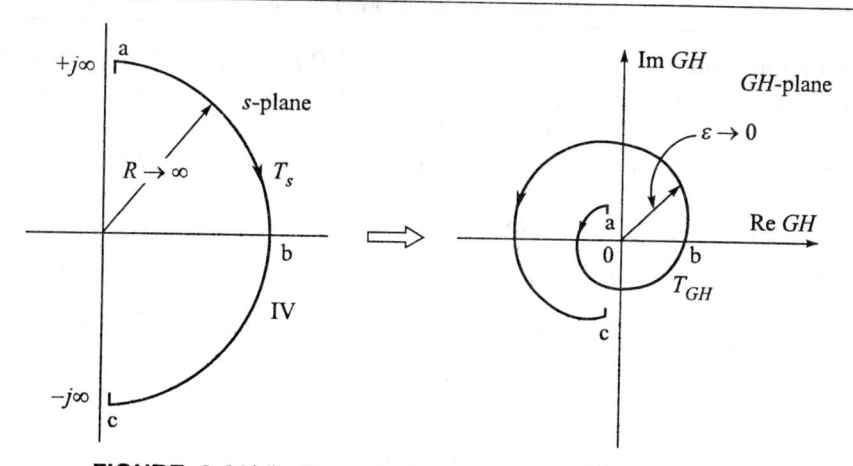

FIGURE 9.44(d) Example 9.9: Nyquist plot of section IV.

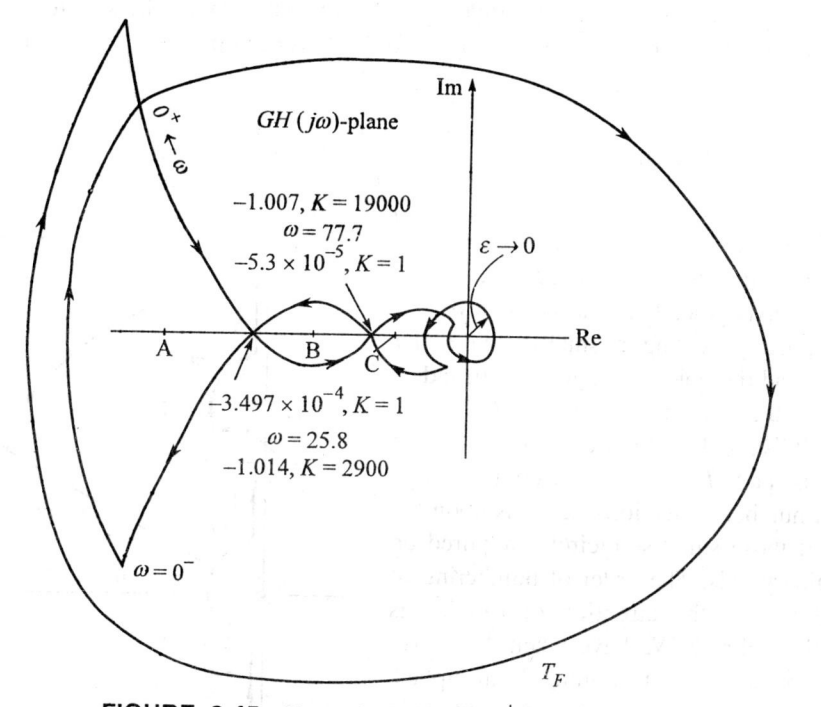

FIGURE 9.45 Example 9.9: complete Nyquist plot.

Hence, the number of positive poles of the closed-loop system, that is $Z = N + P = 2 + 0 = 2$. Therefore the closed-loop system is unstable for the value of gain K in the range $0 < K < 2818$.

For gain K in the range $2818 < K < 18837$, the $(-1, j0)$ point in the GH-plane, that is, the origin of the $F(s)$-plane is encircled once in clockwise direction and second time in the anticlockwise direction, that is, $N = 0$. Then, $Z = N + P = 0 + 0 = 0$. No poles of the closed-loop system lie in the positive-half of s-plane and the closed-loop system is stable for gain in the range $2818 < K < 18837$. In fact, calculations have been given for $K = 19000$ which indicates that the critical point is in the region B and obviously the closed-loop system is stable (as $P = 0$, $N = +1 - 1 = 0$, hence, $Z = N + P = 0$).

Similarly, for gain K in the range $18837 < K < \infty$, the origin of $F(s)$-plane is encircled twice in the clockwise direction; $N = 2$, then, $Z = N + P = 2 + 0$. The closed-loop system is unstable.

The closed-loop system is conditionally stable. The same result is obtained by other methods for the same open-loop transfer function.

9.9 Nyquist Path for Open-Loop Poles on $j\omega$-axis

As per Cauchy's theorem, the Nyquist path should not pass through the singular points of $F(s)$, i.e. the pole at origin or poles on the $j\omega$-axis of the open-loop transfer function $G(s)H(s)$. As an illustration, consider

$$G(s)H(s) = \frac{K}{s(s^2 + \omega_1^2)(s + a)} \quad \text{where } a > 0$$

To study the stability in such cases, the Nyquist contour must be modified so as to by-pass any $j\omega$-axis pole. This can be achieved by indenting, i.e. providing a small semicircular detour, around the pole on the $j\omega$-axis, of radius ε where $\varepsilon \to 0$ as shown in Figure 9.46. For the situation illustrated in Figure 9.46, we divide the Nyquist path T_s into eight sections though the exact number of sections depends upon the number of those small semicircles required on the imaginary axis. The order of numbering of the sections and the direction of rotation is completely arbitrary. We have taken the movement along the Nyquist contour T_s in s-plane clockwise and the corresponding T_F-contour in $F(s)$-plane encirclement of origin in clockwise direction as positive. It may be noted that in some literature counterclockwise movement along the Nyquist contour T_s is considered and accordingly the corresponding T_F-contour encirclement of the origin of $F(s)$-plane is considered positive for counterclockwise encirclement. It may noted that the encirclement of

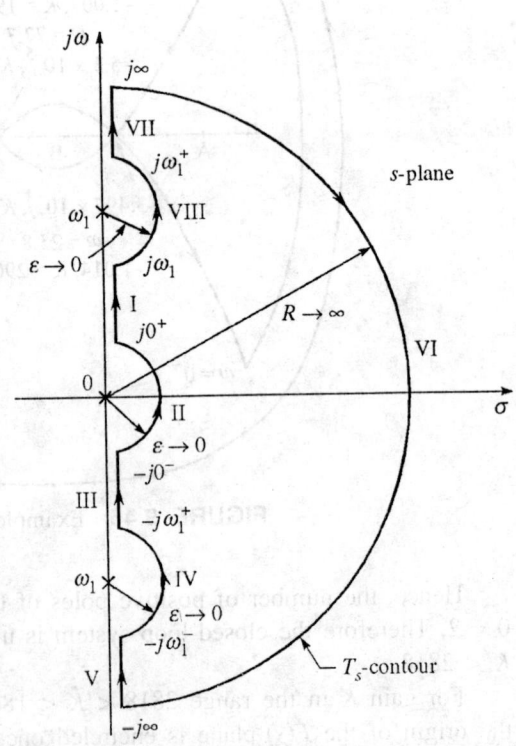

FIGURE 9.46 Nyquist path for open-loop poles on the $j\omega$-axis.

origin of $F(s)$-plane by T_F-contour is similar to the encirclement of $(-1 + j0)$ point in GH-plane by T_{GH}-contour. In Figure 9.46 the eight sections are listed as well as tabulated in Table 9.7 as follows:

Section V : From $\omega = -\infty$ to $\omega = -\omega_1^-$ along the $j\omega$-axis (section V is mirror image of section VII).

Section IV : From $\omega = -\omega_1^-$ to $\omega = -\omega_1^+$ along the small semicircle around $\omega = -\omega_1$ (section IV is mirror image of section VIII).

Section III : From $\omega = -\omega_1^+$ to $\omega = 0^-$ along the $j\omega$-axis (section III is mirror image of section I).

Section II : From $\omega = 0^-$ to $\omega = 0^+$ along the small semicircle around $\omega = 0$.

Section I : From $\omega = 0^+$ to $\omega = \omega_1^-$ along the $j\omega$-axis.

Section VIII : From $\omega = \omega_1^-$ to $\omega = \omega_1^+$ along the small semicircle around $\omega = \omega_1$.

Section VII : From $\omega = \omega_1^+$ to $\omega = +\infty$ along the $j\omega$-axis.

Section VI : From $\omega = +\infty$ to $\omega = -\infty$ along the semicircle of infinite radius moving in clockwise direction covering the entire right-hand side of s-plane.

TABLE 9.7 Mathematical equation for the Nyquist contour of Figure 9.46

Section	Equation	Range of validity
I	$s = j\omega$	$0^+ < \omega < +\omega_1^-$
VIII	$s = \lim_{\varepsilon \to 0} (j\omega_1 + \varepsilon e^{j\theta})$	$-90° \leq \theta \leq +90°$ through $0°$
VII	$s = j\omega$	$+\omega_1^+ < \omega < \infty$
VI	$s = \lim_{R \to \infty} (Re^{j\phi})$	$+90° \geq \phi \geq +90°$ through $0°$
V	$s = j\omega$	$-\infty < \omega < -\omega_1^-$
IV	$s = \lim_{\varepsilon \to 0} (-j\omega_1 + \varepsilon e^{j\theta})$	$-90° \leq \theta \leq +90°$ through $0°$
III	$s = j\omega$	$-\omega_1 < \omega < 0^-$
II	$s = \lim_{\varepsilon \to 0} (\varepsilon e^{j\theta})$	$-90° \leq \theta \leq +90°$ through $0°$

9.10 Closed-Loop Frequency Response

Uptil now we have seen that in frequency response method of analysis, we have got the closed-loop stability information from the open-loop frequency response, be it Bode plot, polar plot or Nyquist plot. We considered the open-loop transfer function $G(j\omega)H(j\omega)$, plotted the open-loop frequency response and from that plot we derived informaton on stability of the closed-loop system. This was the technique that we used althrough. In this section, we will discuss a technique by which the closed-loop frequency response information can be obtained directly from the drawing of the M(magnitude) and N(phase) circles. To begin with, without loss of generality, we restrict ourselves to unity-feedback systems because of simplicity.

9.10.1 *M*-circles: constant magnitude loci

The relationship between the open-loop and the closed-loop frequency response can be established easily for a unity-feedback system of Figure 9.47 as

$$T(j\omega) = M(\omega)e^{j\phi(\omega)} = \frac{G(j\omega)}{1 + G(j\omega)} \qquad (9.54)$$

FIGURE 9.47 Unity-feedback system.

From the polar plot of the open-loop transfer function as shown in Figure 9.48, Eq. (9.54) can be given an interesting graphical interpretation. Since for a given ω, OP represents the open-loop frequency response $G(j\omega)$, the vector AP represents $1 + G(j\omega)$. Hence the closed-loop gain M is obstained as the ratio OP/AP.

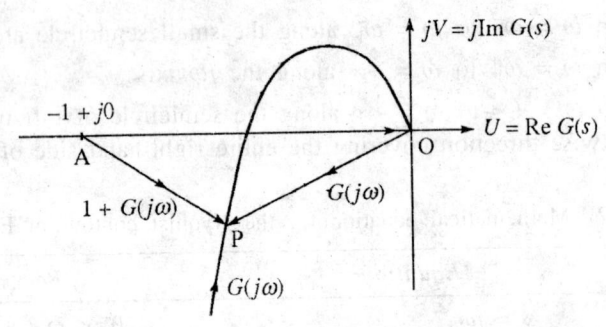

FIGURE 9.48 Graphical representation of closed-loop frequency response of unity-feedback system.

We are going to show that the points representing a constant value of the closed-loop gain, M, lie on a circle in the G-plane. Let

$$G(j\omega) = \text{Re } G(j\omega) + j \text{ Im } G(j\omega) = U(\omega) + jV(\omega)$$

Then,
$$M = \left| \frac{U + jV}{1 + U + jV} \right| = \frac{(U^2 + V^2)^{1/2}}{[(1 + U)^2 + V^2]^{1/2}} \qquad (9.55)$$

or
$$M^2 = \frac{U^2 + V^2}{(1 + U)^2 + V^2}$$

or
$$(1 - M^2)U^2 + (1 - M^2)V^2 - 2M^2 U = M^2 \qquad (9.56)$$

Dividing by the term $(1 - M^2)$ throughout and adding to both sides the term $\left(\dfrac{M^2}{1 - M^2}\right)^2$, we get

$$\left[U^2 - \frac{2M^2}{1 - M^2} U + \left(\frac{M^2}{1 - M^2} \right)^2 \right] + V^2 = \frac{M^2}{1 - M^2} + \left(\frac{M^2}{1 - M^2} \right)^2 \qquad (9.57)$$

which can be rearranged as

$$\left(U - \frac{M^2}{1-M^2}\right)^2 + V^2 = \left(\frac{M}{1-M^2}\right)^2 \tag{9.58}$$

This is the equation of a circle with centre at $U = [-M^2/(M^2 - 1), V = 0]$ and radius $|M/(M^2 - 1)|$ for constant M.

Thus, it in possible to draw a family of constant M-circles in the $G(j\omega)$-plane, as shown in Figure 9.49.

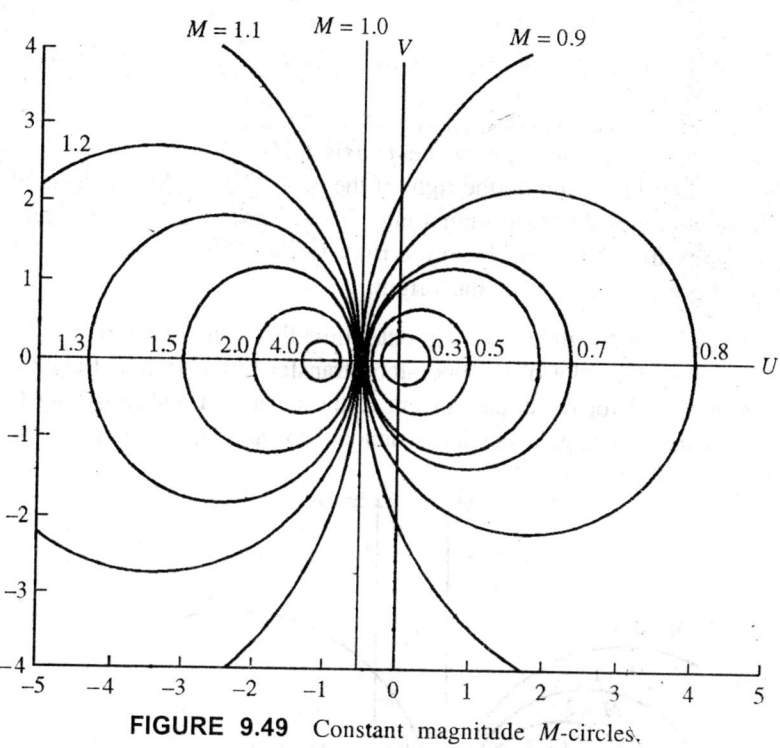

FIGURE 9.49 Constant magnitude M-circles.

It may be observed that:

(a) For $M > 1$: As M increases, the radii of M-circles decrease monotonically and the centre on the −ve real axis progressively shifts towards $(-1 + j0)$ till for $M = \infty$ when radius becomes zero at $U = -1$, that is, the $M = \infty$ circle has a zero radius and centre at $(-1 + j0)$ in the $G(j\omega)$-plane, i.e. U–V plane.

(b) For $M = 1$: The radius becomes ∞ and the centre is at $U = -\infty$. Thus $M = 1$ circle is of infinite radius with centre at infinity on the real axis U. This implies that $M = 1$ is a straight line parallel to the V-axis. Its intercept on the U-axis is $-1/2$. To explain this, the U-axis intercept is obtained by putting $V = 0$ in the equation of the M-circle and we get

$$\left[U + \frac{M^2}{M^2 - 1}\right]^2 = \left(\frac{M}{M^2 - 1}\right)^2$$

which gives $\left[U + \frac{M^2}{M^2 - 1} + \frac{M}{M^2 - 1}\right]\left[U + \frac{M^2}{M^2 - 1} - \frac{M}{M^2 - 1}\right] = 0$

For the first factor to be zero, we need $U = -\frac{M}{M-1}$ which implies that for $M = 1$, $U = -\infty$. For the second factor to be zero, $U = -\frac{M}{M+1}$ which implies that for $M = 1$, $U = -(1/2)$. Thus we conclude that for $M = 1$, the circle is a straight line parallel to the V-axis with finite intercept on the U-axis at $U = -(1/2)$.

(c) For $M < 1$: The circles are on the right of the $M = 1$ circle. As the value of M decreases, the radii of M-circles decrease monotonically, and are located on the positive real axis which shift towards the origin from the right side till for $M = 0$, the radius is zero and obviously the centre is also at the origin.

With the help of these M-circles, we can determine the value of the peak M_p of the closed-loop response from the polar plot of the open-loop transfer function in a straightforward way as the largest value of $M = 2$ for the circle tangent to the polar plot and satisfying Eq. (9.58). The frequency at which maximum gain occurs is denoted by ω_p. This is indicated in Figure 9.50.

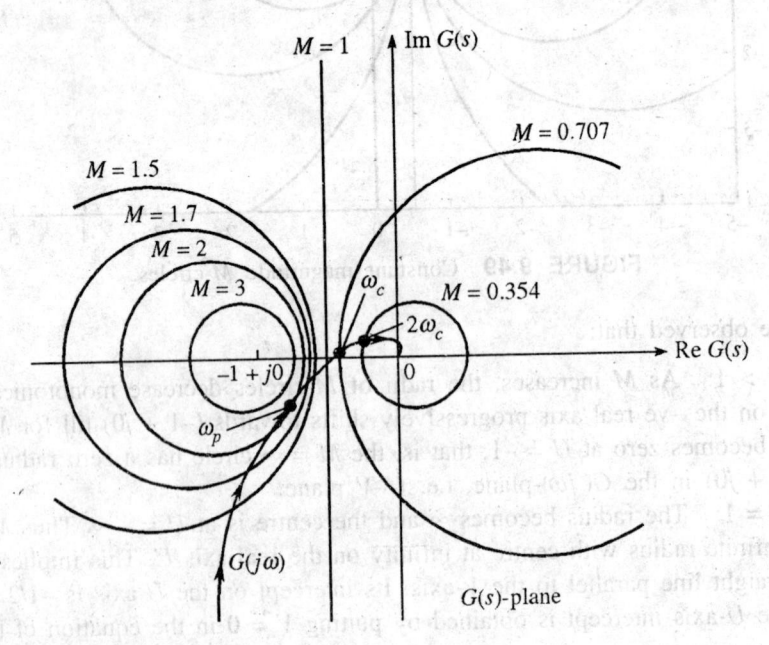

FIGURE 9.50 The $G(j\omega) = \dfrac{K}{s(s+1)(s+2)}$ locus with superimposed M-circles.

Nyquist Stability

It is also possible to use a simple geometric construction to determine the value of the gain-factor K of $G(s)$ of a unity-feedback system that will provide a desired M_p. The various steps are explained in the following. Refer Figure 9.51.

1. Draw the polar plot of $G(j\omega)$ assuming a certain value of $K = K'$.
2. Calculate the angle ϕ given by

$$\phi = \sin^{-1} \frac{1}{M_p} \tag{9.59}$$

and draw a radial line OB from the origin, making an angle ϕ with the negative real axis as shown in Figure 9.51.

3. Draw the M_p-circle with centre on the negative real axis, which is tangent to both the straight line OB and the polar plot of $G(j\omega)$. Let the circle touch OB at C.
4. From C, draw the perpendicular CD to the negative real axis.
5. Then the desired gain is given by dividing the assumed value of $K = K'$ by the length of the line OD; i.e. the required gain-factor K for the specified M_p is given by

$$K_{M_p} = \frac{K'}{\text{length of line OD}}$$

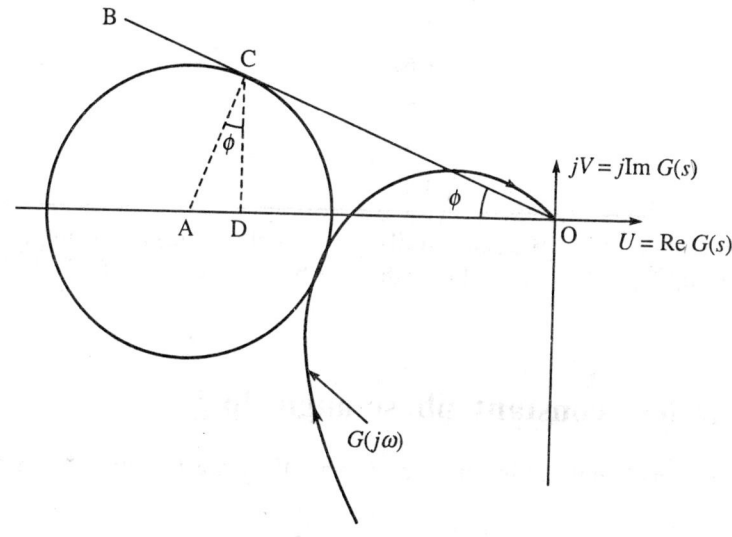

FIGURE 9.51 Construction for determining the open-loop gain for a specified M_p.

Equation (9.58) leads to the circle having radius and centre's coordinate. For the Mth-circle, the radius $AC = \dfrac{M}{M^2 - 1}$ and U coordinate of the centre is OA and is given by $M^2/(M^2 - 1)$.

So,

$$\sin \phi = \frac{AC}{OA} = \frac{\dfrac{M}{M^2 - 1}}{\dfrac{M^2}{M^2 - 1}} = \frac{1}{M} \tag{9.60}$$

and
$$OD = OA - AC \sin \phi = \frac{M^2}{M^2-1} - \frac{M}{M^2-1} \cdot \frac{1}{M} = 1 \qquad (9.61)$$

Thus the location of the point D represents the point $(-1 + j0)$ in the $G(j\omega)$-plane and gives the scale factor by which the loop-gain must be changed to obtain the desired M_p.

EXAMPLE 9.10 The forward transfer function of a unity-feedback system is given by

$$G(s) = \frac{K}{s(s^2 + 2s + 5)} \qquad (9.62)$$

Determine the value of K so that $M_p = 3$ dB.

Solution: Assuming $K = 10$, the frequency response is calculated as shown in Table 9.8.

TABLE 9.8 Frequency response of the transfer function given by Eq. (9.62).

ω	Gain (dB)	Phase shift
0.1	26	$-92.2°$
1.0	7	$-116.6°$
2.0	1.67	$-166°$
2.3	-0.5	$183.6°$
2.4	-1.34	$-189.0°$
2.8	-4.9	$-206.9°$

The gain may be plotted in the M-circles in dB. The system is unstable with M_p approximately infinity. To obtain $M_p = 3$ dB, we must lower the plot by approximately 6 dB. Hence the desired $K = 5$.

9.10.2 N-circles: constant phase-angle loci

Let us obtain the phase angle α in terms of U and V in the $G(j\omega)$-plane. Since

$$\angle e^{j\alpha} = \angle \frac{U + jV}{1 + U + jV}$$

the phase angle α is given by

$$\alpha = \tan^{-1}\left(\frac{V}{U}\right) - \tan^{-1}\left(\frac{V}{1+U}\right)$$

Let us define

$$\tan \alpha = N$$

then,

$$N = \tan\left[\tan^{-1}\left(\frac{V}{U}\right) - \tan^{-1}\left(\frac{V}{1+U}\right)\right]$$

Since
$$\tan(A-B) = \frac{\tan A - \tan B}{1 + \tan A \tan B}$$

we obtain
$$N = \frac{\dfrac{V}{U} - \dfrac{V}{1+U}}{1 + \dfrac{V}{U}\left(\dfrac{V}{1+U}\right)} = \frac{V}{U^2 + U + V^2}$$

or
$$U^2 + U + V^2 - \frac{1}{N}V = 0$$

The addition of $\left[\dfrac{1}{4} + \dfrac{1}{(2N)^2}\right]$ to both sides, yields

$$\left(U + \frac{1}{2}\right)^2 + \left(V - \frac{1}{2N}\right)^2 = \frac{1}{4} + \left(\frac{1}{2N}\right)^2$$

This is an equation of circle with centre at

$$U = -\frac{1}{2}, \; V = 1/2N \text{ and with radius } \sqrt{(1/4) + 1/(2N)^2}$$

For example, if $\alpha = 30°$, then $N = \tan \alpha = 0.577$ and the centre and the radius of the circle corresponding to $\alpha = 30°$ are $(-0.5, 0.866)$ and unity respectively. Since the last equation satisfied for $U = 0 = V$ and $U = -1, V = 0$ regardless of the value of N, this implies that each circle passes through the origin and $-1 + j0$ point. The constant α-loci can be drawn easily, once the value of N is given. A family of constant N-circles is shown in Figure 9.52 with α as parameter, It should be noted that the constant N locus for a given value of α is actually not the entire circle but only an arc. The $\alpha = +30°$ and $\alpha = -150°$ arcs are parts of the same circle.

The use of M and N circles enables us to find the entire closed-loop frequency response from the open-loop frequency response $G(j\omega)$ without calculating the magnitude and phase of the closed-loop transfer function at each frequency. Figure 9.53(a) shows the $G(j\omega)$ locus superimposed on a family of M-circles and Figure 9.53(b) on a family of N-circles. The intersections of the locus of open-loop transfer function $G(j\omega)$ with the M-circles and those with the N-circles give the values of M and N at frequency points on the $G(j\omega)$ locus. Figure 9.53(c) shows the closed-loop frequency response curve for the system. The upper curve is the M versus frequency ω curve and the lower curve is the phase angle α versus frequency ω curve. The resonant peak value is the value of M corresponding to the M-circles of smallest radius that is tangent to the $G(j\omega)$ locus. Thus, in the polar plot, the resonant peak value M_p and the resonant frequency ω_p can be found from the M-circle tangency to the $G(j\omega)$ locus.

FIGURE 9.52 Constant N-circles.

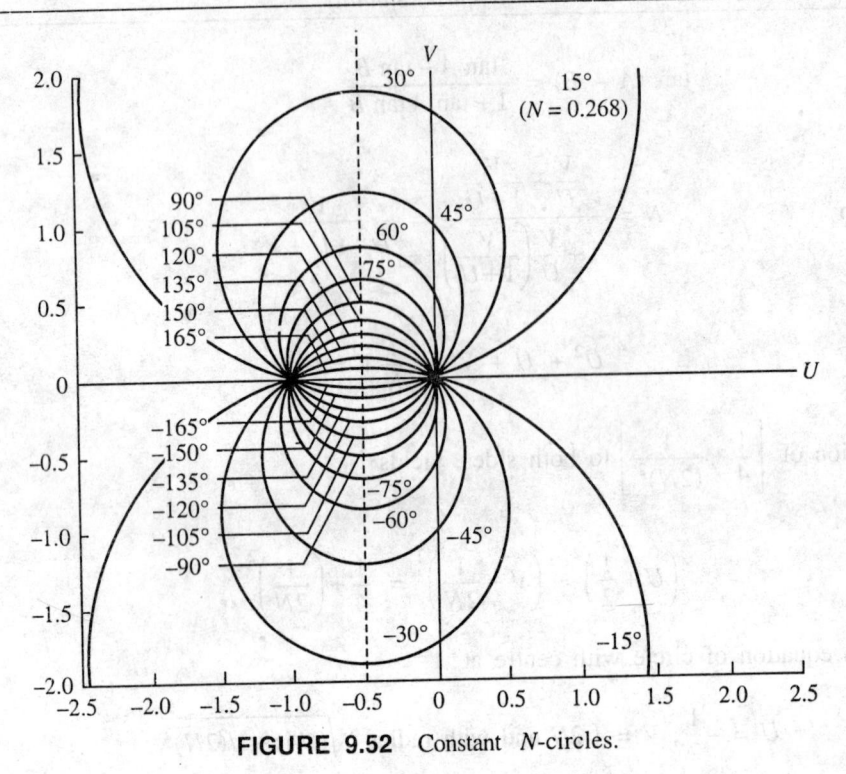

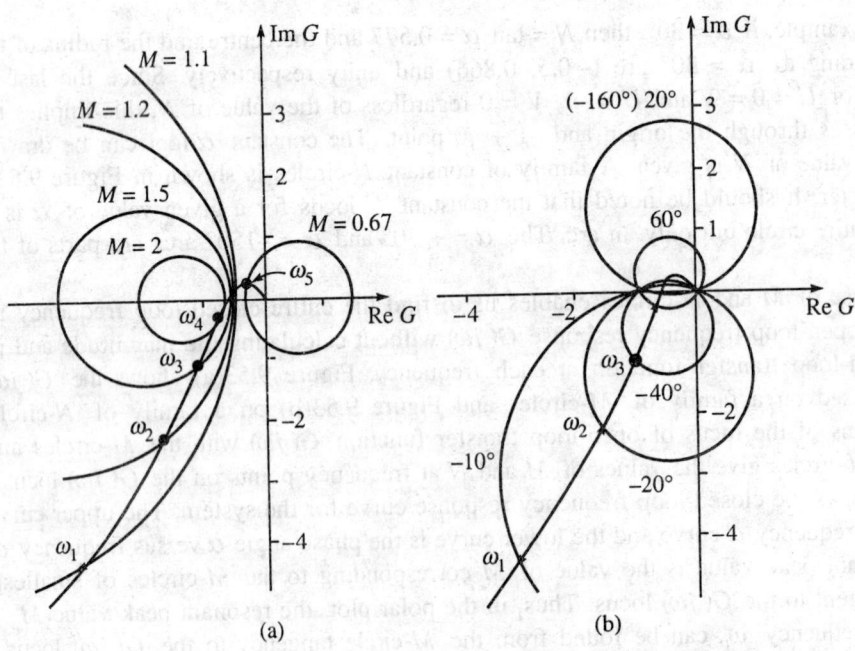

FIGURE 9.53(a) and (b) (a) $G(j\omega)$ locus superimposed on a family of M-circles, (b) $G(j\omega)$ locus superimposed on a family of N-circles.

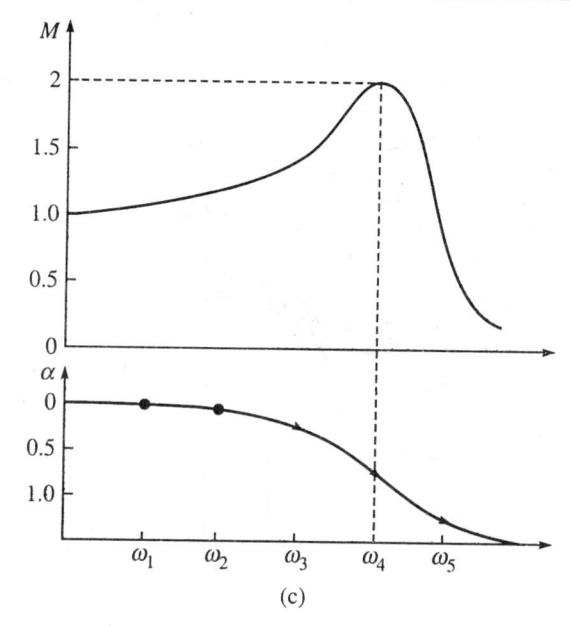

FIGURE 9.53(c) Closed-loop frequency response curves.

9.10.3 Nichols chart

A major disadvantage of working with the polar coordinates for the $G(j\omega)$ plot is that the curve no longer retains its original shape when in a design problem the loop gain is changed. In Bode plot and in magnitude-versus-phase plot, the entire $G(j\omega)$ curve is shifted up or down vertically when the gain is altered and hence it is far more convenient to work in Bode plot or in the magnitude-versus-phase plot domain.

The constant M and N loci in the polar coordinates may be transferred to the magnitude-versus-phase coordinates as in Figure 9.54(a) for $M > 1$. A point on a constant M-circle in the $G(j\omega)$-plane is mapped in the magnitude in dB-versus-phase plane by drawing a vector directly from the origin of the $G(j\omega)$-plane to the particular point on the constant M-circle; the length of the vector in dB and the phase angle in degrees give the corresponding point in the magnitude-versus-phase plane [Figure 9.54(b)]. The critical point $(-1 + j0)$ in the $G(j\omega)$-plane corresponds to the point with 0-dB and -180 degrees in magnitude-versus-phase plane. Further, for $0 < M < \infty$, the M-circles mapped in the magnitude in dB-versus-phase in degrees have been shown in Figure 9.55.

Using the same procedure, the constant N-loci can be mapped into the magnitude-versus-phase plane.

The constant M and N-loci in the magnitude-versus-phase coordinates were devised by Nichols and known as Nichols chart which is commercially available. A typical Nichols chart is shown in Figure 9.56.

The Nichols chart is a nomograph consisting of curves that are maps of the M- and N-circles on a new coordinate system; the M- and N-loci in dB-magnitude-versus-phase diagram. The

476   *Modern Control Engineering*

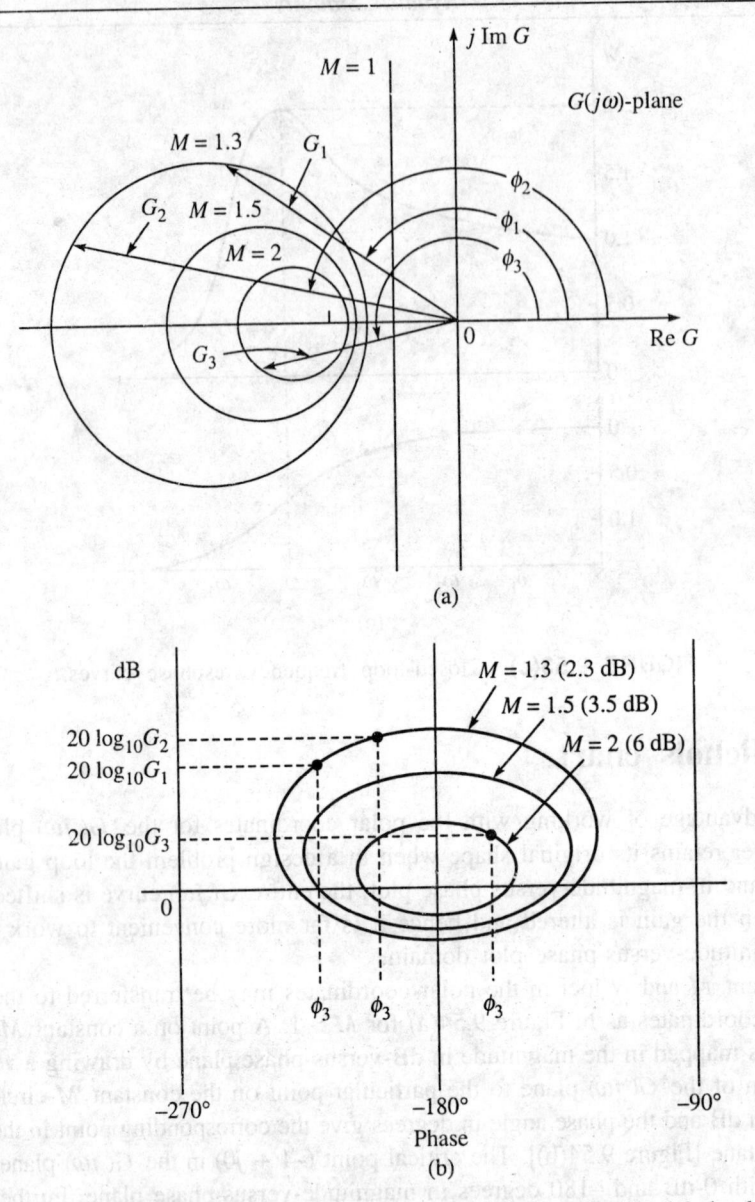

FIGURE 9.54 (a) Constant M-circles ($M > 1$) in the $G(j\omega)$-plane and (b) Nichols chart for $M > 1$ in the magnitude-versus-phase coordinates.

Nichols coordinate system plots $|G|$ in dB as ordinate versus $\angle G$ as abscissa, with origin at $|G| = 0$-dB and $\angle G = -180°$. Note that the critical point $(-1 + j0)$ of the polar coordinate plot is mapped to the Nichols chart as the point (0-dB, $-180°$) and becomes the origin of the Nichols chart. The Nichols chart is symmetric about the $-180°$ axis. The M-loci are centred about the critical point (0-dB, $-180°$). The M- and N-loci repeat for every $360°$ and there is a symmetry at every $180°$ interval.

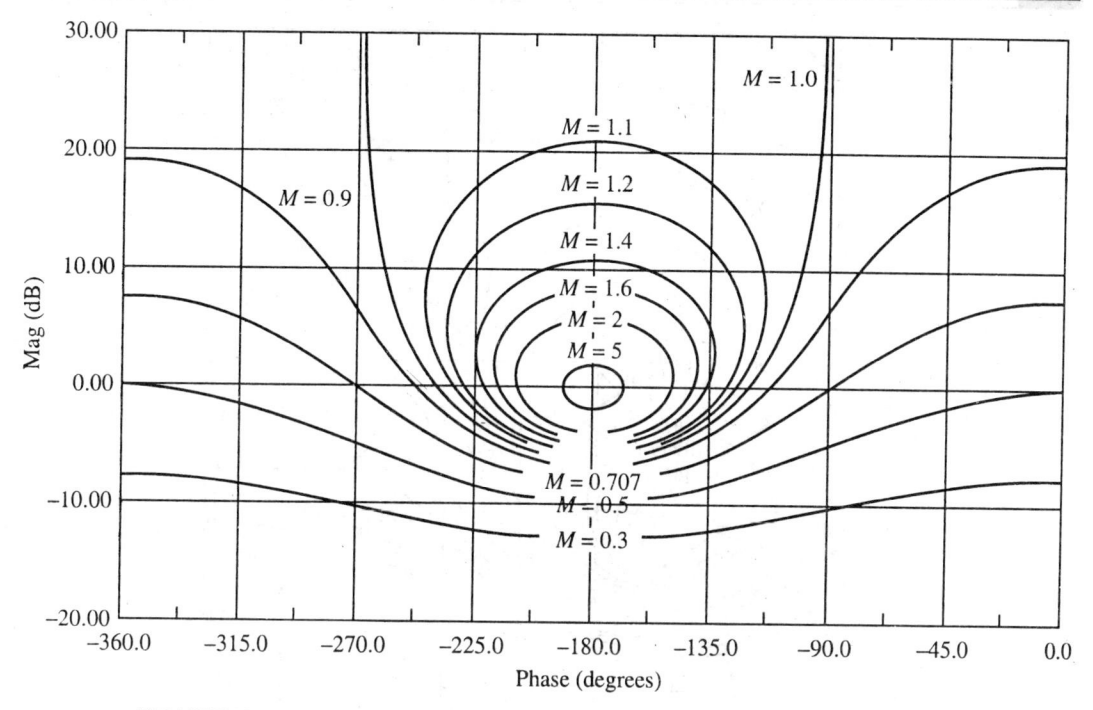

FIGURE 9.55 Nichols chart for $0 < M < \infty$ (for phase from $-180°$ to $0°$).

The Nichols chart is useful for determining the closed-loop frequency response from that of the open-loop. If the open-loop frequency response curve is superimposed on the commercially available Nichols chart, the intersection of the open-loop frequency response curve $G(j\omega)$ and the M and N loci give the value for the magnitude M and phase angle α ($= \tan^{-1}N$) of the closed-loop frequency response at each frequency point. The $G(j\omega)$ locus tangent to $M = M_p$ locus gives the resonant peak value of M of the closed-loop frequency response and is given by M_p. The resonant peak frequency ω_p is given by the frequency at the point of tangency. Information on the gain and phase margins together with the various closed-loop specifications is summarized in Figure 9.57.

The bandwidth of the closed-loop system is found from the intersection of the $G(j\omega)$ locus with the -3 dB M-contour. Unlike the Bode plots, the Nichols chart does not allow asymptotic approximations to be made. Therefore the data points must be calculated directly or transferred from the Bode to the Nichols chart.

Correlation of Bode and Nichols charts is illustrated in Figures 9.58 and 9.59 for determining the gain and phase margins.

It may be mentioned that a change in gain does not alter the shape of the $G(j\omega)$ locus on the Nichols chart but shifts it vertically. This is helpful for the design of compensators.

As an illustration, tracing the $[KG(j\omega)]$ curve (Figure 9.58) on a transparent sheet and sliding it up and down, the controller gain K varies. Nichols grid is an easy method of examining the effect of changing the gain K of a controller as shown in Figure 9.59.

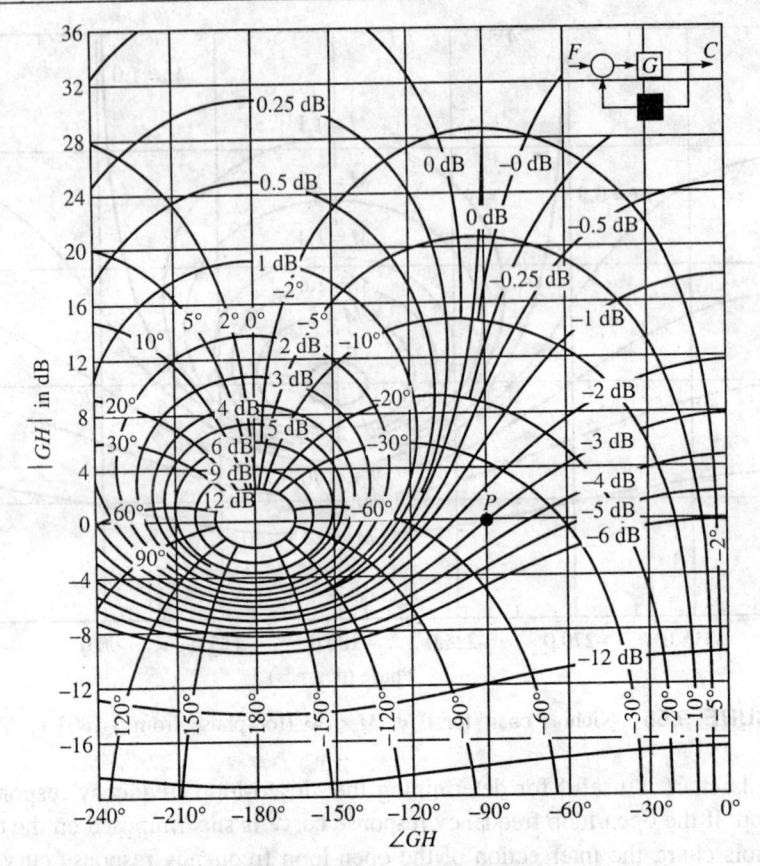

FIGURE 9.56 Nichols chart.

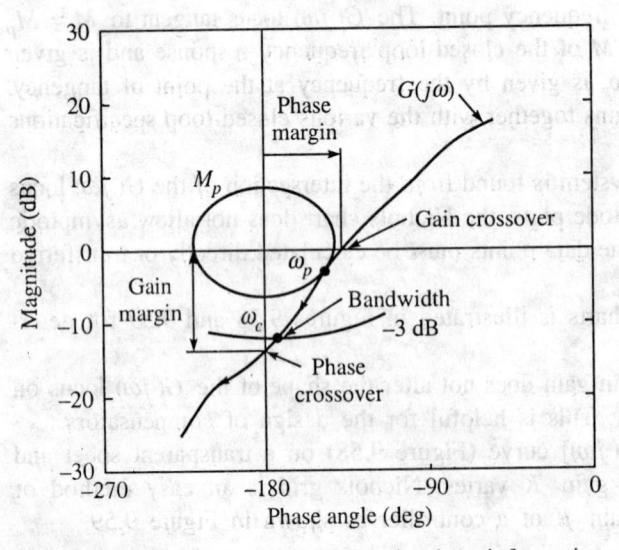

FIGURE 9.57 Summary of Nichols chart information.

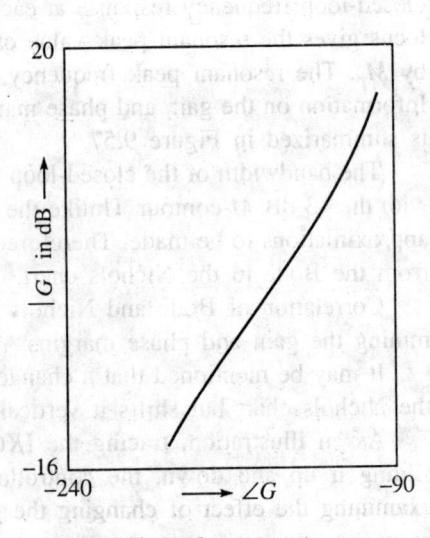

FIGURE 9.58 $KG(j\omega)$ curve.

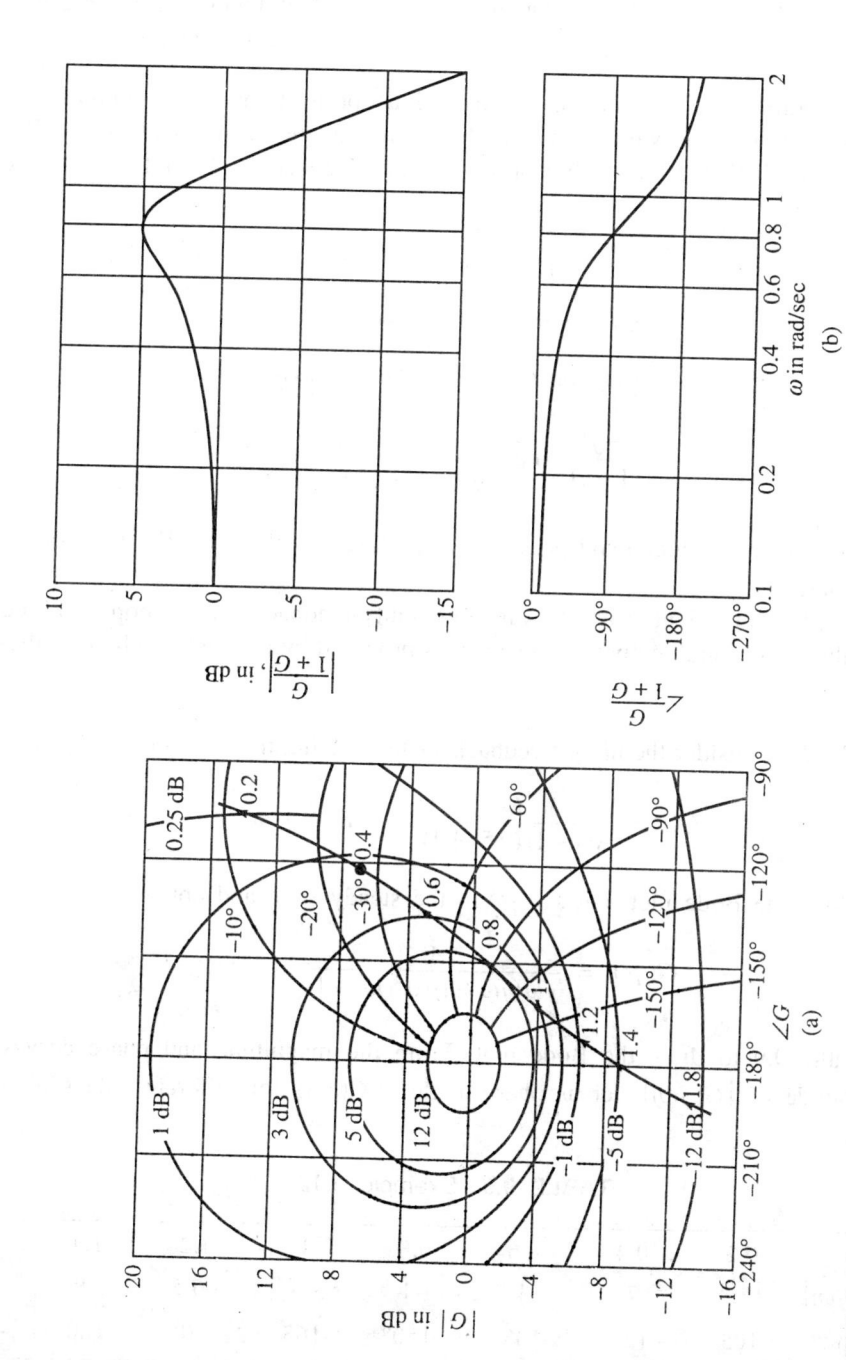

FIGURE 9.59 (a) Plot of $KG(j\omega)$ of Figure 9.58 superimposed on Nichols chart and (b) closed-loop frequency-response curves.

480 *Modern Control Engineering*

EXAMPLE 9.11 The open-loop amplitude ratio and phase shift of a unity-feedback system are 1(0-dB) and –90°, respectively. Using the Nichols chart of Figure 9.56, write down the closed-loop amplitude ratio (in dB) and the phase. Check the result algebraically.

Solution: The 0-dB, –90° point can be located as point P on the rectangular grid of Figure 9.56. The closed-loop values can then be read from the curved grid as –3 dB, –45°.

In order to check the result algebraically, the unity-feedback transfer function here is

$$\frac{G(j\omega)}{1+G(j\omega)}$$

we have
$$G(j\omega) = 1\angle{-90°} = -j1$$

Hence the closed-loop transfer function reduces to the expression

$$\frac{-j1}{1-j1}, \text{ i.e. } \frac{1\angle{-90°}}{\sqrt{2}\angle{-45°}}$$

Hence the closed-loop amplitude and phase are $1/\sqrt{2}$ (i.e. –3dB) and –45° as predicted from the Nichols chart.

A particularly convenient way of evaluating unity-feedback closed-loop amplitude and phase, given the corresponding open-loop values, is provided by the Nichols chart, as illustrated in Example 9.12.

EXAMPLE 9.12 Consider the unity-freedback system having the open-loop transfer function

$$G(s) = \frac{K}{s(s+1)(0.5s+1)}; \quad K = 1$$

In order to draw the Bode plot, put $s = j\omega$ in the steady state and obtain

$$G(j\omega) = \frac{1}{j\omega(j\omega+1)(0.5j\omega+1)}$$

Formulate Table 9.9 to draw the Bode plot. From the magnitude and phase curves, draw the dB-magnitude of $|G(j\omega)|$ versus phase angle $\angle G(j\omega)$ for different values of ω as in Figure 9.58.

TABLE 9.9 Example 9.12

ω	0.2	0.4	0.6	0.8	1	1.2	1.4	1.8
$20 \log |G(j\omega)|$	14.2	7	3	–1	–4	–6.5	–9	–13.5
$\angle G(j\omega)$	–105°	–123°	–138°	–150°	–163°	–170°	–180°	–192°

Figure 9.59(a) shows the $G(j\omega)$-locus of Figure 9.58 superimposed on the commercially available Nichols chart. The closed-loop frequency-response curves may be constructed by reading the magnitudes and phase angles at various frequency points on the $G(j\omega)$-locus from the Nichols chart as per Table 9.10 and are drawn as shown in Figure 9.59(b). We obtain from this figure $M_p = 5$ dB $\omega_p = 0.8$ rad/s; $\omega_{gc} = 1.16$ rad/s $\omega_{pc} = 1.4$ rad/s; GM = 9 dB; PM = 35°.

TABLE 9.10 Example 9.12

ω (rad/s)	Closed-loop amplitude ratio (dB), M	Closed-loop phase shift in degrees, ϕ
0.2	0.25	−105
0.4	2	−27
0.6	3	−45
0.8	5	−90
1.0	3	−135
1.2	−1	−165
1.4	−5	−180
1.8	−11	−190

The procedure for obtaining the closed-loop response from the Nichols chart is as follows:

1. The open-loop frequency response is superimposed on the Nichols chart.
2. From the intersection of this curve with the contours at various frequency points, values of M and ϕ_m are read from the plot.
3. The closed-loop frequency response curves are obtained from the above values.
4. The resonant peak values M_p and the corresponding ω_p are given at a point where $G(j\omega)$ is tangent to an M-circle.
5. The bandwidth ω_B is obtained by noting the frequency at which the $G(j\omega)$ curve intersects the $M = 0.707$ locus.

In order to use the Nichols chart for non-unity feedback systems, the block diagram must be rearranged to obtain an equivalent unity feedback system.

9.10.4 Non-unity feedback system

For a non-unity feedback system having forward path transfer function $G(s)$ and feedback path transfer function $H(s)$, the closed-loop transfer function becomes

$$T(s) = \frac{C(s)}{R(s)} = \frac{G(s)}{1 + G(s)H(s)}$$

This may be rewritten as

$$T(s) = \frac{1}{H(s)}\left[\frac{G(s)H(s)}{1 + G(s)H(s)}\right] = \frac{1}{H(s)}\left[\frac{G_0(s)}{1 + G_0(s)}\right] = \frac{1}{H(s)}T_0(s) \qquad (9.63)$$

where

$$G_0(s) = G(s)H(s) \tag{9.64}$$

and

$$T_0(s) = \frac{G_0(s)}{1+G_0(s)} \tag{9.65}$$

Obviously $T_0(s)$ is the usual standard form of a unity-feedback closed-loop system. Constant M- and N-circles are used for $G_0(s)$ to obtain $T_0(s)$. However, $T(s)$ can then be obtained by multiplying $T_0(s)$ by $1/H(s)$. This multiplication can be carried out through the Bode plot of $T_0(j\omega)$ and $H(-j\omega)$. Then graphically, subtracting log-magnitude and phase-angle curves of $H(j\omega)$ from those of $T_0(j\omega)$, the resulting Bode plot gives the closed-loop frequency response of $T(j\omega)$.

9.10.5 Frequency domain specifications for design

The open-loop systems of types-0, 1 and 2 having a gain margin > 1 combined wth a phase margin > 0° will guarantee the closed-loop stability. As a simple thumb rule, phase margin in degrees divided by 100 gives the closed-loop damping ratio. With unstable open-loop systems or systems having any other type number, alternative methods for checking the closed-loop stability must be employed.

For a second-order system that has the prototype closed-loop transfer function

$$\frac{\omega_n^2}{s^2 + 2\zeta\omega_n s + \omega_n^2}$$

the maximum magnitude ratio usually referred to as 'M-peak' (M_p) gives an indication of system's relative stability to step response and is given as

$$M_p = \frac{1}{2\zeta\sqrt{1-\zeta^2}}; \quad 0 < \zeta < 0.707 \tag{9.66}$$

and

$$M_p = 1 \text{ for } \zeta \geq 0.707.$$

The relationship of underdamped ζ with M_p is given as

$$\zeta^2 = \frac{1}{2} - \frac{1}{2}\sqrt{\left(1-\frac{1}{M^2}\right)}; \quad \text{for } M_p > 1 \tag{9.67}$$

Normally, a large M_p corresponds to a large peak overshoot for a step input. The optimum value of M_p within $1.1 < M_p < 1.5$ corresponds to the damping factor of second-order underdamped stable system as $0.54 < \zeta < 0.36$.

Resonant frequency ω_p

The resonant frequency ω_p is the frequency at which M_p occurs. For the second-order system, ω_p is given by

$$\omega_p = \omega_n\sqrt{1-2\zeta^2} \qquad (9.68)$$

and indicates the system's speed of response.

Bandwidth

The bandwidth, BW, is the frequency as which the magnitude of $|M(j\omega)|$ drops to 70.7% or 3-dB down from the zero frequency value.

For a prototype second-order system the bandwidth is

$$BW = \omega_n\left[(1-2\zeta^2)+\sqrt{4\zeta^4-4\zeta^2+2}\right]^{1/2} \qquad (9.69)$$

A large bandwidth corresponds to faster rise time, because the higher frequency signals are passed out easily on to the output. Conversely, a small bandwidth corresponds to slow and sluggish time response because only the signals of relatively low frequencies are passed. The bandwidth of a system is also an indicator of its noise filtering characteristics: an unnecessarily wide bandwidth would produce a system with poor noise rejection characteristics.

Note that these equations may not yeild accurate results for higher-order systems depending on whether a pair of dominant poles truly exists or not.

Summary

This chapter presented the basic philosophy of Nyquist stability criterion for determining the stability of the closed-loop system from the open-loop frequency response without finding the closed-loop poles. It has been demonstrated that the stability of a single-loop control system can be investigated by studying the behaviour of the Nyquist plot of the open-loop transfer function $G(s)H(s)$ for $\omega = -\infty$ to $\omega = \infty$ with respect to the number of encirclements of the $(-1, j0)$ point. The chapter also presented several illustrative examples of stability analysis of control systems using the Nyquist stability criterion.

An example of a conditionally stable system was given—a system that is stable only for limited ranges of values of the open-loop gain.

Continuing with the discussion on frequency response method of stability analysis, another technique by which the closed-loop frequency response information can be obtained directly from M (magnitude) circles and N (phase) circles was presented. It was then shown how to construct the M- and N-loci in the log-magnitude-versus-phase plane, i.e. Nichols chart. The usefulness of the Nichols chart for dealing with design problems was demonstrated.

Problems

9.1 By use of the Nyquist criterion, determine whether the closed-loop systems having the following open-loop transfer functions are stable or not. If not, how many closed-loop poles lie in the right-half of s-plane?

(a) $G(s)H(s) = \dfrac{1+4s}{s^2(1+s)(1+2s)}$ (b) $G(s)H(s) = \dfrac{1}{s(1+2s)(1+s)}$

(c) $G(s)H(s) = \dfrac{1}{s^2+100}$

9.2 Check the stability of the systems by completing the corresponding Nyquist plots for the given polar plots in Figure P.10.2.

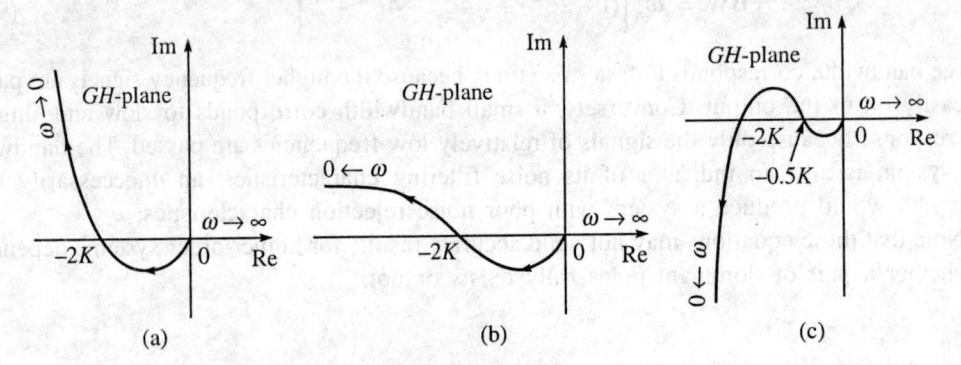

FIGURE P.10.2

9.3 Sketch a Nyquist plot for a system with the open-loop transfer function

$$\dfrac{K(1+0.5s)(s+1)}{(1+10s)(s-1)}$$

Determine the range of values of K for which the system is stable.

9.4 Consider a feedback system having the characteristic equation

$$1 + K = 0 \quad \dfrac{K}{(s-1)(s+1.5)(s+2)}$$

It is desired that all the roots of the characteristic equation have real parts less than -1. Extend the Nyquist stability criterion to find the largest value of K, satisfying this condition.

9.5 For the unity-feedback system having open-loop transfer function

$$\dfrac{100K(s+5)(s+40)}{s^3(s+100)(s+200)}$$

draw the Nyquist plot and find the stability taking $K = 1$.

Further verify the stability from Bode plot taking $K = 1$, and also from root-loci plot for $0 < K < \infty$.

Compensation Techniques

10.1 Introduction

The performance of a control system may be described either in terms of the time-domain performance-measures or in terms of the frequency-domain performance-measures. The performance of a system may be specified in terms of the time-domain performance-measures by specifying a certain peak time—for getting maximum overshoot and settling-time for a step input. Furthermore, it is usually necessary to specify the maximum allowable steady-state error for several standard test signal inputs. These performance specifications may be related in terms of the desirable locations of the poles and zeros of the closed-loop system transfer function. Thus the location of the closed-loop poles and zeros may be specified as per the performance-measures desired, for which the root locus approach may be considered. However, when the locus of the roots does not result in a suitable root configuration, we must add a compensating network in order to be able to alter the locus of the roots with the variation of one parameter.

Alternatively, we may describe the performance in terms of the frequency-domain performance-measures such as the peak of the closed-loop frequency response $M_{p\omega}$, the resonant frequency ω_r, the bandwidth, the gain and phase margin of the closed-loop system. We may add a suitable compensation network $G_c(s)$ and develop the design of the network in terms of the frequency response as portrayed on the polar

OBJECTIVE

The objective of this chapter is to present procedures for the design and compensation of single-input-single-output, linear, time-invariant control systems by the frequency-response approach. The root-locus method gives direct information on transient response and is useful in reshaping the transient response of closed-loop control systems whereas the frequency-response approach gives the information indirectly. The designer has to know both the approaches very well, in order to understand design specifications for realizing a successful system as different specifications leades to different approach to design.

CHAPTER OUTLINE

Introduction
Cascade Compensating Networks
Lead Compensating Networks
Characteristics of Lead Networks
Lag Compensation
Lag-Lead Compensation
Compensation of Operational Amplifier

plot, Bode diagram, or Nichols chart. A cascade transfer function is readily accounted for, on a Bode plot by adding the frequency response of the network, hence the Bode diagram in frequency response approach is preferred.

The frequency response and root locus may be used as tools for designing any of the compensation schemes, except that the root locus is not convenient when more than one or possibly two variable parameters are introduced.

It is the purpose of this chapter to further describe the addition of several compensation networks to a feedback control system. First, we shall consider the addition of a so-called phase-lead compensating network and describe the design of the network by root locus and frequency response techniques. Then, using both the root locus and frequency response techniques, we shall describe the design of the lag compensation networks in order to obtain a suitable system performance. Finally, we shall describe a lag–lead compensator in order to get a suitable performance criterion both in low and high frequency range.

The design aspect of the state-feedback compensator will be discussed in the concerned chapter on state variable approach.

While the design of linear control systems is carried out in the frequency domain, the frequency response design provides information on the steady-state response, stability margin and system bandwidth. The transient response performance can be estimated indirectly in terms of the phase margin, the gain margin and resonant peak magnitude. The percent overshoot is reduced with an increase in the phase margin, and the speed of response is increased with an increase in the bandwidth. Thus, the gain-crossover frequency, the resonant frequency and the bandwidth give a rough estimate of the speed of transient response.

A common approach to the frequency response design is to adjust the open-loop gain so that the requirement on the steady-state accuracy is achieved. If the specifications concerning the phase margin and gain margin are not satisfied, then it is necessary to reshape the open-loop transfer function by adding an additional controller to the open-loop transfer function. When only the gain is varied, the phase angle plot will not be affected. The Bode magnitude curve is shifted up or down to correspond to the increase or decrease in K_p, the gain of the proportional controller. Similarly, the effect of changing K_p on the polar plot is to enlarge or reduce it, but the shape of the polar plot cannot be changed.

Several compensation schemes are shown in Figure 10.1 for a simple single-loop control system. When the compensator $G_c(s)$ is placed in the forward path as in Figure 10.1(a), it is called a *series* or *cascade* compensator. The other compensation schemes are *feedback* compensation as in Figure 10.1(b), *output* or *load* compensation as in Figure 10.1(c), and *input* compensation as in Figure 10.1(d). The choice of the compensation scheme depends upon the specification, power level at various signal modes in the system and the availability of components. Suppose we want to alter the damping coefficient as per the requirement of our specification. We can put (i) a high-pass *RC* circuit or an electronic circuit using operational amplifiers or mechanical spring-dashpot systems in the forward path, or (ii) a tachogenerator in the feedback path. Obviously, a power amplifier would be required in the forward path because of the low level of error signal. On the contrary, tachogenerator feedback does not require any amplification because of the high level of output signal. Both the forward path high-pass *RC* circuit (derivative control) and tachogenerator feedback control do change the damping ratio; but the use of a proper compensation circuit depends on the availability of components and signal level.

Compensation Techniques

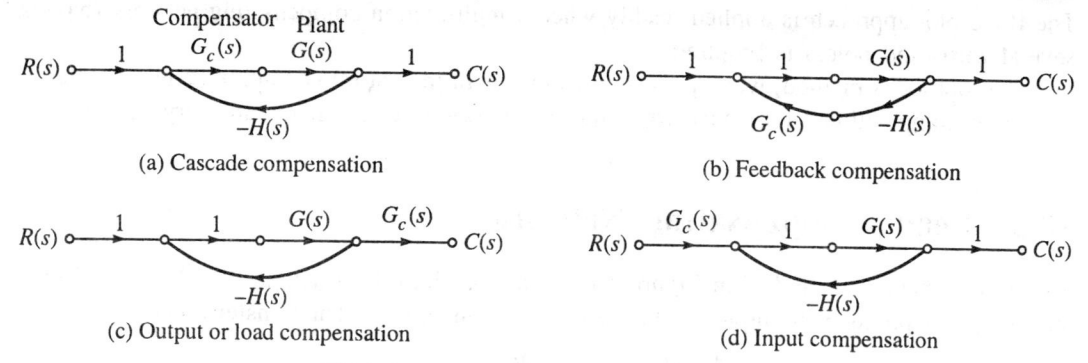

FIGURE 10.1 Compensation schemes.

(a) Cascade compensation
(b) Feedback compensation
(c) Output or load compensation
(d) Input compensation

10.2 Cascade Compensation Networks

A system is cascade compensated when the controller (compensator) is placed in the main forward transmission path as shown in Figure 10.2. The compensation network $G_c(s)$ is cascaded with the unalterable process $G(s)$ in order to provide a suitable loop transfer function $G_c(s)G(s)$. Clearly, the compensator $G_c(s)$ may be chosen to alter the shape of the root locus or the frequency response.

Mathematically, $G_c(s)$ is a ratio of two polynomials, i.e.

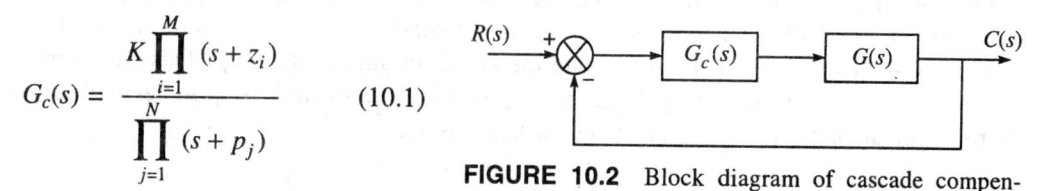

$$G_c(s) = \frac{K \prod_{i=1}^{M} (s + z_i)}{\prod_{j=1}^{N} (s + p_j)} \quad (10.1)$$

FIGURE 10.2 Block diagram of cascade compensation scheme.

Then the problem reduces to the judicious selection of the poles and zeros of the compensator. When graphical design methods are used, the problem of selecting suitable locations of the poles and zeros is solved by trial and error.

To illustrate the properties of the compensation network, we shall consider a first-order compensator. The compensation approach developed on the basis of a first-order compensator may then be extended to higher-order compensators.

To understand the use of compensators, it is both necessary and desirable that the effects of the compensating networks on both the Bode curves and the root locus be clearly understood.

Consider
$$G_c(s) = K \frac{s+z}{s+p} \quad (10.2)$$

where both z and p are real. This transfer function may have:

$|z| < |p|$ = high-pass filter, also called the phase-lead compensator (by taking the phase of the numerator and denominator polynomials).

$|z| > |p|$ = low-pass filter, also called the phase-lag compensator.

The Bode plot approach is applied readily when a multisection compensating network (having several zeros and poles) is required.

The root-locus method, while providing an excellent insight into complex system problems, is a good computational tool only when the compensator requirements are very modest.

10.3 Lead Compensating Networks

The arrangement of R and C in Figure 10.3 causes a phase-lead action. An alternating input signal v_{in} may produce an output v_{out} that leads v_{in} by an angle ϕ. The transfer function becomes

$$\frac{V_{out}(s)}{V_{in}(s)} = \frac{RCs}{RCs+1} \tag{10.3}$$

The phase angle $\phi = 90° - \tan^{-1} \omega RC$ is a positive acute angle. However, this simple circuit is not used in a servo system, for its output v_{out} becomes zero as the signal frequency ω is decreased towards zero, that is, the steady-state condition; a loop containing this circuit has no gain at standstill and therefore has low accuracy.

Zero at the origin is not permitted in a cascade compensator because that would block the zero-frequency (dc) component of the error signal and the system would then be incapable of reaching the desired steady state.

If another resistor R_1 is added across C as shown in Figure 10.4, this becomes a useful phase-lead circuit. At low frequency (or when $\omega = 0$), the current in C is negligible; R_1 and R_2 act as a simple voltage divider so that $v_{out} = v_{in}(R_2/(R_1 + R_2))$. The loss of steady-state gain may be offset (compensated) by increasing the amplifier gain in the loop. This circuit provides no leading angle when ω is very large or very small. At very high frequencies, C becomes a short-circuit around R_1, so v_{out} becomes as large as v_{in}.

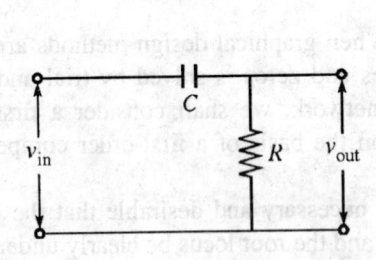

FIGURE 10.3 Phase-lead circuit having no gain at low frequencies.

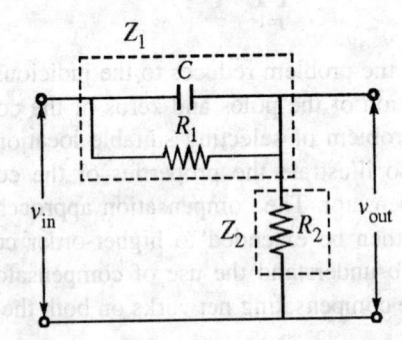

FIGURE 10.4 A useful RC phase-lead circuit.

Using the symbols defined in Figure 10.4, we find that the complex impedances Z_1 and Z_2 are

$$Z_1 = \frac{R_1}{R_1 Cs + 1}; \quad Z_2 = R_2$$

Compensation Techniques

The transfer function between the output $V_{out}(s)$ and the input $V_{in}(s)$ becomes

$$\frac{V_{out}(s)}{V_{in}(s)} = \frac{Z_2}{Z_1 + Z_2} = \frac{R_2}{R_1 + R_2} \frac{R_1 Cs + 1}{\frac{R_1 R_2}{R_1 + R_2} Cs + 1} \tag{10.4}$$

Let us define:

$$R_1 C = T; \quad \frac{R_2}{R_1 + R_2} = \alpha < 1 \tag{10.5}$$

Then the transfer function becomes

$$\frac{V_{out}(s)}{V_{in}(s)} = \alpha \left(\frac{Ts + 1}{\alpha Ts + 1} \right) = \frac{s + 1/T}{s + 1/\alpha T} \tag{10.6}$$

$$= K_c \left(\frac{s + 1/T}{s + 1/\alpha T} \right) = K_c \alpha \left(\frac{Ts + 1}{\alpha Ts + 1} \right)$$

with $K_c = 1$ to make the uniformity with Eq. (10.8).

The phase angle is

$$\phi = \tan^{-1} \omega T - \tan^{-1} \alpha \omega T = \text{+ve acute angle} \quad \text{for } 0 < \alpha < 1 \tag{10.7}$$

It is thus obvious that the output leads the input as $0 < \alpha < 1$.

There are many ways to realize a continuous time compensator. Let us consider an electronic circuit using operational amplifiers as in Figure 10.5. The transfer function for this circuit is obtained as follows.

$$\frac{E_{out}(s)}{E_{in}(s)} = \left(-\frac{R_4}{R_3} \right)\left(-\frac{Z_2}{Z_1} \right) = \left(\frac{R_4 C_1}{R_3 C_2} \right)\left(\frac{s + 1/R_1 C_1}{s + 1/R_2 C_2} \right)$$

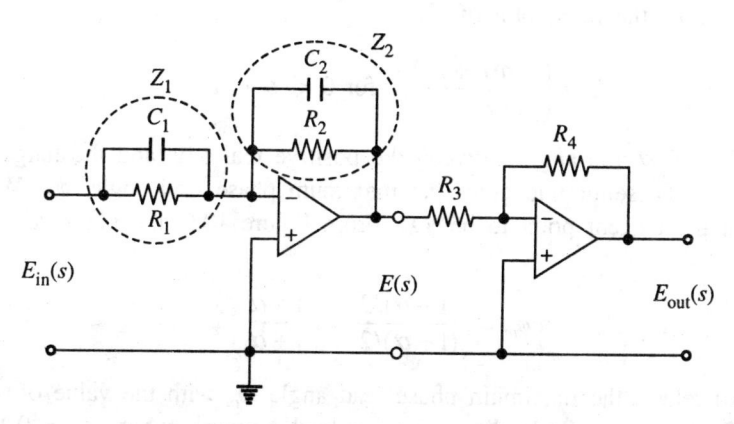

FIGURE 10.5 Electronic circuit which is a lead network if $R_1 C_1 > R_2 C_2$ and a lag network if $R_1 C_1 < R_2 C_2$.

where
$$Z_2 = \frac{R_2}{R_2 C_2 s + 1}; \quad Z_1 = \frac{R_1}{R_1 C_1 s + 1}$$

or
$$\frac{E_{out}(s)}{E_{in}(s)} = K_c \alpha \left(\frac{Ts+1}{\alpha Ts+1} \right) = K_c \left(\frac{s+1/T}{s+1/\alpha T} \right) \tag{10.8}$$

where
$$T = R_1 C_1, \quad \alpha T = R_2 C_2, \quad K_c = \frac{R_4 C_1}{R_3 C_2}$$

Note that,
$$K_c \alpha = \frac{R_4 C_1}{R_3 C_2} \cdot \frac{R_2 C_2}{R_1 C_1} = \frac{R_2 R_4}{R_1 R_3}; \quad \alpha = \frac{R_2 C_2}{R_1 C_1}$$

The electronic op-amp. circuit has a dc gain of $K_c \alpha$. We see that this network is a lead network if $R_1 C_1 > R_2 C_2$, i.e. $0 < \alpha < 1$. It is a lag network if $R_1 C_1 < R_2 C_2$, i.e. $\alpha > 1$.

10.4 Characteristics of Lead Networks

A lead network as per Eq. (10.8) has a zero at $s = -1/T$ and a pole at $s = -1/\alpha T$. Since $0 < \alpha < 1$, we see that the zero is always located to the right of the pole in the complex s-plane. For a small value of α (which is in practice) the pole is located far to the left. The minimum value of α is limited by the physical construction of the lead network and is usually taken to be about 0.05. If the value of α is small, it is necessary to cascade an amplifier in order to compensate for the attenuation caused by the lead network.

Figure 10.6 show the polar plot of

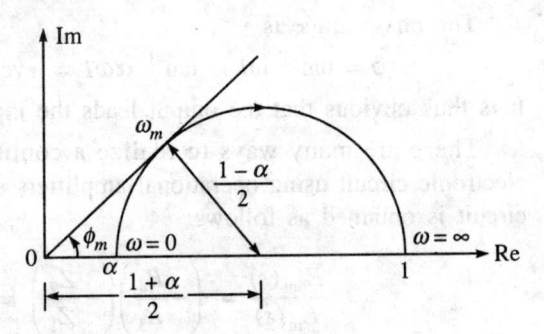

FIGURE 10.6 Polar plot of a lead network.

$$K_c \alpha \left(\frac{j\omega T + 1}{j\omega \alpha T + 1} \right); \quad \text{for } 0 < \alpha < 1 \tag{10.9}$$

For a given value of α, the angle between the positive real axis and the tangent line drawn from the origin to the semicircle gives the maximum phase lead angle ϕ_m. We will denote the frequency at the tangent point to be ω_m. From Figure 10.6, the phase angle at $\omega = \omega_m$ is

$$\sin \phi_m = \frac{(1-\alpha)/2}{(1+\alpha)/2} = \frac{1-\alpha}{1+\alpha} \tag{10.10}$$

Equation (10.10) relates the maximum phase lead angle ϕ_m with the value of α.

Figure 10.7 shows the Bode diagram of a lead network when $\alpha = 0.1$. The corner frequencies for the lead network are $\omega = 1/T$ and $\omega = 1/\alpha T = 10/T$. By examining this figure,

Compensation Techniques

we see that ω_m is the geometric mean of the two corner frequencies, i.e.

$$\log \omega_m = \frac{1}{2}[\log (1/T) + \log (1/\alpha T)]$$

or
$$\omega_m = \frac{1}{\sqrt{\alpha} T} \qquad (10.11)$$

As seen from Figure 10.7, the lead network is basically a high-pass filter. The high frequencies are passed but the low frequencies are attenuated. Therefore, an additional gain elsewhere is needed to increase the low frequency gain.

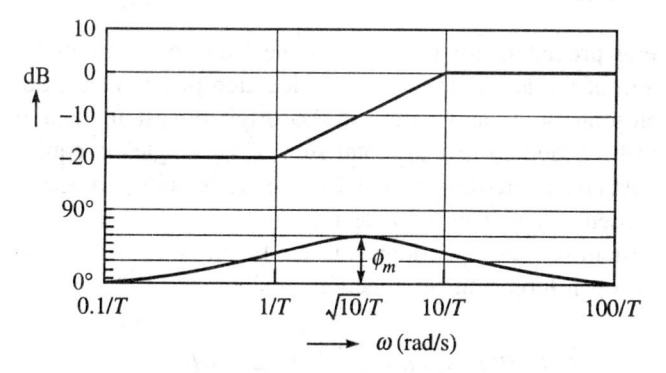

FIGURE 10.7 Bode diagram of a lead compensator $\alpha(j\omega T + 1)/(j\omega\alpha T + 1)$, where $\alpha = 0.1$.

EXAMPLE 10.1 Consider the system shown in Figure 10.8. It is desired to modify the closed loop so that an undamped natural frequency $\omega_n = 4$ rad/s is obtained without changing the value of the damping ratio. What type of compensation would you propose?

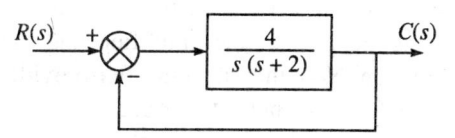

FIGURE 10.8 Example 10.1.

Solution: The characteristic equation of the unmodified closed-loop system is

$$1 + G(s)H(s) = 1 + \frac{1}{s(s+2)} = s^2 + 2s + 4 = 0$$

Then, $\omega_n = 2$ and $\zeta = 0.5$. We know $\zeta = \cos\theta = 0.5$, which gives $\theta = \pm 60°$. A damping ratio of 0.5 implies that the complex poles lie on the lines drawn through the origin making angles of $\pm 60°$ with the negative real axis.

The desired locations of the closed-loop poles are

$$s = -2 \pm j2\sqrt{3}$$

as the given $\omega_n = 4$ and $\zeta = 0.5$.

In the present system, the angle of $G(s)H(s)$ at the desired closed-loop poles is

$$\angle \frac{4}{s(s+2)}\bigg|_{s=-2+j2\sqrt{3}} = -210°$$

Hence we propose to have a lead compensator in the forward path to provide a leading angle of at least $+30°$.

EXAMPLE 10.2 In continuation of Example 10.1 where we concluded that a lead compensator is required in order to satisfy the given specifications, let us now design a suitable lead compensator for the same.

Solution: A general procedure for determining the lead compensator is as follows:
First, find the sum of the angles at the desired location of one of the dominant closed-loop poles with the open-loop poles and zeros of the original system, and then determine the necessary angle ϕ to be added so that the total sum of the angles is equal to $\pm 180°(2k+1)$. The lead network must contribute this angle. (If the angle is quite large, then two or more lead networks may be needed rather than a single one.)

If the original system has the open-loop transfer function $G(s)H(s)$, then the compensated system will have the open-loop transfer function

$$G_1(s)H(s) = \left(\alpha \frac{Ts+1}{\alpha Ts+1}\right) K_c G(s)H(s)$$

where the first term on the right-hand side corresponds to the lead network, the second term K_c is the gain of the amplifier, and the last term $G(s)H(s)$ is the original open-loop transfer function. Note that the amplifier provides the desired impedance matching as well as the desired gain K_c. Also note, there are many possible values for T that will yield the necessary angle contribution at the desired closed-loop poles. The next step is to determine the locations of the pole and zero of the lead network; in other words, the value of T. In choosing the value of T, we shall introduce a procedure to obtain the largest possible value for α so that the additional gain required of the amplifier is as small as possible. First draw a horizontal line passing through point P, the desired location for one of the dominant closed-loop poles at $-2 \pm j2\sqrt{3}$. This is shown as line PA in Figure 10.9. Draw also a line, connecting the point P and the origin. Bisect the angle between the lines AP and PO. Draw two lines PC and PD which make angles $\pm\phi/2$ with the bisector PB. The intersections of PC and PD with the negative real axis give the necessary location for the pole and zero of the lead network. The compensator thus designed will make point P, a point on the root locus of the compensated system as shown in Figure 10.10. The open-loop gain is determined by means of the magnitude condition.

$$\text{Transfer function} = \frac{400}{745}\left(\frac{0.345s+1}{0.185s+1}\right) = \frac{s+2.9}{s+5.4}.$$

We determine the pole and zero of the lead network as in Figure 10.10 to be

pole at $s = -5.4$ and   zero at $s = -2.9$

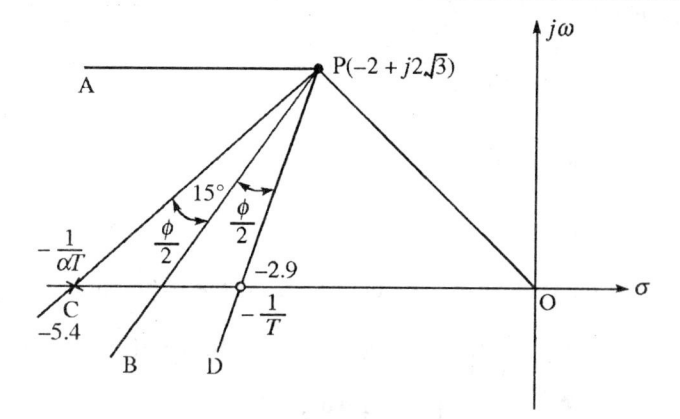

FIGURE 10.9 Example 10.2.

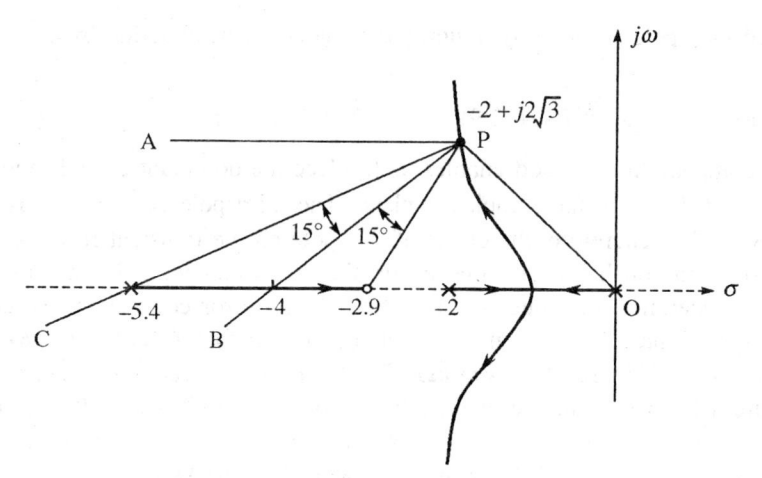

FIGURE 10.10 Example 10.2: root locus of the compensated system.

Hence the transfer function of the lead compensator $G_c(s)$ is

$$G_c(s) = \frac{(s+2.9)}{(s+5.4)} K_c$$

where the gain K_c of the amplifier is for gain adjustment for satisfying the magnitude condition of the root locus approach.

Thus the open-loop transfer function of the compensated system becomes

$$G_c(s)G(s) = \frac{s+2.9}{s+5.4} K_c \frac{4}{s(s+2)} = \frac{K(s+2.9)}{s(s+2)(s+5.4)} \quad \text{where} \quad K = 4K_c$$

The root-locus plot for the compensated system is shown in Figure 10.10. The gain K is evaluated from the magnitude condition as follows:

$$\left| \frac{K(s+2.9)}{s(s+2)(s+5.4)} \right|_{s=-2+j2\sqrt{3}} = 1 \quad \text{or} \quad K = 18.7$$

It therefore follows that

$$G_c(s)G(s) = \frac{18.7(s+2.9)}{s(s+2)(s+5.4)}$$

The gain constant K_c of the amplifier is

$$K_c = \frac{K}{4} = 4.68$$

The static velocity error coefficient K_v is obtained from the expression

$$K_v = \lim_{s \to 0} sG_c(s)G(s)H(s); \quad H(s) = 1$$

$$= \lim_{s \to 0} \frac{18.7(s+2.9)s}{s(s+2)(s+5.4)} = 5.02 \text{ s}^{-1}$$

The third closed-loop pole is found by dividing the characteristic equation by the known factors as follows:

$$s(s+2)(s+5.4) + 18.7(s+2.9) = (s+2-j2\sqrt{3})(s+2+j2\sqrt{3})(s+3.4)$$

The foregoing compensation method enables us to place the dominant closed-loop poles at the desired points $-2 \pm j2\sqrt{3}$ in the complex s-plane. The third pole at $s = -3.4$ is close to the added zero at $s = -2.9$. Therefore, the effect of this pole on the transient response is relatively small. Since no restriction has been imposed on the non-dominant pole and no specification has been given concerning the value of the static velocity error coefficient, we conclude that the present design of the lead compensating network is satisfactory either by RC network of Figure 10.4 with $R_1 = 345$ kΩ, $R_2 = 400$ kΩ, $C = 1$ µF or by electronic operational amplifier circuit of Figure 10.5 with values either (i) first choice: $R_1 = 345$ kΩ, $R_2 = 400$ kΩ, $C_1 = 1$ µF, $C_2 = 0.47$ µF, $R_4 = 10$ kΩ and $R_3 = 4.7$ kΩ; or (ii) second choice: $R_1 = 34.5$ kΩ, $R_2 = 18.5$ kΩ, $C_1 = C_2 = 10$ µF, $R_4 = 46.8$ kΩ and $R_3 = 10$ kΩ.

EXAMPLE 10.3 Consider the unity-feedback system having open-loop transfer function $\frac{4K}{s(s+2)}$. It is desired to find a compensator for the system so that the static velocity error coefficient K_v is 20 s^{-1}, the phase margin is at least 50°, and the gain margin is at least 10 dB.

Solution: In the example, the phase and gain margins have been specified. We shall therefore employ Bode diagrams. Adjusting the gain K to meet the steady-state performance specification or providing the required static velocity error coefficient which is given as 20 s^{-1}, we obtain

$$K_v = \lim_{s \to 0} sG(s)H(s) = \lim_{s \to 0} \frac{4Ks}{s(s+2)} = 2K = 20$$

or

$$K = 10$$

With $K = 10$ the given system satisfies the steady-state requirement.

We next plot the Bode diagram of

$$G(j\omega)H(j\omega) = \frac{40}{j\omega(j\omega+2)} = \frac{20}{j\omega(0.5j\omega+1)}$$

Figure 10.11 shows the magnitude and phase-angle curves of Bode diagram of $G(j\omega)H(j\omega)$. From this plot, the phase and gain margins of the system are found to be 17° and $+\infty$ dB respectively. The specification calls for a phase margin of at least 50°. We thus find that the additional phase lead necessary to satisfy the relative stability requirement is 33°. In order to achieve a phase margin of 50° without decreasing the value of K, it is necessary to insert a suitable lead compensator into the system.

Noting that the addition of a lead compensator modifies the magnitude curve in the Bode diagram, we realize that the gain-crossover frequency will be shifted to the right. We must offset the increased phase lag of $G(j\omega)H(j\omega)$ due to this increase in the gain-crossover frequency. Considering the shift of the gain crossover frequency, we may assume that ϕ_m, the maximum phase lead required, as approximately 38° with margin of tolerance as 5° added to the relative stability requirement of 33°.

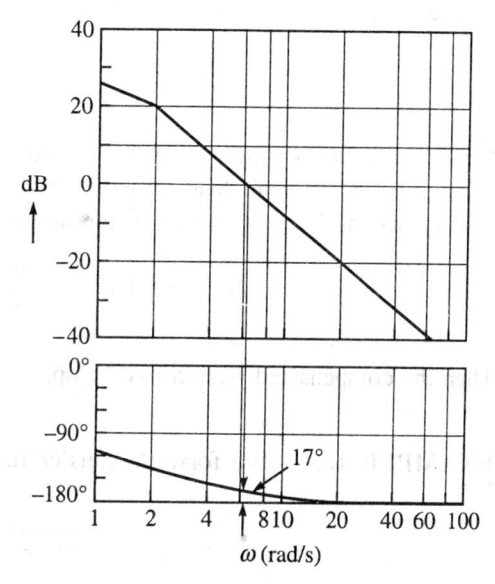

FIGURE 10.11 Example 10.3: Bode diagram for $G(j\omega)H(j\omega)$.

Since,
$$\sin \phi_m = \frac{1-\alpha}{1+\alpha} \tag{10.12}$$

$\phi_m = 38°$ corresponds to $\alpha = 0.24$.

Once the attenuation factor α has been determined on the basis of the required phase lead angle, the next step is to determine the corner frequencies $\omega = 1/T$ and $\omega = 1/(\alpha T)$ of the lead network. To do so, we first note that the maximum phase lead angle ϕ_m occurs at the geometric mean of the two corner frequencies which makes $\omega = 1/(T\sqrt{\alpha})$. The amount of the modification in the magnitude curve at $\omega = 1/(T\sqrt{\alpha})$ due to the inclusion of the term $(Ts + 1)/(\alpha Ts + 1)$ is

$$\left|\frac{1+j\omega T}{1+j\alpha\omega T}\right|_{\omega=1/\sqrt{\alpha}T} = \frac{1}{\sqrt{\alpha}}$$

where
$$\frac{1}{\sqrt{\alpha}} = \frac{1}{\sqrt{0.24}} = \frac{1}{0.49} = 6.2 \text{ dB}$$

Note that $|G(j\omega)| = 6.2$ dB corresponds to $\omega = 9$ rad/s. We shall select this frequency to be the new gain crossover frequency ω_c. Noting that this frequency corresponds to $1/(T\sqrt{\alpha})$, or $\omega_c = 1/(T\sqrt{\alpha})$, we obtain

$$\frac{1}{T} = \sqrt{\alpha}\,\omega_c = 4.41 \quad \text{and} \quad \frac{1}{\alpha T} = \frac{\omega_c}{\sqrt{\alpha}} = 18.4$$

The lead network thus determined is

$$\frac{s+4.41}{s+18.4} = \frac{0.24(0.227s+1)}{0.054s+1}$$

We increase the amplifier gain by a factor of $1/0.24 = 4.17$ in order to compensate for the attenuation due to the lead network. Then the transfer function of the compensator which consists of the lead network and the amplifier becomes

$$G_c(s) = (4.17)K\,\frac{s+4.41}{s+18.4} = 10\,\frac{0.227s+1}{0.054s+1}$$

Then the compensated system has the open-loop transfer function as $41.7\left(\dfrac{s+4.41}{s+18.4}\right)\left(\dfrac{4}{s(s+2)}\right)$.

EXAMPLE 10.4 The forward transfer function of a unity-feedback system is given by

$$\frac{28}{s(s+3)(s+6)}$$

It is desired that the real part of the dominant poles of the closed-loop system be not less than 4, keeping the damping ratio ζ unchanged. Also the static velocity error coefficient K_v must be at least 10. Design a suitable compensator as needed by this system.

Solution: Given the open-loop transfer function as

$$G(s)H(s) = \frac{28}{s(s+3)(s+6)}$$

the closed-loop poles are obtained from the characteristic equation $1 + GH(s) = 0$, as -7, $-1 \pm j\sqrt{3}$.

The damping coefficient of the dominant closed-loop poles $-1 \pm j\sqrt{3}$ is $\zeta = 0.5$. Hence the desired complex conjugate closed-loop poles are $-4 \pm j4\sqrt{3}$.

In the present system, the angle $G(s)H(s)$ at the desired closed-loop pole is

$$\angle \frac{28}{s(s+3)(s+6)}\bigg|_{s=-4+j4\sqrt{3}} = -292.11°$$

Hence, we need a lead compensator in the forward path to provide a leading angle of at least $(292.11° - 180°) = 112.11°$.

In order to satisfy the criterion for velocity error coefficient $K_v \geq 10$, we need an amplifier in cascade of gain $K \geq 6.43$ as

$$K_v = 10 = \lim_{s \to 0} s[KG(s)H(s)] = \frac{28}{18}K, \text{ i.e. } K = 6.43$$

EXAMPLE 10.5 Let us consider a single-loop compensated feedback control system as shown in Figure 10.12 where $G(s) = K/s^2$ and $H(s) = 1$.

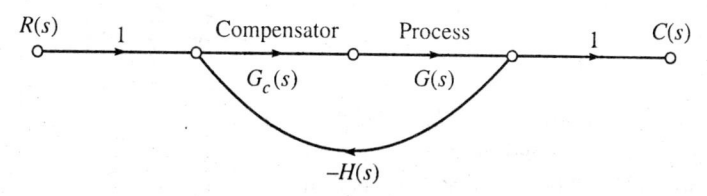

FIGURE 10.12 Example 10.5.

A lead compensating network has to be designed to meet the following specifications:

(i) Settling time, $t_s \leq 4$ s
(ii) Percent overshoot for a step input < 20 %
(iii) Acceleration error coefficient, $K_a \geq 2$

Solution: From Figure 10.13, the percent overshoot versus damping ratio curve of the second-order underdamped system, we get for percent overshoot $\leq 20\%$, the damping ratio as $\zeta \geq 0.45$. The settling time (2% criterion) requirement is, $t_s = 4 = 4/\zeta\omega_n$. Hence

$$\omega_n = 1/\zeta = 1/0.45 = 2.22.$$

Therefore, the desired dominant complex conjugate roots for $\zeta = 0.45$ and $\omega_n = 2.22$ are $-1 \pm j2$.

We have to design a lead compensator

$$G_c(s) = \frac{s+z}{s+p} \quad \text{where } |z| < |p|$$

such that the points $-1 \pm j2$ lie on the root loci of the compensated system whose open-loop transfer function is

$$G_c(s)G(s)H(s) = \frac{K(s+z)}{s^2(s+p)}$$

Place the zero of the compensator directly below the desired location at $s = -z = -1$ as shown in Figure 10.14. We have to place the

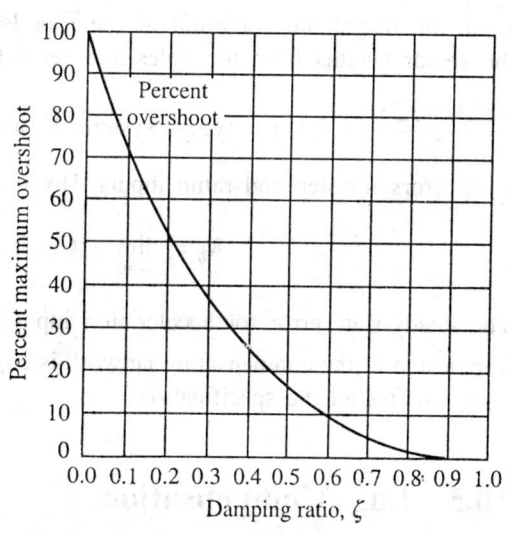

FIGURE 10.13 Example 10.5: percent overshoot vs. damping ratio ζ for a second-order underdamped system.

undetermined pole location of the compensator so that the angle condition of the root loci is satisfied for the compensated system. Hence the angle condition for

$$\left| \frac{K(s-1)}{s^2(s+p)} \right|_{s=-1+j2} = -180°$$

i.e. $90° - 2(116.56°) - \theta_p = -180°$

or $\theta_p = 36.88°$

Then a line is drawn at an angle $\theta_p = 36.88°$ intersecting the desired root location at the real axis as shown in Figure 10.14. The point of intersection with the real axis is then $s = -p = -3.6$. Therefore, the compensator is

$$G_c(s) = \frac{s+1}{s+3.6}$$

The open-loop transfer function of the compensated system is

$$G(s)H(s)G_c(s) = \frac{K(s+1)}{s^2(s+3.6)}$$

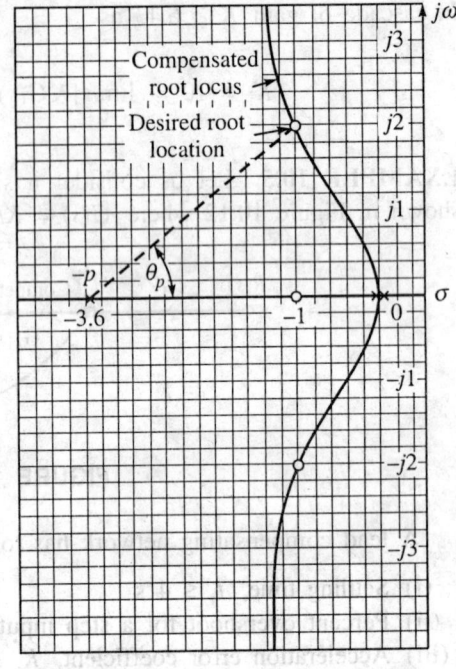

FIGURE 10.14 Example 10.5: phase-lead compensation.

From the magnitude condition of the root loci, we get the value of gain K by measuring the vector lengths from the poles and zeros to the desired root location at $-1 + j2$ as $K = \frac{(2.23)^2 (3.25)}{2} = 8.1$. For type-2 system, K_p and K_v are infinity, hence it results in zero steady-state errors for step and ramp inputs. However, the acceleration constant K_a is

$$K_a = \lim_{s \to 0} s^2 G(s)H(s)G_c(s) = \frac{8.1}{3.6} = 2.25 > 2$$

The steady-state error for acceleration input is finite. The design of the compensating network is complete as it satisfies all the specifications.

10.5 Lag Compensation

Figure 10.15 shows an electrical lag network. The name "lag network" comes from the fact that when the input voltage v_{in} is sinusoidal, the output voltage

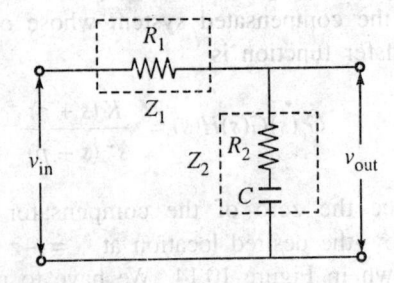

FIGURE 10.15 Electrical lag network.

v_{out} is sinusoidal but lags the input by an angle which is a function of the frequency of the input sinusoid. The complex impedances Z_1 and Z_2 are

$$Z_1 = R_1, \quad Z_2 = R_2 + \frac{1}{sC}$$

The transfer function between the output voltage $V_{out}(s)$ and the input voltage $V_{in}(s)$ is given by

$$\frac{V_{out}(s)}{V_{in}(s)} = \frac{Z_2}{Z_1 + Z_2} = \frac{R_2 Cs}{(R_1 + R_2)Cs + 1}$$

Let us define

$$R_2 C = T; \quad \frac{R_1 + R_2}{R_2} = \beta > 1$$

Then the transfer function becomes

$$\frac{V_{out}(s)}{V_{in}(s)} = \frac{Ts + 1}{\beta Ts + 1} = \frac{1}{\beta}\left(\frac{s + \frac{1}{T}}{s + \frac{1}{\beta T}}\right) \qquad (10.13)$$

10.5.1 Characteristics of lag networks

An RC-lag network of Figure 10.15 has the following transfer function

$$\frac{Ts + 1}{\beta Ts + 1} = \frac{1}{\beta}\left(\frac{s + \frac{1}{T}}{s + \frac{1}{\beta T}}\right); \quad (\beta > 1)$$

In the complex plane, a lag network has a pole at $s = -1/\beta T$ and a zero at $s = -1/T$, i.e., the pole is located to the right of the zero.

Figure 10.16 shows the polar plot of a lag network. Figure 10.17 shows the Bode diagram of a lag network when $\beta = 10$. The corner frequencies of the lag network are $\omega = 1/T$ and $\omega = 1/\beta T$. The lag network is essentially a low-pass filter.

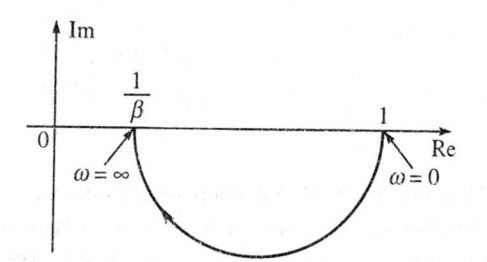

FIGURE 10.16 Polar plot of a lag network.

One can use the electronic op-amp lag circuit of Figure 10.5 just to make it different. In the RC circuit, β has been used for the lag circuit whereas the electronic circuit, with proper choice of values, can be used as lag circuit where $\alpha > 1$.

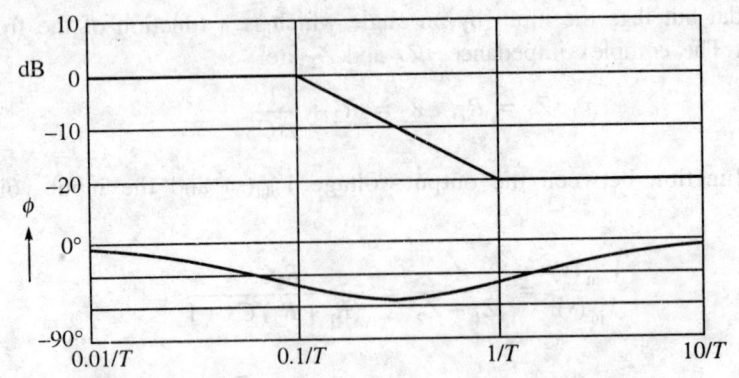

FIGURE 10.17 Bode plot of a lag network with $\beta = 10$.

10.5.2 Lag compensation techniques based on the root-locus approach

Consider the problem of finding a suitable compensation network for the case where the system exhibits satisfactory transient response characteristics but unsatisfactory steady-state characteristics. Compensation in this case essentially consists of increasing the open-loop gain without appreciably changing the transient-response characteristics. This means that the root locus in the neighbourhood of the dominant closed-loop poles should not be changed appreciably, but the open-loop gain should be increased as much as needed.

To avoid an appreciable change in the root loci, the angle contribution of the lag network should be limited to a small amount, say 5°. To assure this, we place the pole and zero of the lag network relatively close together and near the origin of the s-plane. Then the closed-loop poles of the compensated system will be shifted only slightly from their original locations. Hence the transient response characteristics will essentially be unchanged.

Note that if we place the pole and zero of the lag network very close to each other, then $(s_1 + 1/T)$ and $(s_1 + 1/\beta T)$ are almost equal, where s_1 is the closed-loop pole. Thus

$$\left| \frac{1}{\beta} \left(\frac{s + \frac{1}{T}}{s + \frac{1}{\beta T}} \right) \right| \approx \left| \frac{1}{\beta} \left(\frac{s_1 + \frac{1}{T}}{s_1 + \frac{1}{\beta T}} \right) \right| \approx \frac{1}{\beta} \tag{10.14}$$

This implies that the open-loop gain can be increased approximately by a factor of β without altering the transient-response characteristics. If the pole and zero are placed very close to the origin, the value of β can be made large. Usually, $1 < \beta < 15$, and $\beta = 10$ is a good choice.

An increase in gain means an increase in the static error coefficients. If the open-loop transfer function of the uncompensated system is $G(s)$, then the static velocity error coefficient K_v is

$$K_v = \lim_{s \to 0} sG(s)H(s)$$

Compensation Techniques

If the compensator is chosen as

$$G_c(s) = K_c \frac{Ts+1}{\beta Ts+1} = \frac{K_c}{\beta}\left(\frac{s+\frac{1}{T}}{s+\frac{1}{\beta T}}\right) \qquad (10.15)$$

Then for the compensated system with the open-loop transfer function $G_c(s)G(s)H(s)$, the static velocity error coefficient $\hat{K}_v$ becomes

$$\hat{K}_v = \lim_{s\to 0} sG_c(s)G(s)H(s) = \lim_{s\to 0} G_c(s) K_v; \qquad \text{as } H(s) = 1$$

$$= K_c K_v$$

Thus if the compensator is given by Eq. (10.15), then the static velocity error coefficient is increased by a factor of K_c.

EXAMPLE 10.6 Consider the system shown in Figure 10.18. It is desired that the dominant pole of the closed-loop system should have the damping ratio of 0.5 and the magnitude of the real part of the pole be less than unity. Also, the velocity error coefficient should be at least 10. Design a suitable compensator.

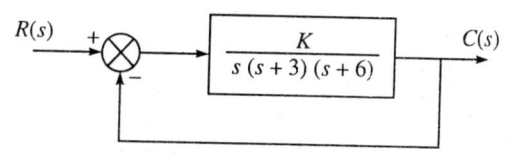

FIGURE 10.18 Example 10.6.

Solution: Let us draw the root loci for the open-loop transfer function

$$G(s)H(s) = \frac{K}{s(s+3)(s+6)}; 0 < K < \infty$$

The root loci is drawn as shown in Figure 10.19. Draw $\zeta = \cos\theta = 0.5$ or $\theta = 60°$ line. The intersection of root loci in quadrant II with $\theta = 60°$ line gives the value of $K = 28$. The characteristic equation of the closed-loop system at this value of $K = 28$ is

$$1 + GH(s) = 1 + \frac{28}{s(s+3)(s+6)}$$

$$= \frac{s(s+3)(s+6)+28}{s(s+3)(s+6)} = \frac{(s+7)(s^2+2s+4)}{s(s+3)(s+6)}$$

The roots of the closed-loop system are

$$(s+7)(s^2+2s+4) = 0$$

$$(s+7)(s+1+j\sqrt{3})(s+1-j\sqrt{3}) = 0$$

The real part of the complex poles is –1, which just meets our specification. Further, for

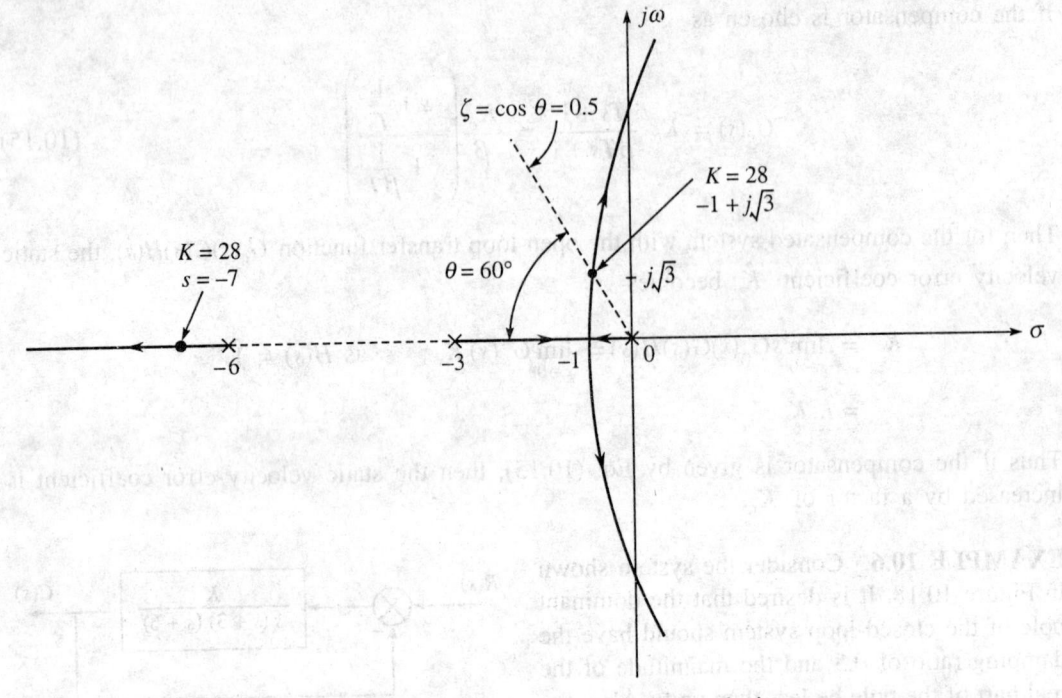

FIGURE 10.19 Example 10.6: root loci.

$K = 28$, the static velocity error coefficient K_v becomes

$$K_v = \lim_{s \to 0} sG(s)H(s) = \lim_{s \to 0} \frac{28s}{s(s+3)(s+6)} = \frac{28}{18} = 1.55$$

which is less than the desired value of $K_v = 10$.

In order to increase the value of K_v to 10, keeping in mind that without appreciably changing the location of the dominant closed-loop poles (i.e. keeping the constraints intact), let us insert a lag compensator which consists of a lag network and an amplifier, in cascade with the given feed-forward transfer function.

Let us place the pole and zero of the lag network at $s = -0.01$ and $s = -0.1$ respectively. Then the structure of the lag network is

$$\frac{sT+1}{s\beta T+1}; \quad \beta > 1$$

from which we get as

$$\frac{1}{\beta}\frac{\left(s + \frac{1}{T}\right)}{\left(s + \frac{1}{\beta T}\right)} = \frac{1}{10}\left(\frac{s+0.1}{s+0.01}\right)$$

For the attenuation due to the lag network, we cascade an amplifier of gain K_c. The feed-forward transfer function of the compensated system would then be

$$G_1(s) = \frac{1}{10}\left(\frac{s+0.1}{s+0.01}\right)(K_c)\frac{28}{s(s+3)(s+6)} = \frac{K(s+0.1)}{s(s+0.01)(s+3)(s+6)}; \text{ where } K = 2.8K_c$$

It may be noted that the distances of the compensator pole and zero from the origin of the s-plane are chosen to be small compared with the distance of the dominant poles from the origin so that the compensator will not affect the locus near the dominant poles.

The lag compensator is only to modify the static error coefficient, in this case K_v. The block diagram of the compensated system is shown in Figure 10.20.

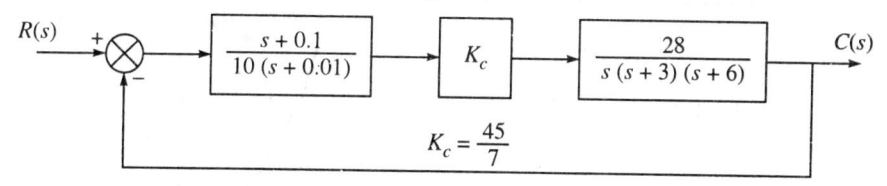

FIGURE 10.20 Example 10.6: block diagram of the compensated system.

The desired static velocity error coefficient K_v of the compensated unity-feedback system is 10, then by definition the modified static velocity error coefficient is given by

$$\lim_{s \to 0} sG_1(s)H(s) = 10 \quad \text{or} \quad \lim_{s \to 0}\left[\frac{s(2.8K_c)(s+0.1)}{s(s+0.01)(s+3)(s+6)}\right] = 10$$

which leads to

$$K_c = \frac{45}{7}$$

A plot of the root loci for the system with the lag compensator is shown in Figure 10.21. It may be seen that the root-locus plot is almost unchanged near the dominant poles. Although we are adding a pole and a zero with the transfer function of the forward path, their contribution to the argument near the dominant poles is negligible, so that the angle condition is still satisfied at that point.

EXAMPLE 10.7 For a unity-feedback system having forward transfer function

$$G(s) = \frac{100}{s(s+1)(0.1s+1)}$$

design a suitable compensator so that the phase margin is at least 45°.

Solution: The Bode plot (with error correction) of the uncompensated system

$$G(s)H(s) = \frac{100}{s(s+1)(0.1s+1)}$$

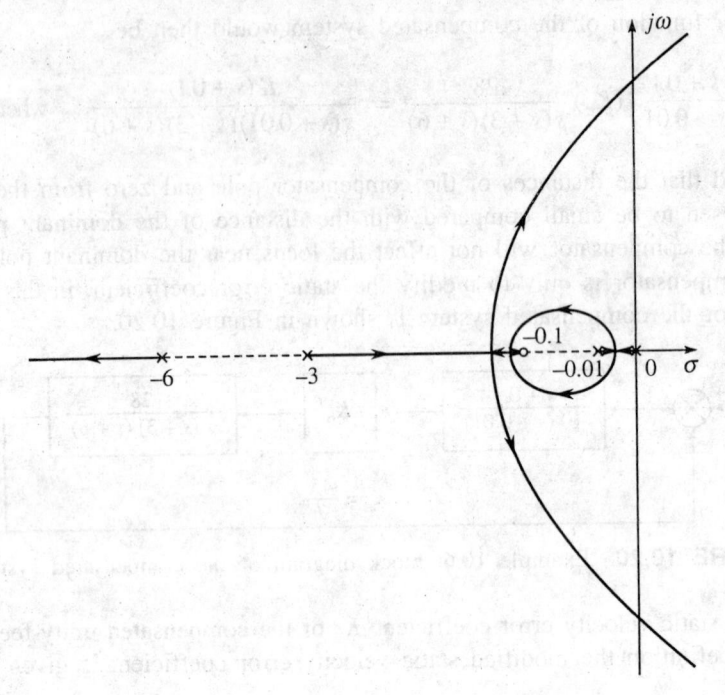

FIGURE 10.21 Example 10.6: root loci of the compensated system of Figure 10.20.

is drawn in Figure 10.22. The gain-crossover frequency ω_g is 8.5 rad/s. The phase-crossover frequency ω_p is 3.2 rad/s. The gain margin (GM) is −20 dB and the phase margin (PM) is −35°. The system is unstable. If the gain-crossover frequency were moved to a new value $\omega_{gd} = 0.7$ by providing the required gain reduction which comes out to be 43 dB from the Bode plot of Figure 10.22, the phase margin would have been 50°. A phase lag network can provide the required attenuation (gain reduction) of 43 dB which comes out to be gain reduction by a factor 140 [i.e. 20 log 140 = 42.92 dB ≈ 43 dB] without affecting the phase curve of the uncompensated system. The usual lag compensating network structure is

$$G_c(s) = \frac{1}{\alpha}\left(\frac{s+(1/T)}{s+(1/\alpha T)}\right)$$

that is, a pole at $-1/(\alpha T)$ and a zero at $-1/T$ and $1/\alpha$ is the attenuation. The proposed lag network should have $\alpha = 140$. Then the lag compensator network with pole at $-1/140T$ and zero at $-1/T$ is separated by slightly more than two decades.

The compensator zero is introduced at $\omega_{gd}/10 = 0.07$ and the phase curve is corrected. From the Bode diagram, it can be seen that the net phase margin is about 50° which is in accordance with the demand.

The compensator is then

$$G_c(s) = \frac{1}{140}\left(\frac{s+0.07}{s+0.0005}\right) = \frac{\frac{s}{0.07}+1}{\frac{s}{0.0005}+1} = \frac{14.29s+1}{2000s+1}$$

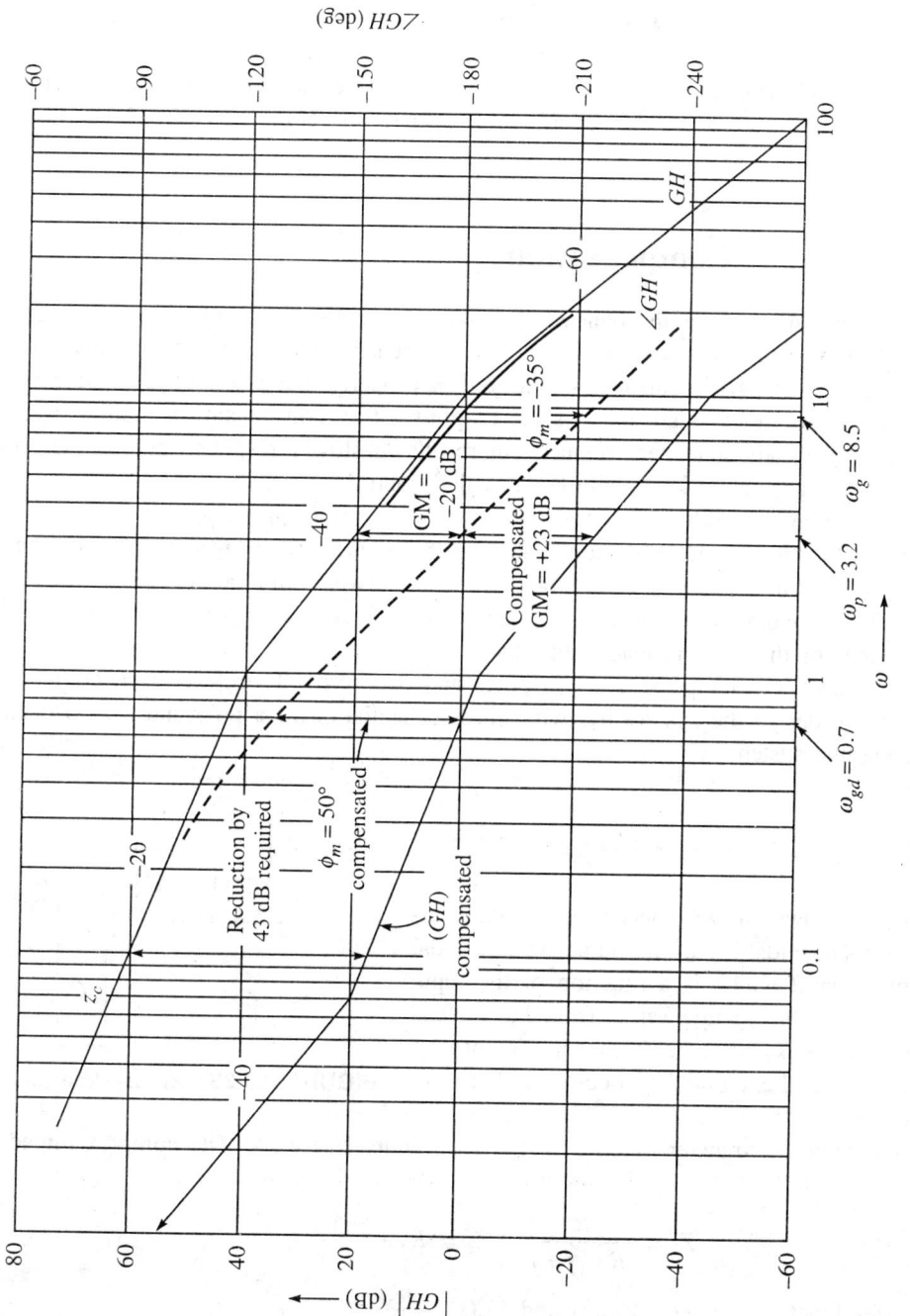

FIGURE 10.22 Example 10.7: Bode plots.

The compensated open-loop transfer function is therefore given by

$$G_c(s)G(s) = \frac{100(14.29s+1)}{s(s+1)(0.1s+1)(2000s+1)}$$

The magnitude and phase plots of the compensated system are shown in Figure 10.22. The phase margin is about 50° and as per the specification the gain margin is +23 dB. The compensated system is stable.

10.6 Lag–lead Compensation

Lead compensation increases the bandwidth, which improves the speed of response, and also reduces the amount of overshoot. However, improvement in steady-state performance is rather small. Basically, derivative control has this type of property that is effective in transient part and ineffective in steady-state part of the response. Lag compensation results in a large improvement in steady-state performance but results in slower response due to the reduced bandwidth. This lag network has similarity with I-control.

If improvements in both transient and steady-state response (namely, large increases in the gain and bandwidth) are desired, then both a lead network and a lag network may be used simultaneously. Rather then introducing both lead network and lag networks as separate elements, it is economical to use a single lag–lead network. The lag–lead network combines the advantages of the lag and lead networks.

The lag–lead network possesses two poles and two zeros. Therefore, such compensation increases the order of the system by two, unless cancellation of a pole and a zero occurs in the compensated system.

10.6.1 Lag–lead networks

Figure 10.23 shows an RC electrical lag–lead network. For a sinusoidal input, the output is sinusoidal with a phase shift which is a function of the input frequency. This phase angle varies from lag to lead as the frequency is increased from zero to infinity. Note that phase lead and lag occur in different frequency bands.

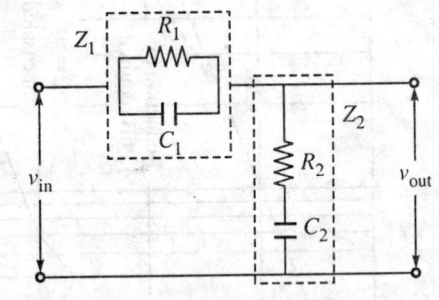

FIGURE 10.23 RC lag–lead network.

Let us obtain the transfer function of the RC lag–lead network. The complex impedances Z_1 and Z_2 are

$$Z_1 = \frac{R_1}{R_1C_1s+1}, \quad Z_2 = R_2 + \frac{1}{C_2s}$$

The transfer function between $V_{out}(s)$ and $V_{in}(s)$

$$\frac{V_{out}(s)}{V_{in}(s)} = \frac{Z_2}{Z_1+Z_2} = \frac{(R_1C_1s+1)(R_2C_2s+1)}{(R_1C_1s+1)(R_2C_2s+1)+R_1C_2s}$$

The denominator of this transfer function can be factored into two real terms. Let us define

$$R_1C_1 = T_1 \quad \text{and} \quad R_2C_2 = T_2$$

$$R_1C_1 + R_2C_2 + R_1C_2 = \frac{T_1}{\beta} + \beta T_2 \quad (\beta > 1)$$

Then $V_{out}(s)/V_{in}(s)$ can be simplified to

$$\frac{V_{out}(s)}{V_{in}(s)} = \frac{(T_1s+1)(T_2s+1)}{\left(\frac{T_1}{\beta}s+1\right)(\beta T_2 s+1)} = \frac{\left(s+\frac{1}{T_1}\right)\left(s+\frac{1}{T_2}\right)}{\left(s+\frac{\beta}{T_1}\right)\left(s+\frac{1}{\beta T_2}\right)} \quad (10.16)$$

The electronic circuit using operational amplifiers is shown in Figure 10.24. The transfer function for this compensator is obtained as follows:

$$Z_1 = R_3 \| (R_1 + C_1) = \frac{(R_1C_1s+1)R_3}{(R_1+R_3)C_1s+1}$$

$$Z_2 = R_4 \| (R_2 + C_2) = \frac{(R_2C_2s+1)R_4}{(R_2+R_4)C_2s+1}$$

Again, $\quad \dfrac{V(s)}{V_{in}(s)} = -\dfrac{Z_2}{Z_1} \quad$ and $\quad \dfrac{V_{out}(s)}{V(s)} = -\dfrac{R_6}{R_5}$

Lag–lead network   Sign inverter

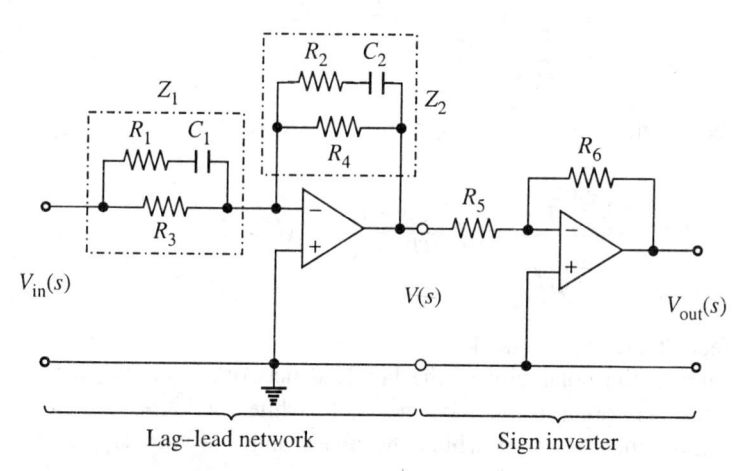

FIGURE 10.24 Lag–lead compensator.

Hence, the transfer function of the circuit is

$$\frac{V_{out}(s)}{V_{in}(s)} = K_c \frac{\beta}{\gamma} \left(\frac{T_1s+1}{\frac{T_1}{\gamma}s+1}\right)\left(\frac{T_2s+1}{\beta T_2 s+1}\right)$$

$$= K_c \frac{(s+1/T_1)(s+1/T_2)}{(s+\gamma/T_1)(s+1/\beta T_2)} \qquad (10.17)$$

where
$$\beta = \frac{R_2 + R_4}{R_2} > 1, \quad \gamma = \frac{R_1 + R_3}{R_1} > 1$$

and
$$K_c = \left(\frac{R_2 R_4 R_6}{R_1 R_3 R_5}\right)\left(\frac{R_1 + R_3}{R_2 + R_4}\right)$$

Note that usually $\gamma = \beta$ is the normal choice. The transfer functions for the *RC* circuit and the electronic circuit with op-amps have the same structure as is evident from Eqs. (10.16) and (10.17).

10.6.2 Characteristics of lag–lead networks

Consider the simplified version of transfer function of the lag–lead network with $K_c = 1$, i.e.

$$\frac{\left(s+\dfrac{1}{T_1}\right)\left(s+\dfrac{1}{T_2}\right)}{\left(s+\dfrac{\beta}{T_1}\right)\left(s+\dfrac{1}{\beta T_2}\right)}$$

The first term
$$\frac{s+\dfrac{1}{T_1}}{s+\dfrac{\beta}{T_1}} = \frac{1}{\beta}\frac{T_1 s + 1}{\dfrac{T_1 s + 1}{\beta}} \qquad (\beta > 1)$$

produces the effect of the lead network and the second term

$$\frac{s+\dfrac{1}{T_2}}{s+\dfrac{1}{\beta T_2}} = \beta\frac{T_2 s + 1}{\beta T_2 s + 1} \qquad (\beta > 1)$$

produces the effect of the lag network.

If the Bode plot or the polar plot of the lag–lead network is drawn, it would be seen that for $0 < \omega < \omega_1$ the network acts as a lag network, while for $\omega_1 < \omega < \infty$ it acts as a lead network, where the frequency ω_1 at which the phase angle is zero is given by

$$\omega_1 = \frac{1}{\sqrt{T_1 T_2}}$$

The magnitude curve will have the value 0-dB at the low-frequency and high-frequency regions. This is because the transfer function of the lag–lead network as a whole does not contain β as a factor.

Compensation Techniques

EXAMPLE 10.8 Consider a unity-feedback control system whose forward transfer function is

$$G(s) = \frac{10K}{s(s+2)(s+8)}$$

Design a compensator so that $K_v = 80$ s^{-1} and the dominant closed-loop poles are located at $-2 \pm j2.3$.

Solution: The characteristic equation is: $s^3 + 10s^2 + 16s + 10 = 0$

$$K_v = \lim_{s \to 0} sG(s)H(s) = \frac{10K}{(2)(8)} = 80$$

Hence, $K = 128$

If we use the lag–lead compensator given be Eq. (10.16), then the open-loop transfer function of the compensated system becomes

$$G_c(s)G(s) = \frac{\left(s+\dfrac{1}{T_1}\right)\left(s+\dfrac{1}{T_2}\right)}{\left(s+\dfrac{\beta}{T_1}\right)\left(s+\dfrac{1}{\beta T_2}\right)} \cdot \frac{1280}{s(s+2)(s+8)}$$

Now if T_2 is chosen large enough so that

$$\frac{s+\dfrac{1}{T_2}}{s+\dfrac{1}{\beta T_2}} \approx 1$$

then for closed-loop poles to lie at $s = -2 \pm j2.3$, the magnitude condition becomes unity. The phase lead portion of the lag–lead network must contribute 136.102° to make the root locus of the compensated system pass through the $-2 \pm j2.3$ points, i.e. the

$$\text{phase condition is } \angle \frac{s+\dfrac{1}{T_1}}{s+\dfrac{\beta}{T_1}}\bigg|_{s=-2+j2.3} = 136.102°$$

This correction cannot be done with a single lead network, so we must connect two identical lead networks and hence conditions for each of them become

$$\left|\frac{s+\dfrac{1}{T_1'}}{s+\dfrac{\beta'}{T_1'}}\right|_{s=-2+j2.3} = \frac{1}{13.33} = 0.274 \quad \text{and} \quad \angle \frac{s+\dfrac{1}{T_1'}}{s+\dfrac{\beta'}{T_1'}}\bigg|_{s=-2+j2.3} = 68.051°$$

So the transfer function of the lead network after detailed calculations becomes

$$\frac{s+2.39}{s+14.411} \cdot \frac{s+2.39}{s+14.411}$$

Now for each of the two lag networks, we require that

$$\left. \frac{s+\dfrac{1}{T_2}}{s+\dfrac{1}{\beta T_2}} \right|_{s=-2+j2.3} = 1$$

and the compensator transfer function becomes

$$G_c(s) = \frac{s+0.05}{s+0.0083} \cdot \frac{s+2.39}{s+14.411}$$

Hence the compensated transfer function becomes

$$G_c(s)G(s) = \frac{1280}{s(s+1)(s+2)} \cdot \frac{s+0.05}{s+0.0083} \cdot \frac{s+2.39}{s+14.411}$$

10.7 Compensation of Operational Amplifier

A compensation scheme for operational amplifiers has been implemented using the root-locus and Bode plot techniques. A practical case has been taken for the 741 op-amp.

The high frequency model of an op-amp with a single corner frequency is shown in Figure 10.25. The open-loop voltage gain is then obtained as follows:

$$V_o = \frac{-jX_C}{R_o - jX_C} A_{OL} V_d \qquad (10.18)$$

or $\quad A = \dfrac{V_o}{V_d} = \dfrac{A_{OL}}{1+j2\pi f R_o C}$

or $\quad A = \dfrac{A_{OL}}{1+j(f/f_1)} \qquad (10.19)$

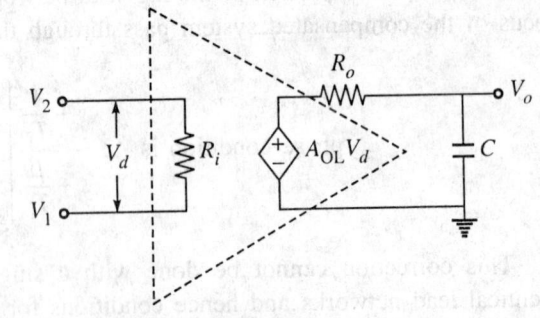

where $\quad f_1 = \dfrac{1}{2\pi R_o C} \qquad (10.20)$

FIGURE 10.25 High frequency model of an op-amp with a single corner frequency.

is the corner frequency of the op-amp.

The voltage transfer function in s-domain can be written as

$$G(s) = A(s) \frac{A_{OL}}{1+j(f/f_1)} = \frac{A_{OL}}{1+j(\omega/\omega_1)}$$

$$= \frac{A_{OL} \cdot \omega_1}{j\omega + \omega_1} = \frac{A_{OL} \cdot \omega_1}{s + \omega_1}$$

where

A_{OL} is the open-loop voltage gain

V_o is the output voltage $= A_{OL}V_d = A_{OL}(V_1 - V_2)$

R_i is the input impedance

R_o is the output impedance

$V_d = (V_1 - V_2)$ is the differential input voltage

V_1 is the non-inverting input signal

V_2 is the inverting input signal

The magnitude and the phase angle of the open-loop voltage transfer function are functions of frequency and can be written as

$$|A| = \frac{A_{OL}}{\sqrt{(1+(f/f_1)^2)}} \qquad (10.21)$$

$$\phi = -\tan^{-1}(f/f_1) \qquad (10.22)$$

The magnitude and phase characteristics from Eq. (10.21) and Eq. (10.22) are shown in Figures 10.26(a) and (b) respectively. It can bee seen that:

(i) For frequency $f \ll f_1$, the magnitude of the gain is $20 \log A_{OL}$ in dB.
(ii) At frequency $f = f_1$, the gain is 3-dB down from the dc value of A_{OL} in dB. This frequency f_1 is called the corner frequency.
(iii) For $f \gg f_1$, the gain rolls off at the rate of -20 dB/decade or -6 dB/octave.

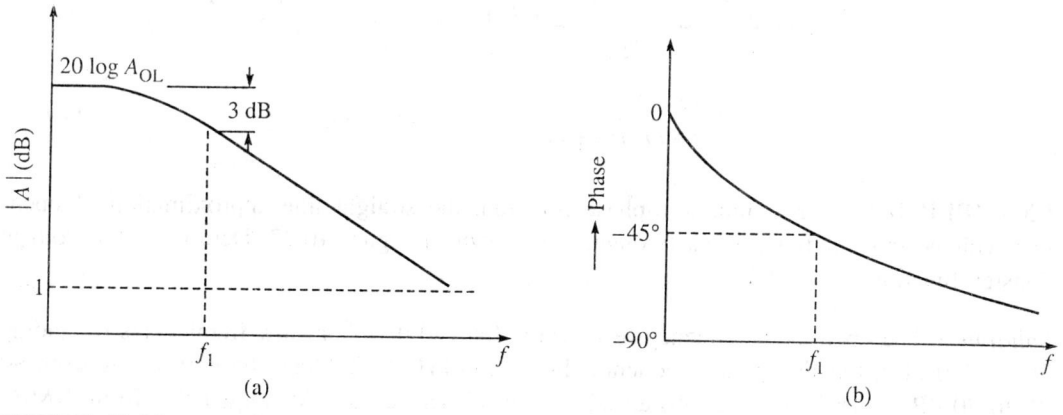

FIGURE 10.26 (a) Open-loop magnitude characteristic of Figure 10.25 and (b) open-loop phase characteristic of Figure 10.25.

It can further be seen from the phase characteristics that the phase angle is zero at frequency $f = 0$. At corner frequency f_1 the phase angle is $-45°$ (lagging) and at infinite frequency the phase angle is $-90°$. This shows that a maximum of $90°$ phase change can occur in an op-amp with a single capacitor. It may be mentioned here that zero frequency does not occur in log scale. For all practical purposes, the zero frequency is taken as one decade below the corner frequency and the infinite frequency is taken one decade above the corner frequency.

10.7.1 Transfer function of a practical op-amp

Ideally, an op-amp should have an infinite bandwidth. This means that if its open-loop gain is 90 dB with dc signal, its gain should remain the same 90 dB through audio and on to high radio frequencies. The practical op-amp gain, however, dcreases (i.e. rolls off) at higher frequencies.

What causes the gain of the op-amp to roll off after a certain frequency is reached? Obviously, there must be a capacitive component in the equivalent circuit of the op-amp. This capacitance is due to the physical characteristics of the device (BJT or FET) used and the internal construction of op-amp. For an op-amp with only one break (corner) frequency, all the capacitor effects can be represented by a single capacitor C as shown in Figure 10.25. This figure represents the high frequency model of the op-amp with a single corner frequency. It may be observed that the high frequency model of Figure 10.25 is a modified version of the low frequency model with a capacitor C at the output. There is one pole due to RC and obviously one -20 dB/decade roll-off comes into effect.

A practical op-amp, however, has a number of stages and each stage produces a capacitive component. Thus due to a number of RC-pole pairs, there will be a number of different break frequencies. The transfer function of an op-amp with three break frequencies can be assumed as

$$A = \frac{A_{OL}}{\left(1+j\frac{f}{f_1}\right)\left(1+j\frac{f}{f_2}\right)\left(1+j\frac{f}{f_3}\right)}; \quad 0 < f_1 < f_2 < f_3 \quad (10.23)$$

or

$$A = \frac{A_{OL} \cdot \omega_1 \cdot \omega_2 \cdot \omega_3}{(s+\omega_1)(s+\omega_2)(s+\omega_3)}; \quad 0 < \omega_1 < \omega_2 < \omega_3 \quad (10.24)$$

EXAMPLE 10.9 For a minimum-phase function, the straight line approximation of open-loop voltage transfer function vs frequency is shown in Figure 10.27. Determine the voltage transfer function.

Solution: The open-loop frequency response is flat (90 dB) from low frequencies (including dc) to 200 kHz, the first break frequency. From 200 kHz to 2 MHz, the gain drops from 90 dB to 70 dB which is at a -20 dB/decade or -6 dB/octave rate. At frequencies from 2 MHz to 20 MHz, the roll-off rate is -40 dB/decade or -12 dB/octave. Accordingly, as the frequency increases, the cascading effect of RC pairs (poles) comes into effect and the roll-off rate increases successively by -20 dB/decade at each corner frequency. Each RC pole pair also introduces a lagging phase of maximum up to $-90°$. Hence the corner frequencies are

$$f_1 = 200 \text{ kHz}; \quad f_2 = 2 \text{ MHz}; \quad \text{and} \quad f_3 = 20 \text{ MHz}$$

Compensation Techniques

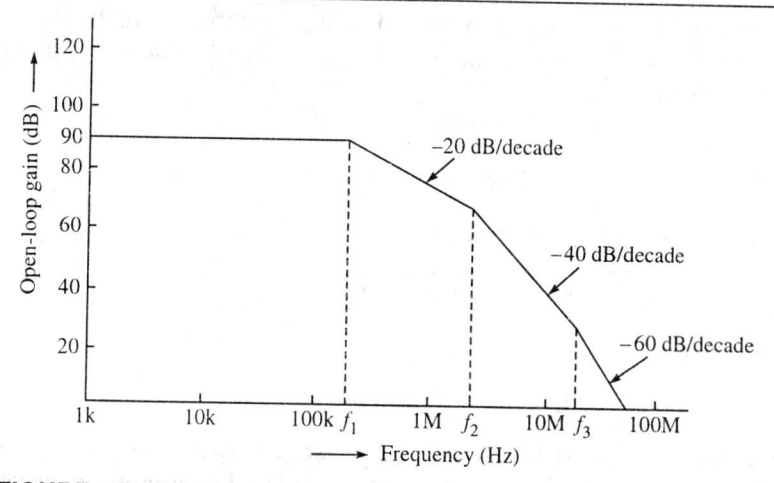

FIGURE 10.27 Approximation of open-loop gain vs frequency curve.

At each successive corner frequency the voltage gain falls by -20 dB/decade. The dc gain is 90 dB which is equal to

$$90 \text{ dB} = 20 \log A_{OL} \quad \text{or} \quad A_{OL} = 31623$$

Therefore, the voltage transfer function

$$A(s) = \frac{A_{OL}\, \omega_1\, \omega_2\, \omega_3}{(s+\omega_1)(s+\omega_2)(s+\omega_3)} = \frac{(31623)(0.2 \times 10^6)(2 \times 10^6)(20 \times 10^6)}{(s+0.2 \times 10^6)(s+2 \times 10^6)(s+20 \times 10^6)}$$

EXAMPLE 10.10 For the circuit of Figure 10.28, determine the closed-loop transfer function, where the open-loop frequency response for minimum-phase function of the op-amp is shown in Figure 10.27.

Solution: From the given open-loop frequency response of Figure 10.27, the open-loop voltage transfer function is

$$A(s) = \frac{A_{OL}\, \omega_1\, \omega_2\, \omega_3}{(s+\omega_1)(s+\omega_2)(s+\omega_3)}; \quad 0 < \omega_1 < \omega_2 < \omega_3$$

where $A_{OL} = 31623$; $f_1 = 0.2$ MHz, $f_2 = 2$ MHz, $f_3 = 20$ MHz and $\omega_i = 2\pi f_i$; $i = 1, 2, 3$.

From the negative feedback concepts, we may write the closed-loop transfer function of the circuit of Figure 10.29 as

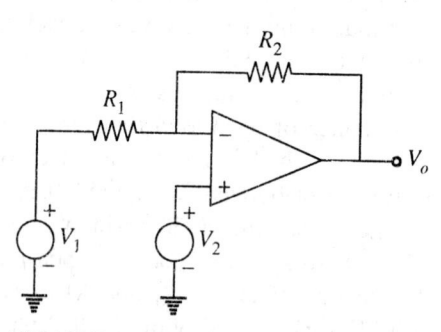

$$A_{CL} = \frac{A}{1+A\beta} \quad (10.25)$$

FIGURE 10.28 Resistive feedback in op-amp.

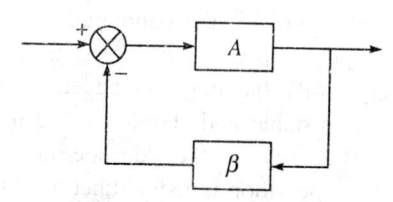

FIGURE 10.29 Feedback loop.

where A is the open-loop voltage gain and β is the feedback ratio. In Eq. (10.25), if the characteristic equation $(1 + A\beta) = 0$, the circuit will become just unstable, that is it will lead into sustained oscillations.

Rewriting the characteristic equation as, $1 + A\beta = 0$ leads to

$$\text{loop gain, } A\beta = -1 \tag{10.26}$$

Since $A\beta$ is a complex quantity, the magnitude condition become

$$|A\beta| = 1 \tag{10.27}$$

and the phase condition is

$$\angle A\beta = \pi \text{(or odd multiple of } \pi) \tag{10.28}$$

In the given circuit, the feedback network is a resistive network, so it does not provide any phase shift. The op-amp is used in the inverting mode and hence provides negative feedback. At high frequencies, due to each corner frequency, an additional phase shift of maximum $-90°$ can take place in open-loop gain A. So for two corner frequencies, a maximum of phase shift that can be associated with gain $A\beta$ is $-180°$ which makes the total phase shift $\angle A\beta$ equal to odd multiples of π. In this case, there is every possibility that the amplifier may begin to oscillate as both the magnitude and phase conditions laid down by Eqs. (10.27) and (10.28) are satisfied. It may be noted that oscillation is just the starting point of instability, or, to be more precise, it is just at the verge of instability. The instability means unbounded output; which can arise from Eq. (10.25), when

$$(1 + A\beta) < 1 \quad \text{or} \quad A\beta < 0 \text{ (i.e. negative)}$$

and then $A_{CL} > A$, i.e. the closed-loop gain increases and leads to instability. The phase contribution by the resistive feedback network is zero. At low frequencies, the additional phase contribution of A is zero, so $A\beta > 0$ and obviously $A_{CL} < A$ and the system is stable. But at high frequencies, the system A having three corner frequencies or three RC-pole pairs, there is a chance of open-loop transfer function $A\beta$ to contribute a maximum of $-270°$ phase shift and for which $A\beta$ may become negative and instability would occur at high frequencies. This is further elaborated in Figure 10.30.

Let us say that a closed-loop gain of 80 dB ($|(A_{OL})| = 10,000$) is desired. The projection of the 80-dB curve upon the open-loop frequency response curve intersects it at a -20 dB/decade rate of closure (point A) as shown in Figure 10.30. The bandwidth is approximately 600 kHz and a maximum of $-90°$ phase shift is added to the open-loop transfer function $A\beta$. The amplifier will remain stable.

Now, if the feedback resistors are so chosen that the op-amp has a closed-loop gain of 1,000 or 60 dB, the bandwidth is about 3.5 MHz. However, now the 60-dB projection on to the open-loop curve intersects at a -40 dB/decade rate of closure (point B). The maximum phase shift that may get added to is now $(-90° - 90°)$, that is, $-180°$. This circuit is likely to be unstable and should not be used without modification. Similarly, a closed-loop gain of 20 dB causes a -60 dB/decade rate of closure (point C). A maximum $-270°$ phase shift is added to the open-loop transfer function $A\beta$ to cause unstable operation. Thus, we may conclude that

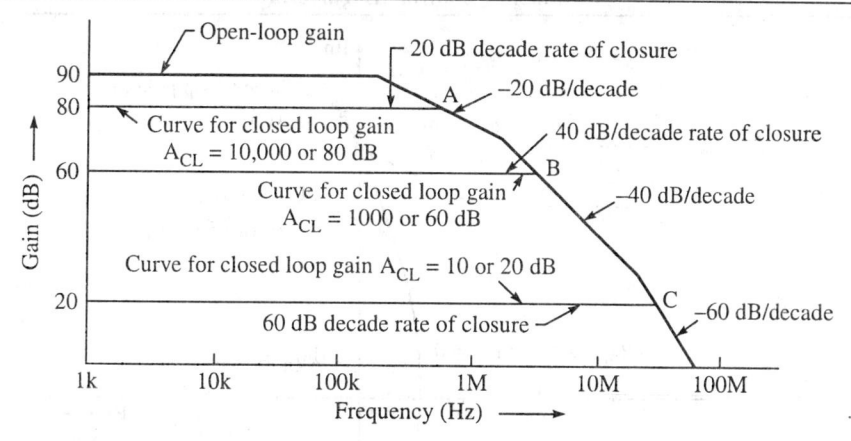

FIGURE 10.30 Effect of feedback on open-loop gain vs. frequency curve.

for stable operation, the rate of closure between the closed-loop gain projection and the open-loop curve should not exceed −20 dB/decade. At higher frequencies for lower closed-loop gains, the feedback becomes significant and regenerative, and may result in sustained oscillations.

So far, we have discussed stability of an op-amp qualitatively. To provide a quantitative discussion on stability, let us rewrite the transfer function of an op-amp characterized by three poles, as

$$A = \frac{A_{OL}\,\omega_1\,\omega_2\,\omega_3}{(s+\omega_1)(s+\omega_2)(s+\omega_3)}; \qquad 0 < \omega_1 < \omega_2 < \omega_3$$

Obviously the poles of the open-loop transfer function are at $-\omega_1, -\omega_2$ and $-\omega_3$. The closed-loop poles, that is, the poles of A_{CL} in Eq. (10.25) will be given by the roots of the characteristic equation

$$1 + A\beta = 0$$

Putting the value of A from Eq. (10.24), we get

$$1 + \frac{\beta A_{OL}\,\omega_1\,\omega_2\,\omega_3}{(s+\omega_1)(s+\omega_2)(s+\omega_3)} = 0$$

or $\qquad (s+\omega_1)(s+\omega_2)(s+\omega_3) + \beta A_{OL}\,\omega_1\omega_2\omega_3 = 0$

or $\qquad s^3 + s^2(\omega_1+\omega_2+\omega_3) + s(\omega_1\omega_2+\omega_1\omega_3+\omega_2\omega_3) + \omega_1\omega_2\omega_3(1+\beta A_{OL}) = 0 \qquad (10.29)$

The roots of this cubic equation depend upon βA_{OL}, the dc loop gain and therefore, βA_{OL} becomes the critical parameter that determines the new pole location. Further βA_{OL} can take any value between zero for no feedback ($\beta = 0$) and A_{OL} for maximum feedback ($\beta = 1$). For variable β in the range $0 < \beta A_{OL} < \infty$, the root loci is shown in Figure 10.31. When $\beta A_{OL} = 0$, the roots are at $-\omega_1, -\omega_2$ and $-\omega_3$ and lie on the negative real axis. For small values of βA_{OL}, the roots still lie on the left-half of the s-plane with one real root and two complex conjugate roots (a, a^*). If βA_{OL} is increased further beyond a critical value $(\beta A_{OL})_c$, the two roots will move to the right half of the s-plane causing instability. We will find out the critical

$$(\beta A_{OL})_c = (2 + \frac{\omega_1}{\omega_2} + \frac{\omega_1}{\omega_3} + \frac{\omega_2}{\omega_1} + \frac{\omega_2}{\omega_3} + \frac{\omega_3}{\omega_1} + \frac{\omega_3}{\omega_2})$$

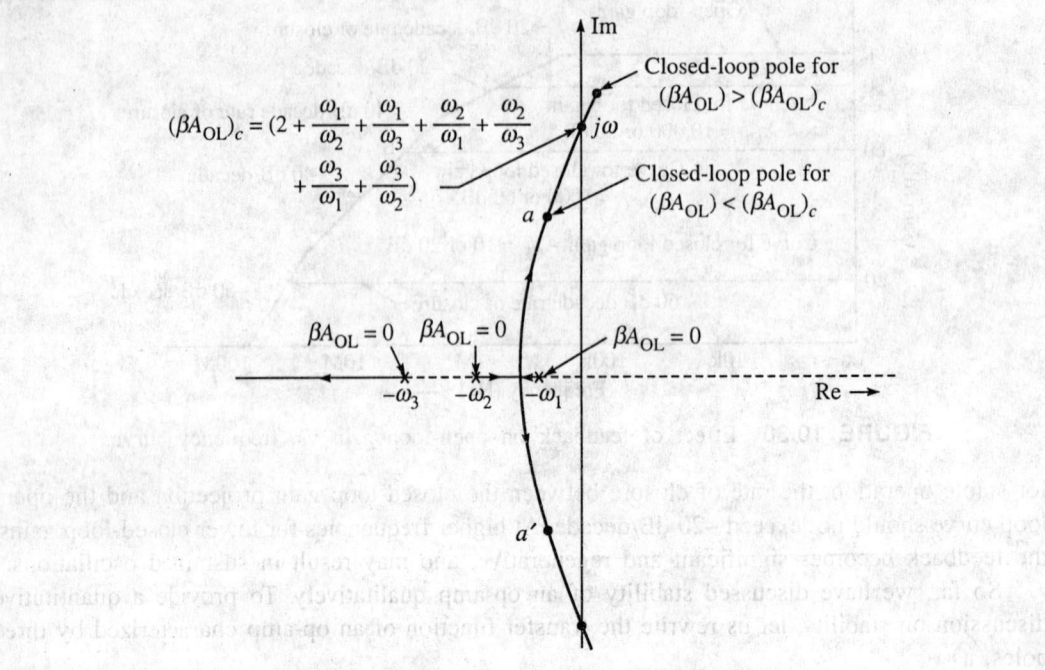

FIGURE 10.31 Root loci as a function βA_{OL}.

value of βA_{OL} if increased further beyond a critical value (βA_{OL}); the two roots will now move to the right-half of the s-plane causing instability. We will find out the critical value of βA_{OL} for which the closed-loop system becomes just unstable. Rewrite Eq. (10.29) as

$$a_3 s^3 + a_2 s^2 + a_1 s^1 + a_0 = 0 \qquad (10.30)$$

where

$a_3 = 1$; $a_2 = \omega_1 + \omega_2 + \omega_3$; $a_1 = \omega_1\omega_2 + \omega_1\omega_3 + \omega_2\omega_3$; and $a_0 = \omega_1\omega_2\omega_3(1 + \beta A_{OL})$

In order to find the critical value of βA_{OL}, apply Routh's stability criterion to Eq. (10.30), that is

(i) All the coefficients a_3, a_2, a_1 and a_0 should be positive.
(ii) $a_2 a_1 - a_3 a_0 \geq 0$ \qquad (10.31)

Putting $s = j\omega$ in Eq. (10.30)

$$a_3(j\omega)^3 + a_2(j\omega)^2 + a_1(j\omega) + a_0 = 0$$

or
$$(a_0 - a_2\omega^2) + j\omega(a_1 - a_3\omega^2) = 0$$

Equating the real and imaginary parts to zero, we get

$$a_0 - a_2\omega^2 = 0 \qquad (10.32)$$
$$a_1 - a_3\omega^2 = 0 \qquad (10.33)$$

Compensation Techniques

Thus, the frequency of oscillations is given by

$$\omega_{osc} = \pm\sqrt{\frac{a_0}{a_2}} = \pm\sqrt{\frac{a_1}{a_3}} \tag{10.34}$$

Putting the values of coefficients

$$\omega_{osc} = \sqrt{\frac{a_1}{a_3}} = \sqrt{\omega_1\omega_2 + \omega_1\omega_3 + \omega_2\omega_3} \tag{10.35}$$

Also from Eq. (10.31), we get

$$a_0 = \frac{a_2 a_1}{a_3} \tag{10.36}$$

or $\quad \omega_1\omega_2\omega_3\{1 + (\beta A_{OL})_c\} = (\omega_1 + \omega_2 + \omega_3)(\omega_1\omega_2 + \omega_1\omega_3 + \omega_2\omega_3)$

or $\quad (\beta A_{OL})_c = 2 + \dfrac{\omega_1}{\omega_2} + \dfrac{\omega_1}{\omega_3} + \dfrac{\omega_2}{\omega_1} + \dfrac{\omega_2}{\omega_3} + \dfrac{\omega_3}{\omega_1} + \dfrac{\omega_3}{\omega_2} \tag{10.37}$

It is obvious that $(\beta A_{OL})_c$ depends upon the ratio of the open-loop pole locations. The minimum value of $(\beta A_{OL})_c$ will occur when all the poles are located at the same place giving $(\beta A_{OL})_c = 8$.

As an example, if $A_{OL} = 10^5$ and $\omega_1 = \omega_2 = \omega_3 = 10^7$ rad/s, then the circuit will oscillate at a frequency of

$$\omega_{osc} = \omega_1\sqrt{3} \text{ rad/s} = 10^7\sqrt{3} \text{ rad/s}$$

On the other hand, if $1000\,\omega_1 = \omega_2 = \omega_3$, the critical loop gain is approximately

$$(\beta A_{OL})_c \approx 2\left(\frac{\omega_2}{\omega_1}\right) \approx 2000 \quad \text{and} \quad \beta_c < \frac{2000}{A_{OL}} \approx \frac{2000}{10000} = 0.2$$

In Figure 10.28, $\beta = R_1/(R_1 + R_2)$

For $\quad \beta < 0.2 \quad \dfrac{R_1 + R_2}{R_1} > \dfrac{1}{0.2} > 5 \quad$ or $\quad \dfrac{R_2}{R_1} \geq 4$

This means that if the op-amp is used as an inverting amplifier in Figure 10.28, the inverting gain magnitude should be greater than 4 and if used as a non-inverting amplifier, the non-inverting gain should be greater than 5 for oscillations to sustain.

If it is desired that the amplifier should remain stable for any resistive network, that is, $0 < \beta < 1$, then A_{OL} must satisfy the most stringent condition for $\beta = 1$, that is,

$$A_{OL} < \left(2 + \frac{\omega_1}{\omega_2} + \frac{\omega_1}{\omega_3} + \frac{\omega_2}{\omega_1} + \frac{\omega_2}{\omega_3} + \frac{\omega_3}{\omega_1} + \frac{\omega_3}{\omega_2}\right)$$

10.7.2 Frequency compensation

Two types of frequency compensating techniques are used: (i) external compensation and (ii) internal compensation.

External frequency compensation

Two common methods for accomplishing this goal are: Dominant-pole compensation and Pole-zero (lag) compensation.

Dominant-pole compensation: Suppose A is the uncompensated transfer function of the op-amp in the open-loop condition as given by Eq. (10.24). Let us introduce a dominant pole by adding an RC network in series with the op-amp as in Figure 10.32 or by connecting a capacitor C from a suitable high resistance point to ground. The compensated transfer function A' becomes

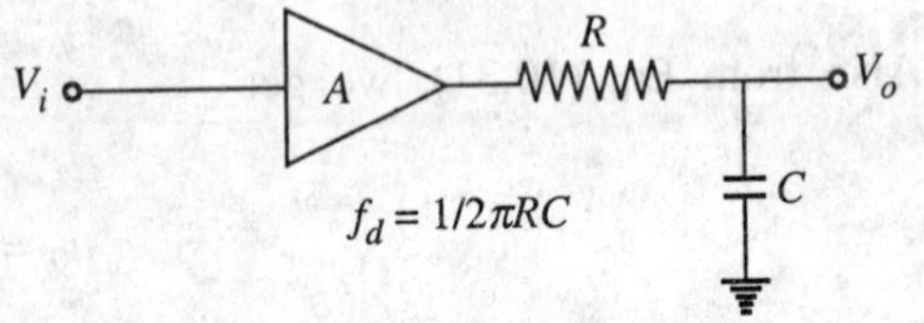

FIGURE 10.32 Dominant pole compensation.

$$A' = \frac{V_o}{V_i} = A \cdot \frac{\frac{-j}{\omega C}}{R - \frac{j}{\omega C}} = \frac{A}{1 + j\frac{f}{f_d}} \quad (10.38)$$

where

$$f_d = \frac{1}{2\pi RC}$$

using Eq. (10.23), we get

$$A' = \frac{A_{OL}}{\left(1 + j\frac{f}{f_d}\right)\left(1 + j\frac{f}{f_1}\right)\left(1 + j\frac{f}{f_2}\right)\left(1 + j\frac{f}{f_3}\right)}; \quad f_d < f_1 < f_2 < f_3$$

The capacitance C is chosen so that the modified loop gain drops to 0-dB with a slope of -20 dB/decade at a frequency where the poles of the uncompensated transfer function A contribute negligible phase shift. Usually $f_d = \omega_d/2\pi$ is selected so that the compensated transfer function A' passes through the 0-dB line at the pole f_1 of the uncompensated A. The frequency can be found graphically by pole f_1 of the uncompensated A. The frequency can be found graphically by having A' pass through the 0-dB line at the frequency f_1 with a slope of -20 dB per decade as shown in Figure 10.33. The value of capacitor C now can be calculated since $f_d = 1/2\pi RC$. The dominant-pole compensation technique reduces the open-loop bandwidth drastically. But the noise immunity of the system is improved.

Compensation Techniques

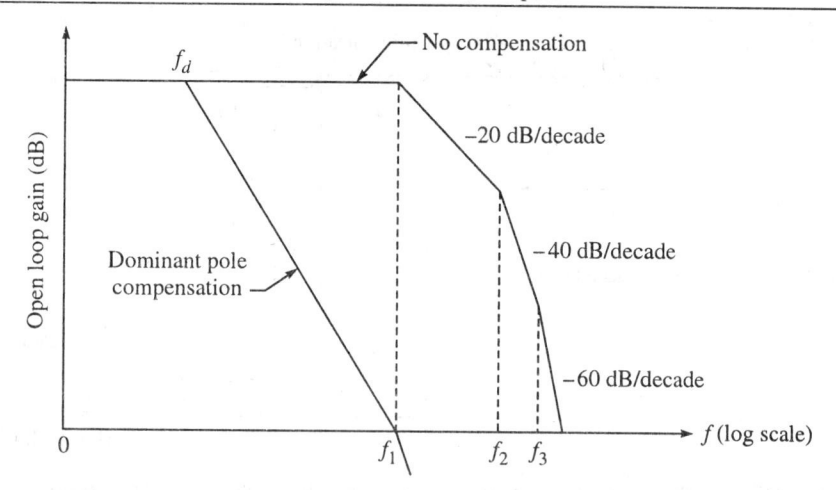

FIGURE 10.33 Gain vs. frequency curve for dominant pole compensation.

Pole-zero compensation: Here the uncompensated transfer function A is altered by adding both pole and zero as shown in Figure 10.34. The zero should be at a higher frequency than the pole. The transfer function of the compensating network alone is

$$\frac{V_o}{V_2} = \frac{Z_2}{Z_1 + Z_2} = \left(\frac{R_2}{R_1 + R_2}\right) \frac{1 + j\frac{f}{f_1}}{1 + j\frac{f}{f_0}} \quad (10.39)$$

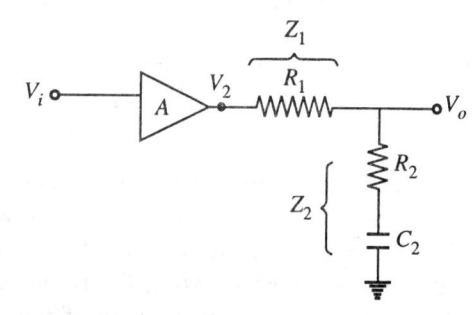

FIGURE 10.34 Pole-zero compensation.

where $Z_1 = R_1$; $Z_2 = R_2 + \dfrac{1}{j\omega C_2}$; $f_1 = \dfrac{1}{2\pi R_2 C_2}$; $f_0 = \dfrac{1}{2\pi (R_1 + R_2) C_2}$

The compensating network is designed to produce a zero at the first corner frequency f_1 of the uncompensated transfer function A. This zero will cancel the effect of the pole at f_1. The pole of the compensating network at $f_0 = \omega_0/2\pi$ is selected so that the compensated transfer function A' passes through 0-dB at the second corner frequency f_2 of the uncompensated transfer function A in Eq. (10.23). The frequency can be found graphically by having A' pass through 0-dB at the frequency f_2 with a slope of −20 dB/decade as shown in Figure 10.35. Assuming that the compensating network does not load the amplifier, i.e. $R_2 \gg R_1$, then the overall transfer function becomes

$$A' = \frac{V_0}{V_2} = \frac{V_0}{V_2} \cdot \frac{V_2}{V_i} = A\left(\frac{R_2}{R_1 + R_2}\right) \frac{1 + j\frac{f}{f_1}}{1 + j\frac{f}{f_0}} \quad (10.40)$$

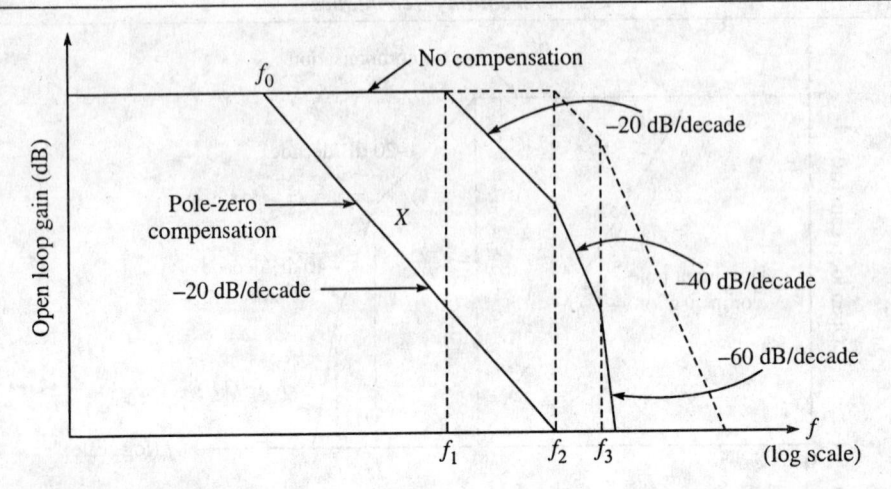

FIGURE 10.35 Open-loop gain vs. frequency for pole-zero compensation.

and note that $R_2 \gg R_1$, so that $R_2/(R_1 + R_2) \approx 1$, then

$$A' = \frac{A_{OL}}{\left(1 + j\dfrac{f}{f_0}\right)\left(1 + j\dfrac{f}{f_2}\right)\left(1 + j\dfrac{f}{f_3}\right)} \quad ; \quad 0 < f_0 < f_1 < f_2 < f_3$$

Consider again the frequency response (Bode plot) for the uncompensated op-amp having three poles at frequencies f_1, f_2 and f_3. Now select R_2 and C_2 so that the zero of the compensating network is equal to the pole at the frequency f_1 (lowest). If there had been no pole added by the compensating network, the response would have changed to that of the dotted curve in Figure 10.35. However, because of the predominance of the pole of the compensating network at f_0, the rate of closure will be -20 dB/decade throughout as shown in the curve X of Figure 10.35. The pole at f_0 should be selected so that the -20 dB/decade fall should meet the 0-dB line at f_2 which is the second pole of A.

Internally compensated operational amplifier

The type 741 op-amp is compensated and has an open-loop gain vs. frequency response as shown in Figure 10.36. The op-amp IC 741 contains a capacitance C_1 of 30 pF (see the manual), that internally shunts off the signal current and thus reduces the available output signal at higher frequencies. This internal capacitance, which is an internal compensating component, causes the open-loop gain to roll off at -20 dB/decade rate and thus assures for a stable circuit. The 741 op-amp has a 1 MHz gain bandwidth product. This means that the product of the coordinates, gain and

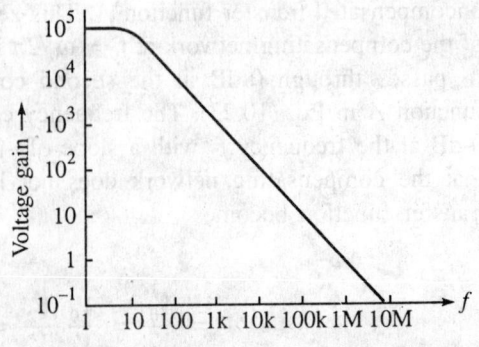

FIGURE 10.36 Frequency response of internally compensated MA741 op-amp.

frequency, of any point on the open-loop gain vs. frequency curve is about 1 MHz. If the 741 op-amp is wired for a closed-loop gain of 10^4 or 80 dB, its bandwidth is 100 Hz as can be seen by projecting to the right from 10^4 in the curve of Figure 10.36. For gain of 10^2, the bandwidth increases to 10 kHz and for gain 1, the bandwidth is 1 MHz. For 741 op-amp, unity gain-bandwidth product is specified as 1 MHz in the data sheet. This simply means that op-amp 741 has 1 MHz bandwidth with unity gain as seen in Figure 10.36. Some internally compensated op-amps are Fairchild's μA741, National Semiconductor's LM741, LM107 and LM112, and Motorala's MC1558. Internally-compensated op-amp 741 is widely used and its compensation can be well understood only through the control systems point of view. That is why we have discussed the compensation of this op-amp in detail.

Summary

As we mentioned earlier, the design and compensation is limited to the classical approach and is applicable to the single-input-single-output linear time-invariant system. One has to be concerned about the environmental changes and accordingly take into account the margin of tolerances in the design. In actual design problems, we must choose the hardware where the design constraints such as cost, size, weight and reliability factors are addressed. For the multivariable system, we apply the modern control approaches, one of which (observability design) is discussed in Chapter 12.

Problems

10.1 Determine the transfer function of a lead compensator that will provide a phase lead of 45° and gain of 10 dB at $\omega = 8$ rad/s.

10.2 Determine the transfer function of a lead compensator that will provide a phase lead of 50° and gain of 8 dB at $\omega = 5$ rad/s.

10.3 While designing a suitable control system for a missile, it was found necessary to introduce a lead of 35° and gain 6.5 dB at $\omega = 2.8$ rad/s. What will be the transfer function of the lead compensator that will satisfy the above requirements?

10.4 Design a lag compensator that will provide a phase lag of 50° and an attenuation of 15 dB at $\omega = 2$ rad/s.

10.5 Design a lag compensator that will provide a phase lag of 45° and antenuation of 10 dB at $\omega = $ rad/s.

10.6 A lag compensator required for a position control system must provide a phase lag of 35° and attenuation of 8 dB at $\omega = 1.9$ rad/s. Determine the transfer function.

10.7 Determine the transfer function of a lag–lead compensator that will provide a phase lead of 50° and attenuation of 15 dB at $\omega = 6$ rad/s.

10.8 Determine the transfer function of a lag–lead compensator that will provide a phase lead of 55° and attenuation of 20 dB at $\omega = 4$ rad/s.

State-Variable Formulation

11.1 Introduction

Over the past few decades a vast amount of research has been expended on linear feedback control theory and its application to the design of control systems. If we look back over the development of feedback control theory, we can see that the chronological development falls into the following two clearly-defined phases:

    Classical approach: up to 1956
    State-variable approach: 1956–todate

The transfer function method is very convenient in the frequency-domain analysis of the system. However, this method suffers a major disadvantage, in that all initial conditions of the system are neglected. Therefore when one is interested in time-domain solution, the transfer function method is inadequate. Further, we cannot apply the transfer function models to non-linear time-varying systems. The transfer function models cannot be used efficiently for multivariable systems (i.e. systems with many inputs and outputs). The transfer function approach confines to input-output behaviour of linear systems only. On the contrary, the state-space representation gives information about the internal behaviour of the system as well as information about its the input-output behaviour. In state-variable approach, inaccessible states can be accessed and for perfect control, all states should be accessible. Hence through the state-variable approach, perfect control of the system is possible. The beauty of the state-variable approach is that it can convert a high-order differential equation to a first-order vector-

OBJECTIVE

The drawbacks of the classical or transfer function approach facilitated the evolution of state variable approach for the analysis and design of control systems. In frequency domain, that is in steady sate analysis the transfer function method is good enough if the system is not of high order. The transient part is missing as it is the steady-state analysis. Further, by its definition, the transfer function approach does not take initial conditions into consideration. This approach, therefore, does not give a clear picture of the system response and its behaviour. Assuming dominant pole concept, the analysis of higher-order systems by transfer function approach is done where you get only the approximate ideas of the system response and analysis. For higher-order systems, the state variable approach is suitable as the system dynamics is converted into first-order vector-matrix differential equation. The important point to be noted is, that for the perfect design of control systems, all states should be accessible, which may not be the reality. But the inaccessible state can be generated under certain constraints for perfect control of the system. And that is why, state variable analysis is synonymous with modern control approach.

CHAPTER OUTLINE

Introduction
Concept of State
State Space Representation
Dual Representation
Similiarty Transformation
Diagonalization
Time-Domain Solution
State Transition Matrix
Algorithm for Power of Matrix
STM of Linear Time Varying System

matrix differential equation. The state-space representation allows us to study multivariable systems. The state-space model handles the time-domain solution. It takes care of the initial conditions of the systems. The state-space model is particularly suitable for use on the digital computer. The design of control systems using optimal control theory is possible only for state-space models. State-space representation is applicable to linear, nonlinear, time-invariant and time-varying systems. Stability analysis using Lyapunov's technique is the only method for stability analysis of nonlinear systems. It requires state-space formulation. The state of a system refers to the past, present and future behaviour of the system. It represents all the information that one cares to know about the future behaviour of the system.

11.2 Concept of State

Consider a single-input-single-output (SISO) linear *RLC* network subjected to an input $u(t)$. The output of the network is a time function $y(t)$. Since the network is known, complete knowledge of input $u(t)$ over the time interval $-\infty$ to t is sufficient to determine the output $y(t)$ over the same time interval. However, if the input is known only over the time interval t_0 to t, then the current through the inductor and the voltage across the capacitor at some time t_1 (usually $t_1 = t_0$) must be known in order to determine the output $y(t)$ over the time interval t_0 to t. These current and voltage constitute the state of the network which is related to the memory of the network. For a purely resistive network (zero memory), only the present input is required to determine the present output. Heuristically, the state of a system separates the future from the past, so that the state contains all the relevant information concerning the past history of the system required to determine the response for any input. The complete solution of an *nth* order linear time-invariant differential equation for $t > t_0$ is obtained in terms of n arbitrary constants. These arbitrary constants can be determined from the fact that the system must satisfy boundary conditions at time t_0. Consider the following differential equation

$$\overset{(n)}{y} + a_{n-1} \overset{(n-1)}{y} + \cdots + a_2 \ddot{y} + a_1 \dot{y} + a_0 y = u(t) \qquad (11.1)$$

Let us rewrite this equation in the form

$$\overset{(n)}{y} = u(t) - a_0 y - a_1 \dot{y} - a_2 \ddot{y} - \cdots - a_{n-1} \overset{(n-1)}{y} \qquad (11.2)$$

It is clear that if we know the forcing function $u(t)$ and also know the value of the variable y and its $(n-1)$ derivatives at some instant of time $t = t_0$, we can calculate the value of the *n*th derivative of $y(t)$ at $t = t_0$ and can proceed to calculate the values for all of the future time $t > t_0$. We may choose to call each of the variable y and each of the first $(n-1)$ derivatives, a *state variable*. Then the number of state variables or states is equal to the order of the differential equation which is normally equal to the number of energy storage elements in the system. Usually, the current in the inductor and the voltage across the capacitor are chosen as the state variables.

In order to provide a systematic mathematical approach to analyze the characteristics of the system, it is convenient to describe the system by a set of simultaneous first-order differential equations with each equation defining one state. This set of equations is called the *state equations*.

Before we proceed further, we must define state, state variable, state vector and state-space.

State: The state of a dynamic system is the minimal amount of information required, together with the initial state at time $t = t_0$ and input excitation, to completely specify the future behaviour of the system for any time $t > t_0$.

State variables: These are the smallest set of variables which determine the state of the dynamic system. If at least n variables $x_1(t), x_2(t), \ldots, x_n(t)$ are needed to completely describe the future behaviour of the system, together with the initial state and input excitation, then these n variables $[x_1(t), x_2(t), \ldots, x_n(t)]$ are a set of state variables. Note that the state variables need not be physically measurable or observable quantities.

State vector: The n state variables can be considered the n components of the state vector $X(t)$ described in n-dimensional vector-space called the state-space, i.e. the state vector

$$X(t) = [x_1(t) \; x_2(t) \; \ldots \; x_n(t)]^T$$

Consider the multiple-input-multiple-output (MIMO) system shown in Figure 11.1, where $y_1(t)$ to $y_m(t)$ are the output signals and $u_1(t)$ to $u_r(t)$ are the input signals. A set of state variables $[x_1(t), x_2(t), \ldots, x_n(t)]$ for the system of Figure 11.1 is a set such that the knowledge of initial values of the state variables $[x_1(t_0), x_2(t_0), \ldots, x_n(t_0)]$ at the initial time t_0 and of the input signals $u_1(t)$ and $u_r(t)$ for all $t > t_0$, suffices to determine the future values of the outputs and state variables completely.

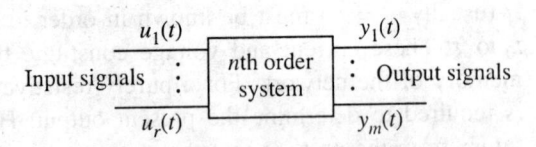

FIGURE 11.1 System block diagram.

State-space equations: The dynamic system must involve elements that memorize the values of the input for $t \geq t_0$, such as inductors and capacitors acting as energy storing devices in electrical systems. The initial conditions of these elements are required for complete representation of the dynamics. The dynamics of any system represented by the integro-differential equation and its simulation can be done by the integrators. The output of such integrators can serve as state variables. These variables can define the internal state of the system. The number of state variables to completely define the dynamics is usually equal to the number of integrators or the energy storing elements in the electrical circuit. For the MIMO system of Figure 11.1, the $(n \times 1)$ state vector $X = [x_1, x_2, \ldots, x_n]^T$, the $(r \times 1)$ input vector $U = [u_1, u_2, \ldots, u_r]^T$ and the $(m \times 1)$ output vector $Y = [y_1, y_2, \ldots, y_m]^T$. Then the system may be described by

$$\begin{aligned}
\dot{x}_1(t) &= f_1(x_1, x_2, \ldots, x_n; u_1, u_2, \ldots, u_r; t) \\
\dot{x}_2(t) &= f_2(x_1, x_2, \ldots, x_n; u_1, u_2, \ldots, u_r; t) \\
&\vdots \\
\dot{x}_n(t) &= f_n(x_1, x_2, \ldots, x_n; u_1, u_2, \ldots, u_r; t)
\end{aligned} \quad (11.3)$$

and the outputs by

$$y_1(t) = g_1(x_1, x_2, \ldots, x_n; u_1, u_2, \ldots, u_r; t)$$
$$y_2(t) = g_2(x_1, x_2, \ldots, x_n; u_1, u_2, \ldots, u_r; t)$$
$$\vdots$$
$$y_m(t) = g_m(x_1, x_2, \ldots, x_n; u_1, u_2, \ldots, u_r; t)$$
(11.4)

Equations (11.3) and (11.4) become

$$\dot{x}(t) = f(x, u, t) \tag{11.5}$$
$$y(t) = g(x, u, t) \tag{11.6}$$

where

$$f(x, u, t) = \begin{bmatrix} f_1(x_1, x_2, \ldots, x_n; u_1, u_2, \ldots, u_r; t) \\ f_2(x_1, x_2, \ldots, x_n; u_1, u_2, \ldots, u_r; t) \\ \vdots \\ f_n(x_1, x_2, \ldots, x_n; u_1, u_2, \ldots, u_r; t) \end{bmatrix}$$

and

$$g(x, u, t) = \begin{bmatrix} g_1(x_1, x_2, \ldots, x_n; u_1, u_2, \ldots, u_r; t) \\ g_2(x_1, x_2, \ldots, x_n; u_1, u_2, \ldots, u_r; t) \\ \vdots \\ g_m(x_1, x_2, \ldots, x_n; u_1, u_2, \ldots, u_r; t) \end{bmatrix}$$

Equation (11.5) is the state equation and Eq. (11.6) is the output equation. If vector functions f and/or g involve time t explicitly then the system is called a time varying system, otherwise time-invariant.

For a time-invariant system, if Eqs. (11.5) and (11.6) are linearized about the operating state, then we have the following linearized state equation and output equation:

$$\dot{X}(t) = AX(t) + Bu(t) \tag{AB}$$
$$y(t) = CX(t) + Du(t) \tag{CD}$$

where A is called the system matrix, B the input coupling, C the output coupling matrix, and D the input-output coupling or direct transmission matrix.

Let us now consider the spring–mass–damper system of Figure 11.2, whose differential equation that represents the dynamics of the system can be written as

$$M\frac{d^2y}{dt^2} + f\frac{dy}{dt} + Ky = u(t)$$

Let the choice of the states be

$$x_1 = y$$
$$x_2 = \dot{y} = \dot{x}_1$$

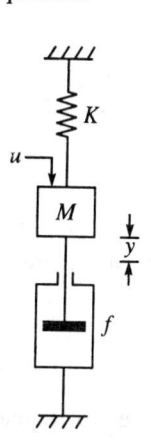

FIGURE 11.2 Spring–mass–damper system.

To represent the system dynamics in terms of state variables, we substitute the state variables in the system equation and obtain

$$M \frac{dx_2}{dt} + f x_2 + K x_1 = u(t)$$

Representing by a set of two first-order differential equations as

$$\dot{x}_1 = x_2 = 0 \cdot x_1 + 1 \cdot x_2 + 0 \cdot u(t)$$

then from the original equation, we get

$$\dot{x}_2 = -\frac{K}{M} x_1 - \frac{f}{M} x_2 - \frac{1}{M} u(t)$$

The above two equations can be written in matrix form as

$$\dot{X} = AX + Bu \tag{AB}$$
$$y = CX + Du \tag{CD}$$

and output

where $\quad A = \begin{bmatrix} 0 & 1 \\ -\frac{K}{M} & -\frac{f}{M} \end{bmatrix}, \quad B = \begin{bmatrix} 0 \\ -\frac{1}{M} \end{bmatrix}, \quad C = [1 \quad 0], \quad D = [0]$

Suppose the dynamics of the spring–mass–damper system is described by

$$\ddot{y} + 3\dot{y} + 2y = u$$

Now, let the of choice of the states be

$$x_1 = y \quad \text{and} \quad x_2 = \dot{x}_1 = \dot{y}$$

Then the above differential equation becomes

$$\ddot{x}_1 + 3\dot{x}_1 + 2x_1 = u$$

or
$$\frac{d}{dt}(\dot{x}_1) + 3\dot{x}_1 + 2x_1 = u$$

or
$$\dot{x}_2 + 3x_2 + 2x_1 = u$$

or
$$\dot{x}_2 = -2x_1 - 3x_2 + u$$

Now we can rewrite these derivatives of the state as

and
$$\dot{x}_1 = x_2 = 0.x_1 + 1.x_2 + 0.u$$
$$\dot{x}_2 = -2.x_1 - 3.x_2 + 1.u$$

The above two equations can written in matrix form as

$$\begin{bmatrix} \dot{x}_1 \\ \dot{x}_2 \end{bmatrix} = \begin{bmatrix} 0 & 1 \\ -2 & -3 \end{bmatrix} \begin{bmatrix} x_1 \\ x_2 \end{bmatrix} + \begin{bmatrix} 0 \\ 1 \end{bmatrix} u$$

State-Variable Formulation

that is
$$\dot{X} = AX + Bu \quad \text{(AB)}$$

where
$$A = \begin{bmatrix} 0 & 1 \\ -2 & -3 \end{bmatrix}, \quad B = \begin{bmatrix} 0 \\ 1 \end{bmatrix}, \quad X = \begin{bmatrix} x_1 \\ x_2 \end{bmatrix}$$

and as we have chosen output $y = x_1$, that is $y = 1.x_1 + 0.x_2 + 0.u$; the output in matrix form can be written as

$$y = \begin{bmatrix} 1 & 0 \end{bmatrix} \begin{bmatrix} x_1 \\ x_2 \end{bmatrix}$$

that is
$$y = CX + Du \quad \text{(CD)}$$

where
$$C = \begin{bmatrix} 1 & 0 \end{bmatrix}, \quad D = [0]$$

Hence the dynamics of the spring–mass–damper system represented by the vector–matrix differential equations (AB) and (CD) where X is the state vector of order (2×1); u is the input of order (1×1), a scalar; y is the output of order (1×1), a scalar; A is the system matrix of order (2×2); B is the input-coupling matrix of order (2×1); C is the output-coupling matrix of order (1×2) and D is the input-output coupling matrix.

The simulation of the system is shown in Figure 11.3.

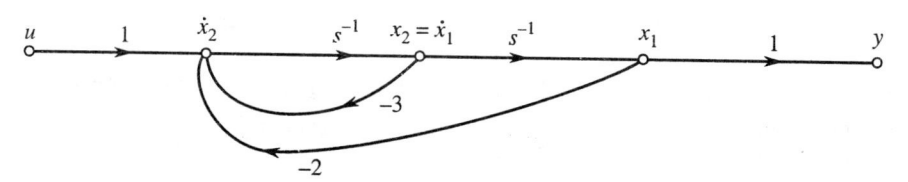

FIGURE 11.3 Simulation of the spring–mass–damper system.

Let us look the other way round. Suppose the simulation of the system is given by the signal flow graph (SFG) of Figure 11.3, and we have to write the state-variable formulation. We then proceed as follows:

From the SFG, write the equation at different nodes in state-variable form as

$$\dot{x}_1 = x_2$$

and
$$\dot{x}_2 = -2x_1 - 3x_2 + u$$

which leads to the vector-matrix differential equation in state-variable form as

$$\begin{bmatrix} \dot{x}_1 \\ \dot{x}_2 \end{bmatrix} = \begin{bmatrix} 0 & 1 \\ -2 & -3 \end{bmatrix} \begin{bmatrix} x_1 \\ x_2 \end{bmatrix} + \begin{bmatrix} 0 \\ 1 \end{bmatrix} u$$

that is,
$$\dot{X} = AX + Bu$$

where
$$A = \begin{bmatrix} 0 & 1 \\ -2 & -3 \end{bmatrix}, \quad B = \begin{bmatrix} 0 \\ 1 \end{bmatrix}, \quad X = \begin{bmatrix} x_1 \\ x_2 \end{bmatrix}$$

and at the output node
$$y = x_1$$

which in matrix form can be written as
$$y = CX, \quad \text{where} \quad C = [1 \; 0]$$

Drill Problem 11.1

Write the state-variable formulation of the system with the SFG as given in Figure 11.4.

$$\text{Ans.} \quad A = \begin{bmatrix} 0 & 1 & 0 \\ 0 & 0 & 1 \\ -1 & -7 & -5 \end{bmatrix}, \quad B = \begin{bmatrix} 0 \\ 0 \\ 1 \end{bmatrix}, \quad C = [3 \; 4 \; 0]$$

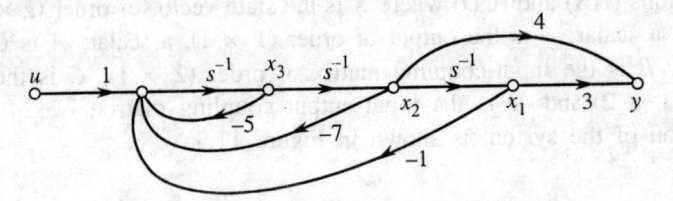

FIGURE 11.4 Drill Problem 11.1: state flow graph.

State-space representation: Consider the dynamics of the system represented by the differential equation as

$$\overset{(n)}{y} + a_{n-1} \overset{(n-1)}{y} + \cdots + a_2 \ddot{y} + a_1 \dot{y} + a_0 y = u(t)$$

Let us choose the state variables as

$$\left. \begin{aligned} x_1 &= y \\ x_2 &= \dot{y} \\ x_3 &= \ddot{y} \\ &\vdots \\ x_{n-1} &= \overset{(n-2)}{y} \\ x_n &= \overset{(n-1)}{y} \end{aligned} \right\} \quad (11.7)$$

Rewriting Eq. (11.7) as

$$\left. \begin{aligned} \dot{x}_1 &= x_2 \\ \dot{x}_2 &= x_3 \\ &\vdots \\ \dot{x}_{n-1} &= x_n \end{aligned} \right\} \quad (11.8)$$

State-Variable Formulation

and from the original equation, we can write

$$\dot{x}_n = -a_0 x_1 - a_1 x_2 - \cdots - a_{n-2} x_{n-1} - a_{n-1} x_n + u \tag{11.9}$$

The output
$$y = x_1 \tag{11.10}$$

Equations (11.8) and (11.9) can be written in matrix form to get

$$\dot{X} = AX + Bu \tag{AB}$$

and output Eq. (11.10) can be written in matrix form to get

$$y = CX + Du \tag{CD}$$

where
$$X = \begin{bmatrix} x_1 \\ x_2 \\ \vdots \\ x_n \end{bmatrix}, \quad A = \begin{bmatrix} 0 & 1 & 0 & \cdots & 0 \\ 0 & 0 & 1 & \cdots & 0 \\ \vdots & \vdots & \vdots & \cdots & \vdots \\ 0 & 0 & 0 & \cdots & 1 \\ -a_0 & -a_1 & -a_2 & \cdots & -a_n \end{bmatrix}, \quad B = \begin{bmatrix} 0 \\ 0 \\ \vdots \\ 0 \\ 1 \end{bmatrix}$$

and
$$C = [1 \ 0 \ 0 \ \cdots \ 0], \quad D = [0] \tag{11.11}$$

Hence the dynamics of an nth order single-input-single-output (SISO) system can be represented by the vector-matrix differential equations (AB) and (CD) where

- A is the system matrix of order $(n \times n)$
- B is the input coupling matrix of order $(n \times 1)$
- C is the output coupling matrix of order $(1 \times n)$
- D is the input-output coupling matrix of order (1×1)
- X is the state vector of order $(n \times 1)$
- u is the scalar input of order (1×1)
- y is the scalar output of order (1×1)

In general, for the nth order multivariable system having m inputs and p outputs, the dynamics can be represented by the vector-matrix differential equation as

$$\dot{X} = AX + BU \tag{AB}$$

and output
$$Y = AX + DU \tag{CD}$$

where the system matrix A is of the order $(n \times n)$, the input-coupling matrix B is of order $(n \times r)$, the output-coupling matrix C is of order $(n \times n)$, the input-output coupling matrix D is of order $(m \times r)$, the state vector X is of order $(n \times 1)$, the output vector Y is of the order $(m \times 1)$ and input vector U is of order $(r \times 1)$.

In expanded form, these preceding equations (AB) and (CD) become

$$\begin{bmatrix} \dot{x}_1(t) \\ \dot{x}_2(t) \\ \vdots \\ \dot{x}_n(t) \end{bmatrix} = \begin{bmatrix} a_{11} & a_{12} & \cdots & a_{1n} \\ a_{21} & a_{22} & \cdots & a_{2n} \\ \vdots & \vdots & & \vdots \\ a_{n1} & a_{n2} & \cdots & a_{nn} \end{bmatrix} \begin{bmatrix} x_1(t) \\ x_2(t) \\ \vdots \\ x_n(t) \end{bmatrix} + \begin{bmatrix} b_{11} & b_{12} & \cdots & b_{1r} \\ b_{21} & b_{22} & \cdots & b_{2r} \\ \vdots & \vdots & & \vdots \\ b_{n1} & b_{n2} & \cdots & b_{nr} \end{bmatrix} \begin{bmatrix} u_1(t) \\ u_2(t) \\ \vdots \\ u_r(t) \end{bmatrix} \quad (11.12)$$

and

$$\begin{bmatrix} y_1(t) \\ y_2(t) \\ \vdots \\ y_m(t) \end{bmatrix} = \begin{bmatrix} c_{11} & c_{12} & \cdots & c_{1n} \\ c_{21} & c_{22} & \cdots & c_{2n} \\ \vdots & \vdots & & \vdots \\ c_{m1} & c_{m2} & \cdots & c_{mn} \end{bmatrix} \begin{bmatrix} x_1(t) \\ x_2(t) \\ \vdots \\ x_n(t) \end{bmatrix} + \begin{bmatrix} d_{11} & d_{12} & \cdots & d_{1n} \\ d_{21} & d_{22} & \cdots & d_{2n} \\ \vdots & \vdots & & \vdots \\ d_{m1} & d_{m2} & \cdots & d_{mr} \end{bmatrix} \begin{bmatrix} u_1(t) \\ u_2(t) \\ \vdots \\ u_r(t) \end{bmatrix} \quad (11.13)$$

Note that this representation is quite general, as it allows for multiple inputs, r of them, and multiple outputs, m of them. Thus the general system representation of Eqs. (AB) and (CD) is adequate for multiple-input-multiple-output systems (multivariable systems). The block diagram representation of the state-variable formulation for multivariable systems is shown in Figure 11.5.

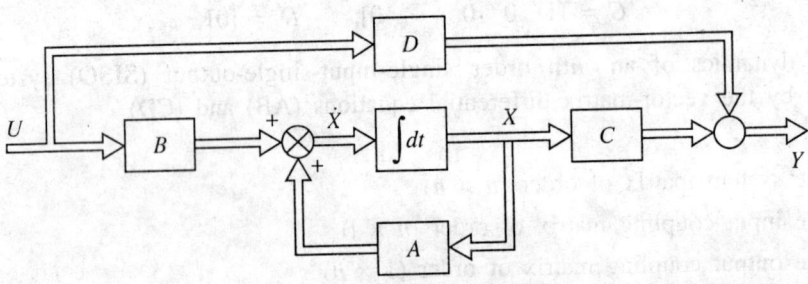

FIGURE 11.5 Block diagram representation of the state-variable formulation for multivariable systems.

Equation (AB) is a set of n first-order differential equations and usually referred to as the plant equation, while Eq. (CD) represents a set of m linear algebraic equations and is referred to as the output expression that appears as

$$Y(t) = CX + DU(t)$$

where the added $Du(t)$ term indicates a direct coupling of the input to the output. The direct coupling of the input to the output is rare in control systems, where power amplification is generally desired.

In general, as there is no direct input-output coupling, that is, $D = 0$, a system can be represented by the vector-matrix differential equation as

$$\dot{X}(t) = AX(t) + BU(t) \quad \text{(AB)}$$

and output

$$Y(t) = CX(t) \quad \text{(C)}$$

State-Variable Formulation

The time-domain state-variable representation of a multiple-input-multiple-output system should be contrasted with the frequency domain transfer function approach of classical control theory. In the latter form, the system discussed above would become simply

$$Y(s) = G(s)U(s)$$

The matrix $G(s)$ is known as the transfer function matrix of order $(m \times r)$. It is usual to represent graphically the transfer function by means of a block diagram.

11.3 Transfer Function

Let us consider first the problem of determining the transfer function of a system having state variable representation as in Eqs. (AB) and (CD). Assuming as usual that in transfer function definition, the initial conditions on X are all zero and taking the Laplace transform, we get

$$sX(s) = AX(s) + BU(s)$$

and

$$Y(s) = CX(s) + DU(s)$$

Grouping the two $X(s)$ terms, we get

$$[sI - A]X(s) = BU(s)$$

where the identity matrix I has been introduced to allow the indicated factoring as the Laplace operator s is a scalar. If both sides of this equation are now pre-multiplied by $[sI - A]^{-1}$, then

$$X(s) = [sI - A]^{-1}BU(s)$$

which may be substituted to obtain

$$Y(s) = C[sI - A]^{-1}BU(s) + DU(s)$$
$$= C[sI - A]^{-1}B + D]U(s)$$

By definition

$$\text{Transfer function} = \frac{\text{Laplace transform of output}}{\text{Laplace transform of input}}\bigg|_{\text{Initial conditions}=0}$$

The transfer function matrix $G(s)$ for a multivariable system is given by

$$G(s) = \frac{Y(s)}{U(s)} = C[sI - A]^{-1}B + D$$

In the case of the single-input-single-output (SISO) system, $G(s)$ is a scalar and is termed transfer function.

If there is no direct coupling between input and output (which is the usual case), i.e. $D = 0$, then in that case.

$$G(s) = C[sI - A]^{-1}B = C\frac{\text{Adjoint }[sI - A]}{|sI - A|}B \tag{11.14}$$

The matrix $[sI - A]^{-1}$ is commonly referred to as the *resolvent matrix* and is designated by $\phi(s)$. Thus,

$$\phi(s) = [sI - A]^{-1}$$

EXAMPLE 11.1 Consider the system represented by

$$\dot{X} = \begin{bmatrix} 0 & 1 \\ -2 & -3 \end{bmatrix} X + \begin{bmatrix} 0 \\ 1 \end{bmatrix} u$$

and output

$$y = \begin{bmatrix} 1 & 0 \end{bmatrix} X$$

In this case, the matrix $[sI - A]$ becomes

$$[sI - A] = \begin{bmatrix} s & -1 \\ 2 & s+3 \end{bmatrix}$$

and its inverse is

$$\phi(s) = [sI - A]^{-1} = \frac{\text{Adj}\,[sI - A]}{\det\,[sI - A]} = \frac{\begin{bmatrix} s+3 & 1 \\ -2 & s \end{bmatrix}}{s^2 + 3s + 2}$$

The transfer function

$$G(s) = C\phi(s)B = \frac{\begin{bmatrix} 1 & 0 \end{bmatrix} \begin{bmatrix} s+3 & 1 \\ -2 & s \end{bmatrix} \begin{bmatrix} 0 \\ 1 \end{bmatrix}}{s^2 + 3s + 2} = \frac{1}{s^2 + 3s + 2}$$

To determine the transfer function of a system from a state-variable representation of the system, we must invert the matrix $[sI - A]$. The inversion of a matrix is never an easy task, and also not easy to program on a computer. Because of this problem it is often easier to obtain the transfer function by carrying out the block diagram reductions or equivalently by using the signal flow graph techniques on the block diagram of the state-variable representation. However, an algorithm for obtaining the transfer function matrix for a general form of system matrix will be presented later.

The transfer function is a ratio of two polynomials of s. The denominator polynomial, that is, the determinant of the matrix $[sI - A]$ gives the characteristic equation, $|sI - A| = 0$.

The nth degree characteristic equation $|sI - A| = 0$ for the nth order system having $n \times n$ system matrix A has n roots or eigenvalues, which may be distinct or repetitive.

For a multivariable system having $m \times 1$ output vector Y and $r \times 1$ input vector U for the nth order system, we get the transfer matrix $G(s)$ of order $(m \times r)$ as

$$G(s) = C[sI - A]^{-1}B + D$$

$$= \begin{bmatrix} G_{11} & G_{12} & \cdots & G_{1r} \\ G_{21} & G_{22} & \cdots & G_{2r} \\ \vdots & \vdots & & \vdots \\ G_{m1} & G_{m2} & \cdots & G_{mr} \end{bmatrix} \quad (11.15)$$

That is, the input-output relationship can be written as

$$Y(s) = G(s)U(s)$$

or

$$\begin{bmatrix} Y_1(s) \\ Y_2(s) \\ \vdots \\ Y_m(s) \end{bmatrix} = \begin{bmatrix} G_{11}(s) & G_{12}(s) & \cdots & G_{1r}(s) \\ G_{21}(s) & G_{22}(s) & \cdots & G_{2r}(s) \\ \vdots & \vdots & & \vdots \\ G_{m1}(s) & G_{m2}(s) & \cdots & G_{mr}(s) \end{bmatrix} \begin{bmatrix} U_1(s) \\ U_2(s) \\ \vdots \\ U_r(s) \end{bmatrix} \quad (11.16)$$

Drill Problem 11.2

Determine the transfer function of the system represented by the vector-matrix differential equation

$$\dot{X} = \begin{bmatrix} -2 & 0 & 1 \\ 1 & -2 & 0 \\ 1 & 1 & -1 \end{bmatrix} X + \begin{bmatrix} 1 \\ 0 \\ 10 \end{bmatrix} u \quad \text{and output} \quad Y = [2 \ 1 \ -1]X$$

Ans. $\dfrac{-8s^2 - 14s + 12}{s^3 + 5s^2 + 7s + 1}$

Drill Problem 11.3

Determine the transfer function if matrices A, B, C of Drill Problem 11.2 remain unchanged and $D = 1$.

Ans. $\dfrac{s^3 - 3s^2 - 7s + 13}{s^3 + 5s^2 + 7s + 1}$

11.3.1 Computation of transfer function using Leverrier's algorithm

The calculation of the transfer function from the state equation requires the inversion of the matrix $[sI - A]$. This can be conveniently done using Leverrier's algorithm (also called Souriau-Frame algorithm). Let

$$[sI - A]^{-1} = \frac{\text{Adj}[sI - A]}{\det[sI - A]} = \frac{P_{n-1}s^{n-1} + P_{n-2}s^{n-2} + \cdots + P_1 s + P_0}{s^n + a_{n-1}s^{n-1} + \cdots + a_1 s + a_0} \quad (11.17)$$

where P_i are $(n \times n)$ matrices and a_i's are scalar. The algorithm can be written as follows:

$$\left.\begin{aligned}
P_{n-1} &= I, \text{ the identity matrix} \\
a_{n-1} &= -\text{tr}[A], \text{ where 'tr' represents the trace of a matrix} \\
P_{n-2} &= P_{n-1}A + a_{n-1}I \\
a_{n-2} &= -\frac{1}{2}\text{tr}[P_{n-2}A] \\
&\vdots \\
P_k &= P_{k+1}A + a_{k+1}I \\
a_k &= -\frac{1}{(n-k)}\text{tr}(P_k A)
\end{aligned}\right\} \quad (11.18)$$

Also, for checking purposes, we can verify that

$$P_0 A + a_0 I = 0$$

It may be noted that the denominator of the transfer function is $|sI - A|$. The necessity of Leverrier's algorithm is to determine $[sI - A]^{-1}$ which a digital computer fails because the Laplace operator s is a scalar. However, this algorithm can be performed by the digital computer easily to obtain $[sI - A]^{-1}$ as illustrated by an example below.

EXAMPLE 11.2 Consider

$$\dot{X} = \begin{bmatrix} -2 & 0 & 1 \\ 1 & -2 & 0 \\ 1 & 1 & -1 \end{bmatrix} X + \begin{bmatrix} 1 & 0 \\ 0 & 1 \\ 1 & 0 \end{bmatrix} \begin{bmatrix} u_1 \\ u_2 \end{bmatrix}$$

and output

$$\begin{bmatrix} y_1 \\ y_2 \end{bmatrix} = \begin{bmatrix} 2 & 1 & -1 \\ 0 & 1 & 0 \end{bmatrix} \begin{bmatrix} x_1 \\ x_2 \\ x_3 \end{bmatrix}$$

In this case, $n = 3$. Hence

$$P_2 = I = \begin{bmatrix} 1 & 0 & 0 \\ 0 & 1 & 0 \\ 0 & 0 & 1 \end{bmatrix}$$

$$a_2 = -\text{tr}[A]$$
$$= -[(-2) + (-2) + (-1)] = 5$$

$$P_1 = P_2A + a_2I = \begin{bmatrix} 3 & 0 & 1 \\ 1 & 3 & 0 \\ 1 & 1 & 4 \end{bmatrix}$$

$$a_1 = -\frac{1}{2}\text{tr}[P_1 A] = -\frac{1}{2}\text{tr}\begin{bmatrix} -5 & 1 & 2 \\ 1 & -6 & 1 \\ 3 & 2 & -3 \end{bmatrix} = 7$$

$$P_0 = P_1 A + a_1 I = \begin{bmatrix} 2 & 1 & 2 \\ 1 & 1 & 1 \\ 3 & 2 & 4 \end{bmatrix}$$

$$a_0 = -\frac{1}{3}\text{tr}[P_0 A] = -\frac{1}{3}\text{tr}\begin{bmatrix} -1 & 0 & 0 \\ 0 & -1 & 0 \\ 0 & 0 & -1 \end{bmatrix} = 1$$

Also check,

$$P_0 A + a_0 I = \begin{bmatrix} -1 & 0 & 0 \\ 0 & -1 & 0 \\ 0 & 0 & -1 \end{bmatrix} + \begin{bmatrix} 1 & 0 & 0 \\ 0 & 1 & 0 \\ 0 & 0 & 1 \end{bmatrix} = 0$$

Hence, $\text{Adj}[sI - A] = P_2 s^2 + P_1 s + P_0 = \begin{bmatrix} s^2 + 3s + 2 & 1 & s + 2 \\ s + 1 & s^2 + 3s + 1 & 1 \\ s + 3 & s + 2 & s^2 + 4s + 4 \end{bmatrix}$

The characteristic polynomial is

$$|sI - A| = s^3 + a_2 s^2 + a_1 s + a_0$$

Putting the values of a_i's, we get

$$|sI - A| = s^3 + 5s^2 + 7s + 1$$

Now the transfer function matrix for the multivariable system becomes

$$G(s) = C[sI - A]^{-1} B = \frac{C\,\text{Adj}[sI - A]B}{s^3 + 5s^2 + 7s + 1}$$

$$= \frac{1}{s^3 + 5s^2 + 7s + 1} \begin{bmatrix} s^2 + 4s + 3 & s^2 + 2s + 1 \\ s + 2 & s^2 + 3s + 1 \end{bmatrix}$$

As a check, let us find by the usual procedure the value of

$$[sI - A]^{-1} = \frac{1}{s^3 + 5s^2 + 7s + 1} \begin{bmatrix} s^2 + 3s + 2 & 1 & s + 2 \\ s + 1 & s^2 + 3s + 1 & 1 \\ s + 3 & s + 2 & s^2 + 4s + 4 \end{bmatrix}$$

Hence by pre- and post-multiplying $[sI - A]^{-1}$ by C and B respectively, the transfer function matrix becomes the same as obtained from that using the Leverrier's algorithm.

The MATLAB program leverrier.m is given in the following:

MATLAB Program 11.1

```
%leverrier.m
clear;
pack;
clc;
A = [-2 0 1;1 -2 0;1 1 -1];
I = [1 0 0; 0 1 0; 0 0 1];
p2 = I
a2 = -trace (A)
    p1 = p2*A + a2*I
a1 = [-1/2]*trace (p1*A)
p0 = p1*A+a1*I
a0 = [-1/3]*trace(p0*A)
p2 =
    1  0  0
    0  1  0
    0  0  1
a2 =
    5
p1 =
    3  0  1
    1  3  0
    1  1  4
a1 =
    7
p0 =
    2  1  2
    1  1  1
    3  2  4
a0 =
    1
```

Drill Problem 11.4

Using the Leverrier's algorithm, calculate the transfer function of the system given in Drill Problem 11.2 and verify the result.

EXAMPLE 11.3 For the *RLC* circuit of Figure 11.6, write the state-variable formulation.

Solution: Applying KCL at node *n*, we get

$$i_R + i_C + i_L = \hat{i}(t)$$

or
$$\frac{e_C}{R} + C\frac{de_C}{dt} + i_L = \hat{i}(t) \quad (11.19)$$

Again
$$i_L = \frac{1}{L}\int e_C \, dt$$

or
$$\frac{di_L}{dt} = \frac{e_C}{L} \quad (11.20)$$

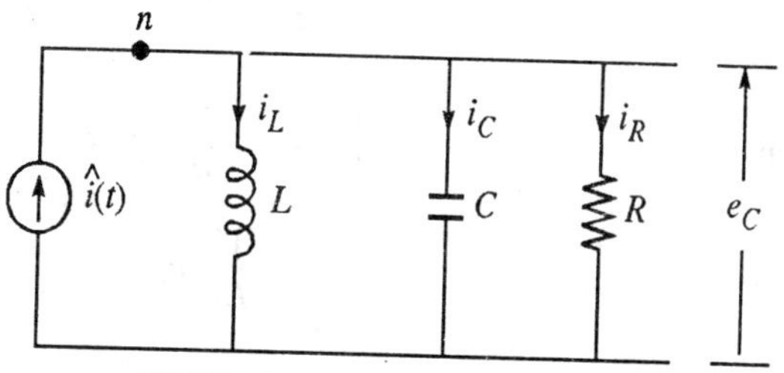

FIGURE 11.6 Example 11.3.

Equations (11.19) and (11.20) respectively can be rewritten as

$$\frac{de_C}{dt} = -\frac{e_C}{RC} - \frac{i_L}{C} + \frac{\hat{i}(t)}{C} \quad (11.21)$$

and
$$\frac{di_L}{dt} = \frac{e_C}{L} - 0.i_L + 0.\hat{i}(t) \quad (11.22)$$

Equations (11.21) and (11.22) can be written in vector-matrix differential equation form as

$$\frac{d}{dt}\begin{bmatrix} e_C \\ i_L \end{bmatrix} = \begin{bmatrix} -1/RC & -1/C \\ 1/L & 0 \end{bmatrix}\begin{bmatrix} e_C \\ i_L \end{bmatrix} + \begin{bmatrix} 1/C \\ 0 \end{bmatrix}\hat{i}(t) \quad (11.23)$$

Usually the voltage across the capacitor and the current through the inductor are chosen as the state variables. Hence let us put the state variables as $x_1 = e_C$ and $x_2 = i_L$.
    Then the state-variable formulation is

$$\dot{X} = AX + Bu$$

and output
$$y = CX$$

where

The state vector, $\quad X = [e_C \quad i_L]^T = [x_1 \quad x_2]^T$

The input scalar, $\quad u = \hat{i}(t)$

The output scalar, $\quad y = e_C$

The system matrix, $\quad A = \begin{bmatrix} -1/RC & -1/C \\ 1/L & 0 \end{bmatrix}$

The input coupling matrix, $\quad B = [1/C \quad 0]^T$

The output coupling matrix, $\quad C = [1 \quad 0]$

The choice of state is **not unique**. For the same circuit, we may choose the voltage across the capacitor e_C and flux linkage ϕ as the state variables. Then for the circuit of Figure 11.6, we get

$$\phi = Li_L \tag{11.24}$$

and

$$\frac{d\phi}{dt} = L\frac{di_L}{dt} = e_C = 1 \cdot e_C + 0 \cdot \phi + 0 \cdot \hat{i}(t) \tag{11.25}$$

as

$$i_L = \frac{1}{L}\int e_C\, dt$$

From KCL, we get

$$i_R + i_C + i_L = \hat{i}(t)$$

or

$$\frac{e_C}{R} + C\frac{de_C}{dt} + i_L = \hat{i}(t) \tag{11.26}$$

Putting Eq. (11.24) in Eq. (11.26), we get

$$\frac{de_C}{dt} = -\frac{1}{RC}e_C - \frac{1}{LC}\phi + \frac{1}{C}\hat{i}(t) \tag{11.27}$$

From Eqs. (11.27) and (11.25), we get the state equation as

$$\frac{d}{dt}\begin{bmatrix} e_C \\ \phi \end{bmatrix} = \begin{bmatrix} -1/RC & -1/LC \\ 1 & 0 \end{bmatrix}\begin{bmatrix} e_C \\ \phi \end{bmatrix} + \begin{bmatrix} 1/C \\ 0 \end{bmatrix}\hat{i}(t) \tag{11.28}$$

where the state vector is $[e_C \;\; \phi]^T$

The output equation remains the same.

EXAMPLE 11.4 Write the state-variable formulation of the circuit shown in Figure 11.7.

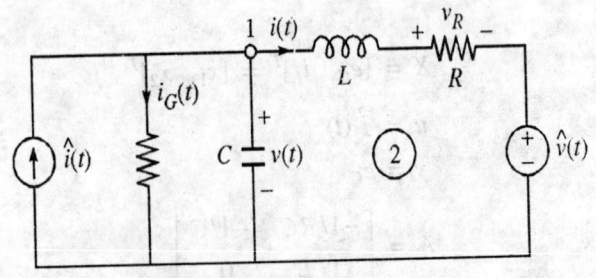

FIGURE 11.7 Example 11.4.

Solution: To write the state-variable form of equation of the given circuit, we may apply Kirchhoff's current law at node 1 and Kirchhoff's voltage law around mesh 2 respectively to

State-Variable Formulation

obtain the following pair of coupled first-order linear time-invariant differential equations

$$C\frac{dv(t)}{dt} = -Gv(t) - i(t) + \hat{i}(t) \quad (11.29)$$

and

$$L\frac{di}{dt} = v(t) - Ri(t) - \hat{v}(t) \quad (11.30)$$

And the output equations as

$$i_G(t) = Gv(t) \quad (11.31)$$

and

$$v_R(t) = Ri(t) \quad (11.32)$$

We can write these set of equations (11.29) to (11.32) to write the state variable formulation as Eqs. (AB) and (C) where

$$A = \begin{bmatrix} -G/C & -1/C \\ 1/L & -R/L \end{bmatrix}, \quad B = \begin{bmatrix} 1/C & 0 \\ 0 & -1/L \end{bmatrix}, \quad C = \begin{bmatrix} G & 0 \\ 0 & R \end{bmatrix}$$

and the state vector, $X = [v(t) \; i(t)]^T$

output vector, $y = [i_G(t) \; v_R(t)]^T$

input vector, $u = [\hat{i}(t) \; \hat{v}(t)]^T$

EXAMPLE 11.5 Consider the network shown in Figure 11.8 and obtain the state variable formulation.

Solution: At node A, by Kirchhoff's current law,

$$\hat{i} = i_L + i_C = i_L + C\frac{dv_C}{dt}$$

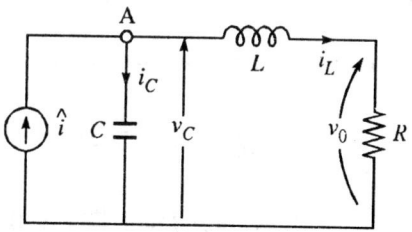

FIGURE 11.8 Example 11.5.

or

$$\dot{v}_C = 0 \cdot v_C + \left(-\frac{1}{C}\right)i_L + \left(\frac{1}{C}\right)\hat{i} \quad (11.33)$$

Using mesh equation (KVL),

$$v_C = L\frac{di_L}{dt} + Ri_L$$

or

$$\frac{di_L}{dt} = \left(\frac{1}{L}\right)v_C + \left(-\frac{R}{L}\right)i_L + (0)\hat{i} \quad (11.34)$$

Combining Eqs. (11.33) and (11.34), the state variable formulation can be written as

$$\frac{d}{dt}\begin{bmatrix} v_C \\ i_L \end{bmatrix} = \begin{bmatrix} 0 & -1/C \\ 1/L & -R/L \end{bmatrix}\begin{bmatrix} v_C \\ i_L \end{bmatrix} + \begin{bmatrix} 1/C \\ 0 \end{bmatrix}\hat{i}$$

and the output
$$v_R = Ri_L = \begin{bmatrix} 0 & R \end{bmatrix} \begin{bmatrix} v_C \\ i_L \end{bmatrix}$$

Note that, usually in the electric circuit problems, the current through the inductor and the voltage across the capacitor are chosen as the state variables.

EXAMPLE 11.6 Derive the state-space representation of the network in Figure 11.9.

Solution: Let the current through the inductor L be $x_1 = i_1 - i_2$, the voltage across the capacitor be x_2 and $e(t)$ be the input voltage to the circuit. Let us choose the current through the inductance and the voltage across the capacitance as the state variables x_1 and x_2 of the system.

We can write the mesh equations as

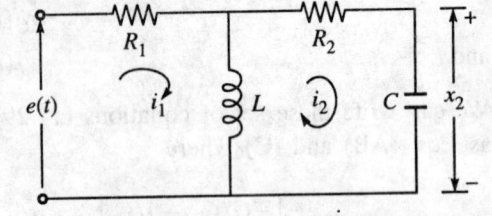

FIGURE 11.9 Example 11.6.

$$L\frac{d}{dt}(i_1 - i_2) + R_1 i_1 = e(t)$$

and
$$\frac{1}{C}\int i_2 \, dt + R_2 i_2 + L\frac{d}{dt}(i_2 - i_1) = 0$$

In terms of x_1 and x_2, we can rewrite these equations as

$$L\frac{dx_1}{dt} + R_1(x_1 + i_2) = e(t)$$

$$x_2 + R_2 i_2 - L\frac{dx_1}{dt} = 0 \qquad \left(\because x_1 = i_1 - i_2 \text{ and } x_2 = \frac{1}{C}\int i_2 \, dt \right)$$

or
$$i_2 = \frac{L\dot{x}_1}{R_2} - \frac{x_2}{R_2}$$

Therefore, $\quad L\dot{x}_1 + R_1 x_1 + \left(\dfrac{R_1 L}{R_2}\right)\dot{x}_1 - \left(\dfrac{R_1}{R_2}\right)x_2 = e(t)$

or
$$\dot{x}_1 = -\frac{R_1 R_2}{L(R_1 + R_2)} x_1 + \frac{R_1}{L(R_1 + R_2)} x_2 + \frac{R_2}{L(R_1 + R_2)} e(t) \qquad (11.35)$$

Again as $\quad x_2 = \dfrac{1}{C}\int i_2 \, dt$, we get

$$\dot{x}_2 = \frac{1}{C} i_2 = \frac{1}{C}\left(\frac{L}{R_2}\dot{x}_1 - \frac{x_2}{R_2}\right) = \frac{L}{CR_2}\dot{x}_1 - \frac{1}{CR_2} x_2$$

$$= \frac{L}{CR_2}\left[\left(-\frac{R_1 R_2}{L(R_1+R_2)}\right)x_1 + \frac{R_1}{L(R_1+R_2)}x_2 + \frac{R_2}{L(R_1+R_2)}e(t)\right] - \frac{1}{CR_2}x_2$$

or

$$\dot{x}_2 = -\frac{R_1}{C(R_1+R_2)}x_1 - \frac{1}{C(R_1+R_2)}x_2 + \frac{1}{C(R_1+R_2)}e(t) \quad (11.36)$$

From Eqs. (11.35) and (11.36), we can write the state-space representation in vector-matrix differential equations as

$$\frac{d}{dt}\begin{bmatrix}x_1 \\ x_2\end{bmatrix} = \begin{bmatrix} \dfrac{-R_1 R_2}{L(R_1+R_2)} & \dfrac{R_1}{L(R_1+R_2)} \\ \dfrac{-R_1}{C(R_1+R_2)} & \dfrac{-1}{C(R_1+R_2)} \end{bmatrix}\begin{bmatrix}x_1 \\ x_2\end{bmatrix} + \begin{bmatrix} \dfrac{R_2}{L(R_1+R_2)} \\ \dfrac{1}{C(R_1+R_2)} \end{bmatrix}e(t)$$

and output $\quad y = [0\ 1]\begin{bmatrix}x_1 \\ x_2\end{bmatrix}$

EXAMPLE 11.7(a) Write the state-variable formulation of the network shown in Figure 11.10 where all component values are of unity magnitude.

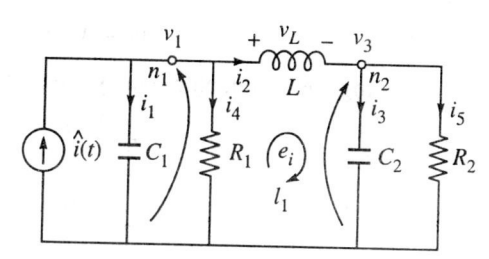

FIGURE 11.10 Example 11.7(a).

Solution: The given component values are $R_1 = R_2 = 1\ \Omega$, $L_1 = 1$ H, $C_1 = C_2 = 1$ F.

Choose v_1, i_2 and v_3 as state variables, where v_1, v_3 are the voltages across the capacitors C_1 and C_2 and i_2 is the current through the inductor L. Write the independent node and loop equations as

Node n_1: $\quad \hat{i} = i_4 + i_1 + i_2 = \dfrac{v_1}{1} + 1 \cdot \dfrac{dv_1}{dt} + i_2$

Loop l_1: $\quad v_1 = v_L + v_3 = 1 \cdot \dfrac{di_2}{dt} + \dfrac{v_3}{1}$

Node n_2: $\quad i_2 = i_3 + i_5 = 1 \cdot \dfrac{dv_3}{dt} + \dfrac{v_3}{1}$

Rearranging these equations, we get

$$\frac{dv_1}{dt} = -v_1 - i_2 + \hat{i}$$

$$\frac{di_2}{dt} = v_1 - v_3$$

$$\frac{dv_3}{dt} = i_2 - v_3$$

The state equation of the state-variable representation in vector-matrix differential equation is

$$\frac{d}{dt}\begin{bmatrix} v_1 \\ i_2 \\ v_3 \end{bmatrix} = \begin{bmatrix} -1 & -1 & 0 \\ 1 & 0 & -1 \\ 0 & 1 & -1 \end{bmatrix} \begin{bmatrix} v_1 \\ i_2 \\ v_3 \end{bmatrix} + \begin{bmatrix} 1 \\ 0 \\ 0 \end{bmatrix} \hat{i}(t)$$

EXAMPLE 11.7(b) Consider the same network as in Figure 11.10. Choose charges and fluxes as the state variables. Write the state-variable representation.

Solution: Let q_1 and q_3 be the charges across the capacitors and ϕ_2 be the flux through the inductor.

Writing the independent node and loop equations as

Node n_1: $\quad\hat{i} = i_4 + i_1 + i_2$

Node n_2: $\quad i_2 - i_3 - i_5 = 0$

Loop l_1: $\quad v_1 = v_L + v_3$

After putting the parameter values, we also have

$$v_1 = \frac{q_1}{C_1} = q_1; \quad i_1 = \frac{dq_1}{dt}$$

$$i_2 = \frac{\phi_2}{L} = \phi_2; \quad i_3 = \frac{dq_3}{dt}$$

$$v_3 = \frac{q_3}{C_2} = q_3; \quad v_3 = \frac{d\phi_2}{dt}$$

$$i_4 = \frac{v_1}{R_1} = v_1; \quad i_5 = \frac{v_3}{R_2} = v_3$$

By algebraic manipulation, eliminate all variables except the state variables. After rearranging, we get the state-variable formulation in vector-matrix differential equation as

$$\frac{d}{dt}\begin{bmatrix} q_1 \\ \phi_2 \\ q_3 \end{bmatrix} = \begin{bmatrix} -1 & -1 & 0 \\ 1 & 0 & -1 \\ 0 & 1 & -1 \end{bmatrix} \begin{bmatrix} q_1 \\ \phi_2 \\ q_3 \end{bmatrix} + \begin{bmatrix} 1 \\ 0 \\ 0 \end{bmatrix} \hat{i}$$

Note that, instead of the voltage across the capacitor, charge q across the capacitor is chosen as the state-variable because of the relation $v = q/C$ (that is, v is proportional to q). Similarly, instead of the current through the inductor, the flux through the inductor is chosen as the state variable because of the relation $i_2 = \phi_2/L$ (i.e. i_2 is proportional to ϕ_2).

EXAMPLE 11.8 Consider a series *RLC* network driven by a voltage source. Write the state-variable representation.

Solution: The dynamic behaviour of the system is completely defined for $t \geq t_0$, if the initial values of the current $i(t_0)$ and capacitor voltage $v_C(t_0)$, together with the input excitation voltage

$u(t)$ for $t \geq t_0$ are known. Thus, the state of the network for $t \geq t_0$ is completely determined by $i(t_0)$ and $v_C(t_0)$, together with the input excitation $u(t)$.

Let us choose the current $i(t)$ through the inductor and the voltage $v_C(t)$ across the capacitor as the state variables.

The equations describing the dynamics of the system are

$$L\frac{di}{dt} + Ri + v_C(t) = u(t)$$

and

$$C\frac{dv_C}{dt} = i$$

After simplifying, we get the state variable formulation in matrix form as

$$\frac{d}{dt}\begin{bmatrix} i \\ v_C \end{bmatrix} = \begin{bmatrix} -R/L & -1/L \\ 1/C & 0 \end{bmatrix}\begin{bmatrix} i \\ v_C \end{bmatrix} + \begin{bmatrix} 1/L \\ 0 \end{bmatrix}u(t)$$

Let us choose the set of variables $[x_1 \ x_2]^T$ as the state variables, where

$$x_1 = i \quad \text{and} \quad x_2 = v_C$$

The output is

$$y = v_C$$

which can be written in matrix form as

$$y = [0 \ 1]\begin{bmatrix} x_1 \\ x_2 \end{bmatrix}$$

Drill Problem 11.5

Write the state variable formulation of the networks shown in Figure 11.11.

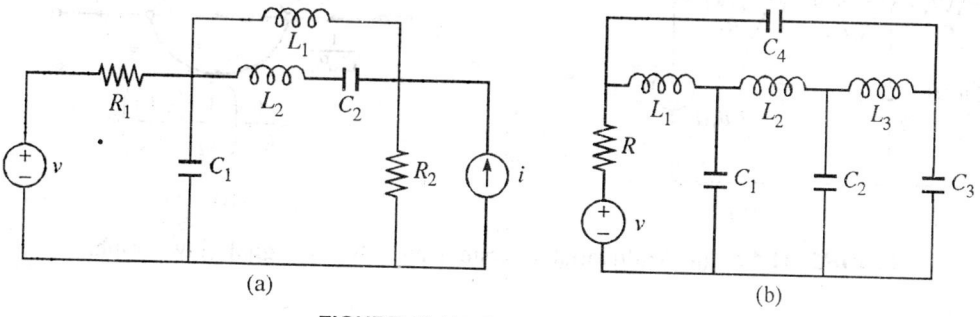

FIGURE 11.11 Drill Problem 11.5.

11.4 State Space Representation of Multivariable Systems

Consider the system shown in Figure 11.12(a). The two node equations (KCL) are:

At node x_1:
$$\frac{x_1 - x_2}{R_3} + \frac{x_1 - u_1}{R_1} + C_1 \frac{dx_1}{dt} = 0$$

And at node x_2:
$$\frac{x_2 - x_1}{R_3} + \frac{x_2 - u_2}{R_2} + C_2 \frac{dx_2}{dt} = 0$$

In matrix form the above two equations can be written as

$$\frac{d}{dt}\begin{bmatrix} x_1 \\ x_2 \end{bmatrix} = \begin{bmatrix} -\frac{1}{C_1}\left(\frac{1}{R_1} + \frac{1}{R_3}\right) & \frac{1}{C_1 R_3} \\ \frac{1}{C_2 R_3} & -\frac{1}{C_2}\left(\frac{1}{R_2} + \frac{1}{R_3}\right) \end{bmatrix}\begin{bmatrix} x_1 \\ x_2 \end{bmatrix} + \begin{bmatrix} \frac{1}{C_1 R_1} & 0 \\ 0 & \frac{1}{C_2 R_2} \end{bmatrix}\begin{bmatrix} u_1 \\ u_2 \end{bmatrix}$$

The output equation is

$$\begin{bmatrix} y_1 \\ y_2 \end{bmatrix} = \begin{bmatrix} x_1 \\ x_2 \end{bmatrix} = \begin{bmatrix} 1 & 0 \\ 0 & 1 \end{bmatrix}\begin{bmatrix} x_1 \\ x_2 \end{bmatrix}$$

The signal flow graph of the system is shown in Figure 11.12(b).

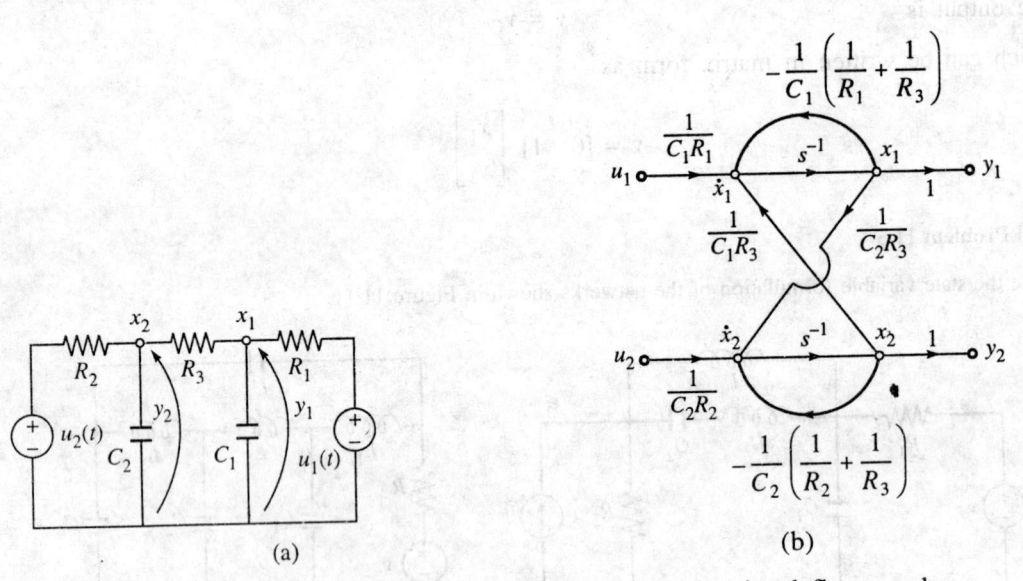

FIGURE 11.12 (a) Multivariable system and (b) its signal flow graph.

11.5 Set of Minimal State Variable Representation

In the earlier examples of state-variable formulation of electric circuits consisting of energy storage elements, the current or flux through the inductor and the voltage or charge across the capacitor are chosen as the state variables. Hence the order of the system matrix A becomes

State-Variable Formulation

the same as that of the number of energy storage elements in the circuit. This is not necessarily true always. A systematic procedure has been developed for state-variable formulation with minimal set of state variables. This has been illustrated with an example.

Let us write the minimal set of state variable formulation of a network shown in Figure 11.13(a). This network has six number of energy storage elements. It has nodes n_1 to n_5 with n_6 as the datum node, and it has four loops l_1 to l_4, and it is driven by the voltage source u. The graph of the given network is depicted in Figure 11.13(b).

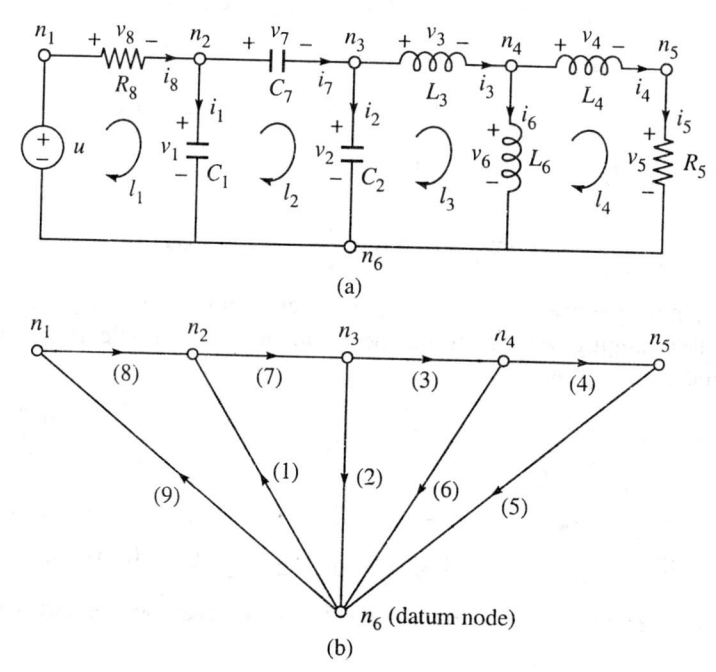

FIGURE 11.13(a) and (b) (a) Network and (b) it graph.

Procedure for writing state equation is:

1. Choose a normal tree of the given network.
2. Take either the voltages or charges across the capacitors, which are in the tree branches (twigs), and either the currents or fluxes through the inductors, which are in the co-tree (chords), as the state variables.
3. Write the independent KVL, KCL and branch voltage-current relations (VCR).
4. Reduce to retain only the state variables by eliminating all the non-state variables.
5. Rearrange to get the vector-matrix differential equation for state-variable representation.

The tree is drawn as continuous lines and the corresponding co-tree is drawn as dotted lines in Figure 11.13(c). Following step 2, the minimal

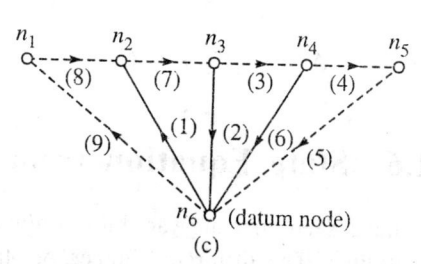

FIGURE 11.13(c) Tree twigs and chords.

state variables are the tree branch capacitor voltages ($v_1 = x_1$ and $v_2 = x_2$) and the currents through the co-tree chords ($i_3 = x_3$ and $i_4 = x_4$). Hence the state vector $X(t)$ is given by

$$X(t) = [x_1(t) \quad x_2(t) \quad x_3(t) \quad x_4(t)]^T = [v_1 \quad v_2 \quad i_3 \quad i_4]^T$$

Writing KCL at nodes:

node n_2: $i_8 = i_7 + C_1 \dot{x}_1$

node n_3: $i_7 = x_3 + C_2 \dot{x}_2$

node n_4: $x_3 = i_6 + x_4$

Writing KVL for loops:

loop l_1: $u = v_8 + x_1$

loop l_2: $x_1 = v_7 + x_2$

loop l_3: $x_2 = v_6 + L_3 \dot{x}_3$

loop l_4: $v_6 = R_5 x_4 + L_4 \dot{x}_4$

Writing the voltage-current relations (VCR) as

$$i_7 = C_7 \dot{v}_7; \quad v_6 = L_6 \frac{di_6}{dt}; \quad v_8 = R_8 i_8$$

Following step 4 more precisely, $i_8, i_7, i_6, v_8, v_7, v_6$ are eliminated using the KCL, KVL, and VCR relations. Rearranging we can write the minimal state-variable formulation in vector-matrix differential equation as

$$\begin{bmatrix} C_7 & -(C_7+C_2) & 0 & 0 \\ C_7+C_1 & -C_7 & 0 & 0 \\ 0 & 0 & L_3+L_6 & -L_6 \\ 0 & 0 & L_6 & -(L_6+L_4) \end{bmatrix} \begin{bmatrix} \dot{x}_1 \\ \dot{x}_2 \\ \dot{x}_3 \\ \dot{x}_4 \end{bmatrix} = \begin{bmatrix} 0 & 0 & 1 & 0 \\ -\frac{1}{R_8} & 0 & 0 & 0 \\ 0 & 1 & 0 & 0 \\ 0 & 0 & 0 & R_5 \end{bmatrix} \begin{bmatrix} x_1 \\ x_2 \\ x_3 \\ x_4 \end{bmatrix} + \begin{bmatrix} 0 \\ \frac{1}{R_8} \\ 0 \\ 0 \end{bmatrix} u$$

Pre-multiplying both sides by the inverse of the matrix on the right of left-hand side, we get

$$\frac{d}{dt}\begin{bmatrix} x_1 \\ x_2 \\ x_3 \\ x_4 \end{bmatrix} = \begin{bmatrix} \frac{-(C_7+C_2)}{\Delta_1 R_8} & 0 & -\frac{C_7}{\Delta_1} & 0 \\ \frac{-C_7}{\Delta_1 R_8} & 0 & \frac{-(C_1+C_7)}{\Delta_1} & 0 \\ 0 & \frac{L_4+L_6}{\Delta_2} & 0 & \frac{R_5 L_6}{\Delta_2} \\ 0 & \frac{-L_6}{\Delta_2} & 0 & \frac{R_5(L_3+L_4)}{\Delta_2} \end{bmatrix} \begin{bmatrix} x_1 \\ x_2 \\ x_3 \\ x_4 \end{bmatrix} + \begin{bmatrix} \frac{C_2+C_7}{\Delta_1} \\ \frac{C_7}{\Delta_1} \\ 0 \\ 0 \end{bmatrix} u$$

where $\Delta_1 = C_1 C_2 + C_2 C_7 + C_1 C_7$ and $\Delta_2 = -(L_3 L_4 + L_3 L_6 + L_4 L_6)$

11.6 State Equation from Transfer Function

We have seen in state-variable formulation of physical systems that the choice of state is not unique. The different choices of state variables lead to different forms of state variable realizations, though the transfer function remains invariant. In a similar way there can be

State-Variable Formulation

different forms of simulation resulting from the unique transfer function. On the other hand, the different forms of simulation lead to different forms of system matrices whereas the transfer function remains same and unique. Now we will discuss the systematic development of different forms of state variable formulations through simulation.

The rational transfer function is the ratio of two polynomials in s as

$$G(s) = \frac{N(s)}{D(s)}$$

When the degree of polynomial $N(s)$ is less than that of $D(s)$, then the rational transfer function is called *strictly rational*. We will get the state-variable formulation for strictly rational transfer functions as

$$\dot{X} = AX + Bu$$

and output
$$y = CX$$

and the responding transfer function will be

$$g(s) = C[sI - A]^{-1}B$$

For a transfer function which is not strictly rational, i.e. the degree of the numerator polynomial $N(s)$ is equal to that of the denominator polynomial $D(s)$, then by division, we will break up the rational transfer function as

$$G(s) = D + g(s)$$

where $g(s)$ is strictly rational.

Hence the rational transfer function $G(s)$ from the state-variable formulation will be

$$G(s) = C[sI - A]^{-1}B + D$$

These will be illustrated with examples.

Consider the control system represented by the transfer function

$$g(s) = \frac{Y(s)}{U(s)} = \frac{K}{s^3 + a_2 s^2 + a_1 s + a_0}$$

Let
$$g(s) = \frac{Y(s)}{U(s)} = \frac{X_1(s)}{U(s)} \frac{Y(s)}{X_1(s)}$$

where
$$\frac{X_1(s)}{U(s)} = \frac{K}{s^3 + a_2 s^2 + a_1 s + a_0} \quad \text{and} \quad \frac{Y(s)}{X_1(s)} = 1$$

By cross multiplying, we get

$$(s^3 + a_2 s^2 + a_1 s + a_0) X_1(s) = K U(s) \quad \text{and} \quad Y(s) = X_1(s)$$

Since, by definition, a transfer function assumes zero initial conditions, this is simply the Laplace transform of a third-order differential equation with zero initial conditions. It is a simple matter to reconstruct the original differential equation by identifying the derivatives with powers of s, so that in the time domain we have

$$\dddot{x}_1(t) + a_2 \ddot{x}_1(t) + a_1 \dot{x}_1(t) + a_0 x_1(t) = Ku(t)$$

and the output
$$y(t) = x_1(t)$$

In order to express the in phase variables as $x_1 = y(t)$, and then according to the definition of the phase variables,

$$x_2 = \dot{x}_1(t) \quad \text{and} \quad x_3 = \dot{x}_2(t)$$

If these substitutions are made in the previous equation, the result is

$$\dot{x}_3 = -a_2 x_3 - a_1 x_2 - a_0 x_1 + Ku$$

And the three first-order differential equations are

$$\dot{x}_1 = x_2$$
$$\dot{x}_2 = x_3$$
$$\dot{x}_3 = -a_0 x_1 - a_1 x_2 - a_2 x_3 + Ku$$

where the output
$$y = x_1$$

These equations are of the same form as (AB) and (C), where

$$A = \begin{bmatrix} 0 & 1 & 0 \\ 0 & 0 & 1 \\ -a_0 & -a_1 & -a_2 \end{bmatrix}, \quad B = \begin{bmatrix} 0 \\ 0 \\ K \end{bmatrix}, \quad C^T = \begin{bmatrix} 1 \\ 0 \\ 0 \end{bmatrix}$$

and the desired system representation has been achieved. Here phase variables are chosen as state variables. Here the system matrix A is in companion form. Further, the state equations for a given transfer function are not unique although the transfer function can be uniquely determined from the state equations.

EXAMPLE 11.9 For the single-input-single-output case, consider the strictly rational transfer function

$$g(s) = \frac{Y(s)}{U(s)} = \frac{1}{s^3 + 7s^2 + 14s + 8}$$

Rewrite the above transfer function as

$$\frac{Y(s)}{X_1(s)} \cdot \frac{X_1(s)}{U(s)} = 1 \cdot \frac{1}{s^3 + 7s^2 + 14s + 8}$$

State-Variable Formulation

Let
$$\frac{X_1(s)}{U(s)} = \frac{1}{s^3 + 7s^2 + 14s + 8}$$

which leads to the differential equation in time domain as

$$\dddot{x}_1 + 7\ddot{x}_1 + 14\dot{x}_1 + 8x(t) = u(t)$$

Now let

$$\dot{x}_1 = x_2 = 0.x_1 + 1.x_2 + 0.x_3 + 0.u$$
$$\dot{x}_2 = x_3 = 0.x_1 + 0.x_2 + 1.x_3 + 0.u$$
$$\dot{x}_3 = -8x_1 - 14x_2 - 7x_3 + 1.u$$

which in matrix form can be written as

$$\frac{d}{dt}\begin{bmatrix} x_1 \\ x_2 \\ x_3 \end{bmatrix} = \begin{bmatrix} 0 & 1 & 0 \\ 0 & 0 & 1 \\ -8 & -14 & -7 \end{bmatrix} \begin{bmatrix} x_1 \\ x_2 \\ x_3 \end{bmatrix} + \begin{bmatrix} 0 \\ 0 \\ 1 \end{bmatrix} u$$

Now from
$$\frac{Y(s)}{X_1(s)} = 1$$

it leads to the output equation in time domain as

$$y(t) = x_1(t) = 1.x_1 + 0.x_2 + 0.x_3$$

or, in matrix form can be written as

$$y = \begin{bmatrix} 1 & 0 & 0 \end{bmatrix} \begin{bmatrix} x_1 \\ x_2 \\ x_3 \end{bmatrix}$$

The analog simulation is shown in Figure 11.14.

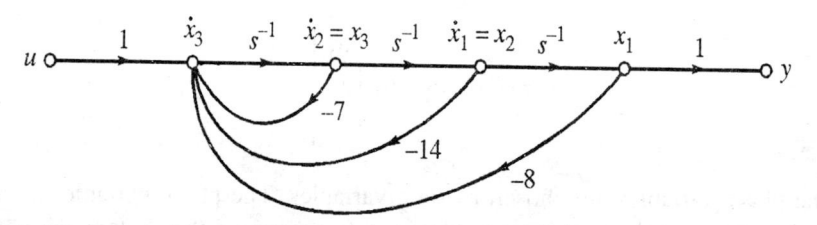

FIGURE 11.14 Example 11.9.

We focus our attention on the question of how the transfer function approach and the state variable approach are related. Consider the strictly proper rational transfer function $g(s)$ having poles and zeros as

$$g(s) = \frac{Y(s)}{U(s)} = \frac{K(c_1 s + c_0)}{s^3 + a_2 s^2 + a_1 s + a_0}$$

Let the transfer function $g(s)$ be divided into two parts in the following manner,

$$g(s) = \frac{Y(s)}{U(s)} = \frac{Y(s)}{X_1(s)} \cdot \frac{X_1(s)}{U(s)}$$

where $\dfrac{X_1(s)}{U(s)} = \dfrac{K}{s^3 + a_2 s^2 + a_1 s + a_0}$ and $\dfrac{Y(s)}{X_1(s)} = c_1 s + c_0$

From the expression $\dfrac{X_1(s)}{U(s)}$, the time domain equation is

$$\dddot{x}_1 + a_2 \ddot{x}_1 + a_1 \dot{x}_1 + a_0 x_1 = Ku$$

Let the state-variable representation be

$$\dot{x}_1 = x_2$$
$$\dot{x}_2 = x_3$$

then $\dot{x}_3 = -a_0 x_1 - a_1 x_2 - a_2 x_3 + Ku$

which can be written in vector-matrix differential equation as

$$\begin{bmatrix} \dot{x}_1 \\ \dot{x}_2 \\ \dot{x}_3 \end{bmatrix} = \begin{bmatrix} 0 & 1 & 0 \\ 0 & 0 & 1 \\ -a_0 & -a_1 & -a_2 \end{bmatrix} + \begin{bmatrix} 0 \\ 0 \\ K \end{bmatrix} u$$

The second expression $\dfrac{Y(s)}{X_1(s)}$ in time-domain can be written as

$$y(t) = c_1 x_2(t) + c_0 x_1(t) = c_0 x_1(t) + c_1 x_2(t) + 0 \cdot x_3(t)$$

which can be written in matrix form as

$$y = \begin{bmatrix} c_0 & c_1 & 0 \end{bmatrix} \begin{bmatrix} x_1 \\ x_2 \\ x_3 \end{bmatrix}$$

Here also the phase variables are chosen as state variables. The phase variables are defined as those particular state variables which are obtained from one of the system variables and its $(n-1)$ derivatives.

State-Variable Formulation

The complete dynamics of the system represented by the given transfer function can be written in state space model as:

$$\dot{X} = AX + Bu$$

and output
$$y = CX$$

where
$$A = \begin{bmatrix} 0 & 1 & 0 \\ 0 & 0 & 1 \\ -a_0 & -a_1 & -a_2 \end{bmatrix}, \quad B = \begin{bmatrix} 0 \\ 0 \\ K \end{bmatrix}, \quad C^T = \begin{bmatrix} c_0 \\ c_1 \\ 0 \end{bmatrix}$$

The state variable formulation consists of two equations. The vector-matrix differential equation $\dot{X} = AX + Bu$ is called the input equation as the state is related with the input. And the equation $Y = CX$ is called the output equation as the state is related with the output. Here both the equations are referred with respect to state vector X.

EXAMPLE 11.10 Consider the following transfer function where the degree of the numerator polynomial is less than that of the denominator polynomial, i.e. the transfer function is a strictly rational function:

$$g(s) = \frac{Y(s)}{U(s)} = \frac{3s^2 + 7s + 15}{s^3 + 7s^2 + 14s + 8}$$

Rewrite the transfer function as

$$\frac{Y(s)}{X_1(s)} \cdot \frac{X_1(s)}{U(s)} = (3s^2 + 7s + 15) \cdot \frac{1}{s^3 + 7s^2 + 14s + 8}$$

Now let
$$\frac{X_1(s)}{U(s)} = \frac{1}{s^3 + 7s^2 + 14s + 8}$$

which leads to the state-variable representation (input equation) as given in Example 11.9.

Let
$$\frac{Y(s)}{X_1(s)} = 3s^2 + 7s + 15$$

which in time domain can be written as

$$y(t) = 3\ddot{x}_1 + 7\dot{x}_1 + 15x_1 = 15x_1(t) + 7x_2(t) + 3x_3(t)$$

and from which the output equation in matrix form can be written as

$$y = \begin{bmatrix} 15 & 7 & 3 \end{bmatrix} \begin{bmatrix} x_1 \\ x_2 \\ x_3 \end{bmatrix}$$

The state equations for the transfer function lead to simple analogue simulation as shown in Figure 11.15.

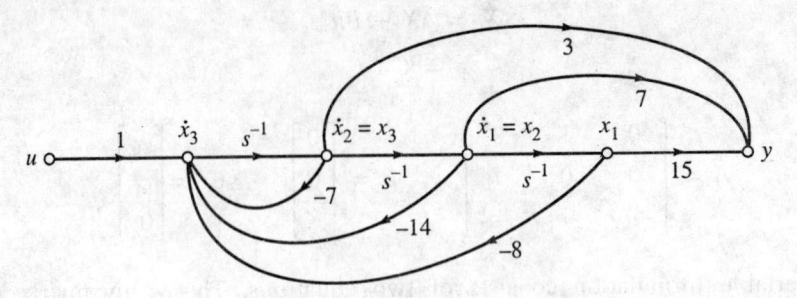

FIGURE 11.15 Example 11.10.

Drill Problem 10.6

Write the state-variable formulation with the choice of phase variable as state-variable for the given transfer function

$$\frac{s^2 + 4s + 3}{s^3 + 5s^2 + 7s + 1}$$

and then draw the SFG. Verify the transfer function applying the Mason's gain formula to the SFG obtained.

EXAMPLE 11.11 Consider the following rational transfer function where the degree of the numerator polynomial is same as that of the denominator polynomial:

$$G(s) = \frac{Y(s)}{U(s)} = \frac{2s^3 + 10s^2 + 21s + 23}{s^3 + 5s^2 + 7s + 10}$$

This can be written after division as

$$\frac{Y(s)}{U(s)} = 2 + \frac{7s + 3}{s^3 + 5s^2 + 7s + 10} = 2 + \frac{Y_1(s)}{U(s)}$$

where the strictly rational part is

$$g(s) = \frac{Y_1(s)}{U(s)} = \frac{7s + 3}{s^3 + 5s^2 + 7s + 10}$$

And obviously, $\quad Y(s) = Y_1(s) + 2U(s)$

Now the strictly rational part $g(s)$ leads to the state-variable formulation as

$$\dot{X} = \begin{bmatrix} 0 & 1 & 0 \\ 0 & 0 & 1 \\ -10 & -7 & -5 \end{bmatrix} X + \begin{bmatrix} 0 \\ 0 \\ 1 \end{bmatrix} u$$

and

$$y_1 = [3 \quad 7 \quad 0] X$$

but $G(s)$ leads to a generalized output equation as

$$y = CX + Du$$

where $\qquad C = [3 \ 7 \ 0] \quad$ and $\quad D = 2$

and in addition $G(s)$ leads to the same state equation, i.e.

$$\dot{X} = \begin{bmatrix} 0 & 1 & 0 \\ 0 & 0 & 1 \\ -10 & -7 & -5 \end{bmatrix} X + \begin{bmatrix} 0 \\ 0 \\ 1 \end{bmatrix} u$$

Now let us take the transfer function of Example 11.11 and see how the MATLAB command

$$[A, B, C, D] = \text{tf2ss(numden)}$$

will give a state-space representation.

MATLAB Program 11.2

```
echo off;
num= [2  10  21  23];
den= [1  5  7  10];
Tr=tf(num,den)
[a,b,c,d] =tf2ss(num,den)
>>
Transfer function:
 2s^3 + 10s^2 + 21s + 23
 ───────────────────────
    s^3 + 5s^2 + 7s + 10

a =

       -5    -7   -10
        1     0     1
        0     1     0

b =

        1
        0
        0

c =

        0     7     3

d =

        2
```

It may be noted that usually in control systems, for all practical purposes there is no direct coupling between input and output, i.e. usually $D = 0$. The simulation is shown in Figure 11.16. It may be noted that if the degree of the numerator polynomial is equal to that of the denominator polynomial, then the state equations are as

$$\dot{X} = AX + Bu$$

and output
$$y = CX + Du \; ; \; D \neq 0$$

Then the transfer function becomes
$$C[sI - A]^{-1}B + D$$

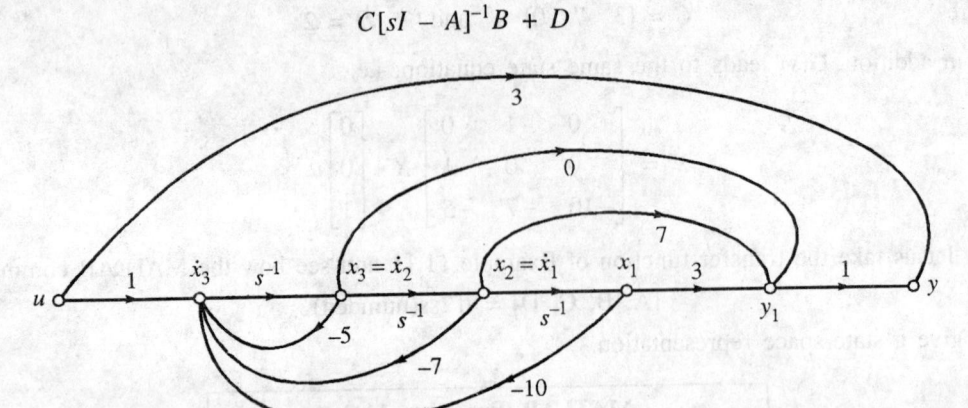

FIGURE 11.16 Example 11.11: simulation.

11.7 Simulation

To simulate a linear time-invariant system in an analog computer, only integrators are used, not the differentiators. The question that may arise is why not use differentiators? In practice, all signals are corrupted by noise and when such a signal is differentiated, the derivative of the usually rapidly varying noise will "drown out" the derivative of the signal. In reality, the derivative of a noise step after differentiation becomes an impulse and the practical amplifier gets saturated, and linearity is lost, that means, the closed-loop system becomes an open-loop one. For these reasons, the simulation in reality is done by using integrators only instead of differentiators.

First approach: using superposition

Consider a third-order system (strictly rational transfer function case) dynamics as

$$\dddot{y} + a_2 \ddot{y} + a_1 \dot{y} + a_0 y = b_2 \ddot{u} + b_1 \dot{u} + b_0 u$$

with zero initial conditions or the transfer function

$$G(s) = \frac{Y(s)}{U(s)} = \frac{b_2 s^2 + b_1 s + b_0}{s^3 + a_2 s^2 + a_1 s + a_0}$$

or
$$\frac{Y(s)}{U(s)} = \frac{Y(s)}{X_1(s)} \cdot \frac{X_1(s)}{U(s)} = b_2 s^2 + b_1 s + b_0 \cdot \frac{1}{s^3 + a_2 s^2 + a_1 s + a_0}$$

where
$$\frac{X_1(s)}{U(s)} = \frac{1}{s^3 + a_2 s^2 + a_1 s + a_0}$$

State-Variable Formulation

The simulation is shown in Figure 11.17.

Again
$$\frac{Y(s)}{X_1(s)} = b_2 s^2 + b_1 s + b_0$$

or
$$Y(s) = (b_2 s^2 + b_1 s + b_0) X_1(s)$$

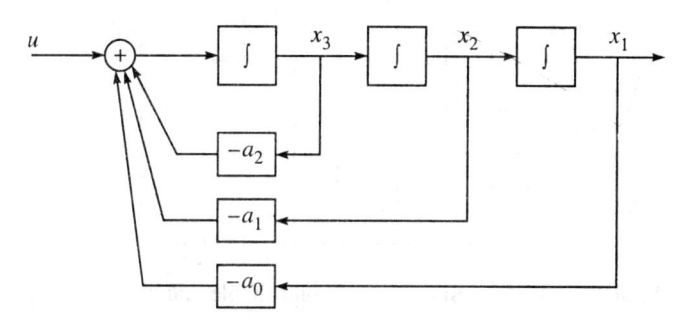

FIGURE 11.17 Simulation of $X_1(s)/U(s)$.

In time domain, we get
$$y(t) = b_2 \ddot{x}_1 + b_1 \dot{x}_1 + b_0 x_1$$

The simulation (from the logical point of view) is shown in Figure 11.18. Now eliminate the differentiators by noticing that in effect integrators and differentiators are in series. Hence move the lines with differentiators over the requisite numbers of integrators, thereby eliminating the need for differentiators. The complete simulation is as shown in Figure 11.19. This is the *controllable canonical form* of realization having the state variable formulation as

$$\dot{X} = \begin{bmatrix} 0 & 1 & 0 \\ 0 & 0 & 1 \\ -a_0 & -a_1 & -a_2 \end{bmatrix} X + \begin{bmatrix} b_0 \\ b_1 \\ b_2 \end{bmatrix} u$$

and output
$$y = \begin{bmatrix} 1 & 0 & 0 \end{bmatrix} X$$

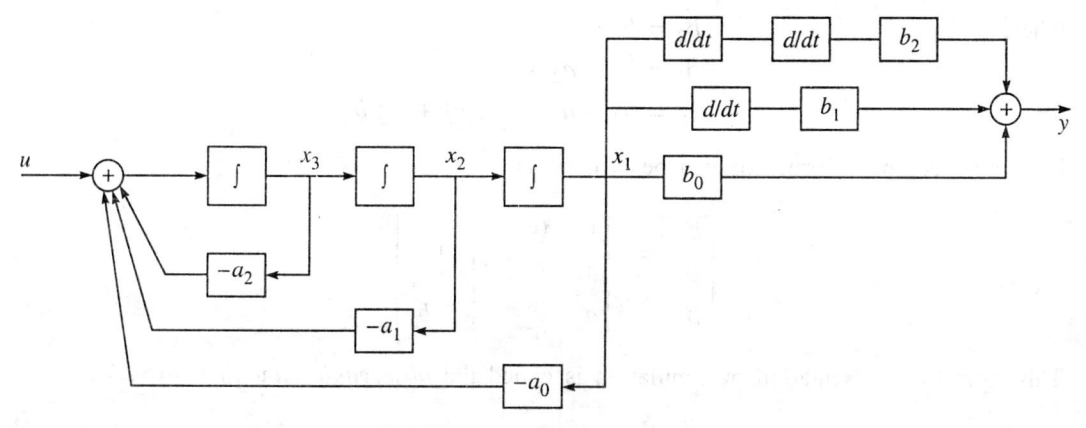

FIGURE 11.18 Simulation.

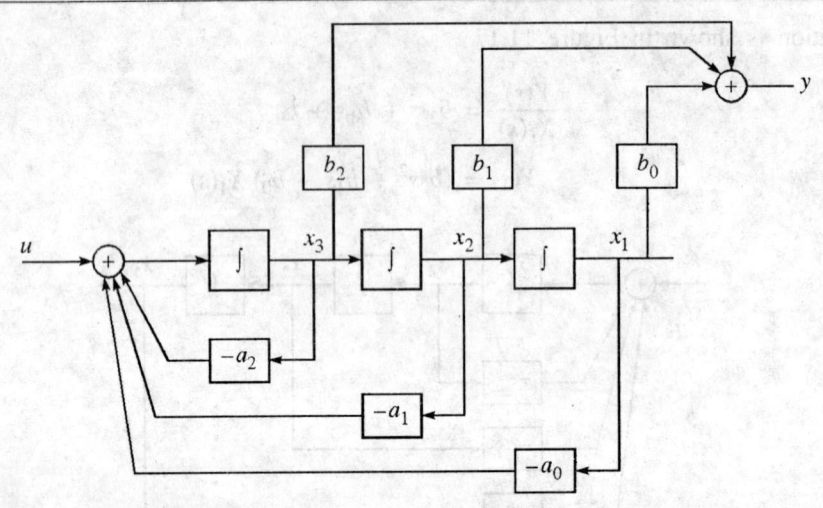

FIGURE 11.19 Simulation: controllable canonical form.

A different realization

Let us first look at the previous procedure in a slightly different way. We rewrite

$$Y(s) = \frac{P(s)}{s^3 + a_2 s^2 + a_1 s + a_0}$$

where
$$P(s) = (b_2 s^2 + b_1 s + b_0)U(s)$$

In time domain,
$$p(t) = b_2 \ddot{u} + b_1 \dot{u} + b_0 u$$

and $y(t)$ becomes
$$\dddot{y} + a_2 \ddot{y} + a_1 \dot{y} + a_0 y = p(t)$$

The time-domain simulation is shown in Figure 11.20.

Now we try to eliminate the differentiators by moving the input lines over the integrators, then the quantities in the feedback lines are affected. However, by careful manipulation, we can obtain the form shown in Figure 11.21.

where
$$\beta_2 = b_2$$
$$\beta_1 = b_1 - a_2 b_2$$
$$\beta_0 = b_0 - a_2 b_1 - a_1 b_2 + a_2^2 b_2$$

In a more compact form, this can be written as

$$\begin{bmatrix} \beta_2 \\ \beta_1 \\ \beta_0 \end{bmatrix} = \begin{bmatrix} 1 & 0 & 0 \\ a_2 & 1 & 0 \\ a_1 & a_2 & 1 \end{bmatrix}^{-1} \begin{bmatrix} b_2 \\ b_1 \\ b_0 \end{bmatrix}$$

This form of representation of simulation is called the *observable canonical form*.

State-Variable Formulation

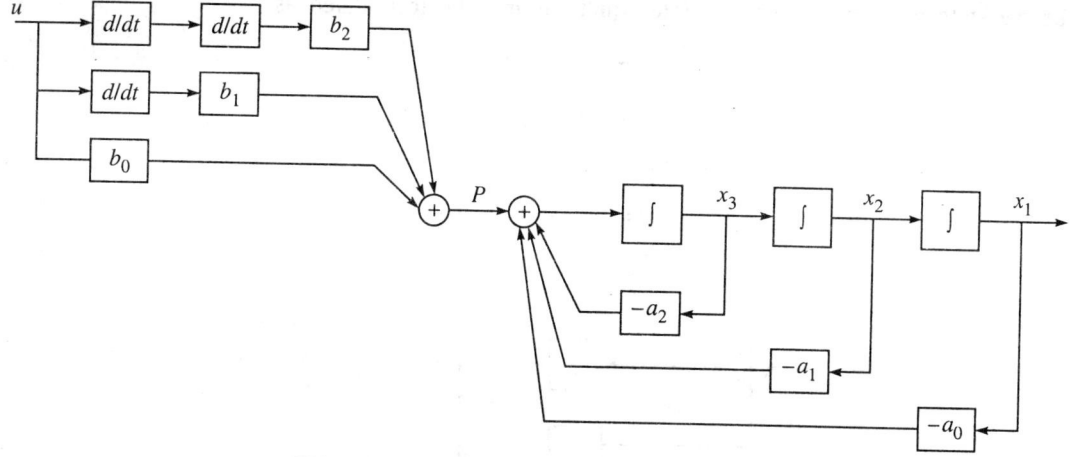

FIGURE 11.20 Time-domain simulation.

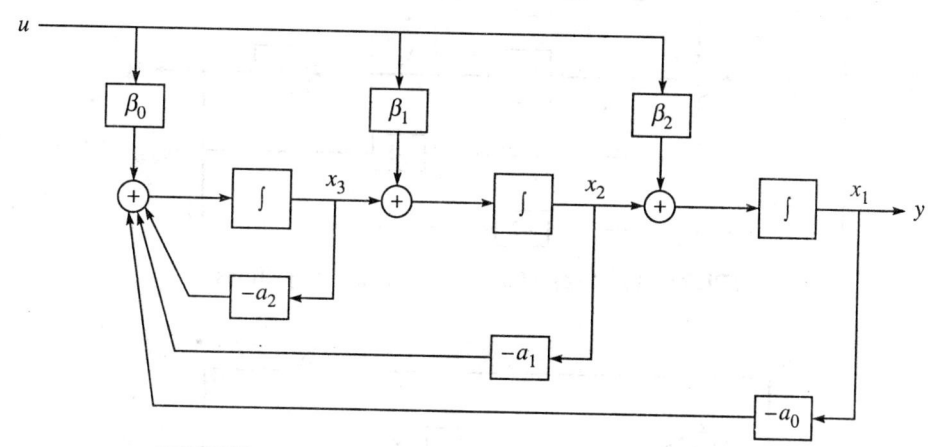

FIGURE 11.21 Simulation: observable canonical form.

11.8 Simultaneous Equations

The simultaneous equations relating the two input controls, $u_1(t)$ and $u_2(t)$, and the two output variables, $y_1(t)$ and $y_2(t)$, are given by

$$\ddot{y}_1 + 3\dot{y}_1 + 2y_2 = u_1$$
$$\ddot{y}_2 + \dot{y}_1 + y_2 = u_2$$

The block diagram of simulation is shown in Figure 11.22(a) and the choice of output of the integrators as the state variables is shown in Figure 11.22(b).

Let
$$y_1 = x_1 \text{ and } y_2 = x_3$$
$$\dot{y}_1 = x_2 \text{ and } \dot{y}_2 = x_4$$

be the state variables. Then the state equations may be formulated as

$$\dot{x}_1 = x_2$$
$$\dot{x}_2 = -3x_2 - 2x_3 + u_1$$
$$\dot{x}_3 = x_4$$
$$\dot{x}_4 = -x_2 - x_3 + u_2$$

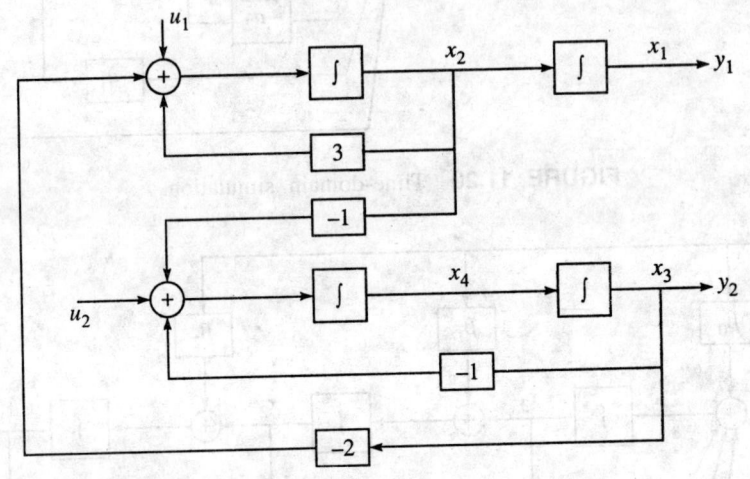

FIGURE 11.22(a) Simulation of state equations.

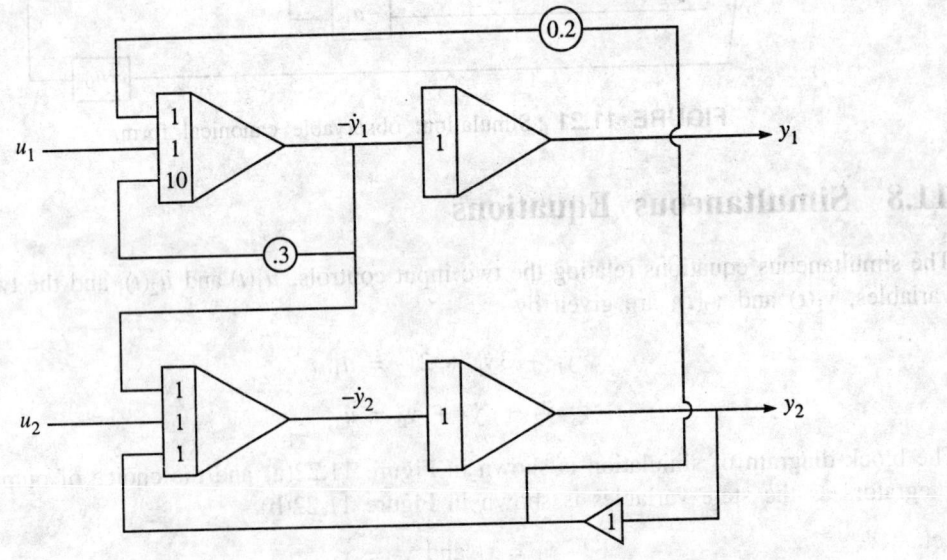

FIGURE 11.22(b) Simulation showing output of integrators as state variables.

The output equations are

$$y_1 = x_1$$

and

$$y_2 = x_3$$

In matrix notation, the above equations may be written as

$$\dot{X} = AX + BU$$

and output

$$\dot{Y} = CX + DU$$

where

the state vector, $X = [x_1 \ x_2 \ x_3 \ x_4]^T$

the output vector, $Y = [y_1 \ y_2]^T$

the input vector, $U = [u_1 \ u_2]^T$

and

$$A = \begin{bmatrix} 0 & 1 & 0 & 0 \\ 0 & -3 & -2 & 0 \\ 0 & 0 & 0 & 1 \\ 0 & -1 & -1 & 0 \end{bmatrix}, \quad B = \begin{bmatrix} 0 & 0 \\ 1 & 0 \\ 0 & 0 \\ 0 & 1 \end{bmatrix}$$

$$C = \begin{bmatrix} 1 & 0 & 0 & 0 \\ 0 & 0 & 1 & 0 \end{bmatrix}, \quad D = \begin{bmatrix} 0 & 0 \\ 0 & 0 \end{bmatrix}$$

11.9 Dual Representation

Further, it may be noted that for the scalar case (single-input-single-output) we get the transfer function as

$$G(s) = C(sI - A)^{-1}B$$

for the dynamics having state variable formulation as

$$\dot{X} = AX + Bu \quad \text{(AB)}$$

and output $\quad y = CX \quad \text{(C)}$

Making the transpose, we get

$$G(s) = [C(sI - A)^{-1}B]^T = B^T(sI - A^T)^{-1}C^T$$

for the dynamics having state variable formulation

$$\dot{X} = A^T X + C^T u \quad (A^T C^T)$$

and output $\quad y = B^T X \quad (B^T)$

Then the pair of Eqs. (AB) and (C) representing the system is dual to the system represented by the pair of Eqs. ($A^T C^T$) and (B^T).

Now the transfer function of Example 11.9 can also be represented by the following state equations

$$\dot{X} = \begin{bmatrix} 0 & 0 & -8 \\ 1 & 0 & -14 \\ 0 & 1 & -7 \end{bmatrix} X + \begin{bmatrix} 15 \\ 7 \\ 3 \end{bmatrix} u$$

and output
$$y = [0 \ 0 \ 1] X$$

The validity of this representation is evident if we note that the system matrix A for this case is the transpose of the system matrix A in Example 11.9.

Similarly the B and C matrices for this representation are the transpose of C, i.e. C^T and the transpose of B, i.e. B^T respectively of Example 11.10.

Let us recheck the transfer function from state-variable formulation and rewrite as

$$\dot{x}_1 = -8x_3 + 15u$$
$$\dot{x}_2 = x_1 - 14x_3 + 7u$$
$$\dot{x}_3 = x_2 - 7x_3 + 3u$$

and output
$$y = x_3$$

Differentiating y, we get

$$\dot{y} = \dot{x}_3 = x_2 - 7x_3 + 3u = x_2 - 7y + 3u$$

Differentiating further, we get

$$\ddot{y} = \dot{x}_2 - 7\dot{y} + 3\dot{u} = (x_1 - 14x_3 + 7u) - 7\dot{y} + 3\dot{u}$$

or
$$\ddot{y} = x_1 - 14y + 7u - 7\dot{y} + 3\dot{u}$$

Differentiating $\ddot{y}$ further, and after rearranging, we get

$$\dddot{y} + 7\ddot{y} + 14\dot{y} + 8y = 3\ddot{u} + 7\dot{u} + 15u$$

After taking the Laplace transform on both the sides with all initial conditions as zero, the transfer function becomes

$$G(s) = \frac{Y(s)}{U(s)} = \frac{3s^2 + 7s + 15}{s^3 + 7s^2 + 14s + 8}$$

The state equation of the transfer function of Example 11.9 can be represented by a set of equations. Again the state equation of the same transfer function is being represented by a set of another equations as shown above. These two representations are dual to each other.

Drill Problem 11.7

Write the state-variable formulation with the choice of phase variables as the state variables from the signal flow graph of Figure 11.23. Then determine the transfer function. Verify your result using Mason's gain formula

$$\text{Ans.} \quad A = \begin{bmatrix} -5 & -3 & -1 \\ 1 & 0 & 0 \\ 0 & 1 & 0 \end{bmatrix}, \quad B = \begin{bmatrix} 1 \\ 0 \\ 0 \end{bmatrix}, \quad C = [0 \ 0 \ 10]$$

$$\text{Transfer function} = \frac{10}{s^3 + 5s^2 + 3s + 1}$$

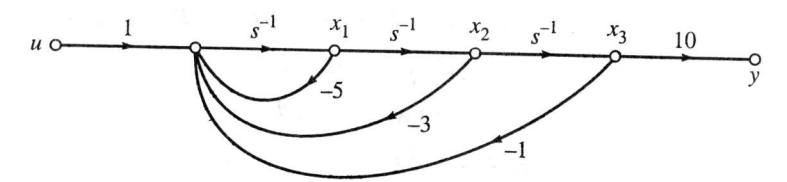

FIGURE 11.23 Drill Problem 11.7.

Drill Problem 11.8

In Drill Problem 11.7, Figure 11.23, interchanging the order of the state variables x_1 and x_3, write the state-variable formulation. Determine the transfer function. Give your comments.

$$\text{Ans.} \quad A = \begin{bmatrix} 0 & 1 & 0 \\ 0 & 0 & 1 \\ -1 & -3 & -5 \end{bmatrix}, \quad B = \begin{bmatrix} 0 \\ 0 \\ 1 \end{bmatrix}, \quad C = [10 \ 0 \ 0]$$

The transfer function remains the same as that of the Drill Problem 11.7.

11.10 Linear Transformation

The choice of state is not unique. Different forms of state-variable representations are possible though the transfer functions remain the same. One form of state-variable representation may have advantages over other forms. Hence there is a need for the transformation matrix. Modal matrix may transform the system matrix to the diagonal matrix either to canonical or to Jordan canonical form. Similarly, transformation to companion form of system matrix can also be possible provided the system is completely controllable and observable. In this chapter we will discuss the transformation matrix.

11.11 Similarity Transformation

The choice of state is not unique for a given system. Suppose that there exists a set of state variables

$$X = [x_1 \ x_2 \ x_3 \ \ldots \ x_n]^T$$

so that a linear or similarity transformation

$$X = P\overline{X}; \quad |P| \neq 0, \quad P^{-1} \text{ exists}$$

i.e.
$$\overline{X} = P^{-1}X \qquad (11.37)$$

transforms to another set of state variables

$$\overline{X} = \begin{bmatrix} \bar{x}_1 & \bar{x}_2 & \bar{x}_3 & \cdots & \bar{x}_n \end{bmatrix}^T$$

Differentiating Eq. (11.37) and using Eq. (AB), we get

$$\dot{\overline{X}} = P^{-1}\dot{X} = P^{-1}[AX + BU] = P^{-1}AX + P^{-1}BU$$
$$= P^{-1}A[P\overline{X}] + P^{-1}BU$$
$$= [P^{-1}AP]\overline{X} + [P^{-1}B]U$$
$$= \overline{A}\,\overline{X} + \overline{B}\,U$$

i.e.
$$\dot{\overline{X}} = \overline{A}\,\overline{X} + \overline{B}\,U \qquad (11.38)$$

and output
$$Y = CX = CP\overline{X} = \overline{C}\,\overline{X}$$

i.e.
$$Y = \overline{C}\,\overline{X} \qquad (11.39)$$

where
$$\left.\begin{array}{l} \overline{A} = P^{-1}AP \\ \overline{B} = P^{-1}B \\ \overline{C} = CP \end{array}\right\} \qquad (11.40)$$

with P as a non-singular transformation matrix.

Hence by similarity transformation, the transformed system can be represented by the vector-matrix differential equation as

$$\dot{\overline{X}} = \overline{A}\,\overline{X} + \overline{B}\,U \qquad \text{(AB)}$$

and output
$$Y = \overline{C}\,\overline{X}$$

where
$$\left.\begin{array}{l} \overline{A} = P^{-1}AP \\ \overline{B} = P^{-1}B \\ \overline{C} = CP \end{array}\right\} \qquad (11.41)$$

It may be observed that the characteristic equation and hence eigenvalues of A and $\overline{A}$ are invariant under similarity transformation. Further, note that the transfer function remains invariant under similarity transformation.

EXAMPLE 11.12 Consider the system described by the differential equation

$$\dddot{y} + 6\ddot{y} + 11\dot{y} + 6y = 6u \qquad (11.42)$$

where u is the input and y the output. Obtain the state-variable formulation.

State-Variable Formulation

Solution: Let us choose the state variables as the phase variables, as

$$x_1 = y$$
$$x_2 = \dot{y}$$
and
$$x_3 = \ddot{y}$$
then we get
$$\dot{x}_1 = x_2$$
$$\dot{x}_2 = x_3$$
$$\dot{x}_3 = -6x_1 - 11x_2 - 6x_3 + 6u$$

which can be written in matrix form as

$$\begin{bmatrix} \dot{x}_1 \\ \dot{x}_2 \\ \dot{x}_3 \end{bmatrix} = \begin{bmatrix} 0 & 1 & 0 \\ 0 & 0 & 1 \\ -6 & -11 & -6 \end{bmatrix} \begin{bmatrix} x_1 \\ x_2 \\ x_3 \end{bmatrix} + \begin{bmatrix} 0 \\ 0 \\ 6 \end{bmatrix} u \quad (11.43)$$

i.e.
$$\dot{X} = AX + Bu$$

and the output equation can be written as

$$y = x_1 = 1.x_1 + 0.x_2 + 0.x_3$$

Writing the output equation in matrix form, we get

$$y = [1 \quad 0 \quad 0] X$$

The block diagram representation of the state equations is shown in Figure 11.24 and it is evident from the simulation that the states are coupled. With this choice of states, the states are coupled, i.e. as one varies one parameter of any of the states, the other states also vary.

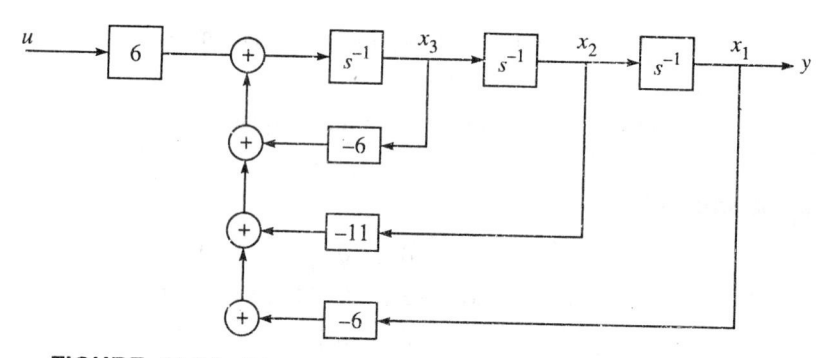

FIGURE 11.24 Block diagram representation of state equations.

The characteristic equation becomes

$$|sI - A| = s^3 + 6s^2 + 11s + 6 = 0$$

Note that the system matrix A has the special form called the companion form of matrix with elements in the last row as the negative of the coefficients of the characteristic equation.

In general, for an $n \times n$ companion form of system matrix A, the element a_{ij} of the companion matrix is defined as

$$a_{ij} = 0; \quad j \neq i+1$$
$$= 1; \quad j = i+1 \tag{11.44}$$

for $i = 1, 2, \ldots, n-1$

and the elements of the last row, i.e. nth row are defined as

$$a_{nj} = -a_{j-1}; \quad j = 1, 2, \ldots, n$$

which are the negative coefficients of the characteristic equation.

EXAMPLE 11.13 The choice of state is not unique. This can be illustrated with the transfer function of Example 11.2, as

$$\frac{Y(s)}{U(s)} = \frac{6}{s^3 + 6s^2 + 11s + 6} = \frac{6}{(s+1)(s+2)(s+3)}$$

$$= \frac{3}{s+1} + \frac{-6}{s+2} + \frac{3}{s+3}$$

Now, $$Y(s) = \frac{3}{s+1} U(s) + \frac{-6}{s+2} U(s) + \frac{3}{s+3} U(s) \tag{11.45}$$

Let us choose:

$\dfrac{3U(s)}{s+1} = Z_1(s)$, which leads to the time domain equation as $\dot{z}_1 = -z_1 + 3u$

$\dfrac{-6U(s)}{s+2} = Z_2(s)$, which leads to the time domain equation as $\dot{z}_2 = -2z_2 - 6u$

$\dfrac{3U(s)}{s+3} = Z_3(s)$, which leads to the time domain equation as $\dot{z}_3 = -3z_3 + 3u$

and these can be written in matrix form as

$$\begin{bmatrix} \dot{z}_1 \\ \dot{z}_2 \\ \dot{z}_3 \end{bmatrix} = \begin{bmatrix} -1 & 0 & 0 \\ 0 & -2 & 0 \\ 0 & 0 & -3 \end{bmatrix} \begin{bmatrix} z_1 \\ z_2 \\ z_3 \end{bmatrix} + \begin{bmatrix} 3 \\ -6 \\ 3 \end{bmatrix} u \tag{11.46}$$

i.e. $\dot{Z} = \Lambda Z + B^* u$

The transformed output $Y(s)$ can be rewritten as

$$Y(s) = Z_1(s) + Z_2(s) + Z_3(s)$$

which in time domain can be written as

$$y(t) = z_1 + z_2 + z_3$$

and the output in matrix form can be written as

$$y = \begin{bmatrix} 1 & 1 & 1 \end{bmatrix} \begin{bmatrix} z_1 \\ z_2 \\ z_3 \end{bmatrix} \qquad (11.47)$$

i.e.
$$y = C * Z$$

The block diagram representation is shown in Figure 11.25. It is obvious from this state diagram that the states are decoupled, that is, as you vary the parameter of any one state, the other states will not vary. Here the states are chosen as Z just to differentiate from the other set of state X; similarly the other matrix.

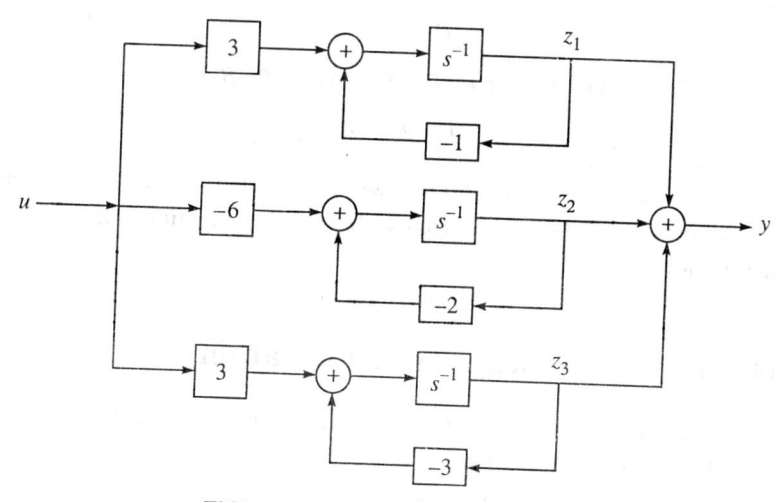

FIGURE 11.25 Block diagram.

Inavariance of eigenvalues

To prove the invariance of the eigenvalues under similarity transformation, we must show that the characteristic polynomials of the original system, i.e. $|sI - A|$ and that of the transformed system, i.e. $|sI - \bar{A}|$ are identical.

Let us begin with the characteristic polynomial of the transformed system as

$$|sI - \bar{A}| = |sP^{-1}P - P^{-1}AP| = |P^{-1}sIP - P^{-1}AP| = |P^{-1}[sI - A]P|$$
$$= |P^{-1}||sI - A||P| = |P^{-1}||P||sI - A|$$

Since the product of determinants $|P^{-1}P|$ is unity, hence

$$|sI - \bar{A}| = |P^{-1}||P||sI - A| = |P^{-1}P||sI - A|$$
$$= |sI - A|$$

that is, the characteristic equations are identical. As eigenvalues are roots of the characteristic polynomial, it is quite obvious that the eigenvalues are invariant under similarity transformation.

11.12 Invariance of Transfer Function

The transfer function of the transformed system can be written as

$$\overline{C}[sI - \overline{A}]^{-1}\overline{B} = CP[sP^{-1}P - P^{-1}AP]^{-1}P^{-1}B$$
$$= CP[P^{-1}sP - P^{-1}AP]^{-1}P^{-1}B$$
$$= CP[P^{-1}(sI - A)P]^{-1}P^{-1}B$$

Since
$$[XY]^{-1} = Y^{-1}X^{-1}$$

We can write
$$[P^{-1}(sI - A)P]^{-1} = P^{-1}[P^{-1}(sI - A)]^{-1} = P^{-1}(sI - A)^{-1}P$$

Therefore,
$$\overline{C}[sI - \overline{A}]^{-1}\overline{B} = CP[P^{-1}(sI - A)^{-1}P]P^{-1}B$$
$$= C[sI - A]^{-1}B \qquad (11.48)$$

Hence the transfer function of the transformed system, i.e. $\overline{C}[sI - \overline{A}]^{-1}\overline{B}$ is identical to that of the original system, i.e. $C[sI - A]^{-1}B$. Hence the transfer function remains invariant under similarity transformation.

11.13 Properties of Linear Transformation

A non-singular transformation matrix P of a system having system matrix A such that $Q = P^{-1}AP$ has the following properties:

(i) $P^{-1}[A]^2P = Q^2$, $P^{-1}[A]^mP = Q^m$, and so on.

(ii) $P^{-1}(A^{-1})P = Q^{-1}$; provided A^{-1} exists

(iii) $\text{trace}[Q] = \text{trace}[A]$

(iv) $\det[sI - A] = \det[sI - Q]$

(v) $\det[Q] = \det[A]$

(vi) The eigenvalues of Q are identical to those of A.

(vii) The transfer function and eigenvalues are invariant under similarity transformation.

11.14 Diagonalization

Earlier we talked about the similarity transformation. Diagonalization is a special form of similarity transformation where the transformation matrix becomes the modal matrix. Diagonalization has the following important features:

(i) If the system matrix is diagonalized, the states will be decoupled.

State-Variable Formulation

(ii) For the diagonal form of system matrix, the state transition matrix becomes diagonal having diagonal elements as $e^{\lambda_i t}$ where $\lambda_i (i = 1, 2,..., n)$ are the n distinct eigenvalues of the $n \times n$ system matrix.

(iii) For the diagonal form of system matrix, the controllability test will be trivial.

Let us consider the linear equation

$$\dot{X} = AX \qquad (11.49)$$

where the $(n \times n)$ order system matrix A has distinct eigenvalues $\lambda_i(i = 1, 2, ..., n)$ of the characteristic equation $|sI - A| = 0$.

Let the non-singular transformation matrix be M known as the modal matrix such that

$$X = MZ \qquad (11.50)$$

transforms

$$\dot{X} = AX$$

to

$$\dot{Z} = \Lambda Z$$

where

$$\Lambda = M^{-1}AM \qquad (11.51)$$

and Λ is a diagonal matrix having only diagonal elements $\lambda_i(i = 1, 2, ..., n)$. The transformed state equation is known as the *canonical form* of representation.

In general, there are several methods available for determining the modal matrix M. It can be shown that the modal matrix M can be formed by the eigenvectors of A, that is,

$$M = [v_1 \; v_2 \; ... \; v_n]$$

where v_i $(i = 1, 2, ..., n)$, a column vector denotes the eigenvector associated with the eigenvalue λ_i and is written with the ith column as

$$v_i = [v_{1i} \; v_{2i} \; ... \; v_{ni}]^T$$

The $n \times 1$ vector v_i satisfies the matrix equation

$$[\lambda_i I - A]v_i = 0$$

where λ_i is the ith eigenvalue of A and v_i is called the eigenvector of A associated with the ith eigenvalue λ_i. This can be rewritten as

$$\lambda_i v_i = A v_i ; \qquad i = 1, 2, ..., n$$

Now forming the $n \times n$ matrix

$$[\lambda_1 v_1 \; \lambda_2 v_2 \; ... \; \lambda_n v_n] = [Av_1 \; Av_2 \; ... \; Av_n] = A[v_1 \; v_2 \; ... \; v_n]$$

or

$$\begin{bmatrix} \lambda_1 v_{11} & \lambda_2 v_{12} & ... & \lambda_n v_{1n} \\ \lambda_1 v_{21} & \lambda_2 v_{22} & ... & \lambda_n v_{2n} \\ \vdots & \vdots & & \vdots \\ \lambda_1 v_{n1} & \lambda_2 v_{n2} & ... & \lambda_n v_{nn} \end{bmatrix} = \begin{bmatrix} a_{11} & ... & a_{1n} \\ a_{21} & ... & a_{2n} \\ \vdots & & \vdots \\ a_{n1} & ... & a_{nn} \end{bmatrix} \begin{bmatrix} v_{11} & ... & v_{1n} \\ v_{21} & ... & v_{2n} \\ \vdots & & \vdots \\ v_{n1} & ... & v_{nn} \end{bmatrix}$$

or

$$[v_1 \; v_2 \; ... \; v_n]\Lambda = A[v_1 \; v_2 \; ... \; v_n]$$

where Λ is a diagonal matrix having the diagonal elements λ_i ($i = 1, 2, ..., n$). Then

$$M\Lambda = AM$$

or

$$\Lambda = M^{-1}AM$$

which is the desired transformation.

EXAMPLE 11.14 Now let us take an example to determine the modal matrix M for transforming any generalized form of an $n \times n$ system matrix A to the canonical form Λ through the transformation $\Lambda = M^{-1}AM$.

Let

$$A = \begin{bmatrix} 0 & 1 & -1 \\ -6 & -11 & 6 \\ -6 & -11 & 5 \end{bmatrix}$$

The characteristic polynomial

$$|sI - A| = s^3 + 6s^2 + 11s + 6 = (s + 1)(s + 2)(s + 3),$$ whose eigenvalues are $-1, -2, -3$.

The eigenvector associated with $\lambda_1 = -1$ is represented as

$$v_1 = [v_{11} \quad v_{21} \quad v_{31}]^T$$

then

$$[\lambda_1 I - A]v_1 = 0$$

or

$$\begin{bmatrix} \lambda_1 & -1 & 1 \\ 6 & \lambda_1 + 11 & -6 \\ 6 & 11 & \lambda_1 - 5 \end{bmatrix} \begin{bmatrix} v_{11} \\ v_{21} \\ v_{31} \end{bmatrix} = 0$$

which gives (putting $\lambda_1 = -1$)

$$-v_{11} - v_{21} + v_{31} = 0$$
$$6v_{11} + 10v_{21} - 6v_{31} = 0$$
$$6v_{11} + 11v_{21} - 6v_{31} = 0$$

From the last two equations, we get $v_{21} = 0$, $v_{11} = v_{31}$. Let us choose arbitrarily $v_{11} = v_{31} = 1$ and get

$$v_1 = [1 \quad 0 \quad 1]^T$$

Now the eigenvector associated with $\lambda_2 = -2$ must satisfy the equation

$$[\lambda_i I - A]v_i = 0$$

Putting the value of $\lambda_2 = -2$, we get

$$-2v_{12} - v_{22} + v_{32} = 0$$
$$6v_{12} + 9v_{22} - 6v_{32} = 0$$
$$6v_{12} + 11v_{22} - 7v_{32} = 0$$

Arbitrary choice of $v_{12} = 1$ gives $v_{22} = 2$ and $v_{32} = 4$.

Thus, $\quad v_2 = [1 \quad 2 \quad 4]^T$

Finally, eigenvector v_3 associated with the eigenvalue $\lambda_3 = -3$ must satisfy

$$[\lambda_3 I - A] v_3 = 0$$

Putting the value of, $\lambda_3 = -3$, we get

$$-3v_{13} - v_{23} + v_{33} = 0$$
$$6v_{13} + 8v_{23} - 6v_{33} = 0$$
$$6v_{13} + 11v_{23} - 8v_{33} = 0$$

Arbitrary choice of $v_{13} = 1$ gives $v_{23} = 6$, $v_{33} = 9$.

Therefore, $\quad v_3 = [1 \quad 6 \quad 9]^T$

Hence the modal matrix consisting of the eigenvectors $v_i (i = 1, 2, 3)$ can be written as

$$M = [v_1 \quad v_2 \quad v_3] = \begin{bmatrix} 1 & 1 & 1 \\ 0 & 2 & 6 \\ 1 & 4 & 9 \end{bmatrix}$$

Further, $|M| \neq 0$, i.e. M is non-singular so that M^{-1} exists. Diagonalization of the system matrix A can be done as

$$\Lambda = M^{-1} A M = \begin{bmatrix} -1 & 0 & 0 \\ 0 & -2 & 0 \\ 0 & 0 & -3 \end{bmatrix}$$

11.15 Vandermonde Matrix

If the system matrix is of the companion form, that is, in phase variable canonical form, then the modal matrix becomes the Vandermonde matrix which can be written as

$$V = \begin{bmatrix} 1 & 1 & \cdots & 1 \\ \lambda_1 & \lambda_2 & \cdots & \lambda_n \\ \vdots & \vdots & & \vdots \\ \lambda_1^{n-1} & \lambda_2^{n-1} & \cdots & \lambda_n^{n-1} \end{bmatrix}$$

where $\lambda_i (i = 1, 2, \ldots, n)$ are the eigenvalues of the system matrix A.

Proof

Consider an $n \times n$ companion form of system matrix $\overline{A}$ as

$$\overline{A} = \begin{bmatrix} 0 & 1 & 0 & \cdots & 0 \\ 0 & 0 & 1 & \cdots & 0 \\ \vdots & \vdots & \vdots & & \vdots \\ 0 & 0 & 0 & \cdots & 1 \\ -a_0 & -a_1 & -a_2 & \cdots & -a_{n-1} \end{bmatrix}$$

Let the ith column of the eigenvector v_i of $\overline{A}$ be

$$v_1 = [v_{1i} \quad v_{2i} \quad v_{3i} \quad \cdots \quad v_{ni}]^T$$

Then

$$[\lambda_i I - \overline{A}] v_i = 0$$

or

$$\begin{bmatrix} \lambda_i & -1 & 0 & \cdots & 0 \\ 0 & \lambda_i & -1 & \cdots & 0 \\ \vdots & \vdots & \vdots & & \vdots \\ a_0 & a_1 & a_2 & \cdots & \lambda_i + a_{n-1} \end{bmatrix} \begin{bmatrix} v_{1i} \\ v_{2i} \\ \vdots \\ v_{ni} \end{bmatrix} = 0$$

or

$$\lambda_i v_{1i} - v_{2i} = 0$$
$$\lambda_i v_{2i} - v_{3i} = 0$$
$$\cdots \cdots \cdots$$
$$\lambda_i v_{(n-1)i} - v_{ni} = 0$$

$$a_1 v_{1i} + a_2 v_{2i} + \cdots + (\lambda_i + a_{n-1}) v_{ni} = 0$$

Let us choose $v_{1i} = 1$ arbitrarily, then

$$v_{2i} = \lambda_i$$
$$v_{3i} = \lambda_i^2$$
$$\cdots \cdots \cdots$$
$$v_{(n-1)i} = \lambda_i^{n-2}$$
$$v_{ni} = \lambda_i^{n-1}$$

which represents the ith column of the eigenvector v_i of the Vandermonde matrix V as

$$\begin{bmatrix} 1 & 1 & \cdots & 1 & \cdots & 1 \\ \lambda_1 & \lambda_2 & \cdots & \lambda_i & \cdots & \lambda_n \\ \lambda_1^2 & \lambda_2^2 & \cdots & \lambda_i^2 & \cdots & \lambda_n^2 \\ \vdots & \vdots & & \vdots & & \vdots \\ \lambda_1^{n-2} & \lambda_2^{n-2} & \cdots & \lambda_i^{n-2} & \cdots & \lambda_n^{n-2} \\ \lambda_1^{n-1} & \lambda_2^{n-1} & \cdots & \lambda_i^{n-1} & \cdots & \lambda_n^{n-1} \end{bmatrix}$$

State-Variable Formulation

Alternative proof

The modal matrix corresponding to the matrix A having distinct eigenvalues is given by the non-zero column of

$$\text{Adjoint } [\lambda I - A]\big|_{\lambda = \lambda_i}; \quad i = 1, 2, \ldots, n$$

where λ_i's are the eigenvalues of A.

For an $n \times n$ companion matrix $\bar{A}$, the last column of

$$\text{Adjoint } [\lambda I - \bar{A}]\big|_{\lambda = \lambda_i}$$

which is always non-zero, can be written as

$$[1 \quad \lambda_i \quad \lambda_i^2 \quad \cdots \quad \lambda_i^{n-1}]^T$$

Hence, the modal matrix corresponding to the companion matrix $\bar{A}$ becomes the Vandermonde matrix V and is given as

$$V = \begin{bmatrix} 1 & 1 & \cdots & 1 \\ \lambda_1 & \lambda_2 & \cdots & \lambda_n \\ \vdots & \vdots & & \vdots \\ \lambda_1^{n-1} & \lambda_2^{n-1} & \cdots & \lambda_n^{n-1} \end{bmatrix}$$

The diagonal matrix Λ for the companion matrix $\bar{A}$ can be obtained as

$$\Lambda = V^{-1} \bar{A} V$$

Again for general matrix A having n distinct eigenvalues, the diagonal matrix Λ is

$$\Lambda = M^{-1} A M$$

where M is the modal matrix.

Further,

$$\Lambda = V^{-1} \bar{A} V = V^{-1}[P^{-1}AP]V = [V^{-1}P^{-1}]A[PV] = [PV]^{-1}A[PV] = M^{-1}AM$$

where P is the similarity transformation matrix to transform any general matrix A to the corresponding companion matrix $\bar{A}$ by the relation $\bar{A} = P^{-1}AP$.

Hence,
$$M = PV$$

As an illustration, consider

$$A = \begin{bmatrix} -2 & 1 & 0 \\ 0 & -1 & 1 \\ 0 & 0 & -3 \end{bmatrix}$$

having distinct roots -1, -2, and -3. The non-singular similarity transformation matrix P (as explained in the next chapter) can be written as

$$P = \begin{bmatrix} -1 & 0 & 0 \\ 2 & 1 & 0 \\ 2 & 3 & 1 \end{bmatrix} \quad ; \quad |P| \neq 0$$

Then,
$$P^{-1} = \begin{bmatrix} 1 & 0 & 0 \\ -2 & 1 & 0 \\ 4 & -3 & 1 \end{bmatrix}$$

The Vandermode matrix becomes

$$V = \begin{bmatrix} 1 & 1 & 1 \\ -1 & -2 & -3 \\ 1 & 4 & 9 \end{bmatrix}$$

This yields the modal matrix M as

$$M = PV = \begin{bmatrix} 1 & 1 & 1 \\ 1 & 0 & -1 \\ 0 & 0 & 2 \end{bmatrix}$$

Hence
$$\Lambda = M^{-1}AM = \begin{bmatrix} -1 & 0 & 0 \\ 0 & -2 & 0 \\ 0 & 0 & -3 \end{bmatrix}$$

EXAMPLE 11.15 If the system matrix would have been given in the companion form as

$$\overline{A} = \begin{bmatrix} 0 & 1 & 0 \\ 0 & 0 & 1 \\ -6 & -11 & -6 \end{bmatrix}$$

the characteristic equation

$$|sI - \overline{A}| = s^3 + 6s^2 + 11s + 6 = 0$$

gives the eigenvalues as -1, -2, and -3, because of the fact that the choice of state is not unique, though the roots are same. In this case the roots are same as those of Example 11.14. The Vandermonde matrix can be written as

$$V = \begin{bmatrix} 1 & 1 & 1 \\ -1 & -2 & -3 \\ 1 & 4 & 9 \end{bmatrix}$$

Obviously V^{-1} exists as V is non-singular. Hence the canonical form can be written as

$$\Lambda = V^{-1}\bar{A}V = \begin{bmatrix} -1 & 0 & 0 \\ 0 & -2 & 0 \\ 0 & 0 & -3 \end{bmatrix}$$

This is to note that any system matrix A can be transformed to its companion form $\bar{A}$ by the similarity transformation

$$\bar{A} = P^{-1}AP$$

Again, $\Lambda = V^{-1}\bar{A}V = V^{-1}[P^{-1}AP]V = [V^{-1}P^{-1}]A[PV] = [PV]^{-1}A[PV] = M^{-1}AM$
where the modal matrix
$$M = PV$$

11.15.1 Vandermonde matrix for multiple eigenvalues

If the companion form of the system matrix involves multiple eigenvalues, then we adopt a different procedure. However, the Jordan canonical form is obtainable. For example, for the 4×4 matrix $\bar{A}$, where

$$\bar{A} = \begin{bmatrix} 0 & 1 & 0 & 0 \\ 0 & 0 & 1 & 0 \\ \vdots & \vdots & \vdots & \vdots \\ 0 & 0 & 0 & 1 \\ -a_0 & -a_1 & -a_2 & -a_3 \end{bmatrix}$$

has the eigenvalues $\lambda_1, \lambda_1, \lambda_1, \lambda_4$, then the transformation matrix, i.e. the Vandermonde matrix

$$V = \begin{bmatrix} 1 & 0 & 0 & 1 \\ \lambda_1 & 1 & 0 & \lambda_4 \\ \lambda_1^2 & 2\lambda_1 & 1 & \lambda_4^2 \\ \lambda_1^3 & 3\lambda_1^2 & 3\lambda_1 & \lambda_4^3 \end{bmatrix}$$

$$\quad\quad\quad\quad\quad \downarrow \quad\quad\quad \downarrow$$

$$\quad\quad\quad \leftarrow \frac{1}{1!}\frac{d}{d\lambda}\text{(1st column)}\Big|_{\lambda=\lambda_1} \quad \frac{1}{2!}\frac{d^2}{d\lambda^2}\text{(1st column)}\Big|_{\lambda=\lambda_1}$$

will yield the Jordan canonical form as

$$J = V^{-1}\bar{A}V = \left[\begin{array}{ccc|c} \lambda_1 & 1 & 0 & 0 \\ 0 & \lambda_1 & 1 & 0 \\ 0 & 0 & \lambda_1 & 0 \\ \hline 0 & 0 & 0 & \lambda_4 \end{array}\right]$$

It may be noted that if the companion form of the system matrix $\overline{A}$ has multiple eigenvalues, then the Vandermonde matrix will be somewhat different as illustrated in Example 11.16.

EXAMPLE 11.16 Suppose a 3×3 matrix

$$\overline{A} = \begin{bmatrix} 0 & 1 & 0 \\ 0 & 0 & 1 \\ -a_0 & -a_1 & -a_2 \end{bmatrix}$$

has the eigenvalues $\lambda_1, \lambda_1, \lambda_3$, that is, repeated root λ_1 of multiplicity 2, then the Vandermonde matrix becomes

$$M = V = \begin{bmatrix} 1 & 0 & 1 \\ \lambda_1 & 1 & \lambda_3 \\ \lambda_1^2 & 2\lambda_1 & \lambda_3^2 \end{bmatrix}$$

which yields
$$\Lambda = M^{-1}\overline{A}M = V^{-1}\overline{A}V$$

$$= \begin{bmatrix} \lambda_1 & 1 & 0 \\ 0 & \lambda_1 & 0 \\ \hline 0 & 0 & \lambda_3 \end{bmatrix}$$

Such a form is called the Jordan canonical form.

EXAMPLE 11.17 Consider the companion form of system matrix $\overline{A}$ as

$$\begin{bmatrix} 0 & 1 & 0 \\ 0 & 0 & 1 \\ -2 & -5 & -4 \end{bmatrix}$$

The characteristic equation $|sI - \overline{A}| = 0$ gives, $(s + 1)^2(s + 2) = 0$. The eigenvalues are $-1, -1, -2$.

The Vandermonde matrix is

$$V = \begin{bmatrix} 1 & 0 & 1 \\ \lambda_1 & 1 & \lambda_3 \\ \lambda_1^2 & 2\lambda_1 & \lambda_3^2 \end{bmatrix} = \begin{bmatrix} 1 & 0 & 1 \\ -1 & 1 & -2 \\ 1 & -2 & 4 \end{bmatrix}$$

We can see that V is non-singular, i.e. V^{-1} exists.

State-Variable Formulation

Hence the Jordan canonical form of the matrix becomes

$$\Lambda = V^{-1}\overline{A}V = \begin{bmatrix} -1 & 1 & 0 \\ 0 & -1 & 0 \\ 0 & 0 & -2 \end{bmatrix}$$

Now, for an $n \times n$ system matrix A having eigenvalues as

$$\lambda_1, \lambda_2, ..., \lambda_{j-1}, \lambda_j, ..., \lambda_j, \lambda_{j+1}, \lambda_{j+2}, ..., \lambda_n$$
$$\leftarrow \text{ } m\text{-repetitive roots } \rightarrow$$

where λ_j is of multiplicity m, the complete transformed matrix in Jordan canonical form can be written as

$$\begin{bmatrix} \begin{bmatrix} \lambda_1 & & 0 \\ & \lambda_2 & \\ & \ddots & \\ 0 & & \lambda_{j-1} \end{bmatrix}_{(j \times j)} & 0 & 0 \\ 0 & \begin{bmatrix} \lambda_j & 1 & 0 & 0 & 0 \\ 0 & \lambda_j & 1 & 0 & 0 \\ \vdots & \vdots & \vdots & \vdots & \vdots \\ 0 & 0 & 0 & \lambda_j & 1 \\ 0 & 0 & 0 & 0 & \lambda_j \end{bmatrix}_{(m \times m)} & 0 \\ 0 & 0 & \begin{bmatrix} \lambda_{j+1} & & 0 \\ & \lambda_{j+2} & \\ & \ddots & \\ 0 & & \lambda_n \end{bmatrix}_{(n-m-j) \times (n-m-j)} \end{bmatrix}$$

Again, for the system matrix A having repetitive roots λ_j of multiplicity m, the mth order Jordan block for repeated eigenvalues λ_j of multiplicity m can be written as

$$\Lambda = \begin{bmatrix} \lambda_j & 1 & 0 & \cdots & 0 & 0 \\ 0 & \lambda_j & 1 & \cdots & 0 & 0 \\ \vdots & \vdots & \vdots & & \vdots & \vdots \\ 0 & 0 & 0 & \cdots & \lambda_j & 1 \\ 0 & 0 & 0 & \cdots & 0 & \lambda_j \end{bmatrix}$$

Then the following relation holds

$$M\Lambda = AM$$

or

$$[v_1 \; v_2 \; \cdots \; v_m] \begin{bmatrix} \lambda_j & 1 & 0 & \cdots & 0 & 0 \\ 0 & \lambda_j & 1 & \cdots & 0 & 0 \\ \vdots & \vdots & \vdots & & \vdots & \vdots \\ 0 & 0 & 0 & \cdots & \lambda_j & 1 \\ 0 & 0 & 0 & \cdots & 0 & \lambda_j \end{bmatrix} = A[v_1 \; v_2 \; \cdots \; v_m]$$

$$\lambda_j v_1 = Av_1$$
$$v_1 + \lambda_j v_2 = Av_2$$
$$v_2 + \lambda_j v_3 = Av_3$$
$$\cdots\cdots\cdots\cdots\cdots\cdots$$
$$v_{m-1} + \lambda_j v_m = Av_m$$

The eigenvectors v_i, $i = 1, 2, \ldots, m$ can be obtained from these equations. Further these equations can be rewritten as

$$\begin{aligned}(\lambda_j I - A)v_1 &= 0 \\ (\lambda_j I - A)v_2 &= -v_1 \\ (\lambda_j I - A)v_3 &= -v_2 \\ \cdots\cdots\cdots\cdots\cdots&\cdots \\ (\lambda_j I - A)v_m &= -v_{m-1}\end{aligned} \quad (11.52)$$

EXAMPLE 11.18 Consider the system matrix

$$A = \begin{bmatrix} 0 & 6 & -5 \\ 1 & 0 & 2 \\ 3 & 2 & 4 \end{bmatrix}$$

The characteristic polynomial

$$|sI - A| = (s - 1)^2 (s - 2)$$

Therefore the system matrix A has a distinct eigenvalue $\lambda_1 = 2$ and double eigenvalues $\lambda_2 = 1$ of multiplicity two.

The eigenvector associated with $\lambda_1 = 2$ is determined from

$$[\lambda_1 I - A]v_1 = 0$$

or

$$\begin{bmatrix} 2 & -6 & 5 \\ -1 & 2 & -2 \\ -3 & -2 & -2 \end{bmatrix} \begin{bmatrix} v_{11} \\ v_{21} \\ v_{31} \end{bmatrix} = 0$$

With the arbitrary choice of $v_{11} = 2$, we get $v_{21} = -1$, $v_{31} = -2$. Therefore,

$$v_1 = [2 \; -1 \; -2]^T$$

Now the eigenvector corresponding to multiple eigenvalues is determined from Eq. (11.52) as
$$(\lambda_2 I - A)v_2 = -v_1 \quad (11.53)$$
and
$$(\lambda_2 I - A)v_3 = -v_2 \quad (11.54)$$

Putting the value of $\lambda_2 = 1$ in Eq. (11.53), we get

$$\begin{bmatrix} 1 & -6 & 5 \\ -1 & 1 & -2 \\ -3 & -2 & -3 \end{bmatrix} \begin{bmatrix} v_{12} \\ v_{22} \\ v_{32} \end{bmatrix} = 0$$

With the arbitrary choice of $v_{12} = 1$, we get $v_{22} = -3/7$ and $v_{32} = -5/7$.

Therefore, $v_2 = [1 \quad -3/17 \quad -5/17]^T$

Similarly for the third eigenvector v_3, putting the value of $\lambda_2 = 1$ in Eq. (11.54), we get

$$\begin{bmatrix} 1 & -6 & -5 \\ -1 & 1 & -2 \\ -3 & -2 & -3 \end{bmatrix} \begin{bmatrix} v_{13} \\ v_{23} \\ v_{33} \end{bmatrix} = \begin{bmatrix} -1 \\ 3/7 \\ 5/7 \end{bmatrix}$$

From which, we get $v_3 = [v_{13} \quad v_{23} \quad v_{33}]^T = [1 \quad -22/49 \quad -46/49]^T$

Thus the modal matrix
$$M = [v_1 \quad v_2 \quad v_3] = \begin{bmatrix} 2 & 1 & 1 \\ -1 & -3/17 & -22/49 \\ 2 & -5/17 & -46/49 \end{bmatrix}$$

Further as M is non-singular, i.e. $M \neq 0$, M^{-1} exists. The Jordan canonical form Λ consists of two Jordan blocks and can be written as

$$\Lambda = M^{-1}AM = \begin{bmatrix} 2 & 0 & 0 \\ 0 & 1 & 1 \\ 0 & 0 & 1 \end{bmatrix}$$

EXAMPLE 11.19 Given the system equation as

$$\dot{X} = \begin{bmatrix} 0 & 1 & 0 \\ 0 & 0 & 1 \\ -3 & -7 & -5 \end{bmatrix} X + \begin{bmatrix} 0 \\ 1 \\ -3 \end{bmatrix} u$$

the characteristic equation is: $s^3 + 5s^2 + 7s + 3 = 0$. The roots are $-1, -1, -3$.
The Vandermonde matrix

$$V = \begin{bmatrix} 1 & 0 & 1 \\ \lambda_1 & 1 & \lambda_2 \\ \lambda_1^2 & 2\lambda_1 & \lambda_2^2 \end{bmatrix} = \begin{bmatrix} 1 & 0 & 1 \\ -1 & 1 & -3 \\ 1 & -2 & 9 \end{bmatrix}$$

Then
$$V^{-1} = \begin{bmatrix} 3/4 & -1/2 & -1/4 \\ 3/2 & 2 & 1/2 \\ 1/4 & 1/2 & 3/4 \end{bmatrix}$$

Therefore,
$$\dot{Z} = \begin{bmatrix} -1 & 1 & 0 \\ 0 & -1 & 0 \\ \hline 0 & 0 & -3 \end{bmatrix} Z + \begin{bmatrix} 1/4 \\ 1/2 \\ -1/4 \end{bmatrix} u$$

11.16 Important Properties of Eigenvalues of a Matrix

1. $\lambda = 0$ is an eigenvalue of A if and only if A is singular. Since for $\lambda = 0$,
$$|\lambda I - A| = |-A| = 0$$

2. The eigenvalues of the matrix kA are k times the eigenvalues of A for any scalar k because
$$|k\lambda I - kA| = 0$$
for each value of λ for which
$$|\lambda I - A| = 0$$

3. The eigenvalues of the matrix A are the same as the eigenvalues of $\bar{A}$ where A and $\bar{A}$ are related by similarity transformation as $\bar{A} = P^{-1}AP$; P is the non-singular similarity transformation matrix. Alternatively, the eigenvalues remain invariant under similarity transformation.

4. The eigenvalues of a non-singular matrix A^{-1} are the inverses of the eigenvalues of A. The characteristic equation is
$$0 = |\lambda I - A| = |\lambda A A^{-1} - A| = |A[\lambda A^{-1} - I]| = -|A|\left|\left(\frac{I}{\lambda}\right) - A^{-1}\right|\lambda$$
$$= \left|\left(\frac{1}{\lambda}\right)I - A^{-1}\right|$$

5. The eigenvalues of A^k (k an integer) are the eigenvalues of A raised to the kth power. The characteristic equation is
$$0 = |\lambda I - A|$$
$$= |\lambda I - A||\lambda I + A| = |(\lambda I - A)(\lambda I + A)|$$
$$= |\lambda^2 I - A^2|$$

State-Variable Formulation

6. The eigenvalues of a diagonal matrix are the diagonal elements.
7. The sum of the eigenvalues of an $n \times n$ matrix A with characteristic equation

$$\lambda^n + a_{n-1}\lambda^{n-1} + a_{n-2}\lambda^{n-2} + \cdots + a_1\lambda + a_0 = 0$$

is

$$\lambda_1 + \lambda_2 + \lambda_3 + \cdots + \lambda_n = -a_{n-1} = \text{trace}[A]$$

Drill Problem 11.9

Given the state equations,

$$\frac{d}{dt}\begin{bmatrix} x_1 \\ x_2 \end{bmatrix} = \begin{bmatrix} -2 & 1 \\ 0 & -2 \end{bmatrix} x + \begin{bmatrix} 0 \\ 1 \end{bmatrix} u$$

and output

$$y = [2 \ \ 4] \begin{bmatrix} x_1 \\ x_2 \end{bmatrix}$$

Determine the transfer function without calculation.

Ans. $G(s) = \dfrac{Y(s)}{U(s)} = \dfrac{2}{(s+2)^2} + \dfrac{4}{s+2}$

Drill Problem 11.10

The Jordan canonical form of state equations is given below.

$$\dot{X} = \begin{bmatrix} -2 & 1 & 0 & 0 \\ 0 & -2 & 1 & 0 \\ 0 & 0 & -2 & 0 \\ 0 & 0 & 0 & -1 \end{bmatrix} X + \begin{bmatrix} 0 \\ 0 \\ 1 \\ 1 \end{bmatrix} u$$

and output

$$y = [2 \ \ 4 \ \ 5 \ \ 3] X$$

Determine the transfer function in factored form (without calculation).

Ans. $G(s) = \dfrac{Y(s)}{U(s)} = \dfrac{2}{(s+2)^3} + \dfrac{4}{(s+2)^2} + \dfrac{5}{s+2} + \dfrac{3}{s+1}$

Drill Problem 11.11

Determine the transfer matrix for a two-input-two-output system described by the following state equations

$$\dot{X} = \begin{bmatrix} -1 & 1 & 0 \\ 0 & -4 & 2 \\ 0 & 0 & -10 \end{bmatrix} X + \begin{bmatrix} 1 & -1 \\ 0 & 2 \\ -1 & 1 \end{bmatrix} u$$

and output

$$Y = \begin{bmatrix} 1 & 0 & 1 \\ 2 & 1 & -3 \end{bmatrix} X$$

EXAMPLE 11.20 For the following transfer function, find the suitable state variable formulations.

$$G(s) = \frac{Y(s)}{U(s)} = \frac{s+3}{s^3 + 9s^2 + 24s + 20}$$

Solution:

Case I From the transfer function, the dynamic equation can be written in time domain as

$$\dddot{y} + 9\ddot{y} + 24\dot{y} + 20y = \dot{u} + 3u$$

or

$$\dddot{y} = -9\ddot{y} - 24\dot{y} - 20y + \dot{u} + 3u$$

Simulation is shown in Figure 11.26(a) and finally in Figure 11.26(b).

From Figure 11.26(b), taking output of the integrator as the state, the state variable formulation is

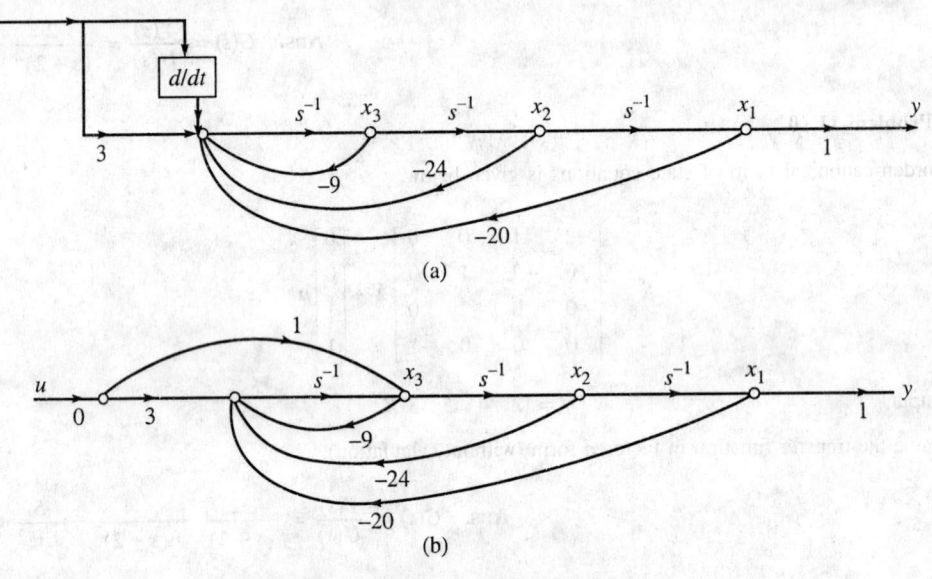

FIGURE 11.26 Example 11.20: case I.

$$\dot{X} = \begin{bmatrix} 0 & 1 & 0 \\ 0 & 0 & 1 \\ -20 & -24 & -9 \end{bmatrix} X + \begin{bmatrix} 0 \\ 0 \\ 1 \end{bmatrix} u$$

and

$$y = [3 \quad 1 \quad 0]X$$

Case II In an alternative way, we can rewrite

$$\dddot{y} = -9\ddot{y} - 24\dot{y} - 20y + \dot{u} + 3u$$

State-Variable Formulation

or

$$y = \int \left\{ \underbrace{-9y + \int \left[\underbrace{-24y + u + \int \underbrace{(-20y + 3u)\,dt}_{x_3}}_{x_2} \right] dt}_{x_1} \right\} dt$$

The simulation diagram is shown in Figure 11.27. The state variable formulation is

$$\dot{X} = \begin{bmatrix} -9 & 1 & 0 \\ -24 & 0 & 1 \\ -20 & 0 & 0 \end{bmatrix} X + \begin{bmatrix} 0 \\ 1 \\ 3 \end{bmatrix} u$$

and

$$y = \begin{bmatrix} 1 & 0 & 0 \end{bmatrix} X$$

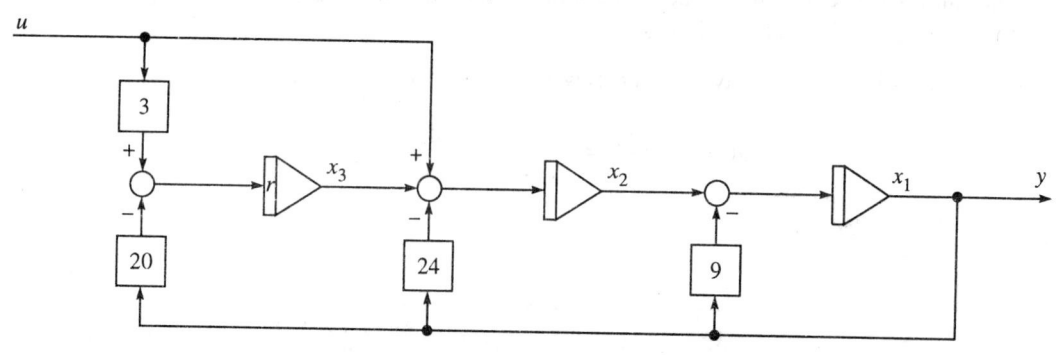

FIGURE 11.27 Example 11.20: case II.

Case III In an another way, we can write by partial fraction expansion as

$$G(s) = \frac{s+3}{(s+2)^2 (s+5)} = \frac{-2/9}{s+5} + \frac{1/3}{(s+2)^2} + \frac{2/9}{s+2}$$

The simulation diagram is shown in Figure 11.28. The state variable formulation from the diagram is

$$\dot{X} = \begin{bmatrix} -5 & 0 & 0 \\ 0 & -2 & 1 \\ 0 & 0 & -2 \end{bmatrix} X + \begin{bmatrix} 1 \\ 0 \\ 1 \end{bmatrix} u$$

and

$$y = \begin{bmatrix} -\frac{2}{9} & \frac{1}{3} & \frac{2}{9} \end{bmatrix} X$$

Here the system matrix is in Jordan canonical form. It may be observed that had separate parallel paths been used for the terms $1/(s+2)^2$ and $1/(s+2)$, one unnecessary integrator would

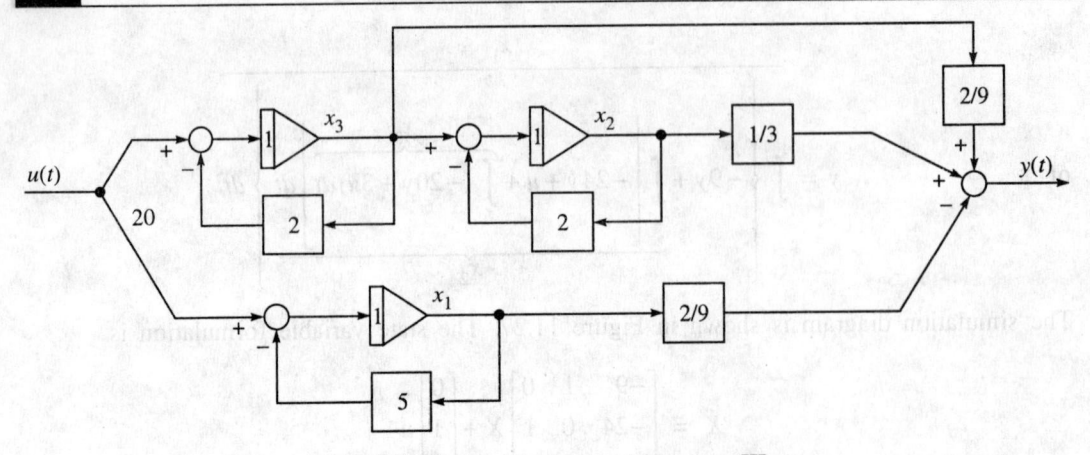

FIGURE 11.28 Example 11.20: case III.

have been used. This must be avoided if the system matrix has to be canonical which means minimal number of components, i.e. integrators. This must be avoided if the integrator outputs are to be taken as the state variables.

Case IV In still another way, we can rewrite the transfer function

$$G(s) = \frac{Y(s)}{U(s)} = \frac{s+3}{s^3 + 9s^2 + 24s + 20}$$

in factored cascaded form as

$$G((s) = \frac{s+3}{(s+2)^2(s+5)} = \frac{s+3}{s+2} \times \frac{1}{s+2} \times \frac{1}{s+5}$$

The simulation diagram is shown in Figure 11.29. The state variable formulation from the diagram is

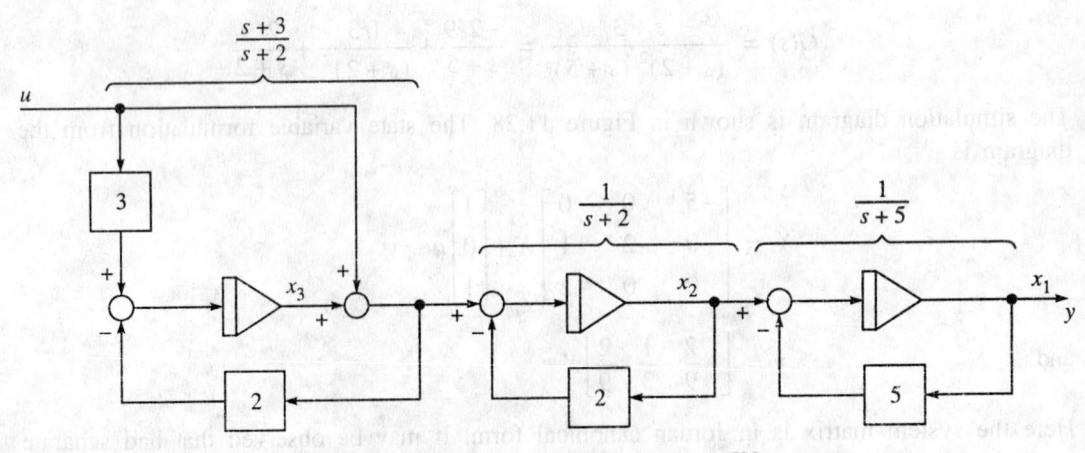

FIGURE 11.29 Example 11.20: case IV.

$$\dot{X} = \begin{bmatrix} -5 & 1 & 0 \\ 0 & -2 & 1 \\ 0 & 0 & -2 \end{bmatrix} X + \begin{bmatrix} 0 \\ 1 \\ 1 \end{bmatrix} u$$

and output
$$y = [1 \quad 0 \quad 0] X$$

For cases III and IV, the diagonal elements of the system matrix are the same, so are the eigenvalues, that is, the system poles.

The choice of state variables is not unique as shown in different forms of representation whereas the eigenvalues, i.e. the transfer function is unique.

11.17 Time-Domain Solution

The dynamics of the system can be represented by the vector-matrix differential equation

$$\dot{X} = AX + BU \qquad \text{(AB)}$$

and the output
$$Y = CX + DU \qquad \text{(CD)}$$

The state variable formulation can be obtained from the transfer function or from the differential equation representing the system. Further, in order to verify, we can get the transfer function from state variable formulation.

It is often desirable to obtain the time response of the state variables of a control system for the analysis of the system and thus examine the performance of the system. The solution can be written as

$$X(t) = \Phi(t) X(0) + \int_0^t \Phi(t-\tau) B u(\tau) d\tau \qquad (11.55)$$

where $\Phi(t)$ is the state transition matrix. The transition from any initial state to any other state for any onward time t can be obtained provided $\Phi(t)$ is known. Hence $\Phi(t)$ plays a paramount role in determining the response. Methods for evaluation of state transition matrix $\Phi(t)$ will be discussed.

Laplace transformed signals

For linear, time-invariant system, the Laplace transform helps to relate the state variable to classical concepts and gives easy derivations of important time-domain results. Taking the Laplace transform of the state and output equation of state variable system model, we get

$$sX(s) - X(0) = AX(s) + BU(s) \qquad (11.56)$$

and
$$Y(s) = CX(s) + DU(s)$$

Rearranging, we get

$$X(s) = [sI - A]^{-1}X(0) \quad + \quad [sI - A]^{-1}BU(s) \quad (11.57)$$
$$\leftarrow \text{zero-input component} \rightarrow \quad \leftarrow \text{zero-state component} \rightarrow$$
$$\text{of state vector} \quad\quad\quad \text{of state vector}$$

where the zero-input and the zero-state components of the state vector are evident. The zero-input component is the solution when the input is zero and the zero-state component is the solution when the initial conditions are zero.

The transform of the system output is then

$$Y(s) = CX(s) + DU(s)$$
$$= C[sI - A]^{-1}X(0) \quad + \quad [C[sI - A]^{-1}B + D]U(s) \quad (11.58)$$
$$\leftarrow \text{zero-input component} \rightarrow \quad \leftarrow \text{zero-state component} \rightarrow$$
$$\text{of output} \quad\quad\quad \text{of output}$$

EXAMPLE 11.21 Given

$$\dot{X} = \begin{bmatrix} -7 & 1 \\ -12 & 0 \end{bmatrix} X + \begin{bmatrix} 2 \\ -1 \end{bmatrix} u$$

and output

$$y = [3 \quad -4]X + [2]u$$

with the input $u(t) = 3e^{-t}$ and initial state $X(0) = \begin{bmatrix} -6 \\ 1 \end{bmatrix}$

we get

$$[sI - A] = \begin{bmatrix} s+7 & -1 \\ 12 & s \end{bmatrix}$$

and

$$(sI - A)^{-1} = \frac{1}{s^2 + 7s + 12} \begin{bmatrix} s & 1 \\ -12 & s+7 \end{bmatrix}$$

Now,

$$X(s) = (sI - A)^{-1}X(0) + (sI - A)^{-1}BU(s)$$

$$= \frac{1}{(s^2 + 7s + 12)(s + 3)} \begin{bmatrix} -6s^2 + s - 2 \\ s^2 + 77s - 14 \end{bmatrix}$$

which could be inverted to find $x_1(t)$ and $x_2(t)$. The system output has the transform

$$Y(s) = CX(s) + BU(s)$$

$$= \frac{-22s^2 - 305s + 50}{(s+1)(s+3)(s+4)} - \frac{6}{s+1}$$

$$= \frac{49.5}{s+1} - \frac{383.5}{s+3} + \frac{306}{s+4}$$

Then
$$y(t) = (49.5)e^{-t} - (383.5)e^{-3t} + (306)e^{-4t} \; ; \; t > 0$$

11.18 State Transition Matrix

In order to obtain the time response of the homogeneous equation

$$\dot{X}(t) = AX(t) \tag{11.59}$$

where the forcing function $u(t)$ has been set equal to zero, taking the Laplace transform, we get

$$X(s) = (sI - A)^{-1}X(0)$$

Taking the inverse Laplace transform, we get

$$X(t) = \mathcal{L}^{-1}X(s) = [\mathcal{L}^{-1}(sI - A)^{-1}]X(0) = [\mathcal{L}^{-1}\Phi(s)]X(0) \tag{11.60}$$

where the resolvent matrix

$$\Phi(s) = (sI - A)^{-1}$$

Further, rewriting $X(s)$ as

$$X(s) = (sI - A)^{-1}X(0) = \left[\frac{1}{s}\left(I - \frac{A}{s}\right)^{-1}\right]X(0)$$

$$= \left[\frac{1}{s}\left(I + \frac{A}{s} + \frac{A^2}{s^2} + \cdots\right)\right]X(0)$$

$$= \left(\frac{I}{s} + \frac{A}{s^2} + \frac{A^2}{s^3} + \cdots\right)X(0) \tag{11.61}$$

Taking the inverse Laplace transform, we get

$$X(t) = \mathcal{L}^{-1}[X(s)]$$

$$= \left[I + At + \frac{A^2 t^2}{2!} + \cdots\right]X(0)$$

$$= \left[\sum_{r=0}^{\infty} \frac{A^r t^r}{r!}\right]X(0) \tag{11.62}$$

$$= e^{At}X(0) = \Phi(t)X(0)$$

Comparing Eqs. (11.60) and (11.62), we get the state transition matrix (STM) as

$$\Phi(t) = \left[\sum_{r=0}^{\infty} \frac{A^r t^r}{r!}\right] = e^{At}$$

$$= \mathcal{L}^{-1}\,[(sI - A)^{-1}] = \mathcal{L}^{-1}\,[\Phi(s)] \qquad (11.63)$$

We derive from this definition, some properties of STM, e^{At} or $\Phi(t)$.

At $t = 0$, $\Phi(t)$ reduces to $e^0 = I$.

Taking the derivative of e^{At} term by term, we have

$$\frac{d}{dt}\left(e^{At}\right) = A + A^2 t + \frac{1}{2!}A^3 t^2 + \cdots + \frac{1}{(n-1)!}A^n t^{n-1} + \cdots$$

$$= A\left(I + At + \frac{A^2 t^2}{2!} + \cdots + \frac{A^{n-1} t^{n-1}}{(n-1)!} + \cdots\right)$$

$$= \left(I + At + \frac{A^2 t^2}{2!} + \cdots + \frac{A^{n-1} t^{n-1}}{(n-1)!} + \cdots\right) A$$

$$= A e^{At} = e^{At} A$$

If $X(t) = e^{At} X(0)$ is the solution of $\dot{X} = AX$, then,

$$\dot{X} = \frac{d}{dt}(X) = \frac{d}{dt}[e^{At} X(0)] = \frac{d}{dt}(e^{At}) X(0)$$

$$= A[e^{At} X(0)]$$

$$= AX$$

Hence it is proved that $X(t) = \Phi(t) X(0)$ is the solution of the dynamic equation

$$\dot{X} = AX$$

The solution to the unforced system (i.e. when $u(t) = 0$) is simply

$$X(t) = \Phi(t) X(0) \qquad (11.64)$$

or

$$\begin{bmatrix} x_1(t) \\ \cdot \\ \cdot \\ \cdot \\ x_n(t) \end{bmatrix} = \begin{bmatrix} \phi_{11}(t) & \cdots & \phi_{1n}(t) \\ \cdot & & \cdot \\ \cdot & & \cdot \\ \cdot & & \cdot \\ \phi_{n1}(t) & \cdots & \phi_{nn}(t) \end{bmatrix} \begin{bmatrix} x_1(0) \\ \cdot \\ \cdot \\ \cdot \\ x_n(0) \end{bmatrix}$$

That is, the term $\phi_{ij}(t)$ is the response of the ith state variable due to a unity initial condition on the jth state variable when there are zero initial conditions on all other states.

From the solution expression in Eq. (11.64), it is obvious that the transition from any initial state $X(0)$ to any other state $X(t)$ for any onward time $t > 0$, is possible provided the $\Phi(t)$ is known and that is why $\Phi(t)$ is called the *state transition matrix* (STM). With these properties, the solution of a non-homogeneous system represented by Eq. (AB) can now be found.

We rewrite Eq. (AB) as

$$\dot{X} - AX = Bu(t)$$

Further,

$$\frac{d}{dt}(e^{-At} X) = \left(\frac{d}{dt} e^{-At}\right) X + e^{-At}\left(\frac{d}{dt} X\right)$$

$$= e^{-At}(-AX + \dot{X}) = e^{-At} Bu(t)$$

Integrating both sides, we can write

$$\int_0^t \frac{d}{dt}(e^{-At} X) dt = \left(e^{-At} X(t)\right)\Big|_0^t = \int_0^t e^{-A\tau} Bu(\tau) d\tau$$

or

$$e^{-At}X(t) - IX(0) = \int_0^t e^{-A\tau} Bu(\tau) d\tau$$

or

$$X(t) = e^{At}\left[X(0) + \int_0^t e^{-A\tau} Bu(\tau) d\tau\right]$$

$$= e^{At} X(0) + \int_0^t e^{A(t-\tau)} Bu(\tau) d\tau$$

$$= \Phi(t) X(0) + \int_0^t \Phi(t-\tau) Bu(\tau) d\tau$$

Hence $\quad X(t) = \Phi(t)X(0) \quad + \quad$ convolution $[\Phi(t), Bu(t)]$
$\qquad\quad \leftarrow$ *complementary* soln. $\rightarrow \qquad \leftarrow$ *particular* soln. $\rightarrow \qquad$ (11.65)
$\qquad\qquad$ *(transient response)* $\qquad\qquad$ *(steady state response)*

For a homogeneous system, i.e. $u(t) = 0$, it is obvious that the solution becomes

$$X(t) = \Phi(t)X(0) \qquad (11.66)$$

If we replace the initial time with t_0 rather than zero at $t = 0$, then the solution of the homogeneous system becomes

$$X(t) = e^{A(t-t_0)} X(t_0) = \Phi(t - t_0) X(t_0)$$

To show that this is in fact the correct result, putting $t = t_0$ in Eq. (11.66), we get

$$X(t_0) = \Phi(t_0) X(0)$$

or

$$X(0) = [\Phi(t_0)]^{-1} X(t_0) = \Phi(-t_0) X(t_0)$$

Substituting in Eq. (11.66), we get

$$X(t) = \Phi(t)\Phi(-t_0)X(t_0) = \Phi(t - t_0)X(t_0) \qquad (11.67)$$

which is the desired result.

11.18.1 Forced response

Our concern here is to find the solution of the vector-matrix differential equation of a forced system

$$\dot{X}(t) = AX(t) + Bu(t) \qquad \text{(AB)}$$

Taking the Laplace transform, we get

$$sX(s) - X(0) = AX(s) + BU(s)$$

or
$$[sI - A]X(s) = X(0) + BU(s)$$

or
$$X(s) = [sI - A]^{-1}X(0) + [sI - A]^{-1}BU(s)$$

or
$$X(t) = \mathscr{L}^{-1}X(s)$$
$$= \mathscr{L}^{-1}\Phi(s)X(0) + \mathscr{L}^{-1}\Phi(s)BU(s)$$
$$= \Phi(t)X(0) + \int_0^t \Phi(t-\tau) B u(\tau) d\tau \qquad (11.68)$$

In Eq. (11.68), $\Phi(t)X(0)$ is the complementary solution and $\int_0^t \Phi(t-\tau) Bu(\tau) d\tau$ is the particular solution for the total solution $X(t)$ of the vector-matrix differential equation (AB). This is, in fact, the correct solution, as can be seen by considering the derivative of $X(t)$, i.e.

$$\dot{X}(t) = \dot{\Phi}(t) X(0) + \int_0^t \dot{\Phi}(t-\tau) Bu(\tau) d\tau + \Phi(0) Bu(\tau)$$

By definition of $\Phi(t)$,

$$\dot{\Phi}(t) = A\Phi(t) \quad \text{and} \quad \Phi(0) = I$$

Therefore,
$$\dot{X}(t) = A\Phi(t) X(0) + A\int_0^t \Phi(t-\tau) Bu(\tau) d\tau + Bu(t)$$
$$= A\left[\Phi(t) X(0) + \int_0^t \Phi(t-\tau) Bu(\tau) d\tau\right] + Bu(t)$$

Using Eq. (11.68), we obtain

$$\dot{X}(t) = AX(t) + Bu(t)$$

and $X(t)$ given by Eq. (11.68) has been shown to be the solution of the forced system represented by Eq. (AB).

State-Variable Formulation

To obtain the solution of the forced system is a more general case where the initial time is taken as t_0 instead of zero, we use Eq. (11.68) to obtain

$$X(t_0) = \Phi(t_0)X(0) + \int_0^{t_0} \Phi(t_0 - \tau) B u(\tau) d\tau$$

or

$$\Phi(t_0)^{-1} X(t_0) = X(0) + \int_0^{t_0} \Phi(t_0)^{-1} \Phi(t_0 - \tau) B u(\tau) d\tau$$

Therefore,

$$X(0) = \Phi(-t_0) X(t_0) - \int_0^{t_0} \Phi(-t_0) \Phi(t_0 - \tau) B u(\tau) d\tau$$

$$= \Phi(-t_0) X(t_0) - \int_0^{t_0} \Phi(-\tau) B u(\tau) d\tau \qquad (11.69)$$

Substituting this in Eq. (11.68), we obtain

$$X(t) = \Phi(t)\left[\Phi(-t_0) X(t_0) - \int_0^{t_0} \Phi(-\tau) B u(\tau) d\tau\right] + \int_0^{t} \Phi(t-\tau) B u(\tau) d\tau$$

$$= \Phi(t - t_0) X(t_0) - \int_0^{t_0} \Phi(t-\tau) B u(\tau) d\tau + \int_0^{t} \Phi(t-\tau) B u(\tau) d\tau$$

$$= \Phi(t - t_0) X(t_0) + \int_{t_0}^{0} \Phi(t-\tau) B u(\tau) d\tau + \int_0^{t} \Phi(t-\tau) B u(\tau) d\tau \qquad (11.70)$$

Therefore, we have the desired result as

$$X(t) = \Phi(t - t_0) X(t_0) \qquad + \qquad \int_{t_0}^{t} \Phi(t-\tau) B u(\tau) d\tau \qquad (11.71)$$

$\leftarrow$ *zero-input component* $\rightarrow$ $\leftarrow$ *zero-state component* $\rightarrow$
or   *or*
complementary solution   *forced or particular solution*
(free vibration)   *(steady state response)*
↓
transient response

From Eq. (11.71), the key role that the STM plays in determining the time response may be seen. We need only three things to obtain the time response of any linear system:

1. The initial state at $t = t_0$, that is, $X(t_0)$
2. The input for $t > t_0$, that is, $u(t)$ for $t > t_0$
3. The STM, $\Phi(t)$

Since a knowledge of the matrix $\Phi(t)$, that is, e^{At} and the initial state of the system allows one to determine the state at any later time, the matrix $\Phi(t)$, i.e. STM plays an important role in obtaining the time response of a system. Because of the key part that it plays in determining the transient response, the evaluation of STM is of paramount importance.

Properties of STM, $\Phi(t)$

The properties of the state transition matrix $\Phi(t)$ are summarized below:

1. $\Phi(t)$ is non-singular for all finite values of time.
2. $\Phi(0) = e^{A0} = I$
3. $\Phi(t) = e^{At} = [e^{-At}]^{-1} = [\Phi(-t)]^{-1}$
4. $\Phi^{-1}(t) = \Phi(-t)$
5. $\Phi(t_1 + t_2) = e^{A(t_1+t_2)} = e^{At_1} \cdot e^{At_2} = \Phi(t_1) \cdot \Phi(t_2) = \Phi(t_2) \cdot \Phi(t_1)$
6. $[\Phi(t)]^n = [e^{At}]^n = e^{Ant} = \Phi(nt)$
7. $\Phi(t_2 - t_1)\Phi(t_1 - t_0) = e^{A(t_2-t_1)} e^{A(t_1-t_0)} = e^{A(t_2-t_0)} = \Phi(t_2 - t_0) = \Phi(t_1 - t_0)\Phi(t_2 - t_1)$
8. $\Phi(t) = \Phi(nT) = [\Phi(T)]^n$
9. $X(t) = \Phi(t)X(0)$, putting $t = nT$; $n = 1, 2, ...$; then

$$X(nT) = \Phi(nT)X(0) = e^{AnT}X(0) = [e^{AT}]^n X(0) = [\Phi(T)]^n X(0)$$

It is very important to remember that

$$e^{(A+B)t} = e^{At}e^{Bt}, \quad \text{if } AB = BA, \text{ that is, commutative}$$
$$e^{(A+B)t} \neq e^{At}e^{Bt}, \quad \text{if } AB \neq BA$$

To prove this, note that

$$e^{(A+B)t} = I + (A+B)t + \frac{(A+B)^2}{2!}t^2 + \frac{(A+B)^3}{3!}t^3 + \cdots$$

and

$$e^{At}e^{Bt} = \left(I + At + \frac{A^2t^2}{2!} + \frac{A^3t^3}{3!} + \cdots\right)\left(I + Bt + \frac{B^2t^2}{2!} + \frac{B^3t^3}{3!} + \cdots\right)$$

$$= I + (A+B)t + \frac{A^2t^2}{2!} + ABt^2 + \frac{B^2t^2}{2!} + \frac{A^3t^3}{3!} + \frac{A^2Bt^3}{2!} + \frac{AB^2t^3}{2!} + \frac{B^3t^3}{3!} + \cdots$$

Hence
$$e^{(A+B)t} - e^{At}e^{Bt} \neq 0$$

The difference between $e^{(A+B)t}$ and $e^{At}e^{Bt}$ vanishes if A and B commute.

11.18.2 Methods for evaluation of STM

The methods for evaluation of the state transition matrix (STM) can be broadly classified into two groups as:

1. Methods which require the knowledge of eigenvalues. Under this category the methods are:

    (a) Laplace inverse transform technique
    (b) Cayley-Hamilton technique
    (c) Sylvester's theorem

State-Variable Formulation

2. Methods which do not require the knowledge of eigenvalues. Under this category the methods are:

   (d) Infinite power series method
   (e) Taylor's series expansion technique.

We will now discuss the methods for the evaluation of STM through examples.

EXAMPLE 11.22 Consider the vector-matrix differential equation describing the dynamics of the system as

$$\dot{X} = \begin{bmatrix} 0 & 1 \\ -6 & -5 \end{bmatrix} X$$

Determine the STM using the Laplace inverse transform technique.

Solution: For the given system,

$$A = \begin{bmatrix} 0 & 1 \\ -6 & -5 \end{bmatrix}, \quad [sI - A] = \begin{bmatrix} s & -1 \\ 6 & s+5 \end{bmatrix}, \quad [sI - A]^{-1} = \frac{1}{s^2 + 5s + 6} \begin{bmatrix} s+5 & 1 \\ -6 & s \end{bmatrix}$$

Therefore, $\Phi(t) = \mathscr{L}^{-1}[sI - A]^{-1} = \begin{bmatrix} \mathscr{L}^{-1}\left[\dfrac{s+5}{(s+2)(s+3)}\right] & \mathscr{L}^{-1}\left[\dfrac{1}{(s+2)(s+3)}\right] \\ \mathscr{L}^{-1}\left[\dfrac{-6}{(s+2)(s+3)}\right] & \mathscr{L}^{-1}\left[\dfrac{s}{(s+2)(s+3)}\right] \end{bmatrix}$

$$= \begin{bmatrix} \phi_{11}(t) & \phi_{12}(t) \\ \phi_{21}(t) & \phi_{22}(t) \end{bmatrix}$$

where $\phi_{11}(t) = \mathscr{L}^{-1}\left[\dfrac{s+5}{(s+2)(s+3)}\right] = \mathscr{L}^{-1}\left[\dfrac{3}{s+2}\right] - \mathscr{L}^{-1}\left[\dfrac{2}{s+3}\right] = 3e^{-2t} - 2e^{-3t}$

$$\phi_{12}(t) = \mathscr{L}^{-1}\left[\dfrac{1}{(s+2)(s+3)}\right] = e^{-2t} - e^{-3t}$$

$$\phi_{21}(t) = \mathscr{L}^{-1}\left[\dfrac{-6}{(s+2)(s+3)}\right] = -6(e^{-2t} - e^{-3t})$$

$$\phi_{22}(t) = \mathscr{L}^{-1}\left[\dfrac{s}{(s+2)(s+3)}\right] = -2e^{-2t} + 3e^{-3t}$$

Hence the STM, $\Phi(t)$, can be written as

$$\Phi(t) = \begin{bmatrix} \phi_{11}(t) & \phi_{12}(t) \\ \phi_{21}(t) & \phi_{22}(t) \end{bmatrix} = \begin{bmatrix} 3e^{-2t} - 2e^{-3t} & e^{-2t} - e^{-3t} \\ -6(e^{-2t} - e^{-3t}) & -2e^{-2t} + 3e^{-3t} \end{bmatrix} \quad (11.72)$$

The MATLAB solution of Example 11.22 is obtained as follows:

MATLAB Program 11.3

```
% Finding State Transition Matrix of a system
% (Example 11.22)
syms t
a = [0 1; -6 -5]
phi = exam (a*t)

>>
a =
    0    1
   -6   -5

Phi =

[ 3*exp(-2*t)-2*exp(-3*t),  -exp(-3*t)+exp(-2*t)]
[ 6*exp(-3*t)-6*exp(-2*t),  -2*exp(-2*t)+3*exp(-3*t)]
```

EXAMPLE 11.23 For Example 11.22, determine $X(t)$ for the given initial values of the state as

$$x_1(0) = 1, \quad x_2(0) = 0$$

Solution: The complementary solution is

$$X(t) = \Phi(t)X(0)$$

Putting the value of $X(0) = \begin{bmatrix} 1 \\ 0 \end{bmatrix}$, we get

$$X(t) = \begin{bmatrix} x_1(t) \\ x_2(t) \end{bmatrix} = \begin{bmatrix} \phi_{11}(t) & \phi_{12}(t) \\ \phi_{21}(t) & \phi_{22}(t) \end{bmatrix} \begin{bmatrix} x_1(0) \\ x_2(0) \end{bmatrix}$$

$$= \begin{bmatrix} 3e^{-2t} - 2e^{-3t} & e^{-2t} - e^{-3t} \\ -6(e^{-2t} - e^{-3t}) & -2e^{-2t} + 3e^{-3t} \end{bmatrix} \begin{bmatrix} 1 \\ 0 \end{bmatrix}$$

$$= \begin{bmatrix} 3e^{-2t} - 2e^{-3t} \\ -6(e^{-2t} - e^{-3t}) \end{bmatrix} \tag{11.73}$$

Further, for a non-homogeneous state equation (AB), i.e. $u(t) \neq 0$ the solution becomes

$$X(t) = \Phi(t)X(0) + \int_0^t \Phi(t-\tau) B u(\tau) d\tau \tag{11.74}$$

and output becomes

$$y(t) = CX(t) + Du(t)$$

For better clarity, the following additional example is given.

EXAMPLE 11.24 Consider the dynamics of a non-homogeneous system as

$$\dot{X}(t) = \begin{bmatrix} 0 & 1 \\ -6 & -5 \end{bmatrix} X(t) + \begin{bmatrix} 0 \\ 1 \end{bmatrix} u(t)$$

and output $\quad y(t) = [1 \ 0] X(t)$

given the initial condition $\quad X(0) = [1 \ 0]^T$

where $u(t)$ is a unit-step input. Determine the output $y(t)$ at $t = 1$ s.

Solution: We obtained STM, $\Phi(t)$, as in Eq. (11.72). Again $\Phi(t)X(0)$ is obtained as in Eq. (11.73). Then from Eq. (11.74), we get

$$X(t) = \begin{bmatrix} x_1(t) \\ x_2(t) \end{bmatrix}$$

$$= \begin{bmatrix} \phi_{11}(t) & \phi_{12}(t) \\ \phi_{21}(t) & \phi_{22}(t) \end{bmatrix} \begin{bmatrix} x_1(0) \\ x_2(0) \end{bmatrix} + \int_0^t \begin{bmatrix} \phi_{11}(t-\tau) & \phi_{12}(t-\tau) \\ \phi_{21}(t-\tau) & \phi_{22}(t-\tau) \end{bmatrix} \begin{bmatrix} 0 \\ 1 \end{bmatrix} 1.d\tau$$

$$= \begin{bmatrix} 3e^{-2t} - 2e^{-3t} \\ -6(e^{-2t} - e^{-3t}) \end{bmatrix} + \begin{bmatrix} (1/6) - (1/2)e^{-2t} + (1/3)e^{-3t} \\ e^{-2t} - e^{-3t} \end{bmatrix}$$

$$= \begin{bmatrix} (1/6) + 2.5e^{-2t} - (5/3)e^{-3t} \\ -5e^{-2t} + 5e^{-3t} \end{bmatrix}$$

Hence the output response

$$y(t) = [1 \ 0] X(t) = \frac{1}{6} + \frac{5}{2} e^{-2t} - \frac{5}{3} e^{-3t}$$

The output at $t = 1$ s is 0.44.

It may be a trivial job to determine the STM for the system given by

$$\dot{X} = \begin{bmatrix} -1 & & 0 \\ & -2 & \\ 0 & & -3 \end{bmatrix} X$$

The STM, $\quad \Phi(t) = \begin{bmatrix} e^{-t} & & 0 \\ & e^{-2t} & \\ 0 & & e^{-3t} \end{bmatrix}$

Similarly, consider the Jordan canonical form of a system matrix given by

$$A = \begin{bmatrix} -1 & 1 & 0 \\ 0 & -1 & 0 \\ 0 & 0 & -2 \end{bmatrix}$$

It can be easily shown that the STM will be

$$\Phi(t) = \begin{bmatrix} e^{-t} & te^{-t} & 0 \\ 0 & e^{-t} & 0 \\ 0 & 0 & e^{-2t} \end{bmatrix}$$

Drill Problem 11.12

Given the transfer function

$$G(s) = \frac{5}{s(s+0.2)(s+5)}$$

(i) Obtain the state variable formulation in observable canonical form.
(ii) Calculate the STM corresponding to this model.

Ans. (i) $\dot{X} = \begin{bmatrix} -5.2 & 1 & 0 \\ -1 & 0 & 1 \\ 0 & 0 & 0 \end{bmatrix} X + \begin{bmatrix} 0 \\ 0 \\ 5 \end{bmatrix} u$

and output $y = \begin{bmatrix} 1 & 0 & 0 \end{bmatrix} X$

Drill Problem 11.13

The following A-matrices are in the Jordan form. Obtain the STM for each of them.

(a) $\begin{bmatrix} -2 & 1 \\ 0 & -2 \end{bmatrix}$ (b) $\begin{bmatrix} -2 & 1 & 0 \\ 0 & -2 & 1 \\ 0 & 0 & -2 \end{bmatrix}$

Ans. (a) $\begin{bmatrix} e^{-2t} & te^{-2t} \\ 0 & e^{-2t} \end{bmatrix}$ (b) $\begin{bmatrix} e^{-2t} & te^{-2t} & \frac{t^2}{2}e^{-2t} \\ 0 & e^{-2t} & te^{-2t} \\ 0 & 0 & e^{-2t} \end{bmatrix}$

EXAMPLE 11.25 The state equation of a linear control system is given as

$$\dot{X} = AX$$

where

$$A = 3\begin{bmatrix} 0 & 1 \\ -6 & -5 \end{bmatrix}$$

and

$$X(0) = \begin{bmatrix} 2 \\ 0 \end{bmatrix}$$

Obtain the STM and the solution of the state equation in terms of STM.

State-Variable Formulation

Solution:

$$\text{STM, } \Phi(t) = \mathcal{L}^{-1}[sI - A]^{-1} = \mathcal{L}^{-1}\begin{bmatrix} s & -3 \\ 18 & s+15 \end{bmatrix}^{-1} = \mathcal{L}^{-1}\left[\frac{1}{\Delta}\begin{bmatrix} s+15 & 3 \\ -18 & s \end{bmatrix}\right]$$

where $\Delta = s^2 + 15s + 54 = (s+6)(s+9)$

Therefore,
$$\Phi(t) = \begin{bmatrix} 3e^{-6t} - 2e^{-9t} & e^{-6t} - e^{-9t} \\ -6(e^{-6t} - e^{-9t}) & -2e^{-6t} + 3e^{-9t} \end{bmatrix}$$

and
$$X(t) = \begin{bmatrix} x_1(t) \\ x_2(t) \end{bmatrix} = \Phi(t)X(0) = \begin{bmatrix} 6e^{-6t} - 4e^{-9t} \\ -12(e^{-6t} - e^{-9t}) \end{bmatrix}$$

11.18.3 Power series method

As an illustration, determine the STM by infinite power series method for the given system matrix

$$A = \begin{bmatrix} 1 & 0 \\ 1 & 1 \end{bmatrix}$$

$$\text{STM, } \Phi(t) = e^{At} = \sum_{r=0}^{\infty} \frac{A^r t^r}{r!}$$

$$A^2 = A \cdot A = \begin{bmatrix} 1 & 0 \\ 1 & 1 \end{bmatrix}\begin{bmatrix} 1 & 0 \\ 1 & 1 \end{bmatrix} = \begin{bmatrix} 1 & 0 \\ 2 & 1 \end{bmatrix}$$

$$A^3 = A^2 \cdot A = \begin{bmatrix} 1 & 0 \\ 3 & 1 \end{bmatrix}; \text{ and so on}$$

Then,
$$\Phi(t) = e^{At} = I + At + \frac{A^2 t^2}{2!} + \frac{A^3 t^3}{3!} + \cdots$$

$$= \begin{bmatrix} 1 & 0 \\ 0 & 1 \end{bmatrix} + \begin{bmatrix} 1 & 0 \\ 1 & 1 \end{bmatrix}t + \begin{bmatrix} 1 & 0 \\ 2 & 1 \end{bmatrix}\frac{t^2}{2!} + \begin{bmatrix} 1 & 0 \\ 3 & 1 \end{bmatrix}\frac{t^3}{6} + \cdots$$

$$= \begin{bmatrix} \left(1 + t + \frac{t^2}{2} + \frac{t^3}{6} + \cdots\right) & 0 \\ \left(t + t^2 + \frac{t^3}{2} + \cdots\right) & \left(1 + t + \frac{t^2}{2} + \frac{t^3}{6} + \cdots\right) \end{bmatrix}$$

$$= \begin{bmatrix} \sum_{r=0}^{\infty} \dfrac{t^r}{r!} & 0 \\ t\sum_{r=0}^{\infty} \dfrac{t^r}{r!} & \sum_{r=0}^{\infty} \dfrac{t^r}{r!} \end{bmatrix} = \begin{bmatrix} e^t & 0 \\ te^t & e^t \end{bmatrix}$$

Now, we will determine the STM for the same system matrix by the inverse Laplace transform method. Given

$$A = \begin{bmatrix} 1 & 0 \\ 1 & 1 \end{bmatrix}$$

Now the resolvent matrix,

$$\Phi(s) = \mathscr{L}^{-1}[sI - A]^{-1} = \begin{bmatrix} s-1 & 0 \\ -1 & s-1 \end{bmatrix}^{-1} = \dfrac{1}{(s-1)^2}\begin{bmatrix} s-1 & 0 \\ 1 & s-1 \end{bmatrix}$$

STM, $\Phi(t) = \mathscr{L}^{-1}[\Phi(s)] = \mathscr{L}^{-1}[sI - A]^{-1}$

$$= \begin{bmatrix} \mathscr{L}^{-1}\left[\dfrac{1}{s-1}\right] & 0 \\ \mathscr{L}^{-1}\left[\dfrac{1}{(s-1)^2}\right] & \mathscr{L}^{-1}\left[\dfrac{1}{s-1}\right] \end{bmatrix} = \begin{bmatrix} e^t & 0 \\ te^t & e^t \end{bmatrix}$$

Obviously both the methods give rise to the same result. The inverse Laplace transform method of evaluation of STM requires the knowledge of eigenvalues which are required for partial fraction of each element ϕ_{ij} of STM, $\Phi(t)$.

In order to find the time-domain solution of the non-homogeneous system whose dynamics is

$$\dot{X} = \begin{bmatrix} 1 & 0 \\ 1 & 1 \end{bmatrix} X + \begin{bmatrix} 1 \\ 0 \end{bmatrix} u$$

and the output

$$y = [1 \quad 0] X$$

and which is subjected to a unit step applied at $t = 0$ and the initial condition

$$X(0) = \begin{bmatrix} 1 \\ 0 \end{bmatrix}$$

we proceed by the power series expansion to obtain the STM, $\Phi(t)$, as

$$\Phi(t) = e^{At} = \begin{bmatrix} \left(1 + t + \dfrac{t^2}{2!} + \dfrac{t^3}{3!} + \cdots\right) & 0 \\ \left(t + t^2 + \dfrac{t^3}{2!} + \cdots\right) & \left(1 + t + \dfrac{t^2}{2!} + \dfrac{t^3}{3!} + \cdots\right) \end{bmatrix}$$

State-Variable Formulation

Now the state equation

$$X(t) = \Phi(t) X(0) + \int_0^t \Phi(t-\tau) Bu(\tau) d\tau = e^{At} X(0) + \int_0^t e^{A(t-\tau)} Bu(\tau) d\tau$$

$$= e^{At} X(0) + e^{At} \int_0^t e^{-A\tau} Bu(\tau) d\tau$$

Let us find for time $t = T$, the STM, e^{-AT} which is

$$e^{-AT} = \begin{bmatrix} \left(1 - T + \dfrac{T^2}{2!} - \dfrac{T^3}{3!} + \cdots\right) & 0 \\ \left(-T + T^2 - \dfrac{T^3}{2!} + \dfrac{T^4}{3!} - \cdots\right) & \left(1 - T + \dfrac{T^2}{2!} - \dfrac{T^3}{3!} + \cdots\right) \end{bmatrix}$$

11.18.4 Cayley–Hamilton theorem

The Cayley–Hamilton theorem states that every matrix satisfies its own characteristic equation. If the characteristic equation of matrix A, that is, $|sI - A| = 0$ is

$$s^n + a_{n-1}s^{n-1} + a_{n-2}s^{n-2} + \cdots + a_1 s + a_0 = 0$$

the matrix A itself satisfies the same equation, namely

$$A^n + a_{n-1}A^{n-1} + a_{n-2}A^{n-2} + \cdots + a_1 A + a_0 I = 0 \qquad (11.75)$$

It may be noted that the nth power and any higher power of A can be expressed in terms of the $(n-1)$th power of A and lower powers, down to and including the zero-th power of A. Hence the state transition matrix which is expressed as power series of A can be evaluated by using the Cayley–Hamilton's theorem.

EXAMPLE 11.26(a) The matrix

$$A = \begin{bmatrix} 0 & 1 \\ -2 & -3 \end{bmatrix}$$

has the characteristic equation

$$|sI - A| = s^2 + 3s + 2 = 0$$

Show that the matrix satisfies its characteristic equation.

Solution: Substituting A for the variable s gives

$$A^2 + 3A + 2I = \begin{bmatrix} -2 & -3 \\ 6 & 7 \end{bmatrix} + 3\begin{bmatrix} 0 & 1 \\ -2 & -3 \end{bmatrix} + 2\begin{bmatrix} 1 & 0 \\ 0 & 1 \end{bmatrix} = \begin{bmatrix} 0 & 0 \\ 0 & 0 \end{bmatrix}$$

That is the null matrix [0]. Hence it is proved.

EXAMPLE 11.26(b) Determine A^2, A^3, A^4 and A^{-1} for the matrix of Example 11.26(a) using the Cayley–Hamilton technique.

Solution: Since A satisfies its own characteristic equation, we get

$$A^2 + 3A + 2I = 0$$

Now, $\quad A^2 = -3A - 2I = -3\begin{bmatrix} 0 & 1 \\ -2 & -3 \end{bmatrix} - 2\begin{bmatrix} 1 & 0 \\ 0 & 1 \end{bmatrix} = \begin{bmatrix} -2 & -3 \\ 6 & 7 \end{bmatrix}$

and $\quad A^3 = A \cdot A^2 = A(-3A - 2I) = -3A^2 - 2A = -3(-3A - 2I) - 2A = 7A + 6I$

$A^4 = [A^2]^2 = (-3A - 2I)^2 = 9A^2 + 12A + 4I = 9(-3A - 2I) + 12A + 4I = -15A - 14I$

$$= \begin{bmatrix} -14 & -15 \\ 30 & 31 \end{bmatrix}$$

From the Cayley–Hamilton theorem, A satisfies its characteristic equation, i.e.

$$A^2 + 3A + 2I = 0$$

or $\quad A + 3I + 2A^{-1} = 0$

or $\quad A^{-1} = -\dfrac{1}{2}A - \dfrac{3}{2}I = \begin{bmatrix} -\dfrac{3}{2} & -\dfrac{1}{2} \\ 1 & 0 \end{bmatrix}$

This is often a convenient way of computing the inverse of a matrix.

EXAMPLE 11.27 Find the polynomial

$$N(A) = A^4 + A^3 + A^2 + A + I$$

if $\quad A = \begin{bmatrix} 0 & 1 \\ -2 & -3 \end{bmatrix}$

Solution: We have already obtained from the characteristic equation using the Cayley–Hamilton technique, the expressions for A^2, A^3, and A^4.

Then $\quad N(A) = A^4 + A^3 + A^2 + A + I$

$= (-15A - 14I) + (7A + 6I) + (-3A - 2I) + A + I$

$= -10A - 9I$

$= \begin{bmatrix} -9 & -10 \\ 20 & 21 \end{bmatrix}$

Hence it can be shown through the example that any polynomial, $N(A)$ of $(n \times n)$ matrix A can be reduced to a linear combination of $I, A, A^2, \ldots, A^{n-1}$ or a polynomial whose highest degree in A is $(n - 1)$.

State-Variable Formulation

An alternative procedure is illustrated. Consider a matrix polynomial $N(A)$ which is of higher degree than the order of A. If $N(s)$ is divided by the characteristic polynomial

$$P(s) = |sI - A|$$

then

$$\frac{N(s)}{P(s)} = Q(s) + \frac{R(s)}{Q(s)} \tag{11.76}$$

where $R(s)$ is the remainder, then

$$N(s) = Q(s)P(s) + R(s) \tag{11.77}$$

Now, if $P(s) = 0$, then we get

$$N(s) = R(s)$$

Correspondingly, since $P(A) = 0$ by the Cayley–Hamilton theorem, the matrix polynomial $N(A)$ is then equal to $R(A)$.

EXAMPLE 11.28 Given

$$N(A) = A^4 + A^3 + A^2 + A + I \quad \text{and} \quad N(s) = s^4 + s^3 + s^2 + s + I$$

The characteristic equation of

$$A = \begin{bmatrix} 0 & 1 \\ -2 & -3 \end{bmatrix}$$

is

$$P(s) = s^2 + 3s + 2 = 0$$

Dividing $N(s)$ by $P(s)$, we get

$$\frac{s^4 + s^3 + s^2 + s + 1}{s^2 + 3s + 2} = (s^2 - 2s + 5) + \frac{-10s - 9}{s^2 + 3s + 2}$$

The remainder is then, $R(s) = -10s - 9$. Hence, $N(A) = R(A) = -10A - 9I$, which is the same result as obtained in Example 11.27.

The preceding technique in which $N(A)$ is a polynomial function of A and when $F(A)$ is desired where $F(s)$ is an analytic function of s, in a region about the origin, then $F(A)$ can be expressed as a polynomial in A of degree $(n - 1)$. Consequently, the remainder $R(s)$ must be a polynomial of degree $(n - 1)$. Then

$$F(s) = Q(s)P(s) + R(s) \tag{11.78}$$

where $P(s)$ is the characteristic polynomial of A and $R(s)$ is a polynomial of the form

$$R(s) = \alpha_0 + \alpha_1 s + \alpha_2 s^2 + \cdots + a_{n-1} s^{n-1} \tag{11.79}$$

The coefficients $\alpha_0, \alpha_1, \ldots, \alpha_{n-1}$ can be obtained successively by substituting the eigenvalues $s_1, s_2, \ldots, s_n$ into Eq. (11.78).

Since
$$P(s_i) = 0$$
the equations
$$\left.\begin{array}{l}F(s_1) = R(s_1) \\ F(s_2) = R(s_2) \\ F(s_3) = R(s_3)\end{array}\right\} \quad (11.80)$$

are obtained. Equation (11.80) describes a set of n linear equations in n unknowns. Therefore, a unique solution can be obtained for all the coefficients of the polynomial $R(s)$. This is valid for distinct roots; for repetitive roots, a modification needs to be done which is not dealt with here.

EXAMPLE 11.29 Find the STM, e^{At}, for the system matrix

$$A = \begin{bmatrix} 0 & -2 \\ 1 & -3 \end{bmatrix}$$

Solution: Here the roots are $-1, -2$. Since A is a second-order matrix, the polynomial $R(s)$ is of first order, i.e.

$$R(s) = \alpha_0 + \alpha_1 s$$

Therefore, two linear equations are obtained by substituting $s_1 = -1$ and $s_2 = -2$. That is,

$$F(s_1) = R(s_1)$$
$$e^{s_1 t} = \alpha_0 + \alpha_1 s_1$$
$$e^{-t} = \alpha_0 - \alpha_1$$

and
$$F(s_2) = R(s_2)$$
$$e^{s_2 t} = \alpha_0 + \alpha_1 s_2$$
$$e^{-2t} = \alpha_0 - 2\alpha_1$$

Solving for α_0 and α_1, we get
$$\alpha_0 = 2e^{-t} - e^{-2t} \quad \text{and} \quad \alpha_1 = e^{-t} - e^{-2t}$$

Hence
$$F(A) = e^{At} = \Phi(t)$$
$$= \alpha_0 I + \alpha_1 A$$
$$= \begin{bmatrix} \alpha_0 & 0 \\ 0 & \alpha_0 \end{bmatrix} + \begin{bmatrix} 0 & -2\alpha_1 \\ \alpha_1 & -3\alpha_1 \end{bmatrix}$$
$$= \begin{bmatrix} 2e^{-t} - e^{-2t} & 2(e^{-2t} - e^{-t}) \\ e^{-t} - e^{-2t} & 2e^{-2t} - e^{-t} \end{bmatrix}$$

EXAMPLE 11.30 Determine the STM for the system having its system matrix of 3×3 Jordan block with eigenvalue λ_1.

State-Variable Formulation

Solution: The characteristic equation is $|sI - J| = (\lambda - \lambda_1)^3$, where

$$J = \begin{bmatrix} \lambda_1 & 1 & 0 \\ 0 & \lambda_1 & 1 \\ 0 & 0 & \lambda_1 \end{bmatrix}$$

The characteristic equation is $(\lambda - \lambda_1)^3 = 0$

and
$$\exp(Jt) = \alpha_0 I + \alpha_1 J + \alpha_2 J^2$$

where
$$\exp(\lambda_1 t) = \alpha_0 + \alpha_1 \lambda_1 + \alpha_2 \lambda_1^2$$

$$\left. \begin{array}{l} \left[\dfrac{d}{d\lambda} \exp(\lambda_1 t)\right]\bigg|_{\lambda=\lambda_1} = t \exp(\lambda_1 t) = \alpha_1 + 2\alpha_2 \lambda_1 \\[2ex] \dfrac{1}{2}\left[\dfrac{d^2}{d\lambda^2} \exp(\lambda_1 t)\right]\bigg|_{\lambda=\lambda_1} = \dfrac{1}{2}t^2 \exp(\lambda_1 t) = \alpha_2 \end{array} \right\} \quad (11.81)$$

Writing Eq. (11.81) in matrix form, we get

$$\begin{bmatrix} 1 \\ t \\ \dfrac{t^2}{2} \end{bmatrix} \exp(\lambda_1 t) = \begin{bmatrix} 1 & \lambda_1 & \lambda_1^2 \\ 0 & 1 & 2\lambda_1 \\ 0 & 0 & 1 \end{bmatrix} \begin{bmatrix} \alpha_0 \\ \alpha_1 \\ \alpha_2 \end{bmatrix}$$

or

$$\begin{bmatrix} \alpha_0 \\ \alpha_1 \\ \alpha_2 \end{bmatrix} = \begin{bmatrix} 1 & \lambda_1 & \lambda_1^2 \\ 0 & 1 & 2\lambda_1 \\ 0 & 0 & 1 \end{bmatrix} \begin{bmatrix} 1 \\ t \\ \dfrac{t^2}{2} \end{bmatrix} \exp(\lambda_1 t)$$

or

$$\begin{bmatrix} \alpha_0 \\ \alpha_1 \\ \alpha_2 \end{bmatrix} = \begin{bmatrix} 1 & -\lambda_1 & \lambda_1^2 \\ 0 & 1 & -2\lambda_1 \\ 0 & 0 & 1 \end{bmatrix} \begin{bmatrix} 1 \\ t \\ \dfrac{t^2}{2} \end{bmatrix} \exp(\lambda_1 t)$$

where

$$\left. \begin{array}{l} \alpha_0 = \left[1 - \lambda_1 t + \lambda_1^2 \dfrac{t^2}{2}\right] \exp(\lambda_1 t) \\[2ex] \alpha_1 = \left[t + \lambda_1 \dfrac{t^2}{2}\right] \exp(\lambda_1 t) \\[2ex] \alpha_2 = \dfrac{t^2}{2} \exp(\lambda_1 t) \end{array} \right\} \quad (11.82)$$

As J is Jordan block of 3×3 order, and $\exp(Jt) = \alpha_0 I + \alpha_1 J + \alpha_2 J^2$ where

$$J = \begin{bmatrix} \lambda_1 & 1 & 0 \\ 0 & \lambda_1 & 1 \\ 0 & 0 & \lambda_1 \end{bmatrix} \quad \text{and} \quad J^2 = \begin{bmatrix} \lambda_1^2 & 2\lambda_1 & 1 \\ 0 & \lambda_1^2 & 2\lambda_1 \\ 0 & 0 & \lambda_1^2 \end{bmatrix}$$

$$\text{STM} = \exp(Jt) = \alpha_0 I + \alpha_1 J + \alpha_2 J^2$$

$$= \begin{bmatrix} \alpha_0 + \alpha_1 \lambda_1 + \alpha_2 \lambda_1^2 & \alpha_1 + 2\lambda_1 \alpha_2 & \alpha_2 \\ 0 & \alpha_0 + \alpha_1 \lambda_1 + \alpha_2 \lambda_1^2 & \alpha_1 + 2\lambda_1 \alpha \\ 0 & 0 & \alpha_0 + \alpha_1 \lambda_1 + \alpha_2 \lambda_1^2 \end{bmatrix}$$

Using Eq. (11.82), we get the value of STM as

$$\text{STM} = \exp(\lambda_1 t) \begin{bmatrix} 1 & t & \frac{1}{2} t^2 \\ 0 & 1 & t \\ 0 & 0 & 1 \end{bmatrix}$$

EXAMPLE 11.31 Use the results of Example 11.30 to find the STM for the given system matrix as 4×4 Jordan block with fourth-order repetitive eigenvalue λ_1.

Solution:

$$J = \begin{bmatrix} \lambda_1 & 1 & 0 & 0 \\ 0 & \lambda_1 & 1 & 0 \\ 0 & 0 & \lambda_1 & 1 \\ 0 & 0 & 0 & \lambda_1 \end{bmatrix}, \quad J^2 = \begin{bmatrix} \lambda_1^2 & 2\lambda_1 & 1 & 0 \\ 0 & \lambda_1^2 & 2\lambda_1 & 1 \\ 0 & 0 & \lambda_1^2 & 2\lambda_1 \\ 0 & 0 & 0 & \lambda_1^2 \end{bmatrix}$$

and

$$J^3 = \begin{bmatrix} \lambda_1^3 & 3\lambda_1^2 & 3\lambda_1 & 1 \\ 0 & \lambda_1^3 & 3\lambda_1^2 & 3\lambda_1 \\ 0 & 0 & \lambda_1^3 & 3\lambda_1^2 \\ 0 & 0 & 0 & \lambda_1^3 \end{bmatrix}$$

$$\text{STM} = \exp(Jt) = \alpha_0 I + \alpha_1 J + \alpha_2 J^2 + \alpha_3 J^3$$

In a similar way, we find a_i's, $i = 0, 1, 2$ and 3 and find STM as

$$\text{STM} = \exp(\lambda_1 t) \begin{bmatrix} 1 & t & \frac{1}{2} t^2 & \frac{1}{3!} t^3 \\ 0 & 1 & t & \frac{1}{2} t^2 \\ 0 & 0 & 1 & t \\ 0 & 0 & 0 & 1 \end{bmatrix}$$

State-Variable Formulation

The pattern illustrated by this example and Example 11.30 can be extended to any $m \times m$ Jordan block.

EXAMPLE 11.32 Find the state transition matrix for the system matrix

$$A = \begin{bmatrix} 0 & -2 \\ 1 & -3 \end{bmatrix}$$

Solution: The power of A can be obtained by successive multiplication by A, so that

$$A^2 = \begin{bmatrix} -2 & 6 \\ -3 & 7 \end{bmatrix}; \quad A^3 = \begin{bmatrix} 6 & 14 \\ 7 & -15 \end{bmatrix}; \quad \text{and so on}$$

Therefore $\Phi(t)$ can be written by power series expansion as

$$\Phi(t) = e^{At} = \sum_{r=0}^{\infty} \frac{A^r t^r}{r!}$$

$$= \begin{bmatrix} 1 & 0 \\ 0 & 1 \end{bmatrix} + \begin{bmatrix} 0 & -2 \\ 1 & -3 \end{bmatrix} t + \begin{bmatrix} -2 & 6 \\ -3 & 7 \end{bmatrix}\frac{t^2}{2!} + \begin{bmatrix} 6 & 14 \\ 7 & -15 \end{bmatrix}\frac{t^3}{3!} + \cdots$$

$$= \begin{bmatrix} 1 - \frac{2t^2}{2!} + \frac{6t^3}{3!} + \cdots & -2t + \frac{6t^2}{2!} + \frac{14t^3}{3!} + \cdots \\ t - \frac{3t^2}{2!} + \frac{7t^3}{3!} + \cdots & 1 - 3t + \frac{7t^2}{2!} + \frac{15t^3}{3!} + \cdots \end{bmatrix}$$

$$= \begin{bmatrix} 2e^{-t} - e^{-2t} & 2(e^{-2t} - e^{-t}) \\ e^{-t} - e^{-2t} & 2e^{-2t} - e^{-t} \end{bmatrix}$$

Note the infinite series for each element. (This is the principal drawback of this method.)

EXAMPLE 11.32(a) For the given system

$$\dot{X} = \begin{bmatrix} 0 & -2 \\ 1 & -3 \end{bmatrix} X + \begin{bmatrix} 0 \\ 1 \end{bmatrix} u$$

and output

$$y(t) = \begin{bmatrix} 1 & 0 \end{bmatrix} X$$

with initial conditions

$$X(0) = \begin{bmatrix} x_1(0) \\ x_2(0) \end{bmatrix}$$

and u as a unit-step function applied at $t = 0$, find the output $y(t)$.

Solution: The STM, $\Phi(t)$ as obtained earlier is

$$\Phi(t) = \begin{bmatrix} 2e^{-t} - e^{-2t} & 2(e^{-2t} - e^{-t}) \\ e^{-t} - e^{-2t} & 2e^{-2t} - e^{-t} \end{bmatrix}$$

The state vector $X(t)$ is obtained as

$$X(t) = \Phi(t)X(0) + \int_0^t \Phi(t-\tau) B u(\tau) d\tau$$

The output $y(t)$ is

$$y(t) = C\Phi(t)X(0) + C\int_0^t \Phi(t-\tau) B u(\tau) d\tau$$

$$= [1\ 0]\begin{bmatrix} \phi_{11}(t) & \phi_{12}(t) \\ \phi_{21}(t) & \phi_{22}(t) \end{bmatrix}\begin{bmatrix} x_1(0) \\ x_2(0) \end{bmatrix} + \int_0^t [1\ 0]\begin{bmatrix} \phi_{11}(t-\tau) & \phi_{12}(t-\tau) \\ \phi_{21}(t-\tau) & \phi_{22}(t-\tau) \end{bmatrix}\begin{bmatrix} 0 \\ 1 \end{bmatrix} d\tau$$

$$= \phi_{11}(t)x_1(0) + \phi_{12}(t)x_2(0) + \int_0^t \phi_{12}(t-\tau) d\tau$$

$$= (2e^{-t} - e^{-2t})x_1(0) + 2(e^{-2t} - e^{-t})x_2(0) + \int_0^t 2\left[e^{-2(t-\tau)} - e^{-(t-\tau)}\right] d\tau$$

$$= (2e^{-t} - e^{-2t})x_1(0) + 2(e^{-2t} - e^{-t})x_2(0) + 2e^{-t} - e^{-2t} - 1\ ;\ t \geq 0$$

11.19 Power of Companion Matrix and Its Applications

If the system has companion form of system matrix, then the state transition matrix can be evaluated by power series expansion. The state transition matrix will be an infinite series.

Let the system dynamics be described by the vector-matrix differential equation

$$\dot{X} = CX \qquad (11.83)$$

where C is the companion form of the system matrix.

Here a special form of system matrix A is taken as companion form of system matrix denoted by C just to avoid confusion. The elements of the companion matrix C are c_{ij}. *Do not mix up with the output coupling matrix.*

The solution to Eq. (11.83) can be written as

$$X(T) = e^{CT}X(0)$$

where the STM e^{CT} is $\phi_C(T)$; the time interval T is chosen which is less than unity for the purpose of convergence of the power series. Thus,

$$\phi_C(T) = \exp(CT) = \sum_{r=0}^{\infty} \frac{C^r T^r}{r!}$$

An algorithm has been developed for finding the powers of companion matrix which can be suitably utilized for the evaluation of STM.

11.19.1 Algorithm: power of companion matrix[†]

The power of any $n \times n$ companion matrix C can be written as

$$C^{r+1} = C \cdot C^r \qquad (11.84)$$
$$= C^r \cdot C \qquad (11.85)$$

It may be noted that for the choice of the phase variables as the state variables, the system matrix assumes the companion form, denoted by C instead of A. The elements c_{ij} of the companion matrix C are denoted by

$$c_{ij} = 1, \; j = i + 1$$
$$= 0, \; j \neq i + 1; \; i = 1, 2, \ldots, n - 1$$

and
$$c_{nj} = -c_{j-1}, \; j = 1, 2, \ldots, n \qquad (11.86)$$

It may be noted that the elements of the last row (nth), that is, $c_{j-1}(j = 1, 2, \ldots, n)$ of companion matrix C of order ($n \times n$) are the coefficients of the characteristic equation

$$|sI - C| = s^n + c_{n-1}s^{n-1} + c_{n-2}s^{n-2} + \cdots + c_2 s^2 + c_1 s + c_0 = 0$$

In order to obtain the first ($n - 1$) rows of C^{r+1}, we proceed as follows:

Partition
$$C = \begin{bmatrix} A_1 & B_1 \\ D_1 & E_1 \end{bmatrix} \qquad (11.87)$$

where

$A_1 = 0$, of order $(n - 1) \times 1$
$B_1 = I$, of order $(n - 1) \times (n - 1)$
D_1 is of order 1×1
E_1 is of order $1 \times (n - 1)$.

Partition
$$C^r = \begin{bmatrix} A_2 & B_2 \\ D_2 & E_2 \end{bmatrix} \qquad (11.88)$$

Let c_{ij}^r be the elements of C^r. The order of A_2 is $1 \times (n - 1)$, B_2 is 1×1, D_2 is $(n - 1) \times (n - 1)$ and E_2 is $1 \times (n - 1)$.

Put Eqs. (11.87) and (11.88) in Eq. (11.84) to get the first ($n - 1$) rows of C^{r+1} as

$$[D_2 \quad E_2] \qquad (11.89)$$

which is the last ($n - 1$) row of C^r and has already been computed. In fact, dimension of Eq. (11.89) is ($n - 1$) × n.

For obtaining the nth row of C^{r+1}, partition C^r as

$$C^r = \begin{bmatrix} A_3 & B_3 \\ D_3 & E_3 \end{bmatrix} \qquad (11.90)$$

[†]D. Roy Chaudhury, "Algorithm for power of companion matrix and its applications," *IEEE Trans. on Automatic Control*, pp. 179–180, 1993.

The dimension of A_3 is $(n-1) \times (n-1)$, B_3 is $(n-1) \times 1$, D_3 is $1 \times (n-1)$ and E_3 is 1×1. Put Eqs. (11.87) and (11.90) in Eq. (11.85) to get the nth row of C^{r+1} as

$$[(D_3A_1 + E_3D_1) \quad (D_3B_1 + E_3E_1)]$$

which simplifies to

$$[E_3D_1 \quad D_3 + E_3E_1] \qquad (11.91)$$

In fact, the dimension of Eq. (11.91) is $1 \times n$.

Putting the elements c_{nj} and c^r_{nj} in Eq. (11.91), we get

$$c^{r+1}_{nk} = c^r_{n(k-1)} + c^r_{nn} c_{nk} \qquad (11.92)$$

where c^{r+1}_{nk} is the element of C^{r+1} of nth row and kth column.

Hence in order to obtain the $(r+1)$th power of companion matrix C, one has to consider only the last row of C. The last row of C^{r+1} is described in Figure 11.30. From Figure 11.30, let us obtain, for example, any element of nth row of C^{r+1}, say c^{r+1}_{n3}, which is obtained by multiplying $(c_{n3} = -c_2)$ by c^r_{nn} and then adding c^r_{n3}, that is,

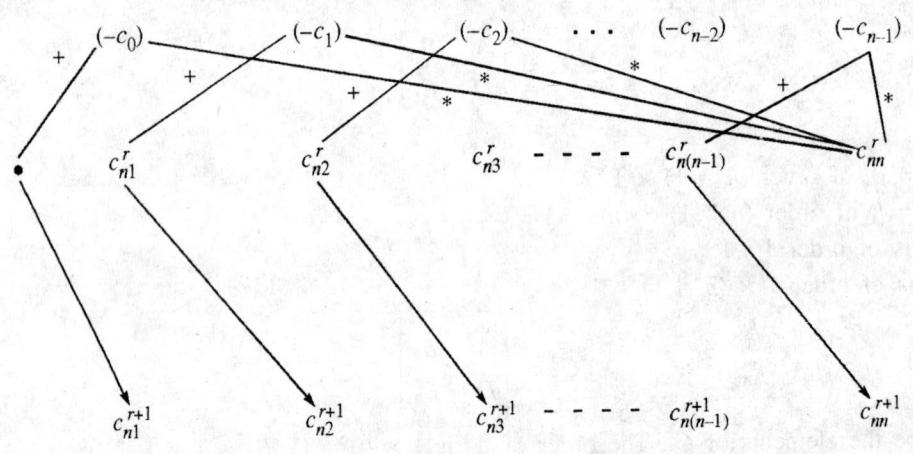

$$c^{r+1}_{n3} = -c_2 * c^r_{nn} + c^r_{n3}$$

FIGURE 11.30 Algorithm of power of companion matrix.

Obviously as c^r_{no} is not there and taken to be zero, hence for obtaining c^{r+1}_{n1} multiply $(c_{n1} = -c_o)$ by c^r_{nn} and then add 0 for c^r_{no}, i.e.

$$c^{r+1}_{n3} = c_{n1} * c^r_{nn} + 0 = c_{n1} c^r_{nn}$$

An algorithm can be developed from Eq. (11.92) for evaluating the successive powers of the companion matrix C as

$$c_{(h+1)k} = c_{nk}c_{hn} + c_{h(k-1)} \qquad (11.93)$$

for
$$h = n, n+1, \ldots; k = 1, 2, \ldots, n$$

State-Variable Formulation

Write the n-rows of the given companion matrix C and the successive rows can be written, with the help of Eq. (11.93) as

$$\begin{matrix} 0 & 1 & 0 & \cdots & 0 \\ 0 & 0 & 1 & \cdots & 0 \\ \cdot & \cdot & \cdot & & \cdot \\ 0 & 0 & 0 & \cdots & 1 \\ c_{n1} & c_{n2} & c_{n3} & \cdots & c_{nn} \\ c_{(n+1)1} & c_{(n+1)2} & c_{(n+2)3} & \cdots & c_{(n+1)n} \\ c_{(n+2)1} & c_{(n+2)2} & c_{(n+2)3} & \cdots & c_{(n+2)n} \\ \cdot & \cdot & \cdot & & \\ \cdot & \cdot & \cdot & & \end{matrix} \qquad (11.94)$$

Thus to obtain any element c_{pq} of the algorithm, the element of the corresponding qth column of the last row of the given companion matrix C is multiplied by the element of the last column of the $(p-1)$th row, and the element $c_{(p-1)(q-1)}$ of the $(p-1)$th row and $(q-1)$th column is added. Naturally, for obtaining the elements of the first column of any new row, the term to be added is zero.

Now the rth power of C can be written straight from the algorithm (11.94) by deleting the first $(r-1)$ rows of the algorithm and thereby forming the matrix C^r with the successive n-rows. So, to obtain the rth power of an $n \times n$ companion matrix, only $(n+r-1)$ rows are to be formed of which the first n-rows are the n-rows of the companion matrix. Hence actually $(r-1)$ rows are to be evaluated for obtaining C^r.

This algorithm for determining the powers of the companion matrix can be suitably utilized for determining the trace and hence for determining the largest pole. Again, this can be used for determining the state transition matrix for any general matrix A. It may be noted that for any general matrix A, the relation $A^r = PC^r P^{-1}$ holds where P is the non-singular similarity transformation matrix.

EXAMPLE 11.33 Consider a companion matrix

$$C = \begin{bmatrix} 0 & 1 & 0 \\ 0 & 0 & 1 \\ -6 & -11 & -6 \end{bmatrix}$$

For obtaining the powers of C, the rows are to be formed and can be written as

$$\begin{matrix} 0 & 1 & 0 \\ 0 & 0 & 1 \\ -6 & -11 & -6 \\ 36 & 60 & 25 \\ -150 & -239 & -90 \\ 540 & 840 & 301 \\ \cdots & \cdots & \cdots \end{matrix}$$

Now suppose C^4 is to be obtained. Then deleting the first three rows of the algorithm and forming the matrix with successive three rows, we get

$$C^4 = \begin{bmatrix} 36 & 60 & 25 \\ -150 & -239 & -90 \\ 540 & 840 & 301 \end{bmatrix}$$

This algorithm for determining the power of the companion matrix can be suitably utilized to determine the STM of any system $\dot{X} = AX$, provided that the similarity transformation matrix $C = P^{-1}AP$ exists where P is the non-singular similarity transformation matrix. The STM, $\Phi(T)$, can be written as

$$\Phi(T) = \exp(AT) = \sum_{r=0}^{\infty} \frac{A^r T^r}{r!} = \sum_{r=0}^{\infty} \frac{(P C^r P^{-1}) T^r}{r!} = P\left(\sum_{r=0}^{\infty} \frac{C^r T^r}{r!}\right) P^{-1}$$

$$= P\Phi_C(T) P^{-1} \quad (11.95)$$

where

$$\Phi_C(T) = \sum_{r=0}^{\infty} \frac{C^r T^r}{r!}$$

is the (state transition matrix) STM corresponding to companion form of the system matrix.

11.20 Trace of a Matrix

We know that transfer function and hence the characteristic function, that is, the roots are invariant under similarity transformation. The trace of a matrix is the sum of the diagonal elements which is equal to the sum of the eigenvalues λ_i's. The trace of a matrix is invariant under similarity transformation.

The traces of the power of a matrix can be conveniently utilized in determining the dominant eigenvalue of the system having magnitude of distinct eigenvalues $\lambda_1 > \lambda_2 > \ldots > \lambda_n$.

Thus,

$$\operatorname{Tr}\left[\bar{A}^r\right] = \sum_{j=1}^{n} \lambda_j^r = \lambda_1^r \left[1 + \left(\frac{\lambda_2}{\lambda_1}\right)^r + \cdots + \left(\frac{\lambda_n}{\lambda_1}\right)^r\right] \approx \lambda_1^r$$

Again

$$\operatorname{Tr}\left[\bar{A}^{r+1}\right] = \sum_{i=1}^{n} \lambda_i^{r+1} = \lambda_1^{r+1} \left[1 + \left(\frac{\lambda_2}{\lambda_1}\right)^{r+1} + \cdots + \left(\frac{\lambda_n}{\lambda_1}\right)^{r+1}\right] \approx \lambda_1^{r+1}$$

because the bracketed series converges for r being very large in the limit $\lim_{r \to \infty}$.

Hence

$$\lambda_1 = \lim_{r \to \infty} \frac{\operatorname{Tr}\left[(\bar{A})^{r+1}\right]}{\operatorname{Tr}\left[(\bar{A})^r\right]}$$

11.21 STM of Linear Time-Varying System

For the case where the system matrix A is not fixed but varying with time, the homogeneous vector-matrix differential equation of a linear time-varying system is

$$\dot{X} = A(t)X \tag{11.96}$$

Let us consider the scalar case where the differential equation is

$$\dot{x} = a(t)x \tag{11.97}$$

The solution is obtained by introducing the integrating factor. The solution becomes

$$x(t) = [\exp b(t)]x(T) \tag{11.98}$$

where T is some fixed initial time instant and

$$b(t) = \int_0^t a(\tau)d\tau$$

The analogous solution to the vector-matrix differential equation would be

$$X(t) = \left[\exp \int_0^t A(\tau)d\tau\right] X(\tau) \tag{11.99}$$

provided
$$\frac{d}{dt}e^{B(t)} = \frac{dB(t)}{dt}e^{B(t)} \tag{11.100}$$

where
$$B(t) = \int_0^t A(\tau)\,d\tau \tag{11.101}$$

which is valid only when the commutativity condition is satisfied, i.e.

$$A(t_1)A(t_2) = A(t_2)A(t_1); \text{ for all } t_1 \text{ and } t_2. \tag{11.102}$$

If the commutativity condition in Eq. (11.102) is satisfied, the STM

$$\Phi(t, T) = \exp \int_0^t A(\tau)\,d\tau \tag{11.103}$$

11.21.1 Matrizant

For the general case where the commutative condition of Eq. (11.102) is not satisfied, the STM is not given by Eq. (11.103). However, the solution of linear, time-varying system in Eq. (11.96) can be obtained by the Peano-Baker method of integration as explained in the following.

Integrating Eq. (11.96) and assuming that the condition $X(T)$ is given, we get

$$X(t) = X(T) + \int_T^t A(\tau) X(\tau)\,d\tau \tag{11.104}$$

This equation is called the Vector Volterra integral equation. It can be solved by repeated substitution of the right-side of the integral equation into the integral for X. For example, the first iteration is

$$X(t) = X(T) + \int_T^t A(\lambda)\left[X(T) + \int_T^t A(s) X(s) ds\right] d\lambda \qquad (11.105)$$

The expression is simplified by introducing the integral operator Q where

$$Q(.) = \int_T^t (.) d\lambda$$

Then Eq. (11.104) can be rewritten as

$$X(t) = X(T) + Q(A)X(T) \qquad (11.106)$$

Continuing the process, $X(t)$ is obtained as the Neuman series as

$$X(t) = [I + Q(A) + Q(AQ(A)) + Q(AQ(AQ(A))) + \cdots]X(T)$$
$$= [G(A)]X(T)$$

where $G(A)$ is called the matrizant and is indeed the STM for the time-varying system and is written as

$$\Phi(t, T) = G(A) = I + Q(A) + Q(AQ(A)) + Q(AQ(AQ(A))) + \cdots \qquad (11.107)$$

Hence the solution of the homogeneous time-varying system of Eq. (11.96) is

$$X(t) = G(A)X(T) = \Phi(t,T)X(T)$$

Clearly, if A is a constant matrix, then

$$\Phi(t,T) = I + (t-T)A + \frac{(t-T)^2}{2!}A^2 + \frac{(t-T)^3}{3!}A^3 + \cdots$$
$$= \exp[A(t-T)] = e^{A(t-T)} \qquad (11.108)$$

The series of Eq. (11.107) is quite lengthy for computation unless the series converges rapidly.

11.21.2 Kinariwala's approach

An interesting alternative solution as proposed by Kinariwala is presented here. Decompose $A(t)$ into two matrices as $A_0(t)$ and $A_1(t)$. The $A_0(t)$ is the symmetric part of $A(t)$ which obviously satisfies the commutative condition and $A_1(t)$ is interpreted as perturbation upon $A_0(t)$. Thus

$$A(t) = A_0(t) + A_1(t) \qquad (11.109)$$

The homogeneous system as defined becomes

$$\dot{X} = [A_0(t) + A_1(t)]X \qquad (11.110)$$

State-Variable Formulation

The solution of the unperturbed equation

$$\dot{X}_0 = A_0(t)X_0$$

becomes

$$X_0(t) = \Phi_0(t, T)X(T)$$

where STM,

$$\Phi(t, T) = \exp\left[\int_T^t A_0(\lambda)d\lambda\right]$$

The perturbation $A_1(t)X(t)$ is equivalent to the forcing function term $Bu(t)$ in case of LTI system. By direct analogy using the principle of superposition, the solution for $X(t)$ is written as

$$X(t) = \Phi_0(t, T)X(T) + \int_T^t \Phi(t,\lambda)A_1(\lambda) X(\lambda) d\lambda \qquad (11.111)$$

This is a Volterra equation which is of the same form as that of Eq. (11.104). Using the same iteration process, the Neumann series for STM, $\Phi(t, T)$ is obtained as

$$\Phi(t, T) = [I + Q(\Phi_0 A_1) + Q(\Phi_0 A_1 \, Q(\Phi_0 A_1)) + \cdots]\Phi_0(t, T) \qquad (11.112)$$

A good first choice of $A_0(t)$ reduces the number of terms required. Further, if $A_1(t)$ is relatively small, then only the first few terms of Eq. (11.112) are necessary for adequate approximation. Needless to say, the solution for the linear time-varying system is quite laborious.

EXAMPLE 11.34 The dynamics of a time-varying system is

$$\dot{X} = \begin{bmatrix} 1 & e^{-t} \\ -e^{-t} & 2 \end{bmatrix} X$$

Obtain the STM, $\Phi(t, 0)$.

Solution:

$$\int_0^t A(\tau)d\tau = \int_0^t \begin{bmatrix} 1 & e^{-t} \\ -e^{-t} & 2 \end{bmatrix} d\tau = \begin{bmatrix} \tau & -e^{-t} \\ -e^{-t} & 2 \end{bmatrix}\bigg|_0^t = \begin{bmatrix} t & -e^{-t}+1 \\ -e^{-t}-1 & t \end{bmatrix}$$

Now

$$\int_0^t \begin{bmatrix} 1 & e^{-T_1} \\ -e^{-T_2} & 2 \end{bmatrix} \left\{\int_0^{T_1} \begin{bmatrix} 1 & e^{-T_2} \\ -e^{-T_2} & 2 \end{bmatrix} dT_2\right\} dT_1$$

$$= \begin{bmatrix} \dfrac{t^2}{2} - \dfrac{e^{-2t}}{2} + e^{-t} - \dfrac{1}{2} & t + e^{-t} - 2te^{-t} - 2e^{-t} + 2 - 1 \\ te^{-t} + e^{-t} - 2e^{-t} - 2t + 2 - 1 & e^{-t} - \dfrac{e^{-2t}}{2} + 4\dfrac{t^2}{2} - \dfrac{1}{2} \end{bmatrix}$$

Thus the STM, $\Phi(t, 0)$ can be written as

$$G(A) = \Phi(t, 0) = \begin{bmatrix} 1 & 0 \\ 0 & 1 \end{bmatrix} + \begin{bmatrix} t & -e^{-t}+1 \\ -e^{-t}-1 & 2t \end{bmatrix} + \begin{bmatrix} \dfrac{t^2}{2} - \dfrac{e^{-2t}}{2} + e^{-t} - \dfrac{1}{2} & t + e^{-t} - 2te^{-t} + 1 \\ te^{-t} + e^{-t} - 2t + 1 & e^{-t} - \dfrac{e^{-2t}}{2} + 2t^2 - \dfrac{1}{2} \end{bmatrix}$$

Summary

This chapter is devoted to state-space analysis of control systems. It explains how the state-variable approach can convert a higher-order differential equation into a first-order vector-matrix differential equation.

The fundamentals on state variables and state equations are first introduced and illustrated through the dynamics of the spring-mass-damper system. The chapter then dealt with the state-space representation of transfer-function systems and calculation of the transfer function from the state equation. The relationship between state equations and transfer functions is thus established. The analog simulation of state equations is discussed.

The need for the transformation matrix is emphasized. Similarity transformation is discussed. It is shown how the transfer function remains invariant under similarity transformation. Diagonalization, a special form of similarity transformation is explained. Similarity transformations to diagonal canonical form and Jordan canonical form are discussed. Characteristic equations and eigenvalues are defined in terms of the state equations and the transfer function.

In order to obtain the time response of the state variables of a control system, the state transition matrix (STM) plays a paramount role. Several methods for evaluation of STM are discussed.

Problems

11.1 (a) For the bridge networks shown in Figure P.11.1, write the state variable formulation of the state equation.

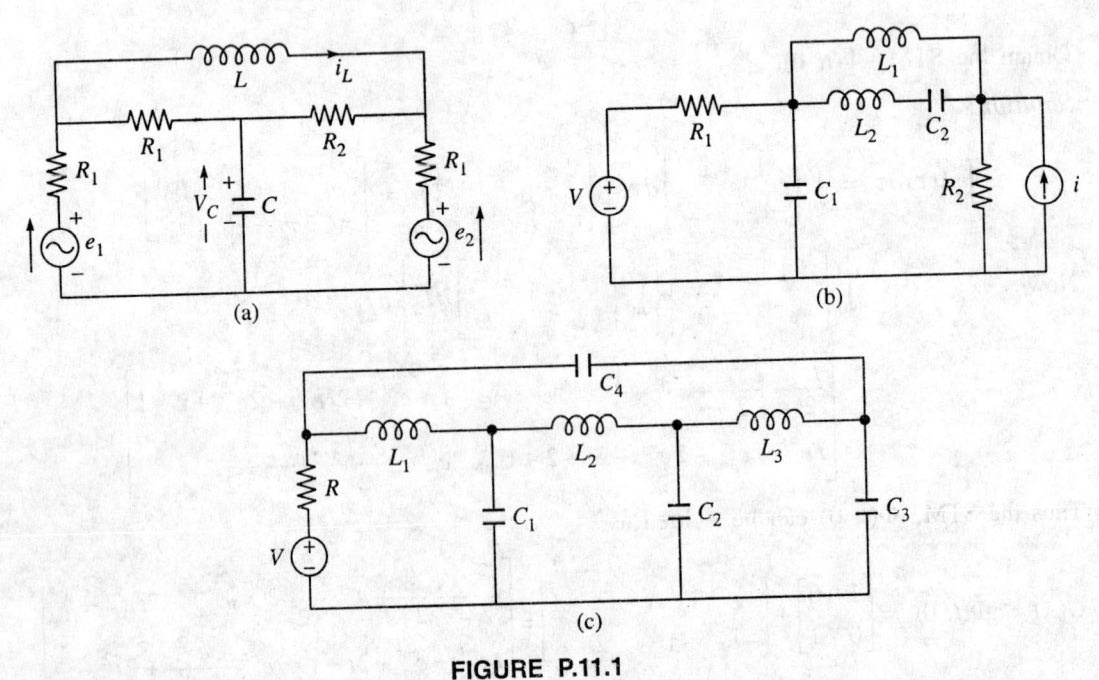

FIGURE P.11.1

(b) The state variable formulation equation of an electric network is given as

$$\frac{d}{dt}\begin{bmatrix} v_C \\ i_L \end{bmatrix} = \begin{bmatrix} -\dfrac{2}{(R_1+R_2)C} & 0 \\ 0 & \dfrac{2R_1R_2}{(R_1+R_2)L} \end{bmatrix}\begin{bmatrix} v_C \\ i_L \end{bmatrix} + \dfrac{1}{R_1+R_2}\begin{bmatrix} \dfrac{1}{C} & \dfrac{1}{C} \\ \dfrac{R_2}{L} & \dfrac{R_2}{L} \end{bmatrix}\begin{bmatrix} e_1 \\ e_2 \end{bmatrix}$$

and output $\qquad Y = \begin{bmatrix} 1 & 0 \end{bmatrix}\begin{bmatrix} v_C \\ i_L \end{bmatrix}$

Determine the transfer matrix from the state variable formulation. Draw the signal flow graph. Verify the transfer matrix using the Mason's gain formula.

11.2 Given the transfer function of the system as

$$\frac{Y(s)}{U(s)} = \frac{s^2 + 3s + 3}{s^3 + 2s^2 + 3s + 1}$$

(a) Write the state variable formulation in controllable canonical form and then draw the signal flow graph.
(b) Write the dual representation and simulate.
(c) Verify that the transfer function remains invariant under similarity transformation.
(d) Determine the similarity transformation matrix which transforms the system representation from (a) to (b).

11.3 Given the transfer function:

$$G(s) = \frac{Y(s)}{U(s)} = \frac{2}{(s+2)^3} + \frac{4}{(s+2)^2} + \frac{5}{s+2} + \frac{3}{s+1}$$

write the state variable formulation in Jordan canonical form.

11.4 Given the canonical form of state variable formulation as

$$\dot{X} = \begin{bmatrix} -2 & 1 & 0 & 0 \\ 0 & -2 & 1 & 0 \\ 0 & 0 & -2 & 0 \\ 0 & 0 & 0 & -1 \end{bmatrix} X + \begin{bmatrix} 0 \\ 0 \\ 1 \\ 1 \end{bmatrix} u$$

and output $\qquad y = \begin{bmatrix} 2 & 4 & 5 & 3 \end{bmatrix} X$

determine the state equation in the controller and observer canonical forms, and simulate the three canonical state-space representations.

11.5 Write the state variable formulation of the system represented by the following differential equations:

(i) $\qquad \dddot{y} + 4\dot{y} + y = 5u(t)$

(ii) $\qquad 2\dddot{y} + 3\ddot{y} + 5\dot{y} + 2y = u(t) + 2\ddot{u}$

(iii) $\qquad \dddot{y} + 5\ddot{y} + 3\dot{y} + y + \displaystyle\int_0^t y(s)\, ds = u(t)$

11.6 For the state variable formulation given below, write the dynamics of the system in terms of the integro-differential equation

$$\dot{X} = AX + Bu$$

and output
$$y = CX$$

where

(a) $A = \begin{bmatrix} -6 & 1 & 0 \\ -11 & 0 & 1 \\ -6 & 0 & 0 \end{bmatrix}$; $B = \begin{bmatrix} 2 \\ 6 \\ 2 \end{bmatrix}$; $C^T = \begin{bmatrix} 1 \\ 0 \\ 0 \end{bmatrix}$

(b) $A = \begin{bmatrix} 0 & 1 & 0 \\ 0 & 0 & 1 \\ -6 & -11 & -6 \end{bmatrix}$; $B = \begin{bmatrix} 0 \\ 0 \\ 1 \end{bmatrix}$; $C^T = \begin{bmatrix} 2 \\ 6 \\ 2 \end{bmatrix}$

(c) $A = \begin{bmatrix} -2 & 1 & 0 \\ 0 & -2 & 0 \\ -1 & -2 & -3 \end{bmatrix}$; $B = \begin{bmatrix} 1 \\ 1 \\ 1 \end{bmatrix}$; $C^T = \begin{bmatrix} 1 \\ 0 \\ 0 \end{bmatrix}$

(d) $A = \begin{bmatrix} 0 & 2 & 0 \\ 1 & 2 & 0 \\ -1 & 1 & 1 \end{bmatrix}$; $B = \begin{bmatrix} 1 \\ 1 \\ 0 \end{bmatrix}$; $C^T = \begin{bmatrix} 1 \\ 0 \\ 1 \end{bmatrix}$

11.7 For the system described in Problem 11.2, find the transformation matrix that transforms the state equation into controllable canonical form (CCF).

11.8 The controllable canonical form of state equations of a time-invariant system has the characteristic equation

$$s^3 + 3s^2 + 2s + 1 = 0$$

and the output equation as

$$y(t) = [1 \ 1 \ 0]X \ ; \quad X^T = [x_1 \ x_2 \ x_3]$$

Transform the state equation to

$$\dot{Z} = AZ + Bu(t)$$

where $Z^T = [x_1 \ \ x_1 + x_2 \ \ \dot{x}_1 + \dot{x}_2]$

11.9 Given the vector-matrix differential equation describing the dynamics of the system as

$$\dot{X} = AX + Bu$$

where

(a) $A = \begin{bmatrix} 0 & 1 & 0 \\ 0 & 0 & 1 \\ -1 & -2 & -3 \end{bmatrix}$; $B = \begin{bmatrix} 0 \\ 0 \\ 1 \end{bmatrix}$; $C = [1 \ 0 \ 0]$

State-Variable Formulation

(b) $A = \begin{bmatrix} 0 & 1 & 0 \\ 0 & 0 & 1 \\ 0 & -1 & -2 \end{bmatrix}$; $B = \begin{bmatrix} 0 \\ 0 \\ 1 \end{bmatrix}$; $C = [1 \ 1 \ 0]$

(c) $A = \begin{bmatrix} 0 & 1 & 0 & 0 \\ 0 & 0 & 1 & 0 \\ 0 & 0 & 0 & 1 \\ 1 & 0 & 0 & 0 \end{bmatrix}$; $B = \begin{bmatrix} 0 \\ 0 \\ 0 \\ 1 \end{bmatrix}$; $C = [1 \ 0 \ 0 \ 0]$

(i) Determine the eigenvalues of A. Then obtain a transformation matrix P such that the system state equation becomes in decoupled form
(ii) Simulate and find the transfer function from the Mason's gain formula.
(iii) Determine the transfer function from the state variable formulation.

11.10 Determine the characteristic equation, the eigenvalues and the state transition matrix of the system represented by the vector-matrix differential equation:

$$\dot{X} = AX + BU$$

where

(a) $A = \begin{bmatrix} 1 & 1 \\ -2 & -1 \end{bmatrix}$, $B = \begin{bmatrix} 0 & 1 \\ 1 & 0 \end{bmatrix}$; (b) $A = \begin{bmatrix} 0 & 1 \\ -4 & -5 \end{bmatrix}$, $B = \begin{bmatrix} 1 \\ 1 \end{bmatrix}$

(c) $A = \begin{bmatrix} 0 & 1 \\ -2 & -3 \end{bmatrix}$; (d) $A = \begin{bmatrix} -3 & 0 \\ 0 & -3 \end{bmatrix}$; (e) $A = \begin{bmatrix} 3 & 0 \\ 0 & -3 \end{bmatrix}$;

(f) $A = \begin{bmatrix} 0 & 2 \\ -2 & 0 \end{bmatrix}$; (g) $A = \begin{bmatrix} -1 & 0 & 0 \\ 0 & -2 & 1 \\ 0 & 0 & -2 \end{bmatrix}$

(h) $A = \begin{bmatrix} -5 & 1 & 0 \\ 0 & -5 & 1 \\ 0 & 0 & -5 \end{bmatrix}$, $B = \begin{bmatrix} 0 \\ 0 \\ 1 \end{bmatrix}$

(i) $A = \begin{bmatrix} -6 & 1 & 0 \\ -11 & 0 & 1 \\ -6 & 0 & 0 \end{bmatrix}$; (j) $A = \begin{bmatrix} 2 & 1 & 0 \\ 0 & 2 & 1 \\ 0 & 0 & 2 \end{bmatrix}$

11.11 Draw the state diagrams for the following systems

$$\dot{X} = AX + BU$$
and output $$Y(t) = CX(t)$$

where

(a) $A = \begin{bmatrix} 0 & 1 & 0 \\ 0 & 0 & 1 \\ -3 & -2 & -1 \end{bmatrix}$; $B = \begin{bmatrix} 0 \\ 0 \\ 1 \end{bmatrix}$; $C = [1 \ 0 \ 0]$

(b) $A = \begin{bmatrix} 0 & 0 & -3 \\ 1 & 0 & -2 \\ 0 & 0 & -1 \end{bmatrix}$; $B = \begin{bmatrix} 1 \\ 0 \\ 0 \end{bmatrix}$; $C = [0 \ 0 \ 1]$

(c) Same A as in part (a), but with $B = \begin{bmatrix} 0 & 1 \\ 1 & 0 \\ 1 & 0 \end{bmatrix}$

11.12 (a) Write the vector-matrix differential equations of the systems shown in Figures P.11.12(a) and (b), and then show that the transfer functions of the two systems are same.

(b) Prove that the two systems in Figures P.11.12(a) and (b) are dual.

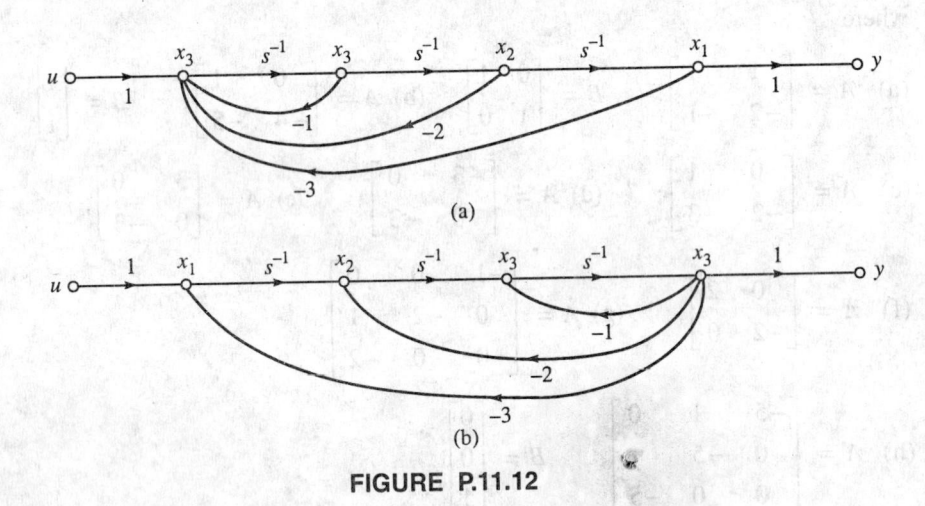

FIGURE P.11.12

11.13 For a linear time-invariant system whose state equations are expressed in controllable canonical form (CCF), show that the last column of

$$[\text{Adjoint}(sI - A)] = \begin{bmatrix} 1 \\ s \\ s^2 \\ \cdot \\ \cdot \\ \cdot \\ s^{n-1} \end{bmatrix}$$

11.14 Assume that the characteristic equation of A has all distinct roots. Then prove that the modal matrix becomes the Vandermonde matrix.

11.15 A linear time-invariant system is described by the differential equation

$$\dddot{y} + 3\ddot{y} + 3\dot{y} + y(t) = u(t)$$

with initial conditions $y(0) = 1$, $\dot{y}(0) = 0$, $\ddot{y}(0) = 0$

(a) Write the state equations of the system in vector-matrix form with the choice of phase variables as state variables.
(b) Determine the characteristic equation and the eigenvalues of A.
(c) Determine the state transition matrix.
(d) Determine the state equation of the system subjected to a unit-step input $u(t)$.
(e) Determine the output $y(t)$ at $t = 0.1$ s.

Rudolf Emil Kalman was born in Budapest, Hungary, on May 19, 1930. He received the bachelor's and the master's degree in electrical engineering, from the Massachusetts Institute of Technology in 1953 and 1954 respectively. He received the doctorate degree from Columbia University in 1957. His major positions include that of Research Mathematician at Research Institute for Advanced Study in Baltimore, between 1958–1964, Professor at Stanford University between 1964–1971, and from 1971 to 1992 Graduate Research Professor, and Director, at the Center for Mathematical System Theory, University of Florida, Gainesville. Moreover, since 1973 he has also held the chair for Mathematical System Theory at Swiss Federal Institute of Technology, Zurich. He is the recipient of numerous awards, including the IEEE Medal of Honor (1974), the IEEE Centennial Medal (1984), the Kyoto Prize in High Technology from the Inamori foundation, Japan (1985), the Steele Prize of the American Mathematical Society (1987), and the Bellman Prize (1997). He is a member of the National Academy of Sciences (USA), the National Academy of Engineering (USA), and the American Academy of Arts and Sciences (USA). He is a foreign member of the Hungarian, French, and Russian Academies of Science, and has received many honorary doctorates. He is married to Constantina nee Stavrou, and they have two children, Andrew and Elisabeth.

Analysis and Design of Modern Control Systems

12.1 Introduction

The concepts of controllability and observability were introduced by Kalman. The controllability and observability play an important role in the design of control systems in state space. The conditions of controllability and observability may govern the existence of a complete solution to the control system design problem. Here we will derive the conditions for controllability and observability. Further, dual representation is quite important from the analysis point of view. The importance of modern control theory lies in its ability to do the perfect design and to accomplish that, state feedback is done. From there we have gone for design by pole placement and its condition of existence. Observer design has been discussed because of the need for inaccessible states for perfect control of the system. Again the condition for observer design has been derived. The similarity transformation matrix for transformation to companion form (which we have studied in Chapter 11) from controllable and observable conditions has been derived. Synthesis of the system has been taken up with the help of examples.

12.2 Controllability

The state-variable formulation of an nth order linear time-invariant multivariable

OBJECTIVE

Controllability and observability are the important concepts of modern control engineering. State space design by arbitrary pole placement method requires that all states should be accessible which is mostly not so in practical systems. The non-accessible states can be made accessible by observer design. Further, If any system is controllable, there exists a similarity transformation matrix by which the system can be transformed to controllable canonical form (CCF).
- If a system is represented in CCF, then the system is controllable.
- If any system is observable, there exists a similarity transformation matrix by which the system can be transformed to observable canonical form (OCF).
- If a system is represented in OCF, then the system is observable.
- Arbitrary pole placement is possible only if the system is controllable.
- Accessibility of inaccessible states through observer design if the system is observable.

In this chapter, we have limited our discussion to single-input-single-output systems only.

The choice of state is not unique. As each form of state-variable formulation has got its own advantages, we have, therefore, discussed different forms of simulation.

CHAPTER OUTLINE

Introduction
Controllability
Observability
Concept of Transfer Function
Duality Property
Similarity Transformation Matrix
Pole-Placement Design
State Observer Design
Realization of Transfer Matrix
Different forms of Simulation

(m-input, p-output) system represented by the vector-matrix differential equation

$$\dot{X} = AX + BU \qquad \text{(AB)}$$

and output
$$Y = CX \qquad \text{(C)}$$

is completely controllable if and only if the rank of the ($n \times mn$) order controllability matrix S is n where S is defined as

$$S = [B \quad AB \quad A^2B \quad \cdots \quad A^{n-1}B] \qquad (12.1)$$

A system is said to be completely state controllable if it is possible to find an input $U(t)$ that will transfer the system from any given initial state $X(t_0)$ to any given final state $X(t_f)$ over a specified interval of time $(t_f - t_0)$.

In the controllability matrix S, since the matrices A and B are involved, sometimes it is said that the pair $[A, B]$ is controllable, which implies that the controllability matrix S is of rank n. If rank of S is not n then the system is not completely controllable or is said to be uncontrollable. Obviously the question will be how to arrive at the result that, for the system to be completely controllable, the controllability matrix S should be of rank n.

Proof
The solution of the vector-matrix differential equation

$$\dot{X} = AX + BU \qquad \text{(AB)}$$

is
$$X(t) = \Phi(t - t_0) X(t_0) + \int_{t_0}^{t} \Phi(t - \tau) BU(\tau) d\tau \qquad \text{for } t > t_0 \qquad (12.2)$$

Without loss of any generality, we can assume that the desired final state $X(t_f) = 0$ for some finite time $t_f > t_0$, given that the initial state $X(t_0) \neq 0$. Then it gives

$$0 = X(t_f) = \Phi(t_f - t_0) X(t_0) + \int_{t_0}^{t_f} \Phi(t_f - \tau) BU(\tau) d\tau$$

or
$$e^{A(t_f - t_0)} X(t_0) = -\int_{t_0}^{t_f} e^{A(t_f - \tau)} BU(\tau) d\tau$$

or
$$X(t_0) = -e^{-A(t_f - t_0)} \int_{t_0}^{t_f} e^{A(t_f - \tau)} BU(\tau) d\tau$$

$$= -\int_{t_0}^{t_f} e^{A(t_f - t_f + t_0 - \tau)} BU(\tau) d\tau = -\int_{t_0}^{t_f} e^{A(t_0 - \tau)} BU(\tau) d\tau$$

$$= -\int_{t_0}^{t_f} \Phi(t_0 - \tau) BU(\tau) d\tau \qquad (12.3)$$

The state transition matrix can be written by Cayley–Hamilton theorem as

$$\Phi(t) = e^{At} = \sum_{k=0}^{n-1} \alpha_k(t) A^k \qquad (12.4)$$

where $a_k(t)$ is a scalar function of t. Then Eq. (12.3) becomes

$$X(t_0) = -\int_{t_0}^{t_f} \sum_{k=0}^{n-1} \alpha_k(t_0 - \tau) A^k BU(\tau) d\tau$$

$$= -\sum_{k=0}^{n-1} A^k B \left[\int_{t_0}^{t_f} \alpha_k(t_0 - \tau) U(\tau) d\tau \right] = \sum_{k=0}^{n-1} A^k B \gamma_k$$

where
$$\gamma_k = \int_{t_0}^{t_f} \alpha_k(t_0 - \tau) U(\tau) d\tau$$

Now,
$$X(t_0) = -[\, B \quad AB \quad A^2B \quad \cdots \quad A^{n-1}B\,]\gamma = -S\gamma \qquad (12.5)$$

where
$$\gamma = [\gamma_0 \quad \gamma_1 \quad \gamma_2 \quad \cdots \quad \gamma_{n-1}]^T \qquad (12.6)$$

and
$$S = [\, B \quad AB \quad A^2B \quad \cdots \quad A^{n-1}B\,] \qquad (12.7)$$

Now the controllability problem may be restated as follows:

Given any initial state $X(t_0)$, find the control vector $U(t)$ so that the final state $X(t_f) = 0$ for finite time $(t_f - t_0)$. This means that given $X(t_0)$ and the controllability matrix S, solve for U, i.e. $\gamma = -S^{-1} X(t_0)$. This means that S should be non-singular so that S^{-1} exists, that is, the rank of the controllability matrix S is n, i.e. there exists a set of n linearly independent column vectors in S.

Therefore the system is completely state controllable if and only if there exists a set of n linearly independent column vectors in S, that is, the matrix S is of rank n.

EXAMPLE 12.1 Consider the system

$$\dot{X} = \begin{bmatrix} -1 & 1 & 0 \\ 0 & -4 & 2 \\ 0 & 0 & -10 \end{bmatrix} X + \begin{bmatrix} 1 \\ 0 \\ -1 \end{bmatrix} u$$

and the output
$$y = [1 \quad 0 \quad 1]X$$

The controllability matrix
$$S = [B \quad AB \quad A^2B] = \begin{bmatrix} 1 & -1 & -1 \\ 0 & -2 & 28 \\ -1 & 10 & -100 \end{bmatrix}$$

That is,
$$|S| \neq 0$$

Hence the rank of S, i.e. Rank$(S) = 3$. The system is completely state controllable. See also MATLAB Program 2.1.

MATLAB Program 12.1

```
% Determine controllability of the system given in Example 12.1:

echo off;
clear;
clf;
clc;
a=[-1 1 0; 0 -4 2; 0 0 -10];
b=[1;0;-1];
c=[1 0 1];
d=[0];
S=ctrb(a,b);
n=det(S);
if abs(n)<eps
    disp('System is not controllable')
else
    disp('System is controllable')
end;

System is controllable
```

EXAMPLE 12.2 Consider the system described by

$$\dot{X} = AX + Bu$$

where

$$A = \begin{bmatrix} 0 & 1 \\ -1 & a \end{bmatrix}; \quad B = \begin{bmatrix} 1 \\ b \end{bmatrix}$$

Determine the region in the *a-b* plane such that the system is completely controllable.

Solution: The condition of controllability is that, the matrix [A AB] should be of rank 2 for the system to be controllable, that is,

$$\begin{vmatrix} 1 & b \\ b & ab-1 \end{vmatrix} \neq 0 \quad \text{or} \quad b^2 - ab + 1 \neq 0$$

or

$$b = \frac{a}{2} \pm \frac{1}{2}\sqrt{a^2 - 4}$$

The region in the *a-b* plane for which the system is controllable is shown in MATLAB Program 12.2.

MATLAB Program 12.2

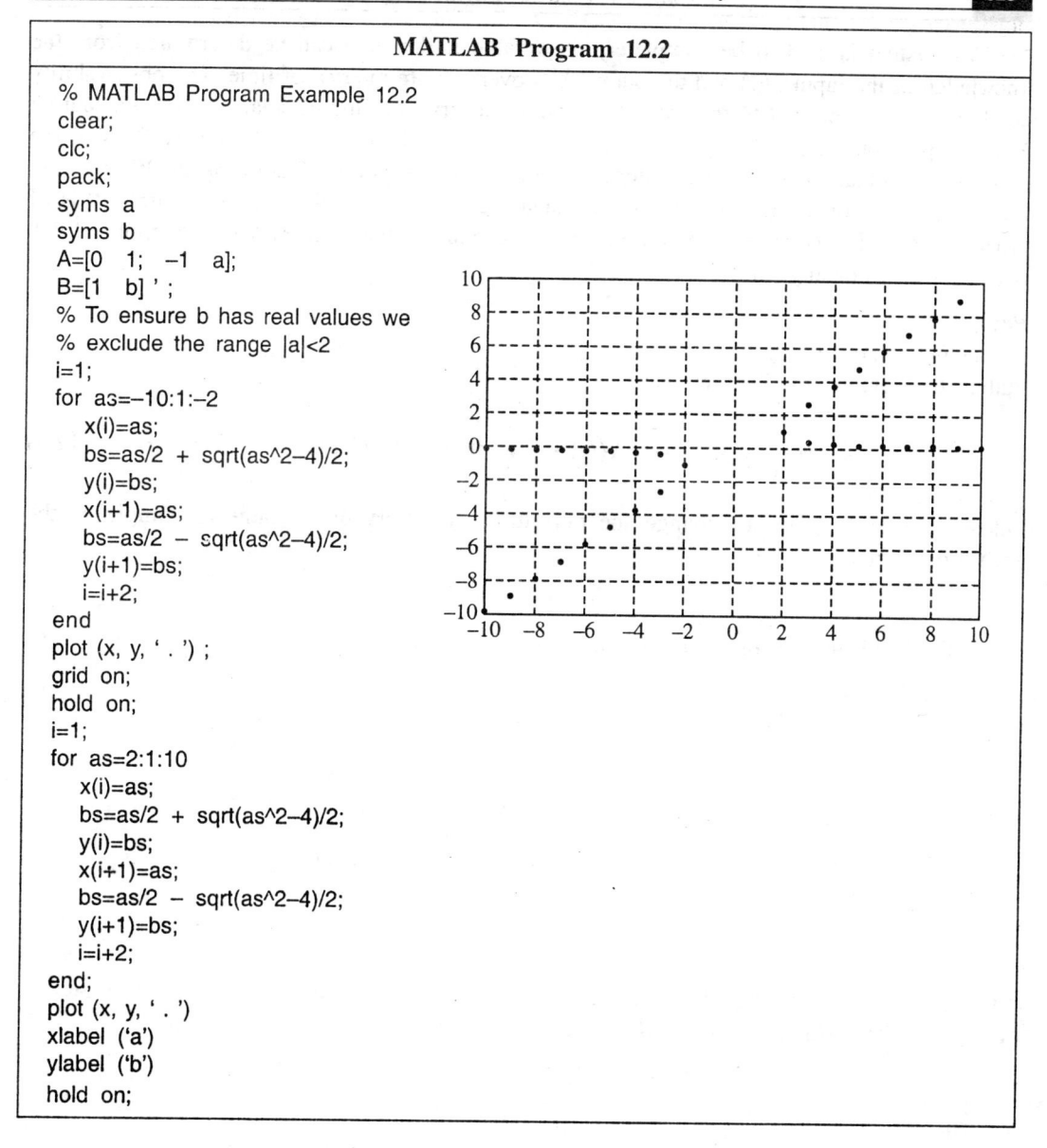

```
% MATLAB Program Example 12.2
clear;
clc;
pack;
syms a
syms b
A=[0 1; -1 a];
B=[1 b]' ;
% To ensure b has real values we
% exclude the range |a|<2
i=1;
for as=-10:1:-2
   x(i)=as;
   bs=as/2 + sqrt(as^2-4)/2;
   y(i)=bs;
   x(i+1)=as;
   bs=as/2 - sqrt(as^2-4)/2;
   y(i+1)=bs;
   i=i+2;
end
plot (x, y, ' . ') ;
grid on;
hold on;
i=1;
for as=2:1:10
   x(i)=as;
   bs=as/2 + sqrt(as^2-4)/2;
   y(i)=bs;
   x(i+1)=as;
   bs=as/2 - sqrt(as^2-4)/2;
   y(i+1)=bs;
   i=i+2;
end;
plot (x, y, ' . ')
xlabel ('a')
ylabel ('b')
hold on;
```

12.3 Observability

The system described by the dynamic equations (AB) and (C) will be completely observable, if and only if the rank of $(n \times np)$ order observability matrix W defined as

$$W = [C^T \quad A^T C^T \quad (A^T)^2 C^T \quad \cdots \quad (A^T)^{n-1} C^T] \tag{12.8}$$

is equal to n. If the matrices are complex then instead of simple transpose, we have to take conjugate transpose. For simplicity we are considering that the matrices are all real until otherwise stated.

The system is said to be completely observable if the state can be determined from the knowledge of the input $U(t)$ and the output $Y(t)$ over a finite interval of time. The observability condition is also referred to the pair $[A, C]$ being observable. In particular, for a scalar output Y the output coupling matrix C is of order $(1 \times n)$, then the observability matrix W is of the order $n \times n$ square matrix. The system is completely observable if the rank of W is n, i.e. Rank $(W) = n$. For a multivariable system having p-outputs, the observability matrix W is of order $(np \times n)$. The columns of W are linearly independent, that is, of rank n; this implies $|W| \neq 0$, i.e. W is nonsingular.

Proof
The output $Y(t) = CX(t)$
Putting the solution of $X(t)$, we get

$$Y(t) = C\Phi(t-t_0)X(t_0) + C\int_{t_0}^{t_f} \Phi(t-\tau)BU(\tau)d\tau \qquad (12.9)$$

Putting $U(t) = 0$ does not change the definition of observability, rather it simplifies the derivation. Then

$$Y(t) = C\Phi(t-t_0)X(t_0)$$

Using Cayley–Hamilton theorem and simplifying, we get

$$\dot{Y}(t) = \sum_{k=0}^{n-1} \alpha_k(t) C A^k X(t_0)$$

$$= [\alpha_0 I \quad \alpha_1 I \quad \ldots \quad \alpha_{n-1} I] \begin{bmatrix} C \\ CA \\ CA^2 \\ \ldots \\ CA^{n-1} \end{bmatrix} X(t_0)$$

Therefore $X(t_0)$ can be uniquely determined for the known (observed) output in a finite interval of time $t_0 \leq t \leq t_f$ if and only if the observability matrix W of order $(np \times n)$ as

$$[C \quad CA \quad CA^2 \quad \ldots \quad CA^{n-1}]^T$$

has rank n.
i.e.
$$W = [C^T \quad A^T C^T \quad (A^T)^2 C^T \quad \ldots \quad (A^T)^{n-1} C^T] \qquad (12.10)$$

has rank n.

Further, for scalar output, the observability matrix W is of order $n \times n$, otherwise of order $(np \times n)$.

The following observations may be noted from the rank conditions of controllability matrix S and observability matrix W.

- Controllability of pair $[A, B]$ implies the observability of the pair $[A^T, B^T]$
- Observability of the pair $[A, C]$ implies the controllability of the pair $[A^T, C^T]$

EXAMPLE 12.3 Consider the system described in Example 12.1 where

$$A = \begin{bmatrix} -1 & 1 & 0 \\ 0 & -4 & 2 \\ 0 & 0 & -10 \end{bmatrix}, B = \begin{bmatrix} 1 \\ 0 \\ -1 \end{bmatrix}, C = \begin{bmatrix} 1 & 0 & 1 \end{bmatrix}$$

The observability matrix

$$W = \begin{bmatrix} C \\ CA \\ CA^2 \end{bmatrix} = \begin{bmatrix} 1 & 0 & 1 \\ -1 & 1 & -10 \\ 1 & -5 & 102 \end{bmatrix}$$

$|W| \neq 0$ and hence its rank is 3. The system is completely observable. See also MATLAB program observable.m

---

MATLAB Program 12.3

```
% observable.m
% Determine the observability of the system given in Example 12.3
clear;
pack;
clc;

a=[-1 1 0;0 -4 2;0 0 -10];
b=[1;0;-1];
c=[1 0 1];
d=[0];
printsys(a,b,c,d)
W=obsv(a,c)
n=det(W)
if abs(n)<eps
%eps is epsilon,a very small quantity ≈ 0 defined in MATLAB
    disp('System is not observable')
else
    disp('system is observable')
end;
system is observable
```

12.4 Controllability Criterion I

The time-invariant system, for which the matrix A has got distinct eigenvalues, is completely controllable if and only if there are no zero rows of $B^* = M^{-1}B$.

EXAMPLE 12.4 A system is described by

$$\dot{X} = \begin{bmatrix} -2 & -2 & 0 \\ 0 & 0 & 1 \\ 0 & -3 & -4 \end{bmatrix} X + \begin{bmatrix} 1 & 0 \\ 0 & 1 \\ 1 & 1 \end{bmatrix} \begin{bmatrix} u_1 \\ u_2 \end{bmatrix}$$

Since A is constant and its eigenvalues are distinct $-1, -2, -3$, by suitable transformation, we get

$$\Lambda = M^{-1}AM = \begin{bmatrix} -1 & 0 & 0 \\ 0 & -2 & 0 \\ 0 & 0 & -3 \end{bmatrix} \quad \text{and} \quad B^* = M^{-1}B = \begin{bmatrix} 1/2 & 2 \\ 3 & 6 \\ 1/2 & 1 \end{bmatrix}$$

We can get the decoupled (diagonal) form. Looking at the transformed input coupling matrix B^* we can conclude (without calculation) that the system is completely controllable because no row of B^* is all zeros as stated in controllability criterion I.

EXAMPLE 12.5 A given system is described by

$$\dot{X} = \begin{bmatrix} 1/2 & 1/2 & 0 \\ 0 & 1 & 0 \\ 5/6 & -13/6 & -1/3 \end{bmatrix} X + \begin{bmatrix} 3 & 1 \\ 2 & 0 \\ -1 & 1 \end{bmatrix} U$$

and output

$$Y = \begin{bmatrix} -1 & 3 & 1 \\ 0 & 1 & 1 \end{bmatrix} X$$

The eigenvalues of A are $1, 1/2$ and $-1/3$. We get the modal matrix M and then

$$\Lambda = M^{-1}AM = \begin{bmatrix} 1 & 0 & 0 \\ 0 & 1/2 & 0 \\ 0 & 0 & -1/3 \end{bmatrix} \quad \text{and} \quad B^* = M^{-1}B = \begin{bmatrix} 2 & 0 \\ 1 & 1 \\ 0 & 0 \end{bmatrix}$$

The system is not completely controllable as one row of B^* contains only zeros, no control can affect the third mode of this system.

12.5 Controllability Criterion II

The time-invariant system for which A does not possess distinct eigenvalues, we can transform the system matrix to Jordan canonical form. Then we can state that the system is completely controllable, if and only if:

(1) no two Jordan blocks are associated with the same eigenvalues,

(2) the elements of any row of transformed input coupling matrix, that is, $B^* = M^{-1}B$, that correspond to the **last row** of each Jordan block are not all zero, and
(3) the elements of each row of B^* that correspond to the distinct eigenvalues are not all zero.

Examples (i) and (ii) of **completely state controllable** systems are given below for better clarity. For the problem (i), the last row of the Jordan block corresponds to the generating vector for mode having an eigenvalue of −2. Hence the second row of B should not have all zero elements for controllability of the mode. For the problem (ii), the last row of the first and second Jordan block corresponds to the generating vector for mode having an eigenvalue −2 and −3 respectively. Hence the third and fifth row of B should not have all zero elements for the controllability of respective modes. Hence problems (i) and (ii) are state controllable. Using the same line of reasoning, problems (iii) and (iv) are not state controllable.

(i) $$\dot{X} = \begin{bmatrix} -2 & 1 & 0 \\ 0 & -2 & 0 \\ 0 & 0 & -3 \end{bmatrix} X + \begin{bmatrix} 0 \\ 4 \\ 5 \end{bmatrix} u$$

(ii) $$\dot{X} = \begin{bmatrix} -2 & 1 & 0 & 0 & 0 \\ 0 & -2 & 1 & 0 & 0 \\ 0 & 0 & -2 & 0 & 0 \\ 0 & 0 & 0 & -3 & 1 \\ 0 & 0 & 0 & 0 & -3 \end{bmatrix} X + \begin{bmatrix} 0 & 1 \\ 0 & 0 \\ 5 & 0 \\ 0 & 0 \\ 2 & 1 \end{bmatrix} U$$

(iii) $$X = \begin{bmatrix} -2 & 1 & 0 \\ 0 & -2 & 0 \\ 0 & 0 & -3 \end{bmatrix} X + \begin{bmatrix} 4 \\ 0 \\ 5 \end{bmatrix} u$$

(iv) $$\dot{X} = \begin{bmatrix} -2 & 1 & 0 & 0 & 0 \\ 0 & -2 & 1 & 0 & 0 \\ 0 & 0 & -2 & 0 & 0 \\ 0 & 0 & 0 & -3 & 1 \\ 0 & 0 & 0 & 0 & -3 \end{bmatrix} X + \begin{bmatrix} 0 & 1 \\ 1 & 0 \\ 5 & 0 \\ 2 & 1 \\ 0 & 0 \end{bmatrix} U$$

12.6 Observability Criterion I

The time-invariant system for which the matrix A has got distinct eigenvalues, is completely observable if and only if there are no zero columns of $C^* = CM$.

EXAMPLE 12.6 A system is described by

$$\dot{X} = \begin{bmatrix} 0 & 1 \\ 8 & -2 \end{bmatrix} X + Bu$$

and output
$$y = [4 \quad 1]X$$

The eigenvalues of A are $-4, 2$. Now $M^{-1}AM = \Lambda$ is obtained and $C^* = CM = [0 \quad 6]$. Since A is a constant, with distinct eigenvalues, the above criterion applies. Column 1 of C^* is zero, so this system is not completely observable.

Again from Example 12.4 with same A which is constant and its eigenvalues are distinct, by suitable transformation, we get

$$\Lambda = M^{-1}AM = \begin{bmatrix} 1 & 0 & 0 \\ 0 & 1/2 & 0 \\ 0 & 0 & -1/3 \end{bmatrix} \quad \text{and} \quad C^* = CM = \begin{bmatrix} 1 & 0 & 1 \\ 0 & 1 & 1 \end{bmatrix}$$

We have got the decoupled (diagonal) form here. Looking at the transformed output coupling matrix C^* we can conclude (without calculation) that the system is completely observable because no column of C^* is all zeros as stated in observability criterion I.

12.7 Observability Criterion II

In case the system matrix A does not possess distinct eigenvalues, we can transform the system matrix to Jordan canonical form. Then we can state that the system is completely observable if and only if:

(1) no two Jordan blocks are associated with the same eigenvalues,

(2) the elements of any column of the transformed output coupling matrix, that is, $C^* = CM$, that correspond to the **first row** of each Jordan block are not all zero, and

(3) the elements of each column of C^* that correspond to the distinct eigenvalues are not all zero.

Examples (i) and (ii) of **completely observable** systems are given below for better clarity.

(i)
$$\dot{X} = \begin{bmatrix} -2 & 1 & 0 \\ 0 & -2 & 0 \\ 0 & 0 & -3 \end{bmatrix} X; \quad Y = \begin{bmatrix} 3 & 0 & 0 \\ 4 & 0 & 1 \end{bmatrix} X$$

(ii)
$$\dot{X} = \begin{bmatrix} -2 & 1 & 0 & 0 & 0 \\ 0 & -2 & 1 & 0 & 0 \\ 0 & 0 & -2 & 0 & 0 \\ 0 & 0 & 0 & -3 & 1 \\ 0 & 0 & 0 & 0 & -3 \end{bmatrix} X; \quad Y = \begin{bmatrix} 0 & 1 \\ 0 & 0 \\ 5 & 0 \\ 2 & 1 \\ 0 & 0 \end{bmatrix}^T X$$

Similarly, examples (iii) and (iv) of not completely observable systems are given below for better clarity.

(iii) $\quad \dot{X} = \begin{bmatrix} -2 & 1 & 0 \\ 0 & -2 & 0 \\ 0 & 0 & -3 \end{bmatrix} X; \quad y = \begin{bmatrix} 0 \\ 4 \\ 5 \end{bmatrix}^T X$

(iv) $\quad \dot{X} = \begin{bmatrix} -2 & 1 & 0 & 0 & 0 \\ 0 & -2 & 1 & 0 & 0 \\ 0 & 0 & -2 & 0 & 0 \\ 0 & 0 & 0 & -3 & 1 \\ 0 & 0 & 0 & 0 & -3 \end{bmatrix} X; \quad Y = \begin{bmatrix} 0 & 0 \\ 1 & 0 \\ 5 & 0 \\ 2 & 1 \\ 0 & 0 \end{bmatrix}^T X$

The criteria for complete controllability and observability given in the preceding sections are useful because of the geometrical insight they provide. However, these criterria are not the most useful because they are restricted to the distinct eigenvalue case.

12.8 Controllable and Observable Systems

For distinct eigenvalues, the system matrix is in diagonal form and the states are decoupled. Each state variable appears in only one equation and the equations are all of first-order. The form of the zero-input response of each of these first-order subsystems is termed its *mode* or the mode associated with the corresponding state variable.

When the system matrix is in diagonal form, any column of the output coupling matrix is zero, the corresponding state variable does not couple to any output, and the mode associated with that state variable is termed *unobservable*, otherwise *observable*.

Similarly, when the system matrix is in diagonal form, any row of the input coupling matrix is zero, the corresponding state cannot be effected by any input and the mode associated with that state variable is termed *uncontrollable*, otherwise *controllable*.

Each mode of a system is either observable or unobservable. If all the modes of a system are observable, then the system is called *completely observable*. Similarly if all the modes of a system are controllable, then the system is called *completely controllable*.

Let us rewrite the transfer function matrix as

$$G(s) = C\,[sI - A]^{-1}\,B + D \qquad (12.11)$$

It may be noted that every element of $G(s)$ shares the denominator polynomial $|sI - A|$. Whenever the system matrix is in diagonal form and there is no coupling to a mode from an input or no coupling from a mode to an output, the transfer function relating that output or input does not have a pole or eigenvalue corresponding to that mode. Thus if any mode is unobservable or uncontrollable, every element of the transfer function matrix will have a pole-zero cancellation.

In general, the system may be decomposed into four parts as indicated in Figure 12.1 and written as

- controllable and observable
- controllable and unobservable
- uncontrollable and observable
- uncontrollable and unobservable

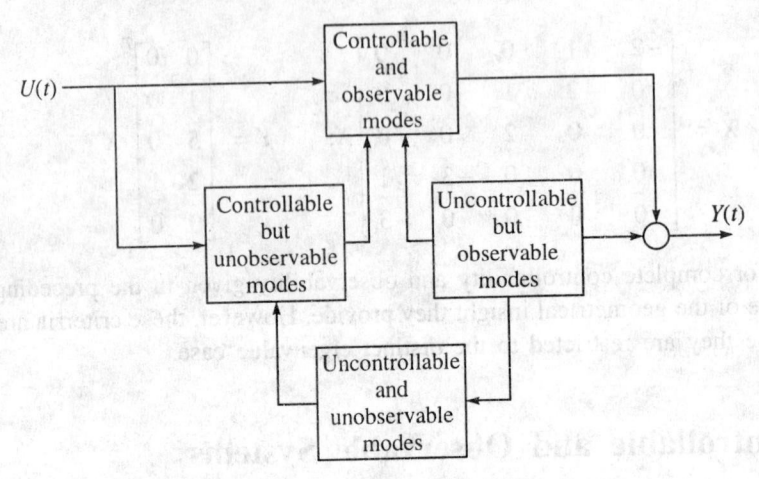

FIGURE 12.1 Decomposition of system.

The transfer function matrix belongs to only the controllable and observable part of the system. Any completely controllable single-input system can be realized in controllable form. A controllable form of representation of a single-input system exists if and only if the system is completely controllable. Similarly, any observable single-output system can be realized in observable form. An observable form representation of a single-output system exists if and only if the system is completely observable.

EXAMPLE 12.7 Subdivide the following system into subsystems as per Figure 12.1.

$$\dot{X} = \begin{bmatrix} -7 & -2 & 6 \\ 2 & -3 & -2 \\ -2 & -2 & 1 \end{bmatrix} X + \begin{bmatrix} 1 & 1 \\ 1 & -1 \\ 1 & 0 \end{bmatrix} U$$

and output

$$Y = \begin{bmatrix} -1 & -1 & 2 \\ 1 & 1 & -1 \end{bmatrix} X$$

The eigenvalues are -1, -3 and -5.

$$M = \begin{bmatrix} 1 & 1 & 1 \\ 0 & 1 & -1 \\ 1 & 1 & 0 \end{bmatrix} \quad \text{and} \quad M^{-1} = \begin{bmatrix} -1 & -1 & 2 \\ 1 & 1 & -1 \\ 1 & 0 & -1 \end{bmatrix}$$

The transformed system becomes

$$\dot{Z} = \Lambda Z + B^*U$$

and output

$$Y = C^*Z$$

where

$$\Lambda = \begin{bmatrix} -1 & 0 & 0 \\ 0 & -3 & 0 \\ 0 & 0 & -5 \end{bmatrix}, \quad B = \begin{bmatrix} 0 & 0 \\ 1 & 0 \\ 0 & 1 \end{bmatrix} \text{ and } C^* = \begin{bmatrix} 1 & 0 & 0 \\ 0 & 1 & 0 \end{bmatrix}$$

The first mode is uncontrollable and the third mode is unobservable. The second mode is both controllable and observable. There is no mode which is both uncontrollable and unobservable. Figure 12.2 illustrates the three subsystems.

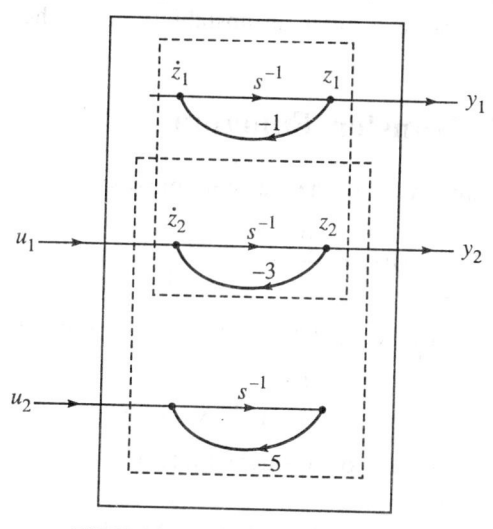

FIGURE 12.2 Example 12.7.

12.9 Pole Cancellation—Stabilization

Consider the system given by the transfer function

$$G(s) = \frac{Y(s)}{U(s)} = \frac{-6+2s}{s^2-9} = \frac{2(s-3)}{(s+3)(s-3)}$$

with the corresponding differential equation

$$\frac{d^2y}{dt^2} - 9y = -6u + 2\frac{du}{dt}$$

A solution to this second-order differential equation, say, for $u(t) = 0$, involves two constants of integration, yielding

$$y(t) = C_1 e^{-3t} + C_2 e^{-3t}$$

If we were to cancel the $(s - 3)$ term in $G(s)$, we would reduce the order of the system, yielding

$$\frac{dy}{dt} + 3y = 2u$$

with a solution, again for $u(t) = 0$, involving only one integration constant, i.e.

$$y(t) = C_1 e^{-3t}$$

Clearly, the two systems are not equivalent. Cancelling a term like $(s - a)$, with $a > 0$, is, therefore, particularly bad because it hides an unstable term in the solution.

12.10 Concept of Transfer Function

Consider the unstable system given by the transfer function

$$G(s) = \frac{Y(s)}{U(s)} = \frac{1}{s-1} \qquad (12.12)$$

In order to stabilize it, we can precede $G(s)$ with a compensator having transfer function

$$G_c(s) = \frac{U(s)}{V(s)} = \frac{s-1}{s+1} \qquad (12.13)$$

to get the overall transfer function of a stable system having overall transfer function as $1/(s + 1)$. See Figure 12.3(a).

But this technique will not work. After a while the system will tend to *burn out* or *saturate*. To see why, the following are the reasons.

$$G_c(s)G(s) = \frac{Y(s)}{V(s)} = \frac{s-1}{s+1} \times \frac{1}{s-1} = \frac{U(s)}{V(s)} \frac{Y(s)}{U(s)}$$

Let
$$\frac{U(s)}{V(s)} = \frac{s-1}{s+1} = 1 + \frac{-2}{s+1} = 1 + \frac{U'(s)}{V(s)}$$

Thus, we get $U(s) = V(s) + U'(s)$

and $u(t) = v(t) + u'(t)$ \hfill (12.14)

and $\dfrac{U'(s)}{V(s)} = \dfrac{-2}{s+1}$ leads to $\dfrac{du'}{dt} = -u' - 2v(t)$ \hfill (12.15)

Further, $\dfrac{Y(s)}{U(s)} = \dfrac{1}{s-1}$ leads to $\dfrac{dy}{dt} = y + u(t)$ \hfill (12.16)

The SFG of blocks $G_c(s)$ and $G(s)$ in cascade is shown in Figure 12.3(b). The integrator outputs are taken as the state variables such as $u' = x_1$ and $y = x_2$. Then we get the state equations as

From Eq. (12.15), $\quad \dot{x}_1 = -x_1 - 2v; \quad x_1(0) = x_{10}$

From Eq. (12.14), $\quad u = x_1 + v$

From Eq. (12.16), $\quad \dot{x}_2 = x_2 + u = x_1 + x_2 + v; \quad x_2(0) = x_{20}$

and output $\quad y = x_2$

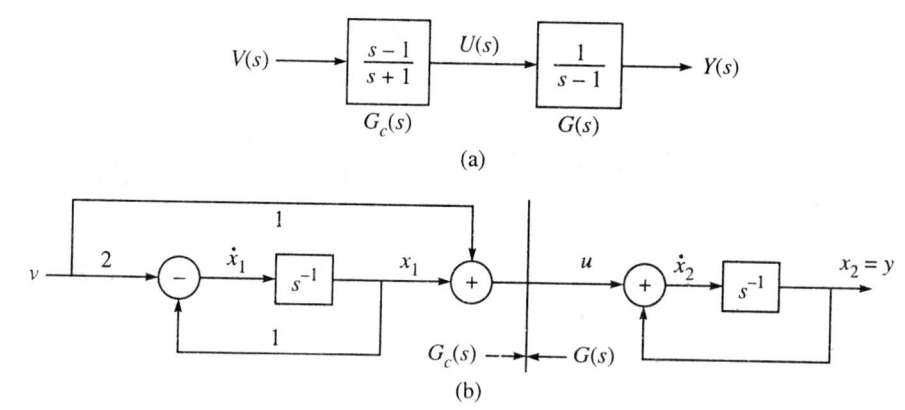

FIGURE 12.3 (a) Cascade compensation and (b) analog simulation of the cascaded system.

Thus the state variable formulation in matrix form is

$$\begin{bmatrix} \dot{x}_1 \\ \dot{x}_2 \end{bmatrix} = \begin{bmatrix} -1 & 0 \\ 1 & 1 \end{bmatrix} \begin{bmatrix} x_1 \\ x_2 \end{bmatrix} + \begin{bmatrix} -2 \\ 1 \end{bmatrix} v; \quad X(0) = \begin{bmatrix} x_1(0) \\ x_2(0) \end{bmatrix} = \begin{bmatrix} x_{10} \\ x_{20} \end{bmatrix} \quad (12.17)$$

and output

$$y = \begin{bmatrix} 0 & 1 \end{bmatrix} \begin{bmatrix} x_1 \\ x_2 \end{bmatrix}$$

The STM is obtained as

$$\Phi(t) = \begin{bmatrix} e^{-t} & 0 \\ \dfrac{e^t - e^{-t}}{2} & e^t \end{bmatrix} \quad (12.18)$$

It yields the time domain solution of states as

$$X(t) = \Phi(t) X(0) + \int_0^t \Phi(t - \tau) Bv(\tau) d\tau$$

which gives

$$x_1(t) = e^{-t} x_{10} - 2 \int_0^t e^{-(t-\tau)} v(\tau) d\tau \quad (12.19)$$

and

$$x_2(t) = e^t x_{20} + \left(\frac{1}{2}\right)(e^t - e^{-t}) x_{10} + \int_0^t e^{-(t-\tau)} v(\tau) d\tau \quad (12.20)$$

The output

$$y(t) = x_2(t) = e^t x_{20} + \left(\frac{1}{2}\right)(e^t - e^{-t}) x_{10} + \int_0^t e^{-(t-\tau)} v(\tau) d\tau \qquad (12.21)$$

The overall transfer function after cancellation of poles becomes $1/(s + 1)$ which by definition of transfer function, has to be calculated with zero initial conditions. Note that, unless all initial conditions are kept zero, the output $y(t)$ grows independently to out-of-bound. Because of the stray voltage, it will be difficult to keep $x_{10} = x_{20} = 0$ and therefore the above method of stabilizing the system is unsatisfactory. The exact cancellation is difficult to achieve because of component tolerances. Further, all initial conditions (except $x_{10} = -2x_{20}$) will excite the unstable mode e^t in Figure 12.3(b).

The system SFG in Figure 12.3(c) is observable (that is, it appears in the output) but it is not controllable (that is, the external input $v(t)$ cannot directly affect it).

FIGURE 12.3(c) System SFG of Figure 12.3(a).

Some further insight into the difficulty is obtained when we take up the same problem where $G(s)$ is followed by $G_c(s)$ as in Figure 12.4(a). The analog simulation is shown in Figure 12.4(b).

The state variable formulation then becomes

$$\dot{X} = \begin{bmatrix} 1 & 0 \\ -2 & -1 \end{bmatrix} X + \begin{bmatrix} 1 \\ 0 \end{bmatrix} v \qquad (12.22)$$

and output

$$y = \begin{bmatrix} 1 & 1 \end{bmatrix} \begin{bmatrix} x_1 \\ x_2 \end{bmatrix} \qquad (12.23)$$

The time domain solution of output is

$$y(t) = (x_{10} + x_{20}) e^{-t} + \int_0^t e^{-(t-\tau)} v(\tau) d\tau \qquad (12.24)$$

In simulation of Figure 12.4(b), the mode e^t is controllable but not observable.

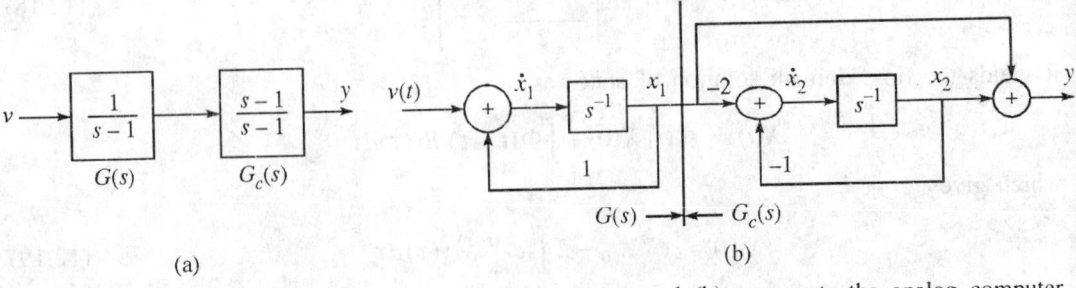

FIGURE 12.4 (a) Compensator follows the given system and (b) represents the analog computer simulation of the cascaded system.

Now the system is stable as far as $y(t)$ goes, even if the initial conditions are non-zero. Are we satisfied then? No, because the realization in Figure 12.4(b) is still internally unstable because $x_1(t)$ and $x_2(t)$ will have terms in them that will grow as e^t, and thus after a while the realization will saturate and/or burn out. The internal behaviour is determined by the natural frequencies of the (undriven) realization, which in this case are $s = +1$ and -1.

However, because of cancellation, not all the corresponding modes of oscillation will appear in the overall transfer function.

Or, to put it in other way, since the transfer function is defined under zero initial conditions, it may not display all the modes of the actual realization of the system. For a complete analysis, we need to keep a track of all the modes, those explicitly displayed by the transfer function and also the 'hidden' ones.

It may be mentioned here that in the simulation in Figure 12.3(b), the unstable mode e^t is observable (it appears in the output) but not controllable (the external input $v(t)$ cannot directly affect it); while in Figure 12.4(b) the mode e^t is controllable but not observable. Hence it may be concluded that the transfer function without any pole-zero cancellation is the controllable and observable part of the system to proceed with for further analysis.

12.11 Duality Property

For the general nth-order multi-input-multi-output linear time-invariant system

$$\dot{X} = AX + BU$$
$$Y = CX$$

the controllability and observability condition due to Kalman is as follows:

The system is completely controllable if and only if the rank of the composite matrix

$$S = [B \quad AB \quad A^2B \quad \ldots \quad A^{n-1}B]$$

is n. This condition is referred to as the pair (A, B) being controllable.

The system is completely observable if and only if the rank of the composite matrix

$$W = [C^T \quad A^T \quad C^T \quad (A^T)^2 C^T \quad \ldots \quad (A^T)^{n-1} C^T]$$

is n. This condition is referred to as the pair (A, C) being observable.

Comparing the composite matrix S with W, the following observations can be made:

The pair (A, B) being controllable signifies that the pair (A^T, B^T) is observable.

The pair (A, C) being observable signifies that the pair (A^T, C^T) is controllable.

Thus the concepts of controllability and observability are dual concepts. It may be noted that the transfer function without pole-zero cancellation is the controllable and observable part of the system.

The principle of duality can be restated clearly as follows:

Consider the system S_1 represented by the vector-matrix differential equation

$$\dot{X} = AX + BU$$

and output
$$Y = CX$$
where

  X is the $(n \times 1)$ state vector

  U is the $(r \times 1)$ input vector

  Y is the $(m \times 1)$ output vector

  A is the $(n \times n)$ system matrix

  B is the $(n \times r)$ input coupling matrix

  C is the $(m \times n)$ output coupling matrix

The dual system S_2 is defined by
$$\dot{Z} = A^T Z + C^T V$$
$$Q = B^T Z$$
and output
where

  Z is the $(n \times 1)$ state vector

  V is the $(m \times 1)$ input vector

  Q is the $(r \times 1)$ output vector

  A^T is the $(n \times n)$ system matrix and is transpose of A

  C^T is the $(n \times m)$ input coupling matrix and is transpose of C

  B^T is the $(r \times n)$ output coupling matrix and is transpose of B

It may be noted here that for any complex matrix, it is the conjugate transpose instead of only transpose. In the text we have considered only the real matrices.

The principle of duality states that the system S_1 is completely state controllable (observable) if and only if the system S_2 is completely observable (state controllable).

To verify the duality principle, let us write down the necessary and sufficient conditions for complete state controllability and complete observability for the dual systems S_1 and S_2.

S_1	S_2
$\dot{X} = AX + BU$ $Y = CX$ Then the conntrollability matrix S is $S = [B \ AB \ \ldots \ A^{n-1}B]$ and the observability matrix W becomes $W = [C^T \ A^T C^T \ \ldots \ (A^T)^{n-1} C^T]$	$\dot{Z} = A^T Z + C^T V$ $Q = B^T Z$ Then the conntrollability matrix S is $S = [C^T \ A^T C^T \ \ldots \ (A^T)^{n-1} \ C^T]$ and the observability matrix W becomes $W = [(B^T)^T \ (A^T)^T \ (B^T)^T \ \ldots \ ((A^T)^T)^{n-1} \ (B^T)^T]$ $= [B \ AB \ \ldots \ A^{n-1} B]$

By comparing these conditions, the truth of this principle is apparent. We can conclude that for dual systems S_1 and S_2, the controllability (observability) condition of system S_1 becomes the observability (controllability) condition of system S_2. By the use of this principal, the observability of a given system can be checked by testing the state controllability of its dual.

As an illustration consider the system as

$$\dot{X} = AX + BU$$

and output

$$Y = CX$$

where

$$A = \begin{bmatrix} 0 & 1 & 0 \\ 0 & 0 & 1 \\ -6 & -11 & -6 \end{bmatrix}; \quad B = \begin{bmatrix} 0 \\ 0 \\ 1 \end{bmatrix}; \quad C^T = \begin{bmatrix} 4 \\ 5 \\ 1 \end{bmatrix}$$

The system is completely controllable as the rank of the controllability matrix $S = [B \ AB \ A^2B]$ is equal to 3. The observability matrix

$$W = [C^T \ A^T C^T \ (A^T)^2 C^T] = \begin{bmatrix} 4 & -6 & 6 \\ 5 & -7 & 5 \\ 1 & -1 & -1 \end{bmatrix}$$

The system is not observable as the rank of W is less than 3.

The transfer function

$$C[sI - A]^{-1} B = \frac{(s+1)(s+4)}{(s+1)(s+2)(s+3)} = \frac{s+4}{(s+2)(s+3)}$$

Clearly, the two factors $(s + 1)$ cancel each other, which means that there are non-zero initial states $X(0)$, which cannot be determined from the measurement of $y(t)$. Hence $\dfrac{s+4}{(s+2)(s+3)}$ is not the transfer function as one mode is hiding because of pole-zero cancellation.

12.12 Similarity Transformation Matrix from Controllability Matrix

We have talked earlier about the similarity transformation by which we can transform the system from any general form to the companion form provided we know the transformation matrix P. After discussing controllability, we will establish the relationship for determining the similarity transformation matrix P as $S\bar{S}^{-1}$.

Let us define $\bar{S}$, the controllability matrix of the transformed system as

$$\dot{\bar{X}} = \bar{A}\bar{X} + \bar{B}u \qquad (\bar{A} \ \bar{B})$$

and

$$y = \bar{C}\bar{X} \qquad (\bar{C})$$

where

$$\bar{A} = P^{-1}AP, \quad \bar{B} = P^{-1}B, \quad \bar{C} = CP$$

It may be noted that the superscript bar corresponds to the phase-variable canonical form of system representation. Thus,

$$\begin{aligned}\overline{S} &= \begin{bmatrix} \overline{B} & \overline{A}\,\overline{B} & \overline{A}^2\overline{B} & \cdots & \overline{A}^{n-1}\overline{B}\end{bmatrix}\\ &= \begin{bmatrix} P^{-1}B & (P^{-1}AP)P^{-1}B & (P^{-1}A^2P)P^{-1}B & \cdots & (P^{-1}A^{n-1}P)P^{-1}B\end{bmatrix}\\ &= \begin{bmatrix} P^{-1}B & P^{-1}AB & P^{-1}A^2B & \cdots & P^{-1}A^{n-1}B\end{bmatrix}\\ &= P^{-1}\begin{bmatrix} B & AB & A^2B & \cdots & A^{n-1}B\end{bmatrix}\\ &= P^{-1}S\end{aligned} \qquad (12.25)$$

Therefore,
$$P = S\overline{S}^{-1}$$

Given the state variable representation of the original system, i.e. when the matrices A and B are given, we can find the controllability matrix S. Further, it may be mentioned that $\overline{S}^{-1}$ of the transformed system can be determined in a short-cut fashion from the characteristic polynomial of the original system, i.e. from

$$|sI - A| = s^n + a_{n-1}s^{n-1} + \cdots + a_1 s + a_0$$

as
$$\overline{S}^{-1} = \begin{bmatrix} a_1 & a_2 & a_3 & \cdots & a_{n-1} & 1 \\ a_2 & a_3 & a_4 & \cdots & 1 & 0 \\ a_3 & a_4 & a_5 & \cdots & 0 & 0 \\ \vdots & \vdots & \vdots & & \vdots & \vdots \\ a_{n-1} & 1 & 0 & \cdots & 0 & 0 \\ 1 & 0 & 0 & \cdots & 0 & 0 \end{bmatrix} \qquad (12.26)$$

Then the similarity transformation matrix P can be determined as

$$P = S\overline{S}^{-1}$$

and then P^{-1} can also be obtained as

$$P^{-1} = \overline{S}\,S^{-1}$$

Now, we get

$$\overline{A} = P^{-1}AP = \begin{bmatrix} 0 & 1 & 0 & \cdots & 0 & 0 \\ 0 & 0 & 1 & \cdots & 0 & 0 \\ \vdots & \vdots & \vdots & & \vdots & \vdots \\ 0 & 0 & 0 & \cdots & 0 & 1 \\ -a_0 & -a_1 & -a_2 & \cdots & -a_{n-2} & -a_{n-1} \end{bmatrix} \qquad (12.27)$$

$$\overline{B} = P^{-1}B = [0 \ 0 \ 0 \ \ldots \ 0 \ 1]^T \qquad (12.28)$$

Hence it has been shown that if a system is state controllable then there always is a similarity transformation matrix by which the system can always be represented in the phase-variable canonical form. The reverse is also true, in that if the system is represented in the phase-variable canonical form, it is always state controllable.

To prove this, the dynamics is given in phase-variable canonical form as

$$\dot{\overline{X}} = \overline{A}\,\overline{X} + \overline{B}u \qquad (12.29)$$

Now find

$$\overline{S} = \begin{bmatrix} \overline{B} & \overline{A}\,\overline{B} & \overline{A}^2\overline{B} & \ldots & \overline{A}^{n-1}\overline{B} \end{bmatrix} \qquad (12.30)$$

and

$$|\overline{S}| = -1$$

that is, the rank of $\overline{S}$ is n.

Hence the system represented in the phase-variable canonical form is controllable. The system represented by Eq. ($\overline{A}\,\overline{B}$) is called the controllable canonical form of representation.

EXAMPLE 12.8 Given

$$A = \begin{bmatrix} -1 & 1 & 0 \\ 0 & -4 & 2 \\ 0 & 0 & -10 \end{bmatrix}; \quad B = [1 \ 0 \ -1]^T; \quad \text{and} \quad C = [1 \ 0 \ 1]$$

Determine the similarity transformation matrix P to transform the system

$$\dot{X} = AX + Bu$$

and output

$$y = CX$$

to the controllable canonical form.

Solution: The controllability matrix

$$S = [B \ AB \ A^2B] = \begin{bmatrix} 1 & -1 & -1 \\ 0 & -2 & 28 \\ -1 & 10 & -100 \end{bmatrix}; \quad |S| \neq 0$$

As S is non-singular, the system is completely controllable. Now the characteristic polynomial of the given system is

$$|sI - A| = s^3 + 15s^2 + 54s + 40$$

Hence

$$\overline{S}^{-1} = \begin{bmatrix} 54 & 15 & 1 \\ 15 & 1 & 0 \\ 1 & 0 & 0 \end{bmatrix}$$

The similarity transformation matrix

$$P = S\bar{S}^{-1} = \begin{bmatrix} 38 & 14 & 1 \\ -2 & -2 & 0 \\ -4 & -5 & -1 \end{bmatrix}, \quad |P| \neq 0, \quad P^{-1} \text{ exists}$$

Now, $\bar{A} = P^{-1}AP = \begin{bmatrix} 0 & 1 & 0 \\ 0 & 0 & 1 \\ -40 & -54 & -15 \end{bmatrix}$; $\bar{B} = P^{-1}B = [0 \ 0 \ 1]^T$; $\bar{C} = CP = [34 \ 9 \ 0]$

Hence the transformed system is

$$\dot{\bar{X}} = \bar{A}\bar{X} + \bar{B}u$$

and output
$$y = \bar{C}\bar{X}$$

The necessity of the state variable approach evolves from the failure of the classical approach in control theory to design the control system perfectly as per the specifications of the user. The advantage of the state variable representation is the arbitrary pole placement as per the requirement of the users, design specification by the state feedback with properly adjustable weight. The state feedback requires that all the states should be accessible which may not be the case always. So there is a need to make the inaccessible states accessible through the observer design. In this chapter we shall discuss the observer design, the condition under which the design is possible, and if so, then how close the observer state variables are to actual state variables.

EXAMPLE 12.9 Consider two carts on frictionless wheels as shown in Figure 12.5. The carts are connected by a damper and a force F is applied to one of the carts.

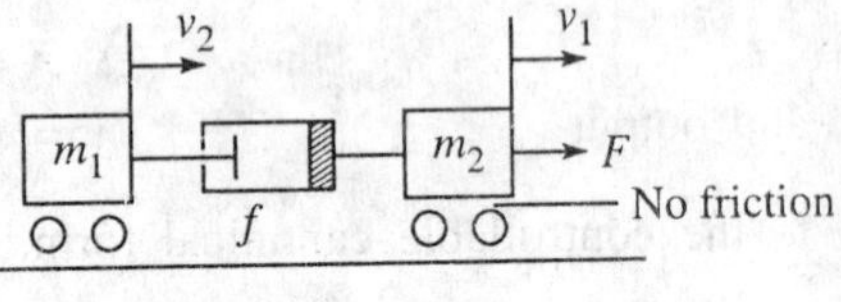

In terms of the speeds v_1 and v_2 of the two carts, the Newton's second law yields

FIGURE 12.5 Example 12.9.

$$m_1\dot{v}_1 = f(v_2 - v_1)$$
$$m_2\dot{v}_2 = F - f(v_2 - v_1)$$

We want to control the system so that the carts move at the same (unspecified) speed, so let us choose the output measurement as the speed difference, i.e.

$$y = v_2 - v_1$$

Suppose it is given that $m_1 = 1$ kg, $m_2 = 0.5$ kg and $f = 1$ N-s/m. Choosing the state variables as v_1 and v_2 and the control input as $u = F$ yields

$$\begin{bmatrix} \dot{v}_1 \\ \dot{v}_2 \end{bmatrix} = \begin{bmatrix} -1 & 1 \\ 2 & -2 \end{bmatrix} \begin{bmatrix} v_1 \\ v_2 \end{bmatrix} + \begin{bmatrix} 0 \\ 2 \end{bmatrix} u$$

and output
$$y = \begin{bmatrix} -1 & 1 \end{bmatrix} \begin{bmatrix} v_1 \\ v_2 \end{bmatrix}$$

To convert to companion form, we first compute the controllability matrix

$$S = [B \quad AB] = \begin{bmatrix} 0 & 2 \\ 2 & -4 \end{bmatrix}; \quad |S| \neq 0$$

The system is controllable. Hence the transformation matrix can be obtained as follows. The characteristic equation is

$$|sI - A| = s^2 + 3s = 0$$

Then
$$\overline{S}^{-1} = \begin{bmatrix} 3 & 1 \\ 1 & 0 \end{bmatrix}$$

The similarity transformation matrix

$$P = S\overline{S}^{-1} = \begin{bmatrix} 2 & 0 \\ 2 & 2 \end{bmatrix}; \quad |P| \neq 0$$

Then
$$P^{-1} = \begin{bmatrix} 0.5 & 0 \\ -0.5 & 0.5 \end{bmatrix}$$

Hence the companion form of the system matrix becomes

$$\overline{A} = P^{-1}AP = \begin{bmatrix} 0.5 & 0 \\ -0.5 & 0.5 \end{bmatrix} \begin{bmatrix} -1 & 1 \\ 2 & -2 \end{bmatrix} \begin{bmatrix} 2 & 0 \\ 2 & 2 \end{bmatrix} = \begin{bmatrix} 0 & 1 \\ 0 & -3 \end{bmatrix}$$

$$\overline{B} = P^{-1}B = \begin{bmatrix} 0.5 & 0 \\ -0.5 & 0.5 \end{bmatrix} \begin{bmatrix} 0 \\ 2 \end{bmatrix} = \begin{bmatrix} 0 \\ 1 \end{bmatrix} \quad \text{and} \quad \overline{C} = \begin{bmatrix} -1 & 1 \end{bmatrix} \begin{bmatrix} 2 & 0 \\ 2 & 2 \end{bmatrix} = \begin{bmatrix} 0 & 2 \end{bmatrix}$$

The companion form of the system is (controllable canonical form)

$$\dot{\overline{X}} = \overline{A}\,\overline{X} + \overline{B}u$$

and
$$y = \overline{C}\,\overline{X}$$

It follows that the observability matrix of this controllable canonical form of system representation is

$$\overline{W} = \begin{bmatrix} \overline{C} \\ \overline{C}\,\overline{A} \end{bmatrix} = \begin{bmatrix} 0 & 2 \\ 0 & -6 \end{bmatrix}; \quad |\overline{W}| = 0$$

As $\overline{W}^{-1}$ does not exist, the observability condition is not satisfied.

12.13 Similarity Transformation Matrix from Observability Matrix

As in the case of a controllable system, we earlier obtained the relationship between the similarity transformation matrix and the controllability matrix. Here also we will find the relationship between the similarity transformation matrix and the observability matrix for a completely observable system. Let $\overline{W}$ be the observability matrix of the transformed system defined as

$$\dot{\overline{X}} = \overline{A}\,\overline{X} + \overline{B}u$$

and output
$$y = \overline{C}\,\overline{X}$$

where
$$\overline{A} = P^{-1}AP; \quad \overline{C} = CP; \quad \overline{B} = P^{-1}B$$

Let us define $\overline{W}$, the observability matrix of the transformed system as

$$\begin{aligned}
\overline{W} &= [\overline{C}\ \overline{C}\,\overline{A}\quad \overline{C}\,\overline{A}^2\ \ldots\ \overline{C}\,\overline{A}^{n-1}]^T \\
&= [CP\ \ CP(P^{-1}AP)\ \ CP(P^{-1}A^2P)\ \ldots\ CP(P^{-1}A^{n-1}P)]^T \\
&= [C\ \ CA\ \ CA^2\ \ldots\ CA^{n-1}]^T P \\
&= WP
\end{aligned}$$

Therefore, $\quad P = W^{-1}\overline{W} \quad$ or $\quad P^{-1} = \overline{W}^{-1}W \quad$ (12.31)

Hence it is shown that if a system is observable, then there exists a similarity transformation matrix by which the system can always be represented in the phase-variable canonical form.

The reverse is always true, in that, if the system is represented in the phase-variable canonical form, it is always observable. This is to note that the superscript 'bar' corresponds to the phase-variable canonical form of system representation.

The later relation $P^{-1} = \overline{W}^{-1}W$ is more suitable as $\overline{W}^{-1}$ can be obtained directly from the characteristic polynomial

$$|sI - A| = s^n + a_{n-1}s^{n-1} + \cdots + a_1 s + a_0$$

by inspection as
$$\overline{W}^{-1} = \begin{bmatrix} a_1 & a_2 & \cdots & a_{n-1} & 1 \\ a_2 & a_3 & \cdots & 1 & 0 \\ \vdots & \vdots & & \vdots & \vdots \\ a_{n-1} & 1 & \cdots & 0 & 0 \\ 1 & 0 & \cdots & 0 & 0 \end{bmatrix}$$

EXAMPLE 12.10 Consider the system of Example 12.3. The observability matrix as obtained earlier is

$$W = \begin{bmatrix} 1 & 0 & 1 \\ -1 & 1 & -10 \\ 1 & -5 & 102 \end{bmatrix}$$

The characteristic polynomial

$$|sI - A| = s^3 + 15s^2 + 54s + 40$$

This gives $\overline{W}^{-1}$ by inspection as

$$\overline{W}^{-1} = \begin{bmatrix} 54 & 15 & 1 \\ 15 & 1 & 0 \\ 1 & 0 & 0 \end{bmatrix}$$

then $P^{-1} = \overline{W}^{-1} W = \begin{bmatrix} 40 & 10 & 6 \\ 14 & 1 & 5 \\ 1 & 0 & 1 \end{bmatrix}$ and $P = \begin{bmatrix} -0.0178 & 0.178 & -0.785 \\ 0.1607 & -0.607 & 2.071 \\ 0.0178 & -0.178 & 1.785 \end{bmatrix}$

which gives the observable canonical form of representation as

$$\overline{A} = P^{-1}AP = \begin{bmatrix} 0 & 0 & -40 \\ 1 & 0 & -54 \\ 0 & 1 & -15 \end{bmatrix}; \quad \overline{B} = P^{-1}B = \begin{bmatrix} 34 \\ 9 \\ 0 \end{bmatrix}; \quad \overline{C} = [0 \ 0 \ 1]$$

12.14 Pole-Placement Design through State Feedback

Knowing the relation between the closed-loop poles and the system performance, we already know that we can carry out the design effectively by specifying the locations of the closed-loop poles of the system. This approach may be described as that of *closed-loop pole placement*. The design methods discussed in classical control theory are all characterized by the necessity that the poles be selected based on the design specification that can be achieved with the fixed controller design configuration and the physical limitations on the range of the controller parameters. Hence pole-placement is the theme of design. A natural question would be: *under what condition would an arbitrary pole-placement of the closed-loop system be possible?* The new approach that we are going to discuss is entirely a new design philosophy and freedom that can be achieved in state variable approach under certain conditions. Again the question would be: what is wrong with classical design, why should we go for a new approach?

The answer is that when we have a control process of third or higher order, the PI, PD, the single-stage phase-lead or phase-lag controller would not be able to control independently all the poles of the system, since there are only two free parameters in each of these controllers. Here lies the necessity of state variable control where arbitrary pole-placement is possible for any higher-order system independently under certain conditions, that is, if the system is completely state controllable, its eigenvalues can be arbitrarily assigned through state feedback.

Design of system with specific eigenvalues

The design of a linear feedback control system with specific eigenvalues is a useful application of state controllability. The linear system described by the state equation is

$$\dot{X} = AX + Bu \tag{AB}$$

The scalar control input is obtained from state feedback as

$$u = -GX + r$$

where G is an $1 \times n$ feedback matrix and r is the noise input. Then, the state equation can be written as

$$\dot{X} = (A - BG)X + Br \tag{12.32}$$

The system under consideration may be represented by the block diagram of Figure 12.6, and the characteristic equation becomes

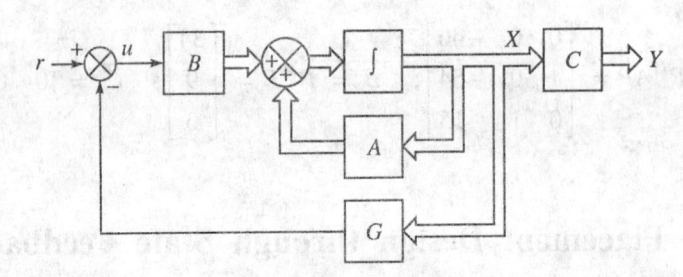

$$|sI - (A - BG)| = 0 \tag{12.33}$$

FIGURE 12.6

In the phase-variable canonical form of state variable representation, the system matrix A and the input coupling matrix B of the state equation (AB) can be written as

$$A = \begin{bmatrix} 0 & 1 & 0 & 0 & \cdots & 0 \\ 0 & 0 & 1 & 0 & \cdots & 0 \\ 0 & 0 & 0 & 1 & \cdots & 0 \\ \vdots & \vdots & \vdots & \vdots & & \vdots \\ 0 & 0 & 0 & 0 & \cdots & 1 \\ -a_0 & -a_1 & -a_2 & -a_3 & \cdots & -a_{n-1} \end{bmatrix} ; \quad B = \begin{bmatrix} 0 \\ 0 \\ 0 \\ \vdots \\ 0 \\ 1 \end{bmatrix}$$

The feedback matrix G can be written as

$$G = [g_1 \quad g_2 \quad \cdots \quad g_n] \quad (12.34)$$

then

$$[A - BG] = \begin{bmatrix} 0 & 1 & 0 & \cdots & 0 \\ 0 & 0 & 1 & \cdots & 0 \\ \vdots & \vdots & \vdots & & \vdots \\ (-a_0 - g_1) & (-a_1 - g_2) & (-a_2 - g_3) & \cdots & (-a_{n-1} - g_n) \end{bmatrix} \quad (12.35)$$

We have shown earlier that arbitrary pole placement is possible with the choice of phase-variable canonical form of state variable representation. Now the question is whether it is always possible to take phase-variable canonical form of state variable representation, that is, system matrix A in companion form and column vector B with all elements zero except the last element? It is possible only when the system is state controllable, or, the matrix

$$S = [B \quad AB \quad A^2B \quad \cdots \quad A^{n-1}B]$$

is of rank n. To show this, we form the following matrices:

$$B = \begin{bmatrix} 0 \\ 0 \\ \vdots \\ \vdots \\ 0 \\ 1 \end{bmatrix}; \quad AB = \begin{bmatrix} 0 \\ 0 \\ \vdots \\ 0 \\ 1 \\ -a_{n-1} \end{bmatrix}; \quad A^2B = \begin{bmatrix} 0 \\ 0 \\ \vdots \\ 1 \\ -a_{n-1} \\ a_{n-1}^2 - a_{n-2} \end{bmatrix}; \quad \cdots \quad A^{n-1}B = \begin{bmatrix} 1 \\ -a_{n-1} \\ -a_{n-1}^2 - a_{n-2} \\ -a_{n-1}^3 + 2a_{n-2}a_{n-1} + a_{n-3} \\ \vdots \\ \vdots \end{bmatrix}$$

It is quite obvious that $|S| \neq 0$ as it is a triangular matrix having diagonal elements as unity. Hence the rank of S is n. S^{-1} exists, so the similarity transformation matrix P can be formulated.

Continuing with the matrix product through $A^{n-1}B$, it will become apparent that regardless of what the values $a_1, a_2, \ldots, a_n$ are, the determinant of

$$S = [B \quad AB \quad A^2B \quad \cdots \quad A^{n-1}B]$$

will always be equal to -1, since S is a triangular matrix with 1's on the main diagonal. Therefore, we have proved that if the system is represented by the *phase-variable canonical form, it is always state controllable.*

The eigenvalues of $[A - BG]$ are then found from the characteristic equation

$$|sI - (A - BG)| = s^n + (a_{n-1} + g_n)s^{n-1} + (a_{n-2} + g_{n-1})s^{n-2} + \cdots + (a_0 + g_1)s^0 = 0 \quad (12.36)$$

Clearly, the eigenvalues can be arbitrarily placed by the proper choice of $g_1, g_2, \ldots, g_n$.

EXAMPLE 12.11 Given the transfer function

$$\frac{10}{s^3 + 3s^2 + 2s}$$

Design a feedback controller so that the eigenvalues of the closed-loop system are at -2, $-1 \pm j1$.

Solution: As there is no pole-zero cancellation, the transfer function is completely state controllable and the observable part of the system, i.e. the system is completely state controllable.

The phase-variable canonical form of state equation can be written as

$$\begin{bmatrix} \dot{x}_1 \\ \dot{x}_2 \\ \dot{x}_3 \end{bmatrix} = \begin{bmatrix} 0 & 1 & 0 \\ 0 & 0 & 1 \\ 0 & -2 & -3 \end{bmatrix} X + \begin{bmatrix} 0 \\ 0 \\ 1 \end{bmatrix} u \tag{12.37}$$

For a third-order system, the feedback matrix G is of the form

$$G = [g_1 \quad g_2 \quad g_3] \tag{12.38}$$

The characteristic equation of the closed-loop system is

$$|sI - (A - BG)| = s^3 + (3 + g_3)s^2 + (2 + g_2)s + g_1 = 0 \tag{12.39}$$

For the desired closed-loop poles at -2, $-1 \pm j1$, the desired characteristic equation becomes

$$s^3 + 4s^2 + 6s + 4 = 0 \tag{12.40}$$

Comparing Eqs. (12.39) and (12.40), we get

$$g_1 = 4, \quad g_2 = 4, \quad g_3 = 1$$

that is
$$G = [4 \quad 4 \quad 1]$$

The state diagram of the overall system is shown in Figure 12.7.

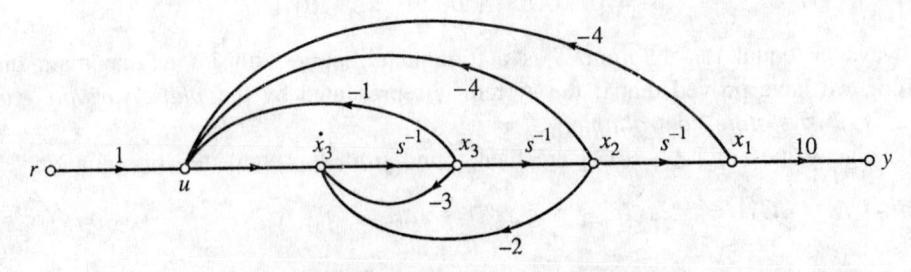

FIGURE 12.7 Example 12.11.

EXAMPLE 12.12 Consider

$$\dot{X} = \begin{bmatrix} 0 & 1 & 0 \\ 0 & 0 & 1 \\ -20 & -6 & 0 \end{bmatrix} X + \begin{bmatrix} 0 \\ 0 \\ 1 \end{bmatrix} u$$

and output $\quad y = [3 \quad 1 \quad 0] X$

The closed-loop transfer function becomes

$$\frac{Y(s)}{U(s)} = \frac{s+3}{s^3 + 6s + 20}$$

The desired closed-loop transfer function

$$\frac{Y(s)}{R(s)} = \frac{40(s+3)}{(s+4)(s^2+8s+30)} = \frac{40(s+3)}{s^3 + 12s^2 + 62s + 120}$$

The desired A_f from Eq. (12.35) can be written as

$$\overline{A}_f = (\overline{A} - \overline{B}G) = \begin{bmatrix} 0 & 1 & 0 \\ 0 & 0 & 1 \\ -120 & -62 & -12 \end{bmatrix}$$

The given system is in controllable canonical form. The characteristic polynomial of the given system is

$$|sI - \overline{A}| = s^3 + 6s + 20$$

Obviously the system is unstable because of the missing term. By proper choice of the feedback matrix, $G = [g_1 \quad g_2 \quad g_3]$, we have to obtain the desired characteristic equation.

We get

$$\overline{A}_f = \overline{A} - \overline{B}GK = \begin{bmatrix} 0 & 1 & 0 \\ 0 & 0 & 1 \\ (-20 - g_1 K) & (-6 - g_2 K) & -g_3 K \end{bmatrix}$$

$$= \begin{bmatrix} 0 & 1 & 0 \\ 0 & 0 & 1 \\ -120 & -62 & -12 \end{bmatrix}$$

and $\quad g_1 K = 100, \ g_2 K = 56, \ g_3 K = 12.$

Hence we have chosen $K = 40$ in order to make dc gain of the closed-loop transfer function unity.

Therefore, $g_1 = 2.5$, $g_2 = 1.4$, $g_3 = 0.3$,
i.e. $G = [2.5 \quad 1.4 \quad 0.3]$

The state diagram of the overall system is shown in Figure 12.8.

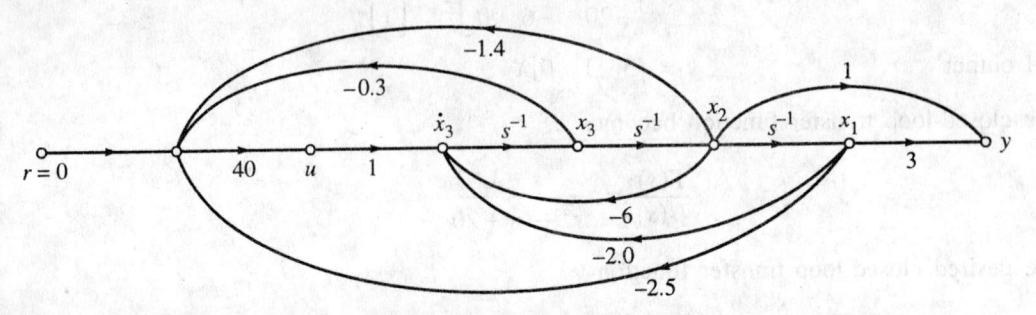

FIGURE 12.8 Example 12.12.

EXAMPLE 12.13 Consider the state equations as

$$\dot{X} = \begin{bmatrix} -1 & 1 & 0 \\ 0 & -4 & 2 \\ 0 & 0 & -10 \end{bmatrix} X + \begin{bmatrix} 1 \\ 0 \\ -1 \end{bmatrix} u$$

and output $y = [1 \quad 0 \quad 1]X$

The system is not in controllable canonical form. If the system is completely controllable, then it can be transformed into controllable canonical form.

The controllability matrix

$$S = [B \quad AB \quad A^2B] = \begin{bmatrix} 1 & -1 & -1 \\ 0 & -2 & 28 \\ -1 & 10 & -100 \end{bmatrix}$$

The rank of S is 3. The system is completely controllable. The characteristic polynomial

$$|sI - A| = s^3 + 15s^2 + 54s + 40$$

gives

$$\bar{S}^{-1} = \begin{bmatrix} 54 & 15 & 1 \\ 15 & 1 & 0 \\ 1 & 0 & 0 \end{bmatrix}$$

From the similarity transformation matrix

$$P = S\bar{S}^{-1} = \begin{bmatrix} 38 & 14 & 1 \\ -2 & -2 & 0 \\ -4 & -5 & -1 \end{bmatrix}$$

P^{-1} is then obtained.

Hence, $\bar{A} = P^{-1}AP = \begin{bmatrix} 0 & 1 & 0 \\ 0 & 0 & 1 \\ -40 & -54 & -15 \end{bmatrix}$; $\bar{B} = P^{-1}B = \begin{bmatrix} 0 \\ 0 \\ 1 \end{bmatrix}$; $\bar{C} = CP = [34 \quad 9 \quad 0]$

Then the transfer function of the closed-loop system is obtained as

$$\bar{C}(sI - \bar{A})^{-1}\bar{B} = \frac{9s + 34}{s^2 + 15s^2 + 54s + 40}$$

The desired transfer function is

$$\frac{32(9s + 34)}{s^2 + 20s^2 + 232s + 1088}$$

to make the dc gain of the closed-loop transfer function unity. Then we have

$$\bar{A}_f = \begin{bmatrix} 0 & 1 & 0 \\ 0 & 0 & 1 \\ -1088 & -232 & -20 \end{bmatrix}$$

Again, $\bar{A}_f = \bar{A} - \bar{B}KG = \begin{bmatrix} 0 & 1 & 0 \\ 0 & 0 & 1 \\ -40 - g_1 K & -54 - g_2 K & -15 - g_3 K \end{bmatrix}$

Comparing, we get

$$Kg_1 = 1048, \quad Kg_2 = 178, \quad Kg_3 = 5$$

A comparison of the numerators gives, $K = 32$

Then $g_1 = 32.75$, $g_2 = 5.5625$, and $g_3 = 0.15625$

i.e. $\bar{G} = [32.75 \quad 5.5625 \quad 0.15625]$

This value of $\bar{G}$ would have been considered if the system matrix of the original system was in the controllable canonical form, but as the original system has been transformed by the similarity transformation matrix P to the controllable canonical form, the proper G would be now

$$G = \bar{G}P^{-1} = [1.093 \quad 2.531 \quad 0.937]$$

The state diagram of the closed-loop system is shown in Figure 12.9.

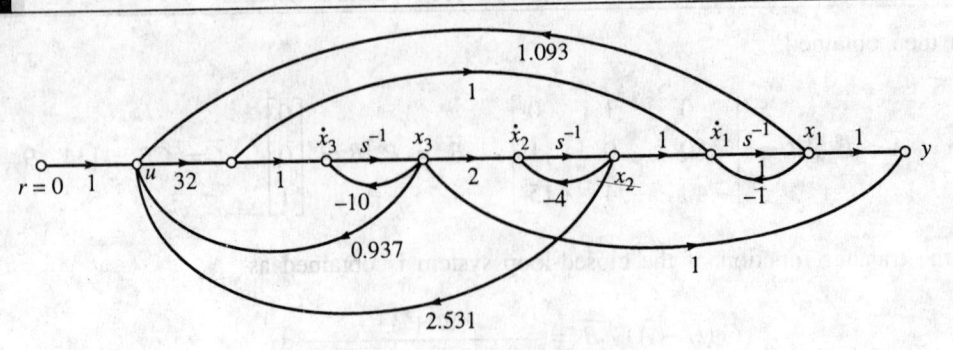

FIGURE 12.9 Example 12.13.

12.15 State Observer

We have shown that if a system is completely controllable, its eigenvalues can be located at any arbitrary desired location by proper state feedback if all the states are accessible. It may be noted that in practical cases some of the states may not be accessible, as for example, in state variable formulation of motor, flux may be chosen as state variable which may not be measurable, i.e. not accessible and hence all state feedback is not possible. Similarly for state variable formulation of a nuclear plant, all states may not be accessible, and hence state feedback is not possible unless there are some means by which the non-accessible states can be generated for being accessible for state feedback. The subsystem that performs the observation of the state variables based on the information received from the measurements of the input $u(t)$ and the output $y(t)$ and their derivatives is called an observer. The output of the observer is the observed state vector $X(t)$. There should be some way by which we can compare how close the approximated state vector $X_e(t)$ is to the exact state vector $X(t)$. But we cannot compare $X_e(t)$ with $X(t)$ as $X(t)$ is not accessible. The next best alternative is to compare the output $y_e(t)$ with $y(t)$ where

$$y_e(t) = CX_e(t) \tag{12.41}$$

and is measurable.

12.15.1 Design of state observer

For arbitrary pole placement as per the design specification, the control input $u(t)$ should be

$$u(t) = -GX(t) + r(t) \tag{12.42}$$

As is evident from Eq. (12.42) that in order to implement the state feedback, all state variables should be accessible. In practice, all the state variables are not accessible. Only the inputs and the outputs are measurable. The subsystem that generates the inaccessible state variables (the estimated one is called $X_e(t)$) based on the information received from the measurements of the input $u(t)$ and the output $y(t)$ and their derivatives is called an *observer*.

Figure 12.10 shows the overall system structure including the observer where the input to the observer are the output $y(t)$ and the control $u(t)$. The output of the observer is the observed state vector $X_e(t)$ (the estimated one). Therefore the actual control input is

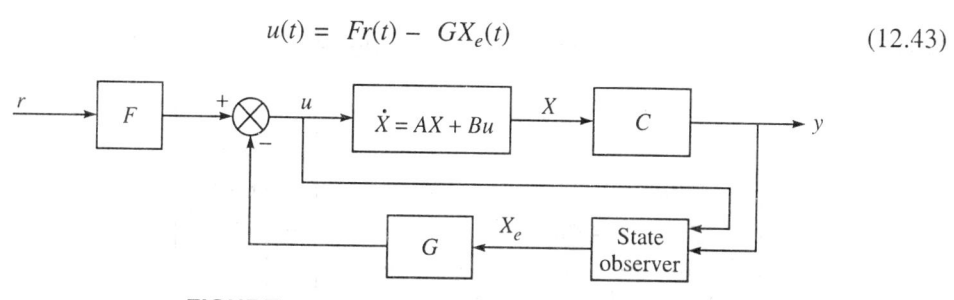

$$u(t) = Fr(t) - GX_e(t) \tag{12.43}$$

FIGURE 12.10 Design of state observer.

The condition for the existence of an observer indicates that the design of observer is closely related to the condition of observability.

12.15.2 Condition for the existence of observer

Consider an nth-order linear time-invariant control system described by the dynamic equations

$$\dot{X}(t) = AX(t) + Bu(t) \tag{12.44}$$

and output

$$y(t) = CX(t) \tag{12.45}$$

with X as state vector of order $(n \times 1)$, u as scalar control input, and y as scalar output.

The inaccessible state vector X_e may be constructed from linear combination of the output y, input u and derivatives of these variables if the system is completely observable.

We will derive the necessary condition for the above statement assuming that the system under consideration is completely observable, i.e. the matrix

$$W = [\, C^T \quad A^T C^T \quad (A^T)^2 \, C^T \quad \ldots \quad (A^T)^{n-1} C^T \,]$$

is of rank n. If the matrix is complex, then its transpose may be replaced by simply the conjugate transpose.

Starting with Eq. (12.45), taking derivatives on both sides with respect to time t, we get after rearranging.

$$\dot{y}(t) - CBu(t) = CAX(t) \tag{12.46}$$

Repeating the time derivative to Eq. (12.46) and after rearranging, we get

$$\ddot{y}(t) - CB\dot{u}(t) = CA\dot{X}(t) = CA[AX(t) + Bu(t)] = CA^2 X(t) + CABu(t)$$

or

$$\ddot{y}(t) - CB\dot{u}(t) - CABu(t) = CA^2 X(t) \tag{12.47}$$

Continuing the process, after taking $(n-1)$ derivatives and rearranging, we get

$$\overset{(n-1)}{y}(t) - CB\overset{(n-2)}{u} - CAB\overset{(n-3)}{u} - \cdots - CA^{n-2}Bu = CA^{n-1}X \qquad (12.48)$$

Arranging in matrix form, Eqs. (12.45), (12.46) through (12.48), we get

$$\begin{bmatrix} y \\ \dot{y} - CBu \\ \ddot{y} - CB\dot{u} - CABu \\ \cdots\cdots\cdots\cdots\cdots\cdots\cdots\cdots \\ \overset{(n-1)}{y} - CB\overset{(n-2)}{u} - CAB\overset{(n-3)}{u} - \ldots - CABu \end{bmatrix} = \begin{bmatrix} C \\ CA \\ CA^2 \\ . \\ . \\ . \\ CA^{n-1} \end{bmatrix} X \qquad (12.49)$$

Therefore, in order to express the state vector in terms of output y, input u and their derivatives, the matrix W should be of rank n, where

$$W = [C \quad CA \quad CA^2 \quad \ldots \quad CA^{n-1}]^T$$
$$= [C^T \quad A^TC^T \quad (A^T)^2C^T \quad \ldots \quad (A^T)^{n-1}C^T]^T \qquad (12.50)$$

which is nothing but the condition for the system to be completely observable.

Hence, the inaccessible state vector can be generated from output, input and their derivatives provided the system is completely observable. If we have the inaccessible state vector (expected) X_e, then the arbitrary pole placement is possible. But who can give the guarantee that the expected one, that is, X_e is identical to the original state of the original system under consideration.

12.16 Eigenvalue Assignment Method for Design of G_e

It is obvious that we have to compare X_e with X, so that the error between the two, i.e. $(X_e - X)$ becomes minimum. But the problem remains because of the inaccessibility of the state vector $X(t)$, we require to generate $X_e(t)$ and hence the question of comparing directly X_e with X does not arise. Since we cannot measure $X(t)$ directly, an alternative is to compare $y(t)$ with $y_e(t)$, where the expected output

$$y_e(t) = CX_e(t) \qquad (12.51)$$

Based on the above argument, a logical arrangement for the state observer is shown in Figure 12.11.

The output error $\qquad E = y_e - y$

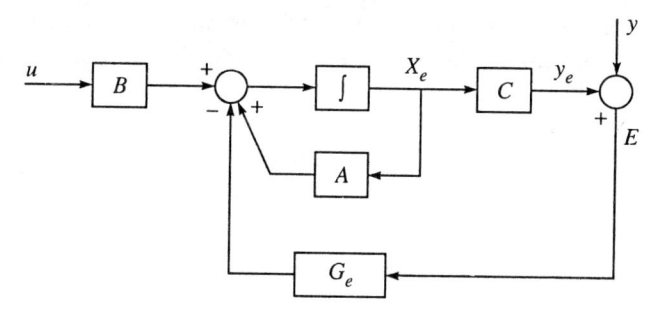

FIGURE 12.11 Block diagram of the state observer.

The dynamics of the observer can be represented by

$$\dot{X}_e(t) = AX_e - G_e E + Bu$$
$$= AX_e - G_e(y_e - y) + Bu$$
$$= AX_e - G_e(CX_e - CX) + Bu$$
$$= AX_e + Bu(t) + G_e C[X(t) - X_e(t)]$$
$$= (A - G_e C)X_e(t) + Bu(t) + G_e CX(t)$$
$$= (A - G_e C)X_e(t) + Bu(t) + G_e y(t)$$
$$= AX_e(t) + Bu(t) + G_e y(t) - G_e CX_e(t)$$
$$= AX_e(t) + Bu(t) + G_e y(t) - G_e y_e(t)$$
$$= AX_e(t) + Bu(t) + G_e [y(t) - y_e(t)] \quad (12.52)$$

Subtracting Eq. (12.52) from Eq. (12.44), we get

$$\frac{d}{dt}[X(t) - X_e(t)] = [A - G_e C][X(t) - X_e(t)] \quad (12.53)$$

which is a homogeneous state equation of linear system with system matrix $[A - G_e C]$. The characteristic equation is

$$|sI - (A - G_e C)| = 0 \quad (12.54)$$

Select the element of G_e such that the response of Eq. (12.53) decays to zero as quickly as possible, that is, eigenvalues of $[A - G_e C]$ should be selected such that $X_e(t)$ approaches $X(t)$ rapidly. Further, it may be noted that by assigning the pole placement, only the denominator polynomial of the transfer function is affected but not the zeros, that is, the numerator polynomial is not affected. However, it may be mentioned that the eigenvalues of $[A - G_e C]$ and those of $[A - G_e C]^T$ are same, as

$$[A - G_e C]^T = [A^T - C^T G_e^T] \quad (12.55)$$

and state vector X_e is same as the original actual state vector X.

The significance of the expression in Eq. (12.53) is that if the initial values of $X(t)$ and $X_e(t)$ are identical, the response of the observer will be identical to that of the original system. Therefore, the design of the feedback matrix G_e for the observer is significant only if the initial conditions to $X(t)$ and $X_e(t)$ are different.

The design objective is to select the feedback matrix G_e such that the expected output $y_e(t)$ will approach the actual output $y(t)$ as fast as possible. When $y_e(t)$ equals $y(t)$ the dynamics of the state observer is described by

$$\dot{X}_e(t) = AX_e(t) + Bu(t) \qquad (12.56)$$

We have proved earlier that arbitrary placement of eigenvalues is possible provided the system is completely state controllable. Further, from the property of duality, controllability and observability conditions are dual to each other. Hence we can rewrite the condition of arbitrary assignment of the eigenvalues of $[A - G_e C]$ or of $[A^T - C^T G_e^T]$ is, that the pair $[A^T, C^T]$ to be completely controllable. This is equivalent to dual system property that the pair $[A, C]$ be required to be completely observable. Therefore, the observability of $[A, C]$ ensures not only that a state observer can be constructed from linear combinations of the output y, input u and the derivatives of these, but also that the eigenvalues of the observer can be arbitrarily assigned by choosing the feedback matrix G_e. The detailed diagram of an observer is shown in Figure 12.12.

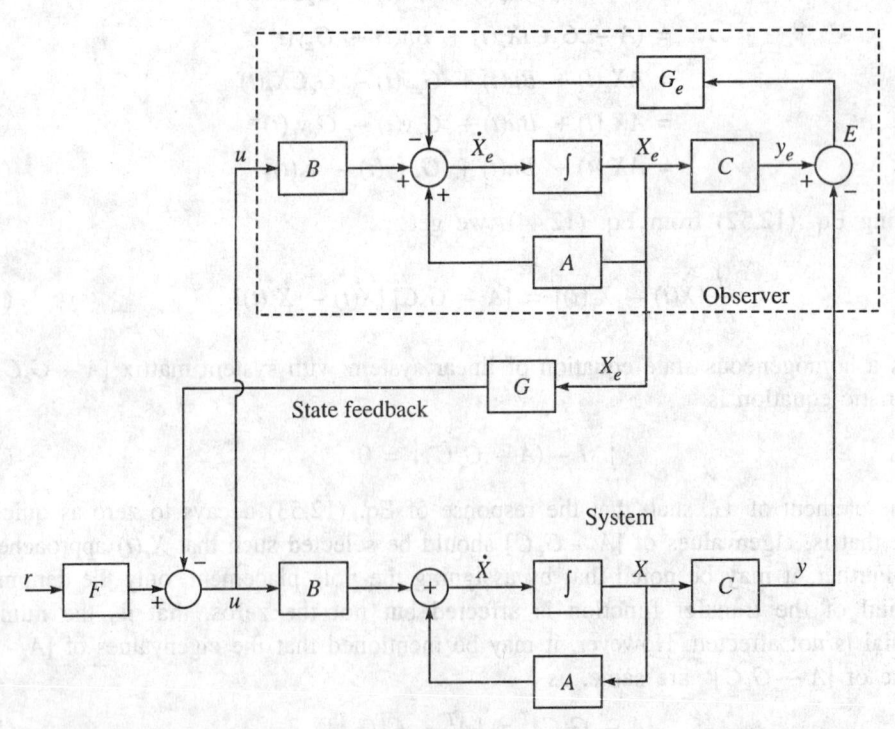

FIGURE 12.12 Block diagram of a linear feedback system with observer.

Hence the design of observer is possible provided the system is completely observable, and arbitrary pole placement is possible if the system is completely controllable. Obviously the question of closeness of the estimated state with the actual state remains. The best possible way to minimize the error is to choose a quadratic cost function of the form

$$J = \frac{1}{2}\int_0^\infty (y - y_e)^T Q(y - y_e)\, dt \qquad (12.57)$$

The matrix Q is symmetric and positive semi-definite; the elements of Q may be selected to give various weightages on the observation of outputs which in turn become the states. This pertains to optimal control theory which is beyond the scope of this book. The only clarification regarding the choice of quadratic cost function is that we have the mathematical tool for testing the sign definiteness of a quadratic function. By Sylvester's criterion, we can test the sign definiteness of a symmetric matrix. The sign of all the principal minors of the symmetric matrix ensure the sign definiteness of the quadratic function.

However, regarding the design of the observer a few points may be noted. If the initial values of $X(t)$ and $X_e(t)$ are identical, the responses of the observer will be identical to those of the original system, the error will be zero and the feedback matrix G_e will have no effect on the responses of the observer whatsoever.

The state observer dynamics becomes slower for pole placement of the observer system closer to the origin and obviously negative.

Arbitrary placement of poles depends on the choice of the feedback matrix G_e. Selecting larger values of G_e matrix to give faster transient response for the observer, does not affect the numerator terms and hence discrepencies may exist in the transient part of the state response of the original system and the observer.

EXAMPLE 12.14 The dynamics of a system is represented by

$$\dot{X} = \begin{bmatrix} 0 & 1 \\ -2 & -3 \end{bmatrix} X + \begin{bmatrix} 0 \\ 1 \end{bmatrix} u$$

and output $\qquad y = [2 \quad 0]X$

with the initial condition $\quad X(0) = [0.5 \quad 0]^T$

It is desired to design a state observer so that the new pole placement will be at $-10, -10$.

Solution: Let the feedback matrix be designated as

$$G_e = [g_{e1} \quad g_{e2}]^T$$

The characteristic equation of the state observer is

$$|sI - (A - G_e C)| = \left| s\begin{bmatrix} 1 & 0 \\ 0 & 1 \end{bmatrix} - \left\{ \begin{bmatrix} 0 & 1 \\ -2 & -3 \end{bmatrix} - \begin{bmatrix} g_{e1} \\ g_{e2} \end{bmatrix}[2 \quad 0] \right\} \right|$$

$$= \begin{vmatrix} s + 2g_{e1} & -1 \\ 2 + 2g_{e_2} & s + 3 \end{vmatrix}$$

$$= s^2 + (2g_{e1} + 3)s + (6g_{e1} + 2 + 2g_{e2}) = 0 \qquad (12.58)$$

For the desired eigenvalues to be placed at –10 and –10, the characteristic equation should be

$$s^2 + 20s + 100 = 0 \tag{12.59}$$

Comparing Eqs. (12.58) and (12.59), after solving we get

$$G_e = \begin{bmatrix} g_{e1} \\ g_{e2} \end{bmatrix} = \begin{bmatrix} 8.5 \\ 23.5 \end{bmatrix}$$

The state diagram for the observer together with the original system is shown in Figure 12.13.

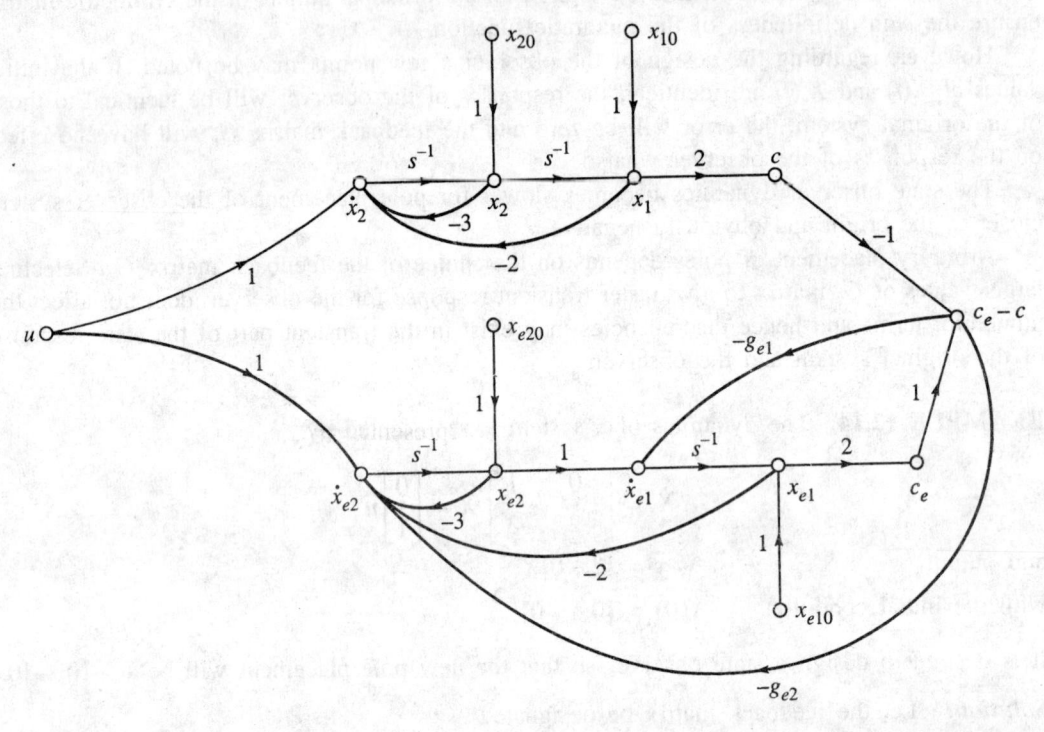

FIGURE 12.13 Example 12.14: state diagram for the observer and the original system.

Now let us see how the observer design is successful by observing the responses of the observer, i.e. $X_e(t)$ is closer to those of the original system, i.e. $X(t)$. It depends on the initial values.

If the initial value of $X(t)$, i.e. $X(0)$ and that of $X_e(t)$, i.e. $X_e(0)$ are identical, then the responses $X(t)$ of the original system and those of $X_e(t)$ of the observer will be identical and the feedback matrix G_e will have no effect on the responses of the observer. But with different initial conditions the observer for $x_1(t)$ and $x_{e1}(t)$ are made. With similar line of action, $x_2(t)$ and $x_{e2}(t)$ can also be made as shown in Figure 12.13.

Case I: For the given system dynamics where the desired eigenvalues are to be $-10, -10$, G_e comes out to be as calculated $[8.5 \quad 23.5]^T$ and with initial conditions $X(0) = [0.5 \quad 0]^T$ and $X_e(0) = [0 \quad 0]^T$, the unit step response of $x_1(t)$ and that of $x_{e1}(t)$ are as shown in Figure 12.14.

Case II: With the same original system having desired pole positions as $-2.5, -2.5$, G_e comes out to be $[1 \quad 1]^T$. The responses $x_1(t)$ and $x_{e1}(t)$ are shown in the same Figure 12.14.

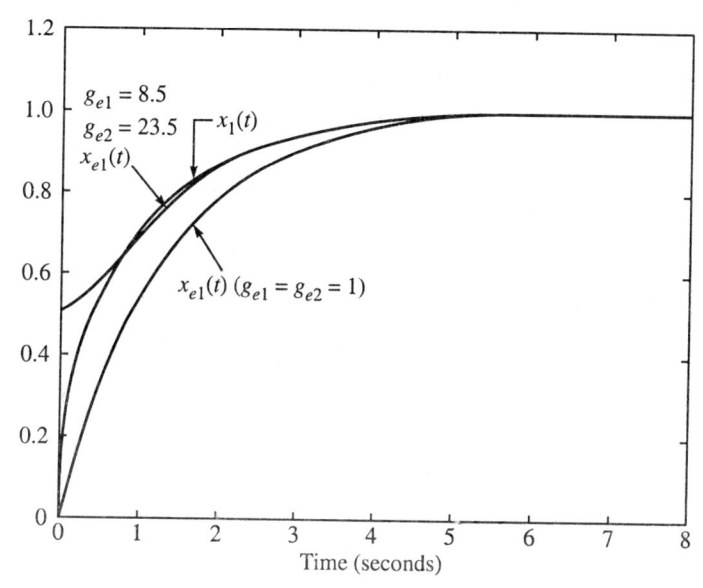

FIGURE 12.14 Example 12.14: state $x_1(t)$ and the observer states of the observer.

Observation: The deviation noticed in case II is of slower dynamics, i.e. eigenvalues. The deviation at steady state is negligible—more or less close to each other—though at initial stage it is high.

EXAMPLE 12.15 Consider a system having transfer function

$$G(s) = \frac{2s + 10}{s^2 + 5s + 6}$$

Write the controllable canonical form of representation of the system.

Solution: The controllable canonical form has the system matrix as

$$A = \begin{bmatrix} 0 & 1 \\ -a_0 & -a_1 \end{bmatrix}$$

where the characteristic equation is

$$|sI - A| = 0$$

and it becomes

$$s^2 + a_1 s + a_0 = 0$$

and $\quad B = [0 \ 1]^T, \quad C = [c_1 \ c_2] \quad$ and $\quad D =$ a scalar,

for single-input-single-output system.

From the transfer function it is obvious that the order of system matrix is two. Hence the structure of the transfer function becomes

$$G(s) = \frac{Y(s)}{U(s)} = C[sI - A]^{-1}B + D$$

$$= \frac{[c_1 \ c_2]\begin{bmatrix} s & -1 \\ a_0 & s+a_1 \end{bmatrix}\begin{bmatrix} 0 \\ 1 \end{bmatrix}}{s^2 + a_1 s + a_0} + D$$

$$= \frac{c_2 s + a_1 c_2 - c_1}{s^2 + a_1 s + a_0} + D = \frac{2s + 10}{s^2 + 5s + 6}$$

By equating both sides of this expression, we get

$$a_0 = 6, \ a_1 = 5, \ c_1 = 0, \ c_2 = 2, \ D = 0$$

The resultant state space formulation is given by

$$\dot{x}_1 = x_2$$
$$\dot{x}_2 = -6x_1 - 5x_2 + u$$

and output $\qquad y = 10x_1 + 2x_2$

12.17 Transfer Function Matrix

For multivariable systems we obtain the transfer function matrix as illustrated with the following example.

EXAMPLE 12.16 Given the system dynamics of three-input-two-output second-order system as

$$\dot{X} = \begin{bmatrix} 0 & 1 \\ -5 & -2 \end{bmatrix} X + \begin{bmatrix} 1 & 0 & -1 \\ 0 & -2 & 3 \end{bmatrix} U$$

and output $\qquad Y = \begin{bmatrix} 0 & -3 \\ 2 & -4 \end{bmatrix} X + \begin{bmatrix} 1 & -2 & 0 \\ 0 & 0 & 0 \end{bmatrix} U$

The transfer function is given by

$$G(s) = C(sI - A)^{-1}B + D$$

$$= \begin{bmatrix} 0 & -3 \\ 2 & -4 \end{bmatrix} \begin{bmatrix} s & -1 \\ 5 & s+2 \end{bmatrix}^{-1} \begin{bmatrix} 1 & 0 & -1 \\ 0 & -2 & 3 \end{bmatrix} + \begin{bmatrix} 1 & -2 & 0 \\ 0 & 0 & 0 \end{bmatrix}$$

$$= \frac{1}{s^2+2s+5}\begin{bmatrix} s^2+2s+20 & -2s^2+2s-10 & -9s-15 \\ 2s+24 & 8s-4 & -14s-18 \end{bmatrix}$$

$$= \begin{bmatrix} G_{11}(s) & G_{12}(s) & G_{13}(s) \\ G_{21}(s) & G_{22}(s) & G_{23}(s) \end{bmatrix}$$

Hence the output can be written as

$$\begin{bmatrix} Y_1(s) \\ Y_2(s) \end{bmatrix} = \begin{bmatrix} G_{11}(s) & G_{12}(s) & G_{13}(s) \\ G_{21}(s) & G_{22}(s) & G_{23}(s) \end{bmatrix} \begin{bmatrix} U_1(s) \\ U_2(s) \\ U_3(s) \end{bmatrix}$$

where

$$G_{11}(s) = Y_1(s)/U_1(s); \quad G_{12}(s) = Y_1(s)/U_2(s); \quad G_{13}(s) = Y_1(s)/U_3(s)$$
$$G_{21}(s) = Y_2(s)/U_1(s); \quad G_{22}(s) = Y_2(s)/U_2(s); \quad G_{23}(s) = Y_2(s)/U_3(s)$$

when the initial conditions and all other inputs are zero.

Alternatively, the signal flow graph simulation can be drawn from the state equations and by Mason's gain formula we can find the transfer function matrix.

12.17.1 Realization of transfer matrix

For an r-input and m-output system the structure of the transfer matrix becomes

$$\begin{bmatrix} Y_1(s) \\ \vdots \\ Y_m(s) \end{bmatrix} = \begin{bmatrix} G_{11}(s) & \cdots & G_{1r}(s) \\ \vdots & & \vdots \\ G_{m1}(s) & \cdots & G_{mr}(s) \end{bmatrix} \begin{bmatrix} U_1(s) \\ \vdots \\ U_r(s) \end{bmatrix}$$

For an nth-order system, i.e. for an n-state vector, the dynamics of the system, dimension-wise looks like the following:

$$\underset{(n\times 1)}{\dot{X}} = \underset{(n\times n)(n\times 1)}{AX} + \underset{(n\times r)(r\times 1)}{BU}$$

$$\underset{(m\times 1)}{Y} = \underset{(m\times n)(n\times 1)}{CX} + \underset{(m\times r)(r\times 1)}{DU}$$

That is,

$$\frac{d}{dt}\begin{bmatrix} x_1 \\ \vdots \\ x_n \end{bmatrix} = \begin{bmatrix} a_{11} & \cdots & a_{1n} \\ \vdots & & \vdots \\ a_{n1} & \cdots & a_{nn} \end{bmatrix}\begin{bmatrix} x_1 \\ \vdots \\ x_n \end{bmatrix} + \begin{bmatrix} b_{11} & \cdots & b_{1r} \\ \vdots & & \vdots \\ b_{n1} & \cdots & b_{nr} \end{bmatrix}\begin{bmatrix} u_1 \\ \vdots \\ u_r \end{bmatrix}$$

and output vector

$$\begin{bmatrix} y_1 \\ \vdots \\ y_n \end{bmatrix} = \begin{bmatrix} c_{11} & \cdots & c_{1n} \\ \vdots & & \vdots \\ c_{n1} & \cdots & c_{nn} \end{bmatrix} \begin{bmatrix} x_1 \\ \vdots \\ x_n \end{bmatrix} + \begin{bmatrix} d_{11} & \cdots & d_{1r} \\ \vdots & & \vdots \\ d_{m1} & \cdots & d_{mr} \end{bmatrix} \begin{bmatrix} u_1 \\ \vdots \\ u_r \end{bmatrix}$$

The transfer matrix G comes out to be dimension-wise as follows:

$$G = \begin{bmatrix} G_{11} & \cdots & G_{1r} \\ \vdots & & \vdots \\ G_{m1} & \cdots & G_{mr} \end{bmatrix} = \underset{(m \times n)}{C} \underset{(n \times n)}{[sI - A]^{-1}} \underset{(n \times r)}{B} + \underset{(m \times r)}{D}$$

$$= \underset{(m \times n)}{\begin{bmatrix} c_{11} & \cdots & c_{1n} \\ \vdots & & \vdots \\ c_{m1} & \cdots & c_{mn} \end{bmatrix}} \underset{(n \times n)}{(sI - A)^{-1}} \underset{(n \times r)}{\begin{bmatrix} b_{11} & \cdots & b_{1r} \\ \vdots & & \vdots \\ b_{n1} & \cdots & b_{nr} \end{bmatrix}} + \underset{(m \times r)}{\begin{bmatrix} d_{11} & \cdots & d_{1r} \\ \vdots & & \vdots \\ d_{m1} & \cdots & d_{mr} \end{bmatrix}}$$

where A is the system matrix. The system matrix A may have different forms which depend upon the choice the states. It may be noted that the choice of states is not unique whereas the transfer matrix for a multivariable system is unique.

The system matrix A may be in normal canonical form, i.e. diagonal form for distinct roots or in Jordan canonical form for repetitive roots and then the states of the system are in decoupled form.

The system matrix A may be in companion form as in controllable canonical form of realization or may be in observable canonical form of realization. The system matrix A may have any other general form as well. In these forms of system matrix, the states of the system are coupled.

Let us illustrate the minimal state variable realization of a single-input-two-output system transfer matrix with the help of the following example.

EXAMPLE 12.17 Consider the 2×1 transfer matrix

$$G(s) = \begin{bmatrix} G_{11}(s) \\ G_{21}(s) \end{bmatrix} = \begin{bmatrix} \dfrac{s+3}{(s+1)(s+2)} \\ \dfrac{s+4}{s+3} \end{bmatrix} = C[sI - A]^{-1}B + D = C[sI - A]^{-1}B + \begin{bmatrix} d_{11} \\ d_{21} \end{bmatrix}$$

$$= \begin{bmatrix} 0 + \dfrac{s+3}{(s+1)(s+2)} \\ 1 + \dfrac{1}{s+3} \end{bmatrix} = \begin{bmatrix} \dfrac{s+3}{(s+1)(s+2)} \\ \dfrac{1}{s+3} \end{bmatrix} + \begin{bmatrix} 0 \\ 1 \end{bmatrix}$$

Now, we get
$$[d_{11} \quad d_{21}]^T = [0 \quad 1]^T$$

and
$$C(sI - A)^{-1}B = \begin{bmatrix} \dfrac{s+3}{(s+1)(s+2)} \\ \dfrac{1}{s+3} \end{bmatrix}$$

It may be noted that $C(sI - A)^{-1}B$ is strictly a proper rational transfer function.

Next we find the least common denominator of G_{ij} for $i = 1, 2$, which comes out to be $(s + 1)(s + 2)(s + 3)$, that is, the characteristic equation $|sI - A| = 0$ is of third-order polynomial and hence the system matrix A is of order 3×3. This expresses G as

$$G(s) = \begin{bmatrix} G_{11}(s) \\ G_{21}(s) \end{bmatrix} = \frac{1}{(s+1)(s+2)(s+3)} \begin{bmatrix} (s+3)^2 \\ (s+1)(s+2) \end{bmatrix} + \begin{bmatrix} 0 \\ 1 \end{bmatrix}$$

$$= \frac{1}{s^3 + 6s^2 + 11s + 6} \begin{bmatrix} s^2 + 6s + 9 \\ s^2 + 3s + 2 \end{bmatrix} + \begin{bmatrix} 0 \\ 1 \end{bmatrix}$$

$$= C[sI - A]^{-1}B + D = \frac{C \, \text{Adj}[sI - A]B}{|sI - A|} + D$$

Further, for the given transfer matrix $G(s)$ of the order (2×1), i.e. for a single-input-two-output system the structure of the system dynamics becomes

$$\underset{(3 \times 1)}{\dot{X}} = \underset{(3 \times 3)(3 \times 1)}{AX} + \underset{(3 \times 1)(1 \times 1)}{Bu}$$

and output
$$\underset{(2 \times 1)}{Y} = \underset{(2 \times 3)(3 \times 1)}{CX} + \underset{(2 \times 1)(1 \times 1)}{Du}$$

Hence a minimal dimensional realization of $G(s)$ in controllable form is given by

$$\begin{bmatrix} \dot{x}_1 \\ \dot{x}_2 \\ \dot{x}_3 \end{bmatrix} = \begin{bmatrix} 0 & 1 & 0 \\ 0 & 0 & 1 \\ -6 & -11 & -6 \end{bmatrix} \begin{bmatrix} x_1 \\ x_2 \\ x_3 \end{bmatrix} + \begin{bmatrix} 0 \\ 0 \\ 1 \end{bmatrix} u$$

and output
$$\begin{bmatrix} y_1 \\ y_2 \end{bmatrix} = \begin{bmatrix} 9 & 6 & 1 \\ 2 & 3 & 1 \end{bmatrix} \begin{bmatrix} x_1 \\ x_2 \\ x_3 \end{bmatrix} + \begin{bmatrix} 0 \\ 1 \end{bmatrix} u$$

Drill Problem 12.1

Determine the transfer function matrix for a two-input-two-output system described by the following state equations.

$$\dot{X} = \begin{bmatrix} -1 & 1 & 0 \\ 0 & -4 & 2 \\ 0 & 0 & 10 \end{bmatrix} X + \begin{bmatrix} 1 & -1 \\ 0 & 2 \\ -1 & 1 \end{bmatrix} U$$

and output
$$Y = \begin{bmatrix} 1 & 0 & 1 \\ 2 & 1 & -3 \end{bmatrix} X$$

Ans. $\dfrac{1}{s^3 + 15s^2 + 54s + 40} \begin{bmatrix} 9s+34 & -7s-14 \\ 5s^2+41s+86 & -(3s^2+15s+26) \end{bmatrix}$

12.18 Diagonal Form of Representation of Transfer Matrix

Let us illustrate the minimal state variable realization of a single-input-two-output system transfer matrix in diagonal form with the help of the following example.

EXAMPLE 12.18 Consider the 2×1 transfer matrix

$$G(s) = \begin{bmatrix} G_{11}(s) \\ G_{21}(s) \end{bmatrix} = \begin{bmatrix} \dfrac{s+3}{(s+1)(s+2)} \\ \dfrac{s+4}{s+1} \end{bmatrix} = \begin{bmatrix} 0 + \dfrac{s+3}{(s+1)(s+2)} \\ 1 + \dfrac{3}{s+1} \end{bmatrix}$$

$$= \begin{bmatrix} \dfrac{s+3}{(s+1)(s+2)} \\ \dfrac{3}{s+1} \end{bmatrix} + \begin{bmatrix} 0 \\ 1 \end{bmatrix} = \begin{bmatrix} \dfrac{2}{s+1} - \dfrac{1}{s+2} \\ \dfrac{3}{s+1} \end{bmatrix} + \begin{bmatrix} 0 \\ 1 \end{bmatrix}$$

The proper rational transfer function $G(s)$ has been broken into $g(s) + D$ where $g(s)$ is the strictly proper rational transfer function, i.e.

$$G(s) = g(s) + D = C(sI - A)^{-1}B + D$$

where
$$g(s) = C(sI - A)^{-1}B$$

$$= \begin{bmatrix} \dfrac{2}{s+1} - \dfrac{1}{s+2} \\ \dfrac{3}{s+1} \end{bmatrix}$$

Hence dimensionally, let us write

$$\underset{(2 \times 1)}{G(s)} \Leftrightarrow \underset{(2 \times 2)}{C} \underset{(2 \times 2)}{[(sI-A)^{-1}]} \underset{(2 \times 1)}{B} + \underset{(2 \times 1)}{D}$$

We want to represent the system matrix A in diagonal form and in fact we have seen after partial fraction expansion that the A matrix should be of order (2×2) and the eigenvalues are -1 and -2. For the $g(s)$, we have two outputs y_1 and y_2 and one input u. Hence from $g(s)$ we can write

$$g(s) = \begin{bmatrix} g_{11}(s) \\ g_{21}(s) \end{bmatrix} = \begin{bmatrix} Y_1(s)/U(s) \\ Y_2(s)/U(s) \end{bmatrix}$$

$$= \begin{bmatrix} \dfrac{2}{s+1} - \dfrac{1}{s+2} \\ \dfrac{3}{s+1} \end{bmatrix} \qquad (12.60)$$

Comparing, we get

$$\dfrac{Y_1(s)}{U(s)} = \dfrac{2}{s+1} - \dfrac{1}{s+2}$$

or
$$Y_1(s) = 2\left[\left(\dfrac{1}{s+1}\right)U(s)\right] - 1\left[\left(\dfrac{1}{s+2}\right)U(s)\right]$$

$$= 2.Z_1(s) - 1.Z_2(s) \qquad (12.61)$$

where
$$\dfrac{Z_1(s)}{U(s)} = \dfrac{1}{s+1} \Rightarrow \dot{z}_1 = -z_1 + u$$

and
$$\dfrac{Z_2(s)}{U(s)} = \dfrac{1}{s+2} \Rightarrow \dot{z}_2 = -2z_2 + u$$

which in matrix form becomes

$$\begin{bmatrix} \dot{z}_1 \\ \dot{z}_2 \end{bmatrix} = \begin{bmatrix} -1 & 0 \\ 0 & -2 \end{bmatrix} \begin{bmatrix} z_1 \\ z_2 \end{bmatrix} + \begin{bmatrix} 1 \\ 1 \end{bmatrix} u \qquad (12.62)$$

and the output
$$\hat{y}_1 = \begin{bmatrix} 2 & -1 \end{bmatrix} \begin{bmatrix} z_1 \\ z_2 \end{bmatrix} = \begin{bmatrix} 2 & -1 \end{bmatrix} Z \qquad (12.63)$$

Similarly from Eq. (12.60), we get

$$\dfrac{Y_2(s)}{U(s)} = \dfrac{3}{s+1}$$

$$Y_2(s) = 3\left[\left(\dfrac{1}{s+1}\right)U(s)\right] + 0.\left[\left(\dfrac{1}{s+2}\right)U(s)\right] \qquad (12.64)$$

$$= 3Z_1(s) + 0.Z_2(s)$$

then the output
$$\hat{y}_2 = \begin{bmatrix} 3 & 0 \end{bmatrix} \begin{bmatrix} z_1 \\ z_2 \end{bmatrix} \qquad (12.65)$$

Combining the set of Eqs. (12.62), (12.63) and (12.65), we get for $g(s)$, the diagonal form of state variable formulation as

$$\dot{Z} = \begin{bmatrix} -1 & 0 \\ 0 & -2 \end{bmatrix} \begin{bmatrix} z_1 \\ z_2 \end{bmatrix} + \begin{bmatrix} 1 \\ 1 \end{bmatrix} u$$

and output
$$\begin{bmatrix} \hat{y}_1 \\ \hat{y}_2 \end{bmatrix} = \begin{bmatrix} 2 & -1 \\ 3 & 0 \end{bmatrix} \begin{bmatrix} z_1 \\ z_2 \end{bmatrix}$$

Hence the diagonal form of state variable representation for the given proper rational transfer function is

$$\begin{bmatrix} \dot{z}_1 \\ \dot{z}_2 \end{bmatrix} = \begin{bmatrix} -1 & 0 \\ 0 & -2 \end{bmatrix} \begin{bmatrix} z_1 \\ z_2 \end{bmatrix} + \begin{bmatrix} 1 \\ 1 \end{bmatrix} u$$

$$\begin{bmatrix} y_1 \\ y_2 \end{bmatrix} = \begin{bmatrix} \hat{y}_1 \\ \hat{y}_2 \end{bmatrix} + \begin{bmatrix} 0 \\ 1 \end{bmatrix} u = \begin{bmatrix} 2 & -1 \\ 3 & 0 \end{bmatrix} \begin{bmatrix} z_1 \\ z_2 \end{bmatrix} + \begin{bmatrix} 0 \\ 1 \end{bmatrix} u$$

EXAMPLE 12.19 A two-input-single-output system has transfer matrix as

$$\begin{bmatrix} \dfrac{1}{s^2+3s+2} & \dfrac{s+3}{s^2+3s+2} \end{bmatrix}$$

Derive the state space formulation in decoupled form. Then from that obtain the controllable canonical form of state space formulation.

Solution: (i) This is a second-order system and hence the order of the system matrix is 2×2. For a two-input system the order of B is 2×2. For a single-output system, the order of C is 1×2 and that of D is 1×2. The state vector X is of order 2×1 and the input vector U is of 2×1 and order of scalar output y is 1×1.

The characteristic equation $s^2 + 3s + 2 = 0$ yields the eigenvalues as $-1, -2$. Thus, the diagonalized state equations will have the A matrix given by

$$A = \begin{bmatrix} -1 & 0 \\ 0 & -2 \end{bmatrix}$$

The condition, $G(s) = C[sI - A]^{-1}B + D$, yields

$$\dfrac{[1\ (s+3)]}{s^2+3s+2} = [c_{11}\ c_{12}] \begin{bmatrix} \dfrac{1}{s+1} & 0 \\ 0 & \dfrac{1}{s+2} \end{bmatrix} \begin{bmatrix} b_{11} & b_{12} \\ b_{21} & b_{22} \end{bmatrix} + [d_{11}\ d_{12}]$$

Thus the elements of B, C and D must statisfy

$$1 = d_{11}s^2 + (c_{11}b_{11} + c_{12}b_{21} + 3d_{11})s + (2c_{11}b_{11} + c_{12}b_{21} + 2d_{11})$$
$$s + 3 = d_{12}s^2 + (c_{11}b_{12} + c_{12}b_{22} + 3d_{12})s + (2c_{11}b_{12} + c_{12}b_{22} + 2d_{12}).$$

Equating the like powers of s and solving the resulting equations yields

$$d_{11} = d_{12} = 0,\ c_{11}b_{11} = 1,\ c_{12}b_{21} = -1,\ c_{11}b_{12} = 2,\ c_{12}b_{22} = -1,\ d_{11} = d_{12} = 0$$

The observability condition requires, $c_{11} \neq 0$. Therefore, these results can be solved for the b_{ij} in terms of c_{11} and c_{12}. The resulting b_{ij} are all nonzero, so the controllability condition is also satisfied. The choice of c_{11} and c_{12} is arbitrary, as long as they are nonzero. For this example, we will choose $c_{11} = c_{12} = 0$, yielding

$$\dot{Z} = AZ + BU$$
$$y = CZ$$

where

$$U = [u_1 \; u_2]^T, \; Z = [z_1 \; z_2]^T, \text{ and}$$

$$A = \begin{bmatrix} -1 & 0 \\ 0 & -2 \end{bmatrix}, \quad B = \begin{bmatrix} 1 & 2 \\ -1 & -1 \end{bmatrix}, \quad C = [1 \; 1]$$

(ii) In order to get the system representation in controllable canonical form, the system matrix should be in companion form. The eigenvalues are -1 and -2. Hence the modal matrix simply becomes as

$$M = \begin{bmatrix} 1 & 1 \\ -1 & -2 \end{bmatrix}$$

As a final step, we transform the diagonalized system by using a coordinate transformation that yields the controllable canonical form of system representation as

$$\dot{X} = \overline{A}X + \overline{B}U$$
$$y = \overline{C}X$$

where, $\overline{A} = MAM^{-1} = \begin{bmatrix} 0 & 1 \\ -2 & -3 \end{bmatrix}$; $\overline{B} = MB = \begin{bmatrix} 0 & 1 \\ 1 & 0 \end{bmatrix}$; and $\overline{C} = CM^{-1} = [1 \; 0]$

EXAMPLE 12.20 For the system (single-input-two-output) represented by the transfer function matrix

$$\begin{bmatrix} Y_1(s) \\ Y_2(s) \end{bmatrix} = \begin{bmatrix} \dfrac{3s^2 - s - 3}{s^2 + (1/3)s - (2/3)} \\ \dfrac{6}{s+1} \end{bmatrix} U(s)$$

write the state variable formulation in controllable form and simulate the same.

Solution: Given: $G_{11}(s) = \dfrac{Y_1(s)}{U(s)} = \dfrac{3s^2 - s - 3}{s^2 + (1/3)s - (2/3)} = 3 + \dfrac{-2s - 1}{s^2 + (1/3)s - (2/3)}$

$$= 3 + \frac{-2s^{-1} - s^{-2}}{1 + (1/3)s^{-1} - (2/3)s^{-2}}$$

and
$$G_{21}(s) = \frac{Y_2(s)}{U(s)} = \frac{6}{s+1} = \frac{6(s-(2/3))}{(s+1)(s-(2/3))} = \frac{6s-4}{s^2 + (1/3)s - (2/3)}$$

$$= \frac{6s^{-1} - 4s^{-2}}{1 + (1/3)s^{-1} - (2/3)s^{-2}}$$

The controllable canonical form of the state variable formulation becomes

$$\begin{bmatrix} \dot{x}_1 \\ \dot{x}_2 \end{bmatrix} = \begin{bmatrix} 0 & 1 \\ (2/3) & -(1/3) \end{bmatrix} \begin{bmatrix} x_1 \\ x_2 \end{bmatrix} + \begin{bmatrix} 0 \\ 1 \end{bmatrix} u$$

and output
$$\begin{bmatrix} y_1 \\ y_2 \end{bmatrix} = \begin{bmatrix} -1 & -2 \\ -4 & 6 \end{bmatrix} \begin{bmatrix} x_1 \\ x_2 \end{bmatrix} + \begin{bmatrix} 3 \\ 0 \end{bmatrix} u$$

EXAMPLE 12.21 The transfer function matrix of a three-input-single-output system is

$$Y(s) = \begin{bmatrix} \dfrac{s^2 + 2s + 2}{s^2 + (1/2)s - (1/2)} & \dfrac{-s}{s-(1/2)} & \dfrac{4}{s^2 + (1/2)s - (1/2)} \end{bmatrix} \begin{bmatrix} U_1(s) \\ U_2(s) \\ U_3(s) \end{bmatrix}$$

Draw the SFG and write the state variable formulation in OCF.

Solution: $\quad G_{11}(s) = \dfrac{Y(s)}{U_1(s)} = \dfrac{s^2 + 2s + 2}{s^2 + (1/2)s - (1/2)} = 1 + \dfrac{(3/2)s + (5/2)}{s^2 + (s/2) - (1/2)}$

$$G_{12}(s) = \frac{Y(s)}{U_2(s)} = \frac{-s}{s-(1/2)} = \frac{-s(s+1)}{(s-1/2)(s+1)} = 1 + \frac{-(1/2)s + (5/2)}{s^2 + (1/2)s - (1/2)}$$

and $\quad G_{13}(s) = \dfrac{4}{s^2 + (1/2)s - (1/2)}$

The state variable representation can be written as

$$\begin{bmatrix} \dot{x}_1 \\ \dot{x}_2 \end{bmatrix} = \begin{bmatrix} (-1/2) & 1 \\ (1/2) & 0 \end{bmatrix} \begin{bmatrix} x_1 \\ x_2 \end{bmatrix} + \begin{bmatrix} (3/2) & (-1/2) & 0 \\ (5/2) & (-1/2) & 4 \end{bmatrix} \begin{bmatrix} u_1 \\ u_2 \\ u_3 \end{bmatrix}$$

and output
$$y(t) = [1\ 0]\begin{bmatrix}x_1\\x_2\end{bmatrix} + [1\ -1\ 0]\begin{bmatrix}u_1\\u_2\\u_3\end{bmatrix}$$

It may be noted that the observable form is especially convenient for the synthesis of multiple-input-single-output systems and the controllable form is convenient for single-input-multiple-output systems.

12.19 State Space Representation in Canonical Form

The choice of state variable is not unique. Many techniques are available for obtaining state space representations. We will show different methods for obtaining state variable representation in the controllable, observable, diagonal or Jordan canonical forms from the same transfer function. There are some standard forms, known as canonical realizations, because these minimize the number of components required in simulation. Consider the rational transfer function of the form

$$G(s) = \frac{Y(s)}{U(s)} = \frac{b_n s^n + b_{n-1} s^{n-1} + \ldots + b_1 s + b_0}{s^n + a_{n-1} s^{n-1} + \ldots + a_1 s + a_0} \tag{12.66}$$

A transfer function is said to be proper when $b_n \neq 0$ and strictly proper when $b_n = 0$ and improper when the degree of the numerator polynomial is greater than that of the denominator polynomial.

Some methods of state space representations in canonical forms are:

1. Direct method for controllable canonical form (CCF) of representation [S1] as in Eqs. (12.70) and (12.71).
2. Nested method for observable canonical form (OCF) of representation [S2] as in Eqs. (12.75) and (12.76).
3. Bush procedure for controllable canonical form (CCF) of representation [S1] as in Eqs. (12.77) and (12.78).
4. Cascaded decomposition method.
5. Partial fraction expansion method—Diagonal or Jordan canonical form.

Direct method for the controllable canonical form

Equation (12.66) can rewritten as

$$G(s) = b_n + g(s) \tag{12.67}$$

where
$$g(s) = \frac{b'_{n-1} s^{n-1} + b'_{n-2} s^{n-2} + \cdots + b'_1 s + b'_0}{s^n + a_{n-1} s^{n-1} + \cdots + a_1 s + a_0} \tag{12.68}$$

and
$$b'_i = (b_i - a_i b_n)\ ;\quad i = 0, 1, \ldots, n-1 \tag{12.69}$$

The system dynamics can be written as

$$\dot{X} = \begin{bmatrix} 0 & 1 & 0 & \cdots & 0 \\ 0 & 0 & 1 & \cdots & 0 \\ \vdots & \vdots & \vdots & & \vdots \\ 0 & 0 & 0 & \cdots & 0 \\ -a_0 & -a_1 & -a_2 & \cdots & -a_{n-1} \end{bmatrix} X + \begin{bmatrix} 0 \\ 0 \\ \vdots \\ 0 \\ 1 \end{bmatrix} U \qquad (12.70)$$

and output
$$y = [\,b_0'\quad b_1'\quad \cdots \quad b_{n-1}'\,]X + b_n u \qquad (12.71)$$

Modification When $b_n = 0$, the transfer function becomes strictly proper. Then the state equation (AB) remains the same and in the output equation, $D = 0$ and C remains the same and $c_i = b_i$ for $i = 0, 1, \ldots, n-1$.

Nested method for the observable canonical form

Consider the rational transfer function of Eq. (12.66) which can be modified into the following form

$$s^n[Y(s) - b_n U(s)] + s^{n-1}[a_{n-1}Y(s) - b_{n-1}U(s)] + \cdots$$
$$+ s[a_1 Y(s) - b_1 U(s)] + s^0[a_0 Y(s) - b_0 U(s)] = 0 \qquad (12.72)$$

Dividing by s^n and rearranging, we get

$$Y(s) = b_n U(s) + \frac{1}{s}\left[(b_{n-1}U(s) - a_{n-1}Y(s)) + \frac{1}{s}[b_{n-1}U(s) - a_{n-2}Y(s)]\right] + \cdots$$
$$+ \frac{1}{s^{n-1}}\left[(b_1 U(s) - a_1 Y(s) + \frac{1}{s}[b_0 U(s) - a_0 Y(s)]\right] \qquad (12.73)$$

Rewriting this as
$$Y(s) = b_n U(s) + X_n(s) \qquad (12.74)$$

with
$$X_n(s) = \frac{1}{s}[b_{n-1}U(s) - a_{n-1}Y(s) + X_{n-1}(s)]$$

$$X_{n-1}(s) = \frac{1}{s}[b_{n-2}U(s) - a_{n-2}Y(s) + X_{n-2}(s)]$$

$$\cdots\cdots\cdots\cdots\cdots\cdots\cdots\cdots\cdots\cdots\cdots$$

$$X_2(s) = \frac{1}{s}[b_1 U(s) - a_1 Y(s) + X_1(s)]$$

$$X_1(s) = \frac{1}{s}[b_0 U(s) - a_0 Y(s)]$$

Taking inverse Laplace transform of the preceding equations and writing them in reverse order, we get the state and output equations in the standard vector-matrix differential equation form as

$$\frac{d}{dt}\begin{bmatrix} x_1 \\ x_2 \\ \vdots \\ x_n \end{bmatrix} = \begin{bmatrix} 0 & 0 & \cdots & 0 & -a_0 \\ 1 & 0 & \cdots & 0 & -a_1 \\ \vdots & \vdots & \vdots & \vdots & \vdots \\ 0 & 0 & \cdots & 1 & -a_{n-1} \end{bmatrix}\begin{bmatrix} x_1 \\ x_2 \\ \vdots \\ x_n \end{bmatrix} + \begin{bmatrix} b_0 & -a_0 b_n \\ b_1 & -a_1 b_n \\ \vdots & \vdots \\ b_{n-1} & -a_{n-1} b_n \end{bmatrix} \quad (12.75)$$

and output

$$y = \begin{bmatrix} 0 & 0 & \cdots & 0 & 1 \end{bmatrix}\begin{bmatrix} x_1 \\ x_2 \\ \vdots \\ x_n \end{bmatrix} + b_n u \quad (12.76)$$

It may be noted that $A^* = A^T$, $B^* = C^T$ and $C^* = B^T$.

The state space representation given by Eqs. (12.75) and (12.76) is said to be in the nested method of observable canonical form (OCF) of representation [S2]. Note that the controllable canonical form and the observable canonical form are dual to each other.

Further, consider the strictly proper transfer function

$$G(s) = C[sI - A]^{-1}B \implies [S1]$$
$$[G(s)]^T = [C[sI - A]^{-1}B]^T = B^T[C[sI - A]^{-1}]^T = B^T[sI - A^T]^{-1}C^T \implies [S2]$$

Now the system [S2] having transfer function $[G(s)]^T$ will lead to the system dynamics in state variable form as

$$\dot{X} = A^T X + C^T u$$
$$y = B^T X$$

and is dual to the original system [S1] having transfer function $G(s)$ with system dynamics as

$$\dot{X} = AX + Bu$$
$$y = CX$$

The systems [S1] and [S2] are dual to each other. It may be noted that the matrices [A, B, C] are not the transpose but truly should be conjugate transpose instead of transpose because of the system parameters being complex.

Hence these state variable representations are dual to each other, i.e. the controllability and observability matrix for both the systems are dual to each other.

This means that the controllability matrix of one representation becomes the observability matrix of the other representation and vice versa. Hence if one is familiar with any one form of state variable formulation, then the other form of representation can be obtained easily by using the duality property.

Bush procedure for controllable canonical form

Let us put the states in the reverse order in the controllable canonical form. Just to differentiate from the Direct Method, we choose a new set of state variables z for the Bush procedure.

Let us rename the states as

$$\dot{z}_1 = \dot{x}_n$$
$$\dot{z}_2 = z_1 = \dot{x}_{n-1} = x_n$$
$$\cdots \cdots \cdots \cdots$$
$$\dot{z}_n = z_{n-1} = \dot{x}_1 = \dot{x}_2$$
$$z_n = x_1$$

Rewriting in terms of z_i's, we get

$$\dot{z}_1 = -a_{n-1}z_1 - a_{n-2}z_2 - \cdots - a_1z_{n-1} - a_0z_n + u$$
$$\dot{z}_2 = z_1$$
$$\dot{z}_3 = z_2$$
$$\cdots \cdots$$
$$\dot{z}_n = z_{n-1}$$

and output
$$y = (b_{n-1} - a_{n-1}b_n)z_1 + (b_{n-2} - a_{n-2}b_n)z_2$$
$$+ \cdots + (b_1 - a_1b_n)z_{n-1} + (b_0 - a_0b_n)z_n + b_nu$$
$$= b'_{n-1}z_1 + b'_{n-2}z_2 + \cdots + b'_1z_{n-1} + b'_0z_n + b_nu$$

where
$$b'_i = b_i - a_ib_n \; ; \; i = 0, 1, \ldots, n-1$$

Writing in matrix form, we get the controllable canonical form CCF following the Bush procedure as

$$\frac{d}{dt}\begin{bmatrix} z_1 \\ z_2 \\ z_3 \\ \vdots \\ z_n \end{bmatrix} = \begin{bmatrix} -a_{n-1} & -a_{n-2} & \cdots & -a_1 & -a_0 \\ 1 & 0 & \cdots & 0 & 0 \\ 0 & 1 & \cdots & 0 & 0 \\ \vdots & \vdots & & \vdots & \vdots \\ 0 & 0 & \cdots & 1 & 0 \end{bmatrix}\begin{bmatrix} z_1 \\ z_2 \\ z_3 \\ \vdots \\ z_n \end{bmatrix} + \begin{bmatrix} 1 \\ 0 \\ 0 \\ \vdots \\ 0 \end{bmatrix}u \quad (12.77)$$

$$y = [b'_{n-1} \; b'_{n-2} \; \cdots \; b'_1 \; b'_0]\begin{bmatrix} z_1 \\ \vdots \\ z_n \end{bmatrix} + b_nu \quad (12.78)$$

Moreover, the duality that we noted earlier for the four canonical realizations has a more general aspect. For any given realization $[A, B, C]$ of a system transfer function, there is a corresponding dual realization $\{A^T, C^T, B^T\}$. Dual realizations will be helpful in saving labour and in suggesting new investigations.

A mnemonic feature to note here is that in the controller and controllability forms, the input either enters each integrator directly or by way of some previous integration (but not other linear operations). Parallel statements may be made for the observer and observability forms if we interchange the roles of input and output.

Parallel and cascade realizations

We have discussed four different forms of simulation of the system described by its input-output relationship. A question may arise as to which form is numerically least sensitive to small changes in parameter values. Such a question is addressed in more specialized books where it is shown that a new form, the so-called sum or parallel or diagonal form, is perhaps the best from the sensitivity point of view. Certain related product or cascade forms are also important in this regard.

Sum or parallel realizations

If we make a partial fraction expansion

$$G(s) = \frac{Y(s)}{U(s)} = \sum_1^n \frac{g_i}{s - \lambda_i}$$

each term on the right can be easily realized and $G(s)$ is then obtained as a parallel combination of these elementary realizations. This is sometimes called a diagonal form representation. Jordan form and the Jordan canonical form have already been discussed is detail.

It may happen that some roots are repeated, then, as in the following example, for the given transfer function,

$$\frac{Y(s)}{U(s)} = \frac{2s^2 + 6s + 5}{s^3 + 4s^2 + 5s + 2} = \frac{1}{(s+1)^2} + \frac{1}{s+1} + \frac{1}{s+2}$$

This is not hard to handle. For complex roots, such as

$$\frac{Y(s)}{U(s)} = \frac{2s^2 + 5s + 7}{s^3 + 3s^2 + 75 + 5} = \frac{\left(1 + \frac{j}{2}\right)}{s + 1 + 2j} + \frac{\left(1 + \frac{j}{2}\right)}{s + 1 - 2j} + \frac{1}{s+1}$$

it is to be noted that this cannot be simulated term by term using real components. The problem can be overcome by working with second-order building blocks. Since $Y(s)/U(s)$ has real coefficients, the complex roots will always occur in conjugate pairs, so we can regroup $Y(s)/U(s)$ as a combination of first-order and second-order terms as

$$\frac{Y(s)}{U(s)} = \frac{2s^2 + 5s + 7}{s^3 + 3s^2 + 7s + 5} = \frac{s+2}{s^2 + 2s + 5} + \frac{1}{s+1}$$

The second-order terms can be simulated in several way.

Product or cascade decomposition

Cascade decomposition may be applied to a transfer function which is in factored form as illustrated by the following example.

Consider the example with

$$G(s) = \frac{s^2 + 6s + 8}{s^3 + 5s^2 + 8s + 6} = \frac{(s+2)(s+4)}{(s+3)(s^2+2s+2)} = g_1(s)g_2(s)$$

where

$$g_1(s) = \frac{s+2}{s+3} = \frac{1+2s^{-1}}{1+3s^{-1}}$$

and

$$g_2(s) = \frac{s+4}{s^2+2s+2} = \frac{s^{-1}+4s^{-2}}{1+2s^{-1}+2s^{-2}}$$

The complete SFG is shown in Figure 12.15. From the SFG, it is observed that

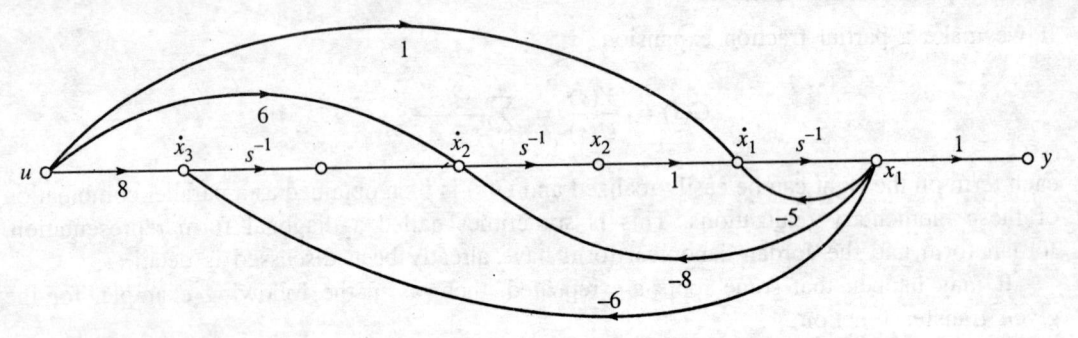

FIGURE 12.15 Observable form of state variable representation under cross-resolution.

$$\dot{x}_1 = x_2$$
$$\dot{x}_2 = 2x_3 + \dot{x}_3 - 2x_2 - 2x_1 = 2x_3 - 3x_3 + u(t) - 2x_1 - 2x_2$$
$$= -2x_1 - 2x_2 - x_3 + u(t)$$
$$\dot{x}_3 = u(t) - 3x_3$$

which in matrix form can be written as

$$\dot{X} = \begin{bmatrix} 0 & 1 & 0 \\ -2 & -2 & -1 \\ 0 & 0 & -3 \end{bmatrix} X + \begin{bmatrix} 0 \\ 1 \\ 1 \end{bmatrix} u(t)$$

and output

$$y(t) = 4x_1 + x_2 = [4 \quad 1 \quad 0]X$$

Drill Problem 12.2

Given the transfer function as

$$\frac{Y(s)}{U(s)} = \frac{A_1}{s+\lambda_1} + \frac{A_2 s + A_3}{s^2 + \alpha s + \beta}$$

Draw the SFG with parallel form of realization. Then write the state variable formulation.

Summary

We have discussed the concept of controllability and observability which is of paramount importance in state variable design. An understanding of *observer* design has been provided and the condition under which the design is possible has been explained. Pole placement is the tool for proper design for which all states should be accessible but this not so in most of the practical and hazardous conditions. So one has to go for observer design to generate the inaccessible states, and that is what constitutes the observer design. Synthesis has been made by realization of state variable formulation from the given transfer function or from the transfer matrix of multivariable systems. Simulation for different forms of state variable representations has been discussed.

Problems

12.1 The vector-matrix differential equation of a system is described by

$$\dot{X} = AX + Bu$$

where

(a) $A = \begin{bmatrix} -2 & 0 \\ 0 & -1 \end{bmatrix}$, $B = \begin{bmatrix} 0 \\ 1 \end{bmatrix}$

(b) $A = \begin{bmatrix} -1 & 1 & 0 \\ 0 & -2 & 0 \\ 0 & -2 & -3 \end{bmatrix}$, $B = \begin{bmatrix} 1 \\ 2 \\ 3 \end{bmatrix}$

(c) $A = \begin{bmatrix} -2 & 1 & 0 \\ 0 & -2 & 0 \\ -1 & -2 & -3 \end{bmatrix}$, $B = \begin{bmatrix} 1 \\ 0 \\ 1 \end{bmatrix}$

(d) $A = \begin{bmatrix} 1 & 2 \\ 1 & 1 \end{bmatrix}$, $B = \begin{bmatrix} 2 \\ \sqrt{2} \end{bmatrix}$

Can the state equations be transformed into controllable canonical form (CCF)? Justify your answer.

12.2 Consider the system

$$\dot{X} = \begin{bmatrix} -1 & 0 & 1 \\ 1 & -2 & 0 \\ 0 & 0 & -3 \end{bmatrix} X + \begin{bmatrix} 0 \\ 0 \\ 1 \end{bmatrix} u$$

and output $\quad y = [1 \ 1 \ 0]X$

Transform the system into (a) controllable canonical form and (b) observable canonical form.

12.3 Consider the system

$$\dot{X} = \begin{bmatrix} -1 & 0 & 1 \\ 1 & -2 & 0 \\ 0 & 0 & -3 \end{bmatrix} X + \begin{bmatrix} 0 \\ 1 \\ 1 \end{bmatrix} u$$

and output $\quad y = [1 \ 1 \ 1]X$

Transform the system into observable canonical form.

12.4 Consider the system

$$\dot{X} = \begin{bmatrix} 0 & 1 & 0 \\ 0 & 0 & 1 \\ -1 & -5 & -6 \end{bmatrix} X + \begin{bmatrix} 0 \\ 1 \\ 1 \end{bmatrix} u$$

By using state-feedback control $u = -KX$, it is desired to have the closed-loop poles at $s = -2 \pm j4$ and $s = -10$. Determine the state feedback gain matrix K.

12.5 Consider the system

$$\dot{X} = \begin{bmatrix} 0 & 1 \\ 0 & 2 \end{bmatrix} X + \begin{bmatrix} 1 \\ 0 \end{bmatrix} u$$

By using state-feedback control $u = -KX$, show that the system cannot be stabilized.

12.6 The dynamics of the system is

$$\dot{X} = \begin{bmatrix} 0 & 1 & 0 \\ 0 & 0 & 1 \\ -6 & -11 & -6 \end{bmatrix} X + \begin{bmatrix} 0 \\ 0 \\ 1 \end{bmatrix} u$$

and output $\quad y = [1 \ 0 \ 0]X$

By using state-feedback control $u = -KX$, it is desired to have the closed-loop poles at $s = -2 \pm j2\sqrt{3}$ and $s = -10$. Determine the state feedback gain matrix K. Verify the result using MATLAB.

12.7 Consider the type-1 servo system described by

$$\dot{X} = \begin{bmatrix} 0 & 1 & 0 \\ 0 & 0 & 1 \\ 0 & -5 & -6 \end{bmatrix} X + \begin{bmatrix} 0 \\ 0 \\ 1 \end{bmatrix} u$$

and output $\quad y = [1 \ 0 \ 0]X$

Determine the feedback gain constants k_1, k_2, k_3 such that the closed-loop poles are located at $s = -2 \pm j4$ and $s = -10$. Obtain the unit-step response and plot the output.

12.8 Write the dynamics of the inverted-pendulum system shown in Figure P.12.8 assuming $M = 2$ kg, $m = 0.5$ kg and $l = 1$ m with the choice of phase variable as state variable. Derive the state space formulation of the system. Determine the state-feedback gain matrix K for the closed-loop system having poles at $-4 \pm j4$ and double poles at -20. Further obtain the state response for the initial condition

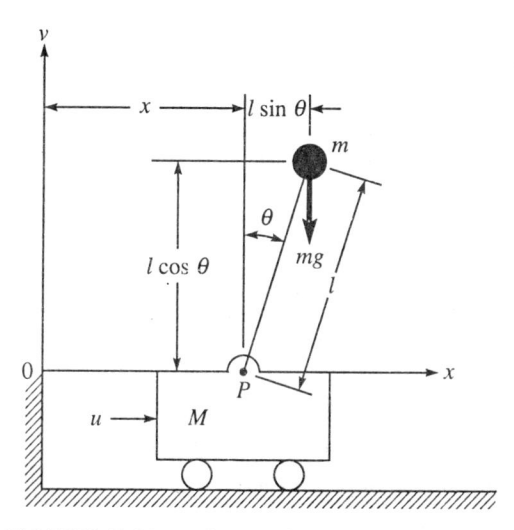

$$X(0) = [0 \ 0 \ 0 \ 1]^T$$

FIGURE P.12.8 Inverted-pendulum system.

12.9 Consider the system defined by

$$\dot{X} = \begin{bmatrix} -1 & 1 \\ 1 & -2 \end{bmatrix} X$$

$$y = [1 \quad 0]X$$

Design a state observer so that the closed-loop system has double poles at -5.

12.10 Consider the system defined by

$$\dot{X} = \begin{bmatrix} 0 & 1 & 0 \\ 0 & 0 & 1 \\ -6 & -11 & -6 \end{bmatrix} X + \begin{bmatrix} 0 \\ 0 \\ 1 \end{bmatrix} u$$

and $\quad y = [1 \quad 0 \quad 0]X$

Design a regular system by the pole-placement method with observer approach assuming that the closed-loop poles are at $-1 \pm j1$ and $s = -5$. The desired observer poles are all located at -6. Also obtain the transfer function of the observer controller.

12.11 Derive the controllability condition of LTI system, show that the controllability and observability conditions are dual to each other.

Digital Control Systems

OBJECTIVE

The advances made in microprocessors, microcomputers and digital signal processors have accelerated the growth of digital control systems theory. Although most contemporary control systems encountered in the industry employ digital control, the fact is that most of the physical processes to be controlled are still analog systems. Therefore, in the earlier part of this book we discussed in detail the continuous-data system theory before the digital control theory in this chapter. Whatever topics we covered in continuous control systems, we have tried to cover almost all of those topics in this chapter on digital control systems to facilitate smooth understanding of the subject keeping in mind the limitation of the volume of the book.

CHAPTER OUTLINE

Introduction
Discrete-Time Systems
Sampled-Data and Digital Control System
Sample-and-Hold Device
z-Transform
Pulse Transfer Function.
Relationship between s-plane and z-plane Poles
Jury's Stability Test
Analysis of Digital Control System
Discrete System Time Response
Steady-State Error
Root Loci
Nyquist Criterion
State-Variable Formulation
Simulation
Solutions to State Equation
Model of National Economy
Direct Decomposition (Phase Variable Form)
Continuous and Discrete State Equation

13.1 Introduction

The growth of digital control systems has been accelerated over the past three decades due to the advent of the microprocessor which has made possible the use of digital computer as a controller. The microprocessor chip can be used as a dedicated digital compensator, which could not be dreamt of three decades ago.

In contrast to the continuous-time systems, the operation of discrete-time systems is described by a set of difference equations. The transform method employed in the analysis of linear time-invariant continuous-time systems is the Laplace transform. In a similar manner, the transform used in the analysis of linear time-invariant discrete-time systems is the z-transform. All the techniques used for the analysis of the continuous time-invariant system where dynamics is expressed in differential equations will now be discussed in case of the discrete system where the system dynamics is expressed in terms of difference equations.

13.2 Discrete-Time Systems

To illustrate the idea of a discrete-time system, for better understanding, the classical linear time-invariant continuous-time feedback control system as in Figure 13.1(a) is shown side-by-side, the digital control system as shown in Figure 13.1(b). The digital computer performs the compensation (or controller) function within the system. The

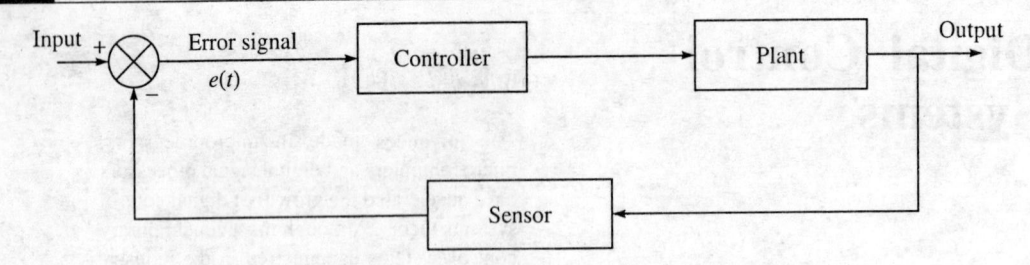

FIGURE 13.1(a) Continuous-time feedback control system.

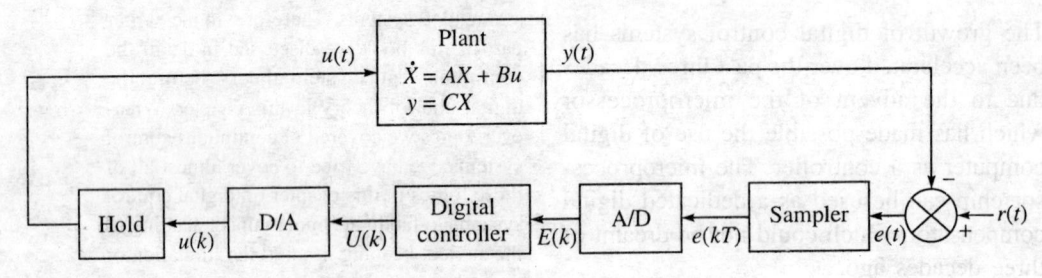

FIGURE 13.1(b) Digital control system.

interface at the input of the computer is an analog-to-digital (A/D) converter, and is required to convert the error signal, which is a continuous-time signal, into a form that can be readily processed by the computer. At the computer output, a digital-to-analog (D/A) converter is required to convert the binary signal of the computer into a form necessary to drive the plant.

The input and output of the digital computer are in digital form whereas those of the plant are in analog form. Hence the data is converted from analog-to-digital form by an A/D converter and from digital-to-analog form by a D/A converter. The digital computer is used to perform the control action (may be PID control or any other compensation) and is termed *digital controller*. Computers are interconnected to the actuator and process data by means of signal converters.

The error signal $e(t)$ is continuous and it cannot be fed in that form to the digital computer. For that we need a *sampler* which will give the value of the continuous signal $e(t)$ at discrete instants of time of sampling T. The sampled signal $e(kT)$ is the impulse of strength as that of the continuous signal at the kth discrete sampling instant of time and the sampled signal is zero in between the sampling instants. Then a data hold is employed. Reconstructing the original signal using zero-order hold (as explained later) and increasing the sampling rate minimizes the error in data reconstruction. A compromise in sampling rate has to be maintained in order to optimize the computational requirement. Otherwise the circuit complexity and realization cost would be enormous if a first or higher-order hold circuit is employed. In practice, the A/D converter itself contains the sample and hold devices. In the digital control system of Figure 13.2(a), the A/D converter, the digital computer, and the D/A converter replace an analog or continuous-time, proportional-integral (PI) controller, such that the digital control system response has essentially the same characteristic as that of the analog system. Since the digital computer can be programmed to multiply, add, and integrate numerically, the controller equation can be realized using a digital computer with the rectangular rule of numerical integration as illustrated in Figure 13.2(b). Of course, the other algorithms of numerical integration may be adopted as well.

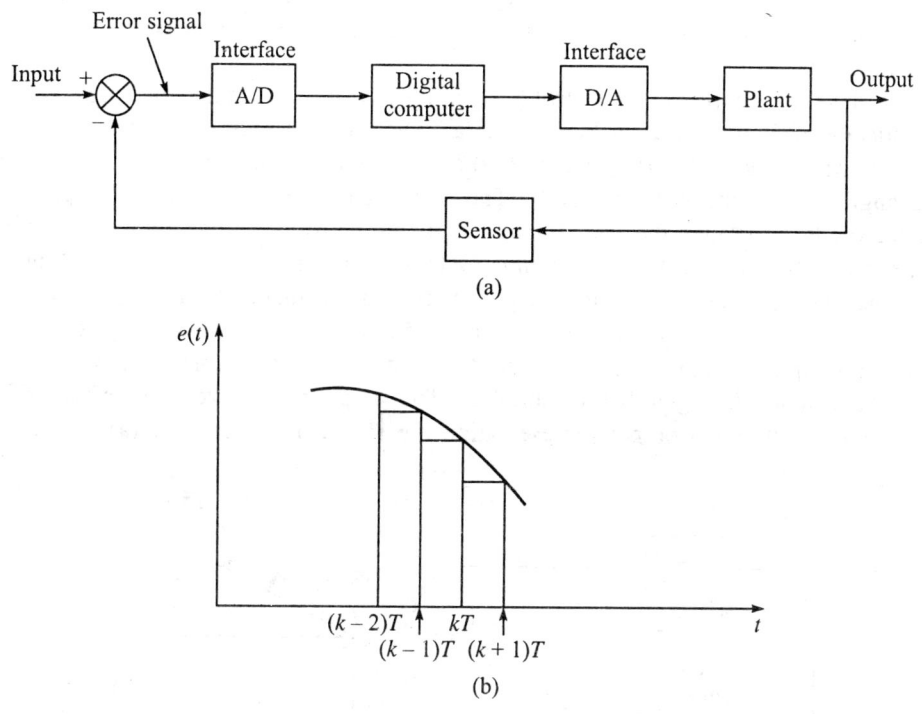

FIGURE 13.2 (a) Block diagram of digital control system and (b) rectangular rule of numerical integration.

The sampling frequency is $1/T$, where T is the sampling period. We sample continuous signal to sampled signal, which is a string of impulses starting at $t = 0$ spaced at T seconds and of amplitude $e(kT)$. We hold the amplitude constant at $e(kT)$ during the following T seconds, i.e. from time interval kT to $(k+1)T$ instant of time, the amplitude remains constant at the previous value $e(kT)$ and this is called a *zero-order hold* (ZOH). During the hold time T, the digital controller can do the processing.

13.3 Sampled-Data and Digital Control System

With this background, we can straightaway go to the mathematical foundation of discrete control systems. Before that, let us broadly introduce the sampled-data control system and the digital control system.

The terms *sampled-data control systems*, *discrete-data control systems*, and *digital-control systems* have all been used loosely and interchangeably in the control system literature and broadly classified as discrete data systems. Strictly speaking, sampled data are pulse-amplitude modulated signals and are obtained by some means of sampling an analog signal. A pulse-amplitude modulated signal is often presented in pulse train form with signal information contained in the amplitude of the pulses. Digital signals are generated by digital transducers or digital computer; often in digitally coded form. The discrete data systems, in broad sense, describe all systems having some form of digital or sampled signals. Furthermore, the existing analytical and design methods are essentially same whether the system contains sampled or digitally coded data.

To introduce the sampled-data control system, let us consider a radar tracking system used to automatically track aircraft as in Figure 13.3(a). The block diagram of such a radar tracking system is shown in Figure 13.3(b), where $\theta_A(t)$ and $\theta_R(t)$ are the angles to the aircraft and the pointing angle of the antenna respectively. The tracking error is $e(t) = \theta_A(t) - \theta_R(t)$. As we will use the digital controller, the error signal $e(t)$ should be followed by sampler and then hold as in Figure 13.3(b). The plant or the antenna system is analog, so only sampler output cannot be fed to the plant because of the high frequency component inherently present in that signal. Therefore, a data-reconstruction device called a *data hold* is inserted into the system directly following the sampler. The purpose of data hold is to reconstruct the sampled signal into a form that closely resembles the signal before sampling. The simplest one is zero-order hold. Sampler and data hold and its input-output representation are shown in Figure 13.4(a).

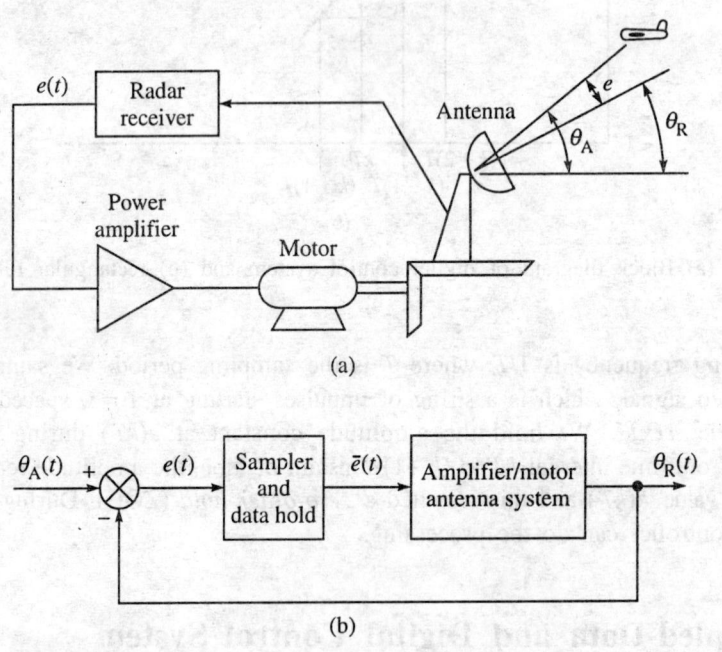

FIGURE 13.3 (a) Radar tracking system and (b) block diagram of radar tracking system.

The digital control system uses digital signals and a digital computer to control a process. The measurement data are converted from analog form to digital form by means of the A/D converter. After processing the inputs, the digital computer provides an output in digital form. This output is then converted to analog form by D/A converter. Digital control systems are used in many applications: machine tools, metal working processes, chemical processes, aircraft control and automobile traffic control.

Digital Control Systems

The signal $\bar{e}(t)$ of Figure 13.4(b) can be expressed as

$$\bar{e}(t) = e(0)[u(t) - u(t - T)] + e(T)[(u(t - T) - u(t - 2T)] + e(2T)[u(t - 2T) - (t - 3T)] + \cdots$$

The Laplace transform of $\bar{e}(t)$ is

$$\bar{E}(s) = e(0)\left[\frac{1}{s} - \frac{e^{-Ts}}{s}\right] + e(T)\left[\frac{e^{-Ts}}{s} - \frac{e^{-2Ts}}{s}\right] + e(2T)\left[\frac{e^{-2Ts}}{s} - \frac{e^{-3Ts}}{s}\right] + \cdots$$

$$= \left[\frac{1 - e^{-Ts}}{s}\right][e(0) + e(T)e^{-Ts} + e(2T)e^{-2Ts} + \cdots]$$

$$= \left[\sum_{n=0}^{\infty} e(nT)e^{-nTs}\right]\left[\frac{1 - e^{-Ts}}{s}\right] = E^*(s)\left[\frac{1 - e^{-Ts}}{s}\right]$$

where $E^*(s)$ is called the starred transformer and is defined as

$$E^*(s) = \left[\sum_{n=0}^{\infty} e(nT)e^{-nTs}\right]$$

and is represented in Figure 13.4(d).

We can proceed directly from the plant transfer function $G_p(s)$ to the pulse transfer function $G(z)$ of the hold, plant and sampler combined as in Figure 13.4(d). In the figure, $G_p(s)$ represents the combined transfer function of the plant and ZOH circuit. The fundamental step

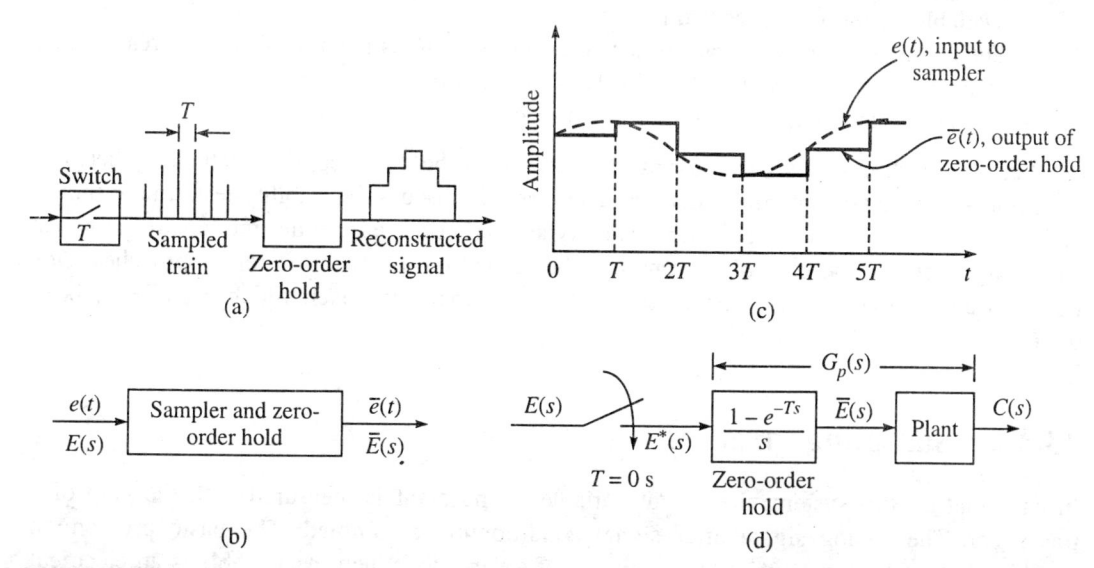

FIGURE 13.4 (a) Sampling a continuous signal, (b) sampler and data hold, (c) input and output signals of sampler/data hold, and (d) a plant with sampler and hold.

in the analysis of pulse transfer function is that we consider the combination of the sampler, hold and plant as a discrete system in its own right. In Figure 13.1(b), the input samples (from the controller algorithm) are processed to give output samples (from the A/D converter). So once again, providing the plant to be modeled as a linear system, the sampler, hold and plant as a whole can be characterized by a linear difference equation of a transfer function in z. This will not give information about the output between the sampling instants.

It should be noted that $E^*(s)$ is not present in the physical system, but appears in the mathematics. The sampler in Figure 13.4(d) does not model the physical sampler and nor does the block model a physical data hold, but the combination does accurately model a physical sampler/data-hold device.

13.3.1 Digital controller vs analog controller

The merits of the digital controller versus the analong controller are enumerated below:

- Digital controllers can perform complex computations with constant accuracy at high speed and with relatively little increase in cost whereas in an analog controller it is not so. Digital control gives improved measurement sensitivity by using digitally-coded signals. Resolution can be enhanced. Digital control has reduced sensitivity to signal noise.
- The use of digitally-coded signals permits the wide application of digital devices and communications.
- Many systems are inherently digital because they send out pulse signals as in the case of a radar tracking system and a space satellite.
- A digital controller is a versatile and flexible device, whereas the same flexibility is not available in an analog controller.
- The mathematics used in case of discrete control systems is z-transform, whereas in case of analog control systems it is the Laplace transform.
- Choice of sampling time may be a cause of problem.

Let us now discuss the z-transforms, the sample and hold device, the relationship between the s-domain and z-domain configurations. Then we will discuss the stability test, time response and steady-state error, root-locus technique, Nyquist criterion, etc. in discrete control systems. The analysis of discrete control systems may be carried out by two different approaches. One is the z-transform approach, and the other is the state space approach which has already been discussed.

13.3.2 Sampling process

In an actual control system, the output variable of the plant is measured with the help of a transducer. The analog signal after signal conditioning is sampled. The basic problem in sampling is the measurement of the amplitude of a variable. When the variable is an electrical signal, the measurement becomes easy. Figure 13.4(a) shows a sampling switch operating at a fixed sampling rate to sample the continuous input signal. The samples are shown as vertical lines and are interpreted as impulses. Almost all physical plants are designed for continuous input and output signals; it is not practical to use impulses as inputs and the interval between the samples is therefore filled with a continuous signal which is generated by a hold circuit. The simplest hold circuit, i.e. ZOH, simply maintains its output at the level of the preceding sample until the next sample is taken as shown in Figure 13.4(a). There are other types of higher-order hold circuits too, such as first-order or higher-order hold, which are more complex in design and adversely affect stability by introducing additional time delay. The time delay introduced by ZOH is approximately one-half the sampling period. As per Shannon's theorem the minimum sampling rate should be at least twice the frequency of the input signal to be sampled.

The resolution of a D/A converter is the smallest change in its output usually expressed as a percentage of full scale. For an n-bit converter, the resolution is $\dfrac{100}{2^n}$ %. Resolution is often specified as dynamic range, i.e.

$$\text{Dynamic range in dB} = 20 \log_{10} 2^n = 6n \text{ dB}$$

EXAMPLE 13.1 Find the resolution and dynamic range of a 12-bit D/A converter?

Solution:

$$\text{Resolution} = \frac{100}{2^{12}} \% = 0.024\%$$

$$\text{Dynamic range} = 20 \log_{10} 2^{12} = 72 \text{ dB}$$

13.4 Sample-and-Hold Device

The role of z-transform in the analysis and design of sampled data systems is similar to that of the Laplace transform in continuous time systems. Discrete time functions arise when continuous time signals are sampled. So we will now discuss sampler and holding devices. A sampler converts a continuous time signal into a train of pulses occurring at the sampling instants 0, T, $2T$, ..., where T is the sampling period. It may be noted that in between the sampling instants the sampler does not transmit any information.

A holding device converts the sampled signal into a continuous signal, which approximately reproduces the signal applied to the sampler. The holding device converts the sampled signal into one which is constant between two consecutive sampling instants. Such a holding device as we now already know is called a zero-order holding device or zero-order hold, and its transfer function G_h is given by

$$G_h = \mathscr{L}[1(t) - 1(t - T)] = \frac{1}{s} - \frac{e^{-Ts}}{s} = \frac{1 - e^{-Ts}}{s} \qquad (13.1)$$

Consider a sequence of impulses of unit strength and period T as shown in Figure 13.5 and defined by

$$\delta_T(t) = \sum_{k=-\infty}^{\infty} \delta(t - kT) \tag{13.2}$$

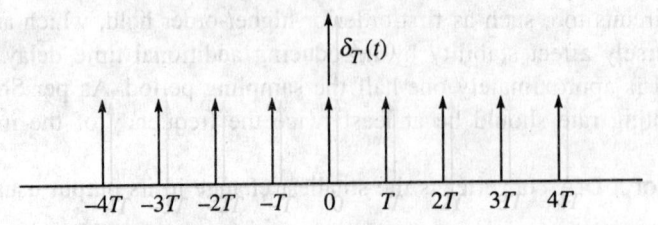

FIGURE 13.5 Sequences of impulses.

The Laplace transform of the sequence of unit impulses is given by

$$I(s) = \mathscr{L}[\delta_T(t)] = \sum_{k=-\infty}^{\infty} e^{-kTs} \tag{13.3}$$

The amplitude of any impulse function is infinite; it is convenient to indicate the strength or area of the impulse function by the length of the arrow, as depicted in Figures 13.6(a) and (b).

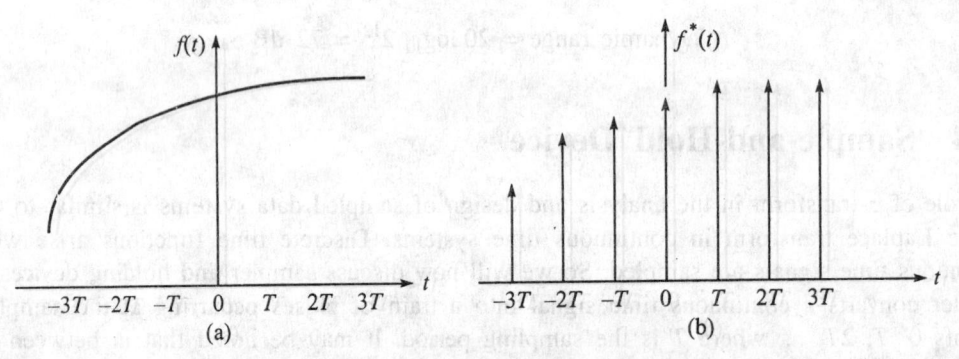

FIGURE 13.6(a) and (b) (a) Continuous time signal and (b) discrete-data output of sampler.

The sampler output is a train of weighted impulses as in Figure 13.6(b). The sampled output $f^*(t)$ can then be written as

$$f^*(t) = \delta_T(t)f(t) = \sum_{k=-\infty}^{\infty} f(kT)\delta(t - kT) \tag{13.4}$$

where $f(t)$ represents the continuous time signal and $\delta_T(t)$ is the train of unit impulses. This sampler of Figure 13.6(c) can be thought of as a modulator circuit having two inputs as in Figure 13.6(d)—one input as continuous signal $f(t)$ and the other input as unit impulse train $\delta_T(t)$.

FIGURE 13.6(c) and (d) (c) Sampler and (d) equivalent modulator circuit.

Most of the standard time functions are zero for $t < 0$, then in Eq. (13.4), the lower limit can be put zero instead of $-\infty$ as the function $f(kT)$ is zero for $t < 0$. Note that both the terms $I(s)$ and $F^*(s)$ involve e^{-Ts}, making them irrational functions of s; it may be however difficult to compute $\mathscr{L}^{-1}[F^*(s)]$. This difficulty can be avoided by defining a transform $z = e^{Ts}$ that will change the irrational function into a rational function $F(s)$. In fact, we could have used e^{-Ts} as the transformation too. Substituting

$$z = e^{Ts} \tag{13.5}$$

the z-transform of the continuous time signal $f(t)$ can be written as

$$F(z) = \mathscr{Z}[f(t)] = \mathscr{L}[f^*(t)]\Big|_{z=e^{Ts}} = \sum_{k=0}^{\infty} f(kT) z^{-k} \tag{13.6}$$

where $\mathscr{Z}$ represents the z-transform and $\mathscr{L}$ represents the Laplace transform.

Basically in the analysis of a sampled data system, it is the mapping from s-plane to z-plane through Eq. (13.6). If $F(s)$ is a rational function of s then $F(z)$ will be a rational function of z. The poles of $F(z)$ will be related to the poles of $F(s)$ through the transformation of Eq. (13.6). We will now obtain the z-transform of some standard input functions as we did in the case of continuous time functions.

A sample-and-hold device with an input and an output buffer amplifier is illustrated in Figure 13.7. During the hold mode, the output voltage of the device may decrease slightly due to the leakage currents associated with the FET switch and the buffer amplifier of the input circuit. The droop in the output of the sample-and-hold device can be greatly reduced by using

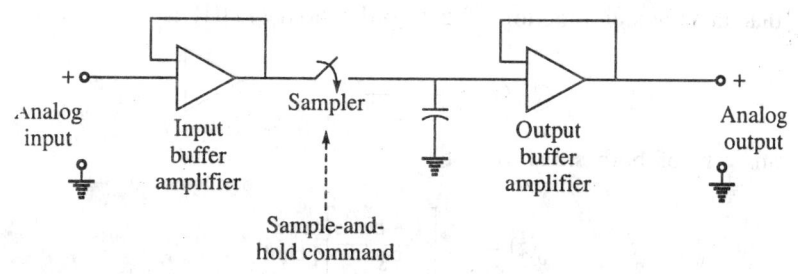

FIGURE 13.7 Sample-and-hold device with input and output buffer amplifiers.

a buffer amplifier with a very high input impedance at the output of the sample-and-hold device. Similarly, the input buffer amplifier keeps the input current of the sample-and-hold device relatively constant.

13.5 Frequency Response of Zero-Order Hold

We already know that the transfer function of a zero-order hold circuit is

$$G_{ho}(s) = \frac{1-e^{-Ts}}{s}$$

For frequency response,

$$G_{ho}(j\omega) = \frac{1-e^{-j\omega T}}{j\omega} \cdot e^{j(\omega T/2)} \cdot e^{-j(\omega T/2)} = \frac{2e^{-j(\omega T/2)}}{\omega}\left[\frac{e^{j(\omega T/2)} - e^{-j(\omega T/2)}}{2j}\right]$$

$$= T \cdot \frac{\sin(\omega T/2)}{\omega T/2} \cdot e^{-j(\omega T/2)}$$

Since $\dfrac{\omega T}{2} = \dfrac{\omega}{2}\left(\dfrac{2\pi}{\omega_s}\right) = \dfrac{\pi\omega}{\omega_s}$, then $G_{ho}(j\omega)$ can be written as

$$G_{ho}(j\omega) = T \cdot \frac{\sin(\pi\omega/\omega_s)}{\pi\omega/\omega_s} \cdot e^{-j(\pi\omega/\omega_s)}$$

Thus, $$|G_{ho}(j\omega)| = T\left|\frac{\sin(\pi\omega/\omega_s)}{\pi\omega/\omega_s}\right|$$

and the phase angle is

$$\angle G_{h0}(j\omega) = -\frac{\pi\omega}{\omega_s} + \theta\,; \qquad \theta = \begin{cases} 0, & \sin\left(\dfrac{\pi\omega}{\omega_s}\right) > 0 \\ \pi, & \sin\left(\dfrac{\pi\omega}{\omega_s}\right) < 0 \end{cases}$$

The amplitude and phase plots are shown in Figure 13.8.

We know that the transfer function of zero-order hold (ZOH) is

$$G_{ho}(s) = \frac{1-e^{-Ts}}{s}$$

Taking the z-transform of both side, we get

$$G_{h_o}(z) = \mathscr{L}\left[\frac{1-e^{-Ts}}{s}\right]$$

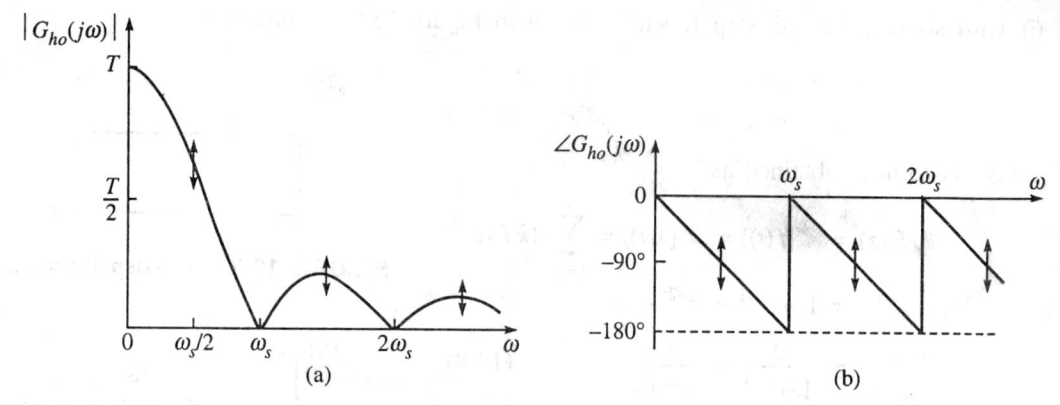

FIGURE 13.8 Frequency response of the zero-order hold: (a) amplitude plot and (b) phase plot.

Since the term e^{-Ts} in the last equation represents a delay of one sampling period, from right-shift theorem, $G_{h_0}(z)$ can be written as

$$G_{h_0}(z) = (1 - z^{-1}) \mathscr{Z}\left[\frac{1}{s}\right] = (1 - z^{-1})\left[\frac{z}{z-1}\right] = 1$$

where the z-transform of unit step is $z/(z - 1)$ as derived in the next section. The results is expected since ZOH simply holds the discrete signal for one sampling period, and the z-transform of the ZOH would revert it to the original sampled signal again.

As an example, let us consider the transfer function $G(s) = K/s(s + a)$.
The z-transform of zero-order hold (ZOH) and $G(s)$ is

$$\mathscr{Z}[G_{h_0}(s)G(s)] = (1 - z^{-1})\mathscr{Z}\left[\frac{K}{s^2(s+a)}\right] = \frac{KT}{a(z-1)} - \frac{K(1-e^{-aT})}{a^2(z-e^{-aT})}$$

Taking the limit as $T \to 0$, and approximating e^{Ts} by $1 + Ts$ and e^{-Ts} by $1 - Ts$, we get

$$\lim_{T \to 0} \mathscr{Z}[G_{h_0}(s)G(s)] = \lim_{T \to 0} \frac{KT}{a(Ts+1-1)} - \lim_{T \to 0} \frac{K(1-1+Ts)}{a^2(1+Ts-1+aT)}$$

$$= \frac{K}{as} - \frac{K}{a(s+a)} = \frac{K}{s(s+a)} = G(s)$$

which is the desired result.

13.6 z-Transform

The z-transforms of some standard functions are derived below using the definition of Eq. (13.6).

(i) **Unit step:** The unit-step function shown in Figure 13.9 is defined as

$$f(t) = 1 \text{ for } t \geq 0$$
$$= 0 \text{ for } t < 0 \quad (13.7)$$

Its z-transform is obtained as

$$F(z) = \mathscr{L}[f(t)] = \mathscr{L}[1(t)] = \sum_{k=0}^{\infty} 1(kT) z^{-k}$$

$$= 1 + z^{-1} + z^{-2} + \cdots$$

$$= \frac{1}{1 - z^{-1}} = \frac{z}{z - 1} \quad (13.8)$$

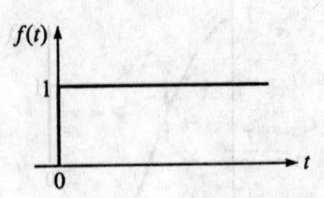

FIGURE 13.9 Unit-step function.

(ii) **Ramp function:** The ramp function shown in Figure 13.10 is defined as

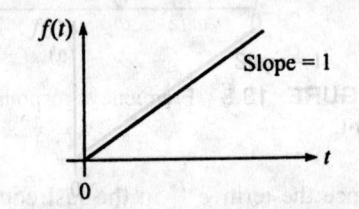

$$f(t) = t,\ t \geq 0$$
$$= 0,\ t < 0 \quad (13.9)$$

FIGURE 13.10 Ramp function.

Its z-transform is obtained as

$$F(z) = \mathscr{L}[t] = \sum_{k=0}^{\infty} f(kT) z^{-k} = \sum_{k=0}^{\infty} (kT) z^{-k} \quad (13.10)$$

$$= 0 + Tz^{-1} + 2Tz^{-2} + 3Tz^{-3} + \cdots$$

$$= Tz^{-1}(1 + 2z^{-1} + 3z^{-2} + \cdots)$$

From the Binomial theorem expansion,

$$(1 - x)^{-n} = 1 + nx + \frac{n(n+1)}{2!} x^2 + \frac{n(n+1)(n+2)}{3!} x^3 + \cdots + \frac{n(n+1)\ldots(n+r-1)}{r!} x^r + \cdots$$

we get

$$(1 - z^{-1})^{-2} = 1 + 2z^{-1} + 3z^{-2} + \cdots$$

Hence

$$F(z) = \mathscr{L}[t] = \frac{Tz^{-1}}{(1 - z^{-1})^2} = \frac{Tz}{(z - 1)^2}$$

Alternatively, multiplying both sides of Eq. (13.10) by z^{-1} and subtracting from Eq. (13.10), we get

$$F(z) - z^{-1} F(z) = \sum_{k=0}^{\infty} (kT) z^{-k} - z^{-1} \sum_{k=0}^{\infty} (kT) z^{-k}$$

or $[1 - z^{-1}]F(z) = [Tz^{-1} + 2Tz^{-2} + 3Tz^{-3} + \cdots] - [Tz^{-2} + 2Tz^{-3} + 3Tz^{-4} + \cdots]$

$$= \sum_{k=0}^{\infty} T z^{-k} \quad (13.11)$$

Next, multiplying both side of Eq. (13.11) by z^{-1} and subtracting from Eq. (13.11), we get

$$[1 - z^{-1}]^2 F(z) = Tz^{-1}$$

Therefore,
$$F(z) = \frac{Tz^{-1}}{(1-z^{-1})^2} = \frac{Tz}{(z-1)^2}$$

(iii) **Exponential function:** The exponential function shown in Figure 13.11 is defined as

$$f(t) = e^{at}, \ t \geq 0$$
$$= 0, \quad t < 0$$

Its z-transform is obtained as

$$\mathscr{Z}[e^{at}] = \sum_{k=0}^{\infty} e^{akT} z^{-k}$$

$$= 1 + e^{aT}z^{-1} + e^{2aT}z^{-2} + \cdots$$

$$= \frac{1}{1 - e^{aT} z^{-1}} \quad \text{[Binomial theorem]}$$

$$= \frac{z}{z - e^{aT}} \qquad (13.12)$$

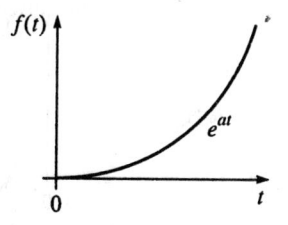

FIGURE 13.11 Exponential function.

Similarly for an exponentially decaying function, e^{-at}, the z-transform will be

$$\mathscr{Z}[e^{-at}] = \sum_{k=0}^{\infty} e^{-akT} z^{-k} = \frac{z}{z - e^{-aT}}$$

As a special case, take $a = 1$, then we get

$$\mathscr{Z}[e^{-t}] = \sum_{k=0}^{\infty} e^{-kT} z^{-k} = \frac{z}{z - e^{-T}}$$

Assuming sampling time $T = 1$ s, we get

$$\mathscr{Z}[e^{-t}] = \frac{z}{z - e^{-1}} = 1 + 0.9512z^{-1} + 0.9048z^{-2} + \cdots$$

(iv) **Sine and cosine functions:** Replacing a in Eq. (13.12) by $j\omega$, we get

$$\mathscr{Z}[e^{j\omega t}] = \mathscr{Z}[\cos \omega t + j \sin \omega t] = \frac{z}{z - e^{j\omega T}}$$

$$= \frac{z}{(z - \cos \omega t) - j \sin \omega T}$$

$$= \frac{z(z - \cos \omega T) + jz \sin \omega T}{(z - \cos \omega T)^2 + (\sin \omega T)^2} \qquad (13.13)$$

Now separating the real and imaginary parts, we get

$$\mathscr{Z}[\cos \omega T] = \frac{z(z - \cos \omega T)}{z^2 - 2z \cos \omega T + 1} \tag{13.14}$$

and
$$\mathscr{Z}[\sin \omega T] = \frac{z \sin \omega T}{z^2 - 2z \cos \omega T + 1} \tag{13.15}$$

(v) **Function with exponential damping:** Let this be defined as

$$f_1(t) = e^{-at} f(t) \tag{13.16}$$

Its z-transform is given by

$$F_1(z) = \sum_{k=0}^{\infty} [f(kT) e^{-akT}] z^{-k} = \sum_{k=0}^{\infty} f(kT)(z e^{aT})^{-k}$$

$$= F(z e^{aT}) \tag{13.17}$$

It may be noted that this is similar to the corresponding shifting theorem in the Laplace transform theory, i.e.

$$\mathscr{L}[e^{-aT} f(t)] = F(s + a) \tag{13.18}$$

For solving the difference equation by the z-transform method, we have to know the following transformation. By definition,

$$\mathscr{Z}[x(k + 1)] = \sum_{k=0}^{\infty} x(k+1) z^{-k} = \sum_{k=1}^{\infty} x(k) z^{-k+1}$$

$$= z \left[\sum_{k=0}^{\infty} x(k) z^{-k} - x(0) \right]$$

$$= z X(z) - z x(0) \tag{13.19}$$

given $\quad X(z) = \mathscr{Z}[x(k)]$

Similarly,
$$\mathscr{Z}[x(k+2)] = z\, x(k+1) - z\, x(1)$$
$$= z[z X(z) - z x(0)] - z x(1)$$
$$= z^2 X(z) - z^2 x(0) - z x(1) \tag{13.20}$$

and so on. The generalized expression becomes

$$\mathscr{Z}[x(k+n)] = z^n X(z) - z^n x(0) - z^{n-1} x(1) - z^{n-2} x(2) - \cdots - z x(n-1) \tag{13.21}$$

where n is a positive integer.

The following examples demonstrate the application of z-transform.

EXAMPLE 13.2 Solve the following difference equation using the z-transform method,

$$x(k+2) + 5x(k+1) + 6x(k) = 0$$

given
$$x(0) = 0, x(1) = 1 \quad (13.22)$$

Solution: Taking the z-transform of each term of the left-hand side of Eq. (13.22), we get

$$z^2 X(z) - z^2 x(0) - z x(1) - 5z X(z) - 5z x(0) + 6 X(z) = 0$$

Substituting the initial conditions and simplifying, we get

$$X(z) = \frac{z}{z^2 + 5z + 6} = \frac{z}{(z+2)(z+3)} = \frac{z}{z+2} - \frac{z}{z+3} \quad (13.23)$$

$$x(k) = \mathscr{Z}^{-1}[X(z)] = \mathscr{Z}^{-1}\left[\frac{z}{z+2}\right] - \mathscr{Z}^{-1}\left[\frac{z}{z+3}\right]$$

or
$$x(k) = (-2^k) - (-3^k); \quad k = 0, 1, 2, \ldots \quad (13.24)$$

EXAMPLE 13.3 Write the z-transform of $\exp(-t)$ sampled at a frequency of 10 Hz.

Solution: The z-transform of $\exp(-t)$ is $\dfrac{z}{z - e^{-T}}$, substituting $T = 0.1$, we get

$$\frac{z}{z - e^{-0.1}} = \frac{z}{z - 0.9}$$

EXAMPLE 13.4 A unit-step function is sampled every T seconds. What would be the z-transform of a sampled step delayed by T seconds?

Solution: The z-transform of unit-step function is $\dfrac{z}{z-1}$ and let it be $X(z)$.
If the sequence is delayed by one sampling interval, then its transform is multiplied by z^{-1}. The transform of the delayed sampled step is, therefore, $z^{-1} X(z) = \dfrac{1}{z-1}$.

EXAMPLE 13.5 Calculate the z-transform of a ramp of slope 2 sampled every second.

Solution: We know the z-transform of unit slope ramp as $\dfrac{Tz}{(z-1)^2}$. Then for a ramp of slope 2, the z-transform is $\dfrac{2Tz}{(z-1)^2}$. Substituting $T = 1$, we get

$$X(z) = \frac{2z}{(z-1)^2}$$

EXAMPLE 13.6 Calculate the z-transform of the system having transfer function $\dfrac{1}{1+2s}$ subjected to a step input sampled at 3 Hz.

Solution: The step response has the Laplace transform

$$X(s) = \frac{1}{s(1+2s)} = \frac{0.5}{s(s+0.5)}$$

Then
$$x(t) = 1 - \exp(-0.5t)$$

The corresponding z-transform is

$$X(z) = \frac{z(1-e^{-0.5T})}{(z-1)(z-e^{-0.5T})}$$

Substituting $T = 0.33$ gives

$$X(z) = \frac{z(1-0.846)}{(z-1)(z-0.846)} = \frac{0.15z}{(z-1)(z-0.846)}$$

The dynamics of continuous time systems can be represented by differential equations. One may require to know, how to convert the continuous time system equation to discrete time system representation in the difference equation form.

EXAMPLE 13.7 Obtain the z-transform of $\dfrac{1}{s(s+1)}$.

Solution: Now by partial fraction expansion, we get

$$x(t) = \mathscr{L}^{-1}\left[\frac{1}{s(s+1)}\right] = \mathscr{L}^{-1}\left[\frac{1}{s} - \frac{1}{s+1}\right] = 1 - e^{-t}$$

Therefore,
$$\mathscr{Z}[x(t)] = X(z) = \mathscr{Z}[1(t)] - \mathscr{Z}[e^{-t}]$$

$$= \frac{z}{z-1} - \frac{z}{z-e^{-T}} = \frac{z(1-e^{-T})}{(z-1)(z-e^{-T})}$$

EXAMPLE 13.8 Obtain the pulse transfer function of the system shown in Figure 13.12.

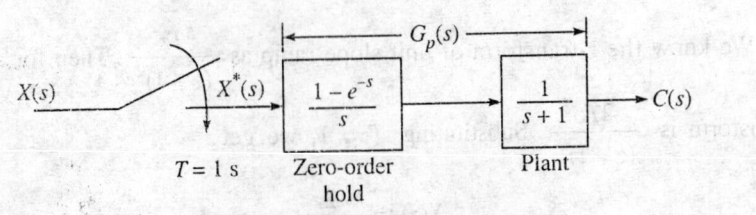

FIGURE 13.12 Example 13.8.

Solution: From the block diagram, we obtain

$$G_p(s) = \frac{C(s)}{X^*(s)} = \frac{1-e^{-s}}{s(s+1)} = (1-e^{-s})\frac{1}{s(s+1)}$$

Taking the result from Example 13.7, we get

$$G_p(z) = (1-z^{-1})\frac{z(1-e^{-T})}{(z-1)(z-e^{-T})} = \frac{1-e^{-T}}{z-e^{-T}}$$

For sampling time $T = 1$ s, we get

$$G_p(z) = \frac{1-e^{-1}}{z-e^{-1}} = \frac{0.632}{z-0.368}$$

EXAMPLE 13.9 Obtain the z-transform of $\frac{1}{s^2(s+1)}$.

Solution: By partial fraction expansion, we get

$$x(t) = \mathscr{L}^{-1}\left[\frac{1}{s^2(s+1)}\right] = \mathscr{L}^{-1}\left[\frac{1}{s^2} - \frac{1}{s} + \frac{1}{s+1}\right] = (t - 1 + e^{-t})1(t)$$

Therefore,

$$X(z) = \mathscr{Z}(t - 1 + e^{-t}) = \frac{Tz}{(z-1)^2} - \frac{z}{z-1} + \frac{z}{z-e^{-T}} = \frac{z(e^{-T}z + 1 - 2e^{-T})}{(z-1)^2(z-e^{-T})}$$

EXAMPLE 13.10 Obtain the pulse transfer function of the system shown in Figure 13.13.

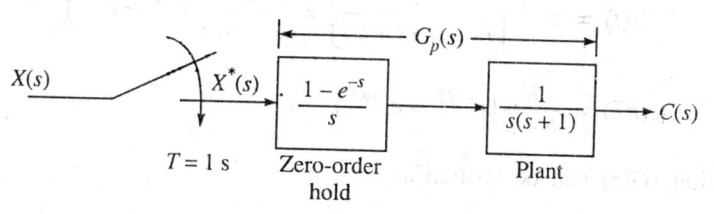

FIGURE 13.13 Example 13.10.

Solution: From the block diagram, we obtain

$$G_p(s) = \frac{C(s)}{X^*(s)} = (1-e^{-s})\frac{1}{s^2(s+1)}$$

Taking the result from Example 13.9, we get

$$G_p(z) = (1-z^{-1})\frac{z(e^{-T}z + 1 - 2e^{-T})}{(z-1)^2(z-e^{-T})} = \frac{e^{-T}z + 1 - 2e^{-T}}{z^2 - (1+e^{-T})z + e^{-T}}$$

For sampling time $T = 1$ s, we get

$$G_p(z) = \frac{e^{-1}z + 1 - 2e^{-1}}{z^2 - (1 + e^{-1})z + e^{-1}} = \frac{0.368z + 1 - 0.736}{z^2 - 1.368z + 0.368}$$

$$= \frac{0.368z + 0.264}{z^2 - 1.368z + 0.368}$$

13.7 Pulse Transfer Function of Cascaded Elements

It may be noted that the pulse transfer functions of the two systems (a) and (b) shown in Figure 13.14 are different.

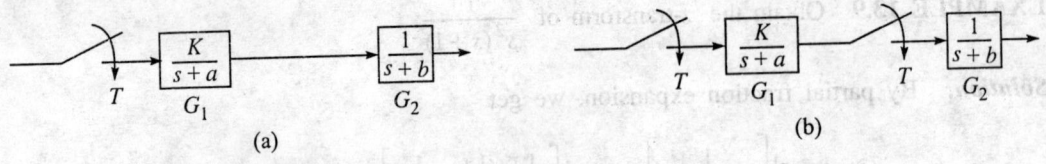

FIGURE 13.14 Pulse transfer function of cascaded elements.

In Figure 13.14(a), the pulse transfer function is

$$G(z) = G_1G_2(z) = \mathscr{Z}\left[\frac{K}{(s+a)(s+b)}\right] = \mathscr{Z}\left[\frac{K}{b-a}\left(\frac{1}{s+a} - \frac{1}{s+b}\right)\right] \quad (13.25)$$

We know that the impulse response function is

$$g(t) = \mathscr{L}^{-1}\left[\frac{K}{(s+a)(s+b)}\right] = \left(\frac{K}{b-a}\right)\left(e^{-at} - e^{-bt}\right) \quad (13.26)$$

Hence, $\qquad g(kT) = \frac{K}{b-a}\left(e^{-akT} - e^{-bkT}\right) \qquad (13.27)$

Then by definition, $G(z)$ can be written as

$$G(z) = \sum_{k=0}^{\infty} \frac{K}{b-a}\left(e^{-akT} - e^{-bkT}\right)z^{-k} \quad (13.28)$$

Now, let us take

$$\sum_{k=0}^{\infty} e^{-akT} z^{-k} = 1 + \sum_{k=1}^{\infty} e^{-akT} z^{-k}$$

$$= 1 + e^{-aT} z^{-1}\left(\sum_{k=0}^{\infty} e^{-akT} z^{-k}\right) \quad (13.29)$$

So, we get

$$\sum_{k=0}^{\infty} e^{-akT} z^{-k} = \frac{1}{1-e^{-aT} z^{-1}} \qquad (13.30)$$

Substituting this in Eq. (13.28), we get

$$G(z) = G_1 G_2(z) = \frac{K}{b-a} \left(\frac{1}{1-e^{-aT} z^{-1}} - \frac{1}{1-e^{-bT} z^{-1}} \right)$$

$$= \frac{K}{b-a} \frac{z\left(e^{-aT} - e^{-bT}\right)}{\left(z-e^{-aT}\right)\left(z-e^{-bT}\right)} \qquad (13.31)$$

The pulse transfer function of the system corresponding to Figure 13.14(b) is

$$G_1(z)G_2(z) = \mathscr{Z}\left[\frac{K}{s+a}\right]\mathscr{Z}\left[\frac{K}{s+b}\right] = \frac{K z^2}{\left(z-e^{-aT}\right)\left(z-e^{-bT}\right)} \qquad (13.32)$$

Obviously the pulse transfer functions of Figures 13.14 (a) and (b) are not identical, i.e.

$$G_1 G_2(z) \neq G_1(z)G_2(z) \qquad (13.33)$$

Therefore, one must carefully observe whether there exists a sampler between the cascaded elements or not.

13.8 Pulse Transfer Function of Closed-Loop Systems

In the closed-loop system shown in Figure 13.15, the actuating error signal is being sampled. From the block diagram,

$$E(s) = R(s) - H(s)C(s)$$

and $\quad C(s) = G(s)E^*(s)$

Hence $\quad E(s) = R(s) - G(s)H(s)E^*(s)$

or $\quad E^*(s) = R^*(s) - \overline{GH}(s)E^*(s)$

or $\quad E^*(s) = \dfrac{R^*(s)}{1+\overline{GH}^*(s)}$

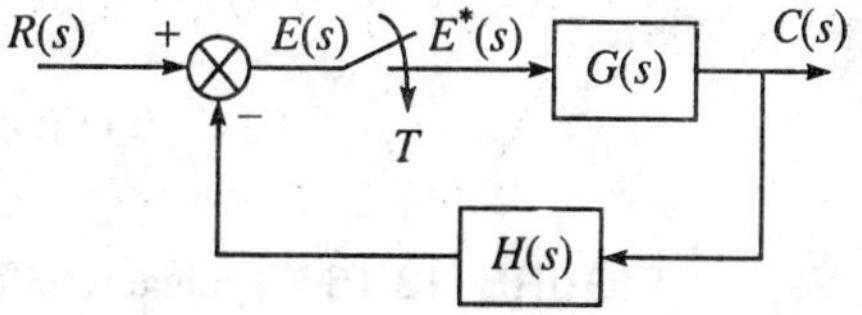

FIGURE 13.15 Single-loop sampled data system.

Since, $\quad C^*(s) = G^*(s)E^*(s)$

we obtain $\quad C^*(s) = \dfrac{G^*(s)R^*(s)}{1+\overline{GH}^*(s)}$

In terms of z-transform, $C(z)$ is given by

$$C(z) = \frac{G(z)R(z)}{1+\overline{GH}(z)} \qquad (13.34)$$

The inverse z-transform of Eq. (13.34) gives the value of the output at the sampling instants. The pulse transfer function of the present closed-loop system is

$$\frac{C(z)}{R(z)} = \frac{G(z)}{1+\overline{GH}(z)}$$

Figure 13.16 shows the five typical configurations of closed-loop discrete-time systems. For each configuration, the corresponding output $C(z)$ is shown.

(a) $C(z) = \dfrac{G(z)R(z)}{1+\overline{GH}(z)}$

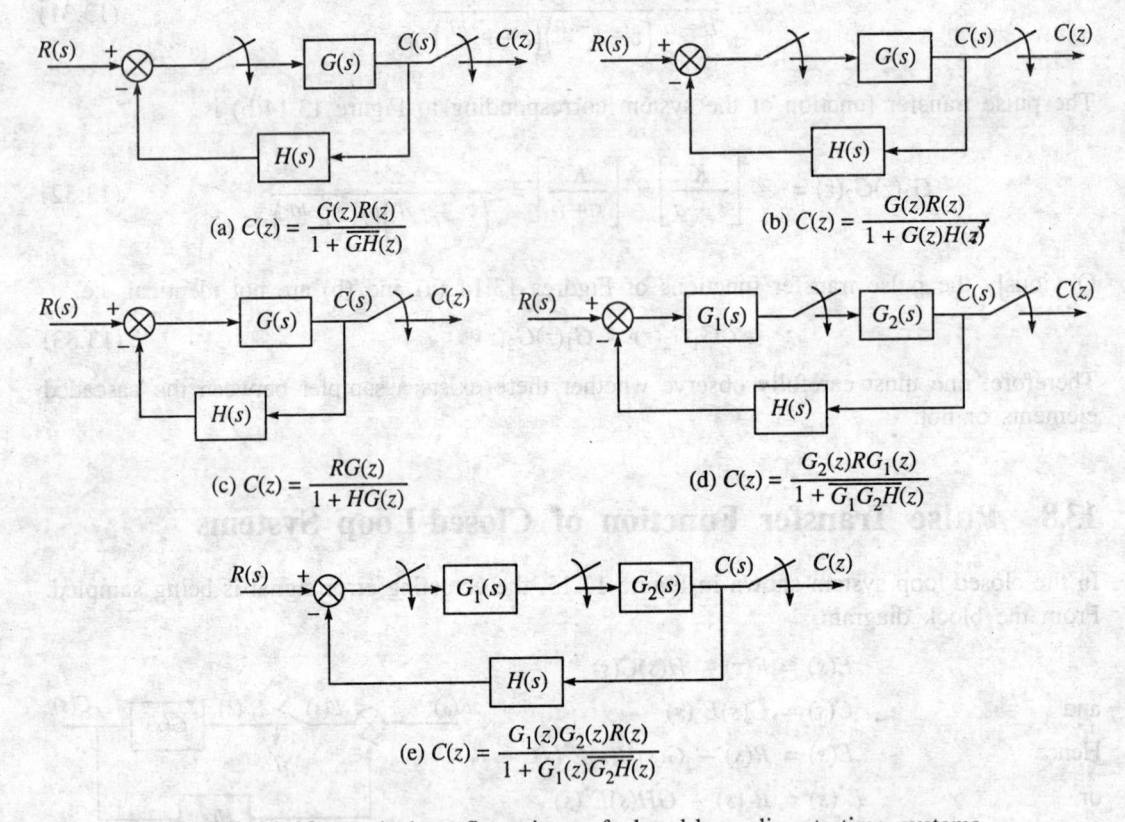

(b) $C(z) = \dfrac{G(z)R(z)}{1+G(z)H(z)}$

(c) $C(z) = \dfrac{RG(z)}{1+HG(z)}$

(d) $C(z) = \dfrac{G_2(z)RG_1(z)}{1+\overline{G_1G_2H}(z)}$

(e) $C(z) = \dfrac{G_1(z)G_2(z)R(z)}{1+G_1(z)\overline{G_2H}(z)}$

FIGURE 13.16 Typical configurations of closed-loop discrete-time systems.

13.8.1 Characteristic equation

Consider the sampled data system as shown in Figure 13.15 where the closed-loop transfer function is

$$\frac{C(z)}{R(z)} = \frac{G(z)}{1+\overline{GH}(z)}$$

and can be written in the form as

$$\frac{C(z)}{R(z)} = \frac{K \prod_{i=1}^{m}(z-z_i)}{\prod_{i=1}^{n}(z-p_i)} \qquad (13.35)$$

Using partial fraction expansion, we can write the output as

$$C(z) = \frac{k_1 z}{z-p_1} + \frac{k_2 z}{z-p_2} + \cdots + \frac{k_n z}{z-p_n} + C_R(z) \qquad (13.36)$$

where $C_R(z)$ contains the terms of $C(z)$ which originates in the poles of $R(z)$. The first n terms are the transient response terms of $C(z)$ which are always present.

The inverse z-transform of the ith term yields

$$\frac{k_i z}{z-p_i} = k_i(p_i)^k \qquad (13.37)$$

Thus it is seen that these p_i terms which are the roots of the polynomial $[1+\overline{GH}(z)]$ can determine the nature or character of the system transient response and hence the equation

$$1 + \overline{GH}(z) = 0 \qquad (13.38)$$

is called the characteristic equation of the system.

EXAMPLE 13.11 Determine the poles of the transfer function, $G(s) = \dfrac{1}{1-e^{-s}}$.

Solution: The denominator $1 - e^{-s} = 0$ gives the poles, that is,

$$e^{-s} = 1 \; ; \; s = \sigma + j\omega$$

or
$$e^{-\sigma}(\cos\omega - j\sin\omega) = 1$$

This means $\sigma = 0$ and $\omega = \pm 2n\pi (n = 0, 1, 2, \ldots)$. Thus, the poles are located at

$$sT = \pm j2n\pi; \qquad (n = 0, 1, 2, \ldots).$$

13.9 Relationship between s-plane and z-plane Poles

We are already aware of the nature of response from our study of the roots of the characteristic equation in case of continuous systems. We should be able to do the same in case of discrete systems too.

Let us now discuss mapping the s-plane into the z-plane and accordingly find the analogous relationship between the two planes. For a decaying function $e(t) = e^{-at}$, we can write

$$E(s) = \frac{1}{s+a} \tag{13.39}$$

and
$$E(z) = \frac{z}{z - e^{-aT}} \tag{13.40}$$

where T is the sampling time. This means that an s-plane pole at $-a$ results in the z-plane pole at $z = e^{-aT}$. Similarly a pole at $s = s_1$ results in a z-plane pole at z_1 where $z_1 = e^{s_1 T}$. A linear continuous feedback control system is stable if all poles of the closed-loop transfer function lie in the left-half of the s-plane. The z-plane is related by the transformation

$$z = e^{sT} = e^{(\sigma + j\omega)T} \tag{13.41}$$

We may also write the magnitude

$$|z| = e^{\sigma T} \tag{13.42}$$

and the phase
$$\angle z = \omega T \tag{13.43}$$

In the left-hand of s-plane, $\sigma < 0$ and therefore the related magnitude of z varies between 0 and 1. The imaginary axis of the s-plane where $\sigma = 0$ corresponds to the unit circle in the z-plane having the centre at origin. The point on the extreme left-half of the s-plane, i.e. $\sigma = -\infty$, is mapped to a point at the centre of z-plane. We can conclude that a sampled system is stable if all the poles of the closed-loop system lie within the unit circle of z-plane. The relative stability increases as the poles are more closer to origin in the z-plane. The region outside the unit circle of z-plane is unstable which corresponds to the poles in the right-hand side of the s-plane. The presence of poles on the unit circle in the z-plane indicates that the system is at the verge of instability.

It is useful to relate certain lines on the s-plane to the corresponding lines on the z-plane:

1. The real axis of the s-plane maps to the real axis on the z-plane as follows:

    (a) At the origin of the s-plane, $s = 0 + j0$, $\Rightarrow z = e^{sT} = e^{(0+j0)T} = e^{0.T} \cdot e^{j0T} = 1 + j0$.
    (b) At $s = -\infty + j0$, $\Rightarrow z = 0 + j0$.

    Thus, the negative real axis of the s-plane maps into the segment $(0 + j0) \le z \le (1 + j0)$. For $\sigma > 0$, $e^{sT} > 1 + j0$ and $z = |e^{\sigma T}|$. Thus the positive real axis of the s-plane maps into the positive real axis of the z-plane for $1 + j0 \le z \le \infty + j0$ and $-\infty + j0 < -z < -1 + j0$; that is, on the real axis outside the unit circle in the z-plane.

2. The imaginary axis of the s-plane maps into the unit circle on the z-plane. It may be noted that, for the sampled data, the sampling frequency is $f_s = 1/T$. Thus,

$$\omega_s = 2\pi f_s = \frac{2\pi}{T} \quad \text{and} \quad T = \frac{2\pi}{\omega_s}$$

For $s = 0 + j\omega$, we have

$$z = e^{j\omega T} = \exp\left[j\omega\left(\frac{2\pi}{\omega_s}\right)\right] = \exp\left[j2\pi\left(\frac{\omega}{\omega_s}\right)\right] = \exp\left[j\pi\frac{\omega}{\omega_s/2}\right]$$

By inspection $|z| = 1$ for all ω and $\angle z = \pi\dfrac{\omega}{\omega_s/2}$, which are the parametric equations of a circle of unit radius with centre at the origin of the z-plane. Figure 13.17 illustrates these results. Note that the unit circle maps frequencies from $-\omega_s/2 \le \omega \le \omega_s/2$.

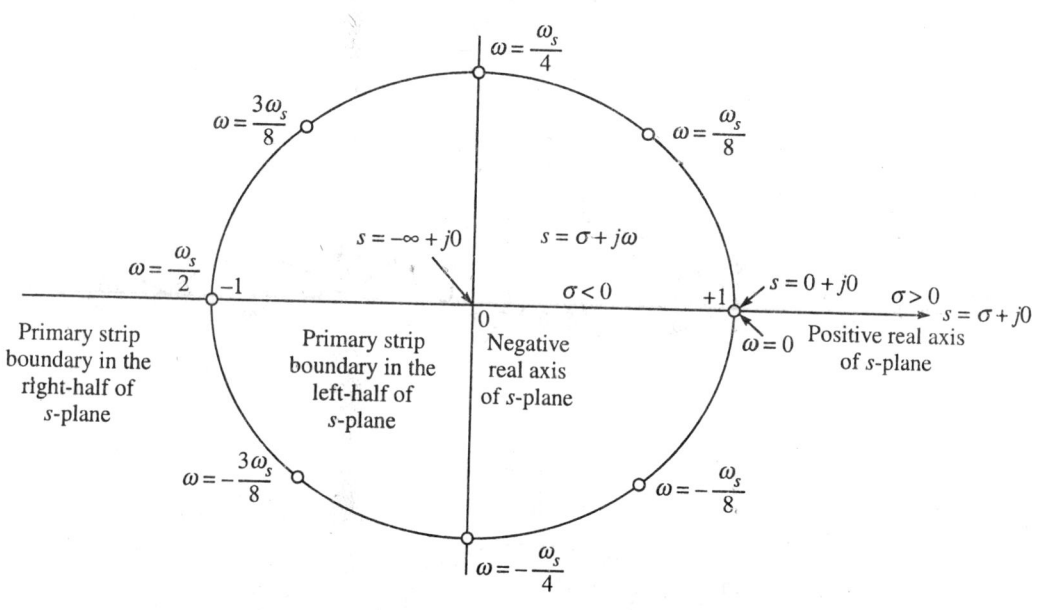

FIGURE 13.17 z-plane map of the s-plane

3. **The primary strip: effect of alias frequencies.**

If the amplitude spectrum of the continuous-time input $f(t)$ is a band-limited signal as shown in Figure 13.18(a), the amplitude spectrum of $F^*(j\omega)$ will be of the general shape of the form as in Figure 13.18(b). We see that the sampling operation retains the fundamental component of $F(j\omega)$, but in addition, the sampler output also contains the harmonic components, $F(j\omega + jn\omega_s)$, for $n = \pm 1, \pm 2, \ldots$. Thus the sampler may be considered as a harmonic generator whose output contains the weighted fundamental component, plus all the weighted complementary components at all frequencies separated by sampling frequency ω_s. The band around zero frequency still carries all the information contained in the continuous input signal, but the same information is also repeated along the frequency axis; the amplitude of each component is weighted by the amplitude of its corresponding Fourier coefficient. The frequency spectrum $F^*(j\omega)$ is shown in Figure 13.18(b) assuming $\omega_s > 2\omega_b$ and obviously, the spectrum of $|F^*(j\omega)|$ around zero frequency bears resemblance to that of original signal. Therefore, theoretically the original signal can be recovered from the spectrum by means of an ideal low-pass filter with a bandwidth that lies between ω_c and $\omega_s - \omega_c$. If $\omega_s < 2\omega_c$, the distortion in the frequency spectrum of $|F^*(j\omega)|$ will appear because of the overlapping of the harmonic components and this effect is known as *aliasing* as shown in Figure 13.18(c). Obviously, the spectrum of $|F^*(j\omega)|$ around zero frequency bears little resemblance to that of original signal and the original signal cannot be recovered from the spectrum.

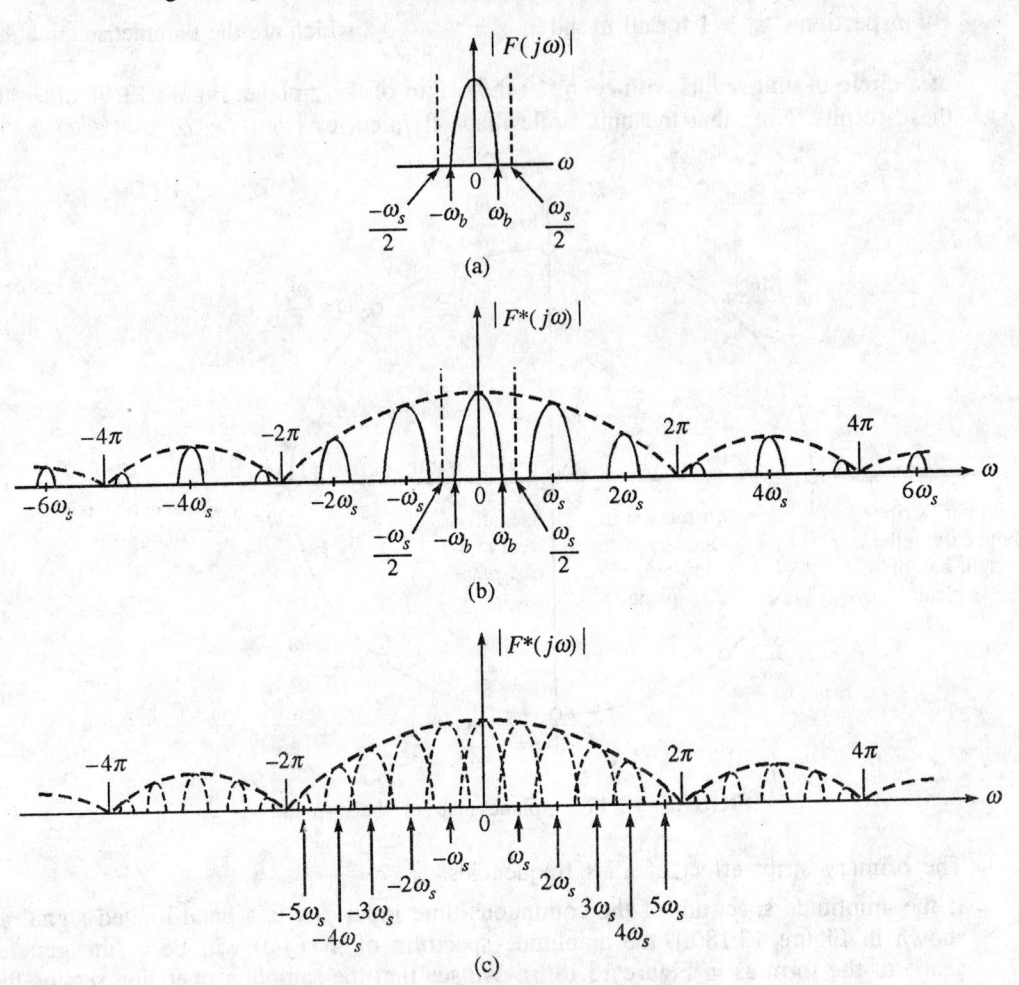

FIGURE 13.18 Amplitude spectra of input and output signals of a finite-pulsewidth sampler: (a) amplitude spectrum of continuous input signal $f(t)$, (b) amplitude spectrum of sampler output ($\omega_s > 2\omega_b$), and (c) amplitude spectrum of sampler output ($\omega_s < 2\omega_b$).

The circle of Figure 13.17 on the z-plane is the map of segment $-\omega_s/2 \leq \omega \leq \omega_s/2$ as shown later in Figure 13.20(a). Horizontal lines at $-\omega_s/2$ and $+\omega_s/2$ bind the region of interest for a system band limited to $\omega_b < \omega_s/2$, and the strip so defined is called the *primary strip*. On the horizontal boundary,

$$s = \mp\sigma \mp j\left(\frac{\omega_s}{2}\right)$$

therefore, $\quad z = e^{sT} = \exp\left(\mp\sigma \mp j\dfrac{\omega_s}{2}\right)T = e^{\mp\sigma T}e^{\mp j\pi}$

from which it is seen that the boundaries of the primary strip map onto the negative real axis of the z-plane. The boundaries in the left-half of the s-plane map on the z-plane between the origin and ± 1 and, those in the right-half of the s-plane map on the z-plane between ± 1 and $\pm\infty$ respectively. This is indicated in Figure 13.17.

Thus we can draw the following conclusions from the general mapping between the s- and z-planes by the z-transform.

- All the points in the left-half of the s-plane correspond to points inside the unit circle in the z-plane.
- All the points in the right-half of the s-plane correspond to points outside the unit circle in the z-plane.
- Points on the $j\omega$-axis in the s-plane correspond to points on the unit circle $|z| = 1$ in the z-plane.

In a like manner, it is readily shown that the imaginary axis segments map on top of the unit circle; also all strip boundaries map on top of each other. The left-half of the s-plane maps inside of the unit circle. Clearly, the unit circle is the stability boundary in the z-plane.

Consider the mapping of the left-half of s-plane in z-plane with the transformation $e^{sT} = z$. Now along the $j\omega$-axis, i.e. $\sigma = 0$,

$$z = e^{sT} = e^{\sigma T} e^{j\omega T} = 1(\cos \omega T + j \sin \omega T) = 1 \angle \omega T \qquad (13.44)$$

The pole location on the unit circle in the z-plane signifies a system with steady-state oscillations in its natural response. This is to note that the $j\omega$-axis between $-j\omega_s/2$ and $+j\omega_s/2$ maps into the unit circle in the z-plane because for $\omega = \omega_s/2$; ωT is equal to π. That means a strip of ω_s in $j\omega$-axis maps into the unit circle in the z-plane. And it repeats after every ω_s strip on the $j\omega$-axis. Mapping from s-plane to z-plane is shown in Figure 13.19.

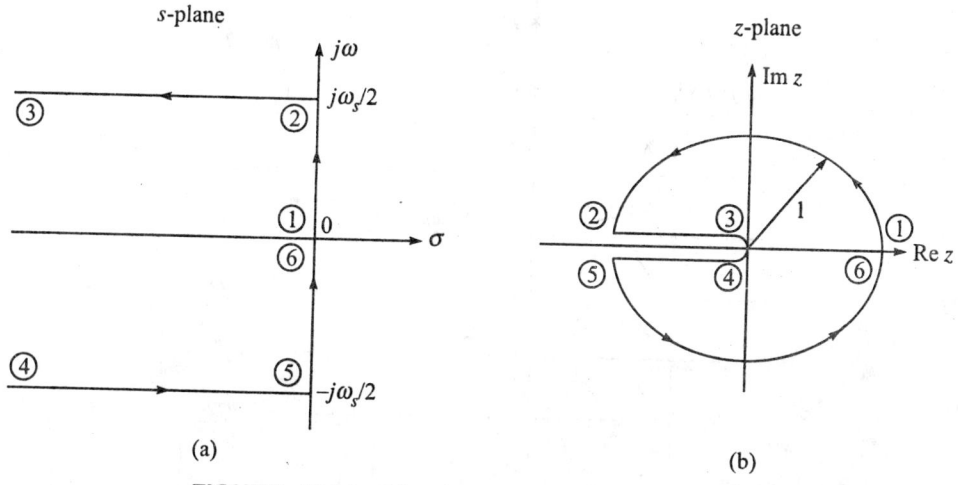

FIGURE 13.19 Mapping from s-plane to z-plane.

We have seen that the sampling operation retains the fundamental component of $F(j\omega)$, but in addition, the sampler output also contains the harmonic components, $F(j\omega + jn\omega_s)$, for $n = \pm 1, \pm 2, \ldots$. Thus, we can write

$$F^*(s) = F^*(s + jm\omega_s) \text{ for any integer } m$$

Hence, for any point $s = s_1$ in the s-plane, the function $F^*(s)$ has the same value at all periodic points $s = s_1 + jm\omega_s$ where m is an integer. In Figure 13.20(a) the s-plane is divided into an infinite number of periodic strips, each with a width of ω_s, the sampling frequency.

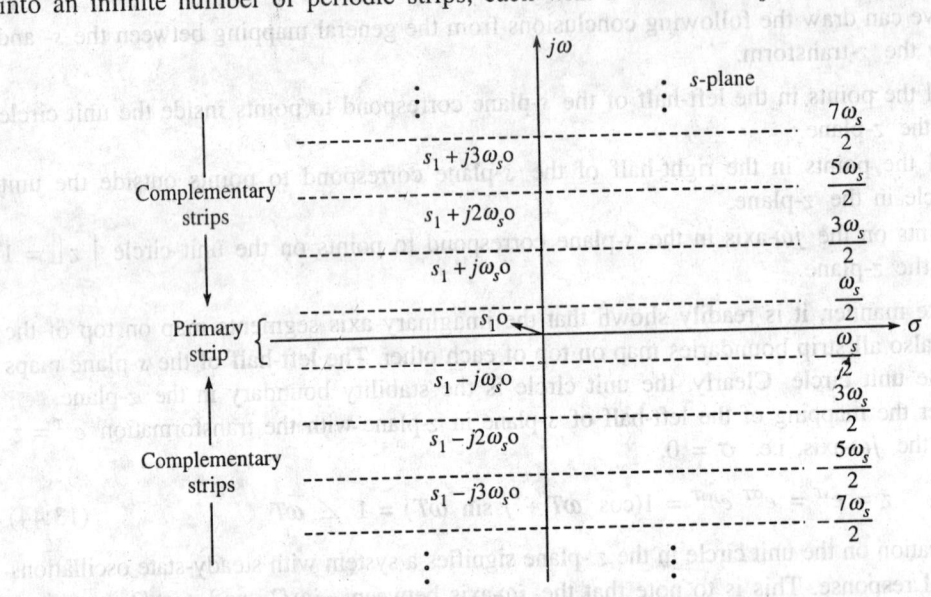

FIGURE 13.20(a) Periodic strips in the s-plane.

FIGURE 13.20(b) Periodicity of the poles of $F^*(s)$.

The strip between $\omega = -\omega_s/2$ and $\omega = \omega_s/2$ is called the *primary strip* and all other occurring at higher frequencies, both positive and negative, are called complementary strips. The function $F^*(s)$ has the same value at all congruent points in the various periodic strips.

If the function $F(s)$ has a pole at $s = s_1$, the $F^*(s)$ has poles at $s = s_1 + jm\omega_s$ where m is an integer from $-\infty$ to ∞ as shown in Figure 13.20(a).

Figure 13.20(b) shows that if $F(s)$ has two real poles in the s-poles, then $F^*(s)$ will have an infinite number of poles located at periodic locations that are spaced by integral multiples of ω_s in the complementary strips.

The constant damping loci in the s-plane (i.e. straight lines with σ constant) map into circles in the z-plane as shown in Figure 13.21 through the relationship

$$z = e^{\sigma_1 T} e^{j\omega T} = e^{\sigma_1 T} \angle \omega T \tag{13.45}$$

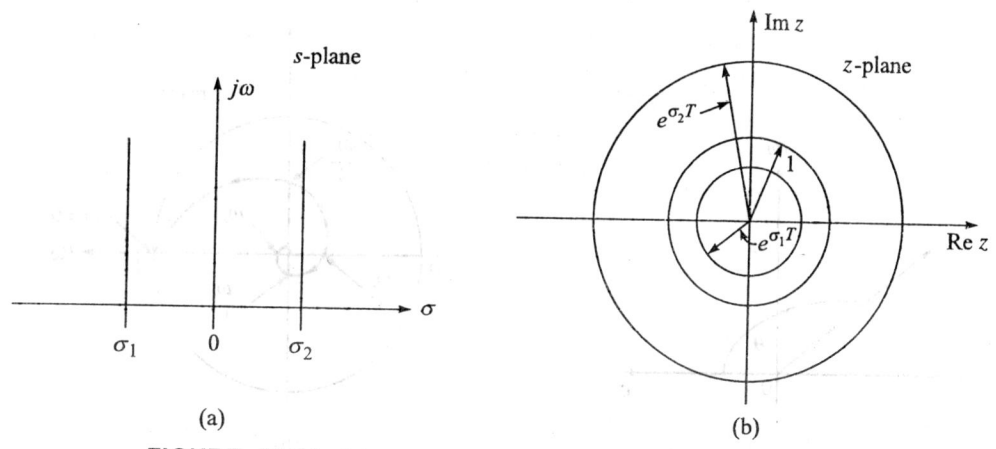

FIGURE 13.21 Mapping of constant damping loci in z-plane.

The constant frequency loci in the s-plane map into rays in the z-plane as shown in Figure 13.22. The constant damping ratio loci in s-plane can be written as $\zeta = \cos\theta$.

where $\quad \tan\theta = \dfrac{\sin\theta}{\cos\theta} = \dfrac{\sqrt{1-\cos^2\theta}}{\cos\theta} = \dfrac{\sqrt{1-\zeta^2}}{\zeta} = \dfrac{\omega_n\sqrt{1-\zeta^2}}{\omega_n\zeta} = \dfrac{\omega_d}{\sigma}$

Therefore, $\quad \omega_d = \sigma \tan\theta \quad$ and $\quad \omega_d T = \sigma T \tan\theta$

Hence $\quad z = e^{sT} = |e^{\sigma T}| \angle \omega T = |e^{\sigma T}| \angle \sigma T \tan\theta \tag{13.46}$

Since σ is negative in the second and third quadrants of the s-plane, Eq. (13.46) describes a logarithmic spiral whose amplitude decreases with σ increasing in magnitude as shown in Figure 13.23.

In a similar way, constant-damping-ratio loci bounded by the primary strip in the s-plane as illustrated in Figure 13.24(a) is mapped into the corresponding loci in the z-plane as shown in Figure 13.24(b).

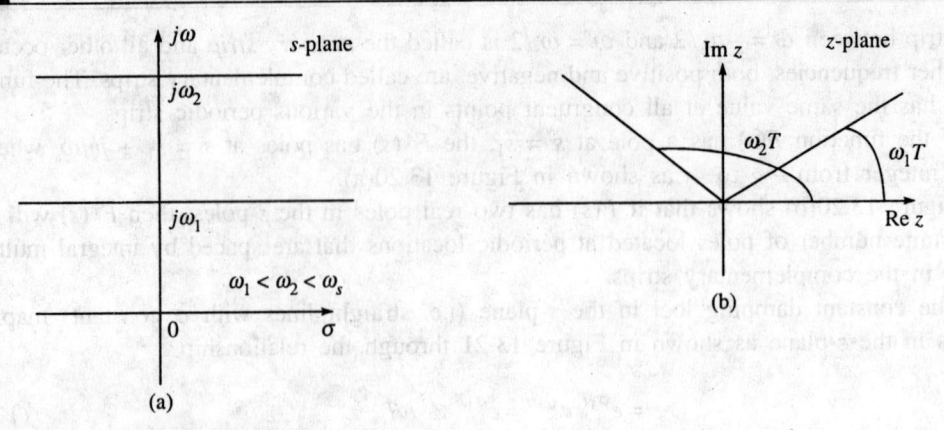

FIGURE 13.22 Mapping of constant frequency loci into z-plane.

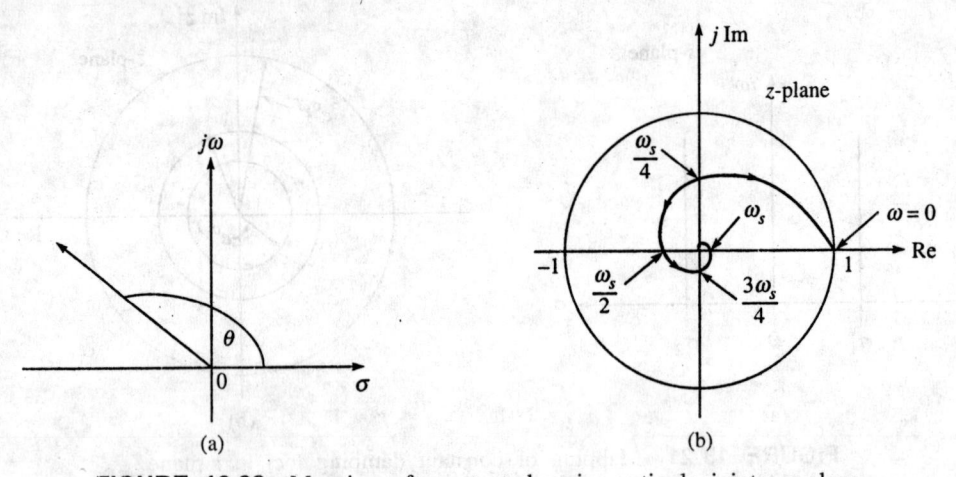

FIGURE 13.23 Mapping of constant damping ratio loci into z-plane.

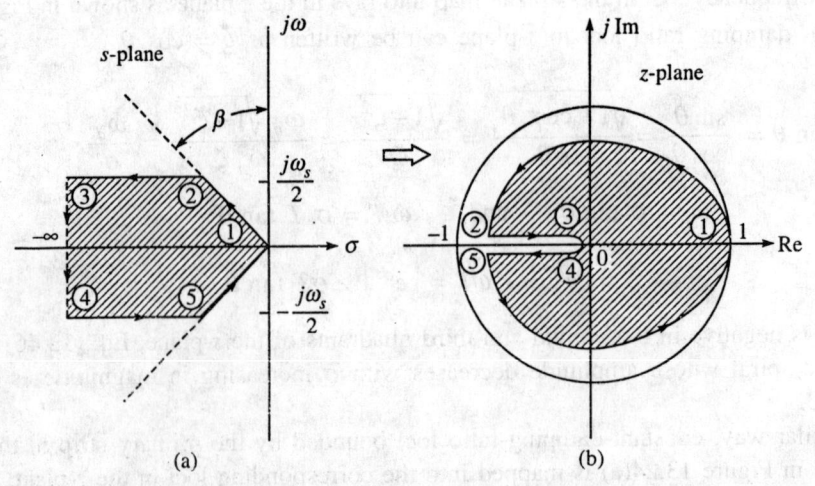

FIGURE 13.24 Constant-damping-ratio loci bounded by the primary strip in the s-plane and the corresponding loci in the z-plane.

Now we should be able to assign time response characteristics from the locations of the roots of the characteristic equation in z-plane. The correspondence of several s-plane and z-plane pole locations is illustrated in Figure 13.25. The time response characteristics of the z-plane pole locations are illustrated in Figure 13.26.

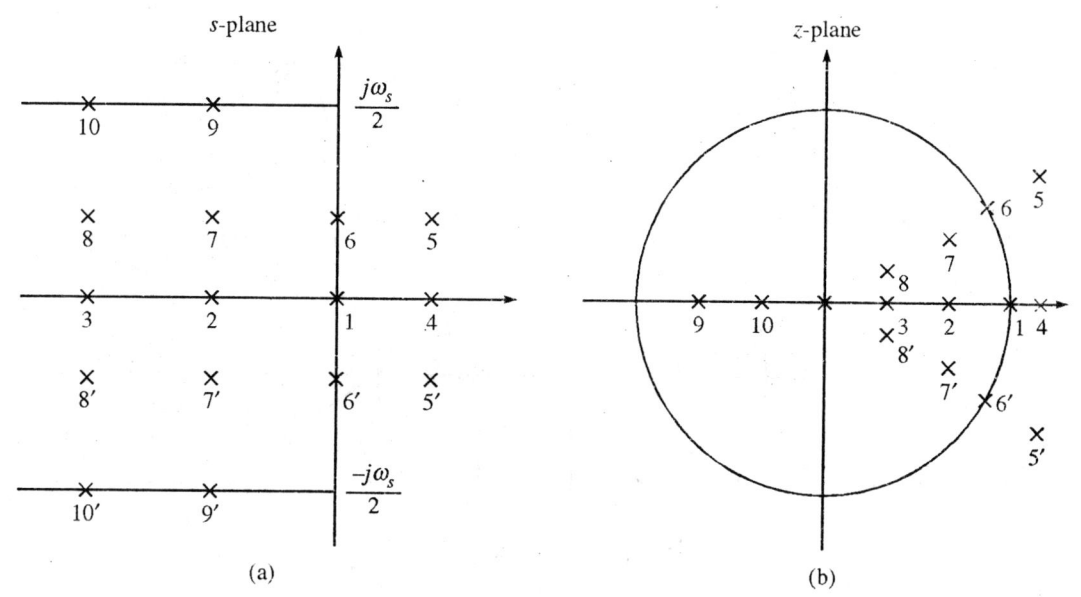

FIGURE 13.25 Corresponding pole locations between the s-plane and the z-plane.

The relationship between s-plane poles and z-plane poles can be written generally as $z = e^{sT}$. For a particular pole location at $s_0 = \sigma_0 + j\omega_0$, we can write

$$z = e^{(\sigma_0 + j\omega_0)T} = e^{\sigma_0 T} e^{j\omega_0 T}$$

Using the usual polar notation for pole position, $z = ce^{j\theta}$, we can identify

$$c = e^{\sigma_0 T} \quad \text{and} \quad \theta = \omega_0 T$$

Figure 13.25 illustrates the relationship and links the stability criterion in the s-plane and z-plane.

Care should be taken to avoid serious misconceptions when interpreting the relationship between the s-plane and z-plane. The expression $z = e^{sT}$ holds between the s-plane poles of a continuous system model and the z-plane poles of the pulse transfer function when combined with a hold at the input and a sampler at the output. It does not necessarily hold for the closed-loop s-plane poles of a continuous system and the closed-loop poles of a discretized version of it. It may be noted that the relationship holds for poles not for zeros.

We can use the relationship $z = e^{sT}$ to characterize complex pole position in the z-plane by means of the concept of damping ratio familiar to us from our analysis of continuous second-order systems. In a continuous system, the constant damping ratio loci are straight lines. Similar loci of constant damping ratio can be constructed in z-plane but owing to the exponential relationship between the s and z-plane pole positions, they are not straight lines. Such lines of

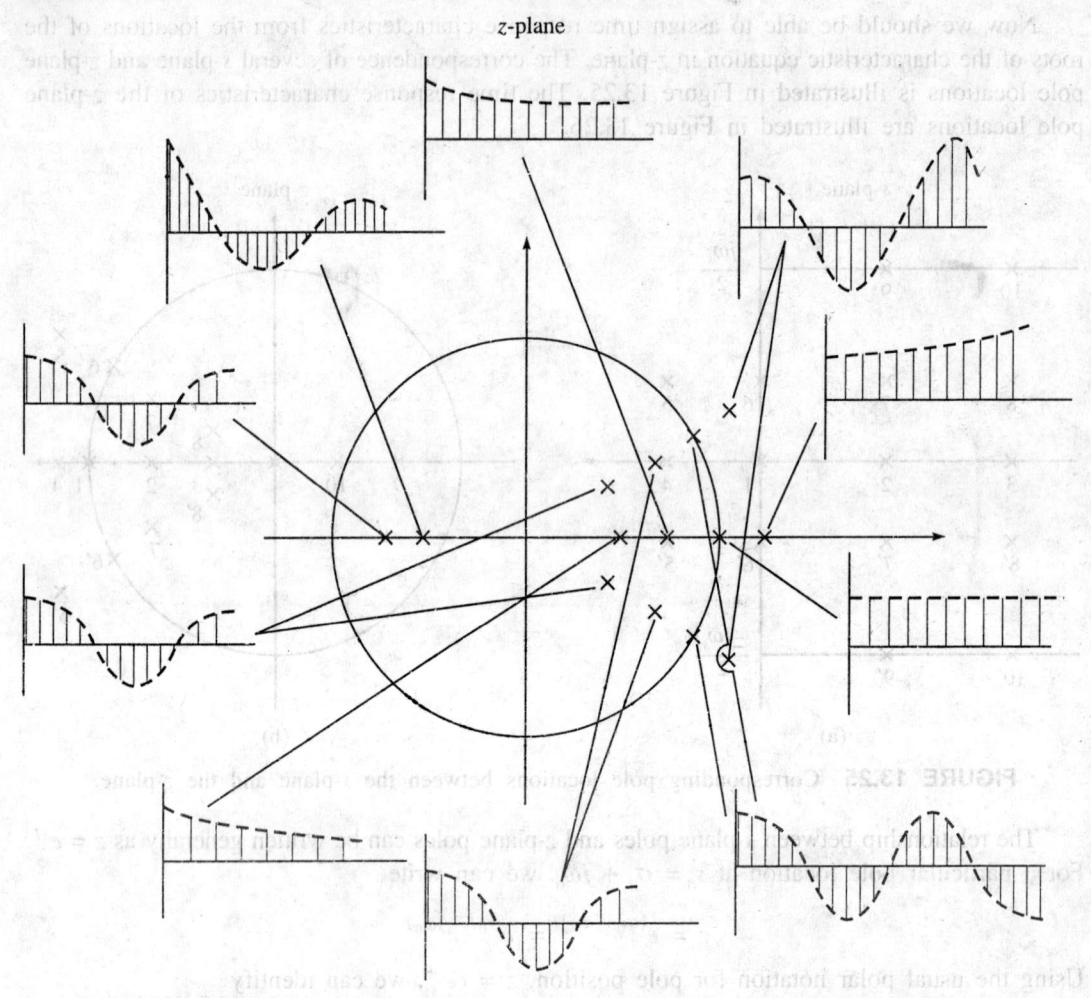

FIGURE 13.26 Transient response characteristics of the z-plane pole locations.

constant damping ratio are shown in Figure 13.23(b) as logarithmic spirals. From this figure, one can easily read the modulus and angle of any given pole, and thus enabling the decay factor and oscillation frequency to be estimated easily.

EXAMPLE 13.12 A unity feedback digital control system of second order has a zero at the origin and a complex conjugate pole-pair. The sampling rate is 50/s. The unit-step response overshoot is 10% and the settling time is 0.5 s for 5% criterion. Determine the approximate z-plane pole position.

Solution: For a pole where $|z| = c$, the transient response envelope decays by a factor of c in the sampling interval. The envelope must be less than 0.05 to its original value after 25 sampling intervals. Hence c is given by $c^{25} = 0.05$.

Taking logarithm on both sides, we get

$$25 \log c = -1.3 \quad \text{or} \quad c = 0.887$$

The 10% overshoot gives damping ratio as 0.6. Hence we get $\theta = 10°$. Therefore, the z-plane poles (underdamped) are at $0.887 \exp(\pm j10)$. Frequency of transient oscillation = 50/36 = 1.4 Hz.

13.10 Jury's Stability Test

A stability criterion for discrete-time systems which is similar to the Routh–Hurwitz criterion and can be applied to the characteristic equation written as a function of z, is the Jury's stability test. Let the characteristic equation of a discrete time system be represented by

$$Q(z) = a_n z^n + a_{n-1} z^{n-1} + \cdots + a_2 z^2 + a_1 z + a_0 = 0 \; ; \; \text{with } a_n > 0 \quad (13.47)$$

We formulate the array as shown in Table 13.1.

TABLE 13.1 Array for Jury's stability test

z^0	z^1	z^2	...	z^{n-k}	...	z^{n-1}	z^n
a_0	a_1	a_2	...	a_{n-k}	...	a_{n-1}	a_n
a_n	a_{n-1}	a_{n-2}	...	a_k	...	a_1	a_0
b_0	b_1	b_2	...	b_{n-k}	...	b_{n-1}	
b_{n-1}	b_{n-2}	b_{n-3}	...	b_{k-1}	...	b_0	
c_0	c_1	c_2	...	c_{n-k}	...		
c_{n-2}	c_{n-3}	c_{n-4}	...	c_{k-2}	...		
$\vdots$	$\vdots$	$\vdots$		$\vdots$			
l_0	l_1	l_2	l_3				
l_3	l_2	l_1					
m_0	m_1	m_2					

The elements of each of the even-numbered rows are the elements of the preceding row in reverse order. The elements of the odd-numbered rows are given as

$$b_k = \begin{vmatrix} a_0 & a_{n-k} \\ a_n & a_k \end{vmatrix}, \quad c_k = \begin{vmatrix} b_0 & b_{n-1-k} \\ b_{n-1} & b_k \end{vmatrix}, \quad d_k = \begin{vmatrix} c_0 & c_{n-2-k} \\ c_{n-2} & c_k \end{vmatrix}, \ldots$$

The necessary and sufficient conditions for the characteristic polynomial $Q(z)$ to have no roots on or outside the unit circle in the z-plane, with $a_n > 0$, are as follows:

$$Q(1) > 0$$

$$(-1)^n Q(-1) > 0$$

$$|a_0| < a_n$$

$$|b_0| > |b_{n-1}|$$
$$|c_0| > |c_{n-2}|$$
$$|d_0| > |d_{n-3}|$$
$$\vdots$$
$$|m_0| > |m_2| \qquad (13.48)$$

Procedure

1. Check the three conditions $Q(1) > 0$, $(-1)^n Q(-1) > 0$ and $|a_0| < a_n$. Stop if any of these conditions are not satisfied.
2. Construct the array, checking the conditions of Eq. (13.48) as each row is calculated. Stop if any condition is not satisfied.

It may be noted that for a second-order system the array consists of only one row. For each additional order, two additional rows are added to the array.

EXAMPLE 13.13 Consider the closed-loop discrete time system as shown in Figure 13.27, where K is the gain of the plant. It is desired to determine the range of K for which the system is stable. Use Jury's stability test. $GH(z)$ is already obtained as in Example 13.10.

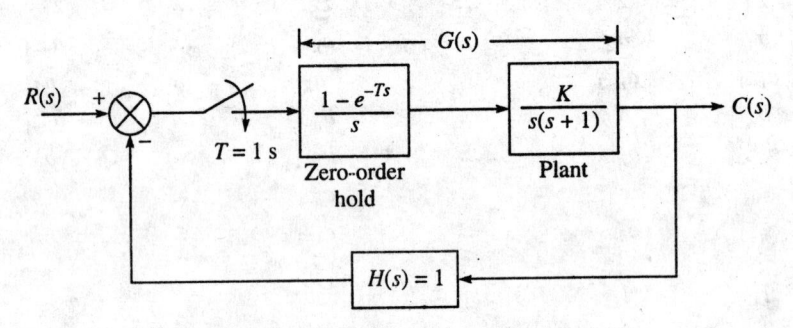

FIGURE 13.27 Example 13.13.

Solution: The closed-loop system characteristic polynomial is

$$Q(z) = 1 + GH(z) = 1 + \frac{0.368K(z + 0.717)}{(z-1)(z-0.368)}$$

$$= z^2 + (0.368K - 1.368)z + (0.368 + 0.264K)$$

The Jury's array is

z^0	z^1	z^2
$0.368 + 0.264K$	$0.368K - 1.368$	1

The constraint $Q(1) > 0$ yields

$$1 + (0.368K - 1.368) + (0.368 + 0.264K) > 0$$

or
$$K > 0$$

The constraint $(-1)^2 Q(-1) > 0$ yields

$$1 - (0.368K - 1.368) + (0.368 + 0.264K) > 0$$

or
$$K < 26.3$$

The constraint $|a_0| < a_2$ yields

$$|0.368 + 0.264K| < 1$$

or
$$K < (1 - 0.368)/0.264 < 2.393$$

Thus the system is stable for $0 < K < 2.393$.

13.11 Shifting Property of z-transform

A sequence $f(k)$ can be shifted to left (advanced) or to right (delayed) and can be written as $f(k + 1)$ or $f(k - 1)$ respectively. Taking z-transform of the shifted sequence reveals interesting interpretations.

The z-transform of $f(k + 1)$, that is,

$$\mathscr{Z}[f(k+1)] = \sum_{k=0}^{\infty} f(k+1) z^{-k} = z \sum_{k=0}^{\infty} f(k+1) z^{-(k+1)}$$

$$= z \sum_{m=1}^{\infty} f(m) z^{-m} \; ; \text{ putting } k+1 = m$$

$$= z \left[\sum_{m=0}^{\infty} f(m) z^{-m} - f(0) \right]$$

$$= zF(z) - z f(0)$$

In a generalized way,

$$\mathscr{Z}[f(k+n)] = z^n F(z) - \sum_{i=0}^{n-1} f(i) z^{n-i} \; ; \; k \geq -n$$

In a nutshell, we can say that, if the sequence $f(k)$ has its z-transform $F(z)$ advanced by n-intervals, then, $f(k + n)$ has its z-transform multiplied by z^n. Similarly, if the sequence $f(k)$ has its z-transform $F(z)$ delayed by n-intervals, then, $f(k - n)$ has its z-transform multiplied by z^{-n}, that is, $\mathscr{Z}[f(k - n)] = z^{-n}F(z)$, which we will derive now.

$$\mathscr{Z}[f(k-1)] = \sum_{k=0}^{\infty} f(k-1) z^{-k} = z^{-1} \sum_{k=0}^{\infty} f(k-1) z^{-(k-1)}$$

Putting $k - 1 = m$ and $f(m) = 0$ for $m < 0$, we obtain

$$\mathscr{Z}[f(k-1)] = z^{-1} \left[\sum_{m=0}^{\infty} f(m) z^{-m} \right] = z^{-1} F(z)$$

In a generalized way,

$$\mathscr{Z}[f(k-n)] = z^{-n} F(z)$$

The above properties are useful in z-transformation of difference equations.

EXAMPLE 13.14 Obtain the impulse response of the system described by the difference equation

$$c(k+1) + 2c(k) = \delta(k); \quad c(0) = 0$$

Solution: Taking the z-transform, we get

$$C(z) = \frac{1}{z+2} = \frac{z^{-1}z}{z+2}$$

We know
$$\frac{z}{z+2} \leftrightarrow (-2)^k$$

Then by shifting property, we get

$$z^{-1}\left(\frac{z}{z+2}\right) \leftrightarrow (-2)^{k-1} \; ; k \geq 1$$

Hence
$$c(k) = (-2)^{k-1} \; ; \; k \geq 1$$

EXAMPLE 13.15 Determine the inverse z-transform, $f(kT)$, in closed-form for the function

$$F(z) = \frac{-10(11z^2 - 15z + 6)}{z^3 - 4z^2 + 5z - 2}; \quad \text{for all } k \geq 0$$

Solution:

$$F(z) = -10\left[\frac{11z^2 - 15z + 6}{(z-1)^2(z-2)}\right] = -10\left[\frac{20}{z-2} + \frac{-9}{z-1} + \frac{-2}{(z-1)^2}\right]$$

Taking inverse z-transform, we get

$$f(k) = \mathscr{Z}^{-1}\left(-10\left[\frac{20}{z-2} + \frac{-9}{z-1} + \frac{-2}{(z-1)^2}\right]\right)$$

$$= (-200)\mathscr{Z}^{-1}\left[\frac{1}{z-2}\right] + (90)\mathscr{Z}^{-1}\left[\frac{1}{z-1}\right] + (20)\mathscr{Z}^{-1}\left[\frac{1}{(z-1)^2}\right]$$

$$= (-200)(2)^{k-1} + (90)(1)^{k-1} + (20)(k-1)(1)^{k-1}$$

$$= (-200)(2)^{k-1} + (90)(1)^{k-1} + (20)(k)(1)^{k-1} - (20)(1)^{k-1}$$

$$= (-200)(2)^{k-1} + (70)(1)^{k-1} + (20)(k)(1)^{k-1}$$

13.12 Analysis of Digital Control Systems

Before going for the design, we must understand how to analyse a digital control system. As in the case of analog control systems, we will evaluate the response of a closed-loop discrete control system, subjected to standard inputs. The root-locus technique for discrete control systems will be discussed with the help of an example. The Nyquist criterion will be discussed with the help of the same example. That is, an attempt has been made in this chapter to illustrate with the help of the same example and to study the different topics of the digital control system in the light of our study of the continuous control system.

13.12.1 Discrete system time response

In this section the time response of a discrete time system is introduced with the help of an example.

EXAMPLE 13.16 Find the unit-step response for the first-order temperature control system shown in Figure 13.28 having the transfer function of the plant as

$$G(s) = \frac{4}{s+2}$$

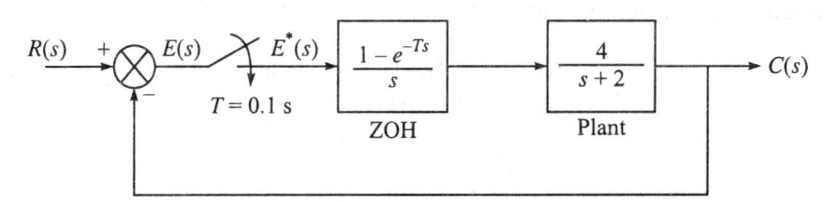

FIGURE 13.28 Example 13.16.

Solution: The system output can be written as

$$C(z) = \frac{G(z)}{1+G(z)} R(z)$$

where $G(z)$ is defined as

$$G(z) = \mathscr{Z}\left[\frac{1-e^{-Ts}}{s} \cdot \frac{4}{s+2}\right] = (1-z^{-1})\mathscr{Z}\left[\frac{4}{s(s+2)}\right]$$

$$= \left(\frac{z-1}{z}\right) X(z)$$

Let us find the z-transform, i.e. $X(z)$ of $\dfrac{4}{s(s+2)}$

By partial fraction expansion, we get

$$x(t) = \mathscr{L}^{-1} \frac{4}{s(s+2)} = \mathscr{L}^{-1}\left[\frac{2}{s} - \frac{2}{s+2}\right] = 2(1 - e^{-2t})$$

Now
$$X(z) = \mathscr{Z}[x(t)] = 2\{\mathscr{Z}[1(t)] - \mathscr{Z}[e^{-2t}]\}$$

$$= 2\left\{\frac{z}{z-1} - \frac{z}{z-e^{-2T}}\right\} = 2\left\{\frac{z(1-e^{-2T})}{(z-1)(z-e^{-2T})}\right\}$$

$$= \left(\frac{2z}{z-1}\right)\left[\frac{1-e^{-2T}}{z-e^{-2T}}\right]$$

Now putting the expression of $X(z)$ in $G(z)$, we get

$$G(z) = \frac{2(1-e^{-2T})}{z-e^{-2T}}$$

For $T = 0.1$ s, we get

$$G(z) = \frac{2(1-e^{-0.2})}{z-e^{-0.2}} = \frac{0.362}{z-0.819}$$

Thus the closed-loop transfer function for sampling period $T = 0.1$ s is

$$G_c(z) = \frac{C(z)}{R(z)} = \frac{G(z)}{1+G(z)} = \frac{0.362}{z-0.457} \qquad (13.49)$$

Since
$$R(z) = \mathscr{Z}\left[\frac{1}{s}\right] = \frac{z}{z-1}$$

$$C(z) = \frac{0.362}{z-0.457} R(z) = \frac{0.362\, z}{(z-1)(z-0.457)}$$

Now
$$\frac{C(z)}{z} = \frac{0.362}{(z-1)(z-0.457)} = \frac{0.667}{z-1} + \frac{-0.667}{z-0.457}$$

Hence
$$C(z) = 0.667\left[\frac{z}{z-1} - \frac{z}{z-0.457}\right]$$

The inverse z-transform gives the response at the sampling instants k as

$$c(kT) = 0.667[1 - (0.457)^k]$$

and this response is listed in Table 13.2.

TABLE 13.2 Response of the discrete system (Example 13.16)

(kT)	c(kT)	c(t)
0.0	0.0	0.0
0.1	0.363	0.3
0.2	0.528	0.466
0.3	0.603	0.557
0.4	0.639	0.606
0.5	0.654	0.634
1.0	0.666	0.665
1.5	0.667	0.667
2.0	0.667	0.667

In order to compare the response of a closed-loop continuous system with the same plant, we get the closed-loop transfer function as

$$\frac{C(s)}{R(s)} = \frac{G(s)}{1+G(s)} = \frac{4}{s+6} \quad (13.50)$$

The unit step response becomes

$$c(t) = \mathscr{L}^{-1}\left[\frac{4}{s+6} \cdot \frac{1}{s}\right] = 0.667(1 - e^{-6t})$$

The analog system response is also listed in Table 13.2 for comparison. The response is shown graphically in Figure 13.29.

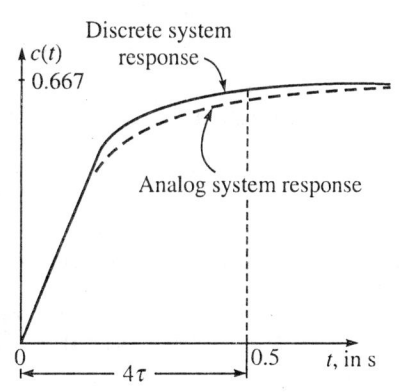

FIGURE 13.29 Example 13.16: output response.

13.12.2 Time constant

The closed-loop transfer function for the system in Eq. (13.49) is

$$\frac{G(z)}{1+G(z)} = \frac{0.362}{z - 0.457}$$

Hence the closed-loop characteristic equation is

$$z - 0.457 = 0 \quad (13.51)$$

Obviously, the pole is at $z_1 = 0.457$ which corresponds to an s-plane pole s_1 on the negative real axis that satisfies

$$z_1 = 0.457 = e^{s_1 T} = e^{0.1 s_1} \quad (\because T = 0.1 \text{ s}) \quad (13.52)$$

Hence, $$s_1 = \frac{\ln(0.457)}{0.1} = -7.848 \qquad (13.53)$$

Since time constant is the reciprocal of the magnitude of a real pole in the s-plane, the closed-loop system has the time constant $\tau = 1/7.848 = 0.127$ s. The settling time (with 2% criterion) is $4\tau = 0.5$ s. This is depicted in Figure 13.29. The time constant of the continuous closed-loop control system is obtained from Eq. (13.50) as 0.1666 s.

EXAMPLE 13.17 Find the time response of the system shown in Figure 13.30.

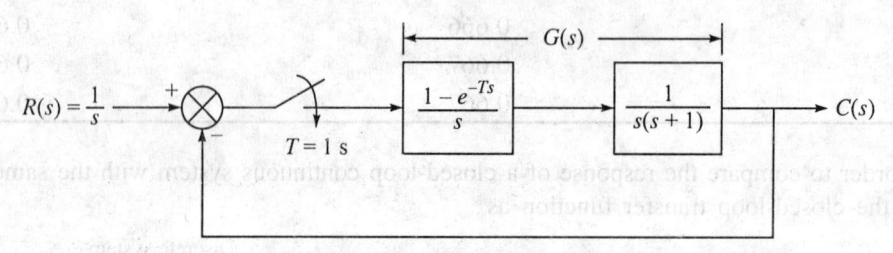

FIGURE 13.30 Example 13.17.

The system output can be expressed as

$$C(z) = \frac{G(z)}{1 + G(z)} R(z)$$

Here, $G(z)$ as already obtained in Example 13.10, is

$$G(z) = \left(\frac{z-1}{z}\right) \mathscr{Z}\left(\frac{1}{s^2(s+1)}\right) = \frac{0.368z + 0.264}{z^2 - 1.368z + 0.368}$$

Then the closed-loop transfer function can be written as

$$\frac{G(z)}{1 + G(z)} = \frac{0.368z + 0.264}{z^2 - z + 0.632}$$

Since $$R(z) = \frac{z}{z-1}$$

Then, $$C(z) = \frac{z(0.368z + 0.264)}{(z-1)(z^2 - z + 0.632)}$$

$$= 0.368z^{-1} + 1.0z^{-2} + 1.4z^{-3} + 1.4z^{-4} + 1.15z^{-5} + 0.9z^{-6} + 0.8z^{-7}$$
$$+ 0.87z^{-8} + 0.99z^{-9} + 1.08z^{-10} + 1.0z^{-11} + 1.00z^{-12} + 0.98z^{-13} + \ldots$$

The final value of $c(nT)$ by final-value theorem is obtained as

$$\lim_{n \to \infty} c(nT) = \lim_{z \to 1} (z-1)C(z) = \frac{0.632}{0.632} = 1$$

Digital Control Systems

The unit step response is plotted in Figure 13.31. Plotted in the same figure is the closed-loop response of the same system subjected to a unit step input, with sampler and data hold removed, that is, nothing but the continuous system for which the closed-loop transfer function becomes $1/(s^2 + s + 1)$. The overshoot of the continuous system is approximately 8%. It may thus be seen that sampling has a destabilizing effect on the system. As a matter of fact, a low sampling frequency has a detrimental effect on the stability of a closed-loop system.

MATLAB program %d_step.m is given in the following:

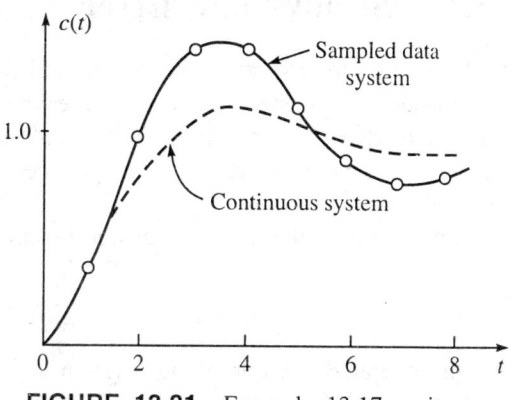

FIGURE 13.31 Example 13.17: unit step response.

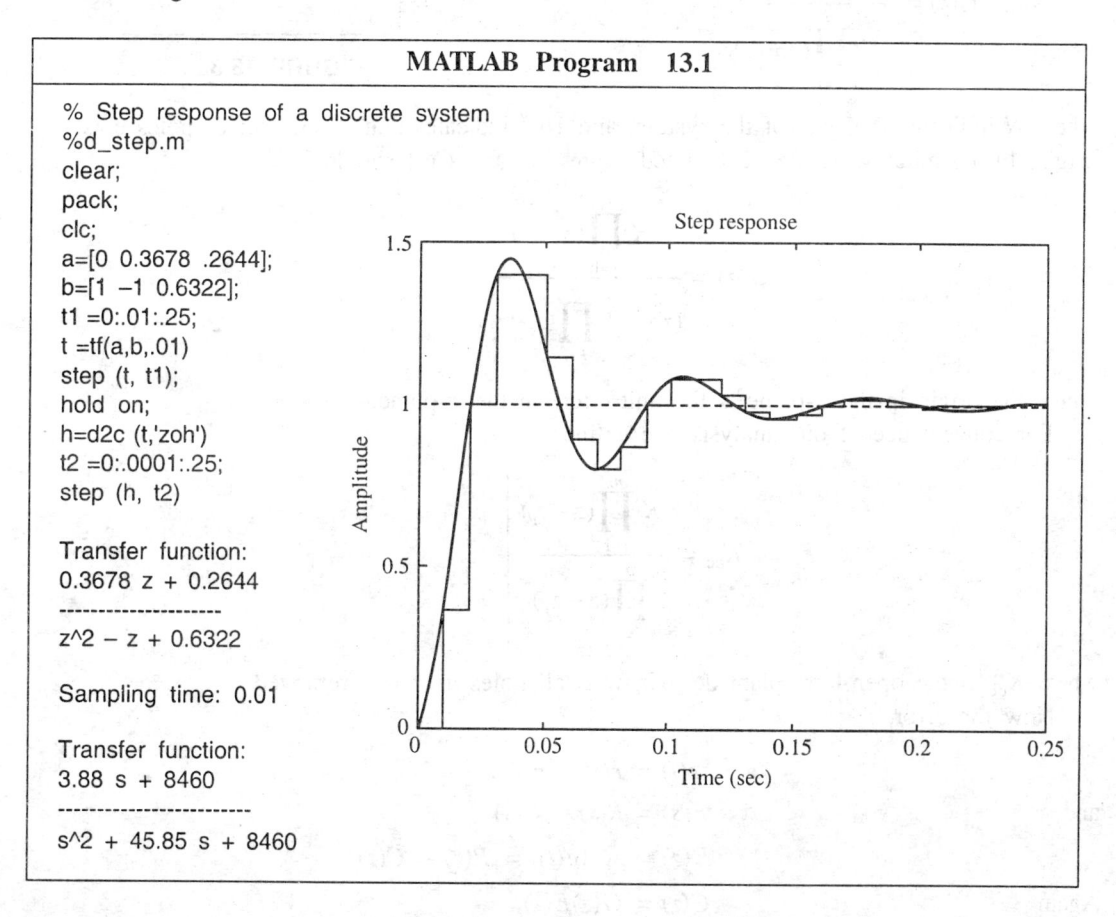

MATLAB Program 13.1

```
% Step response of a discrete system
%d_step.m
clear;
pack;
clc;
a=[0 0.3678 .2644];
b=[1 -1 0.6322];
t1 =0:.01:.25;
t =tf(a,b,.01)
step (t, t1);
hold on;
h=d2c (t,'zoh')
t2 =0:.0001:.25;
step (h, t2)
```

Transfer function:
0.3678 z + 0.2644
----
z^2 − z + 0.6322

Sampling time: 0.01

Transfer function:
3.88 s + 8460
----
s^2 + 45.85 s + 8460

13.13 Steady-State Error

In this section the effects of the system transfer characteristics on the steady-state system errors are considered. The steady-state system errors hint to the ability of the system to follow or track certain standard inputs with a finite minimum error. Here in steady-state error analysis of discrete systems, for convenience, the concerned related expressions for continuous systems are referred.

Consider the closed-loop transfer function of Figure 13.32,

$$\frac{C(z)}{R(z)} = \frac{G(z)}{1+G(z)}$$

In general, for a continuous system

$$G(s) = \frac{K \prod_{i=1}^{m}(s+s_i)}{s^N \prod_{j=1}^{p}(s+s_j)}$$

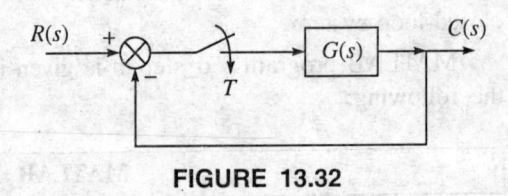

FIGURE 13.32

where N indicates the type of the system, and $1/s^N$ indicates that N number of poles exist at origin. In a similar way, the generalized expression for $G(z)$ should be

$$G(z) = \frac{K \prod_{i=1}^{m}(z-z_i)}{(z-1)^N \prod_{j=1}^{p}(z-z_j)}$$

where the origin in the s-plane is the unit circle in the z-plane.

For convenience of our analysis, we define

$$K_{dc} = \left. \frac{K \prod_{i=1}^{m}(z-z_i)}{\prod_{j=1}^{p}(z-z_j)} \right|_{z=1}$$

where K_{dc} is the open-loop plant dc gain with all poles at $z = 1$ removed.

Now the error,

$$e(t) = r(t) - c(t)$$

and

$$E(s) = R(s) - C(s)$$

$$E(z) = \mathscr{Z}[e(t)] = R(z) - C(z)$$

Again

$$C(z) = G(z)E(z)$$

Hence

$$E(z) = \frac{R(z)}{1+G(z)}$$

For a continuous system, from the final-value theorem, the steady-state error $e_{ss}(t)$ can be rewritten as

$$e_{ss}(t) = \lim_{t \to \infty} e(t) = \lim_{s \to 0} sE(s) = \lim_{s \to 0} (s-0)\frac{R(s)}{1+G(s)}$$

Hence the steady-state error for the discrete system under consideration can be written as

$$e_{ss}(kT) = \lim_{z \to 1} (z-1)E(z) = \lim_{z \to 1} (z-1)\frac{R(z)}{1+G(z)}$$

Case I: Position(Step) input
For the unit-step input,

$$R(z) = \frac{z}{z-1}$$

Then the steady-state error becomes

$$e_{ss}(kT) = \lim_{z \to 1} \frac{(z-1)R(z)}{1+G(z)} = \lim_{z \to 1} \frac{z}{1+G(z)} = \frac{1}{1+\lim_{z \to 1} G(z)} = \frac{1}{1+K_p}$$

where the position error constant, K_p, is

$$K_p = \lim_{z \to 1} G(z)$$

Now for a type-0 (i.e. $N = 0$) system

$$K_p = K_{dc}$$

and

$$e_{ss}(kT) = \frac{1}{1+K_p} = \frac{1}{1+K_{dc}}$$

For $N \geq 1$ (i.e. system of type greater than or equal to 1),

$$K_p = \infty$$

and the steady-state error,

$$e_{ss}(kT) = 0$$

Case II: Velocity(Ramp) input
For the unit-ramp input, i.e. $r(t) = t$,

$$R(z) = \frac{Tz}{(z-1)^2}$$

Then,

$$e_{ss}(kT) = \lim_{z \to 1} \frac{(z-1)R(z)}{1+G(z)} = \lim_{z \to 1} \frac{Tz}{(z-1)+(z-1)G(z)}$$

$$= \frac{T}{\lim_{z \to 1}(z-1)G(z)} = \frac{1}{K_v}$$

where the velocity error constant, K_v, is

$$K_v = \lim_{z \to 1} \frac{1}{T}(z-1)G(z)$$

For a type-0 (i.e. $N = 0$) system, $K_v = 0$, and the steady-state error,

$$e_{ss}(kT) = \frac{1}{K_v} = \infty$$

13.14 Root Loci for Sampled-Data Control Systems

For the sampled-data system of Figure 13.27, the closed loop transfer function is

$$\frac{C(z)}{R(z)} = \frac{G(z)}{1 + \overline{GH}(z)}$$

The characteristic equation is

$$1 + \overline{GH}(z) = 0$$

The root locus is the locus of the root of the closed-loop system obtained from open-loop pole-zero configuration when one parameter, say, gain K, is varied from zero to infinity. The open-loop transfer function of the system is

$$\overline{GH}(z) = \mathscr{Z}\left[\frac{1-e^{-Ts}}{s} \cdot \frac{K}{s(s+1)}\right] = K(1-z^{-1})\mathscr{Z}\left[\frac{1}{s^2(s+1)}\right]$$

$$= K\left(\frac{z-1}{z}\right) \frac{z(1-1+e^{-1}) + z(1-e^{-1}-e^{-1})}{(z-1)^2(z-e^{-1})}$$

Therefore with the sampling time $T = 1$ s,

$$\overline{GH}(z) = \frac{K(0.368z + 0.264)}{z^2 - 1.368z + 0.368} = \frac{0.368K(z + 0.717)}{(z-1)(z-0.368)}$$

The rules for the construction of root loci are:

1. The number of root loci $N = 2$, as number of poles of the open-loop transfer function is $P = 2$, and the number of finite zeros is $Z = 1$.
2. The starting points of root loci are the poles, i.e. $z = 1$ and $z = 0.368$.
3. One terminating point of root loci is the finite zero at $z = -0.717$ and the other is at zero at infinity.
4. The root locus is symmetrical about the real axis.
5. The root locus on the real axis always lies in a section of the real axis to the left of an odd number of poles and zeros on the real axis and is accordingly shown by the dark line on the real axis.

6. The number of asymptotes is $(P - Z)$ with angles

$$\frac{(2k+1)\pi}{P-Z}; \quad k = 0, 1, \ldots, (P - Z - 1)$$

i.e. there is only one asymptote at 180 degree.

7. The intersection of the asymptotes on the real axis is given by

$$= \frac{\sum \text{poles of } \overline{GH}(z) - \sum \text{zeros of } \overline{GH}(z)}{P - Z}$$

The question of intersection of asymptotes on the real axis does not arise in the present example as there is only one asymptote.

8. The break-away/break-in points are the roots of

$$\frac{d}{dz}[\overline{GH}(z)] = 0$$

This gives $z = 0.65$ and $z = -2.08$ as the break-away point and the break-in point respectively.

The value of gain K at the break-away point/break-in point is calculated from the magnitude condition as

$$\overline{GH}(z)\Big|_{z=0.65} = 1$$

which gives $K = 0.196$

and

$$\overline{GH}(z)\Big|_{z=-2.08} = 1$$

gives $K = 15$.

So we can conclude that the root loci for gain K in the range $0 \leq K \leq 0.196$ lie on the real axis, i.e. the roots of the closed-loop system are overdamped, and again in the range $15 \leq K \leq \infty$ the roots are on the real axis and hence the closed-loop system is overdamped, otherwise the closed-loop system would be underdamped as the roots of closed-loop system are complex conjugate for gain K in the range $0.196 \leq K \leq 15$.

9. The point of intersection of root loci with the unit circle can be obtained by Jury's stability test as in Example 13.13 and the value of gain for marginal stability is $K = 2.393$. For this value of K the characteristic equation is

$$[z^2 + (0.368K - 1.368)z + (0.368 + 0.264K)]\Big|_{K=2.393} = z^2 - 0.49z + 1 = 0$$

The roots of this equation are

$$z = 0.0245 \pm j\, 0.97 = 1\angle \pm 75.8° = 1\angle \pm 1.32 \text{ rad}$$
$$= 1\angle \pm \omega T$$
$$= 1\angle \pm \omega \quad (\because T = 1 \text{ s})$$

Since the sampling time $T = 1$ s, the system will oscillate at a frequency of 1.32 rad/s at gain $K = 2.393$. The point of intersection of the root loci with the unit circle is given by $1 \angle \pm 75.8°$ or $1 \angle \pm 1.32$ rad/s.

In the range of gain $0.196 < K < 2.393$, the closed-loop system is underdamped and stable. Further, in the range of gain $2.393 < K < 15$, the closed-loop system is underdamped but unstable.

The root loci is semi-circular for gain $15 > K > 0.196$. The root loci is drawn in Figure 13.33.

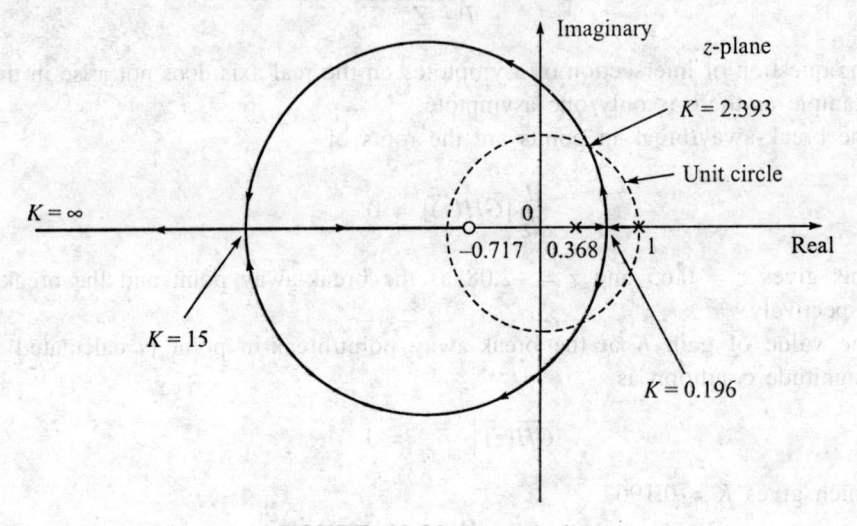

FIGURE 13.33 Root loci.

EXAMPLE 13.18 For the system shown in Figure 13.34(a), the root loci are drawn in Figure 13.34(b). Check the root loci with your own construction.

Solution: The unity-feedback system with the open-loop transfer function is given as

$$KG(z)H(z) = \mathscr{Z}\left[\frac{1-e^{-Ts}}{s}\right]\left[\frac{K(1+10s)}{s^2}\right]$$

$$= K(1 - z^{-1})\mathscr{Z}\left[\frac{K(1+10s)}{s^3}\right] = \left(\frac{z-1}{z}\right)KX(z)$$

Let us find the z-transform of $X(z)$, i.e.

$$\mathscr{Z}\left(\frac{1+10s}{s^3}\right)$$

By partial fraction expansion, we get

$$x(t) = \mathscr{L}^{-1}\left(\frac{1+10s}{s^3}\right) = \mathscr{L}^{-1}\left(\frac{1}{s^3} + \frac{10}{s^2}\right) = \frac{t^2}{2} + 10t$$

Digital Control Systems

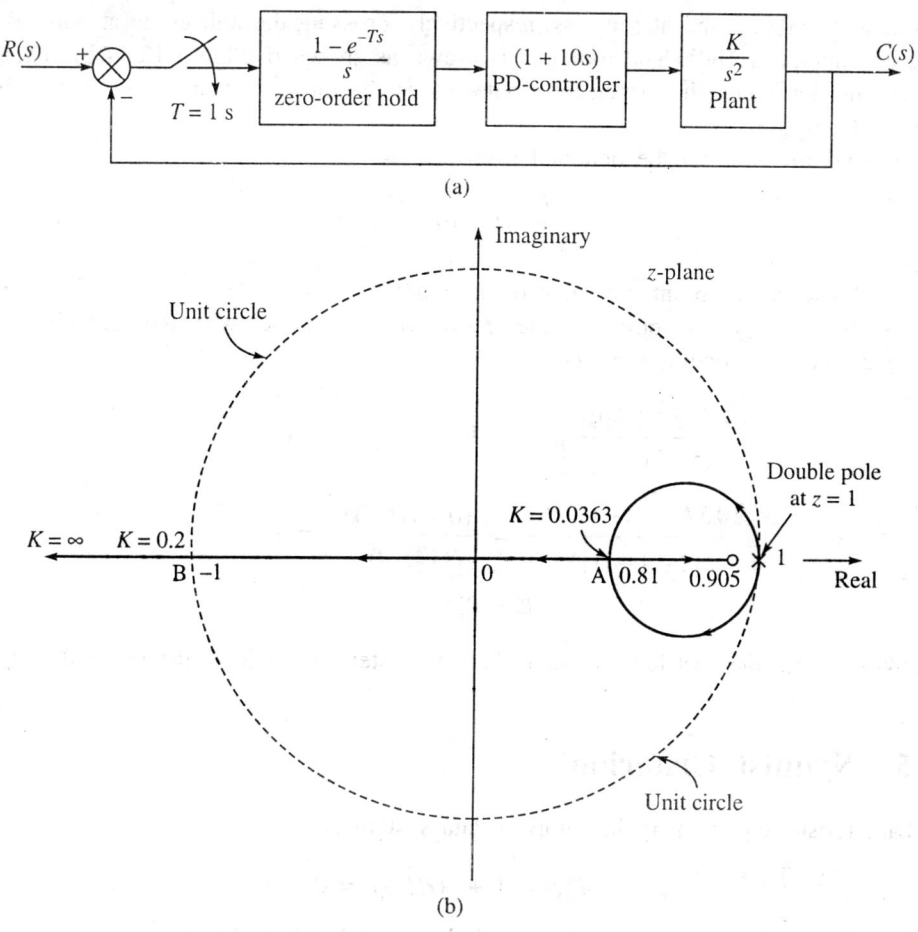

[Figure: (a) system block diagram with R(s), T = 1 s switch, zero-order hold $\frac{1-e^{-Ts}}{s}$, PD-controller $(1 + 10s)$, Plant $\frac{K}{s^2}$, output C(s); (b) root loci in z-plane showing unit circle, double pole at z = 1, zero at z = 0.905, points A at 0.81, B at -1, with K = 0.0363 at A, K = 0.2 at B, K = ∞]

FIGURE 13.34 Example 13.18: (a) system block diagram and (b) its root loci.

Now,
$$X(z) = \mathscr{Z}\,[x(t)] = \mathscr{Z}\left[\frac{t^2}{2}\right] + \mathscr{Z}\,[10t]$$

$$= \frac{T^2 z(z+1)}{2(z-1)^3} + \frac{10Tz}{(z-1)^2}$$

$$= \frac{10.5\, z\,(z - 0.905)}{(z-1)^3} \quad (\because T = 1\text{ s})$$

Putting the value of $X(z)$, we get the z-transform of the open-loop transfer function as

$$KG(z)H(z) = \frac{10.5\,K\,(z - 0.905)}{(z-1)^2}$$

The number of root loci is 2 as the number of finite poles $P = 2$ and is greater than the finite number of zeros $Z = 1$. The two root loci start from the double pole at $z = 1$ and terminate at $z = 0.905$ and at $z = -\infty$, respectively, crossing the unit circle at point B. There is one assymptote at $180°$. The root loci is drawn as shown in Figure 13.34(b). The system becomes unstable when the closed-loop pole leaves the interior of the unit circle at point A shown in the figure.

The break-in point can be obtained from

$$\frac{d}{dz}[KG(z)H(z)] = 0$$

as $z \neq 0.81$ and at this point the value of $K = 0.0363$.

The value of K at the point of intersection B of root loci with the unit circle can be determined from the condition $K|GH| = 1$, i.e.

$$\left.\frac{10.5 K (z - 0.905)}{(z - 1)^2}\right|_{z=-1} = 1$$

or

$$\frac{10.5 K |-1 - 0.905|}{|(-1 - 1)|^2} = \frac{10.5 K(1.905)}{(2)(2)} = 1$$

or

$$K = 0.2$$

It is obvious form the root-loci diagram, that the system is stable in the range $0 < K < 0.2$.

13.15 Nyquist Criterion

The characteristic equation of the sampled data system is

$$F(z) = 1 + \overline{GH}(z) = 0$$

The Nyquist criterion is based on the Cauchy's principle of argument.

Theorem

Let $F(z)$ be the ratio of two polynomials in z. Let the closed curve T_z in the z-plane be mapped into the complex plane through the mapping $F(z)$. If $F(z)$ is analytic within and on T_z contour, except at a finite number of poles (singular points), and if $F(z)$ has neither poles nor zeros on T_z contour, then

$$N = Z - P$$

where Z is the number of zeros of $F(z)$ in T_z, P is the number of poles of $F(z)$ in T_z, and N is the number of encirclements of the origin of F_z contour in $F(z)$-plane, taking the direction of encirclement in the same sense as that of T_z contour in the z-plane.

As for stability, there should not be any zero of the characteristic polynomial, i.e. the poles of the closed-loop system in the right-half of the s-plane for a continuous system, which for the discrete-time system comes out to be the outside of unit circle in the z-plane. In the z-plane, the Nyquist path is the unit circle corresponding to left-hand side of the s-plane and the path of its direction is counterclockwise.

Digital Control Systems

In the continuous time system the Nyquist contour T_s was traversed in a clockwise direction covering the entire right-hand side of the s-plane, which in z-plane comes out to be the contour T_z traversing around the unit circle in the z-plane in counterclockwise direction. This is because of the fact that the right-half of the s-plane maps outside the unit circle in the z-plane and we enclose this region by definition, if it is to our right while we traverse along the Nyquist contour.

To apply Cauchy's Principle of Argument, let

Z_i = number of zeros of $F(z)$ inside the unit circle
P_i = number of poles of $F(z)$ inside the unit circle
Z_o = number of zeros of $F(z)$ outside the unit circle
P_o = number of poles of $F(z)$ outside the unit circle

Then
$$N = Z_i - P_i \tag{13.54}$$

where N is the number of counterclockwise encirclements of origin of $F(z)$-plane or the -1 point made by the Nyquist diagram of $\overline{GH}(z)$. Now in general, the order of numerator and that of denominator of $1 + \overline{GH}(z)$ are the same and let it be n. Then

$$Z_o + Z_i = n \tag{13.55}$$

$$P_o + P_i = n \tag{13.56}$$

Then putting Eqs. (13.55) and (13.56) in Eq. (13.54), we get

$$N = -(Z_o - P_o) \tag{13.57}$$

Thus the Nyquist criterion for sampled-data system is given by Eq. (13.57) with the Nyquist path T_z and N being the number of counterclockwise encirclements of the -1 point by plotting the Nyquist plot $\overline{GH}(z)$ for values of z on the unit circle.

For singular points, we make a small detour around the point of singularity in the z-plane and then proceed further as in case of a continuous time system. This is illustrated below with an example.

EXAMPLE 13.19 Consider the sampled-data system as shown in Figure 13.27 with the value of the gain term $K = 1$. Draw the Nyquist plot and determine the stability of the system.

Solution: The open-loop transfer function of the sampled-data system can be written as

$$\overline{GH}(z) = \frac{0.368z + 0.264}{(z-1)(z-0.368)}$$

The Nyquist contour T_z in the z-plane is shown in Figure 13.35(a). The detour around the $z = 1$ point on the Nyquist contour is necessary since $\overline{GH}(z)$ has a pole at this point. The detour is a small semicircle having radius $\lim_{\varepsilon \to 0} \varepsilon e^{j\theta}$ outside the unit circle. Then on this detour, put

$$z = 1 + \lim_{\varepsilon \to 0} \varepsilon e^{j\theta}$$

and the corresponding mapping in the $\overline{GH}(z)$-plane becomes

$$\overline{GH}(z)\bigg|_{z=1+\lim_{\varepsilon\to 0}\varepsilon e^{j\theta}} = \frac{0.368z + 0.264}{(z-1)(z-0.368)}\bigg|_{z=1+\lim_{\varepsilon\to 0}\varepsilon e^{j\theta}}$$

$$= \lim_{\varepsilon\to 0}\frac{0.632}{\varepsilon e^{j\theta}(0.632)} = \lim_{\varepsilon\to 0}\frac{1}{\varepsilon}e^{-j\theta}$$

$$= (\infty)e^{-j\theta} \tag{13.58}$$

Thus the detour generates the large arc of infinite radius on the Nyquist diagram. The angle θ changes from $-90°$ through $0°$ to $+90°$ in the counterclockwise direction in T_z-contour in the z-plane, while in the corresponding Nyquist plot in $\overline{GH}(z)$-plane θ changes from $+90°$ through $0°$ to $-90°$ with infinite magnitude semicircle as indicated by section I in Figure 13.35(b).

For z on the unit circle,

$$\overline{GH}(z) = \bigg|_{z=e^{j\omega T}} = \frac{0.368 e^{j\omega T} + 0.264}{(e^{j\omega T} - 1)(e^{j\omega T} - 0.368)} \tag{13.59}$$

It may be noted that ω varies from $-\omega_s/2$ (i.e. $-\pi$) to $+\omega_s/2$ (i.e. $+\pi$), where ω_s is the sampling frequency $= 2\pi/T$. Further, it may be noted that $G(e^{j\omega T})$ for $0 > \omega > -\omega_s/2$ is the complex conjugate of $G(e^{j\omega T})$ for $0 < \omega < \omega_s/2$, and hence it is necessary to calculate $G(e^{j\omega T})$ only for $0 < \omega < \omega_s/2$, which is essentially the polar plot for real frequency, i.e. the section II of T_z-contour in z-plane. The Nyquist plot for section II in the z-plane is shown in Figure 13.35(b). It may be noted that $G(-1) = -0.0381$ and the point of intersection with the real axis occurs at -0.418.

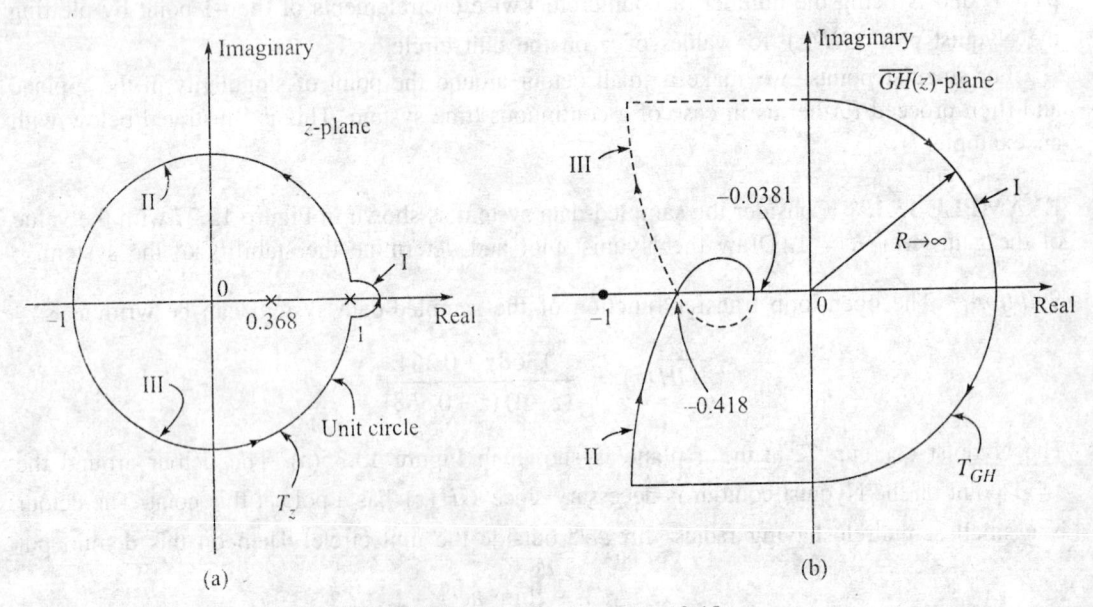

FIGURE 13.35 Example 13.19.

The inverse polar plot of $G(e^{j\omega T})$ for $0 > \omega > -\omega_s/2$ is drawn as shown in section III in Figure 13.35(b).

The entire Nyquist plot in the $\overline{GH}(z)$-plane is shown in Figure 13.35(b).

The number of counterclockwise encirclements of the -1 point in the $\overline{GH}(z)$-plane for $K = 1$ is zero, i.e. $N = 0$. Again $P_o = 0$.

Therefore,
$$Z_o = P_o - N = 0 \tag{13.60}$$

Hence the closed-loop sampled data system is stable for gain $K = 1$. However, it may be noted that the system can be forced to instability by increasing the gain by a factor of $1/0.418$ or 2.39.

13.16 State-Variable Formulation

We have already studied for the continuous linear time-invariant system that the state variable representation of the dynamics of the system can be written by the vector-matrix differential equation as

$$\dot{X} = AX + BU$$

and output
$$Y = CX + DU$$

with $n \times 1$ state vector X, $m \times 1$ output vector Y, $r \times 1$ input vector U, $n \times n$ system matrix A, $n \times r$ input coupling matrix B, $m \times n$ output coupling matrix C, and $m \times r$ input-output coupling matrix D.

For the linear time-invariant discrete multivariable system, the dynamics of the system can be represented by the vector-matrix difference equation as

$$X(k + 1) = AX(k) + BU(k)$$

and output
$$Y(k) = CX(k) + DU(k)$$

The state variable modelling process will now be illustrated by an example.

EXAMPLE 13.20 It is desired to find the state variable representation of the single-input-single-output linear time-invariant discrete system represented by the difference equation

$$y(k + 2) - 1.7y(k + 1) + 0.72y(k) = u(k)$$

Let
$$x_1(k) = y(k)$$
$$x_2(k) = x_1(k + 1) = y(k + 1)$$

Then
$$x_2(k + 1) = y(k + 2) = u(k) + 1.7x_2(k) - 0.72x_1(k)$$

From these equations, we write

$$x_1(k + 1) = x_2(k)$$
$$x_2(k + 1) = -0.72x_1(k) + 1.7x_2(k) + u(k)$$

and output
$$y(k) = x_1(k)$$

We may express these equations in vector-matrix form as

$$X(k+1) = \begin{bmatrix} 0 & 1 \\ -0.72 & 1.7 \end{bmatrix} X(k) + \begin{bmatrix} 0 \\ 1 \end{bmatrix} u(k)$$

$$= AX(k) + Bu(k)$$

and output
$$y(k)] = \begin{bmatrix} 1 & 0 \end{bmatrix} x(k)$$
$$= CX(k) + Du(k)$$

The transfer function of the discrete time system is

$$G(z) = C[zI - A]^{-1}B + D = \frac{1}{z^2 - 1.7z + 0.72}$$

The simulation in signal flow graph representation is shown in Figure 13.36.

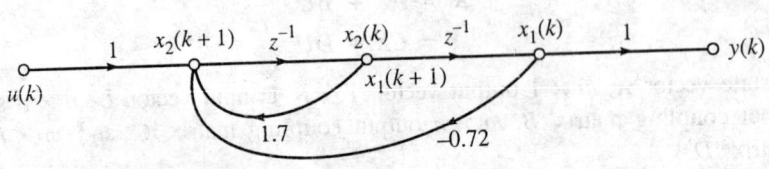

FIGURE 13.36 Example 13.20: signal flow graph.

The system simulation using the delay T for z^{-1} is shown in Figure 13.37.

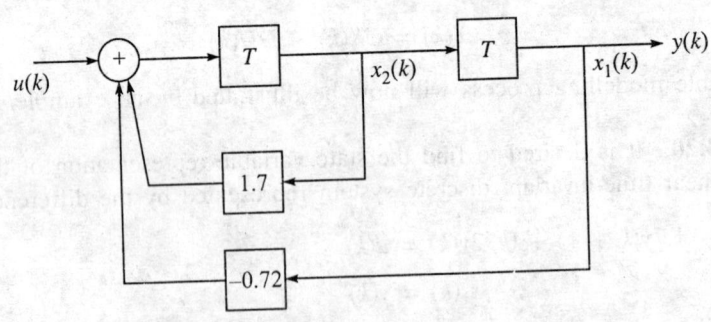

FIGURE 13.37 Example 13.20: system simulation.

It may be noted that the characteristic equation is

$$|zI - A| = 0$$

Further, in generalized terms, for the nth order single-input-single-output discrete linear time-invariant system, the state variable formulation can be written as

$$\begin{bmatrix} x_1(k+1) \\ x_2(k+1) \\ \vdots \\ x_n(k+1) \end{bmatrix} = \begin{bmatrix} 0 & 1 & 0 & 0 & \cdots & 0 \\ 0 & 0 & 1 & 0 & \cdots & 0 \\ \vdots & \vdots & \vdots & \vdots & & \\ -b_0 & -b_1 & -b_2 & -b_3 & \cdots & -b_{n-1} \end{bmatrix} \begin{bmatrix} x_1(k) \\ x_2(k) \\ \vdots \\ x_n(k) \end{bmatrix} + \begin{bmatrix} 0 \\ 0 \\ \vdots \\ 1 \end{bmatrix} u(k)$$

or simply, $\quad X(k+1) = AX(k) + Bu(k)$

and the output
$$y(k) = \begin{bmatrix} a_0 & a_1 & a_2 & \cdots & a_{n-1} \end{bmatrix} \begin{bmatrix} x_1(k) \\ x_2(k) \\ \vdots \\ x_n(k) \end{bmatrix}$$

or simply, $\quad y(k) = CX(k)$

Then the transfer function $G(z)$ becomes

$$G(z) = \frac{Y(z)}{U(z)}$$

$$= C(zI - A)^{-1}B$$

$$= \frac{a_{n-1} z^{n-1} + a_{n-2} z^{n-2} + \cdots + a_1 z + a_0}{z^n + b_{n-1} z^{n-1} + \cdots + b_1 z + b_0} \quad (13.61)$$

A signal flow graph representation is shown in Figure 13.38. The transfer function of a pure delay of T seconds is z^{-1}, when T is the discrete time increment (i.e. T is the sampling interval between the sampling instances k, $k + 1$, $k + 2$, etc.). Using this relationship, the signal flow graph of Figure 13.38 can be converted to the equivalent simulation diagram of Figure 13.39. This representation is called controllable canonical form or phase variable canonical form.

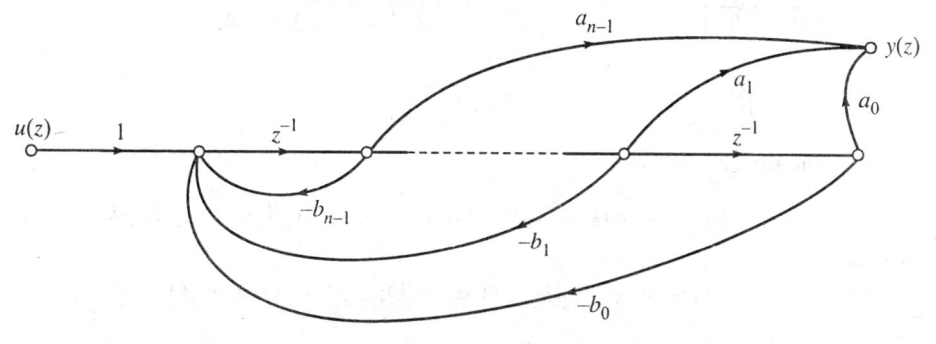

FIGURE 13.38 Signal flow graph.

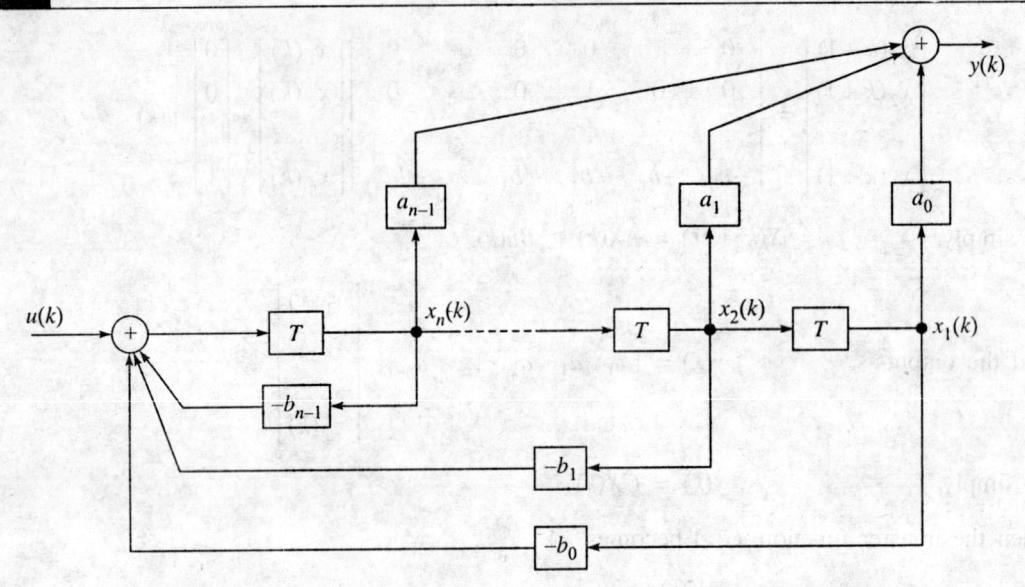

FIGURE 13.39 Simulation diagram of signal flow graph in Figure 13.38.

EXAMPLE 13.21 Consider a fourth-order system with a transfer function given by

$$G(z) = \frac{z^3 - 2z^2 + 2z - 1}{z^4 + z^3 - 3z^2 + 4z - 2}$$

Determine (a) an equivalent state-space form (both state and output equation) for this system and (b) the Jordan canonical form for the system's state equations.

Solution:

(a) $$G(z) = \frac{Y(z)}{U(z)} = \frac{z^3 - 2z^2 + 2z - 1}{z^4 + z^3 - 3z^2 + 4z - 2}$$

or $$\frac{Y(z)}{X_1(z)} \times \frac{X_1(z)}{U(z)} = (z^3 - 2z^2 + 2z - 1) \times \frac{1}{(z^4 + z^3 - 3z^2 + 4z - 2)}$$

Let $$\frac{X_1(z)}{U(z)} = \frac{1}{z^4 + z^3 - 3z^2 + 4z - 2}$$

which in time domain becomes

$$x_1(k+4) = -x_1(k+3) + 3x_1(k+2) - 4x_1(k+1) + 2x_1(k) + u(k)$$

Let us assume

$$x_1(k+1) = x_2(k); \quad x_2(k+1) = x_3(k); \quad x_3(k+1) = x_4(k)$$

Then we get from the previous equation as

$$x_4(k+1) = 2x_1(k) - 4x_2(k) + 3x_3(k) - x_4(k) + u(k)$$

Combining the above equation which in matrix form can be written in controllable canonical form as

$$\begin{bmatrix} x_1(k+1) \\ x_2(k+1) \\ x_3(k+1) \\ x_4(k+1) \end{bmatrix} = \begin{bmatrix} 0 & 1 & 0 & 0 \\ 0 & 0 & 1 & 0 \\ 0 & 0 & 0 & 1 \\ 2 & -4 & 3 & -1 \end{bmatrix} \begin{bmatrix} x_1(k) \\ x_2(k) \\ x_3(k) \\ x_4(k) \end{bmatrix} + \begin{bmatrix} 0 \\ 0 \\ 0 \\ 1 \end{bmatrix} u(k)$$

or $\quad X(k+1) = AX(k) + Bu(k)$

Further, let $\quad \dfrac{Y(z)}{X_1(z)} = z^3 - 2z^2 + 2z - 1$

which in time domain becomes

$$y(k) = x_1(k+3) - 2x_1(k+2) + 2x_1(k+1) - x_1(k)$$
$$= -x_1(k) + 2x_2(k) - 2x_3(k) + x_4(k)$$

which in matrix form can be written as

$$y(k) = [-1 \quad 2 \quad -2 \quad 1]X(k) + Du(k); \quad D = 0$$

or $\quad y(k) = CX(k) + Du(k)$

13.17 Samuelson's Model of National Economy

The national income $y(k)$ for a particular accounting period k (such as a quarter) is given by

$$y(k) = c(k) + i(k) + u(k)$$

where
$\quad c(k)$ = consumer expenditure
$\quad i(k)$ = investment
$\quad u(k)$ = government expenditure

Now according to Samuelson $c(k) = \alpha y(k-1)$

where α is the marginal prosperity to consumer.

Further, $\quad i(k) = \beta[c(k) - c(k-1)] = \alpha\beta[y(k-1) - y(k-2)]$

Finally, government expenditure is a constant for all k, i.e.

$$u(k) = u$$

With these assumptions, the simplified model of national economy is

$$y(k) = \alpha(1+\beta)y(k-1) - \alpha\beta y(k-2) + u$$

Taking z-transform, the pulse transfer function of national economy is

$$\frac{y(z)}{u(z)} = \frac{y(z)}{x(z)} \cdot \frac{x(z)}{u(z)} = 1 \cdot \frac{1}{1-\alpha(1+\beta)z^{-1} - \alpha\beta z^{-2}}$$

It may be noted that the delay of one unit is represented by z^{-1} and delay of two units is represented by z^{-2}, and so on.

Then
$$x(z) = u + \alpha(1 + \beta)z^{-1}x(z) + \alpha\beta z^{-2}x(z)$$

Let
$$z^{-2}x(z) = x_1(z)$$

and
$$z^{-1}x(z) = zx_1(z) = x_2(z)$$

Then in matrix form, we get

$$\begin{bmatrix} x_1(k+1) \\ x_2(k+1) \end{bmatrix} = \begin{bmatrix} 0 & 1 \\ -\alpha\beta & \alpha(1+\beta) \end{bmatrix} \begin{bmatrix} x_1(k) \\ x_2(k) \end{bmatrix} + \begin{bmatrix} 0 \\ 1 \end{bmatrix} u$$

and
$$y(k) = \begin{bmatrix} c_1 & c_2 \end{bmatrix} \begin{bmatrix} x_1(k) \\ x_2(k) \end{bmatrix} + u$$

where
$$c_1 = -\alpha\beta \quad \text{and} \quad c_2 = \alpha(1 + \beta)$$

The simulation is shown in Figure 13.40.

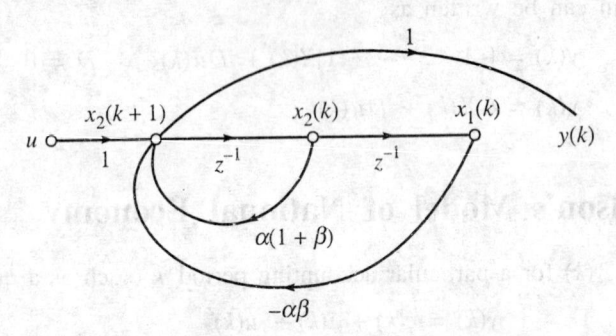

FIGURE 13.40 Model of national economy.

13.18 Solutions to State Equation by Recursive Method

For a linear, single-input-single-output time-invariant discrete system, the state-variable formulation becomes

$$X(k + 1) = AX(k) + Bu(k)$$

and output
$$y(k) = CX(k) + Du(k)$$

and when the given initial condition $X(0)$ and input $u(k)$ for $k = 0, 1, 2, \ldots$ are known, then

$$X(1) = AX(0) + Bu(0)$$

and
$$X(2) = AX(1) + Bu(1)$$
$$= A(AX(0) + Bu(0)) + Bu(1) = A^2X(0) + ABu(0) + Bu(1)$$

In a similar manner, we can get

$$X(3) = A^3X(0) + A^2Bu(0) + ABu(1) + Bu(2)$$

The general solution is given by

$$X(k) = A^k X(0) + \sum_{j=0}^{k-1} A^{(k-1-j)} Bu(j)$$

If we define $\Phi(k) = A^k$

then

$$X(k) = \underbrace{\Phi(k)X(0)}_{\text{zero-input component}} + \underbrace{\sum_{j=0}^{k-1} \Phi(k-1-j) Bu(j)}_{\text{zero-state component}}$$

$$= \Phi(k)X(0) + \text{convolution } [\Phi(k), Bu(k)]$$

and the output solution can be written as

$$y(k) = C\Phi(k)X(0) + \sum_{j=0}^{k-1} C\Phi(k-1-j)Bu(j) + Du(k)$$

where $\Phi(k)$ is called the state transition matrix or the fundamental matrix.

EXAMPLE 13.22 Consider the transfer function

$$G(z) = \frac{z+3}{(z+1)(z+2)}$$

Then state variable formulation is

$$X(k+1) = \begin{bmatrix} 0 & 1 \\ -2 & -3 \end{bmatrix} X(k) + \begin{bmatrix} 0 \\ 1 \end{bmatrix} u(k)$$

and output $y(k) = [3 \quad 1]X(k)$

Assume that the system is initially at rest, i.e. $X(0) = [0 \quad 0]^T$ and the input is a unit step, i.e.

$$u(k) = 1; \; k = 0, 1, 2, \ldots$$

The recursive solution is obtained as

$$X(1) = \begin{bmatrix} 0 & 1 \\ -2 & -3 \end{bmatrix} X(0) + \begin{bmatrix} 0 \\ 1 \end{bmatrix} u(0) = \begin{bmatrix} 0 & 1 \\ -2 & -3 \end{bmatrix}\begin{bmatrix} 0 \\ 0 \end{bmatrix} + \begin{bmatrix} 0 \\ 1 \end{bmatrix}(1) = \begin{bmatrix} 0 \\ 1 \end{bmatrix}$$

and $y(1) = [3 \quad 1]\begin{bmatrix} 0 \\ 1 \end{bmatrix} = 1$

Then, $X(2) = \begin{bmatrix} 0 & 1 \\ -2 & -3 \end{bmatrix} X(1) + \begin{bmatrix} 0 \\ 1 \end{bmatrix} u(1) = \begin{bmatrix} 0 & 1 \\ -2 & -3 \end{bmatrix}\begin{bmatrix} 0 \\ 1 \end{bmatrix} + \begin{bmatrix} 0 \\ 1 \end{bmatrix}(1) = \begin{bmatrix} 1 \\ -2 \end{bmatrix}$

and $y(2) = [3 \quad 1]\begin{bmatrix} 1 \\ -2 \end{bmatrix} = 1$

In a similar manner it can be shown that

$$X(3) = \begin{bmatrix} -2 \\ 5 \end{bmatrix} \text{ and } y(3) = -1$$

Then
$$X(4) = \begin{bmatrix} 5 \\ -10 \end{bmatrix} \text{ and } y(4) = 5, \text{ etc.}$$

Hence the state and output at successive time instants can be determined.

In terms of the convolution formula, the system state at step four is given by

$$X(4) = A^4 X(0) + A^3 Bu(0) + A^2 Bu(1) + ABu(2) + Bu(3)$$

$$= \begin{bmatrix} -14 & -15 \\ 30 & 31 \end{bmatrix} \begin{bmatrix} 0 \\ 0 \end{bmatrix} + \begin{bmatrix} 6 & 7 \\ -14 & -15 \end{bmatrix} \begin{bmatrix} 0 \\ 1 \end{bmatrix}(1) + \begin{bmatrix} -2 & -3 \\ 6 & 7 \end{bmatrix} \begin{bmatrix} 0 \\ 1 \end{bmatrix}(1)$$

$$+ \begin{bmatrix} 0 & 1 \\ -2 & -3 \end{bmatrix} \begin{bmatrix} 0 \\ 1 \end{bmatrix}(1) + \begin{bmatrix} 0 \\ 1 \end{bmatrix}(1) = \begin{bmatrix} 5 \\ -10 \end{bmatrix}$$

Solution by z-transform method

For a homogeneous system, i.e. $u(k) = 0$, the z-transform of

$$X(k + 1) = AX(k)$$

is $\quad zX(z) - zX(0) = AX(z)$

or $\quad X(z) = z[zI - A]^{-1} X(0)$

or $\quad X(k) = \mathscr{Z}^{-1}[X(z)] = \mathscr{Z}^{-1}[(z[zI - A]^{-1})]X(0) = \Phi(k)X(0)$

where STM, $\quad \Phi(k) = \mathscr{Z}^{-1}[(z[zI - A]^{-1})]$

EXAMPLE 13.23 Consider the system

$$X(k) = \begin{bmatrix} 0 & 1 \\ -2 & -3 \end{bmatrix} X(0)$$

Then
$$[zI - A] = \begin{bmatrix} z & -1 \\ 2 & z+3 \end{bmatrix}$$

The determinant

$$|zI - A| = z^2 + 3z + 2 = (z+1)(z+2)$$

Now
$$z[zI - A]^{-1} = \begin{bmatrix} \left(\dfrac{2z}{z+1} + \dfrac{-z}{z+2}\right) & \left(\dfrac{z}{z+1} + \dfrac{-z}{z+2}\right) \\ \left(\dfrac{-2z}{z+1} + \dfrac{2z}{z+2}\right) & \left(\dfrac{-z}{z+1} + \dfrac{2z}{z+2}\right) \end{bmatrix}$$

Then, STM $\Phi(k)$ can be written as

$$\Phi(k) = \mathscr{L}^{-1}[(zI - A)^{-1}] = \begin{bmatrix} 2(-1)^k - (-2)^k & (-1)^k - (-2)^k \\ -2(-1)^k + 2(-2)^k & -(-1)^k + 2(-2)^k \end{bmatrix}$$

The solution of the forced system can be obtained in the same way as for the continuous case.

Controllability and Observability

Further, the condition of controllability and observability will be the same for the discrete system as it was in the case of continuous system.

The controllability condition is the controllability matrix

$$[B \quad AB \quad A^2B \quad \ldots \quad A^{n-1}B]$$

which should be of rank n.

And for observability condition, the observability matrix

$$[C \quad CA \quad CA^2 \quad \ldots \quad CA^{n-1}]^T$$

should be of rank n for nth order system described by

$$X(k + 1) = AX(k) + BU(k)$$

and output

$$Y(k) = CX(k) + DU(k)$$

13.19 Direct Decomposition (Phase Variable Form)

The most versatile way of compensating a discrete-data control system is to use a digital controller. The block diagram in Figure 13.41 can be used to represent a single-input-single-output digital controller.

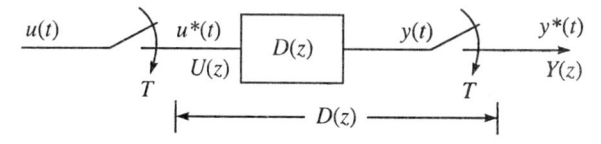

FIGURE 13.41 Block diagram of digital controller.

The input to the controller, $u^*(t)$ is represented as the output of the ideal sampler. The digital controller performs certain linear operations on the sequence $u(kT)$ and delivers the output sequence $y(kT)$ which is portrayed as the output of another sampler $y^*(t)$. The transfer function of the digital controller is described as

$$\frac{Y(z)}{U(z)} = \frac{\beta_0 z^n + \beta_1 z^{n-1} + \cdots + \beta_n}{z^n + \alpha_1 z^{n-1} + \cdots + \alpha_n} \qquad (13.62)$$

Clearly, for the digital controller to be physically realizable, the power-series expansion of $D(z)$ must not contain any positive power of z. Therefore, for the transfer function of Eq. (13.62) to be physically realizable, the highest power of the denominator must be equal to or greater than that of the numerator.

If the digital controller has the same number of poles and zeros, it is expressed as

$$\frac{Y(z)}{U(z)} = \frac{\beta_0 + \beta_1 z^{-1} + \cdots + \beta_n z^{-n}}{1 + \alpha_1 z^{-1} + \cdots + \alpha_n z^{-n}} \tag{13.63}$$

or
$$\frac{Y(z)}{U(z)} = \frac{\beta_0 + \beta_1 z^{-1} + \cdots + \beta_n z^{-n}}{1 + \alpha_1 z^{-1} + \cdots + \alpha_n z^{-n}} \cdot \frac{X_1(z)}{X_1(z)}$$

or
$$\frac{Y(z)}{X_1(z)} \cdot \frac{X_1(z)}{U(z)} = (\beta_0 + \beta_1 z^{-1} + \cdots + \beta_n z^{-n}) \cdot \frac{1}{1 + \alpha_1 z^{-1} + \cdots + \alpha_n z^{-n}}$$

or
$$\frac{X_1(z)}{U(z)} = \frac{1}{1 + \alpha_1 z^{-1} + \cdots + \alpha_n z^{-n}}$$

or
$$X_1(z) = U(z) - \alpha_1 z^{-1} X_1(z) + \cdots + \alpha_n z^{-n} X_1(z) \tag{13.64}$$

and
$$\frac{Y(z)}{X_1(z)} = (\beta_0 + \beta_1 z^{-1} + \cdots + \beta_n z^{-n})$$

or
$$Y(z) = (\beta_0 + \beta_1 z^{-1} + \cdots + \beta_n z^{-n}) X_1(z) \tag{13.65}$$

Choose the state variables as

$$x_1(k) = z^{-n} X_1(z)$$
$$x_2(k) = z^{-(n-1)} X_1(z) = z x_1(k)$$
$$x_3(k) = z^{-(n-2)} X_1(z) = z x_2(k) \tag{13.66}$$
$$\vdots \quad \vdots \quad \vdots \quad \vdots$$
$$x_n(k) = z^{-1} X_1(z) = z x_{n-1}(k)$$

In time the corresponding relation becomes

$$x_1(k + 1) = x_2(k) \tag{13.67}$$
$$\vdots \quad \vdots$$
$$x_{n-1}(k + 1) = x_n(k)$$

Substituting Eq. (13.66) in (13.64) and (13.65) yields

$$z x_n(k) = U(z) - \alpha_1 x_n(k) - \alpha_2 x_{n-1}(k) - \cdots - \alpha_n x_1(k) \tag{13.68}$$

and
$$Y(z) = \beta_0 z x_n(k) + \beta_1 x_n(k) + \cdots + \beta_n x_1(k) \tag{13.69}$$

The corresponding time domain equation

$$x_n(k + 1) = u(k) \alpha_1 x_n(k) - \alpha_2 x_{n-1}(k) - \cdots - \alpha_n x_1(k) \tag{13.70}$$

and
$$y(k) = \beta_0 x_n(k + 1) + \beta_1 x_n(k) + \cdots + \beta_n x_1(k) \tag{13.71}$$

Putting Eq. (13.70) in Eq. (13.71), we get

$$y(k) = c_1 x_1(k) + c_2 x_2(k) + \cdots + c_n x_n(k) \tag{13.72}$$

where
$$c_i = \beta_{n-i} - \alpha_{n-i+1} \beta_0 \tag{13.73}$$

$$X(k+1) = \begin{bmatrix} 0 & 1 & 0 & \cdots & 0 \\ 0 & 0 & 1 & \cdots & 0 \\ \vdots & \vdots & \vdots & & \vdots \\ 0 & 0 & 0 & \cdots & 1 \\ -\alpha_n & -\alpha_{n-1} & -\alpha_{n-2} & \cdots & -\alpha_1 \end{bmatrix} X(k) + \begin{bmatrix} 0 \\ 0 \\ \vdots \\ 0 \\ 1 \end{bmatrix} u(k) \qquad (13.74)$$

$$y(k) = [c_1 \quad c_2 \quad \cdots \quad c_n] X(k) + \beta_0 u(k) \qquad (13.75)$$

The simulation is shown in Figure 13.42.

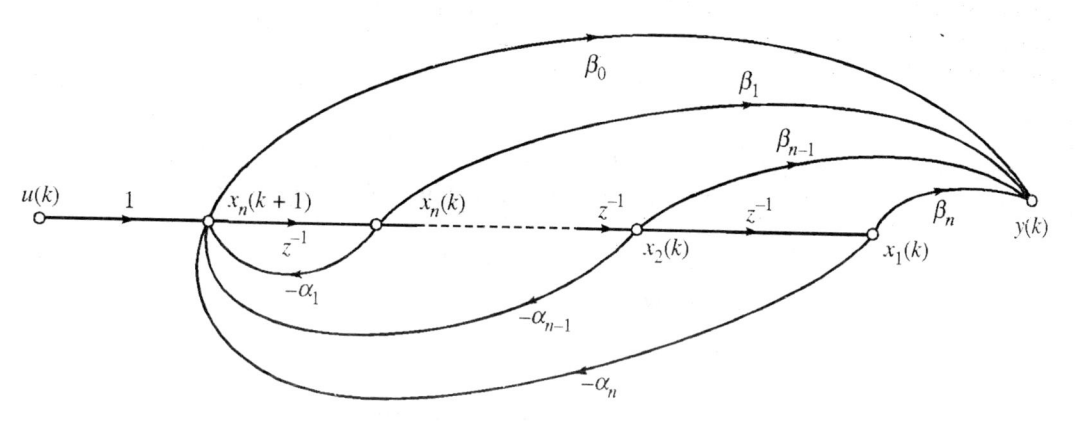

FIGURE 13.42 Direct decomposition.

13.20 Relationship between Continuous and Discrete State Equations

For a single-input-single-output system, the state equations of the continuous portion of the sampled data control system of Figure 13.43 can be written as

$$\dot{V}(t) = A_c V(t) + B_c u(t) \qquad (13.76)$$

and output $\quad y(t) = C_c V(t) + D_c u(t) \qquad (13.77)$

FIGURE 13.43 Continuous controller.

After taking the Laplace transform and manipulating, we get

$$V(s) = \Phi_c(s)V(0) + \Phi_c(s)B_c U(s)$$

where the resolvent matrix

$$\Phi_c(s) = [sI - A_c]^{-1}$$

The time-domain solution is

$$V(t) = \Phi_c(t - t_0)V(t_0) + \int_0^t \Phi_c(t - \tau) B_c u(\tau) d\tau \qquad (13.78)$$

where the state transition matrix (STM)

$$\Phi_c(t - t_0) = \mathcal{L}^{-1}\, \Phi_c(s)$$

The STM $\Phi_c(t - t_0)$ can be calculated by the Laplace inverse transform method if the roots are known, otherwise we have to go for the power-series method specially when the roots are not known, as given below.

$$\Phi_c(t - t_0) = \sum_{r=0}^{\infty} \frac{A_c^r\, (t-t_0)^r}{r!}$$

Put $\quad t_0 = kT \quad \text{and} \quad t = (k + 1)T$ in Eq. (13.78) to get

$$V(k + 1)T = \Phi_c(T)V(kT) + \int_{kT}^{(k+1)T} \Phi_c[(k + 1)T - \tau] B_c m(kT) dT \qquad (13.79)$$

The input $u(t)$ has been replaced by $m(kT)$ since during the time interval $kT \leq t \leq (k + 1)T$, the input $u(t)$ becomes $m(kT)$. The discrete state equation can be represented as

$$X(k + 1) = AX(k) + Bm(k)$$

where
$$X(kT) = V(kT)$$
$$A = \Phi(T)$$
$$B = \int_{kT}^{(k+1)T} \Phi_c[(k + 1)T - \tau] B_c\, d\tau$$

Substituting $\quad kT - \tau = -\sigma,$

$$B = \left[\int_0^T \Phi_c(T - \sigma)\, d\sigma\right] B_c$$

The discrete system matrices A and B may be calculated by determining STM $\Phi_c(t)$ of the continuous system as follows:

$$\Phi_c(T) = \sum_{r=0}^{\infty} \frac{A_c^r\, T^r}{r!}$$

In order to find B, let $\tau = T - \sigma$, then

$$\int_0^T \Phi_c(T - \sigma) d\sigma = \int_0^T \Phi_c(\tau)\, d\tau = \int_0^T \left(\sum_{r=0}^{\infty} \frac{A_c^r\, T^r}{r!}\right) d\tau$$

$$= IT + A_c(T^2/2!) + A_c^2(T^3/3!) + \cdots$$

Hence
$$B = \left[\int_0^T \Phi_c(\tau)\, d\tau\right] B_c$$

Digital Control Systems

For a linear servomotor, the given transfer function is

$$G_p(s) = \frac{10}{s(s+1)}$$

Now referring to this sampled data control system, we proceed as follows:

$$\frac{Y(s)}{U(s)} = \frac{Y(s)}{V_1(s)} \frac{V_1(s)}{U(s)} = 1 \cdot \frac{10}{s(s+1)}$$

Let

$$\frac{V_1(s)}{U(s)} = \frac{10}{s(s+1)}$$

then the time-domain equation becomes

$$\ddot{v}_1 + \dot{v}_1 = 10u(t)$$

Let

$$\dot{v}_1 = v_2$$

then

$$\dot{v}_2 = -v_2 + 10u(t)$$

Then the state equation is

$$\begin{bmatrix} \dot{v}_1 \\ \dot{v}_2 \end{bmatrix} = \begin{bmatrix} 0 & 1 \\ 0 & -1 \end{bmatrix} \begin{bmatrix} v_1 \\ v_2 \end{bmatrix} + \begin{bmatrix} 0 \\ 10 \end{bmatrix} u$$

i.e.

$$\dot{V} = A_c V + B_c u$$

And let

$$\frac{Y(s)}{V_1(s)} = 1$$

which gives the output in time domain in matrix form as

$$y(t) = \begin{bmatrix} 1 & 0 \end{bmatrix} \begin{bmatrix} v_1 \\ v_2 \end{bmatrix}$$

i.e.

$$y(t) = C_c V$$

Then, STM $\Phi_c(t)$ can be written as

$$\Phi_c(t) = \mathcal{L}^{-1}[sI - A_c]^{-1} = \mathcal{L}^{-1}\begin{bmatrix} s & -1 \\ 0 & s+1 \end{bmatrix}^{-1} = \begin{bmatrix} 1 & 1-e^{-t} \\ 0 & e^{-t} \end{bmatrix}$$

Then for $T = 0.1$ s, we get

$$A = \Phi_c(T)\big|_{T=0.1} = \Phi_c(0.1) = \begin{bmatrix} 1 & 0.0952 \\ 0 & 0.905 \end{bmatrix}$$

also

$$\int_0^T \Phi_c(\tau)\, d\tau = \begin{bmatrix} T & T-1+e^{-T} \\ 0 & 1-e^{-T} \end{bmatrix}$$

$$B = \left[\int_0^T \Phi_c(T)\,dT\right]B_c = \begin{bmatrix} 0.1 & 0.00484 \\ 0 & 0.0952 \end{bmatrix}\begin{bmatrix} 0 \\ 10 \end{bmatrix} = \begin{bmatrix} 0.0484 \\ 0.952 \end{bmatrix}$$

Hence the discrete state equation becomes

$$X(k+1) = AX(k) + Bm(k)$$

and output

$$y(k) = CX(k) + Dm(k)$$

where

$$A = \begin{bmatrix} 1 & 0.0952 \\ 0 & 0.905 \end{bmatrix},\ B = \begin{bmatrix} 0.1 & 0.00484 \\ 0 & 0.0952 \end{bmatrix},\ C = [1\ \ 0],\ D = 0$$

The simulation is shown in Figure 13.44.

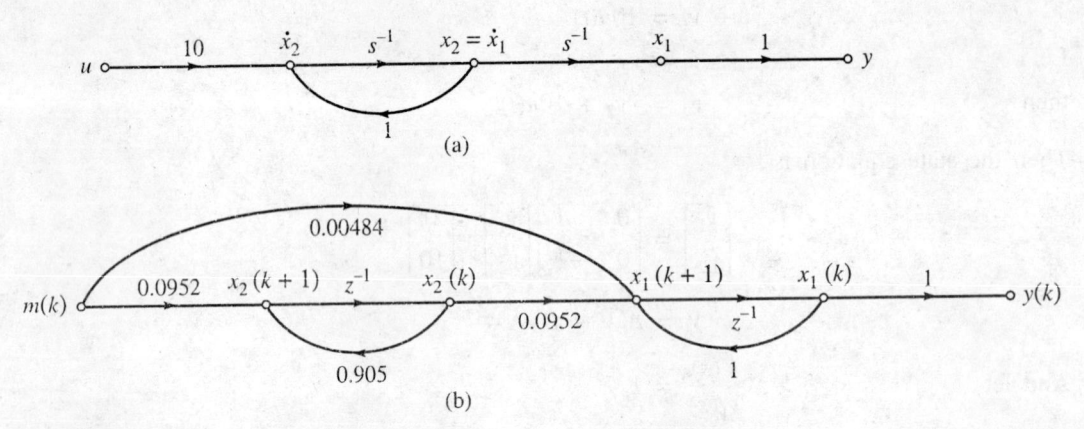

FIGURE 13.44 (a) Simulation of continuous system and (b) simulation of corresponding sampled data system.

13.21 Cayley–Hamilton Method

For discrete time case, the Cayley–Hamilton procedure can be used for computing A^k. Here $F(\lambda_i)$ is used as λ_i^k rather than $e^{\lambda_i t}$ as used for the continuous case.

EXAMPLE 13.24 Compute $\Phi(k)$ for the difference equation

$$y[(k+2)T] + 5y[k+1]T] + 6y(kT) = 0$$

The system matrix A for the system is, $A = \begin{bmatrix} 0 & 1 \\ -6 & -5 \end{bmatrix}$

Assume $x_1(kT) = y(kT)$, $x_2(kT) = y(k+1)T$. The characteristic equation $|\lambda I - A| = 0$ has two characteristic roots $\lambda_1 = -2$ and $\lambda_2 = -3$. Therefore,

$$F(\lambda_1) = \lambda_1^k = (-2)^k = \alpha_0 + \alpha_1\lambda_1 = \alpha_0 - 2\alpha_1$$

$$F(\lambda_2) = \lambda_2{}^k = (-3)^k = \alpha_0 + \alpha_1\lambda_2 = \alpha_0 - 3\alpha_1$$

Solving these two equations, we get

$$\alpha_0 = 3(-2)^k - 2(-3)^k \quad \text{and} \quad \alpha_1 = (-2)^k - (-3)^k$$

Hence
$$F(A) = A^k = \Phi(k) = \alpha_0 I + \alpha_1 A$$

$$= \begin{bmatrix} 3(-2)^k - 2(-3)^k & (-2)^k - (-3)^k \\ -6[(-2)^k - (-3)^k] & -2(-2)^k + 3(-3)^k \end{bmatrix}$$

The problem handles only the distinct roots.

EXAMPLE 13.25 Given $A = \begin{bmatrix} 0 & 1 \\ -1 & -2 \end{bmatrix}$; repetitive roots -1 (double roots). Find the STM.

Solution: In the repetitive roots case, α's are obtained as

$$\left.\frac{d^s F(\lambda)}{d\lambda^s}\right|_{\lambda=\lambda_i} = \left.\frac{d^s F(\lambda)}{d\lambda^s}\left[\sum \alpha_r \lambda^r\right]\right|_{\lambda=\lambda_i}$$

$$= \left.\frac{d^s}{d\lambda^s}[\alpha_0 + \alpha_1 \lambda]\right|_{\lambda=\lambda_i} \quad ; s = 0, 1, 2, \ldots, p_i - 1$$

where p_i is the order of the root.

Hence $\quad (-1)^k = \alpha_0 + \alpha_1 \lambda_1 = \alpha_0 - \alpha_1 \quad \text{and} \quad -k(-1)^k = \alpha_1$

or $\quad \alpha_1 = -k(-1)^k \text{ and } \alpha_0 = (-1)^k (1 - k)$

Therefore, $\quad \Phi(k) = \alpha_0 I + \alpha_1 A = \begin{bmatrix} \alpha_0 & \alpha_1 \\ -\alpha_1 & \alpha_0 - 2\alpha_1 \end{bmatrix} = (-1)^k \begin{bmatrix} 1-k & -k \\ k & 1+k \end{bmatrix}$

EXAMPLE 13.26 Given $A = \begin{bmatrix} 0 & 1 \\ -2 & 2 \end{bmatrix}$; roots $-1 \pm j$. Find the STM.

Solution: In polar form $\sqrt{2}\, e^{\pm j\pi/4}$

$$F(\lambda_1) = (2)^{k/2}\, e^{j k\pi/4} = \alpha_0 + \alpha_1 + j\alpha_1$$
$$F(\lambda_2) = (2)^{k/2}\, e^{-j k\pi/4} = \alpha_0 + \alpha_1 - j\alpha_1$$

Solving both the equations, we get

$$\alpha_1 = (2)^{k/2} \sin(k\pi/4)$$

and $\quad \alpha_0 = (2)^{k/2} [\cos(k\pi/4) - \sin(k\pi/4)]$

Now, STM $\quad\Phi(k) = \alpha_0 I + \alpha_1 A$

$$= \begin{bmatrix} \alpha_0 & \alpha_1 \\ -\alpha_1 & \alpha_0 - 2\alpha_1 \end{bmatrix} = (2)^{k/2} \begin{bmatrix} \left(\cos\dfrac{k\pi}{4} - \sin\dfrac{k\pi}{4}\right) & \sin\dfrac{k\pi}{4} \\ -2\sin\dfrac{k\pi}{4} & \left(\cos\dfrac{k\pi}{4} + \sin\dfrac{k\pi}{4}\right) \end{bmatrix}$$

EXAMPLE 13.27 Determine the unit-step response in closed-form for the system given by

$$X(k+1) = \begin{bmatrix} 0 & 1 \\ 0 & -1 \end{bmatrix} X(k) + \begin{bmatrix} 0 \\ 1 \end{bmatrix} u(k) \text{ for all } k \geq 0 \text{ with } X(0) = 0$$

Solution: $A = \begin{bmatrix} 0 & 1 \\ 0 & -1 \end{bmatrix}$; $|\lambda I - A| = \lambda^2 + \lambda = 0$ gives eigenvalues as $\lambda_1 = 0$ and $\lambda_2 = -1$.

Using Cayley–Hamilton theorem, the STM

$$\Phi(k) = A^k = \alpha_0 I + \alpha_1 A$$

Substituting $\lambda = 0$, we get $\alpha_0 = 0$ and substituting $\lambda = -1$, we get $\alpha_1 = (-1)^{k-1}$

Then $\quad\Phi(k) = \begin{bmatrix} 0 & (-1)^{k-1} \\ 0 & (-1)^k \end{bmatrix} = A_d$

Given $X(0) = 0$, hence

$$X(k) = \sum_{i=0}^{k-1} \Phi(k-i-1)B = \sum_{i=0}^{k-1} \begin{bmatrix} 0 & (-1)^{k-i-2} \\ 0 & (-1)^{k-i-1} \end{bmatrix} \begin{bmatrix} 0 \\ 1 \end{bmatrix} = \sum_{i=0}^{k-1} \begin{bmatrix} (-1)^{k-i-2} \\ (-1)^{k-i-1} \end{bmatrix}$$

or $\quad X(k) = \begin{bmatrix} x_1(k) \\ x_2(k) \end{bmatrix} = \begin{bmatrix} (-1)^{k-2} + (-1)^{k-3} + (-1)^{k-4} + \cdots + (-1)^{-1} \\ (-1)^{k-1} + (-1)^{k-2} + (-1)^{k-3} + \cdots + (-1)^0 \end{bmatrix}$

13.22 Discretization of Continuous Time State Equation

Method-1

A technique is developed for determining the discrete state equation of a sampled-data system directly from the state equation. In fact, the state of continuous model becomes the state of discrete model.

The sampled-data system corresponding to the continuous system is already obtained. The time domain solution of the state equation of the continuous system is written as

$$X(t) = \Phi(t - t_0)X(t_0) + \int_{t_0}^{t} \Phi(t - \tau) Bu(\tau)\, d\tau \qquad (13.80)$$

Digital Control Systems

where t_0 is the initial time. And the state transition matrix $\Phi(t - t_0)$ can be written as

$$\Phi(t - t_0) = e^{A(t-t_0)} = \sum_{k=0}^{\infty} \frac{A^i (t-t_0)^k}{k!} \tag{13.81}$$

To obtain the discrete model, we evaluate Eq. (13.81) at $t = kT + T$ with $t_0 = kT$, that is

$$\Phi(T) = e^{AT} = \sum_{k=0}^{\infty} \frac{A^i T^k}{k!} \tag{13.82}$$

Then from Eq. (13.80), we get

$$X(k + 1) \leftarrow X[(k + 1)T] = \Phi(T)X(kT) + \int_{kT}^{(k+1)T} \Phi[(k + 1)T - \tau] Bu(kT) d\tau$$

$$X(k + 1) \leftarrow X[(k + 1)T] = \Phi(T)X(kT) + u(kT) \int_{kT}^{(k+1)T} \Phi[(k + 1)T - \tau] B d\tau$$

We have replaced $u(t)$ with $u(kT)$ since during the time interval $kT \leq t \leq kT + T$, $u(t) = u(kT)$. This development is vaild only if $u(t)$ is the output of a zero-order hold.

If we let

$$A_d = \Phi(T); \quad B_d = \int_{kT}^{kT+T} \Phi(kT + T - \tau) B d\tau; \quad \text{and with } C_d = C \text{ and } D_d = D$$

then the discrete state equation can be written as

$$X(k + 1) = A_d X(k) + B_d u(k)$$

and output

$$y(kT) = C_d X(kT) + D_d u(kT)$$

Let $kT - \tau = -\sigma$, then the equation of B_d becomes

$$B_d = \left[\int_0^T \Phi(T - \sigma) d\tau \right] B \tag{13.83}$$

The discrete system matrices A_d and B_d may be evaluated by finding $\Phi(T)$ and using Laplace transform approach for which roots need to be known. Otherwise, a computer evaluation of $\Phi(T)$ using power series expansion with $t = T < 1$ and $t = 0$, Eq. (13.82) becomes

$$\Phi(T) = \sum_{k=0}^{\infty} \frac{A^i T^k}{k!} = I + AT + A^2 \frac{T^2}{2!} + A^3 \frac{T^3}{3!} + \dots$$

The series is convergent as the choice of $T < 1$ and the series can be truncated with adequate accuracy in the result.

Now the integral of Eq. (13.83), necessary for computation of B, is as follows. Let $\tau = T - \sigma$, then

$$\int_0^T \Phi(T - \sigma) B d\tau = \int_T^0 \Phi(\tau)(-d\tau) = \int_0^T \Phi(\tau) d\tau$$

$$= \int_T^0 \left(I + A\tau + A^2\frac{\tau^2}{2!} + A^3\frac{\tau^3}{3!} + \cdots\right)d\tau$$

$$= IT + A\frac{T^2}{2!} + A^2\frac{T^3}{3!} + A^3\frac{T^4}{4!} + \cdots$$

Then
$$B_d = \left[IT + A\frac{T^2}{2!} + A^2\frac{T^3}{3!} + A^3\frac{T^4}{4!} + \cdots\right]B$$

If A is nonsingular, then the power series can be written as

$$A^{-1}\left(TA + \frac{T^2}{2!}A^2 + \frac{T^3}{3!}A^3 + \cdots + I - I\right) = A^{-1}(e^{AT} - I)B$$

The use of this formula avoids the need of computing an infinite power series. For the multivariable case only, the matrix D_d will be the matrix and each input must be the output of a ZOH. The above derivation for the SISO case is illustrated with an example below.

EXAMPLE 13.28 The linear time-invariant continuous system dynamics is given in vector-matrix differential equation as

$$\dot{X}(t) = \begin{bmatrix} 0 & 1 \\ 0 & -1 \end{bmatrix} X(t) + \begin{bmatrix} 0 \\ 10 \end{bmatrix} u(t)$$

and output
$$y(t) = [1 \quad 0] X(t)$$

Given sampling time $T = 0.1$ s,

The STM
$$\Phi(t) = \begin{bmatrix} 1 & 1 - e^{-t} \\ 0 & e^{-t} \end{bmatrix}$$

and
$$\int_0^T \Phi(t)d\tau = \begin{bmatrix} T & T - 1 + e^T \\ 0 & 1 - e^{-T} \end{bmatrix}$$

Then $A_d = \Phi(T)\big|_{T=0.1} = \begin{bmatrix} 1 & 0.0952 \\ 0 & 0.905 \end{bmatrix}$ and $B_d = \left[\int_0^T \Phi(\tau)d\tau\right] B = \begin{bmatrix} 0.0484 \\ 0.952 \end{bmatrix}$

Hence the discrete state equation can be written as

$$X(k + 1) = \begin{bmatrix} 1 & 0.0952 \\ 0 & 0.905 \end{bmatrix} X(k) + \begin{bmatrix} 0.0484 \\ 0.952 \end{bmatrix} u(kT)$$

and
$$\text{output} = [1 \quad 0] X(k)$$

The simulation diagram is shown in Figure 13.45.

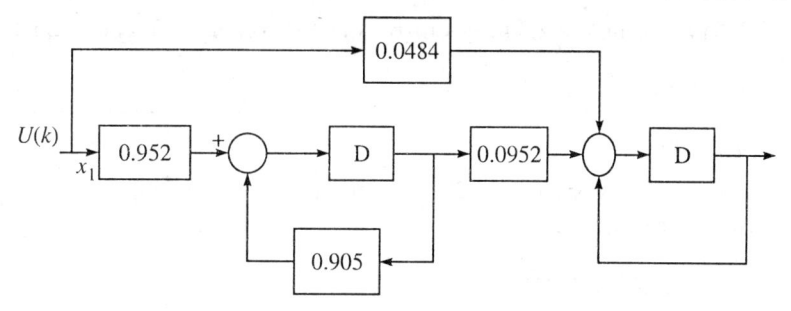

FIGURE 13.45 Example 13.27: simulation diagram.

Mehod-2
Rewrite the linear time-invariant continuous state equations

$$\dot{X}(t) = AX(t) + Bu(t)$$
$$y(t) = CX(t) + Du(t)$$

we know that $\quad \dot{X}(t) = \dfrac{dX(t)}{dt} = \lim\limits_{T \to 0} \dfrac{X(t+T) - X(t)}{T} = AX(t) + Bu(t)$

Then, we get $\quad X(t + T) = X(t) + TAX(t) + TBu(t)$

Putting $t = kT$,

$$X[(k + 1)T] = (I + TA)X(kT) + TBu(kT)$$

and $\quad y(kT) = CX(kT) + DX(kT)$

Defining $\quad A_d = TA + I, \ B_d = TB, \ C_d = C \text{ and } D_d = D$

we arrive at the discrete version of continuous-time system as in Eq. (13.82). Accuracy in this method is improved as T is reduced but computational time will increase. Usually the time interval T is chosen less than one-tenth of the smallest time constant of the system which can be obtained from the knowledge of their eigenvalues.

Between the two methods of discretization presented above, the computation through Method-1 is a closer approximation to continuous system than the other one and means are available in MATLAB too.

Simulation
First consider the LTI system initially at rest by the first-order difference equation

$$y[n] + ay[n - 1] = bx[n]$$

which can be rewritten as

$$y[n] = -ay[n - 1] + bx[n]$$

Realization is in Fig. 13.46 where D is the delay unit which stands for z^{-1}.

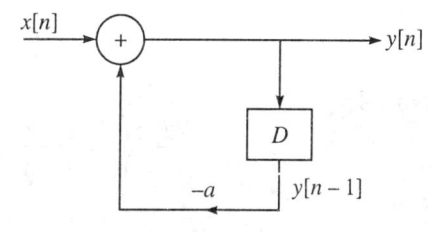

FIGURE 13.46

Consider next the non-recursive LTI system by the first-order difference equation

$$y[n] = b_0 x[n] + b_1 x[n-1]$$

Realization is in Figure 13.47.

Consider next the LTI system initially at rest by the first-order difference equation

$$y[n] + ay[n-1] = b_0 x[n] + b_1 x[n-1] \qquad (13.84)$$

which can be rewritten as

$$y[n] = -ay[n-1] + b_0 x[n] + b_1 x[n-1]$$

FIGURE 13.47

Realization is in Figure 13.48.

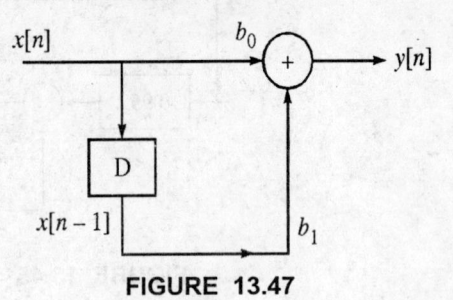

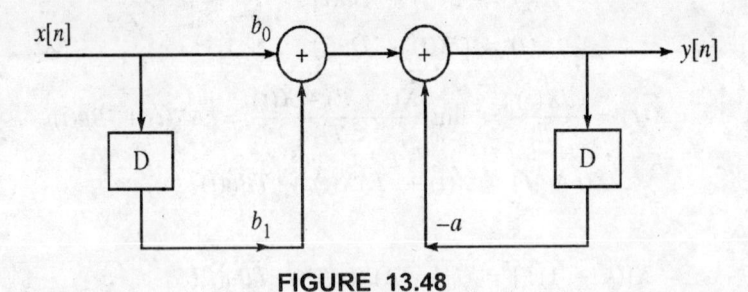

FIGURE 13.48

This is not minimal realization with minimal number of delay units D. For minimal realization, express Eq. (13.84) as

$$Y(z) + az^{-1}Y(z) = b_0 X(z) + b_1 z^{-1} X(z)$$

$$Y(z)[1 + az^{-1}] = X(z)[b_0 + b_1 z^{-1}]$$

the transfer function of the system described by

$$\frac{Y(z)}{X(z)} = \frac{b_0 + b_1 z^{-1}}{1 + az^{-1}}$$

Rewrite this as

$$\frac{Y(z)}{X(z)} = \frac{b_0 + b_1 z^{-1}}{1 + az^{-1}} \frac{P(z)}{P(z)}$$

Now, let us rewrite

$$\frac{Y(z)}{P(z)} \frac{P(z)}{X(z)} = (b_0 + b_1 z^{-1}) \frac{1}{1 + az^{-1}}$$

Now let

$$\frac{P(z)}{X(z)} = \frac{1}{1 + az^{-1}} \Rightarrow p[n] = -ap[n-1] + x[n]$$

and

$$\frac{Y(z)}{P(z)} = (b_0 + b_1 z^{-1}) \Rightarrow y[n] = b_0 p[n] + b_1 p[n-1] \qquad (13.85)$$

Simulation of Eqs. (13.85), that is for system dynamics represented by the difference equation, is shown in Figure 13.49.

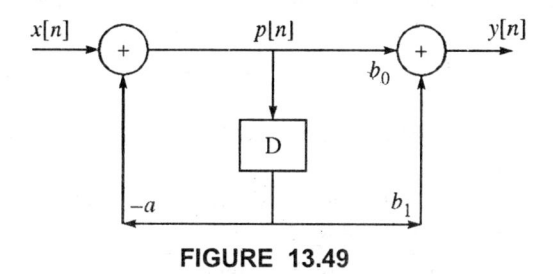

FIGURE 13.49

The same basic idea can be applied to the general recursive equation

$$y[n] = \frac{1}{a_0}\left\{\sum_{k\to 0}^{N} b_k x[n-k] - \sum_{k\to 0}^{N} a_k y[n-k]\right\} \qquad (13.86)$$

Simulation of system dynamics of Eq. (13.86) is shown in Figure 13.50 for minimal realization.

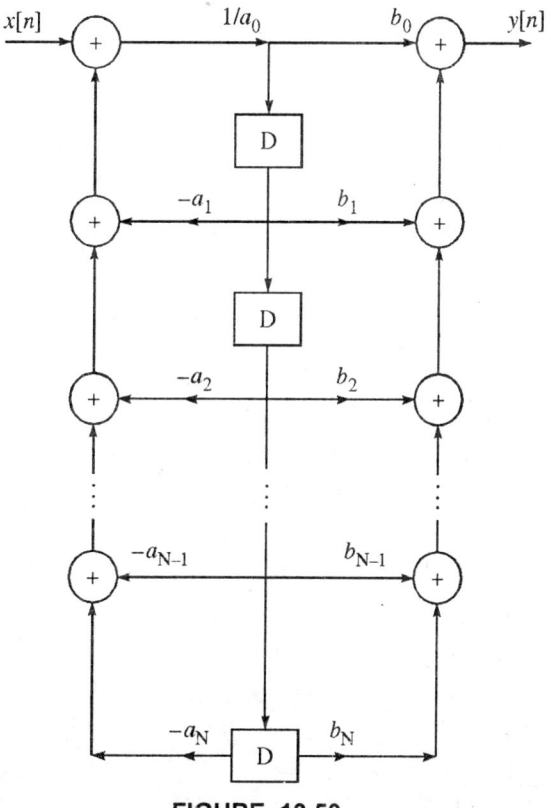

FIGURE 13.50

Summary

The use of digital computer as the compensation device for a closed-loop control system has grown of late due to dramatic reductions in cost and manifold increase in the reliability of computers. A computer can be used to complete many calculations during the sampling interval, T, and to provide the output signal which is used to drive an actuator of a process. Computer control is used for chemical processes, aircraft control, machine tools, robot and many other common processes today.

The z-transform may be used to analyze the stability and response of a sampled system and to design appropriate systems incorporating a computer.

We have presented an overview of sampled-data control systems. We have talked about sampled-data and discrete-data systems and their advantages. In this regard, we have touched upon sample and hold, z-transform, pulse transfer function, characteristic equation, closed-loop system stability, root loci, Nyquist stability criterion, state-variable formulation, and simulation of transfer function of a discrete control system. In a nutshell, whatever we have talked about the continuous-time system has been touched upon in this chapter for better understanding of the digital control system.

Problems

13.1 Determine the z-transforms of the following functions. You may performs partial-fraction expansion, and then use the z-transform table.

(a) $F(s) = \dfrac{5}{s(s^2 + 4)}$

(b) $F(s) = \dfrac{4}{s^2(s + 2)}$

(c) $F(s) = \dfrac{2}{s^2 + s + 2}$

(d) $F(s) = \dfrac{2(s + 1)}{s(s + 5)}$

(e) $F(s) = \dfrac{10}{s(s^2 + 5 + 2)}$

13.2 Given that the z-transform of $g(t)$ for $T = 1$ s is

$$G(z) = \dfrac{z(z - 0.2)}{4(z - 0.8)(z - 1)}$$

find the sequence $g(kT)$ for $k = 0, 1, 2, \ldots, 10$. What is the final value of $g(kT)$ when $K \to \infty$?

13.3 Find the inverse z-transform $f(k)$ of the following functions:

(a) $F(z) = \dfrac{2z + 1}{(z - 0.1)^2}$

(b) $F(z) = \dfrac{2z}{z^2 - 1.2z + 0.5}$

(c) $F(z) = \dfrac{z}{z^2 + 1}$

(d) $F(z) = \dfrac{10z}{z^2 - 1}$

(e) $F(z) = \dfrac{1}{z(z - 0.2)}$

13.4 Solve the following difference equation using the z-transform method:
$$c(k+2) - 0.1c(k+1) - 0.2c(k) = r(k+1) + r(k)$$
where $r(k) = 1(k)$ for $k = 0, 1, 2, \ldots$; $c(0) = 0$ and $c(1) = 0$.

13.5 Solve the following difference equation using the z-transform method:
$$c(k+2) - 1.5c(k+1) + c(k) = 2(k)$$
where $c(0) = 0$ and $c(1) = 1$

13.6 The weighting sequence of a linear discrete-data system is
$$g(k) = \begin{cases} 0.15(0.6)^k - 0.15(0.4)^k & k \geq 0 \\ 0 & k < 0 \end{cases}$$

Find the transfer function $G(z)$ of the system.

13.7 For the system shown in Figure P.13.7, find the output at the sampling instants $c(kT)$. The input is a unit impulse, and the sampling period is 0.1 s. Find the final value of $c(kT)$ as $k \to \infty$.

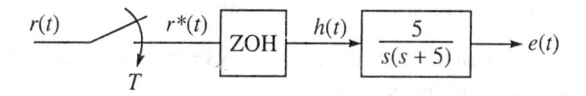

FIGURE P.13.7

13.8 (a) Draw a state diagram for the discrete-data system modeled by the following dynamic equations:
$$X(k+1) = AX(k) + Bu(k)$$
$$C(k) = DX(k)$$

(b) Find the transfer function $C(z)/U(z)$.
(c) Find the characteristic equation of the following systems:

(i) $A = \begin{bmatrix} -1 & 1 \\ -0.5 & 0.2 \end{bmatrix}$ $B = \begin{bmatrix} 0 \\ 1 \end{bmatrix}$ $D = \begin{bmatrix} 1 & 0 \end{bmatrix}$

(ii) $A = \begin{bmatrix} 0 & 2 & -1 \\ 0 & 1 & 1 \\ 3 & 3 & -1 \end{bmatrix}$ $B = \begin{bmatrix} 0 \\ 0 \\ 1 \end{bmatrix}$ $D = \begin{bmatrix} 1 & 0 & 0 \end{bmatrix}$

(iii) $A = \begin{bmatrix} 0 & 1 & 0 \\ 0 & 0 & 1 \\ 0.2 & -1 & 0.5 \end{bmatrix}$ $B = \begin{bmatrix} 1 \\ 0 \\ 1 \end{bmatrix}$ $D = \begin{bmatrix} 1 & 1 & 0 \end{bmatrix}$

13.9 Find the state transition equations of the following systems by means of the state diagram method.

$$X(k + 1) = AX(k) + Bu(k)$$

The initial states are given as $X(0)$, and

$$A = \begin{bmatrix} 0 & 1 \\ 0.5 & 0.3 \end{bmatrix} \quad \text{and} \quad B = \begin{bmatrix} 0 \\ 1 \end{bmatrix}$$

13.10 The input–output transfer functions of some linear discrete-data systems are given below.

(i) $\dfrac{C(z)}{R(z)} = \dfrac{z + 0.5}{z^2 + 0.2z + 0.1}$

(ii) $\dfrac{C(z)}{R(z)} = \dfrac{z^2}{z^3 - z^2 + 0.5z - 0.5}$

(a) Draw the state diagrams for the systems.
(b) Write the dynamic equations of the systems.

13.11 The state diagram of a discrete-data control system is shown in Figure P.13.11. Write the dynamic equations of the system.

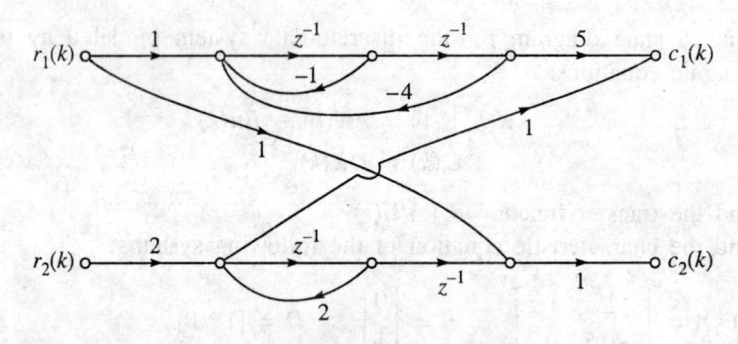

FIGURE P.13.11

13.12 Decompose the following transfer functions by direct decomposition, and draw the state diagrams. Write the discrete state equations in vector-matrix form.

(a) $\dfrac{C(z)}{R(z)} = \dfrac{z + 0.5}{z^3 + 2z^2 + z + 0.5}$

(b) $\dfrac{C(z)}{R(z)} = \dfrac{z(z + 0.5)}{z^3 + z^2 + 2z + 0.5}$

13.13 Given the discrete-data control system

$$X(k+1) = AX(k) + Bu(k)$$
$$c(k) = DX(k)$$

where

$$A = \begin{bmatrix} 0 & 1 \\ -2 & -3 \end{bmatrix} \quad B = \begin{bmatrix} 0 \\ 1 \end{bmatrix} \quad D = \begin{bmatrix} 1 & -1 \end{bmatrix}$$

The control is realized through state feedback,

$$u(k) = -GX(k) = -[g_1 \quad g_2]X(k)$$

where g_1 and g_2 are real constants.

Determine the values of g_1 and g_2 that must be avoided for the system to be completely observable.

13.14 Determine the stability conditions of the discrete-data control systems that are represented by the following characteristic equations:

(a) $z^2 + 0.5z + 0.2 = 0$
(b) $z^3 + z^2 + 3z + 0.2 = 0$
(c) $z^3 - 1.5z^2 + 1.2z - 0.5 = 0$
(d) $z^4 - 1.2z^3 + 0.22z^2 + 0.066z - 0.008 = 0$
(e) $z^3 - 1.4z^2 + 0.53z - 0.04 = 0$
(f) $z^4 - 2z^3 + z^2 - 2z + 1 = 0$
(g) $z^4 - z^3 + z^2 - z + 1 = 0$

13.15 The block diagram of a discrete-data control system is shown in Figure P.13.15. Compute and plot the unit-step response $c^*(t)$ of the system. Find (a) c^*_{max} and the sampling instant at which it occurs, (b) the step-, ramp-, and parabolic-error constants, and (c) the final value of $c(kT)$.

where

(a) $G_p(s) = \dfrac{20}{s(s+5)}$; $T = 0.5$ s

(b) $G_p(s) = \dfrac{2(s+1)}{s(s+2)}$; $T = 0.5$ s

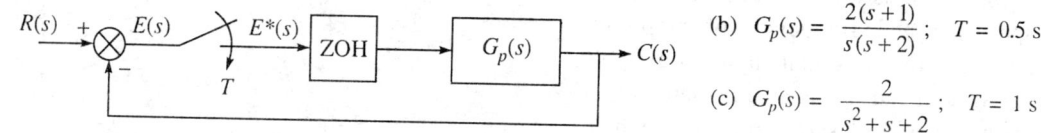

(c) $G_p(s) = \dfrac{2}{s^2+s+2}$; $T = 1$ s

FIGURE P.13.15

Nonlinear Systems

14.1 Introduction

In the preceding chapters we were concerned exclusively with linear time-invariant systems. But systems in practice are nonlinear in nature. Many control systems after linearization, with the application of linear control theory, have produced good results. Many practical systems are sufficiently nonlinear, thus, the important features of their performance can get completely overlooked if they are analyzed and designed through linear techniques. For such systems, great efforts have been expended to evolve some analytical, graphical and numerical techniques which can take care of nonlinearities.

Before going further, we first explain here what a nonlinear system is and how it differs from a usual linear system. The principle of superposition is not applicable to nonlinear systems. In a linear system, altering the size of the input does not change the shape of the response of the system, whereas in the case of nonlinear systems there is considerable change in shape, size and frequency of response. Application of sinusoidal input to a stable linear system causes the steady-state output to be a sinusoid of same frequency, which may differ from the input in phase and magnitude. In nonlinear systems, on the other hand, the steady-state output may contain harmonics of the input as well.

Limit cycle is an unusual feature of nonlinear systems. The nonlinear system may produce oscillations of a certain fre-

OBJECTIVE

We have analyzed linear control systems so far. But systems in nature are nonlinear. Hence the analysis of nonlinear systems is of prime importance as well. There exist the phase-plane method and the describing function method of analysis of nonlinear systems which are graphical in nature. The Liapunov function method of analysis for nonlinear systems exists too. The limitation of the first two methods is that the order of system has to be of second order. Liapunov's method, on the other hand, can be used for asymptotic stability analysis of any order nonlinear system. Needless to say that the methods which are engaged for nonlinear system analysis are also well applicable for linear system analysis. For better understanding we have in our discussion, applied these methods first to linear system analysis and then to nonlinear system analysis.

CHAPTER OUTLINE

Introduction
Common Nonlinearities
Phase-Plane Analysis
Linear Control Systems
Nonlinear Systems
Describing Function Analysis
Liapunov Stability Analysis

quency and amplitude which may not be sinusoidal regardless of the magnitude of the input or the initial conditions.

Jump resonance is another unusual feature of nonlinear systems. This implies jumps in magnitude and phase as the frequency is changed near resonance.

Jump resonance

We are already familiar with the dynamics and response of the spring-mass-damper system as depicted in Figure 14.1(a). The equation of this system when subjected to a sinusoidal forcing function (assuming the components to be linear) can be rewritten as

$$M\ddot{x} + f\dot{x} + Kx = F \cos \omega t$$

The frequency response of this system is given in Figure 14.1(b). Let us now assume that the restoring force of the spring is nonlinear, i.e. of the form $K_1 x + K_2 x^3$, whose characteristics are shown in Figure 14.1(c). The spring is linear for $K_2 = 0$, while it is hard spring when $K_2 > 0$ and soft spring when $K_2 < 0$. The dynamics for the system with nonlinear spring, becomes.

$$M\ddot{x} + f\dot{x} + K_1 x + K_2 x^3 = F \cos \omega t, \quad K_2 \neq 0$$

The frequency response of the system is as shown in Figure 14.1(d) for $K_2 > 0$, that is, hard spring and as shown in Figure 14.1(e) for $K_2 < 0$, that is, soft spring.

The presence of the nonlinear term $K_2 x^3$ ($K_2 > 0$) has caused the resonance peak to bend towards higher frequencies as shown in Figure 14.1(d). The measured response follows the curve through the points A, B, and C, but at C an increment in frequency results in a discontinuous jump down to point D, after which with further increase in frequency, the response curve follows through DE. If the frequency is now decreased, the response follows EDF with a jump up to B occurring at F, and then the response curve moves towards A. That means in a certain range of frequencies, the response function has hysteresis. The same hysteresis occurs in case $K_2 < 0$, i.e. soft spring, as shown in Figure 14.1(e). This phenomenon which is peculiar to nonlinear systems is known as *jump resonance*.

The frequency versus amplitude of oscillation of the spring-mass-damper system with nonlinear spring is depicted in Figure 14.1(f). The explanation is quite obvious. As the amplitude of oscillation, i.e. x tends towards zero the contribution of the nonlinear term $K_2 x^3$ becomes negligible and the frequency of oscillation tends towards $\sqrt{K/M}$, the same as for the linear case. For the linear case, when driven by a sinusoidal input of frequency ω, the output will have the same frequency but differ only in phase. But in a nonlinear case, there are harmonics in the output which are multiples of the driven frequency ω.

Limit cycles

The disturbed nonlinear system, even when staying within its tolerance limits of steady-state oscillation, may exhibit a closed trajectory or limit cycle. The limit cycles describe the oscillation of nonlinear systems. The existence of a limit cycle corresponds to an oscillation of fixed amplitude and period.

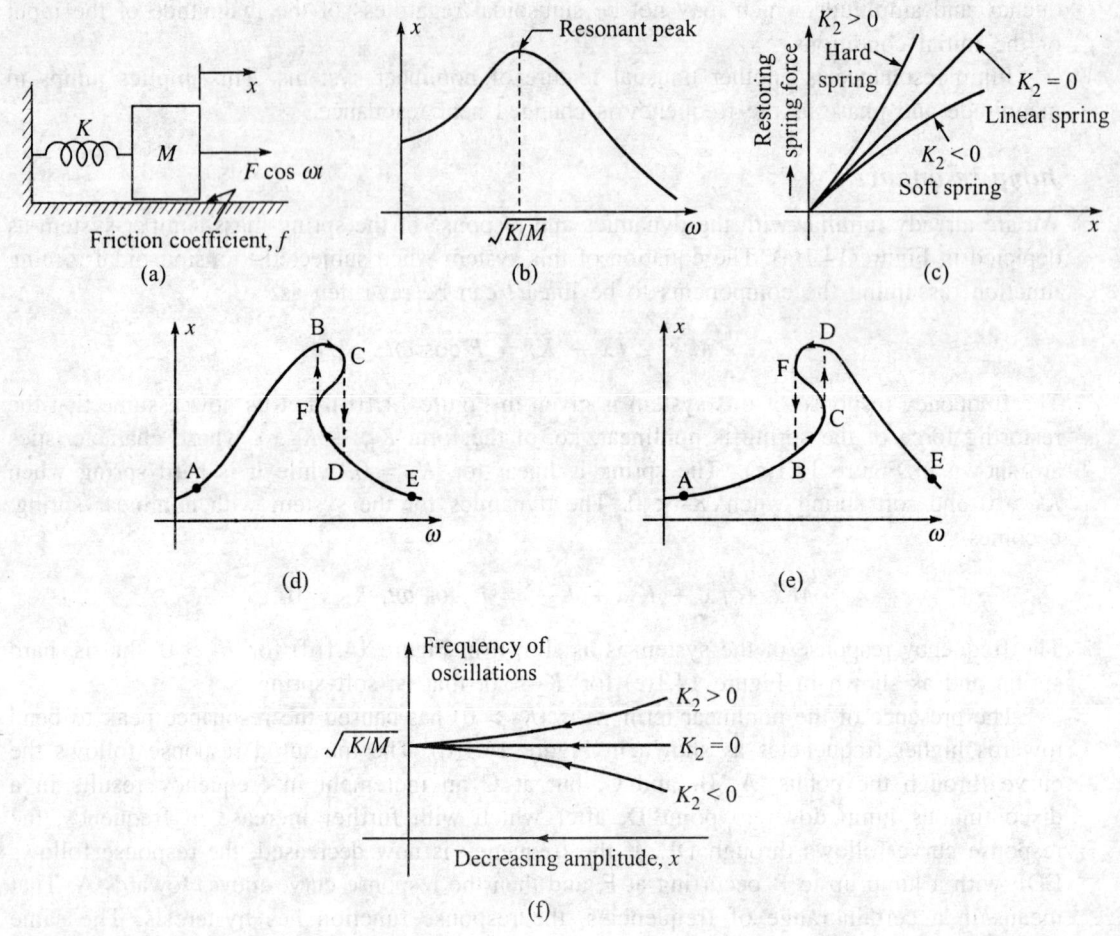

FIGURE 14.1 (a) A spring-mass-damper system, (b) frequency response of spring-mass-damper system, (c) spring characteristics, (d) jump resonance in nonlinear systems (hard spring case), (e) jump resonance in nonlinear systems (soft spring case), (f) frequency versus amplitude for the oscillations of spring-mass-damper system with nonlinear spring.

Let us consider the well-known Vander Pol's differential equation

$$\ddot{x} - \mu(1 - x^2)\dot{x} + x = 0$$

With the choice of phase variable as state variable, the dynamics of the system becomes

$$\dot{X} = AX$$

where the system matrix

$$A = \begin{bmatrix} 0 & 1 \\ -1 & \mu(1 - x_1^2) \end{bmatrix}$$

From the characteristic equation

$$|sI - A| = s^2 - \mu(1 - x_1^2)s + 1$$

we get the damping coefficient as $-\frac{\mu}{2}(1 - x^2)$ that depends on the system output x. We can see that when $|x| \gg 1$, the damping coefficient has large positive values and the system becomes overdamped. Consequently, amplitude x decreases with time which in turn effectively reduces the damping coefficient gradually, and finally the system enters the limit cycle when the damping coefficient becomes zero. See Figure 14.2(a).

On the other hand, when $|x| \ll 1$, the damping coefficient has negative values and the system becomes unstable. Consequently, amplitude x increases with time which in turn effectively increases the damping coefficient gradually from its original negative value and finally the system enters the limit cycle when the damping coefficient becomes zero. See Figure 14.2(a).

Now we consider the Vander Pol's equation with sign of its damping term reversed, that is,

$$\ddot{x} + \mu(1 - x^2)\dot{x} + x = 0$$

Then following the same line of reasoning we get the unstable limit cycle (closed trajectory) diverge away from it. See Figure 14.2(b).

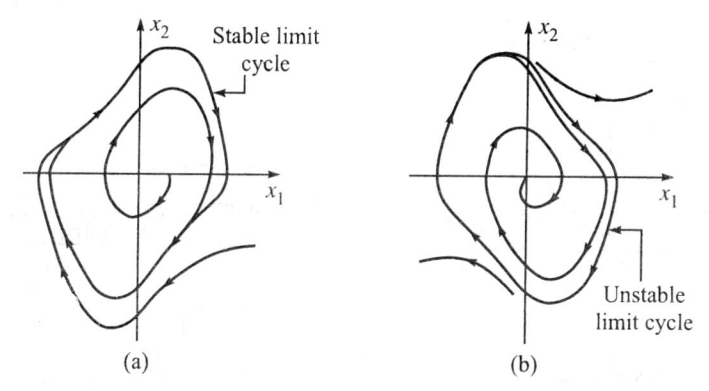

FIGURE 14.2 Limit cycle behaviour of nonlinear systems.

It may be noted that in control system characteristics, in general, the limit cycle is an undesirable phenomenon as it leads to instability of the system.

14.2 Common Nonlinearities

Nonlinearities are classified as incidental and intentional. Incidental nonlinearities are those which are present inherently in the system such as saturation, dead-zone, coulomb friction, stiction, backlash, etc. Intentional nonlinearities (such as relay) are those which are deliberately introduced in the system to modify the system characteristics.

The basic features of commonly encountered nonlinearities are discussed next.

Saturation

There are many components such as amplifiers where the output is proportional to the input for a limited range of operation, and beyond which the output gets saturated. The change from one range to other is rather gradual (see Figure 14.3). Piecewise linear approximation for saturation nonlinearity is used for the analysis of linear systems.

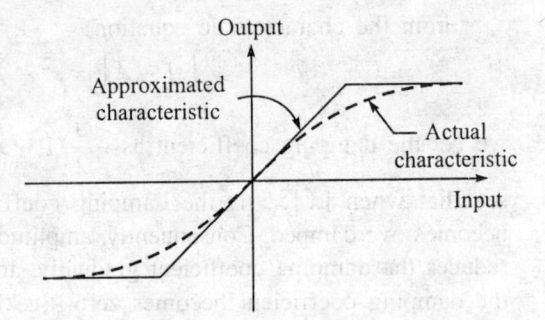

FIGURE 14.3 Piecewise linear approximation of saturation nonlinearity.

Friction

The predominant amongst the frictional forces is the viscous friction which is linearly proportional to the relative velocity of sliding surfaces, that is,

$$\text{Viscous friction force} = f\dot{x}$$

where f is the proportionality constant and $\dot{x}$ the relative velocity [see Figure 14.4(a)].

In addition to viscous friction, the other two friction types are coulomb friction and stiction. Coulomb friction is a constant retarding force always opposing the relative motion. Stiction is the force required to initiate motion. The force of stiction is always greater than that of the coulomb friction [see Figure 14.4(b)].

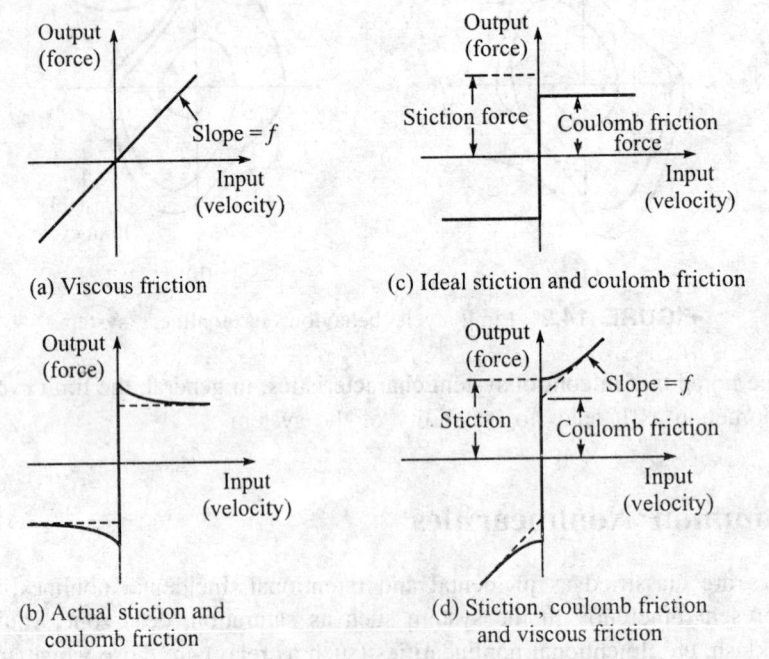

FIGURE 14.4 Characteristics of various types of friction: (a) viscous friction, (b) ideal stiction and coulomb friction, (c) actual stiction and coulomb friction, and (d) stiction, coulomb friction, and viscous friction.

More force is required to move an object from rest than to maintain it in motion because of interlocking of surface irregularities. In actual practice, the stiction force gradually decreases with velocity and changes over to coulomb friction force as depicted in Figure 14.4(c). The composite friction characteristic is shown in Figure 14.4(d).

The other single-valued nonlinearity is the dead-zone nonlinearity as shown in Figure 14.5(a). A relay is a nonlinear power amplifier which can provide large power amplification inexpensively and is therefore introduced deliberately in control systems. The characteristic of such a relay is shown in Figure 14.5(b). Similarly, a relay with dead-zone is shown in Figure 14.5(c).

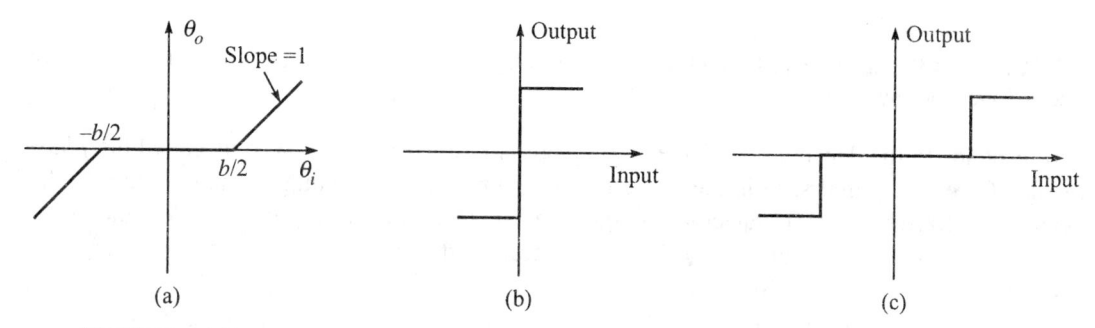

FIGURE 14.5 (a) Dead-zone nonlinearity, (b) ideal relay, and (c) relay with dead-zone.

We can classify nonlinearities, in general, into two categories:

(a) Single-valued nonlinearity such as saturation, dead-zone, viscous friction, and coulomb friction, etc.
(b) Double-valued nonlinearity such as hysteresis, backlash, etc. For a given input, the output is double-valued and out of those two values, which particular output will result, depends upon the history of the input. This type of nonlinearity has thus inherent memory and is referred to as memory type nonlinearity. The nonlinearity commonly referred to as *backlash* as in Figure 14.6(a), occurring in physical systems is hysteresis in mechanical transmission such as gear trains and linkages. Backlash, in fact, is the play between the teeth of the gear drive and those of the driven gear. The other double-valued nonlinearity is relay with pure hysteresis as in Figure 14.6(b). The more practical characteristic with both dead-zone and hysteresis is shown in Figure 14.6(c).

In practice, all physical systems have some nonlinearity; we need some understanding of the basic methods for the analysis of nonlinear systems. To mention a few, these are:

(a) Phase-plane analysis
(b) Describing function method of analysis
(c) Liapunov's stability analysis

14.3 Phase-Plane Analysis

Phase-plane method is basically a graphical method of solving second-order nonlinear systems, obviously the technique can handle linear systems as well. The coordinate plane with axes that

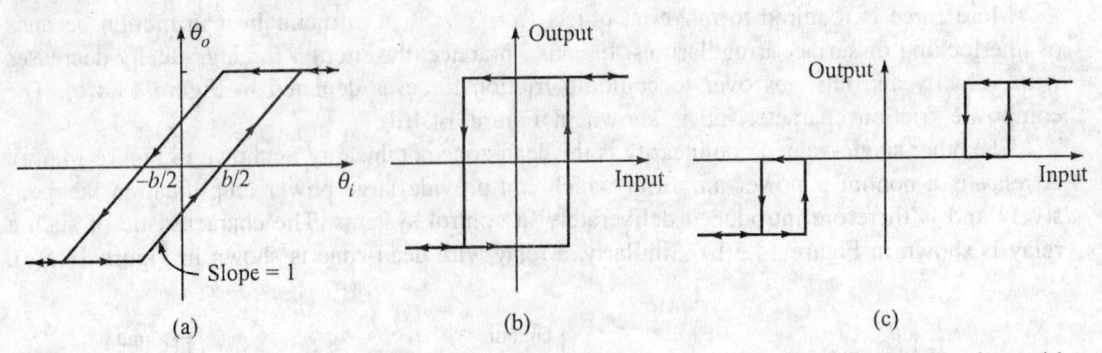

FIGURE 14.6 (a) Backlash nonlinearity, (b) relay with pure hysteresis, and (c) relay with dead-zone and hysteresis.

corresponds to the dependent variable $x_1 = x$ and its first derivative $x_2 = \dot{x}$, is called the *phase-plane*. Given the initial state in the phase-plane with time as the running parameter, this method can be utilized to plot a trajectory in the phase-plane. This trajectory is called the *phase trajectory*. From the set of trajectories for different initial conditions, we can obtain the phase portrait which provides information about the stability and the existence of limit cycle. The phase trajectories are unique with one and only one curve passing through any point except for certain critical points, called *singular points*, through which an infinite number of trajectories pass.

Different initial conditions produce different but geometrically similar trajectories as shown in Figure 14.7 for the undamped case ($\zeta = 0$). The family of trajectories is called *phase portrait*.

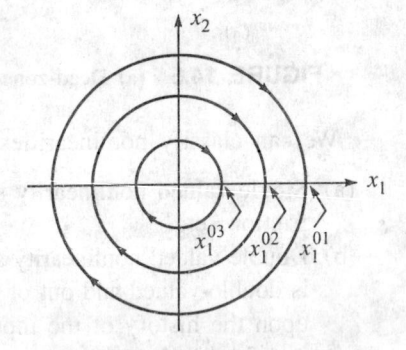

FIGURE 14.7 Phase portrait.

14.3.1 Phase-plane method—basic concept

The dynamics of a linear autonomous spring-mass-damper system is represented by

$$M\ddot{x} + f\dot{x} + Kx = 0 \tag{14.1}$$

which in standard form can be written as

$$\ddot{x} + 2\zeta\omega_n \dot{x} + \omega_n^2 x = 0 \tag{14.2}$$

Let us choose the phase variables as state variables, i.e.

$$x = x_1 \quad \text{and} \quad x_2 = \dot{x}_1$$

Then the dynamics of the system in vector-matrix differential equation in state variable form can be written as

$$\begin{bmatrix} \dot{x}_1 \\ \dot{x}_2 \end{bmatrix} = \begin{bmatrix} 0 & 1 \\ -\omega_n^2 & -2\zeta\omega_n \end{bmatrix} \begin{bmatrix} x_1 \\ x_2 \end{bmatrix}$$

with initial conditions of state vector, $X(0) = [x_1^0 \quad 0]^T$.

Nonlinear Systems

The solution for the state vector can be obtained and plotted for different initial conditions. The phase portrait can be drawn by eliminating time from both the variables x_1 and x_2. However, when the analytical solutions for x_1 and x_2 are available, there is no necessity in practical sense for phase portrait and phase-plane analysis. But the systems in practice are nonlinear in nature and for which no analytical method is available to get the solution of state variables x_1 and x_2. For example, suppose the spring force is nonlinear and is defined as

$$K_1 x + K_2 x^3 \tag{14.3}$$

then the dynamics of the nonlinear spring-mass-damper system can be written as

$$M\ddot{x} + f\dot{x} + K_1 x + K_2 x^3 = 0 \tag{14.4}$$

which in turn is written in state variable form as

$$\frac{d}{dt}\begin{bmatrix} x_1 \\ x_2 \end{bmatrix} = \begin{bmatrix} 0 & 1 \\ -\frac{K_1}{M} - \frac{K_2}{M}x_1^2 & -\frac{f}{M} \end{bmatrix} \begin{bmatrix} x_1 \\ x_2 \end{bmatrix} \tag{14.5}$$

Solving these equations is no more an easy task. For these nonlinear systems, a graphical method known as the phase-plane method is found to be very useful.

14.3.2 Methods for constructing trajectories

In the phase-plane analysis of second-order systems, trajectories may be constructed analytically, graphically, or experimentally.

The analytical method is the most straightforward one for obtaining phase portraits. It is usually not useful from the practical point of view, because we do not need to resort to the phase-plane representation for a solution if the differential equation is integrable.

The graphical methods are useful when it is tedious or impossible to solve the given differential equation analytically. The graphical methods are applicable to both linear and nonlinear equations. All the graphical methods are essentially based on step-by-step procedures.

Method of isocline

The method of isocline is a graphical procedure for determining the phase portrait.

Consider a second order autonomous system represented by the vector-matrix differential equation

$$\dot{X} = \begin{bmatrix} \dot{x}_1 \\ \dot{x}_2 \end{bmatrix} = \begin{bmatrix} f_1(x_1, x_2) \\ f_2(x_1, x_2) \end{bmatrix}$$

The slope m of the trajectory at any point in the phase-plane is given by

$$\frac{dx_2}{dx_1} = m = \frac{f_2(x_1, x_2)}{f_1(x_1, x_2)} \tag{14.6}$$

Every point (x_1, x_2) of the phase-plane, has with it the associated slope of the trajectory except at singular points at which the trajectory slope is indeterminate.

Isoclines are lines in the phase-plane corresponding to slopes of the phase portrait. To construct the phase trajectory, the slope equation must be integrated which is difficult except for a few simple cases. Because the procedure is a numerical approximation, closer spacing of the isoclines increases the accuracy of the resulting trajectory. The direction of phase trajectory at any point (x_1, x_2) can be obtained from the sign of

$$\Delta x_1 = f_1(x_1, x_2)\Delta t$$
$$\Delta x_2 = f_1(x_1, x_2)\Delta t$$

for a small positive increment of time Δt. Further, for a specific trajectory slope m_1, we have

$$f_2(x_1, x_2) = m_1 f_1(x_1, x_2)$$

This equation defines the locus of all such points in the phase-plane of which the slope of the phase trajectory is m_1. Such a locus is called an isocline. See Figure 14.8.

Undamped system

As an illustration, consider a linear undamped normalized second-order system

$$\ddot{\theta} + \theta = 0 \qquad (14.7)$$

Defining the state variables as

$$x_1 = \theta$$
$$x_2 = \dot{\theta} \qquad (14.8)$$

we have

$$\dot{x}_1 = x_2$$
$$\dot{x}_2 = -x_1 \qquad (14.9)$$

Dividing, we get

$$\frac{dx_2}{dx_1} = -\frac{x_1}{x_2} \qquad (14.10)$$

or

$$x_2\, dx_2 + x_1\, dx_1 = 0 \qquad (14.11)$$

Integrating Eq. (14.11), we get

$$x_1^2 + x_2^2 = C \qquad (14.12)$$

where C is an arbitrary constant easily determined from the initial conditions. In the phase-plane $(x, \dot{x})$, or the $(\theta, \dot{\theta})$ plane, the phase portraits are shown in Figure 14.7. These are unique. The initial conditions determine the particular trajectory followed. Here dx_2/dx_1 or $d\dot{\theta}/d\theta$ corresponds to the slope of the trajectories that form the phase portrait. In order to plot the phase portrait on a normalized plane $(\theta, \dot{\theta})$, let

$$\frac{d\dot{\theta}}{d\theta} = -\frac{\dot{\theta}}{\theta} = m \qquad (14.13)$$

where m represents the slope of the trajectory. Isoclines associated with the slopes corresponding to Eq. (14.6) constitute a family of straight lines passing through the origin and are illustrated in Figure 14.8.

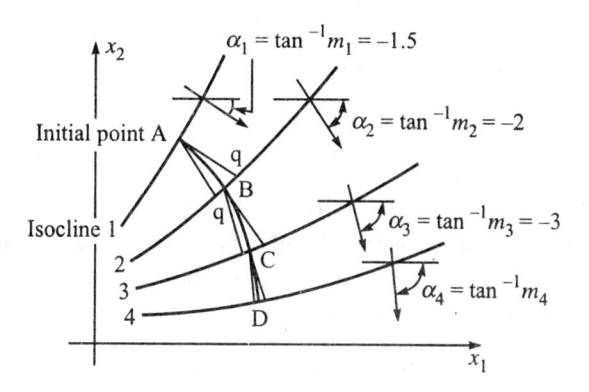

FIGURE 14.8 Construction of trajectory by the isocline method.

Assume that the initial conditions are such that the initial point is located at A on the isocline corresponding to a slope of -1.5. The phase trajectory drawn from point A on the isocline whose slope is -1.5 to that whose slope is -2 would be a straight line whose slope is the average of -1.5 and -2 or -1.75 touching at point B on the isocline to a slope of -2. This is indicated in Figure 14.8 as line segment AB. The constructional procedure is repeated at B and the segment BC whose slope is -2.5 is illustrated in Figure 14.8, and so on. A smooth curve drawn through A, B, C, ... gives the trajectory starting from initial point A.

Underdamped stable system

Now let us consider a linear underdamped second-order stable system as

$$\ddot{\theta} + \dot{\theta} + \theta = 0 \qquad (14.14)$$

Defining the state variables as

$$x_1 = \theta$$
$$x_2 = \dot{\theta} \qquad (14.15)$$

we have

$$\dot{x}_1 = x_2$$

and

$$\dot{x}_2 = -x_1 - x_2 \qquad (14.16)$$

Dividing, we get

$$\frac{dx_2}{dx_1} = -\frac{x_1}{x_2} - 1 = m' \tag{14.17}$$

The new slope of the trajectory is $m' = m - 1$. The plot of isoclines for various values of slope m' can be drawn. With the help of these isoclines the phase trajectory can be constructed.

Singular point

The phase-plane portrait of an autonomous system is a family of non-crossing trajectories which describe the response of the system to all possible initial conditions. Consider the dynamics of the system expressed by second-order differential equation

$$\ddot{x} + a\dot{x} + bx = 0 \tag{14.18}$$

where a, b are functions of x and $\dot{x}$.

The singular point or equilibrium point is the point where the variable and all its derivatives are zero. So the location of the singular point in the x, $\dot{x}$ plane is the origin. Let us examine the behaviour of trajectories near the singular points.

Let us choose the variable $x = x_1$ and let

$$\frac{dx_1}{dt} = x_2 \equiv f_1(x_1, x_2)$$

and from Eq. (14.18), we get

$$\frac{dx_2}{dt} = -bx_1 - ax_2 \equiv f_2(x_1, x_2)$$

where $f_1(x_1, x_2)$ and $f_2(x_1, x_2)$ are analytic functions of the variables x_1 and x_2 in the neighbourhood of the origin.

Expand $f_1(x_1, x_2)$ and $f_2(x_1, x_2)$ in Taylor series in the neighbourhood of the origin, where x_1 and x_2 are very small. Then the system equations become

$$\dot{x}_1 = a_1 x_1 + b_1 x_2 + a_{11} x_1^2 + a_{12} x_1 x_2 + a_{22} x_2^2 + g_1(x_1, x_2)$$
$$\approx a_1 x_2 + b_1 x_2 \tag{14.19}$$
$$\dot{x}_2 = a_2 x_1 + b_2 x_2 + b_{11} x_1^2 + b_{12} x_1 x_2 + b_{22} x_2^2 + g_2(x_1, x_2)$$
$$\approx a_2 x_1 + b_2 x_2 \tag{14.20}$$

where $g_1(x_1, x_2)$ and $g_2(x_1, x_2)$ involve only third and higher order powers of x_1 and x_2. In the neighbourhood of the origin where x_1 and x_2 are very small, the system equations can be approximated by linear terms only, provided they are dominant in the neighbourhood of the origin.

The characteristics of the singular point can be determined by eliminating one of the two variables of Eqs. (14.19) and (14.20) and studying the resulting characteristic equation. Using Laplace transform, we get

$$X_1(s) = \frac{(s - b_2)x_1(0) + b_1 x_2(0)}{s^2 - (a_1 + b_2)s + (a_1 b_2 - b_1 a_2)} \tag{14.21}$$

Nonlinear Systems

and the characteristic equation

$$s^2 - (a_1 + b_2)s + (a_1 b_2 - b_1 a_2) = 0 \tag{14.22}$$

which is modified to the system characteristic equation

$$s^2 + as + b = 0 \tag{14.23}$$

where $\qquad a = -(a_1 + b_2) \qquad$ and $\qquad b = a_1 b_2 - a_2 b_1$

If the roots of the characteristic polynomial in Eq. (14.23) have negative real parts, all trajectories near the origin will approach the singular point or equilibrium point $x_1 = 0$, $x_2 = \dot{x}_1 = 0$, that is, the origin in the $x, \dot{x}$ plane as time t increases indefinitely.

If at least one root is zero (that is, $b = 0$), then the stability cannot be determined from the linearized equation. In this case the behaviour of trajectories near the origin depends on the higher-order terms in Eqs. (14.19) and (14.20).

The nature of the solution of Eq. (14.23) depends on the location of the two roots λ_1 and λ_2 of the characteristic equation

$$\lambda^2 + a\lambda + b = 0 \tag{14.24}$$

where a and b are constants with b not equal to zero. The location of the characteristic roots λ_1 and λ_2 in complex s-plane determines the characteristics of the singular point. Six cases may arise as follows:

1. Stable system with complex roots, that is, λ_1 and λ_2 are complex conjugates having negative real parts (lying in the left-half of s-plane). The transient response is an exponentially damped sinusoid. The system is underdamped, i.e. $0 < \zeta < 1$. The phase trajectory is a logarithmic spiral with origin as singular point which is called a *stable focus*.

2. Unstable system with complex roots, that is, λ_1 and λ_2 are complex conjugate having positive real parts (lying in the right-half of s-plane). The transient is an exponentially increasing sinusoid. The phase trajectory is a logarithmic spiral expanding out of the singular point and is called an *unstable focus*.

   Suppose the state variable formulation of a linear time-invariant system in canonical form is given as

   $$\dot{Z} = \Lambda Z$$

   where $\qquad \Lambda = \begin{bmatrix} \sigma + j\omega & 0 \\ 0 & \sigma - j\omega \end{bmatrix} \tag{14.25}$

   Let the transformation matrix be, $Z = NY$

   where $\qquad N = \begin{bmatrix} 1/2 & -j/2 \\ 1/2 & j/2 \end{bmatrix} \tag{14.26}$

   transforms $\dot{Z} = \Lambda Z$ into $\dot{Y} = PY$

   where $\qquad P = N^{-1} \Lambda N$

such that

$$\dot{y}_1 = \sigma y_1 + \omega y_2 \quad (14.27)$$
$$\dot{y}_2 = -\omega y_1 + \sigma y_2 \quad (14.28)$$

Now,

$$\frac{\dot{y}_2}{\dot{y}_1} = \frac{dy_2}{dy_1} = \frac{\sigma y_2 - \omega y_1}{\sigma y_1 + \omega y_2} = \frac{y_2 - k y_1}{y_1 + k y_2} = \frac{(y_2/y_1) - k}{1 + k(y_2/y_1)} \; ; \; k = \frac{\omega}{\sigma} \quad (14.29)$$

Let us define

$$\frac{dy_2}{dy_1} = \tan \phi \quad \text{and} \quad \frac{y_2}{y_1} = \tan \theta$$

so that we get

$$\tan \phi = \frac{\tan \theta - k}{1 + k \tan \theta}$$

or

$$k = \frac{\tan \theta - \tan \phi}{1 + \tan \theta \tan \phi} = \tan (\theta - \phi) \quad (14.30)$$

This is the equation of a logarithmic spiral.

For $\sigma < 0$, the plot of this equation is a family of equiangular spirals having origin as the singular point which in this case is called a stable focus [see Figure 14.9(a)].

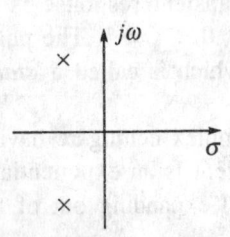

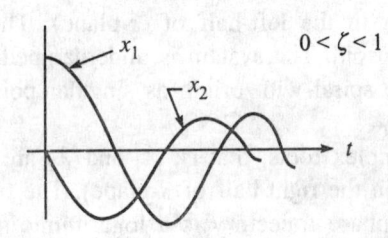

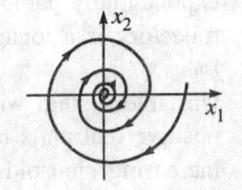

Location of eigenvalues   Response   Phase trajectory of stable focus
(a)

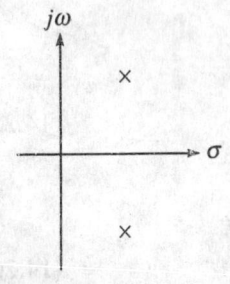

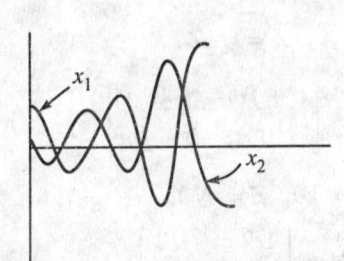

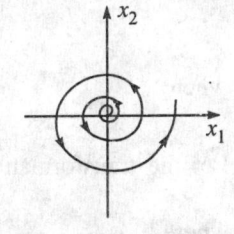

Location of eigenvalues   Response   Phase trajectory of unstable focus
(b)

FIGURE 14.9 Eigenvalues, time response and phase trajectories for linear underdamped (a) stable and (b) unstable system.

For $\sigma > 0$, the plot of this equation consists of expanding spirals as shown in Figure 14.9(b) and the singular point is an unstable focus.

3. Stable system with real, finite and unequal roots, i.e. λ_1 and λ_2 are real and lie in the left-half of s-plane. The system is overdamped, i.e. $\zeta > 1$. The singular point is called a *stable node*. The time responses x_1 and x_2 are monotonically decreasing. The phase portrait contains two trajectories that are exactly straight lines. They are the eigenvectors of the system, and the slopes of these lines are numerically equal to the root values. All trajectories except the fast eigenvector corresponding to the larger root, approach the singular point asymptotic to the slow eigenvector corresponding to the smaller root. The eigenvectors lie in the second and fourth quadrants. See Figure 14.10.

Location of eigenvalues | Response | Stable node in (z_1, z_2) plane | Stable node in (x_1, x_2) plane. Change of coordinate axes by linear transformation $X = MZ$, M = modal matrix.

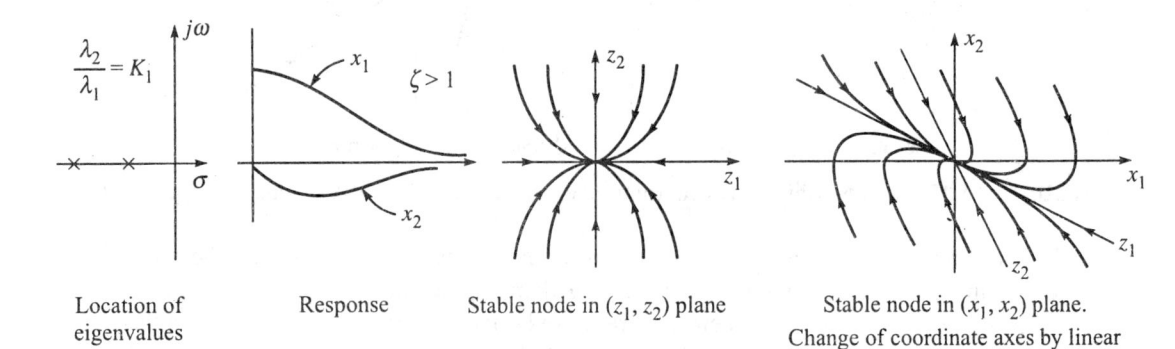

FIGURE 14.10 Relationship between eigenvalues, response and nodes for stable system in original and transformed plane.

4. Unstable system with both roots real, finite, unequal and positive, i.e. λ_1 and λ_2 are real and lie in the right-half of s-plane. This is also called a node but an unstable node. There are two eigenvectors and all trajectories emerge from the singular point and go to infinity. The time response becomes unbounded. The eigenvectors lie in the first and third quadrants. See Figure 14.11.

Location of eigenvalues | Unstable node in (z_1, z_2) plane | Unstable node in (x_1, x_2) plane.

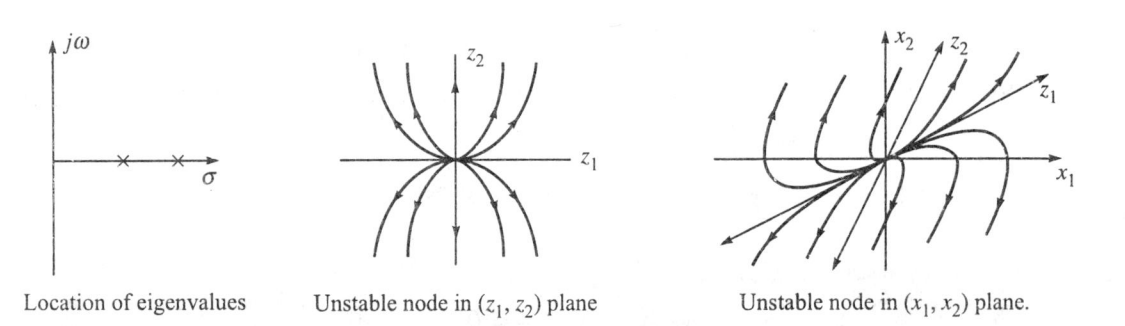

FIGURE 14.11 Relationship between eigenvalues and nodes for unstable system in original and transformed plane.

5. Systems with repeated roots: This happens when the system is critically damped, i.e. $\zeta = 1$. The response monotonically decreases to the singular point. In such cases the two eigenvectors coalesce into a single eigenvector, again with slope determined by the root value and the phase portrait is called a stable node. In this case the roots becomes repetitive, i.e. $\lambda_1 = \lambda_2$. See Figure 14.12.

(a) Location of eigenvalues (b) Stable critically damped system

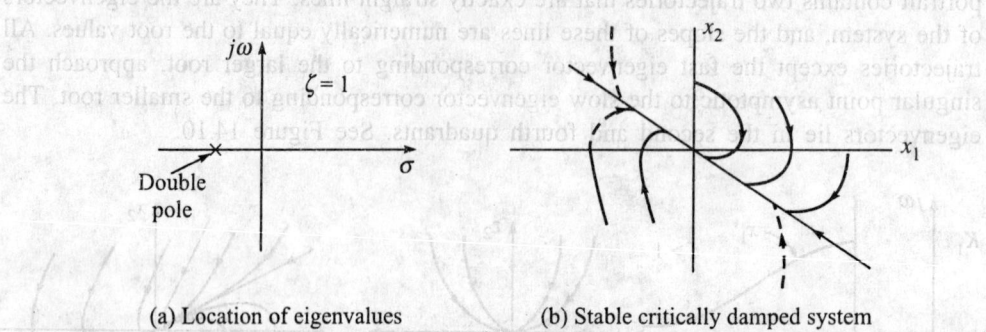

FIGURE 14.12 Relationship between roots and phase trajectory of critically damped system.

6. If the real part of the complex conjugate roots happens to be zero, i.e. the roots lie on the imaginary axis which means that the damping factor $\zeta = 0$. This gives sustained oscillations. The phase trajectories are closed curves concentric with the singular point, which is called a centre. See Figure 14.13.

(a) Location of eigenvalues   (b) Response   (c) Phase trajectory of undamped system

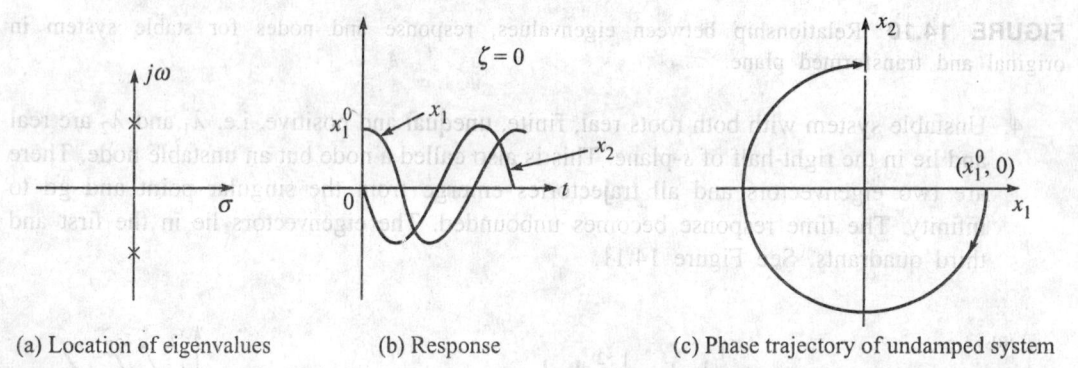

FIGURE 14.13 Relationship between eigenvalues, time response and phase trajectory of undamped system.

From Eqs. (14.27) and (14.28), for $\sigma = 0$, we get

$$\frac{dy_2}{dy_1} = -\frac{y_1}{y_2} \tag{14.31}$$

or
$$y_1\, dy_1 + y_2\, dy_2 = 0$$

Therefore,
$$y_1^2 + y_2^2 = C^2 \tag{14.32}$$

Thus the equation of spiral of Eq. (14.30), becomes that of a circle when $\sigma = 0$, for various values of integration constant C which depend upon initial conditions. The phase portrait is a family of circles or ellipses in the transformed plane with origin as the singular point, which is called a centre.

7. Systems with one positive real root and the other negative real root, that is, λ_1 and λ_2 are real ; λ_1 lies in the right-half and λ_2 lies in the left-half of s-plane. The eigenvector due to the negative root provides a trajectory that enters the singular point, while the trajectory due to the positive root leaves the singular point which is called the *saddle point*. All other trajectories approach the singular point adjacent to the incoming eigenvector, then curve away and leave the vicinity of the singular point, eventually approaching the second eigenvector asymptotically. This configuration of the trajectories is called a *saddle*. See Figure 14.14.

Location of eigenvalues   Saddle point in (z_1, z_2) plane   Saddle point in (x_1, x_2) plane

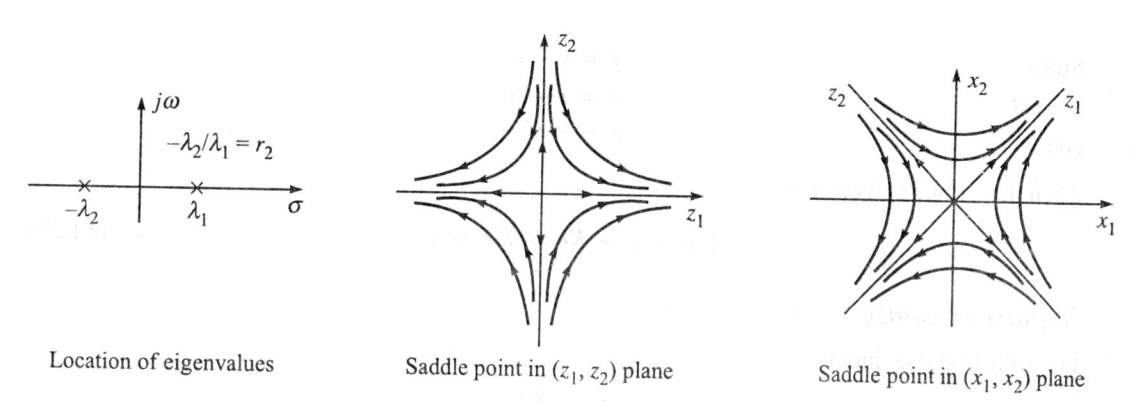

FIGURE 14.14 Location of eigenvalues, response, phase portrait of a second-order system with emphasis on stable and unstable focus, stable and unstable node, critically damped case, undamped system and saddle point in phase-plane (x_1, x_2) and in transformed plane (z_1, z_2) where $X = MZ$ and M is modal matrix.

14.4 Phase-Plane Analysis of Linear Control Systems

Actually, for linear system analysis, there exist a lot many other methods based on the transfer function approach. But for nonlinear systems, a very few methods are there such as Liapunov's method, Nyquist plot, etc. The phase-plane method has a greater role in the analysis of nonlinear control systems. The phase-plane method is quite useful for analyzing second-order systems. Further, it may be noted that the phase-plane method of analysis is limited to second-order systems only and this is its limitation.

Before analyzing a nonlinear control system, let us consider an application of the phase-plane method to the transient response analysis of a stable second-order linear control system shown in Figure 14.15.

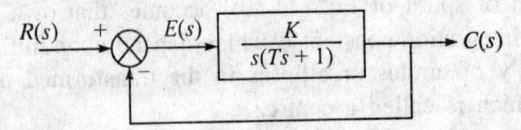

FIGURE 14.15 Second-order control system.

Assume that the system is initially at rest, i.e. $c(0) = \dot{c}(0) = 0$. Then

$$\frac{C(s)}{E(s)} = \frac{K}{s(Ts+1)} \tag{14.33}$$

which in time domain can be written as

$$T\ddot{c} + \dot{c} = Ke \tag{14.34}$$

Since
we get
and

$$e = r - c$$
$$\dot{e} = \dot{r} - \dot{c}$$
$$\ddot{e} = \ddot{r} - \ddot{c}$$

Then Eq. (14.34) becomes

$$T\ddot{e} + \dot{e} + Ke = T\ddot{r} + r \tag{14.35}$$

Impulse response

For unit-impulse input

$$r(t) = \delta(t)$$

So,

$$R(s) = 1$$

Then

$$\dot{r} = \ddot{r} = 0$$

Since

$$\frac{C(s)}{E(s)} = \frac{K}{s(Ts+1)}$$

In the time domain, we get

$$T\ddot{c} + \dot{c} = Ke = K(r-c)$$

or

$$T\ddot{c} + \dot{c} + Kc = Kr \tag{14.36}$$

For impulse input, the dynamics becomes

$$T\ddot{c} + \dot{c} + Kc = K(t) \text{ for } t > 0 \tag{14.37}$$

and from Eq. (14.34) applying the initial-value theorem, we get

$$c(0^+) = \lim_{s \to \infty} s \left[\frac{K}{Ts^2 + s + K} \right] = 0 \tag{14.38}$$

and
$$\dot{c}(0^+) = \lim_{s \to \infty} s^2 \left[\frac{K}{Ts^2 + s + K} \right] = \frac{K}{T} \quad (14.39)$$

Writing in terms of error, we get the system equation as
$$T\ddot{e} + \dot{e} + Ke = T\ddot{r} + \dot{r} \quad (14.40)$$
or
$$T\ddot{e} + \dot{e} + Ke = 0 \quad \text{for } t > 0$$
and the initial conditions are
$$e(0) = 0, \quad \dot{e}(0) = -\frac{K}{T}$$

The dynamics in error plane becomes
$$T\ddot{e} + \dot{e} + Ke = 0 \quad \text{for } t > 0 \quad (14.41)$$
with initial condition as
$$e(0) = 0, \quad \dot{e}(0) = -\frac{K}{T}$$

Suppose the values of T and K are such that the system is underdamped and stable. The singular point in error plane is $(0, 0)$, that is, where the independent variable and all its derivatives are zero. As the system is of second order, the singular point e and its derivative $\dot{e}$ are zero and in general, up to $(n - 1)$ order derivatives for an nth-order system are zero. The phase trajectory in the $e, \dot{e}$ plane starts at $e(0) = 0, \dot{e}(0) = -K/T$ and reaches the singular point $(0, 0)$, that is, $e = \dot{e} = 0$. The trajectory can be obtained through analog simulation.

One can also sense the correctness of the phase trajectory from the following analytical calculations.
$$\ddot{e} = -\frac{\dot{e}}{T} - \frac{K}{T} e$$
or
$$\frac{d}{dt}\dot{e} = -\frac{\dot{e}}{T} - \frac{K}{T} e$$
or
$$\int d(\dot{e}) = \int -\frac{\dot{e}}{T} dt - \int \frac{K}{T} e \, dt$$
or
$$\dot{e}(t) = \int -\left(\frac{\dot{e}}{T}\right) dt - \left(\frac{K}{T}\right) \int e \, dt + \dot{e}(0)$$

Putting the initial values, we get
$$\dot{e}(t) = \text{increasing from } (-K/T) \text{ further}$$
and
$$e(t) = \int \dot{e} \, dt + e(0) = \int -\frac{K}{T} dt + 0$$
$$= -\text{ve quantity}$$

which signifies that the error $e(t)$ is decreasing.

Further, the error will be the damped sinusoid as the described system in error plane is underdamped and stable. Hence $\dot{e}(t)$ will be damped sinusoid as obtained from derivative of $e(t)$ variable. From these considerations, we can get analytically the nature of the phase portrait in the e-$\dot{e}$ plane and obviously after rethinking the phase portrait in the c-$\dot{c}$ plane can also be obtained.

The trajectories indicating unit impulse responses for $t > 0$ are shown in Figure 14.16 for the overdamped system in the c-$\dot{c}$ and e-$\dot{e}$ planes. In the c-$\dot{c}$ plane the starting point corresponding to $t = (0^+)$ is $(0, K/T)$ and in the e-$\dot{e}$ plane the starting point is $(0, K/T)$.

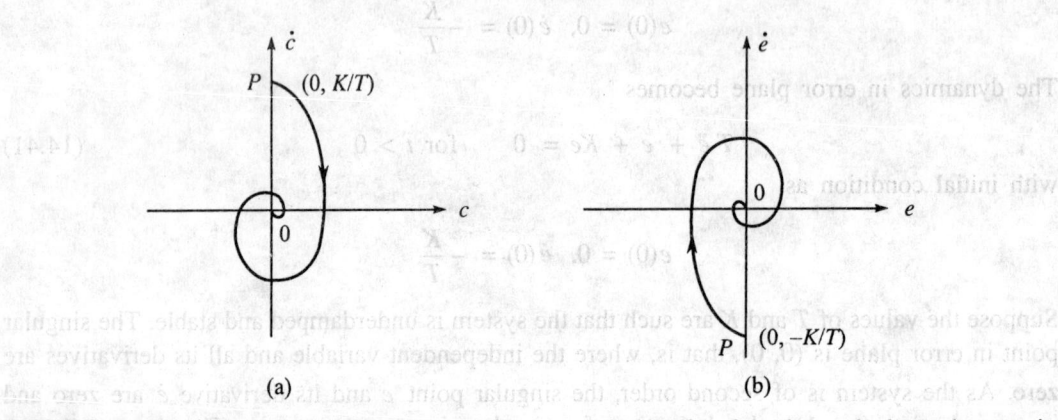

FIGURE 14.16 Phase trajectories corresponding to unit impulse for the underdamped case in (a) c-$\dot{c}$ or output plane and (b) e-$\dot{e}$ or error plane.

Step response

For the step input

$$r(t) = R \qquad (14.42)$$

so

$$\dot{r} = \ddot{r} = 0 \qquad (14.43)$$

Then Eq. (14.35) becomes

$$T\ddot{e} + \dot{e} + Ke = 0 \qquad \text{for } t > 0 \qquad (14.44)$$

with

$$e(0) = R, \ \dot{e}(0) = 0 \qquad (14.45)$$

The trajectory in the e-$\dot{e}$ plane starts at a point $(R, 0)$ and converges to the singular point of the system. Let the gain K be such that the system becomes underdamped, i.e. the roots become complex conjugate, and the transient response of error for step input swings across the zero line and ultimately asymptotically reaches the origin as shown in Figure 14.17(b).

We can get analytically both the time responses $e(t)$ and $\dot{e}(t)$ as shown in Figure 14.17(a). Eliminating time t from both $e(t)$ and $\dot{e}(t)$, we can get the phase portrait as shown in Figure 14.17(b). The phase trajectory in error plane decreases monotonically in clockwise direction.

Nonlinear Systems

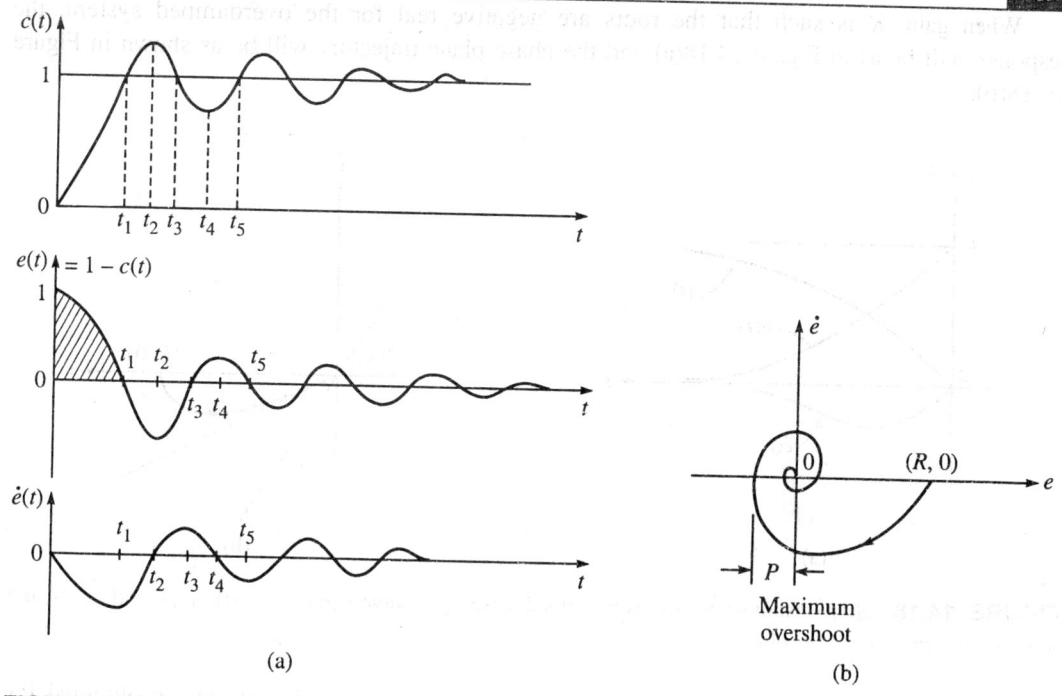

FIGURE 14.17 Step response to an underdamped case: (a) waveforms of $c(t)$, $e(t)$, and $\dot{e}(t)$ and (b) phase-plane portrait in e-$\dot{e}$ plane.

Let us proceed as

$$\ddot{e} = -\frac{\dot{e}}{T} - \frac{K}{T}e$$

Putting the initial values, we get

$$\ddot{e} = -\frac{K}{T}R$$

or

$$\frac{d}{dt}(\dot{e}) = -\frac{K}{T}R$$

or

$$\dot{e}(t) = \int \left(-\frac{K}{T}R\right)dt + \dot{e}(0)$$

or

$$\dot{e}(t) = -\text{ve},$$

That is, $\dot{e}(t)$ is decreasing.

Similarly,

$$\dot{e} = \frac{d}{dt}(e) = e(t) - e(0) = -\text{ve}$$

or

$$e(t) = -\text{ve quantity} + R$$
$$= \text{decreasing}$$

Thus, $e(t)$ is decreasing.

When gain K is such that the roots are negative real for the overdamped system, the response will be as in Figure 14.18(a) and the phase-plane trajectory will be as shown in Figure 14.18(b).

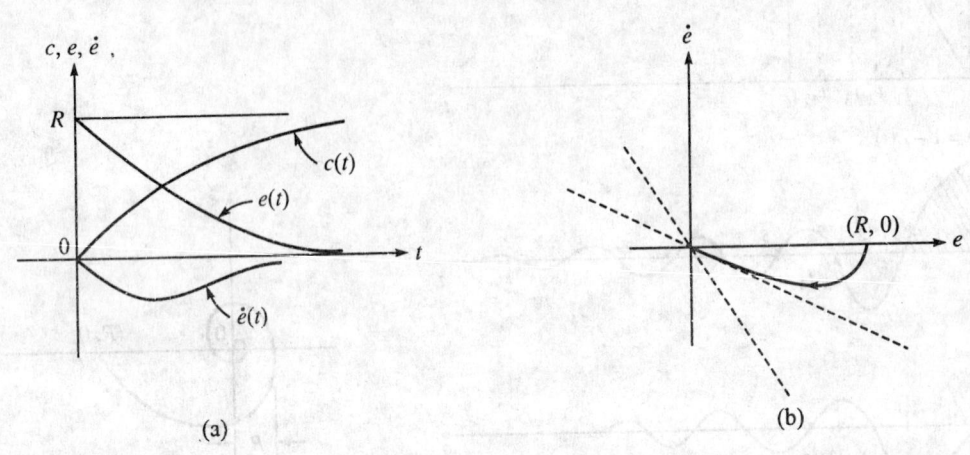

FIGURE 14.18 Step response to an overdamped case: (a) waveforms of $c(t)$, $e(t)$, and $\dot{e}(t)$ and (b) phase-portrait in $e\text{-}\dot{e}$ plane.

It may be noted that for linear systems where an analytical solution can be obtained for c and $\dot{c}$, there is, in fact, no necessity for phase-plane portrait. The phase-plane analysis is effectively applicable purposefully in case of nonlinear systems where analytical solutions are not available. We mention here out of academic interest that analog simulation as in Figure 14.19 for linear systems can be done in an analog computer, and the phase portrait can be obtained with the corresponding error e and the derivative of error $\dot{e}$ applied to the horizontal and vertical plates of a CRO in external time base mode.

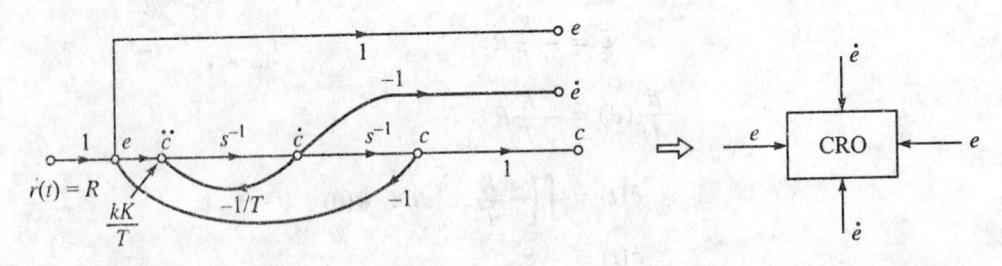

FIGURE 14.19 Analog simulation for a linear system to obtain phase-plane portrait.

Ramp input

For ramp input

$$r(t) = V.t \qquad (14.46)$$

or

$$\dot{r} = V$$

or

$$\ddot{r} = 0 \qquad (14.47)$$

Then Eq. (14.35) becomes

$$T\ddot{e} + \dot{e} + Ke = V$$

or
$$T\ddot{e} + \dot{e} + K\left(e - \frac{V}{K}\right) = 0 \tag{14.48}$$

Let
$$e - \frac{V}{K} = x$$

then
$$\dot{e} = \dot{x}$$

and
$$\ddot{e} = \ddot{x} \tag{14.49}$$

Equation (14.48) now becomes

$$T\ddot{x} + \dot{x} + Kx = 0 \tag{14.50}$$

The phase-plane trajectory in the x-$\dot{x}$ plane is the same as that in the e-$\dot{e}$ plane which is only coordinate shift of x-$\dot{x}$ plane by $(V/K, 0)$. The singular point for Eq. (14.50) is $x = \dot{x} = 0$, that is, in e-$\dot{e}$ plane the singular point is $e = V/K$ and $\dot{e} = 0$ as shown in Figure 14.20. Figure 14.20(a) is for K value such that the system becomes underdamped having complex conjugate roots with −ve real part. Figure 14.20(b) is for K such that the roots are −ve real, that is, the system is overdamped. For both the underdamped and overdamped cases, the trajectory starts initially at point A as shown in Figures 14.20(a) and (b).

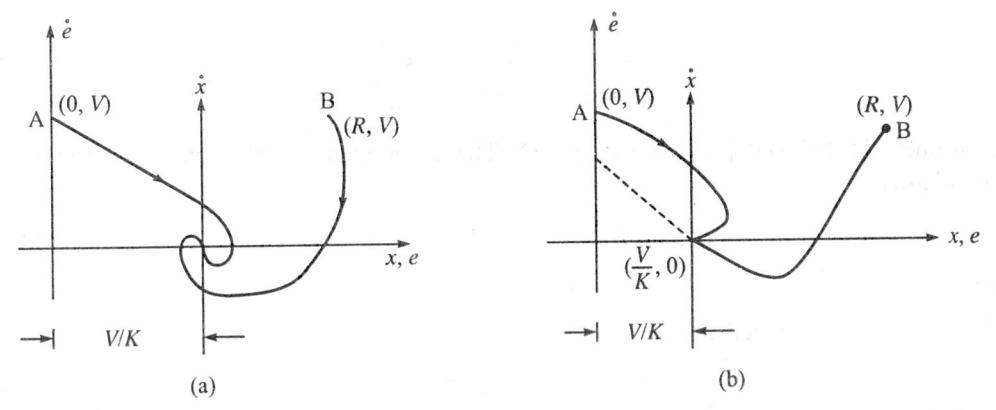

FIGURE 14.20 Trajectories corresponding to ramp response of the system for (a) the underdamped case and (b) the overdamped case.

For ramp plus step input, that is, $r(t) = Vt + R$, similar analysis will hold good. The trajectory in Figure 14.20(a) for the underdamped case starts initially from point B. In both the cases the singular point is $(V/K, 0)$.

14.5 Phase-Plane Analysis of Nonlinear Systems

When nonlinearity can be made piecewise linear, we can analyze nonlinear systems as illustrated next.

14.5.1 Control system with nonlinear gain

Consider the control system shown in Figure 14.21(a). The block G_N is the nonlinear gain element and its input–output characteristic is shown in Figure 14.21(b). The input–output relationship for G_N can be written as

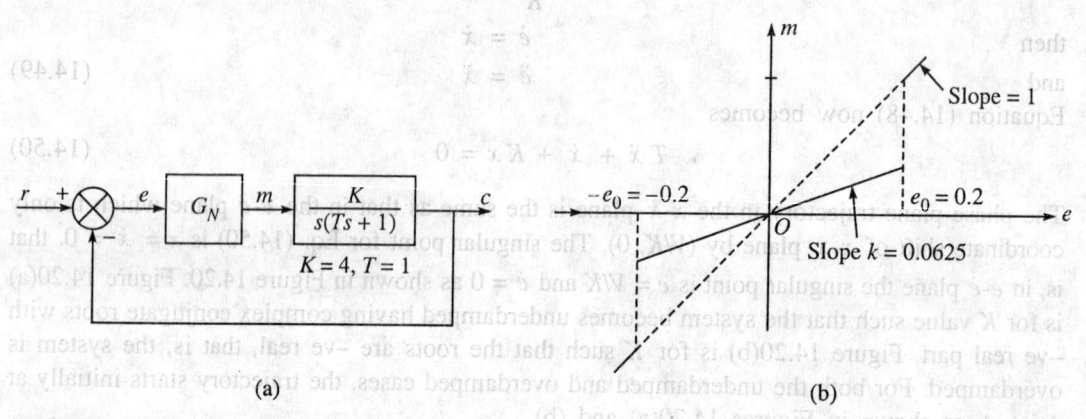

FIGURE 14.21 Control system with nonlinear gain: (a) block diagram and (b) input–output characteristic.

$$m = e \quad \text{for } |e| > e_0 \qquad (14.51)$$
$$ = ke \quad \text{for } |e| < e_0 \qquad (14.52)$$

We assume that the system is initially at rest. The differential equation relating c and m can be written as

$$T\ddot{c} + \dot{c} = Km \qquad (14.53)$$

Since
$$e = r - c$$

Eq. (14.53) can be written as

$$T\ddot{e} + \dot{e} + Km = T\ddot{r} + \dot{r} \qquad (14.54)$$

For a step input

$$\dot{r} = \ddot{r} = 0 \text{ for } t > 0 \qquad (14.55)$$

Eq. (14.54) becomes

$$T\ddot{e} + \dot{e} + Km = 0 \qquad (14.56)$$

From Eqs. (14.51), (14.52), and (14.56), we obtain the following equations

$$T\ddot{e} + \dot{e} + Ke = 0 \quad \text{for } |e| > e_0 \qquad (14.57)$$

and
$$T\ddot{e} + \dot{e} + Kke = 0 \quad \text{for } |e| < e_0 \qquad (14.58)$$

It is clear that the system dynamics represented by Eqs. (14.57) and (14.58) have the origin (0,0) as the singular point in the e-$\dot{e}$ plane. With the parameter values of $K = 4$, $T = 1$ the roots are complex conjugate with negative real parts, i.e. the system represented by Eq. (14.57) becomes linear underdamped. The point (0,0) becomes the stable focus.

The system dynamics represented by Eq. (14.58) is linear and the value of $k = 0.0625$ is such that the damping coefficient becomes unity. Then the singular point (0,0) becomes a stable node. Thus the system becomes linear underdamped for large error and linear overdamped for small error. For the relationship $m = ke$, $k = 0.0625$, the system becomes critically damped governed by Eq. (14.58), having the damping coefficient as unity; the phase-plane plot for this linear system is as shown in Figure 14.22(a). Now for $m = e$, the system dynamics is governed by Eq. (14.57), i.e. the system is underdamped, overshoot occurs in the response, the phase-plane portrait is as shown in Figure 14.22(b).

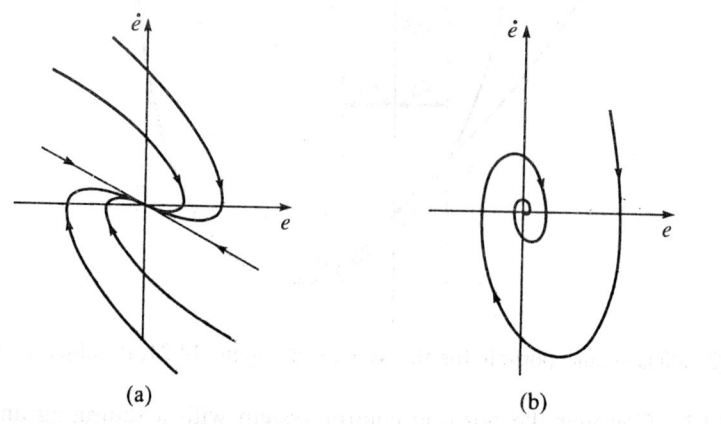

FIGURE 14.22 Phase-plane portraits for linear systems: (a) critically damped case ($m = ke$; $k = 0.0625$) and (b) underdamped case ($m = e$; $k = 1$).

Now for the system of Figure 14.21(a) subjected to unit step input, the e-$\dot{e}$ phase-plane is divided into three regions such as the region bounded by the lines $|e| = e_0$ and a linear operation governed by Eq. (14.58). Outside the region $|e| > e_0$, the governing equation is (14.57). The trajectory in Figure 14.23 starting at point A having $e(0) = 1$, $\dot{e}(0) = 0$ tends to converge to stable focus (0,0). The operation of the system switches at point B on the boundary $e = e_0$ line and at this point the governing equation is (14.58) with $k = 0.0625$, and the system is critically damped. The trajectory tends to converge to stable node (0,0) until the operation of the system switches again at point C at $e = -e_0$ line and the governing equation becomes (14.57) and the system is underdamped and tends to converge to stable focus (0,0) until the point D is reached on the boundary line $e = -e_0$ where the governing equation becomes (14.58) and the system tends to converge to stable node (0,0). Repeating the same process, the trajectory finally converges to the stable node (0,0). At steady state there is no error.

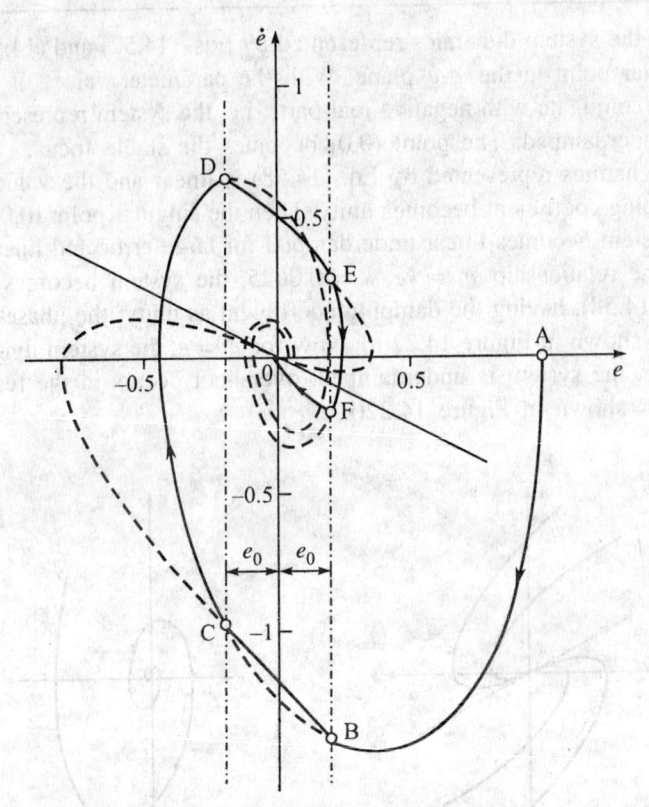

FIGURE 14.23 Phase-plane portrait for the system of Figure 14.21(a) subjected to unit-step input.

EXAMPLE 14.1 Consider the position control system with a saturating amplifier, approximated by straight line segments as shown in Figure 14.24. It is desired to obtain the phase-plane portrait for step input $r(t) = R$. The parameter values are given as $a = 1$, $K = 4$, and zero initial conditions are assumed.

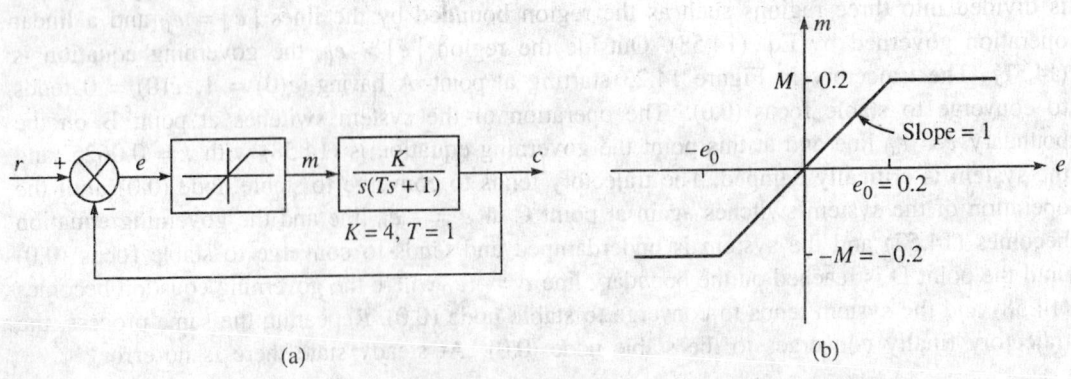

FIGURE 14.24 Example 14.1: (a) position control system with saturating amplifier and (b) input–output characteristic curve of the saturation nonlinearity.

Nonlinear Systems

Solution: From the saturation characteristic curve shown in Figure 14.24(b), we obtain

$$m = e \quad \text{for } |e| \leq e_0 = 0.5$$
$$= M \quad \text{for } |e| > e_0 = 0.5$$
$$= -M \quad \text{for } |e| < -e_0 = -0.5 \tag{14.59}$$

From Figure 14.24(a), we have the forward path transfer function as

$$\frac{C(s)}{m(s)} = \frac{K}{s(Ts+1)} \tag{14.60}$$

which in time domain can be written as

$$T\ddot{c} + \dot{c} = Km \tag{14.61}$$

Given that $\dot{c}(0) = c(0) = 0$

from the given negative feedback system, we get

$$e = r - c \tag{14.62}$$

Then $\dot{c} = \dot{r} - \dot{e} \tag{14.63}$

and $\ddot{c} = \ddot{r} - \ddot{e} \tag{14.64}$

Substituting in Eq. (14.61), we get

$$T\ddot{e} + \dot{e} + Km = T\ddot{r} + \dot{r} \tag{14.65}$$

Step input

$r(t) = R = $ constant, then

$$\dot{r} = \ddot{r} = 0 \text{ for } t > 0 \tag{14.66}$$

Hence Eq. (14.65) becomes

$$T\ddot{e} + \dot{e} + Km = 0 \tag{14.67}$$

For linear operation of the system, i.e. for $|e| \leq e_0$, the output of the saturation type nonlinearity becomes $m = e$, then the above equation becomes

$$T\ddot{e} + \dot{e} + Ke = 0 \tag{14.68}$$

with the initial conditions $e(0) = R$, $\dot{e}(0) = 0$.

The origin of the e-$\dot{e}$ plane, that is, $e = 0$, $\dot{e} = 0$ is the singular point. The singular point (0,0) is either a stable node or a stable focus.

For the nonlinear operation of the system, i.e. for output of the saturation type of nonlinearity

$$m = M \quad \text{for } e > e_0$$

and $$m = -M \quad \text{for } e < -e_0$$

Equation (14.67) becomes

$$T\ddot{e} + \dot{e} + KM = 0 \quad \text{for } e > e_0$$

i.e.
$$\ddot{e} + \frac{\dot{e}}{T} + \frac{KM}{T} = 0 \quad \text{for } e > e_0 \qquad (14.69)$$

and
$$\ddot{e} + \frac{\dot{e}}{T} - \frac{KM}{T} = 0 \quad \text{for } e < -e_0 \qquad (14.70)$$

Let us define,

$$\frac{d\dot{e}}{de} = \alpha$$

i.e.
$$\frac{d\dot{e}/dt}{de/dt} = \alpha$$

or
$$\ddot{e} = \alpha \dot{e}$$

Then, we obtain from Eq. (14.69)

$$\ddot{e} + \frac{\dot{e}}{T} = \frac{-KM}{T}$$

or
$$\dot{e}\left(\frac{1}{T} + \alpha\right) = \frac{-KM}{T}$$

or
$$\dot{e} = \frac{\dfrac{-KM}{T}}{\alpha + \dfrac{1}{T}} \quad \text{for } e > e_0 \qquad (14.71)$$

and from Eq. (14.70) as

$$\dot{e} = \frac{\dfrac{KM}{T}}{\alpha + \dfrac{1}{T}} \quad \text{for } e < -e_0 \qquad (14.72)$$

From Eq. (14.71), it can be seen that for $e > e_0$ all trajectories are asymptotic to the line

$$\dot{e} = -KM \qquad (14.73)$$

which corresponds to $\alpha = 0$.

Similarly, from Eq. (14.72) for $e < e_0$, all trajectories are asymptotic to the line

$$\dot{e} = KM \qquad (14.74)$$

Figure 14.25 shows the phase-plane portrait for the region $|e| > e_0$. Figure 14.26 shows the trajectory when the system is subjected to a step input of magnitude $R = 2$. Then the initial conditions are

$$e(0) = 2, \quad \dot{e}(0) = 0$$

As
$$e = r - c$$

and
$$\dot{e} = \dot{r} - \dot{c}$$

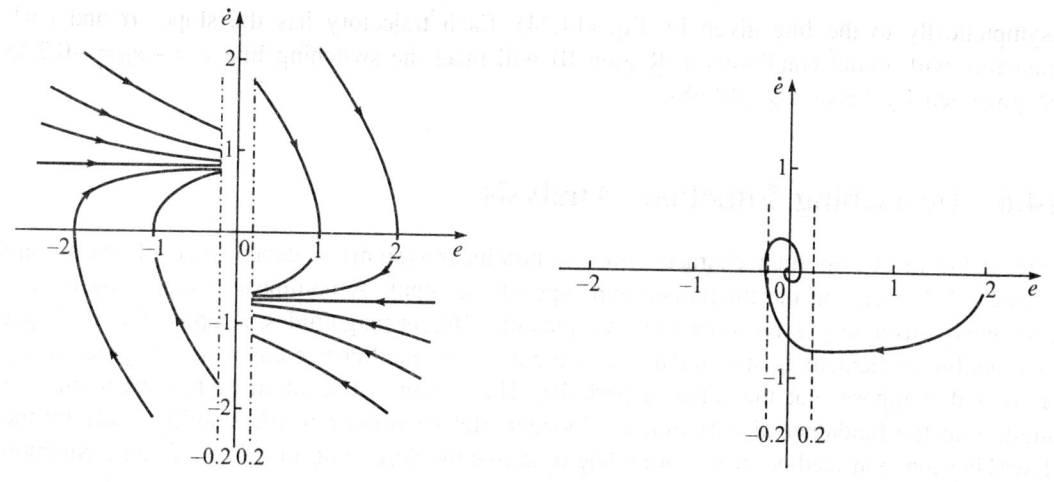

FIGURE 14.25 Phase-plane portrait for the region $|e| = e_0 = 0.2$ for the system shown in Figure 14.24(a).

FIGURE 14.26 Trajectory corresponding to step response of the system shown in Figure 14.24(a).

then for step input

$$e(0) = 2 - c(0) = 2 - 0 = 2$$

and

$$\dot{e}(0) = \dot{r}(0) - \dot{c}(0) = 0 - 0 = 0$$

as the system was initially at rest, i.e. $c(0) = \dot{c}(0) = 0$. The phase-plane trajectory may be divided into three regions:

Region I (defined by $e > e_0 = 0.2$)

In this region, the system characteristic is governed by nonlinear Eq. (14.69). All the trajectories with different initial conditions in Region I will meet asymptotically to the line $\dot{e} = -0.8$ given by Eq. (14.73).

Each trajectory has the slope α. Each trajectory is also unique. The two switching lines are $e = e_0 = 0.2$ and $e = -e_0 = -0.2$. The trajectories with initial conditions in Region I will meet the switching line $e = e_0 = 0.2$ to be governed by linear Eq. (14.68).

Region II (defined by $|e| < |e_0| = |0.2|$, that is, for $-0.2 = -e_0 < e < e_0 = 0.2$)

In this region the system is governed by Eq. (14.68) and moves towards the singular point (0,0). The path of the trajectory in the linear region II depends on striking point of the trajectory on the switching line which becomes the initial condition for the linear Eq. (14.68).

Region III (defined by $e < -e_0 = -0.2$).

In this region the system is governed by nonlinear Eq. (14.70) and follows the similar line as that of Region I. All trajectories with different initial conditions in Region III will meet asymptotically to the line given by Eq. (14.74). Each trajectory has the slope α and each trajectory with initial conditions in Region III will meet the switching line $e = -e_0 = -0.2$ to be governed by linear Eq. (14.68).

14.6 Describing Function Analysis

One of the most important characteristics of nonlinear systems is dependence of the system response behaviour on the magnitude and type of the input. A nonlinear system may behave completely differently in response to step inputs of different magnitudes. Suppose that the input to a nonlinear element is sinusoidal. The output of the nonlinear element is, in general, not sinusoidal. Suppose that the input is periodic. The output will contain higher harmonics, in addition to the fundamental harmonic component. Before coming to the stability study by the describing function method, it is worthwhile to derive the describing functions of some common nonlinearities.

14.6.1 Derivation of describing functions

The describing function of a nonlinear element is defined to be the complex ratio of the fundamental harmonic component of the output to the input, that is,

$$N = (Y_1/X) \angle \phi_1 \qquad (14.75)$$

where
  X is the amplitude of the input sinusoid
  Y_1 is the amplitude of the fundamental harmonic component of the output
  ϕ_1 is the phase shift of the fundamental harmonic component of the output with respect to the input

Therefore for computing the describing function of a nonlinear element, we are simply required to find the fundamental harmonic component of its output for an input

$$x = X \sin \omega t \qquad (14.76)$$

The output $y(t)$ may be expressed as Fourier series as

$$y(t) = A_0 + \sum_{n=1}^{\infty} (A_n \sin n\omega t + B_n \cos n\omega t)$$

$$= A_0 + \sum_{n=1}^{\infty} Y_n \sin(n\omega t + \phi_n) \qquad (14.77)$$

where

$$A_n = \frac{1}{\pi} \int_0^{2\pi} y(t) \sin n\omega t \, d(\omega t) \qquad (14.78)$$

$$B_n = \frac{1}{\pi} \int_0^{2\pi} y(t) \cos n\omega t \, d(\omega t) \qquad (14.79)$$

Nonlinear Systems

$$Y_n = \sqrt{A_n^2 + B_n^2} \tag{14.80}$$

$$\phi_n = \tan^{-1}\left(\frac{B_n}{A_n}\right) \tag{14.81}$$

If the nonlinearity is symmetric, then $A_0 = 0$.

The fundamental component of the output can be written as

$$\begin{aligned} Y_1 &= A_1 \sin \omega t + B_1 \cos \omega t \\ &= Y_1 \sin(\omega t + \phi_1) \end{aligned} \tag{14.82}$$

where the coefficients A_1 and B_1 of the Fourier series are

$$B_1 = \frac{1}{\pi} \int_0^{2\pi} y \cos \omega t \, d(\omega t) \tag{14.83}$$

$$A_1 = \frac{1}{\pi} \int_0^{2\pi} y \sin \omega t \, d(\omega t) \tag{14.84}$$

The amplitude and phase angle of the fundamental component of the output are given by

$$Y_1 = \sqrt{A_1^2 + B_1^2} \tag{14.85}$$

$$\phi_1 = \tan^{-1}\left(\frac{B_1}{A_1}\right) \tag{14.86}$$

Illustrative derivations of the describing functions of some of the commonly encountered nonlinearities are given below:

Dead-zone nonlinearity

The idealized characteristic of a nonlinearity having dead-zone and its response to sinusoidal input are shown in Figure 14.27. In a dead-zone nonlinearity, there is no output for inputs within the dead-zone amplitude.

Referring to Figure 14.27, the output $y(t)$ for $0 \leq \omega t \leq \pi$ is given by

$$\begin{aligned} y(t) &= 0 & \text{for } 0 < \omega t < \omega t_1 \\ &= k(X \sin \omega t - \Delta) & \text{for } \omega t_1 \leq \omega t \leq (\pi - \omega t_1) \\ &= 0 & \text{for } (\pi - \omega t_1) < \omega t \leq \pi \end{aligned}$$

Since the output $y(t)$ is an odd function, A_n terms are zero, and its Fourier expansion has only sine terms. The fundamental harmonic component of the output is given by

$$y_1(t) = A_1 \sin \omega t \tag{14.87}$$

where

$$A_1 = \frac{1}{\pi} \int_0^{2\pi} y(t) \sin \omega t \, d(\omega t) \tag{14.88}$$

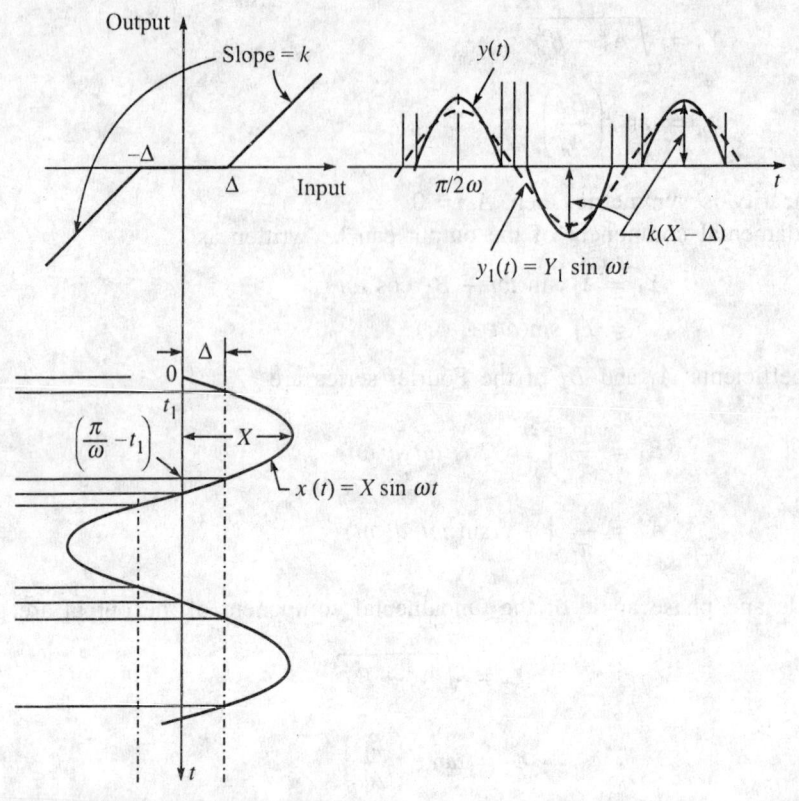

FIGURE 14.27 Sinusoidal response of nonlinearity with dead-zone nonlinearity.

Taking half-wave symmetry into account, we get

$$A_1 = \frac{4}{\pi} \int_0^{\pi/2} y(t) \sin \omega t \, d(\omega t)$$

$$= \frac{4k}{\pi} \int_{\omega t_1}^{\pi/2} (X \sin \omega t - \Delta) \sin \omega t \, d(\omega t) \qquad (14.89)$$

where
$$\Delta = X \sin \omega t_1 \qquad (14.90)$$

or
$$\omega t_1 = \sin^{-1}\left(\frac{\Delta}{X}\right) \qquad (14.91)$$

Hence

$$A_1 = \frac{4Xk}{\pi} \left[\int_{\omega t_1}^{\pi/2} \sin^2 \omega t \, d(\omega t) - \sin \omega t_1 \int_{\omega t_1}^{\pi/2} \sin \omega t \, d(\omega t) \right]$$

$$= \frac{2Xk}{\pi} \left[\frac{\pi}{2} - \sin^{-1}\left(\frac{\Delta}{X}\right) - \frac{\Delta}{X}\sqrt{1-\left(\frac{\Delta}{X}\right)^2} \right] \qquad (14.92)$$

The describing function for an element with dead-zone can be obtained as

$$N = \begin{cases} \dfrac{A_1}{X} \angle 0° \\ \left[k - \dfrac{2k}{\pi}\left[\sin^{-1}\dfrac{\Delta}{X} + \dfrac{\Delta}{X}\sqrt{1-\left(\dfrac{\Delta}{X}\right)^2} \right] \right] \angle 0° \end{cases} \qquad (14.93)$$

Figure 14.28 shows a plot of N/k as a function of Δ/X. Note that for $(\Delta/X) > 1$, the output is zero and the value of the describing function is also zero.

Saturation nonlinearity

An input–output characteristic curve for the saturation nonlinearity is shown in Figure 14.29. For small input signals, the output of the saturation element is proportional to the input. For larger input signals the output will not increase proportionally and finally for very large input signals the output is constant. Figure 14.29 depicts the input–output waveforms of the saturation nonlinearity. Idealized characteristic of saturation type nonlinearity and its response to the sinusoidal input are shown in Figure 14.29. Since the output in Figure 14.29 has odd symmetry, its fundamental component will not have the cosine term. It will therefore be of the form ($B \sin \omega t$). Taking the property of half-wave symmetry, it is seen that for the element with saturation nonlinearity shown in Figure 14.29, the output $y(t)$ for $0 < \omega t < \pi/2$ is given by

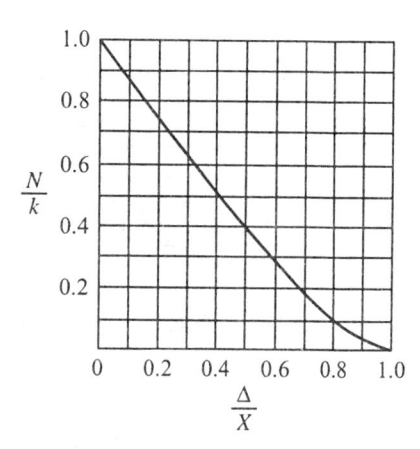

FIGURE 14.28 Describing function plot for the dead-zone nonlinearity.

$$y(t) = kx \sin \omega t \quad \text{for } 0 < \omega t < \beta$$
$$= kS \qquad \text{for } \beta < \omega t < \pi/2 \qquad (14.94)$$

$$B_1 = \dfrac{1}{\pi} \int_0^{2\pi} y(t) \sin \omega t \, d(\omega t)$$

$$= \dfrac{4}{\pi} \int_0^{\pi/2} y(t) \sin \omega t \, d(\omega t)$$

$$= \dfrac{4}{\pi} \int_\beta^{\pi/2} kS \sin \omega t \, d(\omega t)$$

$$= \dfrac{4k}{\pi} \left[\dfrac{X}{2\beta} + S \cos \beta \right] \qquad (14.95)$$

where

$$\beta = \sin^{-1}\left(\dfrac{S}{X}\right) \qquad (14.96)$$

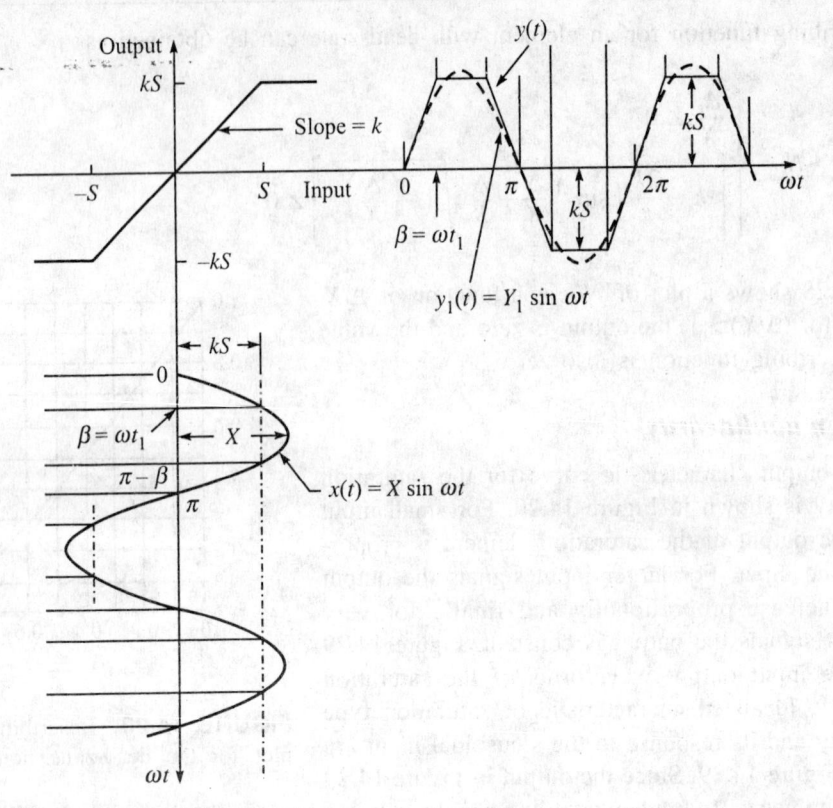

FIGURE 14.29 Sinusoidal response of nonlinearity with saturation nonlinearity.

Hence the describing function is

$$N = \frac{B_1}{X} \angle 0°$$

$$= \frac{2k}{\pi}\left[\sin^{-1}\left(\frac{S}{X}\right) + \frac{S}{X}\sqrt{1-\left(\frac{S}{X}\right)^2}\right] \quad (14.97)$$

Figure 14.30 shows a plot of N/k as a function of S/X. For $(S/X) > 1$, the value of the describing function is unity.

Note that the describing function for the dead-zone nonlinearity and that for the saturation nonlinearity are related as follows.

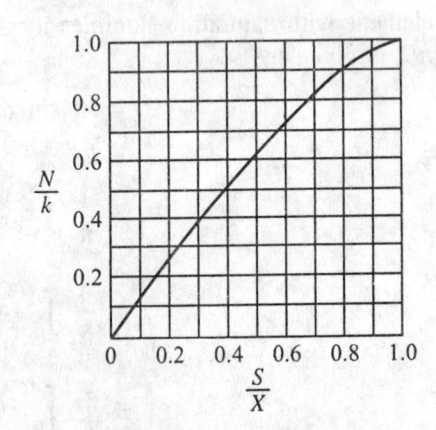

FIGURE 14.30 Describing function plot for the saturation nonlinearity.

$$N_{\text{dead-zone}} = k - N_{\text{saturation}} \quad \text{for } \Delta = S \quad (14.98)$$

On-off nonlinearity

The on-off nonlinearity is often called the two-position nonlinearity, its input–output characteristic is shown in Figure 14.31.

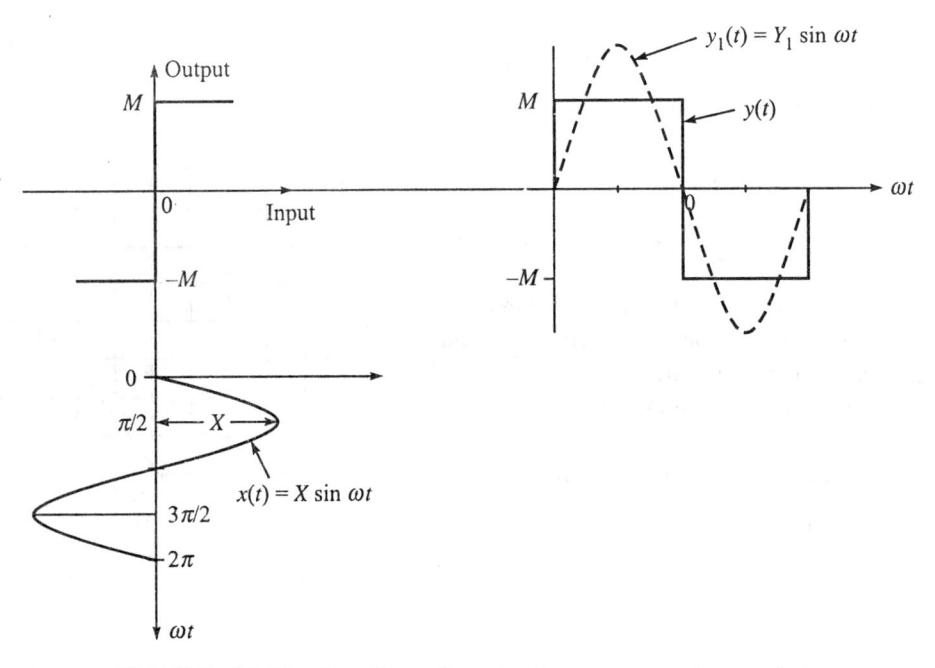

FIGURE 14.31 On-off nonlinearity: input–output characteristic.

The Fourier series expansion of the output $y(t)$ for the on-off nonlinear element is

$$y(t) = A_0 + \sum_{n=1}^{\infty} (A_n \sin n\omega t + B_n \cos n\omega t)$$

The output is an odd function and for that $B_n = 0$ ($n = 0, 1, 2, \ldots$). Then

$$y(t) = \sum_{n=1}^{\infty} A_n \sin n\omega t \qquad (14.99)$$

The fundamental component of $y(t)$ is

$$y_1(t) = A_1 \sin \omega t$$

where

$$A_1 = \frac{1}{\pi} \int_0^{2\pi} y(t) \sin \omega t \, d(\omega t)$$

$$= \frac{2}{\pi} \int_0^{\pi} y(t) \sin \omega t \, d(\omega t) \qquad (14.100)$$

Now,
$$y(t) = M, \quad \text{for} \quad 0 \leq \omega t \leq \pi$$
$$= -M, \quad \text{for} \quad \pi \leq \omega t \leq 2\pi \qquad (14.101)$$

then, we get
$$A_1 = \frac{2M}{\pi} \int_0^\pi y(t) \sin \omega t \, d(\omega t) = \frac{4M}{\pi} \qquad (14.102)$$

Thus,
$$y_1(t) = \frac{4M}{\pi} \sin \omega t \qquad (14.103)$$

The describing function N is given by
$$N = \frac{y_1}{X} \angle 0°$$
$$= \frac{4M}{\pi X} \angle 0° \qquad (14.104)$$

The describing function of the on-off nonlinearity is a real quantity and is only a function of input amplitude X. A plot of the describing function N versus M/X is shown in Figure 14.32.

On-off nonlinearity with dead-zone

The input–output characteristic of on-off nonlinearity with dead-zone is shown in Figure 14.33.

For sinusoidal input, the output of the nonlinear element is

$$y(t) = 0 \quad \text{for} \quad 0 < \omega t < \omega t_1$$
$$= M \quad \text{for} \quad \omega t_1 < \omega t < (\pi - \omega t_1)$$
$$= 0 \quad \text{for} \quad (\pi - \omega t_1) < \omega t < \pi \qquad (14.105)$$

The output waveform is an odd function and hence $A_n = 0$. The fundamental harmonic component of the output $y(t)$ is

$$y_1(t) = A_1 \sin \omega t \qquad (14.106)$$

where
$$A_1 = \frac{1}{\pi} \int_0^{2\pi} y(t) \sin \omega t \, d(\omega t)$$
$$= \frac{4}{\pi} \int_0^{\pi/2} y(t) \sin \omega t \, d(\omega t) \qquad (14.107)$$

Because of half-wave symmetry
$$A_1 = \frac{4}{\pi} \int_{\omega t_1}^{\pi/2} M \sin \omega t \, d(\omega t)$$
$$= \frac{4M}{\pi} \cos \omega t_1 \qquad (14.108)$$

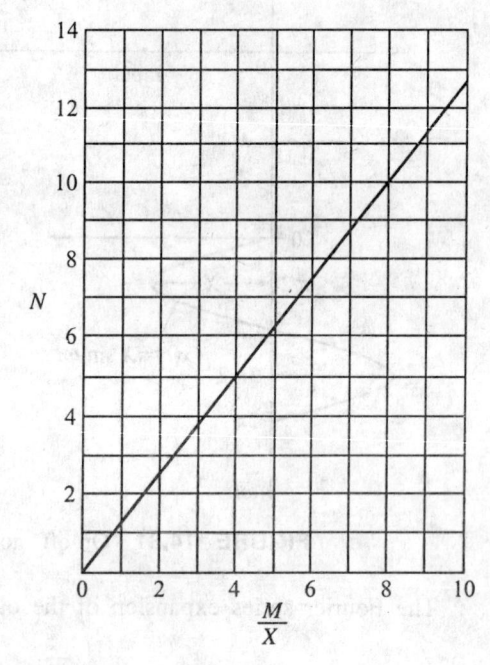

FIGURE 14.32 Describing function for on-off nonlinearity.

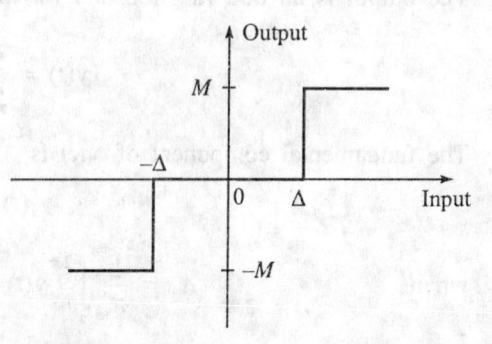

FIGURE 14.33 On-off nonlinearity with dead-zone.

Now,
$$\sin \omega t_1 = \frac{\Delta}{X} \tag{14.109}$$

Hence,
$$\cos \omega t_1 = \sqrt{1 - \left(\frac{\Delta}{X}\right)^2} \tag{14.110}$$

Therefore the describing function N (see Figure 14.34) is

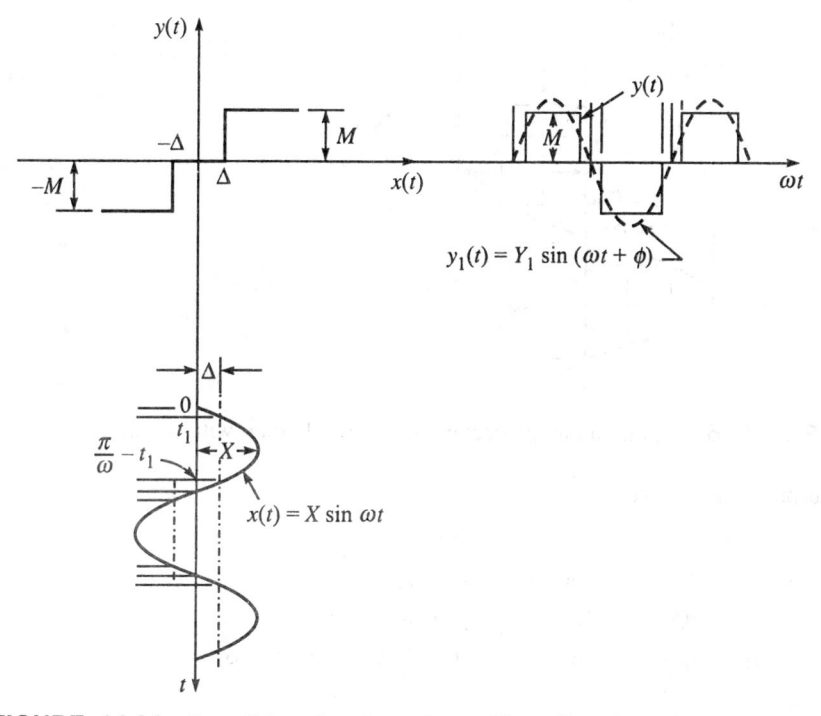

FIGURE 14.34 Describing function of on-off nonlinearity with dead-zone.

$$N = \frac{4M}{\pi X} \cos \omega t_1$$

$$= \frac{4M}{\pi X} \sqrt{1 - \left(\frac{\Delta}{X}\right)^2} \angle 0° \tag{14.111}$$

The describing function plot is shown in Figure 14.35.

Dead-zone and saturation

The input–output characteristic of the dead-zone with saturation nonlinearity is shown in Figure 14.36.

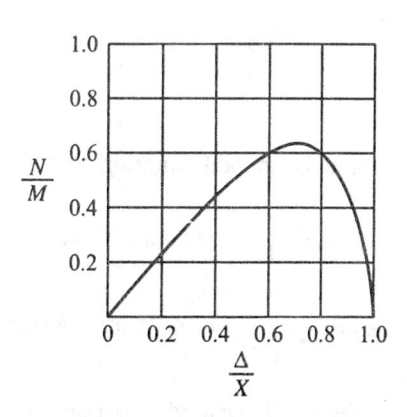

FIGURE 14.35 Describing function plot of on-off nonlinearity with dead-zone.

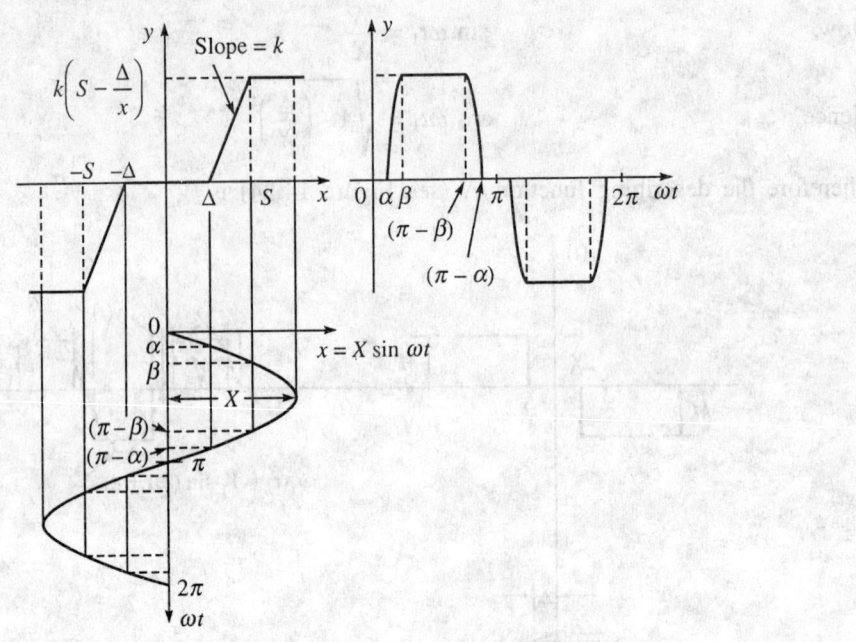

FIGURE 14.36 Input–output characteristic of dead-zone with saturation nonlinearity.

The output can be written as

$$\begin{aligned}
y &= 0 & \text{for} & \quad 0 \leq \omega t \leq \alpha \\
&= k(k - \Delta) & \text{for} & \quad \alpha \leq \omega t \leq \beta \\
&= k(S - \Delta) & \text{for} & \quad \beta < \omega t < \pi - \beta \\
&= k(k - \Delta) & \text{for} & \quad \pi - \beta < \omega t < \pi - \alpha \\
&= 0 & \text{for} & \quad \pi - \alpha \leq \omega t \leq \pi
\end{aligned} \quad (14.112)$$

where

$$\alpha = \sin^{-1} \frac{\Delta}{X}$$

and

$$\beta = \sin^{-1} \frac{S}{X}$$

The output is an odd function with half-wave and quarter-wave symmetry, then $B_1 = 0$

and $\quad A_1 = \dfrac{4}{\pi} \displaystyle\int_0^{\pi/2} y(t) \sin \omega t \, d(\omega t)$

or $\quad A_1 = \dfrac{4}{\pi} \left[\displaystyle\int_0^{\beta} k(X \sin \omega t - \Delta) \sin \omega t \, d(\omega t) + \displaystyle\int_{\beta}^{\pi/2} k(S - \Delta) \sin \omega t \, d(\omega t) \right]$

$\qquad = \dfrac{k}{\pi} [2X(\beta - \alpha) - X(\sin 2\beta - \sin 2\alpha) + 4\{\Delta(\cos \beta - \cos \alpha) + (S - \Delta)\cos \beta\}]$

$\qquad = \dfrac{kX}{\pi} [2(\beta - \alpha) + (\sin 2\beta - \sin 2\alpha)] \quad (14.113)$

Nonlinear Systems

The describing function is

$$\frac{N}{k} = \begin{cases} 0 & ; X < \Delta, \ \alpha - \beta = \pi/2 \\ 1 - \frac{2}{\pi}(\alpha + \sin\alpha \cos\alpha) & ; \Delta < X < S, \ \beta = \pi/2 \\ \frac{1}{\pi}[2(\beta - \alpha) + (\sin 2\beta - \sin 2\alpha)] & ; X > S \end{cases} \quad (14.114)$$

This is the describing function of a generalized single-valued nonlinearity from which special cases are verified with the results derived earlier.

Case I

Saturation nonlinearity ($\Delta = 0$, $\alpha = 0$) has been derived earlier as

$$\frac{N}{k} = \begin{cases} 1 & ;|X| < S \\ \frac{2}{\pi}\left[\sin^{-1}\left(\frac{S}{X}\right) + \frac{S}{X}\sqrt{1 - \left(\frac{S}{X}\right)^2}\right] & ;|X| \geq |S| \end{cases} \quad (14.115)$$

This function is sketched in Figure 14.37.

Case II

Dead-zone nonlinearity ($S \to \infty$, $\beta = \pi/2$) is expressed as

$$\frac{N}{k} = \begin{cases} 0 & ;|X| < \Delta \\ 1 - \frac{2}{\pi}\sin^{-1}\left(\frac{\Delta}{X}\right) + \frac{\Delta}{X}\sqrt{1 - \left(\frac{\Delta}{X}\right)^2} & ;|X| > |\Delta| \end{cases} \quad (14.116)$$

This function is sketched in Figure 14.38.

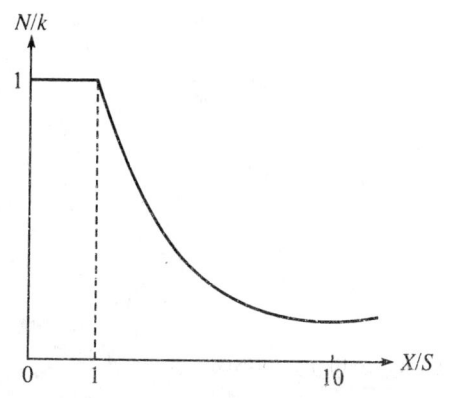

FIGURE 14.37 Describing function of saturation nonlinearity.

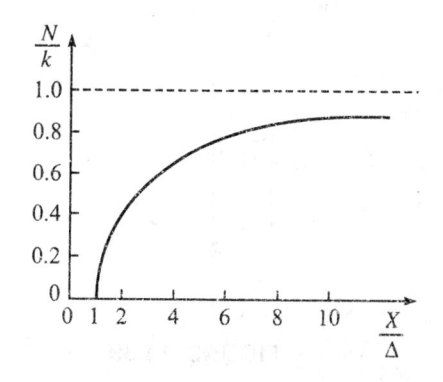

FIGURE 14.38 Describing function of dead-zone nonlinearity.

Case III

The characteristic of the input–output waveform of relay with dead-zone is as shown in Figure 14.33. Letting $S = \Delta$ and slope $k = \infty$ and saturation value as $|M|$ in Eq. (14.114) and then $\alpha = \beta$, we get the expression N/M as

$$\frac{N}{M} = \begin{cases} 0 & ; |X| < \Delta \\ \left(\dfrac{4}{\pi X}\right)\sqrt{1 - \left(\dfrac{\Delta}{X}\right)^2} & ; |X| > \Delta \end{cases} \quad (14.117)$$

It may be noted that for memoryless nonlinearity, the describing function becomes independent of frequency having zero phase shift but amplitude dependent.

On-off nonlinearity with hysteresis

Consider an on-off nonlinear element with hysterisis whose input $x(t)$ and output $y(t)$ are shown in Figure 14.39. Further, $y_1(t)$ is the fundamental harmonic component of output $y(t)$.

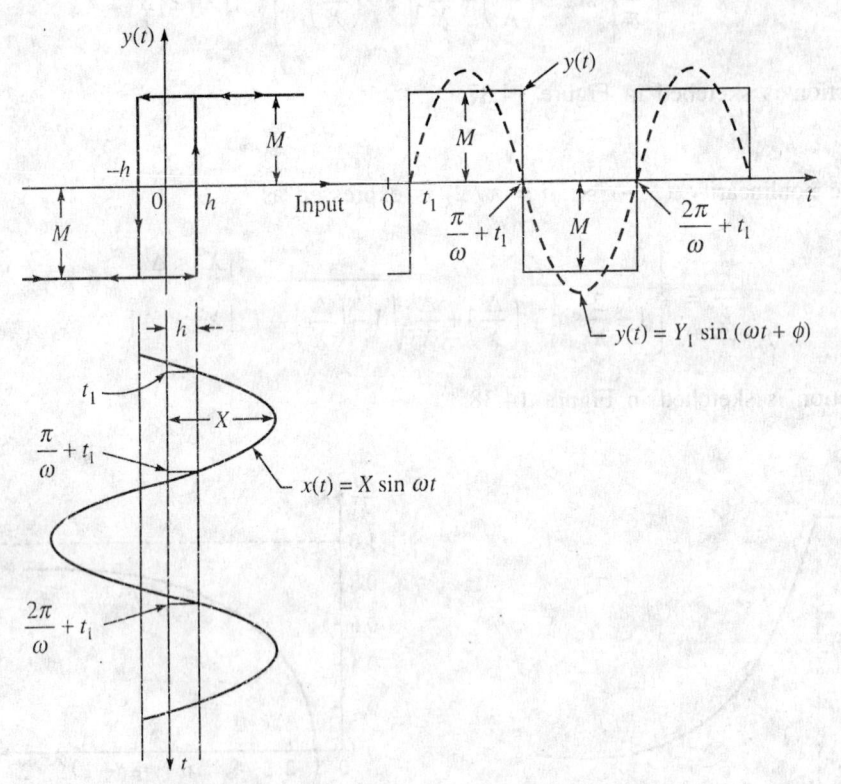

FIGURE 14.39 On-off nonlinearity with hysteresis.

Let input

$$x(t) = X \sin \omega t, \text{ for } X > h \quad (14.118)$$

where h is the dead-zone of on-off nonlinearity with hysteresis. The output $y(t)$ can be described as

$$y = \begin{cases} M & ; \omega t_1 \leq \omega t \leq \pi + \omega t_1 \\ -M & ; \pi + \omega t_1 < \omega t < 2\pi + \omega t_1 \end{cases} \tag{14.119}$$

where

$$\omega t_1 = \sin^{-1}\left(\frac{h}{X}\right) \tag{14.120}$$

Since the output has odd symmetry, its fundamental term would not have a cosine term. It will therefore be of the form $A_1 \sin \omega t$. Taking the property of half-wave symmetry, it is seen that

$$A_1 = \frac{1}{\pi} \int_0^{2\pi} y(t) \sin \omega t \, d(\omega t)$$

$$= \frac{2}{\pi} \int_{\omega t_1}^{\pi + \omega t_1} y(t) \sin \omega t \, d(\omega t)$$

$$= \frac{2}{\pi} \int_{\omega t_1}^{\pi + \omega t_1} M \sin \omega t \, d(\omega t)$$

$$= \frac{4M}{\pi}$$

Hence the magnitude of the describing function for this on-off nonlinearity with hysteresis type nonlinear element is

$$\frac{A_1}{X} = \frac{4M}{\pi X} \tag{14.121}$$

Clearly, the output is a square wave, but it lags the input by

$$\omega t_1 = \sin^{-1}\left(\frac{h}{X}\right)$$

Hence the describing function of on-off nonlinearity with hysteresis is

$$N = \frac{4M}{\pi X} \angle -\sin^{-1}\left(\frac{h}{X}\right) \tag{14.122}$$

or

$$\frac{h}{M} N = \frac{4h}{\pi X} \angle -\sin\left(\frac{h}{X}\right) \tag{14.123}$$

A plot of hN/M versus h/X is shown in Figure 14.40. The plot of (hN/M) versus (h/X) is taken rather than N versus (h/X) because (hN/M) is a function of (h/X) only.

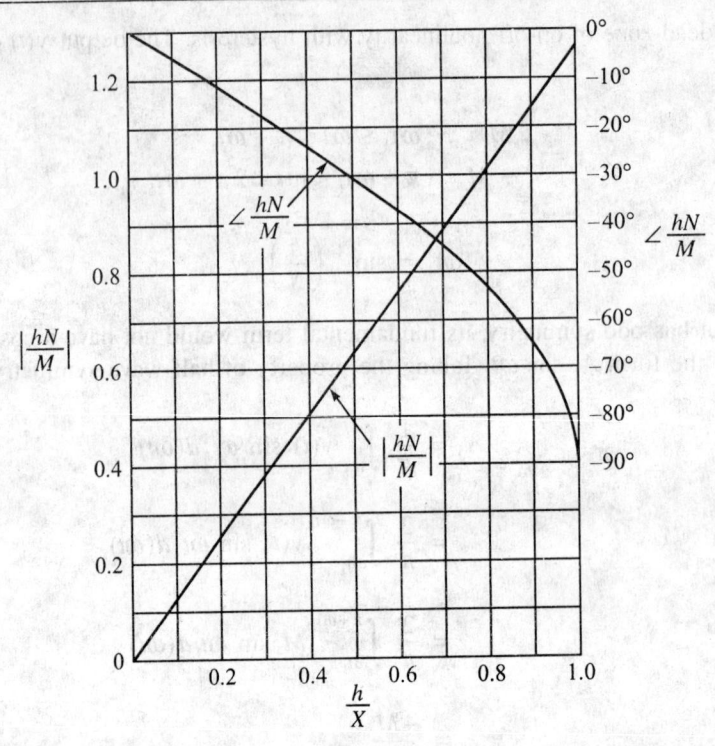

FIGURE 14.40 Plot of hN/M versus h/X.

Relay with dead-zone and hysteresis

The input–output characteristic is shown in Figure 14.41. Half-wave symmetry exists.

The output $y(t)$ may be described as

$$y(t) = \begin{cases} 0 & ; 0 \leq \omega t \leq \alpha \\ M & ; \alpha \leq \omega t \leq (\pi - \beta) \\ 0 & ; (\pi - \beta) \leq \omega t \leq (\pi + \alpha) \\ -M & ; (\pi + \alpha) \leq \omega t \leq (2\pi - \beta) \\ 0 & ; (2\pi - \beta) \leq \omega t \leq 2\pi \end{cases} \quad (14.124)$$

where

$$\alpha = \sin^{-1}\left(\frac{\Delta}{X}\right)$$

$$\beta = \sin^{-1}\left(\frac{\Delta - 2h}{X}\right)$$

we have

$$B_1 = \frac{2}{\pi} \int_0^\pi y \cos \omega t \, d(\omega t) = \frac{2}{\pi} \int_\alpha^{\pi - \beta} M \cos \omega t \, d(\omega t)$$

Nonlinear Systems

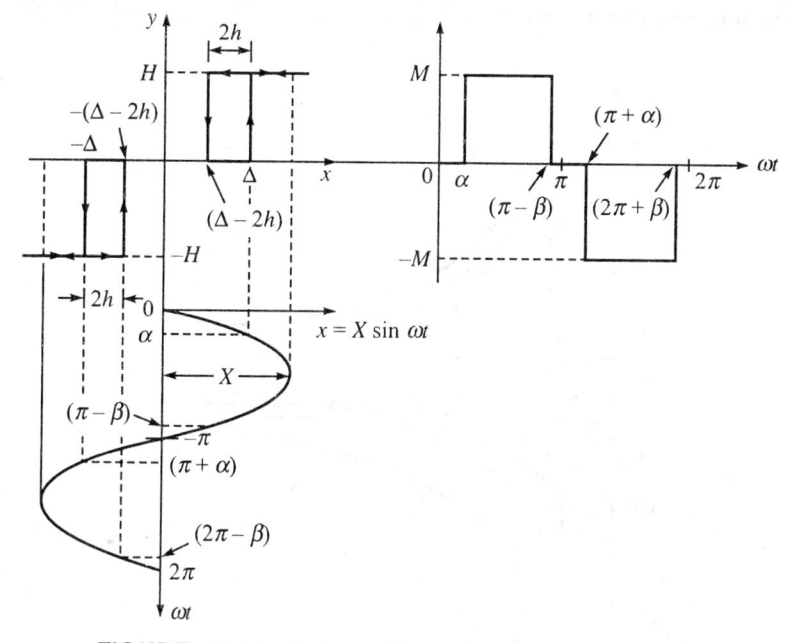

FIGURE 14.41 Relay with dead-zone and hysteresis.

$$= \frac{2M}{\pi}(\sin \beta - \sin \alpha) = \frac{2M}{\pi}\left(\frac{-2h}{X}\right) = -\frac{4Mh}{\pi X}$$

or

$$\frac{B_1}{X} = \frac{2M}{\pi X}\left(-\frac{2h}{X}\right)$$

Further,

$$A_1 = \frac{2}{\pi}\int_{\alpha}^{\pi-\beta} M \sin \omega t \, d(\omega t) = \frac{2M}{\pi}(\cos \alpha + \cos \beta)$$

$$= \frac{2M}{\pi}\left\{\sqrt{\left[1-\left(\frac{\Delta}{X}\right)^2\right]} + \sqrt{\left[1-\left(\frac{\Delta-2h}{X}\right)^2\right]}\right\}$$

or

$$\frac{A_1}{X} = \frac{2M}{\pi X}\left\{\sqrt{\left[1-\left(\frac{\Delta}{X}\right)^2\right]} + \sqrt{\left[1-\left(\frac{\Delta-2h}{X}\right)^2\right]}\right\}$$

Then the describing function is

$$N = \begin{cases} \sqrt{\left(\frac{A_1}{X}\right)^2 + \left(\frac{B_1}{X}\right)^2} \angle \tan^{-1}\left(\frac{B_1}{A_1}\right) & ; \quad |X| > \Delta \\ 0 & ; \quad |X| < \Delta \end{cases} \qquad (14.125)$$

The describing function plot is shown in Figure 14.42.

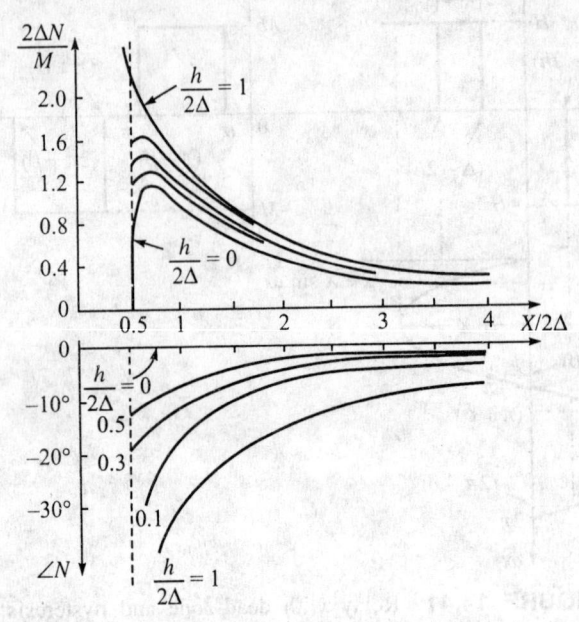

FIGURE 14.42 Describing function plot of relay with dead-zone and hysteresis.

Backlash

The input–output characteristic of backlash nonlinearity is shown in Figure 14.43. The output can be written as

$$
\begin{aligned}
y &= X - b/2 & ; && 0 \le \omega t \le \pi/2 \\
&= X - b/2 & ; && \pi/2 \le \omega t \le (\pi - \beta) \\
&= X + b/2 & ; && (\pi - \beta) \le \omega t \le 3\pi/2 \\
&= -X + b/2 & ; && 3\pi/2 \le \omega t \le (2\pi - \beta) \\
&= X - b/2 & ; && (2\pi - \beta) \le \omega t \le 2\pi
\end{aligned}
\qquad (14.126)
$$

where

$$\beta = \sin^{-1}\left(1 - \frac{b}{X}\right)$$

Half-wave symmetry of output exists. Therefore,

$$
N = \begin{cases} 0 & ; \quad |X| \le b/2 \\ \sqrt{\left(\dfrac{A_1}{X}\right)^2 + \left(\dfrac{B_1}{X}\right)^2} \; \angle \tan^{-1}\left(\dfrac{B_1}{A_1}\right) & ; \quad |X| > b/2 \end{cases}
\qquad (14.127)
$$

The magnitude and phase characteristic of this describing function are shown in Figure 14.44.

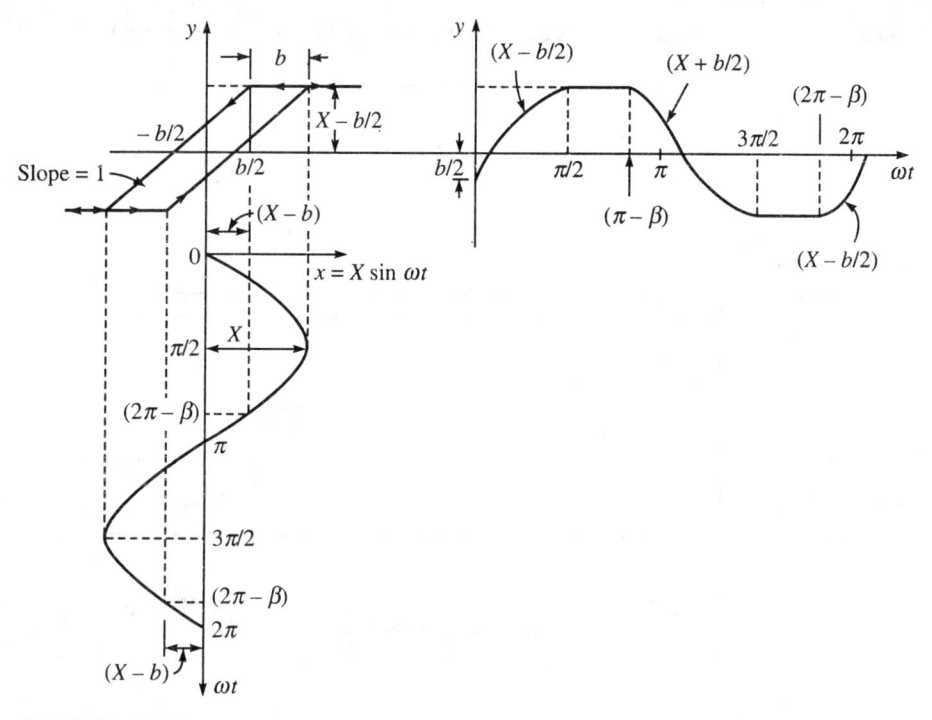

FIGURE 14.43 Characteristic and sinusoidal response of backlash nonlinearity.

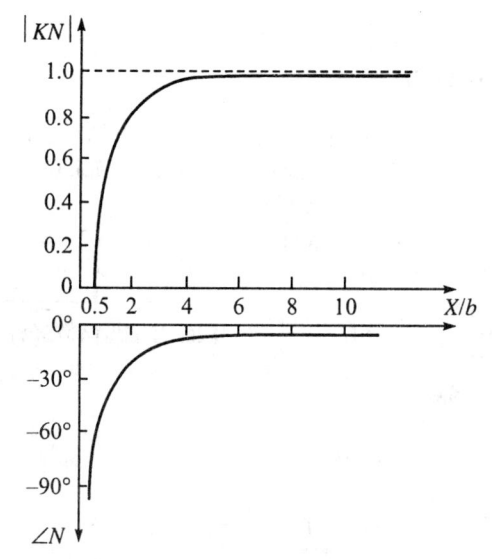

FIGURE 14.44 Describing function of backlash nonlinearity.

Note, when two nonlinearities are connected in cascade, the resulting describing function cannot be obtained by multiplying the describing functions of the individual nonlinearities.

Drill Problem 14.1 Verify for three nonlinearities their describing functions as given below:

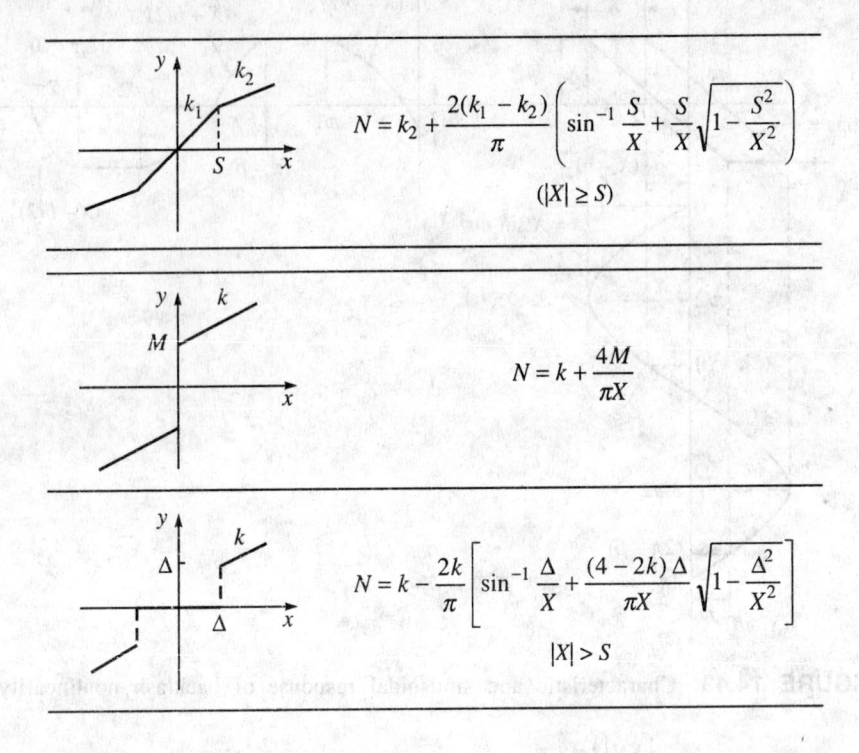

$$N = k_2 + \frac{2(k_1 - k_2)}{\pi}\left(\sin^{-1}\frac{S}{X} + \frac{S}{X}\sqrt{1 - \frac{S^2}{X^2}}\right)$$

$$(|X| \geq S)$$

$$N = k + \frac{4M}{\pi X}$$

$$N = k - \frac{2k}{\pi}\left[\sin^{-1}\frac{\Delta}{X} + \frac{(4-2k)\Delta}{\pi X}\sqrt{1 - \frac{\Delta^2}{X^2}}\right]$$

$$|X| > S$$

14.7 Stability Analysis

Consider the system shown in Figure 14.45(a). The block N denotes the describing function of the nonlinear element. If the higher harmonics are sufficiently attenuated, the describing function N can be treated as a real or complex variable gain. Then the closed-loop frequency response becomes

$$\frac{C(j\omega)}{R(j\omega)} = \frac{N[G(j\omega)]}{1 + N[GH(j\omega)]} \quad (14.128)$$

FIGURE 14.45(a) Unity-feedback nonlinear control system.

The characteristic equation becomes

$$1 + N[GH(j\omega)] = 0$$

or

$$GH(j\omega) = -\frac{1}{N} \quad (14.129)$$

If Eq. (14.29) is satisfied, then the limit cycle exists. Let us plot the locus of $-1/N$ and $GH(j\omega)$. Assume minimum-phase function of $GH(j\omega)$ so that all poles and zeros of $GH(s)$ lie in the left-hand side of the s-plane including the $j\omega$-axis. If $-1/N$ locus is not enclosed by the $GH(j\omega)$ locus, then the system is stable or there is no limit cycle at steady state. On the other hand, if the $-1/N$ locus is enclosed by the $GH(j\omega)$ locus, then the system is unstable. If the $-1/N$ locus and the $GH(j\omega)$ locus intersect, then the system output extends a sustained oscillation, i.e. a limit cycle. This is supported by a few examples.

Stability of oscillations

Let us elaborate further about the stability of sustained oscillations or limit cycle. Assume the unity-feedback nonlinear control system of Figure 14.45(a) and let the loci of $-1/N$ and $G(j\omega)$ intersect as shown in Figure 14.45(b). The points A and B on $-1/N$ locus correspond to small and large value of X, the amplitude of sinusoidal input signal. The value of X increases in the direction from A to B.

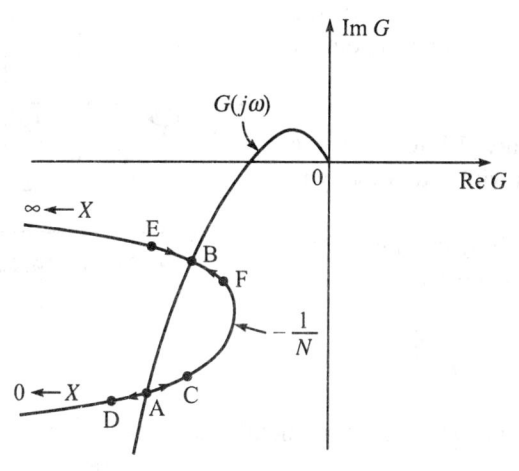

FIGURE 14.45(b) Stability analysis of limit-cycle operations of nonlinear control system.

Assume that the system is originally operated at point A having oscillation amplitude X_A and frequency ω_A, determined from the $-1/N$ locus and the $G(j\omega)$ locus. A slight disturbance at point A moves point A to point C on the $-1/N$ locus so that the amplitude of the input to the nonlinear element is increased slightly. Then the operating point C corresponds to the critical point or to the $(-1 + j0)$ point in the complex plane for linear control systems. Therefore, as seen from Figure 14.45(b) the $G(j\omega)$ locus encloses point C in the Nyquist sense. Since this is similar to the case where the open-loop locus of a linear system encloses the $(-1 + j0)$ point, the amplitude will increase and the operating point moves towards point B. The amplitude of oscillation grows.

Next suppose that a slight disturbance decreases the amplitude of the sinusoidal input to the nonlinear element. Assume that the operating point is moved from point A to point D on the $-1/N$ locus. Point D then corresponds to the critical point. In this case, the $G(j\omega)$ locus does not enclose the critical point and, therefore, the amplitude of the input to the nonlinear element decreases, and the operating point moves further from point D to the left. Thus, point A possesses divergent characteristics and corresponds to an unstable limit cycle.

Consider next the case where a slight disturbance is given to the system operating at point B. Assume that the operating point is moved to point E on the $-1/N$ locus. Then the $G(j\omega)$ locus in this case does not enclose the critical point (point E). The amplitude of the sinusoidal input to the nonlinear element decreases, and the operating point moves towards point B.

Similarly, assume that a slight disturbance causes the system operating point to move from point B to point F. Then the $G(j\omega)$ locus will enclose the critical point (point F). Therefore, the amplitude of oscillation will increase, and the operating point moves from point F towards point B. Thus, point B possesses convergent characteristics, and the system operation at point B is stable; in other words, the limit cycle at this point is stable.

For the system shown in Figure 14.45(a), the stable limit cycle corresponding to point B can be experimentally observed, but the unstable limit cycle corresponding to point A cannot be.

EXAMPLE 14.2 Figure 14.46 shows a nonlinear control system with saturation type of nonlinearity.

We assume that $G(s)H(s)$ is a minimum-phase transfer function given by

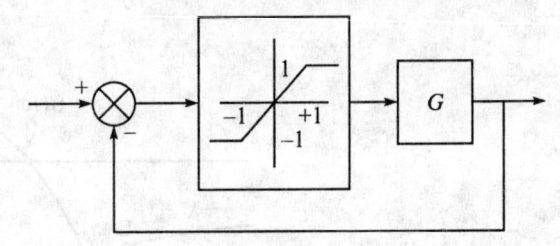

FIGURE 14.46 Example 14.2.

$$G(s)H(s) = \frac{K}{s(Ts+1)}$$

The describing function of saturation type nonlinearity as derived earlier in Eq. (14.97) can be rewritten as

$$N = \begin{cases} k & \text{for } |X| < S \\ \frac{2k}{\pi}\left[\sin^{-1}\left(\frac{S}{X}\right) + \left(\frac{S}{X}\right)\sqrt{1-\left(\frac{S}{X}\right)^2}\right]\angle 0° & \text{for } |X| > S \end{cases} \quad (14.130)$$

Here, in this example $S = 1$, $k = 1$. Then

$$N = \begin{cases} 1 & \text{for } |X| > 1 \\ \frac{2}{\pi}\left[\sin^{-1}\left(\frac{1}{X}\right) + \frac{1}{X}\sqrt{1-\left(\frac{1}{X}\right)^2}\right]\angle 0° & \text{for } |X| > 1 \end{cases} \quad (14.131)$$

The locus of $-1/N$ for the saturation type of nonlinearity is shown in Figure 14.47 in which the arrows indicate the direction of increasing X. The $-1/N$ locus starts from the point $(-1, j0)$ and extends to $-\infty$. Clearly, N is a function only of the amplitude of the input signal $x(t) = X \sin \omega t$.

Figure 14.48 shows the plot of $G(j\omega)$. The $G(j\omega)$ locus is a function of only ω.

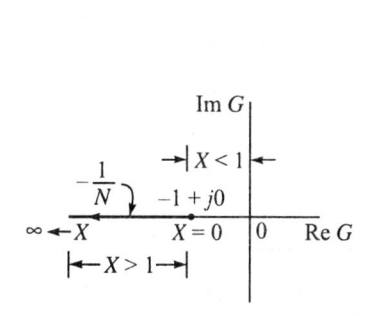

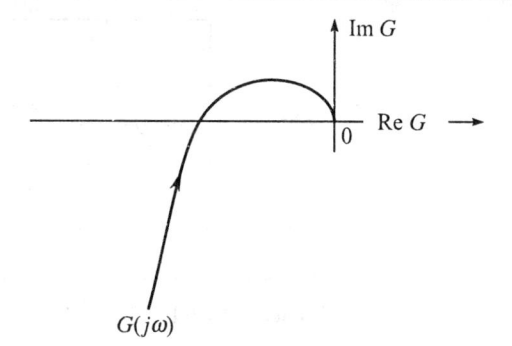

FIGURE 14.47 Example 14.2: locus of $-1/N$.

FIGURE 14.48 Example 14.2: locus of $G(j\omega)$.

Figure 14.49 shows the plot of the $-1/N$ locus and $G(j\omega)$ locus. The two loci intersect. This intersection corresponds to a stable limit cycle. The amplitude of the limit cycle is read from the $-1/N$ locus as $X = X_1$. The frequency of the limit cycle is read from the $G(j\omega)$ locus as $\omega = \omega_1$.

In the absence of any reference input, the output of this system at steady state exhibits a sustained oscillation with amplitude equal to X_1 and frequency equal to ω_1.

If the gain K of $G(j\omega)$ is decreased so that $-1/N$ locus and $G(j\omega)$ locus do not intersect, then the system becomes stable as shown in Figure 14.50. Due to disturbance, oscillation may occur which will die out and no sustained oscillation will exist in steady state. This is because the $-1/N$ locus is to the left of the $G(j\omega)$ locus, i.e. the $G(j\omega)$ locus does not enclose the $(-1, j0)$ point.

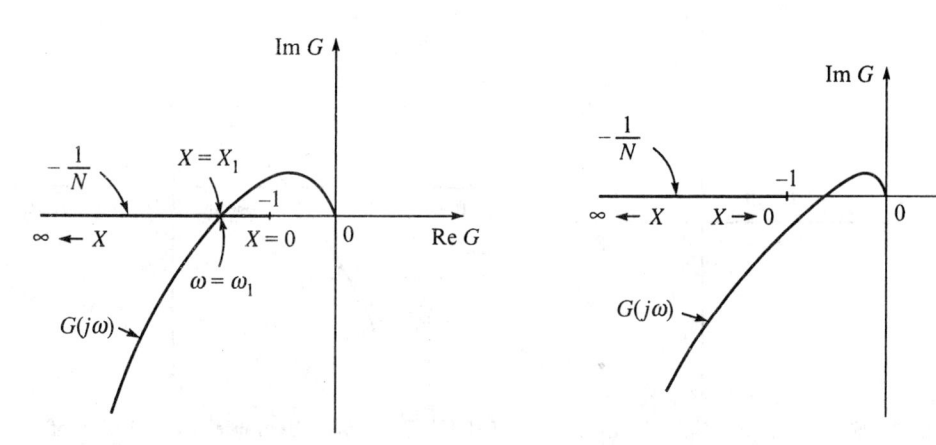

FIGURE 14.49 Example 14.2: plot of $-1/N$ locus and $G(j\omega)$ locus.

FIGURE 14.50 Example 14.2: plot of $-1/N$ locus and $G(j\omega)$ locus with decreased gain.

EXAMPLE 14.3 Figure 14.51 shows a nonlinear control system with dead-zone type of nonlinearity. We assume that $G(s)H(s)$ is a minimum-phase transfer function given by

$$G(s)H(s) = \frac{K}{s(Ts+1)} \quad (14.132)$$

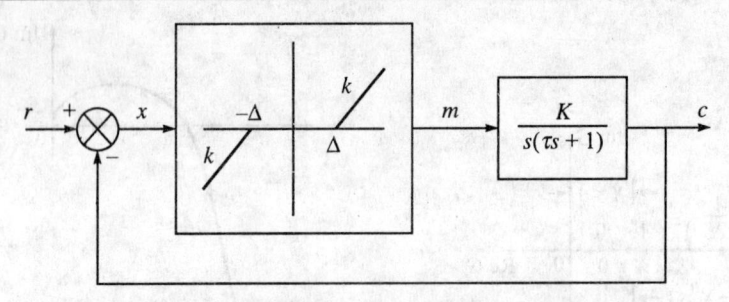

FIGURE 14.51 Example 14.3: nonlinear control system.

The describing function of dead-zone type nonlinearity as derived earlier in Eq. (14.93) can be rewritten as

$$N = \begin{cases} 0 & ; \text{ for } |X| < \Delta \\ k - \dfrac{2k}{\pi}\left[\sin^{-1}\left(\dfrac{\Delta}{X}\right)\right] + \dfrac{\Delta}{X}\sqrt{1 - \left(\dfrac{\Delta}{X}\right)^2} & ; \text{ for } |X| > \Delta \end{cases}$$

Let
$$x(t) = X \sin \omega t$$

The locus of $-1/N$ for the dead-zone type nonlinearity is shown in Figure 14.52 in which the arrows indicate the direction of increasing X. Figure 14.53 shows a plot of $-1/N$ locus for dead-zone nonlinearity and $G(j\omega)$ locus. They intersect. The limit cycle is unstable and the oscillation either dies out or increases indefinitely.

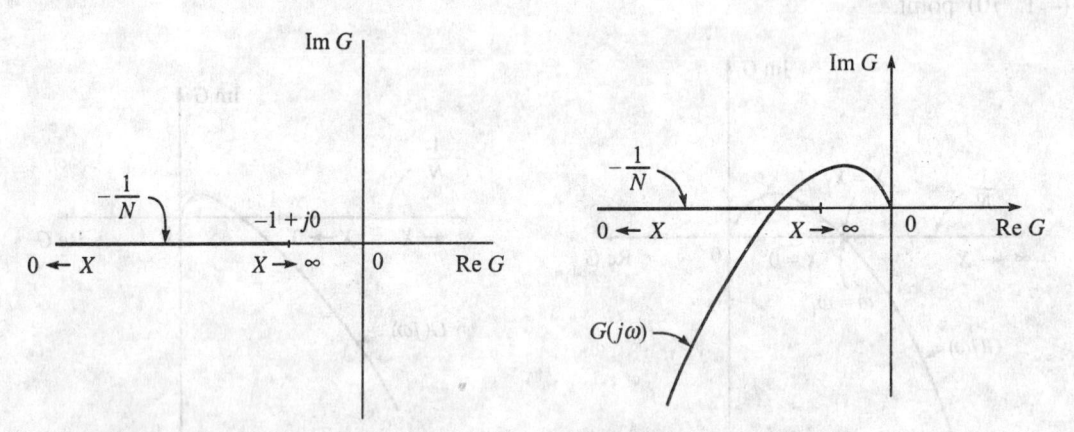

FIGURE 14.52 Example 14.3: locus of $-1/N$.

FIGURE 14.53 Example 14.3: locus of $-1/N$ and $G(j\omega)$.

14.8 Liapunov Stability Analysis

For linear time-invariant systems, many stability criteria are available. Among them are the Nyquist stability criterion, Routh's stability criterion, etc. If the system is nonlinear, or linear but time-varying, however, then such stability criteria do not apply. For a special group of nonlinear systems, Nyquist criterion may be applicable. The describing function approach is an approximate one. Phase-plane analysis is limited to second-order systems.

The second method of Liapunov (which is also called the direct method of Liapunov) is the most general method for determining the stability of nonlinear and/or time-varying systems. It avoids the necessity of solving state equations.

The system is

$$\dot{X} = F(X, t)$$

where X is an n-dimensional state vector and $F(X, t)$ is also an n-dimensional vector. Let the solution be $\phi(t; X_0, t_0)$, where $X = X_0$ at $t = t_0$ and t is the observed time. Thus, $(t_0; X_0, t_0) = X_0$.

Let $F(X_e, t) = 0$ for all t be called an equilibrium state of the system. If the system is linear time-invariant, namely, if $f(X, t) = AX$, then there exists only one equilibrium state if A is singular. For nonlinear systems, there may be one or more equilibrium state. Any isolated equilibrium state can be shifted to the origin of the coordinates, or $F(0, t) = 0$, by a translation of coordinates.

Stability in the sense of Liapunov

In the following, we will denote a spherical region of radius k about an equilibrium state X, as

$$\|X - X_e\| \leq k$$

where $\|X - X_e\|$ is called the Edclidean norm and is defined by

$$\|X - X_e\| = [(x_1 - x_{1e})^2 + (x_2 - x_{2e})^2 + \cdots + (x_n - x_{ne})^2]^{1/2}$$

Let $S(\ )$ consist of all points such that

$$\|X_0 - X_e\| \leq d$$

and let $S(\epsilon)$ consist of all points such that

$$\|\phi(t; X_0, t_0) - X_e\| \leq \epsilon \text{ for all } t \geq t_0$$

An equilibrium state X_e of the system is said to be stable in the sense of Liapunov if, corresponding to each $S(\epsilon)$, there is an $S(\ )$ such that the trajectories starting in $S(\ )$ do not leave $S(\epsilon)$ as t increases indefinitely.

We first choose the region $S(\epsilon)$ and for each $S(\epsilon)$, there must be a region $S(\ )$ such that trajectories starting within $S(\ )$ do not leave $S(\epsilon)$ as t increases indefinitely.

An equilibrium state X_e of the system is said to be asymptotically stable if it is stable in the sense of Liapunov and if every solution starting within $S(\ )$ converges, without leaving $S(\epsilon)$ to X_e as t increases indefinitely.

If asymptotic stability holds for all states (all points in state space) from which trajectories originate, the equilibrium state is said to be asymptotically stable in the large. That is, the equilibrium state X_e of the system is said to be asymptotically stable in the large if it is stable and if every solution converges to X_e as t increases indefinitely. Obviously, a necessary condition for asymptotic stability in the large is that there be only one equilibrium state in the whole state space. For all practical purposes, however, it is sufficient to determine a region of asymptotic stability large enough so that no disturbance will exceed it.

Instability

An equilibrium state X_e is said to be unstable if for some real number $\epsilon > 0$, no matter how small, there is always a state X_e in $S(\ )$ such that the trajectory starting at this state leaves $S(\epsilon)$.

Sign definiteness

Positive definiteness of scalar functions: A scalar function $V(X)$ is said to be positive definite in a region Ω (which includes the origin of the state space) if $V(X) > 0$ for all nonzero states X in the region Ω and $V(0) = 0$.

Negative definiteness of scalar functions: A scalar function $V(X)$ is negative definite if $-V(X)$ is positive definite.

Positive semi-definiteness of scalar functions: A scalar function $V(X)$ is said to be positive semi-definite if it is positive for all states in the region Ω except at the origin and at certain other states, where it is zero.

Negative semi-definiteness of scalar function: A scalar function $V(X)$ is said to be negative semi-definite if $-V(X)$ is positive semi-definite.

Indefiniteness of scalar function: A scalar function $V(X)$ is said to be indefinite if in the region Ω it assumes both positive and negative values, no matter how small the region Ω is.

Here we assume X to be a two-dimensional vector $[x_1 \ x_2]^T$, then the sign definiteness of the following scalar functions will be as follows.

1. $V(X) = x_1^2 + 3x_2^2$ positive definite
2. $V(X) = (x_1 + x_2)^2$ positive semi-definite
3. $V(X) = -x_1^2 - (x_1 + x_2)^2$ negative definite
4. $V(X) = x_1^2 + x_1 x_2$ indefinite
5. $V(X) = x_1^2 + \dfrac{1}{x_2^2}$ positive definite

It may be noted that the V-function that we are choosing is quadratic in nature. Why? This is because the mathematical tool available at our disposal is very limited. The positive definiteness of the quadratic form $V(X)$ can be determined by Sylvester's criterion, which states that the necessary and sufficient condition that a quadratic function $V(X)$ is positive definite where $V(X) = X^T P X$ and all the successive principal minors of real, symmetric matrix P be positive. This is illustrated with the following example.

EXAMPLE 14.4 Determine the sign definiteness of the quadratic function

$$O = 10 x_1^2 + 4x_2^2 + x_3^2 + 2x_2 x_1 - 2x_2 x_3 - 4x_1 x_3$$

Solution: Since Q is function of x_1, x_2 and x_3, writing Q as $Q(x_1, x_2, x_3)$ and writing in the from $X^T P X$, we get

$$Q(x_1, x_2, x_3) = X^T P X = [x_1 \; x_2 \; x_3] \begin{bmatrix} 10 & 1 & -2 \\ 1 & 4 & -1 \\ -2 & -1 & 1 \end{bmatrix} \begin{bmatrix} x_1 \\ x_2 \\ x_3 \end{bmatrix}$$

Now the Sylvester's criterion states that for quadratic form expressed as

$$Q(X) = X^T P X = [x_1 \; x_2 \; \cdots \; x_n] \begin{bmatrix} P_{11} & P_{12} & \cdots & P_{1n} \\ P & P_{22} & \cdots & P_{2n} \\ \vdots & \vdots & & \vdots \\ P & P & \cdots & P_{nn} \end{bmatrix} \begin{bmatrix} x_1 \\ x_2 \\ \vdots \\ x_n \end{bmatrix}$$

to be positive definite is that all leading principal minors of symmetric matrix P be positive, i.e.

$$P_{11} > 0, \quad \begin{vmatrix} P_{11} & P_{12} \\ P_{12} & P_{22} \end{vmatrix} > 0, \quad \ldots, \quad \begin{vmatrix} P_{11} & P_{12} & \cdots & P_{1n} \\ \vdots & \vdots & & \vdots \\ P_{1n} & P_{2n} & \cdots & P_{nn} \end{vmatrix} > 0$$

Applying the above criterion, we get

$$10 > 0; \quad \begin{vmatrix} 10 & 1 \\ 1 & 4 \end{vmatrix} = 39 > 0; \quad \begin{bmatrix} 10 & 1 & -2 \\ 1 & 4 & -1 \\ -2 & -1 & 1 \end{bmatrix} = 17 > 0$$

Here we see that all leading principal minors of symmetric matrix P are positive, which implies that the quadratic function $Q(x_1, x_2, x_3)$ is positive definite.

It may be noted that $V(X) = X^T P X$ would have been positive semi-definite[†] if P was singular and some of the principal minors of P were non-negative.

$V(X)$ would have been negative definite if $-V(X)$ was positive definite. Similarly, $V(X)$ would have been negative semi-definite if $-V(X)$ was positive semi-definite.

The Sylvester's criterion can be used for investigating local stability around the point of equilibrium by simply applying Routh's criterion to the characteristic polynomial of A which is the Jacobian. Let us take an example.

Example: In order to determine the point of equilibrium and stability in the sense of Liapunov for the given nonlinear differential equation

$$\ddot{x} + x^2(\dot{x} - 1) + x = 0$$

the state variable formulation becomes

$$\dot{X} = f(X) = \begin{bmatrix} f_1(X) \\ f_2(X) \end{bmatrix} = \begin{bmatrix} x_2 \\ x_1^2(1 - x_2) - x_1 \end{bmatrix}$$

[†] K.N. Swami, "On Sylvester's criterion for positive semi-definite matrices," *IEEE Trans. on Automatic Control,* June 1973, p. 306.

Then,
$$\begin{bmatrix} \dfrac{\partial f_1}{\partial x_1} & \dfrac{\partial f_1}{\partial x_2} \\ \dfrac{\partial f_2}{\partial x_1} & \dfrac{\partial f_2}{\partial x_2} \end{bmatrix} = \begin{bmatrix} 0 & 1 \\ 2x_1(1-x_2)-1 & -x_1^2 \end{bmatrix}$$

Hence for the equilibrium point $X = [0\ \ 0]^T$, we have

$$A = \begin{bmatrix} 0 & 1 \\ -1 & 0 \end{bmatrix},$$

and the characteristic equation becomes $s^2 + 1 = 0$. The system is oscillatory, roots are on the imaginary axis and hence stable in the sense of Liapunov, which gives limit cycle in the neighbourhood of the equilibrium point (0,0).

Further, for equilibrium point $X = [1\ \ 0]^T$, we have

$$A = \begin{bmatrix} 0 & 1 \\ 1 & -1 \end{bmatrix},$$

and the characteristic equation becomes $s^2 + s - 1 = 0$. The system is unstable in the neighbourhood of the equilibrium point.

Linearization of a nonlinear function $f(x)$ on Taylor series expansion about an operating point x_0 gives

$$f(x) = f(x_0) + \dfrac{df}{dx}\bigg|_{x=x_0}(x-x_0) + \dfrac{d^2 f}{dx^2}\bigg|_{x=x_0}\dfrac{(x-x_0)^2}{2!} + \cdots \qquad (14.133)$$

Let us ignore all higher order terms, only retaining the first-order derivatives, then we get a good linear approximation of Eq. (14.133), if either $(x - x_0)$ is very small or higher-order derivatives are very small.

The state variable representation of an n-dimensional nonlinear time-invariant system is

$$\dot{X} = f(X, U) = \begin{bmatrix} f_1(X, U) \\ f_2(X, U) \\ \vdots \\ f_n(X, U) \end{bmatrix}$$

Ignoring the higher-order terms in Taylor series expansion about $X_0 = 0$ to make origin as the equilibrium point and $U_0 = 0$, we get the linearized version as

$$\dot{X} = AX + BU$$

where $A = \begin{bmatrix} \frac{\partial f_1}{\partial x_1} & \frac{\partial f_1}{\partial x_2} & \cdots & \frac{\partial f_1}{\partial x_n} \\ \frac{\partial f_2}{\partial x_1} & \frac{\partial f_2}{\partial x_2} & \cdots & \frac{\partial f_2}{\partial x_n} \\ \vdots & \vdots & & \vdots \\ \frac{\partial f_n}{\partial x_1} & \frac{\partial f_n}{\partial x_2} & \cdots & \frac{\partial f_n}{\partial x_n} \end{bmatrix}$ and $B = \begin{bmatrix} \frac{\partial f_1}{\partial u_1} & \frac{\partial f_1}{\partial u_2} & \cdots & \frac{\partial f_1}{\partial u_n} \\ \frac{\partial f_2}{\partial u_1} & \frac{\partial f_2}{\partial u_2} & \cdots & \frac{\partial f_2}{\partial u_n} \\ \vdots & \vdots & & \vdots \\ \frac{\partial f_n}{\partial u_1} & \frac{\partial f_n}{\partial u_2} & \cdots & \frac{\partial f_n}{\partial u_n} \end{bmatrix}$

A and B are said to be Jacobian matrices. Their linearized model will be valid for small deviation around the point of equilibrium. Neverthless, for the autonomous system the dynamics is

$$\dot{X} = AX$$

In 1892, A.M. Liapunov of Russia presented two methods for determining the stability of dynamic systems described by ordinary differential equations. The first method consists of all procedures in which the explicit forms of the solutions of the differential equations are used for analysis. This mainly tells us about the bounded output with bounded input.

The second method talks about the asymptotic stability, stability in large, that is, the global stability. This does not require the solutions of the system differential equations. This is applicable to linear as well as nonlinear systems. In this context we may mention that for linear system stability analysis, there are a lot many other methods that are available. But truly speaking, we do not have the methods for the stability analysis of nonlinear systems; the describing function method or phase-plane analysis, or the Nyquist criterion can be applied for the analysis of nonlinear systems in a limited sense. That is why the Liapunov's second method of stability plays an important role in control system study, specially for nonlinear systems.

Before we launch upon the second method of Liapunov, it may be worth mentioning that by this method, we can ascertain the asymptotic stability of the system, not otherwise, that is, we cannot predict the instability of the system. The success of finding suitable V-function depends on the ingenuity of the engineer. With these words, we can go for the proof of the Liapunov's second method.

From the classical theory of mechanics, we know that a vibratory system is stable if its total energy (a positive definite function) is continually decreasing (which means that the time derivative of the total energy must be negative definite) until an equilibrium state is reached.

14.8.1 Asymptotic stability analysis

Consider the linear, time-invariant system

$$\dot{X} = AX \; ; \; |A| \neq 0$$

where X is an n-dimensional state vector and A is an $n \times n$ constant matrix. The only equilibrium state is the origin, i.e. $X = 0$.

Let us choose a possible Liapunov function as

$$V(X) = X^T P X$$

where P is a positive definite Hermitian matrix (that is, P is real symmetric). Then, we get the time derivative of $V(X)$ as

$$\dot{V}(X) = \dot{X}^T P X + X^T P \dot{X}$$
$$= (AX)^T P X + X^T P (AX)$$
$$= X^T A^T P X + X^T P A X$$
$$= X^T (A^T P + P A) X$$

Therefore, we require that $\dot{V} = -X^T Q X$; where

$$Q = -(A^T P + PA) = \text{positive definite}$$

Instead of first specifying a positive definite matrix P and examining whether or not Q is positive definite, it is convenient to specify a positive definite matrix Q first and then examine whether or not P determined from

$$A^T P + PA = -Q$$

is also positive definite.

This necessitates solving $n(n + 1)/2$ linear equations to find out the elements of the symmetric P matrix of order n.

The restriction may be loosened by taking Q as positive semidefinite instead of positive definite, that is if $\dot{V}(X) = -X^T Q X$ does not vanish identically along any trajectory, then Q may be chosen to be positive semidefinite.

EXAMPLE 14.5 Consider the dynamics of the system represented by

$$\begin{bmatrix} \dot{x}_1 \\ \dot{x}_2 \end{bmatrix} = \begin{bmatrix} 0 & 1 \\ -1 & -1 \end{bmatrix} \begin{bmatrix} x_1 \\ x_2 \end{bmatrix}$$

Formulate the Liapunov function to test the asymptotic stability of the system.

Solution: Let the tentative V-function be

$$V(X) = X^T P X$$

where the symmetric matrix P is arbitrarily chosen as

$$P = \begin{bmatrix} 1.5 & 0.5 \\ 0.5 & 1 \end{bmatrix}$$

$[P]$ is positive definite as the principal minors $0.5 > 0$ and $|P| = 1.25 > 0$.

Now the time derivative of V, that is, $\dot{V}(X) = -X^T Q X$ where $= -Q = A^T P + PA$ comes out to be

$$-Q = \begin{bmatrix} 0 & -1 \\ 1 & -1 \end{bmatrix} \begin{bmatrix} 1.5 & 0.5 \\ 0.5 & 1 \end{bmatrix} + \begin{bmatrix} 1.5 & 0.5 \\ 0.5 & 1 \end{bmatrix} \begin{bmatrix} 0 & 1 \\ -1 & -1 \end{bmatrix}$$

or
$$Q = \begin{bmatrix} 1 & 0 \\ 0 & 1 \end{bmatrix}$$

and the matrix Q becomes positive definite as the successive principal minors are positive.

Hence we have got a suitable V-function for which $\dot{V}$ is sign opposite to that of V-function. The equilibrium state is the origin and the system described above is asymptotically stable. The Liapunov function is

$$V(X) = (1.5x_1^2 + x_1x_2 + x_2^2) \quad \text{and} \quad \dot{V}(X) = -(x_1^2 + x_2^2)$$

EXAMPLE 14.6 Consider the same system as in Example 14.5, i.e.

$$\dot{X} = \begin{bmatrix} 0 & 1 \\ -1 & -1 \end{bmatrix} X$$

Determine the asymptotic stability by using the Liapunov's second method.

Solution: Let us choose the Liapunov function $V(X)$ arbitrarily as

$$V(X) = X^T P X = X^T \begin{bmatrix} 2 & 0 \\ 0 & 1 \end{bmatrix} X = 2x_1^2 + x_2^2 = \text{positive definite}$$

Then, $\quad \dot{V}(X) = -X^T(A^T P + PA)X = -X^T \begin{bmatrix} 0 & -1 \\ -1 & 2 \end{bmatrix} X = 2x_1x_2 - x_2^2$

Here $\dot{V}(X)$ is indefinite. So we cannot predict the stability of the system. This implies that this particular $V(X)$ is not a proper Liapunov function. The choice of a proper $V(X)$ function depends on the ingenuity of the designer. If we are unable to get the proper $V(X)$ function, that means $V(X)$ and $\dot{V}(X)$ are not sign opposite to each other, we cannot predict the instability of the system. It may be mentioned that we can predict only the stability of the system by a proper choice of a suitable $V(X)$ function but it is not true otherwise.

Let us choose some other $V(X)$ function for the same system as

$$V(X) = X^T \begin{bmatrix} 1 & 0 \\ 0 & 1 \end{bmatrix} X = 2x_1^2 + x_2^2 = \text{positive definite}$$

Then $\quad \dot{V}(X) = X^T(A^T P + PA)X = X^T \begin{bmatrix} 0 & 0 \\ 0 & -2 \end{bmatrix} X = -2x_2^2$

which is negative semidefinite. Further, the origin $X = 0$ is the equilibrium state. If $\dot{V}(X)$ is to vanish then x_2 must be zero for all $t > 0$. This requires that $\dot{x}_2 = 0$ for $t > 0$. Since from the system equation $\dot{x}_2 = -x_1 - x_2$, x_1 must also be equal to zero for $t > 0$. This means that $\dot{V}(X)$ vanishes only at the origin. Hence the equilibrium state at the origin is asymptotically stable in large.

It is suggested that instead of first specifying a positive definite matrix P and examining whether or not Q is positive definite, it is convenient to specify a positive definite matrix Q first and then examine whether or not P determined from

$$A^T P + PA = -Q$$

is also positive definite.

It may be noted that if $\dot{V}(X) = -X^T Q X$ does not vanish identically along any trajectory, then Q may be chosen to be positive semi-definite instead of positive definite. Then determining P as positive definite will be the necessary and sufficient condition for the asymptotic stability with the equilibrium state as $X = 0$.

EXAMPLE 14.7 Consider the same system as in Example 14.5.

The equilibrium state is the origin. Let $V(X) = X^T P X$ where the symmetric matrix P is to be determined from $(A^T P + PA) = -Q = -1$, i.e.

$$\begin{bmatrix} 0 & -1 \\ 1 & -1 \end{bmatrix} \begin{bmatrix} P_{11} & P_{12} \\ P_{12} & P_{22} \end{bmatrix} + \begin{bmatrix} P_{11} & P_{12} \\ P_{12} & P_{22} \end{bmatrix} \begin{bmatrix} 0 & 1 \\ -1 & -1 \end{bmatrix} = \begin{bmatrix} -1 & 0 \\ 0 & -1 \end{bmatrix}$$

which leads to three simultaneous equations as

$$-2 P_{12} = -1 \; ; \; P_{11} - P_{12} - P_{22} = 0 \text{ and } 2 P_{12} - 2 P_{22} = -1$$

These equations after solving give the values as $P_{11} = 1.5$, $P_{12} = 0.5$ and $P_{22} = 2$. Clearly P is positive definite. Then the Liapunov function is

$$V(X) = 0.5(3 x_1^2 + 2 x_1 x_2 + 2 x_2^2) = \text{positive definite}$$

and
$$\dot{V}(X) = -(x_1^2 + x_2^2) = \text{negative definite}$$

Hence the system is asymptotically stable with origin as the equilibrium state.

EXAMPLE 14.8 A system is described by

$$\dot{x}_1 = -x_1 + x_2 + x_1(x_1^2 + x_2^2)$$
$$\dot{x}_2 = -x_1 - x_2 + x_2(x_1^2 + x_2^2)$$

Detemine the asymptotic stability using the Liapunov's second method while the function

$$V(X) = x_1^2 + x_2^2$$

Solution: Here,

$$V(X) = x_1^2 + x_2^2 = \text{positive definite}$$
$$\dot{V}(X) = 2 x_1 \dot{x}_1 + 2 x_2 \dot{x}_2$$
$$= 2 x_1 [-x_1 + x_2 + x_1(x_1^2 + x_2^2)] + 2 x_2 [-x_1 - x_2 + x_2(x_1^2 + x_2^2)]$$
$$= 2 (x_1^2 + x_2^2)(x_1^2 + x_2^2 - 1)$$

$\dot{V}(X)$ will be negative definite provided $(x_1^2 + x_2^2) < 1$. Thus $V(X)$ is a Liapunov function. From this Liapunov function we can conclude that the origin ($x_1 = 0$, $x_2 = 0$) of the system is asymptotically stable. The system is asymptotically stable within the unit circle $x_1^2 + x_2^2 = 1$.

EXAMPLE 14.9 Test the asymptotic stability using the Liapunov second method for the system dynamics

$$\dot{Y} = \begin{bmatrix} -1 & -2 \\ 1 & -4 \end{bmatrix} Y$$

Solution: To determine the stability of the equilibrium state at the origin, we solve Liapunov equation

$$A^T P + PA = -Q = -\begin{bmatrix} 1 & 0 \\ 0 & 1 \end{bmatrix} \quad \text{(say choosen arbitrarily)}$$

from which we obtain

$$P = \begin{bmatrix} \dfrac{23}{60} & -\dfrac{7}{60} \\ -\dfrac{7}{60} & \dfrac{11}{60} \end{bmatrix}$$

Obviously P is positive definite. Hence the equilibrium state is asymptotically stable.

EXAMPLE 14.10 Determine the stability of the system dynamics

$$\dot{x}_1 = -x_1 - 2x_2 + 2$$
$$\dot{x}_2 = x_1 - 4x_2 - 1$$

Solution: Let us define $y_1 = x_1 + a$ and $y_2 = x_2 + b$
Then, we get

$$\dot{y}_1 = -y_1 - 2y_2 \quad \text{provided} \quad a + 2b + 2 = 0$$
$$\dot{y}_2 = y_1 - 4y_2 \quad \text{provided} \quad -a + 4b - 1 = 0$$

From these two equations, we get $a = -5/3$, $b = -1/6$. With these values of a and b we can translate the equilibrium state to origin in the y_1-y_2 plane. The system dynamics becomes

$$\dot{Y} = \begin{bmatrix} -1 & -2 \\ 1 & -4 \end{bmatrix} Y \, ; \, Y = [y_1 \quad y_2]^T$$

Now stability analysis can be made following the earlier example.

EXAMPLE 14.11 Determine the range of value of K by applying the Liapunov's second method for the given system dynamics

$$\dot{x}_1 = x_2$$
$$\dot{x}_2 = -x_2 + x_3$$
$$\dot{x}_3 = -Kx_1 - 4x_3$$

and the given scalar function
$$V(X) = 5Kx_1^2 + 2Kx_1x_2 + 20x_2^2 + 8x_2x_3 + x_3^2$$

Solution: Here,
$$\frac{dV}{dt} = 10Kx_1\dot{x}_1 + 2Kx_1\dot{x}_2 + 2K\dot{x}_1x_2 + 40x_2\dot{x}_2 + 8x_2\dot{x}_3 + 8\dot{x}_2x_3 + 2x_3\dot{x}_3$$

Putting the value of derivatives of state variable, we get
$$\frac{dV}{dt} = -(40 - 2K)x_2^2$$

For $\dot{V}$ to be negative definite, $K < 20$.

Again, $V(X) = X^T P X = 5Kx_1^2 + 2Kx_1x_2 + 20x_2^2 + 8x_2x_3 + x_3^2$

then,
$$P = \begin{bmatrix} 5K & K & 0 \\ K & 20 & 4 \\ 0 & 4 & 1 \end{bmatrix}$$

Applying Sylvester's criterion for the symmetric matrix P to be positive definite, we need
$$5K > 0 \; ; \; \text{i.e.} \; K > 0$$

$$\begin{vmatrix} 5K & K \\ K & 20 \end{vmatrix} > 0 \; \text{and this gives} \; K < 100$$

and $|P| > 0$ gives $K < 20$

Hence the range of K for which the system is stable is $0 < K < 20$.

EXAMPLE 14.12 For the dynamics described by
$$\dot{x}_1 = x_2$$
$$\dot{x}_2 = -x_1^3 - x_2$$

Prove that the system is globally asymptotically stable using Liapunov's function of the form
$$V(X) = \alpha x_1^4 + \beta x_1^2 + x_1x_2 + x_2^2$$

What values of α and β are appropriate?

Solution: Given Liapunov function
$$V(X) = \alpha x_1^4 + \beta x_1^2 + x_1x_2 + x_2^2$$

$$\frac{dV}{dt} = \left(\frac{dV}{dx_1}\right)\left(\frac{dx_1}{dt}\right) + \left(\frac{dV}{dx_2}\right)\left(\frac{dx_2}{dt}\right)$$

$$= (4\alpha x_1^3 + 2\beta x_1 + x_2)x_2 + (x_1 + 2x_2)(-x_1^3 - x_2)$$

$$= x_1x_2[2x_1^2(2\alpha - 1) + (2\beta - 1)] - x_2^2 - x_1^4$$

Now, dV/dt is negative definite, provided $(x_1 x_2) > 0$
and $(2\alpha - 1) < 0$ which gives $\alpha < 1/2$
and $(2\beta - 1) < 0$ which gives $\beta < 1/2$

The generalized Liapunov's second method of asymptotic stability can be stated as follows: Suppose that the system is described by

$$\dot{X} = f(X, t)$$

where $f(0, t) = 0$ for all t

If there exists a scalar function $V(X, t)$ having continuous, first partial derivatives and satisfying the conditions

$$V(X, t) \text{ is positive definite}$$

and $$\dot{V}(X, t) \text{ is negative definite}$$

then the equilibrium state at the origin is uniformly asymptotically stable.

EXAMPLE 14.13 Consider the system

$$\dot{X} = \begin{bmatrix} -(x_1^2 + x_2^2) & 1 \\ -1 & -(x_1^2 + x_2^2) \end{bmatrix} X$$

Clearly, the origin ($x_1 = 0$, $x_2 = 0$) is the only equilibrium state. Let us choose the scalar function $V(X)$ as

$$V(X) = X^T \begin{bmatrix} 1 & 0 \\ 0 & 1 \end{bmatrix} X = x_1^2 + x_2^2$$

which is positive definite. Then $\dot{V}(X)$ comes out to be

$$\dot{V}(X) = 2x_1 \dot{x}_1 + 2x_2 \dot{x}_2 = -2(x_1^2 + x_2^2)^2$$

which is negative definite. Hence the system is asymptotically stable.

EXAMPLE 14.14 Determine the equilibrium points and investigate the stability for the following systems:

(a) $\begin{cases} \dot{x}_1 = -x_1 + x_2 \\ \dot{x}_2 = (x_1 + x_2) \sin x_1 - 3x_2 \end{cases}$

(b) $\begin{cases} \dot{x}_1 = -x_1^3 + x_2 \\ \dot{x}_2 = -ax_1 - bx_2 \end{cases}$; a and $b > 0$

(c) $\begin{cases} \dot{x}_1 = -x_1 + x_2^2 \\ \dot{x}_2 = -x_2 \end{cases}$

(d) $\begin{cases} \dot{x}_1 = -x_1 + x_1^2 x_2 \\ \dot{x}_2 = -x_2 + x_1 \end{cases}$

(e) $\begin{cases} \dot{x}_1 = -x_1 - x_2 \\ \dot{x}_2 = x_1 - x_2^3 \end{cases}$

Solution: The significant feature of a singular or equilibrium point is that all the derivatives characterizing the system behaviour are zero. In other words, the derivatives of the state variables are zero.

(a) For a singular point all derivatives should be equal to zero, that is, $\dot{x}_1 = 0$ leads to $x_1 = x_2$, and $\dot{x}_2 = 0$ leads to $2x_1 \sin x_1 - 3x_1 = 0 \rightarrow x_1(2 \sin x_1 - 3) = 0 \rightarrow x_1 = 0$ and $\sin x_1 = 3/2 > 1$ which is not possible. Hence there is only one singular point $(x_1 = x_2 = 0$, the origin)

Stability:

Let $\quad V(X) = x_1^2 + x_2^2 \rightarrow$ positive definite

Then, $\quad \dfrac{dV}{dt} = \left(\dfrac{dV}{dx_1}\right)\left(\dfrac{dx_1}{dt}\right) + \left(\dfrac{dV}{dx_2}\right)\left(\dfrac{dx_2}{dt}\right)$

$\qquad = 2x_1 \dot{x}_1 + 2x_2 \dot{x}_2$
$\qquad = 2x_1(-x_1 + x_2) + 2x_2[(x_1 + x_2)\sin x_1 - 3x_2]$
$\qquad = -2x_1^2 - 6x_2^2 + 2x_2[x_1 + (x_1 + x_2)\sin x_1]$

For dV/dt to be negative definite: $x_2[x_1 + (x_1 + x_2)\sin x_1] < 0$; that is, $x_1 = -x_2$ is the condition for dV/dt to be negative definite. The locus for stability is the $-45°$ line in second and fourth quadrants of x_1-x_2 plane.

(b) The singular point is obtained in the same way since $\dot{x}_1 = 0$, and $\dot{x}_2 = 0$ leads to $x_1 = (-b/a)x_2$, from $\dot{x}_1$ to be 0 leads to $x_2 = x_1^3 = (-bx_2/a)^3$ (substituting the value) and this after simplification gives

$$x_2[1 + (b/a)^3 x_2^2] = 0$$

i.e. $x_2^2 = -(a/b)^3$ leads to no real solution and $x_2 = 0$ is another permissible solution. Hence the only permissible singular or equilibrium point is $(x_1 = x_2 = 0$, the origin)

Stability:

Let $\quad V(X) = x_1^2 + x_2^2 \rightarrow$ positive definite

Then, $\quad \dfrac{dV}{dt} = 2x_1 \dot{x}_1 + 2x_2 \dot{x}_2$

$\qquad = 2x_1(-x_1^3 + x_2) - 2x_2(ax_1 + bx_2)$
$\qquad = -2x_1^4 - 2bx_2^2 + 2x_1 x_2(1 - a)$

Nonlinear Systems   811

For dV/dt to be negative definite; $x_1 x_2 (1 - a) < 0$; that is (i) $a < 1$, $x_1 x_2 < 0$ and (ii) $a > 1$, $x_1 x_2 > 0$ are the conditions for dV/dt to be negative definite.

(c) From the given second equation, we get $x_2 = 0$ and from the given first equation, we get $x_1 = x_2^2$.
Hence the singular or equilibrium point is the origin $x_1 = x_2 = 0$.

Stability:

Let $\quad V(X) = x_1^2 + x_2^2 \rightarrow$ positive definite

Then, $\quad \dfrac{dV}{dt} = 2x_1 \dot{x}_1 + 2x_2 \dot{x}_2$

$\qquad = 2x_1(-x_1 + x_2^2) - 2x_2 x_2$

$\qquad = -2x_1^2 - 2x_2^2 + 2x_1 x_2^2$

For dV/dt to be negative definite; $2x_1 x_2^2 < 0$; that is, $x_1 < 0$ is the condition for dV/dt to be negative definite. The range of stability is the negative half of x_1-x_2 plane.

(d) The singular point is obtained in the same way since $\dot{x}_1 = 0$, and $\dot{x}_2 = 0$ leads to $x_1 = x_2$, and $\dot{x}_1$ to be 0 leads to $x_1 = x_1^2 x_2$, gives $x_1(1 - x_1 x_2) = 0$.
So $x_1 = x_2 = 0$ is one set of singular points. Another is $1 = x_1 x_2$, or $1 = x_1^2$ which leads to $x_1 = \pm 1$, and $x_2 = \pm 1$ as the sets of other singular points apart from origin.
Hence the three sets of singular or equilibrium points are ($x_1 = x_2 = 0$, the origin) and ($x_1 = x_2 = 1$) and ($x_1 = x_2 = -1$)

Stability:

Let $\quad V(X) = x_1^2 + x_2^2 \rightarrow$ positive definite

Then, $\quad \dfrac{dV}{dt} = 2x_1 \dot{x}_1 + 2x_2 \dot{x}_2$

$\qquad = 2x_1(-x_1 + x_2 x_1^2) + 2x_2(-x_2 + x_1)$

$\qquad = -2(x_1^2 + x_2^2) + 2x_1 x_2(x_1^2 + 1)$

As $(x_1^2 + 1) > 0$ then for dV/dt to negative definite, $x_1 x_2 < 0$; that is (i) $x_1 < 0$ and $x_2 > 0$; or (ii) $x_1 > 0$ and $x_2 < 0$ are the conditions for dV/dt to be negative definite. The range of stability is in the second and fourth quadrants of x_1-x_2 plane.

(e) The singular point is obtained in the same way since $\dot{x}_1 = 0$, and $\dot{x}_2 = 0$ leads to $x_1 = -x_2$ and $x_1 = x_2^3$ or $-x_2 = x_2^3$ leads to $x_2(1 + x_2^2) = 0$. So $x_2 = 0$ or $x_2^2 = -1$ which is not real solution. Hence the only permissible singular point is the origin, that is, $x_1 = x_2 = 0$ is the only set of singular point.

Stability:

Let $\quad V(X) = x_1^2 + x_2^2 \rightarrow$ positive definite

Then, $\quad \dfrac{dV}{dt} = 2x_1 \dot{x}_1 + 2x_2 \dot{x}_2$

$\qquad = -2x_1(x_1 + x_2) + 2x_2(-x_2^3 + x_1)$

$\qquad = -2x_1^2 - 2x_2^4$

and $dV/dt < 0$ for all values of x_1 and x_2.
The range of stability is therefore for all values of x_1 and x_2.

EXAMPLE 14.15 Consider a time-varying system given by

$$\dot{X}(t) = \begin{bmatrix} 0 & 1 & 0 \\ 0 & 0 & 1 \\ -23 & -15 & -5 \end{bmatrix} x(t) + \begin{bmatrix} e^{-2t} \\ \sin(4t) \\ 1 \end{bmatrix} u^2(t) \quad \text{for all } t \in [0, \infty]$$

with sample time $T = 0.1$ s.

(a) Determine the asymptotic stability of the discrete time system's state equations.
(b) Determine whether the discrete time system's output is asymptotically stable.

$$y(t) = \begin{bmatrix} 1 & -1 & 2 \end{bmatrix} x(t)$$

Solution:

MATLAB Program 14.1

```
%a.
clear;
pack;
clc;
At=[0 1 0; 0 0 1; -23 -15 -5];
string1=sprintf('The matrix At in continuous time system is \n');
disp(string1);
disp(At);
Ateig=eig(At);
T=0.1;
String2=sprintf('The Eigen value(s) of At is (are)\n');
disp(string2);
disp(Ateig);
string3=sprintf('now converting to Discrete A => Ad with sampling time T=%f\n' T);
disp(string3);
Ad=expm(T*At);
string4=sprintf('Ad is \n');
disp(string4)
disp(Ad);
string5=sprintf('Elgen values of Ad are \n');
disp(string5);
disp(eig(Ad));

%b

%%%%%%%%%%%%%%%%%%%
% To ensure that the stability of
% X(k+1)=AX(k)+BU(k)
```

```
% = AX(k) + f(U(k))
%
% is equal to
%
% the stability of
%
% X(k+1)=AX(k)
%
% iff
%
% f(0)=0
%
% where
%
% f(U(k))=BU(k)
%%%%%%%%%%%%%%%%%%%

Adjordan=jordan(Ad);
string6=sprintf('The Jordan canonical form of Ad is\n');
disp(string6);
disp (Adjordan);

% now Phi(k) = Adjordan ^k

syms k;

PhiK=Adjordan.^k;

%As Y(k) = C(k)X(k) = C(k).PhiK.X(0)
%Cd=C(k)
Cd=[k −1 2];
digits(4);
CPhi=vpa(Cd*PhiK,4);
fun1=abs(CPhi(1,1));
fun2=abs(CPhi(1,2));
fun3=abs(CPhi(1,3));
k=double(subs(k,1:100));
fun1a=subs(fun1,k);
fun2a=subs(fun2,k);
fun3a=subs(fun3,k);
plot(k, fun1a, 'r–');
hold
grid
plot(k, fun2a, 'g–');
plot(k, fun3a, 'b–');
xlabel('k');
title('CPhi(1,1)[RED], CPhi(1,2)[GREEN], CPhi(1,3)[BLUE] Vs k');
fun1i=inline('fun1');
fun2i=inline('fun2');
fun3i=inline('fun3');
```

```
[x1,fval1,exitflag1] = fminbnd(fun1i,0,1000);
[x2,fval2,exitflag2] = fminbnd(fun2i,0,1000);
[x3,fval3,exitflag3] = fminbnd(fun3i,0,1000);
if((exitflag1>0)&(exitflag2>0)&(exitflag3>0))
    string7=sprinft('\n THE DISCRETE TIME SYSTEMS OUTPUT IS ASYMPTOTICALLY STABLE \n')
else
    string7=sprinft('\n THE DISCRETE TIME SYSTEMS OUTPUT IS UNSTABLE\n');
end
disp(string7);
Output_____
```

The matrix At in continous time system is

$$\begin{matrix} 0 & 1 & 0 \\ 0 & 0 & 1 \\ -23 & -15 & -5 \end{matrix}$$

The Eigen values(s) of At is (are)

−2.6240

−1.1880 + 2.7118i

−1.1880 − 2.7118i

now converting to Discrete A => Ad with sampling time T = 0.100000

Ad is

$$\begin{matrix} 0.9966 & 0.0977 & 0.0042 \\ -0.0968 & 0.9335 & 0.0767 \\ -1.7634 & -1.2469 & 0.5501 \end{matrix}$$

Eigen values of Ad are

0.7692
0.8555 + 0.2379i
0.8555 − 0.2379i

The Jordan canonical form of Ad is

$$\begin{matrix} 0.7692 & 0 & 0 \\ 0 & 0.8555 + 0.2379i & 0 \\ 0 & 0 & 0.8555 - 0.2379i \end{matrix}$$

Current plot held

THE DISCRETE TIME SYSTEMS OUTPUT IS ASYMPTOTICALLY STABLE

Summary

Phase-plane analysis is restricted to second-order systems only and this is its limitation. Basically, the nonlinearities in the system are approximated by linear segments for different ranges of input and the dynamics of the system is changed for different linear segments and analysed for that range. The describing function analysis is an extension of linear techniques to the study of nonlinear systems. Therefore, the application of the describing function to systems with a low degree of nonlinearity is good enough but the application of the describing function to a high degree of nonlinearity is not good enough as the result will be erroneous. Liapunov's stability analysis in particular, is extremly useful for nonlinear systems, provided a proper V-function can be generated which, however, depends on the ingenuity of the engineer. If a suitable V-function is not generated, one cannot predict anything about the instability of nonlinear systems.

Problems

14.1 The equation of motion of simple pendulum is

$$\ddot{x} + \frac{g}{l} \sin x = 0$$

Draw the phase trajectory.

14.2 Draw the phase trajectory for

$$\ddot{x} + 0.6\dot{x} + x = 0$$

with $x = 1$, $\dot{x} = 0$ as the initial conditions.

14.3 Draw the phase-plane trajectory for the system

$$\dot{x}_1 = x_1 + x_2$$
$$\dot{x}_2 = 2x_1 + x_2$$

14.4 Draw the phase-plane portrait for the system

$$\ddot{x} + \dot{x} + |x| = 0$$

14.5 Draw the phase-plane portrait for the system

$$\ddot{\theta} + \dot{\theta} + \sin \theta = 0$$

14.6 Draw the phase-plane portrait for the following systems:

(i) $\ddot{x} + \dot{x} = 0$
(ii) $\ddot{x} + |x|\dot{x} + x = 0$

14.7 Determine the type of singular point and draw the phase-plane portrait for the given Van der Pol equation

$$\ddot{x} - (1 - x^2)\dot{x} + x = 0$$

14.8 The system response is shown in Figure P.14.8. From the system response, obtain the phase-plane portrait when the system is subjected to the input function

$$r(t) = R_1 1(t) + R_2 1(t - 2) + R_3 1(t - 3)$$

Assume that the system is at rest initially.

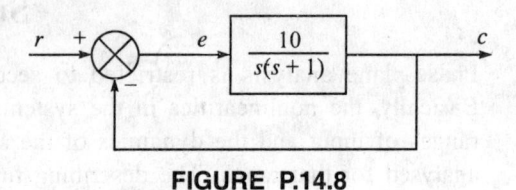

FIGURE P.14.8

14.9 For the system shown in Figure P.14.9, draw the phase-plane portrait in the e-$\dot{e}$ plane when $K = 0$ and $K = 1$. Assume that the input $r(t)$ is zero for $t > 0$, and the system is subjected only to the initial conditions.

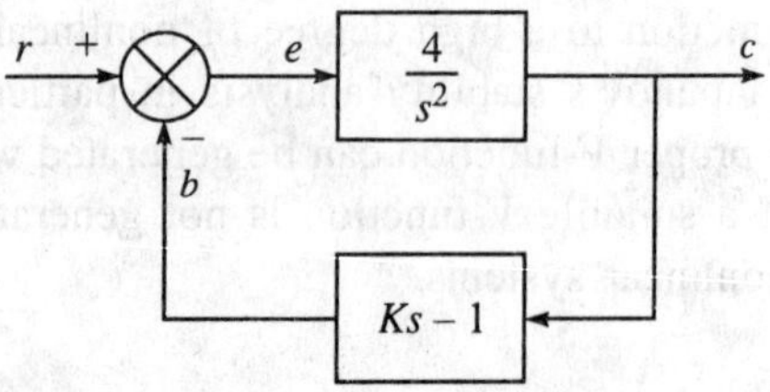

FIGURE P.14.9

14.10 Draw the phase-plane diagrams for the system shown in Figure P.14.10 when $\Delta = 0$ and $\Delta = 0.1$. The input r is a unit step function. Take e and $\dot{e}$ as the coordinate.

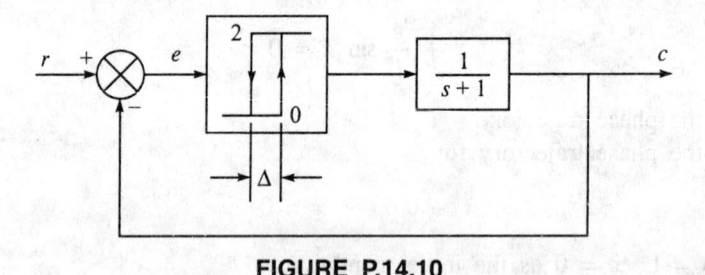

FIGURE P.14.10

14.11 Consider the system shown in Figure P.14.11. The input r is a unit ramp function. Draw the phase-plane portrait in the e-$\dot{e}$ plane.

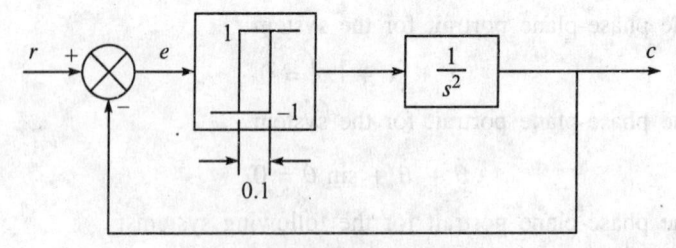

FIGURE P.14.11

14.12 Consider the system shown in Figure P.14.12. Assume that it is subjected only to the initial conditions. Draw the phase-plane portrait in the e-$\dot{e}$ plane.

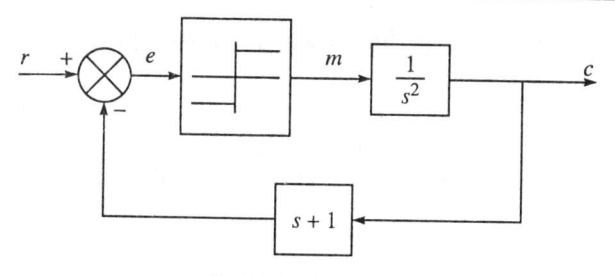

FIGURE P.14.12

14.13 Determine the equilibrium points and investigate the stability for the following systems:
$$\begin{cases} \dot{x}_1 = (x_1 - x_2)(x_1^2 + x_2^2 - 1) \\ \dot{x}_2 = (x_1 + x_2)(x_1^2 + x_2^2 - 1) \end{cases}$$

14.14 Given
$$\begin{cases} \dot{x}_1 = x_2 \\ \dot{x}_2 = -a \sin x_1 - kx_1 - dx_2 - cx_3 \\ \dot{x}_3 = -x_3 + x_2 \end{cases}$$
where all coefficients are positive and $k > a$. Using
$$V(x) = 2a \int_0^{x_1} \sin y\, dy + kx_1^2 + x_2^2 + px_3^2$$
with $p > 0$, show that the origin is a global asymptotical stable equilibrium point.

14.15 Consider the system
$$A_1 \ddot{y} + A_2 \dot{y} + A_3 y = 0$$
where the n by n matrices A_1, A_2 and A_3 are all constant symmetric positive-definite. Show that the origin at $y = 0$ and $\dot{y} = 0$ is the global asymptotical stable equilibrium point.

14.16 An ideal relay is introduced as a nonlinearity in a unity feedback linear servo system with forward transfer function
$$G(s) = \frac{100}{s(0.1s + 1)}$$
The relay has a maximum output of 10 volts. Discuss the stability of the system.

14.17 If the forward transfer function in Problem 14.16 would have been
$$G'(s) = \frac{25}{s(0.1s + 1)(0.025s + 1)},$$
will the system still remain stable?

14.18 Determine the describing function for the nonlinear element described by
$$y = x^3$$

where

    x is the input to the nonlinear element (sinusoidal)

    y is the output of the nonlinear element.

14.19 An ideal on-off relay with output ± 1 drives a unity feedback control system with the $G(j\omega)$ shown in P.14.19. How many periodic solutions for zero references input are obtained from the describing function analysis? Indicate which of these are stable?

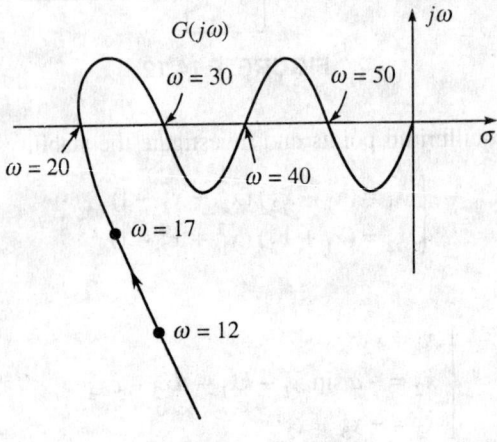

FIGURE P.14.19

14.20 The control system shown in Figure 14.21(a) has a nonlinear element whose output $m(t)$ is the following function of the input $x(t)$

$$m(t) = \left(\frac{dx}{dt}\right)^3 + x^2 \frac{dx}{dt}$$

  (a) Indicate how a describing function would be derived for the above nonlinear element.

  (b) Determine the possibility of a limit cycle in the system shown.

14.21 A system has

$$G(j\omega) = -\frac{10\sqrt{2}}{j\omega(1+0.5j\omega)}$$

and $N = \dfrac{1}{X} \angle -45°$

Find the amplitude and frequency of the possible solution. Is it a stable oscillation? For the system shown in Figure 14.21(a), the nonlinear element N is described as

$$\begin{aligned} m &= x &&\text{for} \quad |x| < 1 \\ &= 1 &&\text{for} \quad x > 1 \\ &= -1 &&\text{for} \quad x < -1. \end{aligned}$$

Given $$G(s) = \frac{1}{s^2 + s + 1}$$

sketch the $\dot{e}$-e plot and also the e-$\dot{e}$ plot for $r = 0$.

14.22 Consider the system described by

$$\ddot{x} - \left(0.1 - \frac{10}{3}\dot{x}^2\right)\dot{x} + x + x^2 = 10$$

(a) Define all singularities.

(b) Sketch the phase-plane trajectories near each singularity.

14.23 Test the sign definiteness of the following quadratic scalar functions $F(x)$.

(i) $x_1^2 + 4x_2^2 + x_3^2 + 2x_1x_2 - 6x_2x_3 - 2x_1x_3$

(ii) $-x_1^2 + 3x_2^2 - 11x_3^2 + 2x_1x_2 - 4x_3x_3 - 2x_1x_3$

14.24 Determine the asymptotic stability using the second method of Liapunov for the system dynamics given as

(a) $\dot{X} = \begin{bmatrix} -1 & 1 \\ 2 & -3 \end{bmatrix} X + \begin{bmatrix} 0 \\ 10 \end{bmatrix} u$

(b) $\dot{X} = \begin{bmatrix} 1 & 3 & 0 \\ -3 & -2 & -3 \\ 1 & 0 & 0 \end{bmatrix} X$

Index

Acceleration error coefficient, 382
Accelerometer, 87
Airy, G.B., 5
Alias frequencies, 699
    primary strip, 703, 704
All-pass network, 386
Analogous
    electrical circuits, 49, 50, 51, 52
    quantities, 50
    system components, 50
Automatic control, 1, 4
    developments in, 4
    system, 2

Block diagram, 95
    procedure for, 95
    reduction techniques, 96
Bode, H.W., 6, 352
Bode plot, 354, 355, 356, 358, 359
    construction of, 359
        constant term, 360
        quadratic poles, 365
        quadratic zeros, 368
        simple pole, 362, 365
        simple zero, 361, 365
        zeros (or poles) at origin, 360
    estimation of transfer function from, 382
    of open-loop transfer function, 354

Cauchy's
    Principle of Argument, 723
    theorem, 424, 430, 431
        conformal mapping, 427, 429
        examples of, 432
Characteristic equation, 19
Classical control theory, 7
Closed-loop frequency response
    curves, 475
    M-circles, 467
    N-circles, 467
Companion matrix, 604, 607, 608
    power of, 605
Compensated system
    root loci, 504

Compensating network, 485
    lag, 498, 503, 504
        lag-lead, 486, 506, 507
        characteristics of, 508, 509
    phase–lead, 486, 488, 506
        Bode diagram, 491
        characteristics of, 490
        design of, 492
        RC circuit, 488
Compensation
    cascade, 633
Compensator
    continuous time, 489
    first-order, 487
    phase-lag, 487
    phase-lead, 487
    output or load, 486
    series or cascade, 486, 487, 488
    state-feedback, 486
Completely state controllable, 627
Conjugate poles, 222
Continuous control, 323
Continuous controller, 323, 330
    designing using root loci, 346, 349
        root loci for P control, 347
        root loci for PI control, 347
        root loci for PID control, 348
    proportional control, 324, 328
        offset, 328
        time response, 341
Control engineering, 1
Control system
    adaptive, 3
    closed-loop, 3, 10
    continuous-time, 4
    discrete-time, 4
    multivariable, 4
    open-loop, 3, 9
    with time delay, 387
        Nyquist criterion, 387
    time-invariant, 4
    time-varying, 4
Controllability, 619
    condition of, 622, 635, 665
    criterion, 626, 627
    matrix, 620, 621, 637, 638
Controllable system, 629
Corner frequency, 362

D'Alembert's principle, 48, 49
Damped natural frequency, 158, 170
Damping coefficient, 169, 170, 171
    effect of roots on, 185
Data hold, 680
Delay time, 147, 162
Derivative control, 196, 199, 200, 201, 202, 330, 333
Diagonalization, 566
Digital
    control, 8
    system, 677 679, 680
        analysis of, 711
Digital controller, 678, 682, 733, 734
    merits vs analog controller, 682
Discrete control system, 682
    Cayley-Hamilton method, 738
    compensating, 733
    controllability and observability, 733
    Jury's stability test, 707
    relationship between s-plane and z-plane poles, 697
    steady-state analysis, 716
    state variable formulation, 725, 730
    time response, 711, 714
Discrete-data control system, 679
Discrete-time system, 677, 708
Disturbance, 2
    rejection, 139
Dominant-pole analysis, 154, 186, 187
Dynamic system, 2
    mathematical model, 46

Electrical analogy, 46
Evans, W.R., 6, 253

Feedback amplifiers, 130, 135
Feedback control, 2
    demerits of, 13
    merits of, 13
    history of, 4
First-order systems, 146
    unit-impulse response, 152
    unit-ramp response, 152
    unit-step response, 146
Forced system, 588
Frequency compensation, 518
    external
        dominant pole, 518
        pole-zero, 519
    internal
        internally compensated op-amp, 520

Frequency response, 353
    of linear time-invariant system, 358
    open-loop, 354
    plot, 378
Friction
    types of, 754
$F(s)$-plane, 433

Gain crossover frequency, 358, 376, 377, 410
Gain margin, 354, 355, 358, 359, 376, 377, 411, 477
$G(H)$-plane, 433

Házen, H.L., 6
Heaviside, O., 5
Heaviside expansion from pole-zero map, 181
Homogeneity, 34, 35
Hurwitz, A., 5, 221

Input signals, test, 145
Integral control, 328
Integral performance criterion, 195
    IAE (integral of the absolute volume of the error), 196
    ITAE (integral of the time multiplied by the absolute value of the error, 196
    ITSE (integral of the time multiplied by the square of the error, 196
Inverse polar plot, 463
Inverse root loci
    rules for the construction of, 298

Jordon canonical for, 575
Jump resonance, 751

Kalman filter, 8
Kharitonov's
    method, 240
    polynomials, 243, 245, 246
    theory behind, 243
Kinariwala's approach, 610

Laplace transform, 14–16
Learning control systems, 3
Liapunov, A.M., 5
Limit cycle, 750, 751, 753
Linear approximation, 34
Linear variable differential transformer, 85–87
Liquid-level systems, 63–65, 148
Load disturbance, 137

M-circles, 468, 469, 470, 472, 473, 474, 476, 482
Magnetic amplifier, 89–93
    B-H magnetization curve, 90
    saturable core reactor, 89, 91
Mason's gain formula, 109, 110
MATLAB
    partial fraction expansion, 21
    step-response to first-order transfer function, 150, 151
    unit-step response to second order system, 172
    zeros and poles of transfer function, 25
Matrix
    eigenvalues, 578
Maxwell, J.C., 5
Mechanical coupling devices, 53
    friction wheels, 53
        f–i analogy, 53
        f–v analogy, 53
    gear train, 54
        T–i analogy, 56
        T–v analogy, 55
Mechanical systems, 56, 57, 58, 59, 60, 61, 62, 63
    f–i analogy, 57, 58
    f–v analogy, 57, 58, 63
Modal matrix, 566, 571
Modern control theory, 9
Multivariable system, 31
    state-variable formulation or representation, 530, 543
    system transfer matrix, 32, 660
    transfer function matrix, 658, 660

N-circles, 472, 473, 474, 482
Negative feedback, 3, 125, 128, 132
    sensitivity, 126, 136
        for the closed-loop system, 128
        definition, 128
        to parameter variations, 127
        to variation in parameter of the feedback element, 131, 133, 134
Nichols, N.B., 6
Nichols plot (chart), 376, 475, 476, 477, 478, 479, 481
Noise, 137, 138
Nonlinear control, 8
Nonlinear systems, 750
    describing, function method, 755, 778–794
    Liapunov's stability analysis, 755, 794, 798
    phase-plane analysis, 755, 756, 765, 771
Nonlinearities
    basic features
        backlash, 755, 756, 792
        dead-zone, 755, 779, 785, 790
        friction, 754

        hysteresis, 755, 756, 788, 790
        saturation, 754, 781, 785
    incidental, 753
    intentional, 753
    on-off, 783
Nyquist, H., 6, 424
Nyquist contour, 430, 437, 440, 441, 442, 443, 444, 445, 446, 448, 449, 450, 451, 452, 453, 454, 455, 456, 457, 459, 462, 464
Nyquist graphical technique, 425
Nyquist plot(s), 434, 435, 436, 438, 440, 441, 442, 443, 444, 445, 446, 448, 449, 450, 451, 452, 453, 454, 455, 456, 457, 460, 462, 463, 464, 465
Nyquist stability criterion, 424, 433, 434
    closed path, 426
    conditionally stable system, 461
    for open-loop poles on $j\omega$-axis, 466
Nyquist stability formula, 450

Observability, 623
    condition, 624, 625, 665
    criterion, 628
    matrix, 624, 642
Observable systems, 629
Observer, 651
    condition for existence, 651
    feedback matrix, 652, 654, 655
Offset, 4(7)
On-off control, 323, 326, 327
Optimal control, 8
Overshoot, maximum, 164

Peak time, 162, 164
Performance indices, 161, 162, 170, 195
Phase crossover frequency, 358, 359, 376, 377, 411
Phase margin, 354, 355, 358, 376, 377, 410, 477
Phase
    isocline method, 757, 759
    portrait, 756
    trajectory, 756
Plant, 2
Polar plot, 393
    conditionally stable system, 417
    construction of, 394
    of type-0, type-1, and type-2 systems, 401
Pole, 33
Polynomial, roots of, 223
Position error coefficient, static, 378
Principle of duality, 636, 637
Principle of superposition, 34
Process, 2

Process control system, 2, 323
    direct action, 326
    reverse action, 326
    Ziegler–Nichols rules, 338, 339, 341, 342, 343
        tuning method, 346
Proportional and integral control (PI), 206, 207, 330, 332, 678
    time response, 341
Proportional-integral-derivative (PID) control, 199, 331, 332, 336
    design, 338
    time response, 341
    transfer function of, 335
    tuning rules, 334
    using op-amps, 336
Pulse transfer function
    of cascaded elements, 694
    of closed-loop discrete-time systems, 695

Rate feedback, 200, 202
Relative stability, 234, 237, 408
Resonant frequency, 367
Rise time, 148, 162
Robust control, 240
Root contour, 306
    MATLAB program for, 308
    for multi-loop system, 313
Root-loci
    basic conditions for, 254
    of characteristic equation, 269
    of the compensated system, 294
    design based on, 315
    effect of
        addition of poles, 300
        addition of zeros, 303
        cancellation of poles and zeros, 305
        varying the pole position, 304
    obtaining with MATLAB, 290, 295
    rules for the construction of, 257
    for second-order system, 251
    for system with transportation lag, 314
Rotational systems, 46
    components of, 47
Routh, E.J., 220, 221
Routh array, 225, 226
Routh–Hurwitz criterion, 221, 222, 232, 233
    rules for, 226

Saddle, 765
    point, 765
Sample-and-hold device, 683, 685
Sampled-data control system, 679, 680
    characteristic equation, 696, 722

discrete state equation of a, 740
Nyquist criterion, 722
root loci for, 718
Sampler, 678, 680, 683
    discrete data output of, 684
Sampler and data hold, 681
Sampling process, 682
Samuelson's Model of National Economy, 729
Sensitivity
    analysis, 139, 140
    derivation of, 126
Sensors, 75
    LVDT (see linear variable differential transformer)
    potentiometers, 75–79
    synchros, 79–83
    tachometers, 83–85
Servomotors, 65
    dc, 68
        armature controlled, 69
        field controlled, 73
        state-space formulation, 71
    two-phase, 66
Setting time, 147, 163, 166, 170
Sign definiteness, 800
Signal Flow Graph, 108
Similarity transformation, 561, 562, 645
    eigenvalues, invariance of, 565
    matrix, 638, 639, 641, 642
Simulation
    controllable canonical form, 555
    observable canonical form, 556
    of state equations, 558
    of time delay, 389
Singular points, 756, 760
Singularity, 33
Spring-mass-damper system, 525
    simulation of, 527
    state-variable formulation, 527
Stable
    focus, 761, 773
    node, 763, 764, 773
Stability, 222, 253
    necessary condition, 224
    under parameter uncertainty, 240
Steady-state error, 188, 191
    unit acceleration input, 190
    unit-ramp input, 189
    unit-step input, 189
State, 26, 524, 538
    concept of, 523
State controllable, 621
State feedback, 644, 650
State observer, 650
    design of, 650

Index

State space, 26
    equations, 524
        from transfer function, 546
State space representation
    in canonical form, 667
    realizations, 671
State transition matrix, 583, 585, 586, 589, 604
    Cayley–Hamilton method, 597
    evaluation of, 590
    infinite power series method, 595
    of linear time-varying system, 609
    properties of, 590
State variable approach, 522
State variable(s), 26, 523, 524
    controllable, 630
    observable, 630
    uncontrollable, 630
    unobservable, 630
State variable formulation
    minimal set, 545
State vector, 26, 524
Stepper motor, variable reluctance, 93, 94

Time-domain behaviour from pole-zero plot, 182
Trace of a matrix, 608
Transducer, 3
Transfer function, 19, 27, 30
    forms of, 33
    invariance of, 566
    minimum phase, 384, 385, 387
    for the multivariable system, 535
    non-minimum phase, 384, 385, 387
    of an op-amp, 512
    properties of, 32

    of a system having state variable representation, 531, 532
        using Leverrier's algorithm, 533
Transfer matrix, 28, 32
    diagonal form of representation, 662
Transformation matrix, 561
    similarity, 571
Transient response, 144, 154, 253
    characteristics of z-plane pole locations, 706
    envelop curves of, 167
Translational systems, 46
    passive elements, 47
    spring-mass-damper, 48, 49

Vander Pol's differential equation, 752, 753
Vandermonde matrix, 569, 570
    for multiple eigenvalues, 573
Velocity error coefficient, static, 380
    from Bode plot, 381
Volterra equation, 611

Zero, 33
Zero-order
    circuit, 681
    frequency response of, 686, 687
    hold, 678, 679, 681, 683
z-transform, 683, 685, 687
    exponential damping, 690
    exponential function, 689
    ramp function, 688
    sine and cosine functions, 689
    shifting property of, 709
    unit step, 688